THE COMMON MARMOSET IN CAPTIVITY AND BIOMEDICAL RESEARCH

American College of Laboratory Animal Medicine Series

Steven H. Weisbroth, Ronald E. Flatt, and Alan L. Kraus, eds.
The Biology of the Laboratory Rabbit, 1974

Joseph E. Wagner and Patrick J. Manning, eds.
The Biology of the Guinea Pig, 1976

Edwin J. Andrews, Billy C. Ward, and Norman H. Altman, eds.
Spontaneous Animal Models of Human Disease, Volume 1, 1979; Volume II, 1979

Henry J. Baker, J. Russell Lindsey, and Steven H. Weisbroth, eds.
The Laboratory Rat, Volume I: Biology and Diseases, 1979; Volume II: Research Applications, 1980

Henry L. Foster, J. David Small, and James G. Fox, eds.
The Mouse in Biomedical Research, Volume I: History, Genetics, and Wild Mice, 1981; Volume II: Diseases, 1982; Volume III: Normative Biology, Immunology, and Husbandry, 1983; Volume IV: Experimental Biology and Oncology, 1982

James G. Fox, Bennett J. Cohen, and Franklin M. Loew, eds.
Laboratory Animal Medicine, 1984

G. L. Van Hoosier, Jr., and Charles W McPherson, eds.
Laboratory Hamsters, 1987

Patrick J. Manning, Daniel H. Ringler, and Christian E. Newcomer, eds.
The Biology of the Laboratory Rabbit, 2nd ed., 1994

B. Taylor Bennett, Christian R. Abee, and Roy Henrickson, eds.
Nonhuman Primates in Biomedical Research, Volume I: Biology and Management, 1995; Volume II: Diseases, 1998

Dennis F. Kohn, Sally K. Wixson, William J. White, and G. John Benson, eds.
Anesthesia and Analgesia in Laboratory Animals, 1997

James G. Fox, Lynn C. Anderson, Franklin M. Loew and Fred W. Quimby, eds.
Laboratory Animal Medicine, 2nd ed., 2002

Mark A. Suckow, Steven H. Weisbroth and Craig L. Franklin, eds.
The Laboratory Rat, 2nd ed., 2006

James G. Fox, Muriel T. Davisson, Fred W. Quimby, Stephen W. Barthold, Christian E. Newcomer and Abigail L. Smith, eds.
The Mouse in Biomedical Research, 2nd ed., Volume I: History, Wild Mice, and Genetics, 2007; Volume II: Diseases, 2007; Volume III: Normative Biology, Husbandry, and Models, 2007; Volume IV: Immunology, 2007

David Backer, ed.
Flynn's Parasites of Laboratory Animals, 2007 (Wiley)

Richard E. Fish, Marilyn J. Brown, Peggy J. Danneman and Alicia Z. Karas, eds.
Anesthesia and Analgesia in Laboratory Animals, 2nd ed., 2008

Jack R. Hessler and Noel D.M. Lehner, eds.
Planning and Designing Animal Research Facilities, 2009

Mark A. Suckow, Karla A. Stevens, and Ronald P. Wilson, eds.
The Laboratory Rabbit, Guinea Pig, Hamster and other Rodents, 2011

J. Harkness, P Turner, S VandeWoude, C Wheler, eds.
Biology and Medicine of Rabbits and Rodents, 5th ed., 2012 (Wiley)

Christian R. Abee, Keith Mansfield, Suzette Tardif and Timothy Morris, eds.
Nonhuman Primates in Biomedical Research, 2nd ed., Volume I: Biology and Management, 2012; Volume II: Diseases, 2012

Kathryn Bayne and Patricia V. Turner, eds.
Laboratory Animal Welfare, 2013

James G. Fox and Robert P. Marini eds.
Biology and Diseases of the Ferret, 3rd ed., 2014 (Wiley)

James Fox, Lynn Anderson, Glen Otto, Kathleen Pritchett-Corning, Mark Whary eds.
Laboratory Animal Medicine, 3rd ed., 2015 (Elsevier)

M. Michael Swindle, ed.
Swine in the Laboratory: Surgery, Anesthesia, Imaging and Experimental Techniques, 3rd ed., 2015 (Taylor & Francis)

Trenton Schoeb and Kathryn Eaton, eds.
Gnotobiotics, 1st ed., 2017 (Elsevier)

David M. Kurtz and Gregory S. Travlos, eds.
The Clinical Chemistry of Laboratory Animals, 3rd ed., 2017 (CRC Press)

THE COMMON MARMOSET IN CAPTIVITY AND BIOMEDICAL RESEARCH

Edited By

James G. Fox (Editor-in-Chief)
Division of Comparative Medicine, Massachusetts Institute of Technology
Cambridge, MA, United States

Robert P. Marini
Division of Comparative Medicine, Massachusetts Institute of Technology
Cambridge, MA, United States

Lynn M. Wachtman
Division of Comparative Medicine, Massachusetts Institute of Technology
Cambridge, MA, United States

Suzette D. Tardif
Southwest National Primate Research Center
Texas Biomedical Research Institute, San Antonio, TX, United States

Keith Mansfield
Discovery and Investigative Pathology, Novartis, Institutes for Biomedical Research
Cambridge, MA, United States

ELSEVIER

ACADEMIC PRESS
An imprint of Elsevier

Academic Press is an imprint of Elsevier
125 London Wall, London EC2Y 5AS, United Kingdom
525 B Street, Suite 1650, San Diego, CA 92101, United States
50 Hampshire Street, 5th Floor, Cambridge, MA 02139, United States
The Boulevard, Langford Lane, Kidlington, Oxford OX5 1GB, United Kingdom

Notices
Knowledge and best practice in this field are constantly changing. As new research and experience broaden our understanding, changes in research methods, professional practices, or medical treatment may become necessary.

Practitioners and researchers must always rely on their own experience and knowledge in evaluating and using any information, methods, compounds, or experiments described herein. In using such information or methods they should be mindful of their own safety and the safety of others, including parties for whom they have a professional responsibility.

To the fullest extent of the law, neither the Publisher nor the authors, contributors, or editors, assume any liability for any injury and/or damage to persons or property as a matter of products liability, negligence or otherwise, or from any use or operation of any methods, products, instructions, or ideas contained in the material herein.

Library of Congress Cataloging-in-Publication Data
A catalog record for this book is available from the Library of Congress

British Library Cataloguing-in-Publication Data
A catalogue record for this book is available from the British Library

ISBN: 978-0-12-811829-0

For information on all Academic Press publications visit our website at https://www.elsevier.com/books-and-journals

Publisher: Andre Gerhard Wolff
Acquisition Editor: Andre Gerhard Wolff
Editorial Project Manager: Pat Gonzalez
Production Project Manager: Sreejith Viswanathan
Designer: Greg Harris

Typeset by TNQ Technologies

Contents

III

RESEARCH APPLICATIONS

Contributors

David H. Abbott Department of Obstetrics and Gynecology, University of Wisconsin, Madison, WI, United States; Wisconsin National Primate Research Center, University of Wisconsin, Madison, WI, United States; Endocrinology-Reproductive Physiology Training Program, University of Wisconsin, Madison, WI, United States

Lyndsey Ace Wisconsin National Primate Research Center, University of Wisconsin, Madison, WI, United States

Arrilton Araujo Laboratory of Behavioral Biology, Psychobiology Graduate Program, Universidade Federal do Rio Grande do Norte, Natal, Brazil

Hayley Ash Wisconsin National Primate Research Center, University of Wisconsin, Madison, WI, United States; School of Natural Sciences, University of Stirling, Stirling, United Kingdom

Bert A. 't Hart Department of Immunobiology, Biomedical Primate Research Centre, Rijswijk, The Netherlands; University of Groningen, University Medical Center, Department of Neuroscience, Groningen, The Netherlands

David G. Baker School of Veterinary Medicine, Louisiana State University, Baton Rouge, LA, United States

Jaco Bakker Animal Science Department, Biomedical Primate Research Centre, Rijswijk, The Netherlands

Victoria K. Baxter Department of Pathology and Laboratory Medicine, University of North Carolina, Chapel Hill, NC, United States

Martina Bleyer German Primate Center, Pathology Unit, Göttingen, Germany

Monika Burns Division of Comparative Medicine, Massachusetts Institute of Technology, Cambridge, MA, United States

Christophe Casteleyn Department of Morphology, Faculty of Veterinary Medicine, Ghent University, Merelbeke, Belgium; Laboratory for Applied Veterinary Morphology, Department of Veterinary Sciences, Faculty of Pharmaceutical, Biomedical and Veterinary Sciences, University of Antwerp, Wilrijk, Belgium

Maria de Fátima Arruda Laboratory of Behavioral Biology, Universidade Federal do Rio Grande do Norte, Natal, Brazil

Jordon Dunham Department of Immunobiology, Biomedical Primate Research Centre, Rijswijk, The Netherlands; University of Groningen, University Medical Center, Department of Neuroscience, Groningen, The Netherlands

Steven J. Eliades Auditory and Communications Systems Laboratory, Department of Otorhinolaryngology: Head and Neck Surgery, University of Pennsylvania Perelman School of Medicine, Philadelphia, PA, United States

Jeffrey Everitt Duke University School of Medicine, Durham, NC, United States

James G. Fox Division of Comparative Medicine, Massachusetts Institute of Technology, Cambridge, MA, United States

Jeffrey A. French Neuroscience Program and Callitrichid Research Center, University of Nebraska at Omaha, NE, United States

Makoto Fukushima Laboratory for Marmoset Neural Architecture, RIKEN Brain Science Institute, Saitama, Japan; Department of Physiology, Keio University School of Medicine, Tokyo, Japan

Caroline M. Garrett PolarityTE Inc., Salt Lake City, UT, United States

Anna Goodroe Wisconsin National Primate Research Center, University of Wisconsin, Madison, WI, United States

Jennifer Haupt Division of Comparative Medicine, Massachusetts Institute of Technology, Cambridge, MA, United States

Noritaka Ichinohe Laboratory for Ultrastructural Research, National Institute of Neuroscience, National Institute of Neurology and Psychiatry, Tokyo, Japan; Laboratory for Molecular Analysis of Higher Cognitive Function, RIKEN Brain Science Institute, Saitama, Japan

Yolanda S. Kap Department of Immunobiology, Biomedical Primate Research Centre, Rijswijk, The Netherlands

Sven Korte Covance Preclinical Services GmbH, Münster, Germany

Liz Koutsos Department of Nutrition Science, Smithsonian National Zoological Park and Conservation Biology Institute, Washington, DC, United States

Joshua A. Kramer Laboratory Animal Sciences Program, Leidos Biomedical Research, Bethesda, MD, United States

Robin Kramer Division of Comparative Medicine, Massachusetts Institute of Technology, Cambridge, MA, United States

Marissa Kraynak Department of Obstetrics and Gynecology, University of Wisconsin, Madison, WI, United States; Endocrinology-Reproductive Physiology Training Program, University of Wisconsin, Madison, WI, United States

Robert E. Lanford Southwest National Primate Research Center, Texas Biomedical Research Institute, San Antonio, TX, United States

Donna Layne-Colon Southwest National Primate Research Center, San Antonio, TX, United States

David A. Leopold Section on Cognitive Neurophysiology and Imaging, Laboratory of Neuropsychology, National Institute of Mental Health, National Institutes of Health, Bethesda, MD, United States; Neurophysiology Imaging Facility, National Institute of Mental Health, National Institute of Neurological Disorders and Stroke, National Eye Institute, National Institutes of Health, Bethesda, MD, United States

Jon E. Levine Department of Neuroscience, University of Wisconsin, Madison, WI, United States; Wisconsin National Primate Research Center, University of Wisconsin, Madison, WI, United States; Endocrinology-Reproductive Physiology Training Program, University of Wisconsin, Madison, WI, United States

Keith G. Mansfield Discovery and Investigative Pathology, Novartis, Institutes for Biomedical Research, Cambridge, MA, United States

Robert P. Marini Division of Comparative Medicine, Massachusetts Institute of Technology, Cambridge, MA, United States

Daniel Marques de Almeida Pessoa Laboratory of Sensory Ecology and Primate Center, Universidade Federal do Rio Grande do Norte, Natal, Brazil

Kerstin Mätz-Rensing German Primate Center, Pathology Unit, Göttingen, Germany

Kelly A. Metcalf Pate Department of Molecular and Comparative Pathobiology, Johns Hopkins University School of Medicine, Baltimore, MD, United States

Andrew D. Miller Department of Biomedical Sciences, Section of Anatomic Pathology, Cornell University College of Veterinary Medicine, Ithaca, NY, United States

Jude F. Mitchell Brain and Cognitive Sciences Department, Meliora Hall, University of Rochester, Rochester, NY, United States

Hideyuki Okano Laboratory for Marmoset Neural Architecture, RIKEN Brain Science Institute, Saitama, Japan; Department of Physiology, Keio University School of Medicine, Tokyo, Japan

Jean L. Patterson Virology and Immunology, Texas Biomedical Research Institute, San Antonio, TX, United States

Ingrid H.C.H.M. Philippens Animal Science Department, Biomedical Primate Research Centre, Rijswijk, The Netherlands

Michael L. Power Conservation Ecology Center, Smithsonian National Zoological Park and Conservation Biology Institute, Washington, DC, United States

Corinna N. Ross Texas A&M University San Antonio, Department of Science and Mathematics, San Antonio, TX, United States

Erika Sasaki Department of Applied Developmental Biology, Central Institute for Experimental Animals, Kawasaki, Japan; Keio Advanced Research Centers (KARC), Keio University, Tokyo, Japan; Laboratory for Marmoset Neural Architecture, RIKEN Brain Science Institute, Wako, Japan

Nancy Schultz-Darken Wisconsin National Primate Research Center, University of Wisconsin, Madison, WI, United States

Gillian C. Shaw School of Veterinary Medicine, University of Wisconsin, Madison, WI, United States

Suzette D. Tardif Southwest National Primate Research Center, Texas Biomedical Research Institute, San Antonio, TX, United States

Joji Tsunada Auditory and Communications Systems Laboratory, Department of Otorhinolaryngology: Head and Neck Surgery, University of Pennsylvania Perelman School of Medicine, Philadelphia, PA, United States

Eric J. Vallender University of Mississippi Medical Center, Jackson, MS, United States; Tulane National Primate Research Center, Covington, LA, United States

Lynn Wachtman Division of Comparative Medicine, Massachusetts Institute of Technology, Cambridge, MA, United States

Maria Emilia Yamamoto Psychobiology Graduate Program, Universidade Federal do Rio Grande do Norte, Natal, Brazil

Preface

In the fall of 2014, investigators at the Massachusetts Institute of Technology approached the Division of Comparative Medicine for assistance in establishing a colony of common marmosets. Their motivation was to develop models of human neurologic conditions that might be met by this small and socially complex nonhuman primate. At that time, the Institute was fortunate that a cohort of common marmosets was available from the pedigreed colony of the New England Primate Research Center. The effort to establish the colony was led by a former staff veterinarian of the primate center and one of the editors of this text, Dr. Lynn Wachtman. As space and resources were set aside, and standard operating procedures and staff training developed, those of us engaged in the task noted that there was no central repository for information on the species. The journal *Comparative Medicine* had devoted one half of its August 2003 edition to the common marmoset, but its focus was, of necessity, somewhat limited and did not contain the myriad bits of information required for a comprehensive understanding of the species or the areas in which it might provide insights in the form of animal models of human physiology and disease. The editorial team began discussing the possibility of creating such a repository in the form of a textbook, one that would serve as a resource to others interested in the species. After further discussion with colleagues from around the world concerning the need for such a resource, we decided to proceed. The American College of Laboratory Animal Medicine (ACLAM), as part of their sustained effort to provide educational materials that promote the humane and responsible use of animals in research, testing, and teaching, agreed to sponsor the text as part of their authoritative Blue Book Series. Elsevier Ltd., a long-trusted party in these endeavors, was engaged as publisher. To help support ACLAM's mission, royalties from the sale of the text will be donated to the College.

The Common Marmoset in Captivity and Biomedical Research contains 29 chapters divided into three sections: Biology and Management, Veterinary Care, and Research Applications. The text has a diverse and international authorship, representing experts from 26 institutions, 6 countries, and 4 continents. The editors are indebted to these individuals, as well as to those who agreed to review their work, for the willingness to share their considerable knowledge and experience and for the gift of their effort and time. It is our hope that these chapters will instruct and inform their readers, but moreover, that they will challenge scientists, veterinarians, technicians, and managers to generate new information that will define further the care and use of these fascinating and engaging animals. We are grateful to our partners at Elsevier, in particular Pat Gonzalez, for the perseverance and professionalism necessary to bring this text into being. We thank as well our indispensable staff at MIT, chiefly Lucy Wilhelm and Alyssa Terestre, but also Parisa Zarringhalam, for their skill, patience, and good cheer. We send our love and recognition to our respective families for indulging the entire enterprise, knowing that efforts of this sort necessarily involve their own and sometimes unacknowledged sacrifice. Finally, the editors thank those engaged in caring for common marmosets at their own institutions and around the world, for their hard work, compassion, and expertise in supporting the worthy goals of humane care, education, and basic and applied research in this extraordinary species.

Robert P. Marini for Lynn Wachtman
Keith Mansfield
Suzette Tardif
James G. Fox
Cambridge, Massachusetts
May 2018

List of Reviewers

Kimberly Ange-Van Heugten North Carolina State University, Raleigh, NC, United States

Karen Bales University of California at Davis, Davis, CA, United States

Saverio Capuano Attending Veterinarian, Wisconsin National Primate Research Center, Madison, WI, United States

Cathy Carlson Professor and Chair, Department of Veterinary Clinical Sciences, College of Veterinary Medicine, University of Minnesota, Minneapolis, MN, United States

Ricki Colman Wisconsin National Primate Research Center, Madison, WI, United States

Gaby Doxiadis Rijswijk, The Netherlands

David Elmore Southwest National Primate Research Center, San Antonio, TX, United States

Paul Flecknell Emeritus Professor at Newcastle University, Newcastle University, University of Cambridge, Newcastle upon Tyne, United Kingdom; Emeritus Prof. Paul Flecknell, Comparative Biology Centre, Medical School, Framlington, Newcastle, United Kingdom

Paul Garber University of Illinois, Urbana-Champaign, IL, United States

Bert't Hart Chairman, Department of Immunobiology, Biomedical Primate Research Center, Professor in Neuroimmunology, University of Groningen, Groningen, The Netherlands

Amber Hoggatt Director and Attending Veterinarian, Center of Comparative Medicine, Brigham and Women's Hospital, Boston, MA, United States

Jon Kaas Gertrude Conaway Vanderbilt Chair and Distinguished Centennial Professor of Psychology, Vanderbilt University, Nashville, TN, United States

Joshua Kramer Senior Animal Program Veterinarian, NCI/Leidos Biomedical Research, Bethesda, MD, United States

Mathias Leblanc Senior Director/Staff Scientist, Animal Resources Department, The Salk Institute for Biological Studies, La Jolla, CA, United States

Steve Lever Principal Scientist, Dstl Porton Down, Salisbury, Wilts, United Kingdom

Gwendalyn Maginnis Nonhuman Primate Specialist, APHIS Center for Animal Welfare, USDA-APHIS-AC, Kansas City, MO, United States

Leah Makaron Veterinarian, University Laboratory Animal Resources, University of Pennsylvania, Philadelphia, PA, United States

Andres Mejia Pathology Services Unit, Wisconsin National Primate Research Center, University of Wisconsin, Madison, WI, United States

Andrew Miller Assistant Professor, Cornell University College of Veterinary Medicine, Department of Biomedical Sciences, Section of Anatomic Pathology, Ithaca, NY, United States

Guy B. Mulder Executive Director, Veterinary and Professional Services, Charles River, Wilmington, MA, United States

Atsushi Okumura Associate Research Scientist, Center for Infection and Immunity, Mailman School of Public Health, Columbia University, New York, NY, United States

Daniele de Paula Faria University of São Paulo, São Paolo, Brazil

Kimberley Phillips Texas Biomedical Research Institute, San Antonio, TX, United States

Christopher Pryce University of Zurich, Zurich, Switzerland

Evan Remington Postdoctoral Associate, McGovern Institute for Brain Research, Massachusetts Institute of Technology, Cambridge, MA, United States

Corinna Ross Texas A&M University at San Antonio, San Antonio, TX, United States

Pritam Sahota Retired from Novartis Pharmaceuticals as Executive Director of Pathology, Kennewick, WA, United States

Jenna Kropp Schmidt Assistant Scientist, Wisconsin National Primate Research Center, University of Wisconsin–Madison, Madison, WI, United States

Nancy Schultz-Darken Associate Scientist, Wisconsin National Primate Research Center, University of Wisconsin–Madison, Madison, WI, United States

Jacintha Shenton Scientific Director Biologics Toxicology, Janssen BioTherapeutics at Janssen R&D, Spring House, PA, United States

Heather Simmons Unit Head, Pathology Services Unit, Wisconsin National Primate Research Center, University of Wisconsin, Madison, WI, United States

Cassandra Turcotte Center for the Advanced Study of Human Paleobiology: George Washington University, Washington, DC, United States

Susan Westmoreland Abbvie Pharamceuticals, Worcester, MA, United States

Ruth Woodward National Institutes of Health, Bethesda, MD, United States

S E C T I O N I

BIOLOGY AND MANAGEMENT

CHAPTER

1

Taxonomy and Natural History

Maria de Fátima Arruda[1], *Maria Emilia Yamamoto*[2], *Daniel Marques de Almeida Pessoa*[3], *Arrilton Araujo*[4]

[1]**Laboratory of Behavioral Biology, Universidade Federal do Rio Grande do Norte, Natal, Brazil;** [2]**Psychobiology Graduate Program, Universidade Federal do Rio Grande do Norte, Natal, Brazil;** [3]**Laboratory of Sensory Ecology and Primate Center, Universidade Federal do Rio Grande do Norte, Natal, Brazil;** [4]**Laboratory of Behavioral Biology, Psychobiology Graduate Program, Universidade Federal do Rio Grande do Norte, Natal, Brazil**

INTRODUCTION

The goal of this chapter is to present a view of the common marmoset in the natural environment, emphasizing the flexibility to variable social and ecological conditions, characteristic of the species. The text describes taxonomy and natural history, patterns of activity and diet, and social and reproductive behavior of the species.

The common marmoset [*Callithrix jacchus* (Linnaeus, 1758)] is a small New World anthropoid (order Primates, suborder Haplorrhini, infraorder Simiiformes, parvorder Platyrrhini) that belongs to the family Cebidae and subfamily Callitrichinae [1]. Callitrichines comprise 41 species of marmosets and tamarins, in addition to Goeldi's monkey (*Callimico goeldii*) [2], accounting for nearly 30% of all living Neotropical primate diversity [3,4]. This subfamily is further subdivided into six [5] different genera (*Callimico*, *Callithrix*, *Cebuella*, *Leontopithecus*, *Mico*, and *Saguinus*) since the genus *Callibella* [6] is considered by some to be invalid [7].

Three of these callitrichines (*Callithrix*, *Cebuella*, and *Mico*) comprise all 22 species of marmosets, short-tusked animals that show dental and behavioral specializations for tree-gouging and exudate-feeding [4]. Recent molecular data indicate that the genus *Callithrix* split from its sister lineage *Mico/Cebuella* between five and six million years ago [8], and, since then, has diversified into six different species [2]: *C. jacchus* (Linnaeus, 1758); *Callithrix penicillata* (É. Geoffroy, 1812); *Callithrix kuhlii* (Coimbra-Filho, 1985); *Callithrix geoffroyi* (É. Geoffroy, 1812); *Callithrix aurita* (É. Geoffroy, 1812); *Callithrix flaviceps* (Thomas, 1903). Estimates also indicate that *C. jacchus* (common marmoset) diverged from its sister species, *C. penicillata* (black-tufted-ear marmoset), less than 1 million years ago [1].

The phylogenetic relationships attributed to cebids [1] corroborate the phyletic dwarfism hypothesis [9]; in other words, the marmoset's small body size should not be interpreted as a primitive trait, but rather one derived from a process of declining body size. The hierarchical branching order among Callitrichinae, from *Leontopithecus*, *Callimico*, *Callithrix* to *Mico*, mirrors decreasing body size and culminates in the smallest anthropoid, *Cebuella pygmaea*, as the most derived [1].

In addition to being small, arboreal, and diurnal, marmosets exhibit a number of taxa-specific anatomical features, such as reproductive twinning, resulting in a heavier overall litter weight (relative to the mother's weight), clawlike nails and reduced hallux, loss of the third molars, uniquely enlarged chisel-like mandibular incisors, loss of lingual enamel and buccal enamel thickening on mandibular incisors, falciform canines expanded in the buccolingual diameter, V-shaped jaw, and large complex cecum [10,11]. These traits have been argued to be related to heavy predation pressure associated with high reproductive output [10,12] and to the exploitation of a new feeding niche by these small primates, which not only cling to large vertical trunks and lianas while exploiting resources [11] but also gouge holes in trunks and large branches to stimulate the production of exudates [13], an exceptionally important food item for *C. jacchus* [14,15].

Callitrichine primates employ crypticity as an important behavioral strategy to thwart predation, moving quietly through the dense foliage [16]. They also use predator-specific alarm calls and show specific reactions

The Common Marmoset in Captivity and Biomedical Research
https://doi.org/10.1016/B978-0-12-811829-0.00001-7

to potential aerial and terrestrial predators, such as remaining motionless or moving to denser vegetation [17]. The compromise between these two selective pressures, predator avoidance, and new feeding niche exploitation, also seems to reflect how marmosets see colors. These animals exhibit X-linked triallelic color vision polymorphism, with 66% of females (heterozygotes) displaying trichromatic color vision (similar to that found in humans with normal color vision), whereas 33% of females (homozygotes) and 100% of males (hemizygotes) show dichromatic color vision (similar to that found in colorblind humans) [18,19]. According to recent suggestions [20], as small-bodied primates (less than 5 kg), such as marmosets, capuchin monkeys (*Cebus* and *Sapajus*), and squirrel monkeys (*Saimiri*), include more insects and fewer leaves in their diets [10], in addition to suffering from higher predation pressure [21], color vision polymorphism is expected to develop in these species because small primates benefit from both the trichromatic advantage of predator detection [20] and the dichromatic advantage of cryptic insect foraging [22]. Ref. [20] proposes that this evolutionary scenario should apply for all anthropoid primates, even when we take into account that color vision among New World primates is not uniform and that it differs from Old World primates' color vision [23].

Marmosets employ a broad array of visual, auditory, olfactory and tactile methods of communication. As mentioned above, the presence of predators is communicated by vocalizations; however, *C. jacchus* individuals also advertise their sex and age through acoustic cues [17,24]. Tactile, visual, and chemical information are also frequently used by common marmosets in sociosexual contexts [25].

Tactile signals play an important role in developing parental care, by enhancing prolactin circulating levels in parents that carry their offspring (Fig. 1.1C) [26–28]. The interchange of tactile stimulation between reproductive pairs (e.g., grooming—Fig. 1.1A), along with visual displays (e.g., genital display—Fig. 1.1E), are strong indicators of pair bonding between males and females in captive [29] and wild [30] common marmosets (see Ref. [31] for alternate views of the role of pair bonding in marmosets and other New World monkeys). Ref. [32] also has shown that infant-licking seems to be associated with high levels of oxytocin.

Common marmosets show rapid tongue-flicking, a visual signal associated with either reproduction or agonistic encounters [33,34]. Other signals frequently observed in agonistic encounters are piloerection and tail raising, although the former may also be displayed to predators, presumably, to appear larger and intimidate the adversary [34]. Chromatic and achromatic signals are also relayed by the sexual skin of pregnant marmosets and may indicate when parturition will occur [35].

Although primates are highly visual animals [36], chemical signals also exert a key role in their communication, especially in a sociosexual context [37]. In *C. jacchus*, scent-marking is a very common behavior that does not differ in frequency between dominant and subordinate individuals [38] but is more frequently observed in females than in males [39]. Common marmosets are able to discriminate between the scents of periovulatory and anovulatory females [40], and between a familiar and novel individual conspecific scent, a cue that may be used to assess individual quality [41].

Common marmosets (*C. jacchus*), which do not exhibit clear sexual dimorphism [42] in body size (approx. 320 g), can be identified by the following diagnostic traits [43]: (1) white tufts circling their ears; (2) a white spot on their forehead; (3) posterior third of their dorsum exhibiting a striated staining pattern; (4) a grizzly dark brown belly; (5) a ring tail, with narrow light gray rings and wider black rings. This species occurs naturally in all northeastern states of Brazil (Maranhão, Piauí, Ceará, Rio Grande do Norte, Paraíba, Pernambuco, Alagoas, Sergipe, and Bahia) [43] and can exploit contrasting environments (e.g., the humid Atlantic Forest and the dry scrub forest of the Caatinga) because they are capable of making behavioral adjustments to their activity budgets according to temperature fluctuations and food availability [44,45].

Marmosets' ancestors dispersed (11 million years ago) from the northwestern to southeastern Amazon and then underwent significant range expansion (6 million years ago) to occupy much of the Amazon and Atlantic Forest [8]. Biogeographic reconstructions also indicate a south–north dispersion through the Atlantic Forest, with occupation of the northern Atlantic Forest by invaders from the southern Atlantic Forest, and posterior invasion of the Cerrado (Brazilian savanna), in central Brazil, by *C. penicillata*, followed by a recent spread (1 million years ago) of *C. jacchus* into the highly seasonal Caatinga habitat (dry scrub forest), in northeastern Brazil [8,46].

Despite being endemic to northeastern Brazil, in the past century common marmosets were massively introduced into other Brazilian states through animal trafficking [47–49]. In Rio de Janeiro, the most affected state [50], common marmosets (*C. jacchus*) and black-tufted-ear marmosets (*C. penicillata*), in addition to their hybrids, are more abundant than the native golden-lion tamarins (*Leontopithecus rosalia*) and frequently compete with them for food [51]. Although often parasitized [52,53], these populations of invader marmosets exhibit a normal rate of reproduction and average body weight exceeding [51] that found in populations of native marmosets [42], reinforcing the view that common marmosets have the ability to adapt and exploit a range of different environments [54].

FIGURE 1.1 Grooming interaction between adults (A); interaction of individuals of the group with a mother and infant (highlighted by *) (B); mother transporting infants on a gumming tree (C); foraging in dry scrub (D); agonistic behavior (E); copulation behavior (F).

DAILY ACTIVITY, DIET, AND HOME RANGE

Daily activities are usually performed by the group as a social unit in their home range, even if some individuals may be apart from others, especially subordinates [14]. The animals leave their sleeping tree soon after dawn, which occurs earlier in the dry season. A difference of about an hour change in day length (light phase) was recorded in an Atlantic Forest study site [55], with a day range lasting from 10 to 12 h. By the end of the afternoon, all the animals move toward the sleeping site and, once there, huddle overnight in the canopy of the sleeping tree. This is an important signal of their social relations since separation from the group for sleeping indicates that a migration is probable within a short interval (Arruda, unpublished data).

Following free-ranging groups in three different sites—a human-impacted area of secondary Atlantic Forest, an area of Atlantic secondary forest and of Caatinga, in northeastern Brazil—showed that foraging was the most prominent activity, accounting for almost 45% of the activity budget throughout the study (Table 1.1). Time spent in foraging, feeding, and social interaction was similar among groups in different environments, each with characteristic floristic composition for fruit and gum sources. Other behaviors varied significantly according to the study site, in particular locomotion and resting. Groups from Caatinga, for example, presented lower levels of locomotion and, as a consequence, more time spent resting (Table 1.1). The temperature in Caatinga can reach up to 37°C and humidity up to 40% at least during 8 months of the year. The high temperature associated with low humidity probably induces marmosets to stay in more comfortable temperature patches and close to exudate sources. In the other environments, the vegetation is more abundant and average temperature is 26°C all year around.

Species of the subfamily Callitrichinae have a diet primarily composed of fruits, insects, and exudates [15,56–59]. Studies conducted by our group in the sites of secondary Atlantic Forest and Caatinga, along with a study at a second Caatinga site [60] show that, based on feeding time, the main food items of *C. jacchus* are exudates (68.5%), fruits (24.3%), and animal prey (6.9%). Other food items such as flowers, eggs, and small vertebrates have also been occasionally recorded.

Fruit consumption occurs whenever it is available in the environment. In an area of secondary Atlantic Forest (Fig. 1.3A), peak fruit consumption occurs from December to March. In the other types of environment, such as the human-impacted area and Caatinga, fruit intake is more evenly distributed throughout the year (Fig. 1.3B and C, respectively). In groups that occupy completely altered areas or those with some exotic fruit species, consumption depends only on their availability, often throughout the year, due to the succession of fruiting species [14].

At our study sites, both in secondary Atlantic Forest and in Caatinga, exotic fruit species were introduced by humans on the edges of natural environments. As such, it was observed that in the Atlantic Forest, marmoset groups ingested 12 native and 8 exotic species, whereas in the Caatinga, 19 native and 5 exotic species of fruit trees were consumed [14,61]. In the area of the Atlantic Forest, 10 species were ingested as sources of fruits, 6 as sources of exudates, and 1 as a source of nectar. It is important to highlight the importance of *Tapirira guianensis* in providing fruit, exudate, and nectar. In the Caatinga, there is a numerical inversion in relation to food sources, with 15 species as sources of exudate, 8 as sources of fruits, and 3 as sources of nectar. Of these 15 exudate species, the most commonly exploited source accounted for 52% of feeding time (Table 1.2). The same type of diet was found in other studies at the Caatinga [44,60] and in the Atlantic Forest [62]. Fruits are rich in simple carbohydrates (76.3%), which are easy to digest, contain 1.3% lipids and 17.4% fiber [61], in addition to low-protein content (5.2%). Based on data from other sites, fruits are poor in minerals compared with exudates and insects [59,63,64] (Fig. 1.3).

Exudates are obtained from the holes of gum tree trunks, mostly (89%) concentrated in the canopy, the uppermost layers containing the largest number of holes [65]. Indeed, both *C. jacchus* and *C. penicillata* depend heavily on exudate and are classified as exudativorous–insectivorous [15,58] (Table 1.2).

A possible explanation for the exploitation of exudates is the need for calcium in the bone development process, particularly in pregnant and nursing females [14,61,64]. This item is a rich source of energy

TABLE 1.1 Percentage of Time/Month (Mean ± SD) of Behavioral Activities in *C. jacchus* Groups From Three Different Environments

	Field Studies			
Behavior	**Secondary Atlantic Forest (SAF)**	**Caatinga (CA)**	**Anthropized Area (AA)**	**Mann–Whitney U Test**
Locomotion	8.4 ± 4.6	3.9 ± 1.4	10.9 ± 6.1	AA > CA $P = 0.01$ SAF > CA $P = 0.03$
Resting	14.6 ± 4.6	21.5 ± 11.8	18.0 ± 4.5	AA > SAF $P = 0.04$
Foraging	45.2 ± 12.8	46.4 ± 11.0	46.4 ± 11.0	$P > 0.05$[a]
Feeding	21.9 ± 11.6	14.8 ± 7.0	15.4 ± 4.6	$P > 0.05$[a]
Social interaction	9.9 ± 5.1	13.5 ± 7.8	16.9 ± 3.8	$P > 0.05$[a]

[a]*For all comparisons.*

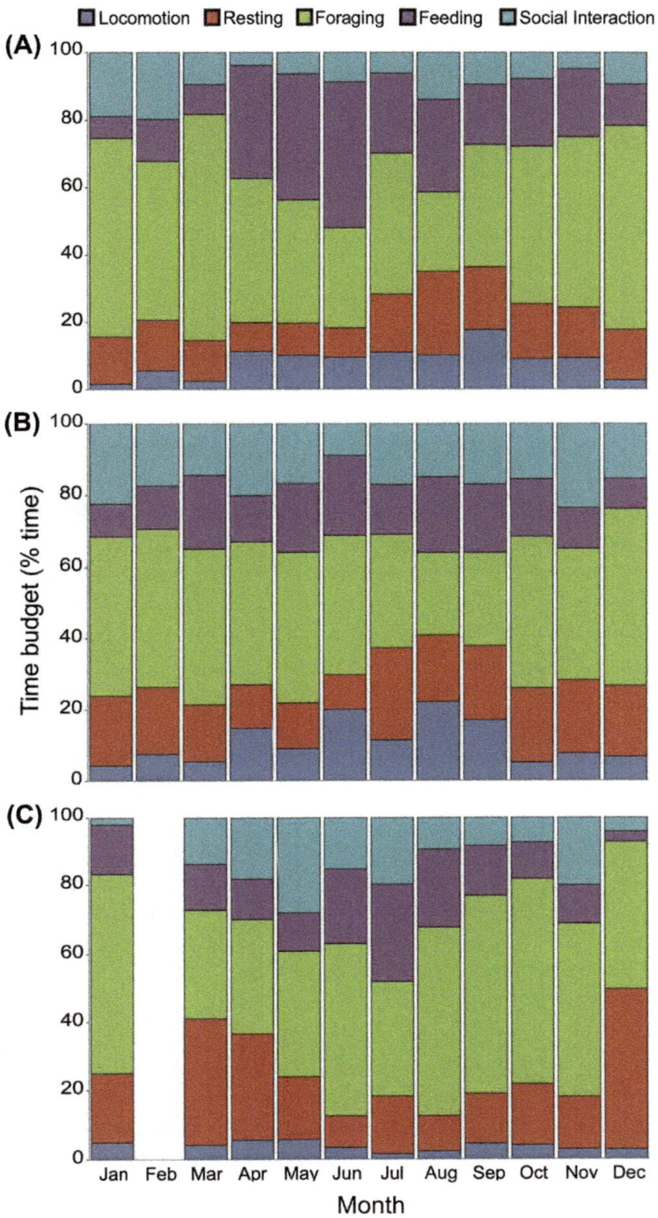

FIGURE 1.2 Daily activity (% time) of wild *C. jacchus* groups in three environments: secondary Atlantic Forest (A); anthropized area (B); Caatinga (C); rainy season in the secondary Atlantic Forest and anthropized area occurs from March to July and in the Caatinga, from February to May.

(carbohydrates) and calcium, which may compensate for the lack or reduction of fruits available during some periods of the year. Chemical analysis of *Anacardium excelsum* exudates shows the presence of several minerals, primarily calcium and phosphorus, at a proportion of 31:1, respectively [64].

In contrast to exudates and fruits, insects, the main prey exploited by common marmosets, are rich in proteins (50%–75%) and lipids (7%–18%) and have a

TABLE 1.2 Tree Species Used as Food Source in Secondary Atlantic Forest (SAF), in Caatinga (CA), and in Anthropized Areas With Exotic Species Both in Atlantic Forest (ESa) and in Caatinga (ESc)

Family/Specie	Forest Type	Food Type
ACANTHACEAE		
sp. 1	CA	Nectar
ANACARDIACEAE		
Anacardium occidentale	SAF/CA	Fruit, exudate
Tapirira guianensis	SAF	Fruit, exudate, nectar
Spondias dulcis	SAF	Fruit, exudate
Spondias purpurea	CA	Fruit, exudate
Mangifera indica	ESa	Fruit
Myracrodruon urundeuva	CA	Exudate
ANNONACEAE		
Annona muricata	ESa	Fruit
Annona crassiflora	ESc	Fruit
BIGNONIACEA		
Tabebuia impetiginosa	CA	Exudate
BOMBACACEAE		
Pseudobombax marginatum	CA	Exudate
BORAGINACEAE		
Cordia superba	SAF	Fruit
BURSERACEAE		
Protium brasiliense	SAF	Fruit
Commiphora leptophloeos	CA	Exudate
CACTACEAE		
sp. 2	CA	Fruit
CAPPARACEAE		
Capparia flexuosa	CA	Nectar
CARICACEAE		
Carica papaya	ESc	Fruit
CECROPIACEAE		
Cecropia sp.	SAF	Fruit
CHRYSOBALANACEAE		
Cupanea sp.	SAF	Fruit
COMBRETACEAE		
Terminalia catappa	ESa	Fruit, exudate
EUPHORBIACEAE		
Sebastiania sp.	CA	Exudate

Continued

TABLE 1.2 Tree Species Used as Food Source in Secondary Atlantic Forest (SAF), in Caatinga (CA), and in Anthropized Areas With Exotic Species Both in Atlantic Forest (ESa) and in Caatinga (ESc)—cont'd

Family/Specie	Forest Type	Food Type
FABACEAE		
Bauhinia cheilantha	CA	Nectar
Caesalpinia pyramidalis	CA	Exudate
Amburana cearensis	CA	Exudate
Enterolobium contortisiliquum	ESa	Exudate
Canavalia brasiliensis	CA	Exudate
Pithecellobium diversifolium	CA	Exudate
Mimosa caesalpiniaefolia	CA	Exudate
Anadenanthera colubrina	CA	Exudate
Leucaena leucocephala	CA	Exudate
MALPIGHIACEAE		
Byrsonima sp.	SAF	Fruit
Malpighia emarginata	ESc	Fruit
MELIACEAE		
Swietenia macrophylla	ESa	Exudate
MORACEAE		
Artocarpus communis	ESa	Fruit
MYRTACEAE		
Eugenia uniflora	ESa	Fruit
Myrciaria tenella	SAF	Fruit
Psidium guayara	ESc	Fruit
Syzygium jambolanum	ESa	Fruit
POLYGONACEAE		
Coccoloba cordifolia	SAF	Fruit
RUBIACEAE		
sp. 3	SAF	Fruit
RUTACEAE		
Citrus aurantifolia	ESc	Exudate
SAPINDACEAE		
Allophylus puberulus	SAF	Fruit
VERBENACEAE		
Lantana camara	CA	Fruit

high proportion of phosphorus in relation to calcium, especially locusts and crickets [64]. In addition, most small primates obtain almost all their nitrogen from insects [64]. Orthoptera (grasshoppers, crickets, and locusts) accounts for most of the insects consumed by common marmosets [14,44]. Prey consumption is relatively constant throughout the year, although in some areas consumption is more intense during the rainy season (Fig. 1.2).

Marmosets search for prey by manipulating branches, trunks, and especially on green or dead leaves trapped in tangled vines. During foraging, individuals spread within a radius of up to 12 m, possibly to avoid intragroup competition [14,61]. However, individuals are also found to remain close during insect foraging, which may reflect social bonds based on age cohorts, hierarchical level, or when caring for infants less than 3 months old [14]. Common marmosets prefer to be alone when they feed and on average are 2 m apart. Depending on the type of environment, groups forage from the ground to the top of the tree canopy but prefer low strata in the understory [14,66,67].

The home range represents the area used by the group throughout the year. In the genus *Callithrix*, the size of the home range increases as the proportion of fruits becomes more important in their diet. Thus, exudativorous species (*C. jacchus* and *C. penicillata*) generally have a smaller home range (Table 1.3). In *C. jacchus*, home range size varies between groups and areas where populations of the species are present. Moreover, the home range is affected by the number and location of sleeping trees and food sources and can vary throughout the year from 0.5 to 6.9 ha.

Exudates, one of the main sources of energy and minerals, are a key resource for *C. jacchus*, accounting for 60%–90% of their diet at certain times of the year (Fig. 1.2). Trees exploited as a source of exudate, as well as sleeping trees, are clumped in the home range, most exploited only by the resident group [14,68]. The fact that they are clustered and located in exclusive areas allows the group to reduce travel, thereby reducing energy expenditure [14,68], especially when fruits are not available [61].

SOCIAL AND REPRODUCTIVE BEHAVIOR

Wild *C. jacchus* live in cooperative breeding groups, usually extended families [69], ranging from 3 to 17 individuals, including adults, subadults, juveniles, and infants from sequential twin sets [14,15]. The empirical basis for our statement is basically behavioral. Nine groups were followed for extended periods (from 20 to 120 months) in intensive observational schedules in two field sites. This allowed us to form a kinship chart of groups based on observed matings, births, and migrations. These observations allowed us to suggest that in most groups (c. 70%) there was only one female that

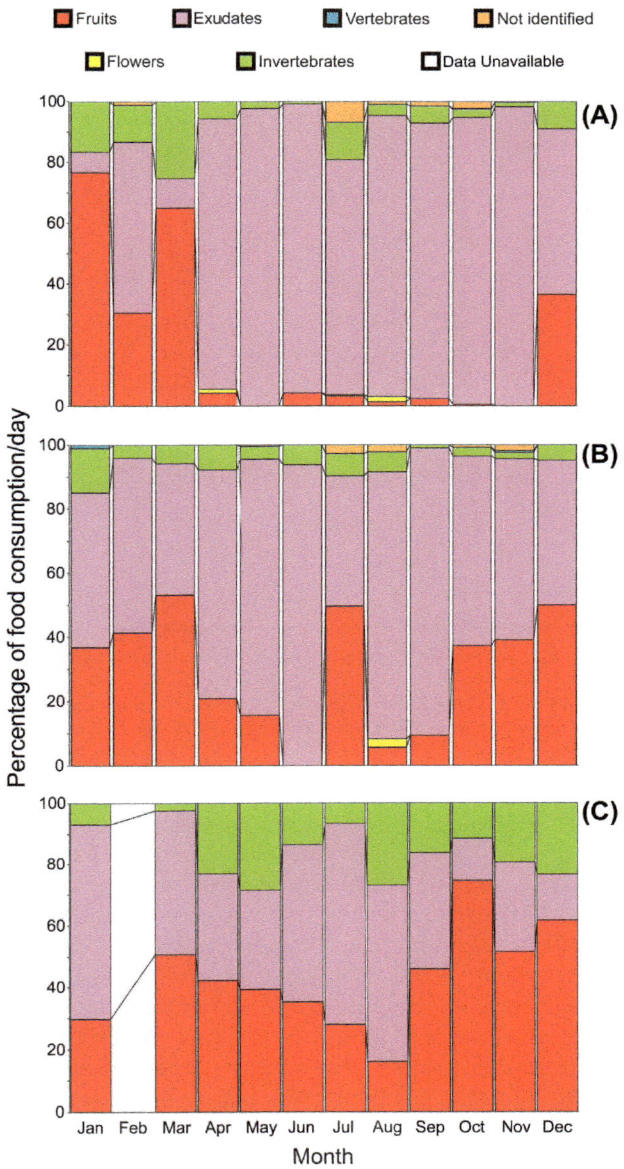

FIGURE 1.3 Feeding activity/pattern (% time) of wild *C. jacchus* groups in three environments: secondary Atlantic Forest (A); anthropized area (B); Caatinga (C). Rainy season in the secondary Atlantic Forest and anthropized area occurs from March to July and in the Caatinga, from February to May.

bred repeatedly and had surviving offspring; these females were never observed mating with another male other than the dominant resident male in their groups; in some of these groups, a second female would give birth to live offspring, but, as a rule, none of those offspring survived, and the females left the groups soon after breeding; some of these females were seen copulating with extra-group males in periods that suggest that the offspring was sired by the extra-group males; in a few occasions the offspring of these females survived probably because they were born almost simultaneously to the dominant female offspring and all four infants were cared for equally by all group members; in circa 30% of the groups two females alternated reproduction [70] and were seen copulating with both the resident and extra-group males [71]; both females bred repeatedly and both had surviving offspring; both were permanent residents and did not leave the group after breeding (see Refs. [70,72–76]).

This information is at odds with Refs. [77,78] as these authors suggest that both genetic and group composition data suggest a nonmonogamous breeding system, with multiple male and female breeders. Although most of their data derive from wild *Saguinus* groups, the authors suggest that the same is probably true for *Callithrix*, as the presence of nonbreeding adults of both sexes from different maternal lineages in the latter is consistent with multiple breeding. Of course, we lack genetic data for these groups, however, behavioral and group composition data suggest that in most groups there is only one pair that is effectively able to raise viable offspring [70,74,76]. This assumption is not incompatible with the genetic diversity in group composition, as replacement of both male and female breeders by migrating individuals has been observed in several groups. Also, the offspring of incoming males and females remain in the group together with the offspring of the previous breeders. Besides, during breeder replacement more than one individual may immigrate and remain in the group, contributing to genetic diversity [76]. In the few groups with two active breeding females, subordinate females were seen copulating with out-group males, suggesting parenting by those males [70,71,74]. Genetic data on one of the populations we followed [69,79] suggest that most group members are close relatives and that groups tend to stabilize on an extended family structure. So, we recognize that a paternity genetic analysis would add valuable data that would help to clarify the breeding systems in *C. jacchus*. However, on the other hand, we suggest that long-term behavioral data offer an alternative hypothesis, at least to the populations that we have been following. We believe that this alternative hypothesis should not be discarded until paternity genetic data can be gathered.

Early reports on Callitrichinae reproductive systems suggested that they were monogamous and formed extended families [80]. These reports were based on the observation of a few species in captivity, and this organization was considered to represent all callitrichine species. However, two papers [59,81] contested this assumption, proposing that Callitrichinae did not form family groups and were not monogamous. Rather, they suggested a flexible mating system and, abiding by a strict definition of monogamy [80], proposed that callitrichines lived in multimale–multifemale groups

TABLE 1.3 Home Range Size for *Callithrix*[a]

Species	Home range (ha)	Groups (N)	Duration of Study	References
Callithrix jacchus	2.5	1	3 weeks	[118][1]
C. jacchus	5.0	1	14 months	[66][1]
C. jacchus	0.72–1.62	5	6 months	[119][2]
C. jacchus	0.5	1	6 months	[15][2]
C. jacchus	3.9–5.2	3	18 months	[120][3]
C. jacchus	3.9–4.4	2	12 months	[14][3]
C. jacchus	3.2–6.9	5	18 months	[62]
Callithrix penicillata	1.25–4.5	2	4 months	[121]
C. penicillata	3.5	5	12 months	[122,123]
Callithrix kuhlii	10.0–12.0	8	3 months	[67,124]
Callithrix aurita	11.0–16.0	2	13 months	[125][b]
C. aurita	11.5	8	5 months	[126,127]
Callithrix flaviceps	35.5	1	12 months	[128]

[a]*Adapted from Rylands AB, Faria DS. Habitats, feeding ecology, and home range size in the genus* Callithrix. *In: Rylands AB, editor. Marmosets and tamarins: systematics, behavior and ecology. Oxford (USA): Oxford University Press; 1993. pp. 262–72.*
[b]*Cited by Rylands AB, Faria DS. Habitats, feeding ecology, and home range size in the genus* Callithrix. *In: Rylands AB, editor. Marmosets and tamarins: systematics, behavior and ecology. Oxford (USA): Oxford University Press; 1993. pp. 262–72; studies developed in: [1]Atlantic Forest Reserve from Federal University of Paraiba, [2]Tapacura Station—Pernambuco, [3]Floresta Nacional de Nisia Floresta—Rio Grande do Norte, Northeastern Brazil.*

that communally reared the dominant female offspring, in a polyandrous breeding system. More recent studies suggest that, although all callitrichines show a flexible mating system, the organization proposed by the papers above [59,81] best describes *Saguinus* species.

Marmosets, and *C. jacchus* in particular, have a flexible mating system that is predominantly restricted to a single female, and as far as behavioral data indicate, infants are sired by a dominant resident male [70,74,75]. In captivity, groups are stable when there is only one breeding pair. They usually breed to form an extended family that helps care for the breeding female offspring [82]. Adult individuals may remain in the group for varying lengths of time, but ovulating subordinate females often suffer aggression from the breeding female and have to be removed from the cage [83,84].

In the wild, breeding strategies are more varied. In a 10-year study of 14 wild breeding groups in northeastern Brazil, only 3 groups (15%) had two breeding females [74]. In the remaining 11 groups, only one female bred repeatedly. In *L. rosalia*, the percentage of wild polygynous groups is similar [85]. In the same marmoset population, it was observed that subordinate females that copulated with extra-group males (extra-group copulation, EGC) left the group when EGC resulted in pregnancy [70]. We suggest that this common occurrence may be an attempt by subordinate ovulating females to scan for breeding vacancies and establish bonds with neighboring group males [74].

Nevertheless, in a short-term or episodic follow-up of the groups, this occurrence could be mistaken as polygyny, as two females from the same group gave birth. However, the interval between expulsion or voluntary dispersal from the group and the acquisition of a breeding position in another group usually takes years [70,86], and therefore was only identified as a single failed attempt in long-term studies. In our long-term monitoring of groups in secondary Atlantic Forest, four females that had left their groups were later identified as the reproductive female in another group in periods varying from 6 to 44 months [86].

Dispersal is an important aspect of marmoset group dynamics for both females and males. As a rule, males and females remain in the natal group after the onset of puberty and emigrate later, singly or in pairs [14,87]. However, females leave their groups more frequently than males [75,88]. Although it might appear to be voluntary rather than forced in some occasions, high levels of cortisol in emigrant females [88] indicated high levels of competition within the social group. This process is usually associated with an increasing level of isolation from other group members, such as during daily activities, or by suffering harassment before leaving the group. Three cases of continuous harassment were recorded in our studies, two of them by a dominant male toward a subordinate one [69] and one by a subordinate female toward the dominant one (Arruda, unpublished data), whereby the animal being chased finally left the area. After leaving the group, the individual might be chased if it encounters the group again [89].

Expulsion of subordinate females in particular may occur not only after a pregnancy, abortion, or infanticide [75], but also when subordinates become sexually mature [88]. Even when it appears to be voluntary, the proximate mechanisms triggering dispersal are not always clear [90]. However, subordinate females may show deference to a dominant female instead of competitive behavior, as demonstrated when paired in the laboratory [91,92]. The pairing may trigger a clear or contested dominance status, demonstrated by aggression versus submission and changes in the level of reproductive hormones, such as progesterone. In uncontested dominance pairs, only dominant females presented progesterone levels indicative of ovulatory cycles, whereas subordinate females exhibited low or inconsistent levels of progesterone [91]. Ref. [92] reported that some, but not all, subordinate females in captive groups showed willingness to act prosocially, which the author suggests is a signal of deference.

Breeding individuals are dominant over all others [30,89,93], showing preferential access to food and sleeping sites and exhibiting high levels of affiliative interactions and mate guarding. Other individuals also show high levels of affiliative interactions with the breeding pair, in particular, adult males [14,72,76]. In a feeding context, in particular, the reproductive female has preferential access to the resources both in the field [14] and in captivity [94]. Agonism is usually subtle, although subordinates leave the place as the dominant female approaches. In captive condition, it was also demonstrated that when peer groups are formed, one female becomes dominant to others and inhibits the subordinate females' ovarian cycling [95].

Among the affiliative interactions, grooming interactions—giving or receiving [76,96], occur mostly during resting periods when the group splits into several smaller subgroups [14,89]. This may occur during the warmest hours of the day [45] and indicate the group's affiliative dynamics. The dominant pair has been reported to be preferential grooming partners [89,96] forming a core group with other adult individuals [76]. However, immatures are reported to receive more grooming from adults than vice versa [89]. In one study, subordinate individuals were groomed more by the dominant female than vice versa during intergroup encounters [38]. Furthermore, in a second study by the end of pregnancy, the dominant female groomed the male more than in the first week postpartum, when she was cycling again and receiving more grooming from the male [97].

Proximity is another important indication of affiliation between individuals, the reproductive couple usually being very close to each other in different contexts, such as moving from one area to another in the home range, foraging, and resting. Subordinate individuals form subgroups [89], although we have observed a decline in grooming interactions and proximity after emergence of the reproductive female vacancy [30] (Arruda, unpublished data).

The need for helpers in the common marmoset breeding system requires cooperation, even considering that competition, mainly among females, and dominance relations are the basis of their social system. However, it also depends on cooperation that is displayed mostly by males [75]. As in other cooperative breeders, the dominant male (who is likely the father) and nonreproductive individuals act as helpers, carrying, sharing food, and protecting the infants of the breeding female from predators [98]. Help from other group members is necessary because reproduction is energetically demanding for marmoset females since they give birth to twins with high combined birth weights and may become pregnant again while still nursing the previous set of infants [82]. However, not all members of the group help equally. A comparison of care in two captive populations showed that the adults, the oldest siblings present in the group, which could be adults and subadults, and singletons of both sexes carried infants more than twins in the group. Males also carried infants more than females, but this difference was not statistically significant [99]. In wild populations, however, nonreproductive helpers, usually males, provide the majority of the care [74,89,100]. In the same population, nonbreeding females were sometimes prevented from carrying infants [72], suggesting that male helpers are generally sufficient to provide the necessary care for infants.

Ref. [101] questions why some helpers help and why they remain in the group without breeding. Postponing reproduction may incur heavy costs to helpers and even prevent them from breeding if they remain long enough in the group as helpers. Moreover, infant care has costs in callitrichines since caregivers have less time to forage, are likely more susceptible to predators, and have decreased mobility while carrying infants [102–104]. These costs may be particularly heavy when resources are low. As such, there must be important benefits for individuals that care for infants.

Four potential benefits for helpers have been suggested [105]: (1) enhanced survivorship through both increased group size and access to the physical and social resources; (2) enhanced future probability of breeding through the budding off of natal territory or the taking over of a vacancy when a breeder dies or leaves the group; (3) increased fecundity as a breeder through previous exposure to parental skills and by recruiting helpers from among the animals that it helped raise; (4) increased production of nondescendent kin, thereby enhancing its inclusive fitness.

There is evidence to support all four of these hypotheses among callitrichines [30,75,76,106–111]. However, Ref. [112] argued that evidence was insufficient, as it is based on correlations between characteristics or behavior and proposed benefits, without full proof of a causal relationship. For instance, benefits accrued to experience, which would improve reproductive performance, may reflect the lessening of a neophobic response to infants [113] and may be gained by simple exposure to infants as opposed to actual participation in infant care [114]. In the field, we have seen supportive evidence for inheritance of the breeding post by subordinate males, particularly when they were the second dominant male in the group's hierarchy, just below the breeding male [76]. These males are usually strong caretakers [72,74,89,104]. Thus, participation in caretaking may benefit some helpers more than others. It is a risk-taking behavior, likely maintained by proximate motivations, such as attraction to infants and pressure from the breeding pair. However, these questions can only be properly answered by experimental studies.

A recent study [115] adds a further potential benefit for helpers, derived from the occurrence of not only hematopoietic chimerism in *Callithrix* [116] but also in many other tissues, including the germ line. As a result, chimeric Callitrichine twins may be more closely related to each other than nonchimeric siblings, an average of $r = 0.574$ for male twins in *C. kuhlii* [115]; however, see Chapter 8 and [117] for controversy on this topic. Chimerism may therefore explain the attraction of males to infants in callitrichines, particularly when caring for their twin offspring, in line with the amount of carrying, provisioning, and guarding they engage in, resulting in an additional potential benefit for helpers. However, in cases in which offspring are either unrelated or not closely related to helpers, other explanations are necessary.

CONCLUDING REMARKS

The wide range of environments where groups of *C. jacchus* can be found demonstrates that using their species-specific adaptive pattern, even when introduced into areas where they are not endemic, they are highly successful. Feeding flexibility, small home ranges, and the exploitation of exudate trees avoids the challenge that other species face when there are drastic changes in seasonality or anthropomorphic interference in their environment. The mating system is characterized by a balance between cooperation, primarily between males, and competition, largely between females. This enables high fertility in breeding females (twin offspring) and help from related and unrelated individuals, some of whom delay their reproduction. Furthermore, the occurrence of chimeric twins may have an impact on care and benefits for helpers.

Given that *C. jacchus* has become a model in biomedical research, it is important to understand the flexible ecology and behavior of this species. A number of aspects of social dynamics are subtle and by understanding their everyday habits and social interactions, researchers can provide a captive environment for the species that preserves its natural behavior and allows the production of better research data. A captive environment that fulfills all the species' needs will also provide the best conditions for studying the species in different areas of research.

References

[1] Perelman P, Johnson WE, Roos C, Seuanez HN, Horvath JE, Moreira MAM, Kessing B, Pontius J, Roelke M, Rumpler Y, Schneider MPC, Silva A, O'Brien SJ, Pecon-Slattery J. A molecular phylogeny of living primates. PLoS Genet 2011;73: e1001342.

[2] Rylands AB, Mittermeier RA. The diversity of the New World primates (Platyrrhini): an annotated taxonomy. In: Garber PA, Estrada A, Bicca-Marques JC, Heymann EW, Strier KB, editors. South American primates: comparative perspectives in the study of behavior, ecology, and conservation. New York: Springer; 2009. p. 23–54.

[3] Rylands AB, Schneider H, Langguth A, Mittermeier RA, Groves CP, Rodríguez- Luna E. An assessment of the diversity of New World primates. Neotrop Primates 2000;8(2):61–93.

[4] Rylands AB, Coimbra-Filho AF, Mittermeier RA. The systematics and distributions of the marmosets (*Callithrix, Callibella, Cebuella,* and *Mico*) and *Callimico* (Callimico) (Callitrichidae, primates). In: Ford SM, Porter LM, Davis LC, editors. The smallest anthropoids: the marmoset/Callimico radiation. New York: Springer US; 2009. p. 25–61.

[5] Rylands AB, Heymann EW, Alfaro JL, et al. Taxonomic review of the new World tamarins (primates: Callitrichidae). Zool J Linn Soc 2016;177(4):1003–28.

[6] Roosmalen MGM, van Roosmalen T. The description of a new marmoset genus, *Callibella* (Callitrichinae, Primates), including its molecular phylogenetic status. Neotrop Primates 2003;11(1): 1–10.

[7] Garbino GST. How many marmoset (Primates: Cebidae: Callitrichinae) genera are there? A phylogenetic analysis based on multiple morphological systems. Cladistics 2015;31:652–78.

[8] Buckner JC, Lynch Alfaro J, Rylands AB, Alfaro ME. Biogeography of the marmosets and tamarins (Callitrichidae). Mol Phylogenet Evol 2015;82:413–25.

[9] Ford SM. Callitrichids as phyletic dwarfs, and the place of the Callitrichidae in Platyrrhini. Primates 1980;21:31–43.

[10] Fleagle JG. Primate adaptation and evolution. 2nd ed. New York: Academic Press; 1998. 596 p.

[11] Garber PA. Vertical clinging, small body size, and the evolution of feeding adaptations in the Callitrichinae. Am J Phys Anthropol 1992;88:469–82.

[12] Caine NG. Cutting costs in response to predatory threat by Geoffroy's marmosets (*Callithrix geoffroyi*). Am J Primatol 1998; 46(3):187–96.

[13] Lacher TE, Fonseca GA, Da, Alves C, Magalhaes-Castro B. Parasitism of trees by marmosets in a central Brazilian gallery forest. Biotropica 1984;16:202–9.

[14] Araújo A. Influence des facteurs écologiques, comportementaux et démografhiques sur la dispersion de *Callithrix jacchus*. Paris: Université Paris-Nord (Paris XIII); 1996. p. 206 [thesis].

[15] Stevenson MF, Rylands AB. The marmosets, genus *Callithrix*. In: Mittermeier RA, Rylands AB, Coimbra-Filho A, Fonseca GAB, editors. Ecology and behavior of neotropical primates, vol. 2. Contagem: Littera Maciel; 1988. p. 131–222.

[16] Terborgh J. Five New World primates: a study in comparative ecology. Princeton: Princeton University Press; 1983. 280 p.

[17] Bezerra BM, Souto AS. Structure and usage of the vocal repertoire of *Callithrix jacchus*. Int J Primatol 2008;29:671–701.

[18] Hunt DM, Williams AJ, Bowmaker JK, Mollon JD. Structure and evolution of the polymorphic photopigment gene of the marmoset. Vision Res 1993;33:147–54.

[19] Pessoa DMA, Tomaz C, Pessoa VF. Color vision in marmosets and tamarins: behavioral evidence. Am J Primatol 2005;67: 487–95.

[20] Pessoa DMA, Maia R, Ajuz RCA, de Moraes PZPMR, Spyrides MH, Pessoa VF. The adaptive value of primate color vision for predator detection. Am J Primatol 2014;76:721–9.

[21] Stanford CB. Avoiding predators: expectations and evidence in primate antipredator behavior. Int J Primatol 2002;23:741–57.

[22] Melin AD, Fedigan ML, Hiramatsu C, Sendall CL, Kawamura S. Effects of colour vision phenotype on insect capture by a free-ranging population of white-faced capuchins, *Cebus capucinus*. Anim Behav 2007;73:205–14.

[23] Jacobs GH. New World monkeys and color. Int J Primatol 2007; 28:729–59.

[24] Norcross JL, Newman JD. Context and gender specific differences in the acoustic structure of common marmosets (*Callithrix jacchus*). Am J Primatol 1993;30:37–54.

[25] Moreira LAA, Pessoa DMA, Sousa MBC. Socio-sexual communication: a review of the sensory modalities used by non-human primates. Oecol Aust 2013;17:303–19.

[26] Dixson AF, George L. Prolactin and parental behavior in a male New-World primate. Nature 1982;299(5883):551–3.

[27] Mota MT, Sousa MBC. Prolactin levels of fathers and helpers related to alloparental care in common marmosets, (*Callithrix jacchus*). Folia Primatol 2000;71:22–6.

[28] Mota MTS, Franci CR, Sousa MBC. Hormonal changes related to paternal and alloparental care in common marmosets (*Callithrix jacchus*). Horm Behav 2006;49:293–302.

[29] Silva HPA, Sousa MBC. The pair-bond formation and its role in the stimulation of reproductive function in female common marmosets (*Callithrix jacchus*). Int J Primatol 1997;18:387–400.

[30] Lazaro-Perea C, Castro CSS, Harrison R, Araujo A, Arruda MF, Snowdon CT. Behavioral and demographic changes following the loss of the breeding female in cooperatively breeding marmosets. Behav Ecol Sociobiol 2000;48(2):137–46.

[31] Díaz-Muñoz SL, Bales KL. "Monogamy" in primates: variability, trends, and synthesis: introduction to special issue on primate monogamy. Am J Primatol 2016;78(3):283–7.

[32] Finkenwirth C, Martins E, Deschner T, Burkart JM. Oxytocin is associated with infant-care behavior and motivation in cooperatively breeding marmoset monkeys. Horm Behav 2016;80:10–8.

[33] Snowdon CT, Soini P. The tamarins, genus Saguinus. In: Mittermeier RA, Rylands AB, Coimbra-Filho AF, da Fonseca GAB, editors. Ecology and behavior of neotropical primates, vol. 2. Washington: World Wildlife Fund; 1988. p. 223–98.

[34] Stevenson MF, Poole TB. An ethogram of the common marmoset (*Callithrix jacchus* jacchus): general behavioural repertoire. Anim Behav 1976;24:428–51.

[35] Moreira LAA, De Oliveira DGR, Sousa MBC, Pessoa DMA. Parturition signaling by visual cues in female marmosets (*Callithrix jacchus*). PLoS One 2015;10:e0129319.

[36] Barton RA. Visual specialization and brain evolution in primates. Proc R Soc Lond B Biol Sci 1998;265:1933–7.

[37] Barton RA. Olfactory evolution and behavioral ecology in primates. Am J Primatol 2006;68:545–58.

[38] Lazaro-Perea C, Snowdon CT, Arruda MF. Scent-marking behavior in wild groups of common marmosets (*Callithrix jacchus*). Behav Ecol Sociobiol 1999;46:313–24.

[39] Sousa MBC, Moura SLM, Menezes AAL. Circadian variation with a diurnal bimodal profile on scent-marking behavior in captive common marmosets (*Callitrhix jacchus*). Int J Primatol 2006;27:263–72.

[40] Smith TE, Abbott DH. Behavioral discrimination between circumgenital odor from peri-ovulatory dominant and anovulatory female common marmosets (*Callithrix jacchus*). Am J Primatol 1998;46:265–84.

[41] Smith T. Individual olfactory signatures in common marmosets (*Callithrix jacchus*). Am J Primatol 2006;68(6):585–604.

[42] Araújo A, Arruda MF, Alencar AI, Aubuquerque F, Nascimetno MC, Yamamoto ME. Body weight in wild and captive common marmosets (*Callithrix jacchus*). Int J Primatol 2000;21:317–24.

[43] Vivo M. Taxonomia de Callithrix Erxleben, 1777 (Callitrichidae, Primates). Belo Horizonte. Fundação Biodiversitas; 1991. 105 p.

[44] Abreu F, De La Fuente MFC, Schiel N, Souto A. Feeding ecology and behavioral adjustments: flexibility of a small neotropical primate (*Callithrix jacchus*) to survive in a semiarid environment. J Mammalian Ova Res 2016;1:1–9.

[45] De la Fuente MFC, Souto A, Sampaio MB, Schiel N. Behavioral adjustments by a small neotropical primate (*Callithrix jacchus*) in a semiarid caatinga environment. Sci World J 2014:1–8.

[46] Faulkes CG, Arruda MF, Monteiro Da Cruz MAO. Matrilineal genetic structure within and among populations of the cooperatively breeding common marmoset, *Callithrix jacchus*. Mol Ecol 2003;12:1101–8.

[47] Malukiewicz J, Boere V, Fuzessy LF, Grativol AD, French JA, Silva IO, Pereira LCM, Ruiz-Miranda CR, Valença YM, Stone AC. Hybridization effects and genetic diversity of the common and black-tufted marmoset (*Callithrix jacchus* and *C. penicillata*) mitochondrial control region. Am J Phys Anthropol 2014;155:522–36.

[48] Malukiewicz J, Boere V, Fuzessy LF, Grativol AD, de Oliveira e Silva I, Pereira LCM, et al. Natural and anthropogenic hybridization in two species of Eastern Brazilian marmosets (*Callithrix jacchus* and *C. penicillata*). PLoS One 2015;10(6):e0127268.

[49] Ruiz-Miranda CR, Affonso AG, Martins A, Beck B. Distribuição do sagui (*Callithrix jacchus*) nas áreas de ocorrência do mico-leão-dourado (*Leontopithecus rosalia*) no estado do Rio de Janeiro. Neotrop Primates 2000;8:98–101.

[50] Ruiz-Miranda CR, Morais Júnior MM, De Paula VR, Grativol AD, Rambaldi DM. O problema dos saguis introduzidos no Rio de Janeiro: vítimas e vilões. Ciência Hoje 2011;48: 44–9.

[51] Ruiz-Miranda C, Affonso A, de Morais M, Verona C, Martins A, Beck B. Behavioral and ecological interactions between reintroduced golden lion tamarins (*Leontopithecus rosalia* Linnaeus, 1766) and introduced marmosets (*Callithrix* spp, linnaeus, 1758) in Brazil's Atlantic coast forest frag- ments. Braz Arch Biol Technol 2006;49:99–109.

[52] Sales IS, Ruiz-Miranda CR, Santos CP. Helminths found in marmosets (*Callithrix penicillata* and *Callithrix jacchus*) introduced to the region of occurrence of golden lion tamarins (*Leontopithecus rosalia*) in Brazil. Vet Parasitol 2010;171:123–9.

[53] Tavela AO, Fuzessy LF, Silva VHD, Da Silva FFR, Carretta Junior M, Silva IO, Souza VB. Helminths of wild hybrid marmosets (*Callithrix* sp.) living in an environment with high human activity. Rev Bras Parasitol Vet 2013;22.
[54] Schiel N, Souto A. The common marmoset: an overview of its natural history, ecology and behavior. Dev Neurobiol 2016;1: 1–59.
[55] Castro CSS, Menezes AAL, Moreira LFS. Locomotor activity rhythm in free-ranging common marmosets (*Callithrix jacchus*). Biol Rhythm Res 2003;34(1):23–30.
[56] Garber PA. Feeding ecology and behaviour of the genus Saguinus. In: Rylands AB, editor. Marmosets and tamarins: systematic, behaviour, and ecology. Oxford University Press; 1993. p. 273–95.
[57] Rylands AB, Coimbra- Filho AF, Mittermeier RA. Systematics, geographic distribution, and some notes on the conservation status of the Callitrichidae. In: Rylands AB, editor. Marmosets and tamarins: systematics, behavior and ecology. Oxford (USA): Oxford University Press; 1993. p. 11–77.
[58] Rylands AB, Faria DS. Habitats, feeding ecology, and home range size in the genus *Callithrix*. In: Rylands AB, editor. Marmosets and tamarins: systematics, behavior and ecology. Oxford (USA): Oxford University Press; 1993. p. 262–72.
[59] Sussman RW, Kinzey WG. The ecological role of the Callitrichidae: a review. Am J Phys Anthropol 1984;64(4):419–49.
[60] Amora TD, Beltrão-Mendes R, Ferrari SF. Use of alternative plant resources by common marmosets (*Callithrix jacchus*) in the semiarid caatinga scrub forests of Northeastern Brazil. Am J Primatol 2013;75:333–41.
[61] Mantilla GR. Ingestão alimentar e nutricional de Calllithrix jacchus: relação com hierarquia social e disponibilidade de alimento. Brazil: UFRN; 2012. p. 63 [Masters thesis].
[62] Thompson CL, Robl NJ, Melo LCdO, Valença-Montenegro MM, Valle YBM, Oliveira MAB d, Vinyard CJ. Spatial distribution and exploitation of trees gouged by common marmosets. Int J Primatol 2013;34:65–85.
[63] Garber PA. Proposed nutritional importance of plant exudates in the diet of the Panamanian tamarin, *Saguinus oedipus* geoffroyi. Int J Primatol 1984;5(1):1–15.
[64] Garber PA. Use of habitat and positional behavior in a Neotropical primate, *Saguinus oedipus*. In: Rodman PS, Cant JGH, editors. Adaptations for foraging in nonhuman primates. New York: Columbia University Press; 1984. p. 112–33.
[65] Francisco TM, Couto DR, Zanuncio JC, Serrão JE, Silva IO, Boere V. Vegetable exudates as food for *Callithrix* spp. (Callitrichidae): exploratory patterns. PLoS One 2014;9:1–7.
[66] Alonso C, Langguth A. Ecologia e comportamento de *Callithrix jacchus* (Primates: Callitrichidae) numa ilha de floresta atlântica. Rev Nord Biol 1989;6:105–37.
[67] Rylands AB. Sympatric Brazilian callitrichids: the black tuffed-ear marmoset, *Callithrix kuhli*, and the golden-headed lion tamarin, *Leontopithecus chrysomelas*. J Hum Evol 1989;18:679–95.
[68] Thompson CL, Valença-Montenegro MM, Melo LDO, Valle YM, d Oliveira MAB, Lucas PW, Vinyard CJ. Accessing foods can exert multiple distinct, and potentially competing, selective pressures on feeding in common marmoset monkeys. J Zool 2014; 294(3):161–9.
[69] Faulkes C, Arruda MF, Monteiro da Cruz MAO. Genetic structure within and among populations of the common marmoset, *Callithrix jacchus*: implications for cooperative breeding. In: Ford SM, Porter LM, Davis LC, editors. The smallest anthropoids: the Marmoset/Callimico radiation. New York: Springer Press; 2009. p. 103–17.
[70] Arruda MF, Araújo A, Sousa MBC, Albuquerque FS, Albuquerque ACSR, Yamamoto ME. Two breeding females within free-living groups may not always indicate polygyny: alternative subordinate female strategies in common marmosets (*Callithrix jacchus*). Folia Primatol 2005;76(1):10–20.
[71] Digby LJ. Sexual behavior and extragroup copulations in a wild population of common marmosets (*Callithrix jacchus*). Folia Primatol 1999;70(3):136–45.
[72] Albuquerque FS. Cuidado cooperativo à prole: dinâmica em ambiente natural. Brazil: Universidade de São Paulo; 1999. p. 202 [Ph.D. dissertation].
[73] Oliveira Terceiro FE. Cuidado parental diferencial em *Callithrix jacchus* em ambiente natural. Brazil: UFRN; 2016. p. 64 [Masters thesis].
[74] Yamamoto ME, Arruda MF, Alencar AI, Sousa MBC, Araújo A. Mating systems and female-female competition in the common marmosets, *Callithrix jacchus*. In: Ford SM, Porter LM, Davis LC, editors. The smallest anthropoids: the marmoset/callimico radiation. New York: Springer Press; 2009. p. 119–33.
[75] Yamamoto ME, Araújo A, Sousa MBC, Arruda MF. Social organization in *Callithrix jacchus*: cooperation and competition. In: Macedo R, editor. Advances in the study of behavior, vol. 42. Academic Press; 2010. p. 259–73.
[76] Yamamoto ME, Araujo A, Arruda MF, Lima AKM, Siqueira JO, Hattori WT. Male and female breeding strategies in a cooperative primate. Behav Process 2014;109(Pt A):27–33.
[77] Deleted in Review.
[78] Garber PA, Porter LM, Spross J, Di Fiore A. Tamarins: insights into monogamous and non-monogamous single female social and breeding systems. Am J Primatol 2016;78(3):298–314.
[79] Nievergelt CM, Digby LJ, Ramiakrishnan U, Woodruff DS. Genetic analysis of group composition and breeding system in a wild common marmoset (*Calltrhix jacchus*) population. Int J Primatol 2000;21:1–20.
[80] Kleiman DG. Monogamy in mammals. Q Rev Biol 1977;52(1): 39–69.
[81] Sussman RW, Garber PA. A new interpretation of the social organization and mating system of the Callitrichidae. Int J Primatol 1987;8:73–92.
[82] Tardif SD, Smucny DA, Abbott DH, Mansfield K, Schultz-Darken N, Yamamoto ME. Reproduction in captive common marmosets (*Callithrix jacchus*). Comp Med 2003;53:364–8.
[83] Saltzman W, Schultz-Darken NJ, Severin JM, Abbott DH. Escape from social suppression of sexual behavior and of ovulation in female common marmosets. Ann NY Acad Sci 1997;807: 567–70.
[84] Saltzman W, Severin JM, Schultz-Darken NJ, Abbott DH. Behavioral and social correlates of escape from suppression of ovulation in female common marmosets housed with the natal family. Am J Primatol 1997;41:1–21.
[85] Dietz JM, Baker AJ. Polygyny and female reproductive success in golden lion tamarins, *Leontopithecus rosalia*. Anim Behav 1993;46: 1067–78.
[86] Tardif SD, Araujo A, Arruda MF, French JA, Sousa MBC, Yamamoto ME. Reproduction and aging in marmosets and tamarins. In: Atsalis S, Margulis SW, Hof PR, editors. Primate reproductive aging. Cross-taxon perpectives, vol. 36. Basel: Karger Publishers; 2008. p. 29–48.
[87] Monteiro da Cruz MAO. Dinâmica reprodutiva em uma população de Sagui-do-Nordeste (*Callithrix jacchus*) na Estação Ecológica do Tapacurá, Pernambuco. Brazil: UFPE; 1988 [Ph.D dissertation].
[88] Sousa MBC, Albuquerque ACSR, Yamamoto ME, Araújo A, Arruda MF. Emigration as a reproductive strategy of the common marmoset (*Callithrix jacchus*). In: Ford SM, Porter LM, Davis LC, editors. The smallest anthropoids. Springer US; 2009. p. 167–82.
[89] Digby L. Infant care, infanticide, and female reproductive strategies in polygynous groups of common marmosets (*Callithrix jacchus*). Behav Ecol Sociobiol 1995;37(1):51–61.

[90] Díaz-Munoz SL. Complex cooperative breeders: using infant care costs to explain variability in callitrichine social and reproductive behavior. Am J Primatol 2016;78:372–87.
[91] Alencar AI, Sousa MBC, Abbott DH, Yamamoto ME. Contested dominance modifies the anovulatory consequences of social subordination in female marmosets. Braz J Med Biol Res 2006;39: 647–58.
[92] Burkart JM. Opposite effects of male and female helpers on social tolerance and proactive prosociality in callitrichid family groups. Sci Rep 2015;5:9622.
[93] Saltzman W, Digby LJ, Abbott DH. Reproductive skew in female common marmosets: what can proximate mechanisms tell us about ultimate causes? Proc R Soc B 2009;276:389–99.
[94] Yamamoto ME, Lopes FA. Effect of removal from the family group on feeding behavior by captive *Callithrix jacchus*. Int J Primatol 2004;25(2):489–500.
[95] Abbott DH, McNeilly AS, Lun SF, Hulme MJ, Burden FJ. Inhibition of ovarian function in subordinate female marmoset monkeys (*Callithrix jacchus* jacchus). J Reprod and fertil 1981;63: 335–45.
[96] Lazaro-Perea C, Arruda MF, Snowdon CT. Grooming as a reward? Social function of grooming between females in cooperatively breeding marmosets. Anim Behav 2004;67:627–36.
[97] Mota MTS, Sousa MBC, Campos MF. The effect of parturition on the interactions within pairs of common marmosets (*Callithrix jacchus*). Braz J Med Biol Res 1995;28:108–12.
[98] Snowdon CT. Infant care in cooperatively breeding species. In: Rosenblatt JS, Snowdon CT, editors. Parental care: Evolution, mechanisms and adaptive significance. San Diego (CA): Academic Press; 1996. p. 643–89.
[99] Yamamoto ME, Box HO. The role of non-reproductive helpers in infant care in captive *Callithrix jacchus*. Ethology 1997;103(9): 760–71.
[100] Albuquerque FS. Distribuição do cuidado à prole em grupos de *Callithrix jacchus* (Callitrichidae: primates), no ambiente natural. Brazil: UFRN; 1994. p. 75 [Masters thesis].
[101] Emlen ST. Predicting family dynamics in social vertebrates. In: Krebs JR, Davies NB, editors. Behavioural ecology: an evolutionary approach. 4th ed. Oxford: Blackwell; 1997. p. 228–53.
[102] Goldizen A. Tamarins and marmosets: communal care of offspring. In: Smuts BB, Cheney DL, Seyfarth RM, Wrangham RW, Struhsaker TT, editors. Primate societies. Chicago: University Chicago Press; 1987. p. 34–43.
[103] Price EC. The benefits of helpers: effects of group and litter size on infant care in tamarins (*Saguinus oedipus*). Am J Primatol 1992; 26:179–90.
[104] Sanchez S, Pelaez F, Gil-Burmann C, Kaumanns W. Costs of infant carrying in the cotton-top tamarin (*Saguinus oedipus*). Am J Primatol 1999;48:99–111.
[105] Emlen ST. Evolution of cooperative breeding in birds and mammals. In: Krebs JR, Davies NB, editors. Behavioral ecology. Oxford: Blackwell; 1991. p. 301–37.
[106] Baker AJ, Dietz JM, Kleiman DG. Behavioural evidence for monopolization of paternity in multi-male groups of golden lion tamarins. Anim Behav 1993;46:1091–103.
[107] Digby LJ, Ferrari SF. Multiple breeding in free-ranging groups of *Callithrix jacchus*. Int J Primatol 1994;3:389–97.
[108] Epple J. Reproductive and social behavior of marmosets with special references to captive breeding. In: Gengozian N, Deinhardt FW, editors. Marmosets in experimental medicine. Basel: S. Karger; 1978. p. 50–62 [Chapters 7, 10].
[109] Garber PA, Moya L, Malaga C. A preliminary field study of the moustached tamarin monkey (*Saguinus mystax*) in northeastern Peru: questions concerned with the evolution of a communal breeding system. Folia Primatol 1984;42(1):17–32.
[110] Tardif SD, Richter CB, Carson RL. Effects of sibling rearing experience on future reproductive success in two species of Callitrichidae. Am J Primatol 1984;6:377–80.
[111] Yamamoto ME. Infant care in callitrichids: cooperation and competition. Ann Rev Biomed Sci 2005;7:149–60.
[112] Tardif S. The bioenergetics of parental behavior and the evolution of alloparental care in marmosets and tamarins. In: Solomon N, French J, editors. Cooperative breeding in mammals. Cambridge: Cambridge University Press; 1997. p. 11–33.
[113] Pryce CR. A comparative systems model of the regulation of maternal motivation in mammals. Anim Behav 1992;43:417–42.
[114] Tardif SD, Carson S, Gangaware BL. Infant-care behavior of nonreproductive helpers in a communal-care primate, the cotton-top tamarin (*Saguinus oedipus*). Ethology 1992;92:155–67.
[115] Ross CN, French JA, Orti G. Germ-line chimerism and paternal care in marmosets (*Callithrix kuhlii*). Proc Natl Acad Sci USA 2007;104:6278–82.
[116] Benirshke K, Brownhill LE. Heterosexual cells in testes of chimeric marmoset monkeys. Cytogenet Genome Res 1963;2:331–41.
[117] Sweeney CG, Curran E, Westmoreland SV, Mansfield KG, Vallender EJ. Quantitative molecular assessment of chimerism across tissues in marmosets and tamarins. BMC Genom 2012; 13(1):98.
[118] Maier W, Alonso C, Langguth A. Field Observations on *Callithrix jacchus jacchus* L. Z Saugetierkunde 1982;47:334–46.
[119] Hubrecht RC. Home range size and use and territorial behaviour in the common marmoset, *Callithrix jacchus jacchus*, at the Tapacura Field Station, Recife, Brazil. Int J Primatol 1985;6:533–50.
[120] Digby LJ. Social organization and reproductive strategies in a wild population of common marmosets (*Callithrix jacchus*). Davis: University of California; 1994 [thesis].
[121] Fonseca GAB, Lacher Jr TE. Exudate-feeding by *Callithrix jacchus penicillata* in semideciduous woodland (Cerradão) in central Brazil. Primates 1984;25(4):441–50.
[122] Faria DS. Tamanho, composição de um grupo social e área de vivência (home-range) do sagui *Callithrix jacchus penicillata* na mata ciliar do córrego Capetinga, Brasília, DF. In: de Mello MT, editor. A Primatologia no Brasil, vol. 2. Brasília: Sociedade Brasileira de Primatologia; 1986. p. 87–105.
[123] Faria DS. O estudo de campo com o mico-estrela no planalto central brasileiro. In: Ades C, editor. Etologia de animais e de homens. Brasília: EDICON/EDUSP; 1989. p. 109–21.
[124] Rylands AB. Exudate-eating and tree-gouging by marmosets (Callitrichidae, Primates). In: Chadwick et AC, Sutton SL, editors. Tropical rain forest: The Leeds Symposium. Leeds: Leeds Philosophical and Literary Society; 1984. p. 155–68.
[125] Torres de Assumpção C. An ecological study of primates in southern Brazil, with a reappraisal of *Cebus apella* races. University of Edinburgh; 1983 [thesis].
[126] Muskin A. Preliminary field observations of *Callithrix aurita* (Callitrichinae, Cebidae). In: de Mello MT, editor. A Primatologia no Brasil, vol. 1. Brasília: Sociedade Brasileira de Primatologia; 1984a. p. 79–82.
[127] Muskin A. Field notes and geographic distribution of *Callithrix aurita* in eastern Brazil. Am J Primatol 1984b;7:377–80.
[128] Ferrari SF. The behaviour and ecology of the buffy-headed marmoset, Callithrix flaviceps (O. Thomas, 1903). London: University College; 1988 [thesis].

CHAPTER

2

The Anatomy of the Common Marmoset

Christophe Casteleyn[1,2], *Jaco Bakker*[3]

[1]Department of Morphology, Faculty of Veterinary Medicine, Ghent University, Merelbeke, Belgium; [2]Laboratory for Applied Veterinary Morphology, Department of Veterinary Sciences, Faculty of Pharmaceutical, Biomedical and Veterinary Sciences, University of Antwerp, Wilrijk, Belgium; [3]Animal Science Department, Biomedical Primate Research Centre, Rijswijk, The Netherlands

INTRODUCTION

The present work is largely based on the monograph entitled *The Anatomy of the Common Marmoset* (Hapale jacchus *Kuhl*) by Beattie [1]. However, this work particularly focuses on the myology and angiology of the common marmoset. As the osteology is rather poorly illustrated in this monograph, we used the work of Casteleyn et al. [2] as a basis for the paragraph on the marmoset's skeleton. Beattie's description of the internal organs of the marmoset was further substantiated by the dissection of a male and a female specimen. In addition, specific anatomical data were obtained by reviewing the literature. Because most veins follow the courses of their arterial counterparts, only the superficial veins that can be accessed for blood drawing or intravenous injection have been substantiated in the present work. Although the common marmoset shows some anatomical similarity with humans, veterinary nomenclature is used throughout the work. When analogous structures are differently named in veterinary and human anatomy, the human term is provided between square brackets.

EXTERNAL CHARACTERISTICS AND INTEGUMENT

Callitrichidae are the most diminutive of the living anthropoids [3]. The body length is 14–18 cm and the tail measures 25–32 cm in length [4]. The weight varies between 300 and 450 g [3].

The head is egg-shaped. The mouth opening is wide. The oval nasal openings are rostrolaterally oriented. The internarial distance is large. The short facial hairs are creamy. Three or four tactile hairs (*vibrissae*) can be observed on the supraorbital ridge. Two or three are implanted bilateral to the nares. Four large vibrissae lie near the corners of the mouth. Six very short vibrissae are seen on the lower lip [1]. The lips of the common marmoset contain large, acinous, ramified sebaceous glands. Their ducts open into the pilary canal of the hair follicles [5]. The remainder of the face is covered with black hairs. The large ears are characterized by a bunch of long black hairs tipped with white (Fig. 2.1). This characteristic has resulted in the fact that the common marmoset is also known as the cotton-eared marmoset. However, this tuft of white hairs is not yet present in the neonatal animal (Fig. 2.2).

A well-defined neck is not present. On cross-section, the thorax is heart-shaped with the apex pointing dorsally. The abdomen is shaped like a cone with the apex being directed toward the tail [1]. The dorsal side of the body is covered with white-tipped black hairs that are golden brown in the middle. Shorter brown hairs are distributed along the ventral neck, thorax, and abdomen. In particular, the medial surfaces of the upper arm and thigh, the axilla, and the flexor surface of the forearm are covered with short brown hairs. The perineum, the skin over the sternum, and the ventral surface of the tail are devoid of hairs. A pair of pectoral nipples is present in the female. The tip of the tail presents a bunch of white-tipped black hairs [1] (Fig. 2.3).

The forelimbs are relatively short. The dorsal side of the hand and fingers is covered with fine brownish-black hairs. A carpal pad (*torus carpeus*) that contains four small vibrissae can be found on the ulnar side of the forearm [1]. The skin of the palms and soles of the marmoset presents well-developed epidermal ridges. The epidermis is composed of a thick *stratum corneum*.

The Common Marmoset in Captivity and Biomedical Research
https://doi.org/10.1016/B978-0-12-811829-0.00002-9

FIGURE 2.1 Facial characteristics of the adult common marmoset. The bunch of white-tipped hairs on the ears is very typical. Notice the infant behind the right adult.

FIGURE 2.3 External appearance of the adult common marmoset, showing in particular the coat pattern. Notice that this marmoset has just been rewarded with a marshmallow.

The *stratum granulosum* contains one to three layers of cells that contain keratohyalin granules. The basal layer harbors many melanin-containing cells [6]. Palmar pads (*torus metacarpeus*) can be observed. Terminal or digital pads (*torus digitalis*) are present on the terminal phalanx of the digits (Fig. 2.4). Each digit of the hand has a sharp, curved claw (*tegula* or *falcula*) [7].

The hind limbs are long and equal the crown–rump length. The dorsal side of the foot is covered similar to the hand. Plantar pads (*torus metatarsus*) and digital pads are present [1] (Fig. 2.5). Each digit of the foot has a sharp, curved claw, except the hallux that bears a flattened nail (*ungula* or *unguis*) [7] (Fig. 2.6). The fact that claws persisted in the common marmoset could be related to the poor manipulative ability of the hand and foot. The marmoset nail on the hallux is probably advantageous in grasping, which plays an important role in the tree gouging and exudate feeding in *Callitrichidae* [3,7].

The palms and soles of the common marmoset contain simple tubular eccrine sweat glands. These are more numerous on the digits [8].

FIGURE 2.2 External appearance of the infant common marmoset. Notice the absence of the bunch of white-tipped hairs on the ears.

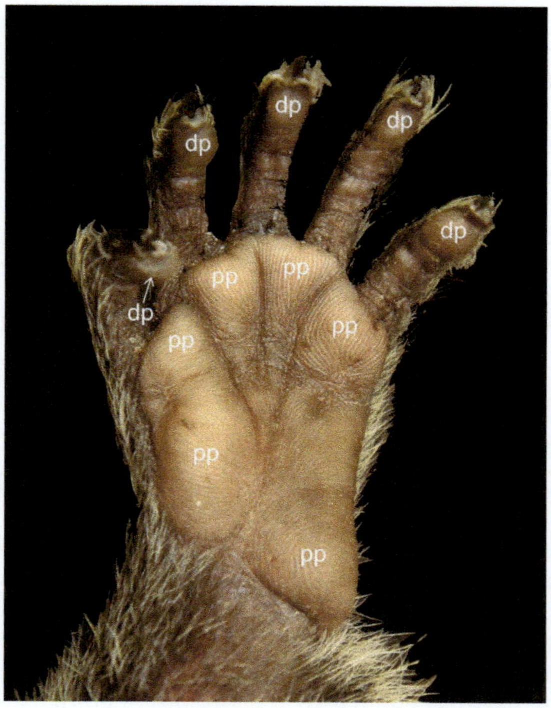

FIGURE 2.4 Palm of the left hand of the common marmoset showing the epidermal ridges, the six palmar pads (pp), and the five digital pads (dp), one on each finger.

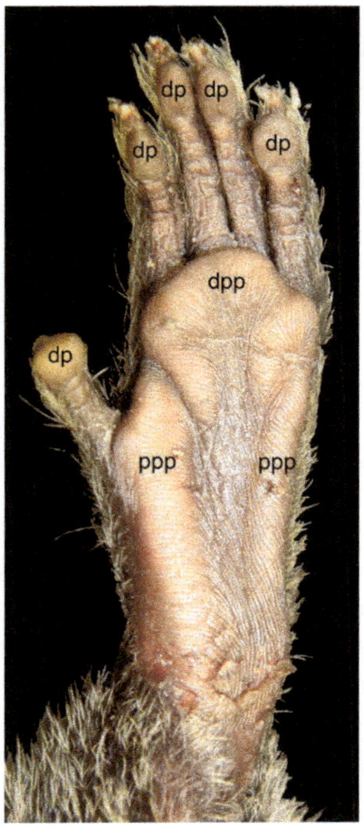

FIGURE 2.5 Sole of the left foot of the common marmoset showing the epidermal ridges, the proximal plantar pads (ppp), the distal plantar pad (dpp), and the five digital pads (dp), one on each toe. Notice the position of the hallux.

Subcutaneous injections can be administered in the interscapular region or at the level of the abdomen. Identification microchips can be implanted at the base of the tail, but the interscapular region is also possible. Furthermore, this region is preferred for injection of anticonception hormone implants in nonbreeding females.

FIGURE 2.6 Medial view of the left foot of the common marmoset showing the presence of a flat nail on the hallux, which contrasts with the claws on the other toes.

LOCOMOTOR SYSTEM

Osteology

Skull

The skull of the adult marmoset measures approximately 45 mm in length and 29 mm in width. The *splanchnocranium* presents a reduced length. The very large conical orbits are almost completely postorbitally closed. The mandible, particularly its body, is relatively large. A prominent mandibular angle can be seen. The symphysis between the left and right mandibles is synostotic [2]. Some major anatomical landmarks of the marmoset skull are presented in Fig. 2.7. More anatomical details can be found in Ref. [1].

The hyoid bone of the marmoset is relatively simple. The basihyoid measures 0.36–0.46 cm in height, 0.18–0.26 cm in length, and 0.52–0.69 cm in width. This structure is considerably different from that in the baboon and rhesus monkey as it is quadrilateral in shape, with a greater horizontal than vertical dimension, and it presents no median air sac [9]. The other hyoid bones comprise the thyrohyoid and the ceratohyoid. The former is well developed, long, and slender. Its lateral surface is convex. The small ceratohyoid is slightly truncated in shape [10] (Fig. 2.8).

Vertebral Column

A mounted skeleton of the common marmoset is depicted in Fig. 2.9. The vertebral column consists of 7 cervical vertebrae, 12 or 13 thoracic vertebrae, 6 or 7

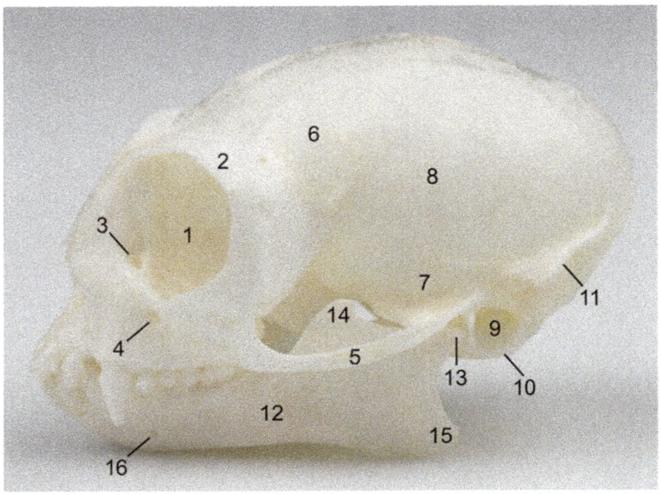

FIGURE 2.7 Left lateral view of the skull of an adult common marmoset with indication of the orbit (1), the superciliary ridge (2), the lacrimal canal (3), the infraorbital foramen (4), the zygomatic arch (5), the temporal line (6), the temporal fossa (7), the parietal bone (8), the external acoustic pore (9), the tympanic bulla (10), the nuchal crest (11), the mandible (12), the condylar process (13), the coronoid process (14), the angular process (15), and the mental foramen (16).

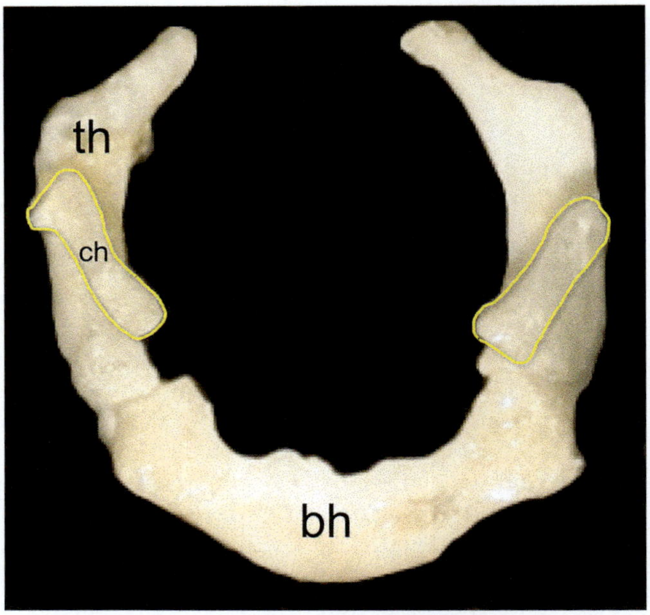

FIGURE 2.8 Craniodorsal view of the hyoid bone of the common marmoset. *bh*, basihyoid; *ch*, ceratohyoid; *th*, thyrohyoid.

lumbar vertebrae, 2 or 3 sacral vertebrae, and 26–29 caudal vertebrae. The sum of the thoracic and lumbar vertebrae always amounts to 19. The cervical, thoracic, lumbar, sacral, and caudal regions measure 23.18 ± 3.08 (SD) mm, 56.47 ± 4.34 mm, 55.81 ± 4.41 mm, 15.02 ± 2.15 mm, and 297.65 ± 50.82 mm in length, respectively [2].

The bodies of the cervical vertebrae are short. The atlas is characterized by small, rectangular wings and large, lateral vertebral foramina. The crest of the axis is very prominent. The spinous processes increase in length from the third to seventh cervical vertebra. On the last two cervical vertebrae, the tips of the spinous processes form heavy tuberosities. The sixth cervical vertebra can easily be recognized by its ventral lamina [2].

The bodies of the thoracic vertebrae elongate toward the lumbar region, whereas the spinous processes shorten and become broader. These processes point in a caudal direction up to the eighth or ninth thoracic vertebra. The spinous processes of the last three or four thoracic vertebrae, and those of the lumbar vertebrae, point in a cranial direction. As a result, the anticlinal vertebra is T9 or T10 [2].

Although the number of lumbar vertebrae is only half that of the thoracic vertebrae, the lengths of the thoracic and lumbar regions are similar. The very long lumbar vertebrae possess prominent spinous and cranioventrally inclined transverse processes that become larger toward the sacrum. The last lumbar vertebra is always shorter than the other lumbar vertebrae. Only a narrow lumbosacral interarcual space is present between the most caudal lumbar vertebra and the sacrum [2].

In most cases the sacrum is composed of three fused vertebrae. However, Casteleyn et al. [2] has described anatomical variations. In one animal, only two sacral vertebrae were observed. In another marmoset, a transitional vertebra was located between the last lumbar vertebra and the sacrum and was composed of two fused sacral vertebrae. A transitional vertebra between the second sacral vertebra and the tail was seen once.

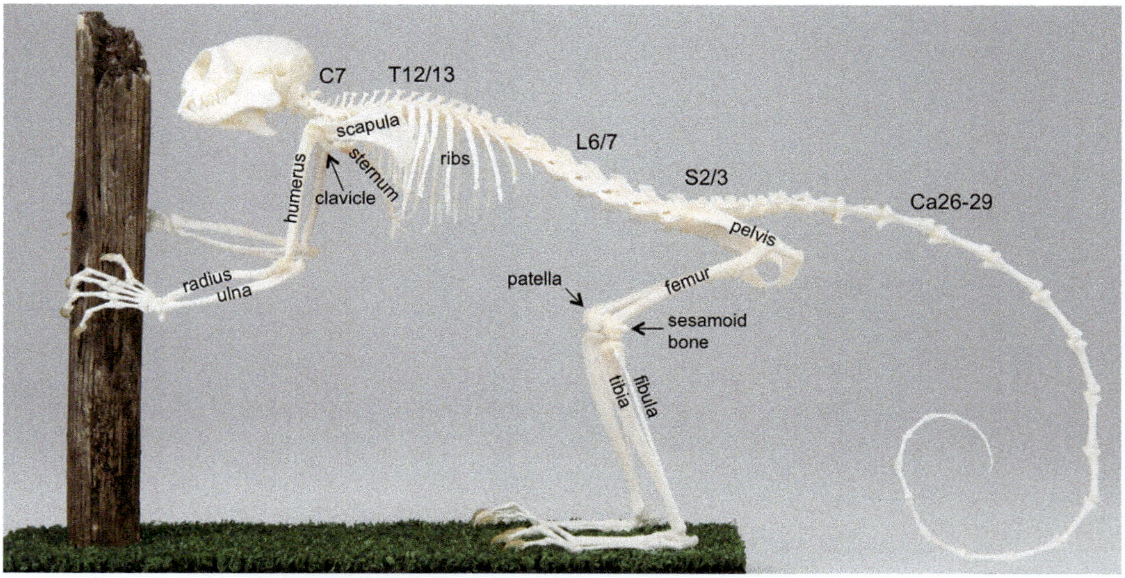

FIGURE 2.9 Left lateral view of the skeleton of the common marmoset. The vertebral formula, i.e., the number of vertebrae in each region, is indicated (*C*, cervical vertebrae; *Ca*, caudal vertebrae; *L*, lumbar vertebrae; *S*, sacral vertebrae; *T*, thoracic vertebrae).

Although the bodies of the sacral vertebrae are fused, the transverse processes of the first can be recognized individually. These processes, together with the cranial third of the sacral wings that are formed by the transverse processes of the subsequent sacral vertebrae, are connected to the auricular surface of the ilium [2].

The caudal vertebrae elongate toward the middle of the tail and then start to shorten again. Closed neural arches are present on the first three to four vertebrae, whereas closed hemal arches can be recognized from the fourth to eighth caudal vertebra [2].

The number of ribs depends on the number of thoracic vertebrae and varies between 12 and 13 pairs. However, having 13 thoracic vertebrae is predominant because only one of the seven female marmosets and one of the three male marmosets that were investigated by Castelyn [2] had 12 thoracic vertebrae. The seventh or eighth pair of ribs, depending on the number of thoracic vertebrae, is the last sternal pair that is directly attached to the sternum by costal cartilage (*costae verae*). The caudal pairs are asternal ribs that have indirect connections with the sternum, as their costal cartilages are attached to that of the previous rib (*costae spuriae*). The last pair of ribs is very short and floating (*costae fluctuantes*), thus lacking any connection with the sternum [2].

The sternum is composed of a broad manubrium, four or five cuboidal sternebrae related to the number of thoracic vertebrae, and a slender xiphoid process. Articular surfaces for the clavicles are located bilaterally on the craniolateral surfaces of the manubrium. Caudal to these, the costal cartilages of the first pair of ribs attach. Subsequent sternebrae are connected to each other by means of intersternebral cartilages onto which the costal cartilages attach [2].

Penile Bone

A small penile bone or *baculum*, approximately 2.5 mm in length, is present in the tip of the penis. It has a slender cranial process, which extends to the glans of the penis, and a more voluminous caudal part [2] (Fig. 2.10).

Forelimb

The long bones of the forelimb are indicated in Fig. 2.9. These comprise the scapula, humerus, radius, and ulna that measure 28.77 ± 1.64 (SD) mm, 45.46 ± 1.13 mm, 40.21 ± 1.32 mm, and 47.08 ± 1.40 mm in length, respectively. The scapula is characterized by a prominent coracoid process and an acromion that has a connection with the sigmoidal clavicle. The humerus is slender and shows a discrete humeral crest and deltoid tuberosity. In contrast, the medial epicondyle is very well developed and protrudes far medially. Neither supracondylar nor supratrochlear foramen is present.

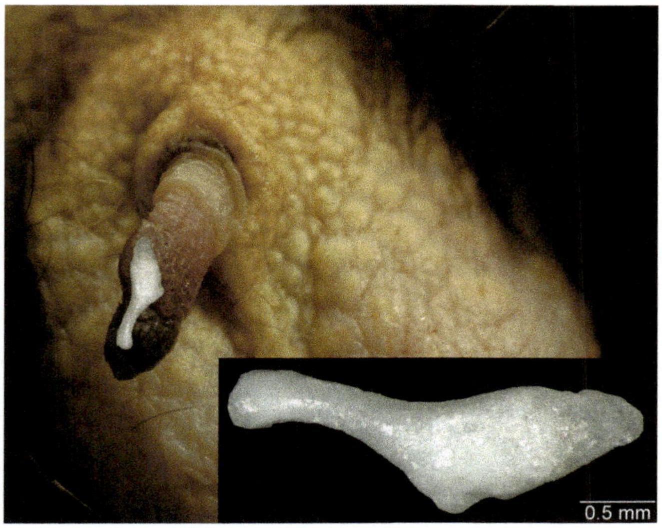

FIGURE 2.10 A small penile bone is present in the tip of the penis in the male common marmoset. The penis has been everted showing the glans penis. Notice the pearly white nodules on the skin of the scrotum and the lateral elevations caused by the presence of the testes.

The radius and ulna are separated by a wide interosseous space and are not fused.

The carpus contains 10 carpal bones, viz., 4 in the antebrachial and metacarpal rows, a central carpal bone that is inserted axially between both rows and a sesamoid bone located in the abductor muscle of the thumb (*os sesamoideum musculi abductoris pollicis longi*). Axial and abaxial ovoid sesamoid bones articulate with the palmar sides of the distal trochleae of each of the five metacarpal bones (*os sesamoideum proximale*). The pollex has only two phalanges, whereas the other four digits contain three phalanges. No distal sesamoid bones are present at the palmar sides of the distal interphalangeal joints [2] (Fig. 2.11).

Hind Limb

The hip bone (*os coxae*, i.e., the fused unilateral *os ilium*, *os ischii*, and *os pubis*) measures 39.92 ± 0.64 (SD) mm from the sacral to the ischial tuberosity. It contains a very large obturator foramen. The acetabulum is deep and contains a lunate articular face that is interrupted by an acetabular notch. Dorsal to the acetabulum, a discrete ischial spine is present. The pelvic symphysis remains syndesmotic in adult animals, and the angle between the left and right hip bones is approximately 70 degrees on craniocaudal view. Sexual dimorphism of the pelvis is absent [2].

The long bones of the hind limb comprise the femur, tibia, and fibula of which the lengths amount to 56.82 ± 2.26 (SD) mm, 60.36 ± 1.98 mm, and 57.94 ± 1.73 mm, respectively (Fig. 2.9). The femur is very slender and shows a well-delineated head, a distinct

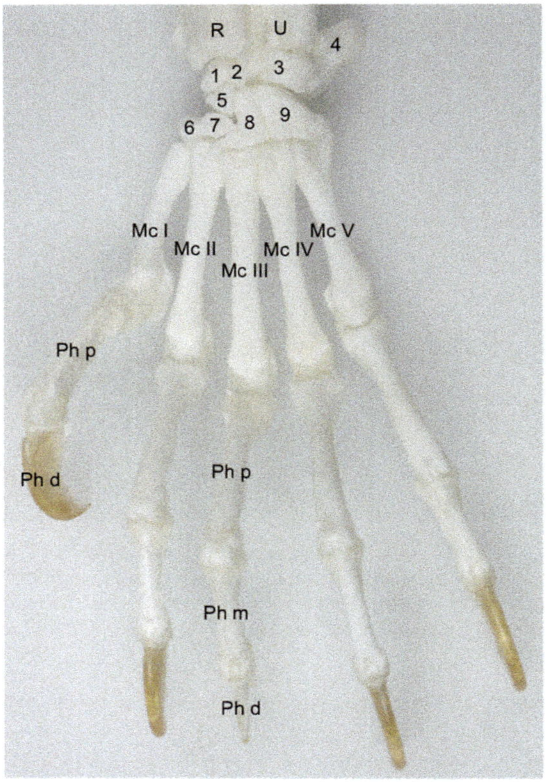

FIGURE 2.11 Dorsal view of the left hand of the skeleton of the common marmoset. The radius (R) and ulna (U) articulate with the carpus that is composed of the radial carpal bone (*os carpi radiale* or *os scaphoideum*) (1), the intermediate carpal bone (*os carpi intermedium* or *os lunatum*) (2), the ulnar carpal bone (*os carpi ulnare* or *os triquetrum*) (3), the accessory carpal bone (*os carpi accessorium* or *os pisiforme*) (4), the central carpal bone (*os carpi centrale*) (5), the first carpal bone (*os carpale primum* or *os trapezium*) (6), the second carpal bone (*os carpale secundum* or *os trapezoideum*) (7), the third carpal bone (*os carpale tertium* or *os capitatum*) (8), and the fourth carpal bone (*os carpale quartum* or *os hamatum*) (9). Distal to the carpus, five metacarpal bones (Mc I to Mc V) can be found. All lie at the base of a digit that consists of a proximal phalanx (*phalanx proximalis*) (Ph p), a middle phalanx (*phalanx media*) (Ph m), and a distal phalanx (*phalanx distalis*) (Ph d), with the exception of the thumb or pollex that lacks the middle phalanx.

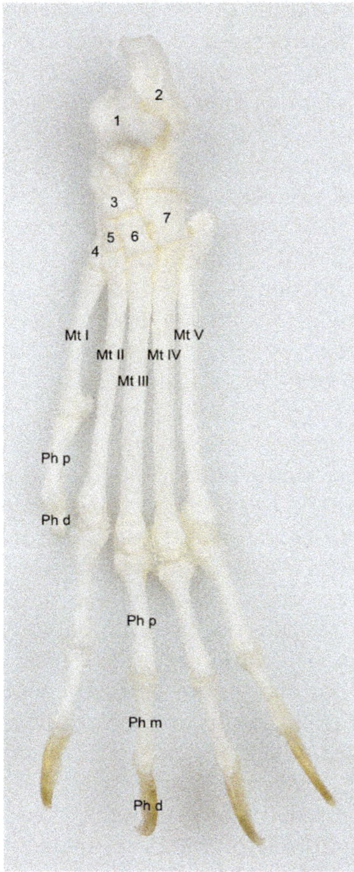

FIGURE 2.12 Dorsal view of the left foot of the skeleton of the common marmoset. The tarsus is composed of the talus (1), the calcaneus (2), the central tarsal bone (*os tarsi centrale* or *os naviculare*) (3), the first tarsal bone (*os tarsale primum* or *os cuneiforme mediale*) (4), the second tarsal bone (*os tarsale secundum* or *os cuneiforme intermedium*) (5), the third tarsal bone (*os tarsale tertium* or *os cuneiforme laterale*) (6), and the fourth tarsal bone (*os tarsale quartum* or *os cuboideum*) (7). Five metatarsal bones (Mt I to Mt V) support the digits that each consist of a proximal phalanx (*phalanx proximalis*) (Ph p), a middle phalanx (*phalanx media*) (Ph m), and a distal phalanx (*phalanx distalis*) (Ph d), except for the hallux that lacks the middle phalanx.

neck, and pronounced major and minor trochanters, with the intertrochanteric fossa in between. Neither the third trochanter nor any obvious gluteal tuberosity is present. The patella has an ovoid shape. A small, round to cuboidal sesamoid bone that is present in the tendon of the gastrocnemius muscle (*fabella* or *os sesamoideum musculi gastrocnemii*) articulates on the proximal surface of each femoral condyle. Like the radius and ulna, the tibia and fibula are separated by an interosseous space and are not fused.

The tarsus contains seven tarsal bones, i.e., the talus and calcaneus in the crural row, the central tarsal bone (forming a reduced middle row), and four bones of the metatarsal row that increase in size from medial to lateral. Five digits are present. The hallux contains two phalanges, whereas the other four digits have three phalanges. Axial and abaxial ovoid sesamoid bones are present at the plantar sides of the distal trochleae of the metatarsal bones. No distal sesamoid bones are present at the plantar sides of the distal interphalangeal joints [2] (Fig. 2.12).

Arthrology

The knee joint has been studied in detail in view of hindlimb malformations that have been encountered in a colony of common marmosets at a primate research center [11]. The small ovoid patella bears a single straight patellar ligament that inserts on the tibial tuberosity. The *fabellae* are embodied in the outer wall of the joint capsule [1]. The incongruence between the femoral

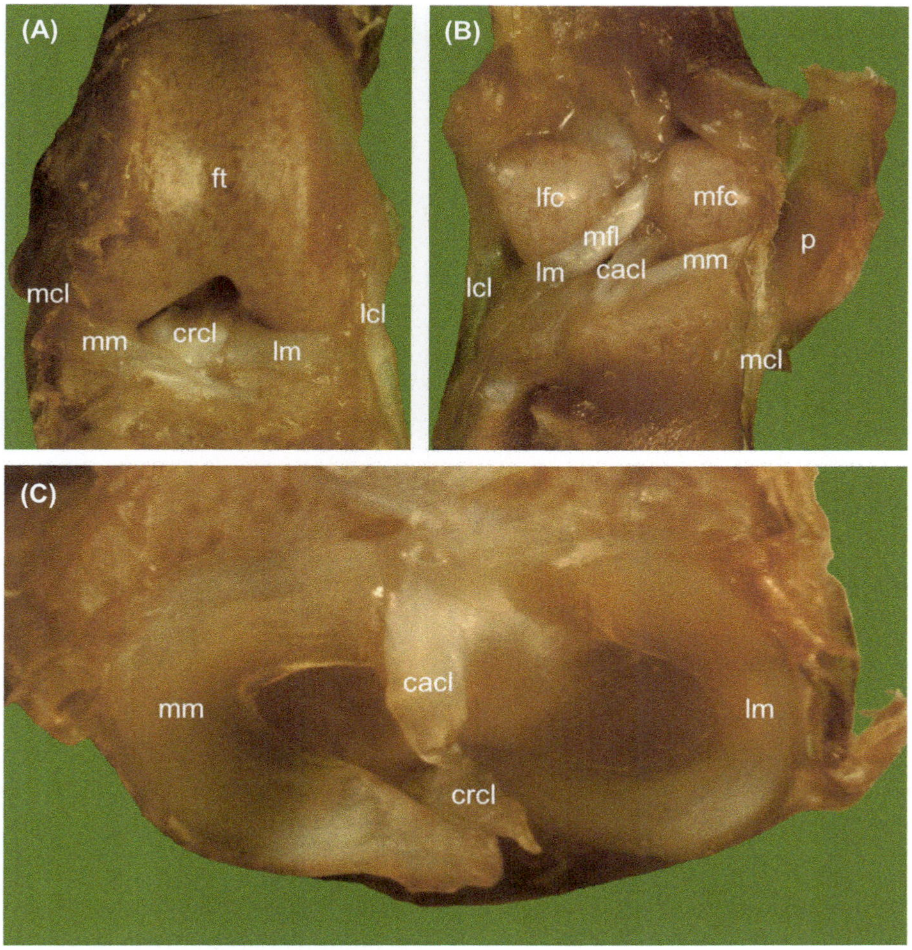

FIGURE 2.13 Arthrological preparation of the left knee showing a cranial view (A), a caudal view (B), and a proximal view (C) on the tibial plateau with the menisci. *cacl*, caudal cruciate ligament; *crcl*, cranial cruciate ligament; *ft*, femoral trochlea; *lcl*, lateral collateral ligament; *lfc*, lateral femoral condyle; *lm*, lateral meniscus; *mcl*, medial collateral ligament; *mfc*, medial femoral condyle; *mfl*, meniscofemoral ligament; *mm*, medial meniscus; *p*, patella.

condyles and the tibial plateau is eliminated by the presence of menisci. The lateral meniscus is C-shaped, whereas the medial meniscus is more crescentic [1]. Both menisci are cranially and caudally attached to the tibia by means of small meniscal ligaments. A cranial, or lateral, and a caudal, or medial, cruciate ligament can be observed between the femoral intercondylar fossa and the tibial intercondylar eminence. In addition, a meniscofemoral ligament or false cruciate ligament inserts on the caudal tip of the lateral meniscus (Fig. 2.13).

Myology

The superficial muscles that can be observed immediately after skinning the animal are illustrated in Figs. 2.14 and 2.15, which are a left lateral and a ventral view, respectively. Below, the musculature of the common marmoset is summarized per region by means of the indication of the origin and insertion of each muscle. Readers are referred to Ref. [1] for more details.

Facial Musculature

- *M. platysma*: This is a very thin superficial muscle that overlies the neck region. The most cranial fibers reach the base of the ear, the orbicularis oculi muscle, the angle of the mouth, and the lower lip.
- *M. sphincter colli profundus*: This muscle lies deep to the platysma muscle. It originates from the base of the ear pinna and runs to the midline of the neck.
- *M. occipitofrontalis*: This muscle originates from the skin and the adjacent bone of the supraorbital margin. It runs over the calvarium and is partly inserted into the nuchal line, whereas other fibers merge with the platysma muscle.

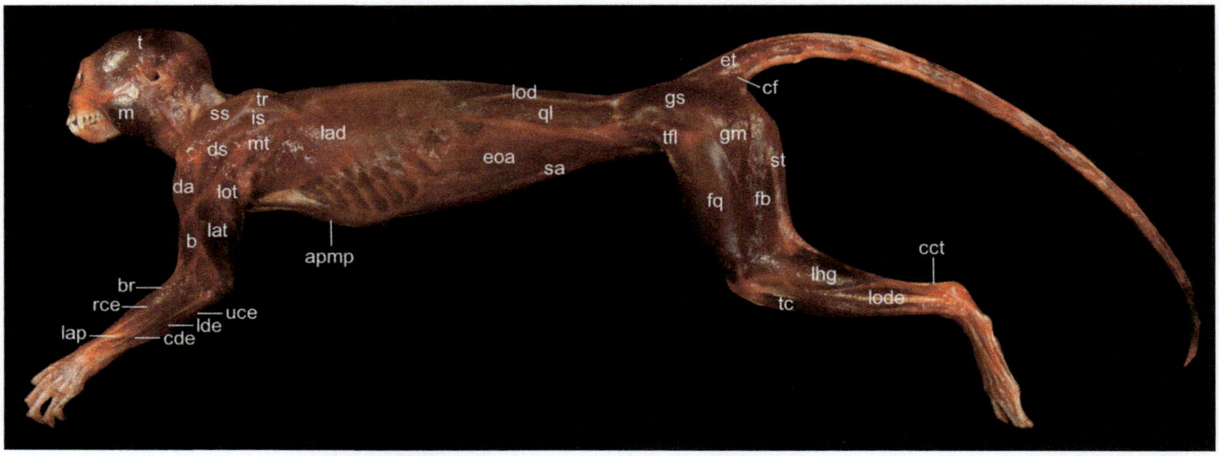

FIGURE 2.14 Left lateral view of the superficial musculature of the common marmoset with indication of the major muscles. *apmp*, abdominal part of the major pectoral muscle; *b*, biceps muscle; *br*, brachialis muscle; *cct*, common calcaneal tendon; *cde*, common digital extensor; *cf*, caudofemoralis muscle; *da*, acromial part of the deltoid muscle; *ds*, scapular part of the deltoid muscle; *eoa*, external oblique abdominal muscle; *et*, extensor muscles of the tail; *fb*, femoral biceps muscle; *fq*, femoral quadriceps muscle; *gm*, gluteus medius muscle; *gs*, gluteus superficialis muscle (*m. gluteus maximus*); *is*, infraspinous muscle; *lad*, latissimus dorsi muscle; *lap*, long abductor of the pollex; *lat*, lateral head of the triceps muscle; *lde*, lateral digital extensor; *lhg*, lateral head of the gastrocnemius muscle; *lod*, longissimus dorsi muscle; *lode*, long digital extensor; *lot*, long head of the triceps muscle; *m*, masseter muscle; *mt*, teres major muscle; *ql*, quadratus lumborum muscle; *rce*, radial carpal extensor; sa, straight abdominal muscle; *ss*, supraspinous muscle; *st*, semitendinosus muscle; *t*, temporal muscle; *tc*, tibialis cranialis muscle; *tfl*, tensor fasciae latae muscle; *tr*, trapezius muscle; *uce*, ulnar carpal extensor.

- *M. orbicularis oculi*: This is a very thin muscle that extends from the orbit to the free margins of the eyelids.
- *M. orbicularis oris*: This is a circular muscle whose fibers insert into the skin around the mouth and the lips.
- *M. auriculolabialis*: The auriculolabialis muscle runs from the base of the ear pinna and inserts into the orbicularis oris and oculi muscles.
- *M. quadratus labii superioris*: This muscle can be found in between the orbicularis oculi muscle and the dorsolateral aspect of the orbicularis oris muscle.

Muscles of Mastication

- *M. masseter*: The masseter originates from the zygomatic arch. It consists of a superficial and a deep part that both insert into the mandible.
- *M. buccinator*: The buccinator is a deep muscle that originates from the rostral part of the zygomatic arch and the maxilla. It is inserted into the mandible.
- *M. temporalis*: This muscle arises from the temporal region of the skull and the temporal line to converge on the coronoid process.
- *M. pterygoideus*: The internal part arises from the pterygoid fossa and inserts into the deep surface of the mandibular angle. The smaller external part originates from the pterygoid lamina and the orbital plate of the malar bone and inserts into the rostral edge of the mandibular condyloid process, the rostral part of the joint capsule, and the intraarticular fibrocartilaginous tissue.
- *M. digastricus*: The rostral and caudal bellies are separated by an intermediate tendon. The digastricus is located in between the mastoid region of the temporal bone and the mandibular symphysis. It is the only masticatory muscle that opens the mouth.

Muscles of the Neck and Tongue

- *M. sternocleidomastoideus*: The lateral portion is the *m. cleidooccipitalis* that arises from the clavicle and inserts into the nuchal line of the skull. The medial portion is the *m. sternomastoideus* that runs from the manubrium of the sternum to the mastoid process of the skull.
- *M. omohyoideus*: This muscle originates from the dorsal border of the scapula and inserts laterally into the hyoid bone. It runs medial to the sternocleidomastoid muscle. The common carotid artery and vagosympathetic trunk lie medial to the omohyoid muscle.
- *M. sternothyroideus*: The sternothyroid muscle lies medial to the former muscle, the common carotid artery, and the vagosympathetic trunk. It has the same origin but inserts laterally into the thyroid cartilage.
- *M. sternohyoideus*: This muscle arises from the manubrium and inserts medially into the hyoid bone. Therefore, this muscle lies in the midline of the neck

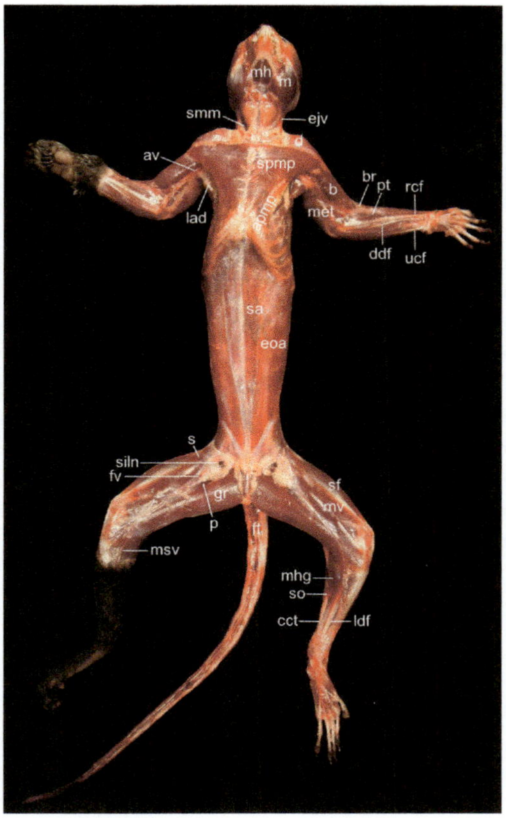

FIGURE 2.15 Ventral view of the superficial musculature of the common marmoset with indication of the major muscles. *apmp*, abdominal part of the major pectoral muscle; *av*, the axillary vein; *b*, biceps muscle; *br*, brachialis muscle; *cct*: common calcaneal tendon; *d*, deltoid muscle; *ddf*, deep digital flexor; *eju*, the external jugular vein; *eoa*, external oblique abdominal muscle; *ft*, flexor muscles of the tail; *fv*, the femoral vein; *gr*, gracilis muscle; *lad*, latissimus dorsi muscle; *ldf*, long digital flexor; *m*, masseter muscle; *met*, medial head of the triceps muscle; *mh*, mylohyoid muscle; *mhg*, medial head of the gastrocnemius muscle; *mv*, medial vastus muscle (head of the femoral quadriceps muscle); *p*, pectineus muscle; *pt*, pronator teres muscle; *rcf*, radial carpal flexor; *s*, sartorius muscle; *sa*, straight abdominal muscle; *sf*, straight femoral muscle (head of the femoral quadriceps muscle); *siln*, the superficial inguinal lymph node; *smm*, sternomastoid muscle; *so*, soleus muscle; *spmp*, sternal part of the major pectoral muscle; *ucf*, ulnar carpal flexor; *vsm*, the medial saphenous vein are also indicated.

adjacent to its contralateral counterpart. Both muscles cover the trachea ventrally.

- *M. thyrohyoideus*: This muscle has its origin on the thyroid cartilage and inserts into the hyoid bone.
- *M. mylohyoideus*: It presents a transverse course with respect to the lingual axis, as this muscle originates from the medial surface of the mandibular body and inserts into the median raphe of the tongue.
- *M. geniohyoideus*: This muscle originates from the mandible near the chin and travels toward the hyoid.
- *M. hyoglossus*: The hyoid is the origin of this muscle that inserts into the tongue.

Muscles of the Cervical Region

- *M. rectus capitis*: The major part arises from the crest of the axis and inserts into the occipital bone. The minor part originates from the dorsal tubercle of the atlas.
- *M. obliquus capitis*: The cranial part runs from the wing of the atlas to the occiput. The caudal part arises from the crest of the axis and inserts into the wing of the atlas.
- *M. trachelomastoideus*: This muscle arises from the second to fourth thoracic vertebrae and is inserted into the occipital bone and the mastoid process.
- *M. complexus*: This muscle arises from the transverse processes of the third to fifth thoracic vertebrae and is inserted into the occipital bone below the nuchal crest near the median plane.
- *M. splenius*: Its origin is from the nuchal ligament, the spinal processes, and interspinal ligaments of the last cervical and first two thoracic vertebrae. The cervical part inserts laterally into the atlas and axis. The cranial part inserts into the nuchal crest.
- *M. longus collis*: This prevertebral muscle can be divided into two parts. The thoracocervical part arises from the first two or three thoracic vertebrae and inserts into the cervical vertebrae. The cervical part originates from the third to fifth cervical vertebrae and inserts into the ventral tubercle of the atlas.
- *M. rectus capitis cranialis*: The major part of this prevertebral muscle arises from the ventral sides of the bodies of the fourth to sixth cervical vertebrae and is inserted into the basioccipital bone. The minor part also inserts into the basiocciput but originates from the atlas.
- *M. scalenus*: This muscle presents a smaller part that arises from the transverse processes of the fifth and sixth cervical vertebrae and inserts into the first rib. The larger part originates from the transverse processes of the second to sixth cervical vertebrae and inserts into the first four ribs.

Muscles of the Spine

The *m. longissimus dorsi* is inserted into the transverse processes of the lumbar and thoracic vertebrae, the articular processes of the thoracic vertebrae, and the tubercles of the 3rd to 12th ribs. The *mm. multifidi* can be recognized in the thoracic region, but the *m. semispinalis* is very weak.

Tail Musculature

The tail of the common marmoset is nonprehensile. The muscles found on the dorsal aspect of the caudal vertebrae are the *mm. interspinales caudae*, the *m. extensor caudae medialis*, the *m. extensor caudae lateralis*, the *m. abductor caudae medialis/internus*, and the *m. abductor caudae lateralis/externus*. The ventral muscles of the tail comprise the *m. flexor caudae brevis*, *m. flexor caudae longus*, and the *mm. intertransversarii caudae* [4] (Figs. 2.14 and 2.15).

Abdominal Muscles

- *M. obliquus abdominis externus*: This abdominal muscle arises from all the ribs and runs in a caudoventral direction to insert into the linea alba by means of an aponeurosis (most fibers) or the iliac crest (more caudal fibers).
- *M. obliquus abdominis internus*: The origin of this muscle is from the lumbodorsal fascia and the cranioventral spine of the ilium. The fibers run in a cranioventral direction to become tendinous (aponeurosis) at the level of the rectus abdominis muscle.
- *M. transversus abdominis*: This muscle arises from the lumbodorsal fascia, the costal cartilages, and the ribs (from the seventh to the last). The fibers run in a dorsoventral direction to insert into the linea alba.
- *M. rectus abdominis*: This muscle originates from the pubis and runs in a cranial direction along the ventral midline to be inserted into the costal cartilages. In contrast to man and domestic mammals, no tendinous intersections are present.
- *M. triangularis sterni*: This muscle arises from the sternum and the xiphoid process and is inserted into the second to seventh ribs.

Forelimb Musculature

Muscles that Connect the Forelimb to the Body

- *M. trapezius*: The cervical and thoracic parts are both triangular muscles that arise from the scapular spine. The former part is inserted into the nuchal ligament as far cranial as the third cervical vertebra. The latter part is inserted into the spines and interspinous ligaments of the 4th to 9th or 10th thoracic vertebrae.
- *M. rhomboideus*: The cervical and thoracic parts arise from the dorsal border of the scapula and insert into the occiput and nuchal ligament and the first three thoracic vertebrae, respectively.
- *M. levator scapulae cranialis*: This muscle runs from the acromion to the atlas.
- *M. levator anguli scapulae*: This muscle connects the cranial angle of the scapula with the transverse processes of the third to fifth cervical vertebrae.
- *M. serratus ventralis*: This muscle arises medially on the scapula and inserts into the first to eighth ribs.
- *M. latissimus dorsi*: This muscle originates from the caudal side of the humeral shaft and is inserted into the spinal processes and interspinal ligaments of the fourth thoracic to third lumbar vertebrae and the lumbodorsal fascia.
- Pectoral muscles:
  - *M. pectoralis superficialis* (*m. pectoralis major*): The sternal part (*pars sternalis*) arises from the sternum over its entire length and is inserted into the humerus. The abdominal part (*pars abdominalis*) arises from the linea alba and from the 9th and 10th costal cartilages to insert deep to the sternal part into the humerus.
  - *M. pectoralis profundus* (*m. pectoralis minor*): This muscle lies deep to the major pectoral muscle. It originates from the costal cartilages of the second to sixth ribs and is inserted into the greater tuberosity of the humerus.
  - *M. subclavius*: This small fusiform muscle arises from the first costal cartilage and is inserted into the clavicle.

Muscles of the Shoulder Region

- *M. supraspinatus*: This muscle arises from the supraspinous fossa and inserts into the greater tuberosity of the humerus.
- *M. infraspinatus*: In contrast to the former muscle, this muscle not only arises from the infraspinous fossa but is also inserted into the greater tuberosity of the humerus, immediately distal to the insertion of the supraspinatus muscle.
- *M. subscapularis*: The origin and insertion of this muscle are the subscapular fossa and the lesser tubercle of the humerus, respectively.
- *M. deltoideus*: The clavicular part (*pars clavicularis*), acromial part (*pars acromialis*), and scapular part (*pars scapularis*) arise from the lateral part of the clavicle, the acromion, and the clavicular spine, respectively. The insertion is at the deltoid tuberosity of the humerus.
- *M. teres minor*: The origin of this muscle is the ventral part of the caudal border of the scapula. The insertion is the greater tuberosity of the humerus, distal to the insertion of the infraspinatus muscle.
- *M. teres major*: This muscle originates at the ventral angle and caudal border of the scapula and inserts medially into the humeral shaft in its proximal third.

Muscles of the Upper Arm

- *M. triceps*: The triceps muscle inserts into the olecranon of the ulna. Its long head (*caput longum*), lateral head (*caput laterale*), and medial head (*caput mediale*) originate from the glenoid lip and the caudal border of the scapula, the greater tuberosity of the humerus, and the proximomedial side of the humeral shaft, respectively. The *m. anconeus* arises from the distal half of the humeral shaft. The *m. dorsoepitrochlearis* arises from the lower margin of the latissimus dorsi muscle.

- *M. biceps brachii*: The long head (*caput longum*) finds its origin at the supraglenoid tuberosity of the scapula, whereas the short head (*caput breve*) arises from the coracoid process of the scapula. The common insertion of both heads is the bicipital tubercle of the radius.
- *M. brachialis*: The origin of this muscle is the humeral shaft, immediately distal to the insertion of the deltoid muscle, and the crest of the lateral epicondyle of the humerus. The muscle fibers insert into the lateral coronoid process of the ulna.
- *M. coracobrachialis*: This muscle arises from the coracoid process of the scapula and inserts into the crest of the medial epicondyle of the humerus.

Muscles of the Forearm (Extensor Musculature [Lateral Side of the Arm])

- *M. extensor carpi radialis (longus et brevis)*: The lateral epicondylar crest of the humerus forms the origin of this muscle. The insertion is into the base of the first metacarpal bone (long part) and second metacarpal bone (short part).
- *M. extensor carpi ulnaris*: This muscle arises from the lateral epicondyle of the humerus and the olecranon and is inserted into the base of the fifth metacarpal bone.
- *M. extensor digitorum communis*: This muscle arises from the lateral epicondyle of the humerus and inserts by means of five tendons into the distal phalanges of each digit.
- *M. extensor digiti minimi* (can be considered as the *m. extensor digitorum lateralis*): This muscle arises immediately caudal to the origin of the common extensor of the fingers. Its tendon fuses with the tendon of the common digital extensor muscle to the fifth finger.
- *M. abductor pollicis longus*: The origin of the muscle is from the upper two-thirds of the shaft of the radius and ulna. The tendon inserts into the base of the first metacarpal bone.
- *M. extensor indicis*: This muscle arises from the upper two-thirds of the dorsal side of the ulnar shaft.

Muscles of the Forearm (Flexor Musculature [Medial Side of the Arm])

- *M. flexor carpi radialis*: This muscle also arises from the medial condyle of the humerus and inserts into the base of the second metacarpal bone.
- *M. palmaris longus*: The origin of this muscle is the medial humeral epicondyle. The muscle fibers attach to the palmar pads.
- *M. flexor carpi ulnaris*: The radial head (*caput radiale*) arises from the medial humeral condyle, whereas the ulnar head (*caput ulnare*) finds its origin on the olecranon and the proximal part of the ulnar shaft. The tendon attaches to the fifth metacarpal bone via the pisiform bone.
- *M. flexor sublimis digitorum*: This very thin muscle originates on the medial epicondyle of the humerus. Its four tendons insert on the base of the second phalanx of digits I to IV.
- *M. flexor digitorum profundus*: This muscle arises from the medial epicondyle of the humerus (*caput humerale*), the proximal half of the ulna (*caput ulnare*), and the upper two-thirds of the radius (*caput radiale*). Four tendons arise, which are inserted into the palmar sides of the terminal phalanges of digits I to IV.
- *M. flexor pollicis longus*: The medial epicondyle of the humerus and the radial shaft are the origins of this muscle. The insertion is into the base of the terminal phalanx of the pollex.

Muscles of the Forearm (the Pronators and Supinator)

- *M. pronator teres*: This muscle finds its origin on the medial epicondyle of the humerus and is inserted into the middle third of the shaft of the radius.
- *M. pronator quadratus*: The fibers of this muscle originate from the proximal radius and ulna and are inserted into the distal third of the radius.
- *M. supinator longus*: This muscle finds it origin at the lateral epicondylar crest of the humerus and is inserted into the base of the styloid process of the radius.

Muscles of the Hand

- *M. palmaris brevis*: It arises from the medial side of the transverse carpal ligament and is inserted into the fascia over the head of the metacarpal bone.
- *Mm. lumbricales*: These muscles, which are four in number, are very well developed. They arise from the medial side of the deep flexor tendons to digits III–V. They are inserted into the base of the proximal phalanx and the metacarpophalangeal joints.
- *Mm. interossei*: These muscles form pairs of muscles that are present in each interdigital cleft. They attach to the sides of the metacarpophalangeal joints.
- *M. abductor pollicis brevis*: This thin muscle arises from the first carpal bone and the transverse carpal ligament. It is inserted into the base of the first phalanx of the pollex.
- *M. abductor digiti minimi*: This abductor of the pollex arises from the transverse carpal ligament and the pisiform bone. It inserts into the base of the first phalanx of the thumb.

- *M. adductor pollicis*: This muscle can be found between the second metacarpal bone and the first phalanx of the pollex.
- *M. flexor brevis pollicis*: This very strong, short flexor arises from the first carpal bone and is inserted into the base of the first phalanx of the pollex.
- *M. flexor brevis minimi digiti*: This small muscle arises from the fourth carpal bone and is inserted into the base of the first phalanx of the thumb.

Musculature of the Hind Limb

Flexor Muscles of the Hip and Hip Joint

- *M. psoas major*: The psoas major muscle arises from the bodies of the second to sixth lumbar vertebrae. The muscle is inserted into the lesser trochanter of the femur.
- *M. psoas minor*: This psoas muscle lies ventral to the *M. psoas major.* It originates from the second and third lumbar vertebrae and is inserted cranially on the pubic bone.
- *M. iliacus*: The origin of this muscle is the medial aspect of the ilium. It runs lateral to the psoas major muscle into which it inserts in addition to the insertion into the lesser trochanter of the femur and its proximal shaft.
- *M. quadratus lumborum*: This muscle finds it origin on the crest and wing of the ilium. This thin quadrilateral muscle is inserted into the last rib and transverse processes of the lumbar vertebrae.

Extensor Muscles of the Hip Joint

- *M. gluteus superficialis* (*m. gluteus maximus*): The superficial head arises from the lumbar fascia, whereas the deep head finds its origin on the first caudal vertebrae. The tendon is inserted into the greater trochanter of the femur.
- *M. gluteus medius*: This part of the gluteus musculature arises from the lateral surface of the wing of the ilium, the sacroiliac joint, and the first caudal vertebra. The large muscle is inserted into the greater trochanter of the femur.
- *M. gluteus profundus (m. gluteus minimus)*: This muscle, which is variable in conformation, has its origin on the dorsal aspect of the ilium and inserts into the greater trochanter of the femur. The most ventral part of this muscle is known as the *m. scansorius.*

The Adductors of the Hind Limb

- *M. sartorius*: This thin muscle arises from the cranioventral spine of the ilium. It inserts into the medial side of the proximal half of the tibia.
- The adductor group of muscles, to which the *m. gracilis, m. pectineus, m. adductor magnus, m. adductor brevis*, and *m. adductor longus* belong, has its origin on the ventral side of the pelvis, near the symphysis. The insertion is into the medial side of the limb, more specifically the femur (*mm. adductores* and the pectineus muscle) or the tibia (*m. gracilis*)

The Supinators of the Hind Limb

- *M. obturatorius externus*: This muscle arises from the obturator membrane and the bone surrounding the obturator foramen, at its dorsal side. The tendon is inserted into the intertrochanteric fossa.
- *M. obturatorius internus*: This muscle also arises from the obturator membrane and the bone surrounding the obturator foramen, albeit at its ventral side. The insertion is at the medial side of the greater trochanter of the femur.
- *Mm. gemelli*: The gemelli originate from the ischium. Their tendons are inserted into the tendon of the *m. obturatorius internus.*
- *M. piriformis*: This muscle runs from the sacrum to the greater trochanter of the femur.
- *M. caudofemoralis*: This muscle runs from the first caudal vertebra to the caudal side of the femoral shaft, in its middle third.

The Extensors of the Knee

- *M. quadriceps*: Intramuscular injections can be administered in this muscle that consists of four parts [12].
  - *M. rectus femoris*: One part of this muscle arises from the caudoventral spine of the ilium. A second head arises just dorsal to the acetabulum. The muscle inserts into the patella.
  - *M. vastus lateralis*: This part of the quadriceps muscle arises from the greater trochanter of the femur and from the lateral surface of the shaft of the femur. The tendon is inserted into the lateral side of the patella.
  - *M. vastus medialis*: This muscle arises from the intertrochanteric crest on the femur. The tendon is inserted into the medial side of the patella.
  - *M. vastus intermedius* (formerly described as the *m. crureus*): This muscle arises from the proximal three-fourths of the shaft of the femur. Its tendon inserts into the cartilage that is present in the blended tendons of the lateral and medial vasti muscles, just proximal to the patella.
- *M. tensor fasciae latae*: The origin of this muscle is the ilium and the fascia overlying the gluteus medius muscle. The muscle is inserted into the fascia lata.

The Caudal Thigh Muscles (Hamstrings)

- *M. biceps femoris*: The biceps femoris muscle arises from the ischial tuberosity. The muscle forms a thin aponeurosis that is inserted into the deep fascia of the proximal hind limb, the head of the fibula, the lateral tibial condyle, and the capsule of the knee joint. This muscle can be used to administer intramuscular injections [12].

- *M. semitendinosus*: This muscle also arises from the ischial tuberosity. However, there is a second head originating from the first caudal vertebra. The tendon lies superficial to the semimembranosus muscle and attaches to the medial surface of the tibial shaft, often in conjunction with the tendon of the gracilis.
- *M. semimembranosus*: The origin of this muscle is the ischial tuberosity. It is inserted into the medial surface of the medial condyle of the tibia.
- *M. popliteus:* This muscle is the only intrinsic flexor muscle of the knee. The fan-shaped muscle is located at the caudal side of the proximal tibial shaft. Its tendon inserts into the popliteal fossa of the femur.

The Flexors of the Tarsal Joint and Extensor of the Digits

- *M. tibialis cranialis*: This muscle arises from the lateral condyle of the tibia and from the upper two-thirds of its shaft. The medial tendon attaches to the medial side of the first tarsal bone, whereas the lateral tendon is inserted into the head of the first metatarsal.
- *M. extensor digitorum longus*: The origins of this muscle are the lateral condyle of the tibia, the interosseous membrane between the tibia and the fibula, and the fibula. Two tendons arise from two muscular bellies, one superficial and one deep. After these tendons have joined together, four slips are inserted into the bases of the terminal phalanges of the second to fifth digits.
- *M. extensor hallucis longus*: This very thin muscle that lies deep to the former muscle obtains its origin from the interosseous membrane in between the *ossa cruris*. The tendon is inserted into the terminal phalanx of the hallux.
- *M. fibularis longus*: This muscle has its origin on the head and proximal half of the shaft of the fibula.
- *M. fibularis brevis*: This muscle arises from the lower two-thirds of the shaft of the fibula.
- *M. fibularis digiti quinti*: This muscle arises from the proximal half of the lateral surface of the fibula.

The fibularis muscles pass behind the lateral malleolus and attach to the dorsolateral side of the calcaneus.

The Extensors of the Tarsal Joint and Flexors of the Digits

- *M. gastrocnemius*: The lateral and medial heads of the gastrocnemius muscle arise from the lateral and medial epicondyle of the femur, respectively. A sesamoid bone is present in each tendon of origin. The *tendo calcaneus* (Achilles tendon) attaches to the tuber calcanei.
- *M. soleus*: This thin muscle arises from the head of the fibula. Its tendon fuses with the gastrocnemius muscle.
- *M. plantaris*: The thin *m. plantaris* has its origin on the lateral condyle of the femur. Its thin tendon lies on the medial side of the Achilles tendon and is inserted into the plantar fascia, hence its name.
- *M. flexor digitorum longus tibialis* (can be considered as the *m. flexor digitorum superficialis*): This muscle arises from the caudal side of the proximal tibia. A thin tendon arises at the medial side. This is the only flexor tendon of the hallux. Two other tendons attach to digits II and V.
- *M. flexor digitorum longus fibularis* (can be considered as the *m. flexor digitorum profundus*): This muscle lies deep to the *M. flexor digitorum longus tibialis.* It arises from the caudomedial aspect of the fibula and the interosseous membrane between the tibia and fibula. It fuses with the *M. flexor digitorum longus tibialis* at the level of the shafts of the metatarsals. The two tendons of the fibularis attach to the third and fourth digits.
- *M. flexor digitorum brevis*: The superficial head has its origin on the tuber calcanei. This head forms the short flexor of digit II. The deep head arises from the flexor tendons, halfway down the metatarsals. The three tendons are inserted into the base of the second phalanx of digits III, IV, and V. The long flexor tendons perforate these three tendons.
- *Muscles of the foot*: The *mm. lumbricales* and *mm. interossei* have a similar conformation and function as those of the hand. As regards the short muscles of the hallux, the *m. adductor hallucis* and *m. abductor hallucis* are present. The *m. abductor digiti quinti* is present as a well-marked muscle mass.

DIGESTIVE SYSTEM

Cranial Structures

Oral Cavity

Dentition

The teeth that are present in the upper and lower jaw are depicted in Figs. 2.16 and 2.17, respectively. The teeth of the marmoset are brachydont, i.e., they have a closed root, a short neck, and a crown that is covered with enamel [13]. The dentition of the marmoset is heterodont, i.e., different types of teeth can be distinguished. 2 incisors (I), a canine tooth (C), 3 premolars (P), and 2 molars (M) can be observed in each oral quadrant of the adult animal. Thus, a total of 32 teeth are present [2]. The incisors are single-rooted teeth that present 1

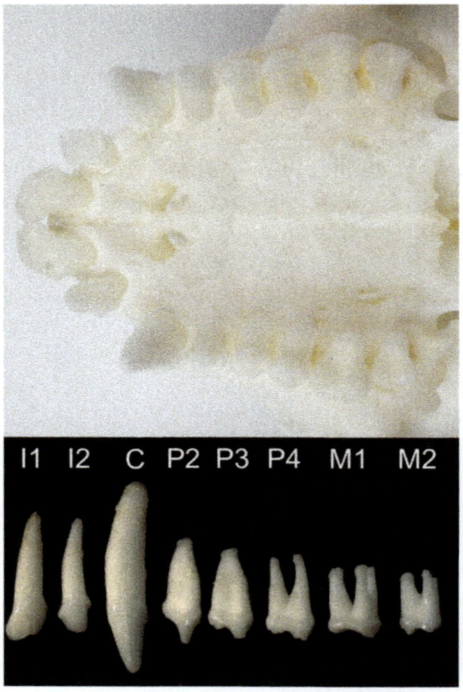

FIGURE 2.16 Dentition of the upper jaw in the common marmoset. The upper panel presents a ventral view of the incisive bone and maxilla with the teeth in position. The lower panel shows the teeth of the left side of the upper jaw after extraction. *C*, canine tooth; *I1*, 1st incisor; *I2*, 2nd incisor; *M1*, 1st molar; *M2*, 2nd molar; *P2*, 2nd premolar; *P3*, 3rd premolar; *P4*, 4th premolar.

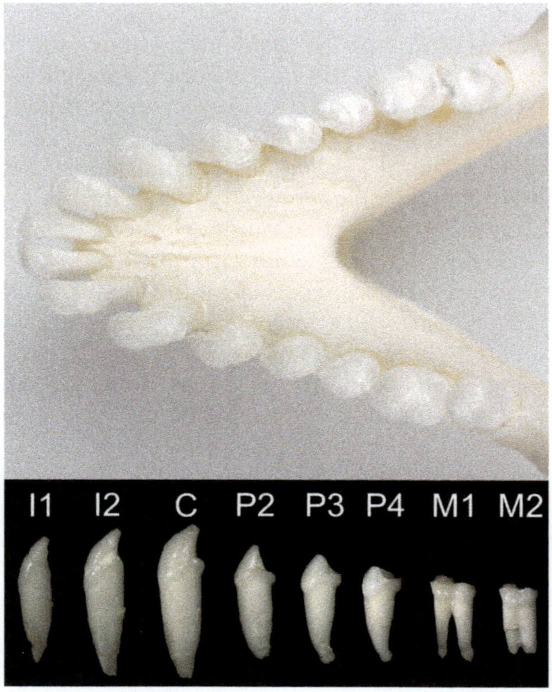

FIGURE 2.17 Dentition of the lower jaw in the common marmoset. The upper panel presents a dorsal view of the mandible with the teeth in position. The lower panel shows the teeth of the left side of the lower jaw after extraction. *C*, canine tooth; *I1*, 1st incisor; *I2*, 2nd incisor; *M1*, 1st molar; *M2*, 2nd molar; *P2*, 2nd premolar; *P3*, 3rd premolar; *P4*, 4th premolar.

cusp. The upper incisors are spade-shaped. I2 is smaller than I1 [2]. The lower incisors are specialized to gouge holes in trunks and large branches [3,14]. To this purpose, they are more elongate and chisel-shaped compared with the upper incisors. In addition, they present a loss of lingual enamel and a thickening of buccal enamel [3,15]. In contrast to the incisors of the upper cheek, I2 is larger than I1 and is almost as large as the canine tooth [2]. The single-rooted canines are short-tusked and falciform [2,3]. The crowns of the upper canines are distinctly longer than those of the adjacent incisors and premolars [16]. In contrast, the mandibular canine teeth are not higher than the incisors [2]. They also present a loss of enamel at their lingual sides [15]. In the maxilla, the 3 premolars, i.e., P2, P3, and P4, have 1 or 2 roots [2,13,16,17]. The sizes of the maxillary premolars are fairly equal. However, the number of cusps differs. P2 possesses 1 cusp, whereas P3 and P4 have 2 cusps [2,16]. The 3 mandibular premolars, i.e., P2, P3 and P4, present 1 root and 1 cusp [2,17]. Marmosets show a loss or reduction of mandibular and maxillary third molars [3]. As a consequence, only M1 and M2 are present [2]. Both maxillary molars bear 3 roots and 3 cusps [2,13,16,17]. The first maxillary molar is markedly larger than P4 and M2. Both molars of the mandibles, of which M1 is distinctly larger than P4 and slightly more voluminous than M2, have 2 roots and 4 cusps.

The dentition of the marmoset is diphyodont, i.e., they have a deciduous and a permanent set of teeth [13]. At birth, only the first deciduous incisors have erupted, whereas the second deciduous incisors can be palpated through the gums. These second incisors erupt at the age of 2 weeks. Very soon, the deciduous canines and the rostral and middle premolars follow. At 4 weeks of age, the caudal deciduous premolars erupt [18]. The deciduous set of teeth, which is devoid of molars, is thus complete in the 1-month-old marmoset. The number of roots of the upper and lower deciduous premolars is 1–3–2 and 1–1–2, respectively [2]. The permanent set of teeth appears from the age of 3–4 months onward when the rostral molars erupt. When the animal is 4–6 months old, the caudal molars become visible. In the 7- to 8-month-old marmoset, the permanent incisors and the caudal premolar breach the gums. During the next 3 months, the rostral and middle premolars erupt. Finally, the permanent canines complement the permanent set of teeth at the age of 11–12 months [18].

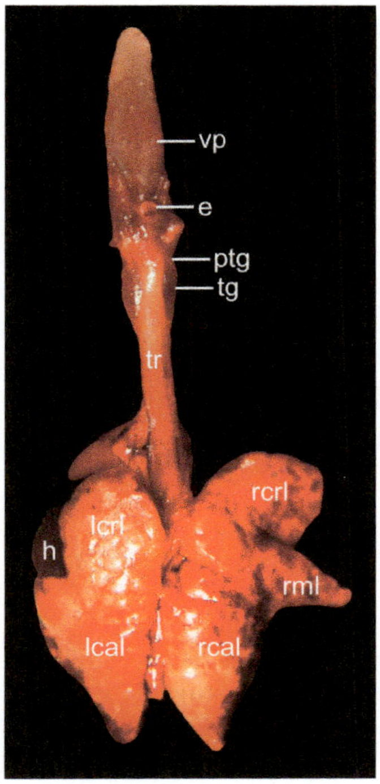

FIGURE 2.18 Dorsal view of the lungs, trachea, larynx, and tongue of the common marmoset. *e*, epiglottis; *h*, heart; *lcal*, left caudal lung lobe; *lcrl*, left cranial lung lobe; *ptg*, parathyroid gland; *rcal*, right caudal lung lobe; *rcrl*, right cranial lung lobe; *rml*, right middle lung lobe; *tg*, thyroid gland; *tr*, trachea; *vp*, lingual vallate papilla.

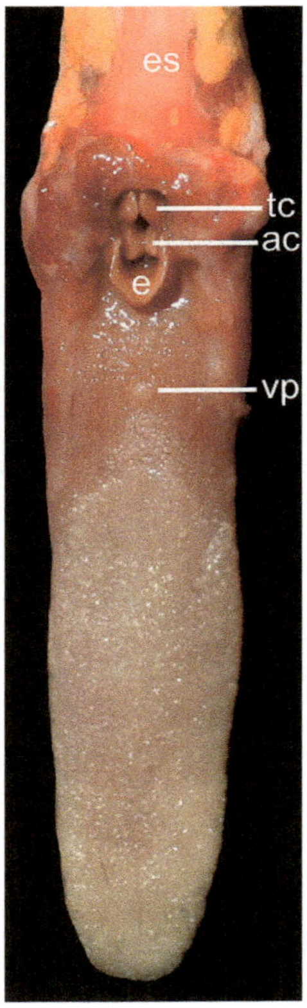

FIGURE 2.19 Craniodorsal view of the root of the tongue and the larynx of the common marmoset. *ac*, arytenoid cartilages; *e*, epiglottis; *es*, esophagus; tc, thyroid cartilage (dorsal horn) *vp*, vallate papilla.

Tongue

In the adult common marmoset, the tongue measures approximately 21 mm in length and 8 mm in width (Fig. 2.18). The tip of the tongue is flat and slightly tuberculated [19]. The ventral surface of the tongue presents a well-marked groove in the midline [1]. Gustatory and mechanical papillae are located at the dorsal lingual surface that is covered with a stratified squamous epithelium. The gustatory papillae comprise the circumvallate, foliate, and fungiform papillae that all contain taste buds. Three circumvallate papillae are arranged in a V-shape with the central, most caudal papilla being located just rostral to the epiglottis (Fig. 2.19). Both more rostral papillae are smaller. The pair of foliate papillae lies at the lateral side of the tongue, rostral to the attachment of the palatoglossal fold. Each foliate papilla consists of about three furrows. The fungiform papillae are relatively few and lie scattered between the filiform papillae. The mechanical filiform papillae can be found throughout the dorsal lingual surface [19].

Palate

The palate consists of the rostral hard palate and the caudal soft palate. The hard palate is characterized by six transverse ridges that are covered with mechanical papillae. Only the rostral two ridges are complete [1]. Taste buds are present in the soft palate. Their number varies from approximately 150 in the newborn animal to 500 in the adult. The taste buds of the soft palate are densely gathered into several groups [20].

Pharynx

The pharynx consists of the nasopharynx, the oropharynx, and the laryngopharynx. The Eustachian tubes (*tuba auditiva*) open in the lateral walls of the nasopharynx. The pharyngeal tonsil or adenoid is absent. However, some aggregated lymphoid follicles can be present in the nasopharyngeal roof. A pair of palatine tonsils is present in the oropharynx. This tonsil is a very simple pocket with lymphoid tissue that is located in the lateral wall of the *fauces*. A well-marked supratonsillar recess is present [1]. The palatoglossal arch (i.e., a mucosal fold between the soft palate and the lateral aspect of the tongue) delineates the *fauces* rostrally, whereas the palatopharyngeal arch (i.e., a muscular fold between the soft palate and the lateral wall of the pharynx) is the caudal border.

Salivary Glands

The (sub)mandibular gland (mean weight 437 mg) is the larger salivary gland in the common marmoset [21].

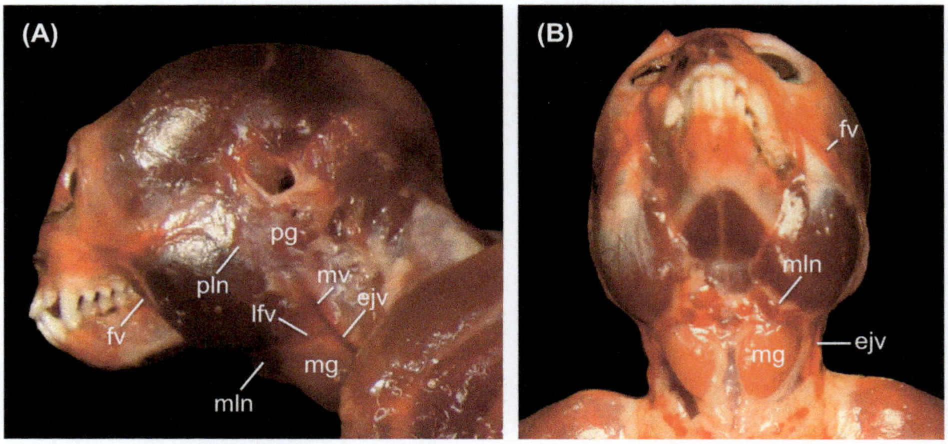

FIGURE 2.20 Left lateral (A) and ventral (B) views of the skinned head of the common marmoset showing the salivary glands, the associated lymph nodes, and the major veins. *ejv*, external jugular vein; *fv*, facial vein; *lfv*, linguofacial vein; *mg*, mandibular gland; *mln*, mandibular lymph node; *mv*, maxillary vein; *pg*, parotid gland; *pln*, parotid lymph node.

It is located near the mandibular angle. It is often divided into a rostral and a caudal lobe [1]. This salivary gland is encapsulated and lobulated and contains a mixture of serous and mucous acini [21] (Fig. 2.20).

The parotid gland is intermediate in size (mean weight 256 mg). It lies rostroventral to the external auditory meatus and extends rostrally over the caudal border of the masseter muscle. This gland is encapsulated and lobulated and is composed of pure serous acini [21]. The parotid duct arises from the rostral aspect of the gland and perforates the buccinator muscle at the level of the maxillary P4 [1] (Fig. 2.20).

The sublingual gland is the smaller salivary gland (mean weight 6 mg). It consists of an unencapsulated, lobulated mass of pure mucous acini that is located on the floor of the mouth, lateral to the frenum [1,21].

Intestinal Tract

When transecting the abdominal wall in the ventral midline, the gut, and in particular the small intestine, is readily visible (Fig. 2.21). The stomach lies cranially in the abdominal cavity and is covered by the thoracic wall. The isolated intestinal tract of the common

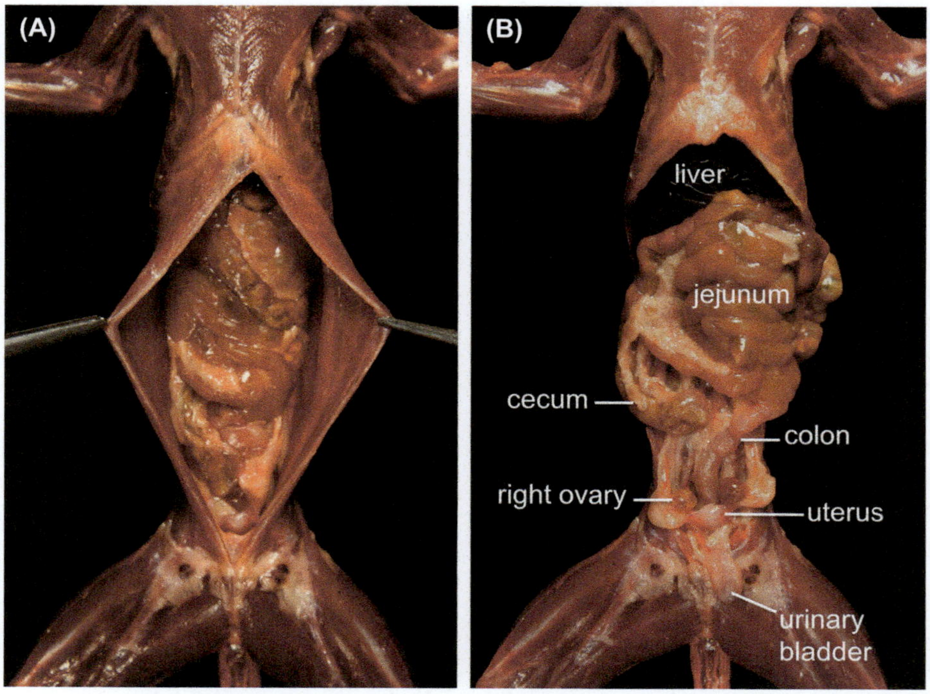

FIGURE 2.21 Topography of the abdominal organs of a female common marmoset after transecting the abdominal wall in the ventral midline (A) and after resecting the abdominal wall and exposing the major organs (B).

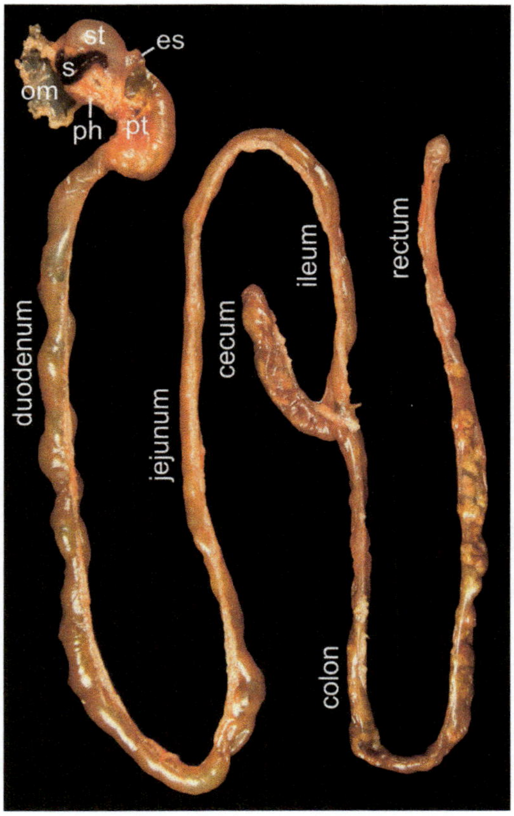

FIGURE 2.22 Isolated gastrointestinal tract of the female common marmoset. The spleen (s) is attached to the stomach (st) by means of the gastrosplenic ligament, which is part of the greater omentum (om). The esophagus (es) has been cut. The pancreas consists of the pancreatic head (ph) that lies in the base of the greater omentum and the pancreatic tail (pt) that is located in the mesoduodenum. Notice the absence of the appendix.

marmoset is depicted in Fig. 2.22. Because marmosets are bark-gouging animals, their intestinal tract is adapted to gummivory. This means that a specialization for the fermentative digestion of plant gums is present [14].

Esophagus

The esophagus measures approximately 10 cm in length. Cranially, the esophagus lies on the midline, dorsal to the trachea (Fig. 2.19). It bends to the right side of the body when reaching the thorax and deviates to the left side again to perforate the diaphragm [1]. Its muscular layer is composed of an outer layer of longitudinally orientated fibers and an inner layer of circular fibers. The transition from striated muscle to smooth muscle begins 2 cm proximal to the stomach [22].

Stomach

The stomach of the common marmoset (Fig. 2.22) consists of the fundus, the body, the pyloric canal, and the pyloric antrum. The fundus is large and extends cranially left of the esophagus. The corpus is continuous with the esophagus and cannot be delineated from the fundus by any anatomical landmark. The pyloric antrum is continuous with the corpus. It can be distinguished from the body by its smaller diameter. The narrow short tube that follows is the pyloric canal that ends at the pyloric sphincter [1]. The surface area of the stomach in *Callitrichidae* is approximately 9 cm^2 [14].

Small Intestine

The small intestine in *Callitrichinae* (Fig. 2.22) has a surface area of approximately 33 cm^2 [14]. The length is almost 40 cm [14]. The small intestine is composed of the duodenum, the jejunum, and the ileum. The duodenum presents a long, descending part that is located at the right side of the abdomen, a short transverse part in which the chyme travels from right to left in the caudal half of the abdominal cavity, and a short ascending part at the left side of the abdomen. The base of the mesentery lies in the middle of the J-shaped duodenum. The common bile duct and the pancreatic ducts (i.e., *ductus pancreaticus* and *ductus pancreaticus accessorius*) enter the descending part of the duodenum at one-third of its length. The accessory pancreatic duct enters the duodenum separately on the minor duodenal papilla, whereas the common bile duct and the principal pancreatic duct join to terminate on the major duodenal papilla [1,9]. Within the mesoduodenum, the tail of the pancreas is found. The jejunum presents three loops and continues as the ileum. This segment travels caudally along the medial side of the ascending part of the duodenum and then forms a small loop before joining the colon [1].

Large Intestine

The cecum can be found at the junction between the ileum and colon at the right side of the abdomen (Fig. 2.21). The cecum is enlarged for the fermentation of plant exudates and lacks an appendix [3,14]. Its length is approximately 4 cm, whereas its surface area is approximately 18 cm^2 [14] (Fig. 2.22). Internal ribbon-like structures (*teniae*) are present [3]. The U-shaped colon consists of the ascending part, the transverse part, and the descending part. Its length is approximately 34 cm, whereas its surface area is ca. 41 cm^2 [14]. The ascending colon lies in the right abdomen and is slightly shorter than the cecum. The ascending colon subsequently turns to the left to form the transverse part. Finally, the descending colon travels along the left side of the abdomen and passes insensibly into the rectum [1] (Figs. 2.21 and 2.22).

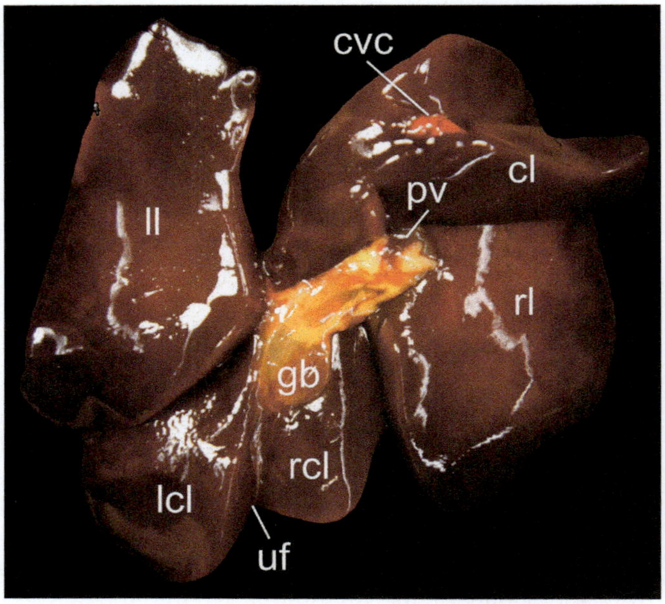

FIGURE 2.23 Visceral side of the liver of the common marmoset, which consists of the left liver lobe (ll), the central lobe that is divided into the left central lobe (lcl), and the right central lobe (rcl) by the umbilical fissure (uf), the right liver lobe (rl), and the caudate lobe (cl). The gallbladder (gb) is lodged between both parts of the central lobe. The portal vein (pv) enters the liver at the hilum. The hepatic veins leave the liver at the dorsal side to flow into the caudal vena cava (cvc).

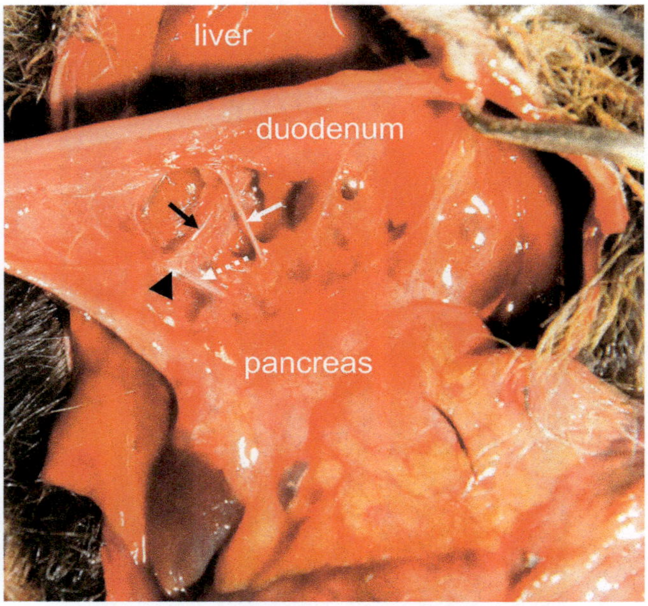

FIGURE 2.24 Ventral view of the opened cranial abdomen of the common marmoset showing the liver, the pancreas, and the mesenteric side of the duodenal wall. The principal pancreatic duct (*dashed white arrow*) joins the common bile duct (*black arrow*) to penetrate the duodenal wall at the major duodenal papilla (*black arrowhead*), whereas the accessory pancreatic duct (*white arrow*) enters the duodenum individually.

Liver

The liver lies most cranial in the abdomen (Fig. 2.21). Its diaphragmatic side is located against the diaphragm, whereas its visceral side faces the viscera, in particular the stomach. The liver of the common marmoset is relatively small [1]. In the male marmoset with a body weight range of 182.9–381.9 g, the liver may weigh 9.182–20.22 g [23]. For the female marmoset that may weigh between 201.2 and 412.1 g, the acceptable range for the liver weight is 9.912–23.213 g [23]. The liver is divided into a left, a central, a right, and a caudate lobe. The right lobe is the larger lobe. The central lobe is divided into a left and a right half by the umbilical fissure. The right half presents a deep fossa in which the gallbladder is lodged. The esophagus runs in a fissure between the left and the caudate lobe [1] (Fig. 2.23). The common bile duct joins the principal pancreatic duct to enter the duodenum on the major duodenal papilla.

Pancreas

The pancreas consists of the vertical head and the transverse tail (Fig. 2.22). The head lies in the mesoduodenum at the right side of the abdominal cavity. The tail traverses the midline in the base of the attachment of the greater omentum to the dorsal abdominal wall. The principal pancreatic duct joins the common bile duct before penetrating the duodenal wall and opening onto the major duodenal papilla. The accessory pancreatic duct opens more proximally onto the minor duodenal papilla (Fig. 2.24).

RESPIRATORY SYSTEM

Nose and Nasal Cavity

The rhinarium is smooth. A shallow median furrow can be observed in the external surface of the upper lip. A cleft (*philtrum*), splitting the upper lip, is absent. Just rostral to the first palatal ridge, the openings of the incisive ducts, which lead to the vomeronasal organ, can be seen. The paired palatine papilla of the marmoset are very typical as they overhang the incisive openings [24]. The vomeronasal organ is present at the rostral base of the nasal septum [24]. The nasal conchae comprise the dorsocaudal ethmoturbinate and the rostroventral maxilloturbinate. The ethmoturbinate arises from the ethmoidal plate and extends rostrally to the first premolar of the upper jaw. The maxilloturbinate has its origin on the maxilla, lateral to the nasal cavity, and reaches the nasal vestibule. The turbinates have no definite ramification [25].

The nasal cavity is lined by four types of epithelium. Stratified squamous epithelium can be found from the nasal inlet to the first upper premolar, i.e., the vestibular portion of the nasal cavity. Nasal transitional epithelium, which is an epithelium containing nonciliated cuboidal or columnar cells and a few goblet cells, is distributed in a narrow area between the squamous and respiratory epithelium that extends to the nasopharyngeal duct. The latter epithelium is pseudostratified and characterized by ciliated columnar cells and goblet cells. Olfactory epithelium lines the dorsocaudal part of the lateral wall of the nasal cavity and the nasal septum [25,26].

The maxillary sinus pneumatizes the medial part of the maxilla in callitrichines [27].

Larynx and Trachea

The laryngeal cavity is cranially bound by the large epiglottis, the aryepiglottidean folds and the arytenoid cartilages (Figs. 2.18 and 2.19). This wide cavity narrows to the level of the false vocal cords. In between these and the true vocal cords lies the laryngeal ventricle [1]. Marmosets have a rich vocal repertoire that is produced by means of the vocal cords. The frequency ranges of the different calls that are recognized in the marmoset vary from near-infrasonic to ultrasonic (80 kHz) [28]. Ultrasonic vocalizations are not used to indicate pain, discomfort, or distress [29].

The cartilages of the larynx comprise the thyroid cartilage, the cricoid cartilage, and the arytenoid cartilages. The thyroid cartilage is formed by two quadrilateral plates that are cranially united. The blunt cranial horn articulates with the hyoid bone, whereas the caudal horn articulates with the cricoid cartilage. This cartilage is ring-shaped. The arytenoid cartilages are small pyramidal structures that are attached to the cricoid by strong ligaments [1].

The trachea, which is approximately 3 cm long in the female and almost 4 cm in the male, is composed of 30 incomplete rings (Fig. 2.18). The dorsal portions of the rings are connected by smooth muscle. The hyaline cartilage of the tracheal rings is generally calcified. The trachea is lined by a pseudostratified columnar epithelium with cilia and scarce goblet cells in the proximal portion. The goblet cells increase in number and volume toward the tracheal bifurcation. Four main branches are given off to the right lung. The left lung receives two main branches [1,26].

Lungs

The lungs have a length of 3.5 cm in an average adult marmoset. The left lung is composed of a cranial and a caudal lobe. A cardiac scissure is present on the cranial lobe. The right lung contains a cranial lobe, a middle lobe, a caudal lobe, and a small medial cardiac lobe [26] (Fig. 2.18). The alveolar surface of the adult marmoset has been estimated to be 7662 ± 1647 cm^2. The lung volume of a marmoset weighing 303 g is 18.1 mL [30].

UROGENITAL TRACT

Urinary System

The bean-shaped kidneys of the common marmoset are unilobar and unipapillate. They lie at the same level in the abdomen, from the second to the third lumbar vertebrae (Fig. 2.25). The cranial renal pole is somewhat

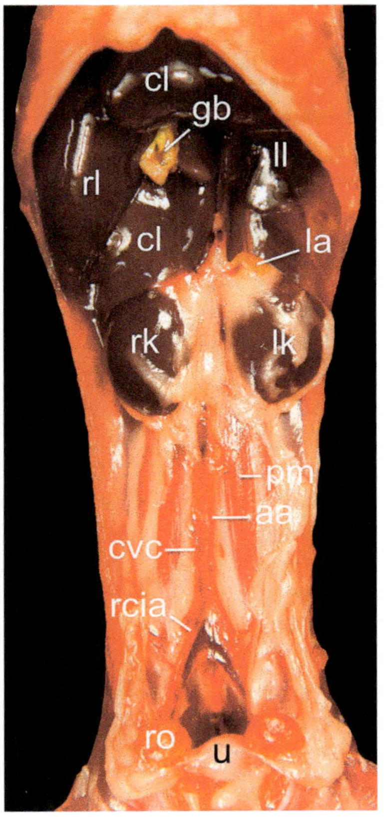

FIGURE 2.25 Ventral view of the abdominal organs of the common marmoset after resection of the gastrointestinal tract. The left kidney (lk) and the right kidney (rk) lie at the same level. The left adrenal gland (la) is visible at the cranial pole of the left kidney. The caudate lobe of the liver (cl) presents an impression in which the cranial pole of the right kidney resides. The right liver lobe (rl), the central liver lobe (cl), the left liver lobe (ll), and the gallbladder (gb) are also visible. The abdominal aorta (aa) and the caudal vena cava (cvc) course in the midline between the psoas major muscles (pm). The bifurcation of the abdominal aorta into the common iliac arteries (rcia, right common iliac artery) lies cranial to the uterus (u) and the ovaries (ro, right ovary).

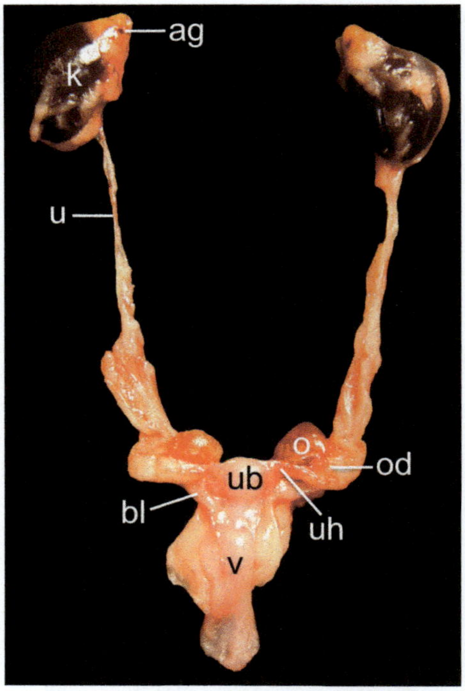

FIGURE 2.26 Dorsal view of the isolated urogenital tract of the female common marmoset. The adrenal gland (ag) is closely attached to the kidney (k) from which the ureter (u) emerges. The urinary bladder is covered by the female genital organs that consist of the ovaries (o) that are connected to the uterine body (ub) by means the oviducts (od) that lead to the uterine horns (uh). The broad ligament of the uterus (bl) contains much adipose tissue, as does the tissue surrounding the vagina (v).

flattened for the adrenal gland [1] (Fig. 2.26). In the male marmoset with a body weight range of 182.9–381.9 g, the kidney weight varies from 0.881 to 2.230 g [23]. In the female marmoset that weighs between 201.2 and 412.1 g, the kidney weight is 1.160–2.441 g [23]. The kidneys are covered by a thin fibrous capsule that contains a small number of elastic fibers [31]. The renal cortex is relatively thick. The pelvis is small. Its apex does not project into the ureter.

The ureters run along the line of the apices of the transverse processes of the lumbar vertebrae toward the trigone of the urinary bladder. In the female, the ureter travels through the base of the broad uterine ligament (Fig. 2.26). In the male, the vas deferens hooks around it [1].

The urinary bladder is conical (Fig. 2.26). Its trigone has three openings, i.e., two ureteric openings and the urethral opening [1].

Male Genital System

In the adult male, the scrotum is a horseshoe elevation of which the skin is hairless and covered with small pearly white nodules. The lateral masses that can be observed within the scrotum are the testes [1] (Fig. 2.10). The testicle of the common marmoset is enveloped by a thick tunica albuginea consisting of collagen fibers intermingled with some elastic fibers. For a male marmoset with an acceptable body weight range of 182.9–381.9 g, the acceptable range for the testicular weight is 0.204–1.365 g [23]. The epididymis consists of a single coiled duct [32]. It is large and is almost one-half the size of the testis. The tail is one-half the size of the entire epididymis [1]. The ductus deferens, which consists of the tunica mucosa, the very thick tunica muscularis, and the tunica adventitia, arises at the medial side of the epididymis and runs toward the inguinal canal [1,32]. After crossing the ureter, the ductus deferens joins the duct of the seminal vesicle that lies lateral to it. Finally, the common ejaculatory duct enters the prostatic urethra [1]. The seminal vesicles are 3 mm in length and 2 mm in width. They are filled with a yellowish glairy fluid [1]. The prostate has the shape of a horse chestnut. It surrounds the proximal segment of the urethra. Its capsule is fibroelastic and contains some smooth muscle cells [32]. In the midline of the scrotum the tip of the penis can be seen [1]. The penis is composed of two *corpora cavernosa penis* and the single ventral *corpus spongiosum urethrae* that terminates at the glans penis [32]. The glans penis presents androgen-dependent keratinized spines. These spines are robust epidermal structures, supported by a papilla of dermal tissue, that measure approximately 180 μm in length. The spines enhance the tactile sensitivity of the glans penis in sexually experienced males [33]. A short penile bone (*os penis* or *baculum*) is located in the septum that joins the two *corpora cavernosa penis* [32] (Fig. 2.10).

Female Genital System

The isolated female urogenital tract is depicted in Fig. 2.26. The pea-shaped ovaries, which are attached to the dorsal layer of the broad uterine ligament by a short mesovarium, measure approximately 5.3 mm × 4.3 mm × 3.8 mm [1,34]. The ovaries of the female marmoset with a weight ranging between 201.2 and 412.1 g can weigh up to 0.347 g [23]. The mean length of the oviduct (fallopian tube or *salpinx*) is 10.5 mm with a width of 1.5 mm. The proximal segments of the oviducts present fimbriae that lie adjacent to the adipose tissue surrounding the ovaries [34]. The female marmoset possesses a single pyriform-shaped uterus that presents two small horns or cornua. The mean uterine size is 8.4 mm (length) × 10.0 mm (width between the horns) × 6.4 mm (thickness) [34]. The uterus is attached to the abdominal wall by the paired broad ligaments (Fig. 2.26). The uterine wall has a thickness of

1–1.5 mm. The cervix is 4–5 mm long and protrudes into the vagina for 2–3 mm. A clear squamocolumnar junction is present within the cervical os. Cervical glands can be found within the walls of the cervical canal [34]. The cranial and caudal parts of the vagina are marked with a vaginal isthmus [34]. Each part measures 17 mm in length [34]. The mucosa of the caudal part is purple, presents many longitudinal rugae, and is lined by a stratified squamous epithelium. The mucosa of the cranial part has occasional pink folds and is also lined by a stratified squamous epithelium. The urethral orifice is located in the vulva [34].

The vulva is pendulous with a small opening located near its most ventral aspect, similar to the preputial orifice of the male [35]. The skin of the perineum is covered with numerous nodules. The urogenital cleft, which is laterally bordered by the labia, is separated by a considerable distance from the anus. The clitoris is completely buried [1] (Fig. 2.27).

Gender Determination

In the neonatal male, the small scrotum is an irregular fold of skin that presents small nodules. It is located at the caudal aspect of the penis, whose presence can be revealed by the small, rounded preputial opening. At 45 days of age, the penis can more easily be exposed in the preputial opening. The small nodules on the scrotal skin now have a pearly white appearance. At 2–3 months of age, the testes can be observed as small subcutaneous swellings craniolateral to the base of the penis. The testicles descend into the scrotum at the age of 8–11 months. In the adult male, the scrotum still covers the penis. Its skin is devoid of hairs and the small pearly white nodules are pronounced. The median raphe is obvious [35].

The perineal skin in the neonatal female resembles the scrotal skin fold of the male neonate. This skin is also characterized by small nodules. The pendulous vulva presents a small, slit-like vestibular opening at the ventral aspect of the perineum. At the age of 45 days, the cutaneous nodules become pearly white and manipulation allows for opening the vulvar cleft [35].

Sex determination of neonatal and young animals must rely on manipulation to demonstrate the round preputial opening and glans penis or the slitlike vaginal opening. When the male animal is stressed, e.g., when handled, it will retract the testes into the inguinal canal [35].

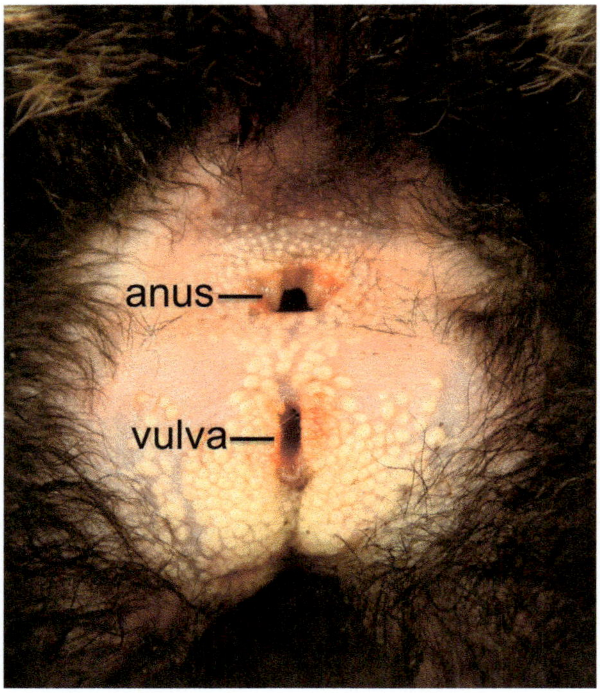

FIGURE 2.27 Caudal view of the perineum of the female common marmoset. Notice the large anogenital distance.

CIRCULATORY SYSTEM

The Heart

The heart lies in the thoracic cavity in the region between the second and sixth ribs (Fig. 2.28). After removal of the left thoracic wall, the blunt apex of the heart, which is formed by the left ventricle, can be observed between the left cranial and caudal lung lobes as the longitudinal axis of the heart presents a deviation of approximately 45 degrees toward the left. The right heart, including the right atrium and ventricle, rests on the sternum as a result of the counterclockwise, quarter rotation of the longitudinal cardiac axis. The right

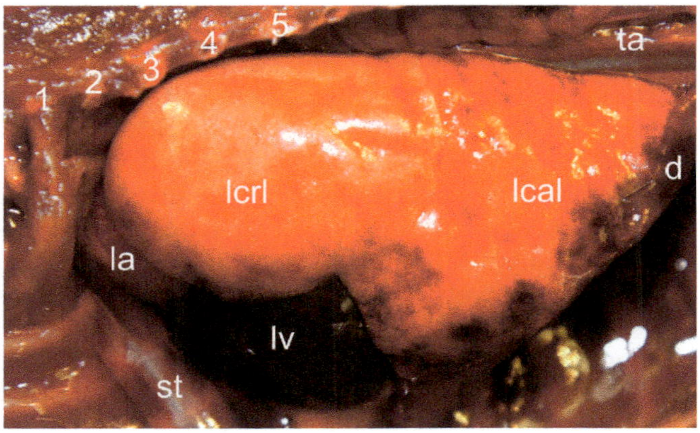

FIGURE 2.28 Left lateral view of the opened thoracic cage. The left lung is composed of the left cranial lobe (lcrl) and the left caudal lobe (lcal). The thoracic aorta (ta) lies dorsal to the lungs. The heart, of which the left atrium (la) and the left ventricle (lv) are visible, rests on the sternum (st). It is positioned in the region that is delineated by the second and sixth ribs (the first five ribs are numbered). The diaphragm separates the thoracic cavity from the abdominal cavity.

atrium receives the systemic venous blood through the cranial and caudal vena cava. The right auriculoventricular valve presents 3 cusps and is therefore known as the tricuspid valve. The wall of the right ventricle is well trabeculated. The right lumen contains a few septomarginal tendinae. The pulmonary valve has the typical arrangement. The left atrium presents three openings. A large opening is connected with the pulmonary vein from the right lung, whereas the two smaller openings are for the veins from the cranial and caudal left lung lobes. The bicuspid mitral valve separates the left atrium from the left ventricle. The latter presents a well-trabeculated wall. The aortic valve contains three flaps that are thicker near the free edge and have a small central nodule. The serofibrous pericardial sac that invests the heart is thin [1].

The cardiac weight of a common marmoset that weighs 209 g is 1.72 g [36]. For a body weight range of 182.9–381.9 g in the male marmoset, the acceptable range for the heart weight is 0.904–2.524 g [23]. For a female marmoset weighing 201.2–412.1 g, the heart weight can vary from 1.110 to 2.474 g [23].

Pulmonary Circulation

The pulmonary trunk that originates from the right ventricle initially lies to the left of the aorta. After crossing this structure dorsally, it runs at the right side of the aorta and branches into the left and right pulmonary arteries. Each pulmonary artery splits up into lobar branches at the hilum of the lungs. The right pulmonary vein enters the left auricle after the four lobar veins of the right lung have joined. The two lobar veins of the left lung do not join but enter the left auricle separately [1].

Systemic Arterial Circulation

The ascending aorta is very short, only 3–4 mm in length. The aortic arch runs backward and to the left. It gives origin to the brachiocephalic trunk that gives off the right subclavian artery and continues its course as the right common carotid artery, the left common carotid artery, and the left subclavian artery. Each common carotid artery travels along the neck to divide into the internal and external carotid arteries at the level of the hyoid bone. The external carotid artery subdivides into the lingual, cranial thyroid, facial, internal maxillary, and superficial temporal arteries. The facial artery and its ramifications were described by Takeda and Okuda [37]. These authors also provide diameters of its different branches.

The intrathoracic descending aorta deviates from the left to the right side of the vertebral column while giving off intercostal branches. The abdominal aorta gives origin to several major branches. The short celiac artery splits into the hepatic and splenic arteries. The hepatic artery supplies the lesser and greater curvature of the stomach, the pancreas, and the descending part of the duodenum, before branching into the liver lobes. The large splenic artery runs through the lienorenal (splenorenal) ligament toward the spleen. Along its course, it gives off branches toward the greater curvature of the stomach through the gastrosplenic ligament. The suprarenal artery is the second branch from the abdominal aorta. It is relatively large in relation to the adrenal gland it supplies. The cranial mesenteric artery, which arises immediately caudal to the suprarenal artery, supplies the ascending part of the duodenum, the jejunum, the ileum, the cecum, and the ascending and transverse parts of the colon. The renal arteries branch subsequently from the abdominal aorta. The renal artery remains a single vessel unto the level of the renal hilum. Caudal to the left renal artery, the caudal mesenteric artery splits from the abdominal aorta. It gives off a large colic artery and several smaller branches to the descending colon and rectum. The ovarian or spermatic arteries arise from the abdominal aorta at the level of the caudal poles of the kidneys. They lie in close association with the ureters. The abdominal aorta finally divides into the two common iliac arteries at the junction of the fifth and sixth lumbar vertebrae (Fig. 2.25). However, it first gives off a pair of iliolumbar arteries that run along the lateral femoral cutaneous nerve into the thigh. The continuation of the abdominal aorta is the middle sacral artery that supplies the ventral muscles of the tail. The common iliac artery divides into the external and internal iliac arteries. The lateral branch of the internal iliac artery supplies the lateral pelvic wall, whereas the medial branch supplies the urinary bladder and, in the female, gives off the uterine artery that runs in the broad uterine ligament. The external iliac artery gives origin to the caudal epigastric artery, the obturator artery, and the medial circumflex artery of the thigh [1].

The arteries of the limbs have been scrutinized previously by Beattie [1]. Here, the major branches are briefly described. After crossing the first rib, the subclavian artery continues as the axillary artery that is accompanied by the axillary vein (Fig. 2.15). The latter artery gives off the superficial brachial artery in the axilla and continues as the deep brachial artery. The superficial brachial artery that lies in the groove between the biceps and triceps muscles on the medial side of the arm first gives off the thoracoacromial and the lateral thoracic arteries. Proximal to the elbow, the transverse cubital artery that lies between the brachialis muscle and the humerus arises. The continuation of this branch is the radial artery that courses along the mediodorsal side of the forearm. It divides into the superficial or palmar branch and the deep or dorsal branch. From the deep brachial artery that runs along the median nerve branch, the circumflex humeral, the subscapular, and, more distally, the *a. profunda brachii*, emerge. At the elbow, the deep brachial artery divides into the ulnar and the interosseus arteries. The former artery forms the superficial palmar arch together with the palmar branch of the radial artery. From this arch arise the common digital arteries. The latter artery divides into palmar branches and supplies the muscles of the forearm.

In the proximal part of the thigh, the external iliac artery continues as the femoral artery that gives off the *a. profunda femoris* that in turn gives origin to the lateral circumflex artery, whose branches supply the vasti muscles. The femoral artery then divides into the saphenous artery and the popliteal artery. Just distal to the medial side of the knee, the saphenous artery divides into the caudal tibial artery and a cranial branch that ramifies into superficial and deep branches. These go on to form superficial and deep arches on the dorsum of the foot. The caudal tibial artery passes into the sole of the foot as the medial plantar artery. The popliteal artery travels to the caudal aspect of the hind limb to reach the popliteal region. Here, the *a. genu lateralis* and *medialis* arise.

Systemic Venous Circulation

The external jugular vein travels along the lateral aspect of the neck and drains the venous blood from the head. Just caudal to the mandibular angle, it branches into the linguofacial vein and the maxillary vein. The mandibular gland is situated between both branches (Fig. 2.20).

The venous circulation of the limbs consists of a deep and a superficial system. The deep system accompanies the arteries, whereas the superficial veins have no arterial counterpart. The cephalic vein, which is located at the cranial side of the antebrachium, is the major superficial vein of the forelimb. The hind limb presents two superficial veins. The medial saphenous vein (*v. saphena magna*) runs on the medial side of the leg. Its cranial and caudal branches join below the knee to travel as an unpaired vein toward the femoral vein. The latter is located in the femoral canal and is covered by the superficial inguinal lymph node. The lateral saphenous vein (*v. saphena parva*) distally runs on the caudal side of the leg. Here, the vein deviates proximally to the lateral side of the knee where it runs adjacent to the popliteal, where it can be accessed for intravenous injection and blood collection (Fig. 2.29A). Larger volumes can be collected from the femoral vein (Fig. 2.15). The lateral saphenous vein accompanies the popliteal lymph node in the popliteal fossa and has a connection with the deep venous system (Fig. 2.29B).

Lymphatic System

The mandibular and parotid lymph nodes are illustrated in Fig. 2.20. The superficial inguinal lymph node can easily be observed after superficial dissection of the femoral canal through which the femoral artery courses (Fig. 2.15). The popliteal lymph node is rather superficially located in the popliteal fossa. Here, it is accompanied by the lateral saphenous vein (Fig. 2.29).

In New World monkeys, connections between the lymphatic and venous system are present in the region of the renal veins. In this way, the interstitial fluid from the abdominal viscera, the pelvic organs, and the caudal half of the body drains into the caudal vena cava. The lymph drainage of the upper half of the body is similar to that of all the other primates [1,9].

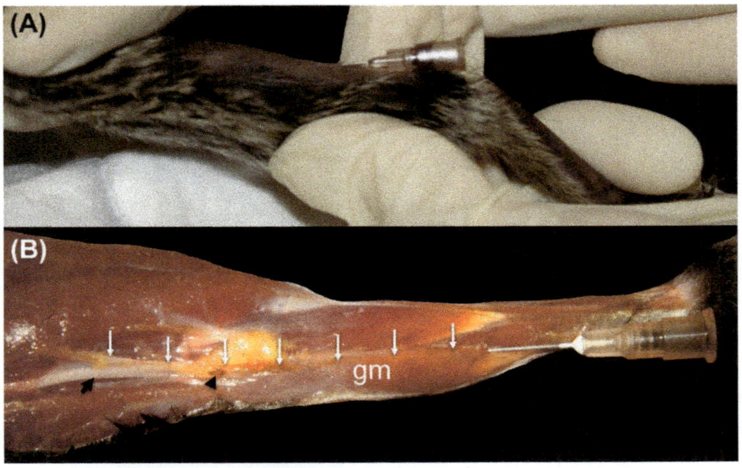

FIGURE 2.29 The caudal branch of the medial saphenous vein can be used for intravenous injection or blood drawing. A: The needle is inserted proximal to the calcaneus at the caudal side of the calf. B: Caudal view of the left skinned hind limb showing the medial saphenous vein (*white arrows*) running distally in the midline between both parts of the gastrocnemius muscle (gm). At the level of the knee, the vein passes by the popliteal lymph node (*black arrowhead*). More proximally, the sciatic nerve (*black arrow*) runs adjacent to the vein.

Spleen

The reddish spleen is tongue-shaped and lies at the left side of the abdomen. It is connected to the stomach by means of the gastrosplenic ligament (Fig. 2.22). The spleen may weigh 0.168–0.949 g in the male marmoset with a body weight range of 182.9–381.9 g [23]. In the female marmoset that has a weight of 201.2–412.1 g, the spleen weight can be 0.190–1.048 g [23].

ENDOCRINE ORGANS

Pancreas

The pancreas is described above as part of the digestive system.

Adrenal Glands

The pale adrenal glands lie on the cranial poles of the kidneys. They are pyramidal in shape with the base resting on the kidney (Figs. 2.25 and 2.26). They may weigh 0.051–0.116 g in the male marmoset that has a body weight range of 182.9–381.9 g [23]. For a female marmoset with an acceptable body weight range of 201.2–412.1 g, the value is 0.069–0.143 g [23].

Thyroid and Parathyroid Gland

The elongated, paired thyroid gland is located at the lateral side of the trachea, just caudal to the larynx (Fig. 2.18). A ventral isthmus can only seldom be observed. The gland has a translucent pale amber to yellow color. It measures approximately 8 mm in length, 3 mm in width, and has a thickness of 1 mm. The average weight is 150 mg. The gland is composed of colloid-filled follicles and is surrounded by a delicate, fibroelastic connective tissue capsule [38]. The parathyroid gland is very small and lies cranially on the thyroid gland (Fig. 2.18).

SENSORY ORGANS

The Eye

The lacrimal gland of the common marmoset lies within the orbit at the dorsolateral and ventral sides near the inner canthus (the glands of Krause). The two, small, glandular masses are not enveloped by a connective tissue capsule. These glands are serous and of the tubuloalveolar type [39]. Their location makes it sometimes hard to sample the gland for examination in toxicological studies. The extraorbital lacrimal gland lies in the temporal fossa just caudal to the lateral margin of the orbit. It is a gray-colored, flat gland measuring approximately 10 mm in length and 2 mm in width. The gland is surrounded by connective tissue that also divides the gland into large lobes with smaller lobules. The gland consists of serous secretory cells that form a tubuloacinar pattern. Well-differentiated excretory ducts can be observed within the connective tissue [40].

The pupil is round and the iris is colored brown (Fig. 2.1). Male marmosets are dichromatic, whereas females are di- or trichromatic. The marmoset possesses a wide range of accommodation. The retina has a well-developed cone-dominated fovea. The paraxial optics of the marmoset eye are remarkably well represented as a scaled-down version of the human eye. The cone topography of the marmoset retina is similar to that of humans and rhesus monkeys. The peak foveal cone density and mean foveal cone spacing are the same in the marmoset and the human [41].

The Ear

The pinna is approximately 2 cm high and 1.5 cm wide. Its caudal surface is covered with dense tufts of white-tipped hair (Fig. 2.1). The pinnae are relatively fixed in the marmoset. Therefore, the animal shows characteristic head movements while localizing the sound source. The hearing range has lower and upper limits of 100 Hz and 25 kHz, respectively [28].

The eardrum is almost circular as it has dimensions of 4–4.3 mm. The malleus has no prominent lateral, rostral, or muscular processes. A small lenticular bone is present between the incus and the stapes [2]. The malleus weighs approximately 1.4 mg. The incus weighs about 1.8 mg and has a stout body and two processes, the longer of which is attached to the stapes [28] (Fig. 2.30).

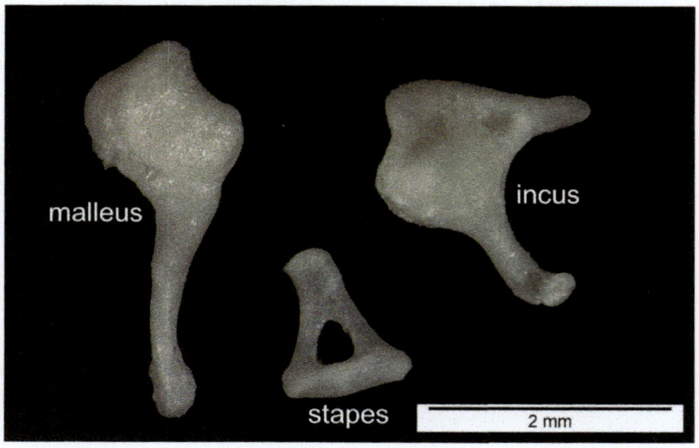

FIGURE 2.30 The three auricular ossicles of the common marmoset.

References

[1] Beattie J. The anatomy of the common marmoset (*Hapale jacchus* Kuhl). Proc Zool Soc Lond 1927;97:593–718.

[2] Casteleyn C, Bakker J, Breugelmans S, Kondova I, Saunders J, Langermans JAM, Cornillie P, Van den Broeck W, Van Loo D, Van Hoorebeke L, Bosseler L, Chiers K, Decostere A. Anatomical description and morphometry of the skeleton of the common marmoset (*Callithrix jacchus*). Lab Anim 2012;46:152–63.

[3] Garber P. Vertical clinging, small body size, and the evolution of feeding adaptations in the *Callitrichinae*. Am J Phys Anthropol 1992;88:469–82.

[4] Lemelin P. Comparative and functional myology of the prehensile tail in new world monkeys. J Morphol 1995;224:351–68.

[5] Miraglia T, dos Santos AJ. Histochemical data on the sebaceous glands in the lips of the marmoset (*Callithrix jacchus*). Acta Anat 1971;78:295–305.

[6] Miraglia T, da Costa NJW, Costa Guedes ML. The skin of the palms and soles of the marmosets (*Callithrix jacchus* and *Callithrix penicillata*). Acta Anat 1976;96:97–114.

[7] Thorndike EE. A microscopic study of the marmoset claw and nail. Am J Phys Anthropol 1968;28:247–62.

[8] de Castro Faria MS, Miraglia T. The eccrine sweat glands of the palms and soles of the common marmoset (*Callithrix jacchus*). Acta Anat 1971;79:466–80.

[9] Swindler DR, Wood CD. An atlas of gross primate anatomy – baboon, chimpanzee, and man. London and Seattle: University of Washington Press; 1973.

[10] Hilloowala RA. Comparative anatomical study of the hyoid apparatus in selected primates. Am J Anat 1975;142:367–84.

[11] Bosseler L, Cornillie P, Saunders J, Bakker J, Langermans J, Casteleyn C, Decostere A, Chiers K. Micromelic dysplasia-like syndrome in a captive colony of common marmosets (*Callithrix jacchus*). Comp Med 2014;64:394–403.

[12] Bakker J, Uilenreef JJ, Pelt ERJ, Brok HPM, Remarque EJ, Langermans JAM. Comparison of three different sedative-anaesthetic protocols (ketamine, ketamine-medetomidine and alphaxalone) in common marmosets (*Callithrix jacchus*). BMC Vet Res 2013;2013(9):113. https://doi.org/10.1186/1746-6148-9-113.

[13] Bertassoli BM, Lessa TB, de Oliveira TD, de Abreu DK, Stunitz de Silva LC, dos Santos AD, Grassi Rici RE, Assis Neto AC. Techniques used for morphological and ultrastructural description from teeth white-tufted-ear-marmosets (*Callithrix jacchus*). In: Méndez-Vilas A, editor. Current microscopy contributions to advances in science and technology. Badajoz: Formatex; 2012. p. 245–50.

[14] Ferrari SF, Aparecida Lopes M, Krause EAK. Gut morphology of *Callithrix nigriceps* and *Sanguinus labiatus* from western Brazilian Amazonia. Am J Phys Anthropol 1993;90:487–93.

[15] Rosenberger AL. Loss of incisor enamel in marmosets. J Mammol 1978;59:207–8.

[16] Hershkovitz P. Living new world monkeys (*Platyrrhini*) - with an introduction to primates, vol. 1. Chicago (London): The University of Chicago Press; 1977.

[17] Shaw JH, Auskaps AM. Studies on the dentition of the marmosets. Oral Surg Oral Med Oral Pathol 1947;7:671–7.

[18] Johnston GW, Dreizen S, Levy BM. Dental development in the cotton ear marmoset (*Callithrix jacchus*). Am J Phys Anthropol 1970; 33:41–8.

[19] Kubota K, Hayama S. Comparative anatomical and neurohistological observations on the tongues of Pigmy and common marmosets. Anat Rec 1964;150:473–86.

[20] Yamaguchi K, Harada S, Kanemaru N, Kasahara Y. Age-related alteration of taste bud distribution in the common marmoset. Chem Senses 2001;26:1–6.

[21] Dreizen S, Goodrich JS, Levy BM. Comparison of the salivary glands and salivary electrolytes in man and marmoset. Arch Oral Biol 1968;13:229–37.

[22] Whitmore I. Oesophageal striated muscle arrangement and histochemical fiber types in Guinea-pig, marmoset, macaque and man. J Anat 1982;134:685–95.

[23] Wadsworth PF, Budgett DA, Forster ML. Organ weight data in juvenile and adult marmosets (*Callithrix jacchus*). Lab Anim 1981;15:385–8.

[24] Hunter AJ, Fleming D, Dixson AF. The structure of the vomeronasal organ and nasopalatine ducts in *Aotus trivirgatus* and some other primate species. J Anat 1984;138:217–25.

[25] Wako K, Hiratsuka H, Katsuta O, Tsuchitani M. Anatomical structure and surface epithelial distribution in the nasal cavity of the common cotton-eared marmoset (*Callithrix jacchus*). Exp Anim 1999;48:31–6.

[26] Surribas JF, von Lawzewitsch I. The respiratory system of the genus Callithrix, the South American Monkey. Primates 1987;28: 393–402.

[27] Nishimura TD, Takai M, Tsubamoto T, Egi N, Shigehara N. Variation in maxillary sinus anatomy among platyrrhine monkeys. J Hum Evol 2005;49:370–89.

[28] Aitkin L, Park V. Audition and the auditory pathway of a vocal new world primate, the common marmoset. Prog Neurobiol 1993;41:345–67.

[29] Bakker J, van Nijnatten TJM, Louwerse AL, Baarends G, Arndt SS, Langermans JAM. Evaluation of ultrasonic vocalizations in common marmosets (*Callithrix jacchus*) as a potential indicator of welfare. Lab Anim 2014;43:313–20.

[30] Barbier A, Bachofen H. The lung of the marmoset (*Callithrix jacchus*): ultrastructure and morphometric data. Respir Physiol 2000;120:167–77.

[31] Miraglia T, Rossi R, Moreira E. Data on the kidneys of the marmoset (*Callithrix jacchus*). Acta Anat 1968;69:274–81.

[32] Miraglia T, Telles Filho M, Lobão Castelo Branco A. The male reproductive system of the common marmoset (*Callithrix jacchus*). Acta Anat 1970;76:594–611.

[33] Dixson AF. Penile spines affect copulatory behaviour in a primate (*Callithrix jacchus*). Physiol Behav 1991;49:557–62.

[34] Cui K-H, Matthews CD. Anatomy of female common marmoset (*Callithrix jacchus*) reproductive system. J Anat 1994;185:481–8.

[35] Stein FJ. Sex determination in the common marmoset. Lab Anim Sci 1978;28:75–80.

[36] de Freitas Burity CH, Mandarim-de-Lacerda CA. The weight of the heart in *Callithrix* Erxleben, 1777. Bulletin de l'Association des Anatomistes 1995;79:21–4.

[37] Takeda Y, Okuda H. The facial artery of the common marmoset (*Callithrix jacchus*). Okajimas Folia Anat Jpn 1990;67:97–110.

[38] Levy BM, Hampton S, Dreizen S, Hampton Jr JK. Thyroiditis in the marmoset (*Callithrix* spp. and *Saguinus* spp. J Comp Pathol 1972;82:99–103.

[39] Marback R, Miraglia T, Ferreira Gomes N. Data on the structure and distribution of polysaccharides in the lacrimal glands of the common marmoset (*Callithrix jacchus*). Acta Anat 1972;83: 106–11.

[40] Katsuta O, Yamaguchi-Onozawa M, Okazaki K, Itoh T, Okazaki Y, Tsuchitani M. Gross and microscopic anatomy of the extraorbital lacrimal gland of the common marmoset (*Callithrix jacchus*). Comp Med 2000;50:609–12.

[41] Troilo D, Howland HC, Judge SC. Visual optics and retinal cone topography in the common marmoset (*Callithrix jacchus*). Vis Res 1993;33:1301–10.

CHAPTER

3

Neuroanatomy of the Marmoset

Makoto Fukushima[1,2], *Noritaka Ichinohe*[3,4], *Hideyuki Okano*[1,2]

[1]Laboratory for Marmoset Neural Architecture, RIKEN Brain Science Institute, Saitama, Japan; [2]Department of Physiology, Keio University School of Medicine, Tokyo, Japan; [3]Laboratory for Ultrastructural Research, National Institute of Neuroscience, National Institute of Neurology and Psychiatry, Tokyo, Japan; [4]Laboratory for Molecular Analysis of Higher Cognitive Function, RIKEN Brain Science Institute, Saitama, Japan

INTRODUCTION

Decades of neuroscience research on nonhuman primates have provided invaluable insights into the neural mechanisms underlying human behavior and cognition. The common marmoset (*Callithrix jacchus*) has recently gained popularity as an attractive animal model for neuroscience research for several reasons [1], among which are its natural social behavior, which is similar to that of humans [2], and short breeding cycle, which facilitates genetic manipulations. Genetic modification techniques have been developed to generate transgenic marmosets using lentiviral-mediated gene transfer [3] as well as gene-knockout marmosets using genome editing [4]. Furthermore, genetic tools have become available to observe and manipulate neuronal populations and circuits in the marmoset brain [5–7].

The neuroanatomical structure of the marmoset is reminiscent of that in humans. The spinal cord of the marmoset is similar to that of other mammals with some deviations. The marmoset spinal cord consists of 34 nerve segments (8 cervical, 12 thoracic, 7 lumbar, 3 sacral, and 4 coccygeal segments) [8], which is slightly less than that of the human spinal cord (31 segments). As a notable deviation, the hand motor neurons reside in the eighth cervical segment (C8) of the marmoset spinal cord, whereas they are located in the first thoracic segment (T1) in other mammals. The spinal cord ends at the coccygeal vertebrae within the spinal canal [9]. Neural tracts of the marmoset spinal cord have been studied with neurophysiological and neuroanatomical methods [10] as well as neuroimaging techniques such as diffusion tensor tractography [11]. In the marmoset, the majority of the corticospinal tract (CST) descends in the contralateral dorsolateral funiculus with considerable individual variability (74.4%–90.7%) [10]. Compared with humans [12], a small number of the CST axons (0.5%–5.0%) descend in the ventromedial funiculus [10]. In contrast to humans and macaques, the direct corticomotoneuronal connection has not been identified in marmosets, which have very sparse corticospinal termination of the CST in lamina IX [10].

The marmoset brain shares a number of similarities with the human brain, such as an expanded temporal lobe and hierarchically structured sensory cortices [13,14] as well as a highly developed prefrontal cortex (PFC) [15,16] (Figs. 3.1A and 3.2). Such similarities increase the attractiveness of marmosets as an animal model for the identification of neuronal mechanisms underlying human behavior and cognition. However, by definition, any animal model will present differences from humans, and assuming a complete one-to-one mapping of cortical areas among human and other primate species is implausible [19]. In this chapter, we review the neuroanatomy of marmosets by emphasizing fundamental similarities and differences between marmosets and other primate species, including macaques and humans.

The Common Marmoset in Captivity and Biomedical Research
https://doi.org/10.1016/B978-0-12-811829-0.00003-0

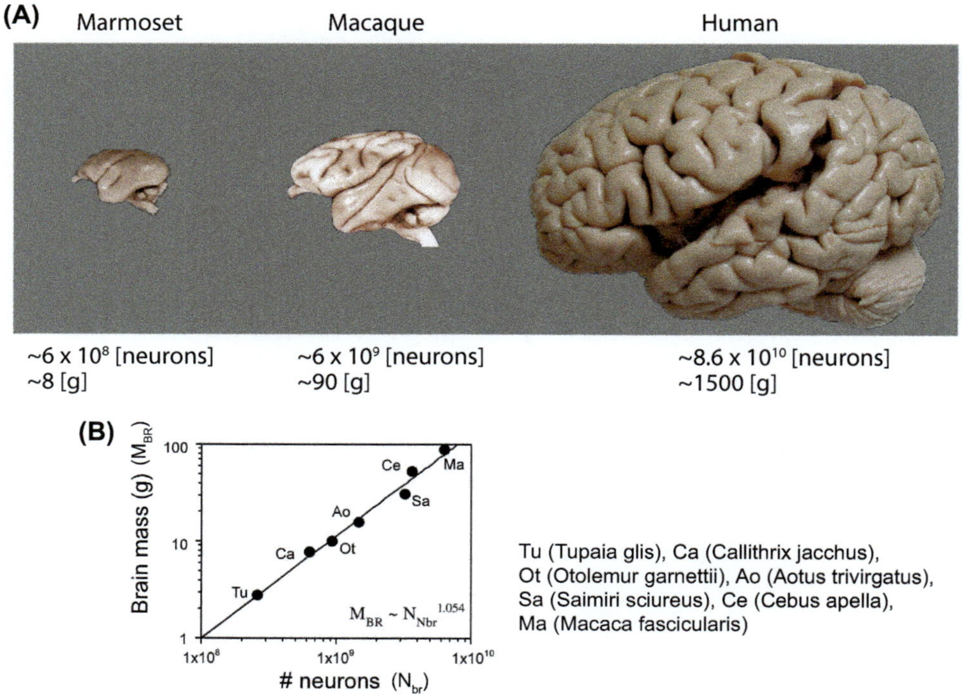

FIGURE 3.1 **Marmoset brain size compared with the brain sizes of other primate species.** (A) Size and number of neurons in the marmoset, macaque, and human brains. (B) Relationship between brain mass (M_{BR}) and neuron number (N_{br}) for six nonhuman primates (Ca, Ot, Ao, Sa, Ce, and Ma) and the tree shrew (Tu). *(A) Adapted from Herculano-Houzel S. The human brain in numbers: a linearly scaled-up primate brain. Front Hum Neurosci 2009;3:1–11. https://doi.org/10.3389/neuro.09.031.2009. (B) Adapted from Herculano-Houzel S, Collins CE, Wong P, Kaas JH. Cellular scaling rules for primate brains. P Natl Acad Sci USA 2007;104:3562–67. https://doi.org/10.1073/pnas.0611396104.*

ORGANIZATION OF THE MARMOSET BRAIN STRUCTURE

Comparison of the Brain of Marmosets With the Brain of Humans and Other Mammals

The common marmoset is a small-sized monkey (300–500 g for adults) that exhibits ease of handling and breeding in laboratory environments, and their small size is one of the advantages of utilizing marmosets in neuroscience research. However, one may question whether the brain structure of this small monkey can be appropriately scaled to represent a good model of the human brain. Moreover, whether the marmoset model is superior to rodent models regarding neuroanatomical structure and the reasons for favoring marmoset models are issues that remain to be discussed.

The marmoset brain and other nonhuman primate brains can be considered isometrically scaled versions of the human brain with regard to weight and neuron number. The marmoset brain weighs approximately 8 g, which is approximately 11 times smaller than the brain of a macaque (approximately 90 g) and approximately 190 times smaller than the brain of a human (approximately 1500 g) [17] (Fig. 3.1A). The isotropic fractionator method has been used to estimate the number of neurons in the brain of many animal species [20]. The number of neurons in the marmoset brain is approximately 60 million, which is approximately 100 times less than that in the macaque brain (6 billion) and 1400 times less than that in a human brain (860 billion) (Fig. 3.1A). Thus, the number of neurons in the brain increases isometrically as a function of weight [18] (Fig. 3.1B), suggesting that the average size of the neuron does not significantly increase with the size of the brain. However, the isometric relationship in rodents is considerably different from that in primates, and a larger brain would be required to support the same number of neurons observed in the primate brain. Following the rodent scaling law, to hold approximately 100 billion neurons, the brain would have to weigh >45 kg [18,21]; however, the human brain holds more than 100 billion neurons at a considerably reduced weight (several kilograms). Indeed, the cellular composition of the human brain follows the isometric relationship derived from *nonhuman primate* species; thus, the human brain can be considered an isometrically scaled-up version of the nonhuman primate brain, including the marmoset brain [22].

Adherence to the scaling law of the primate brain in terms of size and weight does not exclude the possibility that the internal structure of the marmoset brain is organized differently from that of the human brain. Compared with other nonhuman primate species,

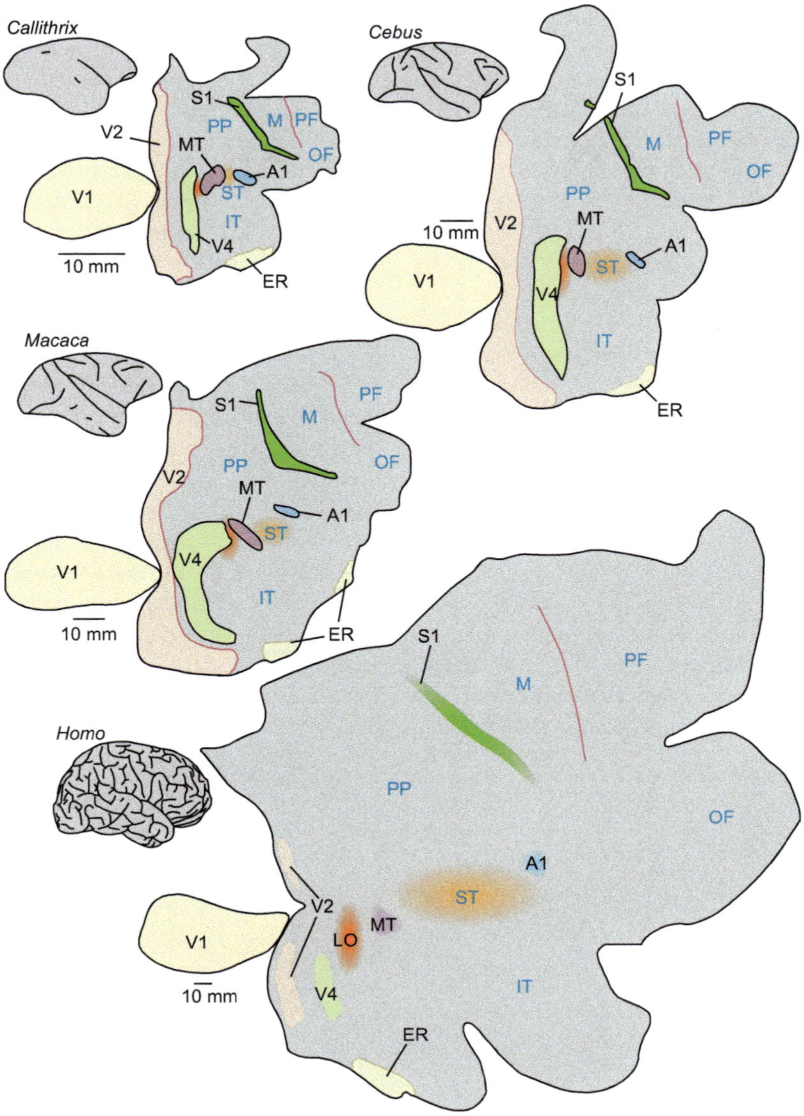

FIGURE 3.2 **Relative size of the visual cortical area varies across primate species.** Flattened maps of the cerebral cortex of four primates are plotted. Note the different scale bars for each species. *Reproduced from Rosa MGP, Tweedale R, Tweedale R. Brain maps, great and small: lessons from comparative studies of primate visual cortical organization. Philos Trans R Soc Lond B Biol Sci 2005;360:665–91. https://doi.org/10.1098/rstb.2005.1626.*

relative increases in the number of human prefrontal neurons compared with other cortical areas are not observed as assessed by the distribution of neurons over the anteroposterior axis [23].[1] This result clearly suggests that the *absolute* number of neurons is one important factor that differentiates humans from nonhuman primates. However, the relative size of the cortical areas differs significantly across primate species. For example, the size of the primary visual cortex (V1) in marmosets is twice as large as that of the secondary visual cortex (V2), whereas the sizes of the V1 and V2 are nearly equivalent in the macaque brain [19,24]. (Fig. 3.2). The proportion of gray matter occupied by the V1 in the entire brain is higher in the marmoset brain (14.5%) than in the human brain (approximately 1.5%) [14,25]. Furthermore, whole-brain comparisons using the cortical surface registration of magnetic resonance (MR) images suggests that three cortical structures present significant expansion across four primate species (marmoset, capuchin, macaque, and human): the temporal parietal junction, the ventrolateral PFC, and the dorsal anterior cingulate cortex [26]. As evident

[1] Note that in the study by Ref. [23], marmosets were not included.

in this study, the marmoset brain has a flat "lissencephalic" cortex with fewer sulci than the brains of other primates, such as humans and macaques. For example, the marmoset brain does not possess the following major sulci of the macaque brain: the central, arcuate, principal, intraparietal, and lunate sulci. However, shallow indentations corresponding to a number of these sulci (e.g., central) could be identified by examining the curvature of the adjacent subcortical white matter [27]. The superior temporal sulcus in marmosets is shorter and shallower than that of macaques. The lateral sulcus is the deepest sulcus of the marmoset brain, although the core areas of the auditory cortex, which are embedded in the lateral sulcus in the macaque brain, are partially exposed on the gyrus.

Annotation of the Cortical Areas: Atlas of the Marmoset Brain

Although cross-species analyses of cortical areas that are identifiable in all species of interest could be straightforward, certain cortical areas are not easily identified in different species. The parcellation of the cortical and subcortical areas is classically based on a combination of cytoarchitectural and topographic mapping of the human brain parcellation [28] (Fig. 3.3A). Knowledge of neuroanatomical connections helps demarcate and categorize cortical area boundaries [29,30]. Recently, functional couplings have been incorporated to enable consistent and novel cortical parcellation in the human brain [31]. Such an approach has not been fully applied for parcellation in marmosets, although it could be useful because of the similarities between marmoset and human cortical structures and the substantial variability across individual marmosets [32].

The current standard atlas for the marmoset brain was published by Paxinos et al. [15] (Fig. 3.3B). The atlas follows stereotaxically precise coordinates and provides coronal sections stained for Nissl, calbindin, acetylcholinesterase, and nonphosphorylated neurofilaments (SMI-32 immunostaining). Cortical and subcortical areas in this atlas were delineated using information regarding the cytoarchitecture as well as results from neuroanatomical studies on marmosets. Each of those areas was named following the nomenclature and abbreviations used to describe the brains of rodents and macaques. In this chapter, we follow the nomenclature described in Ref. [15] unless otherwise noted.

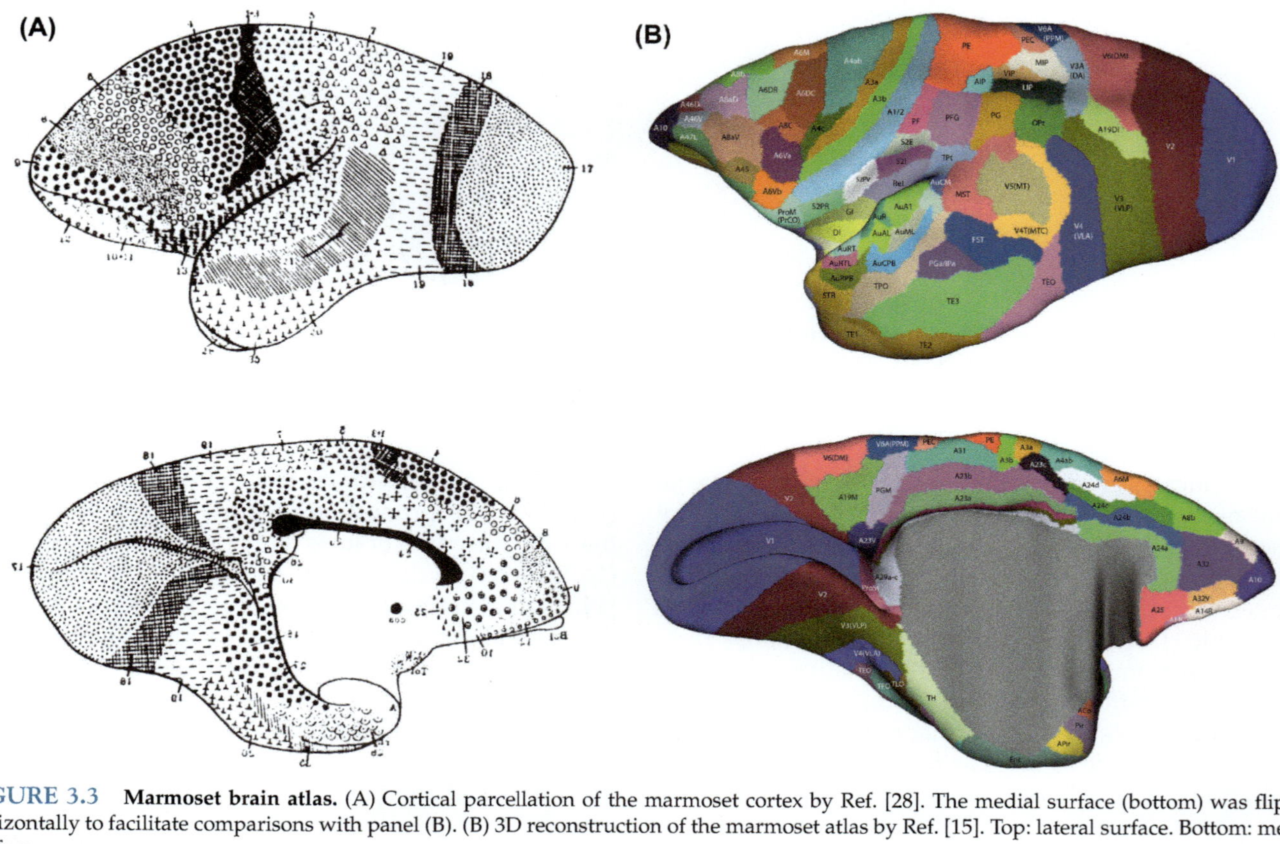

FIGURE 3.3 **Marmoset brain atlas.** (A) Cortical parcellation of the marmoset cortex by Ref. [28]. The medial surface (bottom) was flipped horizontally to facilitate comparisons with panel (B). (B) 3D reconstruction of the marmoset atlas by Ref. [15]. Top: lateral surface. Bottom: medial surface.

Marmoset Cortical Areas: Notable Examples

In the current nomenclature of marmosets, certain cortical areas are uniquely defined. As discussed above, performing simple topographic mapping from one primate to another is often difficult, if not impossible. Next, we describe examples that illustrate the differences between the brain of marmosets and the brains of other primates.

Premotor Cortex: Area 8C, a Newly Defined Field in Marmosets

The marmoset premotor cortex consists of five areas: areas 6DR, 6DC, and 8C in the dorsal premotor cortex and areas 6Va and 6Vb in the ventral premotor cortex. Area 8C was described in Ref. [15] but not in prior studies (e.g., [33]). A homologue of this area in macaques has not been clearly identified. Because of the cytoarchitecture of this area, it was defined as a distinct area, and a tracer injection study subsequently confirmed the connection patterns from area 8C, suggesting its homologous relationship with the premotor cortex [29]. We will discuss this issue in the later section on the premotor cortex.

Missing Area 44 in the Marmoset Brain

Marmosets are highly vocal and present active vocal communication among conspecifics [34–36]. Thus, exploring the evolutional origin of human speech in the marmoset is reasonable. One of the major cortical structures associated with human speech is "Broca's area," which was identified based on Broca's findings in patients with a lesion in the posterior part of the ventrolateral frontal lobe who exhibited severe speech production deficits [37]. The degree to which Broca's area contributes to human speech production has been controversial [38], and the exact definition of Broca's area varies significantly across studies [39]. However, most studies indicate that the major components of Broca's area are areas 44 and 45. Interestingly, area 44 is *missing* in the marmoset brain, although area 45 is defined as the part of the lateral PFC [15,16] (Fig. 3.4). Why has area 44 not been identified in the vocal marmoset?

Petrides and his colleagues defined area 44 in the macaque brain as the cortical area that satisfies the following three characteristics of area 44 in the human brain: (1) dysgranular cortical layer, (2) functional relevance to orofacial function, and (3) topographic mapping of cortical areas in the human PFC [41,42]. The layout of the cortical layers in area 44 can be considered *dysgranular* and is thin and scattered throughout cortical layer 4 in area 44 of the human brain. Layers 4 and 5B contain small to medium pyramidal cells. Layer 5A has distinct large pyramidal cells. Layer 2 is poorly defined, whereas layer 3 contains pyramidal cells of various sizes, with the larger cells at the bottom of the layer [41,42]. These features are distinct from the cortical areas adjacent to area 44: the cortical layer of area 45 has a clear layer 4 that separates layer 3 from layer 5 (i.e., granular), whereas area 6V completely lacks layer 4 and is called the *agranular* cortex in macaques. Although dorsal area 6 (areas 6DR and 6DC) in the marmoset brain may be agranular, the ventral area 6V (6Va and 6Vb) does seem to contain layer 4 (Fig. 41A and B in Ref. [15], for example).

As for the topographical relationship, area 44 should be located immediately caudal to area 45. In the macaque brain, area 44 is located within the posterior bank of the lower ramus of the arcuate sulcus; however,

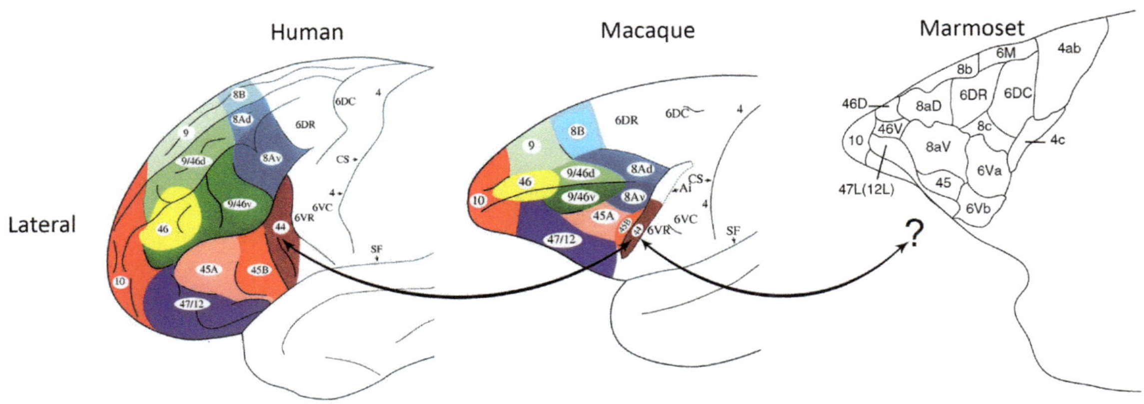

FIGURE 3.4 **Comparison of the lateral prefrontal cortex across three primate species.** The lateral surface of the prefrontal cortex for humans (left), macaques (middle), and marmosets (right). The *arrows* indicate the locations of the Brodmann's area 44 in the human and macaque brain. In the marmoset brain, area 44 is missing. *Middle: adapted from Petrides M. Lateral prefrontal cortex: architectonic and functional organization. Philos Trans R Soc Lond B Biol Sci 2005;360:781–95. https://doi.org/10.1098/rstb.2005.1631 and right: cortical boundaries extracted from Paxinos G, Watson C, Petrides M, Rosa M, Tokuno H. The marmoset brain in stereotaxic coordinates. Academic Press; 2012.*

the arcuate sulcus is not observed in the marmoset brain (or the human brain). Therefore, identifying the area in the marmoset brain that is homologous to area 44 in the macaque brain is not straightforward. Moreover, marmoset studies have not performed cortical electrical stimulation to induce orofacial movements (e.g., jaw and lip movements), which has been performed in macaques [42,43].

Concluding that all possibilities for identifying the area 44 homologue in the marmoset brain have been exhausted may be unreasonable. A potential candidate for this homologous area in the marmoset brain is area 6Vb, which has a scattered ("agranular") layer 4 and is topologically located just caudal to area 45. In the cortical areas dorsal to area 6Vb, such as areas 6Va and 8C, the pattern of neuroanatomical connectivity is similar to that of the premotor area rather than the prefrontal area [29], although the "incipient" layer 4 of area 8C [15] appears to present the dysgranular feature of area 44. Although the cortical inputs to area 6Vb remain to be evaluated, projections from the STS have been found, which has led to the identification of neurons near 6Vb that have "mirror" properties [44] (additional information is included in the section on the premotor cortex below). Thus, the search for the area 44 homologue in marmosets remains ongoing. In such a complex situation, cortical parcellation should be investigated by combining multiple characteristics, such as functional properties and connectivity, rather than focusing on anatomical cytoarchitectural features alone. The case of area 44 represents a notable example showing that cortical parcellation across multiple species is not always straightforward.

NEUROANATOMICAL CONNECTIONS

Next, we describe our current understanding of the neuroanatomical connections among the cortical and subcortical areas in marmosets. Considerable research must still be performed to fully identify the neuroanatomical connections in the marmoset brain; thus, this is an active field of investigation [1].

Auditory Cortex

The auditory cortex is the main cortical structure underlying sound perception. Processing auditory information is crucial in marmosets, which engage actively in vocal communication and demonstrate complex sound perception, such as detecting the pitch of sounds [45–47]. Four published tracer injection studies on the marmoset brain have attempted to identify the neuroanatomical connections to, from, and within the auditory cortex. Here, we summarize these studies regarding the corticocortical connections [48–50] and the subcortical inputs to the auditory cortex [51].[2]

The basic structure of the marmoset auditory cortex is similar to that of the macaque [52–54], and it consists of three concentric fields: core, belt, and parabelt (Fig. 3.5). The core area is further subdivided into three areas (from caudal to rostral): A1, R, and RT. Eight belt areas surround the core areas, and the four medially located belt areas, called the medial belt, are as follows: CM, MM, RM, and RTM. The other belt areas are positioned laterally to the core areas, called the lateral belt: CL ML, AL, and RTL. CM and CL are located caudally to A1, and each of the other belt areas is either positioned laterally or medially to each core field (Fig. 3.5D). In marmosets, the lateral half of the core area and the lateral belt areas (CL, ML, AL, and RTL) are located on the gyrus, which is inconsistent with that in macaques, whereas other auditory areas are buried in the lateral sulcus, which is similar to that in macaques. The superior temporal rostral area (STR) is rostral to RT in the marmoset brain in the gyrus and the lateral sulcus. The STR may correspond to the RTp (inside of the LS) and STGr (gyrus) in the macaque brain. This area has been recently shown to have selective responses to conspecific vocalizations among marmosets [55] and is similar to the RTp or temporal pole in the macaque brain [56–59].

Subcortical Inputs to the Auditory Cortex

An operational definition of the auditory cortex is that the cortex receives inputs from the medial geniculate nucleus (MGN) of the thalamus [13,60]. In the marmoset brain, the thalamic inputs to the auditory cortex have not yet been examined completely. Thus far, two tracer injection studies have reported thalamic inputs to/from the three core areas (A1, R, and RT) and two medial belt areas (RM and CM/MM) [50,51] (Fig. 3.5A and B).

The ventral division of MGN (MGv) provides the major inputs to the core areas. Tracer injections to the two core areas A1 and R showed that the highest percentage of labeled cells occurred in the MGv (approximately 60% for A1 and 70% for R) [51] (Fig. 3.5A). The macaque A1 and R also received a similar proportion of the inputs from MGv [60]. RT is another core area rostral

[2] The distinction between CM and MM has been identified relatively recently; thus, two of the tracer injection studies [48,51] reviewed here do not include this distinction, and the term "CM" in their results should be interpreted as representing either CM or MM. Therefore, we collectively refer to the "CM" in their results as CM/MM.

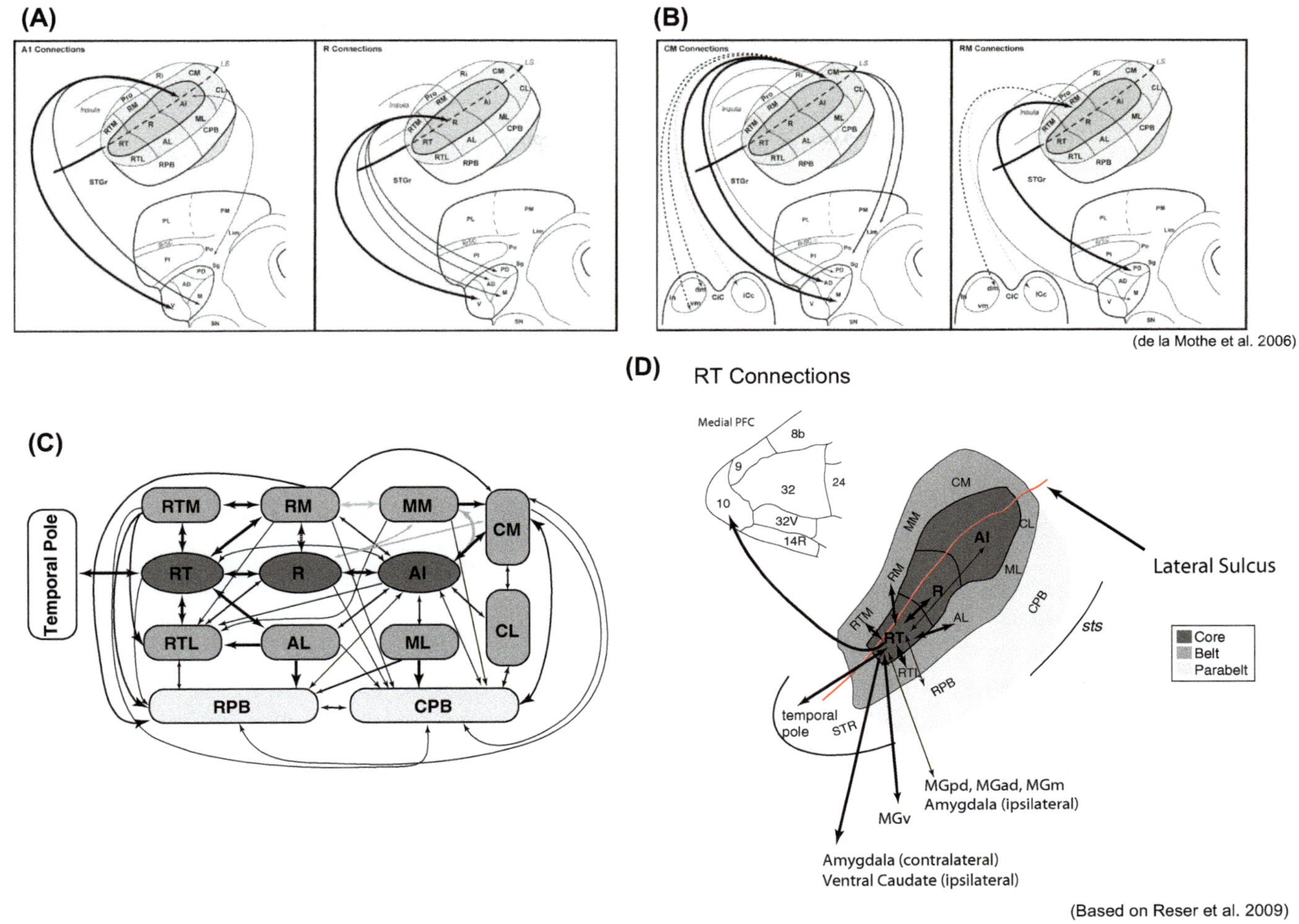

FIGURE 3.5 **Cortical connection revealed by tracer injections in the marmoset auditory cortex.** Thalamocortical connections of (A) two auditory core areas (A1 and R) and (B) belt areas (CM and RM; note: CM refers to CM and MM in the current nomenclature; see the main text). (C) Summary of the corticocortical connections of the auditory cortex. Cortical connections estimated from three tracer injection studies [48–50] were summarized. The thickness of the lines indicates the strength of the connection. (D) Summary diagram for the cortical connections of the rostral core area (RT). A schematic of the medial PFC was delineated from the surface plot of the marmoset atlas by Ref. [15] (Fig. 3.3B). *A1*, primary auditory cortex; *AL*, anterolateral belt; *CL*, caudolateral belt; *CM*, caudomedial belt; *ML*, middle lateral belt; *MM*, middle medial belt; *PFC*, prefrontal cortex; *R*, rostral core; *RT*, rostrotemporal core; *RTL*, rostrotemporal-lateral belt; *RTM*, rostrotemporal-medial belt. *(B) Adapted from La Mothe De LA, Blumell S, Kajikawa Y, Hackett TA. Thalamic connections of the auditory cortex in marmoset monkeys: core and medial belt regions. J Comp Neurol 2006;496:72–96. https://doi.org/10.1002/cne.20924. The strength of the connection was estimated based on Reser DH, Burman KJ, Richardson KE, Spitzer MW, Rosa MGP. Connections of the marmoset rostrotemporal auditory area: express pathways for analysis of affective content in hearing. Eur J Neurosci 2009;30:578–92. https://doi.org/10.1111/j.1460-9568.2009.06846.x.*

to R, and it has dense reciprocal projections with the MGv [50] (Fig. 3.4). For weaker connections than those with the MGv, A1 has reciprocal connections with the magnocellular division of the MGN (MGm) and the posterior nucleus (Po). R is reciprocally connected with MGm as well as the anterior and posterior dorsal divisions of the MGN (MGad and MGpd). The RT also has reciprocal connections with the MGad, MGpd, MGm, and the contralateral ventral caudate. Moreover, RT is bilaterally connected with the amygdala and presents stronger connections with the contralateral side than with the ipsilateral side [50].

The primary inputs to the belt areas are provided from the MGad and MGpd, which is different from the primary inputs to the core areas [51] (Fig. 3.5B). This pattern is similar to that observed in the macaque auditory cortex [13]. The CM/MM receives dominant projections from MGad and MGm and weaker connections from the MGpd. The connection with the caudal part of the CM/MM[3] was largely restricted in the

[3] Putative "CM" in the current definition.

MGad and MGm. Outside of the MGN, the caudal CM/MM is connected with the Po, suprageniculate nucleus, limitans nucleus, and medial pulvinar. However, the rostral part of the CM/MM[4] has weaker connections from the MGpd in addition to those from the MGad and MGm. The CM also has corticotectal projections with the dorsomedial (dm) region of the inferior colliculus (IC). The projection also extends caudally to the pericentral shell of the ventromedial boundary of the IC. Compared with the CM/MM, the RM receives most of the thalamic inputs from the MGN, primarily from the MGpd and secondarily from MGm. The corticotectal connection with the RM is on the dm region of the IC (Fig. 3.4).

Corticocortical Connections in the Auditory Cortex

The subdivisions of the core, belt, and parabelt are highly interconnected with other structures [48—50] and can be summarized in a single diagram that shows the complex and dense patterns of corticocortical connections (Fig. 3.5C). The core areas A1, R, and RT in marmosets approximately confirm the nearest stepwise neighbor connections in which the connection tends to be stronger for the connection to the next core area (i.e., A1 and R, R and RT) [48], which is similar to observations in macaques [54]. The connection can be considered a within-level projection. In the macaque brain, the feedforward input from R is terminated in layer VI and deep layer III of the RT, suggesting that the A1, R, and RT areas do not belong to the same hierarchical level of auditory processing [54]. Tracer injections to the most rostral core, RT, have demonstrated a projection to the dorsal temporal pole and the rostromedial PFC (area 10) [50] (Fig. 3.5D).

A1 and R are topographically connected with the medial and lateral belt regions as well as the infragranular inputs from the parabelt [48]. A1 is strongly connected with the CM area but has weaker connections with the CL and ML areas and the belt fields lateral to R. Both the belt and parabelt regions tend to have topographic connections along the rostrocaudal axis, where injections into the caudal areas have shown stronger connections with caudal areas and injections into the rostral areas have shown stronger connections with rostral areas [49]. RT is also connected with the rostral auditory belt (RTL and RTM) and the rostral parabelt [50]. Overall, the pattern of connections is summarized by stepwise connections between the core and the belt and between the belt and the parabelt, and additional recurrent connections are observed among these structures.

[4] Putative "MM" in the current definition.

Visual Cortex

Although the macaque has primarily been used to investigate the neuroanatomy of the primate visual system, the marmoset brain has also emerged as a useful model in the study of vision. The marmoset brain is equipped with a visual system akin to that of humans, including high acuity foveal vision, trichromacy, and binocular visual fields [14]. The marmoset visual cortex is hierarchically structured and presents similar functional properties to that of macaques, including the V1 [61], V2 [62,63], and area MT [64,65] and face patches [66]. A number of studies have focused on the visual system of primates, and excellent reviews on this topic have recently been published [14,24]. Thus, only a brief overview of the cortical pathways in the marmoset visual system is presented here.

Similarly to other primates, the V1 in marmosets receives thalamic inputs from the lateral geniculate nucleus (LGN). The inputs from the parvocellular and magnocellular layers of the LGN primarily terminate in layers 4 and 6 of the V1, whereas neurons in the koniocellular layer of the LGN tend to project to the supragranular layer [24,67]. Direct inputs from the LGN to MT that bypass the V1 are also observed [68]. These inputs originate from the medial portion of the inferior pulvinar nucleus and the koniocellular layers of the LGN, which receive inputs from the retina [69].

Four Streams From the V1 to V2

Neurons in the V1 project to the V2 via four distinct streams, each of which has spatially segregated targets in the V2 from layer 4B [70]. Cytochrome oxidase (CO) staining has revealed three distinct structures in the V2 (thick, thin, and pale CO stripes) that interleave with each other [62,63,71]. All stripe types receive dominant V1 inputs from layers 2/3 and fewer inputs from layer 4A. Cells in layer 4B of the V1 significantly project to only the thick stripes and the pale stripes positioned laterally to the nearest thick stripe (pale-lateral) (Fig. 3.6A). The projection from layer 4B of the V1 to the pale stripe medially located to the nearest thick stripe (pale-medial) is nearly absent in the marmoset brain and limited in the macaque brain [72]. The thin stripes receive projections from CO blobs in the V1, whereas the pale stripes are projected from the interblob regions of the V1. However, the thick stripes receive projections from the blob/interblob border region. Furthermore, individual neurons in the blob or interblob project predominantly to a single stripe type. In summary, four parallel processing streams are observed from the V1 to V2 (Fig. 3.6A).

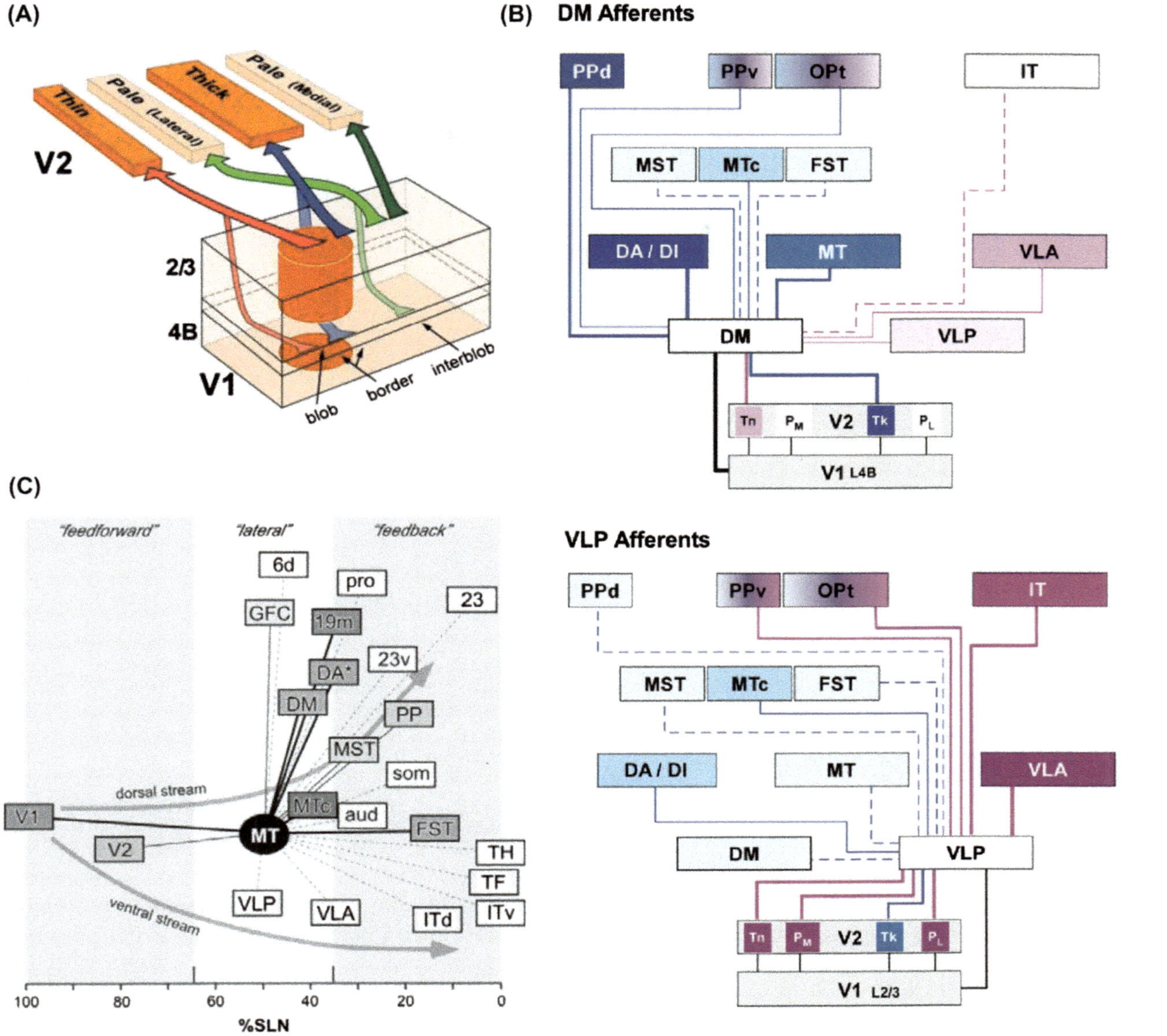

FIGURE 3.6 **Cortical connections of the marmoset visual cortex.** (A) Four pathways from the V1 to V2. The inputs from layer 4A and 5/6 of the V1 are omitted here. (B and C) Major cortical inputs to area V6 (DM) and VLP from the dorsal stream (blue) and the ventral steam (purple). P_L, pale-lateral; P_M, pale-medial; *Tk*, thick stripe; *Tn*, thin stripe. *(A) Adapted from Federer F, Ichida JM, Jeffs J, Schiessl I, McLoughlin N, Angelucci, A. Four projection streams from primate V1 to the cytochrome oxidase stripes of V2. J Neurosci 2009; 29:15455–71. https://doi.org/10.1523/JNEUROSCI.1648-09.2009. (B) Adapted from Palmer SM, Rosa MGP. Quantitative analysis of the corticocortical projections to the middle temporal area in the marmoset monkey: evolutionary and functional implications. Cereb Cortex 2006; 16:1361–75. https://doi.org/10.1093/cercor/bhj078. (C) Adapted from Jeffs J, Federer F, Angelucci A. Corticocortical connection patterns reveal two distinct visual cortical areas bordering dorsal V2 in marmoset monkey. Visual Neurosci 2015;32:E012. https://doi.org/10.1017/S0952523815000097.*

"Third-Tier" Visual Cortex

Visual information is further passed from the V1 and V2 to the "third-tier" visual areas adjacent to the anterior border of the V2. The organization of the third-tier visual cortex has been debated for decades because it originally referred to a structure in owl monkeys [73]. Several models have been proposed for this organization (Fig. 3.7), and comparisons among the models are described in previous review articles [24,74]. Briefly, the simplest model assumes that the visual area anterior to the V2 is the field of V3 (e.g., V3 only in Fig. 3.7A), although the third-tier visual cortex was later found to consist of multiple smaller areas, each of which contains a representation of the visual field with the upper and lower quadrants. Compared with other models, the "revised multiple-areas" model better explains a range of findings, including those from recent tracer injection studies [75–77]. In this model, the DM and VLP occupy

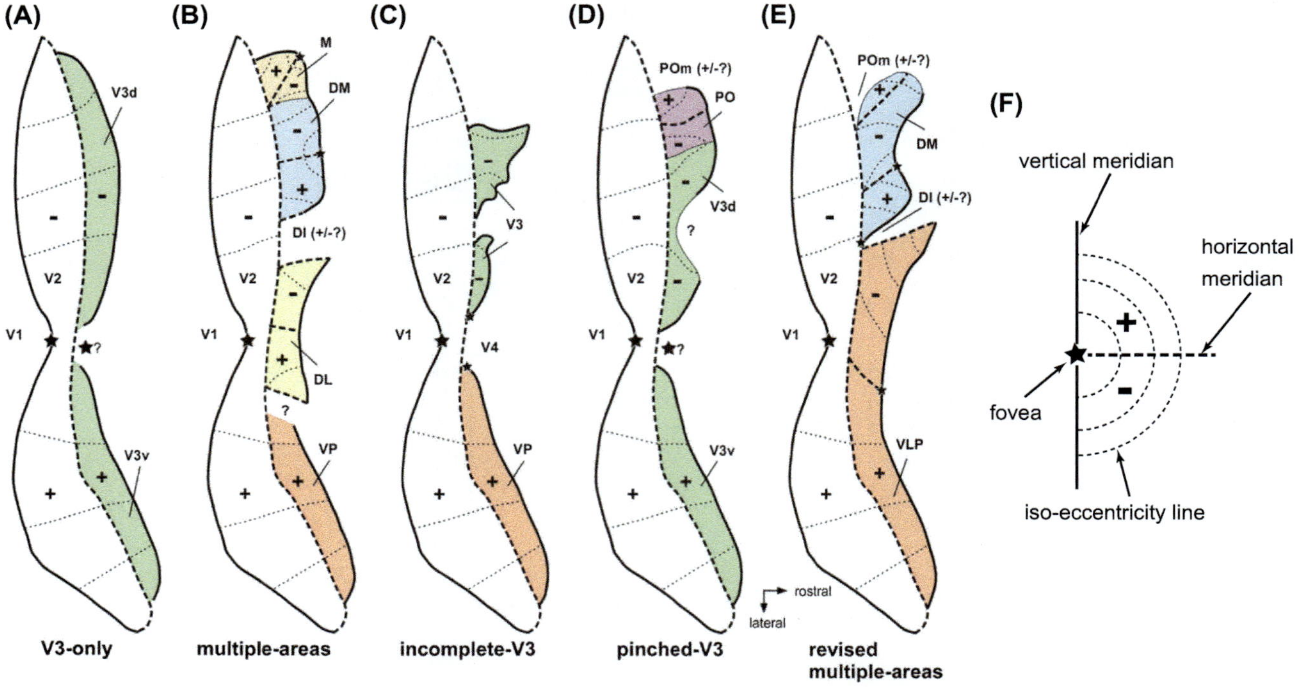

FIGURE 3.7 **Models of the third-tier visual cortex.** *Thick solid line*: vertical meridian of the visual field. *Dashed line*: horizontal meridian of the visual field. *Star*: Foveal representations. *Thin dotted lines*: iso-eccentricity lines. +: Upper visual quadrant representation. −: Lower visual quadrant representation. (A) "V3-only" model, (B) original "multiple-areas" model, (C) "incomplete-V3" model, (D) "pinched-V3" model, and (E) "revised multiple-areas" model. Thin solid contours in (B), (D), and (E) indicate uncertainties of the meridian representation. (F) Visual field mapped to the corresponding visual cortical areas. *(A—E) Reproduced from Angelucci A, Rosa MGP. Resolving the organization of the third tier visual cortex in primates: a hypothesis-based approach. Visual Neurosci 2015;32:E010. https://doi.org/10.1017/S0952523815000073.*

the dorsal aspect of the third-tier visual cortex (Fig. 3.7E).

The DM and VLP also have distinct patterns of corticocortical connections. The DM receives inputs from the thick and thin strips of the V2 as well as layer 4B of the V1, and it primarily projects to the dorsal visual stream, including areas DA, DI, and MT, as well as dorsal subdivisions of the posterior parietal cortex [74,75] (Fig. 3.6B). However, the VLP primarily receives inputs from layer 2/3 of the V1 and all four stripes of the V2. The VLP is connected with cortical areas in the ventral streams, such as the VLA and IT (Fig. 3.6B).

Area MT: A Motion-Sensitive Area

Area MT is an extrastriate visual cortex that is involved in processing visual motion. Compared with most primate species, the entire extent of area MT in the marmoset brain is exposed on the gyrus. Layer 4B of the V1 provides inputs to area MT that terminate in layers 3 and 4 [78,79].[5] As mentioned above, although the neurons in layer 4B of the V1 also project to the DM, a morphological difference is observed between the neurons projecting to the MT and those projecting to the DM [80]. The MT also receives inputs from the V2, other motion-sensitive areas (MTC, MST, and FST) [78,81], and the third-tier cortical areas described above (Fig. 3.7B). However, the connection patterns with the MT are quite different in the region that represents the far peripheral visual field. The portion of the MT for far periphery representations is connected with area 23v and the prostriata [82] (retrosplenial cortex) as well as the V1 and MST, whereas it is not connected to the V2 [83].

Prefrontal Cortex

Similar to other primate species, the PFC of the marmoset brain can be roughly split into the following five divisions: dorsolateral (dlPFC), ventrolateral (vlPFC), orbital (OFC), medial (mPFC), and frontal pole (FP), each of which consists of multiple cortical areas.

Coarse Level Connections of Prefrontal Cortex Divisions

Excitotoxic lesions in marmosets trained on discrimination learning tasks demonstrated a double

[5] Layer 4B is called layer 3C following Hässler's scheme.

dissociation between the location of the lesion on the PFC and the task performance. vlPFC lesions produced a loss of inhibitory control in attentional selection, whereas OFC damage caused a loss of inhibitory control in affective processing [84]. These results suggest that cortical networks present distinct types that involve each of the divisions of the PFC. This possibility was investigated by probing the coarse anatomical connections among the four PFC divisions (OFC, dlPFC, vlPFC, and mPFC) using anterograde and retrograde tracers [16]. At the coarse level, cortical areas in a division of the PFC tend to project to the same area outside of the PFC. The large divisions of the PFC are also interconnected with each other. Areas in the vlPFC tend to be connected with the dlPFC and the OFC; areas in the OFC are connected with the mPFC; and areas in the mPFC are connected with the dlPFC [16] (Fig. 3.8A).

The granular region of the vlPFC has extensive connections with other cortical areas, including areas around the STS (TPO, FST, MT), the somatosensory cortex (S2), the rostral and inferior temporal lobe, the posterior cingulate cortex (area 23), and the posterior parietal cortex (PF, PFG, PG) [16] (Fig. 3.8B). The dlPFC also projects to some of these areas (S2, STS, the posterior parietal cortex, and the posterior cingulate cortex), and it also receives projections from the auditory cortex (CL in particular) and the temporoparietal area (Tpt) [16] (Fig. 3.8B). However, the OFC is connected with the inferior temporal gyrus and areas around the STS (TPO and FSTv), and the mPFC receives inputs from the superior temporal gyrus, inferior temporal lobe, and areas around the STS (Fig. 3.8C). The OFC and mPFC are also connected with limbic structures, such as the cingulate cortex (area 23), amygdala, and hippocampus [16].

The FP has specific long- and short-range connections with several areas of the PFC. Area 10 of the FP is strongly connected with its surrounding regions, including areas 9, 14, and 32, whereas area 46, which is adjacent to area 10, is only sparsely connected. However, although area 13L is not adjacent to area 10, it is strongly connected [85] (Fig. 3.9). Area 10 receives many inputs from the temporal association areas, including the ITr, ITd, STR, and STP, and it also receives projections from the temporal pole. Regarding other sensory inputs, weak inputs are observed from the visual (MST and prostriata) and auditory (parabelt) areas. Overall, the projection pattern to the FP is similar to that of the mPFC and OFC in the sense that the FP receives projections from the mPFC and OFC.

Fine Cortical Network Stemming From Areas Within a Prefrontal Cortex Division

Thus far, we have reviewed the connections among the divisions of the PFC; however, cortical areas from the same division of the PFC could also project to various different cortical areas. Subdivisions of area 8 in the dlPFC project to different cortical areas. Area 8 can be divided into three regions (8b, 8aD, and 8aV) that present different patterns of neuroanatomical connections [86] (Fig. 3.10). The functional dissociation among these areas was suggested from previous studies of the macaque, which found that area 8b was associated with higher cognitive functions and decision-making [87] and area 8a was associated with eye movements [88].

The projection to area 8b is particularly strong from the neighboring areas of the dlPFC (areas 9, 10, and 46) the mPFC (area 32), and the anterior cingulate cortex (area 24a) [86]. Area 8b is also connected with the OFC (areas 11 and 13L) and the vlPFC (areas 12L, 12M, and 12O), and areas 8aD and 8aV present few connections with the mPFC. Areas 8b and 8aD receive strong inputs from the posterior medial temporal lobe, whereas area 8aV receives limited input from this area. In the marmoset brain, area 8aV receives input from the visual system, which is similar to the process in the macaque brain that is responsible for eye movements and visual attention. However, certain cortical areas of the marmoset brain, such as area 6DR, project equally to all three divisions of area 8.Thus, whether the projection from the PFC has a columnar organization should be considered when identifying its contribution to functions [89,90].

Motor and Premotor Cortices

The marmoset motor and premotor cortices are divided into multiple areas [29,15] (Figs. 3.3B and 3.4) that follow the cortical divisions of the macaque brain [91]. The homologous relationship in the cortical divisions between marmosets and macaques has been confirmed for the motor and premotor areas in terms of the cytoarchitecture and neuroanatomical connections. Here, we overview the cortical connections of the motor and premotor cortexes.

Motor Cortex

The primary motor cortex (M1) of marmosets receives strong inputs from the posterior parietal lobe (PE). In addition, the M1 receives projections from the somatosensory areas (3a, 3b, 1/2, and S2) as well as areas in the medial wall (areas 23 and 24) and the parietal cortex (PF and PFG). The premotor areas 6DC and 6M also project to the M1 (Fig. 3.11A and B). Each input to the dorsal region of the M1 tends to originate from dorsal cortical areas, whereas the inputs to the ventral region of M1 tend to originate from ventral cortical areas. In the macaque brain, the M1 is designated F1 and primarily receives projections from the PE and

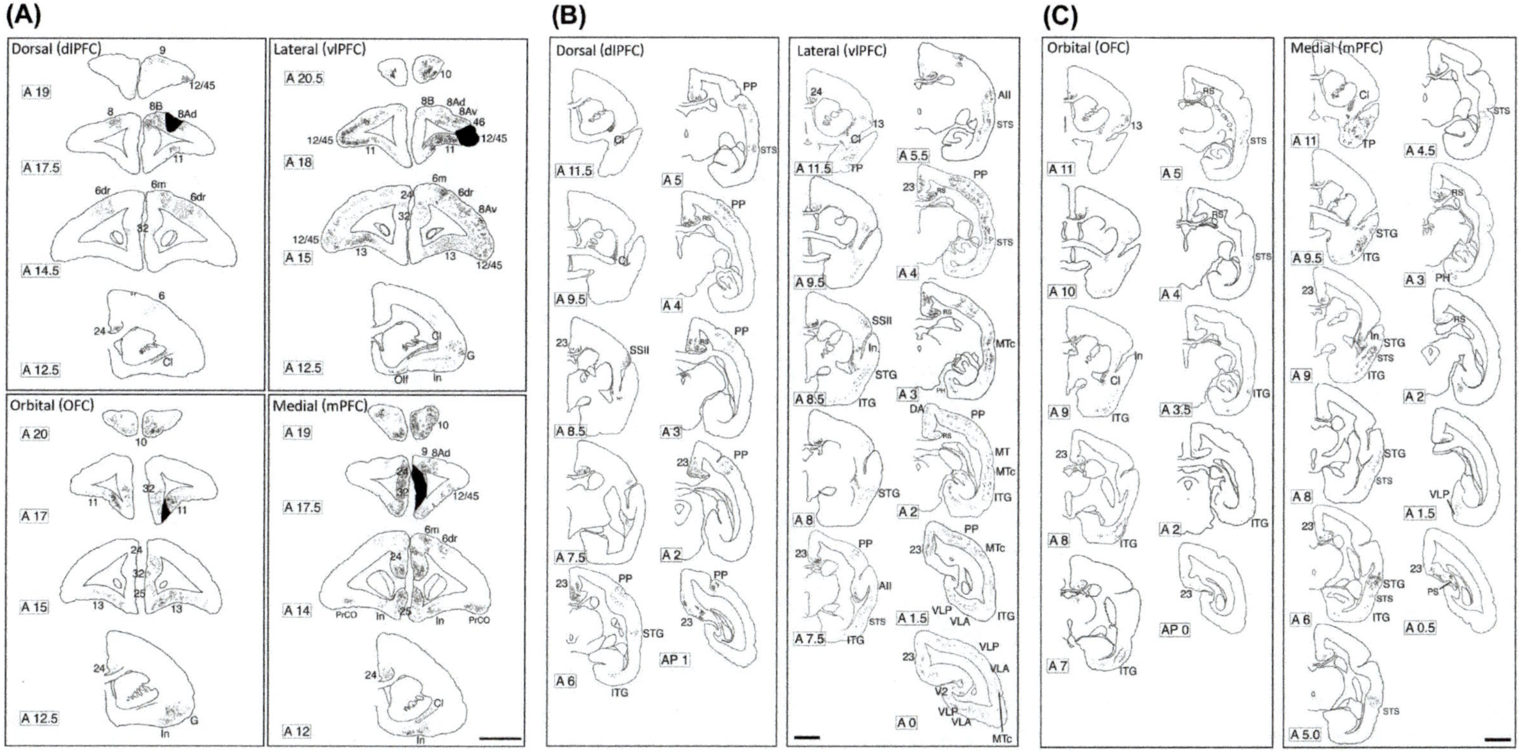

FIGURE 3.8 **Cortical inputs to the four divisions of the PFC (dlPFC, vlPFC, OFC, and mPFC).** (A) Tracer-labeled cells to the dlPFC (top left), vlPFC (top right), OFC (bottom left), and mPFC (bottom right). Injection sites are painted black. AP level: A12–A20.5. (B) Tracer-labeled cells to the dlPFC (left) and vlPFC (right). AP level: A0–A11.5. (C) Tracer-labeled cells to the OFC (left) and mPFC (right). AP level: A0–A11. *Reproduced from Roberts AC, Tomic DL, Parkinson CH, Roeling TA, Cutter DJ, Robbins TW, Everitt BJ. Forebrain connectivity of the prefrontal cortex in the marmoset monkey* (Callithrix jacchus)*: an anterograde and retrograde tract-tracing study. J Comp Neurol 2007;502:86–112. https://doi.org/10.1002/cne.21300.*

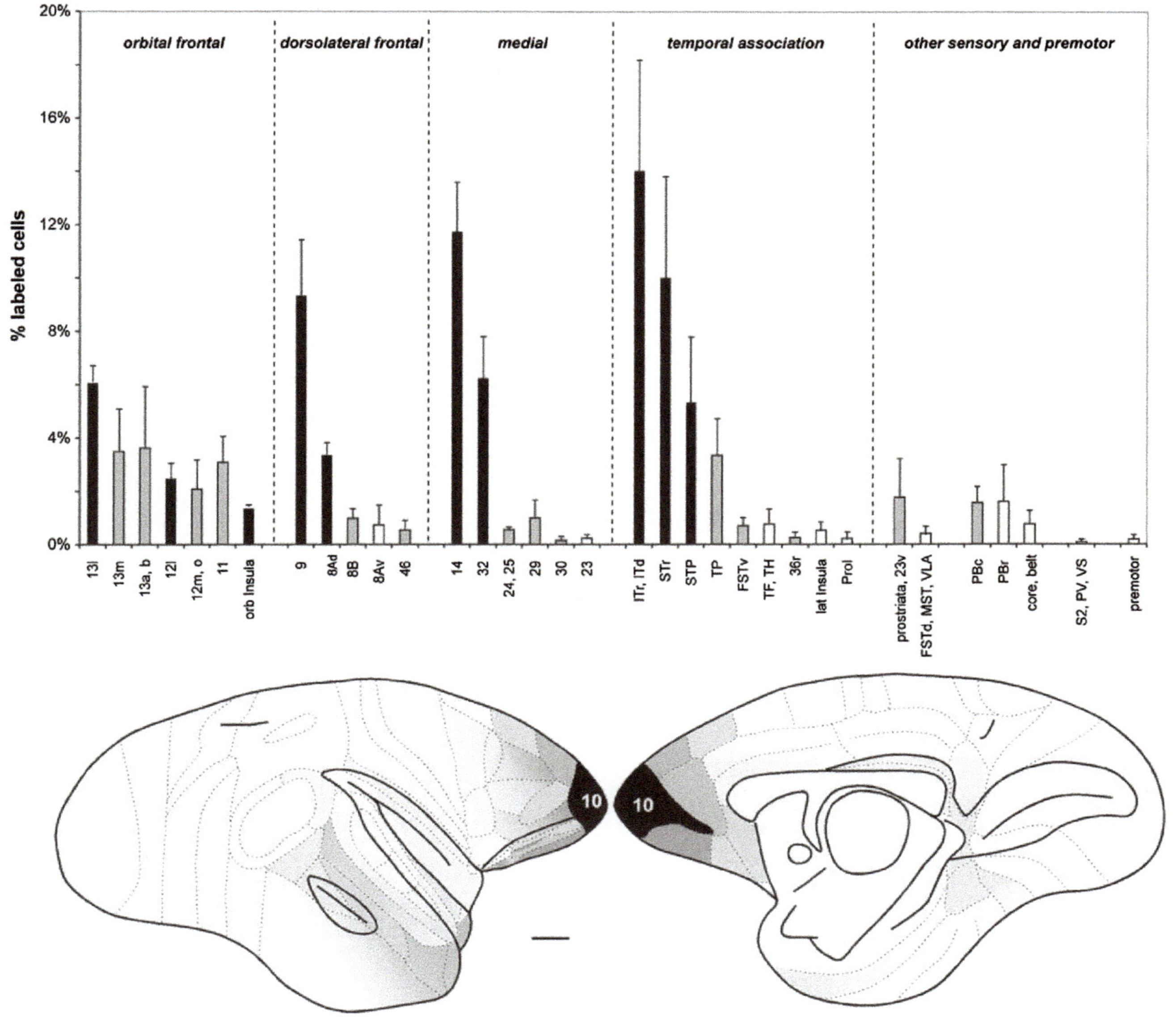

FIGURE 3.9 **Cortical inputs to the frontal pole (area 10).** *Reproduced from Burman KJ, Reser DH, Yu H-H, Rosa MGP. Cortical input to the frontal pole of the marmoset monkey. Cereb Cortex 2011;21:1712–37. https://doi.org/10.1093/cercor/bhq239.*

area 24d (Figs. 3.11C and 3.12B). Thus, the anatomical connectivity supports a homologous relationship between the marmoset M1 and the macaque F1 [91].

Premotor Cortex

The premotor cortex in the marmoset brain is roughly divided into the following three regions: medial (area 6M),[6] dorsal (areas 6DR, 6DC, 8C), and ventral (areas 6Va, 6Vb). Area 6M strongly projects to areas 6DC and 6DR but not to other premotor areas.

The projection patterns from other areas to 6DR are distinct from those to 6DC (Fig. 3.12A) [92]. 6DR is different from 6DC in that the PFC strongly projects to 6DR but weakly projects to 6DC. Moreover, 6DC receives strong input from the primary motor cortex and the somatosensory cortex, whereas 6DR receives limited input from these areas. The anterior cingulate cortex projects to 6DC, whereas the posterior cingulate cortex and the neighboring areas in the medial wall project to 6DR. The PE in the parietal cortex projects

[6] This area can also be referred to as F3 or the supplementary motor area (SMA).

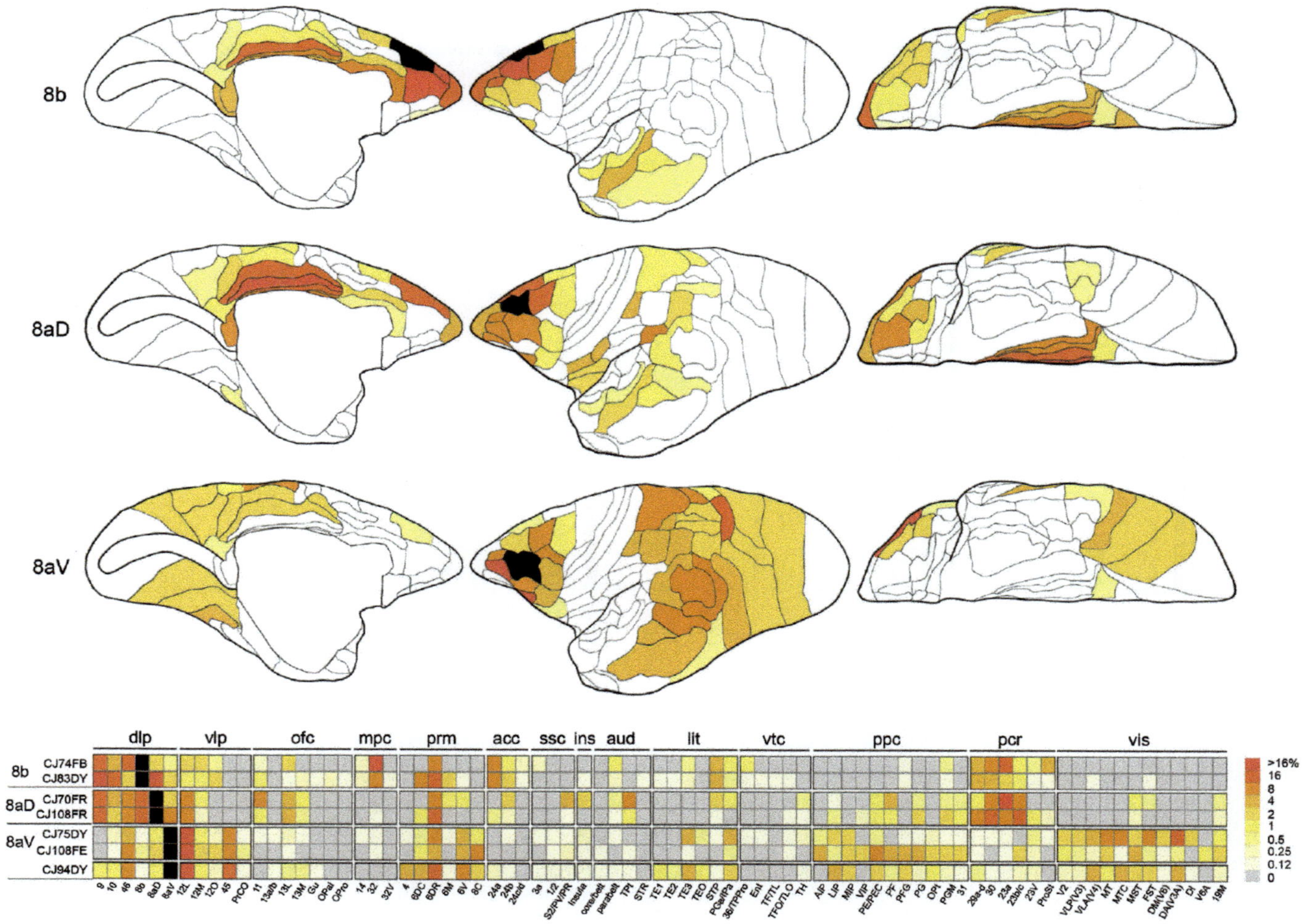

FIGURE 3.10 **Cortical inputs to area 8 in the prefrontal cortex.** The ipsilateral cortical inputs to areas 8b, 8aD, and 8aV are summarized as a matrix. The strength of the connections is shown by the percentage of extrinsic connections. *Reproduced from Reser DH, Burman KJ, Yu H-H, Chaplin TA, Richardson KE, Worthy KH, Rosa MGP. Contrasting patterns of cortical input to architectural subdivisions of the area 8 complex: a retrograde tracing study in marmoset monkeys. Cereb Cortex 2013;23:1901–22. https://doi.org/10.1093/cercor/bhs177.*

very strongly to 6DC but weakly to 6DR. However, 6DR receives a strong projection from the lateral intraparietal cortex. For the ventral parietal lobe, 6DR receives inputs from the caudal areas (PG and Opt), whereas 6DC receives projections from the rostral areas (PF and PFG). Based on these patterns of anatomical connections to areas 6DR and 6DC, 6DC may be involved in the preparation and execution of actions and 6DR may integrate sensory information and execution commands into a plan of action. Moreover, according to the projection patterns of the dorsal premotor cortex (F2 and F7) in the macaque brain (Fig. 3.12C), areas 6DC and 6DR in the marmoset brain are homologous to F2 and F7 in the macaque brain, respectively [92].

Area 8C is a narrow strip of the cortex located between 6DC and 6Va. This area does not have a homologous area in the macaque brain, as mentioned earlier in this chapter. Area 8C was recognized in the marmoset brain atlas by Paxinos et al. [15] based on the cytoarchitecture of an incipient layer 4 that corresponds to a transition between the agranular area 6 and the granular area 8 [15]. Area 8C receives projections from the parietal cortex (PE, PF, and PFG), premotor cortex (6Va, 6Vb, 6DC, and 6DR), PFC (45, 46, 12, and 8aV), and visual cortex (MST), although it does not receive input from the M1 (Fig. 3.12D). The connection patterns to area 8C are roughly similar to those for area 6DR [29,92], which is involved in visuospatial behaviors in macaques.

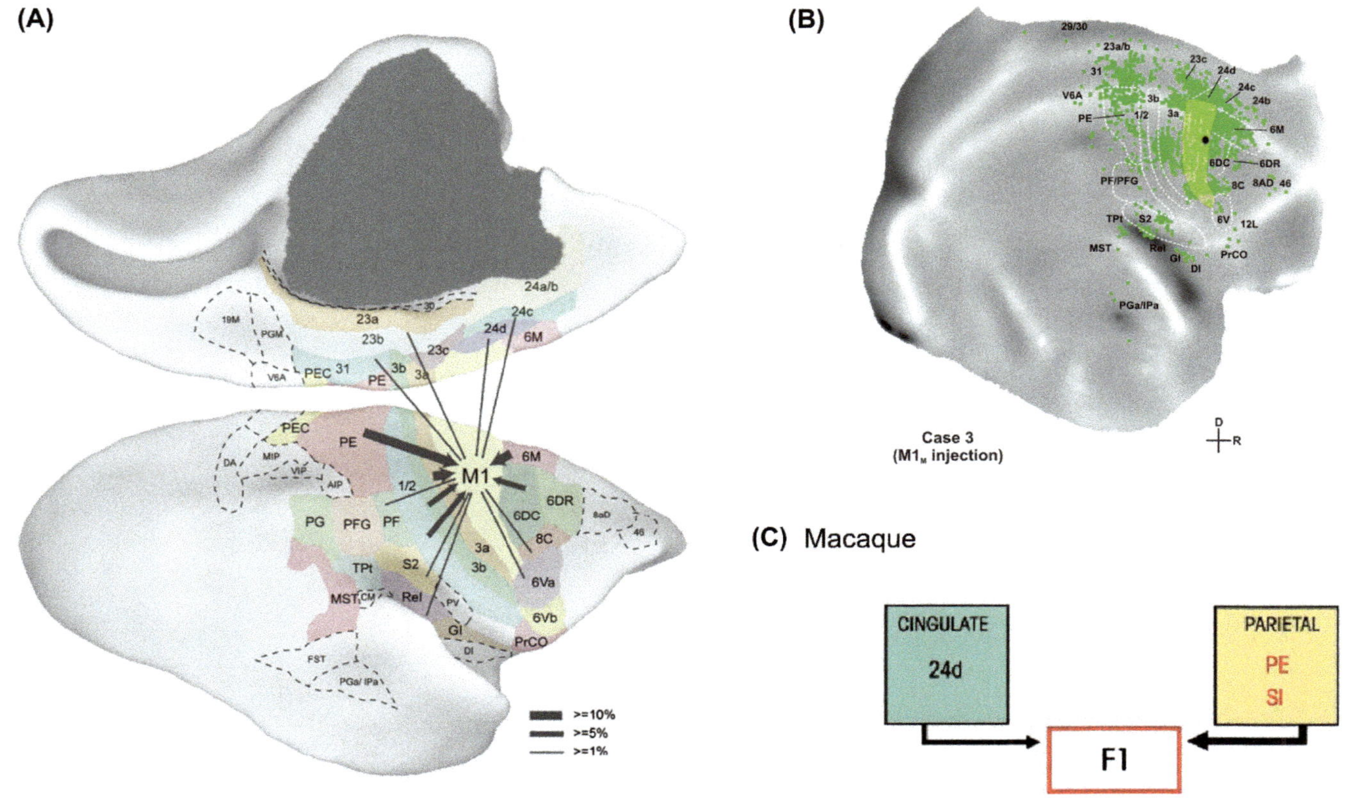

FIGURE 3.11 **Cortical inputs to the primary motor cortex (M1).** (A) Cortical inputs to M1 in the marmoset brain. The *thickness of the arrows* is proportional to the average percentage of labeled neurons by the retrograde tracer injected to M1 among seven cases. (B) Spatial pattern of labeled cells on the flattened cortex from a tracer injection to the medial M1 ($M1_M$). (C) Diagram of the cortical inputs to F1 in the macaque brain. The *thickness of the arrow* represents the strength of the connections. *(A and B) Reproduced from Burman KJ, Bakola S, Richardson KE, Reser DH, Rosa MGP. Patterns of cortical input to the primary motor area in the marmoset monkey. J Comp Neurol 2014;522:811–43. https://doi.org/10.1002/cne.23447. (C) Reproduced from Rizzolatti G, Luppino G. The cortical motor system. Neuron 2001;31:889–901.*

Based on the pattern of corticocortical connections, area 8C may be more similar to the premotor cortex than the PFC [29].

The ventral premotor cortex in the marmoset brain consists of areas 6Va and 6Vb, which could correspond to areas F4/F5 in the macaque brain (Fig. 3.12D and E). Although area 6Va is adjacent to area 8C, the connection pattern of area 6Va is distinct from that of area 8C (Fig. 3.12D) because it receives strong inputs from the M1 and the somatosensory areas (3a, 3b, 1/2, and S2), although 6Va also receives inputs similar to that of area 8C, such as those from the parietal cortex (PE, PF, and PFG). Area 6Vb projects to other premotor areas, including 6Va and 8C [29]. Cortical inputs to area 6Vb in the marmoset brain have not been thoroughly investigated. However, tracer injections to a region of the STS revealed a connection to the ventral part of 6V, which is likely to be area 6Vb [44]. In this study, neurons in the STS region responded to visual stimuli, particularly the grasping action of others, and neurons with "mirror" properties that are projected from the STS have also been found in the region of area 6V.

DISCUSSION

Additional tracer injection studies are required to map out the pattern of connectivity among cortical areas. For example, the frontotemporal connections are not fully identified. In the auditory cortex, although tracer injections have revealed certain major cortical fields, the connections to and from the PFC remain unexplored. Eventually, a systematic attempt to map the connections among cortical areas may produce a whole-brain

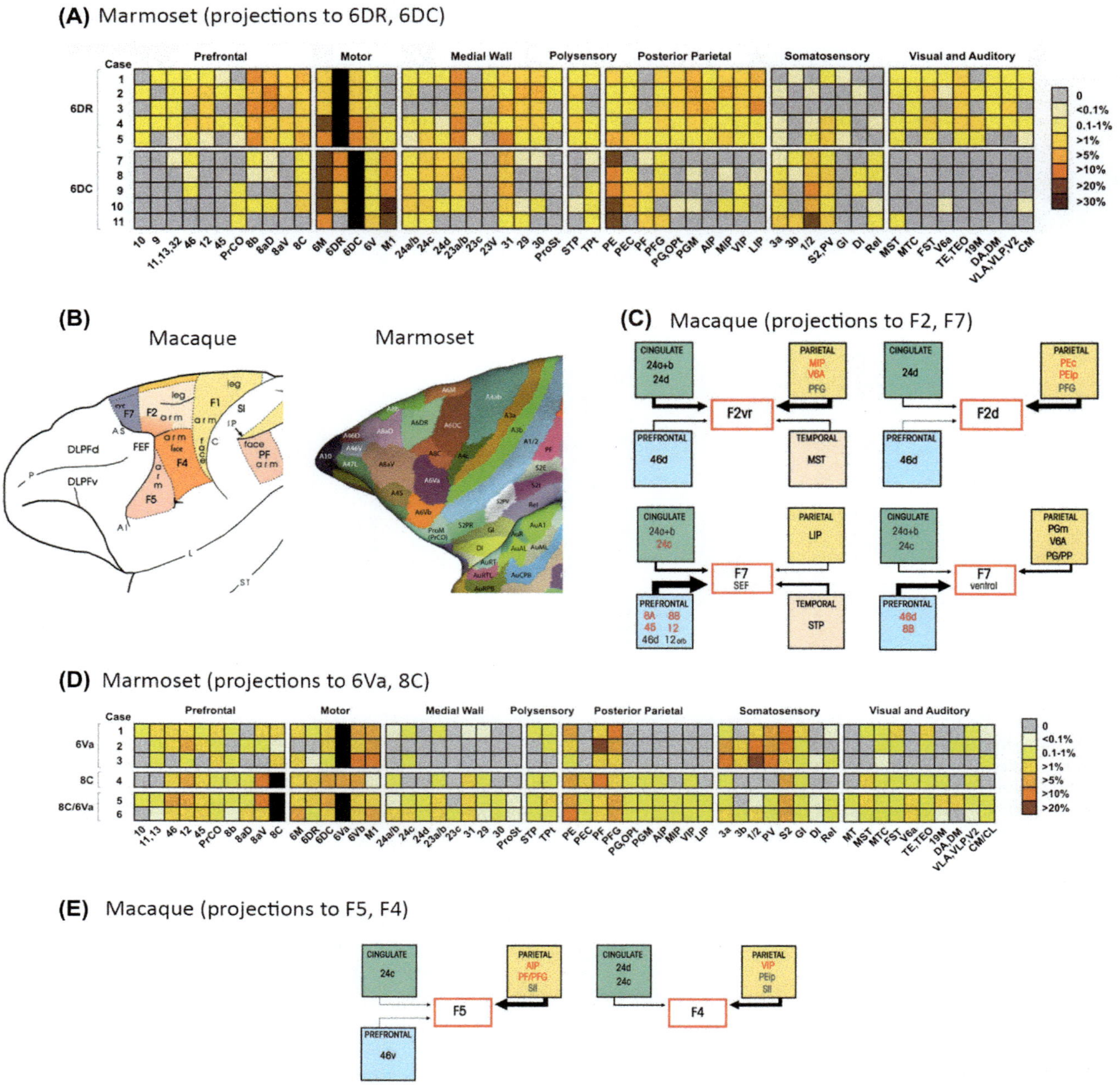

FIGURE 3.12 **Cortical inputs to the premotor cortical areas.** (A) Ipsilateral cortical areas project to areas 6DR and 6DC. Notations for the matrix are the same as that in Fig. 3.10. (B) Lateral frontal cortex of the macaque brain [91] and the marmoset brain [15] showing the motor and premotor cortical areas. (C) Diagram of the cortical inputs to the dorsal premotor cortex (F2 and F7) in macaques. (D) Ipsilateral cortical areas project to areas 6Va and 8C. Notations for the matrix are the same as that in (A). (E) Diagram of cortical inputs to the ventral premotor cortex (F5 and F4) in the macaque brain. *(A) Reproduced from Burman KJ, Bakola S, Richardson KE, Reser DH, Rosa MGP. Patterns of afferent input to the caudal and rostral areas of the dorsal premotor cortex (6DC and 6DR) in the marmoset monkey. J Comp Neurol 2014;522:3683–716. https://doi.org/10.1002/cne.23633. (C and E) Reproduced from Rizzolatti G, Luppino G. The cortical motor system. Neuron 2001;31:889–901. (D) Reproduced from Burman KJ, Bakola S, Richardson KE, Yu H-H, Reser DH, Rosa MGP. Cortical and thalamic projections to cytoarchitectural areas 6Va and 8C of the marmoset monkey: connectionally distinct subdivisions of the lateral premotor cortex. J Comp Neurol 2015;523:1222–47. https://doi.org/10.1002/cne.23734.*

connectome matrix on the mesoscale level [93,94]. Furthermore, to fully identify the cortical connections, the layer- and cell-specific connections must be better understood [95].

To connect available data among different levels of understanding, mesoscale neuroanatomical connectomes must be compared with macroscale connectomes, such as those obtained from the application of diffusion MRI. MRI presents an advantage in probing these connections in vivo, although the degree of precision of diffusion MRI mapping of brain-wide connectivity is still under debate [96,97]. We also note that the flat "lissencephalic" cortex of the marmoset could be suitable for diffusion MR tractography [98]. Another possible method of mapping out the neuronal network is polarized light microscopy, which allows for higher spatial resolution than diffusion MRI [99,100]. Combinations of these methods may help reveal complex neural networks in the marmoset brain and generate insights toward the utilization of marmosets as a biomedical model.

References

[1] Okano H, Sasaki E, Yamamori T, Iriki A, Shimogori T, Yamaguchi Y, Kasai K, Miyawaki A. Brain/minds: a Japanese national brain project for marmoset neuroscience. Neuron 2016;92:582–90. https://doi.org/10.1016/j.neuron.2016.10.018.

[2] Miller CT, Freiwald WA, Leopold DA, Mitchell JF, Silva AC, Wang X. Marmosets: a neuroscientific model of human social behavior. Neuron 2016;90:219–33. https://doi.org/10.1016/j.neuron.2016.03.018.

[3] Sasaki E, Suemizu H, Shimada A, Hanazawa K, Oiwa R, Kamioka M, Tomioka I, Sotomaru Y, Hirakawa R, Eto T, Shiozawa S, Maeda T, Ito M, Ito R, Kito C, Yagihashi C, Kawai K, Miyoshi H, Tanioka Y, Tamaoki N, Habu S, Okano H, Nomura T. Generation of transgenic non-human primates with germline transmission. Nature 2009;459:523–7. https://doi.org/10.1038/nature08090.

[4] Sato K, Oiwa R, Kumita W, Henry R, Sakuma T, Ito R, Nozu R, Inoue T, Katano I, Sato K, Okahara N, Okahara J, Shimizu Y, Yamamoto M, Hanazawa K, Kawakami T, Kametani Y, Suzuki R, Takahashi T, Weinstein EJ, Yamamoto T, Sakakibara Y, Habu S, Hata J-I, Okano H, Sasaki E. Generation of a nonhuman primate model of severe combined immunodeficiency using highly efficient genome editing. Stem Cell 2016:1–51. https://doi.org/10.1016/j.stem.2016.06.003.

[5] MacDougall M, Nummela SU, Coop S, Disney A, Mitchell JF, Miller CT. Optogenetic manipulation of neural circuits in awake marmosets. J Neurophysiol 2016;116:1286–94. https://doi.org/10.1152/jn.00197.2016.

[6] Sadakane O, Masamizu Y, Watakabe A, Terada S-I, Ohtsuka M, Takaji M, Mizukami H, Ozawa K, Kawasaki H, Matsuzaki M, Yamamori T. Long-term two-photon calcium imaging of neuronal populations with subcellular resolution in adult non-human primates. Cell Rep 2015;13. https://doi.org/10.1016/j.celrep.2015.10.050.

[7] Takaji M, Takemoto A, Yokoyama C, Watakabe A, Mizukami H, Ozawa K, Onoe H, Nakamura K, Yamamori T. Distinct roles for primate caudate dopamine D1 and D2 receptors in visual discrimination learning revealed using shRNA knockdown. Sci Rep 2016;6:1–16. https://doi.org/10.1038/srep35809.

[8] Watson C, Sengul G, Tanaka I, Rusznak Z, Tokuno H. The spinal cord of the common marmoset (*Callithrix jacchus*). Neurosci Res 2015;93:164–75. https://doi.org/10.1016/j.neures.2014.12.012.

[9] Tokuno H, Tanaka I, Senoo A, Umitsu Y, Akazawa T, Nakamura Y, Watson C. Internet-based atlas of the primate spinal cord. Neurosci Res 2011;70:128–32. https://doi.org/10.1016/j.neures.2011.01.016.

[10] Kondo T, Yoshihara Y, Yoshino-Saito K, Sekiguchi T, Kosugi A, Miyazaki Y, Nishimura Y, Okano HJ, Nakamura M, Okano H, Isa T, Ushiba J. Histological and electrophysiological analysis of the corticospinal pathway to forelimb motoneurons in common marmosets. Neurosci Res 2015;98:35–44. https://doi.org/10.1016/j.neures.2015.05.001.

[11] Fujiyoshi K, Yamada M, Nakamura M, Yamane J, Katoh H, Kitamura K, Kawai K, Okada S, Momoshima S, Toyama Y, Okano H. In vivo tracing of neural tracts in the intact and injured spinal cord of marmosets by diffusion tensor tractography. J Neurosci 2007;27:11991–8. https://doi.org/10.1523/JNEUROSCI.3354-07.2007.

[12] Donkelaar HJ, Lammens M, Wesseling P, Hori A, Keyser A, Rotteveel J. Development and malformations of the human pyramidal tract. J Neurol 2004;251:1429–42. https://doi.org/10.1007/s00415-004-0653-3.

[13] Hackett TA. Information flow in the auditory cortical network. Hear Res 2011;271:133–46. https://doi.org/10.1016/j.heares.2010.01.011.

[14] Mitchell JF, Leopold DA. The marmoset monkey as a model for visual neuroscience. Neurosci Res 2015;93:20–46. https://doi.org/10.1016/j.neures.2015.01.008.

[15] Paxinos G, Watson C, Petrides M, Rosa M, Tokuno H. The marmoset brain in stereotaxic coordinates. Academic Press; 2012.

[16] Roberts AC, Tomic DL, Parkinson CH, Roeling TA, Cutter DJ, Robbins TW, Everitt BJ. Forebrain connectivity of the prefrontal cortex in the marmoset monkey (*Callithrix jacchus*): an anterograde and retrograde tract-tracing study. J Comp Neurol 2007;502:86–112. https://doi.org/10.1002/cne.21300.

[17] Herculano-Houzel S. The human brain in numbers: a linearly scaled-up primate brain. Front Hum Neurosci 2009;3:1–11. https://doi.org/10.3389/neuro.09.031.2009.

[18] Herculano-Houzel S, Collins CE, Wong P, Kaas JH. Cellular scaling rules for primate brains. P Natl Acad Sci USA 2007;104:3562–7. https://doi.org/10.1073/pnas.0611396104.

[19] Rosa MGP, Tweedale R, Tweedale R. Brain maps, great and small: lessons from comparative studies of primate visual cortical organization. Philos Trans R Soc Lond B Biol Sci 2005;360:665–91. https://doi.org/10.1098/rstb.2005.1626.

[20] Herculano-Houzel S, Lent R. Isotropic fractionator: a simple, rapid method for the quantification of total cell and neuron numbers in the brain. J Neurosci 2005;25:2518–21. https://doi.org/10.1523/JNEUROSCI.4526-04.2005.

[21] Herculano-Houzel S, Mota B, Lent R. Cellular scaling rules for rodent brains. P Natl Acad Sci USA 2006;103:12138–43. https://doi.org/10.1073/pnas.0604911103.

[22] Azevedo FAC, Carvalho LRB, Grinberg LT, Farfel JM, Ferretti REL, Leite REP, Filho WJ, Lent R, Herculano-Houzel S. Equal numbers of neuronal and nonneuronal cells make the human brain an isometrically scaled-up primate brain. J Comp Neurol 2009;513:532–41. https://doi.org/10.1002/cne.21974.

[23] Gabi M, Neves K, Masseron C, Ribeiro PFM, Ventura-Antunes L, Torres L, Mota B, Kaas JH, Herculano-Houzel S. No relative expansion of the number of prefrontal neurons in primate and human evolution. P Natl Acad Sci USA 2016;113:9617–22. https://doi.org/10.1073/pnas.1610178113.

[24] Solomon SG, Rosa MGP. A simpler primate brain: the visual system of the marmoset monkey. Front Neural Circuits 2014;8:96. https://doi.org/10.3389/fncir.2014.00096.

[25] Collins CE, Leitch DB, Wong P, Kaas JH, Herculano-Houzel S. Faster scaling of visual neurons in cortical areas relative to subcortical structures in non-human primate brains. Brain Struct Funct 2012;218:805–16. https://doi.org/10.1007/s00429-012-0430-5.

[26] Chaplin TA, Yu HH, Soares JGM, Gattass R, Rosa MGP. A conserved pattern of differential expansion of cortical areas in simian primates. J Neurosci 2013;33:15120–5. https://doi.org/10.1523/JNEUROSCI.2909-13.2013.

[27] Sawada K, Hikishima K, Murayama AY, Okano HJ, Sasaki E, Okano H. Fetal sulcation and gyrification in common marmosets (*Callithrix jacchus*) obtained by ex vivo magnetic resonance imaging. Neuroscience 2014;257:158–74. https://doi.org/10.1016/j.neuroscience.2013.10.067.

[28] Brodmann K. Vergleichende Lokalisationslehre der Grosshirnrinde in ihren Prinzipien dargestellt auf Grund des Zellenbaues (Brodmann's Localization in the Cerebral Cortex (Smith Gordon, 1994) [transl. Garey, L.J.]. Verlag von Johann Ambrosius Barth; 1909.

[29] Burman KJ, Bakola S, Richardson KE, Yu H-H, Reser DH, Rosa MGP. Cortical and thalamic projections to cytoarchitectural areas 6Va and 8C of the marmoset monkey: connectionally distinct subdivisions of the lateral premotor cortex. J Comp Neurol 2015;523:1222–47. https://doi.org/10.1002/cne.23734.

[30] Rosa MGP, Palmer SM, Gamberini M, Burman KJ, Yu HH, Reser DH, Bourne JA, Tweedale R, Galletti C. Connections of the dorsomedial visual area: pathways for early integration of dorsal and ventral streams in extrastriate cortex. J Neurosci 2009;29: 4548–63. https://doi.org/10.1523/JNEUROSCI.0529-09.2009.

[31] Glasser MF, Coalson TS, Robinson EC, Hacker CD, Harwell J, Yacoub E, Ugurbil K, Andersson J, Beckmann CF, Jenkinson M, Smith SM, Van Essen DC. A multi-modal parcellation of human cerebral cortex. Nature 2016;536:171–8. https://doi.org/10.1038/nature18933.

[32] Majka P, Chaplin TA, Yu H-H, Tolpygo A, Mitra PP, Wójcik DK, Rosa MGP. Towards a comprehensive atlas of cortical connections in a primate brain: mapping tracer injection studies of the common marmoset into a reference digital template. J Comp Neurol 2016;524:2161–81. https://doi.org/10.1002/cne.24023.

[33] Burman KJ, Palmer SM, Gamberini M, Rosa MG. Cytoarchitectonic subdivisions of the dorsolateral frontal cortex of the marmoset monkey (*Callithrix jacchus*), and their projections to dorsal visual areas. J Comp Neurol 2006;495:149–72. https://doi.org/10.1002/cne.20837.

[34] Miller CT, Eliades SJ, Wang X. Motor planning for vocal production in common marmosets. Anim Behav 2009;78:1195–203. https://doi.org/10.1016/j.anbehav.2009.07.038.

[35] Roy S, Miller CT, Gottsch D, Wang X. Vocal control by the common marmoset in the presence of interfering noise. J Exp Biol 2011;214:3619–29. https://doi.org/10.1242/jeb.056101.

[36] Takahashi DY, Narayanan DZ, Ghazanfar AA. Coupled oscillator dynamicsof vocal turn-taking in monkeys. Curr Biol 2013:1–7. https://doi.org/10.1016/j.cub.2013.09.005.

[37] Petrides M. Neuroanatomy of language regions of the human brain. Academic Press; 2013.

[38] Flinker A, Korzeniewska A, Shestyuk AY, Franaszczuk PJ, Dronkers NF, Knight RT, Crone NE. Redefining the role of Broca's area in speech. Proc Natl Acad Sci USA 2015. https://doi.org/10.1073/pnas.1414491112. 201414491.

[39] Tremblay P, Dick AS. Broca and Wernicke are dead, or moving past the classic model of language neurobiology. Brain Lang 2016;162:60–71. https://doi.org/10.1016/j.bandl.2016.08.004.

[40] Petrides M. Lateral prefrontal cortex: architectonic and functional organization. Philos Trans R Soc Lond B Biol Sci 2005; 360:781–95. https://doi.org/10.1098/rstb.2005.1631.

[41] Petrides M, Pandya DN. Comparative architectonic analysis of the human and the macaque frontal cortex. In: Boller F, Grafman J, editors. Handbook of neuropsychology; 1994. p. 17–58.

[42] Petrides M, Cadoret G, Mackey S. Orofacial somatomotor responses in the macaque monkey homologue of Broca's area. Nature 2005;435:1235–8. https://doi.org/10.1038/nature03628.

[43] Coudé G, Ferrari PF, Rodà F, Maranesi M, Borelli E, Veroni V, Monti F, Rozzi S, Fogassi L. Neurons controlling voluntary vocalization in the macaque ventral premotor cortex. PLoS One 2011;6:e26822. https://doi.org/10.1371/journal.pone.0026822.t001.

[44] Suzuki W, Banno T, Miyakawa N, Abe H, Goda N, Ichinohe N. Mirror neurons in a new world monkey, common marmoset. Front Neurosci 2015;9:1806–14. https://doi.org/10.3389/fnins.2015.00459.

[45] Bendor D, Osmanski MS, Wang X. Dual-pitch processing mechanisms in primate auditory cortex. J Neurosci 2012;32: 16149–61. https://doi.org/10.1523/JNEUROSCI.2563-12.2012.

[46] Osmanski MS, Song X, Wang X. The role of harmonic resolvability in pitch perception in a vocal nonhuman primate, the common marmoset (*Callithrix jacchus*). J Neurosci 2013;33: 9161–8. https://doi.org/10.1523/JNEUROSCI.0066-13.2013.

[47] Song X, Osmanski MS, Guo Y, Wang X. Complex pitch perception mechanisms are shared by humans and a New World monkey. Proc Natl Acad Sci USA 2015. https://doi.org/10.1073/pnas.1516120113. 201516120–9.

[48] La Mothe De LA, Blumell S, Kajikawa Y, Hackett T. Cortical connections of the auditory cortex in marmoset monkeys: core and medial belt regions. J Comp Neurol 2006;496:27–71. https://doi.org/10.1002/cne.20923.

[49] La Mothe De LA, Blumell S, Kajikawa Y, Hackett TA. Cortical connections of auditory cortex in marmoset monkeys: lateral belt and parabelt regions. Anat Rec 2012;295:800–21. https://doi.org/10.1002/ar.22451.

[50] Reser DH, Burman KJ, Richardson KE, Spitzer MW, Rosa MGP. Connections of the marmoset rostrotemporal auditory area: express pathways for analysis of affective content in hearing. Eur J Neurosci 2009;30:578–92. https://doi.org/10.1111/j.1460-9568.2009.06846.x.

[51] La Mothe De LA, Blumell S, Kajikawa Y, Hackett TA. Thalamic connections of the auditory cortex in marmoset monkeys: core and medial belt regions. J Comp Neurol 2006;496:72–96. https://doi.org/10.1002/cne.20924.

[52] Hackett TA, Stepniewska I, Kaas JH. Thalamocortical connections of the parabelt auditory cortex in macaque monkeys. J Comp Neurol 1998;400:271–86.

[53] Romanski LM, Averbeck BB. The primate cortical auditory system and neural representation of conspecific vocalizations. Annu Rev Neurosci 2009;32:315–46. https://doi.org/10.1146/annurev.neuro.051508.135431.

[54] Scott BH, Leccese PA, Saleem KS, Kikuchi Y, Mullarkey MP, Fukushima M, Mishkin M, Saunders RC. Intrinsic connections of the core auditory cortical regions and rostral supratemporal plane in the macaque monkey. Cereb Cortex 2015. https://doi.org/10.1093/cercor/bhv277. bhv277–32.

[55] Sadagopan S, Temiz-Karayol NZ, Voss HU. High-field functional magnetic resonance imaging of vocalization processing in marmosets. Sci Rep 2015;5:10950. https://doi.org/10.1038/srep10950.

[56] Fukushima M, Saunders RC, Leopold DA, Mishkin M, Averbeck BB. Differential coding of conspecific vocalizations in the ventral auditory cortical stream. J Neurosci 2014;34: 4665–76. https://doi.org/10.1523/JNEUROSCI.3969-13.2014.

[57] Perrodin C, Kayser C, Logothetis NK, Petkov CI. Voice cells in the primate temporal lobe. Curr Biol 2011;21:1408–15. https://doi.org/10.1016/j.cub.2011.07.028.

[58] Petkov CI, Kayser C, Steudel T, Whittingstall K, Augath M, Logothetis NK. A voice region in the monkey brain 2008;11:367–74. https://doi.org/10.1038/nn2043.

[59] Poremba A, Malloy M, Saunders R, Carson R, Herscovitch P, Mishkin M. Species-specific calls evoke asymmetric activity in the monkey's temporal poles. Nature 2004;427:448–51. https://doi.org/10.1038/nature02268.

[60] Scott BH, Saleem KS, Kikuchi Y, Fukushima M, Mishkin M, Saunders RC. Thalamic connections of the core auditory cortex and rostral supratemporal plane in the macaque monkey. J Comp Neurol 2017;525:3488–513. https://doi.org/10.1002/cne.24283.

[61] Fritsches KA, Rosa MG. Visuotopic organisation of striate cortex in the marmoset monkey (*Callithrix jacchus*). J Comp Neurol 1996;372:264–82. https://doi.org/10.1002/(SICI)1096-9861(19960819)372:2<264::AID-CNE8>3.0.CO;2-1.

[62] Roe AW, Fritsches K, Pettigrew JD. Optical imaging of functional organization of V1 and V2 in marmoset visual cortex 2005;287A:1213–25. https://doi.org/10.1002/ar.a.20248.

[63] Rosa MG, Fritsches KA, Elston GN. The second visual area in the marmoset monkey: visuotopic organisation, magnification factors, architectonical boundaries, and modularity. J Comp Neurol 1997;387:547–67.

[64] Bourne JA, Warner CE, Upton DJ, Rosa MGP. Chemoarchitecture of the middle temporal visual area in the marmoset monkey (*Callithrix jacchus*): laminar distribution of calcium-binding proteins (calbindin, parvalbumin) and nonphosphorylated neurofilament. J Comp Neurol 2006;500:832–49. https://doi.org/10.1002/cne.21190.

[65] Solomon SS, Tailby C, Gharaei S, Camp AJ, Bourne JA, Solomon SG. Visual motion integration by neurons in the middle temporal area of a New World monkey, the marmoset. J Physiol (Lond) 2011;589:5741–58. https://doi.org/10.1113/jphysiol.2011.213520.

[66] Hung C-C, Yen CC, Ciuchta JL, Papoti D, Bock NA, Leopold DA, Silva AC. Functional mapping of face-selective regions in the extrastriate visual cortex of the marmoset. J Neurosci 2015;35:1160–72. https://doi.org/10.1523/JNEUROSCI.2659-14.2015.

[67] Solomon SG. Striate cortex in dichromatic and trichromatic marmosets: neurochemical compartmentalization and geniculate input. J Comp Neurol 2002;450:366–81. https://doi.org/10.1002/cne.10327.

[68] Dick A, Kaske A, Creutzfeldt OD. Topographical and topological organization of the thalamocortical projection to the striate and prestriate cortex in the marmoset (*Callithrix jacchus*). Exp Brain Res 1991;84:233–53.

[69] Warner CE, Goldshmit Y, Bourne JA. Retinal afferents synapse with relay cells targeting the middle temporal area in the pulvinar and lateral geniculate nuclei 2010;4:8. https://doi.org/10.3389/neuro.05.008.2010.

[70] Federer F, Ichida JM, Jeffs J, Schiessl I, McLoughlin N, Angelucci A. Four projection streams from primate V1 to the cytochrome oxidase stripes of V2. J Neurosci 2009;29:15455–71. https://doi.org/10.1523/JNEUROSCI.1648-09.2009.

[71] Lyon DC, Kaas JH. Connectional and architectonic evidence for dorsal and ventral V3, and dorsomedial area in marmoset monkeys. J Neurosci 2001;21:249–61.

[72] Federer F, Williams D, Ichida JM, Merlin S, Angelucci A. Two projection streams from macaque V1 to the pale cytochrome oxidase stripes of V2. J Neurosci 2013;33:11530–9. https://doi.org/10.1523/JNEUROSCI.5053-12.2013.

[73] Allman JM, Kaas JH. The dorsomedial cortical visual area: a third tier area in the occipital lobe of the owl monkey (Aotus trivirgatus). Brain Res 1975;100:473–87.

[74] Angelucci A, Rosa MGP. Resolving the organization of the third tier visual cortex in primates: a hypothesis-based approach. Visual Neurosci 2015;32:E010. https://doi.org/10.1017/S0952523815000073.

[75] Jeffs J, Federer F, Angelucci A. Corticocortical connection patterns reveal two distinct visual cortical areas bordering dorsal V2 in marmoset monkey. Visual Neurosci 2015;32:E012. https://doi.org/10.1017/S0952523815000097.

[76] Rosa MG, Schmid LM. Visual areas in the dorsal and medial extrastriate cortices of the marmoset. J Comp Neurol 1995;359:272–99. https://doi.org/10.1002/cne.903590207.

[77] Rosa MGP, Tweedale R. Visual areas in lateral and ventral extrastriate cortices of the marmoset monkey. J Comp Neurol 2000;422:621–51. https://doi.org/10.1002/1096-9861(20000710)422:4<621::AID-CNE10>3.0.CO;2-E.

[78] Palmer SM, Rosa MGP. Quantitative analysis of the corticocortical projections to the middle temporal area in the marmoset monkey: evolutionary and functional implications. Cereb Cortex 2006;16:1361–75. https://doi.org/10.1093/cercor/bhj078.

[79] Spatz WB. Topographically organized reciprocal connections between areas 17 and MT (visual area of superior temporal sulcus) in the marmoset *Callithrix jacchus*. Exp Brain Res 1977;27:559–72.

[80] vogt Weisenhorn DM, Illing RB, Spatz WB. Morphology and connections of neurons in area 17 projecting to the extrastriate areas MT and 19DM and to the superior colliculus in the monkey *Callithrix jacchus*. J Comp Neurol 1995;362:233–55. https://doi.org/10.1002/cne.903620207.

[81] Krubitzer LA, Kaas JH. Cortical connections of MT in four species of primates: areal, modular, and retinotopic patterns. Visual Neurosci 1990;5:165–204.

[82] Yu H-H, Chaplin TA, Davies AJ, Verma R, Rosa MGP. A specialized area in limbic cortex for fast analysis of peripheral vision. Curr Biol 2012;22:1351–7. https://doi.org/10.1016/j.cub.2012.05.029.

[83] Palmer SM, Rosa MGP. A distinct anatomical network of cortical areas for analysis of motion in far peripheral vision. Eur J Neurosci 2006;24:2389–405. https://doi.org/10.1111/j.1460-9568.2006.05113.x.

[84] Dias R, Robbins TW, Roberts AC. Dissociable forms of inhibitory control within prefrontal cortex with an analog of the Wisconsin Card Sort Test: restriction to novel situations and independence from "on-line" processing. J Neurosci 1997;17:9285–97.

[85] Burman KJ, Reser DH, Yu H-H, Rosa MGP. Cortical input to the frontal pole of the marmoset monkey. Cereb Cortex 2011;21:1712–37. https://doi.org/10.1093/cercor/bhq239.

[86] Reser DH, Burman KJ, Yu H-H, Chaplin TA, Richardson KE, Worthy KH, Rosa MGP. Contrasting patterns of cortical input to architectural subdivisions of the area 8 complex: a retrograde tracing study in marmoset monkeys. Cereb Cortex 2013;23:1901–22. https://doi.org/10.1093/cercor/bhs177.

[87] Eradath MK, Matsumoto M, Matsumoto K, Tanaka K, Ichinohe N. Anatomical inputs to sulcal portions of areas 9m and 8Bm in the macaque monkey. Front Neuroanat 2015;9:30. https://doi.org/10.3389/fnana.2015.00030.

[88] Bruce CJ, Goldberg ME, Bushnell MC, Stanton GB. Primate frontal eye fields. II. Physiological and anatomical correlates of electrically evoked eye movements. J Neurophysiol 1985;54:714–34.

[89] Constantinidis C, Franowicz MN, Goldman-Rakic PS. Coding specificity in cortical microcircuits: a multiple-electrode analysis of primate prefrontal cortex. J Neurosci 2001;21:3646–55.
[90] Tremblay S, Pieper F, Sachs A, Martinez-Trujillo J. Attentional filtering of visual information by neuronal ensembles in the primate lateral prefrontal cortex. Neuron 2015;85:202–15. https://doi.org/10.1016/j.neuron.2014.11.021.
[91] Rizzolatti G, Luppino G. The cortical motor system. Neuron 2001;31:889–901.
[92] Burman KJ, Bakola S, Richardson KE, Reser DH, Rosa MGP. Patterns of afferent input to the caudal and rostral areas of the dorsal premotor cortex (6DC and 6DR) in the marmoset monkey. J Comp Neurol 2014;522:3683–716. https://doi.org/10.1002/cne.23633.
[93] Mitra PP. The circuit architecture of whole brains at the mesoscopic scale. Neuron 2014;83:1273–83. https://doi.org/10.1016/j.neuron.2014.08.055.
[94] Oh SW, Harris JA, Ng L, Winslow B, Cain N, Mihalas S, Wang Q, Lau C, Kuan L, Henry AM, Mortrud MT, Ouellette B, Nguyen TN, Sorensen SA, Slaughterbeck CR, Wakeman W, Li Y, Feng D, Ho A, Nicholas E, Hirokawa KE, Bohn P, Joines KM, Peng H, Hawrylycz MJ, Phillips JW, Hohmann JG, Wohnoutka P, Gerfen CR, Koch C, Bernard A, Dang C, Jones AR, Zeng H. A mesoscale connectome of the mouse brain. Nature 2014;508: 1–21. https://doi.org/10.1038/nature13186.
[95] Jiang X, Shen S, Cadwell CR, Berens P, Sinz F, Ecker AS, Patel S, Tolias AS. Principles of connectivity among morphologically defined cell types in adult neocortex. Science 2015;350:aac9462. https://doi.org/10.1126/science.aac9462.
[96] Azadbakht H, Parkes LM, Haroon HA, Augath M, Logothetis NK, de Crespigny A, D'Arceuil HE, Parker GJM. Validation of high-resolution tractography against in vivo tracing in the macaque visual cortex. Cereb Cortex 2015. https://doi.org/10.1093/cercor/bhu326.
[97] Thomas C, Ye FQ, Irfanoglu MO, Modi P, Saleem KS, Leopold DA, Pierpaoli C. Anatomical accuracy of brain connections derived from diffusion MRI tractography is inherently limited. Proc Natl Acad Sci USA 2014;111:16574–9. https://doi.org/10.1073/pnas.1405672111.
[98] Reveley C, Seth AK, Pierpaoli C, Silva AC, Yu D, Saunders RC, Leopold DA, Ye FQ. Superficial white matter fiber systems impede detection of long-range cortical connections in diffusion MR tractography. Proc Natl Acad Sci USA 2015;112:E2820–8. https://doi.org/10.1073/pnas.1418198112.
[99] Axer M, Strohmer S, Gräbel D, Bücker O, Dohmen M, Reckfort J, Zilles K, Amunts K. Estimating fiber orientation distribution functions in 3D-polarized light imaging. Front Neuroanat. 2016;10:40. https://doi.org/10.3389/fnana.2016.00040.
[100] Zeineh MM, Palomero-Gallagher N, Axer M, Gräβel D, Goubran M, Wree A, Woods R, Amunts K, Zilles K. Direct visualization and mapping of the spatial course of fiber tracts at microscopic resolution in the human Hippocampus. Cereb Cortex 2016. https://doi.org/10.1093/cercor/bhw010. bhw010–16.
[101] Burman KJ, Bakola S, Richardson KE, Reser DH, Rosa MGP. Patterns of cortical input to the primary motor area in the marmoset monkey. J Comp Neurol 2014;522:811–43. https://doi.org/10.1002/cne.23447.

CHAPTER

4

Marmoset Nutrition and Dietary Husbandry

Michael L. Power[1], *Liz Koutsos*[2]

[1]Conservation Ecology Center, Smithsonian National Zoological Park and Conservation Biology Institute, Washington, DC, United States; [2]Department of Nutrition Science, Smithsonian National Zoological Park and Conservation Biology Institute, Washington, DC, United States

INTRODUCTION

Appropriate management of common marmoset colonies requires focused attention on nutrition to meet needs for life stage, health, and well-being, and also to reduce the risk of erroneous or variable research results due to improper nutrition. Nutrition throughout life affects the risks and vulnerabilities of animals to morbidity and mortality from disease processes. Early nutrition can have lasting effects on phenotype through effects on growth and development. The consequences of early life nutrition can include epigenetic processes, indicating the potential for effects on phenotype across generations. Proper dietary husbandry practices are crucial for successful colony management.

In addition, within the context of biomedical research, nutritional considerations are very important for the outcomes and interpretation of experimental data. Evolution has produced regulatory mechanisms that allow animals to survive and reproduce on a range of nutritional planes. That means that two animals on different nutritional planes may be equally healthy; however, their internal physiology, and thus their responses to experimental conditions, may differ. Reducing the variation in nutritional status and in dietary husbandry across individuals and colonies may reduce the variability within a group and improve the ability to compare and test research results across colonies.

There is a long history of concerns over the dietary management of captive common marmosets. Much of what we know about diets in marmosets comes from anecdotal or historic precedence, animal food preferences, or published reports that focus on the role of limited numbers of specific diet components in relatively small numbers of animals. The literature on rigorous nutritional studies of marmosets is relatively thin. Repeated surveys of the membership of the Marmoset Research Group of the Americas (MaRGA) indicate that lack of standardized dietary husbandry is considered a major deterrent of research use. Because we have an incomplete understanding of the dietary requirements of this species, there is wide nutrient and ingredient variability in diets provided to captive marmosets. This inconsistency can result in variation in nutrient intake of marmosets both within and among colonies [1,2], which may contribute unnecessary variation in experimental outcomes and may play a role in some of the more common clinical diseases seen in this species in captivity. The goals of this chapter are to provide a review of the known information relevant to marmoset nutrient requirements, discuss practical aspects of diets and feeding programs, and, finally, to suggest areas of marmoset nutrition and dietary management in which research efforts are needed.

BASICS

It is a truism that although animals eat food, what they require are nutrients. Evolution has provided species with morphological, physiological, metabolic, and behavioral adaptations that allow them to extract nutrients from the food items found in their natural environment. This does not mean, however, that diets for captive marmosets need to mimic or even approximate natural, wild diets. It does mean that the evolved dietary adaptations of marmosets need to be considered when diets are formulated, as well as the altered circumstances presented by captive conditions.

Nutrients are molecules that are required for life that cannot be synthesized by an animal but rather are acquired from the environment, generally by eating food.

The Common Marmoset in Captivity and Biomedical Research
https://doi.org/10.1016/B978-0-12-811829-0.00004-2

They can be chemical elements (e.g., sodium, potassium), components of enzymes (e.g., vitamin C, niacin), or even steroid hormones (e.g., vitamin D). Many of the macronutrients (those nutrients required at the highest concentration in the diet) are classes of molecules (e.g., amino acids as components of proteins), which are sometimes fairly complex to describe (e.g., fiber) and in which specific molecules can have varied functions (e.g., fatty acids). Even vitamins (e.g., vitamins A and E, folate) can consist of suites of related molecules all with varying levels of biological activity. Finally, most nutrients have both minimal levels that animals must obtain and maximal levels above which health can be compromised. In other words, many nutrients are both required and potentially toxic.

Feeding marmosets diets with appropriate levels of nutrients is the obvious necessary condition for successful dietary husbandry. However, captive marmosets exhibit substantial variation in digestive efficiency, and hence nutrient absorption, even when fed identical diets [3,4]. Marmoset dietary husbandry is complicated by this large range of variation in ability to absorb nutrients from the diet. For example, histograms of the apparent digestibility of dry matter for marmosets from the Southwest (SNPRC) and Wisconsin (WNPRC) National Primate Research Centers show not only a significant difference between the two colonies but also substantial variation within each colony (Fig. 4.1). All of the animals used in the digestion trials were ostensibly healthy, with no outward signs of disease. They ranged in age from 2–9 (SNPRC) and 2–10 years (WNPRC) with equal numbers of males and females. The difference between the two colonies can be explained by the fact that WNPRC uses a high-fiber variant [~6% crude fiber, 10% neutral detergent fiber (NDF) vs. ~4% crude fiber, 6.7% NDF] of the marmoset diet used at both sites. However, differences in diet do not explain the approximately 8% points spread in apparent digestibility at SNPRC and the 11%–12% points spread at WNPRC. Those differences are likely due to differences within the animals' digestive tracts, with low digestive efficiency animals potentially suffering from some level of intestinal tract inflammation.

Nutrient requirements for marmosets with healthy digestive tracts are probably little different from those of other monkeys. Based upon the author's experience studying marmoset nutrition, the conservative estimates for these requirements from the 2003 NRC publication on primates likely are adequate for healthy marmosets. Marmosets with poor digestive efficiency will have effectively higher nutrient requirements, as they will gain less from their food. The dietary management concern is how to avoid oversupplementing the healthy animals, while providing sufficient nutrient density to satisfy requirements for animals with lower digestive efficiency. Of course, reducing the proportion of marmosets that suffer from intestinal tract inflammation and hence poor nutrient absorption would be the preferred goal. Research is underway investigating this issue; however, the underlying causes of intestinal inflammation in marmosets are not well understood, despite many years of investigation [3–6].

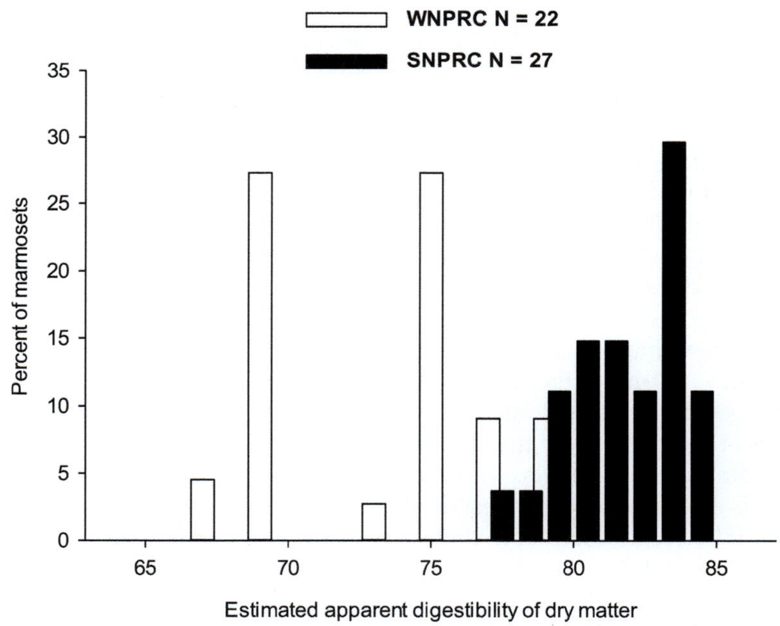

FIGURE 4.1 The range of estimated values for the apparent digestibility of dry matter at the Southwest (SNPRC) and Wisconsin (WNPRC) National Primate Research Centers. Values were the means from two consecutive 4-day digestion trials using the homogeneous base diets for the colonies only, with no supplementary foods given.

GASTROINTESTINAL ANATOMY AND WILD-TYPE DIETS

The common marmoset is a gummivore–frugivore–insectivore. In the wild, marmosets feed heavily on insects and small vertebrates, fruit, and the sap, gum, latex, and other exudates from trees and lianas [7]. This last dietary specialization is somewhat unique among mammals, though fairly common within primates [8–10]. Marmosets have elongated lower incisors [11,12], with a lack of enamel on the lingual side of the mandibular incisors [13]. This produces chisel-like lower incisors adapted for gouging bark to allow consumption of the resulting flow of gums and exudates [14]. In the wild, common marmosets are estimated to spend 6%–60% of their feeding time consuming gums [15], with larger proportions of time devoted to gum feeding when fruit is scarce [16]. Gum is an important food for wild marmosets but is not preferred over fruit or insects.

Few gums have been chemically analyzed; those that have been assayed are rich in complex polysaccharides that would likely require microbial fermentation to be digested [17]. Gum is a problematic food, being difficult to digest and providing mainly energy from the fermentation of polysaccharides and some minerals (e.g., calcium) [18]. Gum is not a complete food, but in theory it would serve to balance a heavily insectivorous diet, as insects are high in phosphorus and low in calcium, whereas gums are generally the reverse.

The gastrointestinal (GI) anatomy of the common marmoset is reflective of its adaptation to a diet prevalent in gums and other exudates [12]. The common marmoset has a well-developed cecum [11,14,19] with an inner structure that produces many small pockets [20] where bacterial populations can flourish and be protected from washout [4,18]. There is some evidence of selective retention of fluid in the cecum, supporting the hypothesis of cecal–colonic fermentation of gums as an evolutionary adaptation of marmosets. Mean retention time (MRT) of a fluid-phase marker was slightly longer than that of particulate matter in commercial-type diets (14 v 12 h, respectively) [21]. In contrast, in marmosets fed a purified gel-type diet, MRT was similar (~17 h) for fluid and particulate markers [4]. In that study, the particulate markers were smaller on average than the ones used by Caton and colleagues (≤2 vs. 3 mm, respectively). A study using polystyrene beads, which were larger than the markers in the other two studies, found MRT to be about 8 h [7]. Thus, marmosets appear to be typical hindgut-fermenting mammals, in that they pass large particles through the GI tract the fastest and fluids and small particles the slowest. This digestive strategy allows the rapid elimination of indigestible material, such as seeds from ingested fruit, while retaining gums within the cecum and colon for a longer time, allowing fermentation [4].

The gut and digestive specializations of the common marmoset are relatively minor and probably are not of much importance for captive management. Marmosets are more capable of digesting soluble fiber, such as in gums, than are tamarins [22]. There are no known contraindications to offering marmosets gum in their diet; however, it should be a minor constituent, as it is deficient and even absent in many nutrients (e.g., vitamins) and will serve to dilute the nutrient content of the rest of the diet. Gum arabic syrup has been used successfully as an enrichment item; however, there are no convincing data that gum provides any essential nutritional or physiological benefits that cannot be met by other constituents of captive diets [1,18].

NUTRIENTS/REQUIREMENTS

Energy

Energy is not a nutrient, rather, energy is derived from the oxidation of nutrients and is the "fuel" for biochemical processes and life. Wild animals may be constrained by the need to meet energy requirements, due to food availability, energy expenditure related to foraging, avoidance of predators, in addition to many other factors. In captively managed situations, however, energy generally needs to be managed to prevent overconsumption and the resulting overweight and obesity. This is in part due to the lower energetic expenditures of animals which are not actively foraging over large geographies, have a reduced need for thermoregulation due to management of environmental conditions, and also to the general high palatability, digestibility, and energy content of food items fed in captivity.

The energy content of food items can be defined in multiple ways. Gross energy (GE) is the amount of energy released through complete oxidation of a food and is determined by combustion. While relatively easy to measure, GE overestimates the amount of energy available to the animal as it does not account for losses during digestion and metabolism. Digestible energy (DE) reflects the proportion of the GE that is digested by the animal and is assessed by subtracting the amount of energy in feces from the GE content of the diet. Further refinement of DE will account for the losses due to metabolism and includes the energy losses in urine and gaseous excretions, which is termed metabolizable energy (ME). The latter two measures of energy require in vivo work and can vary depending on the physiological status of the animal [23]. Estimates of

ME provided by a diet are often used in lieu of animal work and are usually based on the values of 4 kcal/g for dietary protein, 9 kcal/g for fat, and 4 kcal/g for carbohydrate. The factor for carbohydrate assumes low levels of fiber or other carbohydrates that require fermentation in the diet. Carbohydrates that require fermentation will deliver significantly less than 4 kcal/g to the host. Marmoset diets are generally low in fiber, so the estimation is assumed to be reliable for most marmoset diets.

The energy requirements of a marmoset consist of several components, including the energy for basal metabolic functions, voluntary activity, thermoregulation, and synthesis of tissues, in addition to the energy losses that occur during these processes. The basal metabolic rate is the energy needed for the basic functions of life in an animal that is in thermoneutrality, postabsorptive, and awake but resting and unstressed. Resting metabolic rate is a similar measure and these terms are often used interchangeably [23].

Of importance to note is the lower critical temperature for marmosets; this is the lowest temperature that is within the thermal neutral zone for the species below which the animal will increase metabolic rate to maintain body temperature. The thermal neutral zone for marmosets is estimated to be 27−33°C [7]. Marmosets in captive colonies might be maintained at temperatures below the lower critical value of 27°C and may have some additional energy costs of thermal regulation. The thermal regulatory cost will depend on housing conditions, specifically whether animals are singly versus group housed. The value of 27°C for the lower critical temperature is for singly housed animals. Animals housed in pairs or groups can huddle, reducing thermal regulatory costs and lowering the effective lower critical temperature by several degrees C. Attention to the lower critical temperature may be important for marmosets that need to be removed from their group for medical reasons. If an animal is already under stress due to disease or injury, adding additional thermal regulatory costs by keeping the animal at temperatures below 27°C should be avoided, if possible. Ideally, research colonies within and across institutions should provide similar environment temperatures for their animals, for results to be more readily comparable.

The general equation for predicting maintenance energy requirements of captive primates is 140 kcal per day $\times$ BW $(kg)^{0.75}$ [23]. It is important to note that the allometric exponent of 0.75 is a good estimate of maintenance energy requirement differences between two species of different body size but may overestimate the difference in maintenance energy requirement between animals of different masses of the same species. This is especially true for obese animals, as excess adipose tissue is not as metabolically active as lean body mass. The value derived from this equation should be considered a ballpark figure for animals of a healthy weight. A reasonable range for energy requirement for marmosets is 60−75 kcal per day of ME (body mass range 325−425 g). Marmosets with body mass above 435 g have less favorable lipid profiles, indicating potential obesity and susceptibility to obesity-related diseases [24]. Animals with body mass below 300 g tend to have higher rates of morbidity that can extend over years (Tardif, pers. comm.).

A digestibility study of eight singly housed adult marmosets (mean body weight = 355 g, range 295−416 g) found DE intake to average 74 kcal per day (range 62−85 kcal per day), which was approximately twice the resting metabolic rate [7]. These animals were singly housed at temperatures ranging from 22 to 25°C, below the lower critical temperature for marmosets, implying they were expending extra energy for thermal regulation compared with animals housed above 27°C.

Energy requirements will be higher for growing animals, and for lactating females. Gestation does not seem to impart significantly increased energy needs in common marmosets [25]; however, energy restriction during early-to-mid pregnancy leads to rapid pregnancy loss [26]. Lactation was associated with 1.5−2.0× increases in energy intake, with higher intakes during later stages of lactation [25]. Interestingly, maternal energy intake did not vary with number of offspring during lactation [27].

Energy requirements for growth of marmosets are reported to be 30%−50% higher than those of adults at maintenance for Old World and New World primates [28], although it is important to note that data are lacking in this area. Marmoset infants grow at about 1 g per day through the first 4 months. Infants that grew faster than that tended to become obese as young adults [29]. Additionally, earlier onset of eating solid food (before 28 days) is associated with an increased rate of feeding and higher incidence of obesity and impairments in glucose homeostatic mechanisms in common marmosets [30,31].

Protein

Amino acids are the building blocks of proteins. There are 20 amino acids, of which 10 are essential for primates, and the remaining can be synthesized by the animal (assuming there is enough dietary nitrogen). Protein requirements for maintenance are generally less than 10% of the ME in the diet from protein, and requirements for growth, gestation, and lactation are thought to be higher than that of maintenance [23].

Protein sources in marmoset diets will likely consist of purified or semipurified ingredients such as soy protein fractions (isolated soy protein, soy protein concentrate), casein, or lactalbumin or may be derived from less purified ingredients such as grains and seeds (e.g., corn gluten meal, flaxseed meal), legume meals (e.g., soybean meal), or animal products such as eggs, animal or fish protein meals, or insects. There has been some concern over potential allergens in grain-based diets, due to their gluten content, or more specifically, the protein gliadin within gluten [5,32]. Gliadin is, obviously, an evolutionarily novel protein for marmosets, and gluten sensitivity in humans causes intestinal inflammation, a major concern in marmosets, as mentioned above. At present, the limited evidence suggests that some proportion of marmosets may have an antibody reaction to gluten, which may be associated with intestinal tract inflammation and which appears to be ameliorated by removing gluten from the diet [5], though the effect is underwhelming. However, marmosets maintained on gluten-free diets still exhibit a wide range of digestive efficiency [3,4] and exhibit inflammatory lesions in the intestinal tract (Tardif, pers. comm.). Removing gluten from the diet is unlikely to solve this marmoset health issue. This issue will be discussed in more detail at the end of this chapter in the section on research needs.

Essential amino acid levels have not been clearly defined, but levels that appeared to meet or exceed needs of common marmosets have been published [33]. Digestibility of protein from different sources will vary, and current recommendations for dietary protein for marmosets fed purified or semipurified diets are a minimum of 7% protein for maintenance and 12%–18% for growth, gestation, and lactation, and fed diets with "traditional" feed ingredients are 15%–22% crude protein (dry matter basis) [23]. Conventional understanding has suggested that marmosets may benefit from a diet with increased dietary protein [1,95]. This understanding, in part, comes from early literature suggesting that diets with inadequate protein composition are likely to represent causal factors for wasting disease in tamarins, a species closely related to marmosets [34]. While protein amount or source may play a role in intestinal inflammation, the relatively frequent identification of interstitial nephritis calls into question the proper balance of protein in commercial marmoset diets [35]. Many commercial products, particularly those designed for exotic primates in zoo settings, are formulated under the assumption that supplementary, lower-nutrient-density foods will make up as much as 50% of the diet, which would dilute the concentrations of nutrients such as protein. The manufacturer should clarify the expected inclusion rate of their product, and those designed for more supplementation typically have protein levels at or slightly above 20%, which appears to be well above the actual protein requirement for marmosets. The estimated minimal protein requirements of adult common marmosets are estimated to be 6.6% protein (% of dry matter intake; 6% of ME as protein) or 2.5 g protein/kg BW*d [36]. Low dietary protein levels (below 6%) have been associated with weight loss and coprophagy in marmosets [37]. Breeding common marmosets have been successfully maintained on diets containing 15%–16% high-quality protein (lactalbumin) on a dry matter basis for many years [38], a protein level which is likely above minimal requirements but not of concern for excess.

Lipids/Fatty Acids

Primate diets require omega-6 and omega-3 fatty acids for normal growth and development. Current recommendations are for the omega-6 fatty acid, linoleic acid (18:2n6), to be supplied at ≥2% of the diet (dry matter basis), and for the omega-3 fatty acid, linolenic acid (18:3n3), and/or EPA (20:5n3) and DHA (22:6n3) to be supplied at ≥0.5% of the diet (dry matter basis) [23]. Additional lipid in the diet will serve as an energy source. Deficiency of fatty acids is associated with skin lesions, alopecia [39], and impaired brain and nervous system development [23].

Blood cholesterol and triglyceride levels are modified by dietary fat inclusion in marmosets. A higher dietary fat level (12%) was associated with lower blood cholesterol than feeding 4% dietary fat, and diets high in omega-3 fatty acids from fish oil (generally longer-chain PUFA) were associated with lower blood triglyceride levels as compared with diets rich in linoleic acid or saturated fatty acids [40]. As seen in many other species, the dietary fatty acid profile is reflected in the blood and skeletal muscle fatty acid profile [40]. Recent work using high-fat "snacks" (e.g., mealworms and peanuts) has shown improvement in glucoregulatory functions [41], but further work is needed to determine whether fatty acid nutrition, energy nutrition, or some other parameter is responsible for the observed effects.

Carbohydrates

Carbohydrates serve as the primary energy source in primate diets, generally in the form of starch in commercial primate diets (from grains such as corn, wheat, or rice) and sugars (in purified or semipurified diets). Indigestible carbohydrates (fiber, resistant starches) may be supplied from forages (e.g., alfalfa) and by-products of agricultural industries (e.g., beet pulp, soybean hulls, rice hulls). Additionally, soluble fiber fractions such as pectins and gums would be found in wild-type diets of marmosets. Fiber is not essential per se but is important for a healthy GI tract due to multiple factors including physical properties (bulk, water-holding capacity, gut motility), providing substrate for the microbiome in the cecum and colon, and generation of short-chain fatty acids via microbial fermentation (e.g., acetate, propionate, and butyrate). These fatty acids serve as energy sources for the host and may modulate intestinal health directly. For example, butyrate has been shown to be an effective energy substrate for colonic epithelial cells and may improve cell function and health [42].

There is little evidence that marmosets effectively ferment insoluble fiber; however, marmosets have a good ability to ferment soluble fiber such as gum arabic [7,22]. Whether the provision of soluble fibers, such as gum arabic, has health benefits for marmosets is possible but not demonstrated. We consider this issue further in the final section of this chapter.

Vitamins

Vitamins are organic compounds that serve as cofactors, hormones, and regulators of cellular differentiation and which are generally required in trace amounts from the diet to meet needs for growth, reproduction, and health. Vitamins may be categorized as fat-soluble (i.e., vitamins A, D, E, K) or water-soluble (i.e., B vitamins including cobalamin [B12], biotin, folate, niacin, pantothenate [B5], pyridoxine [B6], riboflavin, thiamin [B1], as well as vitamin C [ascorbic acid], and choline). Fat-soluble vitamins have a higher risk of toxicity because they are stored in the liver and adipose and it is harder to excrete excess amounts.

FAT-SOLUBLE VITAMINS

Vitamin D

The primary known biological actions of vitamin D are in support of calcium metabolism and bone health [43]. However, the vitamin D receptor is found on most cells, including not only those cells important in calcium homeostasis (e.g., intestine, kidney, parathyroid gland, and bone) but also active immune cells, pancreatic beta cells, bronchial epithelial cells, skin epithelial cells, testes, and mammary gland [44]. Varying evidence indicates potential roles for vitamin D in immune function, metabolism, and cell differentiation [43,45]. Adequate vitamin D status appears to be beneficial to health in more ways than for calcium metabolism and bone.

Although it is recommended that all captive primates be given a dietary source of vitamin D, under wild, free-living conditions vitamin D is not a required nutrient but rather is an endogenously produced precursor to the active steroid hormone, 1,25-dihydroxyvitamin D ($1,25[OH]_2$ D). When exposed to certain wavelengths of ultraviolet light (UVB), which is prevalent in unfiltered sunlight, most vertebrates, including primates, produce vitamin D in the skin by photoconversion of 7-hydrocholesterol. Vitamin D itself has limited biological activity. Two sequential hydroxylations are required to produce the most biologically active metabolite, 1,25 $[OH]_2$ D. The first is an unregulated hydroxylation that takes place in the liver to produce 25 hydroxyvitamin D or 25-OH-D. This is the primary circulating vitamin D metabolite and serves as the substrate for the regulated hydroxylation in the kidney that produces the active metabolite. Thus, 25-OH-D can be considered the main storage form of vitamin D. Low serum concentrations of 25-OH-D are considered diagnostic of vitamin D deficiency [43].

There is a long history of concern and confusion over vitamin D requirements of marmosets. It has been demonstrated that marmosets do not utilize vitamin D_2 (ergocalciferol) [46], so any dietary supplementation must be done with vitamin D_3 (cholecalciferol). It has been suggested that marmosets have higher vitamin D requirements than do other monkeys because they have circulating levels of $1,25[OH]_2$ D that are 2–4 times higher than rhesus macaques (*Macaca mulatta*) or humans [47]. However, levels of 25-OH-D in healthy, captive black-tufted marmosets (*Callithrix pennicilatta*) with full sunlight exposure (100–140 ng/mL) [48], while higher than is seen in indigenous human populations [49] and free-ranging Old World monkeys [50], were lower than that typically reported in indoor-housed marmosets [46].

Dietary levels of vitamin D_3 vary widely among marmoset colonies and while marmosets appear tolerant of high doses of vitamin D, oversupplementation of this potentially toxic nutrient may have long-term detrimental effects on health. Because of the diverse roles 1,25-dihydroxyvitamin D performs in immune function and cell development, in addition to its calciotropic function, variation in vitamin D status among marmoset colonies also may result in different metabolic and immunologic states between colonies that might affect experimental results.

The vitamin D receptor in callitrichid primates has been shown to bind 1,25-dihydroxyvitamin D as efficiently as does the human vitamin D receptor, implying that any vitamin D resistance is not due to receptor binding [51]. However, in callitrichid primates there is a highly expressed intracellular binding protein that competes with ligand-complexed vitamin D receptor at vitamin D response elements, thus reducing 1,25-dihydroxyvitamin D signaling [52] (reviewed in Chapter 29 of this volume). It is not clear that this translates into a higher vitamin D requirement for these species. Cotton top tamarins (*Saguinus oedipus*) have been successfully maintained on diets with vitamin D at 2500 IU/kg for at least 2 years, with no access to UVB [53]. At the same time, in colonies of cotton top tamarins and common marmosets fed diets with 3–5 times that level of dietary vitamin D there have been instances of vitamin D deficiency. Dietary vitamin D appears to result in variable vitamin D status in captive callitrichid primates, possibly due to individual health conditions relating to the intestinal tract that can affect nutrient absorption. Marmosets with good digestive efficiency have circulating levels of 25-OH-D well above the range exhibited by wild animals, whereas marmosets with poor digestive efficiency have substantially lower circulating 25-OH-D levels, often at deficient levels [3]. Thus, current levels of vitamin D in marmoset diets appear to be above the requirements for healthy animals but may be necessary for marmosets with intestinal inflammation.

Vitamin A

Vitamin A is essential for cellular differentiation and vision in higher-order vertebrates. Vitamin A (retinol) may be derived from animal or microbial origin, and some plant-based carotenoids may serve as provitamin A compounds with varying efficiencies of conversion. Vitamin A deficiency is associated with eye pathology, anorexia, and diarrhea, and can take months to years to manifest while liver stores of vitamin A are depleted. Vitamin A toxicity is also of concern [54,55], and vitamin A tends to be formulated at high levels in commercially prepared diets due to its instability due to heat and light exposure. Liver retinol determination is necessary to evaluate vitamin A status [23], and work in marmosets has demonstrated slightly elevated (as compared with human reference ranges) liver vitamin A concentrations in marmosets fed diets that greatly exceeded recommendations for vitamin A intake (20,000 IU vit A/kg diet; 500 IU/d) [56]. In addition, kidney retinyl esters were also quite high in these animals [55]. Vitamin A status is not generally able to be accurately measured with blood values, although the levels of retinyl esters were elevated in fasted marmosets previously shown to have high liver vitamin A levels and may provide a mechanism by which to assess vitamin A status in blood [57]. From these studies, it has been suggested that less toxic forms of vitamin A be supplied to primates, in particular, β-carotene, for which conversion to vitamin A is regulated by vitamin A status [58,59].

There are some concerns about excess vitamin A and impacts on oxidative status via depletion of vitamin E. In chicks and in rats, high vitamin A increased oxidation of vitamin E in the GI tract prior to absorption and increased vitamin E turnover rate [60,61]. It is unknown if this relationship holds true for primates, but caution in feeding excess vitamin A is warranted. Vitamin A toxicity has been associated with osteoporosis and other manifestations of altered bone metabolism, and there is evidence of antagonism between vitamin A and D in the bone, at the receptor level [62].

Vitamin E

Vitamin E, α-tocopherol, is the most biologically active form of the tocopherols and tocotrienols that make up the family of vitamin E molecules. These molecules serve as antioxidants, stabilize membranes, affect eicosanoid signaling and cellular proliferation, and modulate immune responses. The requirement for vitamin E is generally proportional to the level of dietary PUFA. High-PUFA diets can hasten development of vitamin E deficiency symptoms in marmosets and other primates [63]. Plasma vitamin E levels may be used as an indicator of status. Common marmosets fed a diet with no supplemental vitamin E had plasma α-tocopherol concentrations of ~250 μg/dL, which increased to ~600 μg/dL with supplementation (40 IU/kg) [63] Anemia, dermatitis, and steatitis have been reported in marmosets with plasma α-tocopherol levels <450 μg/dL and conditions resolved after vitamin E supplementation [64]. Other reports support the positive response of marmosets to dietary vitamin E enrichment, in regard to plasma vitamin E levels and potentially oxidative stability (e.g., lowered RBC hemolysis with supplementation) (3–53 mg vitamin E equiv. per day) [65].

Vitamin K

Vitamin K is essential for normal blood clotting and is also involved in bone formation. Menaquinones are one form of vitamin K compounds that are produced by intestinal bacteria, and it is unclear to what extent intestinal flora contribute to vitamin K nutrition of nonhuman primates [23]. Marmosets are not known to be vulnerable to bleeding problems in captivity, implying that dietary levels of vitamin K are sufficient, but little is known about their vitamin K requirements or whether vitamin K is a dietary necessity or endogenously produced. Regardless, all diets should contain vitamin K.

WATER-SOLUBLE VITAMINS

The water-soluble vitamins have a lower toxicity risk and are frequently oversupplemented to ensure optimal levels are met. Thus, research efforts beyond defining minimal requirements are scarce. Of the water-soluble vitamins, vitamin C generally receives the most attention as it is an essential dietary nutrient for primates. Recent research on dietary vitamin C has been enabled by the development of more stable forms of this vitamin. L-ascorbyl-polyphosphate is quite stable [66] and, thus, commercial diets can be used to meet the needs for this vitamin. Early work in marmosets demonstrated that whole-blood ascorbic acid was responsive to dietary supplementation with vitamin C, and that very high doses of vitamin C (200 mg per day) reduced the duration of parainfluenza virus-related illness and mortality [67].

In regard to the other water-soluble vitamins, a study of 50 marmosets (24 females, 26 males) provided some baseline circulating values for future work [68]. Blood riboflavin, biotin, choline, and pantothenate were lower in marmosets than human reference ranges, whereas niacin and pyridoxine were higher. These values were not significantly different due to marmoset age or sex [68]. Levels of these nutrients sufficient for any primate most likely suffice for marmosets.

Minerals

Minerals are often subdivided into macrominerals (calcium [Ca], phosphorus [P]), electrolytes (sodium [Na], potassium [K], chloride [Cl], magnesium [Mg]), trace minerals (copper [Cu], Zn, iron [Fe], zinc [Zn], manganese [Mn], Se), and ultratrace minerals (iodine, cobalt [Co], chromium [Cr], molybdenum [Mo]) based on the general levels in the diet.

Calcium and phosphorus are essential for bone mineralization, and their metabolism is closely related with vitamin D metabolism. The bone can serve as a depot of these minerals and as such deficiency may be slow to manifest. The recommended ratio of dietary calcium:phosphorus is 1:1 to 2:1. The electrolytes (Na, Cl, K, Mg) are unlikely to be deficient if animals are consuming natural ingredient-based diets or purified or semipurified diets with mineral supplementation.

Among the trace minerals, iron is of potential concern due to reports of iron storage disease in marmosets [69]. Vitamin C may enhance iron absorption, thus diets high in iron and vitamin C should be fed cautiously.

A purified diet with Fe at 200 mg/kg on a dry weight basis was successfully used as a sole food for marmosets for over 4 years [27]. This diet was recently reformulated to lower Fe levels to 100 mg/kg dry weight with no apparent ill effects. The concentration of Fe in natural ingredient diets is difficult to regulate due to contamination in the milling process. The bioavailability of Fe in natural ingredient diet is also difficult to ascertain. Natural ingredient diets can be both high in Fe and also have low Fe bioavailability. At present, the concern over Fe toxicity in marmosets is unproven and the most appropriate level of Fe in the diet uncertain. However, purified diets do not need more than 200 mg/kg.

INFANTS

Most of the above discussion and information is pertinent to adult marmosets. Juvenile marmosets appear to do well on the same diets as do adults [27]. There is little solid evidence for nutrient requirements of infants before weaning. Our best estimates must rely on the composition of marmoset milk, for which there are solid data only for water, fat, crude protein, total sugar, and Ca and P content [70, 71]. The mean values for these nutrients in milk from wild and captive marmosets is presented in Table 4.1. Note that in wild marmosets there is a greater amount of energy in the milk from sugar,

TABLE 4.1 Proximate Nutrient Composition of Milk From Wild and Captive Common Marmosets

	Mean ± SEM on an As Is Basis	Mean ± SEM on a Dry Matter Basis	Mean ± SEM as Percent of Gross Energy
Water	Wild: 87.3 ± 0.4% Captive: 86.0 ± 0.4%	–	–
Crude protein	Wild: 2.2 ± 0.2% Captive: 2.7 ± 0.1%	Wild: 17.3 ± 1.6% Captive: 19.4 ± 0.7%	Wild: 19.8 ± 1.1% Captive: 20.8 ± 0.8%
Fat	Wild: 2.3 ± 0.3% Captive: 3.6 ± 0.4%	Wild: 18.1 ± 2.4% Captive: 24.5 ± 2.9%	Wild: 31.0 ± 2.4% Captive: 39.8 ± 2.8%
Sugar[a]	Wild: 8.0 ± 0.2% Captive: 7.4 ± 0.1%	Wild: 63.0 ± 1.6% Captive: 53.2 ± 0.7%	Wild: 49.2 ± 2.9% Captive: 39.3 ± 2.3%
Ca (mg/g)[b]	Captive: 1.03 ± 0.08	Captive: 7.4 ± 0.6	–
P (mg/g)[b]	Captive: 0.42 ± 0.07	Captive: 3.0 ± 0.5	–

[a] *Most sugar likely lactose with only about 5% in the form of other oligosaccharides [97].*

[b] *Data reported in Hinde K, Foster AB, Landis LM, Rendina D, Oftedal OT, Power ML. Daughter dearest: sex-biased calcium concentrations in mother's milk among rhesus macaques. Am J Phys Anthropol 2013;151:144–150.*

*Data from Power ML, Oftedal OT, Tardif SD. Does the milk of callitrichid monkeys differ from that of larger anthropoids? Am J Primatol 2002;56:117–127 [70] and Power ML, Verona CE, Ruiz-Miranda C, Oftedal OT. The composition of milk from free-living common marmosets (*Callithrix jacchus*) in Brazil. Am J Primatol 2008;70:78–83 [71].*

whereas in captive marmosets there is more energy from fat. Of note is that the energy from milk protein is not different between wild and captive marmosets and has the lowest variance. This implies that the relative proportions of fat and sugar in infant diets are not essential, but that protein in any milk replacer should provide about 20% of GE. Because infants will most likely have greater protein requirements than will adult animals, the composition of marmoset milk adds further evidence that adult marmoset diets should meet requirements with protein at less than 20% of energy or less than 17% of dry matter.

FEEDING

Food/Diet Types

In research labs, marmosets are often fed a variety of food items, including fruits, legumes, dairy products, seeds, nuts, insects, and commercially prepared foods included extruded biscuits, canned diet, and gel diets (based on "typical" feed ingredients or based on purified or semipurified ingredients) [72]. It is of importance to note that "cafeteria"-style feeding programs are unlikely to result in appropriate nutrient intakes. Capuchin monkeys (*Cebus apella*) offered a choice of food items versus traditional primate biscuits, had variable intake. For example, biscuits were chosen and consumed less often when familiar food items were offered, particularly if those food items were offered relatively infrequently to the capuchins. In contrast, when novel food items were offered, biscuits were chosen more often. Clearly, these primates were not choosing food items based on nutrition content but rather on novelty or lack thereof [73]. Adult marmoset breeding females are dominant in food situations [74]. They are able to monopolize desired food items. In a cafeteria diet feeding scheme, this can result in the breeding female eating the least of the base, complete diet.

There is some evidence of "specific appetite," or intentional selection for a particular nutrient, for protein in some instances in primates [75,76], and likely for sodium [77,78]. Common marmosets, especially if lactating or with low vitamin D status, displayed a preference for calcium lactate solutions over plain water [78]. There is little evidence for nutritional wisdom as it relates to other nutrients, thus providing balanced diets or food items in appropriate volumes to ensure desired nutrient intake is critical.

Extruded biscuits are commonly fed to primates and are produced by mixing ground ingredients (which may then be further ground to a desired particle size), subjecting the mix to steam and pressure while the mix is forced through an extruder barrel. During this time (generally minutes), some cooking takes place. The mix exits the extruder barrel through precise equipment, which controls the shape and size of the final biscuit. This product is very high in moisture and thus the biscuit must then be dried to a final moisture level of ~8% [79]. The heat associated with the extrusion process results in starch gelatinization, which improves its digestibility by ~ twofold [80], some protein denaturation (which can also improve digestibility), and significant destruction of microorganisms [81]. At the same time, some vitamins are very heat-sensitive (notably thiamine, folate, pyridoxine, and vitamin B12), and losses during extrusion must be accounted for by the manufacturer [79].

Moist/canned diets are manufactured using high-moisture ingredients and/or by the addition of moisture. The canning process requires higher temperatures than that of extrusion, generally 250°F, with some variation due to time and can size and so nutrient stability (generally vitamins and fatty acids) is also a concern for these products, particularly for thiamine [82].

Gel diets are also high-moisture diets but generally sold in a dry meal form, and the end-user adds water to create a semimoist diet. These diets can be used to customize feeding protocols for an individual or group and to provide medications under the care of a veterinarian. It is important to ensure that any additives to gel diets are not deactivated by the high water temperature that gel diets are initially mixed in (or that additives are mixed in only when the temperature of the gel has cooled to appropriate levels). In addition, once prepared, gel diets should be handled and stored like a raw meat item. A purified gel-type diet has been used successfully as a sole diet for many years with common marmosets [38].

For any of the diet items, autoclaving or irradiation may be desired due to experimental constraints. Autoclaving is particularly destructive to vitamins, thus over-fortification prior to autoclaving is required [83]. Irradiation is generally less destructive to vitamins and other sensitive nutrients, unless very high doses of irradiation are used [84]. Marmosets have been maintained on an irradiated purified diet for more than 6 years with normal health and reproduction (Tardif, pers. comm.).

To compare multiple diet items, it is important to assess nutrients on a dry matter basis (i.e., without moisture content). For example, an extruded biscuit containing 20% crude protein and 10% moisture would contain ~22.2% crude protein on a dry matter basis. A canned diet containing 9% crude protein and 60% moisture would contain ~22.5% crude protein on a dry matter basis.

Clearly the diet type affects the animals' physiology. Marmosets fed a mixed diet including fruits, vegetables, and commercially prepared canned diet had significantly different plasma metabolites as compared with when the same animals were fed a purified diet. Specifically, metabolites associated with amino acid metabolism and fatty acid metabolism (carnitine metabolites) varied due to diet type, which may be related to digestibility, overall nutrient profiles, or some other interactive effect of nutrient level and source [72].

Enrichment

Beyond meeting the nutritional needs of the marmoset, generally accomplished with one of the base diet types discussed above, food may also be used as a form of enrichment or for training purposes. Food used for enrichment and/or training can dilute the overall nutrition of the base diet and should be considered in the overall nutrient intake of the animal. The base diet can also be used as enrichment if provided in a novel manner (e.g., puzzle feeders) [85]. The use of gums as enrichment has been considered based on the wild-type feeding strategy of these animals [86,87]. As a benefit in research colonies, gum arabic is available in a purified form, thus reducing the variability in nutrition that may be associated with items such as fruits and vegetables.

When constructing a diet, all ingested food for whatever purposes must be accounted for in the diet analysis. Enrichment foods must be incorporated into the diet analysis to ensure that the enrichment process does not compromise the nutritional plane for the animals. Enrichment foods also have the potential to increase the nutritional variation among animals. As long as enrichment foods make up a small proportion of the ingested diet this should be of little concern.

Dry matter intake by captive marmosets is generally about 4%–6% of body mass [4,7]; for marmosets with healthy weights (range 300–450 g) that corresponds to about 13–25 g of dry matter ingested per day. Enrichment foods should account for only about 10% of the diet or 1.3–2.5 g of dry matter intake per day. That is the approximate equivalent of 1–2 grapes, 3–5 raisins, 1/16 of a medium apple, or 1/15 of a medium banana.

CURRENT ISSUES IN MARMOSET NUTRITION

Wasting Syndrome and Metabolic Bone Disease

Marmoset wasting syndrome and metabolic bone disease are commonly reported nutrition-related diseases in marmoset colonies [6]. A review of necropsy records and examination of bone and GI tissue of 53 marmosets from one institution revealed that 55% of marmosets had both bone and GI disease, 6% had bone disease only, 22% had GI disease only, and 17% had neither bone nor GI disease [88]. Another study found colitis to be the primary cause of death (~22–23% of mortality) [89]. More recent work suggests that the common pathologies in marmosets are associated with similar underlying issues of intestinal inflammation and resulting malabsorption of nutrients [3]. For example, a relationship between energy digestibility and vitamin D status was detected, and animals with higher energy digestibility of their diets had higher blood 25-OH-D levels and greater bone mineral density [3]. Marmosets with bone and GI disease or bone disease alone (diagnosed postmortem) had significantly lower serum albumin levels than their unaffected counterparts, suggesting that this blood component could serve as an antemortem indication of bone disease in marmosets [88,90]. Matrix metalloproteinase 9 (MMP9) has also been identified as a potential biomarker of this condition [90]. In a study of marmosets, a large number in the sample population (10/13) had very low/undetectable populations of *Lactobacillus* spp. in the intestines and very high levels (relative to other mammalian populations) of Gram-negative bacteria, although it is unclear if this GI microflora profile is indicative of any GI inflammation [91].

In marmosets with clinical symptoms of wasting disease, 70% (7/10) had antibodies to gliadin after consuming diets with moderate to high gluten levels (>80 ppm). This gluten challenge was also associated with significant diarrhea in animals with prior clinical symptoms [5]. However, the gluten challenge was not made to healthy animals so the study lacked a control group. In addition, one animal in the study had the highest values for a nonspecific inflammation marker when switched to the gluten-free diet. A study to test whether a gluten-free diet will improve digestive efficiency in outwardly healthy marmosets with poor values for apparent dry matter digestibility is underway. But at present, the evidence of extensive variation in digestive efficiency in marmosets on all diets, whether containing gluten or not, suggests that gluten sensitivity is, at best, a minor component of this problem.

Use of Gum as a Dietary Item

In the wild, many primates feed on gums, a soluble fiber. This is especially true for the small-bodied callitrichid primates. The common marmoset (*Callithrix jacchus*) feeds extensively on gum in its native Brazil. It is unknown if there are nutritional or physiological reasons to feed marmosets gum.

Gums are incomplete foods, providing mainly energy in the form of fermentable soluble fiber and minerals [18]. Gums have been suggested to serve as important calcium sources for galagos (*Galago senegalensis*) [92] and marmosets and other callitrichids [93] in the wild. The levels of calcium in gums that have been assayed range from below 0.5% to around 1% on a dry matter basis. These are fairly high levels for wild foods, but most manufactured primate diets contain 0.8%–1.2% calcium on a dry matter basis. If feeding gum reduces consumption of the nutritionally complete feed that should form the base of the diet, then calcium intake may not be increased. In contrast, substituting gum for fruit may indeed increase calcium consumption but will come at the expense of decreasing the consumption of vitamins in fruit.

At present, there are no demonstrated nutritional reasons for including gum in the diet of captive gum-feeding primates. There are hypothetical advantages, such as providing a fermentable substrate that would increase butyrate production in the colon to enhance colonic health [42]. However, some starches and pectins in captive callitrichid diets probably already reach the colon to be fermented, providing a source of butyrate. Gum added to the diets of common and pygmy marmosets appeared to slow the passage rate of digesta [7,22]. Whether a slower passage rate of digesta would have any positive health benefits in captivity is unknown. Based on the lack of evidence that gum has a nutritional purpose in captive callitrichid diets, it should be treated as an enrichment food. Gum arabic syrup has been used successfully as an enrichment device for marmosets [86].

Obesity

Historically, underweight marmosets (below 300 g) have represented the group at greatest risk for morbidity and mortality, and low-weight individuals are still routinely found in marmoset colonies. However, with the increasing prevalence of obesity in captive marmoset colonies [24,94], we now have two groups of marmosets at higher risk for mortality—those of low weight and those of high weight. Obesity accompanied by dyslipidemia and altered glucose metabolism, including insulin resistance [24,30], is emerging as a frequent clinical finding in marmosets housed in captivity. Obesity has generated increases in pathologies rarely seen previously in marmosets, such as hepatomegaly, hepatic steatosis, diabetes, atherosclerosis, cardiomyopathies, and stroke [24,94]. Marmosets have been thought to require diets high in ME; however, the increasing prevalence of obesity in marmoset colonies may indicate we are overestimating energy requirements and/or providing captive marmosets with food that is too palatable [1,95].

The onset of obesity can develop very early in life. Marmosets judged obese at 1 year of age had higher percent body fat at 1 month of age [29]. One-year-old marmosets already show increased fasting insulin levels and poor glucose control [30]. Obese females are more likely to produce triplets or higher litters, which generally do not do as well as the standard twin litters. Obesity in marmosets has the potential to seriously disrupt the breeding potential of a marmoset colony, as well as causing excessive early mortality and morbidity (see Chapter 8 of this volume).

There is a danger that colony managers and care staff without a longer, historical perspective may believe that 400 + g marmosets are "normal," when in the past a marmoset over 400 g was a rarity. Twenty years ago, a "normal" marmoset weighed between 350 and 400 g, with plenty of healthy marmosets in the low 300 g range. Wild lactating female common marmosets had a mean mass of 383 g [71]. A marmoset over 400 g can be healthy; however, changes in serum lipid values were noticeable in captive marmosets above 435 g [24]. Research is underway to better characterize a healthy marmoset and to develop body condition scores to assist colony managers and care staff to assess their animals. Obesity may be one of the most significant problems facing marmoset colonies in the future.

References

[1] Power ML, Toddes B, Koutsos L. Non-human primate nutrition: nutrient requirements and dietary husbandry principles for captive non-human primates. In: Tardif SD, Abee C, editors. Nonhuman primates in biomedical research. 2nd ed., vol. I. Boston (MA): American College of Laboratory Animal Medicine, Academic Press; 2012. p. 269–86.

[2] Rensing S, Oerke AK. Husbandry and management of new world species: marmosets and tamarins. In: Wolfe-Coote S, editor. The laboratory primate. London: Academic Press; 2005. p. 145–62.

[3] Jarcho MR, Power ML, Layne-Colon DG, Tardif SD. Digestive efficiency mediated by serum calcium predicts bone mineral density in the common marmoset (*Callithrix jacchus*). Am J Primatol 2013;75(2):153–60.

[4] Power ML, Myers EW. Digestion in the common marmoset (*Callithrix jacchus*), a gummivore-frugivore. Am J Primatol 2009;71(12):957–63.

[5] Kuehnel F, Mietsch M, Buettner T, Vervuert I, Ababneh R, Einspanier A. The influence of gluten on clinical and immunological status of common marmosets (*Callithrix jacchus*). J Med Primatol 2013;42(6):300–9.

[6] Ludlage E, Mansfield K. Clinical care and diseases of the common marmoset (*Callithrix jacchus*). Comp Med 2003;53(4):369–82.

[7] Power ML. Digestive function, energy intake and the response to dietary gum in captive callitrichids. University of California at Berkeley; 1991 [Doctoral dissertation].

[8] Burrows AM, Nash LT, editors. The evolution of exudativory in primates. New York: Springer; 2010.

[9] Cabana F, Dierenfeld ES, Donati G, Nekaris K. Exploiting a readily available but hard to digest resource: a review of exudativorous mammals identified thus far and how they cope in captivity. Integr Zool 2017.

[10] Smith AC. Exudativory in primates: interspecific patterns. In: Burrows AM, Nash LT, editors. The evolution of exudativory in primates. New York: Springer; 2010. p. 45–88.

[11] Beattie J. The anatomy of the common marmoset (*Hapale jacchus* Kuhl). Proc Zool Soc Lond 1927 [Wiley Online Library].
[12] Coimbra-Filho AF, Mittermeier RA. Exudate-eating and tree-gouging in marmosets. Nature 1976;262(5569):630.
[13] Rosenberger AL. Loss of incisor enamel in marmosets. J Mammal 1978;59:207–8.
[14] Ferrari SF, Martins ES. Gummivory and gut morphology in two sympatric callitrichids (*Callithrix emiliae* and *Saguinus fuscicollis* weddelli) from western Brazilian Amazonia. Am J Phys Anthropol 1992;88(1):97–103.
[15] Scanlon C, Chalmers N, Montiero da Cruz M. Home range use and the exploitation of gum in the marmoset *Callithrix jacchus* jacchus. Int J Primatol 1989;10(2):123–36.
[16] Rylands AB. Exudate-eating and tree-gouging by marmosets (Callitrichidae, primates). Behaviour 1980.
[17] Power ML. The other side of callitrichine gummivory. In: Norconk MA, Rosenberger AL, Garber PA, editors. Adaptive radiations of neotropical primates. Boston (MA): Springer; 1996. p. 97–110.
[18] Power ML. Nutritional and digestive challenges to being a gum-feeding primate. In: Burrows AM, Nash LT, editors. The evolution of exudativory in primates. New York: Springer; 2010. p. 25–44.
[19] Hill W, Rewell RE. The caecum of primates: its appendages, mesenteries and blood supply. Trans Zool Soc Lond 1948;26(3):198–256.
[20] Coimbra-Filho AF, Rocha NDC, Pissinatti A. Morfofisiologia do ceco e sua correlacao com o tipo odontologico em Callitrichidae (Platyrrhini, Primates). Rev Bras Biol 1980;40:177.
[21] Caton J, Hill D, Hume I, Crook G. The digestive strategy of the common marmoset, *Callithrix jacchus*. Comp Biochem Physiol Physiol 1996;114(1):1–8.
[22] Power ML, Oftedal OT. Differences among captive callitrichids in the digestive responses to dietary gum. Am J Primatol 1996;40: 131–44.
[23] NRC. Nutrient requirements of nonhuman primates: second revised edition. Washington (DC): National Academies Press; 2003.
[24] Tardif S, Power M, Ross C, Rutherford J, Layne-Colon D, Paulik M. Characterization of obese phenotypes in a small nonhuman primate, the common marmoset (*Callithrix jacchus*). Obesity 2009;17:1499–505.
[25] Nievergelt CM, Martin RD. Energy intake during reproduction in captive common marmosets (*Callithrix jacchus*). Physiol Behav 1998;65(4):849–54.
[26] Tardif S, Power M, Layne D, Smucny D, Ziegler T. Energy restriction initiated at different gestational ages has varying effects on maternal weight gain and pregnancy outcome in common marmoset monkeys (*Callithrix jacchus*). Br J Nutr 2004;92(5):841–9.
[27] Tardif SD, Power M, Oftedal OT, Power RA, Layne DG. Lactation, maternal behavior and infant growth in common marmoset monkeys (*Callithrix jacchus*): effects of maternal size and litter size. Behav Ecol Sociobiol 2001;51(1):17–25.
[28] Nicolosi RJ, Hunt RD. Dietary allowances for nutrients in nonhuman primates. Primates Nutr Res 1979:11–37.
[29] Power ML, Ross CN, Schulkin J, Tardif SD. The development of obesity begins at an early age in captive common marmosets (*Callithrix jacchus*). Am J Primatol 2012;74(3):261–9.
[30] Power ML, Ross CN, Schulkin J, Ziegler TE, Tardif SD. Metabolic consequences of the early onset of obesity in common marmoset monkeys. Obesity 2013;21(12):E592–8.
[31] Ross CN, Power ML, Artavia JM, Tardif SD. Relation of food intake behaviors and obesity development in young common marmoset monkeys. Obesity 2013;21(9):1891–9.
[32] Gore MA, Brandes F, Kaup FJ, Lenzner R, Mothes T, Osman AA. Callitrichid nutrition and food sensitivity. J Med Primatol 2001;30: 179–84.
[33] Flurer C, Sappl A, Adler H, Zucker H. Determination of the protein requirement of marmosets (*Callithrix jacchus*) by nitrogen balance with regard to the concentration of essential amino acids in the diet. J Anim Physiol Anim Nutr 1987;57(1–5):23–31.
[34] Barnard D, Knapka J, Renquist D. The apparent reversal of a wasting syndrome by nutritional intervention in *Saguinus mystax*. Lab Anim Sci 1998;38:282–8.
[35] Friedman AN. High-protein diets: potential effects on the kidney in renal health and disease. Am J Kidney Dis 2004;44(6): 950–62.
[36] Flurer C, Krommer G, Zucker H. Endogenous N-excretion and minimal protein requirement for maintenance of the common marmoset (*Callithrix jacchus*). Lab Anim Sci 1988;38(2):183–6.
[37] Flurer C, Zucker H. Coprophagy in marmosets due to insufficient protein (amino acid) intake. Lab Anim 1988;22(4):330–1.
[38] Tardif S, Jaquish C, Layne D, Bales K, Power M, Power R, Oftedal O. Growth variation in common marmoset monkeys (*Callithrix jacchus*) fed a purified diet: relation to care-giving and weaning behaviors. Comp Med 1998;48(3):264–9.
[39] Fiennes R, Sinclair A, Crawford M. Essential fatty acid studies in primates linolenic acid requirements of capuchins. J Med Primatol 1972;2(3):155–69.
[40] Charnock J, Abeywardena M, Poletti V, McLennan P. Differences in fatty acid composition of various tissues of the marmoset monkey (*Callithrix jacchus*) after different lipid supplemented diets. Comp Biochem Physiol Physiol 1992;101(2):387–93.
[41] Ziegler TE, Sosa ME, Peterson LJ, Colman RJ. Using snacks high in fat and protein to improve glucoregulatory function in adolescent male marmosets (*Callithrix jacchus*). J Am Assoc Lab Anim Sci 2013;52(6):756–62.
[42] Wong JMW, de Souza R, Kendall CWC, Emam A, Jenkins DJA. Colonic health: fermentation and short chain fatty acids. J Clin Gastroenterol 2006;40(3):235–43.
[43] Pludowski P, Holick MF, Pilz S, Wagner CL, Hollis BW, Grant WB, Shoenfeld Y, Lerchbaum E, Llewellyn DJ. Vitamin D effects on musculoskeletal health, immunity, autoimmunity, cardiovascular disease, cancer, fertility, pregnancy, dementia and mortality – a review of recent evidence. Autoimmun Rev 2013;12:976–89.
[44] Wang Y, Zhu J, DeLuca HF. Where is the vitamin D receptor? Arch Biochem Biophys 2012;523(1):123–33.
[45] Rosen CJ, Adams JS, Bikle DD, Black DM, Demay MB, Manson JE, Hassan Murad M, Kovacs CS. The nonskeletal effects of vitamin D: an endocrine society scientific statement. Endocr Rev 2012; 33(3):456–92.
[46] Ziegler TE, Kapoor A, Hedman CJ, Binkley N, Kemnitz JW. Measurement of 25-hydroxyvitamin $D_{2\&3}$ and 1, 25-dihydroxyvitamin $D_{2\&3}$ by tandem mass spectrometry: a primate multispecies comparison. Am J Primatol 2015;77:801–10.
[47] Shinki T, Shiina Y, Takahashi N, Tanioka Y, Koizumi H, Suda T. Extremely high circulating levels of 1α, 25-dihydroxyvitamin D 3 in the marmoset, a new world monkey. Biochem Biophys Res Commun 1983;114(2):452–7.
[48] Teixeira DS, Nobrega YKM, Valencia CEU, Gandolfi L, Pratesi R, Castro LCG. Evaluation of 25-hydroxy-vitamin D and parathyroid hormone in *Callithrix penicillata* primates living in their natural habitat in Brazil. J Med Primatol 2012;41:364–71.
[49] Luxwolda MF, Kuipers RS, Kema IP, van der Veer E, Dijck-Brouwer DA, Muskiet FA. Vitamin D status indicators in indigenous populations in East Africa. Eur J Nutr 2013;52: 1115–25.
[50] Power ML, Dittus WPJ. Vitamin D status in wild toque macaques (*Macaca sinica*) in Sri Lanka. Am J Primatol 2017;79:e22655.
[51] Chun RF, Chen H, Boldrick L, Sweet C, Adams JS. Cloning, sequencing, and functional characterization of the vitamin D receptor in vitamin D-resistant New World primates. Am J Primatol 2001;54(2):107–18.

[52] Chen H, Hewison M, Adams JS. Functional characterization of heterogeneous nuclear ribonuclear protein C1/C2 in vitamin D resistance: a novel response element-binding protein. J Biol Chem 2006;281:39114–20.
[53] Ullrey DE, Bernard JB, Peter GK, Zhiren L, Chen TC, Sikarskie JG, Holick M. Vitamin D intakes by cotton-top tamarins (*Saguinus oedipus*) and associated serum 25-hydroxyvitamin D concentrations. Zoo Biol 1999;18:473–80.
[54] Dever JT, Tanumihardjo SA. Hypervitaminosis A in experimental nonhuman primates: evidence, causes, and the road to recovery. Am J Primatol 2009;71(10):813–6.
[55] Mills JP, Penniston KL, Tanumihardjo SA. Extra-hepatic vitamin A concentrations in captive rhesus (*Macaca mulatta*) and marmoset (*Callithrix jacchus*) monkeys fed excess vitamin A. Int J Vitam Nutr Res 2005;75(2):126–32.
[56] Penniston KL, Tanumihardjo SA. Subtoxic hepatic vitamin A concentrations in captive rhesus monkeys (*Macaca mulatta*). J Nutr 2001;131(11):2904–9.
[57] Penniston KL, Thayer JC, Tanumihardjo SA. Serum vitamin A esters are high in captive rhesus (*Macaca mulatta*) and marmoset (*Callithrix jacchus*) monkeys. J Nutr 2003;133(12):4202–6.
[58] Krinsky NI, Mathews-Roth MM, Welankiwar S, Sehgal PK, Lausen NC, Russett M. The metabolism of [14C] beta-carotene and the presence of other carotenoids in rats and monkeys. J Nutr 1990;120(1):81–7.
[59] Lee CM, Boileau AC, Boileau TWM, Williams AW, Swanson KS, Heintz KA, Erdman Jr JW. Review of animal models in carotenoid research. J Nutr 1999;129(12):2271–7.
[60] Blakely SR, Mitchell GV, Jenkins MY, Grundel E, Whittaker P. Canthaxanthin and excess vitamin A alter alpha-tocopherol, carotenoid and iron status in adult rats. J Nutr 1991;121(10): 1649–55.
[61] Sklan D, Donoghue S. Vitamin E response to high dietary vitamin A in the chick. J Nutr 1982;112(4):759–65.
[62] Penniston KL, Tanumihardjo SA. The acute and chronic toxic effects of vitamin A. Am J Clin Nutr 2006;83(2):191–201.
[63] McIntosh GH, Bulman FH, Looker JW, Russell GR, James M. The influence of linoleate and vitamin E from sunflower seed oil on platelet function and prostaglandin production in the common marmoset monkey. J Nutr Sci Vitaminol 1987;33(4):299–312.
[64] Juan-Sallés C, Prats N, Resendes A, Domingo M, Hilton D, Ruiz J, Garner M, Valls X, Marco A. Anemia, myopathy, and pansteatitis in vitamin E-deficient captive marmosets (*Callithrix* spp.). Vet Pathol 2003;40(5):540–7.
[65] Ghebremeskel K, Harbige L, Williams G, Crawford M, Hawkey C. The effect of dietary change on in vitro erythrocyte haemolysis, skin lesions and alopecia in common marmosets (*Callithrix jacchus*). Comp Biochemy Physiol Part A Physio 1991;100(4):891–6.
[66] Gadient M, Fenster R. Stability of ascorbic acid and other vitamins in extruded fish feeds. Aquaculture 1994;124(1–4):207–11.
[67] Murphy B, Krushak D, Maynard J, Bradley D. Ascorbic acid (vitamin C) and its effects on parainfluenza type IH virus infection in cotton-topped marmosets. Lab Anim Sci 1974;24(1 Part II):229–32.
[68] Go YM, Liang Y, Uppal K, Soltow QA, Promislow DE, Wachtman LM, Jones DP. Metabolic characterization of the common marmoset (*Callithrix jacchus*). PLoS One 2015;10(11): e0142916.
[69] Miller GF, Barnard DE, Woodward RA, Flynn BM, Bulte JW. Hepatic hemosiderosis in common marmosets, *Callithrix jacchus*: effect of diet on incidence and severity. Lab Anim Sci 1997;47(2): 138–42.
[70] Power ML, Oftedal OT, Tardif SD. Does the milk of callitrichid monkeys differ from that of larger anthropoids? Am J Primatol 2002;10:117–27.
[71] Power ML, Verona CE, Ruiz-Miranda C, Oftedal OT. The composition of milk from free-living common marmosets (*Callithrix jacchus*) in Brazil. Am J Primatol 2008;70:78–83.
[72] Banton SA, Soltow QA, Liu KH, Uppal K, Promislow DE, Power ML, Tardif SD, Wachtman LM, Jones DP. Plasma metabolomics of common marmosets (*Callithrix jacchus*) to evaluate diet and feeding husbandry. J Am Assoc Lab Anim Sci 2016;55(2): 137–46.
[73] Addessi E, Stammati M, Sabbatini G, Visalberghi E. How tufted capuchin monkeys (*Cebus apella*) rank monkey chow in relation to other foods. Anim Welf 2005;14(3):215.
[74] Tardif SD, Richter CB. Competition for a desired food in family groups of the common marmoset (*Callithrix jacchus*) and the cotton-top tamarin (*Saguinus oedipus*). Lab Anim Sci 1981;31(1): 52–5.
[75] Felton AM, Felton A, Raubenheimer D, Simpson SJ, Foley WJ, Wood JT, Wallis IR, Lindenmayer DB. Protein content of diets dictates the daily energy intake of a free-ranging primate. Behav Ecol 2009;20(4):685–90.
[76] Oftedal OT, Whiten A, Southgate D, Van Soest P. The nutritional consequences of foraging in primates: the relationship of nutrient intakes to nutrient requirements. Philos Trans R Soc Lond B Biol Sci 1991;334(1270):161–70.
[77] Contreras RJ. Gustatory mechanisms of a specific appetite. Neural Mech Taste 1989;494:119.
[78] Power ML, Tardif SD, Layne DG, Schulkin J. Ingestion of calcium solutions by common marmosets (*Callithrix jacchus*). Am J Primatol 1999;47(3):255–61.
[79] Cheftel JC. Nutritional effects of extrusion-cooking. Food Chem 1986;20(4):263–83.
[80] Tran QD, Hendriks WH, van der Poel AF. Effects of extrusion processing on nutrients in dry pet food. J Sci Food Agric 2008;88(9): 1487–93.
[81] Harper JM, Clark JP. Food extrusion. Crit Rev Food Sci Nutr 1979; 11(2):155–215.
[82] Molnar L. Thiamine in a wet pet food application. Kansas State University; 2017 [Master's thesis].
[83] Maerki U, Rossbach W, Leuenberger J. Consistency of laboratory animal food following incubation prior to autoclaving. Lab Anim 1989;23(4):319–23.
[84] Hanis T, Mnukova J, Jelen P, Klir P, Perez B, Pesek M. Effect of gamma irradiation on survival of natural microflora and some nutrients in cereal meals. Cereal Chem 1988;65(5):381–3.
[85] De Rosa C, Vitale A, Puopolo M. The puzzle-feeder as feeding enrichment for common marmosets (*Callithrix jacchus*): a pilot study. Lab Anim 2003;37(2):100–7.
[86] McGrew W, Brennan J, Russell J. An artificial "gum-tree" for marmosets (Callithrix j. jacchus). Zoo Biol 1986;5(1):45–50.
[87] Roberts RL, Roytburd LA, Newman JD. Puzzle feeders and gum feeders as environmental enrichment for common marmosets. J Am Assoc Lab Anim Sci 1999;38(5):27–31.
[88] Baxter VK, Shaw GC, Sotuyo NP, Carlson CS, Olson EJ, Zink MC, Mankowski JL, Adams RJ, Hutchinson EK, Pate KAM. Serum albumin and body weight as biomarkers for the antemortem identification of bone and gastrointestinal disease in the common marmoset. PLoS One 2013;8(12):e82747.
[89] David JM, Dick Jr EJ, Hubbard GB. Spontaneous pathology of the common marmoset (*Callithrix jacchus*) and tamarins (*Saguinus oedipus*, *Saguinus mystax*). J Med Primatol 2009;38(5):347–59.
[90] Yoshimoto T, Niimi K, Takahashi E. Serum matrix metalloproteinase 9 (MMP9) as a biochemical marker for wasting marmoset syndrome. J Vet Med Sci 2016;78(5):837–43.
[91] Bailey MT, Coe CL. Intestinal microbial patterns of the common marmoset and rhesus macaque. Comp Biochem Physiol Mol Integr Physiol 2002;133(2):379–88.

[92] Bearder SK, Martin RD. *Acacia* gum and its use by bushbabies, *Galago* senegalensis (Primates: Lorisidae). Int J Primatol 1980;1:103–28.
[93] Garber PA. Proposed nutritional importance of plant exudates in the diet of the Panamanian tamarin, *Saguinus oedipus* geoffroyi. Int J Primatol 1984;5:1–5.
[94] Wachtman LM, Kramer JA, Miller AD, Hachey AM, Curran EH, Mansfield KG. Differential contribution of dietary fat and monosaccharide to metabolic syndrome in the common marmoset (*Callithrix jacchus*). Obesity 2011;19:1145–56.
[95] Layne DG, Power RA. Husbandry, handling and nutrition for marmosets. Comp Med 2003;53(4):351–9.
[96] Tao N, Wu S, Kim J, An HJ, Hinde K, Power ML, Gagneux P, German JB, Lebrilla CB. Evolutionary glycomics: characterization of milk oligosaccharides in primates. J Proteome Res 2011;10:1548–57.
[97] Hinde K, Foster AB, Landis LM, Rendina D, Oftedal OT, Power ML. Daughter dearest: sex-biased calcium concentrations in mother's milk among rhesus macaques. Am J Phys Anthropol 2013;151:144–50.

CHAPTER

5

Husbandry and Housing of Common Marmosets

Donna Layne-Colon[1], *Anna Goodroe*[2], *Monika Burns*[3]

[1]Southwest National Primate Research Center, San Antonio, TX, United States; [2]Wisconsin National Primate Research Center, University of Wisconsin, Madison, WI, United States; [3]Division of Comparative Medicine, Massachusetts Institute of Technology, Cambridge, MA, United States

INTRODUCTION

The history of husbandry practices used to manage captive marmoset colonies in the United States has been defined by trial and error with colonies derived from wild-caught animals, as well as extrapolation from common husbandry practices used with other species of Callitrichinae such as *Saguinus* spp. Use of the common marmoset (*Callithrix jacchus*) as a research model has increased over the past 30 years, yet resources outlining appropriate husbandry practices are limited. In 1985, Dr. Martin Morin offered a "discouraging analysis" of the then-current state of husbandry practices and model development for tamarins and marmosets [1]. Over 30 years later, information on how to house and care for marmosets when new colonies are established is still often provided by direct contact with management staff of source colonies regarding frequently used but unvalidated practices. Due perhaps to a lack of published reference literature and standardized housing and husbandry practices, there remains substantial variation in management practices across marmoset colonies. This chapter presents an overview of the husbandry practices utilized in several modern marmoset research colonies.

HOUSING

Indoor Environments

Temperature

In research facilities within the United States, marmosets are predominately housed indoors; therefore, the temperature and humidity of the animal housing room determines the ambient climate experienced by marmosets housed within a particular room. While recommended temperature ranges of 64–84°F (18–29°C) for nonhuman primates are outlined in the Guide for the Care and Use of Laboratory Animals [2], marmosets require temperatures at the higher end of the recommended range to maintain thermoneutrality. A thermoneutral range of 81–91°F (27–33°C) for indoor-housed common marmosets has been described previously and is also discussed in Chapter 4 of this text [3,58]. Paired or group-housed marmosets may crowd together within the cage and thus may be able to tolerate housing room temperatures below 81°F (27°C).

Marmosets in the wild are found in diverse habitats that can experience temperature extremes [4–6]. Although they can adapt to periodic temperature fluctuations below 69°F (21°C) or above 87°F (31°C) [4,7,8] they are usually maintained within captive facilities at temperatures between 75 and 81°F (24–27°C); however, a temperature range of 75–81°F (27°C) is below the thermoneutral zone for marmosets established by Power in 1991. A study performed in a different species of Callitrichinae, *Saguinus oedipus* (cotton-top tamarin), suggests that tamarins do not adapt to cooler temperatures in the laboratory housing environment and are metabolically stressed at housing room temperatures of less than 89.6°F (32°C) [9]. Normal body temperature values are discussed in Chapter 6 of this text [59]. Callitrichid species, including common marmosets, enter a state of torpor overnight with a decline in their body temperature and resting metabolic rate [10]. If individuals are isolated temporarily during the postoperative period or for other medical reasons, supplemental heat sources such as heat lamps or warm air circulators may be affixed to the exterior of the cage. It is also important to consider that the thermoregulation of laboratory-housed marmosets is dependent on the caging materials that the animals are in contact with (i.e., wood vs.

The Common Marmoset in Captivity and Biomedical Research
https://doi.org/10.1016/B978-0-12-811829-0.00005-4

stainless steel), not just the ambient room temperature. When supplemental heat sources are used, precautions must be taken to avoid thermal burns or hyperthermia. Small animal intensive care incubators can be used to provide additional thermal support outside of the home cage environment. When animals are transferred to a new facility, consideration as to their former husbandry and housing standards is important; this includes attention to room temperature set points. While preliminary work has provided recommended temperature ranges for marmoset housing rooms, further study is warranted to validate these ranges for use in common marmoset colony management.

Humidity

While the Guide recommends a broad range of 30%–70% for relative humidity in nonhuman primate housing facilities, the European Association of Zoos and Aquaria Husbandry Guidelines recommend a minimum humidity of 60% for indoor-housed Callitrichidae [2,7]. Average humidity within a room, and the ability to control humidity ranges, may be impacted by the geographic location of the facility, heating, ventilation, and air conditioning (HVAC) system, building insulation, and sanitation practices performed within the facility. The number of air exchanges per hour within the animal housing room may also affect temperature and humidity. The recommended rate of air exchanges (10–15 per hour) described in the Guide may overventilate some animal housing rooms to the point of reducing humidity and temperature below desired set points. It is therefore important to evaluate the air exchange rate in addition to previously mentioned variables when determining humidity and temperature set points. Heating, ventilation, and air conditioning settings will need to be adjusted seasonally and in response to increases or decreases in room population to maintain stable room temperature and humidity. HVAC systems that specifically manage both humidity and temperature may be best equipped to provide a higher rate of room air exchanges. Portable humidifiers may be used in animal rooms to maintain adequate room humidity, though the effectiveness of such units may be limited due to the extensive air exchanges typical in an animal room. Portable humidifiers require routine cleaning to ensure no growth of mold, yeast, or bacteria. More success may be had with units that directly humidify the incoming room air. Humidity and temperature values should be monitored continuously and logged daily, with periodic evaluation of records by facility management and veterinary staff to ensure that appropriate ranges are maintained.

Lighting

While wild and outdoor-housed marmosets are exposed to variable photoperiods throughout the year, a consistent 12 h:12 h diurnal light:dark cycle is typically utilized in indoor marmoset housing facilities. This cycle provides adequate illumination to support the expression of normal behaviors during the average active period of the marmoset, which has been documented at 11.4 h/day [8,11]. Automated timers are routinely used to control light cycles, but light cycles should be evaluated regularly to ensure accurate implementation of preset cycles. Rooms with red light capabilities should be assessed during the dark period to ensure that marmosets are not exposed to excessive red light at night. The authors have observed that delayed onset of activity in housing rooms can be a clue that a timer has drifted from its original settings. Light dimmers are available to facilitate a more natural transition between light and dark periods. Some colonies have full or partial exposure to natural sunlight, though the effects of exposure to natural lighting on marmoset colony health have not been defined. As different durations of daily exposure to natural light might confound study results if uneven across animal housing rooms, control of duration of exposure with shades or opaque panels may be necessary for some studies. Changes in photoperiod length may affect behavior—for example, the authors have found that extending the light period to 14 h with access to natural light during summer led to a siesta in the late morning.

While many marmoset colonies use fluorescent bulbs as light sources within animal housing rooms, some colonies use ultraviolet (UV) or full-spectrum light sources, such as Vita-lites, as exclusive light sources within animal rooms (personal comm—all authors). Historically, providing special lighting in the attempt to support adequate vitamin D metabolism was recognized by marmoset colony management staff; the use of mercury-vapor lamps as a source of UV has been recorded for marmosets as early as 1927 ([12,12a]). With the development and use of vitamin D3 fortified diets, and the limitations of such lighting as a source for UV-driven modification of 7-dehydrocholesterol (7-DHC) to vitamin D, it is not necessary to use UV lighting exclusively within marmoset housing rooms. Some limitations include the specific properties of the light, the amount of time that the light is on, and more importantly, the distance from the light source to the animals in the cage determine the amount of radiation accessible to the animals. Further considerations that are important to weigh in the decision whether or not to use full-spectrum or other special lights within animal rooms include the position of lights in the room relative to personnel working within the room, any potential safety issues for personnel, and control of exposure time [58]. The authors caution colony managers who intend to replicate exposure to sunlight with use of full-spectrum lights in large rooms with large cages, as this is highly unlikely. Manufacturer guidelines must be consulted to determine the required frequency of change of full-spectrum light sources. The intensity of a light source may also influence locomotion, levels of activity, and grooming [13–15]. The authors suggest that frequent measurement of light intensity at different areas within the animal housing room would provide more information on history of daily light exposure of individual animals within the colony.

While wild-raised marmosets have a low incidence of refractive errors, captive-raised marmosets have a high incidence of myopia, which has been shown to impact performance during vision studies [16]. This suggests that refractive errors should be considered as a differential diagnosis for poor performance observed in visual tasks. Several studies in laboratory animals have shown that intensity [17–19] and spectral composition [17] of ambient light may influence normal emmetropization. Other studies performed in macaques have found that typical lighting conditions in research housing rooms contribute to a high incidence of myopia when compared to high light level housing conditions [17], and that early life exposure to sunlight may lower the risk of myopia later in life [20]. Interestingly, high levels of outdoor activity have also been shown to reduce the prevalence of myopia in children [21]. While the exact underlying mechanisms are not fully understood, these observations suggest that light intensity and spectrum may influence emmetropization, and that a visually enriched environment with high-level, full-spectrum lighting mimicking outdoors conditions might be optimal for reducing the prevalence of myopia in nonhuman primate colonies. Further study is needed to determine optimal light intensity values for marmoset housing rooms.

Sound

The impact of sound within the primary environment should be taken into account as it may impact marmoset behavior [61]. Marmosets have an extensive vocal repertoire to facilitate communication and several reviews are available [62,63]. Noise production generated by neighboring housing rooms or equipment should be considered during facility design. Marmosets can hear sound frequencies between 125–36,000 Hz [64], and wild populations have been noted to avoid areas with sound production over 55 dB [22]. Availability of additional space for use as temporary housing during extensive repairs generating sound, vibrations, excess activity, or noxious odors is ideal, and such space should be accommodated for during facility design and renovation processes.

Housing Room Density and Environmental Changes

Animal density within a room impacts social interactions and behavior, and increased intra- and/or inter-cage aggression may be noted as room density increases. Common marmosets are a highly territorial species that exist in family groups in the wild, and the space occupied by each free-ranging family is large compared with the cages and rooms in which most animals live in the laboratory animal setting [5,60]. The increased density and constant visual access may induce stress in captive-housed marmosets, which can result in clinical signs such as diarrhea. Changes in visual access may lead to new tensions even without any corresponding increase in animal density within the housing room. Ideally, expanding colonies should be allocated additional space, but when space is not available, increased use of visual barriers may help alleviate tension associated with increasing room density. Females are also more likely to experience reproductive suppression when housed in animal rooms with a high density of other females [23]. The effects of high room density on marmoset vocalization type and rate, scent marking patterns, and sleep cycles are unknown. While facility space constraints and research demands may require periodic changes to animal housing environments and conditions, it is important to consider that environmental changes, including alterations in cage dimension, ambient temperature, humidity, and feeding schedule, may affect social interactions in captive marmoset colonies [24]. The authors note that seemingly minor changes within the cage or the immediate environment, such as novel enrichment items, small changes to cage structure, or new personnel patterns within the room, may impact animal stress level and/or health. Studies have also shown that there is a relationship between high cortisol levels and changes made to environmental and housing conditions in common marmoset colonies [25]. Variation in personnel and management practices on weekends compared with weekdays has also been shown to impact behavioral patterns of marmoset family groups [26].

OUTDOOR ENVIRONMENTS

Marmosets are often found in outdoor environments within zoos and some research animal facilities, including locations in the Netherlands and Australia [27]. The Biomedical Primate Research Centre (BPRC) in the Netherlands maintains marmosets outdoors, with access to an attached heated indoor enclosure. A review of clinical disease incidence, behavior patterns, and reproductive trends showed minimal to no change once the colony was transitioned from indoor enclosures to a combination of indoor and outdoor housing [28]. Inherent risks in housing marmosets outdoors include exposure to additional pathogens through contact with vectors such as insects or wild animals. Ensuring appropriate barriers to wildlife and reviewing geographical information regarding endemic pathogens is strongly advised. Examples of case reports associated with outdoor housing environments include the following: fatal poxvirus infections after suspected exposure to wild rodents [29] and *Trypansoma cruzi* infections secondary to exposure to vector species or insects [30]. During the transition to partial outdoor housing at the BPRC, multiple acute outbreaks of yersiniosis were noted within the marmoset colony [28]. Vaccination of colony animals every 6 months in conjunction with changes in husbandry practices prevented further outbreaks caused by infection with *Yersinia* spp. Rabies virus is also a concern as it is endemic within wild marmoset populations in Brazil [31], and populations housed outdoors may be exposed to reservoir species such as bats. Additionally, native plant species that may be toxic should be removed from outdoor enclosures or a substrate such as gravel utilized to cover any potentially dangerous flora. Review of geographical seasonal temperature and weather extremes should also be considered, and necessary modifications made to animal housing spaces to ensure that marmosets exposed to the outdoors may maintain thermoregulation.

PRIMARY ENCLOSURE

While marmosets have been housed in captivity in research animal facilities and zoos for decades, there are no validated standards in terms of size, design, and materials for marmoset cages. Consideration of the marmosets' natural behaviors and social interactions have led to the current variety of caging systems employed in research animal housing facilities. Cages must be designed with multiple aims in mind. Important considerations include the ability to remove marmosets from the cage at a frequency dictated by preventative health-care program needs and research-related uses, provision of space to encourage species-specific behaviors, and ease of cleaning to maintain an environment that is appropriate and safe for all inhabitants.

Cage Sizes

The Guide provides a table with minimum floor square footage and height requirements for enclosures relative to primate body weight, but specifies that overall cage volume and linear perch space should be considerations for many neotropical arboreal species such as marmosets [2]. The requirements in the United Sates Department of Agriculture's (USDA's) Animal Welfare Regulations describe slightly smaller cage volumes for primates less than 1 kg [32]. Marmosets have been noted to exhibit greater activity level when cage height is doubled from 0.82 to 1.95 m [33]. A comparison of weights and activity levels in marmosets housed in two cage types (0.60 m × 0.43 m × 0.66 m; 1.20 m × 0.75 m × 2.10 m) found an increase in body weight with a reduction in stereotypical pacing behavior for marmosets provided with additional cage furniture, volume, and height [34]. Emphasis should be placed on providing cage spaces that meet physiological and behavioral needs of marmosets as opposed to meeting minimum standards based solely on animal weight set forth in the Guide (2011) and Animal Welfare Regulations (2017) [2]. Similar trends in determining appropriate cage volume and configuration are found in Europe [35–38], Australia [27], and Japan [67] (Table 5.1). The Science Council of Japan advises that "Rather than simply a larger floor area, increased cage height or cage wall area, sheltered area, and complex cage structures are important considerations in designing the best caging. Whether housing space is appropriate can be determined by taking animal traits and behavior as indices. In addition, to referring to literature references, the opinions of specialists, and the necessity

TABLE 5.1 Recommended Minimum Space Guidelines as Reported for the United States and United Kingdom

RECOMMENDED MINIMUM CAGE SIZES FOR* CALLITHRIX JACCHUS *(US GUIDELINES)

NHP Group	Height (m)	Floor area (m^2)	Volume
Group 1 (<1.5 kg)	0.76	0.2/animal	Infers volume of 0.15 m^3/animal
The Guide			

RECOMMENDED MINIMUM CAGE SIZES FOR* CALLITHRIX JACCHUS *(UK GUIDELINES)

Family/group size	Height (m)	Floor area (m^2)	Volume
Breeding pair plus one generation of offspring	1.5	0.55	Infers volume of 0.825 m^3
Group of maximum eight	1.5	1.00	Infers volume of 1.5 m^3
Stock (minimum floor area per animal 0.135 m^2)	1.5	0.55	Infers volume of 0.2 m^3 per animal

Code of practice for the housing and care of animals bred, supplied or used for scientific purposes—Dec 2014. *NHP*, nonhuman primate.

of conducting the research should be taken into consideration." [67].

Cage Construction

The material forming the walls of the cage impacts how the cage is utilized by the marmoset. Climbing, vertical clinging and leaping, and hanging suspended are all species typical postures and modes of locomotion for marmosets [39,40]. Cage walls are typically constructed of metal or plastic. These surfaces are either vertical bars, a mesh grid, or a solid panel. Mesh ceiling and walls provide extensive surface area to climb on and hang from [10,41]. Vertical bars can also provide a surface for ambulation, but care must be taken to ensure that spacing prevents young marmosets from exiting the enclosure. Solid panels provide privacy from neighboring family groups; panels may be a primary component of the cage or an accessory, such as a plastic or cloth panel hung between cages. The cage floor made of mesh, with a grid spacing of less than 2.54 cm, allows excrement and waste food to fall through, but not an infant. Paper or bedding in a waste tray below the floor grate is utilized to catch waste. If there is no waste tray, absorbent paper on the room floor provides for efficient sanitation. If the floor is a solid panel, an absorbent material, such as paper or bedding, may be utilized to soften the surface and absorb waste (Table 5.2).

Examples of Cage Types

Cages commonly used in the United States for housing marmosets fall into the following general categories:

TABLE 5.2 Internal Cage Space Volume Measurements

Facility	Frame Material	Width (m)	Depth (m)	Height (m)	Floor Area (m^2)	Volume (m^3)	Number of Vertical Compartments	Reference
CLEA Japan	Stainless steel	0.39	0.55	0.70	0.21	0.15	2	[66]
Riken Brain Science Institute	Stainless steel	0.60	0.43	0.66	0.26	0.17	2	[34]
Riken Brain Science Institute	Stainless steel	1.20	0.75	2.10	0.90	1.89	2	[34]
WNPRC vertical booth (custom)	Stainless steel	0.60	0.91	1.82	0.56	1.01	1	Author (in use at WNPRC)
SNPRC rectangle (custom)	Aluminum	1.63	0.61	1.3	0.99	1.29	1	Author (in use at SNPRC)
SNPRC one-over-one (custom)	Stainless steel	0.56	0.56	0.70	0.31	0.22	2	Author (in use at SNPRC)
SNPRC single (custom)	Aluminum	0.53	0.53	0.91	0.28	0.26	1	Author (in use at SNPRC)
Barshop one-over-one (custom)	Aluminum	0.61	0.61	0.70	0.37	0.26	2	Author (in use at SNPRC & UTHSA barrier)/C. Ross personal comm.
LP "aviary" (modified)	Stainless steel	1.22	0.71	1.57	0.87	1.36	1	Author (in use at SNPRC, previously at NEPRC; E. Ludlage personal comm.)
Britz and Co. "Marmolade"	Stainless steel	0.76	0.81	1.70	0.62	1.05	1–2	Author (in use at MIT)
Otto Environmental "Marmoset"	Stainless steel	1.22	0.91	2.03	1.11	2.25	1	https://www.ottoenvironmental.com/marmosetsmall-animal-cage-3
Lab Products 1-over-1 (modified)	Stainless steel	0.71	0.71	0.77	0.50	0.39	2	Author (in use at SNPRC)
MRC-HRSU	Stainless steel	0.55	1.10	0.95	0.61	0.57	2	[42]

SNPRC, Southwest National Primate Research Center; *WNPRC*, Wisconsin National Primate Research Center; *LP*, Lab Products; *UTHSA*, Univ of Texas Heath-San Antonio; *CLEA*, Central Laboratory for Experimental Animals; *MRC-HRSU*, Medical Research Council-Human Reproductive Science Unit, Scotland

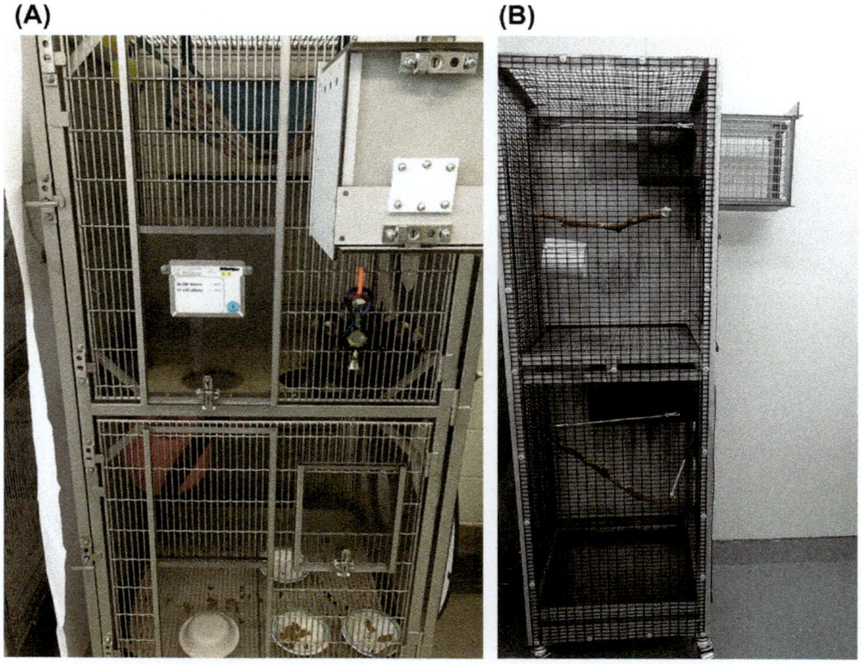

FIGURE 5.1 Examples of one-over-one type cages. (A) Commercial manufactured cage with detachable nest box mounted on the front. Cage card identification and colored plastic tag notification system described in text is visible on the front of the cage. Refer to Table 5.2 Britz & Co. for dimensions. (B) Custom-designed one-over-one type cage with nest box visible in lower section and a balcony insert in place of the top nest box. The balcony is interchangeable with any nest box location. Refer to Table 5.2 Barshop one-over-one custom for dimensions. *(A) Photo courtesy of author Burns, MIT. (B) Photo courtesy of J. Adams SNPRC.*

The two-tiered cage (Fig. 5.1): The two-tiered marmoset cage is available from multiple manufacturers. This two-tier style cage may come in sets of two, four, or six cages. The enclosure walls are typically constructed of metal vertical bars or mesh. A waste pan slides under the mesh floor grate. Manufacturers may add the option to remove horizontal or vertical dividers, providing flexibility in the amount of space available for use. These dividers allow the cages to be connected or control the visual and physical contact provided. Flexibility in configuring cage space is advantageous for new colonies that begin with young families that grow in size. A nest box mounted in or accessible from the front or side of the cage is a useful addition. Disadvantages of this cage design include inconsistent light provided to animals housed in the lower cage relative to animals housed in the upper cages. It has been suggested that lower cages do not permit a marmoset to move above a perceived potential threat (vertical flee response) [42]. The spacing of the vertical bars or size of mesh may allow small marmosets to exit the enclosure.

The aviary cage (Fig. 5.2): As the name implies, this is a large cage based on designs utilized for avian species. These units provide ample space for cage furniture (wooden dowels, plastic pipes, ladders, etc.) to encourage locomotion and horizontal space for quadrupedal running [40,43]. Family groups up to 10 marmosets may be housed successfully in appropriately sized aviary cages. Aviary cages may utilize a combination of vertical and horizontal stainless-steel bars and mesh to form enclosure walls and surfaces. Shelves provide floor space on which food dishes can be placed. A nest box may be integrated into the cage or may be an

FIGURE 5.2 Aviary cage. Each cage contains a nest box, grid shelves, and natural wood branches or large dowels. An example of partial visual screen panels is visible on one side of each cage. Refer to Table 5.2 Aviary for dimensions. *Photo courtesy of E. Ludlage 2002.*

addition. Ideally, methods to divide the cage with panels to facilitate isolation or capture of marmosets are available. Positive reinforcement and acclimation to entering a nest box or restraint device will make removal of marmosets more efficient from larger cages.

Vertical cages (Fig. 5.3): This vertical or booth style of cage measures 0.91 m × 0.61 m × 1.83 m and is custom produced. The cage sides and floor are constructed of stainless steel mesh (12.7 mm by 12.7 mm openings with 1.6 mm wire diameter). One or more cage sides consist of a solid high-density polyethylene panel which provides a visual barrier between neighboring cages. A mobile nest box (stainless steel and aluminum) mounted in the rear of the cage is utilized as a means to remove marmosets from the enclosure. A large front door allows personnel access into the cage to move the marmosets into the nest box prior to removal. Two permanent shelves provide additional floor space and are utilized to hold food receptacles. A sliding polycarbonate window facilitates easy access to provide and remove food receptacles or enrichment without opening the front door of the enclosure. Larger cages are provided for breeding groups (1.22 m × 0.61 m × 1.83 m). Plastic panels may be inserted into larger cages to isolate animals and facilitate removal from the enclosure. This cage design provides ample light throughout the cage and maximizes vertical space.

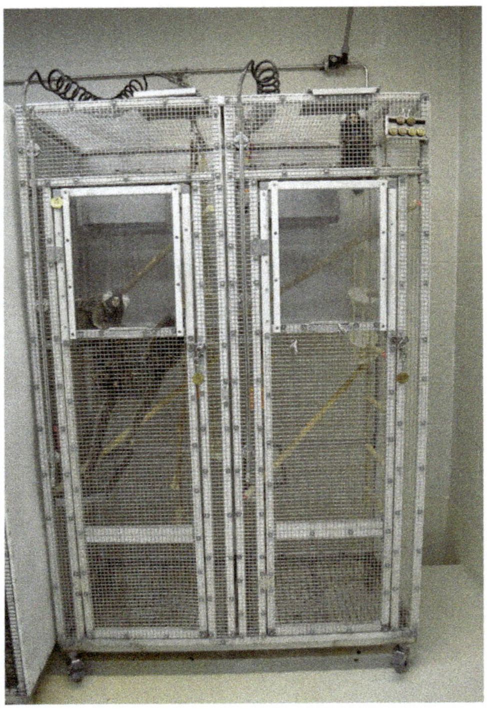

FIGURE 5.3 Vertical-type cage. This example of a vertical cage has dual-use nest boxes mounted inside and is furnished with wooden dowels, rope ladders, and foraging devices. The identification system that uses stamped metal discs corresponding to the marmosets' collars is visible top right. Refer to Table 5.2 vertical for dimensions. *Photo courtesy of author Goodroe, WNPRC.*

FIGURE 5.4 Raised rectangle type cage. This example of a custom-designed cage has dual-use nest boxes and is furnished with natural wood branches, swings made of recycled plastic containers, or T-pipes. One of the cage card identification systems described is visible along the top of the cage. Refer to Table 5.2 Raised rectangle for dimensions. *Photo courtesy of author Layne-Colon, SNPRC.*

Raised rectangle cage (Fig. 5.4): This custom cage is a mesh rectangle that can be divided into two spaces. The frame is aluminum angle, and the ceiling, floor, front and back are plastic-coated mesh. The grid of the mesh is 0.013 m by 0.025 m. There is a center wall that divides the space into two sections. End walls or the center walls are a solid aluminum panel. Within the center wall there is an opening for free passage between the sections that can be closed by inserting a mesh or solid divider panel. Along either side of the center opening is a mesh platform that runs the depth of the cage. The platform supports a portable scale for weighing marmosets in their home cage. The cage floor is raised 0.43 m from the room floor, and there is no waste pan under the mesh floor. Each cage has two nest boxes constructed of polycarbonate plastic or aluminum with a mesh front made from the same mesh as the walls. The nest boxes are located in the top-right corner of each side of the cage, so that if a divider is placed in the center, both sections will have a nest box. Each nest box slides inside the cage on a track, that is a piece of aluminum T-bar, that is welded to the frame. The nest box is always accessible in the cage and can be used to capture a marmoset and remove it from the cage.

Room as the primary enclosure (Fig. 5.5): The room may also serve as the primary enclosure. Considerations include having an anteroom to enter and exit the primary housing room to prevent animal escapes, door construction to help prevent injury of animals by door closure, and implementation of training strategies to remove animals from room for husbandry and clinical

FIGURE 5.5 Use of a room as the primary enclosure. *Photo courtesy of author Goodroe, WNPRC.*

procedures. Absorbent bedding is used to collect excrement.

Nest Boxes

A nest box functions primarily as a place to sleep but can also serve as a visual barrier from cage mates or humans [22]. Nest boxes are typically constructed of plastic polymers or stainless steel [10]. Nest boxes made of wood have been utilized [44]. Use of a material that aids in the conservation of body heat may be beneficial, especially for singly housed marmosets, when they undergo a nightly torpor. Integrated nest boxes are not a common feature but are available through a few cage manufacturers. While nest boxes can be custom made, options include repurposing small mammal boxes such as those designed for guinea pigs or chinchillas as well as using modified gallon plastic jugs, such as those used for juice or milk. Nest boxes help provide hiding spaces and secure areas for marmosets, but they must also allow for adequate passive ventilation, and the interior should be able to be visually inspected by staff. Nest boxes can be placed or suspended within the enclosure or integrated within a wall or door of the enclosure. The position of the nest box relative to branches and platforms within the cage adds to the complexity of the usable space [10,33,45]. It may be ideal to select a stable location within the enclosure and position the nest box at a higher location within the cage [46]. Multiple nest boxes may be provided to larger family groups. If the nest box can be removed from the cage, it can be utilized as a means to remove the marmosets from the cage through positive reinforcement training. A box mounted on the front could also serve as a veranda or test apparatus, depending on the availability of options for clear walls or mesh inserts [45,47]. Some examples of nest boxes and where they are positioned in different types of cages can be seen in previous cage figures, and details on construction and use can be seen in Fig. 5.6.

Provision of Food and Water

An overview of nutrition requirements and diet formulation is provided in Chapter 4 [58]. This section will focus on presentation of food and water relative to husbandry practices. Food is typically provided in metal bowls or disposable trays (material should be safe for digestion as they may be consumed). Depending on the number of marmosets housed within one cage, multiple food bowls may be necessary, especially if food aggression is observed. There is variability in the number of times a day that fresh food is offered. As long as the marmosets have access to appropriate food throughout the day, feeding once per day is acceptable. Advantages of feeding once per day include (1) reduced staff time and effort to prepare and serve; (2) fewer dishes to wash or less disposable trays used;

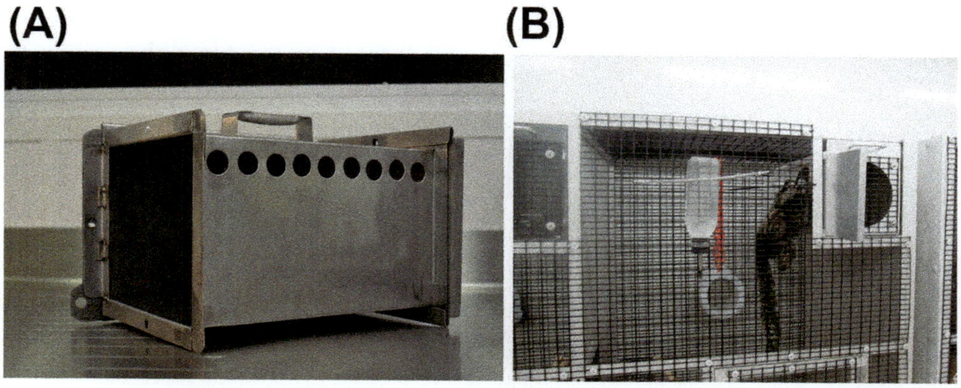

FIGURE 5.6 The nest box. (A) Close up of an interior-mounted nest box from a vertical cage. (B) The use of a nest box as positioned on a raised rectangle cage for capture and transport from the home cage. *(A) Photo courtesy of author Goodroe, WNPRC. (B) Photo courtesy of author Layne-Colon, SNPRC.*

and (3) a limit to individual variation in food intake. The reduction in labor to prepare, feed and wash, along with fewer consumables used translates into lower operating costs. Potential disadvantages to feeding only once per day include (1) more food being offered to ensure all animals have access ad libitum; (2) prolonged exposure of food to warm temperature and higher humidity may facilitate bacterial growth. Removal of uneaten food from the cage at least every 24 h reduces risk of exposure to possibly pathogenic bacterial contamination. More frequent removal of uneaten food from the cage aids in control of pest populations, which tends to be more of a problem in colonies with outdoor access or zoo exhibits. The stabilized forms of vitamin C added to marmoset diets have eliminated the concern about potential vitamin C degradation (M. Power, pers. com).

Water can be provided in bottles or automated watering lines. Selection of water source is dependent on quality of local water and desire for exclusion of solutes or pathogens; sources include tap water, reverse osmosis, UV filtered, or water autoclaved in bottles. When changing facilities, consider current and future water presentation; provide an acclimation period if transitioning to a new watering device.

Cage Furniture and Enrichment

Marmosets exhibit increased activity levels in enriched cages that encourage species-specific behaviors. They are more likely to show a reduction in stereotypical pacing behaviors and increased allogrooming if socially housed in enriched enclosures [33]. Important cage furnishings for any style of cage include wooden branches, shelves, or plastic tubes placed at various heights and angles to increase the use of vertical cage space. Marmosets are arboreal quadrupeds, and branches of various diameters (approximately 1–3 cm) and compliance provide the best substrates for species typical locomotion [7,43,65]. Object orientation in the cage (horizontal or vertical) may influence their use as marmosets engage in scansorial travel and vertical leaping on larger-diameter supports [39]. Additional cage furniture to encourage use of vertical space includes ladders, swings, and other suspended structures such as plastic containers and hammocks. Suspended items should be attached to the enclosure, so that entanglement or entrapment will not occur. Care must be taken with cloth hammocks that marmosets cannot chew holes in the cloth, as this also represents an entrapment risk, particularly for curious infants and juveniles. Plastic or metal links are commonly used. Solid or mesh shelves may be integrated into the enclosure to provide additional surface area. Enclosures may be modified to utilize large-diameter plastic or metal tunnels that facilitate marmoset movement between enclosures. A variety of cage furnishings can be seen in the cages depicted in Figs. 5.1–5.7.

Rotation of cage furniture and novel objects may increase exploratory behavior [47]. Manipulable objects such as puzzle feeders filled with mealworms or gum were shown to increase foraging behavior and decrease stereotypical behaviors such as pacing or inactivity [48,49]. There are countless ways to utilize novel objects to encourage foraging; wood or plastic objects with holes to fill with gum, metal skewers hung from the cage walls with fruit attached, and objects with holes that can be filled with small food items are all simple, reusable foraging opportunities. Overall, there is more flexibility in cage furniture materials and types due to the marmosets' small size and inability to destroy objects as readily as larger primates. Examples of manipulanda and puzzle feeders are shown in Fig. 5.8.

FIGURE 5.7 Example of the indoor–outdoor housing environment. The animals also have access to an indoor housing space (not shown) and can move freely between both environments. Cage furniture includes fixed and swinging branches, ropes, and runways. Deep litter bedding is used in both indoor and outdoor housing spaces. *Photo courtesy of J. Bakker, Biomedical Primate Research Center, Rijswijk, Netherlands.*

FIGURE 5.8 Close up of some examples of manipulada and puzzle feeders. *Photo courtesy of author Goodroe, WNPRC.*

Marmosets are gumnivores with dentition specialized for gnawing on wood [60]; wood in the cage, therefore, provides a natural surface for gnawing and also absorbs scent markings. Natural branches last longer than finished dowels due to their bark covering. Types of wood commonly used include manzanita wood purchased from local suppliers or distributors, unfinished wooden dowels (avoid softwood, pressure-treated, or sealed wood), or wood collected from the local environment (ensure that wood is nontoxic).

Wild populations of marmosets have been observed to spend up to 70% of their time 2–5 m above the ground [22]. Use of the cage floor of an enclosure is influenced by cage complexity and size [33] or the provision of foraging opportunities [50]. Examples of foraging opportunities include adding bedding with small food items to containers that may be placed on the cage floor or adding the mix to the cage pan if accessible from the cage floor. The proximity of the enclosure floor to the ground surface may also play a role; use of the enclosure floor for eating, resting, grooming, and playing has been observed in a cage with a mesh floor that is 0.43 m from the room floor [10].

Cage Identification/Animal Identification

Effective management of captive marmoset colonies relies heavily on the use of accurate animal and cage identification practices. There are three different levels at which identification must be made: (1) individual animal identification, (2) cage labels to indicate the number and IDs of individual animals within the cage, and (3) identification of special diet requests, research restrictions, medications, etc., directly on the cage. The following text will address these identification and cage-front communication practices.

Individual Animal Identification

Identification methods used with many other species may be used to identify individual marmosets within captive marmoset colonies as well. One method of identification is through the use of neck collars comprised of plastic tags stamped with the individual identification number hung on size #3 stainless steel ball chains with stainless steel fasteners. These plastic ID tags are also color coded, with blue tags for males and red tags for females. A second method of animal identification includes the use of microchip implants. Microchips can be implanted subcutaneously at 3–6 months of age. The placement of tattoos represents a more permanent visible method of individual animal identification. Tattoos are often placed on the less-densely haired mid-to-caudal abdomen or on inner thigh, and both places are visible when the animal is encouraged to stretch up at the front of the cage. Tattoo ink may be visible in inguinal lymph nodes after placement of the tattoo. Tattoo sites will need to be reshaved when they become overgrown with hair, and some animals have conflicting dark skin pigmentation that may alter the choice of site for that individual. Tattoos are placed after 6 months to a year of age and should be inspected and reapplied as needed during preventative health exams. During some life stages, temporary identification of individual animals may be required instead of permanent identification. Methods of temporary identification include the dying of ear tufts or dorsal hair coat with nontoxic dye and the placement of shaved markings on the hair coat. Infants and juveniles may have hair clipped from the dorsum, limbs, or tail for identification, though reshaving must be performed consistently to ensure that regrowth of hair does not obscure individual identification markings.

Cage-Front Labels Indicating ID Numbers and Total Animal Numbers

Because marmosets are most often housed in family groups or pairs, the cage-front surface is an important location for communication about the number and identification of animals housed within the individual cage (Fig. 5.3). At one institution, tags stamped with the individual animal ID are attached by velcro to a metal plate on the front of the cage. This system allows for easy modifications when animals are removed from the cage or reunited. Another institution uses custom cage cards for each individual. In breeding groups, a main cage identification card with all of the individual animal ID numbers found in the cage, sex of each animal, and any special markings is placed at the front of the cage in a metal cage card holder (Fig. 5.4). The individual identification card of each animal in that group is placed behind the main card. When an individual is temporarily separated, that individual's card is moved to that side of the group cage and is then returned when the group is reunited. When an individual animal is assigned to a new cage, the corresponding card moves with the animal to the new enclosure. When individuals are assigned to a new study, different colors of cards are used. It is important to have the total number of animals within a particular cage readily visible on the cage-front to facilitate counting during clinical rounds or census.

Identification of Special Diet Requests, Research Restrictions, and Medications Directly on the Cage

The cage-front is an important tool for communication of animal needs and treatments to clinical and research staff. Examples include the use of plastic tags with "lobster claw" hooks hung on the cage to indicate when individual animals or groups are to be food deprived, have special diet needs, require weight checks more frequently than the rest of the colony, require pregnancy watch, and other indications. Color-coded plastic livestock tags can be used for cage front communication with each color assigned to a specific indication. Alternatively, permanent etching or non-permanent marker can be used on plastic tags of any color for general cage front communication. These tags are attached with a quick link made of plastic or stainless steel or a cable tied to a cage to alert clinical staff that an individual animal or group of animals requires special treatment or consideration. A third use of the plastic livestock tags in various colors involves permanent indicators for each specific notification (Fig. 5.1). The tags are attached to the cage with a metal clip and removed as soon as the request has been filled or the notification is no longer relevant. Some facilities may use indicators such as shaving or hair dye applied to ear tufts to specify which individual animals in a cage need medication, procedures such as blood collection, monitoring after recovery from procedures or surgery, or have different dietary needs.

RESEARCH-SPECIFIC CAGING

Biocontainment

Animal biosafety levels (ABSLs) 3 and 4 may require ventilated caging systems which maintain negative pressure relative to the room environment. Detailed animal handling and housing requirements in ABSL 3 and 4 are beyond the scope of this book and "Biosafety in Microbiological and Biomedical Laboratories, fifth Edition" should be referenced [68]. The higher temperature and humidity requirements of marmosets are impeded by high air exchange rates required to maintain negative pressure. In the authors' experience, ceramic heating panels mounted behind the cages help to maintain appropriate temperatures (Fig. 5.9). Humidifiers have been utilized in housing areas with high air exchange rates to maintain appropriate humidity [51]. To reduce bites or scratches constituting an exposure risk, immobilization of the marmoset with an anesthetic is required prior to handling. Hand restraint for injections is typically used but is not ideal in an ABSL 3 or 4 environment. Traditional squeeze back systems utilized in primate caging often fail to fully immobilize a marmoset for injection and present a hazard to the marmoset if an appendage is entrapped during restraint. Modifications used by the authors include incorporation of a squeeze back made of a fabric mesh or fine wire mesh netting that molds to the marmoset and is light enough not to injure. Powell et al. [52] utilized a box attached to the front wall of the enclosure that could be closed without opening the enclosure [52]. Marmosets were trained to enter the box and then were moved into an appropriate biosafety cabinet where inhalant anesthetic was provided to the animal within the box, facilitating immobilization prior to removal from the box.

Metabolic Housing

Urine and feces can be collected by various methods that do not involve removal from the marmosets' home cage (see Chapter 10). Early examples of the use of a metabolism cage for these collections can be referenced [53,54]. The use of a metabolic cage for more intense collections over several hours to days does involve removal from the home cage. Early examples made use of plastic or metal buckets upturned over a rat metabolism unit. The marmosets were acclimated to the cage, and they were maintained in the animal room with familiar marmosets [54]. Marmosets urinate out from the cage. To capture every drop and provide a perch of sufficient height for their normal posture, an upper chamber made of polycarbonate is mounted over the lower funnel pieces of a rat metabolism cage unit. There are two clear, polycarbonate cylinders, each with its own urine and feces separation provided by the rat metabolism cage, mounted on a cart that can be placed in the same location as the home cage. Each chamber has a wire mesh basket suspended in the center of the cylinder, and any urine or feces falling from

FIGURE 5.9 Example of a biocontainment set up for use in ABSL 3–4 with ceramic panels mounted behind the cages to provide a supplemental heat source in a cooler, high air exchange environment. *Photo courtesy of author Layne-Colon, SNPRC.*

this location goes directly into the collection area. The polycarbonate sides do reduce the area the marmoset can climb on, but any urine sprayed out will fall onto a larger funnel and then down into the metabolism unit's collection system. A pair of marmosets from the same cage can be housed in this way to collect from both over the same time period while in a familiar location. The cylinders can also be connected to one another by a tube with a mesh grid just large enough to allow the passing of an infant between parents. In this way, the metabolic sample collections can be done on a lactating female in proximity to her partner. The male cannot contaminate her samples but can still take turns carrying the infant.

Behavioral Evaluation

Behavioral evaluations are key components of many research projects utilizing marmosets. Caging modifications to facilitate behavioral observations include use of plexiglass panels for easy visualization or filming. Cage additions may also be utilized to facilitate training and human interaction [45,47,55] (see Fig. 5.1B). It is important to also consider environmental factors beyond the intracage environment. Location of the cage relative to other cages may impact behavior due to visual, vocalization, or olfactory exposure. Some facilities may use special caging elements for breeding pairs, especially if video recording is employed to monitor dams for parturition. Facility noise and vibrations may also impact behavior. It is prudent to consider acclimation to a novel cage or cage location prior to behavioral observations.

Caging for Transportation

Research facilities are often not centralized and animal housing areas may not be in the immediate vicinity of surgery, imaging, and procedural rooms. The marmosets' small size facilitates easy transport with appropriate cages that may be novel containers or the nest box (Fig. 5.6B). Transport cages typically consist of a small box with an easily manipulated door and carrying handle. Transport cages can also be used to move marmosets between cages during cage sanitation or relocation. Positive reinforcement training to enter and exit the transport box, as well as to prolonged restraint within the box, should be a part of any husbandry or research acclimation plan.

Modifications for Clinical Care

Clinical cases requiring intensive treatment or those undergoing postoperative recovery may require housing in a smaller environment to facilitate frequent capture and restraint of the marmoset as well as reduced disruption to the marmoset's cage mates. If a cage with a divider is not available, a smaller mobile cage can be utilized that can be positioned near the home cage to allow continued visual, auditory, and olfactory contact with cage mates (Fig. 5.10). In such cases, a veterinary cage exemption should be recorded for temporary housing of the marmoset during treatment or postoperative recovery if the animal is housed in a cage that does not meet standard regulatory specifications. If heat support is required, a mobile heat lamp or incubator should be available.

SANITATION

Sanitation of primary enclosures and accessories should ensure that the animal has a clean, dry space to reside. Excreta and food waste must be removed daily from within the enclosure. Debris outside of the enclosure should be removed as often as necessary to prevent excess accumulation (United States Department of Agriculture [USDA]). Many cage designs allow for placement of an absorbent substrate (paper or wood chips, sheet paper, etc.) in the cage pan or underneath the enclosure to catch debris. Alternatively, spraying the bottom of the cage with water can remove debris but may also contribute to aerosolization of pathogens.

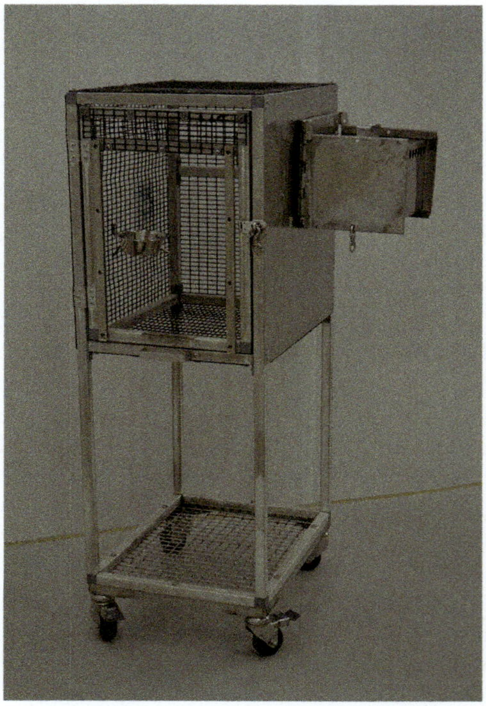

FIGURE 5.10 Small cage for temporary housing during recovery or introductions. *Photo courtesy of author Goodroe, WNPRC.*

Enclosure furniture, including perches, shelves, and next boxes, should be cleared of debris daily (USDA). The room, including the floors and walls, should be sanitized on a regular basis.

The Guide and the Animal Welfare Regulations (AWR) specify that primate primary cages and accessories should be sanitized at least once every 2 weeks. Many cages are able to be sanitized in a conventional cage washer which utilizes hot water, detergents, and disinfectants. The prewash and wash cycle are dependent on the cage materials. The AWR specifies that primary enclosures for scent-marking primates may be sanitized at regular intervals as determined by generally accepted professional practice. It has been suggested that removing all cage elements that are commonly scent marked is stressful; some institutions choose to stagger sanitation of primary cages and cage furniture for this reason, and the AWR provision for alteration of sanitation intervals facilitates this practice. Washing at high temperatures does not remove all scents from cage elements [10]. Movement of washed cage elements with scent markings between family groups has not been shown to cause distress or affect reproduction efficiency (Tardif, pers comm; [56]), but impact on behavior studies should be considered. Porous wood or soft materials such as hammocks may require alternative care as the harsh nature of cage washer detergents and disinfectants may not be appropriate. It is acceptable to leave materials that cannot be sanitized with steam or chemicals within the primary enclosure until they are deemed too contaminated and then dispose of them (AWR). Bottles and sipper tubes are to be sanitized at least once a week per the Guide. Sipper tubes can transfer pathogens quickly throughout a colony if not properly sanitized. Automated drinking-water systems should be flushed routinely to help prevent biofilm accumulation. Routine cleaning of water lines associated with caging should be considered for the same reason. Regular evaluation of the effectiveness of sanitation practices is recommended by the Guide and may be performed via microbiologic culture or ATP bioluminescence testing.

Care should be taken to ensure that detergents or disinfectants that may come into inadvertent contact with animals are nontoxic. An association between chromosomal disorders within blood cells and the use of chemical disinfectant (halogen-substituted hydrocarbon didecyldimethylammonium chloride) has been suggested [57]. Marmosets are often handled with leather gloves to protect handlers from bite wounds; a regular sanitation and replacement program for handling gloves should be considered. The AWR and Guide require a pest control program to be implemented. Investigators and veterinarians should be involved in decisions to utilize chemical pesticides [2].

CLOSING REMARKS

Husbandry practices are in a constant state of refinement as we learn more about the unique physiological and social needs of the marmoset. The impact of the environment, enclosure, and sanitation on research studies must always be balanced with ensuring a safe, stimulating environment for the marmosets. Continued communication through presentations and publications will contribute to establishment of validated, evidence-based husbandry practices.

References

[1] Morin ML. Colony management problems encountered in using marmosets and tamarins in biomedical research. Dig Dis Sci December 1985;30(12):14S–6S.

[2] National Research Council (U.S.), Committee for the Update of the Guide for the Care and Use of Laboratory Animals, Institute for Laboratory Animal Research (U.S.) and National Academies Press (U.S.). Guide for the care and use of laboratory animals, xxv. Washington, D.C.: National Academies Press; 2011. 220 pp.

[3] Power M. Digestive function, energy intake, and the response to dietary gum in captive callitrichids [Dissertation]. 1991.

[4] De la Fuente MF, Souto A, Sampaio MB, Schiel N. Behavioral adjustments by a small neotropical primate (*Callithrix jacchus*) in a semiarid Caatinga environment. Sci World J 2014;2014, 326524.

[5] Schiel N, Souto A. The common marmoset: an overview of its natural history, ecology and behavior. Dev Neurobiol 2017; 77(3):244–62.

[6] Valenca-Silva G, Maciel FG, Zaganini RL, Lucindo AS, Caramaschi S, de Paula HM. Reporting social behaviours of mixed-species troops formed by *Callithrix jacchus* and *Callithrix penicillata* (Primate, Callitrichidae). Braz J Biol 2014;74(3):607–11.

[7] EB R. EAZA husbandry guidelines for Callitrichidae. 2010.

[8] Suchi S, Rothe H. The influence of abiotic factors on the onset and cessation of activity of semi-free *Callithrix jacchus*. Am J Primatol 1999;47(3):241–53.

[9] Stonerook MJ, Weiss HS, Rodriguez MA, Rodriguez JV, Hernandez JI, Peck OC, Wood JD. Temperature-metabolism relations in the cotton-top tamarin (*Saguinus oedipus*) model for ulcerative colitis. J Med Primatol 1994;23(1):16–22.

[10] Layne DG, Power RA. Husbandry, handling, and nutrition for marmosets. Comp Med 2003;53(4):351–9.

[11] Schardt U, Wilhelm I, Erkert HG. Splitting of the circadian activity rhythm in common marmosets (*Callithrix j. jacchus*; primates). Experientia 1989;45(11–12):1112–5.

[12] Lucas NS, Hume EM, Henderson-Smith H. On the breeding of the common marmoset (*Hapale jacchus Linn.*) in captivity when irradiated with ultra-violet rays. J Zool 1927;97(2):447–51.

[12a] Lucas NS, B. M, F. Z. S., Margaret Hume E, Henderson Smith H. On the breeding of the common marmoset (Hapale jacchus Linn.) in captivity when irradiated with ultra-violet rays.-II. A ten years' family history. Proceedings of the Zoological Society London 1937:205–14.

[13] Box HO. Primate responses to environmental change. Springer; 1991.

[14] Lampert RM, de Azevedo CV, Menezes AA. Influence of different light intensities on the daily grooming distribution of common marmosets *Callithrix jacchus*. Folia Primatol 2011;82(3): 131–42.

[15] Scott RC, Corrigan MA, Smith F, Mason H. The influence of skin structure on permeability: an intersite and interspecies comparison with hydrophilic penetrants. J Invest Dermatol 1991;96(6):921–5.
[16] Nummela SU, Coop SH, Cloherty SL, Boisvert CJ, Leblanc M, Mitchell JF. Psychophysical measurement of marmoset acuity and myopia. Dev Neurobiol 2016;77(3):300–13.
[17] Smith 3rd EL, Hung LF, Huang J. Protective effects of high ambient lighting on the development of form-deprivation myopia in rhesus monkeys. Invest Ophthalmol Vis Sci 2012; 53(1):421–8.
[18] Norton TT, Siegwart Jr JT. Light levels, refractive development, and myopia—a speculative review. Exp Eye Res 2013;114:48–57.
[19] Karouta C, Ashby RS. Correlation between light levels and the development of deprivation myopia. Invest Ophthalmol Vis Sci 2014;56:299–309.
[20] Wang Y, Ding H, Stell WK, Liu L, Li S, Liu H, Zhong X. Exposure to sunlight reduces the risk of myopia in rhesus monkeys. PLoS One 2015;10(6):e0127863.
[21] Rose KA, Morgan IG, Ip J, Kifley A, Huynh S, Smith W, Mitchell P. Outdoor activity reduces the prevalence of myopia in children. Ophthalmology 2008;115(8):1279–85.
[22] Duarte MH, Vecci MA, Hirsch A, Young RJ. Noisy human neighbours affect where urban monkeys live. Biol Lett 2011;7(6):840–2.
[23] Tardif S, Hyde K, Digby L. Evidence for suppression of ovulation in singly-housed female common marmosets (*Callithrix jacchus*). Lab Primate Newsletter 1994;33(2).
[24] Mota MTS. Hormonal and behavioral studies of parental and alloparental care and reproductive success in common marmosets. PhD Thesis, Reading University, UK. 1999.
[25] Peters VM. Bem estar animal e os resultados da pesquisa [simpo sio]. Sociedade Brasileira de Etologia J. d. F. Anais do XIX Congresso Brasileiro de Etologia: 44. 2001.
[26] Barbosa MN, Mota MT. Behavioral and hormonal response of common marmosets, *Callithrix jacchus*, to two environmental conditions. Primates 2009;50(3):253–60.
[27] National Health and Medical Research Council. Use of non-human primates for scientific purposes | National Health and Medical Research Council. National Health & Medical Research Council; 2016. www.nhmrc.gov.au/health-ethics/animal-research-ethics/use-non-human-primates-scientific-purposes.
[28] Bakker J, Ouwerling B, Heidt PJ, Kondova I, Langermans JA. Advantages and risks of husbandry and housing changes to improve animal wellbeing in a breeding colony of common marmosets (*Callithrix jacchus*). J Am Assoc Lab Anim Sci 2015;54(3): 273–9.
[29] Mätz-Rensing K, Ellerbrok H, Ehlers B, Pauli G, Floto A, Alex M, Czerny CP, Kaup FJ. Fatal poxvirus outbreak in a colony of New World monkeys. Vet Pathol 2006;43(2):212–8.
[30] Minuzzi-Souza TT, Nitz N, Knox MB, Reis F, Hagström L, Cuba CA, Hecht MM, Gurgel-Gonçalves R. Vector-borne transmission of Trypanosoma cruzi among captive Neotropical primates in a Brazilian zoo. Parasit Vectors 2016;9:39.
[31] Favoretto SR, de Mattos CC, Morais NB, Alves Araujo FA, de Mattos CA. Rabies in marmosets (*Callithrix jacchus*), Ceara, Brazil. Emerg Infect Dis 2001;7(6):1062–5.
[32] United Sates Department of Agriculture (USDA) (2017). Animal Welfare Act and Animal Welfare Regulations, USDA-APHIS. Available online: https://www.aphis.usda.gov/aphis/ourfocus/animalwelfare/SA_AWA. http://www.aphis.usda.gov/animal_welfare/awa_info.shtml.
[33] Kitchen AM, Martin AA. The effects of cage size and complexity on the behaviour of captive common marmosets, *Callithrix jacchus jacchus*. Lab Anim 1996;30(4):317–26.
[34] Yoshimoto T, Takahashi E, Yamashita S, Ohara K, Niimi K. Larger cages with housing unit environment enrichment improve the welfare of marmosets. Exp Anim 2018;67(1):31–9.
[35] Code of practice for the housing and care of animals bred, supplied or used for scientific purposes. London: Stationery Office; 2014.
[36] Code of practice for the housing and care of animals bred, supplied or used for scientific purposes. London: Stationery Office; 2017.
[37] Jennings M, Prescott MJ, Members of the Joint Working Group on R, Buchanan-Smith HM, Gamble MR, Gore M, Hawkins P, Hubrecht R, Hudson S, Jennings M, Keeley JR, Morris K, Morton DB, Owen S, Pearce PC, Prescott MJ, Robb D, Rumble RJ, Wolfensohn S, Buist D. Refinements in husbandry, care and common procedures for non-human primates: ninth report of the BVAAWF/FRAME/RSPCA/UFAW Joint Working Group on Refinement. Lab Anim 2009;43(Suppl. 1):1–47.
[38] Vitale A, Cirulli F, Capone F, Alleva E. Animal welfare issues under laboratory constraints, an ethological perspective: rodents and marmosets. In: Kaliste E, editor. The welfare of laboratory animals; 2007. p. 315–38.
[39] Garber P, Sallenave A, Blomquist GE, Anzenberger G. A comparative study of the kinematics of trunk-to-trunk leaping in *Callimico goeldii*, *Callithrix jacchus*, and *Cebuella pygmaea*. In: Ford SM, Porter LM, Davis LC, editors. The smallest anthropoids. New York (NY): Springer; 2009. p. 259–78.
[40] Sussman RW, Kinzey WG. The ecological role of the callitrichidae: a review. Am J Phys Anthropol 1984;64(4):419–49.
[41] Francis L, Moore RT, Raymond RT, Baskerville A. The housing and handling of marmosets and tamarins infected with AIDS and other retroviruses. In: Beynen AC, Solleveid HA, editors. New developments in biosciences: their implications for laboratory animal science. The Netherlands: Martinus Nijhoff; 1988. p. 99–103.
[42] Buchanan-Smith HM, Shand C, Morris K. Cage use and feeding height preferences of captive common marmosets (Callithrix j. jacchus) in two-tier cages. J Appl Anim Welf Sci 2002;5(2): 139–49.
[43] Young JW, Stricklen BM, Chadwell BA. Effects of support diameter and compliance on common marmoset (*Callithrix jacchus*) gait kinematics. J Exp Biol 2016;219:2659–72. https://doi.org/10.1242/jeb.140939.
[44] Rumble R, Saville M, Simmons L. The preference of the common marmoset for nest boxes made from three different materials: wood, plastic, metal. ATW 2005;4:185–7.
[45] Ely A, Freer A, Windle C, Ridley RM. Assessment of cage use by laboratory-bred common marmosets (*Callithrix jacchus*). Lab Anim 1998;32(4):427–33.
[46] Hosey GR, Jacques M, Burton M. Allowing captive marmosets to choose the size and position of their nest box. Anim Welf 1999;8: 281–5.
[47] Forster FC. Exploratory behavior and learning in laboratory marmosets (*Callithrix jacchus jacchus*): comparison between experimental-cage and home-cage activity. Primates 1995;36:501–14.
[48] Roberts RL, Roytburd LA, Newman JD. Puzzle feeders and gum feeders as environmental enrichment for common marmosets. Contemp Top 1999;38:27–31.
[49] Vignes S, Newman JD, Roberts RL. Mealworm feeders as environmental enrichment for common marmosets. Contemp Top Lab Anim Sci 2001;40(3):26–9.
[50] Stevenson MF, Rylands AB. The marmosets, genus Callithrix. In: Mittermeier RA, Rylands AB, Coimbra-Filho AF, Fonseca GAB, editors. Ecology and behavior of neotropical primates, vol. 2. Washington, DC: World Wildlife Fund; 1988. p. 131–222.

[51] Ross CN, Austad S, Brasky K, Brown CJ, Forney LJ, Gelfond JA, Lanford R, Richardson A, Tardif SD. The development of a specific pathogen free (SPF) barrier colony of marmosets. Aging (Albany NY) 2017;9(12):2544–58. https://www.ncbi.nlm.nih.gov/pubmed/29227963.

[52] Powell DS, Walker RC, Heflin DT, Fisher D, Kosky JB, Homer LC, Reed DS, Stefano-Cole K, Trichel AM, Hartman AL. Development of novel mechanisms for housing, handling, and remote monitoring of common marmosets at animal biosafety level 3. Pathog Dis 2014;71(2):219–26.

[53] Hearn JP, Lunn SF, Burden FJ, Pilcher MM. Management of marmosets for biomedical research. Lab Anim 1975;9(2):125–34.

[54] Lunn SF. Systems for collection of urine in the captive common marmoset, *Callithrix jacchus*. Lab Anim 1989;23(4):353–6.

[55] Crofts HS, Muggleton NG, Bowditch AP, Pearce PC, Nutt DJ, Scott EAM. Home cage presentation of complex discrimination tasks to marmosets and rhesus monkeys. Lab Anim 1999;33:207–14.

[56] French JA. In: Bayne K, editor. Marmosets and tamarins. Enrichment in non-human primates series. APV/ASP Working Group; 2005.

[57] Delimitreva S, Wedi E, Bakker J, Tkachenko OY, Nikolova V, Nayudu PL. Numerical chromosome disorders in the common marmoset (*Callithrix jacchus*)—comparison between two captive colonies. J Med Primatol 2013;42(4):177–85.

[58] Power ML, Koutsos L. Marmoset nutrition and dietary husbandry. In: The common marmoset in captivity and biomedical research; 2018.

[59] Kramer R, Burns M. Normal clinical and biological parameters of the common marmoset (*Callithrix jacchus*). In: The common marmoset in captivity and biomedical research; 2018.

[60] Arruda MF, Yamamoto ME, Pessoa DMA, Araujo A. Taxonomy and natural history. In: The common marmoset in captivity and biomedical research; 2018.

[61] Kaplan G, Pines MK, Rogers LJ. Stress and stress reduction in common marmosets. Appl. Animal Behav. Sci. 2012;137:175–82. https://doi.org/10.1016/j.applanim.2011.04.011.

[62] Eliades SJ, Miller CT. Marmoset vocal communication: Behavior and neurobiology. Behavioral Neurobiology, Mar 2017;77(3):286–99. https://doi.org/10.1002/dneu.22464 (epub oct 2016).

[63] Agamaite JA, Chang CJ, Osmanski MS, Wang XJ. Acoustic Society America 2015;138(5):2906–28. https://doi.org/10.1121/1.4934268.

[64] Osmanski MS, Wang X. Measurement of absolute auditory thresholds in the common marmoset (Callithrix jacchus). Hearing Research 2011;277:127–33.

[65] Dettling A. Physical environment and its influence on behaviour in captive marmosets (*Callithrix jacchus*). In: Pryce C, Scott L, Schnell C, editors. Marmosets and tamarins in biological and biomedical research. Salisbury, UK: DSSD Imagery; 1997. p. 39–46.

[66] Nishijima K, Saitoh R, Tanaka S, Ohsato-Suzuki M, Ohno T, Kitajima S. Life span of common marmoset (Callithrix jacchus) at CLEA Japan breeding colony. Biogerontology 2012;13:439–43.

[67] Science Council of Japan: Guidelines for Proper Conduct of Animal Experiments. 2006. http://www.scj.go.jp/ja/info/kohyo/pdf/kohyo-20-k16-2e.pdf.

[68] Casey CL, Wilson DE, Editors. United States Department of Health and Human Services (HHS): Biosafety in microbiological and biomedical laboratories. HHS Publication No. (CDC) 21–1112. https://www.cdc.gov/labs/pdf/CDC-Biosafety MicrobiologicalBiomedicalLaboratories-2009-P.pdf

Further Reading

[1] Hume EM, Lucas NS, Smith HH. On the absorption of vitamin D from the skin. Biochem J 1927;21(2):362–7.

CHAPTER

6

Normal Clinical and Biological Parameters of the Common Marmoset (*Callithrix jacchus*)

Robin Kramer, Monika Burns

Division of Comparative Medicine, Massachusetts Institute of Technology, Cambridge, MA, United States

INTRODUCTION

Published reports describing normative data on the common marmoset are relatively limited. This chapter intends to compile pertinent information and references to provide a framework for both the clinician and scientist working with the common marmoset as a research model. Though interest in and use of marmosets in research has waxed and waned throughout recent decades, there are multiple advantages to the use of the species as a nonhuman primate disease model. For example, the required volume of test article when marmosets are used in a biomedical research or toxicology setting is approximately 10- to 15-fold less than that of macaques, due largely to a much smaller mean body size. This feature, along with the reduced risk of occupational exposure to pathogens such as *Macacine herpesvirus 1*, makes the species an attractive and resource-efficient research model [1]. As marmoset colonies expand worldwide, the authors urge those involved in their care to compile, analyze, and report data collected on individual colonies. See Table 6.2 for selected normative data in the common marmoset.

PART I: NORMATIVE BIOLOGICAL PARAMETERS

Natural History, Life Stages, and Life Span

Marmosets are sexually monomorphic, with reported mean adult body weight for both sexes ranging from about 300 to 500 g in captivity [2–4]. Reported mean body weight of wild marmoset populations tends to be lower, ranging from about 320 to 335 g [5]. Captive marmosets weigh significantly more than their wild counterparts in all life stages except the infant stage [5]. A likely contributing factor to this difference in mean body weight is that diets provided to callitrichid colonies in biomedical research settings differ significantly from diets consumed by wild marmosets [6]. In addition to differences in diet composition, variation in housing and husbandry practices may contribute to higher mean body weights in captivity. Representative mean body weights for each life stage are presented in Table 6.1.

Sexual maturity occurs at approximately 12–18 months of age, although some marmosets may not be effective breeders when paired with a member of the opposite sex that early in life [7,8]. Gestation lasts an average of 143–144 days in the marmoset, and average age of first conception is 2.49 years [8]. Common marmosets typically produce litters of dizygotic twins; however, triplet litters are born frequently in captivity. A survey of 1219 litters at a national primate center reported that 37.5% of all litters produced during the survey time frame were triplet litters [9]. Singletons, quadruplet, and quintuplet litters occur less frequently but have also been reported in the marmoset literature [9]. Problems associated with triplet litters are discussed in Chapter 8 (Tardif, 2018). Numerical chromosome disorders (polyploidy and chromosomal nondisjunction) have been documented in the marmoset and may be associated with the use of certain chemical disinfectants, as suggested by Delimitreva et al. [10]. Keeping careful records of births, infant survival, and interbirth intervals are important to document breeding success or failure in marmoset colonies.

The maximum life span of captive marmosets is reported to be 16.5 years, with anecdotal reports of rare individuals surviving up to 19 years [7,11]. The average life span of captive marmosets is reported to be

The Common Marmoset in Captivity and Biomedical Research
https://doi.org/10.1016/B978-0-12-811829-0.00006-6

TABLE 6.1 Representative Body Weight of Common Marmosets

Age/Sex Classes (Months of Age)		Setting	
		Captive	Wild
Infant (2–5 m)	Male	144.9 ± 38.5 (49)	125.3 ± 34.4 (10)
	Female	133.7 ± 32.2 (47)	141.6 ± 27.4 (14)
Juvenile (6–10 m)	Male	238.1 ± 45.6 (64)	164.5 ± 41.7 (31)
	Female	244.8 ± 37.6 (53)	179.8 ± 39.5 (30)
Subadult (11–15 m)	Male	302.2 ± 42.5 (48)	244.9 ± 42.8 (31)
	Female	318.2 ± 45.6 (46)	251.4 ± 32.4 (29)
Adult (>15 m)	Male	347.6 ± 30.5 (46)	317.9 ± 34.1 (69)
	Female	359.7 ± 50.4 (38)	322.0 ± 40.4 (86)

Values are presented as mean body weight (g) + standard deviation for captive and wild animals by age and sex.
Reprinted with permission from Araújo A, Arruda MF, Alencar AI, Albuquerque F, Nascimento MC, and Yamamoto ME. Body weight of wild and captive common marmosets (Callithrix jacchus)*. Int J Primatol 2000;21:317–24.*

TABLE 6.2 Selected Normative Data

Parameter	Value
Age at sexual maturity	Male: 382 days (mean), Female: 477 days (mean) [105]
Gestation length	144 days (mean) [107]
Weaning age	62 (mean) [107]; complete by 100 days [105]
Birth weight	29.68 +/- 3.41 g [106]
Chromosome number	23 [108]
Dental formula	2/2, 1/1, 3/3, 2/2 [106]
Vertebral formula	C7, T12–13, L6–7, S2–3, Cd26–29 [106]
Daily dry matter intake	52 g/kg BW [109]
Daily water intake	11.7 +/- 10.2 mL/marmoset [46]
	18–140 mL/kg (median 62 mL/kg) [51]
Daily urine output	5.5–19.7 mL/marmoset [46]
Urine pH	5–8 [53]
Mean arterial blood pressure	95–107 mm Hg [21–23]
Heart rate	150–230 bpm (undisturbed) [21] up to 400 + bpm (manual restraint) [23,27]
Respiratory rate	Approximately 100 breaths/minute (manual restraint) [17]
Blood volume	60–70 mL/kg [17]
Body temperature	36.8–38.6°C [17,26–28]

Unless specified, continuous data in this table are represented as mean and standard deviation.

approximately 5–7 years [7,8,12,13]. A retrospective study performed at a national primate center found that 12% of marmosets that survived to 60 days of age lived to 8 years and only 1.5% survived to 13 years, indicating that the population of animals in some captive colonies surviving past 13 years of age is quite small. This analysis included animals that died of natural causes but did not provide further information about cause of death. It is unclear if animals used in studies were removed from analysis [9]. Greater age at first birth has been associated with increased survivorship in female marmosets; dams that began reproducing at 4 years of age had higher survivorship than dams that began to reproduce prior to 4 years old [13]. A sex-associated difference in life span has been reported and attributed to high burden of reproduction on breeding females in one captive colony of marmosets (Fig. 6.1). [14] Marmosets may be considered "aged" at 8 years [7,14]. Aging changes described in marmosets include fibrous cartilaginous changes in intraarticular disks, β-amyloid deposition in the cerebral cortex, and reduced neurogenesis in the hippocampus. Serum albumin and proximal thigh circumference are significantly lower in aged marmosets than young marmosets [7]. Low serum albumin levels, poor body condition, and low muscle mass have been associated with the development of inflammatory bowel disease that presents most frequently as chronic lymphocytic enterocolitis and is known historically as marmoset wasting syndrome (MWS). Compared with many Old World nonhuman primates, marmosets have a shorter overall life span, which results in the development of aging phenotypes in a relatively compressed time frame.

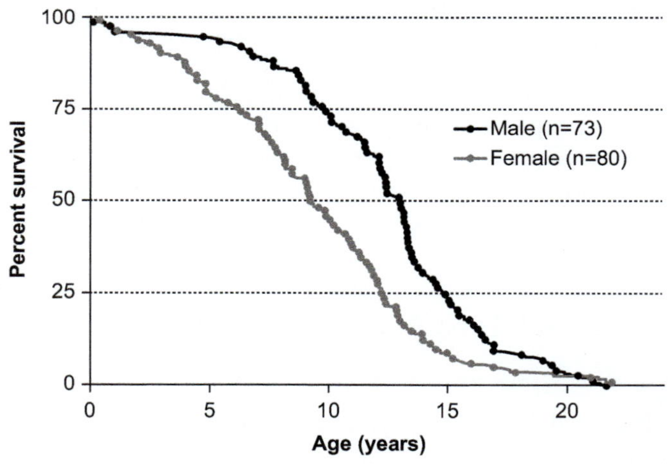

FIGURE 6.1 Survival curve of male and female common marmoset at the Central Laboratory for Experimental Animals (CLEA) Japan breeding colony. Males exceeded females in survival ($P < .01$). *Reprinted with Permission from Nishijima K, Saitoh R, Tanaka S, Ohsato-Suzuki M, Ohno T, Kitajima S. Life span of common marmoset* (Callithrix jacchus) *at CLEA Japan breeding colony. Biogerontology 2012;13: 439–43.*

This characteristic makes the species a unique model for aging research.

Retrospective analyses have found that most commonly reported causes of death or euthanasia in common marmoset colonies tend to be undetermined or nebulous, with "poor condition," "emaciation," or "unknown" frequently documented in medical records [7,15,16]. The rate of spontaneous mortality in one report was 2%–3%, with the presumed causes of death being "digestive problems" such as weight loss and acute infections [17]. Other common causes of death or euthanasia include gastrointestinal disease (colitis, enteritis), amyloidosis, and nephritis. As the use of marmosets increases in biomedical research, efforts should be made to definitively diagnose and publish causes of morbidity and mortality in captive marmoset colonies. The authors recommend that body weight and body condition of all animals be carefully monitored on a monthly basis, and animals that have had a history of weight loss or signs of clinical disease should be tracked on a weekly or even daily basis. Marmosets with poor body condition may appear "normal" to the cage-side observer due to the relatively long hair coat of the species. Because the body condition of an animal may deteriorate without significant change in body weight, the authors emphasize that measurement of both body weight and condition is critical. Further information on body condition evaluation and measurement is provided in Chapter 10 (Wachtman, 2018).

Organ Weights and Morphometry

Due to their increasing use in biomedical research, including toxicology studies, it is important to characterize normal parameters and background lesions in the marmoset. Korte et al. report normal organ weights for healthy adult animals [17]. Mean liver weight of male marmosets was 12.71 g (range 8.95–21.1 g); mean liver weight of female marmosets was 11.94 g (8.42–26.66 g). Mean brain weight of male marmosets was 7.96 g (6.69–9.29 g); mean brain weight of female marmosets was 8.07 g (6.74–9.31 g). Similar to the value in humans of approximately 2%, a marmoset's brain is approximately 2.7% of its body weight, five times larger than that of a similarly sized rat [18]. Other organ weights, including the spleen, adrenal, kidney, thymus, heart, pituitary, thyroid, testes, among others, are also reported by Korte et al. Testicular volume in adult males ranges from 0.26 to 0.88 mL with a mean of 0.53 mL [17]. Crown-rump measurements may be taken after birth for neonates or in utero using ultrasonography [8]. Biparietal diameter growth curves for the fetus in utero have also been reported [19]. Please see Chapter 8 for a full discussion of marmoset reproduction (Tardif, 2018).

Cardiovascular Parameters

Measurement of cardiovascular parameters in marmosets has been accomplished via manual restraint, implanted arterial catheters, electrocardiogram, and remote telemetry [17,20]. Equipment originally designed for rats can be modified for use in this species. Values obtained can vary considerably depending on the combination of habituation, chemical restraint, and manual restraint that is employed. None of these confounders are required to obtain telemetric measurements of normative parameters in the marmoset, which is why measurements obtained via telemetry likely represent the most accurate and unbiased values. Telemetry also allows animals to remain within their family groups, potentially decreasing stress and avoiding false elevation of these parameters. Telemetry has been successfully used for recording periods of up to 11 months in marmosets [21].

Mean blood pressure values were similar as measured via telemetry, indwelling arterial catheter, and tail cuff (noninvasive) at 95, 107, and 96 mmHg, respectively. Importantly, all measurements were performed in awake animals, but only the telemetry measurements reported in this study do not require manual restraint to obtain [21–23].

In contrast, reported heart (HR) rates vary considerably based on the conditions at measurement. Average heart rate for awake, manually restrained marmosets was 405 beats per minute (bpm) as measured via ECG, about twice that of macaques [17]. Michel et al. report an average HR of 348 bpm when measured in manually restrained animals with an indwelling arterial catheter, and Wood et al. found a similar value via the tail cuff method at 312 bpm [22,24]. All of these values are considerably higher than those reported by Schnell and Wood [21]. Using telemetry, they report a heart rate for unstressed animals ranging from 150 bpm at night to 230 bpm during the day.

A predictable diurnal variation in heart rate, blood pressure, and motor activity has been observed, with a rapid increase in blood pressure and heart rate at "lights on" in the morning, a slight trough at midday, relatively steady levels in the afternoon, and a sharp drop at "lights out" in the evening. Slight variations in this pattern will occur depending on the timing of husbandry procedures, in particular timing and frequency of mealtimes. Differences have been observed when animals are fed only once daily versus twice daily [21].

Cardiovascular parameters and motor activity were also recorded during pregnancy, parturition, and lactation via telemetry in a single animal by Gerber et al. [25]. Parturition caused marked changes in heart rate and blood pressure, with smaller changes in blood pressure and heart rate values observed during late gestation and lactation. Respiratory rate is ~100 breaths per minute in awake, manually restrained animals [17].

Body Temperature

Callitrichids develop hypothermic torpor (34°C) during sleep, and the rectal temperature of marmosets oscillates with circadian rhythm [26]. Body temperature ranges from about 38.6°C during daytime hours to about 36.8°C during nighttime hours [27]. Rectal and subcutaneous body temperature measurements in the marmoset have been shown to be in close agreement [28].

PART II: CLINICAL PATHOLOGY

Hematology and Clinical Chemistry

As in other species, evaluation of hematology and serum biochemistry values provides useful information for both the diagnostic and research processes. Several references have been published describing the standard parameters evaluated in many veterinary species, though it is helpful to determine normal values for each individual colony and preferred diagnostic laboratory. Diet and husbandry practices have evolved in the decades since some of these reference values were determined. In addition, there are differences between wild-caught and colony-bred animals, and both age and sex may influence normal reference value ranges [29,30]. Here we summarize the relevant literature and provide in-house data for the marmoset colony at the authors' institution. Serum biochemistry and hematology data are summarized in Tables 6.3 and 6.4, respectively.

Variation in hematological and serum biochemical data by age and sex for a closed colony of animals is described by Yarbrough et al. [30]. This population included both wild-caught and colony-born animals, but the colony was closed for at least 2 years prior to sampling. Sampling was performed without sedation in 55 adult males, 18 adult females, 24 juvenile males (defined as <18 months of age), and 18 juvenile females. Glucose, blood urea nitrogen (BUN), alkaline phosphatase (ALP), lactate dehydrogenase (LDH), phosphorus, and amylase were higher in juveniles than adults, whereas hemoglobin and hematocrit were higher in adults than juveniles. Cholesterol and triglycerides were significantly different between the sexes, with higher values found in males. Davy et al. also evaluated the effects of age and sex on select clinical chemistry parameters in unanesthetized marmosets. Juveniles were defined as 250–300 days old and adults as >500 days; all animals were colony-born [29]. This study also found that ALP was higher in juvenile animals of both sexes, whereas iron was higher in mature males.

The effect of origin (wild- vs. colony-born) on clinical chemistry parameters was investigated by McNees et al. [31]. All animals had been in the colony for at least 2 years at the time of sampling, which was conducted under ketamine sedation. Despite a constant environment for at least 2 years, serum glucose was 20% higher in wild-born animals than those born in the colony. Wild-born animals also had significantly higher serum triglycerides, whereas colony-born animals had higher BUN and serum aspartate aminotransferase (AST). Cholesterol, in contrast, varied with gender, averaging 15% higher in males than females.

More recently, Kuehnel et al. determined mean hematology and chemistry values in 54 adult animals habituated to awake blood collection. Additionally, they, along with an independent study by Ziegler et al., reported normal ranges for serum lipid parameters [32,33]. Normal serum lipid values are summarized in Table 6.5. Lipid values and plasma metabolomic surveys have been undertaken and published in recent years in an effort to better understand the dietary and husbandry requirements of captive marmosets [34].

In addition to factors related to the animal itself, sample size, sampling site, sample handling, and the degree of hemolysis present can affect clinical chemistry parameters. These factors have been shown to influence serum chemistry parameters in rodent studies [35–37], but there are no published reports on the effects of sampling site in the marmoset. Changes in marmoset clinical chemistry values due to the presence of hemolysis were investigated by Davy et al. [29]. This study found that large changes in LDH (+101%) occurred with relatively low levels of hemolysis, whereas a higher degree of hemolysis also affected the measured levels of alanine transferase (ALT), AST, and bilirubin, but to a lesser degree than LDH. The fact that AST was affected by hemolysis indicates that AST may be found in marmoset erythrocytes, as is the case in humans. Total protein, albumin, gamma-glutamyl transferase (GGT), ALP, and plasma iron levels are unlikely to be affected by any level of hemolysis. Sedation used for phlebotomy can also affect clinical chemistry parameters. This is well-documented in macaques, including increases in AST, ALT, and creatine kinase (CK), with reductions in glucose, phosphorus, sodium, and potassium noted under ketamine anesthesia [38,39]. Little has been published on this topic in marmosets, but blood samples are often obtained in awake, acclimated animals in this species [29,30,32]. Further study is merited on the effects of commonly used anesthetic agents, such as ketamine hydrochloride and alfaxalone, on clinical chemistry parameters of the common marmoset.

In comparison to humans, cotton-top tamarins (*Saguinus oedipus*), and white-lipped tamarins (*Saguinus*

TABLE 6.3 Select Clinical Chemistry Reference Ranges

Parameter	Reference 1 [30]	Reference 2 [32]	Reference 3[a]	Reference 4[b]
Alkaline phosphatase (ALP) (IU/L)	Male 22–158 (61) Female 23–88 (58)	44.0–426.3 (115.5)	44.3–114.7 (79.5)	92.8–164.2 (128.5)
Alanine aminotransferase (ALT) (IU/L)	Male 14–99 (55) Female 17–104 (54)	0.8–45.3 (3.0)	3.8–14.0 (8.9)	2.7–19.0 (10.9)
Aspartate aminotransferase (AST) (IU/L)	Male 76–292 (151) Female 70–279 (146)	51.2–316.2 (155.7)	88.0–161.2 (124.6)	76.5–233.3 (155)
Lactate dehydrogenase (LDH) (IU/L)	Male 7–775 (218) Female 115–562 (216)	134.8–1119.7 (279.0)	N/A	N/A
Gammaglutamyl transferase (GGT) (IU/L)	N/A	0.2–13.5(2.8)	0.7–11.3 (6.0)	0.0–10.3 (5)
Total protein (g/dL)	Male 4.9–10.7 (7.2) Female 4.3–7.6 (6.6)	*4.1–7.5 (6.2)*	5.2–6.4 (5.8)	5–6.4 (5.7)
Albumin (g/dL)	Male 2.1–6.8 (5.1) Female 1.7–5.3 (4.4)	*1.1–5.8 (3.3)*	3.1–4.0 (3.5)	3.2–4.2 (3.7)
Calcium (mg/dL)	Male 1.5–15.7 (10.4) Female 1.8–11.2 (9.6)	*7.1–11.8 (9.4)*	7.6–10.5 (9.0)	8.9–10.1(9.5)
Phosphorus (mg/dL)	Male 1.6–10.4 (5.6) Female 2.1–9.1 (5.2)	*1.5–7.5 (3.3)*	2.7–4.5 (3.6)	2.6–4.7 (3.7)
Blood/serum urea nitrogen (BUN/SUN) (mg/dL)	Male 11–43 (22) Female 14–34 (22)	*9.5–52.4 (23.6)*	14.2–25.6 (19.9)	16.0–25.2 (20.6)
Creatinine (mg/dL)	Male 0.1–1.1 (0.6) Female 0–0.9 (0.6)	*0.2–-0.6 (0.3)*	0.1–0.4 (0.2)	0.1–0.3 (0.2)
Glucose (mg/dL)	Male 74–323 (172) Female 97–294 (192)	*72.3–214.2 (124.3)*	70.3–155.8 (113)	82.9–212.5 (147.7)
Iron (μmol/L)	N/A	17.6–39.5 (28.0)	N/A	N/A
Bilirubin (mg/dL) (total)	Male 0.02–4.41 (0.48) Female 0–0.95 (0.25)	*0–0.3 (0.0)*	0.1–0.2 (0.2)	0.1–0.3 (0.2)
Amylase (IU/L)	Male 445–2871 (930) Female 94–2597 (1057)	617.4–1243.2 (924.0)	175.0–287.6 (231.3)	179.6–274.6 (227.1)
Lipase (IU/L)	Male 2.5–150.8 (33) Female 1.7–57.0 (16.8)	10.8–58.8 (27.6)	18.3–47.7 (33.0)	2.6–82.2 (42.4)
Sodium (mM/L)	Male 144–175 (161) Female 152–158 (155)	155.5–174.0(163.0)	147.4–152.5 (149.9)	85.9–181.9 (133.9)
Potassium (mM/L)	Male 3.1–5.5 (4.1) Female 3.7–4.1 (3.9)	2.3–4.7 (3.4)	2.7–3.5 (3.1)	2.6–3.2 (2.9)
Chloride (mM/L)	Male 99–118 (107) Female 100–113 (106)	101.9–119.9 (109.4)	102.4–112.8 (107.6)	102.0–110.6 (106.3)

Italicized values were converted to US Conventional units from the published reference. Data from Reference 1 are represented as ranges with mean values in parentheses. Data from References 3 and 4 include ranges represented as ±2 standard deviations of the mean (in parentheses). Reference 2 ranges are represented as the third to the ninth percentile, with the median in parentheses. All values were rounded to the nearest 10th where applicable.

[a] *Author's institution, Adult (>2 yr) (unpublished) n = 41.*

[b] *Author's institution, Subadult (<2 yr) (unpublished) n = 22.*

TABLE 6.4 Select Hematology Reference Ranges

Parameter	Reference 1 [30]	Reference 2 [32]
Leukocytes	Male 2.7–17.5 (8.1) × 10^3/mm^3	2.7–10.9 (6) (G/L)
	Female 4.5–18.7 (7.4) × 10^3/mm^3	
Neutrophils (%)	Male 17–80 (43)	28.3–78.1 (52.5)
	Female 32–85 (54)	
Lymphocytes (%)	Male 14–82(51)	14.0–62.7 (34.5)
	Female 13–63 (40)	
Monocyte (%)	Male 0–16 (3.3)	2.0–17.1 (6.5)
	Female 0–9 (3.6)	
Eosinophil (%)	Male 0–5 (0.4)	0.0–13.8 (2.5)
	Female 0–3 (0.5)	
Basophil (%)	Male 0–5 (0.8)	0.0–0.0 (0.0)
	Female 0–8 (1.5)	
Red blood cell count (RBC)	Male 3.4–7.2 (5.7) × 10^3/mm^3	3.5–7.9 (6.5) (T/L)
	Female 4.2–9.6 (5.7) × 10^3/mm^3	
Hemoglobin	Male 10.9–19.3 (16.1) (g/dL)	3.4–10.7 (8.9) (mmol/L)
	Female 11.1–17.8 (15.0) (g/dL)	
Hematocrit/packed cell volume	Male 23–52 (43) (%)	0.3–0.5 (0.5) (l/)
	Female 32–56 (41) (%)	
Mean corpuscular volume (MCV)	Male 66–103 (72) (μ^3)	53.3–83.5 (69.0) (fL)
	Female 68–85 (73) (μ^3)	
Mean corpuscular hemoglobin (MCH)	Male 21.2–41.0 (28.0) (pg)	1.2–1.5 (1.4) (fmol)
	Female 17.4–30.5 (26.9) (pg)	
Mean corpuscular hemoglobin concentration (MCHC)	Male 30.0–60.0 (39.0) (%)	18.3–21.2 (20) (mmol/L)
	Female 24.2–42.6 (37.0) (%)	
Platelet count (PLT)	Male 105–465 (281) × 10^3/mm^3	17.6–788.6 (402.5) (G/L)
	Female 93–575 (281) × 10^3/mm^3	

Reference 1 data include ranges and mean values included in parentheses. Reference 2 ranges are represented as the third to the ninth percentile, with the median in parentheses. All values were rounded to the nearest 10th where applicable.

nigricollis), common marmosets have higher glucose, serum urea nitrogen, AST, ALT, LDH, and amylase values [31]. As in other species, measurement of glycosylated hemoglobin values in marmosets can provide information about long-term glucoregulatory control. No studies have evaluated the glycosylated hemoglobin values of marmosets compared with other species. Collecting samples in an awake, unfasted state may explain differences in serum glucose. When compared with cynomolgus macaques (*Macaca fascicularis)*, marmosets have lower GGT and ALT values [40]. Small numbers of nucleated red blood cells are often seen in callitrichid blood smears, and human reticulocyte counts are lower than those reported for marmosets [30,41].

Plasma iron and other iron analytes have also been investigated as indicators of hepatic hemosiderosis, a necropsy finding in callitrichids that has been linked to inflammatory bowel disease in some studies [42–44]. Smith et al. [42] retrospectively evaluated several hematologic iron parameters as markers for hepatic hemosiderosis in 12 species of callitrichids, including the common marmoset (n = 2 of 21 individuals). Blood ferritin levels were positively correlated with both postmortem hepatic iron concentration and histopathologic grades of iron deposition. This study included both serum and plasma samples in their analysis. Interestingly, there was no effect of either age or sex on hepatic iron concentrations in this population.

TABLE 6.5 Lipid Metabolism Parameters

Parameter	Reference 1 [33]	Reference 2 [32]
Total cholesterol (mg/dL)	174.5 ± 10.2	*88.0–196.5 (147.5)*
Total triglycerides (mg/dL)	97.3 ± 14.8	*21.6–101.5 (42.1)*
High-density lipoprotein (HDL) (mg/dL)	90.3 ± 12.1	*21.1–91.1 (51.4)*
Low-density lipoprotein (LDL) (mg/dL)	76.8 ± 13.7	*51-0.7–119.3 (84.9)*
Very low–density lipoprotein (VLDL) (mg/dL)		*0.77–17.4 (4.24)*

Italicized values were converted to US Conventional units from the published reference. Reference 1 ranges are represented as the mean ± the standard error of the mean (SEM). Reference 2 ranges are represented as the third to the ninth percentile, with the median in parentheses. All values were rounded to the nearest 10th where applicable.

The majority of coagulation indices in marmosets are similar to those in humans, including activated partial thromboplastin time (APTT), fibrinogen, and Russell's viper venom time (RVVT) [45]. Mean times, averaged over both males and females, in the marmoset are whole-blood clotting time (CT) = 2.20 min, APTT = 31.0 s, and RVVT = 17.45 s. Marmosets have a lower average prothrombin time than both humans and cynomolgus macaques [40,45], at 8.15 s. In addition, fibrinogen is significantly higher in female marmosets (2.35 g/L) than males (1.62 g/L), and platelet count is about 30% higher in females than males [45]. Other coagulation indices do not differ between genders in the marmoset, though platelet counts are higher in juveniles than adults [30]. The gender difference in platelet count does not exist in humans, and marmoset and human platelets share the same ultrastructural features. Platelet aggregation in the species occurs in response to collagen, adenosine diphosphate, arachidonate, and ristocetin. In contrast to humans, but similar to other nonhuman primates, marmoset platelets do not aggregate in response to epinephrine [45].

Urinalysis

Urine can be collected from marmosets in metabolism cages designed for rats [46]. Urine can also be collected using noninvasive free-catch methods, so that parameters can be evaluated without the confounding influence of stress or sedatives. This eliminates the need for animals to be separated from their family groups for extended periods of time. There are different methods for one-time collections. McKinley et al. describe training marmosets to provide a urine sample using positive reinforcement, whereas Torii et al. entered the cage at the onset of the light cycle to opportunistically collect free-catch samples after a period of acclimation [47,48]. A method of noninvasively and nonstressfully obtaining individual urine samples simultaneously from family groups has also been reported [49]. In this study, marmosets were trained to enter a custom urine collection apparatus immediately after leaving the sleeping box and were given a sweet reward after micturition occurred. Urine samples flowed directly from the collection apparatus into cryotubes.

Urine volumes of 0.2–3 mL per voiding have been described [48]. One study reported a mean water intake over 24 h of 11.7 ± 10.2 mL, which was positively correlated with a urinary output of 12.6 ± 7.1 mL over 24 h [46]. These authors found no difference in urine production between sexes. Additionally, there were no significant differences in urine production in pregnant versus nonpregnant females or gonadectomized versus intact males or females [46]. The range of daily urine production is similar to that reported by Shackleton of 10–25 mL of urine produced per 24 h [50]. Both the median and mean daily water intake for a subset of the colony at the authors' institution is approximately 62 mL/kg, with significant interindividual variation, ranging from 18 to 140 mL/kg daily among individual animals [51].

Little information has been published on normal clinical urinalysis parameters in common marmosets. A study by Drake determined that the average urine specific gravity for marmosets in a laboratory setting was similar to other captive New World primates at 1.026 [52]. Yamada et al. summarized urinalysis and urine cytology findings in animals with various degrees of glomerulonephropathy [53]. Most of the literature relating to urine evaluation in marmosets focuses on measurement of either stress or reproductive hormones. Such measurements are usually normalized against creatinine and expressed as a ratio per milligram of creatinine in the urine. This procedure corrects for variations in water content and concentration of the urine. Please see Part III of this chapter for a discussion of sex hormone measurement in the urine.

Cerebrospinal Fluid

The cerebrospinal fluid (CSF) collection technique in the common marmoset is described by Schultz-Darken [54]. Volumes of 50–150 μL may be collected. As noted in other species, an elevation in white blood cell (WBC) count may be indicative of an infectious or inflammatory process, the presence of red blood cells would suggest contamination of the sample, and elevated protein levels may also indicate inflammation. In cases of suspected meningitis, CSF samples should be submitted for bacterial culture and sensitivity. If another infectious process is suspected, CSF titers should be performed for the suspected or likely organism. Discussion of CSF collection techniques in the marmoset is provided in Chapter 10 (Wachtman, 2018).

Endocrine Hormones

Pituitary and gastrointestinal hormones have been measured in the plasma of marmosets using human reagents [55]. Please see Part III below for a discussion of measuring ovarian and testicular hormones in marmosets.

The pituitary hormones, adrenocorticotropic hormone (ACTH), prolactin, and thyroid-stimulating hormone (TSH), all show circadian variation in this species, whereas only ACTH and prolactin levels are affected by gender [55]. ACTH peaks about 2 h before the start of the active phase, at a level approximately twice that seen at the nadir during the morning (starting ~2 h after the onset of the active phase). This corresponds to a subsequent variation in cortisol levels in the plasma that occurs about 4 h after the changes in ACTH. ACTH is significantly higher in males, varying between 0.04 and 0.1 ng/mL, compared with the reported range in females, which varies from 0.01 to 0.025 ng/mL over the course of a day. In contrast, prolactin levels are significantly higher in females, as has been reported in humans and other mammals [55,56]. In paired but nonbreeding animals, prolactin is relatively constant throughout the day (approximately 2.5 ng/mL in females, 1−1.5 ng/mL in males), with a decrease just after the beginning of the resting phase in both genders to approximately 1 ng/mL in females and 0.75 ng/mL in males. TSH remains relatively constant throughout the circadian cycle (~0.04−0.05 μIU/mL) in both males and females except for a transitory greater than twofold decrease a few hours after the beginning of the active phase in the morning. This is similar to data from rhesus macaques (*Macaca mulatta*) which show no circadian variation in TSH, but in contrast to humans, where TSH is elevated at night [57,58].

Plasma levels of the gut hormones, ghrelin, gastric inhibitory polypeptide, pancreatic polypeptide, and peptide YY, have been reported, although there was no demonstrated effect of time of day or gender [55]. Reported average values are 350, 58, 310, and 507 pg/mL, respectively.

PART III: REPRODUCTIVE PARAMETERS

A complete understanding of marmoset reproductive physiology, normal morphological characteristics of sperm, and female reproductive tract structures is critical to the management of breeding colonies, execution of reproductive research models, and development of effective assisted reproductive technologies for any species. A discussion of sperm morphology and sperm count reference ranges will be provided in this chapter and a thorough discussion of reproductive physiology of the marmoset is provided in Chapter 8 (Tardif, 2018).

Female Reproductive Hormones

Because female marmosets do not menstruate or show other external physiologic indicators of the phase of the reproductive cycle, endocrine monitoring is the most reliable indicator for detecting ovulation and early pregnancy. Plasma progesterone is the gold standard for defining the stage of the estrus cycle, with a rise in progesterone above 10 ng/mL the day after ovulation corresponding to the onset of the luteal phase of the cycle [59,60]. High interindividual fluctuations of serum progesterone and estradiol have been reported, such that absolute values can be difficult to interpret [40]. In general, females that show no elevations of plasma progesterone above 10 ng/mL over the course of 1−2 cycle lengths are not cycling. Cycling females will exhibit sustained, regular progesterone elevations above 10 ng/mL for at least 11 days during the luteal phase and progesterone levels <10 ng/mL for 13 days or less during the follicular phase [61].

The measurement of estrogens in both urine and feces has been investigated [59]. These methods are particularly important for monitoring the female reproductive stage in free-ranging populations or in behavior studies, where the stress of capture and restraint for phlebotomy may confound the experiments. These methods can be useful for the management of breeding populations and family groups in the laboratory setting as well.

Ovarian cyclicity can also be measured in female marmosets by collection of small volumes of urine. Hodges and Eastman found a strong correlation between the hormone-to-creatinine ratio in the urine and plasma hormone levels for several estrogen metabolites [62]. In female marmosets, estrone is the most abundant form of estrogen present in the plasma, whereas 17β-estradiol predominates in the urine [50]. This is somewhat unusual among New World primates, as estrone is the primary urinary metabolite of estrogen in both cotton-top and saddle-back tamarins (*S. fuscicollis*). Another unusual aspect of urinary estrogen metabolism in marmosets is that they excrete mainly sulfate conjugates, whereas cotton-top tamarins, humans, and most Old World monkeys excrete mainly glucuronide conjugates.

It is important to note that while general ovarian cyclicity can be monitored with urinary estrogen metabolites, it is difficult to accurately predict or time ovulation in marmosets with this technique. Very little change in the level of urinary estrogens can be detected until about 4 days after ovulation [63]. Urinary estrogens can be used to support a pregnancy diagnosis after day 20 postovulation, as described in Nievergelt and Pryce [63]. Torii et al. have described changes in urinary and plasma prolactin and luteinizing hormone/chorionic gonadotropin levels over the course of late pregnancy, parturition, and the postpartum period [64].

While urinary hormone analysis remains the predominant method of noninvasive endocrine monitoring, fecal hormone analysis is an alternative technique for monitoring the status of reproductive hormones in marmosets. Heistermann et al. collected paired serum and fecal samples to validate that measurement of unconjugated progesterone, pregnanediol, and estradiol provides reliable information on both luteal and ovarian functions in this species [59]. The interval between the rise in plasma progesterone and fecal progestins averages 48 h. This technique cannot be used for short-term breeding decisions, as the information it provides regarding ovulation is retrospective, but it still provides useful information on effective management of this species.

Telemetric measurement of body temperature has not been shown to be a useful parameter to predict estrous cycling in female marmosets [65]. In a small study done by Gluckman et al., the karyopyknotic index (KPI) of vaginal cytologic samples increased markedly with an increase in progesterone levels and decreased throughout the follicular phase of the cycle, suggesting that vaginal cytology may be useful in reproductive management of common marmosets [66].

Male Reproductive Hormones

Though the benefits are perhaps less obvious than for females, measurement of reproductive hormones can be useful to monitor testicular function in male animals. Plasma testosterone in male marmosets ranges between 0.93 and 178.06 nmol/L, with a mean of 33.61 nmol/L [17]. Mohle et al. characterized testosterone metabolism in urine and feces collected from common marmosets [67]. Using radiolabeled testosterone, they determined that metabolites were excreted primarily into the urine. The predominant metabolite in both the urine and feces was androsterone. This finding contrasts with most other primates, where metabolites such as epiandrosterone and 5β-androstane-3α-ol-17-one/5α-androstan-3,17,dione are more common in the feces.

Sperm

Four semen collection methods have been reported in marmosets: (1) vaginal washing after copulation, (2) electroejaculation, (3) cauda epididymis collection, and (4) penile vibratory stimulation (PVS) [68]. Takabayashi et al. report average sperm concentration of semen collected from 4 + 1.9-year-old male marmosets as 33.4×10^6/mL ± 5.6, with an average motility of 66.1 + 3.5% after centrifugation and swim-up of sperm. A total of 113 attempts of PVS were performed in 27 males, and 94 ejaculate samples were obtained (success rate of semen sample collection was 83.2%).

A comparison of two distinct populations of marmosets revealed that range and variability in sperm concentration was similar across colonies and thus may provide a good reference for marmoset sperm evaluation [69]. PVS was used as the method of semen collection in this study, with a high rate of successful sample provision. In one population of males, samples from the heaviest animals had the highest proportion of sperm major defects. The mean volume of ejaculate of one colony with open genetics and the occasional inclusion of new animals from wild-caught populations was 26.05 ± 2.88 μL. The mean volume of ejaculate of the other colony, which maintained a genetically closed breeding scheme, was 27.05 ± 2.45 μL. The mean concentration of sperm for the open colony was 1096.99 ± 252.37×10^6 spermatozoa/mL and 1032.88 ± 205.56 spermatozoa/mL for the closed colony. Overall health status may influence semen concentration and quality. The range of sample qualities was similar between two independent colonies with different breeding schemes and may therefore be assumed to be typical of captive marmosets. Interestingly, these mean values are greater than the mean concentration of spermatozoa reported by Takabayashi et al. [61]. Both studies used PVS to collect semen samples from male marmosets; however, Takabayashi et al. evaluated only active sperm that were able to swim into a medium layer of supernatant after centrifugation of the diluted sample [68]. It is possible that the difference in semen sample processing protocols could account for the large difference in concentration of spermatozoa between studies.

Sperm morphology, an important parameter predictive of male fertility, has been studied in marmosets. Sperm head abnormalities vary among marmoset donors (from 4% to 22.2%), and the majority of sperm abnormalities present in marmoset semen samples include macrocephalic and microcephalic spermatozoa [69]. The typical marmoset ejaculate represents a heterogeneous population of spermatozoa. Irrespective of the individual marmoset studied, subpopulations of small, average, and large spermatozoa exist. However, depending on the donor, the number of subpopulations within the sample may vary. Semen analysis should be performed for each individual prior to use as a sperm donor to establish morphometric parameters for that individual that may impact fertility. Significant inter- and intramale differences exist in sperm morphometry. Marmosets reportedly have a relatively low level of teratospermy [69].

PART IV: OTHER CLINICAL PARAMETERS

Cortisol

Cortisol is the primary glucocorticoid in primates, and it is released in response to physical and psychological stressors. The measurement of cortisol is useful in laboratory primates to evaluate both acute and chronic stresses, with potential implications for animal welfare. Indeed, measurement of this hormone is often used to categorize procedures (husbandry, experimental, etc.) or experiences as causing higher or lower stress for the animals undergoing manipulation [32,70]. Cortisol in marmosets has been measured in the blood, urine, feces, and saliva. Cortisol in New World primates, including *Callithrix jacchus*, has been reported to be 3–15 times higher than cortisol levels in both Old World primates and humans [71,72]. As with other New World primates, this is due to low levels of cortisol binding globulin in circulation [12]. Previous research indicated that the glucocorticoid receptor in callitrichids has a fairly low affinity for cortisol, but more recent data attribute the relative resistance to the high basal levels of free cortisol in New World primates to high levels of an intracellular cytosolic cortisol binding protein, rather than due to a lack of affinity to the glucocorticoid receptor itself [73]. Regardless of absolute cortisol measurements, the degree of hypothalamic–pituitary–adrenal (HPA) axis activation in marmosets is similar to that in Old World primates, and plasma cortisol does exhibit the same relative circadian variation as seen in humans [12,55].

Bertani et al. measured the circadian variation of plasma cortisol in marmosets and reported that the highest values peaked just after lights turned on, at the beginning of the active phase [55]. These maximum values were approximately 6 times the level of cortisol present in the bloodstream at the lowest point of the day, 2 hours after "lights out". Gender differences were not reported. This cycle is analogous to that reported for corticosterone in rodents relative to each species' active phase (marmosets during the light phase, rodents during the dark phase).

In addition to circadian variation and stressors, social and reproductive status can affect cortisol measurements. The mean peak plasma cortisol, sampled first thing in the morning, in females housed in heterosexual pairs is 214 μg/dL [71]. Consistent with other reports in marmosets, cortisol levels were on average 63% higher in dominant females in the group than in the subordinates, as has been reported in several other species. Saltzman et al. found that subordinate female marmosets did not show significant changes in plasma cortisol following group formation including high levels of agonistic social interactions. This indicates that the difference in plasma cortisol levels in dominant and subordinate females is more likely due to differences in reproductive status within the group, rather than a direct consequence of social status per se. These elevations in plasma cortisol are likely mediated by estradiol. Thus, activation of the hypothalamic pituitary (HPA) axis in this species may not be highly sensitive to psychosocial perturbations related to the formation of new social relationships and subordinate social status.

As in other species, marmosets show suppression of plasma cortisol levels after administration of dexamethasone. Saltzman et al. report a 10-fold decrease in plasma cortisol 24 h after administration of 2 mg of dexamethasone sodium phosphate intramuscularly (mean 129.3 μg/dL at time 0, 13.3 μg/dL at 24 h) [74].

Similar to ovarian steroids, cortisol and its metabolites are excreted primarily in the urine in the marmoset [75]. After intravenous administration of cortisol, peak urinary excretion occurs within 5.5 h, whereas over 91% has been eliminated by 24 h. Bassett et al. found no difference in urinary cortisol levels before and after a mild stressor (removal from the home cage for weighing), but Smith and French describe significantly increased urinary cortisol measurements in Wied's black ear-tufted marmosets (*Callithrix kuhlii*) subjected to more severe stressors that included social isolation for 11 h and manual restraint [70,76]. This work indicates that the method of obtaining urine samples for cortisol measurement must be considered, as some methods may require separation from cage mates for some period of time and thus can confound the interpretation of cortisol levels. Changes in urinary and plasma cortisol levels, for both females and their partners, over the course of late pregnancy, parturition, and lactation, are summarized by Torii et al. [48].

After intravenous administration, cortisol itself is excreted in relatively small amounts in the feces of marmosets (18% of the administered dose) [75]. Unconjugated steroid metabolites (primarily 11-oxoetiocholanolon) represent the majority of fecal excretion of cortisol, which peaks within 26 h after intravenous radiolabeled cortisol administration in the marmoset. The peak fecal excretion time for cortisol metabolites documented by Bahr et al. is less than the approximately 48 h postovulatory rise in plasma progesterone reported for fecal excretion of ovarian steroids in marmosets [59].

There are several advantages to using saliva as the sample for measuring cortisol levels as a means to evaluate acute stress. Saliva samples can be obtained noninvasively, without restraint or sedation, as reported by

Cross et al. [77]. Cortisol diffuses passively into the saliva from the bloodstream, so it is unaffected by the flow rate or the volume of saliva [78]. Thus, salivary cortisol represents the real-time plasma level of cortisol and can be used to evaluate acute stress induced by rapid, short-term exposures. Cross et al. reported that baseline cortisol levels in 8 animals ranged from 464 to 561 nmol/L, a range similar to salivary cortisol values in squirrel monkeys (*Saimiri sciureus*) (500–600 nmol/L) and about 13-fold higher than the value reported in rhesus macaques (34 nmol/L) [77,79,80].

Salivary cortisol levels in marmosets mirror the diurnal variation reported in macaques, squirrel monkeys, and humans. In all of these species, including marmosets, salivary cortisol is highest first thing in the morning, gradually decreasing throughout the day to a nadir just before going to sleep [77]. This mirrors variation in plasma cortisol levels as discussed above. A large amount of individual variation in the magnitude of these circadian changes in salivary cortisol levels has been observed, and a delayed decrease occurs at mealtimes. In contrast, Sousa and Ziegler found higher cortisol levels in fecal samples collected in the afternoon versus the morning [81]. This may be attributable to the length of time between changes in cortisol levels in the bloodstream and their appearance in the feces.

PART V: IMMUNOLOGY AND DRUG METABOLISM

A Brief Overview of the Marmoset Immune System

All marmosets, with the exception of singletons, are born as hematopoietic chimeras due to shared placentation characterized by extensive vascular anastomoses and a fused chorion. Single-cell analysis techniques using fluorescent in situ hybridization (FISH) have been developed for sexing individual cells in the peripheral blood, bone marrow, and other tissues, which has confirmed that cross-sex chimerism in mixed-sex litters exists in both hematopoietic stem cells and peripheral blood leukocytes (see Chapters 14 and 19 for additional discussion on chimerism) [82]. Marmoset lymphocyte populations are described by Wright et al. [83]. Nelson and Loveday report that 63% of lymphocytes in the peripheral blood are T cells, 25% are B cells, and 12% are natural killer (NK) cells, which are comparable to the proportions found in human peripheral blood lymphocytes [84].

Marmoset blood and bone marrow progenitor cells (BMPCs) have also been characterized. CD117+ cells from neonatal cord blood or from bone marrow of adult animals are multipotent hematopoietic stem cells, possessing the ability to differentiate into myeloid lineage, as well as both B and T lymphocytes. Cultured marmoset BMPCs can be stimulated by select human cytokines [85], and marmoset BMPCs have been differentiated into adipocytes, osteocytes, and chondrocytes in vitro [86]. Interestingly, lymphoid maturation is affected by administration of a gonadotropin-releasing hormone (GnRH) antagonist in the neonatal period [87].

Minimal diversity at both major histocompatibility (MHC) Classes I and II loci has been documented in marmosets [88], with an accelerated turnover between generations. The *Callithrix* genus possesses its own unique set of MHC Class I loci, with no loci directly comparable to the human leukocyte antigens (HLA) HLA-A, -B, and -C in humans. HLA-G is the locus that marmoset MHC Class I most closely resembles [12]. Contrary to earlier beliefs but similar to data in rodents, MHC Class I is expressed by a subset of central nervous system neurons in the common marmoset brain [89]. The limited polymorphism of marmoset MHC Class II loci is in contrast to characteristics of MHC Class II loci in cotton-top tamarins. More recently, new alleles have been identified at the marmoset Class II DRB locus by Prasad et al. [90]. The limited variability of marmoset MHC is supportive of a rapid and recent evolution of the species and makes populations vulnerable to viral infections, given the uniform immune responses among individuals within the species to these pathogens [88].

Genes in the marmoset NK complex are collinear and orthologous to the human NK complex, though only a single activating NK receptor gene is encoded. This singularity may correspond to the previously reported low diversity of MHC Class I molecules in these animals. The full genomics and diversity of the complex have been reported [91]. Despite this apparent lack of genetic diversity, phenotypic profiles of marmoset NK cell subpopulations are like those seen in humans, with NKG2D functioning as an activating receptor for marmoset NK cells [92].

Marmosets are host to a gammaherpesvirus in the lymphocryptovirus subgroup, *Callitrichine herpesvirus 3*, which is analogous to the Epstein–Barr virus (EBV) in humans. A prevalence of 79% has been reported in captive colonies. This virus causes transformation of immortalized marmoset PBMCs in vitro, an effect similar to that caused by other gammaherpesviruses. In addition, marmosets experimentally infected with EBV variably develop lymphoproliferative disease similar to syndromes seen in humans [93]. See Chapters 15 (Matz-Rensing, 2018) and 28 (Patterson and Lanford, 2018), respectively, for a full discussion of natural and experimental disease associated with lymphocryptoviruses.

Immunology Research

Though a full discussion of research models is beyond the scope of this chapter, we will touch on select aspects of immunology research here. Please see Chapter 28 (Patterson and Lanford, 2018) for a discussion of marmosets as a model in infectious disease research. General trends in the innate immunological response and immune profile of marmosets after challenge with several acute bacterial pathogens have been reported [84]. Marmosets have also been proposed as a novel preclinical transplant model, where genotyping of select Class II MHC loci may be predictive of future alloreactivity [90].

About 60% of tested human antibodies for IHC are cross-reactive in the marmoset and can be used as reagents in this species [94]. In addition, novel monoclonal antibodies that recognize various subsets of marmoset lymphocytes have been developed [95]. For example, antimarmoset CD8+ mAb has been shown to efficiently deplete marmoset CD8+ T lymphocytes in vivo [96]. An overview of techniques and reagents developed for using marmosets in autoimmune research was published by Jagessar et al., in 2013 [97].

In recent years, marmoset models of experimental autoimmune encephalitis (EAE) have been extensively developed [98–101]. The close evolutionary link between marmosets and humans allows for more relevant models of autoimmune diseases of the nervous system than mouse models provide, and the outbred lineage of marmoset colonies enhances the translational potential of all findings. A more thorough discussion of marmoset EAE models is presented in Chapter 24 (Kap, 2018).

Characteristics of Drug Metabolism in the Common Marmoset

Characteristics of drug metabolism in the common marmoset are not fully understood, though recent work has focused on understanding properties of the cytochrome P450 enzyme system. Marmoset P450 enzymes have high sequence homology to human P450 enzymes, though minor species-specific differences may be significant [102]. Cytochrome P450 3A enzymes are involved in the metabolism of a wide variety of drugs. Marmoset P450 3A enzymes have been shown to have high amino acid sequence identity (88%–90%) to both human and *Macaca fascicularis* orthologues. P450 3A orthologue mRNA is abundant in the marmoset liver and small intestine [103]. Based on four novel P450 4F genes in the marmoset genome, recent work has isolated new P450 4F cDNA in marmoset hepatic tissue. Marmoset P450 4F3B and 4F11 mRNA are predominantly expressed in the liver, whereas marmoset P450 4F2 and 4F12 mRNA are highly expressed in both small intestine and liver [104]. Continued use of marmosets in toxicology research will further elucidate aspects of drug metabolism in this species.

References

[1] Orsi A, Rees D, Andreini I, Venturella S, Cinelli S, Oberto G. Overview of the marmoset as a model in nonclinical development of pharmaceutical products. Regul Toxicol Pharmacol 2011;59:19–27.

[2] Tardif SD, Power ML, Ross CN, Rutherford JN, Layne-Colon DG, Paulik MA. Characterization of obese phenotypes in a small nonhuman primate, the common marmoset (*Callithrix jacchus*). Obesity 2009;17:1499–505. PMC3823549.

[3] Power RA, Power ML, Layne DG, Jaquish CE, Oftedal OT, Tardif SD. Relations among measures of body composition, age, and sex in the common marmoset monkey (*Callithrix jacchus*). Comp Med 2001;51:218–23.

[4] Poole TB, Evans RG. Reproduction, infant survival and productivity of a colony of common marmosets (*Callithrix jacchus jacchus*). Lab Anim 1982;16:88–97.

[5] Araújo A, Arruda MF, Alencar AI, Albuquerque F, Nascimento MC, Yamamoto ME. Body weight of wild and captive common marmosets (*Callithrix jacchus*). Int J Primatol 2000;21: 317–24.

[6] Layne DG, Power RA. Husbandry, handling, and nutrition for marmosets. Comp Med 2003;53:351–9.

[7] Ross CN, Davis K, Dobek G, Tardif SD. Aging phenotypes of common marmosets (*Callithrix jacchus*). J Aging Res 2012;2012: 567143. PMC3312272.

[8] Tardif SD, Smucny DA, Abbott DH, Mansfield K, Schultz-Darken N, Yamamoto ME. Reproduction in captive common marmosets (*Callithrix jacchus*). Comp Med 2003;53:364–8.

[9] Ward JM, Buslov AM, Vallender EJ. Twinning and survivorship of captive common marmosets (*Callithrix jacchus*) and cotton-top tamarins (*Saguinus oedipus*). J Am Assoc Lab Anim Sci 2014;53:7–11. PMC3894641.

[10] Delimitreva S, Wedi E, Bakker J, Tkachenko OY, Nikolova V, Nayudu PL. Numerical chromosome disorders in the common marmoset (*Callithrix jacchus*) – comparison between two captive colonies. J Med Primatol 2013;42:177–85.

[11] Tardif SD, Mansfield KG, Ratnam R, Ross CN, Ziegler TE. The marmoset as a model of aging and age-related diseases. ILAR J 2011;52:54–65. PMC3775658.

[12] Abbott DH, Barnett DK, Colman RJ, Yamamoto ME, Schultz-Darken NJ. Aspects of common marmoset basic biology and life history important for biomedical research. Comp Med 2003;53:339–50.

[13] Smucny DA, Abbott DH, Mansfield KG, Schultz-Darken NJ, Yamamoto ME, Alencar AI, Tardif SD. Reproductive output, maternal age, and survivorship in captive common marmoset females (*Callithrix jacchus*). Am J Primatol 2004;64:107–21.

[14] Nishijima K, Saitoh R, Tanaka S, Ohsato-Suzuki M, Ohno T, Kitajima S. Life span of common marmoset (*Callithrix jacchus*) at CLEA Japan breeding colony. Biogerontology 2012;13:439–43.

[15] David JM, Dick Jr EJ, Hubbard GB. Spontaneous pathology of the common marmoset (*Callithrix jacchus*) and tamarins (*Saguinus oedipus, Saguinus mystax*). J Med Primatol 2009;38:347–59. PMC2740810.

[16] Ash H, Buchanan-Smith HM. Long-term data on reproductive output and longevity in captive female common marmosets (*Callithrix jacchus*). Am J Primatol 2014;76:1062–73.

[17] Korte S, Zuhlke U, Srivastav S. The common marmoset (*Callithrix jacchus*) as a model in toxicology: reference control data comparison with macaques. Baltimore, Maryland USA: abstr Society of Toxicology; 2004.

[18] Abbott DH, Keverne EB, Bercovitch FB, Shively CA, Mendoza SP, Saltzman W, Snowdon CT, Ziegler TE, Banjevic M, Garland Jr T, Sapolsky RM. Are subordinates always stressed? A comparative analysis of rank differences in cortisol levels among primates. Horm Behav 2003;43:67–82.

[19] Tardif SD, Jaquish CE, Toal RL, Layne DG, Power RA. Estimation of gestational ages in the common marmoset (*Callithrix jacchus*) from published prenatal growth curves. J Med Primatol 1998; 27:28–32.

[20] Smith D, Trennery P, Farningham D, Klapwijk J. The selection of marmoset monkeys (*Callithrix jacchus*) in pharmaceutical toxicology. Lab Anim 2001;35:117–30.

[21] Schnell CR, Wood JM. Measurement of blood pressure and heart rate by telemetry in conscious unrestrained marmosets. Lab Anim 1995;29:258–61.

[22] Michel JB, Mahouy G. [The marmoset in biomedical research. Value of this primate model for cardiovascular studies]. Pathol Biol 1990;38:197–204.

[23] Wood JM, Baum HP, Forgiarini P, Gulati N, Jobber RA, Neisius D, Hofbauer KG. Haemodynamic effects of acute and chronic renin inhibition in marmosets. J Hypertens Suppl 1989;7:S37–42.

[24] Wood JM, Baum HP, Jobber RA, Neisius D. Sustained reduction in blood pressure during chronic administration of a renin inhibitor to normotensive marmosets. J Cardiovasc Pharmacol 1987; 10(Suppl 7):S96–8.

[25] Gerber P, Schnell CR, Anzenberger G. Cardiovascular parameters telemetrically measured during pregnancy, parturition, and lactation in a common marmoset (*Callithrix jacchus*). Contemp Top Lab Anim Sci 2000;39:14–7.

[26] Hetherington CM. Circadian oscillations of body temperature in the marmoset, *Callithrix jacchus*. Lab Anim 1978;12:107–8.

[27] Rensing S, Oerke AK. Husbandry and management of new world species:marmosets and tamarins. In: Wolfe-Coote S, editor. The laboratory primate. Amsterdam: Elsevier; 2005. p. 145–62.

[28] Cilia J, Piper DC, Upton N, Hagan JJ. A comparison of rectal and subcutaneous body temperature measurement in the common marmoset. J Pharmacol Toxicol Methods 1998;40:21–6.

[29] Davy CW, Jackson MR, Walker JM. The effect of haemolysis on some clinical chemistry parameters in the marmoset (*Callithrix jacchus*). Lab Anim 1984;18:161–8.

[30] Yarbrough LW, Tollett JL, Montrey RD, Beattie RJ. Serum biochemical, hematological and body measurement data for common marmosets (*Callithrix jacchus jacchus*). Lab Anim Sci 1984;34:276–80.

[31] McNees DW, Lewis RW, Ponzio BJ, Sis RF, Stein FJ. Blood chemistry of the common marmoset (*Callithrix jacchus*) maintained in an indoor-outdoor environment: primate comparisons. Primates 1984;25:103–9.

[32] Kuehnel F, Grohmann J, Buchwald U, Koeller G, Teupser D, Einspanier A. Parameters of haematology, clinical chemistry and lipid metabolism in the common marmoset and alterations under stress conditions. J Med Primatol 2012;41:241–50.

[33] Ziegler TE, Colman RJ, Tardif SD, Sosa ME, Wegner FH, Wittwer DJ, Shrestha H. Development of metabolic function biomarkers in the common marmoset, *Callithrix jacchus*. Am J Primatol 2013;75:500–8. PMC3771328.

[34] Banton SA, Soltow QA, Liu KH, Uppal K, Promislow DE, Power ML, Tardif SD, Wachtman LM, Jones DP. Plasma metabolomics of common marmosets (*Callithrix jacchus*) to evaluate diet and feeding husbandry. J Am Assoc Lab Anim Sci 2016;55: 137–46. PMC4783630.

[35] Schnell MA, Hardy C, Hawley M, Propert KJ, Wilson JM. Effect of blood collection technique in mice on clinical pathology parameters. Hum Gene Ther 2002;13:155–61.

[36] Fernandez I, Pena A, Del Teso N, Perez V, Rodriguez-Cuesta J. Clinical biochemistry parameters in C57BL/6J mice after blood collection from the submandibular vein and retroorbital plexus. J Am Assoc Lab Anim Sci 2010;49:202–6. 2846009.

[37] Neptun DA, Smith CN, Irons RD. Effect of sampling site and collection method on variations in baseline clinical pathology parameters in Fischer-344 rats. 1. Clinical chemistry. Fundam Appl Toxicol 1985;5:1180–5.

[38] Pan X, Zeng X, Hong J, Yuan C, Cui L, Ma J, Chang Y, Hua X. Effects of ketamine on metabolomics of serum and urine in cynomolgus macaques (*Macaca fascicularis*). J Am Assoc Lab Anim Sci 2016;55:558–64. 5029826.

[39] Kim CY, Lee HS, Han SC, Heo JD, Kwon MS, Ha CS, Han SS. Hematological and serum biochemical values in cynomolgus monkeys anesthetized with ketamine hydrochloride. J Med Primatol 2005;34:96–100.

[40] Zuhlke U, Weinbauer G. The common marmoset (*Callithrix jacchus*) as a model in toxicology. Toxicol Pathol 2003;31(Suppl):123–7.

[41] Oerke AK, Einspanier A, Hodges JK. Detection of pregnancy and monitoring patterns of uterine and fetal growth in the marmoset monkey (*Callithrix jacchus*) by real-time ultrasonography. Am J Primatol 1995;36:1–13.

[42] Smith KM, McAloose D, Torregrossa AM, Raphael BL, Calle PP, Moore RP, James SB. Hematologic iron analyte values as an indicator of hepatic hemosiderosis in Callitrichidae. Am J Primatol 2008;70:629–33.

[43] Lowenstine LJ, Munson L. Iron storage in the animal kingdom. In: Fowler ME, Miller RE, editors. Zoo and wild animal medicine CVT IV. Philadelphia: WB Saunders; 1998. p. 260–8.

[44] Miller GF, Barnard DE, Woodward RA, Flynn BM, Bulte JW. Hepatic hemosiderosis in common marmosets, *Callithrix jacchus*: effect of diet on incidence and severity. Lab Anim Sci 1997;47: 138–42.

[45] McIntosh GH, Lawson CA, Rodgers SE, Lloyd JV. Haematological characteristics of the common marmoset (*Callithrix jacchus jacchus*). Res Vet Sci 1985;38:109–14.

[46] Lunn SF. Systems for collection of urine in the captive common marmoset, *Callithrix jacchus*. Lab Anim 1989;23:353–6.

[47] McKinley J, Buchanan-Smith HM, Bassett L, Morris K. Training common marmosets (*Callithrix jacchus*) to cooperate during routine laboratory procedures: ease of training and time investment. J Appl Anim Welf Sci 2003;6:209–20.

[48] Torii R, Moro M, Abbott DH, Nigi H. Urine collection in the common marmoset (*Callithrix jacchus*) and its applicability to endocrinological studies. Primates 1998;39:407–17.

[49] Anzenberger G, Gossweiler H. How to obtain individual urine samples from undisturbed marmoset families. Am J Primatol 1993;31:223–30.

[50] Shackleton CH. The excretion of steroids by the adult marmoset monkey (*Callithrix jacchus*). J Steroid Biochem 1975;6:1429–32.

[51] Bodi-Winn C. Personal Communication. 2017.

[52] Drake P. Species differences in urinary specific gravity of various nonhuman primates. Honors Proj. Grand Valley State University; 2015. https://scholarworks.gvsu.edu/honorsprojects/387/.

[53] Yamada N, Sato J, Kanno T, Wako Y, Tsuchitani M. Morphological study of progressive glomerulonephropathy in common marmosets (*Callithrix jacchus*). Toxicol Pathol 2013;41:1106–15.

[54] Schultz-Darken NJ. Sample collection and restraint techniques used for common marmosets (*Callithrix jacchus*). Comp Med 2003;53:360–3.

[55] Bertani S, Carboni L, Criado A, Michielin F, Mangiarini L, Vicentini E. Circadian profile of peripheral hormone levels in

Sprague-Dawley rats and in common marmosets (*Callithrix jacchus*). In Vivo 2010;24:827–36.
[56] Sassin JF, Frantz AG, Weitzman ED, Kapen S. Human prolactin: 24-hour pattern with increased release during sleep. Science 1972;177:1205–7.
[57] Persani L, Terzolo M, Asteria C, Orlandi F, Angeli A, Beck-Peccoz P. Circadian variations of thyrotropin bioactivity in normal subjects and patients with primary hypothyroidism. J Clin Endocrinol Metab 1995;80:2722–8.
[58] Giannella-Neto D, Quabbe HJ, Witt I. Pattern of thyrotropin and thyroxine plasma concentrations during the 24-hour sleep-wake cycle in the male rhesus monkey. Endocrinology 1981;109: 2144–51.
[59] Heistermann M, Tari S, Hodges JK. Measurement of faecal steroids for monitoring ovarian function in New World primates, Callitrichidae. J Reprod Fertil 1993;99:243–51.
[60] Harlow CR, Hearn JP, Hodges JK. Ovulation in the marmoset monkey: endocrinology, prediction and detection. J Endocrinol 1984;103:17–24.
[61] Saltzman W, Schultz-Darken NJ, Abbott DH. Familial influences on ovulatory function in common marmosets (*Callithrix jacchus*). Am J Primatol 1997;41:159–77.
[62] Hodges JK, Eastman SAK. Monitoring ovarian function in marmosets and tamarins by the measurement of urinary estrogen metabolites. Am J Primatol 1984;6:187–97.
[63] Nievergelt C, Pryce CR. Monitoring and controlling reproduction in captive common marmosets on the basis of urinary oestrogen metabolites. Lab Anim 1996;30:162–70.
[64] Torii R, Utsu S, Tanioka Y. Pituitary-adrenal function in the marmoset monkey (author's transl). Nihon Naibunpi Gakkai Zasshi 1981;57:1177–85.
[65] Drummer C, Konig J, Eildermann K, Fuchs E, Behr R, Hoffmman K. Core body temperature is not a reliable parameter to follow the reproductive cycle in female marmoset monkey (*Callithrix jacchus*). Primate Biol 2014;1:11–22.
[66] Gluckman TL, Walz SE, Schultz-Darken N, Bolton ID. Cytologic assessment of the vaginal epithelium in the common marmoset (*Callithrix jacchus*): a preliminary new approach to reproductive screening. Contemp Top Lab Anim Sci 2004;43:28–31.
[67] Mohle U, Heistermann M, Palme R, Hodges JK. Characterization of urinary and fecal metabolites of testosterone and their measurement for assessing gonadal endocrine function in male nonhuman primates. Gen Comp Endocrinol 2002;129: 135–45.
[68] Takabayashi S, Suzuki Y, Katoh H. Development of a modified artificial insemination technique combining penile vibration stimulation and the swim-up method in the common marmoset. Theriogenology 2015;83:1304–1309 e1302.
[69] Valle Rdel R, Valle CM, Nichi M, Muniz JA, Nayudu PL, Guimaraes MA. Semen characteristics of captive common marmoset (*Callithrix jacchus*): a comparison of a German with a Brazilian colony. J Med Primatol 2014;43:225–30.
[70] Bassett L, Buchanan-Smith HM, McKinley J, Smith TE. Effects of training on stress-related behavior of the common marmoset (*Callithrix jacchus*) in relation to coping with routine husbandry procedures. J Appl Anim Welf Sci 2003;6:221–33.
[71] Saltzman W, Schultz-Darken NJ, Scheffler G, Wegner FH, Abbott DH. Social and reproductive influences on plasma cortisol in female marmoset monkeys. Physiol Behav 1994;56: 801–10.
[72] Coe CL, Savage A, Bromley LJ. Phylogenetic influences on hormone levels across the primate order. Am J Primatol 1992;28: 81–100.
[73] Brandon DD, Markwick AJ, Chrousos GP, Loriaux DL. Glucocorticoid resistance in humans and nonhuman primates. Cancer Res 1989;49:2203s–13s.
[74] Saltzman W, Prudom SL, Schultz-Darken NJ, Abbott DH. Reduced adrenocortical responsiveness to adrenocorticotropic hormone (ACTH) in socially subordinate female marmoset monkeys. Psychoneuroendocrinology 2000;25:463–77.
[75] Bahr NI, Palme R, Mohle U, Hodges JK, Heistermann M. Comparative aspects of the metabolism and excretion of cortisol in three individual nonhuman primates. Gen Comp Endocrinol 2000;117:427–38.
[76] Smith TE, French JA. Psychosocial stress and urinary cortisol excretion in marmoset monkeys (*Callithrix kuhli*). Physiol Behav 1997;62:225–32.
[77] Cross N, Rogers LJ. Diurnal cycle in salivary cortisol levels in common marmosets. Dev Psychobiol 2004;45:134–9.
[78] Kirschbaum C, Hellhammer DH. Salivary cortisol in psychoneuroendocrine research: recent developments and applications. Psychoneuroendocrinology 1994;19:313–33.
[79] Boyce WT, Higley JD, Jemerin JJ, Champoux M, Suomi SJ. Tympanic temperature asymmetry and stress behavior in rhesus macaques and children. Arch Pediatr Adolesc Med 1996;150:518–23.
[80] Tiefenbacher S, Lee B, Meyer JS, Spealman RD. Noninvasive technique for the repeated sampling of salivary free cortisol in awake, unrestrained squirrel monkeys. Am J Primatol 2003;60: 69–75.
[81] Sousa MB, Ziegler TE. Diurnal variation on the excretion patterns of fecal steroids in common marmoset (*Callithrix jacchus*) females. Am J Primatol 1998;46:105–17.
[82] Wedi E, Muller S, Neusser M, Vogt PH, Tkachenko OY, Zimmer J, Smeets D, Michelmann HW, Nayudu PL. Detection of cross-sex chimerism in the common marmoset monkey (*Callithrix jacchus*) in interphase cells using fluorescence in situ hybridisation probes specific for the marmoset X and Y chromosomes. Reprod Fertil Dev 2016;29(5):913–20. https://doi.org/10.1071/RD15321.
[83] Wright J, Falk LA, Wolfe LG, Deinhardt FW. Lymphocyte populations of *Callithrix jacchus* marmosets. Cell Immunol 1978;35:148–57.
[84] Nelson M, Loveday M. Exploring the innate immunological response of an alternative nonhuman primate model of infectious disease; the common marmoset. J Immunol Res 2014; 2014:913632. PMC4129158.
[85] Hibino H, Tani K, Ikebuchi K, Wu MS, Sugiyama H, Nakazaki Y, Tanabe T, Takahashi S, Tojo A, Suzuki S, Tanioka Y, Sugimoto Y, Nakahata T, Asano S. The common marmoset as a target preclinical primate model for cytokine and gene therapy studies. Blood 1999;93:2839–48.
[86] Kanda A, Sotomaru Y, Nobukiyo A, Yamaoka E, Hiyama E. Characterization of common marmoset (*Callithrix jacchus*) bone marrow-derived mesenchymal stem cells. Folia Histochem Cytobiol 2013;51:292–9.
[87] Mann DR, Lunn SF, Akinbami MA, Samuel K, Waterfall M, Fraser HM. Effect of neonatal treatment with a GnRH antagonist on development of the cell-mediated immune response in marmosets. Am J Reprod Immunol 1999;42:175–86.
[88] Antunes SG, de Groot NG, Brok H, Doxiadis G, Menezes AA, Otting N, Bontrop RE. The common marmoset: a new world primate species with limited Mhc class II variability. Proc Natl Acad Sci U S A 1998;95:11745–50. PMC21711.
[89] Rolleke U, Flugge G, Plehm S, Schlumbohm C, Armstrong VW, Dressel R, Uchanska-Ziegler B, Ziegler A, Fuchs E, Czeh B, Walter L. Differential expression of major histocompatibility complex class I molecules in the brain of a New World monkey, the common marmoset (*Callithrix jacchus*). J Neuroimmunol 2006;176:39–50.
[90] Prasad S, Humphreys I, Kireta S, Gilchrist RB, Bardy P, Russ GR, Coates PT. The common marmoset as a novel preclinical transplant model: identification of new MHC class II DRB alleles and prediction of in vitro alloreactivity. Tissue Antigens 2007; 69(Suppl 1):72–5.

[91] Averdam A, Kuhl H, Sontag M, Becker T, Hughes AL, Reinhardt R, Walter L. Genomics and diversity of the common marmoset monkey NK complex. J Immunol 2007;178:7151–61.

[92] Watanabe M, Kudo Y, Kawano M, Nakayama M, Nakamura K, Kameda M, Ebara M, Sato T, Nakamura M, Omine K, Kametani Y, Suzuki R, Ogasawara K. NKG2D functions as an activating receptor on natural killer cells in the common marmoset (*Callithrix jacchus*). Int Immunol 2014;26:597–606.

[93] Jenson HB, Ench Y, Zhang Y, Gao SJ, Arrand JR, Mackett M. Characterization of an Epstein-Barr virus-related gammaherpesvirus from common marmoset (*Callithrix jacchus*). J Gen Virol 2002;83:1621–33.

[94] Kap YS, van Meurs M, van Driel N, Koopman G, Melief MJ, Brok HP, Laman JD, Hart BA. A monoclonal antibody selection for immunohistochemical examination of lymphoid tissues from non-human primates. J Histochem Cytochem 2009;57: 1159–67. 2778089.

[95] Ito R, Maekawa S, Kawai K, Suemizu H, Suzuki S, Ishii H, Tanioka Y, Satake M, Yagita H, Habu S, Ito M. Novel monoclonal antibodies recognizing different subsets of lymphocytes from the common marmoset (*Callithrix jacchus*). Immunol Lett 2008;121: 116–22.

[96] Yoshida T, Suzuki S, Iwasaki Y, Kaneko A, Saito A, Enomoto Y, Higashino A, Watanabe A, Suzuki J, Inoue K, Kuroda T, Takada M, Ito R, Ito M, Akari H. Efficient in vivo depletion of CD8(+) T lymphocytes in common marmosets by novel CD8 monoclonal antibody administration. Immunol Lett 2013;154:12–7.

[97] Jagessar SA, Vierboom M, Blezer EL, Bauer J, Hart BA, Kap YS. Overview of models, methods, and reagents developed for translational autoimmunity research in the common marmoset (*Callithrix jacchus*). Exp Anim 2013;62:159–71. PMC4160941.

[98] Jagessar SA, Kap YS, Heijmans N, van Driel N, van Straalen L, Bajramovic JJ, Brok HP, Blezer EL, Bauer J, Laman JD, t Hart BA. Induction of progressive demyelinating autoimmune encephalomyelitis in common marmoset monkeys using MOG34-56 peptide in incomplete freund adjuvant. J Neuropathol Exp Neurol 2010;69:372–85.

[99] Jagessar SA, Heijmans N, Blezer EL, Bauer J, Weissert R, Hart BA. Immune profile of an atypical EAE model in marmoset monkeys immunized with recombinant human myelin oligodendrocyte glycoprotein in incomplete Freund's adjuvant. J Neuroinflammation 2015;12:169. PMC4574133.

[100] Hart BA, Dunham J, Faber BW, Laman JD, van Horssen J, Bauer J, Kap YS. A B cell-driven autoimmune pathway leading to pathological hallmarks of progressive multiple sclerosis in the marmoset experimental autoimmune encephalomyelitis model. Front Immunol 2017;8:804. PMC5504154.

[101] Mesleh MF, Belmar N, Lu CW, Krishnan VV, Maxwell RS, Genain CP, Cosman M. Marmoset fine B cell and T cell epitope specificities mapped onto a homology model of the extracellular domain of human myelin oligodendrocyte glycoprotein. Neurobiol Dis 2002;9:160–72.

[102] Uno Y, Uehara S, Yamazaki H. Utility of non-human primates in drug development: comparison of non-human primate and human drug-metabolizing cytochrome P450 enzymes. Biochem Pharmacol 2016;121:1–7.

[103] Uehara S, Uno Y, Yuki Y, Inoue T, Sasaki E, Yamazaki H. A new marmoset P450 4F12 enzyme expressed in small intestines and livers efficiently metabolizes antihistaminic drug ebastine. Drug Metab Dispos 2016;44:833–41.

[104] Uehara S, Uno Y, Nakanishi K, Ishii S, Inoue T, Sasaki E, Yamazaki H. Marmoset cytochrome P450 3A4 expressed in liver and small intestine tissues efficiently metabolizes midazolam, alprazolam, nifedipine, and testosterone. Drug Metab Dispos 2017; 45(5):457–67. https://doi.org/10.1124/dmd.116.074898.

[105] Tardif SD, Carville A, Elmore D, Williams LE, Rice K. Reproduction and Breeding of nonuman primates. In: Abbe CR, Mansfield K, Tardif SD, Morris T, editors. Nonhuman Primates in Biomedical Research: Biology and Management, vol. 1. San Diego, CA: Academic Press; 2012. p. 197–247.

[106] Tardif SD, Bales KL. Relations among birth condition, maternal condition, and postnatal growth in captive common marmosets. Am J Primatol 2004;62:83–94.

[107] Saltzman W, Tardif SD, Rutherford JN. Hormones and reproductive cycles in primates. In: Norris DO, Lopez KH, editors. Hormones and Reproduction of Vertebrates, vol. 5. London: Elsevier; 2011. 291e328.

[108] Johnson DR, Levan G, Klein G, Nigida SM, Wolfe LG. Chromosomes and cell surface markers of marmoset lymphocytes and Epstein–Barr virus-transformed marmoset cell lines. Canc Gen & Cyto 1980;3:101–8.

[109] National Research Council. Chapter 2: Energy. Nutrient Requirements of Nonhuman Primates. Second Revised Edition. Washington, DC: The National Academies Press; 2003.

CHAPTER

7

Behavior and Behavioral Management

Nancy Schultz-Darken[1], *Lyndsey Ace*[1], *Hayley Ash*[1,2]

[1]Wisconsin National Primate Research Center, University of Wisconsin, Madison, WI, United States; [2]School of Natural Sciences, University of Stirling, Stirling, United Kingdom

INTRODUCTION

Common marmosets reside in family groups of 3–15 individuals in a variety of forest types in east and central Brazil [1–6] and are characterized by several unique features. Family groups consist of a breeding pair and their offspring. Groups are relatively stable with little immigration [6a], although there are some changes due to immigrations, emigrations, births, and disappearances [7]. Most groups contain only one dominant breeding pair with the dominant female suppressing ovulation in other female group members [8]. However, when a new dominant male joins the group, subordinate females can escape suppression [9]. The dominant male also can suppress sexual behavior in the group subordinate males [10]. However, in the wild, sexual behavior can occur with extragroup mating [11]. Multiple births (most frequently twins) are typical [12], and the intense care of the young involves cooperation of all members of the family [13]. Twinning in the Callitrichidae has been explained as an evolutionary adaptation, allowing small primates to give birth to large-brained offspring [14]. Communal infant rearing may have resulted from this reproductive biology and the high energetic demands of infant care [15], enabling them to successfully rear the twins.

These small-bodied arboreal monkeys are omnivorous, where a primary portion of the diet includes feeding on gum exudate. Because of this, they have a specialized narrow dental formula, claws on all digits except a specialized nail (hallux) on the first digit of each foot, which is the only opposable joint, and specialized scent glands in the chest and genital areas often used to scent mark areas gnawed for gum feeding [16–21].

BEHAVIOR IN CAPTIVITY

Species-Typical Behaviors

Marmosets in captivity will show the complete range of affiliative and agonistic behaviors, intra- and intergroup visual displays, vocalizations, foraging strategies (e.g., gnawing for tree exudate, food capture), and scent marking when housed in a supportive environment (see Chapter 5). These species-typical behaviors are important for maintaining normal marmoset behavior, breeding, and development [22–24]. Additionally, housing marmosets in harmonious social groups allows for species-specific behaviors and buffers the effects of stressful situations [24a].

Although affiliative behaviors such as grooming and huddling are seen in captive marmoset groups, there may also be some agonism observed within the group [24b], most often in the form of visual displays or vocalizations rather than fighting or physical conflict [25]. Captive intergroup interactions can also include visual displays and vocalizations [25] that would normally be present at territorial edges in larger natural areas. This behavior is encouraged when several captive groups are housed within the same room with adequate spacing between the groups. Captive environments may also encourage marmosets to exhibit foraging strategies and scent marking (see the Enrichment section).

Family Behavior

Many detailed observations of captive common marmoset families have found that the behavioral and physical developments of marmosets show a series of

The Common Marmoset in Captivity and Biomedical Research
https://doi.org/10.1016/B978-0-12-811829-0.00007-8

distinct changes from dependence to maturation, which are most evident in the infant phase (reviewed in Refs. [26,27]; see also Chapter 8 of this volume). Throughout infancy, common marmosets also develop characteristically different relationships with each group member [28], as their relative importance changes with age.

Newborns are carried almost continuously for the first 3 weeks of life, with babies returning to the mother for suckling. Caregivers are extremely tolerant of infants at this time, frequently accepting their attempts to climb on [27], and are highly responsive to infant vocal cues [27a]. Various family members take turns in carrying the infants, although some studies have reported the father to be the main caregiver in the first week [28]. In later infancy, caregivers begin to encourage the infants to get off, picking them up less frequently and increasing the amount of rejections. The father may become the main caregiver at this time, with studies finding that there were more successful attempts to climb on to him than the mother [27].

Although general patterns of caregiving are consistent among groups of marmosets, there can be specific family differences [13,27,29], including the amount of care group members are prepared to give and the amount of care that infants seek. Family members could provide little care, and so infants may continue to seek interactions and be highly rejected. However, in other families, caregivers may offer plentiful attention, and so one individual may provide the majority of care an infant requires and will compensate for lower amounts given by other group members.

After week 4, many independent behaviors begin to appear [26,27]. Infants are more mobile and leave their carriers increasingly more to explore and play socially, which is thought to allow young animals to develop complex patterns of adult behavior [30]. Weaning is a process that commences between 3 and 4 weeks of age and is completed between 10 and 12 weeks of age [31,32]. Solid food is often first tasted prior to this through begging or sharing [27,33]. Agonistic behaviors may appear at this stage, as well as scent marking [27], which is known to be an important aspect of communication in the marmoset [34–36].

By the end of the infant phase, there is a marked decrease in the dependence on caregivers [27]. This has also been found in field studies, in which young marmosets were seen to move long distances themselves by the end of the infancy [12]. However, the carrying period may be extended in the wild [12,37]. By 8 weeks, wild infants are still carried up to 50% of the time, whereas captive infants spend approximately 90% of their time off of carriers at this age [13]. Differences in carrying may be due to the larger daily ranges and risk of predation in the wild [37a]. Infants did however gain independence at similar rates. There were few differences in the first appearance of key behaviors between captive and wild common marmosets, except for social play (see Table 7.1).

TABLE 7.1 First Appearance of Developmental Markers in Captive (Reviewed in Ref. [27]) and Wild [88,89] Infant Common Marmosets

Behavior	Mean and Range in Captivity (days)	First day in Wild (days)
Off carriers	15.4 (11–20)	13 [88]
		25 [89]
Foraging	28.6 (25–34)	32 (Fruit) [89]
		38 (Animal prey) [88]
Solitary play/ exploration	19.6 (11–25)	25 [88]
Social play	31.6 (25–49)	61 [89]
Scent mark	62.8 (33–73)	61 [89]

At this time of reduced parental care, interactions with the twin increase significantly. As infant marmosets are adapted to be born as one of a twin pair [14], a principal aspect of their environment is the close presence of an exact age peer. They develop a close proximity during many types of activity, such as playing, exploring, resting, and foraging [38]. Having a twin allows the infants to develop socially together, perhaps being important in promoting security and independence [27]. Support is also found in older siblings [27]. Both captive and field studies have found that the age of siblings may lead to differential interactions. Juveniles rarely help with infant carrying (captive: [38]; wild: [37], with adult helpers having a more active role in infant care), whereas other references suggest that juvenile and young adults are also likely to carry the infants [38a]. Furthermore, the number of helpers in a family group has been associated with differences in infant care ([13,38a,39,40]: *S. oedipus*; [41]: *Callithrix* and *Leontopithecus*; [42]: *S. oedipus*) and survival [43].

MANAGEMENT OF SOCIAL HOUSING

Family Groups

Optimal housing in captivity is in large family groups of up to 8–10 individuals as space allows, giving the offspring the opportunity for good social development. Offspring do not need to be removed unless crowding or agonistic interactions require it, as long as the breeding pair remains intact, as reproduction by these adult offspring in an intact family group is highly unlikely. It is important that the offspring remain in the family group to have experience rearing infants and, if removed too early, they have a much lower likelihood of raising offspring themselves [44].

Management of family group size will be determined by the size of the family enclosure and entails selectively removing adult offspring to form pairs to establish new family groups. When there is loss of one of the breeding adults in a captive family group, management decisions regarding the remaining individuals and the group will depend on specific group characteristics and which adult breeder has been lost. Sex of the adult breeder, sexes of adult children, and ages of the youngest group members need to be taken into account.

Care must be taken if there are any adult same-sex offspring in the group, immediately after the loss and also before considering replacement of another adult breeding partner. The same-sex adult parent may attack the older sons or daughters to eject them from the group (personal observations). This could result in serious injury in captivity where there is no egress as there would be in natural settings [45–47]. In addition, introduction of a new adult male to a family group with older daughters can result in the eldest daughter initiating hormone cycling, becoming pregnant, and resulting infanticide of her infants [48]. A similar description of infanticide directed to an adult subordinate's offspring was noted by Digby [49] in a group of wild common marmosets. Therefore, the best option is to remove adult offspring before adding a new breeding male or female to the family to replace loss of one of the parents. Removed adult offspring can then be paired with an unrelated adult of opposite sex to form a new social group or breeding pair. It is best if the new pair is housed in a room away from their original families [50].

When there are nursing or newly weaned infants present in the family on loss of one of the parents, it is important to consider the makeup of the remaining family before replacing the adult breeder. When introducing a new adult breeder to a family, it is best done with care and includes separating the younger group members into the nest box to provide them time to observe and accept the introduction of the new stepparent to the remaining parent. This will avoid the usual family mobbing that may otherwise occur when the new stepparent is introduced to the remaining family members [50].

It has been known for many years that litter size has been increasing in captive breeding colonies [12], with triplets becoming more common [51] Fig. 7.1. This creates a problem, as marmoset families generally will not care for more than two infants at a time, and so larger litters usually do not survive without human intervention. Tardif et al. [39] found that marmoset mothers actually spent less time carrying and had higher levels of infant harassment when caring for triplet versus twin litters. The likelihood of survival is greatly increased if one infant is partially or completely

FIGURE 7.1 Male marmoset parent carrying triplets. *Photo by Jordana Lenon, Wisconsin National Primate Research Center.*

hand-reared [52], and so early separation from the family is commonly practiced in primate colonies [53]. However, such husbandry practices are often advocated without a sound scientific understanding of their welfare implications [23].

A survey of management practices (at the four main UK marmoset colonies) revealed that various methods of rearing have been developed. Table 7.2 provides full descriptions of each. Complete hand-rearing was carried out at one facility, if infants were abused or rejected, although this was rare. This facility routinely performed partial hand-rearing, in which one infant was removed from the family group for certain periods of the day. Another colony practiced rotational hand-rearing, involving one infant being removed per day on a rotational basis. A third colony carried out supplementary feeding, in which all three of the infants were removed. During this time, infants were kept in incubators, with a woolen pad to cling on to, and hand-fed by care staff at regular intervals. Infants at each facility were occasionally fostered, if an appropriate breeding female was available. The fourth colony routinely euthanized the smallest triplet, to prevent any suffering and avoid the need for hand-rearing, as well as to provide the remaining two infants with a better chance of survival.

TABLE 7.2 Description of Commonly Used Rearing Practices for Infants From Triplet Litters in the Laboratory (Personal Correspondence)

Rearing	Practice	Duration
Complete hand-rearing	One infant is removed from the family group and reared in an incubator, either alone or with peers	Permanent family absence, although the animal is later reintroduced to marmosets
Partial hand-rearing	One infant (usually the smallest or largest triplet) is temporarily removed from the family group for hand feeding	8 h during the day (usually from 8:00 to 16:00), for the first 6–8 weeks of life. After this time, the family remains undisturbed
Rotational hand-rearing	One of the three infants is temporarily removed from the family group per day, on a rotational basis, for hand feeding	8 h during the day (from 8:00 to 16:00), every 3 days, for the first 6–8 weeks of life. After this time, the family remains undisturbed
Supplementary feeding	All three of the infants are temporarily removed from the family group together for hand feeding	2 h twice a day (8:00–10:00 and 16:00–18:00), for the first 6–8 weeks of life. After this time, the family remains undisturbed
Fostering	The infant is introduced to another family, if a suitable dam is available (i.e., has given birth less than a week before and no longer has her own young or has only one infant of a similar age)	The infant remains with the new family group

In marmosets, infants are in full, continuous body contact with their carriers throughout the first several weeks of life [27], and so interruption of this contact for even brief periods is likely to be unnatural. Since Harlow's [54] early experiments of maternal deprivation, more recent studies have looked at the effect of experimentally induced parental separation, e.g., [55–57], finding altered stress physiology/HPA axis activity, impaired cognition, increased fear, and depression-like symptoms in primates that have been separated from their family early in life.

Ash and Buchanan-Smith [58] therefore studied the effects of supplementary feeding using a variety of measures, including tests of cognitive bias and anhedonia. Few differences in affective state were found in those given supplementary feeding, compared with family-reared twins and family-reared infants from triplet litters where only two remain, suggesting that this particular practice to reduce mortality in large litters did not have a substantial impact on welfare and so is recommended if hand-rearing is necessary. The lack of major differences may be because the infants were not isolated during their time away from family, being taken out for short periods with litter mates and reintegrated back into the family as soon as possible. The marmosets were also fed at predictable times, building positive experiences with humans from an early age, and are naturally adapted to being passed between carriers, e.g., [13], which could mean they are less stressed during separation from the mother than other primate species.

The increases in triplet births are in part due to captive marmosets now being significantly heavier than their wild counterparts (marmosets: [58a,59]; tamarins: [60]), leading to higher numbers of ovulations [61]. Maintaining females at lower healthy weights, more similar to their wild counterparts, could therefore be an important management practice, helping to increase the incidence of twin births.

Although there were few differences found between family-reared and supplementary-fed marmoset triplets, separation from the family is not recommended, if it is at all possible to keep the infants with the natal group [3]. Training carriers to allow staff to provide supplementary feeding to the infants while they remain on their backs could therefore be a possible refinement [23,24], mitigating stress to the family and infants. Carriers could be encouraged to take food from a syringe, so they remain in the same position, while the infants are fed. Another practice may be to foster infants to well-experienced surrogate parents [52]. As parents have been found to respond equally to calls from their own and unfamiliar dependent infants [62], fostering is likely to be a successful method, if an appropriate family is available.

Pair Management for Research Subjects

Although captive marmosets are best maintained in captivity in their social family groups, there are problems when the marmosets proliferate beyond the capacity of the laboratory space or when research studies require smaller groups for controlled experiments [63]. Housing marmosets in pairs of the opposite sex works quite well for research studies, but again there needs to be some form of pregnancy control. Two viable options are available for pregnancy control. One option is by administration of a PGF-2α analogue to female marmosets late in the ovulatory cycle to cause luteal regression, either ending the luteal phase or causing abortion of the pregnancy [64] (see Chapter 8 of this volume for more discussion of this topic). The other option is to vasectomize the male marmoset partner to prevent subsequent pregnancies of the dominant pair [65]. Both are frequently used, depending on the colony needs and specific research emphasis with the marmoset pairs.

The procedure for pairing animals usually entails introducing the individuals into either an unfamiliar cage or in a cage previously inhabited by one or both of them (joining two cages together). It is also helpful to pair them within the same room they were previously housed. Pairs are checked often each day in the first week of pairing. A pair is considered compatible if no bidirectional aggression or injuries are observed [65a,66].

Housing with same-sex pairs can be cautiously approached, but these pairs work best when they are related to each other, but always will need close monitoring for any possible aggression between them [67]. Same-sex pairing can be utilized to avoid single housing for a sexually mature adult with no opposite-sex partner available or when research requires paired same-sex subjects. Housing with same-sex pairs can be cautiously approached, but these pairs work best when they are related to each other, but always will need close monitoring for any possible aggression between them [66,67].

Single housing of marmosets is discouraged. And compared to other nonhuman primate species, even infectious disease studies will house the animals in nonreproductive pairs [68–70]. However, there may be certain experimental needs (i.e., metabolism, neuroscience, etc.) that may require separate housing for varying lengths of time [71–73]. In these cases, it is extremely important to allow for visual and auditory contact if possible. The ability to communicate visually either through facial or body displays, olfactory cues, and vocalizations for social contact will alleviate some of the stress of no tactile contact. Although single housing is never the best default housing, marmosets display a much more limited degree of abnormal behavior in this circumstance than do macaques, and self-injurious behavior is extremely rare. It is particularly important to maintain contact with the native social group if the individually housed marmoset will be returned to the group after the brief experimental separation.

ENRICHMENT FOR SPECIES-TYPICAL BEHAVIOR

Although housing marmosets in family groups is the ideal way to encourage species-typical social behavior, there are physical factors that must be included to encourage the full array of species-typical behaviors. Buchanan-Smith [23,24] provides a good summary on marmoset enrichment in captivity.

Wooden Substrates

A wooden branch or branches, or perches, platforms, and other climbing structures provide not only the best surface for resting or locomotion but also the opportunity for gnawing and scent marking [74,75]. Wooden substrates need to be cleaned regularly, however, because chemical communication is important, it is best to infrequently (greater than 2 weeks, if possible) clean or replace wood to retain species scents. In addition, when marmosets are placed into cages with wood surfaces marked by other groups, there are no untoward effects on health, behavior, or reproduction; and the previously marked surfaces may also offer the opportunity for species-typical territorial behavior [50]. Size and orientation of these branches or dowels will also allow a complex three-dimensional environment for normal social interactions [76].

Foraging Opportunities

Adult marmosets in the wild spend a large portion of active hours foraging and visiting gouge holes to feed on tree exudate [77,77a]. They also are known to catch various small prey with a range of strategies [78]. Provision of gum may be important in the marmoset diet [79] and should be considered as a foraging option, even when wooden substrates are provided for gouging. Live food provided as a dietary supplement can also be appropriate environmental stimuli to motivate them to exhibit natural foraging and feeding behaviors [80]. A mealworm-foraging device described by Vignes et al. [81] not only provided dietary enrichment but also led to decreased rates of pacing and sitting, with more overall appropriate use of the other available cage furniture. Such foraging devices provide stimuli-encouraging exploration and manipulation while reducing stereotypical scratching [82].

Climbing and Jumping Opportunities

Marmosets will spend a large portion of active hours leaping and climbing when they are provided a complex three-dimensional environment in captivity. These enclosures should provide enough horizontal space for animals to leap between branches or other furnishings and should offer as much height as possible to exercise a full range of natural activities suited to their arboreal tendencies [76]. The Joint Working Group on Refinement

[3] summarizes that enclosure designs need to provide suitable areas for climbing, resting, vertical clinging, and movement, hidden areas for refuge, as well as for quadrupedal walking and running. These are also important when groups have visual contact to allow for intergroup interactions such as places to "hide" from other groups and horizontal substrates to allow for intergroup visual display, e.g., bristle strut, genital display, etc. [25].

ENRICHMENT—HUMAN INTERACTIONS

Human interactions are a given aspect of captivity, and emphasis should only be of a positive nature. Positive human interactions with marmosets can serve as a buffer to reduce stress reactions by increasing grooming and play while reducing scratching behavior [83]. Waitt et al. [84] describe that nonhuman primates who had friendly relationships with caretakers exhibited lower levels of stress-related behavior during times of high caretaker activity. Therefore, developing a positive relationship with caretakers may help mediate stress-related responses when animals are forced into contact with human personnel during daily animal care routines. Positive reinforcement training (PRT) is a useful method to encourage tractability during husbandry and experimental procedures. Training monkeys to cooperate voluntarily using PRT techniques is one means of significantly reducing the adverse impact of such procedures and husbandry routines [85,86,86a]. The ease and investment of training marmosets can simplify procedures such as weighing, urine collection, elimination of aggressive feeding behaviors, etc., and once training was complete, data collection for trained animals was considerably faster than that collected using standard procedures [87].

CONCLUSIONS

In general, marmoset management should allow for species-typical behaviors. In particular, housing in social family groups allows for behavioral development similar to that observed in natural groups in the wild. Management of triplet births can be decided according to practices of each facility. Nutritional supplementation of triplets is one option but is not required. Adult offspring can be paired with the opposite sex after growing up with the opportunity to participate in the rearing of younger siblings. Short-term, brief single housing is tolerated when certain measurements require this but is discouraged as a default housing condition. In addition to social management, we suggest the provision of physical additions to housing such as wooden substrates, multiple levels for locomotion, and unique foraging opportunities to promote territorial behavior, visual displays, arboreal locomotion, and foraging for food. Finally, the human interactions with marmosets should include PRT of marmosets to various management procedures and provide the least invasive methods for marmoset behavioral management.

References

[1] Digby LJ, Barreto CE. Social organization in a wild population of *Callithrix jacchus*. Folia Primatol 1993;61(3):123–34.

[2] Hubrecht RC. Field observations on group size and composition of the common marmoset (*Callithrix jacchus jacchus*), at Tapacura, Brazil. Primates 1984;25(1):13–21. https://doi.org/10.1007/BF02382291.

[3] Joint Working Group on Refinement (JWRG). Refinements in husbandry, care and common procedures for non-human primates. In: Jennings M, Prescott MJ, editors. Ninth report of the BVAAWF/FRAME/RSPCA/UFAW joint working group on refinement. Laboratory animals, vol. 43; 2009. p. 1–47 (S1).

[4] Pontes ARM, da Cruz MAOM. Home range, intergroup transfers, and reproductive status of common marmosets, *Callithrix jacchus*, in a forest fragment in North-Eastern Brazil. Primates 1995;36(3):335–47.

[5] Rowe N. The pictorial guide to the living primates. New York: Pogonias Press; 1996.

[6] Scanlon CE, Chalmers NR, da Cruz MM. Home range use and the exploitation of gum in the marmoset (*Callithrix jacchus jacchus*). Int J Primatol 1989;10(2):123–36.

[6a] Ferrari SF, Lopes Ferrari M. A re-evaluation of the social organisation of the Callitrichidae, with reference to the ecological differences between genera. Folia Pimatologica 1989;52(3-4):132–380.

[7] Arruda MF, Araujo A, Sousa MBC, Albuquerque FS, Albuquerque ACSR, Yamamoto ME. Two breeding females within free-living groups may not always indicate polygyny: alternative subordinate female strategies in common marmosets (*Callithrix jacchus*). Folia Primatol 2005;76(1):10–20.

[8] Abbott DH. Behavioral and physiological suppression of fertility in subordinate marmoset monkeys. Am J Primatol 1984;6: 169–86. https://doi.org/10.1002/ajp.1350060305.

[9] Saltzman W, Schultz-Darken NJ, Severin JM, Abbott DH. Escape from social suppression of sexual behavior and of ovulation in female common marmosets. Ann N Y Acad Sci 1997;807:567–70. https://doi.org/10.1111/j.1749-6632.1997.tb51970.x.

[10] Baker JV, Abbott DH, Saltzman W. Social determinants of reproductive failure in male common marmosets housed with their natal family. Anim Behav 1999;58(3):501–13.

[11] Digby LJ. Sexual behavior and extragroup copulations in a wild population of common marmosets (*Callithrix jacchus*). Folia Primatol 1999;70(3):136–45.

[12] Stevenson MF, Rylands AB. The marmosets, genus *Callithrix*. Ecol Behav Neotrop Primates 1988;2:131–222.

[13] Ingram JC. Interactions between parents and infants, and the development of independence in the common marmoset (*Callithrix jacchus*). Anim Behav 1977;25:811–27.

[14] Leutenegger W. Maternal-fetal weight relationships in primates. Folia Primatol 1973;20:280–93.

[15] Tardif SD. Relative energetic cost of infant care in small-bodied neotropical primates and its relation to infant-care patterns. Am J Primatol 1994;34:133–43. https://doi.org/10.1002/ajp.1350340205.

[16] Garber PA. Vertical clinging, small body size, and the evolution of feeding adaptations in the *Callitrichinae*. Am J Phys Anthropol 1992;88:469–82.

[17] Natori M, Shigehara N. Interspecific differences in lower dentition among Eastern-Brazilian marmosets. J Mammal 1992;73(3): 668–71.

[18] Canton JM, Hill DM, Hume JD, Crook GA. The digestive strategy of the common marmoset, *Callthrix jacchus*. Comp Biochem Physiol 1996;144:1–8.

[19] Castro CSS, Araujo A. Diet and feeding behavior of marmoset, *Callthrix jacchus*. Rev Bras Ecol 2007;7:14–8.

[20] Magden ER, Mansfield KG, Simmons JH, Abee CR. Nonhuman primates. In: Fox JG, Anderson LC, Otto G, Pritchett-Corning KR, Whary MT, editors. Laboratory animal medicine. 3rd ed. Amsterdam: Elsevier Science; 2015. p. 771–930.

[21] Taylor AB, Vinyard CJ. Comparative analysis of masseter fiber architecture in tree-gouging (*Callithrix jacchus*) and non gouging (*Saguinus oedipus*) callitrichids. J Morphol 2004;261:276–85.

[22] Stevenson MF, Poole TB. An ethogram of the common marmoset (*Calithrix jacchus jacchus*): general behavioural repertoire. Anim Behav 1976;24(2):428–51.

[23] Buchanan-Smith HM. Marmosets and tamarins. In: Hubrecht RC, Kirkwood J, editors. The UFAW handbook on the care and management of laboratory and other research animals. 8th ed. Hoboken, NJ: Wiley Blackwell; 2010a. p. 543–63.

[24] Buchanan-Smith HM. Environmental enrichment for primates in laboratories. Adv Sci Res 2010b;5:41–56. https://doi.org/10.5194/asr-5-41-2010.

[24a] Schaffner CM, Smith TE. Familiarity may buffer the adverse effects of relocation on marmosets (*Callithrix kuhli*): Preliminary evidence. Zoo Biology 2005;24(1):93–100. https://doi.org/10.1002/zoo.20019.

[24b] Sutcliffe AG, Poole TB. Intragroup agonistic behavior in captive groups of the common marmoset *Callithrix jacchus jacchus*. International Journal of Primatology 1984;5(5):473–89.

[25] de Boer RA, Overduin-de Vries AM, Louwerse AL, Sterck EHM. The behavioral context of visual displays in common marmosets (*Callithrix jacchus*). Am J Primatol 2013;75:1084–95. https://doi.org/10.1002/ajp.22167.

[26] Schultz-Darken N, Braun KM, Emborg ME. Neurobehavioral development of common marmoset monkeys. Dev Psychobiol 2016;58(2):141–58. https://doi.org/10.1002/dev.21360.

[27] Yamamoto ME. From dependence to sexual maturity: the behavioral ontogeny of *Callithrichidae*. In: Rylands A, editor. Marmosets and tamarins: systematics, behavior and ecology. Oxford: Oxford University Press; 1993. p. 235–54.

[27a] Sánchez SM, Ziegler TE, Snowdon CT. Both parents respond equally to infant cues in the cooperatively breeding common marmoset, *Callithrix jacchus*. Animal Behaviour 2014;97(1): 95–103.

[28] Locke-Haydon J, Chalmers NR. The development of infant-caregiver relationships in captive common marmosets (*Callithrix jacchus*). Int J Primatol 1983;4(1):63–81.

[29] Arruda MF, Yamamoto ME, Bueno OF. Interactions between parents and infants, and infants-father separation in the common marmoset (*Callithrix jacchus*). Primates 1986;27(2):215–28.

[30] Box HO. Quantitative studies of behaviour within captive groups of marmoset monkeys (*Callithrix jacchus*). Primates 1975a;16(2):155–74. https://doi.org/10.1007/BF02381414.

[31] Ross CN, Power ML, Artavia JM, Tardif SD. Relation of food intake behaviors and obesity development in young common marmoset monkeys. Obesity 2013;21(9):1891–9.

[32] Tardif S, Jaquish C, Layne D, Bales K, Power M, Power R, Oftedal O. Growth variation in common marmoset monkeys (*Callithrix jacchus*) fed a purified diet: relation to care-giving and weaning behaviors. Comp Med 1998;48(3):264–9.

[33] Brown GR, Almond REA, van Bergen Y. Begging, stealing, and offering: food transfers in nonhuman primates. Adv Stud Behav 2004;34:265–95.

[34] Epple G. Quantitative studies on scent marking in the marmoset (*Callithrix jacchus*). Folia Primatol 1970;13:48–62.

[35] Epple G, Belcher AM, Greenfield KL, Scolnick L, Smith AB, Kuderling I, Zeller U. Making sense out of scents: species differences in scent glands, scent-marking behavior, and scent-mark composition in the *Callitrichidae*. Int J Primatol 1987;8(5):434.

[36] Lazaro-Perea C, Snowdon CT, de Fátima Arruda M. Scent-marking behavior in wild groups of common marmosets (*Callithrix jacchus*). Behav Ecol Sociobiol 1999;46(5):313–24.

[37] Yamamoto ME, Box HO, Albuquerque FS, de Fátima Arruda M. Carrying behaviour in captive and wild marmosets (*Callithrix jacchus*): a comparison between two colonies and a field site. Primates 1996;37(3):297–304.

[37a] Yamamoto ME, Box HO, Albuquerque FS, Arruda MF. Carrying behaviorin captive and wild marmosets (*Callthrix jacchus*): A comparison between two colonies and a field site. Primates 1996;37(3):297–304.

[38] Box HO. A social developmental study of young monkeys (*Callithrix jacchus*) within a captive family group. Primates 1975b; 16(4):419–35. https://doi.org/10.1007/BF02382740.

[38a] Fite JE, Patera KJ, French JA, Rukstalis M, Hopkins EC, Ross CN. Opportunistic mothers: female marmosets (*Callithrix khuli*) reduce their investment in offspring when they have to, and when they can. Journal of Human Evolution 2005;49(1):122–42.

[39] Tardif SD, Layne DG, Smucny DA. Can marmoset mothers count to three? Effect of litter size on mother–infant interactions. Ethology 2002;108(9):825–36.

[40] Cleveland J, Snowdon CT. Social development during the first twenty weeks in the cotton-top tamarin (*Saguinus o. oedipus*). Anim Behav 1984;32(2):432–44.

[41] Santos CV, French JA, Otta E. Infant carrying behavior in callitrichid primates: Callithrix and Leontopithecus. Int J Primatol 1997; 18(6):889–907.

[42] McGrew WC. Parental division of infant caretaking varies with family composition in cotton-top tamarins. Anim Behav 1988; 36(1):285–6.

[43] Jaquish CE, Tardif SD, Toal RL, Carson RL. Patterns of prenatal survival in the common marmoset (*Callithrix jacchus*). J Med Primatol 1996;25(1):57–63.

[44] Tardif SD, Richter CB, Carson RL. Effects of sibling-rearing experience on future reproductive success in two species of *Callitrichidae*. Am J Primatol 1984;6:377–80. https://doi.org/10.1002/ajp.1350060408.

[45] Lazaro-Perea C, Castro CS, Harrison R, Araujo A, Arruda MF, Snowdon CT. Behavioral and demographic changes following the loss of the breeding female in cooperatively breeding marmosets. Behav Ecol Sociobiol 2000;48(2):137–46. https://doi.org/10.1007/s002650000215.

[46] Sousa MBC, Albuquerque ACSDR, Albuquerque FDS, Araujo A, Yamamoto ME, Arruda MDF. Behavioral strategies and hormonal profiles of dominant and subordinate common marmoset (*Callithrix jacchus*) females in wild monogamous groups. Am J Primatol 2005;67(1):37–50. https://doi.org/10.1002/ajp.20168.

[47] Yamamoto ME, Araujo A, Sousa MBC, Arruda MF. Social organization in *Callithrix jacchus*: cooperation and competition. Adv Stud Behav 2010;42:259–73. https://doi.org/10.1016/S0065-3454(10)42008-2.

[48] Saltzman W, Pick RR, Salper OJ, Liedl KJ, Abbott DH. Onset of plural cooperative breeding in common marmoset families following replacement of the breeding male. Anim Behav 2004; 68(1):59–73. https://doi.org/10.1016/j.anbehav.2003.07.020.

[49] Digby L. Infant care, infanticide, and female reproductive strategies in polygynous groups of common marmosets (*Callithrix*

jacchus). Behav Ecol Sociobiol 1995;37(1):51–61. https://doi.org/10.1007/BF00173899.

[50] Tardif S, Bales K, Williams L, Ludlage Moeller E, Abbott D, Schultz-Darken N, Mendoza S, Mason W, Bourgeois S, Ruiz J. Preparing new world monkeys for laboratory research. ILAR J 2006;47(4):307–15.

[51] Ash H, Buchanan-Smith HM. Long-term data on reproductive output and longevity in captive female common marmosets (*Callithrix jacchus*). Am J Primatol 2014;76(11):1062–73.

[52] Hearn JP, Burden FJ. Collaborative rearing of marmoset triplets. Lab Anim 1979;13(2):131–3.

[53] Poole TB, Evans RG. Reproduction, infant survival and productivity of a colony of common marmosets (*Callithrix jacchus jacchus*). Lab Anim 1982;16(1):88–97.

[54] Harlow HF. The development of learning in the rhesus monkey. Am Sci 1959;47:459–79.

[55] Dettling AC, Feldon J, Pryce CR. Repeated parental deprivation in the infant common marmoset (*Callithrix jacchus*) primates and analysis of its effects on early development. Biol Psychiatr 2002; 52(11):1037–46.

[56] Dettling AC, Schnell CR, Maier C, Feldon J, Pryce CR. Behavioral and physiological effects of an infant-neglect manipulation in a bi-parental, twinning primate: impact is dependent on familial factors. Psychoneuroendocrinology 2007;32(4):331–49.

[57] Pryce CR, Dettling AC, Spengler M, Schnell CR, Feldon J. Deprivation of parenting disrupts development of homeostatic and reward systems in marmoset monkey offspring. Biol Psychiatr 2004;56(2):72–9.

[58] Ash H, Buchanan-Smith HM. The long-term impact of infant rearing background on the affective state of adult common marmosets (*Callithrix jacchus*). Appl Anim Behav Sci 2016;174: 128–36.

[58a] Tardif SD, Bales KL. Relations among birth condition, maternal condition, and postnatal growth in captive common marmoset monkeys (*Callithrix jacchus*). American Journal of Primatology 2004;62(2):83–94.

[59] Araújo A, Arruda MF, Alencar AI, Albuquerque F, Nascimento MC, Yamamoto ME. Body weight of wild and captive common marmosets (*Callithrix jacchus*). Int J Primatol 2000;21(2):317–24.

[60] Kirkwood JK. Effects of diet on health, weight and litter-size in captive cotton-top tamarins (*Saguinus oedipus oedipus*). Primates 1983;24(4):515–20.

[61] Tardif SD, Jaquish CE. Number of ovulations in the marmoset monkey (*Callithrix jacchus*): relation to body weight, age and repeatability. Am J Primatol 1997;42(4):323–9.

[62] Sánchez SM, Ziegler TE, Snowdon CT. Both parents respond equally to infant cues in the cooperatively breeding common marmoset, *Callithrix jacchus*. Anim Behav 2014;97:95–103.

[63] Sainsbury AW. The humane control of captive marmoset and tamarin populations. Anim Welf 1997;6(3):231–42.

[64] Summers PM, Wennink CJ, Hodges JK. Cloprostenol-induced luteolysis in the marmoset monkey (*Callithrix jacchus*). J Reprod Fertil 1985;73(1):133–8.

[65] Morris TH, David CL. Illustrated guide to surgical technique for vasectomy of the common marmoset. Lab Anim 1993;27:381–4.

[65a] Woodcock AJ. The first weeks of cohabitation of newly-formed heterosecual paris of common marmosets (*Callithrix jacchus*). Folia Primatologica 1982;37(1):228–54.

[66] Majolo B, Buchanan-Smith HM, Morris K. Factors affecting the successful pairing of unfamiliar common marmoset (*Callithrix jacchus*) females: preliminary results. Anim Welf 2003a;12(3): 327–38.

[67] Eckert K. Same-sex pairing of marmosets: a discussion. Lab Primate Newsl 2000;39(2):14–5.

[68] Moncla LH, Ross TM, Dinis JM, Weinfurter JT, Mortimer TD, Schultz-Darken N, Brunner K, Capuano 3rd SV, Boettcher C, Post J, Johnson M, Bloom CE, Weiler AM, Friedrich TC. A novel nonhuman primate model for influenza transmission. PLoS One 2013;8(11):e78750. https://doi.org/10.1371/journal.pone.0078750.

[69] Nelson M, Loveday M. Exploring the innate immunological response of an alternative nonhuman primate model of infectious disease; the common marmoset. J Immunol Res 2014; 2014, 913632. https://doi.org/10.1155/2014/913632.

[70] Smither SJ, Nelson M, Eastaugh L, Nunez A, Salguero FJ, Lever MS. Experimental respiratory infection of marmosets (-*Callithrix jacchus*) with Ebola virus Kikwit. JID (J Infect Dis) 2015;12(Suppl 2):S336–45. https://doi.org/10.1093/infdis/jiv371.

[71] Schultz-Darken NJ, Pape RM, Tannenbaum PL, Saltzman W, Abbott DH. Novel restraint system for neuroendocrine studies of socially living common marmoset monkeys. Lab Anim 2004; 38:393–405.

[72] Tardif SD, Power ML, Ross CN, Rutherford JN, Layne-Colon DG, Paulik MA. Characterization of obese phenotypes in a small nonhuman primate, the common marmoset (*Callithrix jacchus*). Obesity 2009;17(8):1499–505. https://doi.org/10.1038/oby.2009.77.

[73] Walker J, MacLean J, Hatsopoulos NG. The marmoset as a model system for studying voluntary motor control. Dev Neurobiol 2016;77(3):273–85. https://doi.org/10.1002/dneu.22461.

[74] Hearn JP. The common marmoset (*Callithrix jacchus*). In: Hearn JP, editor. Reproduction in new world primates. London, UK: Springer; 1983. p. 183–215.

[75] Layne DG, Power RA. Husbandry, handling, and nutrition for marmosets. Comp Med 2003;53(4):351–9.

[76] Sodaro V. Housing callitrichids and callimico in zoological parks. In: Sodaro V, Saunders N, editors. Callithichid husbandry manual. Chicago, IL: Neotropical Primate Taxon Advisory Group; 1999. p. 1–6.

[77] Francisco TM, Couto DR, Zanuncio J, Serrao JE, Silva IO, Boere V. Exudates as food for *Callithrix* spp. (*Callitrichidae*): exploratory patterns. PLoS One 2014;9:e112321. https://doi.org/10.1371/journal.pone.0112321.

[77a] Abreu F, Castellon De la Fuente MF, Schiel N, Souto A. Feeding ecology and behavioral adjustments: flexibility of a small neotropical primate (*Callithrix jacchus*) to survive in a semiarid environment. Mammal Research 2016;61(3):221–9.

[78] Schiel N, Souto A, Huber L, Bezerra BM. Hunting strategies in wild common marmosets are prey and age dependent. Am J Primatol 2010;72:1039–46. https://doi.org/10.1002/ajp.20860.

[79] Francisco TM, Lopes-Mattos KLB, Picoli EAT, Couto DR, Oliveria JA, Zanuncio JC, Serrao JE, Silva IO, Boere V. Feeding habits of marmosets: a case study of bark anatomy and chemical composition of *Anadenanthera peregrina* gum. Am J Primatol 2016; 79:e22615. https://doi.org/10.1002/ajp.22615.

[80] Lakshminarasimha R, Singh VR. Environmental enrichment for common marmosets (*Callithrix jacchus*) at Mysore Zoo. Zoos' Print 2009;24(10):11–4.

[81] Vignes S, Newman JD, Roberts RL. Mealworm feeders as environmental enrichment for common marmosets. Contemp Top 2001;40(3):26–9.

[82] Majolo B, Buchanan-Smith HM, Bell J. Response to novel objects and foraging tasks by common marmoset (*Callithrix jacchus*) female pairs. Lab Anim 2003b;32(3):32.

[83] Manciocco A, Chiarotti F, Vitale A. Effects of positive interaction with caretakers on the behaviour of socially housed common marmosets (*Callithrix jacchus*). Appl Anim Behav Sci 2009;120: 100–7.

[84] Waitt C, Buchanan-Smith HM, Morris K. The effects of caretaker-primate relationships on primates in the laboratory. J Appl Anim Welf Sci 2002;5(4):303–19. https://doi.org/10.12.7/S15327604JAWS0504-05.

[85] Prescott MJ, Buchanan-Smith HM. Training nonhuman primates using positive reinforcement techniques. J Appl Anim Welf Sci 2003;6(3):157–61. https://doi.org/10.1207/S15327604JAWS0603_01.

[86] Westlund K. Training laboratory primates – benefits and techniques. Primate Biol 2015;2:119–32. https://doi.org/10.5194/pb-2-119-2015.

[86a] Bassett L, Buchanan-Smith HM, McKinley J, Smith TE. Effects of training on stress-related behavior of the common marmoset (*Callithrix jacchus*) in relation to coping with routine husbandry procedures. J Appl Anim Welf Sci 2003;6(3):221–33. https://doi.org/10.1207/S15327604JAWS0603_07.

[87] McKinley J, Buchanan-Smith HM, Bassett L, Morris K. Training common marmosets (*Callithrix jacchus*) to cooperate during routine laboratory procedures: ease of training and time investment. J Appl Anim Welf Sci 2003;6(3):209–20.

[88] Alonso C. Observações de campo sobre o cuidado à prole e o desenvolvimento dos filhotes de Callithrix jacchus. In: de Mello MT, editor. A primatologia no Brasil. I. Brasília, Sociedade Brasileira de Primatologia; 1984. p. 67–78.

[89] Stevenson MF. The behaviour and ecology of the common marmoset (Callithrix jacchus jacchus) in its natural environment. In: Rothe H, Wolters HJ, Hearn JP, editors. Biology and Behaviour of Marmosets. Göttingen: Eigenverlag H. Rothe; 1978. p. 298.

CHAPTER

8

Reproduction, Growth, and Development

Suzette D. Tardif[1], *Corinna N. Ross*[2]

[1]Southwest National Primate Research Center, Texas Biomedical Research Institute, San Antonio, TX, United States; [2]Texas A&M University San Antonio, Department of Science and Mathematics, San Antonio, TX, United States

INTRODUCTION

Marmosets, along with the other callitrichine primates, display a suite of unique reproductive and developmental traits. Understanding these features is critical to management of a healthy and productive marmoset colony and to the development of transgenic, neuroscience, metabolic, and reproduction models.

This chapter provides an overview of reproductive physiology, development, and growth in common marmosets. There is an emphasis on information that will be of likely importance to individuals responsible for the management of breeding populations of this species. In addition, the information presented is relevant to the development of marmosets as models in studies involving reproduction and obesity.

REPRODUCTION

Female Reproductive Physiology

Marmosets typically display a 28-day ovarian cycle. The nature of that cycle differs from that of Old World monkeys and humans in consisting of a relatively short follicular phase (around 8–9 days) and a relatively long luteal phase (around 19–20 days) [1]. Marmosets do not exhibit any external signs of ovulation and while copulations are more frequent during the periovulatory period, they are not limited to that period. Marmosets do not menstruate. Hormonal assays are the most reliable means of determining when ovulation has occurred, but there is at least one report on use of ultrasonographic change in ovarian appearance to estimate day of ovulation [2]. Blood, urine, or fecal samples may be used for hormonal assessments. The most commonly used method to determine phase of the ovarian cycle or pregnancy is arguably concentration of urinary estrone conjugates (e.g., Ref. [3]) or urinary pregnanediol-3-glucuronide (e.g., Ref. [4]) concentrations. Fecal concentrations of steroids have also been successfully used to determine ovulation date [5]. Heisterman [6] provides an excellent overview of the applied aspects of using urinary, fecal, and saliva samples for monitoring reproductive function in nonhuman primates, including marmosets. Measurement of serum progesterone via ELISA is commonly used for genetic engineering projects to determine timing of luteolysis and ovulation [7]. Gluckman et al. [8] describe the use of a karyopyknotic index as adjunct to other means of monitoring stage of ovarian cycle.

A notable difference in endocrine function in marmosets (as well as squirrel monkeys and owl monkeys) when compared with Old World monkeys and humans is the pituitary synthesis and secretion of chorionic gonadotropin (CG), rather than luteinizing hormone (LH). CG secretion appears to function in the same fashion as LH in terms of controlling gonadal functions [9].

As opposed to other anthropoid primates, luteal regression that will either end the luteal phase or will cause abortion of pregnancies can be induced by injection of a single dose of a prostaglandin F-2α (PGF-2α) analogue. Summers et al. [10] detail procedures for use of cloprostenol (Estrumate) to reliably generate luteolysis at 10–17 days postovulation, with normal ovarian cycles and ovulation following induced luteolysis [7]. Some marmoset colonies manage timing of reproduction through this method [11,12]. Nievergelt and Pryce [12] report successful induction of luteolysis up to 64 days after conception, with a failure rate of 13%. Medroxyprogesterone acetate (DMPA) is also reported as a reliable contraceptive, with single injections given every 30 days [11].

The Common Marmoset in Captivity and Biomedical Research
https://doi.org/10.1016/B978-0-12-811829-0.00008-X

While female marmosets undergo the initiation of puberty at 10–11 months of age, they often do not advance to normal, regular ovarian cycles if they are housed in the presence of their mother or another dominant female. This reproductive suppression has been the subject of extensive study over the past 30 years, but the mechanisms behind it are still somewhat unclear. This impaired ovulatory function, which can be maintained for up to several years, is associated with suppressed pituitary secretion of LH/CG, likely associated with enhanced negative-feedback sensitivity to low levels of estrogen or blunted responsiveness to increasing estrogen [13]. Older literature often proposes that female marmosets who are in the presence of their mother are always reproductively suppressed; however, more recent studies reveal that some daughters escape suppression while in the presence of their mothers and do ovulate, though they often display impaired luteal function raising question as to the fertility of these females [14]. However, even if daughters ovulate, as long as the group remains intact, with only the daughter's father and brothers as potential breeding partners, these females only rarely become pregnant (see cooperative infant care, below), resulting in a social structure in which only one female generally reproduces regardless of the number of adult females in the group [13]. Suppression effects may continue after females are removed from their natal group. Studies of cotton-top tamarins, a related species, indicate that continued exposure to the scent marks of mothers will delay onset of ovulation on removal of a daughter from her natal group [15]. In addition, ovarian cycling and ovulation may be affected by cues that individually housed marmosets receive from other females housed in the same room but not in the same cage. Females may change from an acyclic pattern to regular ovarian cyclicity when being moved from a room with a high density of females to a room with a lower female density [16]; therefore, care should be taken in room arrangements for studies in which female reproductive cycling is an important variable.

The callitrichines, including marmosets, are the only anthropoid primates that routinely ovulate more than one ova per cycle. Marmosets are sometimes referred to in the literature as "obligate twinners," but in fact the ovulation number is variable in captive populations [17]. There are no published data on ovulation number in wild marmosets, but observed litter sizes suggest that double ovulations are likely typical in wild marmosets and tamarins. It was long assumed that ovulation of more than two ova per cycle was an artifact of captivity, but at least three studies provide convincing evidence of the presence of triplets in wild callitrichine populations, including marmosets [18–20]. The actual number of triple ovulations in the wild may be even higher given the common occurrence of litter size reduction in utero in marmosets (see Pregnancy). In captivity, ovulations of 1–6 ova per cycle have been reported, with the typical ovulation number being 2–3. Location of ovulation sites between and within cycles is stochastic [21]. Ovulation number is sensitive to maternal condition in this species. Higher ovulation numbers are associated with larger female weight and, within a given female, higher ovulation numbers are associated with weight gain [22,23].

Marmoset females remain fertile until late age. There are age effects on litter size and interbirth interval [24] but the effects are small. Tamarins, species closely related to marmosets, display impaired ovarian cycles that are reflected in acyclical concentrations of urinary estrone conjugates and circulating progesterone. As opposed to humans, late life anovulation is not associated with decreasing circulating steroid concentrations, most likely due to the continued presence of steroidogenic interstitial glands in the ovaries of old tamarin and marmoset females. A thorough review of reproductive aging in marmosets and tamarins is provided by Tardif et al.[25].

Male Reproductive Physiology

Males initiate puberty at around 8–9 months of age. Testosterone concentrations are elevated in infant males from days 15–100, but then decline and remain low until around 8 months of age at which time both testicular volume and testosterone concentration increases, with testosterone concentrations similar to adult males (30–50-ng/mL) [26]. Ejaculation of motile spermatozoa is reported to occur at 11–13 months of age [27]. Abbott and Hearn [26] report the earliest age of successful insemination producing a pregnancy in a 14-month-old male. While males remaining in an intact natal group are unlikely to reproduce, the suppression is behavioral in nature, as opposed to the physiological suppression seen in females.

Tardif et al. [28] provide a review of nonhuman primate male spermatogenesis, sperm maturation, and epididymal function that includes common marmosets. For the most part, these functions are similar among nonhuman primate species. Weinbauer et al. [29] provide a description of spermatogenesis in the common marmoset. Marmosets are similar to apes and humans in displaying a multistage tubular cross section. Marmosets display an acrosome maturation profile similar to humans, with increasing abilities to undergo capacitation and acrosome response when moving from the caput to the caudal epididymis.

Various semen collection methods are reported for marmosets [7]. Morrell et al. [30] compared the efficacy of semen collected by vaginal washes versus electroejaculation, finding that vaginal wash samples were both more reliably produced and more effective in artificial insemination than were samples obtained via electroejaculation. Subsequent to this study, Kuderling et al. [31] reported successful semen collection in unsedated marmosets by penile vibrostimulation, using a modified FertiCare personal vibrator designed for semen collection from humans. Of 10 males who were stimulated 6–12 times, 9 produced ejaculate in at least one session after an average of 49.7 s of stimulation, with the overall success rate per sessions being 35.2%. The total number and concentration of motile sperm was 3–4 times higher than that produced through rectal probe electroejaculation. The first studies successfully producing transgenic marmosets used this semen collection method [32].

Mating and Pregnancy

For breeding purposes, marmosets are typically housed as a mated pair. Selections of animals for mating are generally based on overall health, balanced with the need to maintain an outbred population that maintains as much as possible of the genetic variation represented in the founder pool. Use of genetic management software is essential for proper management as colonies are maintained over long periods of time and become more related. Breeding pairs can be maintained indefinitely, with minimal decline in reproductive performance; however, decisions may be made to retire breeders before the end of their reproductive life if they become overrepresented in subsequent generations. Remating is generally a straightforward process but may require removal of older offspring from the group. Details of cooperative breeding group management are provided in Ref. [33].

There are limited data available on average times from mating to conception or delivery of first offspring, but the available data are relatively consistent, with the average duration from pairing to first delivery of 206 days (median = 168.5, range 143–441) for one US colony and 204 days (range = 143–528) for one UK colony [17]. While males and females may be fertile as early as 13–14 months of age, first successful reproduction is likely to be later and many colony managers recommend delaying pairing until at least 18–24 months of age [17]. Copulation will occur throughout the ovarian cycle and pregnancy, though there are some data to support that copulations are more frequent during the periovulatory period.

Following fertilization, implantation occurs at around day 10–12 [1]. A bidiscoid placenta begins to form following implantation. A thorough description of marmoset placental development is provided by Merker et al. [34]. The multiple embryos are attached to the two discs in a stochastic fashion. There is a rapid expansion of the blastocysts to fill the uterine lumen resulting in a fused chorion, such that all embryos are ultimately enclosed in a single amniotic sac. There is an extensive network of anastomosed vessels between the two discs, such that all litter mates share a single, functional placenta.

The placenta contains hematopoietic foci beginning around day 60. Perhaps the most unusual feature of marmosets stems from this unique embryonic environment—that feature being the fact that litter mates are hematopoietic chimeras. Some studies have suggested that marmosets are chimeric in nonhematopoietic tissues (e.g., Ref. [35]); however, a recent study by Sweeney et al. [36] strongly suggests that chimerism observed in nonhematopoietic tissues is the result of contamination with hematopoietic cells. The hematopoietic chimerism results in immune tolerance among litter mates, a feature that can be used in studies requiring adoptive transfer (e.g., Ref. [37]). The chimerism also creates logistical issues in any genomic studies that are based on blood samples, as the genome of those samples represents an unpredictable percentage of alleles from the individual's litter mates.

Entire or partial pregnancy failure is relatively common in marmosets. Table 8.1 provides data on complete pregnancy loss, as assessed by urinary steroid concentrations, for 212 marmosets monitored over 596 implantation events [38]. Assessment of loss via repeated ultrasonography in a different colony found a similar loss rate, with 22% of pregnancies assessed after ~ day 30 aborting prior to term [39].

A number of factors have been proposed to be associated with such pregnancy loss. Environmental stressors, including construction activities within animal facilities and movement of animals within facilities, have been proposed to increase pregnancy loss. Controlled, modest energy restriction during midpregnancy has predictable effects, with abortion within 11–47 days, whereas the effects of restrictions in late pregnancy are

TABLE 8.1 Total Reproductive Loss in the Marmoset (Values Refer to Percentage of 596 Implantations)

	Percentage at End of Period:	
Pregnancy Stage	**Losses**	**Persistent**
Implantation		100.0
Days 10–29	17.3	82.7
Days 30–39	10.9	71.8
Days 40–49	7.9	63.9
Days 50–79	11.5	52.4
Days 80–142	4.4	48.0

*From Heger W, Merker H-J, Neubert D. Frequency of prenatal loss in marmoset monkeys (*Callithrix jacchus*). In: Neubert D, Merker H-J, Hendrickx A, editors. Nonhuman primates — developmental biology and toxicology. Berlin: Ueberreuter Wissenschaft; 1988. pp. 129–140.*

more variable, with 3/7 females delivering nonviable, preterm infants [40]. These findings suggest that pregnant marmoset females may be particularly sensitive to energetic stressors.

Death of one or more fetuses in a pregnancy that proceeds to live births, with no complications for the dam, is also relatively common. Both laparoscopic data on follicular development and ultrasonographic assessments of pregnancy indicate that singletons births are virtually always the result of twin conceptions in which one conceptus dies in utero and that triplet conceptions resulting in twin births is also relatively common [39,41]. Fig. 8.1 illustrates a macerated fetus found associated with a placenta from a delivery of a live offspring.

A term pregnancy is 143 days long (mean +/ s.d. = 142.4 ± 3.5; median = 143; [38]). Deliveries typically occur overnight. Females found laboring in the morning should be carefully observed for signs of distress or exhaustion. Dystocia is rare in this species. Unpublished observations of normal overnight labor of around 50 deliveries indicate that signs of visible labor (increased activity, visible contraction of the uterus, standing with a raised-tail posture) are usually seen within 2 h of lights going out (2000 h). The time from crowning of the first infant to completed delivery of all infants is generally 20–50 min, depending on the litter size. The placenta is generally expelled at some point around the same time as the last infant or up to an hour later. Labors that are prolonged (greater than 4 h) and include visual and audible signs of contraction with no associated vulvar parting or crowning are indicative of problems. The response of the male to the delivery is highly variable, with some males remaining inactive in the nest box throughout the process while others actively retrieve infants from the female immediately following delivery. The infant's behavior immediately following delivery is critical to its survival. Normal, healthy infants immediately begin to climb upward on the mother and often are rooting within minutes. Live infants that are found the subsequent morning but are not completely cleaned and being transported in a normal posture are likely signs of a problematic delivery or inadequate postdelivery behavior by either mother or infants.

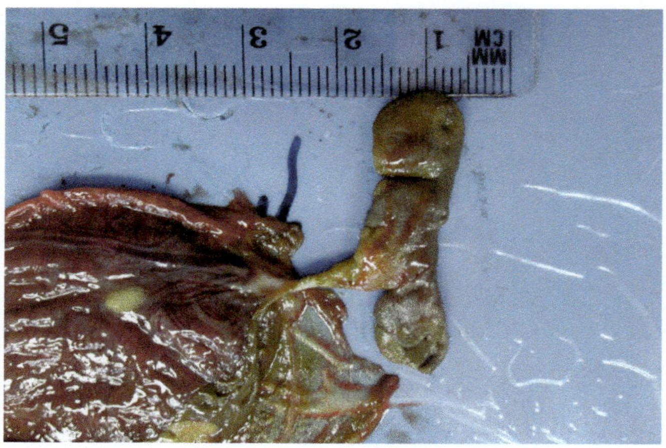

FIGURE 8.1 Macerated fetus delivered attached to the placenta. A normal, term infant was also born from this delivery.

Following delivery, marmosets typically lactate for around 75 days; however, weaning to solid food begins around day 30. Marmoset milk is typical of the milk of other primates in being relatively dilute, with around 14% dry matter and 0.76 kcal/gram of gross energy [42]. While marmoset milk is higher in protein than the milk of other anthropoid primates (2.7% vs. 1.7%; [42]), standard milk substitutes for humans (e.g., Similac) or nonhuman primates (e.g., Primilac) can be successfully used when nursery or supplemental rearing is required.

It is common for marmosets to deliver more than two infants. Triplets are, in many colonies, the most common litter size and quadruplets are not uncommon. However, with two nipples and a cooperative infant care system adapted for the care of two infants, it is extremely rare for marmosets to successfully raise more than two infants at a time (see Ref. [33]). The management of litters larger than two is dependent on the situation at the time of the birth and is ultimately the decision of the experienced colony management staff. Chapter 7 describes the options available and how they have been applied in four marmoset colonies in the United Kingdom. The policy at one US center (SNPRC) follows these basic procedures:

1. Assessment of Infants' Condition. Infants who are small (<24 g) or show signs of significant weakness (cannot cling to an adult without repeatedly falling off) are extremely unlikely to be successfully reared, so these infants may be removed and euthanized shortly following birth to minimize suffering and perhaps increase the chances of successful rearing of the remaining litter.
2. Fostering. If all infants are judged to be in good condition and there is another dam in the colony that is nursing one infant and has given birth within 3 weeks, one of the supernumerary infants may be removed and placed with the group with one infant. This process generally involves placing the infant in the nest box and allowing the foster dam and sire access to it. This process is successful in the majority of cases.
3. Hand-Rearing. Partial or full hand-rearing of supernumerary infants can be successful. Those management decisions are based on production needs balanced with the acknowledgment that these

infants may be phenotypically different from fully marmoset-reared animals, particularly in terms of the impact of early life nutrition. Hand-rearing takes one of two forms:

a. Rotational Rearing: In rotational rearing, one infant per day may be removed from the group and formula fed. In these cases, the infant may be placed in a cage or incubator immediately adjacent to its home cage. The infant is returned to the group at the end of the day, and a different infant removed on the following day. In this way, each infant has access to marmoset milk for at least part of its rearing. This option requires routine removal and replacement of infants from a social group and, therefore, requires staff that are well-trained in handling animals in such situations and should be discouraged if the marmoset group is highly reactive to this process. In our experience, marmosets are quite variable in response, with some animals remaining highly agitated and reactive when infants are removed while other animals are relatively relaxed and will eventually allow staff to remove an infant directly from its back in exchange for a reward.

b. Full Nursery Rearing. For full nursery rearing, an infant may be removed from the group and formula fed. In these cases, the infant may also be placed in a cage or incubator immediately adjacent to its home cage or adjacent to the cage of the group into which it will be fostered on weaning. The infant is generally returned to a social group by 3 months of age.

5. Observation and Euthanasia. If, for management reasons, it is deemed inappropriate to hand-rear any of the infants, then the group should be closely observed and weakened infants removed and humanely euthanized to reduce suffering. An infant found alive but on the floor in the cage should be immediately removed and its condition assessed. If the infant is of acceptable weight and strong enough to cling, it can be warmed then placed into the nest box. The group should be observed 1−2 h later to see if the infant has been retrieved and is clinging to a carrier. Experience has taught us that frequent observations of the group at shorter intervals may actually interfere with the parents' response to the infant, so it is advisable to leave the group alone for a time sufficient to let the group settle and retrieve the infant. If the infant is not retrieved or if the infant is retrieved but is again found on the floor of the cage (likely indicating insufficient strength), then the infant should be immediately removed and euthanized.

Lactation does not prevent or delay ovulation in marmosets, as it does in most anthropoid primates. Typically, marmosets will ovulate 9−10 days following delivery. Production of more than one offspring per delivery, combined with a lack of lactational anestrus, makes marmosets the most fertile anthropoid primates that are commonly used for research. Analyses from a large, multicolony database found average production to be 2.3 young per female per year, contrasting with average production of around 0.44 young per female per year for macaques [17].

Cooperative Infant Care

It is essential that the female's mate remain with her during pregnancy and particularly following delivery. Marmosets practice cooperative infant care, in which the sire and any older offspring who are in the cage play important roles in caring for infants, including physically carrying them and provisioning them with solid food once weaning begins. Typically, the dam will be carrying the infant around 30% of the time during the day. For further information on this topic, see Schultz-Darken [33].

DEVELOPMENT

Prenatal Development and Fetal Growth

Both placental and embryonic developments are notably slower in marmosets than in other primates, including humans. Fig. 8.2 illustrates relative pace of embryonic development in marmosets and humans [34]. This difference should be considered in any studies that involve modeling effects in a given human pregnancy trimester, as the matching phase of development in marmosets will occur at a later relative gestational age.

Gestational age is most reliably determined by identifying the day of ovulation by hormonal assessments. However, a reliable (±4 days) estimate of gestational age can be achieved through ultrasonography or palpated uterine diameter (see also Ref. [43]). Figs. 8.3 and 8.4 provide growth curves for crown-rump length and biparietal diameter [45]. These curves are routinely used by the laboratory in which they were produced to estimate gestational age and term delivery dates for both management and study purposes. The most reliable estimates are provided from crown-rump lengths of 4−12 mm [46,47]. Fig. 8.5 provides growth curves for palpated uterine diameter [1], another method that may be used to estimate gestational age.

Table 8.2 provides normative data for birth weight and birth size. Birth weight has historically been related

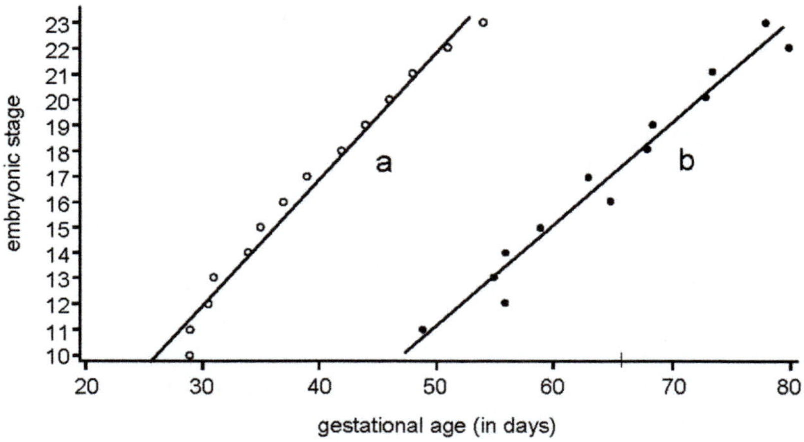

FIGURE 8.2 Comparison of Carnegie embryonic stage × gestational age for marmosets and humans, illustrating the slow pace of marmoset embryonic development in comparison with humans. *Based upon data from Merker H-J, Bremer D, Csato W, Heger W, Gossrau R. Development of the marmoset placenta. In: Neubert D, Merker H-J, Hendrickx A, editors. Nonhuman primates - developmental biology and toxicology. Berlin: Ueberreuter Wissenschaft; 1984. pp. 245–272.*

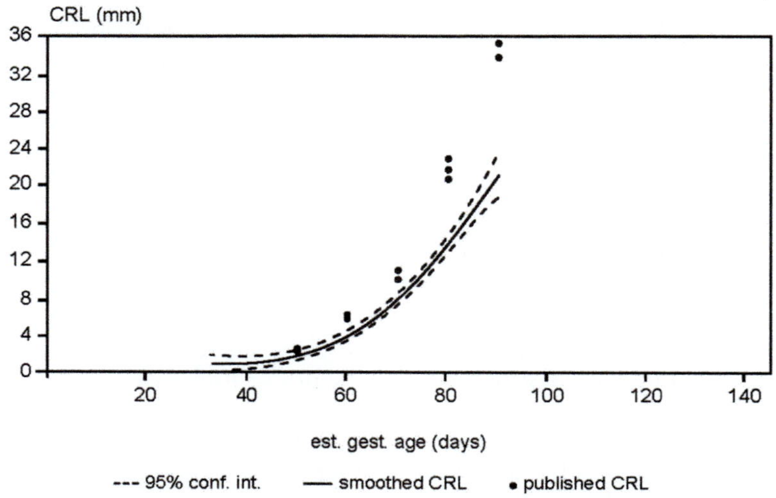

FIGURE 8.3 Cubic spline–fitted curve of the relationship between crown-rump length as measured through ultrasonography and gestational age, based on results from 39 pregnancies. The dots indicate directly measured crown-rump lengths of embryos/fetuses collected via hysterotomy by Chambers and Hearn [44]. *Figure from Jaquish CE, Toal RL, Tardif SD, Carson RL. Use of ultrasound to monitor prenatal growth and development in the common marmoset (*Callithrix jacchus*). Am J Primatol 1995;36(4):259–275. https://doi.org/10.1002/ajp.1350360402; Fig 6b*

to litter size in a predictable fashion, with higher litter sizes having lower average birth weights and also having higher within litter variation in birth weights. However, with the temporal pattern of larger adult weights (see Growth), one may see less difference in birth weights as the birth weights for larger litters increases [48]. Animals with birth weights less than 24 g have a substantially higher risk of preweaning death.

Postnatal Development

Physical

Kohn et al. [49] report on average ages of epiphyseal closures in the marmoset skeleton. The order of closure reflected the general ordered trend in primate skeletal growth, with epiphyses in the elbow and hip closing relatively early (0.55–0.90 years), and epiphyses in the shoulder, wrist, and knee closing relatively late (1.08–1.47 years). With the exception of the ischial tuberosity and the iliac crest, all closures are completed by 1.5 years of age. Abbott and Hearn [26] report an asymptote in knee–heel length at around 1.10 years (400 days). Associated with the marmoset's rapid maturation, deciduous teeth begin to erupt during the first week of life and are completed around week four. The first permanent teeth are visible at 3–4 months of age, and the full complement of 32 permanent teeth has erupted by 11–12 months of age [50]. Body weight can continue to increase through the second year (see Growth).

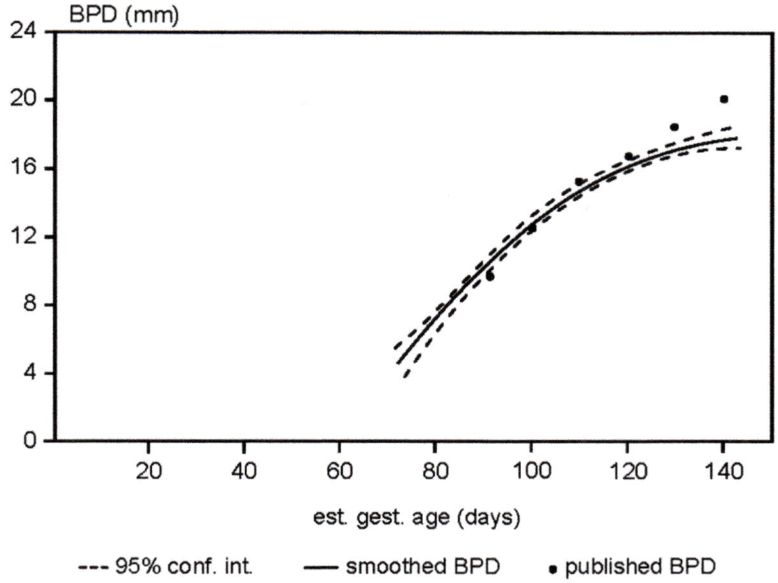

FIGURE 8.4 Cubic spline-fitted curve of the relationship between biparietal diameter x gestational age, based on results from 39 pregnancies. The dots indicate directly measured biparietal diameters of fetuses collected via hysterotomy by Chambers and Hearn [44]. *Figure from Jaquish CE, Toal RL, Tardif SD, Carson RL. Use of ultrasound to monitor prenatal growth and development in the common marmoset* (Callithrix jacchus). *Am J Primatol 1995;36(4):259–275. https://doi.org/10.1002/ajp.1350360402; Fig 7b*

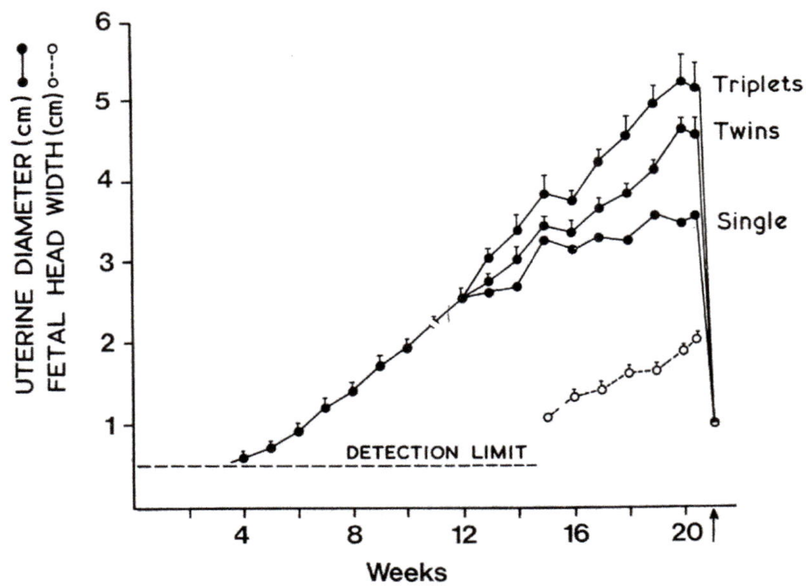

FIGURE 8.5 Diameter of the uterine fundus (*solid lines*) and fetal heads (*dotted line*), measured by transabdominal palpation in marmosets bearing singleton, twins, or triplets. *Figure from Hearn JP. The common marmoset. In: Hearn JP, editor. Reproduction in new world primates. Hingham (MA): MTP Press; 1983. pp. 181–216; Fig 5*

Behavioral

Schultz-Darken et al. [51] provide an excellent overview of what is known regarding behavioral development in marmosets, from the pioneering studies of Epple in the 1970s to the postnatal neurodevelopment assessment scale for marmosets (PPNAS-M) developed by Schultz-Darken and colleagues in 2015 [52]. Fig. 8.6 provides their synthesis of developmental stages as defined by Refs. [53–55]. These studies all present a similar developmental pattern that defines the period from birth to week 4–5 as a period of primary dependence on caregivers with minimal time spent off of those caregivers and little/no consumption of solid food. Weeks 4–10 represent the phase during which infants

TABLE 8.2 Birth Weight Relative to Litter Size and Survival

Litter Size	Outcome	Mean Weight (g)[a]	N
One	Surviving	35.8	8
	Dying	35.6	3
Two	Surviving	31.8	87
	Dying	25.3	29
Three	Surviving	28.4	61
	Dying	26	47

[a] *Weights taken approximately 36–40 h postdelivery.*

establish physical independence, including the onset of grooming, eating solid food, and head cocking behavior—most basic motor skills are acquired by the end of this period. During weeks 12–16, carrying and nursing cease and play interactions intensify.

GROWTH

Normal Weight Gain Trajectories

Kirkwood [56] describes the growth rate of mass in marmosets as a simple decelerating exponential curve, with a growth rate of 1.0–1.4 g per day and growth from 25% to 75% of adult weight taking 234–257 days. This finding is generally reinforced in subsequent studies (e.g., [57,58]). Body weights typically plateau between ~2 and 5 years of age, at which point there is a slight age-related decline. Reports of what is a "typical" body weight for an adult marmoset are highly variable from one colony to another, ranging from as low as 300 g to as high as 500 g. Data from wild marmosets are sparse but suggest that adult weights average around 350 g [59]. Perusal of data on body weights in captive marmoset colonies from the 1970s to present strongly suggests that there has been a secular trend of increasing body weights. Fig. 8.7 provides survival data relative to highest body weight in one large captive marmoset colony (n = 65 naturally occurring deaths between 1987 and 2009), suggesting that animals between greater than 400 g but less than 500 g in body weight have the overall highest age-specific survival. A finer grained analysis of the relationship of weight to survivorship would be a useful addition to the literature on this species.

Fig. 8.8 illustrates growth variation in two marmosets that both subsequently lived into mid-adulthood.

These extreme growth patterns are associated with a variety of environmental factors, which are discussed in subsequent sections. They are likely also associated with genetic factors driving growth and body weight. However, at present there are no published results that shed light on the role of genetic variation in explaining variation in growth and adult body weight in marmosets.

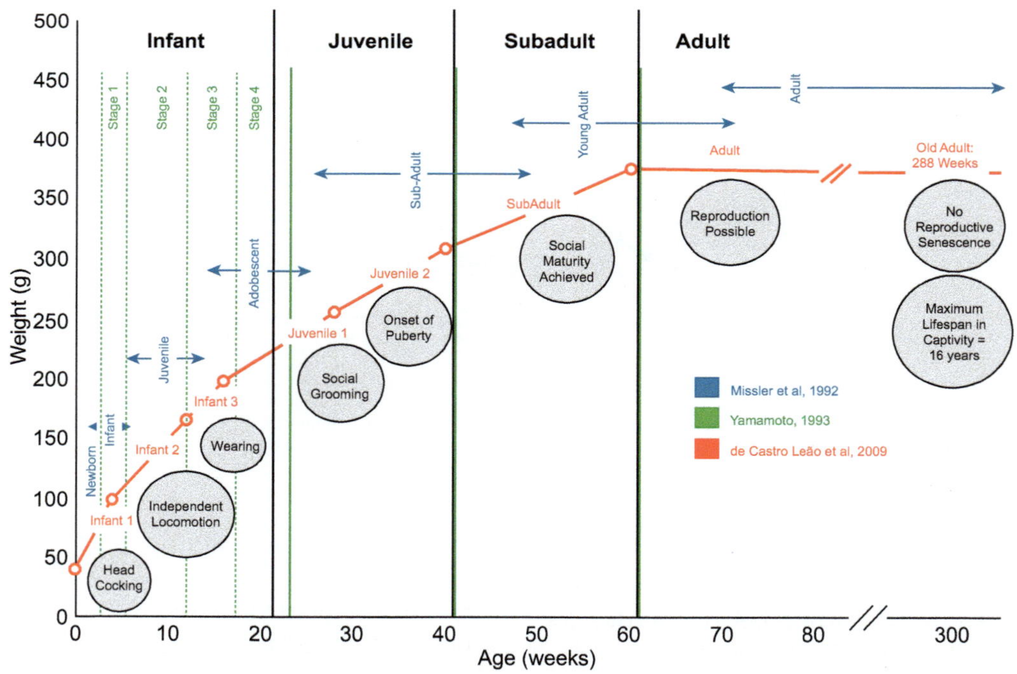

FIGURE 8.6 Graphic representation of common marmoset behavioral development based on descriptions of Missler et al., Yamamoto, and de Castro Leao. Milestones are inserted in the figure along the age timeline below the defined stages. *Figure from Schultz-Darken N, Braun KM, Emborg ME. Neurobehavioral development of common marmoset monkeys. Dev Psychobiol 2016;58(2):141–158. https://doi.org/10.1002/dev.21360, Fig 2*

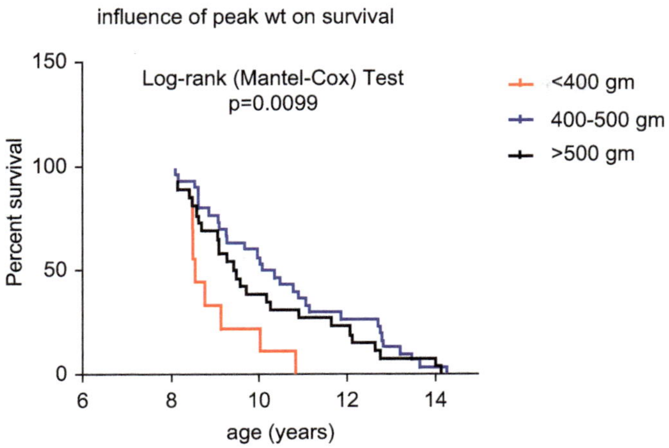

FIGURE 8.7 Relation of peak weight to survival past 6 years of age. *Figure from Tardif SD, Mansfield KG, Ratnam R, Ross CN, Ziegler TE. The marmoset as a model of aging and age-related diseases. ILAR J 2011;52(1): 54–65. https://doi.org/10.1093/ilar.52.1.54; Fig 3c*

Low Weight Gain and Its Causes

Factors that have been associated with poor growth and low adult weight include early life stressors and inflammatory bowel disease. Johnson et al. [61] reported that marmoset infants experiencing abusive caregiver interactions in the first 3 weeks of life were smaller as juveniles and adolescents. Abusive behavior was defined as "serious biting or nipping of the infant's tail by both parents to the extent that the lower third to half of the tail was amputated." This is a behavior that is observed at varying frequencies in virtually all captive marmoset colonies and sometimes requires veterinary intervention, such as tail-tip amputation. There is no evidence of this behavior occurring in the wild. In contrast, Tardif et al. [46,47] report no relationship between this form of abuse and growth outcomes; however, they do report a relationship between birth weight and such abuse, with infants thus abused being, on average, smaller at birth than nonabused infants. Because birth weights were not available in the Johnson study, it is not possible to tell whether disparity in weight and perhaps, therefore, in maturity or general condition at birth may explain the relation between abuse and growth.

Inflammatory bowel disease is one of the most common clinical problems seen in common marmosets. Death from IBD is much more common in young adult marmosets than in middle- or old-aged marmosets, suggesting that onset of these conditions is often an early life event [60]. Unpublished observations from various marmoset colonies suggest that impaired growth and development ("runting") occurs in animals less than 1 year of age. Between January 2006 and August 2009 animal care staff at Southwest National Primate Research Center identified 17 individuals as failure to thrive or runted individuals. These runts were animals that survived the initial critical period of 14 days and were noted to have slower growth at some point during their infant or juvenile phase. The timing of the identification of runting varied from early identification by 30–45 days of age to later identification closer to

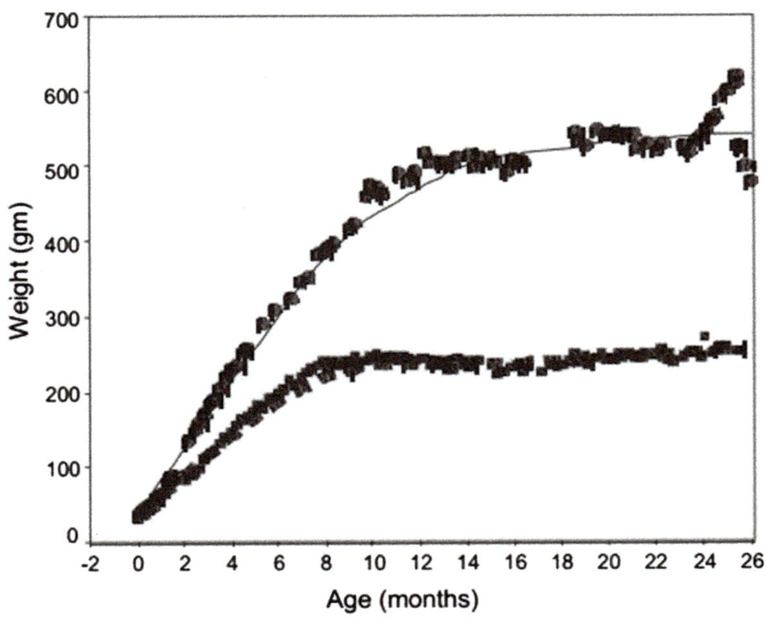

FIGURE 8.8 Growth trajectories for two marmosets, both surviving into early adulthood. *Figure from Tardif S, Power M, Layne D, Smucny D, Ziegler T. Energy restriction initiated at different gestational ages has varying effects on maternal weight gain and pregnancy outcome in common marmoset monkeys* (Callithrix jacchus). *Br J Nutr 2004;92(5):841–849. https://doi.org/10.1079/BJN20041269.*

6 months of age. All of the individuals noted as runts were identified due to being smaller than normal infants and failing to reach normal adult size at any point in their lives. The growth trajectories of these runts varied considerably, but all were smaller than typical marmosets for their age (Fig. 8.9). While the timing of the growth differentiation varied by individual, one notable characteristic of all of the runts is that the weight growth trajectory plateaus significantly prior to the typical age of growth plateau. There also seems to be two types of runts. Type 1 runts exhibited slowed growth and maturation prior to weaning, and these animals typically were not only small in weight but also stature. While we do not have data on long bone length, comments by staff in the records note that 4- to 5-month-old runts were similar in size to 30-day-old infants. These small in stature runts also had delays in pelage change. One animal that died at almost 10 months of age weighing only 80 g was noted to still have its infant coat and coloring as well as the full hair dye that is applied at 1 day of age and is typically lost by 6 months. Type 2 runts were animals that seemed to be near normal growth until approximately 6 months of age when their weight gain stopped and plateaued. These animals were notable for having normal coloration and almost normal stature, but typically being very lean and underweight. The differences in the growth trajectories for the runts may be due to the presence of two growth phases in infant marmosets [58]. The type 1 runts notable for smaller stature and infant pelage seem to have had developmental delays in growth phase 1. The type 2 runts may have had slower trajectories in growth phase 1 but then plateaued early and did not grow during the second developmental growth phase. It is unclear whether there are two distinct causes for these two phenotypic outcomes.

Unfortunately, we do not have body composition data for most of the animals identified as runts. A comparison of body compositions from four runted animals at ages 1, 2, 6, and 12 months to four normal animals not surprisingly reveals that runts had significantly lower body fat ($17.15 + 6.5$ g at 6 months of age and $9.35 + 4.01$ g at 12 months of age; mean +standard deviation) compared with normal age matched animals ($27.28 + 8.6$ g at 6 months of age and $31.35 + 14.1$ g at 12 months of age; $F(1,6) = 9.42$, $P = .02$). Runts also had a lower lean mass ($134.7 + 19.8$ g at 6 months of age and $132.2 + 20.2$ g at 12 months of age) compared with normal age matched animals ($177.6 + 12.6$ g at 6 months of age and $202.1 + 29.9$ g at 12 months of age; $F(1,6) = 10.74$, $P = .017$) (Fig. 8.10).

Many of the runts exhibited substantially higher mortality rates than typical of the colony with age at death ranging from 3 months to 3.5 years. The cause of death could not be determined for all of the individuals, but eight of the runts were noted to have significant alterations in gut morphology due to enteritis and colitis. One animal was noted to have small intestinal lymphoma. One animal had significant lymph node

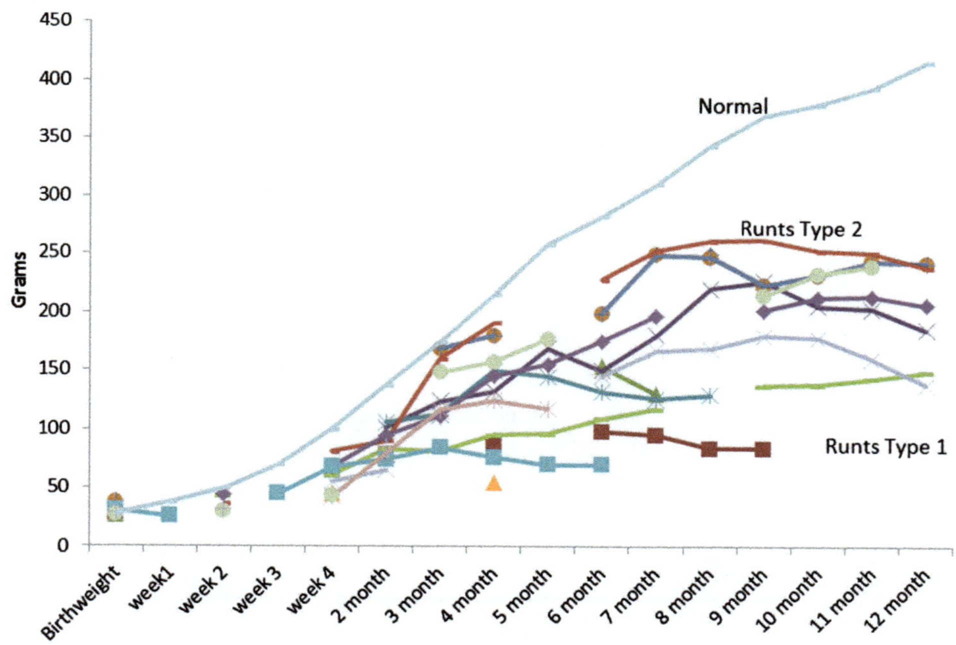

FIGURE 8.9 Growth curves for infants that were identified as runts in the colony between 2006 and 2009, infants are labeled as type and then number of months lived (1–4 is type 1 lived 4 months). Type 1 runts have delayed growth prior to weaning and many never get larger than 100 g. Type 2 runts have slow growth that plateaus around 6 months of age with most individuals not weighing more than 200 g and many lived to at least the age of sexual maturity. The top line represents average normal marmoset infant growth for the colony during the same time frame.

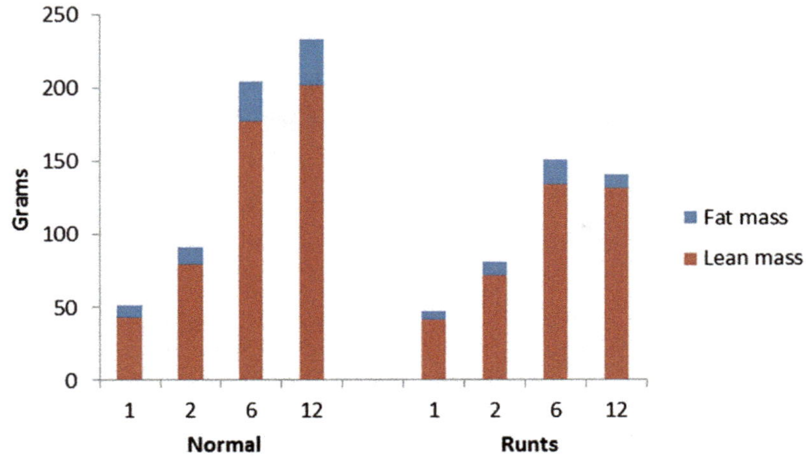

FIGURE 8.10 Body composition data for four normal infants and four infants identified as runts. Fat and lean mass growth differs significantly, with runts having reduced lean and fat mass.

hyperplasia. Two animals were noted to have significant amyloid deposits in the small intestine, colon, cecum, spleen, stomach, pancreas, and adrenal for one individual and small intestine, large intestine, spleen, adrenal, and liver for the second.

Congenital malformations that are not immediately fatal may also contribute to poor growth and development and death during the weaning period. One runted female, for example, only lived for 3 months and at necropsy was noted to be blind with a cloudy appearance to the eyes. At the Wisconsin National Primate Research Center (Nancy Schultz-Darken, personal communication), a 2-month-old was found to have a cleft palate at necropsy. The defect was clearly not enough to completely prevent suckling, as the infant lived for 2 months. However, the infant had no ingesta and was judged to be very dehydrated and malnourished at death.

The etiology of this runting phenomenon in the colony remains unknown. The notation of runts did not occur before 2006 and then seemed to increase in prevalence briefly with 8% of live births in 2006 being noted as runts, 9% in 2007, 12% in 2008, 6% in 2009, and none thereafter. It is possible that this phenomenon was associated with a move of the breeding colony to a new building that had much larger rooms containing more cages. While the density per square foot was not changed by the move, the number of animals living in proximity increased from on average 4–5 groups per room to 12 groups per room. However, there was no change in 2010 to the breeding room to explain the lack of new births of runts. Another possible change that occurred during this time period was the shift of protein source in the base diet from lactalbumin to casein. While a dietary shift is an attractive explanation, there were infants identified as runts prior to the implementation of this diet for all the breeding animals, and the diet was not changed in 2010 when the colony seemed to stop producing runts. For now, this phenomenon remains noteworthy, but unexplained.

HIGH WEIGHT GAIN AND ITS CAUSES

In contrast to the weight loss and cachexia observed in marmosets with progressive inflammatory bowel disease, obesity accompanied by dyslipidemia and altered glucose metabolism [62,63] is also emerging as a frequent clinical finding in marmosets housed in captivity. Obesity in marmosets contrasts with that of other nonhuman primates in being an early life phenomenon. Two large marmoset colonies have reported a high prevalence in subadult marmosets ranging between 1 and 2 years of age with 46%–52% of animals in this age category demonstrating obesity and reduced insulin sensitivity [63–65]. Fig. 8.11 illustrates a young adult female marmoset of a normal weight contrasted with an obese male of the same age. Fig. 8.12 illustrates the growth in fat mass from birth to 6 months of age in animals that subsequently were normal weight (n = 16–19) or obese (n = 13–14), emphasizing the early onset of this condition.

The etiology of pediatric obesity in this species is poorly understood, but some phenotypes associated with subsequent obesity have been identified. Marmoset infants are more likely to become obese if they begin the weaning process sooner and the first day of solid food consumption significantly predicted obesity at 12 months of age [66]. The consumption of solid food was found to be correlated with a number of social behaviors including percent of time spent in independent movement—i.e., not being transported by a parent or

FIGURE 8.11 Two sedated marmosets 17 months of age (equivalent to midteens in humans). Animal on right is of nonobese weight (370 g), whereas the animal on the left weighs 630 g.

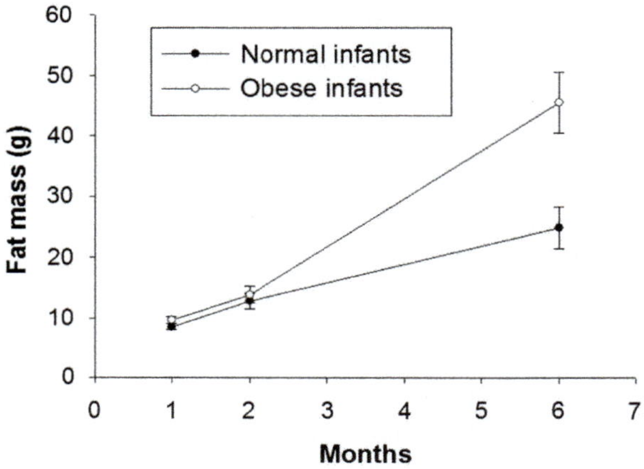

FIGURE 8.12 Differences in fat mass growth in nonobese versus obese marmoset infants.

older sibling. The first day of consumption is neither associated with maternal harassment nor with other negative social interactions, i.e., the infants are initiating their time off of carriers and initiating weaning. Marmosets develop patterns of consumption that are consistent throughout development and animals destined to become obese displayed one interesting and unexpected difference from a nonobese individual that was revealed using a lickometer intake monitoring system. Marmosets that become obese at 12 months were significantly more likely to be consuming more grams with every lick—that is they consumed the same amount of liquid diet with significantly fewer lick contacts, indicating that they were consistently taking in more liquid diet per contact.

Clearly, additional studies are required to define what is a normal and healthy growth trajectory and adult weight in marmosets. These studies will be critical to both defining best practices in clinical management and to understanding how these extremes of growth may serve as useful models of similar human conditions.

References

[1] Hearn JP. The common marmoset. In: Hearn JP, editor. Reproduction in new world primates. Hingham, MA: MTP Press; 1983. p. 181–216.

[2] Oerke A-, Einspanier A, Hodges JK. Noninvasive monitoring of follicle development, ovulation, and corpus luteum formation in the marmoset monkey (*Callithrix jacchus*) by ultrasonography. Am J Primatol 1996;39(2):99–113. https://doi.org/10.1002/(SICI)1098-2345(1996)39:2<99::AID-AJP2>3.0.CO;2-M.

[3] Eastman SA, Makawiti DW, Collins WP, Hodges JK. Pattern of excretion of urinary steroid metabolites during the ovarian cycle and pregnancy in the marmoset monkey. J Endocrinol 1984;102(1):19–26.

[4] French JA, Brewer KJ, Schaffner CM, Schalley J, Hightower-Merritt D, Smith TE, Bell SM. Urinary steroid and gonadotropin excretion across the reproductive cycle in female wied's black tufted-ear marmosets (*Callithrix kuhli*). Am J Primatol 1996;40(3):231–45. https://doi.org/10.1002/(SICI)1098-2345(1996)40:3<231::AID-AJP2>3.0.CO;2-Z.

[5] Ziegler TE, Scheffler G, Wittwer DG, Schultz-Darken N, Snowdon CT, Abbott DH. Metabolism of reproductive steroids during the ovarian cycle in two species of callitrichids, *Saguinus oedipus* and *Callithrix jacchus*, and estimation of the ovulatory period from fecal steroids. Biol Reprod 1996;54(1):91–9. https://doi.org/10.1095/biolreprod54.1.91.

[6] Heistermann M. Noninvasive monitoring of endocrine status in laboratory primates: methods, guidelines and applications. Adv Sci Res 2010;5(1):1–9. https://doi.org/10.5194/asr-5-1-2010.

[7] Sasaki E. Creating transgenic marmosets. In: The common marmoset in captivity and biomedical research; 2018.

[8] Gluckman TL, Walz SE, Schultz-Darken N, Bolton ID. Cytologic assessment of the vaginal epithelium in the common marmoset (*Callithrix jacchus*): a preliminary new approach to reproductive screening. JAALAS 2004;43(2):28–31.

[9] Saltzman W, Tardif SD, Rutherford JN. Hormones and reproductive cycles in primates. In: Hormones and reproduction of vertebrates, vol. 5; 2011. p. 291–327. https://doi.org/10.1016/B978-0-12-374928-4.10013-6.

[10] Summers PM, Wennink CJ, Hodges JK. Cloprostenol-induced luteolysis in the marmoset monkey (callithrix jacchus). J Reprod Fertil 1985;73(1):133–8.

[11] Mustoe AC, Jensen HA, French JA. Describing ovarian cycles, pregnancy characteristics, and the use of contraception in female white-faced marmosets, *Callithrix geoffroyi*. Am J Primatol 2012;74(11):1044–53. https://doi.org/10.1002/ajp.22058.

[12] Nievergelt C, Pryce CR. Monitoring and controlling reproduction in captive common marmosets on the basis of urinary oestrogen metabolites. Lab Anim 1996;30(2):162–70. https://doi.org/10.1258/002367796780865781.

[13] Saltzman W, Digby LJ, Abbott DH. Reproductive skew in female common marmosets: what can proximate mechanisms tell us about ultimate causes? Proc Biol Sci 2009;276(1656):389–99. https://doi.org/10.1098/rspb.2008.1374.

[14] Saltzman W, Schultz-Darken NJ, Abbott DH. Familial influences on ovulatory function in common marmosets (*Callithrix jacchus*). Am J Primatol 1997;41(3):159–77. https://doi.org/10.1002/(SICI)1098-2345(1997)41:3<159::AID-AJP1>3.0.CO;2-W.

[15] Savage A, Ziegler TE, Snowdon CT. Sociosexual development, pair bond formation, and mechanisms of fertility suppression in female cotton-top tamarins (*Saguinus oedipus oedipus*). Am J Primatol 1988;14(4):345–59. https://doi.org/10.1002/ajp.1350140404.

[16] Tardif S, Hyde K, Digby L. Evidence for suppression of ovulation in singly-housed female common marmosets (*Callithrix jacchus*). Lab Primate Newsl 1994;33(2):1–4.

[17] Tardif SD, Smucny DA, Abbott DH, Mansfield K, Schultz-Darken N, Yamamoto ME. Reproduction in captive common marmosets (*Callithrix jacchus*). Comp Med 2003;53(4):364–8.

[18] Bales K, O'Herron M, Baker AJ, Dietz JM. Sources of variability in numbers of live births in wild golden lion tamarins (*Leontopithecus rosalia*). Am J Primatol 2001;54(4):211–21. https://doi.org/10.1002/ajp.1031.

[19] Dixson A, Anzenberger G, Monteriro D. DNA fingerprinting of freeranging groups of marmosets in northeast Brazil. In: Martin R, Dixson A, Wickings E, editors. Paternity in primates: genetic tests and theories. Implications of human DNA fingerprinting. Basel: Karger; 1992. p. 192–202.

[20] Savage A, Soto L, Medina F, Emeris G, Soltis J. Litter size and infant survivorship in wild groups of cotton-top tamarins (*Saguinus oedipus*) in Colombia. Am J Primatol 2009;71(8):707–11. https://doi.org/10.1002/ajp.20696.

[21] Tardif SD, Lacker HM, Feuer M. Follicular development and ovulation in the marmoset monkey as determined by repeated laparoscopic examination. Biol Reprod 1993;48(5):1113–9.

[22] Tardif SD, Jaquish CE. The common marmoset as a model for nutritional impacts upon reproduction. Ann N Y Acad Sci 1994; 709:214–5.

[23] Tardif SD, Jaquish CE. Number of ovulations in the marmoset monkey (*Callithrix jacchus*): relation to body weight, age and repeatability. Am J Primatol 1997;42(4):323–9. https://doi.org/10.1002/(SICI)1098-2345(1997)42:4<323::AID-AJP7>3.0.CO;2-Z.

[24] Smucny DA, Abbott DH, Mansfield KG, Schultz-Darken NJ, Yamamoto ME, Alencar AI, Tardif SD. Reproductive output, maternal age, and survivorship in captive common marmoset females (*Callithrix jacchus*). Am J Primatol 2004;64(1):107–21. https://doi.org/10.1002/ajp.20065.

[25] Tardif S, Araujo A, Arruda M, French J, Sousa M, Yamamoto M. Reproduction and aging in marmosets and tamarins. Primate Reprod Aging 2008;36:29. https://doi.org/10.1159/000137678.

[26] Abbott DH, Hearn JP. Physical, hormonal and behavioural aspects of sexual development in the marmoset monkey, *Callithrix jacchus*. Reproduction 1978;53(1):155–66. https://doi.org/10.1530/jrf.0.0530155.

[27] Nandekar TD, Kholkute SD. Ovarian development and circulating levels of steroid hormones in common marmosets (*Callithrix jacchus*) from birth to senescence. In: Rolland R, editor. Neuroendocrinology of reproduction. proceedings of the VIth reinier de grant symposium. Amsterdam: Excerpta Medica; 1987. p. 271–82.

[28] Tardif S, Carville A, Elmore D, Williams L, Rice K. Reproduction and breeding of nonhuman primates. In: Abee C, Mansfield K, Tardif S, Morris T, editors. Nonhuman primates in biomedical research. Biology and management, vol. 1. London: Elsevier; 2012. p. 197–250.

[29] Weinbauer GF, Aslam H, Krishnamurthy H, Brinkworth MH, Einspanier A, Hodges JK. Quantitative analysis of spermatogenesis and apoptosis in the common marmoset (*Callithrix jacchus*) reveals high rates of spermatogonial turnover and high spermatogenic efficiency. Biol Reprod 2001;64(1):120–6. https://doi.org/10.1095/biolreprod64.1.120.

[30] Morrell JM, Kuderling I, Hodges JK. Influence of semen collection method on ejaculate characteristics in the common marmoset, *Callithrix jacchus*. J Androl 1996;17(2):164.

[31] Kuderling I, Schneider A, Sonksen J, Nayudu P, Hodges J. Non-invasive collection of ejaculates for the common marmoset (*Callithrix jacchus*) using penile vibrostimulation. Am J Primatol 2000; 52:149–54.

[32] Sasaki E, Suemizu H, Shimada A, Hanazawa K, Oiwa R, Kamioka M, Nomura T. Generation of transgenic non-human primates with germline transmission. Nature 2009;459(7246):523–7. https://doi.org/10.1038/nature08090.

[33] Schultz-Darken N. Behavior and behavioral management. In: The common marmoset in captivity and biomedical research; 2018.

[34] Merker H-J, Bremer D, Csato W, Heger W, Gossrau R. Development of the marmoset placenta. In: Neubert D, Merker H-J, Hendrickx A, editors. Non-human primates - developmental biology and toxicology. Berlin: Ueberreuter Wissenschaft; 1984. p. 245–72.

[35] Ross CN, French JA, Orti G. Germ-line chimerism and paternal care in marmosets (*Callithrix kuhlii*). Proc Natl Acad Sci USA 2007;104(15):6278–82. https://doi.org/10.1073/pnas.0607426104.

[36] Sweeney CG, Curran E, Westmoreland SV, Mansfield KG, Vallender EJ. Quantitative molecular assessment of chimerism across tissues in marmosets and tamarins. BMC Genom 2012;13: 98. https://doi.org/10.1186/1471-2164-13-98.

[37] Genain CP, Hauser SL. Experimental allergic encephalomyelitis in the new world monkey *Callithrix jacchus*. Immunol Rev 2001;183: 159–72. https://doi.org/10.1034/j.1600-065x.2001.1830113.x.

[38] Heger W, Merker H-J, Neubert D. Frequency of prenatal loss in marmoset monkeys (*Callithrix jacchus*). In: Neubert D, Merker H-J, Hendrickx A, editors. Non-human primates - developmental biology and toxicology. Berlin: Ueberreuter Wissenschaft; 1988. p. 129–40.

[39] Jaquish CE, Tardif SD, Toal RL, Carson RL. Patterns of prenatal survival in the common marmoset (*Callithrix jacchus*). J Med Primatol 1996;25(1):57–63. https://doi.org/10.1111/j.1600-0684.1996.tb00194.x.

[40] Tardif S, Power M, Layne D, Smucny D, Ziegler T. Energy restriction initiated at different gestational ages has varying effects on maternal weight gain and pregnancy outcome in common marmoset monkeys (*Callithrix jacchus*). Br J Nutr 2004;92(5): 841–9. https://doi.org/10.1079/BJN20041269.

[41] Windle CP, Baker HF, Ridley RM, Oerke A, Martin RD. Unrearable litters and prenatal reduction of litter size in the common marmoset (*Callithrix jacchus*). J Med Primatol 1999;28(2):73–83. https://doi.org/10.1111/j.1600-0684.1999.tb00254.x.

[42] Power ML, Oftedal OT, Tardif SD. Does the milk of callitrichid monkeys differ from that of larger anthropoids? Am J Primatol 2002;56(2):117–27. https://doi.org/10.1002/ajp.1068.

[43] Wachtman L. Physical examination, diagnosis, and common clinical procedures. In: The common marmoset in captivity and biomedical research; 2018.

[44] Chambers PL, Hearn JP. Embryonic, foetal and placental development in the common marmoset monkey (*Callithrix jacchus*). J Zool 1985;207(4):545–61.

[45] Jaquish CE, Toal RL, Tardif SD, Carson RL. Use of ultrasound to monitor prenatal growth and development in the common marmoset (*Callithrix jacchus*). Am J Primatol 1995;36(4):259–75. https://doi.org/10.1002/ajp.1350360402.

[46] Tardif SD, Jaquish CE, Toal RL, Layne DG, Power RA. Estimation of gestational ages in the common marmoset (*Callithrix jacchus*) from published prenatal growth curves. J Med Primatol 1998a; 27(1):28–32.

[47] Tardif S, Jaquish C, Layne D, Bales K, Power M, Power R, Oftedal O. Growth variation in common marmoset monkeys (*Callithrix jacchus*) fed a purified diet: relation to care-giving and weaning behaviors. Lab Anim Sci 1998b;48(3):264–9.

[48] Tardif SD, Power ML, Ross CN, Rutherford JN. Body mass growth in common marmosets: toward a model of pediatric obesity. Am J Phys Anthropol 2013;150(1):21–8. https://doi.org/10.1002/ajpa.22110.
[49] Kohn LA, Olson P, Cheverud JM. Age of epiphyseal closure in tamarins and marmosets. Am J Primatol 1997;41(2):129–39. https://doi.org/10.1002/(SICI)1098-2345(1997)41:2<129::AID-AJP5>3.0.CO;2-Z.
[50] Johnston GW, Dreizen S, Levy BM. Dental development in the cotton ear marmoset (*Callithrix jacchus*). Am J Phys Anthropol 1970; 33(1):41–8. https://doi.org/10.1002/ajpa.1330330107.
[51] Schultz-Darken N, Braun KM, Emborg ME. Neurobehavioral development of common marmoset monkeys. Dev Psychobiol 2016;58(2):141–58. https://doi.org/10.1002/dev.21360.
[52] Braun K, Schultz-Darken N, Schneider M, Moore CF, Emborg ME. Development of a novel postnatal neurobehavioral scale for evaluation of common marmoset monkeys. Am J Primatol 2015;77(4): 401–17. https://doi.org/10.1002/ajp.22356.
[53] Missler M, Wolff JR, Rothe H, Heger W, Merker HJ, Treiber A, Crook GA. Developmental biology of the common marmoset: proposal for a "postnatal staging". J Med Primatol 1992;21(6):285.
[54] Yamamoto M. From dependence to sexual maturity: the behavioural ontogeny of callitrichidae. In: Rylands A, editor. Marmosets and tamarins: systematics, behaviour and ecology. Oxford: Oxford University Press; 1993. p. 235–54.
[55] de Castro Leão A, Duarte Dória Neto A, Cordeiro de Sousa MB. New developmental stages for common marmosets (*Callithrix jacchus*) using mass and age variables obtained by K-means algorithm and self-organizing maps (SOM). Comput Biol Med 2009; 39(10):853–9. https://doi.org/10.1016/j.compbiomed.2009.05.009.
[56] Kirkwood JK. Patterns of growth in primates. J Zool 1985;205(1): 123–36. https://doi.org/10.1111/j.1469-7998.1985.tb05617.x.
[57] Peters VM, Guerra MO. Growth of marmoset monkeys *Callithrix jacchus* in captivity. Folia Primatol 1998;69(5):266–72. https://doi.org/10.1159/000021636.
[58] Tardif SD, Bales KL. Relations among birth condition, maternal condition, and postnatal growth in captive common marmoset monkeys (*Callithrix jacchus*). Am J Primatol 2004;62(2):83–94. https://doi.org/10.1002/ajp.20009.
[59] Araújo A, Arruda M, Alencar A, Albuquerque F, Nascimento M, Yamamoto M. Body weight of wild and captive common marmosets (*Callithrix jacchus*). Int J Primatol 2000;21(2):317–24. https://doi.org/10.1023/A:1005433722475.
[60] Tardif SD, Mansfield KG, Ratnam R, Ross CN, Ziegler TE. The marmoset as a model of aging and age-related diseases. ILAR J 2011;52(1):54–65. https://doi.org/10.1093/ilar.52.1.54.
[61] Johnson EO, Kamilaris TC, Calogero AE, Gold PW, Chrousos GP. Effects of early parenting on growth and development in a small primate. Pediatr Res 1996;39(6):999–1005. https://doi.org/10.1203/00006450-199606000-00012.
[62] Tardif SD, Power ML, Ross CN, Rutherford JN, Layne-Colon DG, Paulik MA. Characterization of obese phenotypes in a small nonhuman primate, the common marmoset. Obesity 2009;17(8): 1499–505. https://doi.org/10.1038/oby.2009.77.
[63] Wachtman LM, Kramer JA, Miller AD, Hachey AM, Curran EH, Mansfield KG. Differential contribution of dietary fat and monosaccharide to metabolic syndrome in the common marmoset. Obesity 2011;19(6):1145–56. https://doi.org/10.1038/oby.2010.303.
[64] Power ML, Ross CN, Schulkin J, Tardif SD. The development of obesity begins at an early age in captive common marmosets (*Callithrix jacchus*). Am J Primatol 2012;74(3):261. https://doi.org/10.1002/ajp.21995.
[65] Power ML, Ross CN, Schulkin J, Ziegler TE, Tardif SD. Metabolic consequences of the early onset of obesity in common marmoset monkeys. Obesity 2013;21(12):E598. https://doi.org/10.1002/oby.20462.
[66] Ross CN, Power ML, Artavia JM, Tardif SD. Relation of food intake behaviors and obesity development in young common marmoset monkeys. Obesity 2013;21(9):1891–9. https://doi.org/10.1002/oby.20432.

CHAPTER

9

Regulatory Considerations

Robert P. Marini

Division of Comparative Medicine, Massachusetts Institute of Technology, Cambridge, MA, United States

The intent of this chapter is threefold: to review regulations specific to the common marmoset, to review currently accepted professional standards or best practices relevant to those regulations, and to identify regulations, guidelines, and best standards that pertain to marmoset acquisition and transport. Regulations refer specifically to statutory requirements promulgated by a local, state, national, or international entity and which therefore have the force of law, whereas industry standards are established standards of care achieved by consensus, experience, and communication of that experience and consensus among experts. Best practices represent an evolutionary stage and comprise incremental improvements in the process of caring for animals. These practices may someday become industry standards as they are adopted and validated by the community. Similarly, industry standards will inform and may one-day be promulgated as regulations. More in-depth treatments of some of this information, especially where it pertains to husbandry, housing, and environmental enrichment, will be presented elsewhere in this text, and an effort has been made to avoid unnecessary repetition of that material.

REGULATIONS

The specific source regulations for this chapter include the following:

International

- Convention on International Trade in Endangered Species of Wild Flora and Fauna (CITES) [1].

United States

- The Animal Welfare Act (AWA), (United States Code (U.S.C.), Title 7 (Agriculture), Chapter 54 (Transportation, Sale, and Handling of Certain Animals), Sections 2131–2159) and the Animal Welfare Regulations (AWR). 9 Code of Federal Regulations (CFR) Chapter 1 [2].
- The Health Research Extension Act of 1985 and the Public Health Service Policy on the Humane Care and Use of Laboratory Animals (Public law 99-158).
- The Centers for Disease Control and Preventions (CDC). (42 CFR 71.53) [3–5].
- The Endangered Species Act (7 U.S.C. § 136, 16 U.S.C. § 1531 et seq.).

European Union

- Directive 86/609/EEC (European Union).
- Appendix A of the Council of Europe Convention ETS 123 [6].

INDUSTRY STANDARDS

The specific source materials for industry standards appearing in this chapter include the following:

- Guide for the Care and Use of Laboratory Animals (US) [26].
- Guidelines for the Veterinary Care of Laboratory Animals: A report of the FELASA-ECLAM-ESLAV (FELASA—the Federation for Laboratory Animal Science Associations; ECLAM—the European College of Laboratory Animal Medicine; ESLAV—the European Society of Laboratory Animal Veterinarians) Joint Working Group on Veterinary Care
- European Association of Zoos and Aquariums (EAZA) Husbandry Guidelines for Callitrichidae 2010 [7,8]
- NIH Nonhuman Primate Management Plan (US)
- Psychological Well-Being of Nonhuman Primates, National Academy Press [9]
- International Air Transport Association (IATA). Live animal transport group

The Common Marmoset in Captivity and Biomedical Research
https://doi.org/10.1016/B978-0-12-811829-0.00009-1

The marmoset is a "covered species" under the AWR (§1.1 AWR 2013). In the United States the two primary regulatory documents are the AWR promulgated by the AWA, and the Public Health Service Policy on the Human Care and Use of Laboratory Animals. Similar information is contained in the Council of Europe [10] document, Appendix A of the Council of Europe Convention ETS 123 [11]. Informing Appendix A is the 2007 publication—Euroguide on the Accommodation and Care of Animals used for Experimental and Other Scientific Purposes—prepared by a FELASA Working Group (2007) [12].

Beyond the United States and Europe, there are country-specific agencies governing the care and use of laboratory animals. An overview of this global regulatory apparatus can be found in the UFAW Handbook on the Care and Management of Laboratory and Other Research Animals: Eighth Edition [13]. Additionally, many institutions voluntarily submit to the review and accreditation process of the Association for the Assessment and Accreditation of Laboratory Animal Care International [14] and similar entities. Regulations for a variety of countries may be found at their website (http://www.aaalac.org/resources/internationalregs.cfm).

PHYSICAL PLANT, HOUSING, AND HUSBANDRY

Housing and husbandry regulations should be considered minimum standards, whereas currently accepted professional standards represent additional refinement of those minimums. Institutions will determine internal standards based on available resources of space and funding and considering the expertise and recommendations of husbandry and veterinary staff, consultants, and their IACUC's or similar bodies. The following is a synthesis of existing regulations and industry standards.

Facilities housing nonhuman primates must be "in good repair, and they must protect the animal from injury, contain the animals securely, and restrict other animals from entering...surfaces must be constructed in a manner and made of materials that allow them to be readily cleaned and sanitized, or removed or replaced when worn or soiled" (§3.75, AWR 2013). More generally, facilities must be constructed with the objective of sealing out vermin, providing access only to appropriate personnel, and creating functional areas for veterinary procedures, aseptic surgery, diet preparation, cage sanitation and storage, and containment or quarantine, animal housing, and necropsy (Guide). Organizations should employ designers, architects, and builders with experience in the design and construction of laboratory animal holding facilities. The AWR contains additional information on storage, drainage and waste disposal, maintenance and replacement of surfaces, water, electric power, heating, ventilation, proper temperature, and lighting (§3.75, §3.76, AWR 2013). Although not a regulatory requirement, full spectrum lighting (providing UV-B radiation) or natural sunlight through UV penetrable glass should be used (EAZA 2010). UV lights should be placed approximately 1–2 m above favorite resting spots to optimize their impact.

Consistent with the statement: "Nonhuman primates may not be housed with other species of animals unless they...are not known to be hazardous to the health and well-being of each other" {§3.81 (a) (3), AWR], marmosets and squirrel monkeys must be physically separate, preferably in different facilities. This precaution exists because of the risk of *Saimirine herpesvirus 1* (formerly *Herpesvirus tamarinus*), which can be carried without clinical signs in squirrel monkeys but can cause fatal disease in marmosets. Work flow should be organized such that care and veterinary staff do not attend to marmosets after visiting squirrel monkeys, and common areas such as surgery should have their scheduling adjusted so that the species do not have contact [15].

PRIMARY ENCLOSURES

Primary enclosures must meet many criteria according to the AWR and cognate documents in other countries. Among these are construction that precludes injury to animals {§3.80 2 (i) (ii) (x)}, prevents accidental opening and escape {§3.80 2(iii)}, provides adequate shelter from extremes of temperature and inclement weather {§3.80 2 (vi)}, allows free access to clean food, water {§3.80 2 (viii)}, and shade {§3.80 2 (vii)}, allows easy cleaning, sanitation, or replacement when worn {§3.80 2 (ix)}, and which provides "sufficient space for the nonhuman primate to make postural adjustments with freedom of movement" {§3.80 2 (xi)}. In both the European Union and the United States, space requirements for primary enclosures of nonhuman primates are determined by weight. In the AWR and the Guide, marmosets and tamarins fall into Group 1, animals under 1 kg body weight, but despite using the identical metric, the requirements

TABLE 9.1A Minimum Space Allowance for Marmosets

	Weight		Floor Area/Animal		Height	
Group	lbs	(kg)	ft^2	(m^2)	in	(cm)
1	Under 2.2	(under 1)	1.6	(0.15)	20	(50.8)

Animal Welfare Regulations – US.

TABLE 9.1 B Minimum Space Allowance for Marmosets

Animals	Weight, kg	Floor Area/Animal, ft^2 (m^2)	Height[a], in cm	Comments
Monkeys[b] (including baboons)				Cage height should be sufficient for the animals to stand comfortably erect with their feet on the floor
Group 1	Up to 1.5	2.1 (0.20)	30 (76.2)	Overall cage volume and linear perch space should be considerations for many neotropical and arboreal species

[a]*From cage floor to cage top.*
[b]*Callitrichidae, Cebidae, Cercopithecidae, and Papio.*
Recommended Minimum Space for Nonhuman Primates Housed in Pairs or Groups. Environment, Housing and Management: The Guide, 2011, (US).

differ slightly (Tables 9.1A and 9.1B). For pair or family-housed marmosets, the floor space required is that required for one animal multiplied by the number of animals. Floor space requirements may be waived in the case of innovative primary enclosures provided that animals have a "sufficient volume of space and the opportunity to express species-typical behavior." Caging that does not meet the minimum engineering standards, whether due to research exemption, veterinary exemption, or because of innovative design, must be approved by the IACUC in the case of research facilities or by the APHIS Administrator in the case of dealers and exhibitors.

European Union guidelines for floor space and cage height are more generous than those in the United States (Table 9.1C). Marmosets value vertical space because of their arboreal nature and the tendency to move vertically when in flight from perceived threats. The Euroguide identifies the following as size determinants for primary enclosures: "The adult size of the animals, sufficient space to provide a complex and challenging environment and; the size of the group to be accommodated." For marmosets and tamarins, they comment on the primary importance of volume of space and vertical height in contrast to floor space for these species. The Euroguide goes on to recommend shade for animals with access to natural light. Guidelines for marmoset enclosures in zoo settings have been set forth by the EAZA (2010). As with the United States and European regulations, the EAZA guidelines state that materials used for enclosures and cage fixtures should be nontoxic and in good repair. Solid materials are preferable for ease of sanitation, pest control, and to preclude predation. Animal doors within enclosures should be 1.5–2 m above substrate, as marmosets are unaccustomed to walking close to the ground. At least two animal doors between contiguous spaces per enclosure are recommended, and these should have an area of 10 cm square and be placed widely enough apart that a dominant animal cannot become a "gatekeeper." Keeper access doors should have "safety porches" to help prevent animal escape. Enclosures should also be designed so that zookeepers can access nest boxes and feeding stations from outside the enclosure. Draft-excluding flaps are used for doors in indoor/outdoor enclosures, and tunnels are suggested for joining enclosures. Sliding doors on either end of a tunnel can be used to isolate animals for identification, observation, or removal. The EAZA minimum enclosure height is 2.5 m for both indoor and outdoor enclosures. The indoor enclosure total floor volume should be 3 square meters versus 10 square meters for outdoor enclosures. Additionally, indoor enclosures should be "configured into two separable areas" to facilitate cleaning, new animal introduction, isolation and other husbandry, and veterinary practices. The combined indoor/outdoor area to which animals have access "80% of their waking hours year round" should be 22.5 cubic meters. For family groups larger than five animals, the guidelines simply state that "more room" should be provided.

SHELTERED AND OUTDOOR HOUSING FACILITIES

The AWR distinguishes between sheltered facilities, those in which animals are provided: "...protection from the elements; and protection from temperature

TABLE 9.1C Minimum Space Allowance for Marmosets—Marmosets and Tamarins: Minimum Enclosure Dimensions and Space Allowance

	Minimum Floor Area of Enclosures for 1[a]-2 Animals Plus Offspring up to 5 months old (m^2)	Minimum Volume per Additional Animal Over 5 months (m^3)[b]	Minimum Enclosure Height (m)
Marmosets	0.5	0.2	1.5
Tamarins	1.5	0.2	1.5

[a]*Animals should only be kept singly under exceptional circumstances.*
[b]*The top of the enclosure should be at least 1.8 m from the floor.*
Non-human primates (marmosets and tamarins, squirrel monkeys, macaca and vervets, baboons) (FELASA and Conv ETS 123 Appendix A) (EU).

extremes at all times." and "outdoor housing," which: "...does not meet the definition of any other type of housing facility provided in the regulations, and in which temperatures cannot be controlled within set limits" (§1.1). For both types of housing, animals must be provided shelter from the elements at all times, shelters must be large enough to accommodate all animals in the shelter, and multiple shelters must be provided so that dominant animals do not preclude subordinate animals from having access {§3.77(d) (e); §3.78 (b) (c)}. In sheltered facilities, ventilation should be adequate to protect and promote animal well-being with respect to minimizing "odors, drafts, ammonia levels, and moisture condensation" {§3.77(b)}. Lighting, either artificial or natural, must be provided in all animal areas, should be uniformly diffuse, and sufficient for observation of the animals and normal housekeeping {§3.77(c)}.

Appendix A of the European Convention not only states similar provisions for shelter from the elements but also specifies that wood within outdoor enclosures be either of "...a type of wood that animals will not chew or protected with mesh and a non-toxic treatment" (Appendix A, ETS 4.3.2). Floors of such enclosures can either be of concrete covered with a "...suitable non-toxic substrate..." or natural vegetation. In agreement with EAZA guidelines, Appendix A also recommends that more than one door be provided to areas with indoor and outdoor components so that subordinate animals are not precluded by more dominant animals from free movement between enclosures. The EAZA recommends that outdoor areas be predator and pest proof.

Perimeter fences must be used for both sheltered and outdoor housing. These fences must be at least 6 feet high and must protect NHPs from incursion by animals "the size of dogs, skunks and raccoons from going through it or under it..." {§3.77(f) (§3.78(d)}. Fences must also be at least 3 feet from the outer boundary of the primary enclosure. Exceptions to these requirements must be approved by the APHIS Administrator. The requirement for perimeter fences is waived if "...the outer walls of primary enclosure are made of a sturdy, durable material...and are high enough and constructed in a manner that restricts contact with or entry by humans or animals that are outside the facility" or alternatively, if a natural barrier exists that both meet these objectives and is approved by the APHIS Administrator {§3.77 (f) (1) (2) (§3.78 (d) (1) (2)}. For zoos, the AWR states: "Fixed public exhibits housing nonhuman primates, such as zoos, must have a barrier between the primary enclosure and the public, at any time the public is present, that restricts physical contact between the public and the nonhuman primates" {§3.77(g) (§3.78(e)}. Similarly, "stand-off" barriers, including moats that are at least 4 m wide and 0.4 m deep, are recommended by the EAZA.

ENVIRONMENTAL ENHANCEMENT

In the United States, the requirement to provide "an appropriate plan for environmental enhancement adequate to promote the psychological well-being of nonhuman primates" was first introduced through amendment of the AWR 1985 (7 U.S.C. § 2143), with details provided in the Animal Welfare Regulations [16] subsequently adopted in 1991 (Title 9, Code of Federal Regulations, Subchapter A—Animal Welfare Part 3 Standards, Subpart D Specifications for the Humane Handling, Care, Treatment, and Transportation of Nonhuman Primates, Section 3.81). A history and general information on environmental enrichment can be found in Nelson and Mandrell [17]; and a template to be used in developing a plan can be found in the NRC text: The Psychological Well-Being of Nonhuman Primates [9]. Dealers, exhibitors, and research facilities are responsible for development, documentation, and implementation of such plans, which are required to be "in accordance with the currently accepted professional journals or reference guides, and as directed by the attending veterinarian."

The plan must include specific provisions to meet social needs, enrichment of the environment, and plans or programs to address special circumstances requiring additional attention (Table 9.2). The language in the AWR emphasizes "providing means of expressing non-injurious species-typical activities...", through provision of "...perches, swings, mirrors, and other increased cage complexities...objects to manipulate; varied food items...foraging or task-oriented feeding methods; and...interaction with the care giver or other familiar and knowledgeable person consistent with personnel safety precautions." Social housing should be considered normative in marmosets, as they are typically housed in family groups, in compatible same-sex pairs or in mixed-sex pairs if the male has been vasectomized or one of the pair, typically the female, has been subjected to pharmaceutical methods of contraception. Exceptions to social housing must be described in the plan and may include overly aggressive animals and those suspected of harboring infectious disease. Alternative methods for meeting social needs include housing marmosets with compatible species of nonhuman primates or other animals—a practice that is likely to be more applicable to zoo settings—as long as it is not prohibited by communicable disease (as is the case with squirrel monkeys)—or by the circumstance of individuals of one species limiting access of others to food,

TABLE 9.2 Plan for Environmental Enrichment: Components of Enrichment Plans for Nonhuman Primates

Social Housing
PHYSICAL ENRICHMENT
Changing size and complexity of enclosure Furniture Manipulanda Visual barriers Substrate and forage Perches, Nest boxes, Swings, Climbing structures
NUTRITIONAL ENRICHMENT
Novel food Novel food delivery Foraging
OCCUPATIONAL ENRICHMENT
Positive Reinforcement Training
SENSORY ENRICHMENT
Visual Tactile Olfactory Auditory
CIRCUMSTANCES REQUIRING SPECIAL ATTENTION
Individuals showing signs of psychological distress Individuals for which an approved study requires restricted activity Individually housed animals without visual access to conspecifics or compatible species

Adapted from Behavioural Management of Nonhuman Primates. https://oacu.oir.nih.gov (US) [18] and the AWR.

water, or shelter. All nonhuman primates should have visual and auditory access to primates of their own or compatible species unless prohibited by the attending veterinarian. Exemptions from the plan can occur if required by an experimental protocol, adequately justified, approved by the IACUC, and reviewed at least yearly. Veterinarians may exempt individual animals from the plan, but such exemptions, unless permanent, must be reviewed every 30 days by the attending veterinarian. Specific examples of enrichment through cage complexity and design can be found in Chapters 5 and 7.

The Euroguide provides additional guidelines on social housing including the age of weaning (8 months), the time of establishment of breeding pairs (2 years), the importance of housing animals in family groups of two unrelated adults and "one or more sets of offspring," and the duration of housing of offspring in their family group so that appropriate parenting skills are learned (13 months). Appendix A of the European Convention document states the need for "some degree of complexity and stimulation." Wooden perches that allow gnawing and scent marking behavior, swings, ropes, and platforms are all recommended, as are secure rest areas. Appendix A recommends the use of foraging devices in the upper reaches of the enclosure. Wood chip substrate and food-containing enrichment devices in the lower part of the enclosure can encourage use of this ordinarily underutilized space. Because marmosets and tamarins spend 50% of their time in the wild in search of food, use of food enrichment that promotes foraging, such as wooden dowels or perches in which drill holes have been filled with gum arabic, is recommended. Similar recommendations are found in the EAZA guidelines; these also provide suggestions specific to zoo settings such as compatible (nontoxic) plants, substrates, and foraging items [7].

HUSBANDRY

There are very few guidelines on husbandry specific to the common marmoset. General comments found in the AWR {§3.82(a-d)} include the following: provision of "clean, wholesome, and palatable" food of sufficient quantity and quality to provide for normal health and weight; a minimum of once daily feeding; ready access to food receptacles; multiple receptacles to preclude competition for food or bullying by dominant animals; observation of the animals as they feed to determine that all animals have access; adoption and proper sanitation of food receptacles at a minimum frequency of once every 2 weeks; and adoption of measures to avoid fouling of food by mold, feces, and excessive moisture. Clean, potable water must be provided at least twice daily for 1 h each time (§3.83), and water receptacles must be sanitized at least once every 2 weeks. The EAZA Guidelines recommend at least two feeding and watering sites per enclosure, 1.5 m from the floor of the enclosure, with perch areas immediately adjacent for "relaxed" food and water consumption.

For sanitation, the AWR requires daily removal of excreta and food waste from primary enclosures {§3.84) (a)}. In those facilities where floors are made of dirt, or covered with bedding material or plantings, spot cleaning of sufficient frequency to avoid animal contact with excreta must be carried out. If cleaning methods include hoses or flush systems, removal of the animals may be required to keep them from being subject to distress, harm, or wetting. As with food and water receptacles, primary enclosures must be sanitized at least once every 2 weeks {§3.84(b) (2)}. This sanitation can either be achieved by live steam under pressure or "...washing with hot water {at least 180°F (82.2°C)} and soap or detergent" {§3.84 (b) (3) (ii)}. For materials that cannot be sanitized by these methods (e.g., dirt, plantings, grass, bedding), contaminated materials must be removed "...as necessary to control odors, diseases, pests, insects, and vermin infestation" {§3.84(b) (4)}. For scent marking species like the common marmoset, "...the primary enclosure must be sanitized at regular intervals determined in accordance with generally accepted professional and husbandry standards" {§3.84(2) (b) (2)}. The Euroguide and Appendix A recommend a staggered sanitation schedule to accommodate the scent marking behavior characteristic of common marmosets. In this schedule, enclosures and enrichment devices are sanitized alternately so that some scent-marked surfaces remain at all times.

The EAZA Guidelines recommend a soft bedding substrate if impervious floors with drains are in use and suggest spot cleaning within enclosures 2–3 times weekly. The substrate can be removed once or twice yearly or more often if required. The Guidelines also reference a "biofloor" composed of a 25-cm thick covering of woodchips beneath which are a filter pad and concrete floor with drain. More information on the system can be accessed through the Apenheul Primate Park website (https://www.apenheul.com).

For indoor and sheltered facilities, the AWR prohibits temperatures from falling below 45°F (7.2°C) or rising above 85°F (29.5°C) for more than 4 h and requires ambient temperatures to be maintained within the range acceptable for a particular species "as directed by the attending veterinarian" {§3.76(a), §3.77(a)}. For outdoor facilities, supplemental heat must be provided such that "the ambient temperature does not fall below 45°F (7.2°C) except as directed by the attending veterinarian" {§3.78(b)}. Appendix A of the European Convention guidelines states that for outdoor enclosures, animals must have "continuous access to adequate heated indoor accommodation" and goes on to highlight the importance of heated accommodations for frostbite-susceptible neonates. The Euroguide and Appendix A recommend a temperature range of 23–28°C but note that slightly higher temperatures are acceptable. For humidity, 40%–70% is acceptable but again the animals may tolerate higher levels. The EAZA guidelines (2010) state that "most zoos keep their indoor callithricid enclosures at a minimum of 18°C with no resultant problems...". The same document recommends relative humidity of at least 60%. Indoor temperatures can vary between 18 and 24°C; zoos may allow animal access to the outdoors to 5°C or even freezing. Outdoor heat lamps and indoor basking areas using heated shelves or heat lamps are sometimes provided. The Euroguide also recommends minimization of exposure to ultrasound, as these frequencies are within the hearing range of marmosets.

ACQUISITION, TRADE, AND INTERNATIONAL TRANSPORT

With regard to acquisition and transport, both the places of origin and of destination are likely to have national, regional, and local ordinances with which to comply. Veterinary authorities in the countries of origin and destination must be consulted for specific requirements of appropriate documentation prior to transport. Attempts to harmonize standards for care are ongoing and reflect the activity of both national and international bodies [13,19]. An excellent resource for individuals purchasing and transporting nonhuman primates is Module 26: Nonhuman Primate Transportation of the National Veterinary Accreditation Program [20] (https://www.aphis.usda.gov/aphis/ourfocus/animalhealth/nvap/ct_aast).

Trade and international travel of common marmosets is regulated and guided by a number of documents and sponsoring organizations in addition to those discussed above. These include the Convention on International Trade in Endangered Species of Wild Flora and Fauna (CITES), the World Organization for Animal Health (OIE: Office International des Epizooties), the US Centers for Disease Control and Prevention (CDC), the Lacey Act (US), the Endangered Species Act (US), and the International Air Transport Association (IATA).

- CITES: The United Nations Environment Programme in Geneva, Switzerland, administers CITES. CITES Appendices I and II cover alive or dead animals and their parts and states: "All import, export, re-export, and introduction of a species has to be authorized through a Licensing system." The common marmoset is an Appendix II species—defined as a species that might become endangered as a consequence of trade. Animals in Appendix II can only be obtained if the animal was legally acquired and if its export will not adversely affect the survival of the species. Importation of a CITES Appendix II species requires an export permit or reexport permit from the management authority of the state of export. In the United States, management authority has been vested in the US Fish and Wildlife Service. Authority in EU member states has been harmonized, permitting movement of animals between member states, subject to regulations contained in Chapter II of the Council Directive 92/65/EEC of 13 July 1992. In fact, the term "international transport" excludes "journeys of less than 50 km and between member states of the European Community" [21].
- The World Organization for Animal Health (OIE [22]; www.oie.int): The OIE (Office International des Epizooties) was established in 1924 to combat animal diseases and promote animal health worldwide. The current designation, World Organization for Animal Health, was adopted in 2003 but the original acronym for the group was preserved. In 2008, the OIE entered

into an agreement with the International Air Transport Association to cooperate on "Issues of Common Interest." These include: mutual provision of general information relating to transport of both live animals and biological specimens; veterinary research into the health and welfare of animals being transported by air; development and revision of international standards for transport of live animals; technical requirements for transport; and communication strategies relative to health and welfare emergencies arising during transport.

- The International Air Transport Association [23] (www.iata.org), Live Animal Transport Group: The IATA is an industry group that promotes safe and reliable air transport for all manner of consumers and services. They publish a text entitled: Live Animal Regulations, currently in its 43rd edition, which contains information on airline and government requirements for air transport of live animals, information on handling, marking, labeling, documentation, container requirements for various species, and guidelines for transport by other means.
- The Lacey Act: Under the Lacey Act (United States), passed in 1900, and amended in 2008, it is unlawful to trade in species of plant or animal that have been taken or collected in violation of state or national laws in the state or country of origin. All species protected under CITES are covered by the law (18 USC 42-43 16 USC 3371-3378 Lacey Act). Its enforcement arm is the US Fish and Wildlife Service, Office of Law Enforcement.
- The Endangered Species Act: In the United States, the Endangered Species Act (7 U.S.C. § 136, 16 U.S.C. § 1531 et seq.) supports CITES and implements national participation in the Convention. The Secretary of the Interior, through the US Fish and Wildlife Service, has management authority over all but marine animals.
- CDC: The Public Health Service of the United States has granted to the CDC the regulatory authority for the prevention of communicable diseases. The regulations for quarantine of nonhuman primates are found in The Code of Federal Regulations (42 CFR 71.53) [24]. The following is a summary of information pertinent to international travel and quarantine of nonhuman primates:
  - Imported nonhuman primates can be used for scientific research, education, teaching, or exhibition. They must not be imported as pets.
  - Importers must register with the director of the CDC and must document the intended legal use of the animals and must not transfer the animals to others without documentation of legal use.
  - A minimum quarantine of 31 days in a CDC-approved quarantine facility is required.
  - Any evidence of an infection in nonhuman primates that endangers the public health may result in a number of actions, from examination of the animal to its destruction at the owner's expense.
  - Quarantine must report within 24 h any primate disease suspected of being yellow fever, monkey pox, or filovirus and must report within 24 h the illness of a staff member.
  - New World primates need not be tested for filovirus.

TRANSPORTATION

A number of national and international bodies regulate or guide the transportation of laboratory animals. These regulations and guidelines cover topics such as the environmental conditions to which transport containers are exposed prior to shipment, the conditions within primary conveyance vehicles, the standards for primary enclosure used in transport, and signage and documentation during travel (Fig. 9.1). There is considerable overlap among these regulations. In the United States, the AWR contains detailed provisions for transport of nonhuman primates from the holding facility of the consigner, to the primary conveyance of the carrier or intermediate handler, to the air or cargo terminal, to the primary conveyance of another carrier or intermediate handler, to the consignee. Transport facilities in which nonhuman primates are held awaiting transport cannot fall below 45°F for more than four consecutive hours. During handling for shipment, as would occur when moving animals into and out of facilities or vehicles, the critical time not to be exceeded is 45 min {AWR (§3.91 and §3.92)} (Guidelines for the Humane Transportation of Research Animals 2006 [25]). These regulations are summarized in Tables 9.3–9.6. For air travel, industry standards have been established by the IATA. The LIVE Animal Transport Working Group of IATA has collaborated with CITES to improve transport standards and regulations, and the IATA Live Animals Regulations have been adopted by CITES, the World Organization for Animal Health, and the European Union. Airlines that are members of IATA must comply with these regulations. The EAZA has its own guidelines for transport of callitrichids but defers to IATA in the case of air transport. Similar issues of materials used, adequacy of ventilation, temperature in the enclosure (between 15 and 25°C), and freedom from contact with urine and feces are discussed in the EAZA

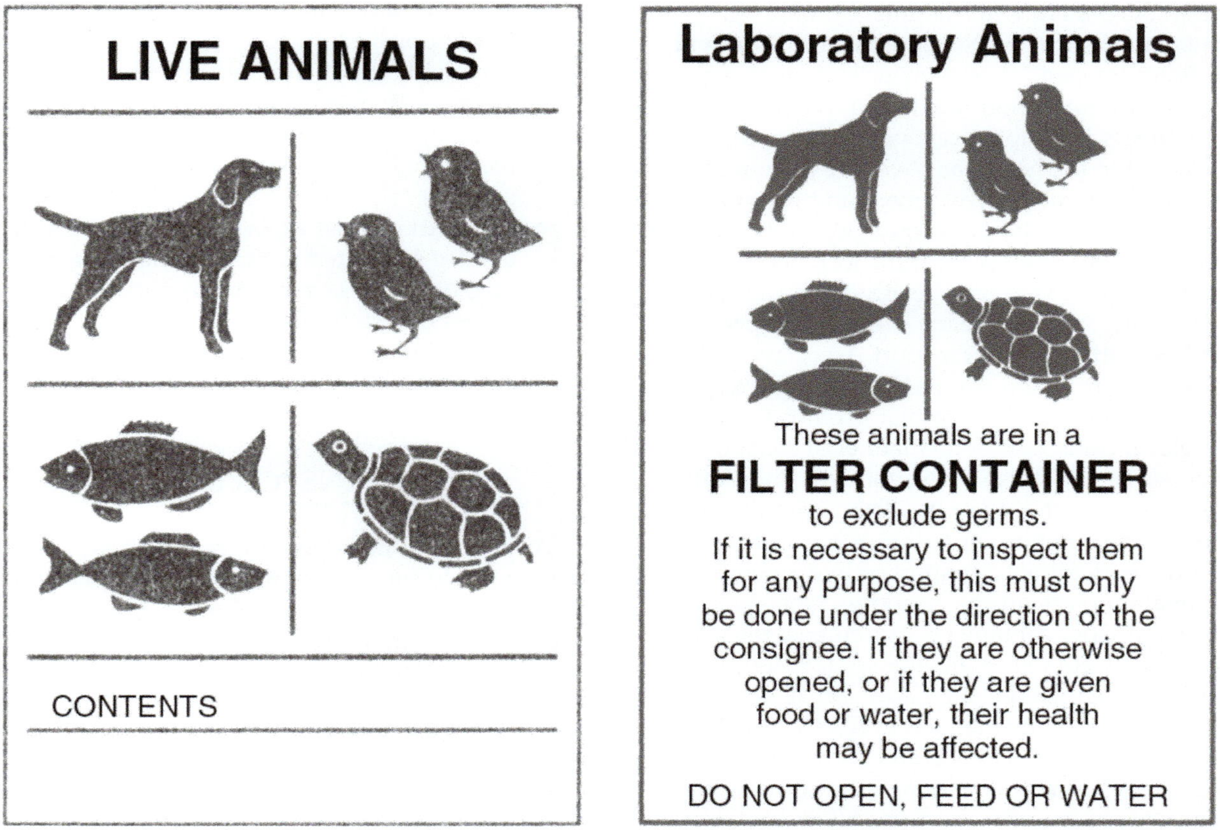

FIGURE 9.1 Examples of IATA-approved signage for shipping live animals. Reproduced with permission from W.J.White, S.T.Chou, C.B.Kole and R.Sutcliffe, Chapter 13 - transportation of laboratory animals, In: R.Hubrecht and J.Kirkwood, (Eds.), The UFAW handbook on the care and management of laboratory and other research animals, 8th ed., 2010, The Universities Federation for Animal Welfare, Wheathamstead, UK, 169-182.

recommendations, but they also invoke specific strategies to relieve stress in callitrichids. These include providing "furniture" or other items from the cage to provide familiar smells, allowing animals a method of supporting themselves on enclosure surfaces in the case of sudden or jarring movements, and the use of two compartments within the enclosure for "movement and seclusion at the choice of the animal." Enclosure

TABLE 9.3 Federal Regulations Relevant to the Transportation of Nonhuman Primates (US)

		Nonhuman Primates	
Agency	**Agency Program or Federal Regulation**	**Interstate**	**Import/ Export**
USDA [27]	Animal Welfare Act	X	X
FWS [28]	CITES [29,30]		X
	Lacey Act	X	X
	Endangered Species Act	X	X
CDC	Division on Global Migration and Quarantine (DGMQ)		X[a]
	Etiologic Agent Import Permit Program (EAIPP)		

[a]*Regulates importation only.*
http://www.nap.edu/catalog/11557.html. Copyright © National Academy of Sciences.

TABLE 9.4 Ventilation Requirements for Primary Enclosures for Transport—AWR (US)

- Bars, wire mesh, or smooth expanded metal having airspaces can be used
- Ventilation to comprise 16% of each wall when located on opposite walls
- Ventilation grills to be located above the vertical midline of the enclosure
- Ventilation to comprise 8% of the surface of each wall when located on all four sides
- Provision of 0.75 in. (1.9 cm) projection rims to prevent obstruction of the ventilation grill (if the cage is not affixed to the primary conveyance)
- When fixed to the primary conveyance, 90% of the surface area of the front wall must be ventilated
- When fixed to the primary conveyance, the front-ventilated area must open to a circumscribed passageway

TABLE 9.5 Documentation for Travel[a] (US)

- Shipping document affixed to the primary enclosure to allow easy removal for examination[b]
- Certification of having offered food and water within 4 h of delivery to the carrier or intermediate handler
- Consignor's name and address, signature, and time certification was signed
- Species of nonhuman primate
- Time and date of last feeding and watering
- Specific instruction for feed and watering for 24 h
- Instructions for medication and special care
- Instruction to withhold food or water (requires instructions by the attending veterinarian)
- Certification by a veterinarian within 10 days of delivery to the carrier or intermediate handler of acclimatization of animals to temperature lower than 50°F (10°C). A minimum temperature to be specified by the veterinarian. Certificate requires date and signature of the veterinarian

[a]*These documents are independent of those required for interstate or international import or export by national or international authorities.*
[b]*Shipping documents may also be held by the carrier while the animal is in ground transport.*

TABLE 9.6 Requirements for Primary Enclosures for Travel—AWR (US)

- Strength to withstand transport
- No interior sharp points or protrusions
- Designed to prevent animals from reaching outside the enclosure
- Animal-proof, secure door
- Constructed with nontoxic materials
- Equipped with exterior handles for "tilt-free portage"
- Designed to preclude contact of the handlers with the animals
- Provided with proper ventilation (see Table 9.4)
- Solid, leak-proof bottom or wire mesh over removable leak-proof bottom
- Food and water receptacles located to preclude contamination by excreta or vermin
- Food and water receptacles attached securely inside and such that refill can occur without opening the door
- Refill apertures of a shape and size to preclude escape
- Space adequate for postural adjustments or to sit in a palms-down position without contact of the head with the top of the cage

size of 30 × 30 × 30 cm for single animals and 40 × 40 × 40 cm for paired animals is suggested.

Documents required for transport of laboratory animals are summarized in Table 9.7, but the country or state of destination should always be contacted for guidance.

TABLE 9.7 Required Documents for Transport of Laboratory Animals, as Specified by Itinerary, Mode of Transport, Route, Species, and Health Status

- AWB number/IATA shipper's certificate for air (this should be on each container)
- Import licenses issued by the state veterinary service
- Crate labels, for example, *This Way Up* (orientation arrows), feeding guide, full consignee address, and 24-h contact telephone number
- CITES permits, where necessary
- Definitive invoices for customs purposes
- Fitness to travel documents
- Individual or group animal records
- Health certificate
- Health screen
- Journey log, if separate
- List of contacts with telephone numbers
- Packing list
- Quarantine labels
- Route plan or animal transport certificate
- Transfer authorizations from specific bodies that regulate laboratory animal use, for example, the UK Home Office, US Animal and Plant Health Inspection Service [32]
- Vehicle registration details and insurance

AWB, Air Waybill; *CITES*, Convention on International Trade in Endangered Species of Wild Flora and Fauna.
Adapted from Swallow J, Anderson D, Buckwell AC, et al. Guidance on the transport of laboratory animals. Lab Anim 2005;39:1–39.

In the United States, requirements for domestic travel of nonhuman primates can be provided by the State Animal Health Official [33] (http://www.usaha.org/Portals/6/StateAnimlaHealthOfficials.pdf) or State Public Health Veterinarian [34] (http://www.nasphv.org/Documents/StatePublicHealthVeterinariansByState.pdf).

References

[1] Convention on International Trade in Endangered Species, 2018. https://www.cites.or.

[2] Animal Welfare Act. Transportation, sale, and handling of certain animals. United States Code, Title 7—Agriculture, chapter 54, Sections 2131– 2159. 1966.

[3] Centers for Disease Control, Prevention (CDC). Animal importation. Bringing a monkey into the United States. 2016. https://www.cdc.gov/importation/bringing-an-animal-into-the-united-states/monkeys.html.

[4] Centers for Disease Control, Prevention. Legal authorities for isolation and quarantine. 2014. https://www.cdc.gov/quarantine/aboutlawsregulationsquarantineisolation.html.

[5] Centers for Disease Control, Prevention. Quarantine station contact list and map. 2017. https://www.cdc.gov/quarantine/quarantinestationcontactlistfull.html.

[6] Council of Europe. European convention for the protection of vertebrate animals used for experimental and other scientific purposes. ETS No. 123, Article 5, Appendix A. Strasbourg: Council of Europe; 2006.
[7] European Association of Zoos and Aquariums, 2015. http://www.eaza.net/assets/Uploads/CCC/2015-Callitrichids-EAZA-Best-Practice-Guidelines-Approved.pdf.
[8] Bairrao Ruivo E, editor. EAZA husbandry guidelines for Callithricidae. 2nd ed. Amsterdam: Beauval Zoo; 2010.
[9] Committee on Well-Being of Nonhuman Primates. Institute for Laboratory Animal Research, Commission on Life Sciences. The psychological well-being of nonhuman primates. National research Council. Washington, D.C: National Academies Press; 1998.
[10] Council of Europe. ETS No. 193. The European Convention on the protection of animals during international transport. 2018. http://conventions.coe.int/treaty/en/Treaties/html/193.htm.
[11] Council of Europe. Appendix A of the European Convention of the protection of vertebrate animals ETS 123. 2018. http://conventions.coe.int?treaty/en/Treaties/Html/123.htm.
[12] Forbes D, Blom H, Kostomitsopoulos N, Moore G, Peretta G. Euroguide: on the accommodation and care of animals used for experimental and other scientific purposes. London UK: Royal Society of Medicine Press; 2007.
[13] Bayne K, Morris TH, France MP. Chapter 8- legislation and oversight of the conduct of research using animals: a global overview. In: Hubrecht R, Kirkwood J, editors. The UFAW handbook on the care and management of laboratory and other research animals. 8th ed. Wheathamstead, UK: The Universities Federation for Animal Welfare; 2010. p. 107−23.
[14] AAALAC International. International regulations. 2018. http://www.aaalac.org/resources/internationalregs.cfm.
[15] Wachtman L, Mansfield K. Viral diseases of nonhuman primates. Chapter 1. In: Abee CR, Mansfield K, Tardif S, Morris T, editors. Nonhuman primates in biomedical research. 2nd ed., vol. 2. London: Elsevier; 2012. p. 1−104.
[16] Animal Welfare Act, Animal Welfare Regulations. Subpart D − specifications for the humane handling, care, treatment, and transportation of nonhuman primates. Title 9- animals and animal products, chapter 1, Subchapter A. Washington, D.C: USDA; 2017.
[17] Nelson RJ, Mandrell TD. Enrichment and Nonhuman Primates: "first, do no harm". ILAR J 2005;46(2):171−7.
[18] United States National Institutes of Health. Regulations and standards for the use of laboratory animals in the United States. 2018. https://oacu.oir.nih.gov/regulations-standards.
[19] White WJ, Chou ST, Kole CB, Sutcliffe R. Chapter 13- transportation of laboratory animals. In: Hubrecht R, Kirkwood J, editors. The UFAW handbook on the care and management of laboratory and other research animals. 8th ed. Wheathamstead, UK: The Universities Federation for Animal Welfare; 2010. p. 169−82.
[20] USDA-APHIS. National import export services contacts. Module 26: Nonhuman Primate Transportation of the National Veterinary Accreditation Program 2018. https://www.aphis.usda.gov/aphis/ourfocus/animalhealth/nvap/ct_aast.
[21] Council of Europe. European convention for the protection of animals during international transport (revised). ETS No. 193. Chisinau: Council of Europe; 2003.
[22] World Organization for Animal Health (OIE), 2018. http://www/oie.int.
[23] International Air Transport Association (IATA), 2018. www.iata.org.
[24] Federal Quarantine Regulations. Subpart F - Importations. Volume 1. Section 71.53-nonhuman primates. Title 42-public health. Chapter I, Subchapter F, Part 71. Washington, D.C: Public Health Service; 2003.
[25] Committee on Guidelines for the Humane Transportation of Laboratory Animals. Guidelines for the Humane transportation of research animals. National Research Council. Washington, D.C: The National Academies Press; 2006.
[26] National Academy of Sciences. Guide for the care and use of laboratory animals. 8th ed. Washington, D.C: National Academies Press; 2011.
[27] USDA-APHIS, 2018. Fillable PDF of APHIS form 7001. www.aphis.usda.gov/library/forms/pdf/APHIS7001.pdf.
[28] U.S. Fish and Wildlife Service (FWS), 2018a. Live Animal Transport Website. http://www/fws.gov/international/travel-and-trade/live-animal-transport.html.
[29] U.S. Fish, Wildlife Service. How CITES works. 2018. https://www.fws.gov/international/cites/how-cites-works.html.
[30] U.S. Fish, Wildlife Service. International affairs. CITES. 2018. http://www/fws.gov/international/cites/index.html.
[31] Swallow J, Anderson D, Buckwell AC, et al. Guidance on the transport of laboratory animals. Lab Anim 2005;39:1−39.
[32] U.S. Department of Agriculture Animal, Plant Health Inspection Service. Final report on environment enhancement to promote the psychological well-being of nonhuman primates. Riverdale, MD: Animal Care; 1999.
[33] Unites States Animal Health Association. State animal health official. 2018. http://www.usaha.org/Portals/6/StateAnimalHealthOfficials.pdf.
[34] National Association of State Public Health Veterinarians, 2018. State Public Health Veterinarian. http://www.nasphv.org/Documents/StatePublicHealthVeterinariansByState.pdf.

Further Reading

[1] IATA Animal Regulations. 33rd ed., 2006. International Air Transport Association, Montreal. http://www.iata.org/ps/publications/9105.htm.
[2] International Air Transport Association (IATA). Live animals regulations (LAR). 2018. http://www.iata.org/whatwedo/cargo/live-animals/Pages/index.aspx.
[3] International Air Transport Association. Live animal regulations. 44th ed. Montreal: International Air Transport Association; 2018.
[4] United States Department of Agriculture, 2018. Animal Care Web site. www.aphis.usda.gov/animal-welfare.
[5] United States Department of Agriculture. Import of Live laboratory mammals and their material (for research purposes). 2018. http://www.aphis.usda.gov/import_export/animals/animal_import/.
[6] United States Department of Agriculture. Live animal export from the United States. 2018. http://www.aphis.usda.gov/regulations/vs/iregs/animals/.
[7] United States Department of Agriculture. USDA-APHIS import export programs. 2018. http://www.aphis.usda.gov/wps/portal/aphis/ourfocus/importexport.

SECTION II

DISEASES AND CLINICAL APPLICATIONS

CHAPTER

10

Physical Examination, Diagnosis, and Common Clinical Procedures

Monika Burns, Lynn Wachtman

Division of Comparative Medicine, Massachusetts Institute of Technology, Cambridge, MA, United States

INTRODUCTION

Common marmosets (*Callithrix jacchus*) are small-bodied, New World nonhuman primates that are easily bred and maintained in captivity. The ease of handling, low risk of zoonotic disease transmission, and tendency to birth twins are factors that have long contributed to the popularity of this species as an animal model. The recent completion of the whole genome sequence and ability to efficiently execute gene editing techniques has resulted in a burgeoning scientific interest and increasing prevalence of the common marmoset in research vivaria. As with any other species, a well-designed preventive medicine program is critical to the veterinarian's ability to monitor the health and well-being of the marmoset colony and thus protect these valuable research assets. As a component of a robust preventative medicine program, physical examination can lead to the early recognition and treatment of a variety of clinical conditions affecting individuals or groups of animals. Each clinician must develop his or her own physical examination method that is complete, efficient, and consistent. The method of physical examination will vary from clinician to clinician and will also depend on whether the animal is alert, sedated, or acclimated to handling by veterinary staff. This chapter describes the marmoset physical examination process and also provides instruction on procedures routinely performed as part of a clinical care program. Common physical examination findings and the diseases most frequently associated with these findings are presented; however, the other chapters in this text will provide a more complete discussion of the etiology and pathogenesis of these diseases.

PART I: THE PHYSICAL EXAMINATION

The physical examination is performed with a set of anticipated findings in mind that are based on age, signalment, clinical and experimental history, as well as individual and colony-wide history. Life stages of the common marmoset are divided into the following groups: infant (0–3 months); juvenile (3–12 months); subadult (13–18 months); sexually mature (18–24 months); prime adult (2–8 years); and aged (>8 years) [1]. The incidence and prevalence of disease processes varies across life stages, and this is reflected in the ranking of possible differential diagnoses. Because marmosets are often housed in family groups of 2–10 animals, it is important to evaluate other group members if any of the clinical observations are likely to result from an infectious process or unsafe cage dynamics.

It is crucial to maintain current and complete medical records on all individual animals in the colony. Whether information is stored within an electronic medical records system or on paper records, a complete medical record will include the following information: animal identification, physical exam results, details of immunizations and prophylactic treatments, documentation and interpretation of diagnostic tests, description of research interventions, treatment plans (including clinical response and follow-up), surgery, anesthesia, analgesia, perioperative care, control of pain and distress, and documentation of euthanasia and necropsy findings [2]. It is a best practice that medical record entries related to the physical examination are entered into the records system as soon as possible after the examination has been performed.

The first step to any examination involves careful cage-side observation of the animal's behavior as it

The Common Marmoset in Captivity and Biomedical Research
https://doi.org/10.1016/B978-0-12-811829-0.00010-8

moves about and interacts with both conspecifics and human observers. In the natural environment, Callitrichids spend about 20% of the daytime traveling, 29% resting, 14% foraging, 13% gouging, and 21% feeding, with the last 3% of time devoted to other activities [3]. These estimations of time allocation may be useful to the clinician when establishing a baseline for normal behavior of colony-housed animals but may vary depending on cage population, cage fixtures, and feeding opportunities. Thus, clinicians should develop a general idea for normal behavior in their facility. Overall alertness and activity level should be noted. Most animals are typically observed in an alert, upright posture in the mid-to-upper tier of the cage (Fig. 10.1). Marmosets in poor health are generally reluctant or slow to move and often have a lethargic or dazed appearance. While it is not necessarily abnormal to observe a marmoset on the bottom of the cage, this finding warrants a closer or prolonged observation to ensure the animal is not ill or subject to aggression by cagemates. Marmosets are a diurnal species, and it is not unusual to observe the group taking an afternoon nap in hammocks or nest boxes after eating a midday meal. While monitoring activity level, any observed deviations from the normal quadrupedal gait and characteristic movement patterns suggest that evaluation of the animal for lameness or neurologic impairment is indicated.

The cage bottom and perches should be evaluated for the presence of diarrhea, vomitus, hematochezia, or hematuria. The presence or absence of food consumption can be confirmed by an inspection of food bowls and food enrichment devices. Water consumption is difficult to evaluate if an automatic watering system is in use. The clinician will need to rely on evidence of urine production and clinical indicators of adequate hydration. The clinician may also want to listen to vocalizations both before and after entering the animal-holding room. This species has a wide range of vocalizations employed in different situations, including chirps, whistles, phees, and screeches. Some vocalizations, such as chirps, are affiliative calls produced in nonstressful situations, whereas other vocalizations only occur when the animal is distressed. Chattering or squeals may indicate intragroup aggression. Interestingly, as marmosets mature, these vocalizations change based on parental feedback in a pattern very similar to that seen in human development [4]. Recordings of marmoset vocalizations can be found online at http://marmosetcare.com/ [5].

FIGURE 10.1 Healthy marmosets observed in the home cage. Animals maintain an upright posture in the mid-to-upper tier of the cage and demonstrate an alert gaze that is oriented on the observer or conspecifics.

For closer evaluation, the marmoset must be moved to a clinic or procedure room. This requires a transfer method. The authors prefer the use of a capture box to isolate the animal from the home cage and move it to a procedure room. This requires both patience and animal acclimation. With training, most marmosets will readily move into a transfer box if offered a food treat. This process will be discussed in more detail in Part III of this chapter. Ideally, the clinical handling space and equipment are dedicated to the examination and treatment of marmosets. Squirrel monkeys (*Saimiri sciureus*) and marmosets should never be handled in the same procedure room or using the same clinical equipment due to the risks of transmission of *Saimirine herpesvirus 1* and *Saimirine herpesvirus 2*. Human handlers should always wear personal protective equipment when handling and examining any nonhuman primate, including marmosets.

Body Weight and Body Condition

After sedative drugs have taken effect and the marmoset is removed from the transfer box, body weight is measured and the current body weight is compared with historical weights. Any increase or decrease of greater than or equal to 10% of the prime body weight should be noted in the medical record and investigated with appropriate diagnostic tests. Reported mean body weights of wild marmosets range from 320 to 335 g, whereas mean body weights of captive marmosets range from 283 to 530 g [6–9]. Detailed information about mean body weight and other normative data is available in Chapter 6 of this text [10]. In addition to objective measurements of body weight, a subjective evaluation of body condition is performed by gentle palpation of the thorax, abdomen, and soft tissues overlying the thoracolumbar dorsal vertebral spinous processes and sacroiliac bones. The body condition scoring system that the authors use is adapted from a scoring system previously validated for use in rhesus macaques (*Macaca mulatta*) [11]. The relatively long haircoat of the marmoset can obscure thin body condition in underweight or emaciated animals. Thorough palpation is therefore essential to the appropriate assignment of body condition score. Visual inspection alone may result in artificially high body condition scores. A body condition score of 1 is reserved for emaciated animals with easily palpable bony structures and indiscernible muscle mass. Animals with an optimal score of 3 have palpable bony structures underneath well-developed muscle mass and subcutaneous fat layers. A score of 5 is assigned to grossly obese animals with obvious and large fat deposits throughout the body, particularly in the axillary or abdominal regions. See Table 10.1 for further guidance on body condition scoring in common marmosets. Due to the subjective nature of body condition scoring, all staff members require training on body-condition scoring criteria to enhance interobserver agreement.

TABLE 10.1 Body Condition Scoring Chart for Common Marmosets

Score	Description
1	**Emaciated**—immediately/easily palpable, possibly visible hip bones. Indiscernible muscle mass palpable over ileum/ischium. Anus appears recessed. No subcutaneous fat.
1+	**Very thin**—immediately/easily palpable, possibly visible hip bones. Very scant muscle palpable over ileum/ischium and epaxial muscle region. Anus may appear recessed. No discernible fat reserves.
2	**Thin**—scant fat reserves, easily palpable hip bones/spinous processes. Scant muscle coverage over ileum/ischium and epaxial muscle region.
2+	**Lean**—obvious muscle mass palpable along spine/hip regions. Thin layer of subcutaneous fat.
3	**Optimum**—good muscle mass is palpable with gentle pressure along spine/hip regions. Thin subcutaneous fat layer. No abdominal, axillary, or inguinal fat pads.
3+	**Slightly overweight**—bony prominences are smooth, small fat pads may be present.
4	**Heavy**—bony prominences are palpable with firm pressure. An abundant subcutaneous fat layer may be present. Fat deposits accumulate in axillary, mammary, and inguinal areas.
4+	**Obese**—prominent fat pads present. Abdomen may appear very round when sitting in cage. Bony contours are smooth and ill-defined and are difficult to palpate. Mammary fat pads may be large, even in males.
5	**Grossly obese**—obvious, large fat deposits are present in many regions. Abdominal palpation is hindered due to a large amount of mesenteric fat. Large fat deposits may alter posture/ambulation. Hip bones/spinous processes are only discernible with deep palpation.

Marmosets are predisposed to diseases that cause weight loss as well as diseases that result in significant weight gain. Differential diagnoses to consider when weight loss has been identified include the following: chronic lymphocytic enterocolitis (CLE, known historically as marmoset wasting syndrome), amyloidosis, chronic renal disease, chronic endoparasitism, and neoplasia. Food hoarding by cagemates has been anecdotally noted to result in decreased diet consumption and resultant weight loss in more submissive animals. Complete blood cell count, serum chemistry, radiography, abdominal ultrasound, and fecal parasite examinations are appropriate initial diagnostic tests to perform following identification of weight loss on physical examination. Of these and depending on the colony, CLE may be the most common. CLE, also known clinically as inflammatory bowel disease (IBD), is a poorly defined and incompletely understood syndrome in marmosets, and the diagnosis is often applied when nonspecific findings such as weight loss, diarrhea, and hypoalbuminemia are identified. These findings may be caused by diverse disease processes. The clinician should make every attempt to arrive at a specific diagnosis before initiating treatment with medications such as oral budesonide, which has been shown to reverse weight losses attributed to IBD/CLE [12]. A body weight of less than 325 g or serum albumin less than 3.5 g/dL has been shown to have a sensitivity of 92% and a specificity of 100% for the identification of chronic lymphoplasmacytic enterocolitis [13]. These findings may prompt the clinician to consider surgical intestinal or colonic biopsy collection. Protocols for these procedures are discussed in Part III of this chapter.

Marmosets easily become obese in captive settings, especially if consistently offered calorically dense foods [9,14]. As an accompaniment to obesity, marmosets are also prone to the development of endocrine disorders such as diabetes mellitus and metabolic syndrome. The precise calorie and nutrient requirements of common marmosets have not been validated to date, and it is possible that many commercial and hand-prepared diets do not optimally match the nutritional needs of captive common marmosets. Further prospective studies are needed to better refine laboratory diets for marmosets. Significant weight gain noted on physical examination should prompt diagnostic bloodwork including complete blood count, serum chemistry, and glycated hemoglobin (HbA_{1c}) testing. Animals classified as obese will likely display dyslipidemia, elevated fasting glucose, and elevated HbA_{1c} [9]. Weaning behaviors of infant marmosets can influence the development of feeding phenotype and the development of juvenile obesity [15]. The likelihood of an individual marmoset developing a disease that leads to either significant weight gain or loss during his or her lifetime is high, and thus, body weight measurement and body condition scoring are important components of every examination.

Body Temperature

Adequate thermal support must be provided to anesthetized marmosets, as they are exquisitely prone to hypothermia due to their large surface area to body mass ratio. Conversely, aggressive thermal support can result in hyperthermia necessitating the need for frequent monitoring during anesthetic episodes. Rectal temperature assessed at the beginning of the physical examination will provide the most direct measurement of the animal's temperature, although stressful capture procedures may increase the temperature obtained at the time of sedation. Normal daytime body temperature acquired using telemetry is reported to be in the range of 38.5°C [16]. Marmosets demonstrate marked diurnal variation with body temperatures as low as 34°C measured during the dark phase of the light cycle [17].

Dermatologic System

During the physical examination, the pelage and skin surfaces are thoroughly inspected for any wounds or indications of dermatitis, and any lesions are treated appropriately. Marmosets often have a yellow/orange tinge to the facial skin that is not associated with an increase in serum liver enzymes and should not be mistaken for icterus. Alopecia, specifically tail alopecia, can be a significant problem in captive marmoset colonies. All regions of alopecia should be documented by an estimation of the total area and location affected, and a plan should be put in place to address the hair loss. Factors contributing to alopecia include poor nutrition, stress, overgrooming by conspecifics, lack of appropriate substrates for scent marking, or systemic disease. Tail alopecia is often noted in animals that have been diagnosed with CLE, but a causal relationship between CLE and the development of tail alopecia has not been proven [18]. Alopecia scoring systems may be developed and employed at each institution to encourage objective assessment of progression or remission of tail alopecia. Other previously reported dermatologic abnormalities in the marmoset include a case report of a subcutaneous mass, ultimately diagnosed as calcinosis circumscripta, in a young male [19].

Occasionally, secretion laden or matted haircoats are noted during the physical examination. Based on the authors' experience, possible causes of this oily appearance include excessive room humidity, inappropriate cage or room density, or lack of appropriate substrates for scent marking. Replacing metal cage structures with wood

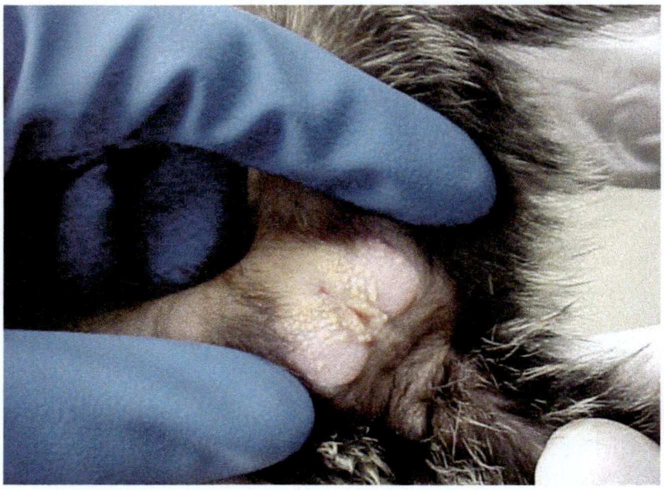

FIGURE 10.2 Circumferential glands distributed over the labia of a female marmoset. The vulva of this female is abnormal and demonstrates a central fusion of the labia.

substrates may help to decrease oily haircoats within the family group, as wood is much more absorbent than metal [20]. If secretion-laden or matted haircoats are noted, individuals can be bathed with warm water and combed to remove matted fur. Occasionally, the mats require removal using clippers. After bathing, room and cage conditions should be evaluated for relative humidity, animal density, and substrate composition.

Scent messages related to sexual or territorial behavior are passed via marmoset scent-marking behaviors [21]. Marmosets have two distinct scent-marking glands: the circumferential and sternal glands. The circumferential glands are sebaceous glands that overlie enlarged apocrine glands covering the labia in the female and the scrotum in the male (Fig. 10.2). Sternal scent glands are principally apocrine glands on the anterior chest and may be focal or diffuse [22]. Individual female marmosets can be identified by the unique ratio of volatile chemicals in their scent marks, and pheromonal cues from the dominant female in the family group suppress LH secretion and ovulation in socially subordinate female marmosets [23,24]. It is not unusual for the sternal scent gland to have a prominent appearance in some individuals. Occasionally, waxy secretions will accumulate around the periphery of the scent glands. Secretions may be orange-to-reddish in color and should not be mistaken for blood or erythema. Secretions may be removed easily with a gauze square soaked in warm water.

Visual and Auditory Systems

The eyes and periorbital structures are evaluated for ocular discharge, symmetry, pupil size, cataracts, palpebral response, conjunctival inflammation, and direct and consensual pupillary light responses. Normative anatomical data regarding the marmoset eye have been reported. The adult marmoset globe is approximately 13 mm in diameter, and the macula is approximately 2.5 disc diameters from the optic disc with a prominent foveal avascular zone [25]. Further information about ocular anatomy and measurements is presented in Guyomard et al. [26]. Ophthalmoscopy and retinal exam are not commonly performed unless indicated by clinical history or future research use. If a marmoset colony is involved in vision research, a more thorough ophthalmic and visual exam is recommended to establish baseline visual capabilities that may affect study design or results, as a high percentage of captive marmosets experience refractive errors after 3 years of age [27,28]. Few abnormalities are routinely encountered during the marmoset ocular exam although hyaloid artery remnants in young animals are reported as one relatively common abnormality [29]. The ears are evaluated for evidence of discharge emanating from the canal and for swelling of the pinnae.

Respiratory System

Examination of the respiratory system begins with an inspection of the nares for symmetry and freedom from discharge or debris. Discharge from the nares may be indicative of upper respiratory tract infection, trauma, presence of a foreign object, benign or neoplastic mass, or parasitism. The color of the discharge, the relative amount, chronicity, and whether it is unilateral or bilateral will impact the rank of possible contributing etiologies. The average respiratory rate of alert, manually restrained marmosets is approximately 100 breaths per minute, whereas animals under sedation will have a lower respiratory rate [30]. The choice of anesthetic protocol may have a profound impact on respiratory rate in the marmoset [31]. Any disease process that causes a mass effect in the thoracic cavity can result in increased respiratory rate, as was noted in a 9-month-old female common marmoset with a mediastinal myeloid sarcoma that filled over 50% of the thoracic cavity [32]. Nasopharyngeal tumors have been documented in the marmoset, and a possible differential for nasal masses is squamous cell carcinoma [33,34]. Pulmonary tumors, including small cell carcinoma, bronchial adenoma, and squamous cell carcinoma, have also been documented in this species [35].

With modern laboratory animal housing facilities and biosecurity practices, the prevalence of infectious disease in marmoset colonies today is certainly lower than historical prevalence. Outbreaks of respiratory disease with both viral and bacterial etiologies have been described in marmoset colonies. An outbreak of *parainfluenza I* virus occurred in marmosets that presented with sneezing, serous to purulent nasal discharge, ocular discharge, depression, and anorexia [36]. *Bordetella bronchiseptica* was isolated from the nasopharynx, trachea, and lungs of a group of 16 young marmosets that died suddenly after presenting with mild rhinitis [37].

Cardiovascular System

An infant or pediatric stethoscope is the most appropriate model of stethoscope to use during cardiac auscultation of the marmoset. The average heart rate of alert, manually restrained marmosets is approximately 405 beats per minute [30]. Though the average marmoset heart rate is extremely rapid during alert examination, it is feasible to auscult heart murmurs or arrhythmias if present. Heart rate and rhythm can be tracked with telemetry [38]. With appropriate equipment, it is also possible to perform electrocardiography and blood pressure measurements. Further discussion of normative cardiovascular system measurements is found in Chapter 6 of this text [10].

A large retrospective study that evaluated causes of mortality for marmosets during a 9-year period at a regional primate center found that 1% of marmosets died as a result of cardiac fibrosis [15]. Another retrospective study found that heart failure was cited as the cause of death in 29% percent of marmoset necropsy reports [39]. Myocardial fibrosis, typically an asymptomatic and incidental finding at necropsy, was noted in 60% of animals who died at greater than 10 years of age and is thought to correlate with increased risk of arrhythmias and congestive heart failure [40]. An experimental study noted that marmosets fed both glucose-enriched and high-fat diets had histopathological evidence of intermediate (Type III) and advanced (Type IV) atherosclerotic lesions, respectively [14]. The presence of lesions correlated with a diagnosis of glucose intolerance in those marmosets fed the glucose-enriched diet. In this study, marmosets with atherosclerosis demonstrated a nonsignificant trend toward elevated total cholesterol. The clinician should be aware that it is possible for atherosclerosis to develop in colony animals with naturally occurring metabolic syndrome. Because the above-referenced, glucose-enriched diet is a commercially available diet designed for routine feeding, clinicians should pay careful attention to diet choice in an effort to limit the availability of simple sugars present in the diet when possible.

Lymphoid/Immune System

Peripheral and mesenteric lymph nodes are palpated during the physical examination to assess for lymphadenomegaly. The inguinal lymph nodes are the easiest to evaluate in the marmoset due to a superficial location. A dark discoloration may be noted in the lymph node tissue if a tattoo has been placed on nearby skin. Splenomegaly is easily palpable during abdominal palpation of the left upper abdominal quadrant. Differentials for lymphadenomegaly and splenomegaly include infectious and neoplastic etiologies. It is not unusual for neoplastic etiologies to be associated with viral infections. Experimental inoculation of marmosets with *Saimirine herpesvirus 2* results in the development of T-cell lymphomas [41,42]. Spontaneous T-cell lymphoma has been documented in a 2-year-old male marmoset that presented with enlargement of submandibular and inguinal lymph nodes and, as noted at necropsy, enlargement of the thymus and spleen [43]. Virus isolation was not attempted in this case. Marmosets with B-cell lymphomas associated with *Callitrichine herpesvirus 3* presented with palpable enlargement of peripheral and mesenteric lymph nodes [44,45]. Enlarged inguinal lymph nodes were also observed in a *Mycobacterium kansasii*-infected marmoset [46].

Digestive System

Oral cavity: A complete dental examination is an important component of the marmoset oral exam, as dental disease represents a frequent clinical problem in captive marmosets. While the prevalence of dental disease in captive marmoset colonies is unreported, the prevalence in wild marmoset populations is approximately 10% [47]. The marmoset dental formula is (I2/2; C1/1; P3/3; M2/2) [16]. Marmosets have procumbent incisors that are the same length as their canine teeth, and allow animals to gnaw into trees and consume their natural diet of gums and exudates [16]. Tooth fractures of the canine teeth, usually the maxillary canines, often occur due to biting of gloves during handling. Tooth fracture can result in pulp exposure, infection of the pulp, and, ultimately, tooth root abscesses, which typically present in the form of facial swelling ventral to the orbit. The swelling can range in size from barely perceptible to quite large, causing reduction in size of the palpebral fissure, and may be erythematous on presentation. Occasionally, draining tracts will appear after the development of facial swelling. Rarely, tooth root abscesses progress to involve other facial tissues and structures and present in the form of cheek enlargement rather than a suborbital swelling. Commonly cultured bacteria include *Staphylococcus aureus* and anaerobic organisms [48]. At the authors' institution, *Neisseria gonorrhoeae* has been isolated from marmoset tooth root abscess cultures; the clinical significance of such isolates is unknown.

The oral examination commences with an evaluation of mucous membrane color and capillary refill time. Pale mucous membrane color is indicative of anemia, jaundice is apparent as a yellow-tinged mucus membrane color, and bright pink/red mucus membranes are observed in the hyperdynamic phase of shock. A slow capillary refill time may suggest the presence of dehydration or shock. As with other nonhuman primate species, care should be taken to thoroughly screen the oral and lingual surfaces for any signs of ulceration or vesicle formation. Fatal infections of *Human herpesvirus 1* (HHV-1) have been reported in captive marmosets with oral, lingual, and facial skin ulcers, and excessive salivation, as well as neurologic signs including prostration, seizure, and aggressive behavior [49,50]. As reviewed in Wachtman and Mansfield, *Saimirine herpesvirus 1* is also a cause of oral ulcerations in the marmoset [42]. If an oral mass is noted, squamous cell carcinoma is one possible differential diagnosis. One report found a 4.9% incidence of invasive oral and nasopharyngeal squamous cell carcinoma in a breeding colony of marmosets, with most cases presenting in males over 4 years of age [33].

Gastrointestinal tract: The perineal area should be inspected for evidence of diarrhea or fecal staining. A thorough abdominal palpation includes the evaluation of the stomach for distension and the examination of the small and large intestines for excess gas or fluid contents. The presence of diarrhea is considered potentially significant. Any persistent diarrhea lasting longer than 24 h, or any diarrhea with frank blood, should be followed up with a thorough diagnostic workup to rule out both infectious and noninfectious causes. Importantly, it is not uncommon for marmosets to vomit shortly after recovery from ketamine sedation.

Documented infectious causes of diarrhea in marmosets include parasitic, viral, and bacterial agents [51–54]. Diarrhea-causing bacteria of particular concern in captive marmoset populations include the following: *Yersinia* spp., Campylobacter spp., *Klebsiella pneumoniae, Shigella flexneri,* and some strains of *Escherichia coli* [54–57]. These organisms are discussed in further detail in Chapter 16 of this volume [58]. *Cryptosporidium parvum* and *Giardia* spp. are protozoal parasites that may also be associated with clinically significant, but potentially less serious, diarrhea.

Helicobacter spp. are isolated from many laboratory animal species, and the marmoset is no exception. One recent study found that 28.2% of a research colony of marmosets was culture positive for a novel *Helicobacter* species, *Helicobacter jaachi* sp. [59]. The role of this organism in clinical gastrointestinal disease in the marmoset is unknown [60,61]. Exposure to clostridial organisms may also result in gastrointestinal disease in marmosets. *C. piliforme* has been documented as the causative agent of spontaneous death in a neonatal marmoset in a zoo facility [62]. Gaseous abdominal distension and sudden death was reported in a female marmoset with nontraumatic gas gangrene caused by *Clostridium perfringens* type A [63].

Gastroenteritis caused by measles, a *morbillivirus* in the *Paramyxoviridae* family, warrants additional discussion. Outbreaks of measles in modern marmoset colonies are rare, likely due to the decreasing prevalence of human disease and increased biosecurity measures, but the risk of devastating disease is great enough that it remains an important differential diagnosis for marmoset gastrointestinal disease. In marmosets, severe gastrointestinal signs and a rapidly progressive course of disease are common clinical signs of measles virus infection, whereas exanthema is characteristically absent [42,64]. The devastating and highly contagious nature of this pathogen in marmoset colonies reinforces the importance of maintaining a closed colony, accessed only by personnel with positive IgG titers to measles virus.

As mentioned previously, marmosets are prone to CLE/IBD [48]. Physical examination findings include evidence of diarrhea combined with a loss of body condition. Histologically, CLE is characterized by altered small intestinal and colonic morphology with marked lymphocytic infiltrates into the lamina propria, crypt cell hyperplasia, villous atrophy and fusion, and increased numbers of intraepithelial lymphocytes [48]. Subclinical gut inflammation is thought to result in reduced digestive efficiency. Even in apparently healthy marmosets, digestive efficiency can vary by up to 8 percentage points between individuals, a result hypothesized to reflect the difference between animals with healthy guts and those with mild-to-moderate intestinal inflammation [65]. In the latter case, prolonged subclinical disease may result in vitamin D deficiency leading to metabolic bone disease [66].

Palpation of an abdominal mass may be indicative of gastrointestinal tract neoplasia. Various gastrointestinal neoplasms have been documented in the marmoset. One retrospective study found that small intestinal adenocarcinoma was the most common neoplastic cause of morbidity and mortality in aged common marmosets. Mean age at necropsy was 6.6 years of age, and no sex predilection was found [67]. Other gastrointestinal masses have been documented in the marmoset, including a report of an inflammatory fibroid polyp that was found incidentally in the duodenal wall of a 32-month-old marmoset during an end-of-study necropsy [68]. Thus, differentials for gastrointestinal masses in the marmoset should include both neoplastic and benign processes.

Liver: During routine abdominal palpation of the marmoset, the liver should not extend beyond the limits of the costal margin. The experienced clinician is able to discern hepatomegaly in the marmoset if firm tissue is palpated in the right cranial quadrant of the abdomen. In one colony of marmosets, palpable hepatomegaly, diagnosed histologically as hepatic lipidosis, was detected in 18% of animals [69]. In this group of animals, marmosets with hepatomegaly had higher body condition scores and increased total body mass compared with those without palpable hepatomegaly. Animals with greater hepatic mass also had increased mean serum triglycerides and plasma insulin levels. Findings from this study suggest that the common marmoset may serve as an animal model for the study of nonalcoholic fatty liver disease and nonalcoholic steatohepatitis.

Another important differential for hepatomegaly in marmosets is amyloidosis [70]. Secondary amyloidosis is the most commonly noted form of the disease in most mammals, with lesions caused by pathologic deposition of the amyloid A (AA) protein. Elevations in the precursor protein, serum amyloid A are typically associated with chronic inflammatory or infectious processes. Amyloidosis is a common diagnosis in captive marmoset colonies; 17.4% of marmosets euthanized at a regional primate center over a 10-year period had evidence of amyloid deposition in at least one tissue [40,70]. In another study, amyloid deposition was noted in 12% of marmosets at necropsy but was listed as a significant morphologic diagnosis in only 3% of animals [39]. Amyloid deposition also causes clinically significant lesions in renal and small intestinal tissue.

Other differentials for hepatobiliary system disease include both infectious and noninfectious etiologies. Lymphocytic choriomeningitis virus infection, known historically as Callitrichid hepatitis, is an important differential for febrile, lethargic, or found-dead animals, especially those with possible exposure to feral rodents [71]. Affected animals present with jaundice and hepatosplenomegaly [72]. A noninfectious differential for liver disease in marmosets is hepatic hemosiderosis [73]. A consequence of iron overload, hepatic hemosiderosis is one of the most common postmortem findings in captive-housed callitrichids [74]. Despite refinements to commercial diets, the levels of dietary iron provided remain high. Captive marmosets lack access to natural tannins obtained from leaves, barks, and fruits, which serve to inhibit iron absorption in free-ranging marmosets. The authors urge careful monitoring of dietary iron access and judicious use of parenteral iron supplementation. Serum iron parameters and liver biopsy may be helpful in ruling out iron overload in the marmoset.

Urinary System

In contrast to larger species of nonhuman primates, both the left and right kidneys are easily palpable in the marmoset. Both kidneys are often quite mobile, and the left kidney is often mobile enough to occasionally be mistaken for an abdominal mass by the inexperienced clinician. Subjective characterization of the relative size, cortical texture, and homogeneity of both kidneys should be appreciated on physical exam. Diagnostic imaging modalities such as abdominal radiography and ultrasound provide a more objective assessment of kidney size and capsular texture. Marmosets will frequently urinate during the process of sedation, and the sample will occasionally be large enough to collect for urine specific gravity analysis. A protocol for training marmosets to provide urine samples on cue has been described [75].

A diagnosis of nephritis was listed as one of the most common morphologic diagnoses in a retrospective analysis of 129 marmoset records [39]. Spontaneous progressive nephropathy with glomerular lesions is commonly reported in marmosets [76]. It is typically characterized histologically by enlargement of glomeruli, an increase in mesangial cells, interstitial infiltration of inflammatory cells, tubular regeneration and hyperplasia, and glomerular sclerosis [77]. Such lesions tend to be seen in older animals, and the prevalence of disease increases with age. The development of chronic renal lesions often occurs concurrently with clinical signs such as weight loss, anemia, and increases in blood urea nitrogen (BUN), creatinine, and alkaline phosphatase (ALP). Imbalances in calcium and phosphorus levels may also be noted, especially in instances of hyperparathyroidism secondary to chronic renal disease. In the authors' experience, changes in serum creatinine values due to renal disease tend to be less severe than those noted in other species; a marmoset with significant renal compromise may still have serum creatinine values of less than 1.0 mg/dL.

Retrospective studies have found that 75% of marmosets aged greater than 8 years of age have some degree of renal pathology at the time of necropsy; however, renal disease was noted as the primary cause of death in only 16.7% of cases [40]. The origin of nephropathy in marmosets may have multiple etiologies, including IgA, IgM, and IgA/IgM antibody deposition [78,79]. Spontaneous glomerular mesangial lesions in the marmoset have been documented as benign nonprogressive glomerulopathies by Collins [80,81]. Animals examined in that study also had low-grade proteinuria with normal renal function. Amyloidosis is another common cause of nephropathy in marmosets [48,70]. A colony of marmosets exposed to bark and leaves from *Eucalyptus viminalis* trees developed renal failure caused by oxalate-induced nephropathy [82].

Infectious agents that commonly cause renal disease in other species, such as *Leptospira* spp, are infrequently cited as the cause of clinical illness in marmosets. Experimental leptospirosis has been induced in *C. jacchus* to model severe pulmonary leptospirosis syndrome seen in humans [83]. Severe tubulointerstitial nephritis was noted in experimentally infected animals, as was hemorrhage surrounding main branches of bronchi. Seroepidemiological surveys have shown that Callitrichids are unlikely to be seropositive for *Leptospira* spp. [84]. Laboratory animal housing environments are typically maintained free of feral rodents, and thus likelihood of exposure of a closed colony of marmosets to leptospirosis, and potentially other pathogens, is low.

Reproductive System

Marmosets are sexually monomorphic, with no characteristics distinguishing males and females, except the external genitalia. In males, the testes are evaluated for symmetry and mobility within the scrotum, and the penis extruded from the prepuce to evaluate for discharge, ulceration, or malformation. Reports of spontaneous diseases of the male genital system in marmosets are lacking. Differentials for observed abnormalities are similar to those reported in other nonhuman primate species and include cryptorchidism, complications of vasectomy, orchitis, prostatitis, seminal vesiculitis, and neoplastic disease. Retrospective studies have documented testicular atrophy, prostatic degeneration, prostatitis, and seminal vesicle impaction in male marmosets [39].

In females, the vulva is examined for evidence of discharge. Females have estrous cycles and do not menstruate. Observation of a hemorrhagic discharge is abnormal and should be evaluated further with additional diagnostic modalities. Rarely, examination of the external genitalia will reveal abnormalities, including small vulvar opening (Fig. 10.2) [85,86]. The clinician should take note of such abnormalities and these animals should be excluded from breeding programs. Endometriosis, a common reproductive tract disorder in female macaques, has been induced experimentally in marmosets but no spontaneous cases have been reported [39,40].

Callitrichids do not exhibit a lactational anestrus, and therefore a marmoset dam may become pregnant after mating during a postpartum ovulation that occurs 7–14 days after parturition. Pregnancy status can be determined with caudal abdominal palpation, or with ultrasound examination. Pregnancy detection and gestational staging is an important component of the physical examination of breeding females. Data on estimated parturition date, when communicated to the husbandry and technical staff, allows for supplemental monitoring in the days leading up to parturition. Dystocia is not common, but will likely occur sporadically in any colony that breeds a significant number of marmosets. When signs of dystocia are observed, medical therapy with oxytocin can be attempted, however, emergency surgical correction is typically required. If not addressed in a timely fashion, dystocia can result in uterine rupture. One case report describes a multiparous female presenting with acute weight loss, abdominal enlargement, and peritoneal effusion that was ultimately diagnosed to be secondary to uterine rupture [87].

Marmosets have one pair of axillary mammary glands. In overweight males, adipose deposition can be mistaken for gynecomastia. Nursing females may develop mastitis, which presents as erythematous, enlarged, firm mammary tissue. Culture of milk samples is indicated if mastitis is suspected. Milk sampling technique is described in the Part III of this chapter.

Musculoskeletal System

The digits and long bones are closely evaluated during the physical examination. Marmosets have claws instead of nails on both front and hind digits; however, the opposable first digit on each foot (the hallux), contains a specialized nail. Careful examination may reveal luxated claws, abrasions, or phalangeal fractures. The clinician should palpate the length of the long bones of the arms and legs as well as check the range of motion of all joints. Long bone fractures are rare in the marmoset, but when they do occur, they require intensive clinical management. Fractures tend to be associated with the process of hand catching animals or with the process of recovering animals from sedation. Animals with fractures generally respond well to immobilization of the bones with a padded bandage and splint. The predisposition of marmosets to bone fractures and bone pathology has been reported in the literature [48,88]. One contributing factor is the marmoset's requirement for high dietary levels of vitamin D3 due to an overexpression of a vitamin D response element binding protein. Despite a normal expression level and function of the vitamin D receptor, the overexpression of the binding protein results in a target cell resistance to vitamin D, the latter being critical to the maintenance of calcium homeostasis and bone mineral density (BMD) [1,89]. Interventions to reduce the risk of bone fractures in colony animals include provision of a nutritionally complete diet and housing of animals under full spectrum lighting.

Measurements of calcium, ionized calcium, and phosphorus levels can be performed as a routine component of the preventative health-care (PHC) program, particularly if bone disease has been identified in colony animals. Parathyroid hormone level (PTH) can also be measured, as low calcium and phosphorus mediate increases in PTH, which allows for mobilization of calcium and phosphorus from bone to ameliorate the deficiency. Annual survey radiography during routine health exams can be used to subjectively monitor changes in bone density. There appears to be a relationship between Chronic lymphocytic enteritis (CLE) and metabolic bone diseases in the marmoset [13]. Often, significant elevations in ALP are seen in animals with metabolic bone disease and/or CLE, and an increase in that parameter may be indicative of disease progression. Olson et al. provides a detailed discussion of bone disease in the common marmoset [90]. This retrospective study characterizes bone disease in two distinct marmoset breeding colonies maintained at two different institutions. Affected marmosets at one institution had bony lesions consistent with diagnosis of rickets, fibrous osteodystrophy (FOD), and osteopenia. Affected animals at another institution had lesions that were not easily categorized. Radiographic findings of this study are discussed in Part III of this chapter.

Tail injuries are not infrequent in captive marmoset colonies. In the authors' experience, tail injuries caused by bites from cagemates do not heal well and are best treated with amputation proximal to the site of injury using standard surgical techniques as described in Chapter 11 of this volume [91]. A tail bandage covering the amputation site that is secured to the proximal tail with nonconstricting tape layers will provide coverage for at least 24–48 h if applied correctly. The amputation site or bandage should be monitored closely, as some animals or their cagemates will continue to irritate the amputation site as it heals.

Other reports of musculoskeletal disease in marmosets include a case report of a 44-month-old male marmoset with bilateral hind limb paralysis that was found to have a large caudal dorsal mass with a definitive diagnosis of rhabdomyosarcoma [92]. In one colony of captive marmosets, a small percentage of adolescent animals developed severely malformed limbs, without lesions associated with rickets, and these cases were ultimately characterized as a micromelic dysplasia-like syndrome [93].

Nervous System

Evaluation of neurologic function begins with an assessment of the animal moving in its home cage. Here, changes in gait, posture, proprioceptive deficits, and other neurologic abnormalities can be observed. Evaluation of many cranial nerves can be accomplished on alert animals through cage-side assessment of functions such as swallowing, pupil size, and facial symmetry, among other parameters. If indicated, assessment of these reflexes and others such as palpebral reflex, pupillary light response, and jaw tone can also be made while the animal is sedated.

While trauma and ischemic injury must be included as a primary differential diagnoses for neurologic abnormalities in marmosets, there are infectious and metabolic differentials to include as well. *Human herpesvirus 1* is an important differential diagnosis for neurologic abnormalities, including seizures [49]. In marmosets that present with nervous system signs such as seizure, another top differential diagnosis is hypoglycemia. As with other small species with high metabolic rates, marmosets are predisposed to the development of hypoglycemia that can progress to a severity that results in seizure activity. Fluid and dextrose resuscitation is indicated in cases of hypoglycemia, as is the provision of thermal support, because hypoglycemic marmosets tend to be hypothermic as well. Interestingly, a high incidence of amyloid beta deposition in the cerebral cortex of aged marmosets has been documented, though the clinical significance of these deposits is unknown [94].

PART II: PREVENTATIVE HEALTH CARE AND QUARANTINE

The physical examination is but one component of a robust PHC program. Such programs are required to protect colony health and to exclude potential sources of research confounding. The components and execution of preventative medicine programs for nonhuman primates have recently been reviewed [95], and these general concepts are directly applicable to the design of programs for the care of marmosets. The frequency at which PHC encounters are performed and the specific components of each encounter depend on several factors including colony profile and history, the clinical and experimental history of the colony, whether breeding is performed, the frequency at which new animals are introduced to the colony, the frequency at which pathologic findings are detected in colony animals, and the utility of the PHC encounter for early detection of those pathologies. At a minimum, marmosets should be examined every 3–6 months. Quarterly examinations have been recommended by some experts [48]. Table 10.2 lists procedures that clinicians may elect to perform at some or all of the scheduled PHC encounters.

In addition to the preventative medicine program for resident animals, a strictly enforced quarantine program should be in place for all newly procured animals [95]. Maintenance of a closed marmoset colony reduces the risk of latent viral or bacterial pathogen introduction; however, additional animals may be required to increase genetic diversity or to expand colony size for experimental needs. Animals should always be obtained from reputable sources. Clinicians responsible for animal procurement should review source colony data including recent and historical disease trends. Depending on the institution, site audits of specific vendors

TABLE 10.2 Procedures and Diagnostic Testing Performed During PHC Encounters and Quarantine Examinations

Procedure	Frequency Performed for PHC Purposes at the Authors' Institution[a]
Physical examination	Biannually
Body weight	Monthly/Biannually[b]
Body condition score	Biannually
Somatometric measurements	As indicated for research purposes
Intradermal tuberculin test	Biannually
Pregnancy palpation	Biannually
Gestational staging via ultrasonography	Biannually
Complete blood cell count	Annually
Serum chemistry	As indicated by PE or clinical history[c]
Serum banking	Annually
Glycated hemoglobin testing	As indicated by PE or clinical history
Fecal floatation and direct smear	Annually
Perianal tape testing	Annually
Giardia ELISA or antigen capture	As indicated by PE or clinical history
Fecal cultures or molecular diagnostics targeting	Annually
Salmonella spp., *Shigella* spp., *Campylobacter* spp., *Klebsiella* spp., *Yersinia enterocolitica*, and Enteropathogenic *Escherichia coli*	Annually or as indicated by PE and clinical history
Viral serology or molecular diagnostics	+/− Once for animals in quarantine[d]
Thoracic radiographs	As indicated by PE or clinical history
Tattoo placement or retouching	Once/Retouching as required

PHC, preventative health care.
[a]*The frequency at which PHC encounters are performed and the specific components of each encounter will vary by institution and depend on several factors including colony profile and history, the clinical and experimental history of the colony, whether breeding is performed, the frequency at which new animals are introduced to the colony, the frequency at which pathologic findings are detected in colony animals, and the utility of the PHC encounter for early detection of those pathologies.*
[b]*Body weights are obtained on sedated marmosets during biannual PHC physical examinations. Body weights are obtained monthly on alert animals shifted into transfer boxes. If weight loss of* $\geq$*10% or progressive weight loss is identified, a physical examination is performed under ketamine sedation.*
[c]*At the authors' institution, total solids are measured annually.*
[d]*Target viruses might include* Saimirine Herpesvirus 1, Saimirine Herpesvirus 2, Callitrichine Herpesvirus 3, *and measles virus. However, validated and commercially available assays to test for these viruses are not yet readily available. There is a critical need for research that will result in best practice recommendations and validated viral testing protocols for PHC and quarantine of marmosets.*

may prove useful to evaluate source colonies. Individual animal health records should be examined with particular attention paid to data such as sex, date of birth, source (bred at the consignor or elsewhere), dam and sire, weight history, clinical and experimental history, reproductive history, and recent and past tuberculin skin testing results. Clinicians responsible for assessing candidate animals may also elect to evaluate complete blood cell counts, serum chemistry panels, fecal parasite tests, fecal cultures, and other clinicopathologic data. Once animals have been procured, they should be maintained in a separate quarantine facility that is physically distant from the routine animal-holding space. The quarantine facility should have dedicated equipment, a dedicated husbandry staff, a separate ventilation system, and a work flow that prevents cross contamination. The duration of the quarantine depends on the perceived risks of integrating the newly sourced animals with the established colony but, at a minimum, should range from 45 to 90 days [95]. The same procedures listed in Table 10.2 are those that clinicians may elect to perform at some or all of the scheduled quarantine encounters.

Surveillance for *Mycobacterium tuberculosis* complex (MTC) organisms is a critical component of any PHC or quarantine program. Historically, New World primates were thought to be relatively resistant to colonization with MTC organisms [96]. Recent experimental work has refuted this hypothesis. Via et al. have demonstrated that marmosets are sensitive to experimental infection with MTC organisms [97]. Animals inoculated via nebulization developed pulmonary cavitation and extrapulmonary spread to the liver, spleen, and lymph nodes. Antemortem symptoms included weight loss, fever, tachypnea, anorexia, and dehydration. Even low dose inoculums of less than 15 colony forming units resulted in strain and host-dependent development of clinical disease and histopathological lesions in this species [98]. Human handlers can also serve as a source of MTC exposure for colony animals [99]. An employee health surveillance program that includes tuberculosis screening should be implemented in all facilities housing marmosets.

The tuberculin skin test (TST) is the primary screening tool for *M. tuberculosis* surveillance in common marmosets. All animals greater than 6 months of age should be tested. A volume of 0.05 mL of old mammalian tuberculin is injected intradermally into the superior palpebra [46,48]. The location of the test site, lot number of tuberculin, and date of injection is recorded in the animal's medical record. Alternate eyelids should be used for each round of testing. Following injection, reactions are observed, graded, and recorded at 24, 48, and 72 h. Table 10.3 details reaction grades, the characteristic appearance of each grade, and interpretation of the test result. Examples of grade 3 and grade 4 reactions are depicted in Fig. 10.3 and 10.4.

TABLE 10.3 Appearance and Interpretation of Tuberculin Skin Test Reactions

TST Reaction Grade	Typical Appearance	Interpretation
Grade 1	Slight bruising of the eyelid	Negative
Grade 2	Erythema of the eyelid without swelling	Negative
Grade 3	Varying degrees of erythema with minimal swelling	Indeterminate
Grade 4	Obvious swelling with ptosis	Positive
Grade 5	Marked swelling and/or necrosis of the eyelid	Positive

FIGURE 10.3 Grade 3 tuberculin skin test reaction in a common marmoset. The animal's left eyelid is erythematous with mild swelling. As is indicative of the subjective nature of this test, some observers may interpret this reaction as demonstrating obvious swelling and assign a grade 4.

Animals that have been given Freund complete adjuvant or those with a history of Bacilli Calmette–Guerin vaccination may demonstrate false positive reactions [100]. Marmosets exposed to environmental, nontuberculous mycobacteria (NTM) can also demonstrate positive TST reactions. In one large marmoset colony, a 3.6% prevalence of equivocal or positive TST reactions observed over an 11-month period was associated with gastrointestinal *M. gordonae* colonization and extraintestinal *M. kansasii* infection [46]. Necropsy of all but one of the grade 4 reactors was performed. Of the animals necropsied, 50% demonstrated multifocal granulomatous lesions in lymph nodes, spleen, liver, and bone. While *M. gordonae* was identified in the gastrointestinal tract, the tissues containing the multifocal granulomatous lesions were culture positive for *M. kansasii* instead. This study indicates that NTM should be a differential for any marmoset demonstrating an equivocal or positive TST reaction. When determining the clinical response to an equivocal or positive TST, the animal's clinical, travel, and experimental history are considered. The response to a

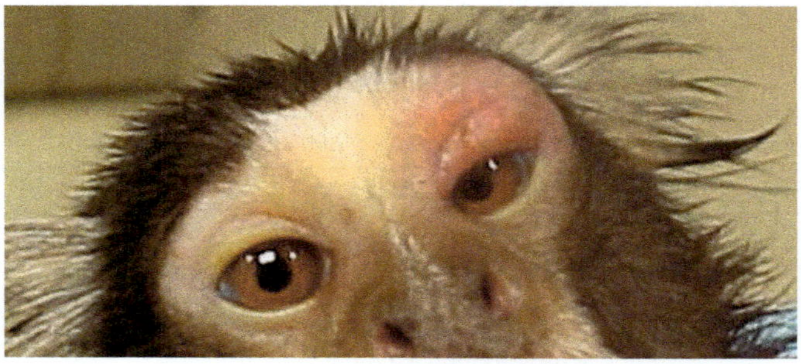

FIGURE 10.4 Grade 4 tuberculin skin test reaction in a common marmoset. The animal's left eyelid is erythematous with obvious swelling and ptosis. As is indicative of the subjective nature of this test, some observers may interpret this reaction as demonstrating marked swelling with slight necrosis and assign a grade 5.

positive test observed in a recently procured animal in quarantine may be quite different from the response to a similar result observed in a longstanding resident of an established colony. Additional diagnostic measures currently available to the clinician include thoracic radiographs, mycobacterial culture of gastric or bronchoalveolar lavage fluid, PCR-based assays, and complete necropsy [46,101].

Although vaccination administration is a routine component of PHC programs for nonhuman primates, few vaccination protocols are described for use in marmoset colonies. Chen et al. used measles vaccination with Attenuvax (Merck and Company, Inc., West Point, PA) to stem an outbreak of measles virus in mustached tamarins (*Saguinus mystax*) and rhesus macaques. At the authors' institution, rather than preventative vaccination of colony marmosets for measles, a rigorous screening program is in place to ensure that all employees with access to animal-holding rooms have demonstrated immunity to measles. There is anecdotal referral in the exotic pet literature to vaccination of marmosets with tetanus toxoid [102]. Chapter authors were not able to find a description of tetanus vaccination protocols for marmosets in the biomedical literature. Rabies infection of free-ranging marmosets has been described, but safety and efficacy of rabies vaccination in marmosets is unproven [103].

PART III: CLINICAL TECHNIQUES AND PROCEDURES

While it is beyond the scope of this chapter to describe all techniques that are employed in the clinical care of marmosets and in the execution of research using this species, the following sections discuss the techniques that are most commonly employed by the veterinary clinician as an adjunct to the physical examination. Therapeutic techniques commonly employed in general supportive care of this species are also discussed.

Removal of Marmosets From the Home Cage

To perform a physical examination of a sedated marmoset or to execute a brief and limited manipulation of an alert animal, the marmoset must first be removed from the home cage. While hand capture of the animal directly from the home cage is possible [20,104], this technique may pose an increased risk for escape of the animal into the holding room, injury to the marmoset, or occupational injury to the human handler. This method may also be more stressful to the animal than other methods of removal. It is preferable to train the animal to enter a small transfer box. There are many designs for marmoset transfer boxes [20,105,106]. It is the authors' preference to use boxes that measure approximately 1 cubic foot, have a smooth stainless steel bottom, three stainless steel mesh sides and top, and a sliding Plexiglas door (Fig. 10.5). An alternative design choice involves two Plexiglas sides and one steel mesh side at the back of the box. Boxes of this design limit the area in which the marmoset can move about during hand capture such that the risk of injury is minimized. Animals that are unaccustomed to the transport box may attempt to escape through the Plexiglas, possibly resulting in injury. The authors recommend the placement of a visual tape marker when transporting naïve marmosets. Once the scheduled procedure has been completed, the marmoset can also be recovered from sedation within the confines of the transfer box. The mesh allows the handler to visualize the animal at all times as well as providing a surface to which the marmoset may cling during the recovery period.

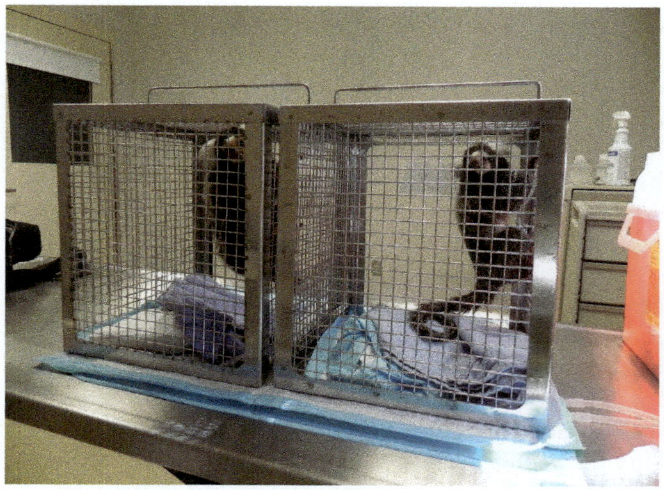

FIGURE 10.5 Transfer boxes containing marmosets in the process of recovering from sedation. The small area of the transfer box limits injury during the sedation and recovery process. The mesh sides provide a surface to which the marmoset can cling during the recovery period.

To capture the animal, the handler raises the Plexiglas door just enough to slip a leather-gloved hand into the transfer box. The other hand is used to block the open space created by raising the door. The marmoset is gently grasped around the thorax and shoulders. Care must be taken that the pressure of the handler's grip is enough to adequately restrain the animal yet not so strong that the animal cannot execute normal chest excursions. The marmoset may cling to the mesh at which point the handler uses his or her other hand to gently extract the animal's clawed hands or feet. The marmoset is then removed from the box, taking care to stabilize the sliding door, and is either sedated or handled as indicated. An alternative technique is to gently hold the marmoset against the mesh wall and inject a sedative or other medication through the mesh openings. This technique avoids the process of removing the alert marmoset from the transfer box, although it requires the handler to have adequate visualization of the quadriceps muscle to safely administer the injection. The authors are most comfortable handling marmosets with leather gloves, although techniques for handling of marmosets without the use of leather gloves have been published [20]. The latter technique should be reserved for animals that are acclimated to handling and performed only by experienced marmoset handlers.

The process of encouraging a marmoset to leave the home cage and enter the transfer box requires patience on the part of the handler and an investment in the development and execution of a training program. Techniques for positive reinforcement training of common marmosets have been well described [75,107]. Such training generally consists of administering a food reward paired with the delivery of a bridge or cue sound to encourage the animal to grasp and hold a target. Once a marmoset is trained to approach and grasp the target, the animal can be habituated to the transfer box and movement of the sliding door. The target is used to direct the animal into the transfer box with gradual habituation to increasing duration and transport within the box (Fig. 10.6). There must be continued opportunity for positive reinforcement training. If the transfer boxes are only used for aversive encounters, the marmoset will quickly learn that there is no benefit to entering the box. Positive reinforcement sessions can be used as an opportunity to obtain serial body weights. Not only do the animals have the opportunity to enter the transfer box without being subjected to an aversive encounter but the routine review of data on individual animals' body weights is an excellent means to monitor health trends.

Sedation

Although other drug combinations have been described, the dissociative anesthetic ketamine hydrochloride is the most commonly used drug for sedation of marmosets [48,91,104,108]. The ketamine dose range is wide with reported doses ranging between 5 and

FIGURE 10.6 Transfer boxes in use and in relation to marmoset housing. The image on the left shows a wire mesh transfer box placed on a shelf and in alignment with a guillotine-style door located on the front of the marmoset caging unit. To transfer a marmoset from the cage to the transfer box, the guillotine style doors of the cage and of the transfer box are simultaneously lifted. The marmoset is encouraged with a treat or using a target to enter the transfer box and, once the marmoset has entered the transfer box, the guillotine doors are slowly lowered. The image in the center shows an integrated nesting box that hangs on the front of the marmoset caging unit. This nesting box also serves as a transfer box. Once the marmoset has entered this nesting box, a panel can be inserted to cover the entrance to the nesting box and secure the marmoset within. The nesting box can then be removed from the front of the cage and transported to a procedure room. If other marmosets will remain within the caging unit, a panel must also be placed to cover the opening that is left after removal of the nesting box. The image on the right shows a marmoset within the integrated nesting box.

50 mg/kg IM [48,104,108]. Selection of dose depends on what must be accomplished during the sedation encounter. For a procedure that does not require complete or prolonged immobilization, the clinician might select a ketamine dose of 5–10 mg/kg. For a physical examination and phlebotomy, one might administer 25–30 mg/kg. For a minor surgical procedure, such as extracting a tooth or suturing a wound, one might use 40–50 mg/kg of ketamine in combination with appropriate analgesic drugs. The clinician's judgment in selecting drugs and doses should always be based on his or her knowledge of the animal's physical condition, concurrently administered medications, and clinical history. For thin, aged, debilitated, or juvenile animals, doses should be lowered. A recent publication by Bakker et al. suggests that alphaxalone administered at a dose of 12 mg/kg IM provides comparable immobilization to ketamine at 50 mg/kg but with fewer side effects [31]. In this study, ketamine was more frequently associated with excess salivation, involuntary muscle twitching, vomiting, and elevated muscle enzymes (aspartate aminotransferase, lactate dehydrogenase, and creatine kinase). The authors suggest that alphaxalone, with a mean immobilization time of 53.72 min, may be used safely for procedures lasting approximately 40 min, whereas ketamine, with a mean immobilization time of 31.23 min, is more appropriate for procedures lasting less than 20 min. Lastly, these authors indicate that, in their experience, a ketamine–medetomidine–atipamezole protocol was associated with unacceptably prolonged recovery periods. Regardless of the sedative drug selected, it is of utmost importance to provide supplemental heat via circulating warm water or warm-air blankets during sedation and recovery. Additional notes on sedation and anesthesia can be found in Chapter 11.

Substance Administration

Standard parenteral and oral routes used for compound delivery in other species are also feasible methods for the administration of drugs to common marmosets. Small volume oral medications can be masked in highly preferred foods such as yogurt, nectars, banana, or marshmallows. Because of the small size of the species, many commercially available oral formulations of veterinary drugs are too highly concentrated to allow for appropriate dosing. Clinicians may wish to establish a relationship with a reputable compounding pharmacy through which palatably flavored, oral suspensions of the most commonly used medications can be purchased. Marmosets can be habituated to receive flavor-masked oral dosing solutions from a blunt tipped syringe [109]. Oral medications can also be delivered via oral gavage using an 8-French infant feeding tube made of flexible polyvinyl chloride. This can be accomplished in alert animals, but because of the risks of aspiration, esophageal injury, and damage of the dental arcade, the use of this procedure should be limited to delivery of unpalatable medications to sedated marmosets, administration of medications or nutritional supplements to critically ill and anorexic patients, or for scientifically justified and IACUC-approved experimental procedures. Maximal oral dosing volumes for marmosets should not exceed 15 mL/kg [110].

Subcutaneous injections of medications and parenteral fluids are administered between the scapulae or in the relatively loose skin of the ventral abdomen. This procedure can be accomplished in alert animals restrained for the clinician by a skilled handler. A maximal subcutaneous dosing volume of 5 mL/kg/site is recommended [110]. Intramuscular medications are routinely administered to the craniolateral surface of the thigh directed into the quadriceps muscle mass. A maximal volume of 0.25–0.5 mL/kg/site is recommended [110]. Intravenous injections can be administered into the lateral tail vein, cephalic vein, or saphenous vein using a 25–27 gauge needle and appropriately sized syringe, usually a 1-mL or 3-mL syringe. Intravenous injections are most readily accomplished in the sedated animal. It is optimal for the clinician to work with an assistant who can stabilize the limb or tail during the injection. Clipping the fur over the vessel and moistening the region with an alcohol swab will improve visualization of the vessel. Maurus et al. describe a technique for tail vein injection and indicate that dilation of the lateral tail vein using gauze sponges soaked in warm water may improve success of this procedure [111]. All three venous access points are superficially located. The vessel is occluded proximal to the injection site to further dilate the vessel. The needle is slowly introduced, bevel up, while applying negative pressure on the syringe plunger until blood is visualized in the needle hub. Vessel occlusion is released and the desired substance is then introduced with a steady hand via a slow bolus. A maximal slow bolus (over 5–10 min) intravenous dosing volume of 2.5–10 mL/kg is recommended [110].

Over the needle, intravenous catheters can be placed in the lateral tail vein, cephalic vein, or saphenous vein. The authors find that the saphenous vein is most readily accessible for this purpose (Fig. 10.7). The technique is similar to that described above for the intravenous injection. In addition to clipping the fur over the vessel, the skin should be aseptically prepared. A 24 gauge over the needle catheter is directed at the dilated vessel, bevel up, using a shallow angle (approximately 10 degrees) that is nearly parallel to the surface of the leg. Once the needle bevel and approximately 1–2 mm

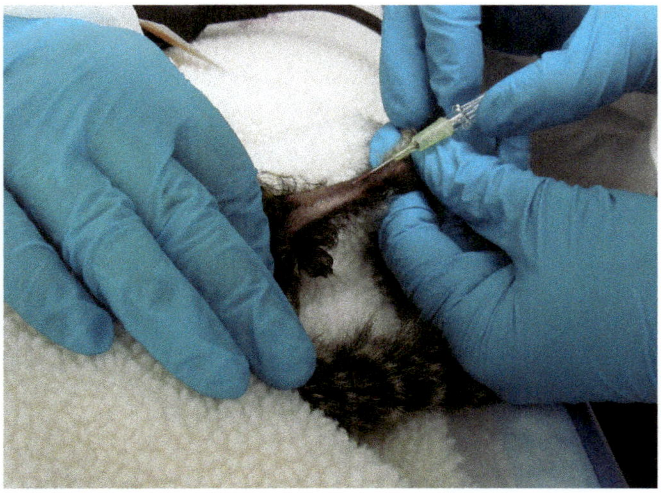

FIGURE 10.7 Placement of an over-the-needle venous catheter in the saphenous vein of a common marmoset. The skin surface has been clipped and aseptically prepared. A 24-guage catheter is directed into the vessel with the bevel up and at a shallow angle of approach.

of the most proximal segment of the catheter are introduced into the vessel, the catheter is then slowly fed off the needle and into the vessel. The needle is withdrawn and adequate placement is confirmed by visualization of venous blood filling the catheter hub. The catheter is then capped with a primed fluid line or injection port and secured to the leg using bandage tape and dressing according to the clinician's preference. Catheter patency is confirmed via a slow flush with a small volume of nonheparinized physiologic saline or other isotonic fluid.

Intravenous catheters are utilized for delivery of substances or intravenous fluids over periods longer than those described above for slow intravenous bolus. Situations that call for the use of intravenous catheters include the delivery of intravenous fluids during surgical procedures or prolonged anesthetic episodes, the delivery of substance volumes that exceed 10 mL/kg or that require a more prolonged dosing period, delivery of agents for which extravasation could cause tissue damage, and the delivery of intravenous fluid support to critically ill or dehydrated patients. When administering intravenous fluid therapy, rate and fluid selection should be based on the estimated degree of dehydration, rate of ongoing fluid losses, assessment of electrolyte measures, and additional components of the animal's clinical history. As in other species, the degree of dehydration is estimated via an assessment of skin turgor, mucus membrane color, capillary refill time, pulse quality, degree of enophthalmos, and measures of the animal's hematocrit, total solids, and renal function values. A calibrated syringe pump is useful for targeted and accurate delivery of small volumes of intravenous fluids.

All but the most critically ill or sedated marmosets are unlikely to leave an intravenous catheter and fluid line undisturbed. The authors have used a technique of stabilizing the catheterized leg of a sedated marmoset with a padded splint. The sedated marmoset is then wrapped with its arms at its side in a padded disposable underpad (chux pad) that is secured around the animal's torso using self-adherent bandaging wrap. Care must be taken not to interfere with the animal's normal chest excursions. Intravenous fluids or other substances can then be delivered for the duration of the sedation and recovery period as well as for a period of time following recovery. The animal must be closely monitored at all times for evidence of distress, escape from the wrap, and normal recovery from sedation. Once the infusion has been completed, a skilled handler can assist the clinician with removal of the wrap, splint, and catheter as well as application of a padded bandage for hemostasis and return of the marmoset to its home cage or transfer box. Unless the animal is habituated, this procedure and associated restraint is likely to be associated with some level of distress. The clinician should determine if the clinical benefit to the animal warrants the stress involved. If this technique is used for experimental delivery of test compounds, the procedure should be justified, described in the IACUC protocol, and associated with a period of training and acclimation to the restraint.

Techniques to facilitate the prolonged administration of compounds to marmosets have been published. O'Byrne describes the technique for implantation of a long-term vascular cannula into the internal jugular vein [112]. Polyvinyl tubing (25 cm in length, I.D. 0.58 mm, O.D. 0.99 mm) was introduced approximately 3 cm into the vessel to allow the cannula tip to lie within the cranial vena cava or atrium. The remainder of the tubing was tunneled subcutaneously and exteriorized between the scapulae with the excess tubing housed in a cloth backpack worn by the marmoset. Cannulas remained patent for up to 7 months. Although this publication describes the use of these cannulas for serial blood sampling, a similar technique could feasibly be used for compound administration. Attachment of the intravascular cannula to a subcutaneously implanted vascular access port (VAP) eliminates the need for cannula exteriorization and the requirement that the marmoset wear a protective garment. Sauer et al. combined VAPs with femoral arterial and femoral venous catheterization in marmosets for serial blood collection, whereas Fitzgerald et al. combined VAPs with femoral venous catheterization for daily administration of test compounds over a 3-week period [113,114]. Selection, implantation, and maintenance of VAPs is reviewed in Swindle et al. [115]. This reference suggests that a 3- to 5-French polyurethane or silicone cannula combined with a small VAP (3 g in weight with a 0.1–0.15 mm port reservoir) is appropriate for nonhuman primates weighing less than 1–2 kg. Care must be taken to ensure that the catheter material selected is compatible with the compound to be administered. A cannula tunneled for approximately 3 cm in the dorsal subcutaneous space and attached to an osmotic mini-pump carried in a cloth backpack has been used for long-term subcutaneous compound infusion [116]. In this study the osmotic pumps were capable of delivering compound for up to 5 days. With replacement of the pump every 10 days, the investigators were able to deliver the compound to marmosets for up to 3 months.

Phlebotomy

Blood samples are collected via the femoral vein, the lateral saphenous vein, or the cephalic vein [117]. The femoral vein is commonly selected for venipuncture when collection of larger blood volumes is required. Venipuncture of the lateral saphenous and cephalic veins is more appropriate for collection of smaller blood volumes (less than 0.1–0.2 mL), although the clinician may elect to preserve the integrity of these vessels for catheter placement. The lateral tail vein can be used for collection of small volumes of blood from an acclimated, alert marmoset. This site is useful for collection of blood for blood glucose or HbA_{1c} testing. The authors have also used dermal puncture of the heel to obtain capillary blood samples for this purpose [14]. Regardless of the vessel selected, clinicians should be mindful of limits on the volume of blood that can be safely procured from this species. The average circulating blood volume of the marmoset is 70 mL/kg [110]. According to Diehl et al., removal of 7.5% of the blood volume (approximately 2.0 mL in a 350 g marmoset) requires at least 1 week recovery time before the marmoset may be resampled. Similarly, removal of 10% (approximately 2.5 mL) or 15% (approximately 3.5 mL) requires at least 2 and 4 weeks recovery time, respectively [110]. The maximum amount of blood collected should be adjusted based on the marmoset's age, body weight, hydration level, and health status.

Equipment required for venipuncture of the femoral vein includes a 26–27 gauge needle of 3/8- or 5/8-inch length attached to a luer lock 1 or 3 mL syringe. The femoral vein is superficial and lays in the groove between the sartorius and vastus medialis muscles cranially and the adductor muscle caudally (Fig. 10.8). Some clinicians, extrapolating from experience with other nonhuman primate species, may elect to perform venipuncture at the femoral triangle, which is demarcated at the base by the abdominal wall, cranially by the sartorius muscle, and caudally by the pectineus and adductor muscles. It is the authors' recommendation that phlebotomists puncture the vein approximately 2 cm distal to the femoral triangle. A blue coloration demarcating the vein is often visible with use of overhead illumination. Visibility is improved by dampening hair and skin with an alcohol swab. The vein may not be visible in well-muscled or obese animals. If the vessel is not visible, the femoral arterial pulse can be palpated and the needle directed just medial to the artery to puncture the vein. To perform the phlebotomy procedure, the needle is inserted parallel to the vessel at a 10–15 degrees angle with the bevel up. The syringe can be balanced on the clinician's thumb, which may be used as a guide to achieve the proper angle of insertion as well as to stabilize the syringe. The syringe plunger is withdrawn just after insertion of the needle to determine the point at which the needle enters the vessel. Once a flash of blood in the needle hub is observed, the plunger is slowly withdrawn until the desired volume of blood is obtained. Care must be taken to avoid collapse of the vein when using a larger syringe, although venipuncture can easily be accomplished as long as back pressure is periodically released. Readjustments should be made slowly because it is easy to pass through the vessel with quick movements. Once the sample is collected, apply pressure to the needle insertion site for approximately 3–4 min to achieve hemostasis. If the femoral artery is punctured, pressure should be applied to the site for at least 10 min.

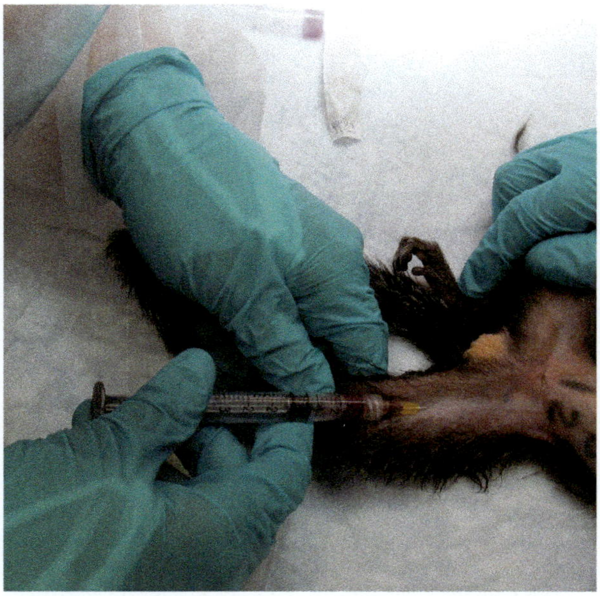

FIGURE 10.8 Femoral vein phlebotomy technique in a common marmoset. The vein is punctured approximately 2 cm distal to the femoral triangle using a 26-gauge needle and a 3 mL syringe. The needle is inserted bevel up, parallel to the vessel, and with a shallow angle of approach.

Although infrequently observed, development of femoral artery hematoma is a clinically significant, adverse event that is associated with the venipuncture procedure [39,48]. Presentation is characterized by swelling over the medial aspect of the thigh, lameness, and, if severe, evidence of anemia and cardiovascular shock. Unless targeted observations are made, symptoms may not be noted for days to months from the phlebotomy encounter. Late-developing hematomas may present as mild-to-severe lameness during cage-side observation. Due to concerns for hematoma formation, it is recommended that clinicians perform frequent rechecks of the phlebotomy site during the animal's recovery from sedation. Following the animal's return to the home cage, a food reward can be used to encourage the animal to present at the front of the cage to recheck the phlebotomy site. If observed in the early stages of formation, application of a pressure bandage to the affected leg may prevent progression. If observed after a prolonged period or in instances where hemorrhage cannot be controlled, surgical ligation or repair of the damaged vessel may be required. Such cases should be considered serious adverse events due to the clinical challenges associated with blood loss and cardiovascular instability.

Although phlebotomy is most easily accomplished in the sedated animal, there may be clinical or experimental requirements to collect blood from alert marmosets. Techniques for venipuncture of marmosets acclimated to hand restraint have been described [118]. Restraint devices to facilitate phlebotomy of alert marmosets have also been designed [119–121]. The most commonly used restraint device consists of a rigid tube that encircles the thorax of the marmoset (Fig. 10.9). This tube is attached to a support post that extends at a 50–60 degree angle from a working-height platform. Velcro straps are used to secure the abdomen and legs. The leg targeted for venipuncture is accessed from below the rigid tube and is stabilized by the phlebotomist's hand or with the Velcro straps. In addition to venipuncture, this restraint device can also be used for semen collection, abdominal ultrasound, substance administration, and other procedures. Positive reinforcement techniques must be used to acclimatize subjects to the device and specific procedures.

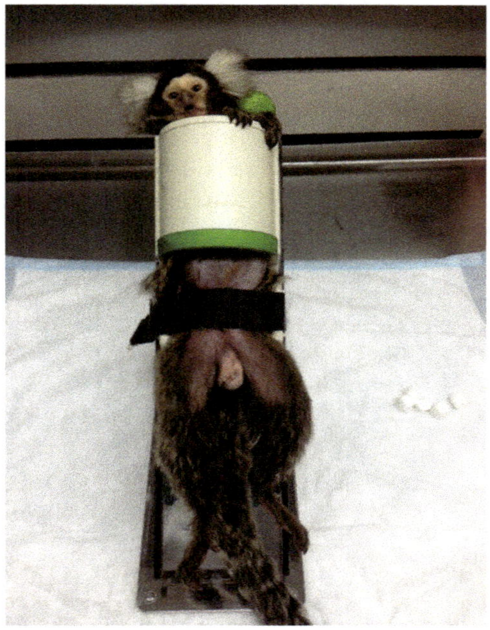

FIGURE 10.9 Common marmoset within restraint device. This restraint device can be used for venipuncture, semen collection, abdominal ultrasound, substance administration, and other procedures.

Blood Transfusion

The administration of blood transfusions to nonhuman primates has recently been reviewed by Bohm and Gilbert [122]. As in other species, indications for blood transfusion in common marmosets include anemia, acute or chronic blood loss, hemolysis, coagulopathy, or thrombocytopenia. In the authors' experience, transfusions are infrequently performed for the clinical support of marmosets, but when they are indicated, are most often required for cases of femoral artery hematoma and obstetric emergencies. Because of this infrequent indication, rather than using banked blood components, whole blood is procured from a healthy donor at the time that the need for a transfusion is recognized.

The donor should be a healthy marmoset of the same specific pathogen-free status as the recipient. A donor marmoset weighing at least 350 g that has not had a phlebotomy procedure in the previous 2 weeks can provide up to 3 mL of whole blood (less than 15% of the circulating blood volume) [110]. Due to the challenges of performing phlebotomy in marmosets, it may be difficult to obtain a spun hematocrit from the donor. Recent PHC data can be consulted. Alternatively, microhematocrit tubes can be filled directly from a small bleb of blood obtained by pricking the lateral tail vein with a small gauge needle. Isotonic parenteral fluids should be administered via the subcutaneous route to replace the blood volume collected from the donor. Citrate phosphate dextrose adenosine (CPDA) at a ratio of 1:7 with whole blood is the recommended anticoagulant. A CPDA-loaded 3 mL syringe with a 26 gauge needle is used for collection of blood from the femoral vein of the donor. Blood will be administered directly to the recipient from this syringe via a 24-gauge catheter inserted into the saphenous or cephalic vein. An in-line, saline-primed hemonate filter is used to remove erythrocyte and platelet microaggregates. Blood can be delivered using a syringe pump or by hand using a slow bolus technique. If more than 3 mL is required for the transfusion, blood from a second donor can be combined with that of the first donor.

There is an older body of literature that describes the study of blood group antigens in nonhuman primates. Old world primates (OWP) do express the same A-B-H antigens that comprise the A-B-O blood group system of humans. Although these antigens are present in gastrointestinal mucus secreting cells, saliva, posterior root ganglia, and vascular endothelium, they are not expressed on the erythrocyte surface of OWP species [123,124]. In contrast, Callitrichids do not express the A-B-H antigens on vascular endothelial cells but do express A-like and B-like antigens on the surface of the erythrocytes [124—127]. Agglutination experiments suggest that these A- and B-like antigens differ significantly from those expressed in humans. Interestingly, unlike OWP, secreted salivary antigens do not correlate with the A-like or B-like antigens expressed on Callitrichid erythrocytes. A more recent publication examined non-A-B-O antigen expression in baboons, rhesus macaques, and squirrel monkeys using commercial microtyping cards [128]. Kell, Lutheran, Kidd, and Duffy antigens were detected in the erythrocytes of all species, Rh antigens were present in squirrel monkey erythrocytes, MNS antigens were present in rhesus macaque and squirrel monkey erythrocytes, and Lewis antigens were present in baboon and rhesus macaque erythrocytes. Although marmosets were not examined in this study, this species also likely expresses multiple antigens, related or unrelated to those observed in humans, on the surface of their red blood cells. This potential repertoire of antigen expression does suggest that there is a risk for transfusion reactions in this species. When possible, a cross-match assay should be performed, although this is often not feasible in emergency situations. The risk of a transfusion reaction is minimal if the recipient has not previously received a blood transfusion.

Collection of Urine Samples

Urine samples can be obtained opportunistically or collected from marmosets trained to urinate for a food reward. Because marmosets will provide a first-void sample shortly after emerging from the nesting box, opportunistic samples are most readily obtained in the early morning just after the room lights turn on. A clean stainless steel cage pan placed below a mesh cage bottom can be used to collect these samples. Several publications describe techniques for training marmosets to urinate on cue and indicate that these techniques are reliable, with a nearly 95% success rate, for collection of samples ranging in volume from 50 to 600 μL [75,129,130]. French et al. describe a training regimen that initially involves provision of a food reward, such as a mealworm or a marshmallow, when the marmoset approaches the cage and then, as training progresses, when the marmoset urinates outside of the cage [131]. Subsequently, the food reward is withheld until the animal provides the desired sample that is caught within a handheld aluminum dish. The use of metabolism cages for collection of 24-h urine samples from marmosets has been described with mean urinary excretion measuring 14.4 mL/24 h and ranging between 2.9 and 34 mL/24 h [132,133].

Collection of Semen

Experimental techniques for the collection of semen have been developed and refined to support the generation of genetically engineered marmosets. Oocyte collection, in vitro fertilization, and embryo transfer techniques are discussed in Chapter 20 of this volume [134]. Clinicians wishing to examine semen samples to evaluate male fertility can procure such samples by electroejaculation, through the use of penile vibrostimulation, or collection of postcopulation vaginal washings. Electroejaculation requires that the animals be sedated. This technique also requires specialized equipment and is perceived as more invasive than other techniques. Some references suggest that electroejaculation has no detrimental effect on seminal quality and sperm production, whereas others have described reduced numbers of total viable sperm, reduced in vitro sperm survivability, and altered sperm velocity and motility parameters [135—137]. It has been suggested that electroejaculation results in a firmer coagulum that is more resistant to dispersion, thus contributing to a greater entrapment of spermatozoa, and possibly resulting in the altered parameters described above [137].

Vibratory stimulation of the penis is performed in alert marmosets that are accustomed to being handheld [137—139]. It is recommended that subjects be separated from their female cagemates for at least 24 h before collection. This technique also requires specialized equipment: a medical vibrator designed to treat male ejaculatory dysfunction (FertiCare personal vibrator, Multicept, Inc., Rungsted, Denmark). Collections are performed in a darkened room that is removed from the animal-holding space. The animal is gently held in a prone or slightly recumbent position. Food rewards are provided during the procedure. The vibrator shaft is fitted with a short piece of vinyl tubing designed to hold a glass or plastic microcentrifuge tube. The tube, which serves both as an artificial vagina and a sample collection vial, is held against the preputial orifice and the vibrator is activated to generate the required stimulation. The initial stimulus is applied at a frequency of

70–80 Hz and amplitude of 1 mm. This stimulation is applied for 2–3 min periods followed by rest periods of 0.5–2 min. Stimulation can be further adjusted in a stepwise fashion to a maximum frequency of 95 Hz and amplitude of 2 mm until ejaculation is achieved. Animals typically ejaculate within the first few minutes of stimulation. Stimulation should be discontinued if ejaculation has not occurred within 15–20 min. Success rates range between 35% and 90% and likely depend on the technique of the collector and characteristics of the subject [137,138]. This technique has been described to produce ejaculates with a mean volume of 31.9 μL, a mean concentration of 1154.2×10^6 sperm/mL, and containing 74.6% live sperm and 59.6% motile sperm [138].

Collection of postcopulation vaginal washings is a noninvasive technique that is performed on alert marmosets and requires no specialized equipment [136,140]. This technique capitalizes on the tendency of marmosets to mate shortly after being first paired or repaired after a separation. Pairs are separated for 12–24 h, reintroduced, and observed for copulation. Copulation is characterized by pelvic thrusting of the male followed immediately by the female seeking to move away from the male. Once these behaviors are observed, the female marmoset is removed from the home cage for the collection procedure. The female is held in a Trendelenburg position. A plastic pipette fitted with a 2-cm long segment of silastic tubing (2 mm O.D.) is used to flush and gently aspirate 100 μL of media or physiologic saline into and from the vagina. Media containing the coagulum is then examined as indicated.

Semen can be analyzed using techniques employed for other species. Parameters examined include percent live sperm, percent motile sperm, percent progressively motile sperm, morphology, and concentration. Sperm counts can be performed using a Neubauer cell counting chamber. Computerized image analysis can be used to collect additional data. Computer-assisted sperm analysis (CASA) uses phase contrast, fluorescence, or bright field microscopy to randomly capture and analyze images of individual spermatozoa and provide great detail on morphological characteristics of the head, nucleus, acrosome, and tail [139]. CASA can also provide greater detail on motility characteristics such as curvilinear velocity, straight line velocity, and linearity [137].

Other Sample Collection

As in other species, cerebrospinal fluid (CSF) is collected from the cisterna magna or from the lumbar subarachnoid space. For collection from either location, marmosets are anesthetized, the hair over the collection site is shaved, and the skin is aseptically prepared. In one published technique for collection from the cisterna magna, animals are placed in ventral recumbency and the head and neck are held in ventroflexion [117]. A 27 gauge needle is inserted perpendicularly, caudal to the occipital protuberance, at the point of intersection of a line drawn between the angles of the mandible and the midline. Gentle suction is applied while slowly advancing the needle until CSF is observed in the needle hub. An alternative technique for collection of CSF from the cisterna magna is described by Geretschläger et al. [141]. Anesthetized animals are held in a sitting position. Using a 25 gauge, winged infusion set, the needle is inserted in an upward direction, on midline, and at a point 8 mm distal to the occipital protuberance until the needle approaches and touches the occipital bone. The needle is then gradually redirected in a ventral direction along the bone to a position that allows puncture of the atlantooccipital membrane. Once the tip of the needle enters the cisterna magna, a 1-mL syringe is used to apply gentle suction and draw the CSF sample into the cannula. Both techniques enable collection of approximately 100 μL of CSF. Wang et al. describe a technique for performing intrathecal injections into the lumbar subarachnoid space [142]. Presumably this technique can be adapted for collection of small volumes of CSF from this site. With the marmoset in ventral recumbency, the space between the fourth and fifth lumbar vertebrae is palpated and a 30 gauge needle inserted into the space until a small sample of CSF fills the hub of the needle. Due to the small size of this species, these procedures require precision and finesse. Clinicians having the opportunity to perfect these techniques using cadavers may wish to do so.

Milk can be obtained from female marmosets during the lactation period, which generally persists for 65–90 days postparturition [143]. Using a technique described by Power et al., females are separated from their infants for 3–4 h to allow milk to accumulate within the mammary glands [144]. Following sedation with ketamine hydrochloride and administration of 2–3 IU oxytocin IM, nipples are cleansed with distilled water and milk is expressed manually into a collection vial.

Tooth Extraction

Tooth root abscesses and fractured maxillary canine teeth are commonly observed in many marmoset colonies. Removal of the affected tooth is the recommended therapeutic measure. Due to the small size of marmoset teeth and the oral cavity itself, dental extractions are technically challenging, but feasible with the appropriate tools, practice, and patience. For maxillary canine extraction, the use of a feline or other small elevator to break down the periodontal ligament is recommended. The tip of the elevator must be sharpened

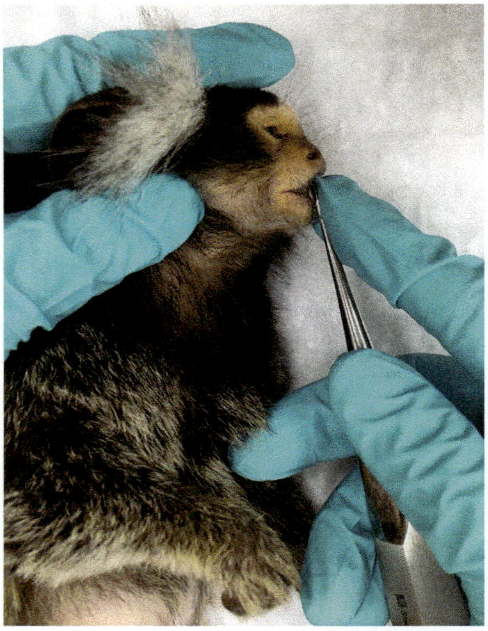

FIGURE 10.10 Positioning of hands and equipment for extraction of a maxillary canine tooth. The marmoset's head is stabilized by the nondominant hand and the elevator is held in the dominant hand. The index finger is positioned close to the sharp edge of the elevator to avoid slipping that could lead to tissue trauma.

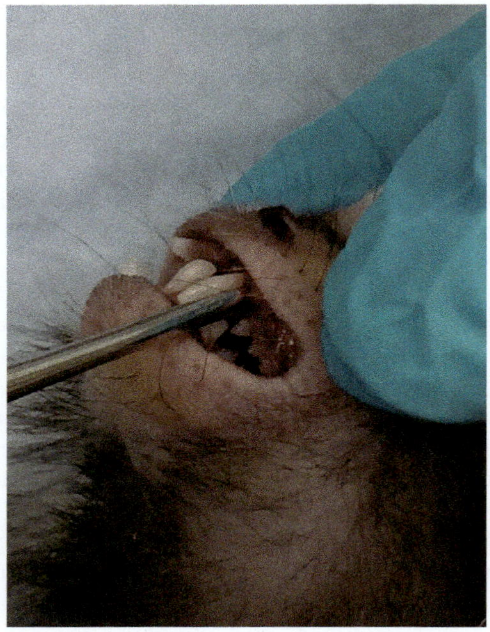

FIGURE 10.11 Introduction of the curved tip of the elevator into the tooth sulcus along the edge of the tooth and perpendicular to the periodontal ligament fibers.

regularly to maintain an edge sharp enough to cut through the periodontal ligament without causing significant trauma to surrounding soft tissues or teeth. With the marmoset's head in the nondominant hand, the elevator is held in a firm but comfortable grip in the dominant hand (Fig. 10.10). The curved tip of the elevator is introduced into the tooth sulcus along the edge of the tooth, perpendicular to periodontal ligament fibers (Fig. 10.11). Gentle pressure is slowly applied with the elevator. Pressure is maintained in one position for about 30 s without any back-and-forth movement. As soon as the periodontal ligament fibers have broken down on one aspect of the tooth, the elevator is repositioned and the process is repeated around the entire circumference of the tooth. When the tooth is mobile, rongeurs are used to gently grasp the tooth and slowly apply rotational force to move it out of the sulcus. If this final step of extraction is attempted prior to full breakdown of periodontal ligament fibers, root tip fracture may occur. The authors do not routinely close tooth extraction sites in marmosets. Antimicrobial therapy based on culture and sensitivity may be warranted after extraction. There is little data regarding the practice and efficacy of prophylactic dentistry in common marmosets, and it is unclear if implementation of a prophylactic dentistry program would reduce the incidence of tooth root abscesses.

Surgical Biopsy Collection

Most surgical biopsies can be performed in marmosets using techniques applied in other species. Clinicians should select instrumentation, suture materials and other consumables, and techniques that are suitable for this species' small size. Sedatives, anesthetics, and pre- and postoperative analgesics appropriate for the procedure and clinical status of the patient must be selected and administered. As in other species, animals should be observed postoperatively at regular intervals for abnormal signs or behavior such as excessive vocalization, decreased appetite, lethargy, bleeding, or incision drainage or dehiscence.

Skin biopsy is routinely performed in marmosets sedated with either ketamine hydrochloride or alphaxalone. An area of the skin, generally the abdomen, dorsum, or lesioned area, is shaved and aseptically prepared. If samples will be utilized for histology, the area should be minimally shaved and not subjected to a surgical scrub. Lidocaine or an equivalent local anesthetic is infiltrated to supplement oral or parenteral analgesics. Using sterile technique, a disposable biopsy punch (2–3 mm) is used to obtain the sample. A small snip biopsy may be performed with sharp Metzenbaum scissors if the biopsy is to be used for fibroblast culture or another purpose that does not depend on maintenance of tissue architecture. The resulting defect is closed with appropriate suture material placed in the subcutis and skin. The authors have administered the nonsteroidal anti-inflammatory agent meloxicam (0.1–0.2 mg/kg, SQ or PO SID) as an analgesic when performing this procedure [145]. Antibiotics and additional or alternative analgesics are used as is appropriate for the patient or if any complications arise. Possible complications include infection or dermatitis at biopsy sites.

The lymph node biopsy procedure is similar to that performed in other species. The most readily accessible lymph nodes are those in the inguinal region. If the inguinal lymph node is selected for biopsy, the clinician should be mindful of the prominent vasculature that courses through this same region. Biopsies can be performed with ketamine hydrochloride, alphaxalone, or under general anesthesia depending on the clinical picture of the patient and skill of the surgeon. Hair is clipped from the surgical site, the area is aseptically prepared, and lidocaine or equivalent local anesthetic is infiltrated along the planned incision. Using surgical draping and sterile technique, a small incision is made and the lymph node is isolated using blunt dissection. Blood vessels supplying the lymph node are ligated or cauterized to prevent bleeding. The subcutaneous tissue and skin is then closed with appropriate suture. The incision should be monitored for any signs of dehiscence or infection. Buprenorphine (5–20 μg/kg, IM) administered perioperatively, combined with meloxicam administered postoperatively, is one analgesic regimen that may be considered when performing this procedure [104,145]. Clinicians should be aware that administration of buprenorphine doses at the higher end of this published range may result in sedation, prolonged anesthetic recovery, and respiratory depression [145]. Antibiotics and alternative analgesics are used as appropriate. Possible complications of lymph node biopsy include hemorrhage, hematoma formation, infection, or incision dehiscence.

CLE is a pathological syndrome of significant import that frequently affects the health of captive-housed common marmosets and serves as a cofounder to research projects employing this species as an animal model. As mentioned previously, this syndrome presents as enteritis and/or colitis with evidence of chronic diarrhea, progressive weight loss, and hypoalbuminemia [13]. While the diagnosis is often one of exclusion, the veterinarian may elect to perform colonic or small intestinal biopsies to aid in clinical decision-making and to distinguish among differential diagnoses such as gastrointestinal lymphoma, small intestinal adenocarcinoma, and systemic amyloidosis [45,67,70]. Colonic biopsies can be performed blindly in animals sedated with ketamine hydrochloride, alphaxalone, or under general anesthesia [146,147]. Briefly, a pediatric endoscopic biopsy forceps (approximately 1.8 mm outer diameter) is inserted into the colon, and an appropriate number of mucosal pinch biopsies are obtained. Postprocedural complications, when encountered, usually only involve transient hematochezia. Very rarely, perforation resulting in peritonitis could result. To perform biopsy collection from marmosets and tamarins under visualization of the colon, Clapp et al. describe a technique using a pediatric bronchoscope with a 4.8-mm insertion tube and 2.0-mm biopsy channel to view the colon from the rectum to the cecum. Small intestinal biopsies must be performed via an open abdomen procedure due to the current lack of appropriately-sized, commercially-available endoscopic equipment. Ludlage and Mansfield describe a technique for performing small intestinal biopsies using endoscopic biopsy forceps inserted, through a 2–3 mm incision, into a segment of small intestine that is accessed via a keyhole laparotomy [48].

Hepatomegaly is another clinical finding commonly encountered in the common marmoset [69]. Liver biopsy can assist clinicians in distinguishing among differential diagnoses that include hepatic steatosis, systemic amyloidosis, and lymphoma [43,45,69,70]. Because hepatic dysfunction can significantly alter production of coagulation factors, clinicians may elect to survey the function of the intrinsic and extrinsic coagulation systems before performing a liver biopsy. Liver biopsy is best performed under general anesthesia. The authors have used buprenorphine as a preemptive and postoperative analgesic combined with a local anesthetic block consisting of 2% lidocaine infiltrated at the incision site. Using surgical draping and sterile technique, collection of small samples can be accomplished using a 20 gauge Tru-Cut biopsy needle inserted through the abdominal musculature and into the liver under ultrasound guidance. A keyhole laparotomy allows for collection of more substantial samples in a controlled manner [148]. Following aseptic preparation of the site and surgical draping, a 2-cm, right paracostal abdominal incision is made. The caudal edge of the right lateral liver lobe is identified, gently stabilized with a sterile, moistened cotton swab, and Metzenbaum scissors are used to excise a tissue sample. Gel foam is placed on the biopsy site to minimize bleeding before returning the lobe to the abdomen. The abdominal musculature, subcutaneous, and cutaneous layers are closed with appropriate suture material and pattern. Patients should be monitored for presence of pallor, abdominal distension, lethargy, and anorexia.

Metabolic Profiling

Marmosets maintained in captivity demonstrate a propensity toward obesity [149]. Obesity in this species is frequently accompanied by hallmarks of the metabolic syndrome including changes in body composition, alterations of glucose metabolism, and dyslipidemia [9,14,69]. For these reasons, veterinarians managing common marmoset colonies may decide to employ techniques for metabolic phenotyping as a routine component of PHC protocols.

Collection of somatometric measurements is a simple means to monitor changes in body composition without the need for specialized equipment. These measurements, taken during routine physical exams using a tape measure, include abdominal circumference, thoracic circumference, and crown-rump length. Abdominal circumference is measured at the level of the umbilicus. Because of this species' tendency to store adipose in the axillary region, thoracic circumference is measured at the level of the mammary glands. Crown-rump length is measured from the occipital crest to the base of the tail. Consistency in collection of measurements is important. As mentioned previously, marmosets should also be assigned a body condition score during any physical examination encounter.

More advanced techniques for assessing body composition exist. Numeric measures of body fat and lean mass can be collected using quantitative magnetic resonance (QMR; Echo Medical Systems, Houston, TX) [9,14,69]. This is an internally-shielded device with a small footprint and can easily be housed within the clinical space of the vivarium. The QMR procedure is performed on sedated or alert animals. For measurement of alert animals, a tube-shaped Plexiglas holder is placed into the animals cage and maneuvered such that the animal places its head within the tube entrance (Fig. 10.12). Conditioned animals will freely move forward into the body of the tube. With some animals, the handler may need to reach behind and gently direct the animal into the body of the tube. Once the marmoset has entered into the holder, the smaller diameter internal tube is secured behind the animal. The tube should be of sufficient size to allow the animal to safely turn yet limit wide-ranging movements that interfere with accurate measurements. The animal is transported within the tube from the cage-side to the QMR machine. Once the tube containing the animal is inserted into the QMR machine, the scan takes approximately 60 s to complete. Once the scan is completed, the animal is transported within the tube to the home cage. The cap is removed and the animal is able to back out of the tube to exit. A treat is provided to the animal following return to the home cage. Data output includes measures of lean mass, fat mass, free water, and total water. The ease of this technique has allowed QMR to largely replace other means of measuring body composition such as total body electroconductivity and labeled-water dilution [8]. QMR does not quantitate bone mass. Dual-energy X-ray absorptiometry has been used in sedated common marmosets to measure BMD (bone mineral density) [66].

Measures of fasting blood glucose, HbA_{1c}, and glucosuria comprise the minimum dataset used by veterinary clinicians to diagnose diabetes mellitus in common marmosets. Data obtained from an intravenous glucose tolerance test can augment this clinical dataset, particularly in cases where the determination of a diabetic phenotype based on typical clinical assays is not clear [14]. Dextrose (50%; 300–750 mg/kg IV) is administered as a slow bolus via a 24-gauge catheter placed in the saphenous vein. Blood glucose is measured in heel stick blood samples using a handheld glucometer at the following time points: −5, −1, 5, 10, 20, 30, 40, and 60 min. Data are plotted as a curve with glucose level on the y-axis and time on the x-axis. Calculated summary measures include peak blood glucose level, 60-min blood glucose level, area under the curve, and

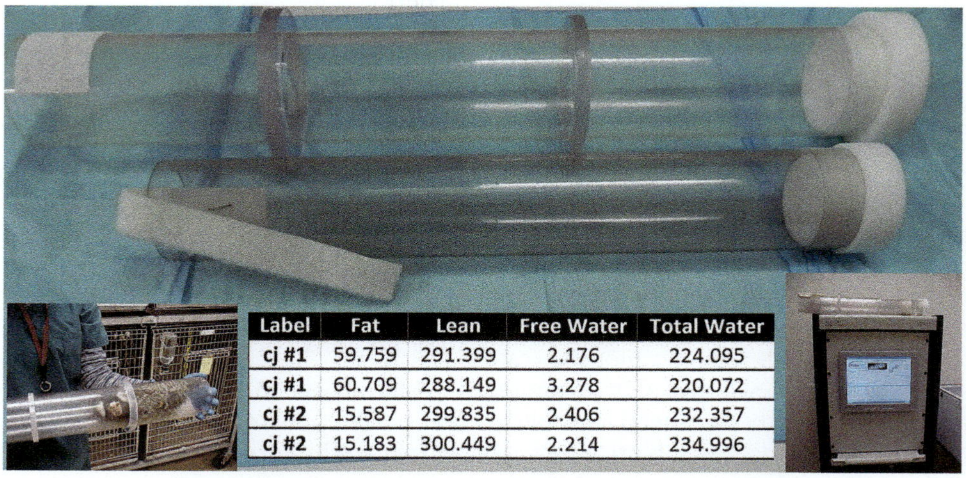

Label	Fat	Lean	Free Water	Total Water
cj #1	59.759	291.399	2.176	224.095
cj #1	60.709	288.149	3.278	220.072
cj #2	15.587	299.835	2.406	232.357
cj #2	15.183	300.449	2.214	234.996

FIGURE 10.12 Performance of body composition measurements in the common marmoset using quantitative magnetic resonance (QMR). The alert or sedated marmoset is placed in a tube-shaped Plexiglas holder with a smaller diameter tube secured behind the animal to prevent escape. Representative data quantifying lean mass, fat mass, free water, and total water is depicted (below, center). The QMR device is internally shielded with a small footprint to allow housing within the clinical space of the marmoset vivarium (below, right).

glucose disappearance rate [Eq. (1): ($\log_e$Glucose$_{5\ min}$ − $\log_e$glucose$_{30\ min}$)/25]. Additionally, measurement of biomarkers such as insulin, adiponectin, leptin, and ghrelin may aid the clinician in completing the clinical picture for a given patient [150].

Imaging Modalities

With the advent of novel gene editing techniques, completion of the whole genome sequence, and an increased understanding of the unique aspects of the behavioral repertoire and social structure of the common marmoset, this species has recently garnered a great deal of interest from neuroscientists [134,151,152]. It is anticipated that the common marmoset's close genetic and physiologic homology to humans will increase the translational utility of neuropsychiatric research performed in this species. In addition to recent studies of genetically modified marmosets, groundbreaking work continues using classical models such as the experimental allergic encephalomyelitis model of multiple sclerosis and the MPTP model of Parkinson's disease [153]. Concurrent with this continued and growing interest, significant advancements in experimental techniques for imaging the central nervous system have been made. Anatomical magnetic resonance imaging (MRI) correlated with histological examinations has allowed the publication of detailed brain atlases, and MR spectroscopy has allowed the in vivo measurement of neurometabolites [154–157]. MRI-based diffusion tensor imaging enables detailed mapping of the structure of neuronal circuits, whereas functional MRI provides great insight into the activation of visual, auditory, and other sensory pathways [158–160]. Lastly, positron emission tomography has been used to quantitate and localize the expression of biological substances such as bound serotonin transporters, dopamine receptors and transporters, monoamine transporters, glucose, and numerous other molecules in both sedated and alert marmosets [157,161]. It is beyond the scope of this chapter to discuss the intricacies of these advanced imaging techniques. Techniques of routine clinical utility such as ultrasound and radiography are discussed in more detail below.

As ultrasound machines become more ubiquitous in the laboratory animal setting, the clinician should develop familiarity with species-specific normal and pathologic appearance of anatomical structures on ultrasound images. An array of ultrasound transducers and scanners used for imaging the common marmoset have been evaluated, and imaging performance of these instruments have been described [162]. Choice of transducer is important; the ideal transducer has high resolution combined with a depth of field that is adequate to visualize structures of interest [162]. High accuracy transducers with a narrow beam width and a depth of field <7 mm may be ideal for imaging superficial structures such as the uterus and ovaries but may not be sufficient to image those structures with a greater distance from the transducer focal region [162]. At the authors' institution, a 21-mHz probe is most frequently used for clinical and experimental purposes although occasionally a 30-mHz probe is used to obtain greater image detail of small structures. Table 10.4 lists select publications, the type of imaging performed, and available details of the transducers used. Pulsed wave Doppler and color Doppler capabilities are useful in assessing parameters such as fetal heart rate and evaluation of blood flow through various tissues. The ability to take measurements of structures on stored ultrasound images and videos is also a valuable feature.

Ultrasonographic techniques and standards for confirmation of pregnancy and reproductive tract evaluation in the marmoset have been described [168–170]. To perform an ultrasonographic examination of the uterine body, ovaries, and other reproductive structures in the marmoset, a high-frequency probe is placed in a transverse orientation at the level of the pelvic bone region and is slowly scanned cranially until the uterine body is visualized. The authors have found the 1995 manuscript published by Oerke et al. to be the most useful reference published to date for performing gestational staging in marmosets, and all of the ultrasonographic indicators of gestational progression described below source to this reference [170]. The ultrasonographic appearance of the nongravid uterus is ovoid in shape with a single bright line transecting the uterine body horizontally (Fig. 10.13). This line represents the apposed layers of the endometrium. At approximately day 15 of gestation, these layers begin to separate and appear as two bright lines (Fig. 10.14). This uterine lumen becomes measurable after approximately day 21 of gestation and individual embryos can be counted starting at approximately day 33. From approximately day 38–75 of gestation, ultrasonographic measurements of uterine diameter and lumen diameter can be used to estimate gestational age. The beating fetal heart becomes visible at approximately day 54. The authors of this chapter have routinely used pulsed wave Doppler settings to measure the fetal heart rate starting at this embryonic stage. This function is of high clinical importance in assessing fetal viability if a dam presents with clinical signs of dystocia. The biparietal diameter of the fetus can be measured starting at approximately day 82 (Fig. 10.15). Ultrasound is also useful for estimation of the number of fetuses. Accuracy, as determined by instances in which a greater number of infants were born relative to the number of fetuses counted, is greater than 92% [171]. After day 110 of gestation, radiographs

TABLE 10.4 Publications Describing the Types of Ultrasound Transducers Used for Select Imaging Purposes in Marmoset Species

Publication	Purpose of Ultrasonography	Description of Transducer Utilized
[163]	Examine *C. penicilliata* with acute abdominal pain	6.6–14 MHz linear transducer
[164]	Guide transcervical cannulation for embryo collection	7.2–14 MHz linear transducer
[162]	Image ovarian structure	20–60 MHz real-time microvisualization transducer (highest rated for this purpose)
		8–16 MHz linear transducer
		5–17 MHz linear transducer
		7–15 MHz linear transducer
		10–22 MHz linear transducer
		6–13 MHz linear transducer (lowest rated for this purpose)
[165]	Echocardiography	13 MHz sector transducer
[166]	Abdominal imaging	Linear array multifrequency transducer (operated at 9 MHz)
[167,168]	Determine embryonic and fetal losses	10 MHz mechanical probe—early-stage exams
		7.5 MHz mechanical probe—late-stage exams
[169]	Gestational staging	7.5 MHz sequential linear array transducer
		7.5 MHz linear sector array transducer
		7.5 MHz annular array transducer
[170]	Gestational staging	7.5 MHz intraoperative sector transducer—early-stage exams
		7.5 MHz linear transducer—late-stage exams

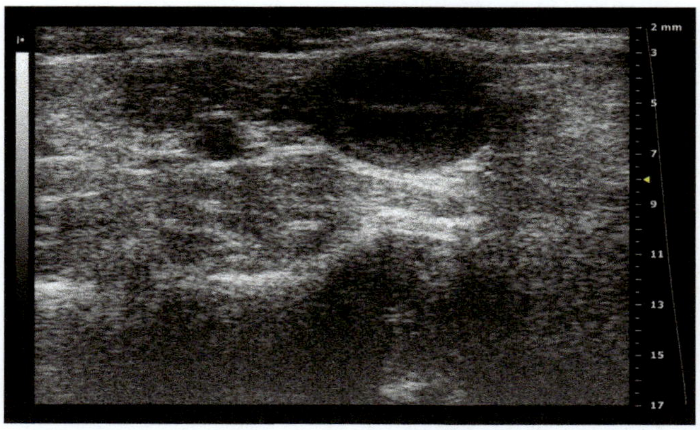

FIGURE 10.13 Ultrasound image depicting a nongravid uterus of a common marmoset. The uterus is ovoid in shape. The apposed layers of the endometrium appear as a single bright, horizontal line transecting the uterine body.

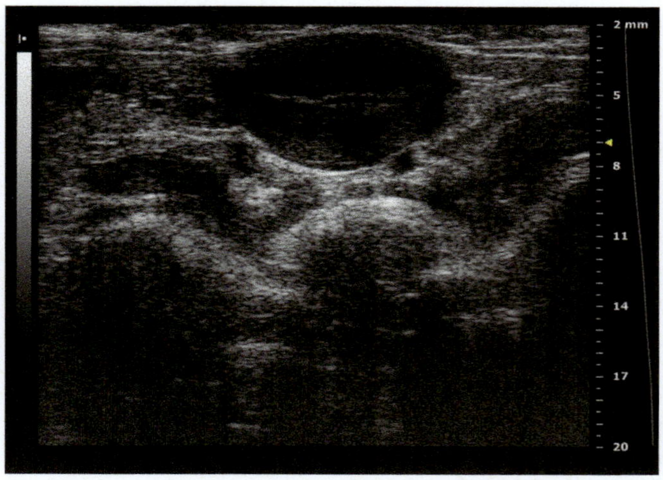

FIGURE 10.14 Ultrasound image depicting day 15 of gestation. The uterus is ovoid in shape. The separating layers of the endometrium appear as two bright, horizontal lines transecting the uterine body.

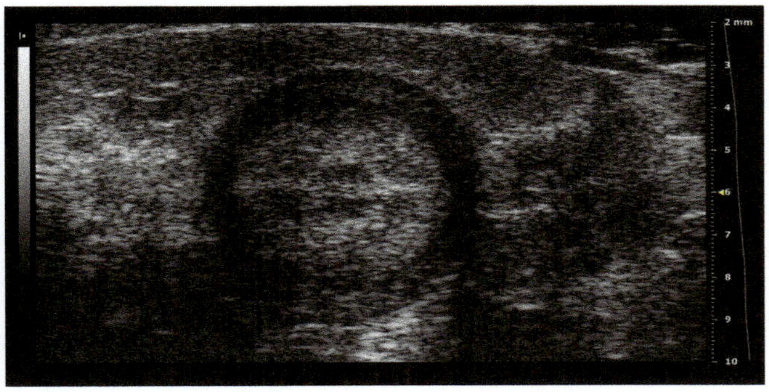

FIGURE 10.15 Ultrasound image depicting measurement of the biparietal diameter (BPD) of the marmoset fetus. BPD can be measured starting at day 82 of gestation. To measure BPD the ultrasound probe is oriented to image the fetal skull in a transverse orientation. To ensure accurate measurement, the cerebral falx should be visible as a central, parallel line, whereas the ocular structures should not be visible in the image. Measurements are obtained at the widest part of the skull.

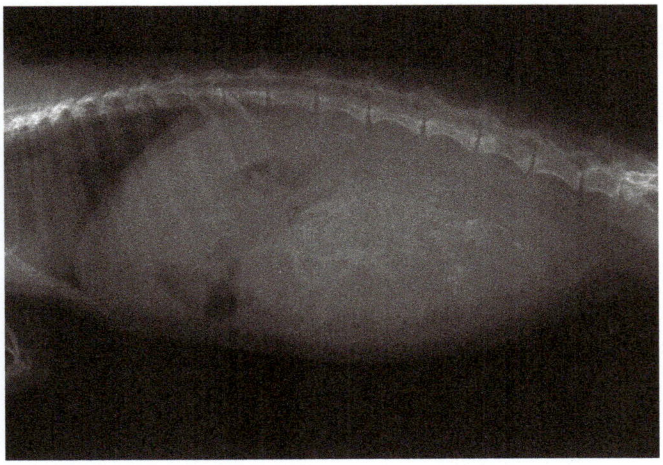

FIGURE 10.16 Radiograph of a pregnant common marmoset. Fetal structures are visible on radiographs starting at approximately day 110 of gestation. Three fetal skulls are evident in this image. The skull that is positioned closest to the pelvis is not fully mineralized.

can be selectively used as an adjunct means to determine the number of fetuses [172] (Fig. 10.16). Highly skilled ultrasonographers are able to count follicles and corpora lutea with a 92% and 78% accuracy, respectively [173]. The disappearance of longitudinally monitored follicles is consistent with the preovulatory LH surge indicating ovulation [173].

A 2005 publication used a linear array multifrequency transducer to scan 17 healthy animals and generated mean measurements for normal abdominal structures in the marmoset [166]. While this study provides a basis for standard measurements of abdominal structures in the marmoset, further evaluation involving larger sample sizes and animals of all life stages will be necessary to validate these measurements. The authors of that study found that the application of canine and feline standards to marmoset ultrasound images could result in misdiagnosis, as marmoset echoanatomy has significant differences from companion animal species. Clinicians are encouraged to develop their own standard practices and measurements using available ultrasound machines and animal populations. Each model of ultrasound machine is unique, and thus caution must be applied when using standards developed on different machines to interpret diagnostic measurements. As with other small species, a common clinician error is the application of too much pressure with the transducer so that structures are compressed and images distorted. Light pressure applied with the transducer and adequate contact with ultrasound gel is optimal during ultrasonographic exam. Accelerated development of imaging technology in recent years will certainly augment the clinical utility of ultrasonography for use in captive marmoset populations, as well as increase accessibility to clinicians for both diagnostic and research use.

As in other species, radiography is a useful modality for the diagnosis of disease in common marmosets. Digital radiographic equipment is becoming increasingly available in the laboratory animal setting and allows for significant improvements in the ease of acquisition of high-quality images. Clinicians should develop radiographic techniques and procedures to maximize the utility of the equipment available to them. A study, run in parallel to the 2005 work described above, was performed using 17 marmosets to generate normative radiologic data [174]. These investigators used a 100-speed mammography system with a source-to-image distance of 105 cm, kVp of 44, mAs of 10, and table top technique to collect right lateral and ventrodorsal images at the end of inspiration. Using this technique, the heart, lungs, liver, kidneys, and gastric axis were readily visible. Measurements for each of these structures are provided in the publication. An interstitial lung pattern was routinely observed. Abdominal detail was poor, with the intestinal tract often discernible only by gas pattern. This lack of detail did not appear to correlate with the amount of body fat. Investigators did note that regional collimation did not improve abdominal detail but did increase the quality of thoracic images. Investigators also stated that the selected radiographic technique may have contributed to the poor abdominal detail, and that further investigation is required to determine optimal radiographic techniques for this species.

Several studies have examined the marmoset skeletal system using radiographic techniques [13,88,90,175]. In one study, collection of radiographs as a component of the PHC program revealed a respective 12.8% and 8.6% prevalence of bone disease over a 9-year time span at two independent facilities [90]. Radiographic findings are summarized as follows. In the first colony, young animals with a histologically confirmed diagnosis of rickets demonstrated radiographic evidence of thickened physes and metaphyseal flaring. Animals with a histologically confirmed diagnosis of FOD demonstrated a "moth-eaten" radiographic appearance of the cortex and medulla of long bones, mandibles, maxillae, and vertebrae. Lastly, animals with a histologically confirmed diagnosis of osteopenia demonstrated a diffuse decrease in radiodensity of the cortex and medulla. In animals from this colony, radiographic and histologic lesions were not mutually exclusive; animals with a primary diagnosis of rickets often demonstrated evidence of FOD and osteopenia and animals with a primary diagnosis of FOD also often displayed evidence of osteopenia. Unique radiographic findings associated with an idiopathic bone disease were observed in the second colony. Lesioned animals demonstrated severe, locally extensive, well-demarcated areas of radiolucency that were accompanied by loss of cortical bone. Occasionally, these lytic lesions coalesced to induce a multilocular expansion of the bone. These findings suggest that routine radiography may be a good complement to a marmoset preventative health program.

In addition to the qualitative findings described in the above study, radiographic techniques can be used for quantitative assessment of BMD. Computed tomography (CT) is considered the gold standard for quantitation of BMD and has the ability to detect bone loss earlier than standard radiographic images [88]. Grohmann et al. used CT to measure BMD and trabecular area ratio in marmosets and found that these measures correlated with measures of ALP and, in males, calcium [175]. In the absence of CT, as described in a study by Baxter et al. [13], a mammographic aluminum step wedge detailing grades of radiolucency was included in digital radiographic images of 10 marmosets. Using image analysis software, the bone radiodensity fraction (BRF; ranging from 0 to 1) was determined by comparison of the distal femur to the step wedge scale. Marmosets with a BRF<0.5 demonstrated an increased risk of bone disease evidenced by one or more of the following histological findings: abnormal growth plates, abnormal cortices, evidence of fracture, increased bone resorption, or myelofibrosis.

Ophthalmic Procedures

As is a limitation with other body systems, this species' small size contributes to difficulties in examining the eyes. Quantification of aqueous tear production is difficult to perform using Schmirmer's tear test (STT) strips, even when they are cut to fit the palpebral fissure. Instead, phenol red impregnated thread, which indicates tear production by a pH-induced color change when moistened, or endodontic absorbent paper points (size 30) are recommended [176]. As with the STT, the thread or paper points are placed in the conjunctival fornix and tear production is measured by the amount that these items are moistened over a set time frame. The anterior and posterior ocular segments can be examined with a standard ophthalmoscope. Mydriasis is achieved with topical 1% tropicamide and 2.5%–10% phenylephrine hydrochloride [25,26]. Guyomard et al. describe a technique to image the fundus and ciliary body of the marmoset using a 2.7-mm diameter, 10-cm long rigid endoscope probe linked to a light source, camera, and monitor [26]. Chronopoulos et al. used optic coherence tomography to image and quantify changes to the retina of marmosets with a diet-induced diabetic phenotype [25].

CONCLUSION

Working with common marmosets is at times not only challenging but also extremely rewarding. To provide the best health care to this species, clinicians should develop consistent and thorough examination procedures. The execution of novel clinical techniques may require a combination of creativity, thorough review of the literature, and discussion with colleagues. To advance the standard of care, the authors of this chapter encourage those working with marmosets to publish the clinical techniques they develop for use in this species.

References

[1] Abbott DH, Barnett DK, Colman RJ, Yamamoto ME, Schultz-Darken NJ. Aspects of common marmoset basic biology and life history important for biomedical research. Comp Med 2003;53:339–50.

[2] Field K, Bailey M, Foresman LL, Harris RL, Motzel SL, Rockar RA, Ruble G, Suckow MA. Medical records for animals used in research, teaching, and testing: public statement from the American College of Laboratory Animal Medicine. ILAR J 2007;48:37–41.

[3] Passamani M. Activity budget of Geoffroy's marmoset (*Callithrix geoffroyi*) in an Atlantic forest in southeastern Brazil. Am J Primatol 1998;46:333–40.

[4] Takahashi DY, Fenley AR, Teramoto Y, Narayanan DZ, Borjon JI, Holmes P, Ghazanfar AA. LANGUAGE DEVELOPMENT. The developmental dynamics of marmoset monkey vocal production. Science 2015;349:734–8.

[5] Watson CFI, Buchanan-Smith HM. Marmoset care website. University of Stirling. http://marmosetcare.com/.

[6] Araujo A, Arruda MF, Alencar AI, Albuquerque F, Nascimento MC, Yamamoto ME. Body weight of wild and captive common marmosets (*Callithrix jacchus*). Int J Primatol 1999;21:317–24.

[7] Poole TB, Evans RG. Reproduction, infant survival and productivity of a colony of common marmosets (*Callithrix jacchus jacchus*). Lab Anim 1982;16:88–97.

[8] Power RA, Power ML, Layne DG, Jaquish CE, Oftedal OT, Tardif SD. Relations among measures of body composition, age, and sex in the common marmoset monkey (*Callithrix jacchus*). Comp Med 2001;51:218–23.

[9] Tardif SD, Power ML, Ross CN, Rutherford JN, Layne-Colon DG, Paulik MA. Characterization of obese phenotypes in a small nonhuman primate, the common marmoset (*Callithrix jacchus*). Obesity 2009;17:1499–505.

[10] Burns M, Kramer R. Normal clinical and biological parameters. In: Marini R, Wachtman L, Tardif S, Mansfield K, Fox JG, editors. The common marmoset in captivity and biomedical research. Academic Press; 2018.

[11] Clingerman KJ, Summers L. Validation of a body condition scoring system in rhesus macaques (*Macaca mulatta*): inter- and intrarater variability. J Am Assoc Lab Anim Sci 2012;51:31–6.

[12] Otovic P, Smith S, Hutchinson E. The use of glucocorticoids in marmoset wasting syndrome. J Med Primatol 2015;44:53–9.

[13] Baxter VK, Shaw GC, Sotuyo NP, Carlson CS, Olson EJ, Zink MC, Mankowski JL, Adams RJ, Hutchinson EK, Metcalf Pate KA. Serum albumin and body weight as biomarkers for the antemortem identification of bone and gastrointestinal disease in the common marmoset. PLoS One 2013;8:e82747.

[14] Wachtman LM, Kramer JA, Miller AD, Hachey AM, Curran EH, Mansfield KG. Differential contribution of dietary fat and monosaccharide to metabolic syndrome in the common marmoset (*Callithrix jacchus*). Obesity 2011a;19:1145–56.

[15] Ross CN, Davis K, Dobek G, Tardif SD. Aging phenotypes of common marmosets (*Callithrix jacchus*). J Aging Res 2012;2012: 567143.

[16] Magden ER, Mansfield KG, Simmons JH, Abee CR. Nonhuman primates. In: Fox JG, Anderson L, Otto G, Prichett-Corning K, Whary M, editors. Laboratory animal medicine. Elsevier; 2015.

[17] Hetherington CM. Circadian oscillations of body temperature in the marmoset, *Callithrix jacchus*. Lab Anim 1978;12:107–8.

[18] Logan AC, Khan KN. Clinical pathologic changes in two marmosets with wasting syndrome. Toxicol Pathol 1996;24:707–9.

[19] Wachtman LM, Pistorio AL, Eliades S, Mankowski JL. Calcinosis circumscripta in a common marmoset (*Callithrix jacchus jacchus*). J Am Assoc Lab Anim Sci 2006;45:54–7.

[20] Layne DG, Power RA. Husbandry, handling, and nutrition for marmosets. Comp Med 2003;53:351–9.

[21] Epple G. Quantitative studies on scent marking in the marmoset (*Callithrix jacchus*). Folia Primatol 1970;13:48–62.

[22] Epple G. The role of pheromones in the social communication of marmoset monkeys (*Callithricidae*). J Reprod Fertil Suppl 1973;19: 447–54.

[23] Barrett J, Abbott DH, George LM. Extension of reproductive suppression by pheromonal cues in subordinate female marmoset monkeys, *Callithrix jacchus*. J Reprod Fertil 1990;90:411–8.

[24] Smith TE, Tomlinson AJ, Mlotkiewicz JA, Abbott DH. Female marmoset monkeys (*Callithrix jacchus*) can be identified from the chemical composition of their scent marks. Chem Senses 2001;26:449–58.

[25] Chronopoulos A, Roy S, Beglova E, Mansfield K, Wachtman L. Hyperhexosemia-induced retinal vascular pathology in a novel primate model of diabetic retinopathy. Diabetes 2015;64:2603–8.

[26] Guyomard JL, Rosolen SG, Paques M, Delyfer MN, Simonutti M, Tessier Y, Sahel JA, Legargasson JF, Picaud S. A low-cost and simple imaging technique of the anterior and posterior segments: eye fundus, ciliary bodies, iridocorneal angle. Invest Ophthalmol Vis Sci 2008;49:5168–74.

[27] Boisvert C, Mitchell J, Adams SC, Reynolds J, Leblanc M. Ophthalmologic assessment of the common marmoset (*Callithrix jacchus*) involved in visual neuroscience. JAALAS 2016;55.

[28] Nummela SU, Coop SH, Cloherty SL, Boisvert CJ, Leblanc M, Mitchell JF. Psychophysical measurement of marmoset acuity and myopia. Dev Neurobiol 2017;77:300–13.

[29] Orsi A, Rees D, Andreini I, Venturella S, Cinelli S, Oberto G. Overview of the marmoset as a model in nonclinical development of pharmaceutical products. Regul Toxicol Pharmacol 2011;59:19–27.

[30] Korte S, Zuhlke U, Srivastav S, Korte U, Weinbauer G, Muller W, Vogel F. The common marmoset (*Callithrix jacchus*) as a model in toxicology: reference control data comparison with macaques. Baltimore, Maryland USA: Society of Toxicology; 2004.

[31] Bakker J, Uilenreef JJ, Pelt ER, Brok HP, Remarque EJ, Langermans JA. Comparison of three different sedative-anaesthetic protocols (ketamine, ketamine-medetomidine and alphaxalone) in common marmosets (*Callithrix jacchus*). BMC Vet Res 2013;9:113.

[32] Morosco DT, Cline CR, Owston MA, Kumar S, Dick Jr EJ. Spontaneous mediastinal myeloid sarcoma in a common marmoset (*Callithrix jacchus*) and review of the veterinary literature. J Med Primatol 2017;46(2).

[33] Betton GR. Spontaneous neoplasms of the marmoset (*Callithrix jacchus*). Oral and nasopharyngeal squamous cell carcinomas. Vet Pathol 1984;21:193–7.

[34] McIntosh GH, Giesecke R, Wilson DF, Goss AN. Spontaneous nasopharyngeal malignancies in the common marmoset. Vet Pathol 1985;22:86–8.

[35] Brack M, Schwartz P, Heinrichs T, Schultz M, Fuchs E. Tumors of the respiratory tract observed at the German Primate Center, 1978-1994. J Med Primatol 1996;25:424–34.

[36] Sutherland SD, Almeida JD, Gardner PS, Skarpa M, Stanton J. Rapid diagnosis and management of parainfluenza I virus infection in common marmosets (*Callithrix jacchus*). Lab Anim 1986; 20:121–6.

[37] Baskerville M, Wood M, Baskerville A. An outbreak of *Bordetella bronchiseptica* pneumonia in a colony of common marmosets (*Callithrix jacchus*). Lab Anim 1983;17:350–5.

[38] Schnell CR, Wood JM. Measurement of blood pressure and heart rate by telemetry in conscious unrestrained marmosets. Lab Anim 1995;29:258–61.

[39] David JM, Dick Jr EJ, Hubbard GB. Spontaneous pathology of the common marmoset (*Callithrix jacchus*) and tamarins (*Saguinus oedipus*, *Saguinus mystax*). J Med Primatol 2009;38:347–59.

[40] Tardif SD, Mansfield KG, Ratnam R, Ross CN, Ziegler TE. The marmoset as a model of aging and age-related diseases. ILAR J 2011;52:54–65.

[41] Hunt RD, Melendez LV, King NW, Gilmore CE, Daniel MD, Williamson ME, Jones TC. Morphology of a disease with features of malignant lymphoma in marmosets and owl monkeys inoculated with Herpesvirus saimiri. J Natl Cancer Inst 1970;44: 447–65.

[42] Wachtman LM, Mansfield K. Viral diseases of nonhuman primates. In: Abee CR, Mansfield K, Tardif S, Morris T, editors. Nonhuman primates in biomedical research. Diseases, vol. 2. Acadmic Press; 2012. p. 1–104.

[43] Yamaguchi I, Myojo K, Sanada H, Sudo E, Ootsuka S, Okumura H, Takami A, Yoneshige T, Suzuki Y, Imaizumi M, Takada C, Kimoto N, Saeki K, Takaba K. Spontaneous malignant T cell lymphoma in a young male common marmoset (*Callithrix jacchus*). J Toxicol Pathol 2013;26:301–7.

[44] Cho Y, Ramer J, Rivailler P, Quink C, Garber RL, Beier DR, Wang F. An Epstein-Barr-related herpesvirus from marmoset lymphomas. Proc Natl Acad Sci U S A 2001;98:1224–9.

[45] Ramer JC, Garber RL, Steele KE, Boyson JF, O'Rourke C, Thomson JA. Fatal lymphoproliferative disease associated with a novel gammaherpesvirus in a captive population of common marmosets. Comp Med 2000;50:59–68.

[46] Wachtman LM, Miller AD, Xia D, Curran EH, Mansfield KG. Colonization with nontuberculous mycobacteria is associated with positive tuberculin skin test reactions in the common marmoset (*Callithrix jacchus*). Comp Med 2011b;61:278–84.
[47] Hershkovitz P. Dental and periodontal diseases and abnormalities in wild-caught marmosets (primates—*Callithricidae*). Am J Phys Anthropol 1970;32:377–94.
[48] Ludlage E, Mansfield K. Clinical care and diseases of the common marmoset (*Callithrix jacchus*). Comp Med 2003;53:369–82.
[49] Casagrande RA, Pannuti CS, Kanamura C, Freire WS, Grespan A, Matushima ER. Fatal *Human herpesvirus 1* (HHV-1) infection in captive marmosets (*Callithrix jacchus* and *Callithrix penicillata*) in Brazil: clinical and pathological characterization. Pesq Vet Bras 2014;34:1109–14.
[50] Matz-Rensing K, Jentsch KD, Rensing S, Langenhuyzen S, Verschoor E, Niphuis H, Kaup FJ. Fatal Herpes simplex infection in a group of common marmosets (*Callithrix jacchus*). Vet Pathol 2003;40:405–11.
[51] Hahn NE, Capuano 3rd SV. Successful treatment of cryptosporidiosis in 2 common marmosets (*Callithrix jacchus*) by using paromomycin. J Am Assoc Lab Anim Sci 2010;49:873–5.
[52] Izzi JMS, Garret C. *Salmonellosis* outbreak in a captive marmoset colony. JAALAS 2015;55:101–14.
[53] Kramer JA, Hachey AM, Wachtman LM, Mansfield KG. Treatment of giardiasis in common marmosets (*Callithrix jacchus*) with tinidazole. Comp Med 2009;59:174–9.
[54] Hayashimoto N, Inoue T, Morita H, Yasuda M, Ueno M, Kawai K, Itoh T. Survey and experimental infection of enteropathogenic *Escherichia coli* in common marmosets (*Callithrix jacchus*). PLoS One 2016;11:e0160116.
[55] Bakker J, Kondova I, de Groot C, Remarque EJ, Heidt PJ. A report on Yersinia-related mortality in a colony of New World monkeys. Lab Primate Newsl 2007;46:11–6.
[56] Cooper JE, Needham JR. An outbreak of shigellosis in laboratory marmosets and tamarins (Family: *Callithricidae*). J Hyg 1976;76: 415–24.
[57] Pisharath HR, Cooper TK, Brice AK, Cianciolo RE, Pistorio AL, Wachtman LM, Mankowski JL, Newcomer CE. Septicemia and peritonitis in a colony of common marmosets (*Callithrix jacchus*) secondary to *Klebsiella pneumoniae* infection. Contemp Top Lab Anim Sci 2005;44:35–7.
[58] Mansfield K, Fox JG. Bacterial diseases. In: Marini R, Wachtman L, Tardif S, Mansfield K, Fox JG, editors. The common marmoset in captivity and biomedical research. Academic Press; 2018.
[59] Shen Z, Feng Y, Sheh A, Everitt J, Bertram F, Paster BJ, Fox JG. Isolation and characterization of a novel Helicobacter species, *Helicobacter jaachi* sp. nov., from common marmosets (*Callithrix jaachus*). J Med Microbiol 2015;64:1063–73.
[60] de Mello MF, Monteiro AB, Fonseca EC, Pissinatti A, Ferreira AM. Identification of Helicobacter sp. in gastric mucosa from captive marmosets (*Callithrix* sp.; *callitrichidae*, primates). Am J Primatol 2005;66:111–8.
[61] Won YS, Vandamme P, Yoon JH, Park YH, Hyun BH, Kim HC, Itoh T, Tanioka Y, Choi YK. *Helicobacter callitrichis* sp. nov., a novel *Helicobacter* species isolated from the feces of the common marmoset (*Callithrix jacchus*). FEMS Microbiol Lett 2007;271: 239–44.
[62] Yoshida K, Nibe K, Nakamura T, Takahashi T, Komatsu M, Ogawa H, Shirota K, Chambers JK, Nakayama H, Uchida K. Spontaneous Tyzzer's disease with the central nerve involvement in a newborn common marmoset. J Vet Med Sci 2013;75: 1119–21.
[63] Yasuda M, Inoue T, Ueno M, Morita H, Hayashimoto N, Kawai K, Itoh T. A case of nontraumatic gas gangrene in a common marmoset (*Callithrix jacchus*). J Vet Med Sci 2016;77:1673–6.
[64] Chen PH, Miller GF, Powell DA. Colitis in a female tamarin (*Saguinus mystax*). Contemp Top Lab Anim Sci 2000;39:47–9.
[65] Power M, Myers E. Digestion in the common marmoset (*Callithrix jacchus*), a gummivory-frugivore. Am J Primatol 2009;71: 957–63.
[66] Jarcho MR, Power ML, Layne-Colon DG, Tardif SD. Digestive efficiency mediated by serum calcium predicts bone mineral density in the common marmoset (*Callithrix jacchus*). Am J Primatol 2013;75:153–60.
[67] Miller AD, Kramer JA, Lin KC, Knight H, Martinot A, Mansfield KG. Small intestinal adenocarcinoma in common marmosets (*Callithrix jacchus*). Vet Pathol 2010;47:969–76.
[68] Yokouchi Y, Imaoka M, Sayama A, Jindo T, Sanbuissho A. Inflammatory fibroid polyp in the duodenum of a common marmoset (*Callithrix jacchus*). Toxicol Pathol 2013;41:80–5.
[69] Kramer JA, Grindley J, Crowell AM, Makaron L, Kohli R, Kirby M, Mansfield KG, Wachtman LM. The common marmoset as a model for the study of nonalcoholic fatty liver disease and nonalcoholic steatohepatitis. Vet Pathol 2015;52:404–13.
[70] Ludlage E, Murphy CL, Davern SM, Solomon A, Weiss DT, Glenn-Smith D, Dworkin S, Mansfield KG. Systemic AA amyloidosis in the common marmoset. Vet Pathol 2005;42:117–24.
[71] Asper M, Hofmann P, Osmann C, Funk J, Metzger C, Bruns M, Kaup FJ, Schmitz H, Gunther S. First outbreak of callitrichid hepatitis in Germany: genetic characterization of the causative lymphocytic choriomeningitis virus strains. Virology 2001;284: 203–13.
[72] Montali RJ, Scanga CA, Pernikoff D, Wessner DR, Ward R, Holmes KV. A common-source outbreak of callitrichid hepatitis in captive tamarins and marmosets. J Infect Dis 1993;167:946–50.
[73] Miller GF, Barnard DE, Woodward RA, Flynn BM, Bulte JW. Hepatic hemosiderosis in common marmosets, *Callithrix jacchus*: effect of diet on incidence and severity. Lab Anim Sci 1997;47: 138–42.
[74] Smith KM, McAloose D, Torregrossa AM, Raphael BL, Calle PP, Moore RP, James SB. Hematologic iron analyte values as an indicator of hepatic hemosiderosis in *Callitrichidae*. Am J Primatol 2008;70:629–33.
[75] McKinley J, Buchanan-Smith HM, Bassett L, Morris K. Training common marmosets (*Callithrix jacchus*) to cooperate during routine laboratory procedures: ease of training and time investment. J Appl Anim Welf Sci 2003;6:209–20.
[76] Isobe K, Adachi K, Hayashi S, Ito T, Miyoshi A, Kato A, Suzuki M. Spontaneous glomerular and tubulointerstitial lesions in common marmosets (*Callithrix jacchus*). Vet Pathol 2012;49: 839–45.
[77] Yamada N, Sato J, Kanno T, Wako Y, Tsuchitani M. Morphological study of progressive glomerulonephropathy in common marmosets (*Callithrix jacchus*). Toxicol Pathol 2013;41:1106–15.
[78] Brack M, Schroeder C, Fooke M, Schlumberger W. IgM/IgA nephropathy in callitrichids: antigen studies. Nephron 1999;82: 221–31.
[79] Brack M, Weber M. Ultrastructural and histochemical mesangial alterations in Callitrichid IgM nephropathy (primates: *platyrrhina*). Nephron 1995;69:286–92.
[80] Buckwold VE, Collins B, Hogan P, Rippeon S, Wei J. Investigation into the ability of GB virus B to replicate in various immortalized cell lines. Antiviral Res 2005;66:165–8.
[81] Collins MG, Rogers NM, Jesudason S, Kireta S, Brealey J, Coates PT. Spontaneous glomerular mesangial lesions in common marmoset monkeys (*Callithrix jacchus*): a benign nonprogressive glomerulopathy. J Med Primatol 2014;43:477–87.
[82] Vanselow BA, Pines MK, Bruhl JJ, Rogers LJ. Oxalate nephropathy in a laboratory colony of common marmoset monkeys (*Callithrix jacchus*) following the ingestion of *Eucalyptus viminalis*. Vet Rec 2011;169:100.

[83] Pereira MM, Da Silva JJ, Pinto MA, Da Silva MF, Machado MP, Lenzi HL, Marchevsky RS. Experimental leptospirosis in marmoset monkeys (*Callithrix jacchus*): a new model for studies of severe pulmonary leptospirosis. Am J Trop Med Hyg 2005;72:13–20.

[84] Molina CV, Catao-Dias JL, Ferreira Neto JS, Vasconcellos SA, Gennari SM, do Valle Rdel R, de Souza GO, de Morais ZM, Vitaliano SN, Strefezzi Rde F, Bueno MG. Sero-epidemiological survey for brucellosis, leptospirosis, and toxoplasmosis in free-ranging *Alouatta caraya* and *Callithrix penicillata* from Sao Paulo State. Brazil J Med Primatol 2014;43:197–201.

[85] Isachenko EF, Nayudu PL, Isachenko VV, Nawroth F, Michelmann HW. Congenitally caused fused labia in the common marmoset (*Callithrix jacchus*). J Med Primatol 2002;31:350–5.

[86] Sanchez-Morgado JM, Haworth R, Morris TH. XY female marmoset (*Callithrix jacchus*). Comp Med 2003;53:539–44.

[87] Divincenti Jr L, Miller AD, Knoedl DJ, Mitchell JF. Uterine rupture in a common marmoset (*Callithrix jacchus*). Comp Med 2016;66:254–8.

[88] Grohmann J, Taetzner S, Theuss T, Kuehnel F, Buchwald U, Einspanier A. The conclusiveness of less-invasive imaging techniques (computer tomography, X-ray) with regard to their identification of bone diseases in a primate model (*Callithrix jacchus*). J Med Primatol 2012b;41:130–7.

[89] Gacad MA, Chen H, Arbelle JE, LeBon T, Adams JS. Functional characterization and purification of an intracellular vitamin D-binding protein in vitamin D-resistant new world primate cells. Amino acid sequence homology with proteins in the hsp-70 family. J Biol Chem 1997;272:8433–40.

[90] Olson EJ, Shaw GC, Hutchinson EK, Schultz-Darken N, Bolton ID, Parker JB, Morrison JM, Baxter VK, Pate KA, Mankowski JL, Carlson CS. Bone disease in the common marmoset: radiographic and histological findings. Vet Pathol 2015;52:883–93.

[91] Marini R, Haupt J. Anesthesia and common surgical procedures. In: Marini R, Wachtman L, Tardif S, Mansfield K, Fox JG, editors. The common marmoset in captivity and biomedical research. Academic Press; 2018.

[92] Tochitani T, Matsumoto I, Hoshino K, Toyosawa K, Kouchi M, Koujitani T, Kimura J, Funabashi H. Spontaneous rhabdomyosarcoma in a common marmoset (*Callithrix jacchus*). J Toxicol Pathol 2013;26:187–91.

[93] Bosseler L, Cornillie P, Saunders JH, Bakker J, Langermans JA, Casteleyn C, Decostere A, Chiers K. Micromelic dysplasia-like syndrome in a captive colony of common marmosets (*Callithrix jacchus*). Comp Med 2014;64:394–403.

[94] Palazzi X, Switzer R, George C. Natural occurrence of amyloid-Aβ deposits in the brain of young common marmosets (*Callithrix jacchus*): a morphological and immunohistochemical evaluation. Vet Pathol 2006;43:777–9.

[95] Kramer JA, Ford EW, Capuano S. Preventative medicine in nonhuman primates. In: Abee CR, Mansfield K, Tardif S, Morris T, editors. Nonhuman primates in biomedical research. Biology and management, vol. 1. Academic Press; 2012. p. 293–321.

[96] Montali RJ, Mikota SK, Cheng LI. Mycobacterium tuberculosis in zoo and wildlife species. Rev Sci Tech 2001;20:291–303.

[97] Via LE, Weiner DM, Schimel D, Lin PL, Dayao E, Tankersley SL, Cai Y, Coleman MT, Tomko J, Paripati P, Orandle M, Kastenmayer RJ, Tartakovsky M, Rosenthal A, Portevin D, Eum SY, Lahouar S, Gagneux S, Young DB, Flynn JL, Barry 3rd CE. Differential virulence and disease progression following *Mycobacterium tuberculosis* complex infection of the common marmoset (*Callithrix jacchus*). Infect Immun 2013;81:2909–19.

[98] Cadena AM, Klein EC, White AG, Tomko JA, Chedrick CL, Reed DS, Via LE, Lin PL, Flynn JL. Very low doses of *Mycobacterium tuberculosis* yield diverse host outcomes in common marmosets (*Callithrix jacchus*). Comp Med 2016;66:412–9.

[99] Michel AL, Huchzermeyer HF. The zoonotic importance of *Mycobacterium tuberculosis*: transmission from human to monkey. J S Afr Vet Assoc 1998;69:64–5.

[100] Lin PL, Yee J, Klein E, Lerche NW. Immunological concepts in tuberculosis diagnostics for non-human primates: a review. J Med Primatol 2008;37(Suppl. 1):44–51.

[101] Housman G, Malukiewicz J, Boere V, Grativol AD, Pereira LC, Silva Ide O, Ruiz-Miranda CR, Truman R, Stone AC. Validation of qPCR methods for the detection of Mycobacterium in new world animal reservoirs. PLoS Negl Trop Dis 2015;9:e0004198.

[102] Wissman MA. Husbandry and medical care of Callitrichids. J Exot Pet Med 2014;23:347–62.

[103] Favoretto SR, de Mattos CC, Morais NB, Alves Araujo FA, de Mattos CA. Rabies in marmosets (*Callithrix jacchus*), Ceara, Brazil. Emerg Infect Dis 2001;7:1062–5.

[104] Buchanan-Smith H. Marmosets and tamarins. In: Hubrecht R, Kirkwood J, editors. The UFAW handbook on the care and management of laboratory and other research animals. Wiley-Blackwell; 2010. p. 543–63.

[105] Powell DS, Walker RC, Heflin DT, Fisher D, Kosky JB, Homer LC, Reed DS, Stefano-Cole K, Trichel AM, Hartman AL. Development of novel mechanisms for housing, handling, and remote monitoring of common marmosets at animal biosafety level 3. Pathog Dis 2014;71:219–26.

[106] Williams PT, Poole MJ, Katos AM, Hilmas CJ. A new device for the capture and transport of small nonhuman primates in scientific research. Lab Anim (NY) 2008;37:116–9.

[107] Prescott M, Bowell V, Buchanan-Smith H. Training laboratory-housed non-human primates, part 2: resources for developing and implementing training programmes. ATW 2005:133–48.

[108] Tardif S, Bales K, Williams L, Moeller EL, Abbott D, Schultz-Darken N, Mendoza S, Mason W, Bourgeois S, Ruiz J. Preparing New World monkeys for laboratory research. ILAR J 2006;47:307–15.

[109] Donnelly MJ, Wickham LA, Kulick AA, Rogers IT, Stribling DS, Strack AM, Doerning BJ, Feeney WP. A refinement of oral dosing in the common marmoset (*Callithrix jacchus*). In: AALAS national meeting, Charlotte, NC; 2007.

[110] Diehl KH, Hull R, Morton D, Pfister R, Rabemampianina Y, Smith D, Vidal JM, van de Vorstenbosch C, European Federation of Pharmaceutical Industries Association and European Centre for the Validation of Alternative Methods. A good practice guide to the administration of substances and removal of blood, including routes and volumes. J Appl Toxicol 2001;21:15–23.

[111] Maurus C, Lange A, Simkins M. Blood collection and intravenous injection in the common marmoset (*Callithrix jacchus*). Indianapolis, IN: AALAS National Meeting; 2008.

[112] O'Byrne KT. Long-term blood sampling technique in the marmoset. Lab Anim 1988;22:151–3.

[113] Fitzgerald AL, Dillon LM, Altrogge DM, Bleavins MR, Breider MA. Use of subcutaneous vascular access ports in common marmosets (*Callithrix jacchus*). Contemp Top Lab Anim Sci 1996;35:57–9.

[114] Sauer M, Dulac H, Aborn S, Doherty J, Batchelder M. Surgical placement of vascular access ports in common marmosets, lessons learned. In: Association of primate veterinarians, 35th annual workshop, Charlotte, NC; 2007.

[115] Swindle MM, Nolan T, Jacobson A, Wolf P, Dalton MJ, Smith AC. Vascular access port (VAP) usage in large animal species. Contemp Top Lab Anim Sci 2005;44:7–17.

[116] Ruiz de Elvira MC, Abbott DH. A backpack system for long-term osmotic minipump infusions into unrestrained marmoset monkeys. Lab Anim 1986;20:329–34.

[117] Schultz-Darken NJ. Sample collection and restraint techniques used for common marmosets (*Callithrix jacchus*). Comp Med 2003;53:360–3.
[118] Hearn JP, Lunn SF, Burden FJ, Pilcher MM. Management of marmosets for biomedical research. Lab Anim 1975;9:125–34.
[119] Greig I, Morris KD, Mathiesen E, Mathiesen R, Buchanan-Smith HM. An improved restraint device for injections and collection of samples from marmosets. Lab Primate Newsl 2006;45:1–5.
[120] Hearn JP. Restraining device for small monkeys. Lab Anim 1977; 11:261–2.
[121] Schultz-Darken NJ, Pape RM, Tannenbaum PL, Saltzman W, Abbott DH. Novel restraint system for neuroendocrine studies of socially living common marmoset monkeys. Lab Anim 2004; 38:393–405.
[122] Bohm RP, Gilbert MH. Emergency medicine and critical care for nonhuman primates. In: Abee C, Mansfield K, Tardif S, Morris T, editors. Nonhuman primates in biomedical research. Biology and management, vol. 1. Academic Press; 2012. p. 359–89.
[123] Chen S, Wei Q, Li J, Xiang Y, Guo H, Ichim TE, Chen G. A simple and reliable method to blood type monkeys using serum samples. Transpl Int 2009;22:999–1004.
[124] Mollicone R, Davies DR, Evans B, Dalix AM, Oriol R. Cellular expression and genetic control of ABH antigens in primary sensory neurons of marmoset, baboon and man. J Neuroimmunol 1986;10:255–69.
[125] Gengozian N. Human A- and B-like antigens on red cells of marmosets. Proc Soc Exp Biol Med 1964;117:858–61.
[126] Merritt CB, Hampton JK. Further evidence for human A-like antigens on marmoset red cells. Proc Soc Exp Biol Med 1967;124: 134–7.
[127] Wiener AS, Moor-Jankowski J, Gordon EB. Blood groups of apes and monkeys. V. Studies on human blood group factors A, B, H and LE in old and new world monkeys. Am J Phys Anthropol 1964;22:175–87.
[128] Ramis G, Martínez-Alarcon L, Quereda JJ, Mrowiec A, Funes C, Ríos A, Ramírez P, Muñoz A, Majado MJ. Non-ABO blood group systems phenotyping in non-human primates for blood banking laboratory and xenotransplantation. Lab Anim 2013;47:100–5.
[129] Smith TE, French JA. Psychosocial stress and urinary cortisol excretion in marmoset monkeys (*Callithrix kuhli*). Physiol Behav 1997;62:225–32.
[130] Smith TE, McCallister JM, Gordon SJ, Whittikar M. Quantitative data on training new world primates to urinate. Am J Primatol 2004;64:83–93.
[131] French JA, Brewer KJ, Schaffner CM, Schalley J, Hightower-Merritt D, Smith TE, Bell SE. Urinary steroid and gonadotropin excretion across the reproductive cycle in female Wied's black tufted-ear marmosets (*Callithrix kuhli*). Am J Primatol 1996;40: 231–45.
[132] Lunn SF. Systems for collection of urine in the captive common marmoset, *Callithrix jacchus*. Lab Anim 1989;23:353–6.
[133] Seltzer LJ, Ziegler TE. Non-invasive measurement of small peptides in the common marmoset (*Callithrix jacchus*): a radiolabeled clearance study and endogenous excretion under varying social conditions. Horm Behav 2007;51:436–42.
[134] Sasaki E. Creating transgenic marmosets. In: Marini R, Wachtman L, Tardif S, Mansfield K, Fox JG, editors. The common marmoset in captivity and biomedical research. Academic Press; 2018.
[135] Cui KH, Flaherty SP, Newble CD, Guerin MV, Napier AJ, Matthews CD. Collection and analysis of semen from the common marmoset (*Callithrix jacchus*). J Androl 1991;12:214–20.
[136] Morrell JM, Küderling I, Hodges JK. Influence of semen collection method on ejaculate characteristics in the common marmoset, *Callithrix jacchus*. J Androl 1996;17:164–72.
[137] Schneiders A, Sonksen J, Hodges JK. Penile vibratory stimulation in the marmoset monkey: a practical alternative to electro-ejaculation, yielding ejaculates of enhanced quality. J Med Primatol 2004;33:98–104.
[138] Kuederling I, Schneiders A, Sønksen J, Nayudu PL, Hodges JK. Non-invasive collection of ejaculates from the common marmoset (*Callithrix jacchus*) using penile vibrostimulation. Am J Primatol 2000;52:149–54.
[139] Valle RR, Nayudu PL, Leal CL, García-Herreros M. Sperm head morphometry in ejaculates of adult marmosets (*Callithrix jacchus*): a model for studying sperm subpopulations and among-donor variations. Theriogenology 2012;78:1152–65.
[140] Kuederling I, Morrell JM, Nayudu PL. Collection of semen from marmoset monkeys (*Callithrix jacchus*) for experimental use by vaginal washing. Lab Anim 1996;30:260–6.
[141] Geretschläger E, Russ H, Mihatsch W, Przuntek H. Suboccipital puncture for cerebrospinal fluid in the common marmoset (*Callithrix jacchus*). Lab Anim 1987;21:91–4.
[142] Wang H, Yang B, Qiu L, Yang C, Kramer J, Su Q, Guo Y, Brown RH, Gao G, Xu Z. Widespread spinal cord transduction by intrathecal injection of rAAV delivers efficacious RNAi therapy for amyotrophic lateral sclerosis. Hum Mol Genet 2014;23: 668–81.
[143] Tardif SD, Smucny DA, Abbott DH, Mansfield K, Schultz-Darken N, Yamamoto ME. Reproduction in captive common marmosets (*Callithrix jacchus*). Comp Med 2003;53:364–8.
[144] Power ML, Oftedal OT, Tardif SD. Does the milk of Callitrichid monkeys differ from that of larger anthropoids? Am J Primatol 2002;56:117–27.
[145] Murphy KL, Baxter MG, Flecknell PA. Anesthesia and analgesia in nonhuman primates. In: Abee CR, Mansfield K, Tardif S, Morris T, editors. Nonhuman primates in biomedical research. Biology and management, vol. 1. Academic Press; 2012. p. 403–35.
[146] Mansfield KG, Lin KC, Xia D, Newman JV, Schauer DB, MacKey J, Lackner AA, Carville A. Enteropathogenic *Escherichia coli* and ulcerative colitis in cotton-top tamarins (*Saguinus oedipus*). J Infect Dis 2001;184:803–7.
[147] Saunders KE, Shen Z, Dewhirst FE, Paster BJ, Dangler CA, Fox JG. Novel intestinal Helicobacter species isolated from cotton-top tamarins (*Saguinus oedipus*) with chronic colitis. J Clin Microbiol 1999;37:146–51.
[148] Manickam C, Rajakumar P, Wachtman L, Kramer JA, Martinot AJ, Varner V, Giavedoni LD, Reeves RK. Acute liver damage associated with innate immune activation in a small nonhuman primate model of hepacivirus infection. J Virol 2016;90:9153–62.
[149] Klimentidis YC, Beasley TM, Lin HY, Murati G, Glass GE, Guyton M, Newton W, Jorgensen M, Heymsfield SB, Kemnitz J, Fairbanks L, Allison DB. Canaries in the coal mine: a cross-species analysis of the plurality of obesity epidemics. Proc Biol Sci 2011;278:1626–32.
[150] Ziegler TE, Colman RJ, Tardif SD, Sosa ME, Wegner FH, Wittwer DJ, Shrestha H. Development of metabolic function biomarkers in the common marmoset, *Callithrix jacchus*. Am J Primatol 2013;75:500–8.
[151] *Marmoset Genome Sequencing and Analysis Consortium*. The common marmoset genome provides insight into primate biology and evolution. Nat Genet 2014;46:850–7.
[152] Izpisua Belmonte JC, Callaway EM, Caddick SJ, Churchland P, Feng G, Homanics GE, Lee KF, Leopold DA, Miller CT, Mitchell JF, Mitalipov S, Moutri AR, Movshon JA, Okano H, Reynolds JH, Ringach D, Sejnowski TJ, Silva AC, Strick PL, Wu J, Zhang F. Brains, genes, and primates. Neuron 2015;86: 617–31.

[153] Mansfield K. Marmoset models commonly used in biomedical research. Comp Med 2003;53:383–92.
[154] Hashikawa T, Nakatomi R, Iriki A. Current models of the marmoset brain. Neurosci Res 2015;93:116–27.
[155] Newman JD, Kenkel WM, Aronoff EC, Bock NA, Zametkin MR, Silva AC. A combined histological and MRI brain atlas of the common marmoset monkey, *Callithrix jacchus*. Brain Res Rev 2009;62:1–18.
[156] Paxinos G, Watson C, Petrides M, Rosa M, Tokuno H. The marmoset brain in stereotaxic coordinates. San Diego: Academic Press; 2012.
[157] Yun JW, Ahn JB, Kang BC. Modeling Parkinson's disease in the common marmoset (*Callithrix jacchus*): overview of models, methods, and animal care. Lab Anim Res 2015;31:155–65.
[158] Belcher AM, Yen CC, Stepp H, Gu H, Lu H, Yang Y, Silva AC, Stein EA. Large-scale brain networks in the awake, truly resting marmoset monkey. J Neurosci 2013;33:16796–804.
[159] Hung CC, Yen CC, Ciuchta JL, Papoti D, Bock NA, Leopold DA, Silva AC. Functional MRI of visual responses in the awake, behaving marmoset. Neuroimage 2015;120:1–11.
[160] Okano H, Mitra P. Brain-mapping projects using the common marmoset. Neurosci Res 2015;93:3–7.
[161] Yokoyama C, Onoe H. Positron emission tomography imaging of the social brain of common marmosets. Neurosci Res 2015;93: 82–90.
[162] Moran CM, Pye SD, Ellis W, Janeczko A, Morris KD, McNeilly AS, Fraser HM. A comparison of the imaging performance of high resolution ultrasound scanners for preclinical imaging. Ultrasound Med Biol 2011;37:493–501.
[163] de Souza AC, Lange RR, Guerios SD, Froes TR. Ultrasonographic examination in non-human primates with acute abdomen signs. J Med Primatol 2013;42:336–42.
[164] Ishibashi H, Motohashi HH, Kumon M, Yamamoto K, Okada H, Okada T, Seki K. Ultrasound-guided non-surgical embryo collection in the common marmoset. Reprod Biol 2013;13:139–44.
[165] Hanton G, Sobry C, Daguès N, Rochefort GY, Bonnet P, Eder V. Cardiovascular toxicity of minoxidil in the marmoset. Toxicol Lett 2008;180:157–65.
[166] Wagner WM, Kirberger RM. Transcutaneous ultrasonography of the abdomen in the normal common marmoset (*Callithrix jacchus*). Vet Radiol Ultrasound 2005b;46:251–8.
[167] Windle CP, Baker HF, Ridley RM, Oerke AK, Martin RD. Unrearable litters and prenatal reduction of litter size in the common marmoset (*Callithrix jacchus*). J Med Primatol 1999;28:73–83.
[168] Nubbemeyer R, Heistermann M, Oerke AK, Hodges JK. Reproductive efficiency in the common marmoset (*Callithrix jacchus*): a longitudinal study from ovulation to birth monitored by ultrasonography. J Med Primatol 1997;26:139–46.
[169] Tardif SD, Jaquish CE, Toal RL, Layne DG, Power RA. Estimation of gestational ages in the common marmoset (*Callithrix jacchus*) from published prenatal growth curves. J Med Primatol 1998;27:28–32.
[170] Oerke A-K, Einspanier A, Hodges J. Detection of pregnancy and monitoring patterns of uterine and fetal growth in the marmoset monkey (*Callithrix jacchus*) by real-time ultrasonography. Am J Primatol 1995;36.
[171] Jaquish CE, Tardif SD, Toal RL, Carson RL. Patterns of prenatal survival in the common marmoset (*Callithrix jacchus*). J Med Primatol 1996;25:57–63.
[172] Phillips IR. Skeletal development in the foetal and neonatal marmoset (*Callithrix jacchus*). Lab Anim 1976;10:317–33.
[173] Oerke A-K, Einspanier A, Hodges J. Noninvasive monitoring of follicle development, ovulation, and corpus luteum formation in the marmoset monkey (*Callithrix jacchus*) by ultrasonography. Am J Primatol 1996;39:99–113.
[174] Wagner WM, Kirberger RM. Radiographic anatomy of the thorax and abdomen of the common marmoset (*Callithrix jacchus*). Vet Radiol Ultrasound 2005a;46:217–24.
[175] Grohmann J, Kuehnel F, Buchwald U, Koeller G, Habla C, Einspanier A. Analysis of the bone metabolism by quantitative computer tomography and clinical chemistry in a primate model (*Callithrix jacchus*). J Med Primatol 2012a;41:1–10.
[176] Lange RR, Lima L, Montiani-Ferreira F. Measurement of tear production in black-tufted marmosets (*Callithrix penicillata*) using three different methods: modified Schirmer's I, phenol red thread and standardized endodontic absorbent paper points. Vet Ophthalmol 2012;15:376–82.

CHAPTER

11

Anesthesia and Select Surgical Procedures

Robert P. Marini, Jennifer Haupt

Division of Comparative Medicine, Massachusetts Institute of Technology, Cambridge, MA, United States

Anesthesia in marmosets must be considered in the context of their size and the physiologic and anatomic consequences thereof. Body mass will determine the volumes that can be safely administered via various sites, the sites and methods of intravenous catheterization and orotracheal intubation, and the relative importance of maintaining heat, energy, and hydration in comparison with larger animals. Metabolic rate, which is scaled to body mass, is considerably higher per unit mass (the specific metabolic rate) than in larger species, and this biological feature has implications in the practice of anesthesia and surgery [1], especially in lengthy procedures. Another notable issue is the lack of information on the pharmacokinetics of agents used in sedation, anesthesia, and analgesia. Drug doses have been derived empirically, and with regard solely to efficacy and untoward effects, so that doses only approximate the ideal. Allometric scaling, which predicts drug dosage in a target species based on its known dose in another species, and which uses the relationship of body mass to metabolic rate, is limited by this dearth of information in basic pharmacology. An example is buprenorphine. As discussed later, this agent was likely responsible for death in three of six marmosets to which it was administered before complete recovery from anesthesia at 0.03 mg/kg, a dose which should have been reasonable based on efficacy and pharmacokinetic studies in other nonhuman primates [2,3]. At this dose in marmosets, however, it led to apnea and cardiovascular collapse.

ANESTHESIA

Despite the challenges of small size and limited drug efficacy and pharmacology data, the common marmoset can be reliably and safely anesthetized by a variety of agents and their combination. As in other species, a check of all anesthetic equipment before induction and careful calculation of anesthetic doses is recommended. An emergency drug sheet with precalculated doses for each patient prior to anesthesia induction is also a prudent measure.

Preanesthetic Assessment and Considerations

Marmosets should be observed for activity and demeanor while in their home cage. Olberg and Sinclair [4] recommend fasts of 6–8 h for animals under 1 kg body weight to avoid perioperative hypoglycemia. Once the animal is sedated or anesthetized, a full physical examination can be performed. If the animal is adequately immobilized, and depending on the procedure, the saphenous or lateral tail vein can be catheterized with a 22–24 gauge, over-the-needle catheter. A point-of-care analyzer such as the I-STAT (Abbott Point of Care Inc., Princeton, NJ) is useful for evaluating hematological and serum chemistry variables prior to anesthesia. A dedicated anesthetist should monitor the animal at all times and attend to its heat, energy, and hydration through use of supplemental heating devices and administration of warm intraperitoneal, subcutaneous, or intravenous fluids. Temperature monitoring is essential for assuring normothermia, as hyperthermia can also occur due to robust heat-preservation measures. During surgery, intravenous fluids should be administered at a maintenance rate of 4 mL/kg/h [5]. The administration rate can be adjusted as the procedure and clinical signs dictate and based on preexisting and contemporary losses. Anesthetists must guard against overhydration and its consequences, volume overload and pulmonary edema. The use of precision fluid pumps, and in longer procedures, monitoring of urine production, packed cell volume, electrolytes, and central venous pressure, with occasional adjustment of the infusion rate, will help avoid these complications.

The Common Marmoset in Captivity and Biomedical Research
https://doi.org/10.1016/B978-0-12-811829-0.00011-X

As in other species, a range of crystalloids can be used (lactated Ringer's solution; 0.9% saline; Normosol; 5% dextrose in lactated Ringer's solution). The use of colloids in marmosets has not been described. The anticholinergics atropine (0.02−0.1 mg/kg, [6]; 0.15 mg/kg [7]) or glycopyrrolate (0.005−0.01 mg/animal [8,9]), can be administered to control salivary and tracheobronchial secretions, especially during inhalant use. In our surgical laboratory, we use atropine at 0.01−0.02 mg/kg SC and IV or glycopyrrolate at 0.005−0.01 mg/kg SC if indicated to help control tracheobronchial secretions or in the treatment of bradycardia. The use of these agents in the absence of a specific indication is not as common as it once was, and many anesthetists avoid their use in the absence of bradycardia, ptyalism, or issues with airway maintenance.

Anesthetic Agents

Injectables

It is common in marmoset medicine and surgery to administer the same agents for sedation, induction, and maintenance. Moreover, the small size of the animal and the relatively large volume of the injectable can result in muscle injury and lead to postoperative complications, including lameness or paralysis. Large volumes should be divided into several sites and be administered with the smallest gauge needle that allows the material injected to pass. The quadriceps group is the largest muscle mass available. The most widely used agents are the dissociative ketamine, the steroid alphaxalone, the benzodiazepines diazepam and midazolam, the alkyl phenol agent propofol, and the α-2 agonist xylazine. The most commonly used regimens combine ketamine or alphaxalone with other agents.

Ketamine

The dissociative agent and NMDA-receptor antagonist ketamine is used in doses from 5 to 15 mg/kg IM for mild sedation and chemical restraint to 50 mg/kg for surgical procedures [10]. Doses at the lower end of the range can be used to prolong anesthesia induced by other agents (e.g., 5 mg/kg ketamine IM after alfaxalone/diazepam (Table 11.1) [11]). The use of additional agents in anesthetic cocktails can prolong anesthesia, increase anesthetic depth, and preclude or mitigate the many undesirable consequences of ketamine when used as a sole agent. These include hypersalivation, myoclonus, poor muscle relaxation, spontaneous movement, prolonged recovery with higher doses, and, potentially, emergence delirium [4]. An additional concern with the intramuscular use of ketamine in marmosets is myotoxicity [12]. The administration of 10 mg (35 mg/kg for a 350 g animal) into the gastrocnemius muscle caused significant elevation in AST, CK, and LDH-5 isoenzyme with consequent myofibrillar shrinkage, loss of striation, necrosis, and infiltration of red and white blood cells. Additional trials using solutions with the low pH and high osmolality of ketamine showed that most, though not all, of the changes, were attributable to ketamine itself [12].

Alfaxalone

The steroidal agent alfaxalone (Alfaxan, Jurox Animal Health) has replaced the combination agent Saffan (alfaxalone 0.9% w/v; alphadalone 0.3% w/v), abandoned because of side effects associated with its diluent cremaphor EL. Saffan is no longer commercially available, but references of its use in marmosets are included for completeness [10,13,14]. Anesthetic effects identified in these references provide a useful index of the effects of alfaxalone alone, in that this agent was the more potent of the steroids in the Saffan combination. Alfaxalone's mechanism of action is chloride ion modulation associated with interaction with the neuronal membrane GABA A receptor. Alfaxalone is safe and effective at 12−15 mg/kg IM for routine noninvasive procedures and lasts approximately 40−45 min [15,16]. The most common side effects are dose-dependent but transient hypotension (15 min), hypoventilation, and apnea. Higher doses of alfaxalone (12 mg/kg) may cause mild hypoxemia, which can be corrected by administration of supplemental oxygen via face mask [17]. Muscle twitching occurs toward the end of immobilization and can be used as a sign of impending emergence. In the surgical laboratory of the authors, alfaxalone is used at a lower concentration (5−7 mg/kg IM) for procedures such as blood work or radiographs; 10−12 mg/kg IM is used for minor surgical procedures. The injection volume at higher doses can be quite large; no more than 0.2 mL per site should be administered. Intravenous doses are used after placement of a catheter; a 2−5 mg/kg bolus IV followed by 0.17 mg/kg min IV CRI can be used for advanced imaging or surgery. IV bolus and CRI are often preceded by an initial IM dose of ketamine or alfaxalone [9]. Alfaxalone is most commonly used in combination with various doses of diazepam (Table 11.1). In this context, it is used for anesthetic induction and endotracheal intubation or stereotaxic fixation with subsequent delivery of an inhalant via face mask [18], as well as for sedation for short imaging procedures such as magnetic resonance imaging (MRI) [19].

Several studies compare combinations of agents containing ketamine or alfaxalone. Bakker et al. [15] compared ketamine (50 mg/kg), alfaxalone (12 mg/kg), and two combinations of ketamine (25 mg/kg) and medetomidine (0.5 mg/kg or 0.05 mg/kg, both with subsequent atipamezole reversal). Variables evaluated included induction, immobilization, and recovery time, pulse and respiratory rate, systolic, diastolic, and mean blood pressure, and pulse oximetry. Alfaxalone had significantly longer induction and immobilization times

TABLE 11.1 Injectable Agents for Use in Sedation, Anesthesia, and Analgesia of the Common Marmoset

Sedative or Anesthetic Drug	Dose	Indications and Comments	References
Ketamine	5–15 mg/kg IM	Chemical restraint	Buchanan-Smith [10]
	5 mg/kg IM	Prolonging anesthesia	Bakker et al. [15]
	50 mg/kg IM	Surgery	Buchanon-Smith [10]
Alfaxalone	10 mg/kg IM	IV catheterization	Thomas et al. [22]
Diazepam	1–3 mg/kg IM	Sedation, premedication	Bourne and Rosa [6] Valverde Salzmann et al. [26]
Alfaxalone	5–7 mg/kg IM	Physical examinations, blood collection, radiographs	Johns Hopkins Formulary 2017, Garrett C. [9]
Alfaxalone	12–15 mg/kg IM	Noninvasive procedures	Ansel et al. [16] Bakker et al. [15]
Propofol	Bolus to effect IV followed by 0.9 mg/kg min CRI	Evaluate for apnea. Useful for functional magnetic resonance imaging studies when NMDA antagonists and/or isoflurane cannot be used	Johns Hopkins Formulary 2017 Garrett C [9].
KETAMINE COMBINATIONS			
Ketamine/ butorphanol/ midazolam	10 mg/kg; 0.015 mg/kg; 0.3 mg/ kg IM	Comparison study on anesthetic effect	Ishibashi et al. [19]
Ketamine/atropine/ Diazepam	10 mg/kg; 0.15 mg/kg; 0.75 mg/ kg IM	IP implantation of osmotic minipumps	Schell and Wood [7]
Ketamine/xylazine	10–15 mg/kg; 1.5 mg/kg IM	Surgery	Buchanan-Smith [10]
	45 mg/kg; 3 mg/kg IM	Stereotaxic surgery (two drops of doxapram sublingually OR 0.2 mg/kg yohimbine if apnea occurs)	Bourne and Rosa [6]
Ketamine/xylazine/ atropine cocktail	50 mg/mL; 50 mg/mL; 0.1 mg/ mL; 1 mL cocktail/kg bw IM	Endotracheal intubation	Hoffmann et al. [23]
Ketamine/ butorphanol/ midazolam	10–30 mg/kg; 0.005–0.015 mg/ kg; 0.1–0.3 mg/kg IM	Ultrasound-guided transcervical embryo collection	Ishibashi et al. [55]
Ketamine/midazolam	10 mg/kg; 1 mg/kg IM	30–45 min effect	Olberg and Sinclair [4]
ALFAXALONE COMBINATIONS			
Alfaxalone/diazepam/ glycopyrolate	10 mg/kg; 0.3 mg/kg; 0.01 mg/ animal IM	Endotracheal intubation in preparation for stereotaxic surgery	Garea-Rodriquez et al. [8] and Helms et al. [11]
Alfaxalone/diazepam	12 mg/kg; 3 mg/kg IM	Anesthesia induction (atropine 0.05 mg/kg IM) and dexamethasone (0.4 mg/kg IM after induction). Author statement: better induction, recovery with fewer complications compared with ketamine/ xylazine	Mundinano et al. [18]
ANALGESICS			
Ketoprofen	2–3 mg/kg IM		Allen et al. [33]
Meloxicam	0.01 mg/kg PO q 24 h	0.5 mg/mL administered by syringe. Loading dose of 0.2 mg/kg is recommended **NOT FOR USE IN ANIMALS UNDER 2.5 MONTHS**	Mundinano et al. [18]
Carprofen	4 mg/kg SQ q 24 h		Buchanan-Smith [10]
Buprenorphine	0.005–0.01 mg/kg IM q 8–12 h	Dilute to 0.03 mg/mL. Administer after surgery	Allen et al. [33] Mundinano et al. [18]

TABLE 11.2 Induction, Immobilization, and Recovery Times for Ketamine and Alfaxalone

	Protocol 1 (Ketamine 50 mg/kg IM)	Protocol 2 (Alfaxalone 12 mg/kg IM)
Induction time	1.24 ± 0.55	2.54 ± 0.65
Immobilization time	31.23 ± 22.39	53.72 ± 13.08
Recovery time	135.84 ± 39.19	55.79 ± 11.02
Total procedure time	168.32 ± 45.60	112.05 ± 15.30

*Adapted from Bakker JJ, Uilenreef JJJ, Pelt ERR, Brok HPP, Remarque EJJ, et al. Comparison of three different sedative-anaesthetic protocols (ketamine, ketamine-medetomidine and alphaxalone) in common marmosets (*Callithrix jacchus*). BMC Vet Res 2013;9:113.*

and a shorter recovery time than ketamine (Table 11.2). Alfaxalone was also superior in its ability to diminish muscle tone and ablate pedal withdrawal and palpebral reflexes. All regimens caused a significant increase in CK concentration, but the ketamine-containing regimens also increased AST significantly. Both ketamine/medetomidine protocols had unacceptably long recovery times, suggesting that the doses were inappropriate. There were no significant differences among regimens for the cardiorespiratory variables.

Ishibashi [19] compared anesthetic induction using ketamine (30 mg/kg) or the cocktail, ketamine (10 mg/kg), midazolam (0.3 mg/kg), and butorphanol (0.015 mg/kg), followed, after loss of the righting reflex, by 5% sevoflurane in 80%–100% O_2 by face mask. With absence of the pedal withdrawal reflex as a measure of surgical anesthesia, 7 of 99 animals did not reach a surgical anesthetic plane when administered ketamine alone, whereas only 2 of 273 failed to do so with ketamine/midazolam/butorphanol.

Xylazine

The α-2 adrenoceptor agonist xylazine has been used in the common marmoset. In contrast to other species, there is limited information on the use of newer, more selective α-2 agonists in this species. Xylazine (3 mg/kg IM) has been used with ketamine (45 mg/kg IM) for anesthesia induction after atropine (0.2 mg/kg IM) and diazepam (3 mg/kg IM) premedication and prior to tracheotomy and stereotaxic surgery [6]. Atropine was used to avoid the bradycardic and ventricular arrythmogenic effects of the agent. The authors also made special note to warn against additional doses of the agent. Xylazine-associated apnea can be reversed with two drops of doxapram hydrochloride applied sublingually. The α-2 adrenoceptor antagonist yohimbine (0.2 mg/kg IM) can be given as well [6]. Alpha-2 adrenoceptor agonists are not used for marmosets in our surgical laboratory.

Propofol

The alkyl phenol agent propofol is used in many species for anesthetic induction and maintenance via constant rate infusion for minimally invasive procedures like imaging. It is a GABA receptor agonist and a glutamate antagonist [20]. The drug is available as a liquid emulsion, has rapid onset, and animals recover rapidly if the drug is not administered in repeated doses. Among its many physiologic effects are reduction of cerebral metabolic rate, intracranial pressure, and cerebral blood flow. Its principal shortcomings are the need to administer it intravenously and its tendency to produce apnea if administered too rapidly. Reports of its use in marmosets are limited to the imaging setting and are described in the balanced anesthetic technique section below.

Inhalant Agents

Induction of anesthesia with inhalant anesthetics can be achieved by using a rodent induction chamber of adequate size. Inhalant induction without premedication is recommended for infants under 3 months of age [18]. The chamber should not be precharged. Inhalant concentration should be 1%–4% isoflurane or 2%–5% sevoflurane, both administered in oxygen. The practice in our surgical laboratory is to start with lower concentrations, increasing by 1% increments every 30–60 s. Adult animals are best premedicated to avoid breath holding prior to induction chamber or mask induction.

Masks of reasonably appropriate size can be purchased commercially. Custom masks to accommodate the very small size of the marmoset face have been described by Mundinano et al. [18] and are generated by MRI (9.4T) followed by 3-D printing. The mask described by these authors incorporates a palatine bar for stereotaxic surgery. Masks can also be constructed from modified 3–6 mL syringe cases or polyvinylchloride tubing, tethered to the anesthetic circuit by cable tie or tape. They can be constructed as needed for single use only.

Endotracheal Intubation

As in other species, airway control is critical during anesthesia, and an endotracheal tube of adequate diameter allows for effective positive-pressure ventilation during periods of apnea. The advantages of the endotracheal tube include ready access to the tracheal mucosa for administration of agents during cardiopulmonary resuscitation, the ability to ventilate mechanically, and reduced operating theater pollution during inhalant anesthetic administration. Placement does require time, however, and surgeons and anesthetists must consider the value of intubation in light of the cost of time, especially in short or routine procedures. When a face mask is used, chest compression can be used to ventilate apneic animals. Tracheotomy can also be used in acute studies during long-duration anesthesia [6].

Endotracheal intubation may be attempted under direct visualization with the animal restrained vertically and the mouth held open by using the thumb and index finger to grasp the zygomatic arch on either side of the head [21]. A cotton tip applicator is used to depress the base of the tongue, which is drawn forward, facilitating visualization of the larynx. A drop of 2% lidocaine without epinephrine is placed on the laryngeal cartilages, and a customized endotracheal tube with stylet is inserted. Tubes are premeasured to avoid improper placement [21]. A Macintosh laryngoscope blade size 0 or Miller blade size 00 works well for visualization.

Alternatively, a modification of the rodent intubation technique using a tilt table (Tilting Workstand, Hallowell Inc., Pittsfield, MA; Fig. 11.1) and Rodent Intubation Pack can be used (http://www.hallowell.com/index.php?pr=Video_Presentations&doc=3) [22]. In this technique, the marmoset is placed in dorsal recumbency on the workstand and secured by its canine teeth and adjustable side bars. Once the animal is secured, the workstand is tilted to a 45-degrees position with the animal's head up. This maneuver extends the neck of the animal. The mouth is opened by deflection of the mandible by a Miller O laryngoscope blade with appropriate handle. The tip of the blade is placed between the tip of the epiglottis and the base of the tongue. Upward deflection of the tip brings the larynx into view. The operator sprays it with 2 mg of lidocaine without epinephrine. The endotracheal tube comprises a 12 or 14 gauge, over-the-needle, intravenous catheter, from which the stylet has been removed. The junction of the catheter and its luer tip is reinforced with a piece of siliconized rubber (Silastic) tubing of appropriate inner diameter and extends from the luer tip to 3 cm from the free end of the catheter. The tip is beveled to 40 degrees, and a Murphy eye added to the larger, 12 gauge catheter. The Rodent Intubation Pack (Hallowell Inc.) is equipped with a syringe whose plunger has a Seldinger wire attached. The catheter is placed over the wire, and is used to intubate the larynx, after which the endotracheal tube is advanced over the wire until the Silastic tubing is positioned immediately rostral to the larynx. Rotating the tube 180 degrees can assist intubation and help overcome the natural 140-degree angle created by the longitudinal axes of the larynx and trachea (Fig. 11.2). Intubation of the trachea is confirmed by demonstrating condensation with each exhalation using a dental mirror and observation of normal thoracic and abdominal excursions with expansion of the rebreathing bag. Confirmation by capnography with placement of an infant/pediatric CO_2 sample line or sensor is helpful. The chest wall can be compressed to help confirm intubation in the apneic patient. The tube can be secured using an umbilical tape or Velcro tie.

Thomas et al. [22] have determined adequate tube length to be a multiple of crown-to-rump length. Proper tube length precludes both inadvertent displacement and bronchial intubation. Endotracheal tubes cut to a

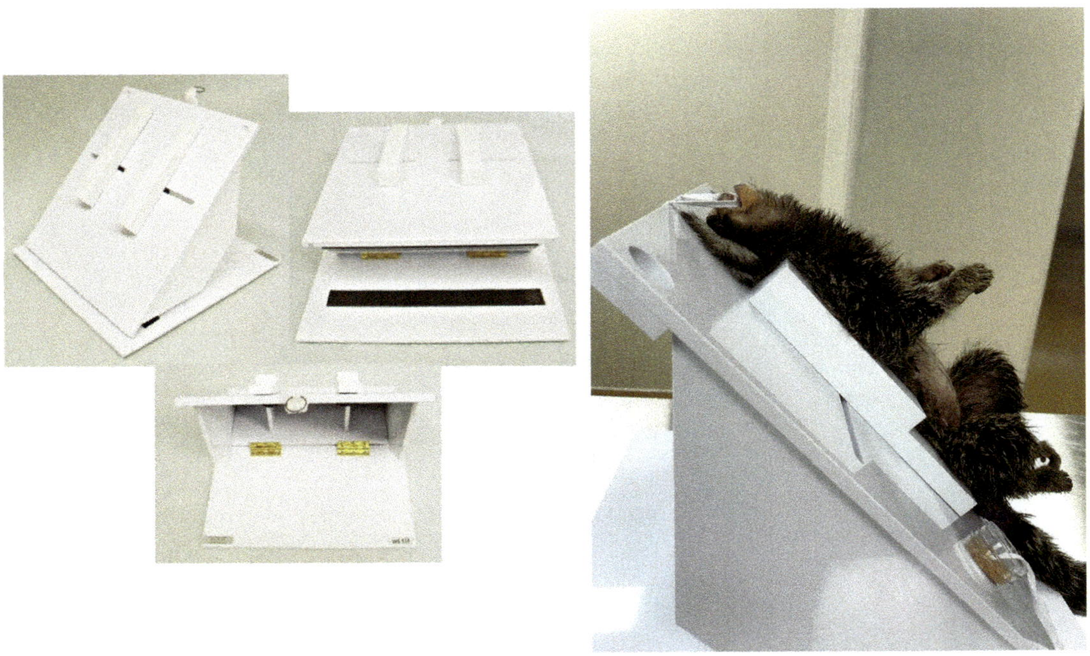

FIGURE 11.1 The Rodent WorkStand (Hallowell Engineering and Manufacturing Corporation) can be used to facilitate the endotracheal intubation of common marmosets.

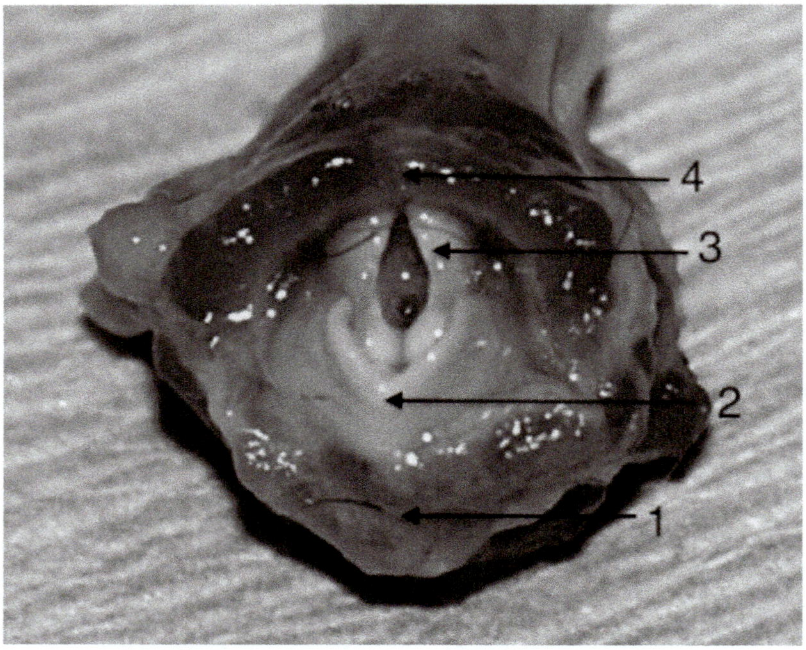

FIGURE 11.2 En face and sagittal view of the common marmoset larynx. (1) Base of the tongue, (2) epiglottis, (3) arytenoid cartilage, and (4) dorsal pharyngeal wall. *Reprinted with permission from Thomas AA, Leach MC, Flecknell PA. An alternative method of endotracheal intubation of common marmosets (*Callithrix jacchus*). Lab Anim 2012;46:71–6.*

length that is the product of 0.42 and the craniosacral length (0.42 × CS)[1] will approximate the nose to tracheobronchial bifurcation distance. A variety of endotracheal tubes have been described, including 2-0 to 3-0 id Portex tubes (uncuffed), modified 1.5 mm id Cole endotracheal tubes, five French red rubber urinary catheters with suitable adapters, 1.2/2.2 id/od polyethylene endotracheal tubes, and 12–14 gauge reinforced IV catheters with beveled tips (Fig. 11.3) [8,21–23]. In our surgical laboratory, adult marmosets are intubated with 1.0–1.5 mm ID silicon endotracheal tubes (Surgivet. Smiths Medical, Waukesha, WI). These are autoclavable and equipped with their own stylets. Tracheotomies will accommodate 3 mm od tubes.

Anesthetists should deliver inhalants to marmosets using nonrebreathing circuits such as the T-piece or Bain coaxial circuit equipped, if possible, with low dead space connectors. Flow rates of 2–2.2 times the minute respiratory volume will preclude rebreathing [24]. A rate of 500–800 mL/min is typically cited. If there is an orotracheal or tracheostomy tube in place, artificial ventilation can be used. A respiratory rate of 40–50 per minute and a tidal volume of 7–10 mL/kg are appropriate. Many ventilators are available, but those

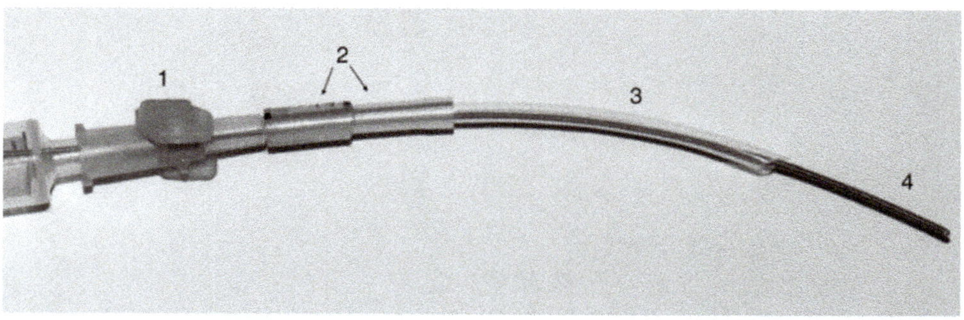

FIGURE 11.3 Custom-made endotracheal catheter created from a 12-gauge over-the-needle catheter: (1) catheter head, (2) silastic tubing, (3) catheter body and (4) syringe-mounted blunt metal stylet. *Reprinted with permission from Thomas AA, Leach MC, Flecknell PA. An alternative method of endotracheal intubation of common marmosets (*Callithrix jacchus*). Lab Anim 2012;46:71–6.*

[1] Top of the parietal bone to the base of the tail.

available from Kent Scientific (www.kentscientific.com; Torrington, CT) have been recommended [24]. Artificial ventilation is essential in regimens incorporating neuromuscular blockade and is useful in longer procedures to avoid the development of atelectasis and to maintain normocapnia. Two issues associated with intermittent positive-pressure ventilation are potential hypotension due to decreased venous return and the necessity to wean animals from ventilation until they are able to breathe spontaneously. The latter can be achieved by gradual reduction of ventilation rate or total air flow while assuring adequate oxygenation and ventilation through pulse oximetry and capnography.

The agents commonly in use in marmosets are isoflurane and sevoflurane. Minimum alveolar concentrations of these gases have not been determined for marmosets. These agents must therefore be used to effect: typically 1%–4% for isoflurane and 2%–5% for sevoflurane. Carrier gases can vary from oxygen to various concentrations of oxygen and medical air or carbon dioxide or nitrogen. The use of room air or nitrogen avoids the high PaO_2 which commonly results from high inspired oxygen in prolonged procedures (e.g., >16 h). Nitrous oxide is used in some neuroscience settings to permit decreased inhalant exposure while still maintaining adequate anesthetic depth. This mitigates the depressive effect of inhalants on many neurologic responses. As in other species, nitrous oxide should not exceed 70% of the carrier gas and should be terminated 10 min prior to the end of anesthesia to preclude the development of diffusion hypoxia.

Isoflurane and sevoflurane will likely cause dose-dependent depression of cardiovascular function. Anecdotally, isoflurane is more likely to cause breath holding than sevoflurane. Those administering inhalants to marmosets must continuously evaluate the breathing of their patients to assure adequacy of respiration. A nurse or veterinarian dedicated to anesthesia delivery and monitoring is essential.

Inhalants can be used for anesthesia induction following premedication with ketamine (4–15 mg/kg IM; [10,25]), diazepam (0.75–3 mg/kg; [6,26]) or other agents [6,26,27]. Sevoflurane has a lower blood solubility than isoflurane; the former is therefore associated in other species with a decrease in induction and recovery times. Sevoflurane is also less potent than isoflurane and so is likely to be used at higher concentrations. Valverde et al. [26] state their preference for sevoflurane. In the United States, the purchase cost of the agent is currently four times that of isoflurane.

Balanced Anesthetic Techniques

A number of balanced anesthetic techniques have been developed in marmosets for use in neuroscience research. These techniques employ several injectable agents, either administered via continuous-rate infusion or intermittent bolus, often with subanesthetic concentrations of inhalants with or without nitrous oxide as a carrier gas. Agents often found in these regimens include opioid analgesics, neuromuscular blocking agents, the alkyl phenol agent propofol, and benozodiazepines. Various agents within the regimen provide the individual components of the anesthetic tetrad: analgesia, muscle relaxation, loss of consciousness, and amnesia. The balanced anesthetic technique serves to maintain neuronal activity depressed by agents such as isoflurane and ketamine [6]. Independent of their neuronal depression-sparing activity, balanced techniques using opioid infusion are also valuable in other settings. After intubation, opioid infusion allows reduction of inhalant concentration (MAC sparing) or a lower rate of infusion of propofol or alfaxalone than would otherwise be needed.

Liu et al. [25] evaluated the influence of anesthetic technique on blood oxygen level-dependent (BOLD) response to electrical stimulation of the forearm. Marmosets were sedated with ketamine (4 mg/kg IM), mask induced with isoflurane (2%), intubated orotracheally, and catheterized via the lateral tail vein. Isoflurane was then discontinued, and artificial ventilation with 30% O_2 in air was initiated. One of two CRIs were begun: low-dose propofol (0.3 mg/kg min) and fentanyl (10–20 μg/kg h) OR high-dose propofol (0.6–1 mg/kg min). Doses were adjusted within these ranges to achieve immobility, then remained unchanged for 6 h. Electrical stimulation of the forearm of awake marmosets resulted in BOLD functional magnetic resonance imaging (fMRI) response in several areas: primary and secondary somatosensory cortex (SI and SII), caudate nucleus, and ipsilateral and contralateral thalamus. Compared with anesthetized marmosets, the BOLD response was stronger and more immediate in awake marmosets independent of stimulation amplitude. Anesthesia eliminated stimulus-evoked responses in the caudate and ipsilateral SI. In other regions, response amplitudes were barely detectable at low frequencies but increased with stimulus frequencies above 16 Hz. Propofol suppression of the BOLD response was dose-dependent (Fig. 11.4). Other examples of balanced anesthetic techniques are described in Table 11.3.

Anesthesia monitoring in the context of balanced techniques using neuromuscular-blocking agents is complicated by the elimination of muscle twitching or palpebral, corneal, and pedal withdrawal reflexes as indices of anesthetic depth. Various methods for assuring adequate depth have been developed and include the use of autonomic indices (heart rate and blood pressure), EEG, administration of inhalants at "MAC-multiples," and periodic removal of the neuromuscular blockade to assess depth via the pedal withdrawal reflex.

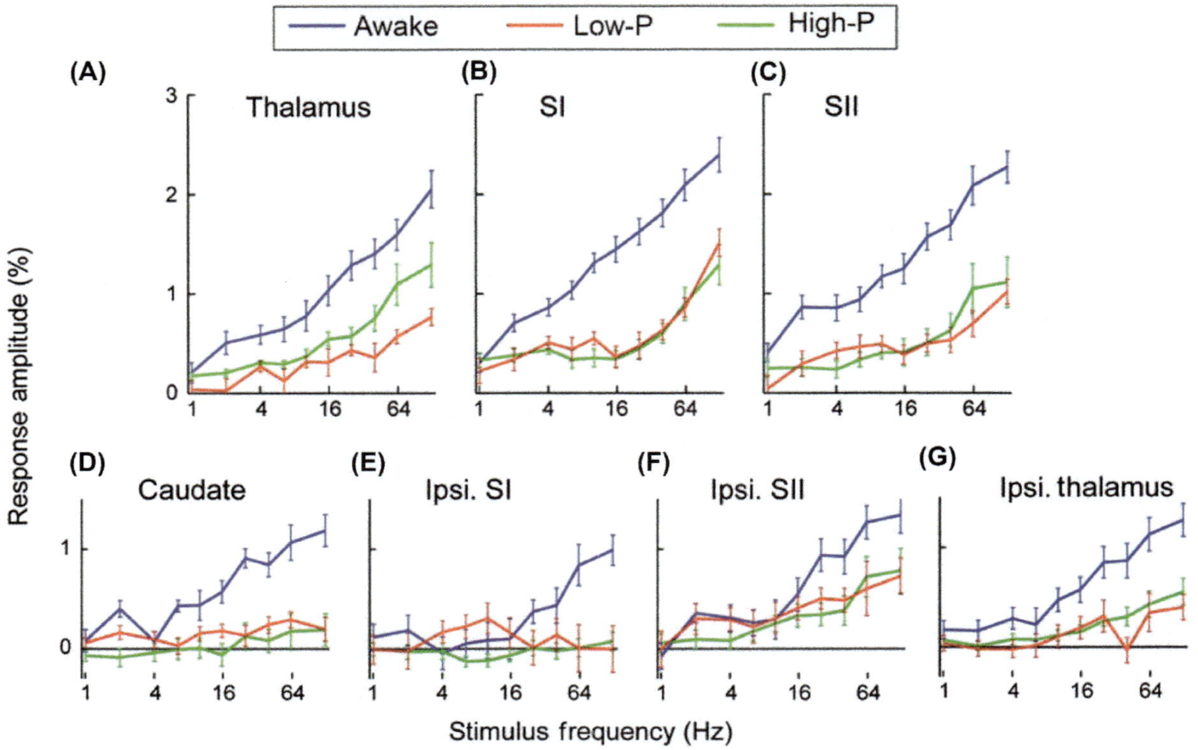

FIGURE 11.4 Bold response elicited by electrical stimulation of the forearm using different anesthetic regimens. (A) Thalamus. (B) SI. (C) SII. (D) Caudate. (E) Ipsi. SI. (F) Ipsi. SII. (G) Ipsi. thalamus. *Reprinted with permission from Liu JV, Hirano Y, Nascimento GC, Stefanovic B, Leopold DA, et al. fMRI in the awake marmoset: somatosensory-evoked responses, functional connectivity, and comparison with propofol anesthesia. Neuroimage 2013;78: 186–95.*

Each of these methods has shortcomings, as recently reviewed by Flecknell [28]. For example, administering inhalants at 1.3–1.5 MAC (1.3 MAC can be considered the ED^{95} of the inhalant) is hampered by the absence of available MAC values in the marmoset and is not likely to be used in neuroscience or other balanced anesthetic settings where subanesthetic doses of inhalants are combined with other drugs to avoid neuronal depression [28,29]. Autonomic index monitoring is the most widely used of these techniques in the author's experience; an increase of 10%–15% in the heart rate or blood pressure from baseline values obtained before the start of neuromuscular blockade prompts the administration of additional anesthetic. Performing all surgical manipulations prior to neuromuscular blockade, when anesthetic depth is more easily measured, is useful and provides a baseline of values associated with adequate depth.

Anesthesia Monitoring

Monitoring anesthesia in marmosets requires constant vigilance by a dedicated anesthetist, especially because of the frequent occurrence of apnea. As in other species, pulse, heart, and respiratory rate, percent saturation of hemoglobin with oxygen, electrocardiogram, noninvasive blood pressure, and end-tidal carbon dioxide values can be measured. In practice, the small size of the animal and attendant risks of hypothermia as time passes during optimization of probe, cuff, and electrode placement have discouraged complete monitoring. At a minimum, pulse oximetry should be used to monitor the pulse rate and oxygen saturation, and constant visual examination of breathing should be used to evaluate respiration. Normal values for the common marmoset can be found in Chapter 6.

Temperature probes are essential for continuously evaluating body temperature, thereby avoiding the physiologic consequences of either hypo- or hyperthermia. Probes designed for rodents work well as they are generally smaller and have a shorter insertion length than probes used in other species. An intraoperative rectal temperature of 100–101°F should be maintained. Supplemental heat sources for use intraoperatively include circulating warm water blanket systems (Stryker T/Pump Localized Therapy System, Stryker Kalamazoo, MI), forced air systems (Bair Hugger Normothermia System) (3M, St. Paul MN) or Baja Breeze Warm Air Blower and Convective Air Blankets (Animal Hospital Supply Inc. Flowery Branch, Georgia), and fabric warming systems (HotDog Patient Warming Systems, Eden prairie, MN). Hyperthermia from aggressive heat

TABLE 11.3 Select Techniques for Long-Term and Balanced Anesthesia

Sedation/Induction	Maintenance	Indications/Comments	References
Ketamine 4 mg/kg IM and 2% isoflurane by face mask	0.3 mg/kg/min propofol with 10-20 μg/kg/hr fentanyl OR 0.6–1 mg/kg/min propofol	IPPV with 30% O_2 in air. Recovery within 15 min	Liu et al. [25]
Alfaxalone	5–7 mg/kg bolus IV followed by 0.17 mg/kg min IV CRI	For advanced imaging or in general surgery. CRI used as maintenance anesthetic for up to 8 h	Johns Hopkins Formulary, Garrett C [9]
Sedation: Atropine 0.2 mg/kg IM; diazepam 2 mg/kg IM *Induction:* Ketamine 45 mg/kg Xylazine 3 mg/kg Incision sites infiltrated with xylocaine (0.5% with 1:200,000 epinephrine) SQ Dexamethasone 0.3 mg/kg IM Procaine penicillin G 25 mg/kg IM Ketamine 20 mg/kg as needed	1.6 mL/kg of an anesthetic cocktail* as an initial bolus IV, followed by 6 μg/kg h sufentanil adjusted to maintain heart rate between 240 and 300 beats/min. After 12 h, (0.3 mg/kg h) xylazine is administered IV	Stereotaxic procedure and single-cell recordings under anesthesia: nonsurvival procedure. IPPV with 50:50 N_2O:O_2 initially, followed by ratio of 70:30 N_2O:O_2	Bourne and Rosa [6]
*Sedation:*0.1 mg/kg atropine IM; 2 mg/kg diazepam IM *Induction:* 4%–5% sevoflurane in balance O_2	Remifentanyl 0.06–0.1 μg/kg min Mivacurium Cl 1 mg/kg h IV 2%–4% sevoflurane in 30% O_2	Stereotaxic procedure Optical imaging, visual stimulation and recording, vessel mapping: nonsurvival procedure	Valverde Salzmann et al. [26]
Ketamine 25 mg/kg IM Atropine 0.15 μg/kg IM	0.1 μg/kg min remifentanil IV 13 μg/kg min rocuronium	Electrophysiology recordings	Suzuki et al. [56]

* Sufentanil 4.8 μg/kg; dexamethasone 0.32 μg/kg; pancuronium 96 μg/kg administered as 1.6 mL/kg of the following: sufentanil (0.15 mg); dexamethasone (10 mg); pancuronium bromide 93 mg) in 15 mLs synthamin, 15 mLs 0.9% sodium chloride, and 15 mLs sodium chloride/glucose (0.18%; 4%, respectively).
IPPV, *intermittent positive-pressure ventilation.*

conservation measures using the systems mentioned above and surgical draping materials can develop rapidly and is treated by removing thermal support and increasing conductive heat loss. Incubators should be used during the immediate postoperative and post-procedural period until normal body temperature is achieved. The temperatures of both marmosets and of incubators must be monitored during postoperative recovery.

Pulse oximeters that are capable of faithfully registering the high pulse rate of marmosets include the Nonin 8600 F–O (Nonin Medical, Plymouth, MA); the DRE Waveline EZ Portable Patient monitor (www.dremed.com; Louisville, KY); and the SystemVet Vetrends 6400 series monitor (Tampa Florida: www.SystemVet.com). Rodent pulse oximetry systems can also be used.

The heart rate can be obtained using a Doppler probe applied directly to the sternum. Electrocardiogram can be evaluated using software-enabled monitors attached to 22G hypodermic needles inserted subcutaneously across the heart (e.g., in each axilla) [6]. Adhesive pads or alligator clips used for cats are also appropriate. The DRE and SystemVet monitors come with three lead electrode sets that are positioned on the left and right arms and left thigh.

Blood pressure measurements have been studied in awake-restrained, anesthetized, and awake, telemetrically monitored animals [7,16,30–32]. A potentially useful monitoring modality in the anesthetic setting is the volumetric tail cuff method, which measures tail blood volume instead of arterial wall oscillations (CODA Surgical Monitor, Kent Scientific, Torrington CT). Ansel et al. [16] tested the repeatability (same day) and reproducibility (next week) of measurements and evaluated the number required to achieve a low standard deviation. Measurements were obtained during alphaxalone anesthesia (15 mg/kg IM) which produced a duration of immobilization of 47.3 ± 13.2 min ($x \pm SD$). The least variability of standard deviation for both systolic and diastolic blood pressure was at 11 min after anesthetic injection. Mean pressure measurements varied from 69.9 ± 4.1 mmHg at 10 min to 98.8 ± 16.8 mmHg 30 min after injection. Using this system, four readings were required for repeatable measures, and there was both repeatability (1.98%) and reproducibility (14.5%) for systolic blood pressure, with similar but slightly elevated values for diastolic blood pressure. Blood pressure measurement using Doppler probes and neonatal size 1 and 2 cuffs on the tail, leg, or arm can be used and at the very least will allow the anesthetist to monitor changes in pressure.

ANALGESIA AND PAIN ASSESSMENT

A number of agents are used to provide analgesia for marmosets, despite a lack of analgesiometry data in the species. Polymodal and preemptive analgesia has become the standard of care in laboratory animals and should be used in common marmosets unless exemptions are warranted on a scientific basis or in the case of buprenorphine, as discussed below. As in other species, the major classes of agents used include nonsteroidal antiinflammatory drugs (NSAIDs), opioids, and local anesthetics.

Nonsteroidal Antiinflammatory Drugs

These agents are traditionally used for mild-to-moderate pain and in polymodal analgesia using opioids and/or local blocks. They are typically administered before the surgical insult, the timing designed to "preempt" the flow of nociceptive impulses to the spinal cord and brain. Most agents in current use in marmosets inhibit cyclooxygenase 2 (COX-2) preferentially, and therefore avoid to some extent the COX-1 inhibition responsible for the untoward effects of earlier agents. Meloxicam (0.1 mg/kg) has more selectivity for COX-2 than does carprofen (4 mg/kg), which weakly inhibits both isozymes. These agents have been used in marmosets (Table 11.1), and are administered daily per OS [18] or subcutaneously [10]. A loading dose of 0.2 mg/kg meloxicam followed by 0.1 mg/kg SID for up to 3 days is used in the surgical laboratory of the authors. Ketoprofen (2–3 mg/kg) use is also reported [33].

Opioids

Buprenorphine is the opioid agent most commonly used in marmosets and has partial opioid agonist activity at the mu (μ) opioid receptor. Intramuscular, intravenous, and subcutaneous administration are used for moderate-to-severe pain. Buprenorphine's popularity stems from its long duration of action and good safety profile. Higher doses are associated with respiratory depression, sometimes in association with other pharmaceuticals metabolized via the hepatic cytochrome p450 system [34]. Allen et al. [33] report a series of three of six deaths in marmosets in which craniotomies were being performed under isoflurane anesthesia after ketamine (15–20 mg/kg IM) induction. All had received 0.03 mg/kg SQ buprenorphine pre- or intraoperatively and had required artificial ventilation after administration. Necropsy and histopathology demonstrated acute centrilobular degeneration suggestive of circulatory shock. After elimination of buprenorphine from the protocol, nine other similarly used animals survived. Based on their experience, the authors recommended buprenorphine administration to marmosets only after surgery and at a lower dosage of 0.005–0.01 mg/kg IV, IM, SQ q 8–12 h. Oral transmucosal (OTM) delivery of buprenorphine has been used by our surgical laboratory at a dosage of 0.01 mg/kg PO q 12 h. OTM buprenorphine use is based upon recommendations in the feline and is only utilized as part of a multimodal analgesic approach [35,36]. Naloxone is a competitive antagonist at the mu, kappa, and sigma receptors and can be used for reversal of buprenorphine in case of inadvertent overdose and respiratory depression. Naloxone can be administered intravenously, intramuscularly, or subcutaneously at a dose of 0.1 mg/kg.

Local Anesthetics

Lidocaine and bupivacaine can be used for line blocks to preempt the transmission of nociceptive impulses along nerves innervating areas of surgical injury. The mechanism of action is inhibition of Na channels. The small size of the marmoset precludes the use of large volumes of these agents.

Mundinano et al. [18] recommend restriction of bupivacaine (0.25% v/v) with adrenaline (1:200,000) to less than 0.4 mL/500 g of body weight. Doses for local anesthesia of up to 4 mg/kg for lidocaine and 2 mg/kg for bupivacaine are extrapolated from other species. Line blocks and ring blocks can be used. Regional blocks, Bier blocks, wound infusion catheters using local anesthetics, and spinal anesthetic techniques have not been reported in the marmoset.

SURGERY

The marmoset poses the same challenges to surgeons as do other small animals. Chief among these is the physiologic cost of long procedures, a limited blood volume, and a smaller scale of structures and organs with attendant issues of limited exposure. The physiologic cost is most manifest in the risk of hypothermia, and specific strategies must be in place to minimize heat loss. These include heated surgical surfaces, the use of fleece or fabric barriers on bare surfaces, circulating hot water blankets, radiant heat sources, and fabric warming (Hot Dog) and forced air systems (BAIR Hugger). Additional strategies against hypothermia include reducing prep and recovery time by proper planning and choice of anesthetic regimen, using warm fluids for infusion, and warming materials used in aseptic preparation. Warm sterile saline can replace alcohol in skin preparation, sparing the animal from subsequent evaporative heat loss after povidone-iodine or chlorhexidine prep [37].

The blood volume of a 350-g marmoset is likely to be approximately 25 mL (7% of lean body weight). This limits acceptable blood loss and necessitates strict adherence to hemostasis. Electrocautery, direct pressure, radiofrequency coagulation, hemostats, hemoclips, and encircling ligatures can all be used to limit blood loss and preserve blood volume. With regard to electrocautery, bipolar electrocautery is recommended to avoid the potential of surface burns from the use of contact plates in monopolar systems [37]. Battery-operated, hand-held cautery pens are also useful. Topical hemostatics such as gelatin sponge (Gelfoam; Pfizer, Pharmacia and Upjohn, Kalamazoo, MI), oxidized regenerated cellulose (Surgicel; Ethicon, Johnson & Johnson, Sommerville, NJ), and Thrombin (Thrombin-JMI; GenTrac, Middleton, WI) can be used on oozing mucosal surfaces. The latter agent can be used alone or with Gelfoam or Surgicel. Blood loss can be estimated by counting blood-soaked gauze (3–5 mL per 2 × 2 square) and cotton-tipped applicators (0.1 mL). Alternatively, these items can be weighed to provide a more accurate measure of blood loss. Intravenous fluids should be administered warm at an initial rate of 4 mL/kg h via syringe pump, fluid pump, or traditional pediatric drip set. Blood transfusion, if required, can be performed according to protocols developed for dogs and cats [38]. More information on blood transfusion can be found in Chapter 10.

The small size of organs and limitations of exposure necessitate the use of microsurgical instruments in addition to smaller versions of conventionally sized instruments. The authors recommend the use of a combination of ophthalmic and microsurgical instruments which can be purchased through standard instrument vendors (e.g., Fine Science Tools—http://www.finescience.com). The contents of surgical packs are a personal preference. High-quality instruments or German grade are recommended. Table 11.4 is a list of recommended instruments for a marmoset surgical pack.

Mehler [37] recommends the following instruments for inclusion in a microsurgical pack: long jeweler's forceps (#3); Castroviejo pen needle holder (5.25 inches), and Tew-Barraquer scissors (curved 7 inches). These are available from Fine Science Tools (Foster City, CA, USA) or RICA Surgical (Schiller Park, IL; http://www.ricasurgical.com). Microsurgery instruments are extremely delicate and should be packed in a specialized microsurgical case to be sterilized separate from other instruments.

TABLE 11.4 Recommended Instruments for Use in Marmoset Surgery

Scalpel Blade Handle	#3 Blade Handle for Use of 15 Blade
Scissors	1. General purpose operating scissors: straight, blunt/blunt tips 2. Iris scissors: 4 or 4-1/2″ long, straight 3. Tissue separating scissors: straight, blunt/blunt tips 4. Metzenbaum dissecting scissors: 5-1/2″ long, straight, blunt/blunt tips 5. Strabismus scissors: 4″ long, straight, blunt/blunt tips
Forceps	1. Eye dressing forceps: 4″ long, half curved, standard, 0.8 mm serrated tips × 2 pairs 2. Iris tissue forceps: 4″ long, 1 × 2 teeth, straight, 0.8 mm tips 3. Brown-Adson tissue forceps 4. Adson tissue forceps: 1 × 2 teeth
Hemostats	Straight and curved hemostats are recommended with fine tips: Extra delicate Halstead mosquito forceps or micromosquito hemostats
Needle holder	1. Halsey needle holders, 5″ long, serrated tip 2. Mayo-Hegar needle holders, 5 or 6″ long 3. Derf needle holders, 4-3/4″ long
Retractors	1. Alm retractor with blades 2. Colibri retractors
Instrument tip protectors	1. Instrument tip protectors, material: plastisol, autoclave safe 2. Bone pin point covers, autoclave safe
Instrument tray	Small sterilizing tray, JorVet J0664T, autoclave safe

Retraction of the abdominal wall can be achieved by a surgical assistant, by use of a number of self-retaining retractors, or with stay sutures attached to instruments with weight adequate to the purpose. The Alm Retractor with Blades (Fine Science Tools) works well and has a low profile.

Surgical drapes should be customized to the species. Adhesive drapes are particularly useful in stereotaxic surgery, especially if the distinctive ear tufts are left unshaved. A 6 × 8″ adhesive surgical drape with povidone-iodine is especially useful for surgeries of all types (JorVet; Jorgenson Laboratories, Inc. Loveland CO). Clear drapes are also of benefit in that they allow both surgeon and assistants to monitor the respiration of the anesthetized marmoset and can help retain heat as compared with fabric or composite drapes. Clear drapes can be custom designed with adhesive portions by Steris Healthcare (Steris, Mentor, Ohio, USA). Surgical towels, if used, can be cut to one quarter their usual size so as not to overwhelm the immediate surgical space. Fenestrated fluid-resistant drapes or half-sheet drapes can be used over surgical towels and should provide a continuous sterile field to the instrument table or tray. Finally, a foot-operated surgical microscope or eye loups are desirable for finer procedures. Many models of loupes of varying costs and sophistication are available. Especially important considerations are focal distance, magnification, and the attached light source.

Postoperative Recovery and Monitoring

Thermal support: In the immediate postoperative period, thermal support is vital for rapid recovery from anesthesia. In our surgical laboratory, animals are maintained on a circulating warm water blanket or in a Bair hugger until they are active, then transitioned to a prewarmed (80–86°F) neonatal care unit incubator (Lyon ProCare Neo-Natal Intensive Care Unit, Lyon Technologies Inc. Chula Vista, California, USA). Rectal temperature is checked at 15, 30, and 60 min postoperatively, then at 30–60-min intervals until in a normal range. Supplemental oxygen can be used in the incubator as well, as some animals may deoxygenate after particularly long procedures (Mattias LeBlanc—personal communication).

Pain management: As described earlier in the chapter, a polymodal analgesic approach is recommended for surgical cases. As pharmokinetics are lacking in the literature for marmosets, many analgesic protocols are extrapolated from other nonhuman primates and other veterinary species. Routinely for the surgical cases in our laboratory, we employ a polymodal approach including preemptive administration of meloxicam (0.2 mg/kg) subcutaneously, and including a local anesthetic block applied during the surgical scrub. Buprenorphine (0.005 mg/kg given subcutaneously) is reserved for recovery when the animal is active in the incubator or capture box. Following the administration of buprenorphine the animals will generally become sedate, so one should continue monitoring them in a recovery area prior to return to the home cage.

The subsequent postoperative analgesic plan is dictated by the invasiveness of the surgical procedure and assessment of postoperative pain. In our surgical laboratory, animals are administered with meloxicam (0.1 mg/kg PO SID) and OTM delivery of buprenorphine (0.01 mg/kg PO BID) [35,36].

A descriptive analysis of pain assessment in marmosets is lacking in the literature. However, one can apply general principles from other veterinary species to assess potential discomfort postoperatively. In our experience, animals experiencing pain are less active in their home cage. For this reason, we advocate modifying the home environment (decreasing cage size with dividers, adding perches or wooden branches) to ease access to food and water. Animals in pain are less interested in consuming their standard diet, so addition of appealing supplements (Ensure or Boost beverage or pudding) assists in caloric consumption. Thermal support may be indicated for postoperative animals that are less active due to pain. For these cases, we recommend a heating lamp on a fluid pole directed toward a corner of the cage so that the animals have distinct thermal zones within their home cage.

Bandaging

Bandages are advocated for all surgical procedures, as marmosets are generally very inclined to remove skin sutures. A light bandage, including a small piece of nonadherent dressing over the incision covered with an adhesive bandage (Hypafix, BSN Medical, Charlotte, NC), may work for small incisions or minor surgical procedures. For larger incisions or procedures where dehiscence may be catastrophic, a layered bandage is advisable. The base layer includes a nonadherent dressing secured over the surgical site with an adhesive bandage, which is covered with absorbent cast padding (2″) and adherent self-adhesive wrap (Vetrap, 3M, St. Paul, MN; Co-Flex, Andover Healthcare, Salisbury, MA or similar product). For animals that are aggressive chewers of bandages, we employ self-adhesive wrap impregnated with bitterants (PetFlex No Chew, Andover Healthcare, Salisbury, MA), followed with elastic tape (Elastikon, Johnson & Johnson Consumer Inc., New Brunswick, NJ). Bandages should be changed every 2–3 days but should be reassessed daily at a minimum.

Select Surgical Procedures: Stereotaxic Surgery

Mundinano and coworkers [18] describe an anesthetic and surgical protocol for stereotaxic surgery born of a decade's experience [18]. This protocol employs an MRI-guided approach using an Agilent 9.4 T scanner for infants and adults. A custom-built MRI-compatible stereotaxic frame is used, as are 3D-printed custom facemasks into which the palatine bar has been incorporated (Fig. 11.5). Key features of the protocol include preoperative and postoperative antibiotics and analgesics; strict attention to body temperature, including the use of forced hot air (37°C) in the scanner, and the use of heat pads during induction, surgery, imaging, and recovery; surgery within the stereotaxic frame prior to imaging (as the pressure exerted by drilling may cause shifting of the skull within the frame); the use of five fiducial cranial markers for orientation before and after surgery and imaging (ear bars, eye bars, and a cranial midline marker to be placed after craniotomy); saline irrigation of craniotomy sites during drilling; and assiduous hemostasis throughout

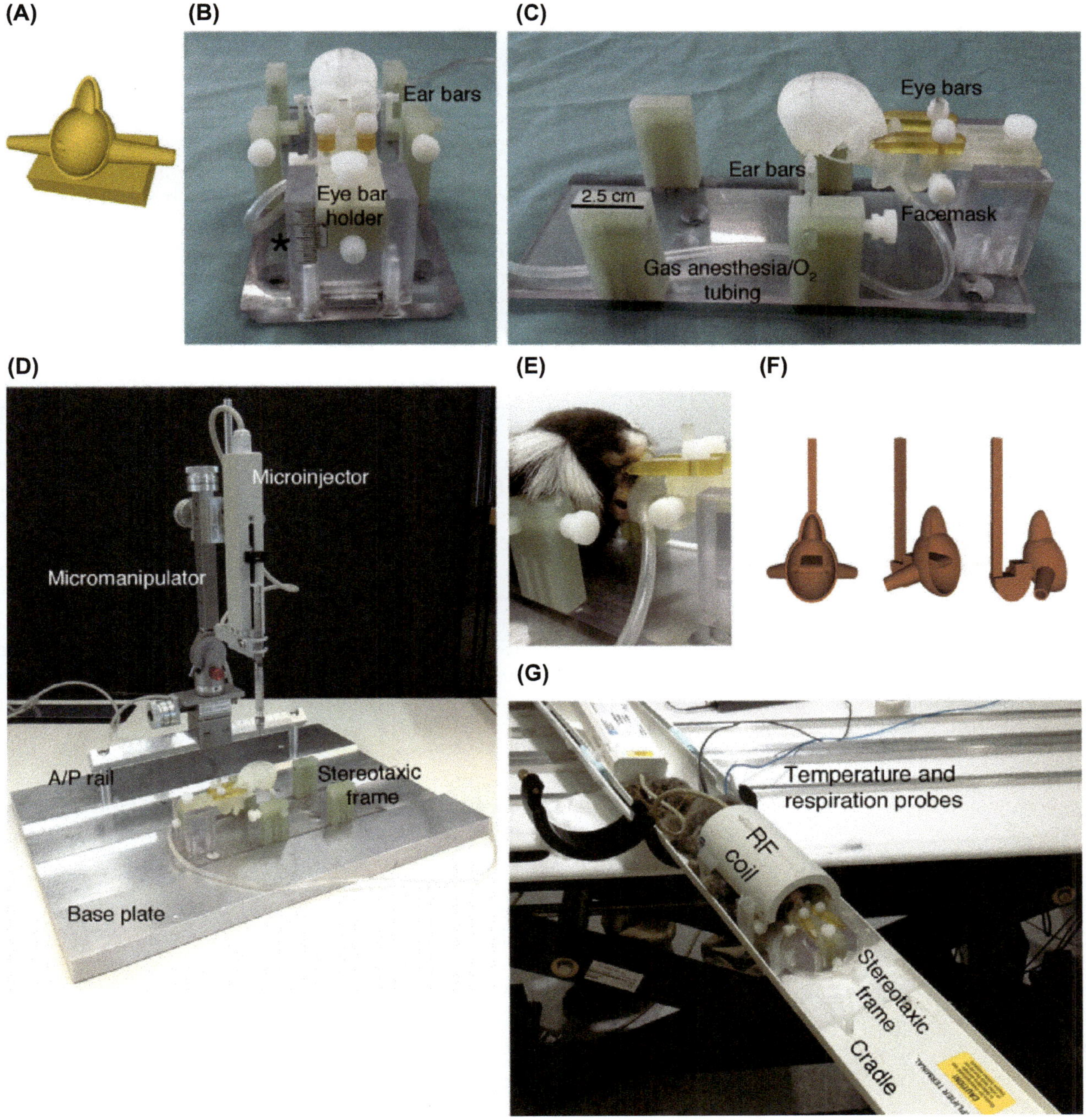

FIGURE 11.5 Custom-built MRI-compatible stereotaxic frame, injection, and MRI equipment, and face mask for marmoset neurosurgery. (A) 3D design of a bench face mask. (B and C) Photographs of the MRI-compatible stereotaxic frame holding a 3D-printed skull of an adult marmoset. The asterisk indicates the Vernier scale for modifying the elevation of the eye bar holder. (D) Example of an injection setup, with stereotaxic frame loaded onto the base plate and secured in place with screws. The MRI-stereotaxic frame is attached to the A/P base plate, which supports one micromanipulator holding an injection device. (E) Photograph of an adult anesthetized marmoset with the face mask within the stereotaxic frame. (F) 3D CAD drawings of custom-designed gas anesthesia face mask. The 3D-printed face mask has a palatine bar embedded in it. (G) Photograph of a marmoset in a stereotaxic frame and positioned in the cradle, ready for MRI scan. The radiofrequency (RF) receiver coil is placed on top of the head. Temperature, ECG, and respiration probes allow monitoring of the animal during the scan. Gas anesthesia is supplied during the scan through the face mask (E and F). *Reprinted with permission from Macmillan Publishers Ltd: Nature Protocols. 2016. Mundinano IC, Flecknell PA, Bourne JA. MRI-guided stereotaxic surgery in the infant and adult common marmoset. 11: 1299-308.*

the procedure. Monitoring of animals during induction, surgery, imaging, and recovery is critical to survival. Heart rate (240–300 bpm), SpO_2 >90%, and body temperature suffice for monitoring; body temperature preservation is assisted by use of a homeothermic blanket (Harvard Apparatus: model no. AH-50-7061).

With the animal fixed within the stereotaxic frame, and after a midsagittal line block with 0.1 mL of bupivacaine, a skin incision is made on the cranium. The superficial fascia is removed to expose the skull. If reflection of the temporalis muscle is required, an injection of 0.1 mL of bupivacaine should precede manipulation by 15 min. After this time, the temporalis is reflected sharply by incision of the muscle at its attachment to the cranium. Bleeding is controlled by cautery. An electrode tower is used to identify the site of craniotomy, previously calculated from the MR image. Microdrills with 0.5-mm diameter bits are used to create the craniotomy. Saline irrigation dissipates the heat generated by drilling, clears bone debris, and assists in determining drill depth. Once the craniotomy is complete, the subjacent dura mater is kept moist with sterile gauze soaked in Sofradex (Sanofi, Surrey, UK). If a bone flap is removed, it should be kept moist for subsequent replacement. The animal is now ready for intracranial injection.

Closure is achieved anatomically. The temporalis muscle is advanced to the approximate location of its cranial attachment and is secured there with tissue adhesive. Sterile gelatin soaked in buffer is used to cover the cortex if an area greater than 1 mm^2 of dura is opened. Bone wax is applied along the border of the craniotomy, and the bone flap is replaced. Tissue adhesive may also be applied. The site is cleaned of debris with saline, and skin is approximated with monofilament absorbable suture (4-0) using a subcuticular or simple interrupted pattern with sutures at 2 mm intervals. In infants, the daily application of dimethyl phthalate to the sutures with a cotton tip helps preclude removal by cage mates. Details of the anesthetic and drug administration protocol can be found in Table 11.2.

Bourne and Rosa [6] describe the application of a stainless steel restraint pedestal for in vivo recording of visual cortex neurons in anesthetized marmosets. Because of the thin and delicate nature of the marmoset skull, 6–8 burr holes are used in lieu of cortical screws to bind dental acrylic to the cranium and restraint pedestal. Layers of acrylic were also used to fashion a recording chamber of adequate depth to hold silicone oil, which is used in these procedures to keep the dura and cortical surface from drying. In contrast, Nummela et al. [39] describe the use of titanium screws to secure a cephalic restraint pedestal. After dissection of the skin and removal of tissue overlying the cranium, the implant area is cleaned with dilute (1.5%) hydrogen peroxide after which Copalite dental varnish is applied. Eight titanium screws (6 mm length, 1.5 mm diameter; Synthes) are inserted into the calvarium to a depth of greater than 1 but less than 1.5 mm, and thin laminae of adhesive dental acrylic (C&B Metabond. Parkell Inc., Edgewood, NJ) are used to create a peripheral ring around the exposed area. The adhesive acrylic impedes the growth of granulation tissue. The interior of the implant is then formed by successive laminae of a clear fMRI-compatible dental acrylic (Ortho-Jet, Lang Dental Manufacturing Company). After flushing the implant tissue interface with saline, tissue adhesive is used to secure the skin to the base of the implant.

A marmoset stereotaxic device has been designed by Narashige Group (Tokyo, Japan) in collaboration with Japanese investigators [40]. The device and accessories (SR-5C-HT and SR-6C-HT) are available commercially and feature ear and orbital bars created to minimize damage to the animal (Fig. 11.6). Head-holding adapters that can be used with the company's SR series of stereotaxic devices for mice are also available.

Hoffman et al. [23] describe the use of cortical screws for fixation of the Neurologger, a data-logging system developed for rodents. A 3-cm midsagittal skin incision is used to expose the cranium, after which it is dried with 3% hydrogen peroxide. A microdrill is used to create four 1.0-mm holes, and 1.2-mm diameter, gold-plated brass screws are inserted to serve as electrodes. These are connected to silver-plated elastic cables, connected in turn to a pin connector. The latter is then secured to the cranium with dental acrylic. A customized protective cover is attached via two screw nuts

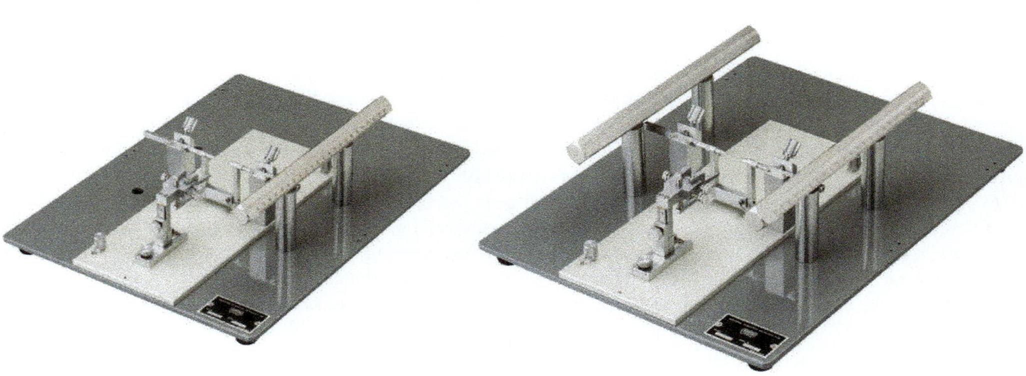

FIGURE 11.6 Narishige stereotaxic apparatus for common marmosets (SR-5C-HT; SR-6C-HT; SRS-AC; CA-6). *Reprinted with permission from Narashige Group (Tokyo, Japan).*

embedded in the acrylic. The skin is sutured flush with the implant using 5-0 monofilament absorbable sutures. Investigators used the device to evaluate sleep parameters. Acute optical imaging using a cephalic chamber secured with dental acrylic has also been described [26].

As marmosets become more commonly used in neuroscience research, the use of cephalic restraint devices, subdural recording arrays, cephalic recording chambers, and cranial windows is likely to increase. These procedures will drive refinement of techniques currently in use. Such refinements should include newer, more robust tissue adhesives, strategies to encourage osteointegration, progressive miniaturization of implants, and the use of materials with greater tissue compatibility than traditionally used titanium alloys and dental acrylic [41–45].

Select Surgical Procedures: Vasectomy

Vasectomy is typically performed in common marmosets for the purpose of birth control. Either hair plucking or electric clippers can be used to remove hair. After aseptic preparation of the caudal abdomen, inguinum, and genital area, a 1-cm ventral midline incision is made 1 cm cranial to the scrotum [46]. Alternatively, two paramedian skin incisions can also be used to access the vas on either side. The subcutaneous tissue and spermatic fascia are dissected bluntly to reveal the parietal vaginal tunic. The tunic is not as substantial as in larger species and can be entered by blunt dissection. The ductus deferens is identified as an off-white cord with an associated small artery. This is likely the artery of the vas deferens, and it can be separated or ligated. The ductus is dissected free of the spermatic vessels and nerve and is double-ligated with 4-0 nonabsorbable suture such that a 1-cm length is isolated. The intervening portion of the duct is excised. Mucosal cautery with handheld cautery pens is recommended [47]. The ends of the duct are allowed to drop back into the spermatic cord and the procedure is repeated on the contralateral side. Closure of the vaginal tunic is not performed, but subcuticular or intradermal sutures (4-0 or 5-0 Vicryl) followed by interrupted skin sutures with 4-0 Monocryl or nylon can be used. Either tissue adhesive (Vetbond) or simple interrupted or cruciate sutures of 3-0 monofilament nylon or PDS can be used for the skin [46]. Monofilament absorbables can also be used (e.g., Monocryl). A figure-of-8 bandage with cast padding and Vetwrap can be applied temporarily to preclude premature removal of sutures. Icing during recovery may help decrease swelling. Hypafix adhesive bandages (BSN Medical, Charlotte, NC) can also be applied. Postoperatively, the bandages are removed 48 h after vasectomy unless visibly soiled. Animals routinely receive approximately three doses of oral buprenorphine (one dose during recovery and two additional doses orally at 12-h intervals), in addition to three doses of meloxicam.

Recommendations regarding vasectomy techniques are derived from experience in our surgical laboratory, and recommendations from the American Urological Association [47]. The goal of vasectomy is a permanent form of contraception, but it is important to note that vasectomy does not immediately produce sterility. In humans, an additional form of contraception is required until a first postvasectomy semen analysis (PVSA) at 8–16 weeks postoperatively. To the author's knowledge, PVSA has not been used as routine evaluation postvasectomy in nonhuman primates, so additional forms of contraceptives such as intramuscular injection of cloprostenol are recommended for 2–3 months after vasectomy [48]. Vasectomy failure and/or recanalization have not been observed in our laboratory, and based on evidence from human literature, the rates after mucosal cautery should be $<1\%$. Vasectomy should be considered a failure if any motile sperm are seen on PVSA at 6 months after vasectomy, in which case repeat vasectomy should be considered.

Select Surgical Procedures: Cesarean Section

Cesarean section is indicated for dystocia, removal of nonviable feti, and for harvest of prenatal infants for experimental purposes. Surgery should be initiated soon after induction of anesthesia to minimize the neonatal respiratory depression associated with inhalant anesthesia. After anesthetic induction and aseptic preparation of the ventral abdomen, a midline abdominal skin incision is made from immediately cranial to the umbilicus to the cranial-most aspect of the pubis. Generally, a large incision (5 cm) is required to exteriorize the gravid uterus. The linea alba is incised sharply but carefully to avoid the uterus and bladder, and the incision is extended using blunt-tipped scissors. Importantly, the body wall is very thin in late pregnancy animals, so extreme care should be taken while opening the abdomen. Once the peritoneal cavity is accessed, the uterus is exteriorized and packed from the adjacent abdominal contents with moistened gauze. The uterus is incised sagittally, avoiding blood vessels, for whatever length is required to allow unimpeded exteriorization of the infants. The amnion is entered as the uterus is incised. For each fetus, the umbilical cord is clamped with either two hemostatic forceps, or bulldog clamps, and transected. The infants can be removed and given to an assistant for resuscitation or collection, depending on the indication. If resuscitation is planned, assistants should be ready with warm gauze and a warm incubator. The uterus is explored digitally to assure that all infants have been removed. The placenta generally separates easily from the uterus with minimal bleeding. Placental remnants can be removed using long cotton-tipped applicators, a spatula, or gauze fastened to a hemostat. Once the uterus is empty it should be lavaged with warm saline inside and out to eliminate tissue debris and blood. Lavage helps avoid adhesions which can potentially affect fertility. The hysterotomy is closed with 4-0 monofilament absorbable suture on a taper-point needle. Due to the significant amount of uterine involution, faster-absorbing suture such as Monocryl is preferred to avoid potential bowel entrapment within loops of suture that may form after involution. It is important to tighten the uterine suture while closing the hysterotomy site. A two-layer uterine closure using a simple continuous appositional first layer and continuous inverting second layer (Cushing or Lembert pattern) is recommended. Suture should engage submucosa but not penetrate the uterine lumen. Once the uterus is closed, 0.5 IU of oxytocin can be injected directly into the uterus to assist involution. In cases of dystocia, placental remnants protruding through the cervix, or autolyzed feti, surgical gloves should be changed prior to abdominal lavage and the closure of the abdomen. Prior to closure, the abdominal organs are visually inspected. The peritoneal cavity is thoroughly lavaged with 100–200 mL/kg of warm saline or balanced electrolyte solution. The abdominal musculature can be closed routinely with simple interrupted sutures of 3-0 PDS or Vicryl. A buried intradermal pattern is generally preferred for skin apposition with 4-0 Monocryl or Vicryl. Skin sutures are optional with simple interrupted or cruciate sutures of 4-0 nylon, Prolene, or Monocryl.

Infants should be resuscitated as in other species by clearing away fetal membranes and secretions using a small bulb syringe, DeLee aspirator, or cotton swabs [49]. Cleaning and drying of the infant will stimulate respiration. Apnea is addressed by providing positive-pressure ventilation or stimulating the Jen Chung acupuncture point [50]. Neonatal hypoxemia is managed by oxygen therapy using a face mask or incubator. Drugs are rarely indicated in the resuscitation of a newborn infant [49,51]. Bradycardia is generally a result of inadequate lung inflation, hypoxemia, and hypothermia. Establishing adequate ventilation and providing supplemental thermal support are vital to neonatal resuscitation. Thermal support during the resuscitation period is achieved using warm dry towels and a neonatal intensive care incubator set to 85–90°F. Surviving infants can be returned to the dam after the latter recovers from anesthesia. If dams reject an infant, fostering to a lactating surrogate dam or euthanasia of the infant is required.

In the postoperative setting, dams should be returned to the home cage when fully awake. Staff may decide to make part of the entire cage inaccessible to the dam and sire. Dams should receive an antibiotic (e.g., cefazolin 25/mg kg bid IM) and analgesics (meloxicam SID for 3 days and buprenorphine bid for 24–48 h postoperatively). Oral cephalexin (20 mg/kg PO BID) or orbifloxacin (7.5 mg/kg PO SID) can be substituted for an injectable agent after a day.

Select Surgical Procedures: Ovariectomy

Ovariectomized marmosets have been used to evaluate the effects of sex steroids on cognitive and brain aging, as well as menopausal symptoms [52–54]. Other indications for ovariectomy include sterilization, neoplasia, and chronic ovarian cysts. After aseptic preparation of the caudal abdomen, a 1–2-cm midline abdominal skin incision is made over the palpable uterus. The linea alba is incised sharply but carefully to avoid the uterus and bladder, and the incision is extended using blunt- tipped scissors. The ovary is exteriorized through the incision for visualization and identification of the ovarian anatomy and vasculature. Using either the blunt end of a curved suture needle or a delicate tip hemostat, a ligature is placed circumferentially around the proper ligament of the ovary and the ovarian branch of the uterine artery. Two ligatures of an absorbable braided suture (such as 4-0 Vircyl) are placed at this location. The remaining ovarian pedicle (containing ovarian artery and vein) is ligated with an additional two ligatures of absorbable braided suture. Dissection of the mesovarium is generally not indicated. The ovary is excised using scissors or electrocautery. Hemostasis is assured prior to routine closure of the abdomen. An adhesive bandage is applied for recovery.

Select Surgical Procedures: Digit Amputations

Digit amputations are most commonly performed for bite wounds to the digit or for traumatic degloving, shearing, or crush injuries. Initial management following an injury should be conservative, as many injuries will resolve with treatment with NSAIDs and antibiotics, and marmosets can function well with osteoarthritis and ankylosis of the interphalangeal joints. Digital wounds with severe soft tissue injury and cellulitis, however, may require amputation. Prior to surgery, local anesthetic is applied within the joint at the site of amputation and in the soft tissues. The level of amputation is determined by the extent of the injury and health of the soft tissue, as these will determine if closure of the wound can be achieved. Due to the small size of the digits, dissection of the deeper tissues (flexor and extensor tendons) is limited. For amputation, a circumferential skin incision should be made at a point distal to the joint to allow for disarticulation of the interphalangeal joint via transection of the flexor and extensor tendons, as well as collateral ligaments and joint capsule. Digital arteries and veins can be difficult to identify and ligate, so hemostasis is generally managed by compression and cautery. Closure of the surgical site can include an interrupted subcutaneous suture with simple interrupted skin sutures. A light bandage can be applied over the surgical site but must allow for use of the remaining digits for ambulation around the cage. The authors recommend a small bandage of a nonadherent dressing (Telfa, Coviden, Mansfield, MA, USA) trimmed to cover only the immediate surgical site followed by an adherent dressing retention sheet (Hypafix, BSN Medical, Charlotte, NC, USA or Tegaderm, 3M Health care, St. Paul, MN, USA). Analgesia should include an NSAID to help reduce inflammation at the surgical site.

Select Surgical Procedures: Tail Amputation

Tail amputation is indicated for traumatic tail injuries and their potential consequences, indolent ulcers, or paralysis. Tail injury is common in this species and is usually inflicted by a family member. The injured tail may be bandaged to attempt to keep cage mates from it, or the tail can be amputated. As in other species, the tail is prepared for aseptic surgery, and a site proximal to the injury is chosen for amputation. In addition to anesthesia or sedation, application of lidocaine/prilocaine topical cream followed by a small subcutaneous dose of injectable lidocaine before amputation is recommended. Prolonged general anesthesia is avoided in neonatal infants as their tails are generally too small and soft for detailed tissue dissection, and the animals tolerate well distal tail amputations with a local block alone. The key to successful healing is to preserve adequate skin for closure of the stump after disarticulation. A circumferential skin incision is made, followed by another incision on the long axis of the ventral surface of the tail. This second incision should reach to the level of the caudal vertebrae that will be disarticulated. Blunt dissection is used to liberate the skin from the underlying fascia and musculature, a site for disarticulation is chosen, and sharp dissection with a scalpel blade or Mayo scissors is used to amputate the tail. Hemostasis can be achieved by cautery or crushing bleeding vessels with a hemostat. For infants, an en bloc amputation using a scalpel blade with cautery and tissue glue is adequate and well tolerated. The ventral, longitudinal incision is closed with simple interrupted sutures of 3-0 or 4-0 nylon or Prolene, followed by similar closure of the circumferential incision. The latter is closed either with a horizontal mattress suture or with simple interrupted sutures. An alternative method is to make two skin incisions on either lateral surface of the tail, joining them ventrally and dorsally with U-shaped extensions such that the convexity of the U faces the tail tip. After amputation, simple interrupted sutures are used to close the resulting defect. Animals should be treated with antibiotics and analgesics at the discretion of the clinician. In our surgical laboratory, tail bandages are applied postoperatively for as long as is necessary to keep cage mates from reinjuring the tail or removing sutures prematurely.

References

[1] Sharma V, McNeill JH. To scale or not to scale: the principles of dose extrapolation. Br J Pharmacol 2009;157:907–21.

[2] Nunamaker EA, Halliday LC, Moody DE, Fang WB, Lindeblad M, et al. Pharmacokinetics of 2 formulations of buprenorphine in macaques (*Macaca mulatta* and *Macaca fascicularis*). J Am Assoc Lab Anim Sci 2013;52:48–56.

[3] Kelly KR, Pypendop BH, Christe KL. Pharmacokinetics of buprenorphine following intravenous and intramuscular administration in male rhesus macaques (*Macaca mulatta*). J Vet Pharmacol Ther 2014;37:480–5.

[4] Olberg R, Sinclair M. Monkeys and gibbons. In: West G, Heard D, Caulkett N, editors. Zoo animal and wildlife immobilization and anesthesia. Ames (IA): Wiley; 2014. p. 561–73.

[5] Davis H, Jensen T, Johnson A, Knowles P, Meyer R, et al. 2013 AAHA/AAFP fluid therapy guidelines for dogs and cats. J Am Anim Hosp Assoc 2013;49:149–59.

[6] Bourne JA, Rosa MG. Preparation for the in vivo recording of neuronal responses in the visual cortex of anaesthetised marmosets (*Callithrix jacchus*). Brain Res Brain Res Protoc 2003;11:168–77.

[7] Schnell CR, Wood JM. Measurement of blood pressure and heart rate by telemetry in conscious, unrestrained marmosets. Am J Physiol 1993;264:H1509–16.

[8] Garea-Rodriguez E, Schlumbohm C, Czeh B, Konig J, Helms G, et al. Visualizing dopamine transporter integrity with iodine-123-FP-CIT SPECT in combination with high resolution MRI in the brain of the common marmoset monkey. J Neurosci Methods 2012;210:195–201.

[9] Garrett C. Johns Hopkins University marmoset drug formulary. Baltimore (MD): Johns Hopkins University; 2017. Caroline Garrett cgarre12@jhmi.edu.

[10] Buchanan-Smith HM. Marmosets and tamarins. In: Hubrecht RC, Kirkwood J, editors. The UFAW handbook on the care and management of laboratory and other research animals. 8th ed. Ames (IA): Wiley-Blackwell; 2010.

[11] Helms G, Garea-Rodriguez E, Schlumbohm C, Konig J, Dechent P, et al. Structural and quantitative neuroimaging of the common marmoset monkey using a clinical MRI system. J Neurosci Methods 2013;215:121–31.

[12] Davy CW, Trennery PN, Edmunds JG, Altman JF, Eichler DA. Local myotoxicity of ketamine hydrochloride in the marmoset. Lab Anim 1987;21:60–7.

[13] Pearce PC, Crofts HS, Muggleton NG, Scott EA. Concurrent monitoring of EEG and performance in the common marmoset: a methodological approach. Physiol Behav 1998;63:591–9.

[14] Whelan G, James MF, Samson NA, Wood NI. Anaesthesia of the common marmoset (*Callithrix jacchus*) using continuous intravenous infusion of alphaxalone/alphadalone. Lab Anim 1999;33:24–9.

[15] Bakker J, Uilenreef JJ, Pelt ER, Brok HP, Remarque EJ, et al. Comparison of three different sedative-anaesthetic protocols (ketamine, ketamine-medetomidine and alphaxalone) in common marmosets (*Callithrix jacchus*). BMC Vet Res 2013;9:113.

[16] Ansel TV, Nour AK, Benavente-Perez A. The effect of anesthesia on blood pressure measured noninvasively by using the tail-cuff method in marmosets (*Callithrix jacchus*). J Am Assoc Lab Anim Sci 2016;55:594–600.

[17] Konoike N, Miwa M, Ishigami A, Nakamura K. Hypoxemia after single-shot anesthesia in common marmosets. J Med Primatol 2017;46:70–4.

[18] Mundinano IC, Flecknell PA, Bourne JA. MRI-guided stereotaxic brain surgery in the infant and adult common marmoset. Nat Protoc 2016;11:1299–308.
[19] Ishibashi H. More effective induction of anesthesia using midazolam-butorphanol-ketamine-sevoflurane compared with ketamine-sevoflurane in the common marmoset monkey (*Callithrix jacchus*). J Vet Med Sci 2016;78:317–9.
[20] Feng AY, Kaye AD, Kaye RJ, Belani K, Urman RD. Novel propofol derivatives and implications for anesthesia practice. J Anaesthesiol Clin Pharmacol 2017;33:9–15.
[21] Morris TH, Jackson RK, Acker WR, Spencer CK, Drag MD. An illustrated guide to endotracheal intubation in small non-human primates. Lab Anim 1997;31:157–62.
[22] Thomas AA, Leach MC, Flecknell PA. An alternative method of endotracheal intubation of common marmosets (*Callithrix jacchus*). Lab Anim 2012;46:71–6.
[23] Hoffmann K, Coolen A, Schlumbohm C, Meerlo P, Fuchs E. Remote long-term registrations of sleep-wake rhythms, core body temperature and activity in marmoset monkeys. Behav Brain Res 2012;235:113–23.
[24] Flecknell PA. Basic principles of anaesthesia. In: Flecknell PA, editor. Laboratory animal anaesthesia. 4th ed. Waltham (MA): Academic Press; 2015. p. 1–75.
[25] Liu JV, Hirano Y, Nascimento GC, Stefanovic B, Leopold DA, et al. fMRI in the awake marmoset: somatosensory-evoked responses, functional connectivity, and comparison with propofol anesthesia. Neuroimage 2013;78:186–95.
[26] Valverde Salzmann MF, Wallace DJ, Logothetis NK, Schuz A. Multimodal vessel mapping for precise large area alignment of functional optical imaging data to neuroanatomical preparations in marmosets. J Neurosci Methods 2011;201:159–72.
[27] Crofts HS, Wilson S, Muggleton NG, Nutt DJ, Scott EA, et al. Investigation of the sleep electrocorticogram of the common marmoset (*Callithrix jacchus*) using radiotelemetry. Clin Neurophysiol 2001;112:2265–73.
[28] Flecknell PA. Special techniques. In: Flecknell PA, editor. Laboratory animal anaesthesia. 4th ed. Waltham (MA): Academic Press; 2015. p. 109–40.
[29] Ko JC, Marini RP. Anesthesia. In: Fox JG, Marini RP, editors. Biology and diseases of the ferret. Ames (IA): John Wiley and Sons, Inc.; 2014. p. 259–84.
[30] Horii I, Kito G, Hamada T, Jikuzono T, Kobayashi K, et al. Development of telemetry system in the common marmoset—cardiovascular effects of astemizole and nicardipine. J Toxicol Sci 2002;27:123–30.
[31] Shiba Y, Santangelo AM, Braesicke K, Agustin-Pavon C, Cockcroft G, et al. Individual differences in behavioral and cardiovascular reactivity to emotive stimuli and their relationship to cognitive flexibility in a primate model of trait anxiety. Front Behav Neurosci 2014;8:137.
[32] Mietsch M, Baldauf K, Reitemeier S, Suchowski M, Schoon HA, et al. Blood pressure as prognostic marker for body condition, cardiovascular, and metabolic diseases in the common marmoset (*Callithrix jacchus*). J Med Primatol 2016;45:126–38.
[33] Allen PS, Liechty ER, Howland MH, Bergin IL. Severe respiratory depression following buprenorphine administration in *Callithrix jacchus*. (AALAS National Meeting, Baltimore, MD). JAALAS 2013;52:646.
[34] Megarbane B, Hreiche R, Pirnay S, Marie N, Baud FJ. Does high-dose buprenorphine cause respiratory depression?: possible mechanisms and therapeutic consequences. Toxicol Rev 2006;25:79–85.
[35] Steagall PV, Monteiro-Steagall BP, Taylor PM. A review of the studies using buprenorphine in cats. J Vet Intern Med 2014;28:762–70.
[36] Robertson SA, Lascelles BD, Taylor PM, Sear JW. PK-PD modeling of buprenorphine in cats: intravenous and oral transmucosal administration. J Vet Pharmacol Ther 2005;28:453–60.
[37] Mehler SJ. Surgery. In: Fox JG, Marini RP, editors. Biology and diseases of the ferret. Ames (IA): John Wiley and Sons, Inc.; 2014. p. 285–310.
[38] Ford RB, Mazzaferro E. Kirk and Bistner's handbook of veterinary procedures and emergency treatment. 9th ed. Philadelphia (PA): W.B. Saunders; 2012.
[39] Nummela SU, Coop SH, Cloherty SL, Boisvert CJ, Leblanc M, et al. Psychophysical measurement of marmoset acuity and myopia. Dev Neurobiol 2017;77:300–13.
[40] Tokuno H, Moriya-Ito K, Tanaka I. Experimental techniques for neuroscience research using common marmosets. Exp Anim 2012;61:389–97.
[41] Shah NJ, Hyder MN, Moskowitz JS, Quadir MA, Morton SW, et al. Surface-mediated bone tissue morphogenesis from tunable nanolayered implant coatings. Sci Transl Med 2013;5:191ra183.
[42] Mulliken GH, Bichot NP, Ghadooshahy A, Sharma J, Kornblith S, et al. Custom-fit radiolucent cranial implants for neurophysiological recording and stimulation. J Neurosci Methods 2015;241:146–54.
[43] Frydman GH, Marini RP, Bakthavatchalu V, Biddle KE, Muthupalani S, et al. Local and systemic changes associated with long-term, percutaneous, static implantation of titanium alloys in rhesus macaques (*Macaca mulatta*). Comp Med 2017;67:1–11.
[44] Samuel RE, Shukla A, Paik DH, Wang MX, Fang JC, et al. Osteoconductive protamine-based polyelectrolyte multilayer functionalized surfaces. Biomaterials 2011;32:7491–502.
[45] Addison O, Davenport AJ, Newport RJ, Kalra S, Monir M, et al. Do 'passive' medical titanium surfaces deteriorate in service in the absence of wear? J R Soc Interface 2012;9:3161–4.
[46] Morris TH, David CL. Illustrated guide to surgical technique for vasectomy of the common marmoset. Lab Anim 1993;27:381–4.
[47] Sharlip ID, Belker AM, Honig S, Labrecque M, Marmar JL, et al. Vasectomy: AUA guideline. J Urol 2012;188:2482–91.
[48] Mustoe AC, Jensen HA, French JA. Describing ovarian cycles, pregnancy characteristics, and the use of contraception in female white-faced marmosets, *Callithrix geoffroyi*. Am J Primatol 2012;74:1044–53.
[49] Johnson CA, Casal ML. Chapter 6: neonatal resuscitation: canine and feline. In: Lopate C, editor. Management of pregnant and neonatal dogs, cats, and exotic pets. Ames, IA; 2012.
[50] Davidson AP. Approaches to reducing neonatal mortality in dogs. Recent advances in small animal reproduction. Ithaca (NY): International Veterinary Information Services; 2003.
[51] Wyckoff MH, Aziz K, Escobedo MB, Kapadia VS, Kattwinkel J, et al. Part 13: neonatal resuscitation: 2015 American Heart Association Guidelines update for cardiopulmonary resuscitation and emergency cardiovascular care. Circulation 2015;132:S543–60.
[52] LaClair M, Lacreuse A. Reversal learning in gonadectomized marmosets with and without hormone replacement: are males more sensitive to punishment? Anim Cogn 2016;19:619–30.
[53] Gervais NJ, Viechweg SS, Mong JA, Lacreuse A. The middle-aged ovariectomized marmoset (*Callithrix jacchus*) as a model of menopausal symptoms: preliminary evidence. Neuroscience 2016;337:1–8.
[54] Lacreuse A, Chang J, Metevier CM, LaClair M, Meyer JS, et al. Oestradiol modulation of cognition in adult female marmosets (*Callithrix jacchus*). J Neuroendocrinol 2014;26:296–309.
[55] Ishibashi H, Motohashi HH, Kumon M, Yamamoto K, Okada H, Okada T, Seki K. Ultrasound-guided non-surgical embryo collection in the common marmoset. Reprod Biol 2013 Jun;13(2):139–44. https://doi.org/10.1016/j.repbio.2013.02.002. Epub 2013 Mar 13.
[56] Suzuki W, Banno T, Miyakawa N, Abe H, Goda N, Ichinohe N. Mirror Neurons in a New World Monkey, Common Marmoset. Front Neurosci 2015 Dec 10;9:459. https://doi.org/10.3389/fnins.2015.00459. eCollection 2015.

CHAPTER

12

Diseases of the Urogenital System

Robert P. Marini

Division of Comparative Medicine, Massachusetts Institute of Technology, Cambridge, MA, United States

Details of the anatomy of the urinary and reproductive systems of the common marmoset can be found in Chapter 2 of this text. There are few specific and well-documented conditions of the urogenital system of common marmosets. The most commonly seen urinary tract condition is glomerulonephropathy, whereas for the reproductive system, infant, fetal, and embryonal loss is the most common. Diagnosis of these and other conditions requires a thorough history and complete physical examination, as well as abdominal imaging studies and clinical laboratory evaluation of blood and urine. This chapter will discuss these diagnostic elements and the conditions they assist in diagnosing.

IMAGING

The two imaging modalities commonly used to assist in the diagnosis of urogenital conditions are radiography and ultrasonography. The appearance of the kidneys and bladder using both modalities has been described [1,2]. Using a 100-speed mammography system, a source to image distance of 105 cm, tabletop technique, and settings of 44 kVP and 10 mAs, Wagner and Kirberger evaluated the abdomens of 17 pet marmosets (8 males and 9 nongravid females [1]). The beam was centered on the last rib. They identified both kidneys in 9 of 17 in the ventrodorsal (VD) and dorsoventral (DV) views and 3 of 17 in lateral views. In contrast to many animals, the right kidney was caudal to the left, and the mean left and right kidney length on VD and DV views was significantly different between males and females, with those of females being longer (Table 12.1). Urinary bladder, uterus, ureters, and ovaries could not be identified. Contrast studies of the urinary and reproductive tracts have not been described.

Wagner and Kirberger [2] also describe an abdominal ultrasonographic examination using a linear array, multifrequency transducer operated at 9 MHz. They describe traditional locations of the kidneys, with the right kidney associated with the liver, and the left associated with the spleen. The kidneys are oval-shaped with the right kidney occasionally presenting a "tear-shaped caudal pole." The renal cortex was hyperechoic to the spleen and occasionally isoechoic to the liver. The renal medulla was thin in comparison and accentuated in the transverse view. The corticomedullary difference was indistinct on sagittal and parasagittal planes, evidently a physiologic characteristic of marmosets, in contrast to dogs and cats in which such a finding suggests disease. No medullary "rim sign," a hyperechoic line of the outer medulla which parallels the corticomedullary junction, was identified [3]. There was no correlation between body weight and kidney size, and renal width was longer than the height. As on the radiographic study, a significant difference was found in kidney length between males and females, and the right kidney was also shorter than the left (Table 12.1). Doppler ultrasonography revealed an interlobar artery flow profile with a steep systolic peak, a clear spectral window, and a smooth downslope to diastole. The renal resistive index (RRI), the difference between the peak systolic and end diastolic velocity of intrarenal vessels indexed to the peak systolic velocity (peak systolic and end diastolic velocity/peak systolic velocity), was measured and found to be 0.75 ± 0.04 for the left kidney and 0.72 ± 0.05 for the right. The RRI is a reflection of cardiovasculorenal change and is a "kidney organ damage marker" [4]. The utility of the RRI as a marker of renal disease in marmosets has not been evaluated.

The bladder, round and anechoic, was observed in 12 of 17 animals. The impression of sludge within the organ was attributed to multiple artifacts [2].

Evaluation of ovarian and uterine morphology and pregnancy determination and estimates of parturition date are commonly achieved through ultrasonography ([5,6]: see also Chapters 8 and 10 of this text). There is a

The Common Marmoset in Captivity and Biomedical Research
https://doi.org/10.1016/B978-0-12-811829-0.00012-1

TABLE 12.1 Radiographic and Ultrasonographic Measurements of Kidneys of Common Marmosets

Variables	Number of Marmosets	Mean (cm)	SD (±) (cm)	Range	
				Minimum (cm)	Maximum (cm)
Left kidney length on ventrodorsal (VD)/dorsoventral (DV)	12	1.91	0.16	1.60	2.10
Males	5	1.78	0.15	1.60	2.00
Females	7	2.00	0.08	1.90	2.10
Right kidney length on VD/DV	9	1.96	0.11	1.80	2.10
Males	3	1.87	0.12	1.80	2.00
Females	6	2.00	0.09	1.90	2.10

Variables	Number of Marmosets	Mean (mm)	SD (±) (mm)
LEFT KIDNEY			
Length	17	19.04	1.95
Height	17	10.01	1.18
Width	14	12.75	2.07
Average Resistive Index (RI)	15	0.75	0.04
RIGHT KIDNEY			
Length	17	18.21	2.09
Height	17	10.14	1.22
Width	8	13.00	1.68
Average Resistive Index (RI)	17	0.72	0.05
BLADDER			
Length	12	12.13	9.12
Height	12	5.48	3.74
Wall thickness	12	0.49	0.25

Adapted from Wagner WM, Kirberger RM. Radiographic anatomy of the thorax and abdomen of the common marmoset (Callithrix jacchus)*. Vet Radiol Ultrasound 2005a;46:217–224; Wagner WM, Kirberger RM. Transcutaneous ultrasonography of the abdomen in the normal common marmoset* (Callithrix jacchus)*. Vet Radiol Ultrasound 2005b;46:251–258.*

single reference describing contrast ultrasonography in the differentiation of ovarian structures [7]. The commercially available contrast agent SonoVue (Bracco UK Ltd., High Wycombe, UK) was administered by intracardiac administration during an acute procedure to 11 marmosets which were subsequently evaluated by both ultrasonography and histology. The authors were able to distinguish 9 of 11 relatively vascular corpora lutea from adjacent, and relatively avascular, ovarian parenchyma and follicles, as well as a large cyst on one ovary.

CLINICAL CHEMISTRY

The use of glomerular filtration rate (GFR) as an index of functional renal mass has only been reported in the experimental setting. Humke et al. [8] determined a number of renal function variables, including GFR and effective renal plasma flow (ERPF) from inulin and para-aminohippuric acid clearance, respectively, in studies of renin antagonism in conscious marmosets (Table 12.2). As in other species, surrogates of GFR such as blood urea nitrogen (BUN) and serum creatinine (sCr) concentrations can be used. Unfortunately, these surrogates are neither specific nor sensitive measures of renal damage [9]. In the case of BUN, concentrations are nonspecific because they are influenced by extrarenal factors such as dehydration or hemorrhage (prerenal), or urinary tract obstruction (postrenal), and are insensitive because considerable damage to the kidney must occur before elevations are observed [10]. Similarly, sCr can be influenced by age, diet, muscle mass, and liver function, and will not increase for several days after ischemic renal injury. Urinary creatinine clearance values are not

TABLE 12.2 Renal Function and Associated Variables in Conscious Common Marmosets

	Clearance Period						
	1	2	3	4	5	6	7
MAP, mmHg	97 ± 3	99 ± 2	97 ± 3	95 ± 2	96 ± 3	96 ± 2	95 ± 3
ERPF, ml min^{-1} 100 g body wt^{-1}	2.5 ± 0.4	2.7 ± 0.4	2.9 ± 0.4	2.9 ± 0.4	2.9 ± 0.4	2.8 ± 0.4	3.0 ± 0.4
ERBF, ml min^{-1} 100 g body wt^{-1}	5.4 ± 0.8	5.5 ± 1.0	5.7 ± 0.8	5.6 ± 0.9	5.6 ± 0.8	5.4 ± 0.9	5.5 ± 0.8
GFR, ml min^{-1} 100 g body wt^{-1}	0.52 ± 0.05	0.45 ± 0.05	0.46 ± 0.04	0.45 ± 0.04	0.47 ± 0.04	0.44 ± 0.04	0.46 ± 0.05
EFF, %	19 ± 1	18 ± 1	17 ± 1	18 ± 2	18 ± 2	18 ± 2	17 ± 2
UV, μl min_{-1} 100 g body wt^{-1}	7.1 ± 1.3	8.2 ± 2.3	7.5 ± 1.2	7.3 ± 1.2	7.5 ± 1.1	6.4 ± 1.1	6.2 ± 1.0
$U_{Na}V$, μmol min^{-1} 100 g body wt^{-1}	0.33 ± 0.14	0.37 ± 0.016	0.40 ± 0.13	0.41 ± 0.15	0.46 ± 0.15	0.38 ± 0.15	0.47 ± 0.15
U_KV, μmol min^{-1} 100 g body wt^{-1}	0.65 ± 0.06	0.67 ± 0.08	0.63 ± 0.06	0.64 ± 0.07	0.69 ± 0.08	0.64 ± 0.07	0.66 ± 0.05
U_{osm}, mosmol/kgH_2O	782 ± 91	705 ± 100	682 ± 88	672 ± 105	718 ± 87	767 ± 97	783 ± 84
Plasma Na, meq/l	142 ± 2	145 ± 1		142 ± 3		144 ± 1	145 ± 2
Plasma K, meq/l	2.9 ± 0.1	2.9 ± 0.2		2.8 ± 0.1		2.8 ± 0.1	2.9 ± 0.1
Plasma osmolality, mosmol/kgH_2O	298 ± 2	297 ± 2		298 ± 3		298 ± 1	297 ± 1
Hematocrit, %	51 ± 1	49 ± 1	49 ± 2	48 ± 1	48 ± 1	47 ± 1	46 ± 1

Values are means ± SE from 10 animals. *EFF*, effective filtration fraction; *ERBF*, effective renal blood flow; *ERPF*, effective renal plasma flow; *GFR*, glomerular filtration rate; *MAP*, mean arterial pressure; U_KV, urine potassium excretion; $U_{Na}V$, urine sodium excretion; U_{osm}, urine osmolality; *UV*, urine volume. Clearance *periods 1* and *2* were measured over 30 min. All other clearance periods were measured over 15 min. The average of the first two control periods was compared with the following clearance periods by repeated measures analysis of variance. No significant changes were determined over time in any of the measured parameters.
Reproduced with permission from Humke U, Levens N, Wood J, Hofbauer K. Responses to renin inhibition in conscious primates. Am J Physiol 1991;261(1 Pt 2):F179–86.

reported for common marmosets but would be of value in the diagnosis of acute renal injury. Moreover, serum and urinary biomarkers of kidney injury such as serum cystatin C, urinary Il-18, and glutathione-S-transferase have not been evaluated in marmosets.

Normal sCr concentrations are considerably lower in marmosets than in many other species (see Chapter 6), with mean concentrations in most surveys of 0.2–0.3 mg/dL. BUN has a wide range of normal values depending on the study but in general is in the range of approximately 20–22 mg/dL. As suggested in Chapter 6, establishment of normal values for each colony is recommended.

URINALYSIS

Published urinalysis results are summarized in Chapter 6. Yamada et al. [11] reported the results of dipstick and cytologic examinations of urine collected directly by cystocentesis at necropsy from 21 common marmosets and indexed the findings to renal lesions of glomerulopathy (Table 12.3). All marmosets in which dipstick analysis of urine was performed had proteinuria, with no apparent relationship of renal lesion grade to degree of proteinuria. Similarly, most animals were positive for occult blood and many (9/12) had red cells present on the cytological exam. In a separate study in which witnessed voided samples were collected from plastic or foil sheets, Drake [12] identified average urine-specific gravity for marmosets as 1.026. Glucosuria, defined as greater than or equal to 100 mg/dL using Clinistix strips (Siemens Diagnostics, Tarrytown, NY), was identified in marmosets, especially those shown to be glucose-intolerant, being used in a high-fat or glucose-enriched diet study [12a].

URINARY TRACT DISORDERS

Congenital Disease

Malignant Nephroblastoma

- *Etiology:* Nephroblastoma is a tumor of embryonic origin, derived from metanephric blastema. Expansion of these blastemal elements, which are considered pluripotent, can occur to create the tumor. It is included here as a congenital disease to reflect this etiopathogenesis.

  The disease in humans has been associated with chromosomal abnormalities and various syndromes. It is referred to as Wilm's tumor and is usually seen in children. In veterinary medicine it is most commonly observed as a naturally occurring

TABLE 12.3 Urinalysis and Urine Cytology of Common Marmosets

				Urinalysis (Dipstick Method)			Urine Cytology				
Animal Number	Sex	Age	Renal Lesion Grade	Protein[a]	Occult Blood[b]	pH	Crystal[c]	RBC[d]	WBC[d]	Epithelial Cell[e]	Cast[f]
1	M	3	1	NE	NE	NE	NE	NE	NE	NE	NE
2	M	3	1	NE	NE	NE	NE	NE	NE	NE	NE
3	M	4	1	++	++	5	−	++	−	−	−
4	M	4	2	++	±	5	−−	−	−	−	−
5	M	4	2	NE	NE	NE	NE	NE	NE	NE	NE
6	M	7	3	NE	NE	NE	NE	NE	NE	NE	NE
7	M	9	3	++	+	7	−	−	−	+	−
8	M	9	2	++	+	8.5	+	+	−	−	−
9	M	10	3	++	−	5.5	−	−	−	−	−
10	F	2	1	NE	NE	NE	NE	NE	NE	NE	NE
11	F	2	1	NE	NE	NE	NE	NE	NE	NE	NE
12	F	3	1	NE	NE	NE	NE	NE	NE	NE	NE
13	F	4	2	+++	+++	6	−	++	−	−	+
14	F	5	1	NE	NE	NE	NE	NE	NE	NE	NE
15	F	9	3	NE	NE	NE	−	+++	−	−	−
16	F	11	4	++	+++	8	−	+++	−	−	−
17	F	11	3	++	++	6	−	+	−	−	+
18	F	11	3	++	++	5.5	−	+++	−	−	−
19	F	11	3	NE	NE	NE	NE	NE	NE	NE	NE
20	F	11	4	+++	+	7.5	−	+	−	−	−
21	F	11	4	+++	±	7	−	+	+	+	+

Note. *BUN*, blood urea nitrogen; *NE*, not examined; *RBC*, red blood cell; *WBC*, white blood cell.

[a] *Urinalysis, protein: (++) 30–100 mg/dL; (+++) 100–300 mg/dL.*

[b] *Urinalysis, occult blood: 1+ was equivalent to hemoglobin 0.06 mg/dL.*

[c] *Urine cytology, crystal: (−) 0 per 10 fields; (+) 1–10 per 10 fields.*

[d] *Urine cytology, RBC and WBC: (−) 0–1 per 10 fields; (+) 2–5 per 10 fields; (++) 6–10 per 10 fields; (+++) >11 per 10 fields.*

[e] *Urine cytology, epithelial cell: (−) 0–2 per 10 fields; (+) 3–10 per 10 fields.*

[f] *Urine cytology, cast: (−) 0 per 10 fields; (+) 1–2 per 10 fields.*

Adapted with permission from Yamada N, Sato J, Kanno T, Wako Y, Tsuchitani M. Morphological study of progressive glomerulonephropathy in common marmosets (Callithrix jacchus). *Toxicol Pathol 2013;41:1106–15.*

condition of swine and chickens. The tumor described by Zoller et al. [13], in a common marmoset, most closely resembles a cystic variant of nephroblastoma. A handful of reports document its occurrence in other nonhuman primates [13].

- *Clinical signs:* A male, 17-month-old, common marmoset presented with increased abdominal girth and hematuria [13].
- *Epizootiology:* This is an isolated case. Renal neoplasms are infrequently reported in nonhuman primates.
- *Pathology:* At necropsy, the left kidney presented as an edematous mass, which had been greatly expanded by multiple cysts. The renal capsule contained approximately 45 mL of brown fluid (Fig. 12.1). Solidly cellular and cystic portions of the neoplasm effaced normal renal parenchyma. Cysts were filled with a proteinaceous eosinophilic fluid, along with cellular debris, and occasional calcareous deposits. They were lined with a flattened epithelium. Solid portions were composed of blastemal, epithelial, and stromal arrangements. Blastemal cells were in various morphologies—densely packed, clustered, palisading—and were supported by a fibrous connective tissue stroma. Scattered multifocal necrosis, dystrophic calcification, and immature epithelium resembling glomeruli and

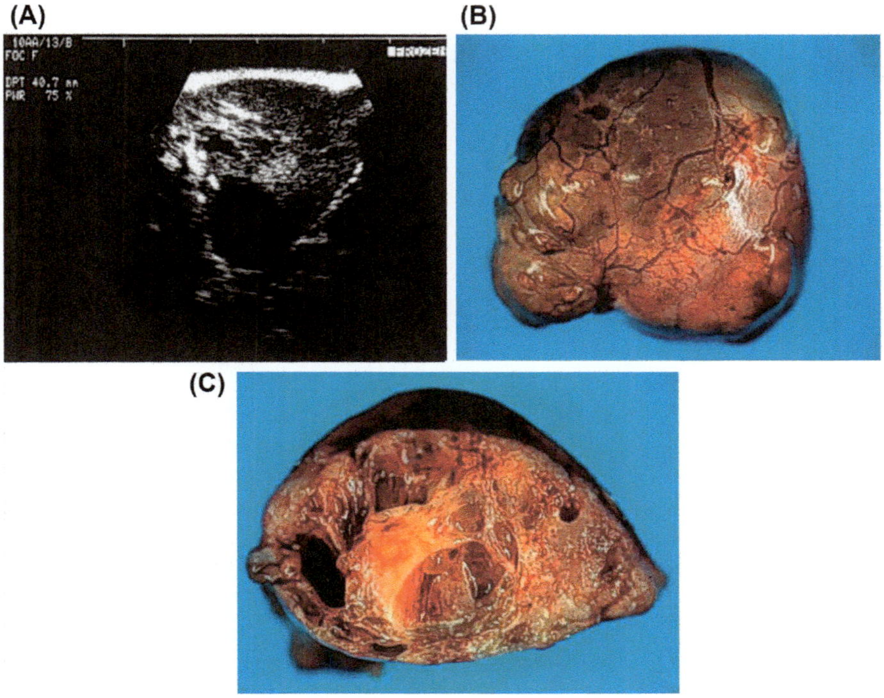

FIGURE 12.1 Nephroblastoma in a common marmoset. (A) Ultrasonographic image of affected kidney. (B and C) Gross necropsy images of nephroblastoma. *Reprinted with permission from Zoller M, Matz-Rensing K, Fahrion A, Kaup F.J. Malignant nephroblastoma in a common marmoset* (Callithrix jacchus). *Vet Pathol 2008;45:80–4.*

tubules were observed. A variety of special stains were also used to more fully characterize the origin of cell types within the tumor.

The right kidney had circumscribed subcapsular and extensive ureteral hemorrhage. The bladder was filled with a "bloody fluid."

- *Diagnosis:* Abdominal ultrasonography identified multiple hypoechogenic structures suggestive of cysts within the left kidney. Euthanasia and complete necropsy were performed.
- *Differential diagnosis:* Polycystic renal disease; renal abscessation, granuloma, and hematoma; hydronephrosis; and neoplasms other than nephroblastoma are differential diagnoses for renomegaly associated with nephroblastoma.
- *Prevention and control:* As malignant nephroblastoma is a rare tumor, prevention and control seem unwarranted. One could presumably remove parents of an affected animal from active breeding, as there have been chromosomal loci implicated in the disease in humans [13]. In the zoo setting, surgical removal might be a consideration.
- *Research complications:* Animals with nephroblastoma are unsuitable for research of any kind, but especially that involving excretory, endocrine, or metabolic functions of the kidneys.

Experimental Disease

Leptospirosis

- *Etiology: Leptospira interrogans* serovar Copenhageni. The isolate used for experimental inoculation of marmosets was from a fatal case of the severe pulmonary form of leptospirosis in a human [14].
- *Clinical signs:* Weight loss, most severe between days 6 and 12 postinoculation (PI), and dehydration were seen.
- *Epizootiology*: Leptospirosis is a potentially severe infectious disease of humans and animals. There are no reports of spontaneous (natural) cases of leptospirosis in common marmosets.
- *Experimental design:* Six female and five male, young adult, wild-caught common marmosets, were inoculated with 2×10^6 organisms intraperitoneally, and euthanized at 3-day intervals until day 28 PI. Blood cultures, complete necropsy, and histopathologic examination of hematoxylin and eosin (H&E) and Giemsa-stained tissue sections, serum chemistry evaluation (BUN, creatinine, AST, ALT, bilirubin), serology, and immunofluorescence staining were performed.
- *Pathology:* Gross pathology included pleural and peritoneal hemorrhage and diffuse jaundice of the

viscera and subcutaneous tissue. Histopathologic examination of the kidneys showed a progression of lesions from acute tubular necrosis to tubular dilatation with intraluminal casts (Day 6) to severe tubulointerstitial nephritis (Day 6 and onward). Minor changes were observed in the glomeruli, with the most striking pathology involving the tubules and interstitium.

Extrarenal lesions included severe hemorrhage and interstitial pneumonitis; mild hepatic vacuolization, hemorrhage, biliary stasis, and focal degeneration and necrosis; and hemorrhage and mild focal necrosis of myocytes. Histopathology was associated with a progressive increase in serologic response to the leptospira antigens and leptospira antigen localization in tissues.

Clinical pathology findings included conjugated hyperbilirubinemia, elevated aminotransferase concentrations, and azotemia peaking at days 6–12 PI.

- *Prevention and control*: Leptospirosis is extremely unlikely in modern laboratory settings. In zoological parks, exposure to wild rodents and water contaminated with rodent urine are potential sources of infection.

Spontaneously Occurring Renal Disease

Renal Failure in the Setting of Vitamin E Deficiency in a Common Marmoset

- *Etiology:* Anemia, degenerative myopathy, and steatitis are highly suggestive of vitamin E deficiency. The etiologies of the comorbidities reported in the animals in the case series described below are uncertain. See subsequent sections of this chapter for a discussion of glomerulonephropathy and Chapter 13 of this volume for a discussion of lymphocytic enterocolitis.
- *Clinical signs, physical examination, and laboratory findings*: Juan-Salles et al. [15] reported a series of five pet callitrichids presenting with signs of vitamin E deficiency. Four of the five animals, three common marmosets, and a black tufted-ear marmoset were without signs of concomitant renal disease. The remaining animal, a two and a half year old, intact female, pet marmoset, paired with a male conspecific, presented with weight loss, impaired locomotion, polyuria, polydipsia, vomiting, and diarrhea. Physical examination revealed hindlimb paresis, hypotrichosis of the tail, and muscle atrophy. Hematology and serum chemistry findings included anemia, leukopenia, marked thrombocytosis, hypercholesterolemia, hypoalbuminemia, azotemia, and hyperphosphatemia. The animal was negative on rectal swab culture for enteric pathogens including *Campylobacter* spp. Urine dipstick evaluation several hours before death revealed proteinuria (30 mg/dL), hematuria (250 red blood cells per μL), and glucosuria (500 mg/dL). Serum vitamin E concentrations were low normal compared to controls and below normal when corrected for hypercholesterolemia.
- *Epizootiology:* Vitamin E deficiency has been described in common marmosets [16]. Inadequate dietary levels of vitamin E or fat malabsorption can both contribute to hypovitaminosis.
- *Pathology:* Renal lesions included white, renal cortical foci, mesangioproliferative glomerulopathy, tubular casts (hyaline and giant cell), and phagocytosis of hyaline casts. There were numerous extrarenal lesions including severe diffuse degenerative myopathy with Ziehl–Neelsen and periodic acid-Schiff (PAS)-positive pigment in esophageal skeletal muscle fibers and cardiomyocytes; diffuse catarrhal enterocolitis; pyogranulomatous steatitis of perirenal and mesenteric fat; hyperplasia of pancreatic islets; hemosiderosis of hepatocytes, reticuloendothelial cells and macrophages; erythrophagocytosis; and ulcerative duodenitis. Pancreatic insulin hyperexpression was demonstrated immunohistochemically.
- *Diagnosis:* Degenerative myopathy, steatitis, and anemia are characteristic of vitamin E deficiency. Chronic lymphocytic enterocolitis, wasting, and renal disease are common comorbidities in marmosets. Moreover, the combination of enterocolitis and renal disease may contribute to anemia through various mechanisms, including iron deficiency, poor absorption of fat-soluble vitamins including vitamin E, and anemia of chronic disease. Aging marmosets frequently have mesangioproliferative glomerulonephropathy and nephritis identified as incidental findings [17]. In this case, renal lesions were associated with failure as manifested by azotemia, hyperphosphatemia, and urinalysis findings.
- *Prevention and control:* Marmoset diets used by modern vivaria should preclude the development of vitamin E deficiency. It is noteworthy that this case and others in the same case series occurred in pet marmosets. Prevention of the renal, enteric, and colonic lesions is a topic of current investigations.
- *Research complications:* Animals in renal failure, with or without comorbidities, would be unsuitable for research use.

Oxalate Nephropathy

- *Etiology:* Oxalate poisoning has been described as a result of ingestion of bark, sap, and leaves of the *Eucalyptus viminalis* Labill tree (the ribbon gum or manna gum tree) [18].
- *Clinical signs:* Animals were presented with depression, inappetence, and weight loss. Several were found dead.
- *Epizootiology:* Seven of 17 common marmosets in a laboratory colony died, six from renal failure, 1 month after introduction of the bark and leaves of the eucalyptus tree. Deaths continued for 5 months after removal of the offending plant parts.
- *Pathology:* Gross lesions were variable, with kidneys displaying a spectrum of gross appearances, from normal to swollen in shape and from normal to dark, mottled or pale in color. Histologic findings revealed severe renal pathology consisting of tubular lesions, glomerulosclerosis, interstitial fibrosis, and nephritis. Tubules were atrophied, distorted, and sometimes flattened, and contained either proteinaceous casts or crystals arranged in a radial pattern. Crystals were typical of calcium oxalate, showing birefringence in cross-polarized light and positivity in both Von Kossa- and Pizzolato-stained sections (Fig. 12.2). Two animals showed additional, unrelated lesions.
- *Diagnosis:* Diagnosis was achieved via histopathology. Partial urinalysis of the surviving animals in the colony of 17 was performed 80 and 122 days after first contact with the plant. All animals had proteinuria (with a mean value of 5 g/dL) and some had ketones. The samples had normal glucose and pH. Blood samples for hematology or serum chemistry evaluation were not taken. Oxalate was identified within the branches and leaves of *E. viminalis* using chemical and gas chromatographic methods.
- *Differential diagnosis:* Renal failure in the common marmoset is most often associated with glomerulonephropathy. Bacterial pyelitis and pyelonephritis are additional differentials.
- *Prevention and control*: Prevention can be achieved by eliminating access to the *E. viminalis* and other trees and shrubs high in oxalates. The European Association of Zoos and Aquaria (EAZA) publishes a list of plants that are safe for captive callitrichids [EAZA Best Practice Guidelines for Callitrichidae—3.1 Edition—2017 Appendix 1 (https://www.eaza.net/assets/Uploads/CCC/2017-Callitrichids-EAZA-Best-Practice-Guidelines-Approved.pdf)].
- *Research complications:* Renal disease would complicate studies involving any of the excretory, hormonal, or metabolic functions of the kidney. Neuroscience studies using water restriction and subsequent water reward as a motivator for performance would also be affected.

Spontaneous Progressive Glomerulonephropathy and Other Renal Lesions

- *Etiology:* This lesion is of unknown etiology, although immunoglobulins of both the IgM and IgA classes are involved. Immunoglobulin deposition may be associated with dietary or other environmental allergens. Investigators have likened spontaneous progressive glomerulonephropathy in marmosets to both IgM and IgA nephropathy of humans.
- *Clinical signs:* In the majority of cases, lesions, regardless of histologic grade, have not been associated with clinical signs. In a study characterizing aging of the common marmoset, Tardif et al. [17] report that 75% of animals over 8 years of age had renal lesions (interstitial nephritis, amyloidosis, and glomerulopathy) at necropsy but

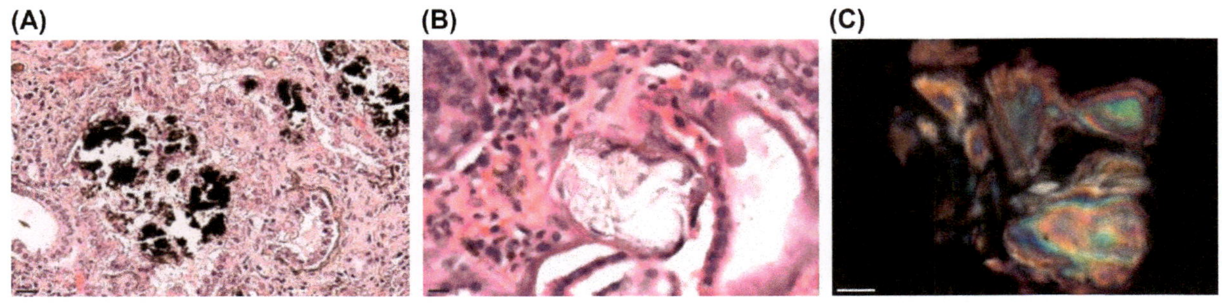

FIGURE 12.2 Oxalate nephropathy in a common marmoset. (A) Von Kossa-stained calcium oxalate crystals within renal tubules. (B) Unstained oxalate crystals within a renal tubule. Renal tubular atrophy and interstitial fibrosis and infiltration with mononuclear cells are evident. (C) Crystal birefringence under cross-polarized light of the section depicted in (B). *Reprinted with permission from Vanselow BA, Pines MK, Bruhl JJ, Rogers LJ. Oxalate nephropathy in a laboratory colony of common marmoset monkeys* (Callithrix jacchus) *following the ingestion of Eucalyptus viminalis. Vet Rec 2011;169:100–3.*

only 17% had renal disease as the primary cause of death. Collins et al. [19] corroborated these findings in their study of 19 marmosets euthanized as part of experimental protocols and six animals euthanized as a consequence of progressive weight loss. All of the animals in the Collins study had normal renal function. Weight loss, polyuria/polydipsia, and anorexia are all signs of renal disease.

- *Epizootiology:* Lesions are found in both male and female common marmosets ranging in age from young adult (18 months) to aged (over 8 years). Yamada et al. [11] characterized lesions of the kidneys in 21 common marmosets, 9 males and 11 females, and correlated lesions with age of the animal (between 2 and 11 years). They created a system in which a total histopathological score and renal lesion grade (1 through 4) were assigned based on glomerular lesions, regeneration tubuli, interstitial fibrosis, and interstitial infiltration (Table 12.4). Higher histopathology scores and grades were observed as the age of the animal increased, and there was no difference between the sexes. Brack et al. [19a] identified a far higher prevalence of glomerular lesions in animals over 4 years of age compared with younger animals, whereas Collins et al. [19] found only moderate mesangial expansion of glomeruli and no significant tubular or interstitial lesions in animals aged 2 to over 12 years old.
- *Pathology:* Based on light microscopy, Yamada et al. [11] characterized renal lesions into four grades of increasing severity. Grade 1 lesions consisted of sporadically occurring foci of lymphocytic infiltration of the interstitium, regenerative tubules, and hyaline casts. No glomerular lesions were observed. In Grade 2, these tubulointerstitial changes progressed, and foci of enlarged glomeruli were seen. Mesangial proliferation and mesangial cells at the glomerular hilus were observed, but neither these nor capillary loops were seen in PAS, periodic acid methenamine silver (PAM), and Masson trichrome (MT)-stained sections. Grade 3-scored sections had additional progression of tubulointerstitial changes, multifocal to diffuse glomerular involvement, mesangial cell matrix proliferation along the glomerular tuft, partial dilation of glomerular capillary loops, and an array of changes seen with special stains. These included red deposits beneath the glomerular basement membrane in MT-stained sections and thickening of capillary loops in PAM- and MT-stained sections. Many of these changes persisted to a greater degree in Grade 4 (Fig. 12.3). Additional lesions included collapse of some capillaries, dilation of Bowman's capsule and thickening of its basement membrane, partial adherence of glomeruli to Bowman's capsule, diffuse tubular lesions, hyperplastic regeneration of tubules, interstitial fibrosis, double loop capillary profiles, and progressive deposition of red material in PAM and MT-stained sections.

Glomerular lesions progressing through Yamada's renal lesion grades (1–4), as identified by transmission electron microscopy, are as follows: effacement of podocytes and thickening of the glomerular basement membranes (Grade 1); progressive capsular thickening and podoctye effacement, mesangial matrix and cellular proliferation at the glomerular hilum, and fraying of

TABLE 12.4 Criteria for Histopathological Score of Progressive Glomerulonephropathy

Score	Glomerular Lesion	Regeneration Tubuli, Hyaline Cast	Interstitial Cell Infiltration	Interstitial Fibrosis
0	–	No change	No change	No change
1	No change	Few (within four areas)	Sporadically in the hilar area of the kidney or the perivascular area	Focal[b]
2	Increase in mesangial matrix predominantly in the hilar area in the glomerulus or segmental[a] mesangial proliferation (focal/multifocal[b])	Many (up to five areas) or diffuse[b]	Few (within four areas) and focal/multifocal[b]	Diffuse (more severe than score 1)[b]
3	Increase in the mesangial matrix predominantly in the hilar area in the glomerulus or global[a] mesangial proliferation (diffuse[b])	–	Many (up to five areas) or diffuse[b]	–

[a]*Segmental, a lesion involving a part of the glomerulus; global, a lesion involving the whole glomerulus.*
[b]*Focal/multifocal, a lesion involving a part of the kidney (<80%); diffuse, a lesion involving all or nearly all area (>80%).*
Reprinted with permission from Yamada N, Sato J, Kanno T, Wako Y, Tsuchitani M. Morphological study of progressive glomerulonephropathy in common marmosets (Callithrix jacchus)*. Toxicol Pathol 2013;41:1106–15.*

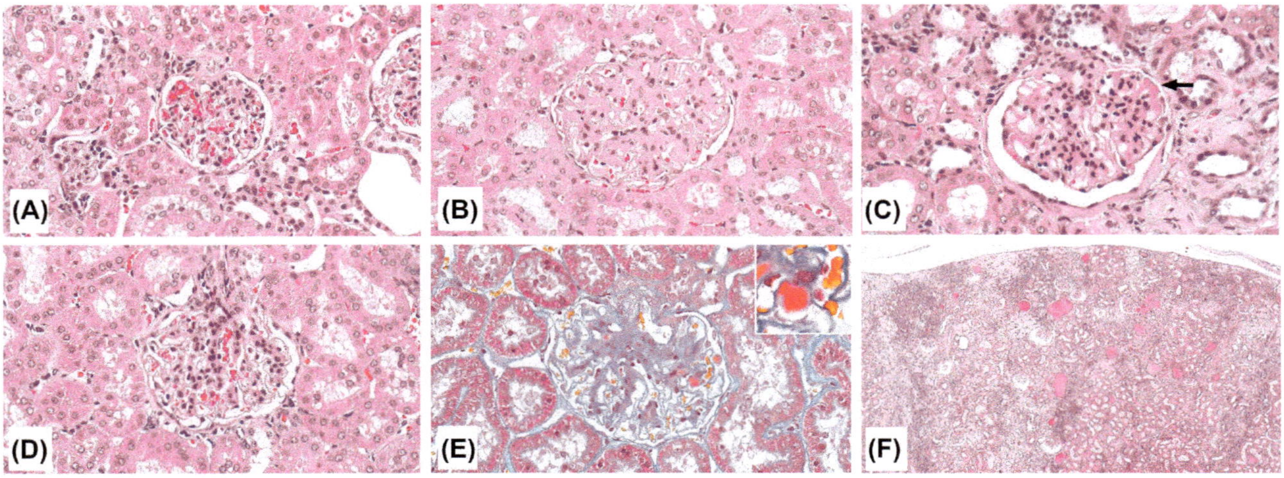

FIGURE 12.3 Histology of kidneys with lesions of glomerulonephropathy corresponding to Grades 1 through 4 in the scheme presented by Yamada et al. [11]. (A) Grade 1 lesion. No remarkable change in glomerular appearance; mild cellular infiltrate in the interstitium. (D) Grade 2 lesion. Mild hilar mesangial cell proliferation with mononuclear cell infiltration of the interstitium. (B and E) Grade 3 lesion. Increased mesangial matrix and glomerular enlargement (B); increased mesangium and red deposits in paramesangial areas (Masson's trichrome stain) (E). (C and F) Grade 4 lesion. Increased glomerular mesangium, thickening of the capsular basement membrane, and "eosinophilic mesangial deposit" (arrow) (C). Interstitial fibrosis and nephritis with tubular casts (F). Magnification objectives 40 × (A–E); 4 × (F). *Reprinted with permission.*

the trilaminar structure of the glomerular basement membrane (GBM) (Grade 2); mesangial matrix proliferation encompassing the mesangium with concomitant appearance of electron dense deposits and progression of podocyte effacement (Grade 3); and continued expansion of the mesangium and marked irregularity of the GBM (Grade 4 [11]; Fig. 12.4). The authors suggest that podocyte foot process effacement and GBM thickening may be the first lesions to occur in glomerulopathy, in that they were not evident by light microscopy of Grade 1 lesions.

There is evidence that the glomerulonephropathy of marmosets is immune-mediated. Brack [19b] evaluated the immunological basis of glomerulonephropathy in a retrospective, tissue-based study using 42 common marmosets

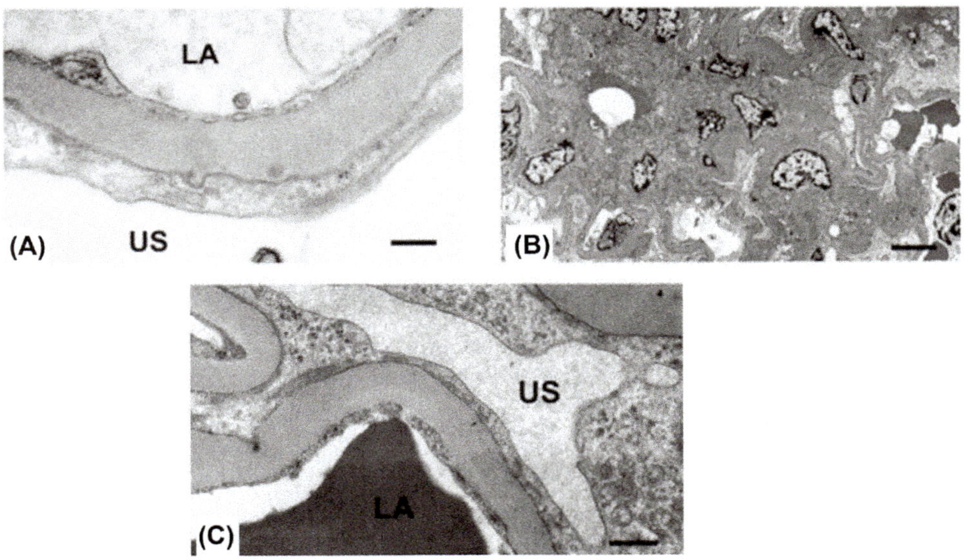

FIGURE 12.4 Transmission electron microscopic images of glomerulonephropathy in common marmosets. (A) Mild thickening of basement membrane and effacement of podocyte foot processes in a Grade 1 glomerulus. Bar 250 nm. *LA*, luminal area; *US*, urinary space. (B) Proliferation of mesangial cells and increased mesangial matrix in Grade 2 glomerulus. Bar 5 μm. (C) Basement membrane thickening and partial irregularity of trilaminar arrangement in a Grade 2 glomerulus. Bar 667 nm. *Reprinted with permission from Yamada N, Sato J, Kanno T, Wako Y, Tsuchitani M. Morphological study of progressive glomerulonephropathy in common marmosets* (Callithrix jacchus). *Toxicol Pathol 2013;41:1106–15.*

as young as 6 months of age, referring to the condition as callitrichid IgM mesangial nephropathy. IgM was identified in 100% of specimens. The immunoglobulins were not only located predominantly in the mesangium, but also along glomerular capillary walls and in hyaline casts. IgA was found in one-third of specimens, again, most frequently in the mesangium. IgG was occasionally found in the interstitium. Subsequent studies attempted to identify nutritional antigens or antinutritional antigen antibodies responsible for antibody deposition, but with inconclusive results [19a]. Gliadins, cereal grain-derived proteins which comprise one of the protein groups in gluten, have been considered a likely candidate antigen due to their presence in foods fed to marmosets and the role of these proteins in celiac disease of humans. Shroeder et al. [20] evaluated 72 common marmosets for a possible association between IgA antibodies to gliadin, glomerular IgA antibody deposition, and wasting marmoset syndrome (chronic lymphocytic enteritis [CLE] or inflammatory bowel disease). Twelve marmosets with CLE and 60 without were evaluated for IgA-antigliadin antibodies (IgA-AGA), IgA containing circulating immune complexes (IgA-CIC), and IgA-antibody deposition in glomeruli. The study showed that glomerular IgA deposition was significantly associated with higher IgA-AGA titer; animals with CLE had significantly higher titers of IgA-AGA and IgA-CIC; and significantly more animals with glomerular IgA deposition had CLE compared with those without such deposits. Yamada et al. [11] suggested that the electron-dense deposits and those identified in MT-stained sections of animals with Stage 3 and 4 renal lesions might be composed of immunoglobulin and have recently identified glomerular IgM deposition in association with the early lesion of podocyte effacement [21]. IgA and IgG were deposited as the lesions progressed to glomerular basement membrane thickening and reticulation. Collins et al. [19] identified IgM by immunofluorescence in all 11 animals evaluated, with some animals also staining positive for IgA, IgG, and C3.

- *Diagnosis*: Diagnosis is made by histopathologic examination of kidneys.
- *Differential diagnosis:* Differentials for renal failure include heritable conditions such as polycystic kidneys and acute kidney disease due to toxins, and bacterial and viral agents.
- *Prevention and control:* The lesions, while suspected of being immune-mediated, are of unknown etiology.

Renal Amyloidosis

- *Etiology:* Amyloidosis in marmosets is of the secondary type and is characterized by infiltration of organs by the N-terminal fragment of the apolipoprotein precursor molecule, serum amyloid A [22]. Although secondary amyloidosis in other species is typically associated with inflammation, genetic factors have been suggested to play a role in the common marmoset.
- *Clinical signs:* Lesions of renal amyloidosis may be associated with signs of chronic renal disease including weight loss, inappetence, and polyuria or oliguria.
- *Epizootiology:* Ludlage et al. [22] evaluated amyloidosis in archival tissue from marmosets over 1 year of age which had been euthanized because of "morbidity or failure to thrive." Fifteen of 86 (17%) animals had infiltration of various organs with amyloid. Nine of the 15 (60%) had infiltration of the renal glomeruli and 7 (47%) had infiltration of the renal interstitium. Three animals had both glomerular and interstitial lesions. When evaluating the group of animals with and without amyloidosis, there were no differences in the serum concentration of serum amyloid A or in the pathology observed between the cohorts.
- *Pathology:* In the renal cortex, the histopathologic lesion consisted of marked accumulation of homogeneous eosinophilic material between tubules and within the glomerulus [22]. The material was congophilic, birefringent under polarized light, and identified by chemical analysis of hepatic extracts and immunohistochemistry as amyloid.
- *Diagnosis*: Diagnosis is made by histopathologic examination of kidneys.
- *Differential diagnosis:* Differentials for chronic renal disease in the marmoset include polycystic renal disease and spontaneous progressive glomerulonephropathy.
- *Prevention and control:* Ludlage et al. [22] concluded that amyloidosis in the common marmoset might have a genetic basis, in that they found both a high prevalence of amyloidosis and a lack of correlation of amyloidosis with inflammatory lesions. Should such genetic elements be identified in the future, whole genome analysis could provide a method of reducing the prevalence of involved alleles through selective breeding.

REPRODUCTIVE DISORDERS

Congenital Diseases

Fused Labia

- *Etiology:* The fused labia defect is most likely recessive and heritable from both parents. Identification of responsible genes or alleles has not been accomplished.
- *Clinical signs*: Isachenko et al. [23] describe 12 female marmosets, 14–40 months old, presenting with labia fused for their entire length, with the exception of 1.5–2.5 mm around the urethral opening (Fig. 12.5). Females were unable to copulate and were functionally sterile.
- *Epizootiology:* All affected females were offspring of parents which had one or both partners come to the German Primate Center from one of two genetically related colonies in Munich, Germany, or another colony in Basel, Switzerland. Forty-five percent of female offspring were affected when both parents came from one of these three colonies, while only 26% had fused vulva when only one parent had so originated.
- *Pathology:* The labia were joined by a thin fibrous band. When surgically opened, the vaginal tract appeared to be of normal morphology. All animals, but one, had normal cyclicity as reflected by progesterone profiles. One animal had persistently elevated progesterone concentrations and cystic ovaries. Histopathology was not reported.
- *Diagnosis:* Physical examination and measurement of the vulva was performed, as was ultrasound examination of the reproductive tract. Plasma progesterone concentrations were determined twice weekly. Photographs of the reproductive tracts of all affected animals were taken at laparotomy or necropsy.
- *Differential diagnosis*: Imperforate vagina [24], influence of a teratogenetic agent of environmental or experimental origin such as androgen administration, XY female, and aromatization consistent with intersex conditions (disorders of sexual differentiation) are all potential differentials.
- *Prevention and control:* Specific protocols to avoid inbreeding and to eliminate affected females from the breeding colony should be preventative. Eventual identification of the gene coupled with genomic evaluation of potential breeding pairs should eliminate the defect from colonies.
- *Research complications:* Presumptive congenital anomalies, especially those which escape detection until the animals are adults, would complicate toxicological investigations.

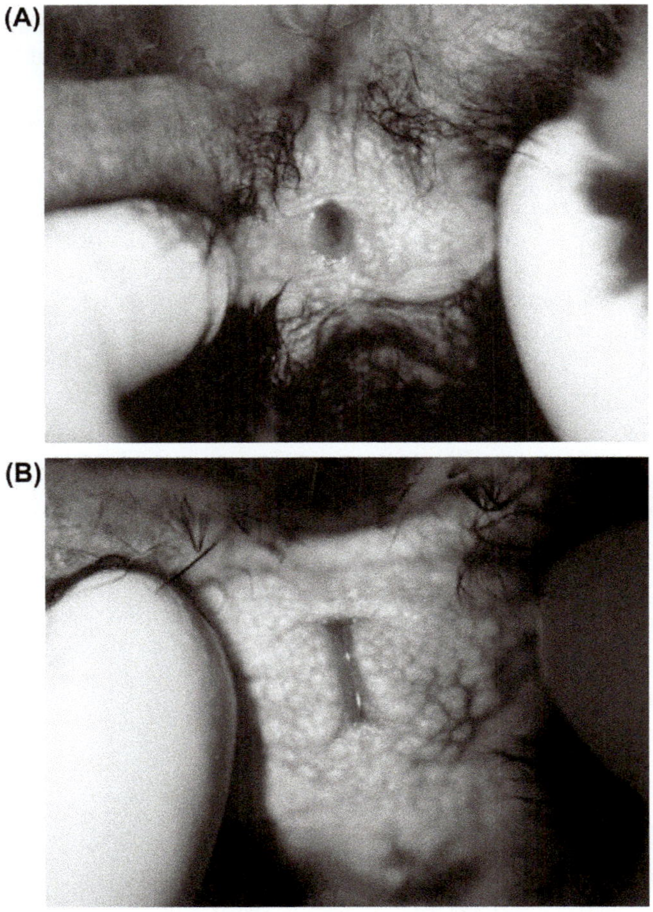

FIGURE 12.5 Affected (A) and normal (B) vulvas of a common marmoset with fused labia. *Reprinted with permission from Isachenko EF, Nayudu PL, Isachenko VV, Nawroth F, Michelmann HW. Congenitally caused fused labia in the common marmoset* (Callithrix jacchus). *J Med Primatol 2002;31:350–5.*

Imperforate Vagina

- *Etiology:* Congenital anomaly.
- *Clinical signs:* Niimi et al. [24] describe a single case of imperforate vagina in a 24-month-old, female, common marmoset. The animal was part of a family group with an actively breeding, dominant female.
- *Epizootiology:* A single animal was affected. Two female siblings of the affected animal had normal vulvar and vaginal morphology, as did all other females in a colony of approximately 200 marmosets.
- *Pathology:* The animal had a normal reproductive tract with the exception of the imperforate vagina. The vulvar skin and subcutaneous glandular tissue, as well as the stratified squamous vaginal epithelium, were continuous and intact (Fig. 12.6).
- *Diagnosis*: Physical examination revealed imperforate vagina, distinguished by the authors from labial fusion by the presence of normal vaginal and vulvar

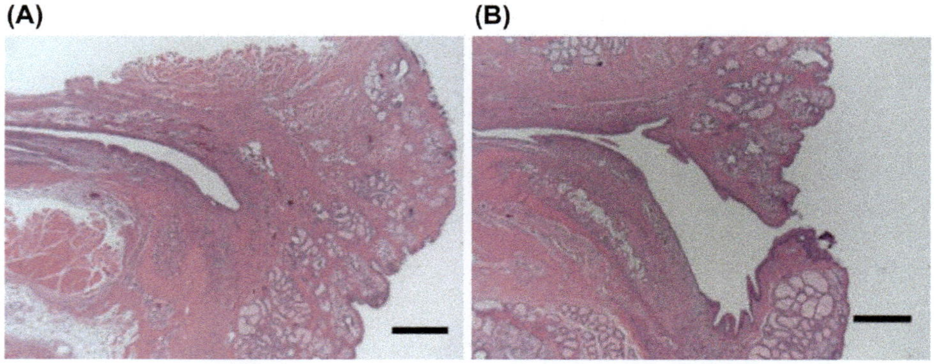

FIGURE 12.6 Histology of the vagina and perivaginal skin surface of an affected (A) and normal (B) marmoset. *Reprinted with permission from Niimi K, Oguchi A, Nishio K, Okano Y, Takahashi E. Congenital malformation of the vaginal orifice, imperforate vagina, in the common marmoset (*Callithrix jacchus*). J Vet Med Sci 2015;77:345–8.*

epithelium as opposed to the "thin fibrous band" described by Isachenko et al. [23]. Plasma progesterone concentrations identified the animal as noncycling, presumably from suppression by the dominant female in the family group.

- *Differential diagnosis:* Labial fusion [23], influence of a teratogenetic agent of environmental or experimental origin, XY female, and aromatization consistent with intersex conditions (disorders of sexual differentiation) are all potential differentials.
- *Prevention and control:* Specific protocols to avoid inbreeding and to eliminate affected females from the breeding colony should be preventative.
- *Research complications:* Presumptive congenital anomalies would complicate toxicological investigations.

XY Female Marmoset

- *Etiology:* Genes normally contained on the Y chromosome in a phenotypically female animal presumably directed differentiation of the embryologic anlage of the external genitalia into a phenotype intermediate between male and female (Fig. 12.7 [25]). The authors showed Y-derived sequences in the mammalian zinc finger protein gene and in the sex-determining region Y gene (SRY) in this female, establishing it as an XY female chimera.
- *Clinical signs:* A female marmoset presented with a single opening for its urethra in the center of the genital area. It had no observable vulva or testicles. The female was the fraternal triplet of both a male and a female marmoset.

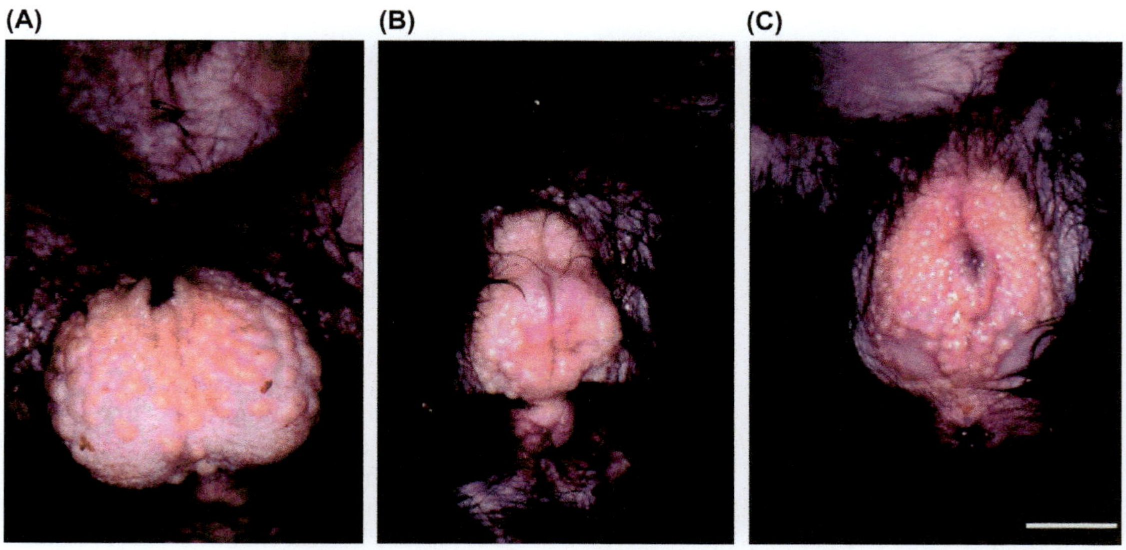

FIGURE 12.7 External genitalia of normal male (A), normal female (B), and affected XY (C) female marmoset. Scale bar = 1 cm. *Reprinted with permission from Sanchez-Morgado JM, Haworth R, Morris TH. XY female marmoset (*Callithrix jacchus*). Comp Med 2003;53:539–44.*

- *Epizootiology*: A single adult female was identified with this abnormality, suggesting aberrant differentiation as opposed to a heritable genetic defect.
- *Pathology:* The reproductive tract was grossly and microscopically female; a histologically normal vagina was seen and there were ovarian structures reflective of normal cyclicity. The skin superficial to the vagina was normal and had modified tubular and sebaceous glands within the dermis. No residual testicular tissue was observed in any section.
- *Diagnosis:* Physical examination and necropsy were used for diagnosis. Plasma hormone analysis identified normal estradiol and testosterone concentrations for a female animal. Amelogenin, zinc finger protein (ZFP), and SRY genes were polymerase chain reaction (PCR)-amplified and the products were sequenced to determine differences in sequence between X- and Y-derived genes. Amelogenin and ZFP have homologues on X and Y chromosomes, whereas SRY only exists on the Y chromosome. The amelogenin gene product could not distinguish between X and Y homologues, but a Y-derived ZFP (ZFY), and the SRY genes identified the animal as an XY chimera.
- *Differential diagnosis:* Labial fusion [23], imperforate vagina [24], influence of a teratogenetic agent of environmental or experimental origin, and aromatization consistent with intersex conditions (disorders of sexual differentiation) are all potential differentials.
- *Prevention and control:* Specific protocols to avoid inbreeding and to eliminate affected females from the breeding colony should be preventative.
- *Research complications:* Presumptive congenital anomalies would complicate toxicological investigations.

GENITAL DISEASE

Experimental Disease

Endometriosis

- *Etiology:* Endometriosis is a disease characterized by the implantation of endometrial cells outside the uterine cavity. Retrograde reflux of these cells from the uterus to the peritoneal space via the oviduct is a leading etiologic hypothesis (retrograde menstruation), although the etiopathogenesis is not well understood [26]. The disease is estrogen-dependent. This syndrome has been experimentally induced in the common marmoset and is not reported to occur naturally.
- *Clinical signs:* Endometriosis causes chronic pain, dysmenorrhea, and infertility in premenopausal women. In macaques it not only causes similar signs but also anorexia and constipation from mechanical obstruction due to chocolate cysts (large cystic structures filled with brown-red fluid [27]). Two marmosets from the study reported here experienced irregular cycles as determined by plasma progesterone profiles. In a later study, Arnold et al. [28] demonstrated decreased grooming and social interaction (established via videotaping), as well as poor habituation to cognitive tests (Wisconsin General Test Apparatus), in marmosets with endometriosis compared with unaffected controls. Treatment for endometriosis returned these behaviors to those of control animals [29].
- *Epizootiology:* Endometriosis is a common disease of women and female, Old World, nonhuman primates. There are no reports of spontaneous (natural) cases of endometriosis in common marmosets [17].
- *Experimental design*: In the study reported by Arnold et al. [26], 29 female marmosets between 2 and 10 years of age had endometriosis induction by retrograde uterine flushing employing a repeated, noninvasive (transcervical) technique or an invasive (transuterine) technique. In the latter technique, the uterus was accessed via laparotomy. Animals failing to develop endometriosis using the noninvasive technique were then subjected to the invasive technique.

  Animals were evaluated via ultrasound and Doppler scanning before and after induction of endometriosis. Once endometriotic foci had been identified by Doppler scanning, laparotomies were performed to confirm the findings and to access tissue (both ectopic foci and uterine endometrial cells) for histology and immunohistochemistry. These tissues were evaluated by both immunohistochemistry and reverse transcriptase (RT)-PCR for endocrine receptors (estrogen, progesterone), steroidogenic enzymes [17 B hydroxysteroid dehydrogenase type 1,2,7; aromatase (RT-PCR only)], and the endometriosis marker CD10 (immunohistochemistry only). The experimental period lasted 18 months.
- *Pathology and results*: Endometriosis was first identified by increased blood flow at 4 months; in total, 21 of 29 animals (72%) developed endometriosis. Animals developing endometriosis with either technique showed no significant increase in the number of lesions after 7 months. Lesions were always associated with the bladder, ovaries, and uterus and were not seen on the parietal peritoneum. Endometriosis was established with both methods but disease derived from the two induction methods differed in histologic grade, time course, and

consistency of induction progression. Foci of endometriosis were more consistently produced using the invasive technique, and endometriosis was of a more advanced histologic grade comprising both endometrial glands and stroma, as opposed to stroma only. The foci of endometriosis were also confirmed by immunohistochemistry and RT-PCR of messenger RNA.

Uterine Rupture

- *Etiology:* There is no known etiology in common marmosets. In women, risk factors are uterine scarring from previous cesarean section, obstetric manipulation, or other uterine surgery. In a minority of cases, approximately 12% in one survey, uterine rupture in women occurred in the absence of uterine scarring. Maternal age and parity are considered risk factors in these latter cases [30]. Changes in uterine tone and activity may also be factors [31].
- *Clinical signs:* Weight loss, weakness, hunched posture, or absence of clinical signs have been documented in cases of uterine rupture of marmosets [30,32]. In women, signs include acute abdomen, excessive vaginal bleeding, absence of uterine tone, and fetal distress.
- *Epizootiology:* There have been two cases of uterine rupture of common marmosets described in the literature. Both animals were pluriparous (one had had four previous pregnancies, the other eight) and presented soon after a successful parturition (2 weeks and 1 day, respectively).
- *Pathology:* In case 1 the uterus was actively bleeding and gray in color at laparotomy. Histopathology revealed the absence of fibrosis and uterine scarring, and the presence of normal endometrium and placental tissue that had herniated from the rupture was in contact with the serosal surface of the uterus. These findings indicated a recent uterine rupture.
- *Diagnosis:* In case 1, two physical exams were performed over the course of several days. Physical exam revealed a palpable abdominal mass that persisted over both examinations. Subsequent radiographs showed a lack of serosal detail: additional clinical signs of weakness and altered posture prompted exploratory laparotomy. Hemoperitoneum and uterine rupture of the posterior uterine wall were diagnosed, ovariohysterectomy was elected, and the animal was returned to her family group. In case 2, no clinical signs were apparent, but abdominal palpation during physical examination a day after delivery of a normal infant revealed a potential twin, after which exploratory laparotomy was performed [32]. An abdominally born twin and posterior uterine rupture were identified, and "subtotal hysterectomy and bilateral ovariectomy" were performed.
- *Differential diagnosis*: Uterine neoplasia, abscess, granuloma, metritis, retained placenta, mummification, or uterine foreign body could present with similar signs.
- *Prevention and control:* Sporadically occurring conditions are difficult to either prevent or control. Institutional policies limiting the number of pregnancies or the duration of breeding life of females might be effective.
- *Research complications:* Uterine rupture has implications for individual and colony reproductive productivity and performance.

Dystocia

- *Etiology:* Uterine atony, fetal oversize, fetal death, and maternal exhaustion are potential etiologies [33,34].
- *Clinical signs:* Lethargy around the expected date of parturition is seen most commonly. Inappetence, hunched posture, huddling in the nest box or hammock, abdominal press or straining, increased respiratory rate and effort, and weakness can also occur. Vaginal bleeding may occur and is abnormal but should be considered in the context of other signs and the time of day during which it occurs.
- *Epizootiology:* Females of breeding age and near the end of gestation, especially primiparous animals, are susceptible.
- *Diagnosis:* Parturition in the common marmoset takes place in the evening. Accordingly, signs consistent with dystocia that occur during the day warrant immediate veterinary assessment and potential intervention (Lynn Wachtman—personal communication). In a series of three cases of cesarean section in common marmosets, two exhibited "abdominal contractions," and one had apathy and anorexia during daylight hours [35]. In each case, these abnormal behaviors prompted ultrasonographic evaluation of the animals and led to a diagnosis of fetal death or fetal oversize.

  At the author's institution, veterinary intervention in the context of vaginal bleeding occurs in the context of the time of day at which clinical signs are observed. If an animal is within its presumptive due date, as determined by biparietal diameter (BPD) measurement during pregnancy (see Chapter 8), is behaving normally in the late afternoon or early evening, but has vaginal bleeding, it can be left overnight to see if it will deliver. If delivery has not occurred by the following morning, the animal is sedated with low-dose ketamine (0.05 mg/kg IM) or alfaxalone (5 mg/kg IM) and ultrasonographic and/or radiographic evaluation of the fetus or feti is performed. Dams that are conditioned to utrasonography can be evaluated without sedation, but sedation with alfaxalone is used when radiographs are taken (Monika Burns—personal communication). If the feti are viable and the heartbeats are strong, the dam can continue to be evaluated every 24 h for continued vaginal bleeding, fetal viability, and the progression of clinical signs. A new BPD can also be obtained for re-assessment of the due date. If an animal has vaginal bleeding during the day but is behaving normally, it should be sedated and examined as described above. Medical or surgical intervention should occur if other signs of dystocia, particularly significant lethargy, are observed.

Medical therapy is as described for other species but is unlikely to have the desired outcome (Wachtman—personal communication). Cesarean section is the preferred treatment. When a cesarean section is elected, it should be done expeditiously. See Chapter 11 for details.

- *Differential diagnosis:* Differentials for lethargy and inappetence during late pregnancy include any serious systemic illness (e.g., cardiomyopathy, sepsis, renal disease, chronic lymphocytic enteritis, or inflammatory bowel disease).
- *Prevention and control*: Selection of dams with successful pregnancy and parturition records may diminish the incidence of dystocia over the preceding history of a colony. Control consists simply of close observation of periparturient females and adequate nutritional and social support throughout pregnancy.
- *Research complications:* Dystocia has implications for individual and colony reproductive productivity and performance.

Pregnancy Loss, Embryonic and Fetal Death, Stillbirth

- *Etiology:* Poor reproductive outcomes of all kinds can generally be considered to derive from fetal, maternal, or environmental factors. Pregnancy loss is poorly understood but presumably occurs because of environmental stressors such as noise, body condition change, and change of social structure [34]. The only etiology for embryonic death established experimentally is modest (25%) energy restriction, which the authors, demonstrating a reduction in urine chorionic gonadotropin, estrogen, and cortisol concentrations from restricted animals, suggest was the consequence of endocrine-mediated depression of placental function [35a]. Intrauterine influences have also recently been identified to have a role in the reproductive success of females that were born as either a twin or a triplet [35b]. As adults, females born as a triplet lose far more offspring before birth than do those born as twins. Another striking finding is that females born with twin brothers lose significantly more offspring as adults in comparison with females born with a female twin. The presumptive influence of the androgenizing effect of male littermates while in utero is contradicted by the absence of such an effect on triplet females, regardless of the number of males with which they have shared a uterus. Additional research on these in utero effects, collectively called developmental programming theory, is clearly necessary [35b].

  The possibility exists that embryonic and fetal loss is a natural mechanism developed evolutionarily by marmosets to allow adjustment of eventual litter size to match available resources. Windle et al. [36] evaluated embryonic and fetal loss of marmosets both by ultrasonic evaluation from conception to birth and by comparison of litter size derived by hysterotomy versus litter size at term birth. They demonstrated progressive reduction in litter size during the embryonic phase (average litter size = 2.81), fetal phase (2.67), and at term (2.47), as established by hysterectomy of dams at specific gestational ages. Comparing litter size at hysterotomy and litter size at term during natural birth, the authors suggested that embryonic and fetal loss, referred to by the authors as "litter size reduction," occurs to reduce the frequency of "unrearable" births; the term "unrearable" being defined as litters of three or more marmosets. They justified this designation by observation of the low frequency with which triplet litters were successfully weaned at the institutions involved in their study (see the next section for further discussion).
- *Clinical signs:* There are typically no signs of embryonic or fetal loss. A possible sign of pregnancy loss is a female's premature return to breeding at its first postpregnancy estrus.
- *Epizootiology:* Reproductive loss, the difference in number of potential offspring between conception and term delivery, has been documented to be as high as 50% in common marmosets [34,36]. Litter size has a negative impact on birth weight and perinatal survival. Ward et al. [36a], in a retrospective study of marmoset breeding-colony data at the New England Primate Research Center over a 25-year period, demonstrated that twin and triplet litters had

significantly more liveborn infants than single or quadruplet litters. In a large multicenter study of factors involved in reproductive variation in common marmosets, mean litter size, survival age of the dam, age of the dam at first parturition, and mean interbirth interval were all determined to have an effect on reproductive production [37]. These factors, however, accounted for less than 50% of total reproductive variance, leaving the authors to suggest that other direct factors such as infant survival and reproductive lifespan, as well as a score of indirect factors (e.g., maternal condition, family group size, reproductive and behavioral experience, and pedigree) could also be involved.

- *Pathology:* Stillborn animals are near-term infants delivered dead and are distinguished at necropsy by failure of their lungs to float in fixative solution. These animals have minimal histopathologic evidence of autolysis. Feti that are dead appear macerated or "mushy" and can exhibit "skin slippage" [35b]. Such animals would have significant autolysis at histopathology. Delivered placentae can be evaluated histologically for evidence of infectious agents, but studies of this kind have not been reported in marmosets.
- *Diagnosis*: Pregnancy, embryonic, and fetal loss can best be demonstrated through serial ultrasonic examination. Laparoscopic and ultrasonographic studies have demonstrated that many single and twin births result from death of a twin or triplet, respectively, in utero [36,38]. Discovery of dead-term (stillborn), near-term, or earlier feti in the family's home cage is diagnostic, but fetophagy and placentophagy likely lead to underestimates of fetal loss (Fig. 12.8). Pathology can help in distinguishing stillborn infants from those that have died earlier in utero.

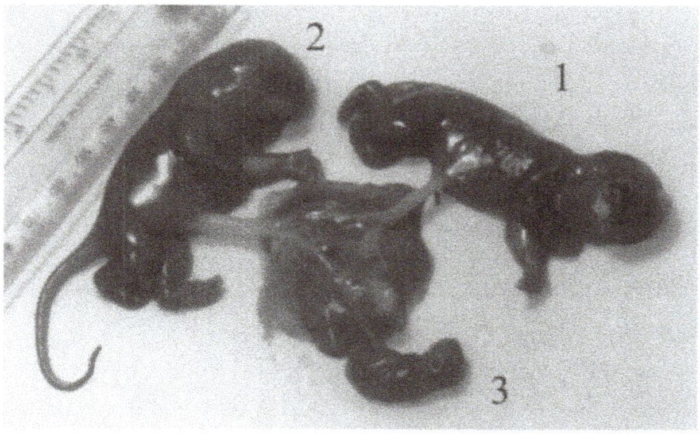

FIGURE 12.8 Aborted feti and placenta. *Reprinted with permission from Windle CP, Baker HF, Ridley RM, Oerke AK, Martin RD. Unrearable litters and prenatal reduction of litter size in the common marmoset (*Callithrix jacchus*). J Med Primatol 1999;28:73–83.*

- *Differential diagnosis:* Pregnancy loss, as well as embryonic and fetal loss, should be distinguished from abortion due to toxins, infectious agents, or virulence factors produced by infectious agents. Toxin analysis or demonstration of causative abortifacients in fetal or maternal tissues would be required for diagnosis of the latter conditions.
- *Prevention and control:* Assuring an adequate nutritional plane, especially energy level, is essential for successful reproduction. Tardif et al. [35a] demonstrated termination of pregnancy in marmosets within 47 days of a 25% reduction in nutritional energy levels compared to controls.
- *Research complications:* Embryonic and fetal loss has implications for individual and colony reproductive productivity and performance.

Infant Morbidity and Mortality

- *Etiology:* Inadequate maternal care, increased litter size, impaired infant mobility, and compromised infant condition—sometimes in association with congenital anomalies—are considered causative [34, 35b, 36a]. One or two infants typically die in litters with three or more marmosets [36]. In a multicenter study of over 1600 litters, the mean litter size was 2.22 ± 0.56, while the mean weaned litter size at day 90 postpartum was 1.14 ± 0.68 [37]. Infants may also be injured by the parents or siblings. Such injuries typically take the form of bites of the tail and distal limbs. Additionally, inflammatory bowel disease and subsequent runting are observed in animals under a year of age and can cause early mortality (see Chapter 8).
- *Clinical signs:* Injured animals are likely to guard the area of injury. Lameness may be observed in animals with limb injuries. Signs may be difficult to discern because infants are typically carried by parents or siblings. Anecdotally, weaker infants often do not curl their tails while being carried and may be found clinging more caudally on juveniles or adults than do the more robust siblings. Weaker infants may also fall off adults or be found abandoned on cage surfaces. See Chapter 8 for strategies to be used with abandoned infants.
- *Epizootiology:* [34] Litter size has an impact on early survival, with singletons and twins being more likely to survive to postnatal day 60 than triplets or quadruplets ([36a], Fig. 12.9). Interestingly, after 60 days, triplets have a longer life expectancy than twins. Most mortality occurs within the first 10 days of life.
- *Pathology:* Necropsy is not typically performed but may provide insights into etiology and pathogenesis.

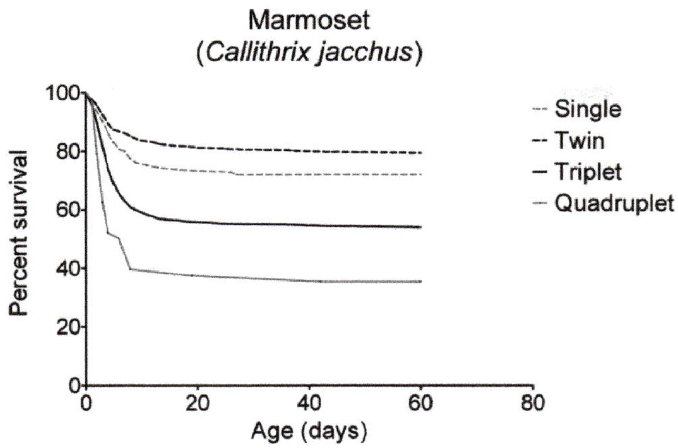

FIGURE 12.9 Litter size and survival in the first 60 days of life. *Reprinted with permission from Ward et al. (2014).*

- *Diagnosis:* Physical examination of injured infants is all that is required for diagnosis. Injuries can be treated medically (clipping, flushing with dilute chlorhexidine solutions and saline, and subsequent bandaging) or by amputation of the affected part (see Chapter 11 for surgical details). Runted individuals, those which survive the first 14 days of life but which then fail to thrive, can be diagnosed by early plateau of their weight gain trajectories (Chapter 8). A subset of these animals also are of smaller stature than their age-matched littermates and maintain their infant pelage.
- *Differential diagnosis:* It may be important from a management perspective to distinguish between parental neglect or rejection, lactational failure, and normal selection.
- *Prevention and control:* Various institutions have adopted different strategies to improve the survival of litters with more than two infants. At the author's institution, rotational rearing, in which a different infant in the litter is supplemented on each successive day, has dramatically increased the number of triplet litters successfully brought to weaning age. Chapter 8 has a full discussion of strategies to improve infant survival.
- *Research complications:* Infant loss has implications for colony reproductive productivity and performance.

References

[1] Wagner WM, Kirberger RM. Radiographic anatomy of the thorax and abdomen of the common marmoset (*Callithrix jacchus*). Vet Radiol Ultrasound 2005;46:217–24.

[2] Wagner WM, Kirberger RM. Transcutaneous ultrasonography of the abdomen in the normal common marmoset (*Callithrix jacchus*). Vet Radiol Ultrasound 2005;46:251–8.

[3] Biller DS, Bradley GA, Partington BP. Renal medullary rim sign: ultrasonographic evidence of renal disease. Vet Radiol Ultrasound 2005;33(5):286–90.

[4] Lubas A, Kade G, Stanisław N. Renal resistive index as a marker of vascular damage in cardiovascular diseases. Int Urol Nephrol 2014;46:395–402.

[5] Oerke A-K, Einspanier A, Hodges JK. Detection of pregnancy and monitoring patterns of uterine and fetal growth in the marmoset monkey (*Callithrix jacchus*) by real-time ultrasonography. Am J Primatol 1995;36:1–13.

[6] Oerke A-K, Einspanier A, Hodges JK. Noninvasive monitoring of follicle development, ovulation, and corpus luteum formation in the marmoset monkey (*Callithrix jacchus*) by ultrasonography. Am J Primatol 1996;39:99–113.

[7] Hastings JM, Morris KD, Allan D, Wilson H, Millar RP, Fraser HM, Moran CM. Contrast imaging ultrasound detects abnormalities in the marmoset ovary *(Callithrix jacchus)*. Am J Primatol 2012;74:1–9.

[8] Humke U, Levens N, Wood J, Hofbauer K. Responses to renin inhibition ent in the marmoset (*Callithrix jacchus*). J Physiol 1991; 261(1 Pt 2):F179–86.

[9] De Loor J, Daminet S, Smets P, Maddens B, Meyer E. Urinary biomarkers for acute kidney injury in dogs. J Vet Intern Med 2013;27: 998–1010.

[10] Krstic D, Tomic N, Radoslavljevic B, Avramovic N, Dragutinovic V, Skodric SJ, Colovic M. Biomarkers of renal function. Curr Med Chem 2016;23:2018–40.

[11] Yamada N, Sato J, Kanno T, Wako Y, Tsuchitani M. Morphological study of progressive glomerulonephropathy in common marmosets (*Callithrix jacchus*). Toxicol Pathol 2013;41:1106–15.

[12] Drake P. Species differences in urinary specific gravity of various nonhuman primates. Honors projects. Grand Valley State University; 2015. http://scholarworks.gvsu.edu/honorsprojects/387.

[12a] Wachtman LM, Kramer JA, Miller AD, Hachey AM, Curran EH, Mansfield KG. Differential contribution of dietary fat and monosaccharide to metabolic syndrome in the common marmoset (Callithrix jacchus). Obesity (Silver Spring) 2011 Jun;19(6):1145–56. https://doi.org/10.1038/oby.2010.303. Epub 2010 Dec 16.

[13] Zoller M, Matz-Rensing K, Fahrion A, Kaup FJ. Malignant nephroblastoma in a common marmoset (*Callithrix jacchus*). Vet Pathol 2008;45:80–4.

[14] Pereira MM, da Silva JJP, Pinto MA, da Silva MC, Machado MP, Lenzi HL, Marchevsky RS. Experimental leptospirosis in marmoset monkeys (*Callithrix jacchus*): a new model for studies of severe pulmonary leptospirosis. Am J Trop Med Hyg 2005;72: 13–20.

[15] Juan-Salles C, Prats N, Resendes A, Domingo M, Hilton D, Ruiz JM, Garner MM, Valls X, Marco AJ. Anemia, myopathy, and pansteatitis in vitamin E–deficient captive marmosets (*Callithrix* spp.). Vet Pathol 2003;40:540–7.

[16] Chalmers DT, Murgatroyd LB, Wadsworth PF. A survey of the pathology of marmosets (*Callithrix jacchus*) derived from a marmoset breeding unit. Lab Anim 1983;17:270–9.

[17] Tardif SD, Mansfield KG, Ratnam R, Ross CN, Ziegler TE. The marmoset as a model of aging and age-related diseases. ILAR J 2011;52:54–65.

[18] Vanselow BA, Pines MK, Bruhl JJ, Rogers LJ. Oxalate nephropathy in a laboratory colony of common marmoset monkeys (*Callithrix jacchus*) following the ingestion of *Eucalyptus viminalis*. Vet Rec 2011;169:100–3.

[19] Collins MG, Rogers NM, Jesudason S, Kireta S, Brealey J, Coates PT. Spontaneous glomerular mesangial lesions in common marmoset monkeys (*Callithrix jacchus*): a benign nonprogressive glomerulopathy. J Med Primatol 2014;43:477–87.

[19a] Brack M, Schroeder C, Fooke M, Schlumberger W. IgM/IgA nephropathy in callitrichids: antigen studies. Nephron 1999;82(3): 221–31.
[19b] Brack M. IgM-mesangial nephropathy in callithricids. Vet Pathol 1988 Jul;25(4):270–6.
[20] Schroeder C, Osman AA, Roggenbuck D, Mothes T. IgA-Gliadin antibodies, IgA-containing circulating immune complexes, and IgA glomerular deposits in wasting marmoset syndrome. Nephrol Dial Transplant 1999;14:1875–80.
[21] Yamada N, Hashimoto N, Kamiie J, Doi T, Sato J, Inoue T, Shirota K, Tsuchitani M. Relationship between immunoglobulin deposition and early lesions of progressive glomerulonephropathy in young common marmosets. Vet Pathol 2018;55:173–6.
[22] Ludlage E, Murphy CL, Davern SM, Solomon A, Weiss DT, Glenn-Smith D, Dworkin S, Mansfield KG. Systemic AA amyloidosis in the common marmoset. Vet Pathol 2005;42: 117–24.
[23] Isachenko EF, Nayudu PL, Isachenko VV, Nawroth F, Michelmann HW. Congenitally caused fused labia in the common marmoset (*Callithrix jacchus*). J Med Primatol 2002;31:350–5.
[24] Niimi K, Oguchi A, Nishio K, Okano Y, Takahashi E. Congenital malformation of the vaginal orifice, imperforate vagina, in the common marmoset (*Callithrix jacchus*). J Vet Med Sci 2015;77: 345–8.
[25] Sanchez-Morgado JM, Haworth R, Morris TH. XY female marmoset (*Callithrix jacchus*). Comp Med 2003;53:539–44.
[26] Einspanier A, Lieder K, Brüns A, Husen B, Thole H, Simon C. Induction of endometriosis in the marmoset monkey (*Callithrix jacchus*). Mol Hum Reprod 2006;12:291–9.
[27] Cline JM, Brignolo L, Ford EW. Chapter 10. Urogenital system. In: Abee CR, Mansfield K, Tardif S, Morris T, editors. Nonhuman primates in biomedical research, diseases. 2nd ed. London: Elsevier Inc.; 2012. p. 510–4.
[28] Arnold C, Lamp J, Lamp O, Einspanier A. Behavioral tests as indicator for pain and distress in a primate endometriosis model. J Med Primatol 2011;40:317–26.
[29] Arnold C, Einspanier A. Medical treatment improves social behavior in a primate endometriosis model (*Callithrix jacchus*). J Med Primatol 2013;42:112–9.
[30] DiVincenti Jr L, Miller AD, Knoedl DJ, Mitchell JF. Uterine rupture in a common marmoset (*Callithrix jacchus*). Comp Med 2016;66:254–8.
[31] Vlemminx MW, de Lau H, Oei SG. Tocogram characteristics of uterine rupture: a systematic review. Arch Gynecol Obstet 2017;295(1):17–26.
[32] Lunn SF. Uterine rupture during labour in a common marmoset (*Callithrix jacchus*). Vet Rec 1985;116:266–7.
[33] Hobson BM, Hobbs KR. Abnormal fetal development in the marmoset (*Callithrix jacchus*). J Reprod Fertil 1974;44:323–4.
[34] Tardif SD, Smucny DA, Abbott DH, Mansfield K, Schultz-Darken N, Yamamoto ME. Reproduction in captive common marmosets (*Callithrix jacchus*). Comp Med 2003;53:364–8.
[35] Prestes NC, Pinheiro JC, Ferraz FMC, Garofalo NA, Simões CRB, Sabino FA, Lea LS, Oba E. Cesarean sections in marmosets: white-tufted marmoset (*Callithrix jacchus*). Vet e Zootec 2014;21:92–7.
[35a] Tardif SD, Ziegler TE, Power M, Layne DG. Endocrine changes in full-term pregnancies and pregnancy loss due to energy restriction in the common marmoset (Callithrix jacchus). J Clin Endocrinol Metab 2005 Jan;90(1):335–9. Epub 2004 Oct 13.
[35b] Rutherford JN, deMartelly VA, Layne Colon DG, Ross CN, Tardif SD. Developmental origins of pregnancy loss in the adult female common marmoset monkey (Callithrix jacchus). PLoS One 2014 May 28;9(5):e96845. https://doi.org/10.1371/journal.pone.0096845. eCollection 2014.
[36] Windle CP, Baker HF, Ridley RM, Oerke AK, Martin RD. Unrearable litters and prenatal reduction of litter size in the common marmoset (*Callithrix jacchus*). J Med Primatol 1999; 28:73–83.
[36a] Ward JM, Buslov AM, Vallender EJ. Twinning and survivorship of captive common marmosets (Callithrix jacchus) and cotton-top tamarins (Saguinus oedipus). J Am Assoc Lab Anim Sci. 2014 Jan;53(1):7–11.
[37] Smucny DA, Abbott DH, Mansfield KG, Schultz-Darken NJ, Yamamoto ME, Alencar AI, Tardif SD. Reproductive output, maternal age, and survivorship in captive common marmoset females (*Callithrix jacchus*). Am J Primatol 2004;64: 107–21.
[38] Jaquish CE, Tardif SD, Toal RL, Carson RL. Patterns of prenatal survival in the common marmoset (callithrix jacchus). J Med Primatol 1996;25:57–63.

Further Reading

[1] David JM, Dick Jr EJ, Hubbard GB. Spontaneous pathology of the common marmoset (*Callithrix jacchus*) and tamarins (*Saguinus oedipus, Saguinus mystax*). J Med Primatol 2009;38:347–59.
[2] Okazaki Y, Kurata Y, Makinodan F, Kidachi F, Yokoyma M, Wako Y, Yamagishi Y, Katsuta O, Takechi M, Tsuchitani M. Spontaneous lesions detected in the cotton-eared marmosets. J Vet Med Sci 1996; 58:181–90.
[3] Ross CN, French JA, Orti G. Germline chimerism and paternal care in marmosets (*Callithrix kuhlii*). Proc Natl Acad Sci USA 2007;104: 6278–82.
[4] Ross CN, Davis K, Dobek G, Tardif SD. Aging phenotypes of common marmosets (*Callithrix jacchus*). J Aging Res 2012. Article ID 56714.
[5] Sweeney CG, Curran E, Westmoreland SV, Mansfield KG, Vallender EJ. Quantitative molecular assessment of chimerism across tissues in marmosets and tamarins. BMC Genom 2012; 13:98.
[6] Tardif SD, Richter CB, Carson RL. Reproductive performance of 3 species of Callitrichidae. Lab Anim Sci 1984;34:272–5.
[7] Zhou LT, Lv LL, Pan MM, Cao YH, Liu H, Feng Y, Ni HF, Liu BC. Are urinary tubular injury markers useful in chronic kidney disease? A systematic review and meta analysis. PLoS One December 1, 2016. https://doi.org/10.1371/journal.pone.0167334.

CHAPTER

13

Diseases of the Gastrointestinal System

Joshua A. Kramer

Laboratory Animal Sciences Program, Leidos Biomedical Research, Bethesda, MD, United States

INTRODUCTION

In captivity, gastrointestinal (GI) diseases are widespread and the most common clinical finding in common marmosets [1–6]. Clinical signs may be due to a host of etiologies and typically manifest as weight loss, diarrhea, changes in eating habits, and/or abdominal distension. Proper recognition of clinical signs and identification of the underlying cause(s) is important to maintain the health of the individual animal and colony as a whole.

Marmosets are broadly characterized as frugivores, insectivores, and exudivores; and their GI system is adapted to meet the needs of this diverse diet [7]. Their teeth and bone structure allow them to gouge tree bark to eat edible plant exudates including gum, sap, and latex [8]. This gouging behavior not only exposes the sappy interior but also stimulates the plant to produce gum, thus resulting in a long-term food source [9,10]. To better digest the gums, marmosets have a large cecum for their body size, which enables efficient fermentation. They may also be able to preferentially retain liquids (i.e., gums) in the cecum while sending solid food to the colon, and their cecum has extensive folds and pockets that may allow for retention of bacteria to repopulate the upper colon when needed [11].

Marmosets in captivity are generally fed a commercial marmoset diet supplemented with fruit, vegetables, and insects or other sources of protein, such as eggs. Due to limitations in caging and diet formulation, the diet fed to captive marmosets likely does not recapitulate their natural diet, which can lead to nutrient deficiencies, secondary metabolic disorders, or inflammatory disease of the GI system. Care should be taken to avoid feeding diets high in simple sugars, such as glucose. While palatable, these diets can lead to secondary clinical problems, as noted below. While it was historically common to feed mouse pups to marmosets, this should be avoided due to the risk of disease transmission.

Additional detail on marmoset nutrition, including vitamin requirements, is discussed elsewhere in this volume. However, it should be noted here that marmosets have a dietary requirement for vitamin D3 due to a reported end-organ resistance [12]. Inadequate D3 levels in feed can result in osteomalacia or other bone pathology. This is covered in additional detail elsewhere in this volume.

Because clinical signs of GI disease are readily apparent to caretakers, veterinary staff are often tasked with identifying the potential causes of diarrhea in marmosets. Every investigation should begin with a thorough cageside examination to better understand the animal's temperament, activity level, and overall appearance. From there, a physical examination should be performed and include evaluation of the oral cavity and abdominal palpation. Veterinary staff should become familiar with the normal characteristics of marmosets including the thickness of the gut loops, liver size, and mesenteric lymph node size. Given their small size, it may be possible to perform a basic examination on awake, restrained animals. However, sedation enhances the ease through which abnormalities can be detected on abdominal palpation and reduces the risk of injuries to staff or animals.

In some cases a physical examination may allow proper diagnosis of a condition. However, additional diagnostics may be required. Radiography and ultrasonography are both readily available and easy to perform and can provide useful diagnostic information. Again, the clinician should become familiar with what is normal utilizing healthy animals as a baseline. There are also published resources available to describe normal radiographic and ultrasonographic anatomy [13,14].

Infectious causes of GI disease are not uncommon. Fecal bacterial culture is important when considering bacterial causes of GI diseases and many viruses can be identified through PCR or other techniques. Fecal

The Common Marmoset in Captivity and Biomedical Research
https://doi.org/10.1016/B978-0-12-811829-0.00013-3

flotation, sediment analysis, or direct examination are useful to identify parasites and protozoa.

More advanced diagnostics such as colonoscopy and biopsy may also be considered and can provide useful diagnostic information for some bacterial, viral, neoplastic, or inflammatory conditions. Intestinal biopsy with or without colonoscopy is easily performed utilizing pinch biopsy forceps with an opening measuring approximately 4 mm. Liver biopsy can be obtained percutaneously or via a small paracostal incision.

In those institutions maintaining colonies of marmosets, consideration should be given to longitudinal tracking of colony health. Tracking body weight and body condition scores is a simple method that can provide useful long-term data on colony health. More complex analyses of body composition, such as EchoMRI or DEXA scans, can also be used and may be simple to perform in marmosets trained to cooperate. This is covered in more detail elsewhere in this book. These modalities can provide useful information to track changes in fat, lean mass, and bone composition over time to evaluate changes in colony health and may be able to identify subtler changes.

Finally, necropsy of index cases also serves as a useful diagnostic tool, especially in larger colonies. A complete and thorough necropsy provides valuable diagnostic information that may not be obtained through other methods. Necropsy examination should be performed by a trained pathologist or person familiar with normal marmoset anatomy and abnormal tissues should be submitted for histopathologic evaluation. Additionally, in the event of an epizootic, euthanasia of affected animals may help limit the spread of the agent, thereby preserving the health of the remainder of the colony.

The GI tract is large and encompasses different organ systems, each of which has its own spectrum of common diseases. Each system will be covered individually below. In addition, specific detail on the multisystemic manifestation of some diseases may be found in other chapters of this manuscript.

DISEASES OF THE ORAL CAVITY AND ESOPHAGUS

Examination of the oral cavity should be a part of every physical examination. Traumatic injuries to the teeth are common and should be identified and treated appropriately to prevent continued oral pain or ongoing pathology.

Teeth

Marmosets have a symmetrical upper and lower tooth arcade with two incisors, one canine, three premolars, and two molars. Detailed descriptions of the order of tooth eruption have been published [15,16]. They are specially adapted to eat plant exudates and in the wild may spend up to 30%–67% of their time foraging for exudates. To aid in this endeavor, they use specially adapted teeth [8].

Dental Fractures and Abscesses

Tooth damage is a common source of morbidity in captive marmosets. Any damage that exposes the root cavity may lead to bacterial infection and a resultant abscess. A common cause of damage is tooth fracture as a result of handling, but wear from enrichment devices, cage material, or other sources may also lead to dental wear and eventual exposure of pulp cavity.

Marmosets with damage to upper teeth present with facial swelling on their cheek, lateral to the nares and below the eye (Fig. 13.1). This is sometimes associated with oculonasal discharge. Damage to lower teeth is less common but can result in abscess formation on the chin or upper neck. Swelling often begins as a red discoloration, progressing rapidly to abscess formation and purulent exudate. Bacterial isolates from the lesions vary, and bacteria may invade the maxilla which will be evident radiographically [17]. Treatment often requires removal of the affected tooth to allow drainage (Chapter 10 of this volume, figures 10 and 11) with subsequent antibiotic treatment to prevent osteomyelitis, damage to the orbit or globe, or sepsis.

Periodontal Disease

Periodontal disease is also a common problem in captive marmosets. Animals have varying clinical signs of gingivitis, hyperemia, pocket formation, and tooth loss [5]. The exact etiology is not described in the

FIGURE 13.1 Tooth root abscesses often present with facial swelling on their cheek, lateral to the nares and below the eye.

literature, but one would presume that dietary factors including reduced mechanical cleaning of teeth due to reduced tree gouging behavior may play a role. A variety of bacterial isolates may also be present. Periodic dental examinations during routine preventive health assessments along with prophylactic cleaning should be performed to prevent gingivitis and periodontal disease.

Viral Infections

While marmosets may be susceptible to a wide range of oral viral pathogens, by far the most important are herpes infections. These agents can cause significant morbidity and mortality, which can greatly affect captive colonies. Care should be taken to avoid infection by strictly separating callitrichids from other species of New World monkeys.

Herpes Simplex

Etiology: Human herpesvirus 1 (HHV-1) (herpes simplex virus) has been identified as the cause of outbreaks of vesicular stomatitis in common marmosets. In all cases, animals have been exposed to a caretaker with recrudescent herpesvirus lesions several days earlier.

Clinical signs: HHV-1 infection is rapidly progressive and animals may die within 1–4 days [18]. Clinical signs include ulcerative lesions on the tongue, gingiva, and lips, which are associated with anorexia, weakness, and hypersalivation. More rarely, signs may progress to include more fulminant signs such as seizures [18,19].

Pathology: Pathology findings are consistent with the clinical signs and include vesicular and ulcerative lesions of the oral and esophageal mucosa. Inflammatory exudate and intranuclear inclusion bodies are associated with these lesions. Animals that live longer may develop meningoencephalitis [19].

Prevention and control: Due to the potential for this virus to spread from animal to animal within a colony, suspected cases should be euthanized immediately and submitted for pathologic examination. Contact animals should be quarantined and observed closely for clinical signs to prevent the spread of disease to other animals in the group. Caretakers with suspect clinical signs should be assigned alternative work duties to prevent exposure of the colony.

Herpes Tamarinus

Etiology: Herpesvirus tamarinus is an alphaherpesvirus related to *Herpes simplex* whose natural host is the squirrel monkey [20]. Spider monkeys and some animals in the genus *Cebus* may also be latent carriers [21]. Though rarely a source of disease in squirrel monkeys, it causes ulcerative and vesicular stomatitis in aberrantly infected callitrichids [22].

Clinical signs: The disease course and clinical signs are similar to that noted for *Herpes simplex*, with a short incubation period and relatively rapid progression to death. In addition to stomatitis, infection is associated with hepatosplenomegaly, ulcerative gastroenteritis, and multifocal necrosis in the adrenal glands, liver, and lymphoid tissue.

Pathology and diagnosis: Clinically apparent oral vesiculation and ulceration should cause a high level of suspicion, and diagnosis can be achieved by biopsy or through viral isolation from saliva. Biopsy will show ulcerative and vesicular lesions with suppurative inflammation, viral syncytia, and intranuclear viral inclusion bodies.

Prevention and control: Treatment with antiherpetic drugs has not been reported, and consideration should be given to euthanize infected animals, and potentially their contacts, to limit the spread of the agent and prevent an epizootic. Because of the risk of this virus, there should be strict separation of callitrichids from other New World primate species. Suspected cases should be euthanized for a diagnostic evaluation including necropsy and histopathology, and contact animals should be quarantined and closely observed.

Fungal Disease

Candida albicans can infect marmosets but is generally only associated with disease in debilitated animals in whom it may cause thrush [23,24]. Thrush is characterized by a white discoloration and proliferation on the tongue and in the esophagus. Lesions are most common in the upper GI tract but may be present in other regions as well. Oral nystatin wash or fluconazole may be effective, and treatment aimed at eliminating the cause of debilitation or immunosuppression should be initiated.

Neoplastic Diseases

In general, neoplastic diseases of the oral cavity are rare in nonhuman primates, including common marmosets. There is, however, a report of oral and nasopharyngeal squamous cell carcinoma affecting eight animals (4.9%) in a single colony [25]. These tumors were locally invasive with expansion into the jaw and/or palate, retrobulbar space, or base of the cranium. While a similar incidence of this specific neoplasm has not been reported elsewhere, it should be included as a differential diagnosis for oral proliferative lesions.

Nutritional Diseases

As in other primate species, marmosets should be provided a balanced diet containing essential nutrients and vitamins. With respect to oral health, vitamin C and niacin are both essential to maintain the integrity of the oral mucosa.

Vitamin C is required to cross-link collagen and therefore an essential dietary nutrient in species lacking gulonolactone oxidase, including marmosets. Deficiency (scurvy) affects multiple species of primates. Marmosets have been reported to have a relatively high requirement for vitamin C and to develop clinical signs after approximately 10 weeks of absolute restriction.

Clinical signs include weight loss, altered gait, and gingival hemorrhage. At necropsy, animals may have hemorrhages around joints and in muscles. Histologically, there was prominent hemorrhage around the periodontal ligament despite not having the characteristic gingivitis that plagues other species of primates [26–29].

To prevent this disease, animals should be fed a commercial diet with adequate levels of vitamin C. Care should be taken to monitor the shelf life and refrigerate diets according to the manufacturer's recommendations to prevent loss of vitamin C over time and secondary nutritional deficiency. Animals that develop clinical signs should be treated with dietary supplementation of at least 500 ppm and by injection of vitamin C, but marmosets may be refractory to treatment [27].

Other nutrient deficiencies may also manifest as oral disease including vitamin A [30], folic acid [31], and niacin [32] deficiencies. Vitamin A deficiency causes aberrant epithelialization and keratinization of the oral mucosa. Folic acid deficiency causes impairment of keratinization and secondary ulceration, which results in weight loss and anorexia. When the related species *Saguinus oedipus* was fed a diet deficient in niacin, animals developed ulcerative stomatitis, ulcerative enterocolitis, and ulcerative and atrophic glossitis [32]. Clinical signs included anorexia and weight loss as a result of the extensive oral and intestinal lesions. Prevention of all of these syndromes can be accomplished through provision of a well-balanced diet.

Congenital Malformations

Though rare, clinicians should be cognizant of potential congenital malformations in marmosets that are aborted near full gestational age or that die soon after birth. Recently, one marmoset breeding colony in the United States had three cases of cleft palate. In all cases, animals developed to full term. One was still-born, one died perinatally, and one lived to 4 days old but eventually died. No other malformations were present [33]. There has also been a single case report of cleft palate in an aborted brown-mantled tamarin fetus [34]. In another report, a single animal was born with cleft lip and cleft palate, microphthalmia, and polydactyly. The animal was born at full term but euthanized after being rejected by its parents. Together, this triad of congenital malformations suggests 13 trisomy [35].

DISEASES OF THE STOMACH AND INTESTINE

Diseases affecting the stomach and intestine of marmosets are the cause of significant morbidity and mortality in marmoset colonies around the world. Often the presenting complaint is diarrhea. Utilization of a scoring system to track diarrhea longitudinally may prove useful, and such systems have been published [36]. Clinicians should become familiar with the common causes of diarrhea and weight loss and pursue diagnostic tests as appropriate to each case.

Bacterial Diseases

Scientific reports of bacterial enterocolitis in common marmosets are surprisingly rare. Marmosets are likely as susceptible to a similar variety of GI pathogens as other primates, including humans. If a bacterial disease is identified, care should be taken to ensure overall colony health by identifying and treating subclinical carriers in addition to those animals with recognizable clinical signs. In addition, many of these pathogens are zoonotic risks, highlighting the need for personal protective equipment and strict sanitation.

Escherichia coli

Etiology: Escherichia coli is a Gram-negative bacillus that includes a diverse array of serotypes and pathogenic subtypes. Enteropathogenic *E. coli* positive for the attaching and effacing locus have been responsible for outbreaks of hemorrhagic diarrhea in common marmosets and related species [37–40]. In two reports, *E. coli* was positive for the attaching and effacing (*eae*) gene, negative for shiga toxin production, and were identified as serogroup O26. This serogroup is associated with outbreaks of diarrhea and hemolytic uremic syndrome in man and thus represents a potentially serious zoonotic threat to caretakers [38].

Clinical signs: Clinical signs in infected animals include weakness, lethargy, dehydration, and voluminous watery diarrhea that progresses rapidly to include hemorrhage [39,40]. Individual cases, or, commonly, outbreaks of diarrhea, should cause a high index of suspicion. Anecdotally, some animals may present with red-tinged urine as well.

Pathology: Intestinal biopsies or samples of intestine from necropsy will show mucosal erosion, adherent bacillary bacteria, and a cobblestone appearance to the mucosa. This attaching and effacing lesion is pathognomonic and may aid in the rapid diagnosis of infection.

Diagnosis: Diagnosis can be made through fecal culture on MacConkey agar, and *E. coli* culture should be specifically requested from the diagnostic lab. If *E. coli* is cultured, serotyping can be pursued to identify the exact etiology and relative risk to humans.

Prevention and control: Prevention of the introduction and spread of disease should be accomplished through washing of fruits and vegetables fed to animals, hygiene of caretakers, and strict use of personal protective equipment. Treatment should be initiated based on bacterial sensitivity results, but empirical selection of enrofloxacin dosed orally or parenterally is normally effective.

Klebsiella pneumoniae

Etiology: Klebsiella pneumoniae is a Gram-negative bacillus that has been associated with outbreaks of fatal peritonitis and GI disease in common marmosets [1,2,41,42] and other New World primates [43].

Clinical signs: In marmosets, clinical signs include anorexia, mesenteric lymphadenopathy, diarrhea, and peritonitis.

Pathology: At necropsy, animals generally have ulcerative colitis and fibrinopurulent peritonitis with bacterial septicemia and dilated colonic crypts (Fig. 13.2).

Diagnosis: Diagnosis is readily made through culture, and sensitivity testing should be performed as bacteria may harbor resistance genes.

Prevention and control: Treatment with enrofloxacin should be rewarding unless the particular strain is resistant. Consideration should be given to screening contact animals with nasopharyngeal and/or rectal cultures and treating any animals identified as carriers with an appropriate antibiotic.

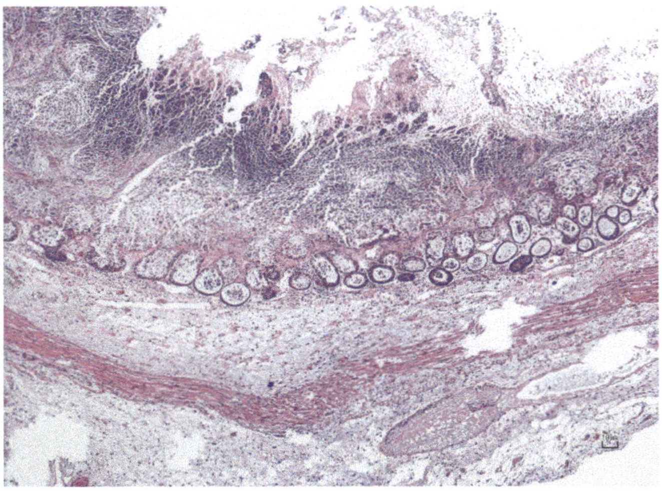

FIGURE 13.2 *Klebsiella* causes severe disease with ulcerative colitis, fibrinopurulent peritonitis, and septicemia. Hematoxylin and eosin, 40x. *Image courtesy of Charles Bailey, Johns Hopkins University.*

Shigella *sp.*

Etiology: Shigella sp. are common pathogens of laboratory primates. Shigellosis is highly contagious and inoculation with as few as 10 organisms may cause infection [44]. Shigellosis is less commonly reported in marmosets than in other nonhuman primate species, but infection with *Shigella sonnei* has been reported [23,37].

Clinical signs: Clinical signs include depression, lethargy, and dehydration due to ongoing hemorrhagic diarrhea. Facial edema, likely due to fluid imbalance, may also be seen.

Pathology: At necropsy, animals will have necrotizing, hemorrhagic, and ulcerative typhlocolitis, similar to the pathology of other nonhuman primates with shigellosis.

Diagnosis: Diagnosis is made through fecal culture. However, laboratory misdiagnosis is not uncommon, and there is some difficulty in distinguishing *E. coli* and *Shigella* [44]. In an outbreak situation, consideration should be given to utilizing alternative diagnostic modalities such as PCR to properly identify the etiologic agent.

Prevention and control: In the cited report, treatment with neomycin was effective; however, one should consider utilizing enrofloxacin or trimethoprim–sulfamethoxazole, which is utilized with success in other species including man [44]. Due to the risk to other animals and zoonotic potential for this pathogen, care should be taken to isolate infected animals, utilize rigid personal protective equipment practices, and properly sanitize the environment. Treated animals should be cultured or examined by PCR following cessation of therapy to identify any residual carriers. These animals should undergo further treatment until cultures and/or fecal PCR show elimination of the carrier state.

Helicobacter *sp.*

Various species of *Helicobacter* have been isolated from marmosets including *Helicobacter jaachi, Helicobacter callitrichis*, and others [45–47]. Due to the nature of the surveys conducted, a correlation with any specific clinical sign or disease syndrome is not possible. In one survey [45], identification of *Helicobacter* sp. histologically was not associated with pathologic changes in the stomach.

Clostridium perfringens

Etiology: Clostridium perfringens is an anaerobic spore-forming, rod-shaped bacterium associated with gas gangrene in a variety of species. There are scant reports of *Clostridium perfringens* causing anorexia, diarrhea, gas gangrene, and death in marmoset colonies including one in which 29 animals died due to gastric dilation following antibiotic treatment for shigellosis [48,49].

Clinical signs: Animals may develop facial swelling and facial erythema. Death is often rapid and sudden, and the GI tract is distended by gas.

Pathology: The intestines, in addition to gas, contain watery brown-gray fecal material and may also have areas of necrosis and mucosal sloughing. Bacteria can be identified in GI contents. In addition, organisms may be identified circulating in the blood and within various tissues, including muscle, bile ducts, liver, lungs, and kidney.

Prevention and control: Given the rapid clinical course, animals may die without premonitory clinical signs. If a diagnosis is made, treatment with antibiotics such as clindamycin, rifampin, or tetracycline should be considered. Prophylactic treatment of other potentially exposed animals may help prevent additional deaths due to this disease.

Clostridium difficile

Clostridium difficile may also be associated with disease but is rarely reported. A high index of suspicion should occur if animals develop diarrhea following antibiotic therapy. Affected marmosets develop bloody diarrhea, weakness, and collapse. Antibiotic therapy alters the normal gut microbiota and allows proliferation of toxin-producing *C. difficile*. At necropsy, lesions range from mild mucosal edema to pseudomembranous colitis. Transmural cecal and colonic necrosis may also be seen [50–52].

Yersinia *sp.*

Yersinia enterocolitica is a Gram-negative coccobacillus that causes gastroenteritis in humans. *Y. enterocolitica* and its relative *Y. pseudotuberculosis* have been reported in other species of primates; and while rarely reported in marmosets, it can cause diarrhea [3,53,54]. The most prominent pathologic change is hepatic necrosis, but there may also be necrosis in the small and large intestines. Diagnosis can be achieved through histologic findings of index cases, fecal culture, or PCR. Treatment may be unrewarding but should include supportive care and aggressive antibiotic therapy with fluoroquinolones or trimethoprim–sulfamethoxazole.

Mycobacteria *sp.*

Atypical mycobacterial species, including *Mycobacterium kansasii* and *Mycobacterium gordonae*, can be isolated from the GI tract including stomach fluid and feces [55]. While not associated with clinical GI disease, bacteria are associated with mesenteric lymphadenitis, and enlarged abdominal lymph nodes can sometimes be palpated. Perhaps most importantly, atypical mycobacteria can cause positive reactions to tuberculin skin tests in marmosets and therefore must be distinguished from *Mycobacterium tuberculosis* infection.

Parasitic Diseases

While parasitic diseases were very common historically, changes in husbandry including the use of antiparasitic drugs and effective pest control have reduced their burden in many marmoset colonies.

Giardia *sp.*

Etiology: Giardia is a flagellated protozoan parasite.

Clinical signs: Giardia sp. has been identified as a common organism in the feces of marmosets but is not generally associated with immediate clinical signs. It may be more common in animals <1 year of age [36,56].

Diagnosis: Diagnosis can be accomplished through ELISA of fecal specimens, and testing of colonies may show a high prevalence of infection despite a lack of appreciable clinical signs. Direct examination of fecal flotations can also be used but is less sensitive than immunoassays. Organisms can also be identified histologically as small crescent-shaped or teardrop-shaped organisms with paired nuclei in the lumen (Fig. 13.3). Regardless of the diagnostic methodology chosen, multiple samples should be sampled over the course of several days due to intermittent shedding.

Prevention and control: Eradication with two doses of oral tinidazole was effective at eliminating the disease and this was associated with weight gain 1 year later. This suggests that elimination of *Giardia* sp. from colonies is worthwhile to improve the overall colony health and may help reduce the incidence of unexplained

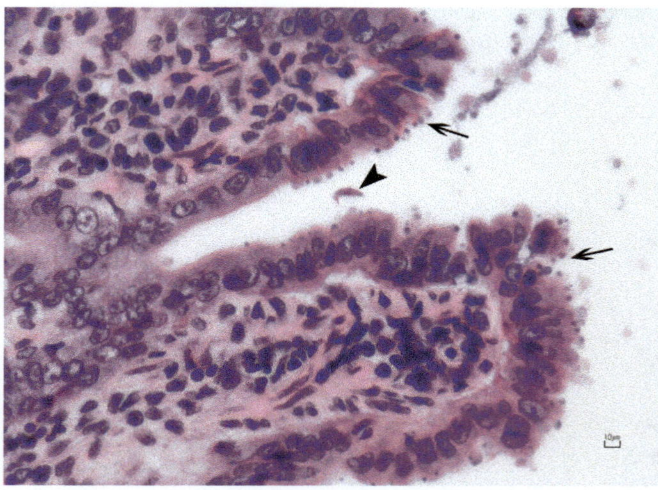

FIGURE 13.3 *Cryptosporidium* sp. and *Giardia* sp. can both cause diarrhea or may be seen in asymptomatic animals. *Cryptosporidium* can be visualized as small basophilic organisms at the luminal surface of the small or large intestine (arrows). Giardia often appears as small crescent-shaped organisms within the lumen (arrowhead). Hematoxylin and eosin, 600x. *Image courtesy of Charles Bailey, Johns Hopkins University.*

weight loss, which is generally associated with inflammatory bowel disease (IBD) [36].

Coccidia

Isospora arctopitheci can infect marmosets. Parasites invade the small intestinal epithelium and, while most infections remain subclinical, may cause clinical signs including bloody diarrhea. In severe infections, animals may die between 3 and 7 days post infection. Oocysts can be identified on fecal flotation or sedimentation. At necropsy animals will have small intestinal lesions characterized by sloughing of the intestinal epithelium. Sporocysts, macrogamonts, and microgamonts are usually visualized in epithelial cells at the distal end of intestinal villi, especially the jejunum. Mice serve as an intermediate host; and therefore, feeding of mouse pups to marmosets should be avoided [57–59].

Prosthenorchis elegans

Prosthenorchis elegans attaches to the wall of the ileum where it forms a fibrous nodule and is associated with inflammation near the site of attachment. In some animals, there may be adhesions between the ileum and adjacent structure, abscess formation, or septic peritonitis resulting from partial rupture of the ileum [60,61]. Depending on the degree of inflammation and adhesion formation, animals may present with no clinical signs or they may present with abdominal turgor, lethargy, weakness, or signs of sepsis.

Other

Entamoeba coli may be present in the GI tract but is not often associated with disease and is considered, like in other primate species, nonpathogenic in most cases.

Cryptosporidium parvum is a known pathogen of a variety of species and has been reported in symptomatic and asymptomatic marmosets [56,62]. On histology, small, round, blue organisms are seen at the luminal surface (Fig. 13.3). Treatment with paromomycin has proven effective.

Viral Disease

Morbilliviruses

Etiology: Morbilliviruses are RNA viruses associated with disease in a variety of species. The most common *Morbillivirus* is measles virus, which is the cause of significant morbidity and mortality in primates.

Clinical signs: In Old World monkeys, it is associated with a multisystemic disease centered on the skin, lymphoid system, brain, and lung. Unlike Old World primates and man, callitrichids infected with either measles or the closely related *Paramyxovirus saguinus* usually develop a GI syndrome characterized clinically by diarrhea, dehydration, and progressive lethargy. However, upper eyelid edema, a skin rash, and pneumonia can also be seen [63].

Experimentally infected mustached tamarins generally succumbed within 2 weeks of infection [64]. Outbreaks have been reported in several species of callitrichids with up to a 100% mortality in some species [63,65]. In these situations, death may occur as soon as 8 h after onset of clinical signs and not all animals will show clinical signs.

Pathology: Gross pathology findings include hemorrhagic gastroenteritis and colitis and variable pneumonia. Histologically, there is multifocal necrotizing gastroenterocolitis and/or bronchointerstitial pneumonia depending on the tropism of the specific virus. Large syncytial cells containing intranuclear and intracytoplasmic viral inclusions may be present in the gut, bile ducts, pancreatic ducts, renal tubules, bronchial epithelium, or hepatic cords; and there may be necrosis of lymphoid tissue, which is typical of morbillivirus infection.

Prevention and control: Based on these reports and clinical knowledge of the disease in man, every effort should be made to prevent entry of measles virus into research colonies. Contact with personnel should be limited to those who have been vaccinated and/or screened for positive antibody titers. If an outbreak does occur, affected animals should be quarantined and consideration should be given to culling affected animals early to help limit the spread of disease.

Vaccination is a common practice in colonies of macaques and other Old World primates. Marmosets have been successfully vaccinated [64]. While routine vaccination does not appear to be a common practice, vaccination with the human measles or canine distemper-measles vaccine should be considered to protect colony health in an epizootic. Treatment of affected animals, if attempted, should include rehydration, supportive care, and antibiotics to treat and prevent secondary infections.

Herpesvirus ateles *and* Herpesvirus saimiri

Etiology: Ateline herpesvirus 2 and 3 *(Herpesvirus ateles)* is a gammaherpesvirus in the genus *Rhadinovirus* related to Epstein–Barr virus (EBV) whose natural host is the spider monkey [66–68]. When experimentally inoculated in common marmosets, it causes Hodgkin's-like lymphoma, which may manifest as GI disease due to proliferation of neoplastic lymphocytes in the gut and abdominal lymph nodes.

Saimiriine herpesvirus 2 *(Herpesvirus saimiri)* is a closely related herpesvirus whose natural host is the squirrel monkey. In squirrel monkeys, it is not generally associated with disease but can persistently infect marmoset lymphocytes and cause malignant transformation [66,67,69–71].

Clinical signs: For both agents, disease has been reported to start approximately 2–3 weeks postinoculation and may prove fatal by 3–4 weeks; but longer courses with slower-growing lymphomas have also been described. Disease is characterized by a mild-to-moderate leukocytosis, circulating neoplastic cells visible on blood smears, generalized lymphadenopathy, and hepatosplenomegaly. Animals may also have upper respiratory disease including nasal discharge.

Pathology: At necropsy, there is diffuse lymphoma affecting multiple organs that can include those noted above plus kidney, adrenal gland, tonsils, and lung. Aggressive cases may also be associated with extensive hemorrhage and necrosis in vital organs. Treatment has not been described though animals have been vaccinated using an attenuated viral strain. Vaccination successfully protected animals from subsequent challenge [72].

Prevention and control: Because of the risk of this virus, there should be strict separation of New World primate species. Suspected cases should be euthanized for a diagnostic evaluation including necropsy and histopathology, and contact animals should be quarantined and closely observed.

Other Gammaherpesviruses

Human herpesvirus 4 (EBV) causes mononucleosis, lymphoma, and nasopharyngeal carcinoma in people. While infection to callitrichids has been documented, common marmosets seem relatively resistant to lymphoma development following EBV infection [73,74]. However, some strains may be more or less oncogenic, and EBV/HHV-4 should be considered as a differential diagnosis for malignant lymphoma in the marmoset [69]. Animals that do progress to lymphoma have a more protracted course than with *Herpesvirus ateles* or *saimiri* with lymphoma developing 4–8 weeks post infection.

Finally, *Callitrichine herpesvirus 3* (CalHV3) has also been associated with GI lymphoma, although an etiologic relationship has not been established. Animals develop clinical signs including inappetence, diarrhea, and weight loss and often have palpable abdominal masses. At necropsy, B cells invade the gut and cause enlargement of mesenteric lymph nodes. While seroprevalence may be as high as 60% in some colonies, the incidence of GI lymphoma may be low [75–77]. Currently it is not clear whether CalHV3 is the etiologic agent responsible for these lymphomas or if it is detected in these lesions because of their high B-cell content.

Other

Experimental infection with the SARS coronavirus has been reported and resulted in diarrhea associated with mild diffuse colitis. At necropsy, animals had colonic crypt hyperplasia [78]. Experimental yellow fever infection has been associated with hemorrhagic diarrhea and melena, with findings of hemorrhagic gastroenteritis at necropsy [79,80].

Neoplastic and Proliferative Disease

Small Intestinal Adenocarcinoma

The most common GI tumor of common marmosets is small intestinal adenocarcinoma [81–83]. Tumors generally originate in the proximal jejunum but may also be present in the distal jejunum or ileum. Metastasis to the mesenteric lymph nodes is common and has also been reported in the lung and kidney.

Not surprisingly, the most common clinical signs are diarrhea and weight loss, though animals may also present with abdominal bloat. At necropsy, there is focal thickening of the small intestine with or without stricture. In man, gastric carcinomas have been associated with EBV infection and with *Helicobacter* sp. infection. Neither was identified as a potential cause for small intestinal adenocarcinomas in the largest report to date [83]. Based on these results, small intestinal carcinoma should be considered as a differential diagnosis in cases of diarrhea for which another underlying cause cannot be identified.

Gastrointestinal Lymphoma

GI lymphoma is not uncommon and may be associated with a viral cause including *Callitrichine herpesvirus 3*, *Herpesvirus saimiri*, *Human herpesvirus 4* (EBV), or *Herpesvirus ateles* as described above.

Other

Reports of other proliferative diseases in common marmosets are exceedingly rare. There is a single case report of an inflammatory fibroid polyp in the duodenum of a common marmoset [84].

Other Noninfectious Diseases

Inflammatory Bowel Disease

Etiology: No discussion of GI disease in common marmosets would be complete without the discussion of IBD. In marmosets, this condition has historically been called marmoset wasting syndrome (MWS). In reality, the historical MWS is most likely a group of distinct disease entities that present with similar clinical signs but diverse etiologies and associated pathology [41]. Because MWS lacks diagnostic specificity, its use should be discouraged and marmoset diseases causing significant weight loss or wasting should be referred to by their etiology or associated morphologic changes. IBD is a known entity and will be described further.

Clinical signs: Clinical signs for both small and large intestinal IBDs are apparently similar in marmosets. Clinically, IBD animals present with diarrhea, weight loss, alopecia, and a failure to thrive. Bloodwork may show hypoproteinemia, anemia, and elevated liver enzymes [85]. As the condition progresses, animals may develop muscle atrophy and weakness, including a characteristic facial appearance with prominence of the cranium and ears.

Pathology: Simplistically, there are two general pathologic patterns of IBD. Historically, IBD in callitrichids has been associated with chronic colonic inflammation in which the lamina propria is expanded by lymphocytes [1,4,60,85,86]. While most associated with tamarins [1,60], this has also been shown to commonly affect adult common marmosets [1,2,4]. Other reports discuss a chronic lymphocytic enteritis (CLE) centered on the small intestine (Fig. 13.4). Histologically, the lamina propria of the jejunum is expanded by lymphocytes and villi are blunted [2,41,86,87].

A specific entity of IBD has been recognized in some common marmoset colonies characterized by a CLE accompanied by progressive weight loss. In this form of IBD lymphocytic infiltrates comprising CD3/CD56-positive cells efface the normal architecture resulting in fusion and blunting of villous tips accompanied by crypt hyperplasia. Intense lymphocytic infiltration of the epithelium is observed consistent with an immune-mediated response to environmental or potentially host antigens [41]. Treatment with prednisone and dietary withdrawal of gluten had no effect on the clinical disease. Elimination of *Giardia* within the colony was associated with a marked reduction in incidence of the disease, suggesting that environmental triggers may initiate a self-perpetuating immune-mediated disease.

Differential diagnosis: Despite the high prevalence of the disease and associated morbidity and mortality, the definitive cause of IBD/CLE has yet to be identified. Weight loss and wasting are nonspecific signs in marmosets and may be associated with a number of clinical entities. These findings have been associated with parasitic infection [36,88–90], gluten sensitivity [91], dietary protein deficiency [92,93], vitamin E deficiency [23], stress, and gliadin hypersensitivity [94,95]. In one survey of zoos, no fewer than 10 causes were proposed by survey respondents [96]. In some cases, subsequent investigations have contradicted previous assertions [97].

As noted earlier, wild marmosets obtain a large percentage of their calories from tree exudates and neither this method of foraging nor the nutritional content of the exudate are exactly recapitulated in current marmoset diets. Additional investigations into dietary factors would be worthwhile. Recent reports of gluten sensitivity including elevations in anti-*Gliadin* antibodies and positive response to food trials eliminating gluten indicate that dietary factors, specifically incorporation of cereals into the diet, may play a role in the development of IBD [91,94,95]. Furthermore, in one

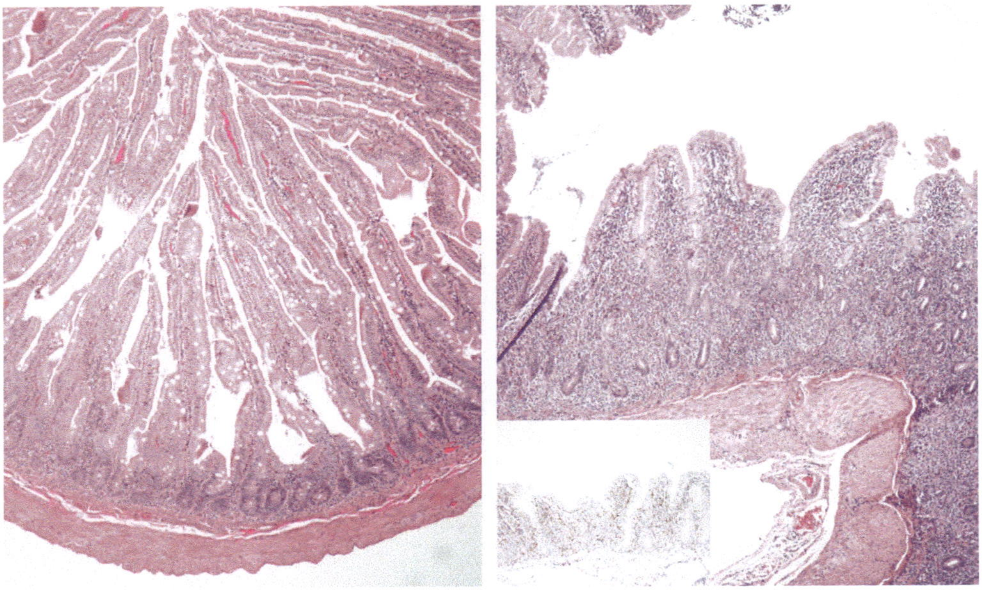

FIGURE 13.4 Compared with normal marmosets (left), animals with chronic lymphocytic enteritis (right) have villous blunting and fusing with expansion of the lamina propria by large numbers of CD3+ lymphocytes (inset). Hematoxylin and eosin, 100x. *Images courtesy of Keith Mansfield, Novartis Institute for Biomedical Research.*

experiment, wild-caught *Saguinus* sp. monkeys were biopsied at baseline then placed in a standard cage and fed a commercial diet. They developed characteristic colonic inflammation during the course of captivity, which would seem to implicate food and/or stress [98].

Diagnosis: Regardless of the cause, IBD poses a unique threat to captive marmoset colonies and represents a significant cause of mortality [99]. Diagnosis is often made through identification of clinical signs but gut biopsies can also prove useful. Recently, calprotectin and matrix metalloproteinase 9 have been identified as possible biomarkers of disease in marmosets [100,101]. Additionally, hypoalbuminemia (<3.5 g/dL) and weight loss (<325 g) together are negatively correlated with outcome [102].

Prevention and control: Treatment should be supportive. Some success has also been reported in treating animals with glucocorticoids. Due to the possible side effects of this and effects on research, long-term glucocorticoid therapy may not be possible in many situations but could be considered for individual cases [103]. Elimination of other causative factors such as parasitic infections should be pursued and animals should be fed a nutritious and balanced diet formulated for their species.

Amyloidosis

While IBD is a major source of mortality in animals <6 years of age, amyloidosis is more common in older animals [99] and in one survey affected 17% of animals overall. Amyloid was identified as AA or reactive amyloid that originates from the N-terminal fragment of the serum amyloid A (SAA) protein [104]. Animals with reactive amyloidosis tend to be older (6.75 years of age) compared with animals that die of other causes and present with weight loss with or without diarrhea and hepatosplenomegaly. Organs affected include, in decreasing order of frequency, the small intestine, liver (Fig. 13.5), adrenal glands, kidney, stomach, colon, and spleen. In some animals, diagnosis can be confirmed through use of rectal biopsy or percutaneous liver biopsy.

DISEASES OF THE LIVER

Bacterial Diseases

Bacterial liver disease in marmosets is uncommon but has been reported both naturally and experimentally.

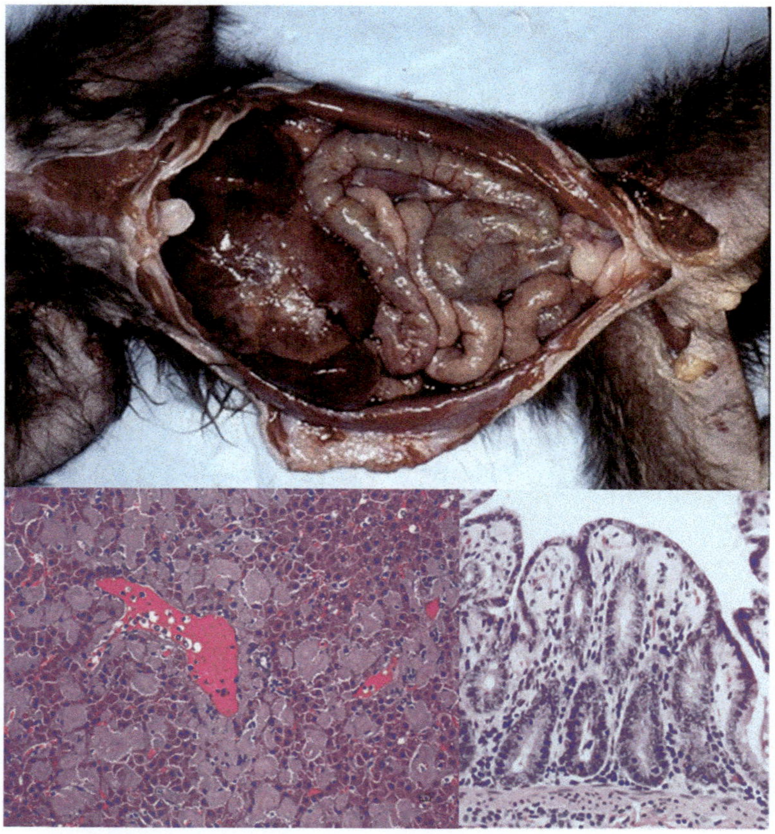

FIGURE 13.5 Amyloidosis commonly affects the liver and intestine. Livers appear large, pale, and waxy. The gut is thickened and tortuous. Histologically, amyloid is identified as extracellular amorphous eosinophilic material, which expands hepatic sinusoids and/or the lamina propria of the small or large intestine. Hematoxylin and eosin, 400x. *Images courtesy of Keith Mansfield, Novartis Institute for Biomedical Research and Charles Bailey, Johns Hopkins University.*

Clostridium piliforme

Etiology: Clostridium piliforme is a rod-shaped, flagellated, anaerobic, spore-forming bacillus that is the causative agent of Tyzzer's disease. Tyzzer's disease can infect a wide range of laboratory animals and has been reported in both marmosets and tamarins.

Clinical signs: Animals appear to be most susceptible at an early age and may die suddenly without premonitory clinical signs.

Pathology: Affected animals have classic lesions of Tyzzer's disease including dissemination of bacteria to heart, liver, and/or GI tract with resultant multifocal hemorrhagic necrosis in these locations [105,106].

Diagnosis: Unfortunately, antemortem diagnosis is difficult because the organisms do not grow in culture.

Prevention and control: Treatment has not been reported but is often unrewarding in other species. Tetracyclines may be attempted.

Francisella tularensis

Etiology: Francisella tularensis, the causative agent of tularemia, has been described as both a natural and experimental infection [107–109].

Clinical signs: Clinical signs of infection may include lethargy, ataxia, and fever, but animals may also present more acutely or die suddenly. Following experimental aerosol exposure, animals generally succumbed within 1 week of infection.

Pathology: At necropsy, animals will have multifocal hepatic necrosis, necrotizing enteritis, bronchopneumonia, lymphadenitis, and splenitis, which are consistent with disease in other species.

Diagnosis: In suspected cases, serology may be beneficial to help diagnose the disease but necropsy and histopathology may be beneficial as well.

Prevention and control: Limiting contact of free-ranging marmosets to possible hosts such as rodents and other small mammals should limit spread of this disease.

Leptospira *sp.*

Leptospira borgpetersenii has rarely been reported in Wied's marmosets and presumably could infect common marmosets as well. Expected clinical signs include diarrhea, weight loss, and jaundice, and bloodwork shows hyperbilirubinemia. Diagnosis can be confirmed through serology or at necropsy where animals will have renal and/or hepatic lesions. Leptospiral organisms may be noted with a silver stain or immunohistochemistry [110].

Other Bacterial Agents

Other potential pathogens include *Yersinia enterocolitica,* which causes miliary hepatic necrosis [53], and *Burkholderia mallei* (glanders), which induces hepatic necrosis when introduced experimentally [111].

Parasitic Disease

As in the stomach and intestines, parasitic disease of the liver and gallbladder are rare in marmosets. The most common parasites are *Platynosomum* sp. and *Toxoplasma gondii*. While modern pest control measures and use of antiparasitic agents have reduced the likelihood of transmission by reducing contact with host species, infection is still possible; and one should be familiar with these agents as potential causes of clinical disease.

Platynosomum *sp.*

Etiology: The trematode flukes *Platynosomum amazonensis, Platynosomum marmoseti*, and *Athesmia foxi* may infect callitrichids including common marmosets [61,90,112,113].

Clinical signs: These parasites reside in the gallbladder and bile ducts, and clinical signs manifest primarily as weight loss. With a severe infection, one would expect to see hyperbilirubinemia and icterus.

Pathology: At necropsy, there is dilation of intrahepatic bile ducts, thickening of the gallbladder, and black discoloration of the bile. Flukes may be grossly visible as small black foreign bodies within the gallbladder or bile ducts. Histologically flukes are associated with portal fibrosis and significant cholestasis.

Prevention and control: This parasite can be treated with 60 mg/kg praziquantel, and care should be taken to control cat, snail, lizard, and toad populations, which can all serve as intermediate or paratenic hosts [90].

Toxoplasma gondii

Etiology: Toxoplasma gondii has also been reported in callitrichids and other New World monkeys [114–117].

Clinical Signs: Experimental infection results in rapid death between 1 and 12 days post infection. Clinically, one would expect to see fever, along with elevations in liver enzymes associated with infection. The organism also commonly infects the kidneys, muscles, heart, lymph nodes, spleen, and lung, which can result in kidney failure, pneumonia, and myocarditis, which may be presenting complaints as well.

Pathology: Infection is prominent in the liver where it is associated with multifocal to coalescing inflammation and necrosis, which can be extensive.

Diagnosis: Diagnosis may be made from blood smears, though organisms may be present in blood only late in the disease. Due to the acute nature of the disease, treatment in marmosets has not been described but includes trimethoprim–sulfamethoxazole or pyrimethamine + sulfadiazine + leucovorin in other species.

Viral Diseases

Lymphocytic Choriomeningitis Virus

Etiology: Lymphocytic choriomeningitis virus (LCMV) is a member of the family *Arenaviridae*. Disease has been reported in a variety of species, and marmosets are highly susceptible to infection. Originally termed callitrichid hepatitis virus, it was later discovered that this transmissible agent is actually a variant of LCMV [118–121].

Clinical signs: Marmosets are most commonly exposed following ingestion of infected rodents, and disease has been associated with the feeding of newborn mice to marmosets [19], though wild mice may also serve as a reservoir [122]. Approximately 1–2 weeks following exposure, animals may present as weak and anorexic, jaundiced, or with unexplained hemorrhage. Sudden death without identification of premonitory clinical signs may also occur. Animals often have elevated liver enzymes.

Pathology: The most common and prominent pathologic finding is multifocal hepatic necrosis with Councilman (acidophilic) bodies and mononuclear inflammation. Animals may also have necrosis in the spleen, lymph nodes, adrenal cortex, and intestinal tract.

Diagnosis: Diagnosis can be made through serology or, more likely, based on characteristic histopathologic findings and use of immunohistochemistry or other molecular techniques to identify infection.

Prevention and control: Care should be taken to avoid infection by controlling the wild rodent population and avoiding the use of mice as food enrichment for marmosets. If disease is suspected, animals should be isolated to avoid spread within the colony, and euthanasia of infected animals and contacts should be considered. Thorough sanitation of the environment is required, and an evaluation of pest control measures should be undertaken to reduce or eliminate the rodent population.

GB Virus

Hepatitis C is a cause of significant morbidity in human populations and is responsible for hundreds of thousands of human deaths annually. Marmosets are susceptible to infection with GB virus B, the most closely related virus to hepatitis C virus, and therefore serve as a useful animal model of hepatitis C [123].

Infected animals develop hepatic pathology similar to that seen in humans with hepatitis C, which include multifocal nonsuppurative hepatitis, formation of lymphocytic nodules, and piecemeal necrosis, which progress to portal fibrosis over the course of approximately 5–6 months. Steatosis may also be present. Progression occurs despite clearance of the virus from plasma [124,125]. These changes are associated with concomitant changes in plasma insulin and glucagon levels consistent with metabolic dysregulation [126]. While acute experimental infection is not associated with notable clinical signs, one may see an elevation in serum ALT, ALP, and/or LDH clinically.

Yellow Fever

Etiology: Marmosets are susceptible to yellow fever, and the virus may occur naturally in wild populations [127–129]. Experimental inoculation of *Callithrix jacchus* may cause disease depending on the specific strain used for infection. In an experiment in which multiple strains were inoculated into common marmosets, the African-origin Asibi strain resulted in low mortality, whereas South American strains such as O.C., J.Z., J.F., Martinez, and A.C. Bolivian strains resulted in high mortality with an average survival time of 5–9 days post infection [79,128].

Clinical signs: Clinically one would expect to see weakness, lethargy, and potentially icterus. Depending on the strain, liver enzymes may be elevated and clotting times may be abnormally long.

Pathology: Gross pathologic changes are somewhat dependent on strain but generally include icterus, splenomegaly, and congestion or hemorrhage of the stomach. Histologically, there is steatosis and multifocal random hepatic necrosis, which may be associated with intranuclear viral inclusion bodies in hepatocytes. Councilman bodies may also be present.

Prevention and control: Unless a facility was importing primates from South America, introduction of this agent into a facility would not be expected. However, as global temperatures rise, the endemic range of mosquitoes harboring this virus (*Aedes aegypti*) could change and lead to an enhanced probability of transmission in some captive settings. As such, measures to control the mosquito population and to limit contact with other infected animals or humans should be implemented to prevent transmission of this virus.

Other Viruses

In addition to these important viruses, marmosets may also be infected experimentally with other pathogens that result in hepatic disease.

Not surprisingly, marmosets are susceptible to *Lassa virus*, the causative agent of Lassa fever. This virus is an arenavirus closely related to LCMV and results in lesions similar to those seen in this important human disease, including hepatocellular necrosis with lesions similar to those described for LCMV above [130].

Marmosets are also experimentally susceptible to Junin virus, the causative agent of Argentine hemorrhagic fever, another arenavirus. Animals present with anorexia, weight loss, and/or neurologic symptoms. Anemia develops over the course of 1–3 weeks, and leukopenia and thrombocytopenia may also be evident [131,132]. Gross lesions include multifocal hemorrhage in the mouth, pharynx, and esophagus. Histologically, there is multifocal hepatic necrosis along with meningoencephalitis, interstitial pneumonia, and lymphocytic depletion [133].

Marmosets have been used as models to study Marburg virus and Ebola virus. Infection with both results in weight loss, anorexia, fever, and depression. Animals may show hemorrhage from mucosal surfaces and dyspnea. Animals generally succumb to disease within 7–10 days. Pathologic findings are typical of filovirus infection and include hepatocellular necrosis along with a host of other pathologic changes such as splenic necrosis, lymphoid necrosis, and disseminated intravascular coagulation [134–136].

Other species of callitrichids, including several species of *Saguinus*, have been experimentally infected with hepatitis A virus. Animals developed elevated liver enzymes, histopathologic changes in the liver consistent with hepatitis A virus infection, and developed circulating antibody titers. The incubation period was approximately 3–5 weeks [137].

Noninfectious Diseases

Marmosets commonly develop nonalcoholic fatty liver disease (NAFLD), which sometimes progresses to nonalcoholic steatohepatitis (NASH). In one report, hepatomegaly was present in 18% of marmosets and was identifiable on physical examination and radiographs [138]. NAFLD and NASH are important human diseases, and marmosets may serve as a useful animal model for these conditions.

Marmosets with fatty liver disease are more likely to have high total body weight and body condition scores and are generally more likely to be female than male. Bloodwork findings are reported to include an elevated GGT and triglycerides.

At necropsy, animals will have markedly enlarged livers which, in severe cases, can extend down to the pelvic region. Histologically, there can be both macrovesicular and microvesicular steatosis, which is associated in some cases with inflammation and fibrosis (Fig. 13.6). Management of NAFLD/NASH is not discussed in the literature, but dietary intervention may prove worthwhile (Table 13.1).

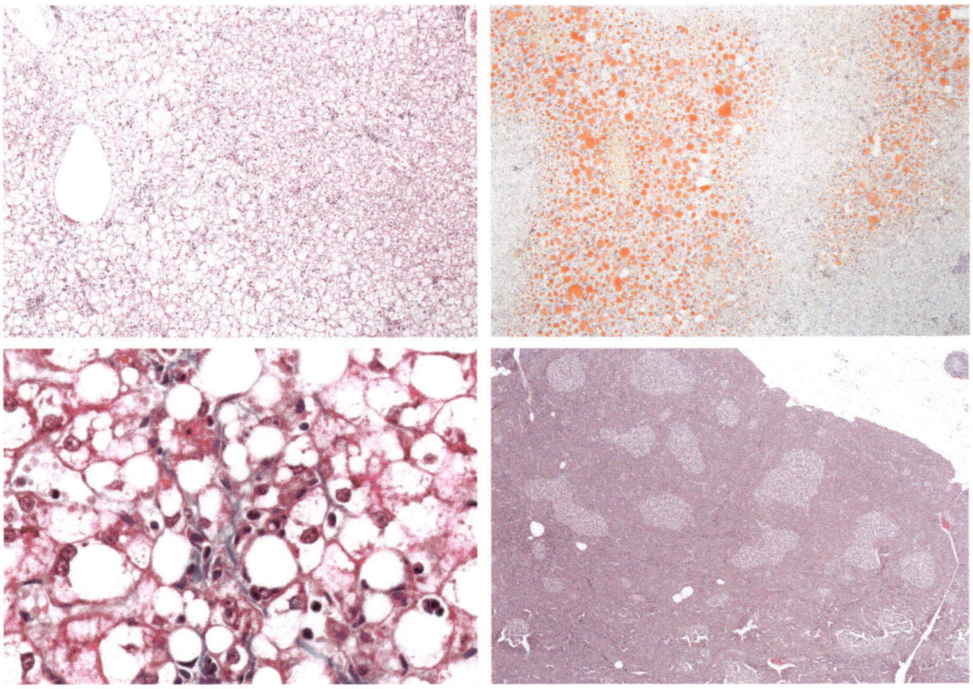

FIGURE 13.6 Marmosets develop fatty liver, which sometimes progresses to nonalcoholic steatohepatitis. Fatty liver is often characterized histologically as macrovesicular steatosis in which hepatocytes contain large, clear, intracytoplasmic fatty inclusions on HE staining (top left). These vacuoles stain with oil red O indicating they are fat (top right). Sometimes, marmosets progress toward nonalcoholic steatohepatitis (NASH), which is characterized by interstitial fibrosis and a ballooning degeneration (Trichrome stain, bottom left). Some of these marmosets also develop diabetes, and histologically animals have large pancreatic islets (bottom right).

TABLE 13.1 Infectious Etiologies of Multifocal Hepatic Necrosis in the Common Marmoset

Etiology	Key Characteristics
HSV, HVT	Intranuclear inclusion bodies and viral syncytial cells; coalescing regions of hepatocellular necrosis
Lymphocytic choriomeningitis virus	Multifocal hepatic necrosis with Councilman bodies and mononuclear cell inflammation. Lesions in spleen, adrenal cortex, and intestines
GB virus	Multifocal nonsuppurative hepatitis. Piecemeal necrosis, lymphocytic nodules, steatosis
Yellow fever	Steatosis, multifocal random or midzonal necrosis, intranuclear viral inclusion bodies in hepatocytes, and Councilman bodies
Yersinia enterocolitica	Mutlifocal hepatic abscesses and necrosis. Associated with mesenteric lymphadenitis and ulcerative enterocolitis
Clostridium piliforme	Multifocal hemorrhagic necrosis of liver, heart, and intestine with bacteria at leading edge of necrosis visible using silver stain as long slender rods
Francisella tularensis	Multifocal necrotizing and granulomatous hepatitis along with enteritis, lymphadenitis, and splenitis
Burkholderia mallei	Multifocal microabscesses and necrosis, intralesional bacilli visible. Also affecting spleen, lung, and nasopharynx with similar histology
Toxoplasmosis	Multifocal necrosis with hemosiderosis with or without PAS + organisms. Also lesions in intestine, heart, lung, kidneys, brain, spleen, and/or lymph nodes

DISEASES OF THE PANCREAS

Parasitic Disease

Trichospirura leptostoma

Etiology: Though far less common than it once was, the nematode *Trichospirura leptostoma* causes chronic pancreatitis [139,140].

Clinical signs: Infected animals have clinical signs including weight loss, muscle wasting, and anemia. Though not reported, one would expect elevated amylase on serum chemistry. Amylase >800 IU/L should raise the level of suspicion for pancreatitis [141].

Pathology: The parasite infects the pancreatic ducts where it is associated with moderate-to-severe ductular fibrosis and blockage of the ducts. Fibrosis expands into exocrine pancreatic tissue.

Diagnosis: Parasite eggs can be identified in feces using formalin sedimentation.

Prevention and control: Treatment of suspected cases is best accomplished with 50 mg/kg fenbendazole for 14 days. Ivermectin at 200–500 μg/kg may also successfully kill the parasite but is less effective [139].

The German cockroach (*Blattella germanica*) and brown-banded cockroach (*Supella longipalpa*) are intermediate hosts. Initial reports of infection in wild-caught marmosets suggested that this parasite did not cause high morbidity, whereas later reports of laboratory samples indicated a higher incidence of disease. One possible explanation for this dichotomy is that some animals experience a rapid infection due to cockroach infestation, which overloads the body's ability to manage the infection. As such, pest control measures should be implemented to prevent infestation and reduce the risk of transmission to marmoset colonies.

Successfully treated animals may show remarkable weight gain and improvement in clinical signs after treatment. The clinical signs together with marked weight gain post treatment have led some to speculate about this parasite's relationship to MWS [88,89]. While it is clear that the parasite causes wasting and characteristic clinical signs that are readily reversible with treatment, it is unlikely that it is the cause of MWS which, as noted above, is associated with chronic inflammation of the intestines and is still present in colonies despite eradication of this agent.

Noninfectious Diseases

Diabetes

Marmosets can develop insulin-resistant diabetes, and this condition is partially dependent on the diet they are fed. Marmosets fed a common, commercially available diet, which has glucose as a main ingredient (32% added glucose), readily develop obesity and prolonged hyperglycemia. Animals fed a high-fat diet also developed hyperglycemia and obesity, though these changes were both less dramatic and delayed by ~6 months [142]. Animals in both groups developed vascular pathology consistent with atherosclerosis and islet hyperplasia consistent with insulin resistance. As noted above, diet is critically important in maintaining healthy marmosets. While palatability is obviously important, the selected diet should be well balanced without unnecessary added sugar.

Animals with insulin resistance or overt diabetes may present with clinical signs including polyuria/polydipsia, anorexia, vomiting, and weakness [143]. Animals may be obese but begin to lose weight as the disease progresses.

Laboratory findings are characteristic of diabetes in other species and include hyperglycemia, hypertriglyceridemia, and glycosuria. Glycosylated hemoglobin (HgbA1c) is also a useful diagnostic tool and is elevated in animals with persistent hyperglycemia. Animals that are euthanized with diabetes will initially show islet hyperplasia at necropsy (Fig. 13.5), but this may progress to islet atrophy as disease progresses.

Dietary management is possible to delay the progression of disease but may prove unrewarding in the long term. Anecdotally, oral hypoglycemic agents such as metformin may be used to treat animals with some success. Insulin therapy could be considered on a case by case basis but is likely not feasible in a research setting due to the intensive care and frequent injections that are required and secondary physiologic changes that would alter the animal's usefulness experimentally.

In addition to developing diabetes naturally, marmosets should also be considered as an animal model for diabetes given the readiness with which they develop insulin resistance and hyperglycemia following a change in diet. Interestingly, marmosets are resistant to diabetes induction with streptozotocin due to altered expression of the GLUT2 glucose transporter in their islets limiting the usefulness of this specific model [144]. However, their islet anatomy and structure are similar to those of humans, making them a potentially useful animal model for an important human disease [145,146].

References

[1] Clapp NK, Henke ML, Lushbaugh CC, Humason GL, Gangaware BL. Effect of various biological factors on spontaneous marmoset and tamarin colitis. A retrospective histopathologic study. Dig Dis Sci 1988;33:1013–9.

[2] David JM, Dick Jr EJ, Hubbard GB. Spontaneous pathology of the common marmoset (*Callithrix jacchus*) and tamarins (*Saguinus oedipus, Saguinus mystax*). J Med Primatol 2009;38:347–59.

[3] Potkay S. Diseases of the Callitrichidae: a review. J Med Primatol 1992;21:189–236.

[4] Sainsbury AW, Kirkwood JK, Appleby EC. Chronic colitis in common marmosets (*Callithrix jacchus*) and cotton-top tamarins (*Saquinus oedipus*). Vet Rec 1987;121:329–30.

[5] Shaw JH, Auskaps AM. Studies on the dentition of the marmoset. Oral Surg Oral Med Oral Pathol 1954;7:671–7.

[6] Tucker MJ. A survey of the pathology of marmosets (*Callithrix jacchus*) under experiment. Lab Anim 1984;18:351–8.

[7] Chivers DJ, Hladik CM. Morphology of the gastrointestinal tract in primates: comparisons with other mammals in relation to diet. J Morphol 1980;166:337–86.

[8] Hogg RT, Ravosa MJ, Ryan TM, Vinyard CJ. The functional morphology of the anterior masticatory apparatus in tree-gouging marmosets (Cebidae, Primates). J Morphol 2011;272: 833–49.

[9] Ferrari SF, Martins ES. Gummivory and gut morphology in two sympatric callitrichids (*Callithrix emiliae* and *Saguinus fuscicollis* weddelli) from western Brazilian Amazonia. Am J Phys Anthropol 1992;88:97–103.

[10] Francisco TM, Couto DR, Zanuncio JC, Serrão JE, Silva IdO, Boere V. Vegetable exudates as food for Callithrix spp. (Callitrichidae): exploratory patterns. PLoS One 2014;9:e112321.

[11] Power ML, Myers EW. Digestion in the common marmoset (*Callithrix jacchus*), a gummivore-frugivore. Am J Primatol 2009;71: 957–63.

[12] Takahashi N, Suda S, Shinki T, Horiuchi N, Shiina Y, Tanioka Y, Koizumi H, Suda T. The mechanism of end-organ resistance to 1 alpha,25-dihydroxycholecalciferol in the common marmoset. Biochem J 1985;227:555–63.

[13] Wagner WM, Kirberger RM. Radiographic anatomy of the thorax and abdomen of the common marmoset (*Callithrix jacchus*). Vet Radiol Ultrasound 2005;46:217–24.

[14] Wagner WM, Kirberger RM. Transcutaneous ultrasonography of the abdomen in the normal common marmoset (*Callithrix jacchus*). Vet Radiol Ultrasound 2005;46:251–8.

[15] Goss AN. A comparison of tooth eruption patterns between two colonies of young marmosets (*Callithrix jacchus*). J Dent Res 1984; 63:44–6.

[16] Johnston GW, Dreizen S, Levy BM. Dental development in the cotton ear marmoset (*Callithrix jacchus*). Am J Phys Anthropol 1970;33:41–8.

[17] Baskerville M. Canine tooth root infection as a cause of facial abscess in the common marmoset (*Callithrix jacchus*). Lab Anim 1984;18:115–8.

[18] Matz-Rensing K, Jentsch KD, Rensing S, Langenhuyzen S, Verschoor E, Niphuis H, Kaup FJ. Fatal Herpes simplex infection in a group of common marmosets (*Callithrix jacchus*). Vet Pathol 2003;40:405–11.

[19] Juan-Salles C, Ramos-Vara JA, Prats N, Sole-Nicolas J, Segales J, Marco AJ. Spontaneous herpes simplex virus infection in common marmosets (*Callithrix jacchus*). J Vet Diagn Invest 1997;9: 341–5.

[20] Desrosiers RC, Falk Jr LA. Herpesvirus tamarinus and its relation to herpes simplex virus. J Gen Virol 1981;56:119–30.

[21] Jones TC, Hunt RD, King NW. Veterinary pathology. 6th ed. Baltimore, MD: Williams and Wilkins; 1997.

[22] King NW, Hunt RD, Daniel MD, Melendez LV. Overt herpes-T infection in squirrel monkeys (*Saimiri sciureus*). Lab Anim Care 1967;17:413–23.

[23] Chalmers DT, Murgatroyd LB, Wadsworth PF. A survey of the pathology of marmosets (*Callithrix jacchus*) derived from a marmoset breeding unit. Lab Anim 1983;17:270–9.

[24] Hill AC, Turton JA, Bleby J. Bacterial and mycoplasma flora of a laboratory colony of the common marmoset (*Callithrix jacchus*). Vet Rec 1978;103:284–7.

[25] Betton GR. Spontaneous neoplasms of the marmoset (*Callithrix jacchus*). Oral and nasopharyngeal squamous cell carcinomas. Vet Pathol 1984;21:193–7.

[26] Dreizen S, Levy BM, Bernick S. Studies on the biology of the periodontium of marmosets. VII. The effect of vitamin C deficiency on the marmoset periodontium. J Periodontal Res 1969; 4:274–80.

[27] Flurer CI, Kern M, Rambeck WA, Zucker H. Ascorbic acid requirement and assessment of ascorbate status in the common marmoset (*Callithrix jacchus*). Ann Nutr Metab 1987; 31:245–52.

[28] Line AS, Eisele PH, Markovits JE. Eruption gingivitis associated with scorbutism in macaques. Lab Anim Sci 1992;42:96–7.

[29] Ratterree MS, Didier PJ, Blanchard JL, Clarke MR, Schaeffer D. Vitamin C deficiency in captive nonhuman primates fed commercial primate diet. Lab Anim Sci 1990;40:165–8.

[30] Dreizen S, Levy BM, Bernick S. Studies on the biology of the periodontium of marmosets. XI. Histopathologic manifestations of spontaneous and induced vitamin A deficiency in the oral structures of adult marmosets. J Dent Res 1973;52:803–9.

[31] Dreizen S, Levy BM, Bernick S. Studies on the biology of the periodontium of marmosets: 8. The effect of folic acid deficiency on the marmoset oral mucosa. J Dent Res 1970;49:616–20.

[32] Dreizen S, Levy BM, Bernick S. Studies on the biology of the periodontium of marmosets. XIII. Histopathology of niacin deficiency stomatitis in the marmoset. J Periodontol 1977;48:452–5.

[33] Personal communication: Mejia A. Cleft palate in marmosets. 2017.

[34] Kraus BS, Garrett WS. Cleft palate in a marmoset: report of a case. Cleft Palate J 1968;5:340–3.

[35] Pugsley SL. Congenital malformations in a common marmoset (*Callithrix jacchus*) similar to human 13-trisomy syndrome. Lab Anim 1985;19:123–4.

[36] Kramer JA, Hachey AM, Wachtman LM, Mansfield KG. Treatment of giardiasis in common marmosets (*Callithrix jacchus*) with tinidazole. Comp Med 2009b;59:174–9.

[37] Cooper JE, Needham JR. An outbreak of shigellosis in laboratory marmosets and tamarins (Family: Callithricidae). J Hyg 1976;76: 415–24.

[38] Durso LM, Bono JL, Keen JE. Molecular serotyping of *Escherichia coli* O26:H11. Appl Environ Microbiol 2005;71:4941–4.

[39] Mansfield KG, Lin KC, Xia D, Newman JV, Schauer DB, MacKey J, Lackner AA, Carville A. Enteropathogenic *Escherichia coli* and ulcerative colitis in cotton-top tamarins (*Saguinus oedipus*). J Infect Dis 2001;184:803–7.

[40] Thomson JA, Scheffler JJ. Hemorrhagic typhlocolitis associated with attaching and effacing *Escherichia coli* in common marmosets. Lab Anim Sci 1996;46:275–9.

[41] Ludlage E, Mansfield K. Clinical care and diseases of the common marmoset (*Callithrix jacchus*). Comp Med 2003;53:369–82.

[42] Pisharath HR, Cooper TK, Brice AK, Cianciolo RE, Pistorio AL, Wachtman LM, Mankowski JL, Newcomer CE. Septicemia and peritonitis in a colony of common marmosets (*Callithrix jacchus*) secondary to *Klebsiella pneumoniae* infection. Contemp Top Lab Anim Sci 2005;44:35–7.

[43] Gozaol A, Montoyo E. *Klebsiella pneumoniae* infection in a New World nonhuman primate center. Lab Primate Newsl 1991;30:13–20.

[44] Dekker JP, Frank KM. Salmonella, Shigella, and yersinia. Clin Lab Med 2015;35:225–46.

[45] de Mello MF, Monteiro AB, Fonseca EC, Pissinatti A, Ferreira AM. Identification of Helicobacter sp. in gastric mucosa from captive marmosets (Callithrix sp.; callitrichidae, primates). Am J Primatol 2005;66:111–8.

[46] Shen Z, Feng Y, Sheh A, Everitt J, Bertram F, Paster BJ, Fox JG. Isolation and characterization of a novel Helicobacter species, Helicobacter jaachi sp. nov., from common marmosets (Callithrix jaachus). J Med Microbiol 2015;64:1063–73.

[47] Won YS, Vandamme P, Yoon JH, Park YH, Hyun BH, Kim HC, Itoh T, Tanioka Y, Choi YK. Helicobacter callitrichis sp. nov., a novel Helicobacter species isolated from the feces of the common marmoset (*Callithrix jacchus*). FEMS Microbiol Lett 2007;271: 239–44.

[48] Stein FJ, Lewis DH, Stott GG, Sis RF. Acute gastric dilatation in common marmosets (*Callithrix jacchus*). Lab Anim Sci 1981;31: 522–3.

[49] Yasuda M, Inoue T, Ueno M, Morita H, Hayashimoto N, Kawai K, Itoh T. A case of nontraumatic gas gangrene in a common marmoset (*Callithrix jacchus*). J Vet Med Sci 2016;77:1673–6.

[50] Keel MK, Songer JG. The comparative pathology of Clostridium difficile-associated disease. Vet Pathol 2006;43:225–40.

[51] Rolland RM, Chalifoux LV, Snook SS, Ausman LM, Johnson LD. Five spontaneous deaths associated with *Clostridium difficile* in a colony of cotton-top tamarins (*Saguinus oedipus*). Lab Anim Sci 1997;47:472–6.

[52] Togerson R, Chalifoux L, King NW, Ringler D, Snook SS. Antibiotic associated pseudomembranous enterocolitis in new world primates (abstract). Vet Pathol 1992;29:454.

[53] Fredriksson-Ahomaa M, Naglic T, Turk N, Seol B, Grabarevic Z, Bata I, Perkovic D, Stolle A. Yersiniosis in zoo marmosets (Callithrix jacchus) caused by Yersinia enterocolitica 4/O:3. Vet Microbiol 2007;121:363–7.

[54] Poelma FG, Borst GH, Zwart P. Yersinia enterocolitica infections in non-human primates. Acta Zool Pathol Antverp 1977:3–9.

[55] Wachtman LM, Miller AD, Xia D, Curran EH, Mansfield KG. Colonization with nontuberculous mycobacteria is associated with positive tuberculin skin test reactions in the common marmoset (*Callithrix jacchus*). Comp Med 2011b;61:278–84.

[56] Kalishman J, Paul-Murphy J, Scheffler J, Thomson JA. Survey of Cryptosporidium and Giardia spp. in a captive population of common marmosets. Lab Anim Sci 1996;46:116–9.

[57] Duszynski DW, Wilson WD, Upton SJ, Levine ND. Coccidia (Apicomplexa: eimeriidae) in the primates and the scandentia. Int J Primatol 1999;20:761–97.

[58] Lindsay DS, Dubey JP, Blagburn BL. Biology of Isospora spp. from humans, nonhuman primates, and domestic animals. Clin Microbiol Rev 1997;10:19–34.

[59] Olcott A, Speer C, Hendricks L. Endogenous development of Isospora arctopitheci Rodham, 1933 in the marmoset Saguinus geoffroy. Proc Helminthol Soc Wash 1982;49:118–26.

[60] Chalifoux LV, Bronson RT, Escajadillo A, McKenna S. An analysis of the association of gastroenteric lesions with chronic wasting syndrome of marmosets. Vet Pathol Suppl 1982;19(Suppl. 7):141–62.

[61] Hunt RD, Anderson MP, Chalifoux LV. Spontaneous infectious diseases of marmosets. Primates Med 1978;10:239–53.

[62] Hahn NE, Capuano 3rd SV. Successful treatment of cryptosporidiosis in 2 common marmosets (*Callithrix jacchus*) by using paromomycin. J Am Assoc Lab Anim Sci 2010;49:873–5.

[63] Levy BM, Mirkovic RR. An epizootic of measles in a marmoset colony. Lab Anim Sci 1971;21:33–9.

[64] Albrecht P, Lorenz D, Klutch MJ, Vickers JH, Ennis FA. Fatal measles infection in marmosets pathogenesis and prophylaxis. Infect Immun 1980;27:969–78.

[65] Fraser CE, Chalifoux L, Sehgal P, Hunt RD, King NW. A paramyxovirus causing fatal gastroenterocolitis in marmoset monkeys. Primates Med 1978;10:261–70.

[66] Ablashi DV, Pearson G, Rabin H, Armstrong G, Easton J, Valerio M, Cicmanec J. Experimental infection of Callithrix Jacchus marmosets with herpesvirus ateles, herpesvirus saimiri, and Epstein Barr virus. Biomedicine 1978;29:7–10.

[67] Laufs R, Steinke H, Gisela S, Petzold D. Latent infection and malignant lymphoma in marmosets (*Callithrix jacchus*) after infection with two oncogenic Herpesviruses from primates. J Natl Cancer Inst 1974;53:195–9.

[68] Melendez LV, Hunt RD, Daniel MD, Fraser CE, Barahona HH, King NW, Garcia FG. Herpesviruses saimiri and ateles—their role in malignant lymphomas of monkeys. Fed Proc 1972;31: 1643–50.

[69] Wolfe LG, Deinhardt F. Overview of viral oncology studies in Saguinus and Callithrix species. Primates Med 1978;10:96–118.

[70] Wright J, Falk LA, Wolfe LG, Ogden J, Deinhardt F. Susceptibility of common marmosets (*Callithrix jacchus*) to oncogenic and attenuated strains of Herpesvirus saimiri. J Natl Cancer Inst 1977a;59: 1475–8.

[71] Wright J, Falk LA, Wolfe LG, Ogden J, Deinhardt F. Susceptibility of common marmosets (*Callithrix jacchus*) to oncogenic and attenuated strains of herpesvirus saimiri. J Natl Cancer Inst 1977b;59: 1469–73.

[72] Wright J, Falk LA, Wolfe LG, Deinhardt F. Herpesvirus saimiri: protective effect of attenuated strain against lymphoma induction. Int J Canc 1980;26:477–82.

[73] Cox C, Chang S, Karran L, Griffin B, Wedderburn N. Persistent Epstein-Barr virus infection in the common marmoset (*Callithrix jacchus*). J Gen Virol 1996;77(Pt 6):1173–80.

[74] Felton SC, Hoffmann CC, Kreier JP, Glaser R. Hematologic and immunologic responses in common marmosets (*Callithrix jacchus*) infected with Plasmodium knowlesi and Epstein-Barr virus. Lab Anim Sci 1984;34:164–8.

[75] Cho Y, Ramer J, Rivailler P, Quink C, Garber RL, Beier DR, Wang F. An Epstein-Barr-related herpesvirus from marmoset lymphomas. Proc Natl Acad Sci USA 2001;98:1224–9.

[76] Ramer JC, Garber RL, Steele KE, Boyson JF, O'Rourke C, Thomson JA. Fatal lymphoproliferative disease associated with a novel gammaherpesvirus in a captive population of common marmosets. Comp Med 2000;50:59–68.

[77] Rivailler P, Cho YG, Wang F. Complete genomic sequence of an Epstein-Barr virus-related herpesvirus naturally infecting a new world primate: a defining point in the evolution of oncogenic lymphocryptoviruses. J Virol 2002;76:12055–68.

[78] Greenough TC, Carville A, Coderre J, Somasundaran M, Sullivan JL, Luzuriaga K, Mansfield K. Pneumonitis and multiorgan system disease in common marmosets (*Callithrix jacchus*) infected with the severe acute respiratory syndrome-associated coronavirus. Am J Pathol 2005;167:455–63.

[79] Davis NC. The susceptibility of marmosets to yellow fever virus. J Exp Med 1930;52:405–16.

[80] Noguchi H. Etiology of yellow fever: III. Symptomatology and pathological findings in animals experimentally infected. J Exp Med 1919;29:585–96.

[81] Brack M. Gastrointestinal tumors observed in nonhuman primates at the German primate center. J Med Primatol 1998;27: 319–24.

[82] Chalifoux LV. Adenocarcinoma of the small intestine in a common marmoset (*Callithrix jacchus*). Lab Anim Sci 1990;40:210–2.

[83] Miller AD, Kramer JA, Lin KC, Knight H, Martinot A, Mansfield KG. Small intestinal adenocarcinoma in common marmosets (*Callithrix jacchus*). Vet Pathol 2010;47:969–76.

[84] Yokouchi Y, Imaoka M, Sayama A, Jindo T, Sanbuissho A. Inflammatory fibroid polyp in the duodenum of a common marmoset (*Callithrix jacchus*). Toxicol Pathol 2013;41:80–5.

[85] Logan AC, Khan KN. Clinical pathologic changes in two marmosets with wasting syndrome. Toxicol Pathol 1996;24:707–9.

[86] Okazaki Y, Kurata Y, Makinodan F, Kidachi F, Yokoyama M, Wako Y, Yamagishi Y, Katsuta O, Takechi M, Tsuchitani M. Spontaneous lesions detected in the common cotton-eared marmosets (*Callithrix jacchus*). J Vet Med Sci 1996;58:181–90.

[87] Ross CN, Davis K, Dobek G, Tardif SD. Aging phenotypes of common marmosets (*Callithrix jacchus*). J Aging Res 2012;2012, 567143.

[88] Beglinger R, Illgen B, Pfister R, Heider K. The parasite Trichospirura leptostoma associated with wasting disease in a colony of common marmosets, *Callithrix jacchus*. Folia Primatol 1988;51:45–51.

[89] Pfister R, Heider K, Illgen B, Beglinger R. Trichospirura leptostoma: a possible cause of wasting disease in the marmoset. Z Versuchstierkd 1990;33:157–61.

[90] Sousa MB, Leao AC, Coutinho JF, de Oliveira Ramos AM. Histopathology findings in common marmosets (*Callithrix jacchus* Linnaeus, 1758) with chronic weight loss associated with bile tract obstruction by infestation with Platynosomum (Loos, 1907). Primates 2008;49:283–7.

[91] Kuehnel F, Mietsch M, Buettner T, Vervuert I, Ababneh R, Einspanier A. The influence of gluten on clinical and immunological status of common marmosets (*Callithrix jacchus*). J Med Primatol 2013;42:300–9.

[92] Brack M, Rothe H. Chronic tubulointerstitial nephritis and wasting disease in marmosets (*Callithrix jacchus*). Vet Pathol 1981;18:45–54.

[93] Shimwell M, Warrington BF, Fowler JS. Dietary habits relating to 'wasting marmoset syndrome' (WMS). Lab Anim 1979;13: 139–42.

[94] Gore MA, Brandes F, Kaup FJ, Lenzner R, Mothes T, Osman AA. Callitrichid nutrition and food sensitivity. J Med Primatol 2001; 30:179–84.

[95] Schroeder C, Osman AA, Roggenbuck D, Mothes T. IgA-gliadin antibodies, IgA-containing circulating immune complexes, and IgA glomerular deposits in wasting marmoset syndrome. Nephrol Dial Transplant 1999;14:1875–80.

[96] Ialeggio D, Baker A. Results of a preliminary survey into wasting marmoset syndrome in callitrichid collections. Toronto, Ontario, Canada. 1995. p. 148–58.

[97] Gutteridge JM, Taffs LF, Hawkey CM, Rice-Evans C. Susceptibility of tamarin (*Saguinus labiatus*) red blood cell membrane lipids to oxidative stress: implications for wasting marmoset syndrome. Lab Anim 1986;20:140–7.

[98] Wood J, Peck O, Sharma H, Mekhjian H, Stone D, Stonerook M, Weiss H, Hernandez J, Rodriguez J, MA R. Captivity promote colitis in the cotton-top tamarin (*Saguinus oedipus*). Gastroenterology 1990;98:A180.

[99] Tardif SD, Mansfield KG, Ratnam R, Ross CN, Ziegler TE. The marmoset as a model of aging and age-related diseases. ILAR J 2011;52:54–65.

[100] Nakashima E, Okano Y, Niimi K, Takahashi E. Detection of calprotectin and apoptotic activity in the colon of marmosets with chronic diarrhea. J Vet Med Sci 2013;75:1633–6.

[101] Yoshimoto T, Niimi K, Takahashi E. Serum matrix metalloproteinase 9 (MMP9) as a biochemical marker for wasting marmoset syndrome. J Vet Med Sci 2016;78:837–43.

[102] Baxter VK, Shaw GC, Sotuyo NP, Carlson CS, Olson EJ, Zink MC, Mankowski JL, Adams RJ, Hutchinson EK, Metcalf Pate KA. Serum albumin and body weight as biomarkers for the antemortem identification of bone and gastrointestinal disease in the common marmoset. PLoS One 2013;8:e82747.

[103] Otovic P, Smith S, Hutchinson E. The use of glucocorticoids in marmoset wasting syndrome. J Med Primatol 2015;44:53–9.

[104] Ludlage E, Murphy CL, Davern SM, Solomon A, Weiss DT, Glenn-Smith D, Dworkin S, Mansfield KG. Systemic AA amyloidosis in the common marmoset. Vet Pathol 2005;42: 117–24.

[105] Sasseville VG, Simon MA, Chalifoux LV, Lin KC, Mansfield KG. Naturally occurring Tyzzer's disease in cotton-top tamarins (*Saguinus oedipus*). Comp Med 2007;57:125–7.

[106] Yoshida K, Nibe K, Nakamura T, Takahashi T, Komatsu M, Ogawa H, Shirota K, Chambers JK, Nakayama H, Uchida K. Spontaneous Tyzzer's disease with the central nerve involvement in a newborn common marmoset. J Vet Med Sci 2013;75: 1119–21.

[107] Nelson M, Lever MS, Dean RE, Savage VL, Salguero FJ, Pearce PC, Stevens DJ, Simpson AJ. Characterization of lethal inhalational infection with Francisella tularensis in the common marmoset (*Callithrix jacchus*). J Med Microbiol 2010;59:1107–13.

[108] Nelson M, Lever MS, Savage VL, Salguero FJ, Pearce PC, Stevens DJ, Simpson AJ. Establishment of lethal inhalational infection with Francisella tularensis (tularaemia) in the common marmoset (*Callithrix jacchus*). Int J Exp Pathol 2009;90:109–18.

[109] Splettstoesser WD, Matz-Rensing K, Seibold E, Tomaso H, Al Dahouk S, Grunow R, Essbauer S, Buckendahl A, Finke EJ, Neubauer H. Re-emergence of Francisella tularensis in Germany: fatal tularaemia in a colony of semi-free-living marmosets (*Callithrix jacchus*). Epidemiol Infect 2007;135:1256–65.

[110] Baitchman EJ, Calle PP, James SB, Linn MJ, Raphael BL. Leptospirosis in Wied's marmosets (Callithrix kuhlii). J Zoo Wildl Med 2006;37:182–5.

[111] Jelesijevic T, Zimmerman SM, Harvey SB, Mead DG, Shaffer TL, Estes DM, Michel F, Quinn FD, Hogan RJ, Lafontaine ER. Use of the common marmoset to study Burkholderia mallei infection. PLoS One 2015;10:e0124181.

[112] Kingston N, Cosgrove G. Two new species of Platynosomum (Trematoda: Dicrocoeliidae) from South American monkeys. Proc Helm Soc Wash 1967;34:147–51.

[113] Melo A, Pereira L. Sobre o parasitismo por Primasubulura jacchi em *Callithrix penicillata* (Primates, Callitrichidae). In: Melo MT, editor. A primatologia no Brasil. 2nd ed. 1986. p. 483–8.

[114] Benirschke K, Richart R. Spontaneous acute toxoplasmosis in a marmoset monkey. Am J Trop Med Hyg 1960;9:269–73.

[115] De Rodaniche E. Susceptibility of the marmoset, Marikina geoffroyi, and the night monkey, Aotus zonalis, to experimental infection with Toxoplasma. Am J Trop Med Hyg 1954;3:1026–32.

[116] Epiphanio S, Sinhorini IL, Catao-Dias JL. Pathology of toxoplasmosis in captive new world primates. J Comp Pathol 2003;129: 196–204.

[117] Molina CV, Catao-Dias JL, Ferreira Neto JS, Vasconcellos SA, Gennari SM, do Valle Rdel R, de Souza GO, de Morais ZM, Vitaliano SN, Strefezzi Rde F, Bueno MG. Sero-epidemiological survey for brucellosis, leptospirosis, and toxoplasmosis in free-ranging *Alouatta caraya* and *Callithrix penicillata* from Sao Paulo State. Brazil J Med Primatol 2014;43:197–201.

[118] Montali RJ, Connolly BM, Armstrong DL, Scanga CA, Holmes KV. Pathology and immunohistochemistry of callitrichid hepatitis, an emerging disease of captive New World primates caused by lymphocytic choriomeningitis virus. Am J Pathol 1995;147:1441–9.

[119] Montali RJ, Scanga CA, Pernikoff D, Wessner DR, Ward R, Holmes KV. A common-source outbreak of callitrichid hepatitis in captive tamarins and marmosets. J Infect Dis 1993;167:946–50.

[120] Ramsay EC, Montali RJ, Worley M, Stephensen CB, Holmes KV. Callitrichid hepatitis: epizootiology of a fatal hepatitis in zoo tamarins and marmosets. J Zoo Wildl Med 1989;20:178–83.

[121] Stephensen CB, Jacob JR, Montali RJ, Holmes KV, Muchmore E, Compans RW, Arms ED, Buchmeier MJ, Lanford RE. Isolation of an arenavirus from a marmoset with callitrichid hepatitis and its serologic association with disease. J Virol 1991;65:3995–4000.

[122] Asper M, Hofmann P, Osmann C, Funk J, Metzger C, Bruns M, Kaup FJ, Schmitz H, Gunther S. First outbreak of callitrichid hepatitis in Germany: genetic characterization of the causative lymphocytic choriomeningitis virus strains. Virology 2001;284: 203–13.

[123] Bright H, Carroll AR, Watts PA, Fenton RJ. Development of a GB virus B marmoset model and its validation with a novel series of hepatitis C virus NS3 protease inhibitors. J Virol 2004;78: 2062–71.

[124] Jacob JR, Lin KC, Tennant BC, Mansfield KG. GB virus B infection of the common marmoset (*Callithrix jacchus*) and associated liver pathology. J Gen Virol 2004;85:2525–33.

[125] Manickam C, Rajakumar P, Wachtman L, Kramer JA, Martinot AJ, Varner V, Giavedoni LD, Reeves RK. Acute liver damage associated with innate immune activation in a small nonhuman primate model of Hepacivirus infection. J Virol 2016;90:9153–62.

[126] Manickam C, Wachtman L, Martinot AJ, Giavedoni LD, Reeves RK. Metabolic dysregulation in Hepacivirus infection of common marmosets (*Callithrix jacchus*). PLoS One 2017;12: e0170240.

[127] Galindo P, Derodaniche E. Surveillance for sylvan yellow fever activity in Panama (1957–1961). Am J Trop Med Hyg 1964;13: 844–50.

[128] Laemmert HJ. Susceptibility of marmosets to different strains of yellow fever virus. Am J Trop Med Hyg 1944;s1–24:71–81.

[129] Laemmert HJ, de Castro Ferreira L. The isolation of yellow fever virus from wild-caught marmosets. Am J Trop Med Hyg 1945; s1–25:231–2.

[130] Carrion Jr R, Brasky K, Mansfield K, Johnson C, Gonzales M, Ticer A, Lukashevich I, Tardif S, Patterson J. Lassa virus infection in experimentally infected marmosets: liver pathology and immunophenotypic alterations in target tissues. J Virol 2007;81: 6482–90.

[131] Frigerio MJ, Rondinone SN, Callelo MA, Paradisi ER, Weissenbacher MC. Junin virus infection of Calithrix jacchus: haematological findings. Acta Virol 1982;26:270–8.

[132] Weissenbacher MC, Calello MA, Colillas OJ, Rondinone SN, Frigerio MJ. Argentine hemorrhagic fever: a primate model. Intervirology 1979;11:363–5.

[133] Gonzalez PH, Laguens RP, Frigerio MJ, Calello MA, Weissenbacher MC. Junin virus infection of *Callithrix jacchus*: pathologic features. Am J Trop Med Hyg 1983;32:417–23.

[134] Carrion Jr R, Ro Y, Hoosien K, Ticer A, Brasky K, de la Garza M, Mansfield K, Patterson JL. A small nonhuman primate model for filovirus-induced disease. Virology 2011;420:117–24.

[135] Smither SJ, Nelson M, Eastaugh L, Laws TR, Taylor C, Smith SA, Salguero FJ, Lever MS. Experimental respiratory Marburg virus haemorrhagic fever infection in the common marmoset (*Callithrix jacchus*). Int J Exp Pathol 2013;94:156–68.

[136] Smither SJ, Nelson M, Eastaugh L, Nunez A, Salguero FJ, Lever MS. Experimental respiratory infection of marmosets (*Callithrix jacchus*) with Ebola virus Kikwit. J Infect Dis 2015; 212(Suppl. 2):S336–45.

[137] Ebert JW, Maynard JE, Bradley DW, Lorenz D, Krushak DH. Experimental infection of marmosets with hepatitis A virus. Primates Med 1978;10:295–9.

[138] Kramer JA, Grindley J, Crowell AM, Makaron L, Kohli R, Kirby M, Mansfield KG, Wachtman LM. The common marmoset as a model for the study of nonalcoholic fatty liver disease and nonalcoholic steatohepatitis. Vet Pathol 2015;52:404–13.

[139] Hawkins JV, Clapp NK, Carson RL, Henke MA, McCracken MD, Faulkner CT, Patton S. Diagnosis and treatment of Trichospirura leptostoma infection in common marmosets (*Callithrix jacchus*). Contemp Top Lab Anim Sci 1997;36:52–5.

[140] Smith WN, Chitwood MB. Trichospirura leptostoma gen. et sp. n. (Nematoda: Thelazioidea) from the pancreatic ducts of the white-eared marmoset *Callithrix jacchus*. J Parasitol 1967;53: 1270–2.

[141] Gatesman TJ. Serum amylase values in callitrichids. Lab Anim Sci 1992;42:46–50.

[142] Wachtman LM, Kramer JA, Miller AD, Hachey AM, Curran EH, Mansfield KG. Differential contribution of dietary fat and monosaccharide to metabolic syndrome in the common marmoset (*Callithrix jacchus*). Obesity 2011a;19:1145–56.

[143] Juan-Salles C, Marco A, Ramos-Vara JA, Resendes A, Verges J, Valls X, Montesinos A. Islet hyperplasia in callitrichids. Primates 2002;43:179–90.

[144] Kramer J, Moeller EL, Hachey A, Mansfield KG, Wachtman LM. Differential expression of GLUT2 in pancreatic islets and kidneys of New and Old World nonhuman primates. Am J Physiol Regul Integr Comp Physiol 2009a;296:R786–93.

[145] Mohanasundaram D, Drogemuller C, Brealey J, Jessup CF, Milner C, Murgia C, Lang CJ, Milton A, Zalewski PD, Russ GR, Coates PT. Ultrastructural analysis, zinc transporters, glucose transporters and hormones expression in New world primate (*Callithrix jacchus*) and human pancreatic islets. Gen Comp Endocrinol 2011;174:71–9.

[146] Plentz RR, Palagani V, Wiedemann A, Diekmann U, Glage S, Naujok O, Jorns A, Muller T. Islet microarchitecture and glucose transporter expression of the pancreas of the marmoset monkey display similarities to the human. Islets 2012;4:123–9.

CHAPTER

14

Bone, Muscle, and Skeletal Disease

Victoria K. Baxter[1], Gillian C. Shaw[2], Caroline M. Garrett[3], Kelly A. Metcalf Pate[4]

[1]Department of Pathology and Laboratory Medicine, University of North Carolina, Chapel Hill, NC, United States; [2]School of Veterinary Medicine, University of Wisconsin, Madison, WI, United States; [3]PolarityTE Inc., Salt Lake City, UT, United States; [4]Department of Molecular and Comparative Pathobiology, Johns Hopkins University School of Medicine, Baltimore, MD, United States

THE NORMAL MUSCULOSKELETAL SYSTEM

Anatomy and Histology of Normal Bone

Bone is composed of a mineralized matrix of collagenous connective tissue (osteoid), noncollagenous proteins known as ground substance, and a number of different cell types. Collagen I and collagen V make up the majority of the osteoid and account for most of the weight of bone. The mineral is made up of hydroxyapatite crystal containing calcium and phosphorus. Osteoprogenitor cells give rise to osteoblasts, which secrete the extracellular matrix (osteoid) of bone. Once osteoblasts are surrounded by matrix, they are called osteocytes and reside within channels known as lacunae. Neighboring lacunae are connected via canaliculi through which processes of osteocytes grow and make contact with each other, allowing cell-to-cell contact and communication. Whereas osteoblasts create bone, osteoclasts, which are multinucleated phagocytic cells derived from hematopoietic cells of the bone marrow, break down bone, and play an important role in releasing calcium from mature bone to maintain blood calcium levels. In releasing calcium and causing bone resorption, osteoclasts create scalloped cavities in bone surfaces called Howship's (or erosion/resorption) lacunae. Other cells of importance are those lining the inner and outer structural surfaces of the bone, named periosteal and endosteal cells, respectively, which are derived from osteoblasts.

Immature bone is known as woven bone and lacks the characteristic lamellar organization of mature bone, known as lamellar bone. It is more highly cellular than mature bone, and the component collagen fibers are arranged haphazardly, a feature that is best visualized using polarized light. Woven bone contains a higher content of ground substance than mature bone, imparting a more basophilic appearance when stained with hematoxylin and eosin stains. Two types of mature/lamellar bone are compact (dense) and cancellous (spongy). Compact bone makes up the cortex and is composed of columnar subunits called osteons or Haversian systems. These are cylinders composed of concentric layers of lamellae that are oriented parallel to the longitudinal axis of the bone and contain centrally located vessels and nerves. Cancellous bone is composed of anastomosing plates or rods (trabeculae) and is found within the medullary canal.

Most bones are composed of a cortex, medullary canal (houses bony trabeculae with bone marrow), and articular surfaces. The major subdivisions of a long bone are the epiphysis (proximal and distal ends of long bones; usually covered by articular cartilage), the metaphysis (short section between the epiphysis and diaphysis), and the diaphysis (central shaft of long bone). Most bones elongate by a highly ordered process called endochondral ossification, which occurs in the growth plate/physis (plate of epiphyseal cartilage located between the epiphysis and metaphysis in growing individuals) and involves the proliferation,

The Common Marmoset in Captivity and Biomedical Research
https://doi.org/10.1016/B978-0-12-811829-0.00014-5

hypertrophy, and mineralization of cartilage that eventually ossifies and is remodeled into outer compact and medullary trabecular bone. The growth plate has several layers in which the matrix, cell morphology, and arrangement differ.

Development of the Musculoskeletal System

The vertebrate musculoskeletal system develops entirely from the mesodermal layer of the developing embryo and includes bone, muscle, tendons, ligaments, articular cartilage, joint capsules, and synovium. The development of the marmoset's musculoskeletal system is similar to that of other primates and begins in utero and continues after birth. Specifically, final maturation of the skeleton, evidenced by closure of growth plates, typically occurs around sexual maturity, but there is great variation in which growth plates close at each chronological age within a given species. The exact age of growth plate closure in the major long bones has been studied in several species of callitrichids including *Callithrix jacchus*, *Saguinus fuscicollis*, and *Saguinus oedipus* [1]. Growth plate closure times for *Callithrix jacchus* range from the earliest closure of the distal humerus (0.55 years) to the proximal humerus and distal ulna (1.47 years) [1]. Specific anatomic features of the common marmoset skeleton have been reported in detail [2].

Bone Physiology

Bone constantly undergoes remodeling throughout a vertebrate organism's life to maintain structural integrity and respond to the body's needs for calcium and phosphorus, which are primarily stored in bone [3]. Both bone formation and resorption are tightly controlled, with calcium-regulating hormones promoting metabolic functions, systemic hormones controlling skeletal growth, and cytokines and growth factors governing bone remodeling at the local level. Two calcium-regulating hormones, parathyroid hormone (PTH) and 1,25-dihydroxyvitamin D [$1,25(OH)_2D$], primarily regulate the metabolic activity of bone. PTH predominantly regulates serum calcium levels and is involved in both bone formation and resorption; $1,25(OH)_2D$ is covered in more depth later in this chapter. A third hormone, calcitonin, does not play a major role in calcium homeostasis but is a powerful inhibitor of bone resorption. Other systemic hormones involved in bone remodeling include growth hormone, glucocorticoids, and estrogen. Local factors that affect bone remodeling include cytokines that promote bone loss (IL-1, TNF-α, osteoclast differentiation factor), cytokines that inhibit bone resorption (IL-4, IFN-γ, osteoprotegerin), colony-stimulating factors, prostaglandins, leukotrienes, nitric oxide, and growth factors (insulin-like growth factor, TGF-β). Balancing the interplay between the metabolic obligations of the skeleton and its need to properly respond to mechanical forces is a constant and continuous challenge, with imbalance resulting in pathophysiologic changes that can cause clinical disease. Additional information on the biology and physiology of normal bone can be found in multiple excellent reviews [4–6].

Chimerism in Bone Marrow

New World nonhuman primates in the Callitrichidae family (marmosets and tamarins) are uniquely capable of producing dizygotic twins that exhibit chimerism in the blood and other cells of the hematopoietic lineage (bone marrow, spleen, and thymus). Chorionic fusion of the placentas, occurring from gestational day 19–29, leads to a single chorion with anastomoses connecting the embryos. This fusion process is associated with a delay in embryonic development that allows for the exchange of embryonic stem cells via blood flow between the twins [154,155]. The result of this phenomenon is genetically chimeric infants with tissues derived from embryonic cell lineages from self as well as from their sibling.

Chimerism outside of the hematopoietic system has also been investigated using high-throughput microsatellite markers in addition to traditional karyotyping methods [156]. While both techniques are capable of successfully detecting chimeric tissue, neither is able to quantitate the level of chimerism within an individual or various tissue types [157]. Most recently, however, quantitative PCR (qPCR) has been utilized to compare and quantify levels of chimerism between individuals as well as between tissue types within an individual animal. Results of qPCR-based studies have determined that although chimerism can be detected across a variety of tissue types, levels are significantly higher in tissues derived primarily from the hematopoietic lineage, such as the bone marrow. Furthermore, while nonquantitative methods found levels of chimerism to correlate within individuals, tissue variation along germ layer lines was not identified, suggesting ultimately that events occurring early in development were not responsible for the chimerism detected in the adult tissues analyzed. The primary hypothesis for detection of chimerism within nonhematopoietic tissues is infiltration by blood or lymphocytes [157].

An understanding of hematopoietic chimerism is imperative to the interpretation of their natural biology and provides an understanding of limitations for the marmoset in biomedical research. In particular, studies that involve genotyping and immunology must consider the source of genomic DNA and potential confounders posed by chimeric blood and lymphocytes [157].

Vitamin D Metabolism

Vitamin D_3 (cholecalciferol) can be produced through exposure to ultraviolet light or is introduced through dietary acquisition (Fig. 14.1) [7,8]. 7-Dehydrocholesterol (7-DHC), found in the skin, is converted into vitamin D_3 by ultraviolet B (UVB) radiation through a two-step nonenzymatic process that is dependent on both UVB intensity and skin pigmentation. Vitamin D_3 acquisition through diet generally comes from vitamin D_2 (ergocalciferol), which is produced by UVB irradiation of ergosterol in plants and fungi. However, marmosets utilize vitamin D_2 very poorly, necessitating that they receive high amounts of dietary vitamin D_3, particularly in captive situations where exposure to UVB is minimal [9,10]. Following consumption, vitamin D_3 is carried in the blood by vitamin D-binding protein (DBP) to the liver. It is then hydroxylated by cytochrome P450 enzymes and converted into 25-hydroxyvitamin D_3 [25(OH)D, calcifediol]. DBP then transports 25(OH)D to the kidney, where it is once again hydroxylated in the proximal renal tubules and converted into 1,25-dihydroxyvitamin D_3 [1,25 $(OH)_2D$, calcitriol]. Hydroxylation of 25(OH)D to 1,25$(OH)_2D$ is tightly regulated by PTH, fibroblast growth factor 23, and 1,25$(OH)_2D$ itself [7].

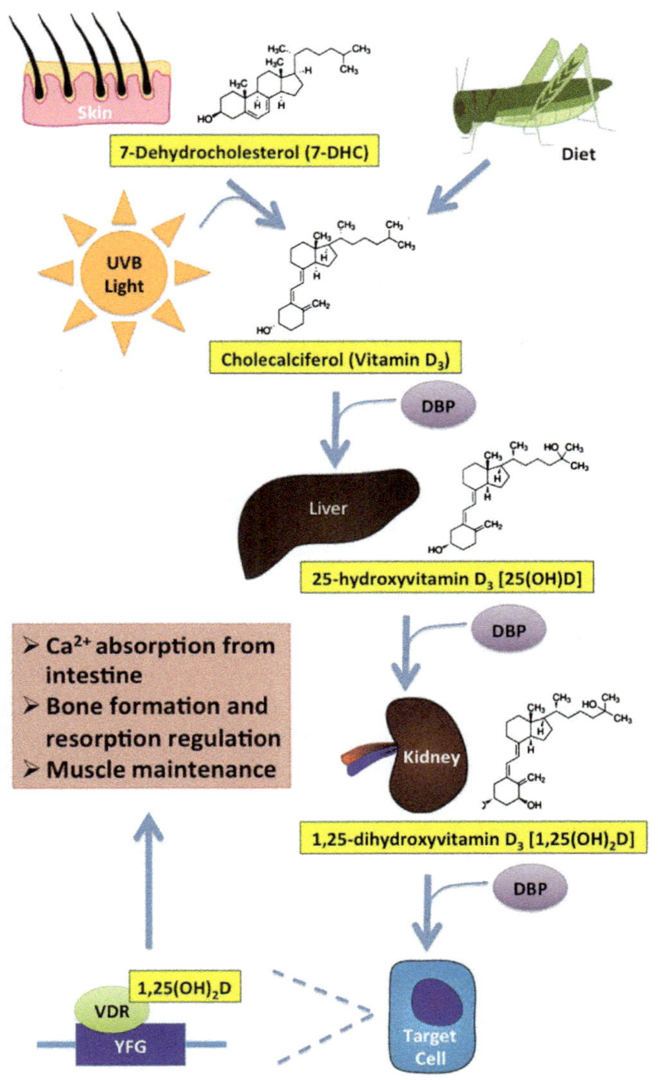

FIGURE 14.1 Vitamin D metabolism. Cholecalciferol (vitamin D_3) is obtained either by conversion of 7-dehydrocholesterol (7-DHC) by UVB light or through dietary consumption. After binding to vitamin D-binding protein (DBP), vitamin D_3 is then transported via the blood to the liver, where it is converted into 25-hydroxyvitamin D_3 [25(OH)D]. 25(OH)D is again bound by DBP and transported via the blood to the kidney, where it is converted into 1,25-dihydroxyvitamin D_3 [1,25$(OH)_2D$]. 1,25$(OH)_2D$ is then once again bound to DBP and transported to target cells, where it binds to vitamin D receptor (VDR), activating its transcription factor activity. The 1,25$(OH)_2D$–VDR complex then regulates gene *YFG* (= Your Favorite Gene) transcription, resulting in a multitude of effector actions.

1,25$(OH)_2D$ represents the hormonally active form of vitamin D_3 and carries out most of its biological functions, including calcium homeostasis and bone metabolism [11]. Once synthesized, 1,25$(OH)_2D$ is again transported by DBP throughout the body to various organs. It then binds to vitamin D receptor (VDR) in target cells, whose now-activated transcription factor activity mediates the expression of several proteins [7]. Arguably 1,25$(OH)_2D$'s principal function in calcium homeostasis involves promoting calcium absorption from the intestine, where it mediates calcium translocation across intestinal enterocytes and paracellular diffusion [12]; this absorption of calcium is balanced by the action of PTH to maintain serum calcium levels in homeostasis [13]. 1,25$(OH)_2D$'s role in bone metabolism includes promoting bone formation through regulating the balance of calcium and phosphorus levels, altering bone responsiveness to growth hormone and expression of skeletal growth factors, and mediating chondrocyte differentiation and function; it also regulates bone resorption through stimulating osteoclast formation and activity [11]. 1,25$(OH)_2D$ possesses multiple functions not related to bone and calcium homeostasis, including maintaining muscle performance and balance [14,15], and can also be synthesized by monocytes and macrophages, where it stimulates the innate immune system [16].

Availability of vitamin D metabolites to perform their vital calcium and bone homeostatic functions is directly related to availability of their precursors [13]. Hydroxylation of vitamin D_3 to 25(OH)D in the liver is directly related to the concentration of vitamin D_3 in the blood, making serum measurements of 25(OH)D an accurate estimate of vitamin D level in humans, and the amount of hydroxylase conversion of 25(OH)D into 1,25$(OH)_2D$ in the kidney depends on the serum concentrations of PTH, phosphate, and calcium. However, despite having extremely high levels of circulating 1,25$(OH)_2D$ in the blood, marmosets are not hypercalcemic and are prone to osteomalacia associated with hypophosphatemia [17]. Historically, it was thought that marmosets have fewer $1{,}25(OH)_2D$ receptors and low $1{,}25(OH)_2D$-receptor complex activity compared with Old World monkeys, suggesting that they possess an end-organ resistance to $1{,}25(OH)_2D$ in contrast to humans and Old World primates [18]. However, a series of subsequent studies have shown that vitamin D receptors are functionally competent and sufficiently numerous to allow for import of vitamin D_3 into New World primate cells at a rate comparable to that for Old World primates, and that the end-organ resistance is rather due to the presence of high levels of an intranuclear protein that binds to the vitamin D response element (VDRE) that inhibits the transcription of the vitamin's downstream targets [19–21]. This unique characteristic further complicates bone metabolism and calcium homeostasis in the marmoset, making them susceptible to musculoskeletal pathology when not provided with sufficient vitamin D.

Marmoset Diet in the Wild and Nutritional Requirements in Captivity

Compared with other anthropoid primates, marmosets and other callitrichids have higher metabolic rates and thus require a higher energy intake. In the wild, marmosets are omnivores and consume a varied diet consisting of high-energy foods including insects, fruits, reproductive plant parts, and small vertebrates. Grasshoppers represent the most common insects eaten by callitrichids, and a large amount of their time is spent hunting and foraging for insect prey [22]. Marmosets possess specialized lower incisors that allow them to gouge holes in trees and extract sap, gum, or resin exudates, which comprise a significant portion of their diet [23]. In contrast to insects, these exudates are high in calcium and low in phosphorus and are therefore thought to help balance the calcium-to-phosphorus ratio in the marmoset's diet [22].

While most commercial diets strive to mimic the nutritional content found in wild marmoset feedstuffs, they are generally significantly different. Commercial callitrichid diets often possess a high amount of protein to mimic the high protein content found in insects, up to 25% [24]. Due to their relatively high energy densities, fat content in commercial New World monkey diets is also often relatively high, around 9%–10%. Negative reactions to soy, milk, gluten, and wheat products have been reported in captive marmosets, suggesting they should be excluded to avoid intestinal problems leading to nutrient malabsorption [25,26]. Most research institutions feed a commercially available canned or biscuit-based diet and supplement with fruits, vegetables, and vitamin C and protein sources [23,24]. However, when given the choice, marmosets will often preferentially eat sugary treats, and so enrichment should be closely monitored to ensure animals are receiving a nutritionally balanced diet [27,28].

Providing adequate dietary vitamin D_3 represents the biggest challenge in captive marmoset diet formulations, as insect chitin and UVB light are thought to provide most of the vitamin D_3 to wild marmosets [23,24,27]. The National Research Council's Nutrient Requirements in Nonhuman Primates currently recommends a diet consisting of 1000 to 3000 IU of vitamin D_3/kg of dry diet, but the vitamin D_3 content in most callitrichid diets is much higher, up to 8000 IU/kg dry diet [24,27]. Despite these high levels, most osteomalacic bone changes seen in marmosets are thought to be due to the unique vitamin D metabolism found in these animals and our incomplete understanding of their fundamental nutritional requirements.

DIAGNOSING DISEASE OF THE MUSCULOSKELETAL SYSTEM

Histologic Lesions of Bone Disease

Knowing what is normal for a given anatomic location of an animal of a given age is essential for being able to evaluate bone disease histologically. Having matched normal controls (age, sex, reproductive status, anatomic location) is very useful in identifying and interpreting lesions. Although techniques to evaluate bone microscopically without decalcification exist, most routine bone pathology is done on decalcified specimens to facilitate sectioning. The decalcification process, however, makes evaluation of mineralization status impossible. There are various techniques and solutions used to decalcify specimens, and these may be employed immediately after fixation in 10% neutral buffered formalin or at a later time. Removal of skin and muscle will facilitate decalcification. For general sampling of bone, it is important to collect bone from a variety of locations, as certain processes will affect certain sites differently than others. At a minimum, skull, lumbar vertebrae, femur, rib, and humerus should be collected. Callitrichids are small enough to allow the entire skeleton to be preserved, thus allowing the choice of samples for evaluation to be done at a later date.

Radiographic evaluation of an animal's skeleton before postmortem examination will also facilitate the decision of which bones to evaluate and will greatly enhance the utility of histologic bone examination. Radiographic evaluation of formalin-fixed (but NOT decalcified) bones is also perfectly acceptable.

Histologic evaluation should include an assessment of all parts of the bone (articular surfaces, epiphysis,

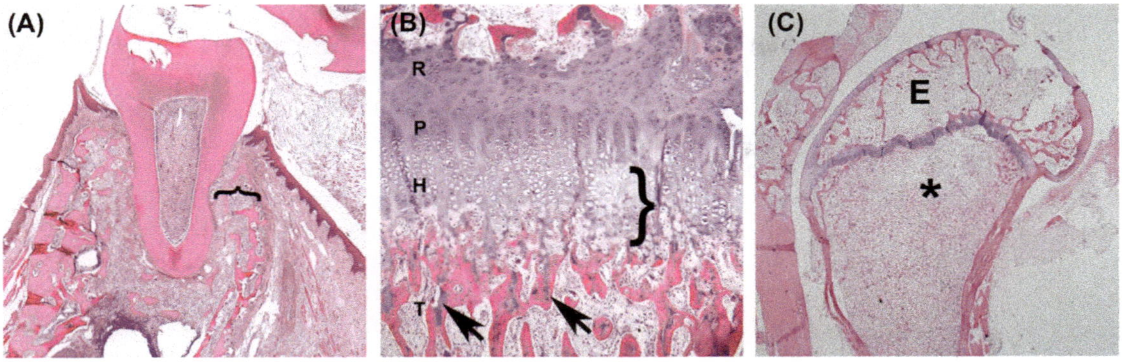

FIGURE 14.2 Typical histopathological lesions associated with bone disease in marmosets. All pictures depict cross-sections through hematoxylin- and eosin-stained sections. (A) Fibrous osteodystrophy (FOD), mandible with gingival epithelium and tooth at 40× magnification. The *bracket* denotes an area where both the bony cup (alveolus) surrounding the tooth and a solid bony mandible should exist, but both have been replaced by dense fibrous connective tissue, accompanied by small spicules of bone. (B) Rickets, the growth plate of a young marmoset at 100× magnification. The epiphyseal side of the plate is at the top of the image, and the metaphyseal side of the plate is at the bottom. The *bracket* denotes an expansion of the zone of hypertrophy of the growth plate, which is characteristic of rickets. Additionally, retained cartilaginous cores are shown by the tongues of cartilage that extend into the trabeculae of primary spongiosa of the marrow cavity (*black arrows*). The trabeculae in this image are thickened and irregularly shaped, another feature of the disease. *H*, hypertrophic zone; *P*, proliferative zone; *R*, resting zone; *T*, trabecular bone. (C) Osteopenia, femoral condyle of a young marmoset. The growth plate is still present in this individual. There is almost a complete lack of trabecular bone within the metaphysis (*) and epiphysis (E) of this bone and the cortices are extremely thin. Hematoxylin- and eosin-stained section, 12.5× magnification.

medullary cavity, metaphyseal and cortical bone, and the cells present in each). The growth plates of young animals also must be evaluated as they are commonly affected by metabolic and nutritional derangements and also are common sites of infection. The thickness and density of cortical bone and the thickness, density, and number of medullary trabeculae should be evaluated. Viable osteocytes should occupy lacunae, and their absence can be interpreted as osteonecrosis. The appearance and number of osteoblasts provides an estimate of the rate of bone formation, whereas the number of osteoclasts provides an estimate of the rate of bone resorption. A detailed histomorphometric evaluation of bone turnover requires administration of fluorochrome labels in vivo and processing of undecalcified bone samples into plastic. Articular cartilage and subchondral bone thickness and quality are used to characterize the severity of osteoarthritis. Both osteoarthritis and inflammatory arthritis (infectious or immune-mediated) can also involve the periarticular soft tissues (joint capsule, synovium, tendons, and ligaments). The presence of inflammation in the marrow spaces, with or without accompanying infectious organisms, is used to diagnose osteomyelitis.

Nutritional bone diseases are often categorized into fibrous osteodystrophy, rickets, and osteopenia/osteoporosis, although they may occur simultaneously in the same individual. Fibrous osteodystrophy can occur at any age and is characterized histologically as increased numbers of osteoclasts within erosion lacunae, myelofibrosis, cortical trabeculation, endosteal resorption, periosteal new bone formation, and incomplete dental alveoli/tooth sockets (Fig. 14.2A). Rickets is a disease of young animals in which there is abnormal growth of long bones characterized histologically as a widened zone of hypertrophic chondrocytes, with or without disorganization of the growth plate and associated widening of the growth plate (Fig. 14.2B). Clinically, these lesions often are accompanied by angular limb deformities and joint enlargement. Osteopenia/osteoporosis is characterized histologically as an overall decrease in the thickness of cortical bone and reduced number and thickness of trabeculae in cancellous bone (Fig. 14.2C) [29].

Imaging Modalities for the Evaluation of Bone Pathology in Marmosets

Imaging is a powerful tool for ante- and postmortem evaluation of the musculoskeletal system in marmosets. Available imaging modalities range from standard radiographic evaluation to advanced techniques such as MRI and computed tomography (CT) to highly specialized technologies for assessment of bone mineral density such as dual-energy X-ray absorptiometry (DEXA or DXA) [30–34]. It is important to note that though bone mineral density is the gold standard measurement when screening for bone pathology such as osteoporosis in human medicine, propensity to fracture is determined by bone quality, a multifactorial measure that takes into account bone architecture, cellularity, mineralization, and turnover [35]. This holds true in marmosets as well, where bone mineral density (BMD), as measured by CT, was found to not correlate with postmortem breaking strength of bones, with the bones most able to withstand stress not having BMDs appreciably different from the most flexible bones, whereas if BMD were truly predictive, one would expect to see that the least flexible bones have the highest BMD [32]. However, imaging remains a valuable screening tool for bone pathology of marmosets.

Two-Dimensional Techniques

Radiography uses X-ray radiation to generate a two-dimensional image that superimposes cortical on trabecular bone. The resulting image demonstrates relatively high spatial resolution (40 μm) and can be used for qualitative or quantitative analysis of trabecular architecture [35]. It is readily available and hence one of the most commonly used imaging modalities for the evaluation of bone [36]. Digital radiographs are replacing more specialized radiographic techniques, including direct magnification radiography, mammographic techniques, and fine-detail radiography, for the diagnosis of bone pathology [37]. Though radiography when paired with advanced digital image analysis methods can correlate with the results obtained by the most sensitive imaging modality, microCT [38], radiography is only able to detect pathology following loss of 30%–40% of bone density [36], making radiography the least sensitive imaging modality reviewed herein. When compared with CT for measurement of plane (2D) bone mineral density (BMD) in marmosets, standard radiographs allow for diagnosis of pathological lesions at 0.08 g/cm^2 [32]. In humans, the sensitivity of radiographs has been significantly increased through the use of standardized acquisition and analysis techniques to quantify lesions. Digital X-ray radiogrammetry (DXR) can detect lesions at 0.0012–0.0028 g/cm^2 and has been used to reliably track changes over time in patients with osteoporosis but cannot be directly extrapolated to marmosets because it relies on human skeletal landmarks [39]. An analogous standardized technique has been developed for use in marmosets with digital radiography, by using a step-wedge bone density standard and a commercially available image analysis software (Nikon Elements) to quantify bone radiodensity fraction (BRF). In one study, this technique was successful in identifying marmosets with bone disease (fibrous osteodystrophy, rickets, and osteopenia, as identified by histopathology) compared with normal controls [40].

Despite its limitations in sensitivity, radiography can allow for specific diagnosis of many common bone diseases in marmosets, including fibrous osteodystrophy, rickets, and osteopenia, following a simple qualitative assessment (Fig. 14.3, see Ref. [29] for review). Fibrous osteodystrophy in marmosets is characterized by multifocal areas of radiolucency and a lacelike appearance of the cortex and medulla of long bones, jaw, and vertebrae with loss of corticomedullary distinction (Fig. 14.3B) [29]. Though not yet reported in marmosets, in humans these lesions are often accompanied by the appearance

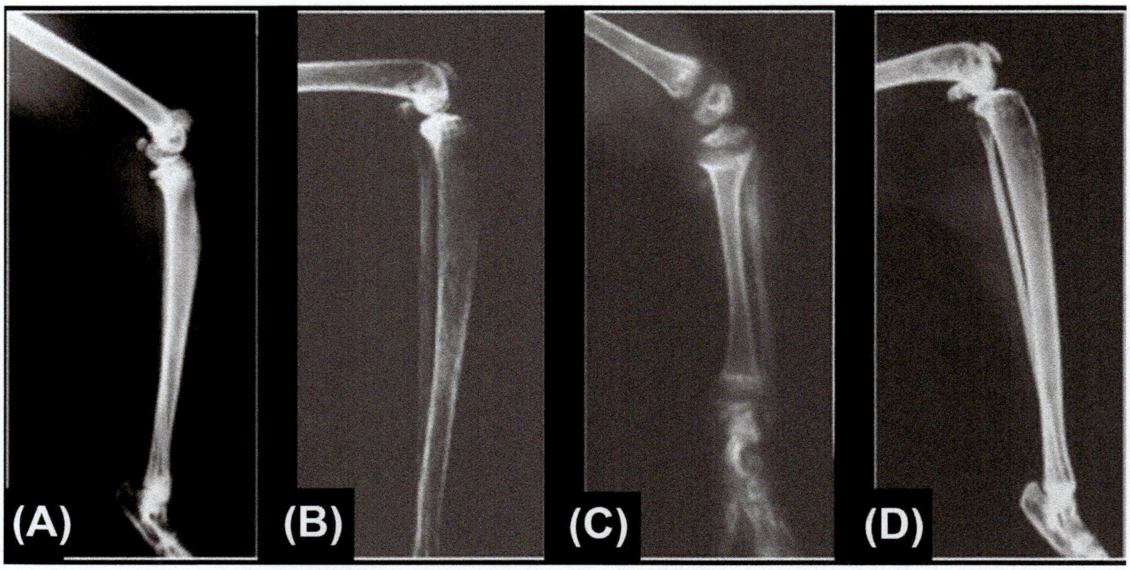

FIGURE 14.3 Typical radiographic lesions associated with bone disease in marmosets. All pictures depict digital radiographs of the stifle area taken at consistent settings on the same machine of animals that had histopathologic diagnoses of bone disease. (A) Normal bone, 4-year-old male. (B) Fibrous osteodystrophy (FOD), 3-year-old female. Multifocal areas of radiolucency and a lacelike appearance are most apparent at the tibial condyles and along the tibial crest. (C) Rickets, 1-year-old female. Subtle physeal flaring is observed at the proximal tibia, and bowing of the tibial shaft is noted. (D) Osteopenia, 4-year-old male. The cortex is notably thin compared with the other animals depicted in this figure. Figure adapted from Olson et al. [29].

of a ground-glass appearance ("salt and pepper calvarium") characteristic of global bony resorption in the skull, and pseudowidening of joint spaces, especially around the pelvic bones [41]. One report also documented 24 cases of atypical radiographic lesions in marmosets characterized by well-circumscribed cystlike lesions, which occasionally involved entire long bones. These were demonstrated histologically to occur secondary to marked osteoclastic resorption of bone. These lesions appeared to be an early stage of a proliferative bone process that is characterized radiographically by locally extensive to disseminated expansion of the bone diameter with loss of a clearly defined cortex. Histologically, this lesion appears to represent marked thinning of the cortical bone due to chronic endosteal resorption and concurrent or subsequent marked periosteal new bone production [29]. In this condition, the early lesions are reminiscent of the bony cysts seen with osteoclastomas in some cases of osteodystrophy in humans; however, the late lesions have no clear analogy to a human condition associated with osteodystrophy [41]. Rickets in marmosets is also characterized by a decrease in bone density and loss of corticomedullary distinction [37] but is accompanied by widening of the physes of long bones with or without metaphyseal flaring and bowing of the long bone axis (Fig. 14.3C) [29]. Osteopenia/osteoporosis in marmosets is characterized by a diffuse decrease in the radiodensity of the cortex and medulla of all bones (Fig. 14.3D) [29]. In humans, fracture without callus formation additionally raises the index of suspicion for osteoporosis [42], and this appears to apply to marmosets as well.

DEXA is another two-dimensional projectional imaging technique that uses dual-energy X-ray spectra to measure relative tissue absorption as area density in g/cm^2 [43]. The resultant T-score (a direct comparison to age- and sex-matched normal controls) or Z-score (an indirect comparison of the standard deviations) is used by the World Health Organization as the basis for their definitions of osteopenia and osteoporosis [44]. The technique is widely accepted, rapid, economical, and poses a low radiation risk to the patient, and commercially available portable DEXA machines designed for the evaluation of BMD in human hands are ideal in size for the imaging of marmosets [42]. However, DEXA is not perfect, suffering from artifacts of positioning, bone curvature, interindividual variation in muscle and fat, and concurrent disease or injury, and as such is most powerful for single time point evaluation rather than longitudinal monitoring of subtle change [35]. It is also important to note that multiple scanners are available, all with their own software and analysis paradigms, so caution should be used when comparing results from different systems [45–47]. As a two-dimensional modality, it cannot differentiate between cortical and trabecular bone [48], and the output of areal BMD represents only 60%–70% of bone strength [49]. Despite the fact that it is the gold standard technique for detection of osteoporosis, DEXA yielded normal results for 60% of postmenopausal women with osteoporosis who suffered insufficiency fractures [50]. Even with its shortcomings, its wide availability and acceptance for the measurement of BMD make DEXA a logical choice for the evaluation of BMD in marmosets, and it has been used for this purpose in at least one study [33].

Quantitative ultrasound uses ultrasound waves at a frequency of 1 kHz (compared with 2–10 MHz for diagnostic ultrasound) to image bone [51] and has been used to screen for osteoporosis in women [52]. Quantitative ultrasound is fast and presents no radiation risk. However, it lacks sensitivity compared with all other modalities [51]. Its use in marmosets for the evaluation of bone pathology has not been reported.

Three-Dimensional Techniques

CT uses a combination of multiple X-ray images taken at different angles to create a three-dimensional image. Quantitative CT (qCT) uses a mineral-equivalent calibration phantom that is scanned simultaneously with the patient and special software to report BMD as true volumetric density (mg/cm^3) [35] and thus is more sensitive than DEXA, especially when finite element analysis modeling (FEA) is used [53]. As a three-dimensional modality, qCT can differentiate between cortical and trabecular bone [48] and can reveal subregional differences in trabecular pattern and density [53]. It can reliably be used for longitudinal monitoring of changes in BMD within a patient [54,55]. Specialized peripheral qCT systems and systems specifically marketed for animal research allow for the imaging of small animals such as marmosets [56]. However, qCT is costly, occupies significant space within a facility, and delivers a high dose of radiation [42]. qCT is furthermore subject to movement artifact due to the long scan acquisition time [49], and its relatively low spatial resolution presents a barrier to measuring thin bone cortices (as may be seen in marmosets with bone disease) or animals with certain bone marrow infiltrates [57,58]. qCT measurement of BMD has been shown to correlate with histopathology results for BMD of the third and fourth lumbar vertebrae in marmosets [32] and is the imaging modality represented in most publications to date [31,32]. It should be noted that plane (2D) BMD as measured by CT increases with age in marmosets until 96 months [31].

MicroCT is a variation of the CT modality that can provide resolution up to 2 μm in three dimensions, allowing for detailed analysis of trabeculae that is nearly equivalent to that achieved by histopathology [59]. In fact, when stereometric sampling techniques are used, microCT has proven to be even more sensitive than histopathology in detecting bone disease [60], and sensitivity can be further increased through the substitution of synchrotron radiation for standard X-rays [35]. These machines are typically limited to research settings for the imaging of small animals and biopsy specimens, due to their small bore size. Though a microCT unit could accommodate a live marmoset, imaging studies that evaluate marmoset BMD using microCT to date have focused on postmortem specimens [61,62].

MRI uses magnetic fields and radio waves to generate a three-dimensional image of soft and hard tissues throughout the body. MRI can resolve structures at 150 μm when obtaining a minimum slice thickness of 500 μm, allowing for the evaluation of larger trabeculae [63], and small machines with high-resolution coils that are designed for distal limbs or research with small animal models can accommodate marmosets. MRI has the advantage of no radiation risk to the animal or operator, and this modality allows for concurrent evaluation of nonstructural aspects of the bone like the bone marrow [42]. However, MRI is costly and similarly susceptible to movement and positional artifacts [42]. Furthermore, bony trabeculae are indirectly imaged as an extrapolation of areas of lack of signal in the fatty marrow [42]. MRI evaluation of the structural trabecular parameters of peripheral limbs with a 3 T magnet correlates well with high-resolution peripheral qCT in human patients [63], and MRI has proven superior to DEXA at identifying osteoporosis in women [64,65]. MRI is not commonly reported in the evaluation of bone pathology in marmosets in the published literature.

Biomarkers of Disease of the Musculoskeletal System

Biomarkers in the serum, plasma, or urine are established for human musculoskeletal disease and represent a relatively noninvasive means for the diagnosis of musculoskeletal disease in marmosets. Utility of these assays is limited by the small volume of biosamples available from marmosets and issues with cross-reactivity of marmoset samples with available assays.

Biomarkers of Bone Disease

Specific biomarkers of bone disease can be categorized as markers of bone formation and bone resorption and have been extensively used in human medicine. Recent reviews determined that serum levels of the osteoblast enzyme bone alkaline phosphatase (bALP), the osteoblast-secreted hormone osteocalcin (OC), and the by-product of collagen synthesis procollagen type I N-terminal propeptide (P1NP) are commonly measured as sensitive and specific markers of bone formation, whereas serum or urine levels of cross-linked telopeptides of type I collagen (serum C-terminal cross-linked telopeptides or urine N- or C-terminal cross-linked telopeptides) are commonly measured as markers of bone resorption [66,67]. Of these, a recent metaanalysis found that elevations in P1NP and serum C-terminal cross-linked telopeptides (also known as procollagen type 1 C-terminal portion, or CTX) are most commonly associated with clinical disease [66], though one study demonstrated that urine CTX and NTX can predict changes in bone mineral density by DEXA [68,69] and another that serum osteocalcin can predict vertebral fracture risk better than DEXA [70]. As there can be significant variation among individuals due to sex, age, and other factors, these markers are most useful in monitoring the progression of bone disease over time rather than diagnosing disease at a given time point [67], or when used in combination with imaging modalities [71].

The validity of a number of these biomarkers for the diagnosis and monitoring of bone disease in marmosets has been evaluated through controlled studies. Similar to humans, serum CTX and P1NP were the most sensitive biomarkers of bone resorption and formation, respectively, following treatment with bisphosphonate ibandronate to suppress bone turnover; other markers, including serum osteocalcin and parathyroid hormone (PTH), also changed appropriately but not significantly [72]. However, when serum CTX and osteocalcin were used to monitor the development of osteopenia in male marmosets following orchidectomy, the increase was transient and not statistically significant, though it preceded the detection of a decrease in cancellous bone density using qCT by 9 months [34]. Serum levels of the hormone responsible for maintaining calcium homeostasis, PTH, also show promise as a biomarker for the diagnosis of osteopenia, rickets, and fibrous osteodystrophy; one study showed significantly higher values in marmosets with bone disease compared with healthy controls and defined a cutoff of values greater than 600 pg/mL as suggestive for bone disease [40].

It should be noted that these biomarkers are likely most sensitive when used longitudinally to monitor the progression of disease, as there is wide variation in baseline values for CTX, P1NP, and osteocalcin in marmosets though no differences by age or sex [72]. This was demonstrated when CTX was evaluated as a potential biomarker for osteopenia, rickets, and fibrous osteodystrophy in a retrospective study that compared serum biomarker results with histopathology for diagnosis; no difference in serum CTX level was found between animals with bone disease and normal controls [40]. In addition, the ELISA for osteocalcin failed to cross-react with marmoset samples, though ELISA assays from other manufacturers had cross-reacted in previous studies [34,40,72].

A serum chemistry panel can also provide valuable information on bone health. Alkaline phosphatase (ALP) can originate from the liver, bone, intestines, or kidneys, and levels will rise in the serum following increased production by osteoblasts in a state of high bone turnover [73]; thus, ALP is used as a biomarker for osteomalacia in humans. ALP has also been extensively studied as a biomarker for bone disease in marmosets. In marmosets, it is elevated in animals with low breaking strength of bones [32], is negatively correlated with both plane and total BMD as measured by qCT [31], and was increased in one marmoset with multifocal myelofibrosis [74]. It should be noted that ALP is higher in juvenile males than females and decreases with age in both sexes [75]. Similarly, calcium is an integral part of the bony matrix and has been positively correlated with BMD in two studies [31,33]. However, calcium is bound to albumin in the blood, and the value reported in serum chemistry panels represents both the bound and unbound fractions. Ideally and especially in the presence of hypoalbuminemia, calcium should be measured as an unbound fraction, or the value of total calcium should be calculated by mathematically correcting for serum albumin level. One study that initially identified differences in serum calcium in bone disease observed that those differences disappeared once a mathematical correction for serum albumin was applied [40]. This is important because hypoalbuminemia is commonly present in animals with bone disease. In fact, a serum albumin value of less than 3.5 g/dL was identified as a biomarker for bone disease in one study but should be interpreted with caution as low albumin can be indicative of a multitude of disease processes [40]. Serum levels of inorganic phosphorus or vitamin D metabolites have not been established as biomarkers for bone disease in marmosets.

Biomarkers of Muscle Disease

Specific biomarkers of muscle disease are readily available in a standard serum chemistry panel. Creatine kinase (CK) is the enzyme that reversibly phosphorylates creatine to phosphocreatine in an ATP-dependent process [76]. Elevations are specific for muscle damage, and it is commonly used in the diagnosis of muscular dystrophy in human patients but is not elevated in all myopathies and can be affected by sex, muscle mass, exercise, and some drugs [76–79]. CK has similarly been shown to be specific for skeletal muscle damage in marmosets [80].

Lactate dehydrogenase (LDH) is the enzyme that reversibly converts pyruvate into lactate. It is used as a marker for muscle damage but is readily affected by mild hemolysis, demonstrates significant interindividual variability, and can also be elevated in liver disease [76,81,82]. Although aspartate aminotransferase (AST) and alanine aminotransferase (ALT) are two other enzymes that are present in myocytes and that will be elevated following muscle injury, both are abundant in the liver and other tissues and are thus nonspecific for muscle disease [83]. In marmosets, CK, AST, ALT, and LDH are consistently increased following treatment with myotoxins [81,84], and CK, AST, and LDH have also been used for monitoring muscle damage secondary to intramuscular administration of injectable anesthetic cocktails [85,86].

Specialized markers of muscle disease in the serum include troponin and xin. Though both proteins are present in the cytoplasm of skeletal and cardiac muscle and increase in serum following injury of either, distinct isoforms are present in each tissue, allowing for differentiation between skeletal and cardiac muscle damage [87,88]. Troponin further has specific isoforms for slow-and fast-twitch skeletal muscle [89]. Though cardiac troponin has been used as a biomarker of cardiac toxicity in marmosets [80], the skeletal muscle isoform is not commonly utilized for the detection of muscle damage in this species.

Urine represents an additional biofluid that can be sampled in the diagnosis of muscle disease. Myoglobin is a structural protein in the cytoplasm of skeletal and cardiac myocytes [90]. Free myoglobin in the blood is excreted in the urine [90] and results in a red-brown discoloration that can be distinguished from hemoglobin by precipitating the hemoglobin with ammonium sulfate [91]. It increases within 30 min after exertion or injury, and elevated levels can persist for up to 5 days [92,93], but samples must be analyzed promptly as myoglobin deteriorates quickly in urine [90]. Myoglobin has yet to be reported as a marker for muscle damage in marmosets.

Biomarkers of Joint Disease

The classic marker for autoimmune (rheumatoid) arthritis in human patients is the presence of rheumatoid factor (RF) autoantibodies; however, this test lacks sensitivity and is negative early in disease [94]. Pyridinoline (PYR, also known as hydroxylysylpyridinoline) and deoxypyridinoline (DPYR, also known as lysyl pyridinoline) are collagen breakdown products that are elevated in the urine in the presence of ongoing joint erosion in both active rheumatoid arthritis and osteoarthritis [95]. Cyclic citrullinated peptides (CCPs) are formed secondary to joint inflammation, and serum levels of both citrullinated peptides and anti-CCP antibody are increased in rheumatoid arthritis, with the rise in peptides preceding the rise in antibody [96,97]. Combinatorial evaluation of citrullinated peptides, anti-CCP antibody, and the bone turnover marker hydroxyproline can distinguish between rheumatoid and osteoarthritis in humans [96].

In marmosets, biomarker use for the diagnosis of arthritis has been limited to the use of urine PYR and DPYR normalized to urine creatinine [98,99]. Arthritis is more commonly diagnosed by clinical signs, though biomarkers of inflammation, such as serum amyloid A, and bone remodeling, such as ALP, are elevated in the collagen-induced arthritis model [98]. Though C-reactive protein has been used as a marker of the acute-phase inflammation that occurs following induction of this model in macaques and to characterize GI disease in marmosets, assays for this protein may have variable cross-reactivity with marmoset samples [40,99].

Emerging Biomarkers

MicroRNAs (miRNAs) are emerging as a potential biomarker of musculoskeletal health in human medicine but have yet to be evaluated in the context of disease in marmosets. Specifically, miR503, miR138, miR705, and miR3077/Sp are sensitive and specific for osteoporosis in humans [100]. Plasma miR133a was found to be more sensitive than CK in detecting the effects of skeletal muscle toxins [101], and 24 specific miRNAs have been identified for the detection of osteoarthritis [100]. The value of miRNAs for the detection of musculoskeletal disease in marmosets warrants further investigation.

DISEASES OF THE MUSCULOSKELETAL SYSTEM

Nutritional Bone Disease

Metabolic Bone Disease and Related Disorders

Metabolic bone disease (MBD) refers to one of many pathologies of the bone that result from disturbances in calcium metabolism. It has long been recognized in nonhuman primates, with a detailed description of "cage paralysis" dating back to 1904 by the New York Zoological Society [102]. Initially thought of as a nervous system disorder, the condition was eventually determined to be osteomalacic in nature, with affected animals showing decreased ambulation, "stiffness" of the extremities, and pathologic fractures; on necropsy, the skeletal system, especially the weight-bearing bones, was reported to be severely deformed with accompanying muscle wasting. Since then, multiple types of MBD, including rickets, fibrous osteodystrophy (FOD), and osteopenia/osteoporosis, have been described, with marmosets especially predisposed due to their unique vitamin D metabolism.

Rickets, which is a disease of growing animals characterized by poorly mineralized bone and cartilage, and osteomalacia, which occurs in mature animals and is characterized by failure of osteoid to properly mineralize, are two related skeletal diseases with a metabolic etiology that are frequently reported in marmosets. Historically, the most commonly observed bone disease in marmosets is vitamin D-dependent rickets type II [103]. Radiographic findings include thickened physes with variable metaphyseal flaring, kyphosis, reduced bone density, pathologic fractures, and resorption of the subperiosteal bone in the hands and in the alveoli of the tooth sockets (Fig. 14.3C) [29,103]. On histopathologic examination, osteoid volume and osteoclast number are considerably increased in affected marmosets compared with unaffected animals, and growth plates have cellular disorganization with a widened zone of hypertrophy due to an increased number of hypertrophic chondrocytes (Fig. 14.2B). Rickets in humans is caused by vitamin D deficiency, and the marmoset's end-organ resistance to $1,25(OH)_2D$ is thought to contribute to the high incidence of the condition, though the exact mechanism has not been elucidated.

Another skeletal pathology reported in adult marmosets is fibrous osteodystrophy (FOD). On radiographs, the skeleton of marmosets with FOD exhibits a "moth-eaten" appearance, where multifocal areas of radiolucency are found in not only the cortex and medulla primarily involving long bones but also occasionally the mandible, maxilla, and vertebrae (Fig. 14.3B) [29,104]. Histologic lesions include increased osteoclast number, peritrabecular fibrosis (particularly in the skull), cortical trabeculation, endosteal resorption, and periosteal new bone formation (Fig. 14.2A). Absence of alveolar bone in the mandible or maxilla is a variable finding but is striking when it occurs. Similar to the human condition, PTH is often elevated in cases of FOD [40], and serum calcium levels are usually low [33,40]. In contrast to rickets and osteomalacia, FOD is predominantly found in marmosets with concurrent gastrointestinal (GI) disease [29].

In the common marmoset, inflammatory bowel disease (IBD) resulting in malabsorption (historically referred to as marmoset wasting syndrome) and MBD in its various forms are associated, with marmosets over seven times more likely to have concurrent bone and GI disease as opposed to disease in only one organ system [40]. Low serum calcium and vitamin D have been recognized to contribute to low BMD [105]; poor digestive efficiency has been directly associated with low serum calcium and vitamin D levels and correlated with bone mineral density via DEXA in animals with low serum calcium levels [33]. These studies support the theory that IBD and MBD share a pathologic process; GI inflammation characteristic of IBD results in poor absorption of nutrients that are critical to bone homeostasis, leading to MBD [33,40].

Osteopenia/osteoporosis is another MBD described in marmosets and is often concurrently diagnosed with rickets and/or FOD in histologic sections [29]. Osteoporosis is a systemic skeletal condition characterized by reduced bone mass, leading to bone fragility and increased risk of pathologic fractures. In marmosets diagnosed with osteopenia, bones have diffusely decreased radiodensity of both the cortex and medulla and reduced bone mineral density (Fig. 14.3D) [29,34]. Histologically, cortical thickness is reduced with variable endosteal resorption and cortical trabeculation (Fig. 14.2C) [29]. The quality of the osteoid remains normal and because calcium–phosphorus balance does not influence the condition, classic therapies for MBD are not effective [104].

To summarize the difference in histopathological lesions: rickets is diagnosed by an abnormally widened and disorganized growth plate in a young animal; fibrous osteodystrophy is diagnosed by loss of cortical and trabecular bone replacement by fibrous tissue; and osteopenia is diagnosed as an overall decrease in cortical and trabecular bone. It is not unusual to diagnose more than one of these lesions simultaneously in the same individual, which occasionally makes the histologic differentiation of these diseases difficult.

Antemortem diagnosis of MBD is most frequently achieved by imaging. While conventional radiography represents the most commonly available modality in most laboratories and clinics, because 40% of mineral must generally be resorbed before lesions become identifiable, it is often not diagnostic until late in the course of MBD [104]. Despite this, by using conventional radiography paired with a digital analysis paradigm, marmosets with bone disease have been found to have significantly reduced bone radiodensity fraction in the distal femur compared with unaffected animals [40]. Other, more sensitive imaging modalities used to evaluate MBD in marmosets include computed tomography and DEXA; in each of these studies, BMD has been found to be reduced in marmosets with MBD [32–34]. Other antemortem diagnostics found to suggest the presence of bone disease in marmosets include high PTH levels, low serum albumin levels, and low body weight [40]. Additionally, serum ionized calcium may be decreased, and the calcium-to-phosphorus ratio inverted [106].

Treatment of MBD focuses on addressing the imbalance in calcium/vitamin D_3 metabolism. Oral salmon calcitonin, a synthetic form of the hormone calcitonin prescribed for osteoporosis treatment in postmenopausal women, combined with injectable vitamin D_3 and injectable and oral calcium supplementation, has proven successful in treating juvenile cases of MBD but is less successful in adult marmosets [107]. In another colony, vitamin D_3 supplementation starting at 1000 IU/animal/day and then decreased to 500 IU/animal/day for maintenance proved to be highly successful in treating cases of MBD [108]. Because many cases of MBD are concomitant with (and potentially secondary to) IBD, additional treatments that address intestinal malabsorption are often warranted [33]. One case report described reversing IBD in a cohort of *Saguinus mystax* tamarins through replacement of a commercial closed (proprietary) formula diet with a more nutritionally appropriate open (publically available) formula diet [109]. In another report, the normal colony diet was transiently replaced with a "recuperation" diet with added human milk substitute [110]. Recent therapies for IBD in marmosets that have proved promising include administration of budesonide, a GI-specific glucocorticoid, or tranexamic acid, a plasmin inhibitor [111,112]. Secondary clinical consequences of MBD, specifically pathologic fractures, should also be addressed with symptomatic treatment. Despite these recent advances, the prognosis for IBD and associated MBD remains poor.

Despite the availability of commercial diets with adequate amounts of vitamin D_3 and a proper calcium-to-phosphorus ratio, MBD continues to persist in marmoset colonies, albeit at a considerably decreased frequency. Daily intake of at least 100 IU/kg vitamin D_3 and 250 mg/kg calcium and a dietary calcium-to-phosphorus ratio of 1.5–2:1 are recommended to help prevent MBD [107]. Additionally, supplemental oral calcium administration may prevent osteomalacia and clinical hypocalcemic tetany in lactating female marmosets [106]. If possible, access to natural sunlight or full-spectrum lighting indoors to provide a nondietary form of vitamin D_3 is highly recommended to prevent development of rickets and osteomalacia [107].

Metastatic and Dystrophic Mineralization

Calcinosis circumscripta is a condition in which focal mineral deposition, particularly calcium salts, occurs in the soft tissue. While well described in humans and dogs and recognized in rhesus macaques, only one case report describes the condition in marmosets [113]. In this case of dystrophic mineralization, a subcutaneous calcified mass was found and successfully excised from the interscapular region of a common marmoset near where a microchip had been placed. While unexpected due to the marmoset's propensity for developing osteomalacic lesions, calcinosis circumscripta represents a possible differential diagnosis for subcutaneous masses in the species.

Metastatic calcification, a form of inappropriate tissue mineralization that occurs secondary to excessively high circulating levels of calcium, has also been reported in the common marmoset. A privately owned marmoset was found to have marked focal calcification of the epicardium and myocardium concomitant with MBD due to calcium and vitamin D_3 deficiency [105]. This unusual case of simultaneous metastatic mineralization and MBD has not been reported in marmosets elsewhere, though concurrent soft tissue calcification and bone resorption have been reported in humans with chronic renal disease and in rodent models of secondary hyperparathyroidism [114,115]. In a report that extensively examined spontaneous lesions of marmosets by histopathology, renal mineralization was found frequently, a possible consequence of high circulating levels of $1,25(OH)_2D$ in the presence of this species' inherent end-organ resistance [116].

Zinc Deficiency

Zinc is an essential trace element that serves a variety of structural, regulatory, and catalytic functions in humans and animals. Structural roles are attributed to the ability of zinc to stabilize and contribute to the function of zinc finger proteins, one of the most abundant proteins in eukaryotic organisms. Regulatory properties include the binding of zinc to metal response element transcription factors (MTFs) that regulate gene expression in the presence of metal ions and oxidative stress. Zinc is also a key component of catalytic sites for enzymes such as matrix metalloproteinases, carbonic anhydrase, and alkaline phosphatase [117].

Zinc is required for normal musculoskeletal, integumentary, immune, reproductive, and cognitive development and function. The skeletal system acts as the body's reservoir for zinc, releasing and reincorporating the ion from the bone to maintain homeostasis. Physical and physiological conditions associated with zinc deficiency in nonhuman primates correlate closely with the concentration of dietary zinc consumed. States of severe dietary zinc deprivation have been reported in squirrel monkeys [118], tamarins [119], rhesus (*Macaca mulatta*), and bonnet macaques (*Macaca radiata*) [120]. Clinical findings included anorexia, apathy, weight loss, alopecia, dermatitis, reproductive complications, and low levels of zinc in the plasma and tissues. In the majority of these studies, zinc was heavily restricted in the diet (<1 ppm). Marginal diet-induced zinc deficiency (<4 ppm) has been evaluated in pregnant female rhesus macaques [121]. Although alterations in lymphocyte mitogen responsiveness and taste sensitivity were reported, clinical complications such as alopecia and weight loss were not observed. To date, there are no reports of spontaneous or experimentally induced zinc deficiency in common marmosets.

Standard commercially available callitrichid diets contain 50–70 ppm of zinc (zinc sulfate, zinc oxide, zinc proteinate, etc.). It should be noted that phytates found in grains have the potential to bind to zinc and reduce bioavailability. Therefore, purified diets may provide higher concentrations of zinc compared with natural ingredient diets [122].

Vitamin A Toxicity

Vitamin A (all-trans retinol or simply retinol) is a fat-soluble micronutrient essential to normal vision, growth, and immune function. Biological activity is measured in international units (IU) where 1 μg of all-trans retinol is equivalent to 3.33 IU of vitamin A. Vitamin A must be acquired through the diet because no animal species is capable of de novo synthesis. Nonhuman primates are capable of metabolizing several provitamin A plant carotenoids (α-carotenoid, β-carotenoid) into retinol-containing compounds. β-Carotenoid in particular is a major source of vitamin A [27]. Typically, commercial laboratory diets for nonhuman primates contain primarily preformed sources of vitamin A such as retinyl acetate or palmitate. These sources are generally less expensive to manufacture and considerably more stable and bioavailable within the gastrointestinal tract compared with provitamin A carotenoids [123].

The health implications associated with chronic, preformed vitamin A over supplementation through diet fortification and use of multivitamins have received considerable attention in human and veterinary medicine over the past few decades [123]. Excessive preformed vitamin A is capable of antagonizing vitamin D absorption and activity and has been linked to reduced skull and long bone density and higher incidences of fracture in humans and several animal species [124–126]. Consequences of vitamin A antagonism are particularly concerning for marmosets that are exclusively housed indoors and are expected to receive 100% of their required vitamin D3 (cholecalciferol) through dietary sources.

Despite the known risks of oversupplementation in commercial diets, studies comparing vitamin A status between wild and captive nonhuman primate populations are limited. Several studies have evaluated serum and organ concentrations of vitamin A in captive rhesus macaques, marmosets, and African Green monkeys fed concentrations up to four times higher than those recommended by the National Research Council [127–129]. All three genera of nonhuman primates were found to have high organ (liver, lung, and kidney) levels of vitamin A, but serum concentrations within the normal range. These findings are consistent with previous reports suggesting serum vitamin A concentrations are tightly regulated and do not serve as good indicators of whole-body vitamin A status [129]. Strategies to reduce the potentially detrimental effects of chronic hypervitaminosis A in nonhuman primates include reduction of diet-based preformed vitamin A to current NRC recommended levels, reformulation of diets to achieve higher amounts of the provitamin A carotenoid and establishment or revision of species-specific dietary recommendations [123].

Vitamin C Deficiency

Vitamin C (ascorbic acid) is a water-soluble cofactor required for multiple essential biochemical reactions driven by hydroxylases within the body, including the synthesis of procollagen, carnitine, and catecholamine neurotransmitters [130]. Vitamin C can be synthesized from glucose in many mammals, but most primates, including marmosets, lack the gulonolactone oxidase enzyme required to catalyze this reaction and therefore require a dietary source. A marmoset should consume 20 mg/kg of metabolic body weight per day spread across multiple feedings to prevent vitamin C deficiency [131]. Vitamin C is highly labile in the presence of oxygen, heat, or ultraviolet radiation. The fact that it is vulnerable to environmental deactivation is the rationale behind federal regulations that require that guinea pig feed not be used more than 90 days from milling date, although encapsulation technologies are able to significantly extend both the practical and regulated shelf life of vitamin C-enriched diets. Supplementation with fresh citrus fruits and other natural sources of vitamin can also be effective in meeting daily nutritional requirements [27].

Vitamin C deficiency is the cause of the multisystemic disease "scurvy." In humans, it initially presents with nonspecific symptoms such as weakness, weight loss, and anorexia. As disease progresses, a number of musculoskeletal signs can occur, including osteoporosis, joint and muscle pain, and bleeding into the joint and periosteal space [132,133], the latter is the cause of the classic cephalhematomas reported in squirrel monkeys (*Saimiri* spp.) [134]. Alveolar bone is dramatically affected, and sloughing of the teeth and bleeding of the gums is common [132,133]. Radiographs demonstrate cortical thinning, osteoporosis, osteonecrosis, and a radiopaque metaphyseal band in long bones, knobby costochondral junctions, and biconcave vertebral deformity. Histopathologic examination may reveal the presence of hemorrhagic bone infarctions. Electron microscopy will show loss of mature collagen fibers in vessel walls and the interstitial tissues [132]. Definitive diagnosis is through the measurement of low serum concentrations of vitamin C, and it should be noted that normal blood vitamin C levels in marmosets (2.5 mg/100 mL) are approximately five times higher than tamarins [135].

The majority of reports of vitamin C deficiency in marmosets are historical and experimentally induced. Onset of signs can take 10 weeks, but disease is fulminant, fast, and fatal. Affected marmosets demonstrate decreased mobility and anorexia accompanied by anemia [136], with the predominant musculoskeletal sign of periodontal bony lysis accompanied by hemorrhage [137]. Once clinical signs have appeared, therapy with vitamin C is ineffective in marmosets but provision of sufficient dietary vitamin C is preventative [136].

Marmosets as a Translational Model for Osteoporosis

Marmosets have been proposed as a highly translational model for osteoporosis using two paradigms as an alternative to rats or macaques. Male and female aged marmosets demonstrate decreased BMD compared with younger marmosets. One study tested the translational value of an aged marmoset model by treating a subset of 7- to 9-year-old male and female marmosets with an osteoclast inhibitor, alendronate, that is commonly used in the treatment of osteoporosis in human patients [30]. Alendronate treatment for 60 days resulted in increased cancellous bone volume in the lumbar vertebrae and distal femur compared with untreated controls when evaluated with microCT. The ability of alendronate to halt the ongoing bone remodeling that is a hallmark of osteoporosis was demonstrated by decreased serum osteocalcin levels in treated marmosets compared with controls. It should be noted that both male and female marmosets demonstrated a similar response to alendronate.

In aging humans, waning levels of sex hormones play a role in the development of osteoporosis, and gonadectomy is a key element in bone loss in rodent models of osteoporosis. Female marmosets, similar to female rodents, do not undergo menopause. Attempts to model postmenopausal osteoporosis in marmosets by surgical ovariectomy have proven ineffective [138]. However, orchidectomy of male marmosets results in decreased BMD within 12 months and represents a second paradigm for modeling osteoporosis [34]. In this model, increased bone turnover is evidenced by elevated serum osteocalcin and CTX levels. In young adult marmosets (4- to 6-years old), the magnitude of loss of BMD fell short of the WHO definition for osteoporosis but did qualify as clinical osteopenia and was limited to the peripheral long bones. However, in two older male marmosets (10 and 11 years old), the magnitude of BMD was marked enough to be categorized as osteoporosis and was observed in both the peripheral long bones and lumbar vertebrae similar to the expected findings in elderly humans. Further study is needed to determine if aged orchidectomized male marmosets represent a translational model for osteoporosis in humans.

Adult marmosets, with their bipedal stance, closed growth plates, lamellar bone structure including Haversian canals, and persistent bone remodeling, present an intriguing animal model for osteoporosis. However, findings in such models should be interpreted with some caution as marmosets can have circulating serum vitamin D levels in excess of 4- to 10-fold that of other primates [139]. Such end-organ resistance not only represents a distinct difference in bone metabolism in marmosets compared with other species but also may be evidence of a more global resistance to select actions of steroid hormones. If this is the case, such end-organ resistance could minimize the translational applicability of osteoporosis models in marmosets [140].

Developmental and Congenital Bone Disease

Cleft Palate

Facial development is a complex process that involves the median frontonasal prominence, paired maxillary prominences, and paired mandibular prominences. Upper lip formation, one of the first facial features to be completed, requires precise fusion between the maxillary prominences and the medial nasal tissues, an outgrowth of the frontonasal prominence. Palatal fusion then proceeds anteriorly to posteriorly ending with fusion of the uvula. Palatal development involves primary and secondary palatal structures. The primary palate (medial palatine process), derived from the fused medial nasal tissues, forms the philtrum, part of the alveolus, and the most anterior segment of the hard palate. The secondary palate, derived from swellings of the maxillary prominence (the lateral palatine processes), forms the uvula, and hard and soft palates [141].

Integration of several complex embryonic processes is necessary for normal facial and oral cavity development. Disruptions in either timing or positioning at any stage of this process can result in cleft lip alone without cleft palate (CL), cleft lip with or without cleft palate (CL/P), which includes cleft lip only and cleft lip with cleft palate, cleft lip with cleft palate (CLP), or posterior cleft palate without cleft lip (CPO). These conditions may arise as isolated birth defects or as components of syndromic or nonsyndromic patterns of malformation [142].

Orofacial clefts are among the most frequent congenital anomalies reported in nonhuman primates [142–150]. Discovery of cleft palate following spontaneous abortion in common marmosets has been documented twice [145,149]. Proposed etiologies included abnormal uterine morphology, compromised fetal–placental blood supply, and environmental stressors with secondary elevations in maternal serum cortisol; physical impediments to the migration of the palate secondary to decreased fetal motion may also contribute [151]. Orofacial clefting in viable neonatal marmosets is extremely rare. One case of complex malformations including CLP, microphthalmia, ocular hypertelorism, and polydactyly has been reported in a full-term triplet with two unaffected siblings [150]. The authors noted that the combination of malformations included the 13-trisomy triad and several of the other abnormalities associated with the human condition. Cytogenetic assessment to compare this nonhuman complex to 13-trisomy syndrome was, however, not possible; hence this hypothesis was based solely on phenotypic observations.

Micromelic Dysplasia

Skeletal dysplasias of long bones are named according to the location of the primary abnormality (diaphyseal, epiphyseal, or metaphyseal) and, if present, bone shortening (rhizomelic = humerus and femur, mesomelic = radius, ulna, tibia, and fibula, or micromelic = all long bones) [152]. Although such conditions are relatively rare in human and nonhuman primates, an idiopathic micromelic dysplasia-like syndrome characterized by severe, progressive bending and shortening of the scapula, humerus, radius, ulna, tibia, and fibula in the mid-to-distal diaphyseal region has been reported in four captive marmosets [153]. Affected animals exhibited no evidence of pain, discomfort, or abnormal behavior despite being unable to fully extend the distal portions of their limbs. Although lesions associated with vitamin D deficiency (i.e., rickets or osteomalacia) were not appreciable, serum vitamin D was decreased in one of the four affected animals (the remaining three were not analyzed). Proposed etiologies for what was concluded to be a primary growth abnormality included genetic defects, metabolic disorders (vitamin C or D deficiency), and toxicities (fluoride, hypervitaminosis A).

Infectious Etiologies for Bone Disease

Reports of infectious causes of musculoskeletal disease in *C. jacchus* are scant, though infectious etiologies have been deemed the culprit for periodontal disease and noma in *Callithrix penicillata* and *Saguinus oedipus*, respectively. Captive-born *C. penicillata* has been used as a model for periodontal disease in humans. The contribution of the oral microbiome, specifically of the virulence factors of *Fusobacterium* spp. and *Actinobacillus* spp., to the pathology of the dentition and jaw has been extensively examined in this naturally occurring model [158,159], but does not reliably translate to the human condition [160]. *S. oedipus* has also been evaluated as a potential model for periodontal disease in humans. However, in wild-caught animals, the incidence of gingival disease was 18.2% and bony involvement less than 0.5% [161]. Furthermore, the mononuclear phagocytic nature of the associated inflammatory response failed to mimic the lymphoplasmacytic nature of the response in humans [162], so induced models were developed. When occlusive pressure is applied to the dentition or when complete Freund's adjuvant is used to model periodontal disease, marked destruction of the periodontal bone is noted [163,164]. One report of two cases of facial noma, a polymicrobial, necrotizing condition, in *S. oedipus* described involvement of the maxillary bone in addition to the surrounding soft tissues [165]. Similar occurrences of muscle or bone disease of infectious etiology may occur in *C. jacchus* but have yet to be reported in the literature.

Joint Disease

Though spontaneous joint disease in marmosets is only rarely documented in the literature, it likely occurs as in any other species. One study reports the prevalence of osteoarthritis, calcium pyrophosphate deposition disease, and spondyloarthropathy in several species of New World primates [166]. None of these conditions were diagnosed in the callitrichids examined in this particular study; however, the authors recognized that the small number (40) of animals examined likely influenced this outcome. Interestingly, they concluded that being raised/housed in confinement (as is typical in research colonies) was a significant risk factor for the development of osteoarthritis and spondyloarthropathy for other New World primates. Another report details spondyloarthropathy in the common marmoset [166] and describes the finding of erosive disease with closely associated reactive new bone formation in the spine in addition to frequent involvement of the appendicular skeleton of adult animals only. Erosive lesions of the zygapophyseal joints of the spine and sacroiliac joint with or without syndesmophyte formation (calcification of the annulus fibrosus) were used to diagnose spondyloarthropathy in this study. Perierosional new bone formation, recognized in spondyloarthropathy, was used to distinguish bone loss caused by other erosive arthritides, such as rheumatoid arthritis, which are considered to be purely bone-resorbing. Osteopenia was not a feature of disease.

A marmoset model of arthritis was developed in the 1980s by sensitization to various antigens (ovalbumin, methylated-BSA, or human fibrin) with or without *Bordetella pertussis* organisms followed by intraarticular injection of a similar antigen [167]. The methylated-BSA and human fibrin sensitization and challenge resulted in joint inflammation, whereas ovalbumin did not.

Marmosets have been used recently as an experimental model of collagen-induced arthritis (CIA), which mimics autoimmune (rheumatoid) arthritis of humans [99]. The disease is typically induced by immunization of a marmoset with collagen type II isolated from a cow or chicken, which causes sensitization and thus periarticular inflammation [99]. Unlike a similarly induced macaque model, which only develops an acute and more destructive form of the disease, the marmoset develops a semiacute or chronic form, making it a more accurate model of chronic rheumatoid arthritis in humans. This model could provide the opportunity to test treatments being developed to help humans suffering from rheumatoid arthritis.

Treatment of Pain Associated With Arthritis

Treatment of musculoskeletal pain in marmosets can be accomplished using a variety of pharmacological agents. One report concerned with providing analgesia that does not interfere with study outcomes in macaque models of arthritis concluded that opioids adequately controlled joint pain, while simultaneously allowing for the evaluation of mechanisms of arthritis development [168].

Skeletal Muscle Disease

Arguably the most important etiology contributing to myopathy in marmosets is what is known historically as marmoset wasting syndrome (MWS). Terminology for this syndrome has evolved to focus on the underlying pathology of chronic lymphocytic enteritis/enterocolitis that results in a clinical syndrome similar to inflammatory bowel disease (IBD). IBD is a syndrome characterized by marked weight loss, alopecia, muscle atrophy, anemia, and colitis with diarrhea. Histologic examination of muscle shows profound type II fiber atrophy, hyaline degeneration, variability in muscle fiber diameter, interstitial fibrosis, and focal necrosis [169]. Despite increased knowledge of the nutritional requirements of marmosets, prevalence of IBD remains high in captive colonies [40,170]. While the etiology of muscle wasting seen during IBD is unknown, nutritional deficiencies secondary to intestinal malabsorption due to inflammation have been postulated, specifically for vitamin E [28,119,171]. Diets deficient in vitamin E have been reported to induce the extragastrointestinal signs found with IBD, and supplemental administration of vitamin E and selenium is commonly administered to prevent the muscle pathology associated with IBD [171].

Myopathy directly resulting from vitamin E deficiency has been reported in marmosets. In a survey of a breeding colony of marmosets, Heinz body hemolytic anemia and skeletal muscle myopathy were the most common pathological findings [172]. The observed myopathy was characterized by hyaline degeneration and necrosis with evidence of regeneration, replacement fibrosis, and muscle atrophy. Vitamin E levels were undetectable in affected animals, and supplementation of marmosets with vitamin E was reported to decrease the incidence of anemia in the colony. In another report, five privately owned marmosets presented with weight loss and anemia in the presence of low serum vitamin E, and on necropsy were diagnosed with degenerative myopathy, systemic lipofuscin deposition, increased erythrophagocytosis, and hemosiderosis [173]. IBD-associated myopathy was thought to have been influenced by vitamin E deficiency in these marmosets and likely contributed to the muscle pathology.

References

[1] Kohn LA, Olson P, Cheverud JM. Age of epiphyseal closure in tamarins and marmosets. Am J Primatol 1997;41:129–39.

[2] Casteleyn C, Bakker J, Breugelmans S, Kondova I, Saunders J, Langermans JA, Cornillie P, Van Den Broeck W, Van Loo D, Van Hoorebeke L, Bosseler L, Chiers K, Decostere A. Anatomical description and morphometry of the skeleton of the common marmoset (*Callithrix jacchus*). Lab Anim 2012;46:152–63.

[3] Raisz LG. Physiology and pathophysiology of bone remodeling. Clin Chem 1999;45:1353–8.

[4] Buck DW, Dumanian GA. Bone biology and physiology: Part I. The fundamentals. Plast Reconstr Surg 2012;129:1314–20.

[5] Dusso AS, Brown AJ, Slatopolsky E. Vitamin D. Am J Physiol Ren Physiol 2005;289:F8–28.

[6] Florencio-Silva R, Sasso GR, Sasso-Cerri E, Simões MJ, Cerri PS. Biology of bone tissue: structure, function, and factors that influence bone cells. BioMed Res Int 2015;2015:421746.

[7] Bikle DD. Vitamin D metabolism, mechanism of action, and clinical applications. Chem Biol 2014;21:319–29.

[8] Christakos S, Ajibade DV, Dhawan P, Fechner AJ, Mady LJ. Vitamin D: metabolism. Endocrinol Metab Clin N Am 2010;39: 243–53.

[9] Hunt RD, Garcia FG, Hegsted DM. A comparison of vitamin D2 and D3 in new world primates. I. Production and regression of osteodystrophia fibrosa. Lab Anim Care 1967a;17:222–34.

[10] Hunt RD, Garcia FG, Hegsted DM, Kaplinsky N. Vitamins D2 and D3 in new world primates: influence on calcium absorption. Science 1967b;157:943–5.

[11] Bikle DD. Vitamin D and bone. Curr Osteoporos Rep 2012;10: 151–9.

[12] Christakos S, Dhawan P, Porta A, Mady LJ, Seth T. Vitamin D and intestinal calcium absorption. Mol Cell Endocrinol 2011;347: 25–9.

[13] Nussey S, Whitehead S. Endocrinology: an integrated approach. Oxford; 2001.

[14] Dawson-Hughes B. Vitamin D and muscle function. J Steroid Biochem Mol Biol 2017;173:313–6. See PMID 28341251.

[15] Rojas-Rivera J, De LA Piedra C, Ramos A, Ortiz A, Egido J. The expanding spectrum of biological actions of vitamin D. Nephrol Dial Transplant 2010;25:2850–65.

[16] Liu PT, Stenger S, Li H, Wenzel L, Tan BH, Krutzik SR, Ochoa MT, Schauber J, Wu K, Meinken C, Kamen DL, Wagner M, Bals R, Steinmeyer A, Zugel U, Gallo RL, Eisenberg D, Hewison M, Hollis BW, Adams JS, Bloom BR, Modlin RL. Toll-like receptor triggering of a vitamin D-mediated human antimicrobial response. Science 2006;311: 1770–3.

[17] Shinki T, Shiina Y, Takahashi N, Tanioka Y, Koizumi H, Suda T. Extremely high circulating levels of 1 alpha,25-dihydroxyvitamin D3 in the marmoset, a new world monkey. Biochem Biophys Res Commun 1983;114:452–7.

[18] Takahashi N, Suda S, Shinki T, Horiuchi N, Shiina Y, Tanioka Y, Koizumi H, Suda T. The mechanism of end-organ resistance to 1 alpha,25-dihydroxycholecalciferol in the common marmoset. Biochem J 1985;227:555–63.

[19] Adams JS, Chen H, Chun RF, Nguyen L, Wu S, Ren SY, Barsony J, Gacad MA. Novel regulators of vitamin D action and metabolism: lessons learned at the Los Angeles zoo. J Cell Biochem 2003;88:308–14.

[20] Chun RF, Chen H, Boldrick L, Sweet C, Adams JS. Cloning, sequencing, and functional characterization of the vitamin D receptor in vitamin D-resistant new world primates. Am J Primatol 2001;54:107–18.

[21] Gacad MA, Adams JS. Endogenous blockade of 1,25-dihydroxyvitamin D-receptor binding in new world primate cells. J Clin Invest 1991;87:996–1001.

[22] Sussman RW, Kinzey WG. The ecological role of the callitrichidae: a review. Am J Phys Anthropol 1984;64:419–49.

[23] Council NR. The psychological well-being of nonhuman primates. Washington, D.C., USA: National Academies Press; 1998.

[24] Layne DG, Power RA. Husbandry, handling, and nutrition for marmosets. Comp Med 2003;53:351–9.

[25] Gore MA, Brandes F, Kaup FJ, Lenzner R, Mothes T, Osman AA. Callitrichid nutrition and food sensitivity. J Med Primatol 2001; 30:179–84.

[26] Kuehnel F, Mietsch M, Buettner T, Vervuert I, Ababneh R, Einspanier A. The influence of gluten on clinical and immunological status of common marmosets (*Callithrix jacchus*). J Med Primatol 2013;42:300–9.

[27] Council NR. Nutrient requirements of nonhuman primates. Washington, D.C., USA: National Academies Press; 2003.

[28] Shimwell M, Warrington BF, Fowler JS. Dietary habits relating to 'wasting marmoset syndrome' (WMS). Lab Anim 1979;13: 139–42.

[29] Olson EJ, Shaw GC, Hutchinson EK, Schultz-Darken N, Bolton ID, Parker JB, Morrison JM, Baxter VK, Pate KA, Mankowski JL, Carlson CS. Bone disease in the common marmoset: radiographic and histological findings. Vet Pathol 2015;52:883–93.

[30] Bagi CM, Volberg M, Moalli M, Shen V, Olson E, Hanson N, Berryman E, Andresen CJ. Age-related changes in marmoset trabecular and cortical bone and response to alendronate therapy resemble human bone physiology and architecture. Anat Rec 2007;290:1005–16.

[31] Grohmann J, Kuehnel F, Buchwald U, Koeller G, Habla C, Einspanier A. Analysis of the bone metabolism by quantitative computer tomography and clinical chemistry in a primate model (*Callithrix jacchus*). J Med Primatol 2012a;41:1–10.

[32] Grohmann J, Taetzner S, Theuss T, Kuehnel F, Buchwald U, Einspanier A. The conclusiveness of less-invasive imaging techniques (computer tomography, X-ray) with regard to their identification of bone diseases in a primate model (*Callithrix jacchus*). J Med Primatol 2012b;41:130–7.

[33] Jarcho MR, Power ML, Layne-Colon DG, Tardif SD. Digestive efficiency mediated by serum calcium predicts bone mineral density in the common marmoset (*Callithrix jacchus*). Am J Primatol 2013;75:153–60.

[34] Seidlová-Wuttke D, Schlumbohm C, Jarry H, Dullin C, Wuttke W. Orchidectomized (orx) marmoset (*Callithrix jacchus*) as a model to study the development of osteopenia/osteoporosis. Am J Primatol 2008;70:294–300.

[35] Griffith JF, Engelke K, Genant HK. Looking beyond bone mineral density: imaging assessment of bone quality. Ann NY Acad Sci 2010;1192:45–56.

[36] Harris WH, Heaney RP. Skeletal renewal and metabolic bone disease. N Engl J Med 1969;280:303–11.

[37] Yochum TR, Rowe LJ. Essentials of skeletal radiology. Baltimore, MD, USA: Lippincott Williams & Wilkins; 2004.

[38] Apostol L, Boudousq V, Basset O, Odet C, Yot S, Tabary J, Dinten JM, Boiler E, Kotzki PO, Peyrin F. Relevance of 2D radiographic texture analysis for the assessment of 3D bone microarchitecture. Med Phys 2006;33:3546–56.

[39] Rosholm A, Hyldstrup L, Backsgaard L, Grunkin M, Thodberg HH. Estimation of bone mineral density by digital X-ray radiogrammetry: theoretical background and clinical testing. Osteoporos Int 2001;12:961–9.

[40] Baxter VK, Shaw GC, Sotuyo NP, Carlson CS, Olson EJ, Zink MC, Mankowski JL, Adams RJ, Hutchinson EK, Metcalf Pate KA. Serum albumin and body weight as biomarkers for the antemortem identification of bone and gastrointestinal disease in the common marmoset. PLoS One 2013;8.

[41] Murphey MD, Sartoris DJ, Quale JL, Pathria MN, Martin NL. Musculoskeletal manifestations of chronic renal insufficiency. Radiographics 1993;13:357–79.

[42] Patel AA, Ramanathan R, Kuban J, Willis MH. Imaging findings and evaluation of metabolic bone disease. Adv Radiol 2015;2015.

[43] Schuit SC, Van Der Klift M, Weel AE, De Laet CE, Burger H, Seeman E, Hofman A, Uitterlinden AG, Van Leeuwen JP, Pols HA. Fracture incidence and association with bone mineral density in elderly men and women: the Rotterdam Study. Bone 2004;34:195–202.

[44] Organization, W. H. Prevention and management of osteoporosis: report of a WHO Scientific Group. Geneva, Switzerland: World Health Organization; 2003.

[45] Dasher LG, Newton CD, Lenchik L. Dual X-ray absorptiometry in today's clinical practice. Radiol Clin 2010;48:541–60.

[46] El Maghraoui A, Roux C. DXA scanning in clinical practice. QJM 2008;101:605–17.

[47] Watts NB. Fundamentals and pitfalls of bone densitometry using dual-energy X-ray absorptiometry (DXA). Osteoporos Int 2004; 15:847–54.

[48] Guglielmi G, Muscarella S, Bazzocchi A. Integrated imaging approach to osteoporosis: state-of-the-art review and update. Radiographics 2011;31:1343–64.

[49] Adams JE, Bishop N. Dual energy X-ray absorptiometry (DXA) in adults and children. Washington, D.C., USA: American Society for Bone and Mineral Research; 2009.

[50] Sornay-Rendu E, Munoz F, Duboeuf F, Delmas PD. Rate of forearm bone loss is associated with an increased risk of fracture independently of bone mass in postmenopausal women: the OFELY study. J Bone Miner Res 2005;20:1929–35.

[51] Adams JE. Advances in bone imaging for osteoporosis. Nat Rev Endocrinol 2013;9:28–42.

[52] Moyad MA. Osteoporosis: a rapid review of risk factors and screening methods. Urol Oncol 2003;21:375–9.

[53] Guglielmi G, Lang TF. Quantitative computed tomography. Semin Muscoskel Radiol 2002;6:219–27.

[54] Genant HK, Engelke K, Fuerst T, Gluer CC, Grampp S, Harris ST, Jergas M, Lang T, Lu Y, Majumdar S, Mathur A, Takada M. Noninvasive assessment of bone mineral and structure: state of the art. J Bone Miner Res 1996;11:707–30.

[55] Haugeberg G. Imaging of metabolic bone diseases. Best Pract Res Clin Rheumatol 2008;22:1127–39.

[56] Pistoia W, Van Rietbergen B, Lochmuller EM, Lill CA, Eckstein F, Ruegsegger P. Estimation of distal radius failure load with micro-finite element analysis models based on three-dimensional peripheral quantitative computed tomography images. Bone 2002;30:842–8.

[57] Lang TF. Quantitative computed tomography. Radiol Clin 2010; 48:589–600.

[58] Prevrhal S, Engelke K, Kalender WA. Accuracy limits for the determination of cortical width and density: the influence of object size and CT imaging parameters. Phys Med Biol 1999;44:751–64.

[59] Phan CM, Matsuura M, Bauer JS, Dunn TC, Newitt D, Lochmueller EM, Eckstein F, Majumdar S, Link TM. Trabecular bone structure of the calcaneus: comparison of MR imaging at 3.0 and 1.5 T with micro-CT as the standard of reference. Radiology 2006;239:488–96.

[60] Jiang Y, Zhao JJ, Mitlak BH, Wang O, Genant HK, Eriksen EF. Recombinant human parathyroid hormone (1-34) [teriparatide] improves both cortical and cancellous bone structure. J Bone Miner Res 2003;18:1932–41.

[61] Johnson LA, Della Santina CC, Wang X. Temporal bone characterization and cochlear implant feasibility in the common marmoset (*Callithrix jacchus*). Hear Res 2012;290:37–44.

[62] Smith TD, Rossie JB, Cooper GM, Schmieg RM, Bonar CJ, Mooney MP, Siegel MI. Comparative microcomputed tomography and histological study of maxillary pneumatization in four species of new world monkeys: the perinatal period. Am J Phys Anthropol 2011;144:392–410.

[63] Krug R, Carballido-Gamio J, Burghardt AJ, Kazakia G, Hyun BH, Jobke B, Banerjee S, Huber M, Link TM, Majumdar S. Assessment of trabecular bone structure comparing magnetic resonance imaging at 3 Tesla with high-resolution peripheral quantitative computed tomography ex vivo and in vivo. Osteoporos Int 2008;19:653–61.

[64] Chesnut 3rd CH, Majumdar S, Newitt DC, Shields A, Van Pelt J, Laschansky E, Azria M, Kriegman A, Olson M, Eriksen EF, Mindeholm L. Effects of salmon calcitonin on trabecular microarchitecture as determined by magnetic resonance imaging: results from the QUEST study. J Bone Miner Res 2005;20: 1548–61.

[65] Krug R, Banerjee S, Han ET, Newitt DC, Link TM, Majumdar S. Feasibility of in vivo structural analysis of high-resolution magnetic resonance images of the proximal femur. Osteoporos Int 2005;16:1307–14.

[66] Cabral HW, Andolphi BF, Ferreira BV, Alves DC, Morelato RL, Chambo A, Borges LSS. The use of biomarkers in clinical osteoporosis. Rev Assoc Med Bras (1992) 2016;62:368–76.

[67] Eastell R, Hannon RA. Biomarkers of bone health and osteoporosis risk. Proc Nutr Soc 2008;67:157–62.

[68] Greenspan SL, Resnick NM, Parker RA. Early changes in biochemical markers of bone turnover are associated with long-term changes in bone mineral density in elderly women on alendronate, hormone replacement therapy, or combination therapy: a three-year, double-blind, placebo-controlled, randomized clinical trial. J Clin Endocrinol Metab 2005;90:2762–7.

[69] Greenspan SL, Rosen HN, Parker RA. Early changes in serum N-telopeptide and C-telopeptide cross-linked collagen type 1 predict long-term response to alendronate therapy in elderly women. J Clin Endocrinol Metab 2000;85:3537–40.

[70] Garnero P, Dargent-Molina P, Hans D, Schott AM, Breart G, Meunier PJ, Delmas PD. Do markers of bone resorption add to bone mineral density and ultrasonographic heel measurement for the prediction of hip fracture in elderly women? The EPIDOS prospective study. Osteoporos Int 1998;8:563–9.

[71] Sarkar S, Reginster JY, Crans GG, DIEZ-Perez A, Pinette KV, Delmas PD. Relationship between changes in biochemical markers of bone turnover and BMD to predict vertebral fracture risk. J Bone Miner Res 2004;19:394–401.

[72] Angeliewa A, Budde M, Schlachter M, Hoyle NR, Bauss F. Biochemical bone turnover markers are useful tools to assess changes in bone metabolism in marmosets. J Bone Miner Metabol 2004;22:192–7.

[73] Woitge HW, Seibel MJ, Ziegler R. Comparison of total and bone-specific alkaline phosphatase in patients with nonskeletal disorder or metabolic bone diseases. Clin Chem 1996;42: 1796–804.

[74] Khan KN, Logan AC, Blomquist EM. Idiopathic myelofibrosis (agnogenic myeloid metaplasia) in a marmoset (*Callithrix jacchus*): hematologic and histopathologic changes. Vet Pathol 1997;34:341–5.

[75] Davy CW, Jackson MR, Walker J. Reference intervals for some clinical chemical parameters in the marmoset (*Callithrix jacchus*): effect of age and sex. Lab Anim 1984a;18:135–42.

[76] Brancaccio P, Lippi G, Maffulli N. Biochemical markers of muscular damage. Clin Chem Lab Med 2010;48:757–67.

[77] Amelink GJ, Koot RW, Erich WB, Van Gijn J, Bar PR. Sex-linked variation in creatine kinase release, and its dependence on oestradiol, can be demonstrated in an in-vitro rat skeletal muscle preparation. Acta Physiol Scand 1990;138:115–24.

[78] Dale KM, White CM, Henyan NN, Kluger J, Coleman CI. Impact of statin dosing intensity on transaminase and creatine kinase. Am J Med 2007;120:706–12.

[79] Swaiman KF, Sandler B. The use of serum creatine phosphokinase and other serum enzymes in the diagnosis of progressive muscular dystrophy. J Pediatr 1963;63:1116–9.

[80] Hanton G, Sobry C, Daguès N, Rochefort GYY, Bonnet P, Eder V. Cardiovascular toxicity of minoxidil in the marmoset. Toxicol Lett 2008;180:157–65.

[81] Davy CW, Fulleylove M, Edmunds JG, Eichler DA, Rushton B, Tudor RJ, Walker JM. The diagnostic usefulness of isocitrate dehydrogenase (ICDH) in the marmoset (*Callithrix jacchus*). J Appl Toxicol 1989;9:33–7.

[82] Davy CW, Jackson MR, Walker JM. The effect of haemolysis on some clinical chemistry parameters in the marmoset (*Callithrix jacchus*). Lab Anim 1984b;18:161–8.

[83] Shin K-AA, Park KD, Ahn J, Park Y, Kim Y-JJ. Comparison of changes in biochemical markers for skeletal muscles, hepatic metabolism, and renal function after three types of long-distance running: observational study. Medicine 2016;95.

[84] Owen K, Pick CR, Libretto SE, Adams MJ. Toxicity of a novel HMG-CoA reductase inhibitor in the common marmoset (*Callithrix jacchus*). Hum Exp Toxicol 1994;13:357–68.

[85] Bakker J, Uilenreef JJ, Pelt ER, Brok HP, Remarque EJ, Langermans JA. Comparison of three different sedative-anaesthetic protocols (ketamine, ketamine-medetomidine and alphaxalone) in common marmosets (*Callithrix jacchus*). BMC Vet Res 2013;9:113.

[86] Davy CW, Trennery PN, Edmunds JG, Altman JF, Eichler DA. Local myotoxicity of ketamine hydrochloride in the marmoset. Lab Anim 1987;21:60–7.

[87] Chapman DW, Simpson JA, Iscoe S, Robins T, Nosaka K. Changes in serum fast and slow skeletal troponin I concentration following maximal eccentric contractions. J Sci Med Sport 2013;16:82–5.

[88] Nilsson MI, Nissar AA, Al-Sajee D, Tarnopolsky MA, Parise G, Lach B, Furst DO, Van Der Ven PF, Kley RA, Hawke TJ. Xin is a marker of skeletal muscle damage severity in myopathies. Am J Pathol 2013;183:1703–9.

[89] Rebalka IA, Hawke TJ. Potential biomarkers of skeletal muscle damage. Biomarkers Med 2014;8:375–8.

[90] Chen-Levy Z, Wener MH, Toivola B, Daum P, Reyes M, Fine JS. Factors affecting urinary myoglobin stability in vitro. Am J Clin Pathol 2005;123:432–8.

[91] Blondheim SH, Margoliash E, Shafrir E. A simple test for myohemoglobinuria (myoglobinuria). J Am Med Assoc 1958;167:453–4.

[92] Ascensao A, Rebelo A, Oliveira E, Marques F, Pereira L, Magalhaes J. Biochemical impact of a soccer match – analysis of oxidative stress and muscle damage markers throughout recovery. Clin Biochem 2008;41:841–51.

[93] Neubauer O, Konig D, Wagner KH. Recovery after an Ironman triathlon: sustained inflammatory responses and muscular stress. Eur J Appl Physiol 2008;104:417–26.

[94] Raza K, Breese M, Nightingale P, Kumar K, Potter T, Carruthers DM, Situnayake D, Gordon C, Buckley CD, Salmon M, Kitas GD. Predictive value of antibodies to cyclic citrullinated peptide in patients with very early inflammatory arthritis. J Rheumatol 2005;32:231–8.

[95] Takahashi M, Kushida K, Hoshino H, Suzuki M, Sano M, Miyamoto S, Inoue T. Concentrations of pyridinoline and deoxypyridinoline in joint tissues from patients with osteoarthritis or rheumatoid arthritis. Ann Rheum Dis 1996;55:324–7.

[96] Ahmed U, Anwar A, Savage RS, Costa ML, Mackay N, Filer A, Raza K, Watts RA, Winyard PG, Tarr J, Haigh RC, Thornalley PJ, Rabbani N. Biomarkers of early stage osteoarthritis, rheumatoid arthritis and musculoskeletal health. Sci Rep 2015;5:9259.

[97] Van Venrooij WJ, Van Beers JJ, Pruijn GJ. Anti-CCP antibodies: the past, the present and the future. Nat Rev Rheumatol 2011;7:391–8.

[98] Van Pelt J, Kamerling JP, Bakker HD, Vliegenthart JF. A comparative study of sialyloligosaccharides isolated from sialidosis and galactosialidosis urine. J Inherit Metab Dis 1991;14:730–40.

[99] Vierboom MP, Breedveld E, Kondova I, T Hart BA. Collagen-induced arthritis in common marmosets: a new nonhuman primate model for chronic arthritis. Arthritis Res Ther 2010;12:R200.

[100] Seeliger C, Balmayor ER, VAN Griensven M. miRNAs related to skeletal diseases. Stem Cells Dev 2016;25:1261–81.

[101] Laterza OF, Lim L, Garrett-Engele PW, Vlasakova K, Muniappa N, Tanaka WK, Johnson JM, Sina JF, Fare TL, Sistare FD, Glaab WE. Plasma microRNAs as sensitive and specific biomarkers of tissue injury. Clin Chem 2009;55:1977–83.

[102] Brooks H, Blair WR. Osteomalacia of primates in captivity: a clinical and pathological study of "cage paralysis". Ninth Annu Rep NY Zool Soc 1905.

[103] Yamaguchi A, Kohno Y, Yamazaki T, Takahashi N, Shinki T, Horiuchi N, Suda T, Koizumi H, Tanioka Y, Yoshiki S. Bone in the marmoset: a resemblance to vitamin D-dependent rickets, type II. Calcif Tissue Int 1986;39:22–7.

[104] Fowler ME. Metabolic bone disease. Zoo and wildlife medicine. Philadelphia, PA, USA: W.B. Saunders Comapny; 1986.

[105] Hatt JM, Sainsbury AW. Unusual case of metabolic bone disease in a common marmoset (*Callithrix jacchus*). Vet Rec 1998;143: 78–80.

[106] Ludlage E, Mansfield K. Clinical care and diseases of the common marmoset (*Callithrix jacchus*). Comp Med 2003;53:369–82.

[107] Wissman MA. Nutrition and husbandry of callitrichids (marmosets and tamarins). Vet Clin North Am Exot Anim Pract 1999;2: 209–40.

[108] Hampton Jr JK, Hampton SH, Landwehr BT. Observations on a successful breeding colony of the marmoset, Oedipomidas oedipus. Folia Primatol 1966;4:265–87.

[109] Barnard D, Knapka J, Renquist D. The apparent reversal of a wasting syndrome by nutritional intervention in *Saguinus mystax*. Lab Anim Sci 1988;38:282–8.

[110] Crook G. A nutritional reversal of marmoset wasting syndrome. Australian Primatol 1989;4:21.

[111] Otovic P, Smith S, Hutchinson E. The use of glucocorticoids in marmoset wasting syndrome. J Med Primatol 2015;44:53–9.
[112] Yoshimoto T, Niimi K, Takahashi E. Tranexamic acid and supportive measures to treat wasting marmoset syndrome. Comp Med 2016;66:468–73.
[113] Wachtman LM, Pistorio AL, Eliades S, Mankowski JL. Calcinosis circumscripta in a common marmoset (*Callithrix jacchus* jacchus). J Am Assoc Lab Anim Sci 2006;45:54–7.
[114] Nikodimopoulou M, Liakos S. Secondary hyperparathyroidism and target organs in chronic kidney disease. Hippokratia 2011; 15:33–8.
[115] Tamagaki K, Yuan Q, Ohkawa H, Imazeki I, Moriguchi Y, Imai N, Sasaki S, Takeda K, Fukagawa M. Severe hyperparathyroidism with bone abnormalities and metastatic calcification in rats with adenine-induced uraemia. Nephrol Dial Transplant 2006; 21:651–9.
[116] Kaspareit J, Friderichs-Gromoll S, Buse E, Habermann G. Background pathology of the common marmoset (*Callithrix jacchus*) in toxicological studies. Exp Toxicol Pathol 2006;57:405–10.
[117] Daniels PJ, Bittel D, Smirnova IV, Winge DR, Andrews GK. Mammalian metal response element-binding transcription factor-1 functions as a zinc sensor in yeast, but not as a sensor of cadmium or oxidative stress. Nucleic Acids Res 2002;30: 3130–40.
[118] Macapinlac MP, Barney GH, Pearson WN, Darby WJ. Production of zinc deficiency in the squirrel monkey (*Saimiri sciureus*). J Nutr 1967;93:499–510.
[119] Chadwick DP, May JC, Lorenz D. Spontaneous zinc deficiency in marmosets, saguinus mystax. Lab Anim Sci 1979;29:482–5.
[120] Swenerton H, Hurley LS. Zinc deficiency in rhesus and bonnet monkeys, including effects on reproduction. J Nutr 1980;110: 575–83.
[121] Golub MS, Gershwin ME, Hurley LS, Hendrickx AG, Baly DL. Induction of marginal zinc deficiency in female rhesus monkeys. Am J Primatol 1982;3:299–305.
[122] Abee CR, Mansfield K, Tardiff S, Morris T. Nonhuman primates in biomedical research: biology and management. Orlando, FL: Academic Press; 2016.
[123] Dever JT, Tanumihardjo SA. Hypervitaminosis A in experimental nonhuman primates: evidence, causes, and the road to recovery. Am J Primatol 2009;71:813–6.
[124] Feskanich D, Singh V, Willett WC, Colditz GA. Vitamin A intake and hip fractures among postmenopausal women. J Am Med Assoc 2002;287:47–54.
[125] Lind T, Ohman C, Calounova G, Rasmusson A, Andersson G, Pejler G, Melhus H. Excessive dietary intake of vitamin A reduces skull bone thickness in mice. PLoS One 2017;12: e0176217.
[126] Michaelsson K, Lithell H, Vessby B, Melhus H. Serum retinol levels and the risk of fracture. N Engl J Med 2003;348: 287–94.
[127] Mills JP, Penniston KL, Tanumihardjo SA. Extra-hepatic vitamin A concentrations in captive Rhesus (*Macaca mulatta*) and Marmoset (*Callithrix jacchus*) monkeys fed excess vitamin A. Int J Vitam Nutr Res 2005;75:126–32.
[128] Mills JP, Tanumihardjo SA. Vitamin A toxicity in wild-caught African green vervet monkeys (*Chlorocebus aethiops*) after 2 years in captivity. Comp Med 2006;56:421–5.
[129] Penniston KL, Thayer JC, Tanumihardjo SA. Serum vitamin A esters are high in captive rhesus (*Macaca mulatta*) and marmoset (*Callithrix jacchus*) monkeys. J Nutr 2003;133:4202–6.
[130] Levine M. New concepts in the biology and biochemistry of ascorbic acid. N Engl J Med 1986;314:892–902.
[131] Flurer CI, Zucker H. Ascorbic acid in a new world monkey family: species difference and influence of stressors on ascorbic acid metabolism. Z Ernahrungswiss 1989;28:49–55.
[132] Fain O. Musculoskeletal manifestations of scurvy. Joint Bone Spine 2005;72:124–8.
[133] Hodges RE, Hood J, Canham JE, Sauberlich HE, Baker EM. Clinical manifestations of ascorbic acid deficiency in man. Am J Clin Nutr 1971;24:432–43.
[134] Blackwell CA, Manning PJ, Hutchinson TC, Fisk SK. Cranial hyperostosis of squirrel monkeys (*Saimiri sciureus*). Lab Anim Sci 1974;24:541–4.
[135] Flurer CI, Zucker H. Difference in serum ascorbate in two species of Callithricidae. Int J Vitam Nutr Res 1987;57:297–8.
[136] Flurer CI, Kern M, Rambeck WA, Zucker H. Ascorbic acid requirement and assessment of ascorbate status in the common marmoset (*Callithrix jacchus*). Ann Nutr Metab 1987;31:245–52.
[137] Driezen S, Levy BM. Histopathology of experimentally induced nutritional deficiency cheilosis in the marmoset (*Callithrix jacchus*). Arch Oral Biol 1969;14:577–82.
[138] Colman RJ, Saltzman W, Abbott DH, Kemnitz JW, Binkley N. Absence of estrogen depletion bone loss in female common marmosets. J Bone Miner Res 1997;12:S342.
[139] Hunt RD, Garcia FG, Hagsted DM. Hypervitaminosis D in new world monkeys. Am J Clin Nutr 1969;22:358–66.
[140] Abbott DH, Barnett DK, Colman RJ, Yamamoto ME, Schultz-Darken NJ. Aspects of common marmoset basic biology and life history important for biomedical research. Comp Med 2003;53:339–50.
[141] Shkoukani MA, Chen M, Vong A. Cleft lip – a comprehensive review. Front Pediatr 2013;1:53.
[142] Goldschmidt B, Lopes CA, Moura M, Nogueira DM, Goncalves MA, Fasano DM, Andrade MC, Nascimento LW, Marinho AM. Cleft lip and palate associated with other malformations in a neotropical primate (*Saimiri ustus*). J Am Assoc Lab Anim Sci 2010;49:357–60.
[143] Baker CA, Hendrickx AG, Cooper RW. Spontaneous malformations in squirrel monkey [*Saimiri sciureus*] fetuses with emphasis on cleft lip and palate. J Med Primatol 1977;6:13–22.
[144] Benirschke K, Kumamoto AT, Bogart MH. Congenital anomalies in Lemur variegatus. J Med Primatol 1981;10:38–45.
[145] Ferraz MC, Matos AVR, Prestes NC, Ferreira JCP, Francelino LC, Cunha FS, Oba E. Fetal malformation in a marmoset (*Callithrix jacchus*): case report. Arq Bras Med Vet Zootec 2014;66:1401–5.
[146] Hill WC, Sabater Pi J. Notes on two anomalies in mandrills (*Mandrillus sphinx* Linn.). Folia Primatol 1970;12:290–5.
[147] Hoopes CW, Jerome CP. Oral-facial clefts and associated malformations in the squirrel monkey (*Saimiri sciureus*). J Med Primatol 1987;16:203–9.
[148] Koford CB, Farber PA, Windle WF. Twins and teratisms in rhesus monkeys. Folia Primatol 1966;4:221–6.
[149] Kraus BS, Garrett WS. Cleft palate in a marmoset: report of a case. Cleft Palate J 1968;5:340–3.
[150] Pugsley SL. Congenital malformations in a common marmoset (*Callithrix jacchus*) similar to human 13-trisomy syndrome. Lab Anim 1985;19:123–4.
[151] Panter KE, Bunch TD, Keeler RF, Sisson DV, Callan RJ. Multiple congenital contractures (MCC) and cleft palate induced in goats by ingestion of piperidine alkaloid-containing plants: reduction in fetal movement as the probable cause. J Toxicol Clin Toxicol 1990;28:69–83.
[152] Rothenberg AB, Berdon WE, Woodard JC, Cowles RA. Hypervitaminosis A-induced premature closure of epiphyses (physeal obliteration) in humans and calves (hyena disease): a historical review of the human and veterinary literature. Pediatr Radiol 2007;37:1264–7.
[153] Bosseler L, Cornillie P, Saunders JH, Bakker J, Langermans JA, Casteleyn C, Decostere A, Chiers K. Micromelic dysplasia-like syndrome in a captive colony of common marmosets (*Callithrix jacchus*). Comp Med 2014;64:394–403.

[154] Benirschke K, Anderson JM, Brownhill LE. Marrow chimerism in marmosets. Science 1962;138:513–5.
[155] Gengozian N, Batson JS, Eide P. Hematologic and cytogenic evidence for chimerism in the marmoset, tamarinus nigricollis. Sam-Tdr-64–61. AMD TR Rep 1964:1–10.
[156] Ross CN, French JA, Orti G. Germ-line chimerism and paternal care in marmosets (Callithrix kuhlii). Proc Natl Acad Sci U S A 2007;104:6278–82.
[157] Sweeney CG, Curran E, Westmoreland SV, Mansfield KG, Vallender EJ. Quantitative molecular assessment of chimerism across tissues in marmosets and tamarins. BMC Genom 2012;13:98.
[158] Lucia LF, Farias FF, Eustaquio CJ, Auxiliadora M, Carvalho R, Alviano CS, Farias LM. Bacteriocin production by Actinobacillus actinomycetemcomitans isolated from the oral cavity of humans with periodontal disease, periodontally healthy subjects and marmosets. Res Microbiol 2002;153:45–52.
[159] Oliveira AA, Farias LM, Nicoli JR, Costa JE, Carvalho MA. Bacteriocin production by Fusobacterium isolates recovered from the oral cavity of human subjects with and without periodontal disease and of marmosets. Res Microbiol 1998;149:585–94.
[160] Saddi-Ortega L, Carvalho MA, Cisalpino PS, Moreira ES. Actinobacillus actinomycetemcomitans genetic heterogeneity: amplification of JP2-like ltx promoter pattern correlated with specific arbitrarily primed polymerase chain reaction (AP-PCR) genotypes from human but not marmoset Brazilian isolates. Can J Microbiol 2002;48:602–10.
[161] Ammons WF, Schectman LR, Page RC. Host tissue response in chronic periodontal disease. 1. The normal periodontium and clinical manifestations of dental and periodontal disease in the marmoset. J Periodontal Res 1972;7:131–43.
[162] Schectman LR, Ammons WF, Simpson DM, Page RC. Host tissue response in chronic periodontal disease. 2. Histologic features of the normal periodontium, and histologic and ultrastructural manifestations of disease in the marmoset. J Periodontal Res 1972;7:195–212.
[163] Dreizen S, Friedman LA, Levy BM, Bernick S. Studies on the biology of the periodontium of marmosets. XIV. The effect of experimentally produced osteomyelitis on the marmoset periodontium. J Periodontol 1972;43:347–51.
[164] Levy BM, Nelms DC, Dreizen S, Bernick S. Pathogenesis of adjuvant-induced destructive periodontitis in marmosets. Primates Med 1978;10:205–14.
[165] Brack M. Noma in *Saguinus oedipus*: a report of 2 cases. Lab Anim 1982;16(4):361–3.
[166] Rothschild BM. Arthritis of the spondyloarthropathy variety in *Callithrix jacchus*. J Med Primatol 1993;22:313–6.
[167] Hunneyball IM. Investigations into the induction of chronic experimental arthritis in the common marmoset (*Callithrix jacchus*). Rheumatol Int 1983;3:69–74.
[168] Vierboom MP, Breedveld E, Keehnen M, Klomp R, Bakker J. Pain relief in nonhuman primate models of arthritis. Meth Mol Biol 2017;1559:411–7.
[169] Murgatroyd LB, Chalmers DT. A histological and histochemical study of normal and diseased skeletal muscle in marmosets *Callithrix jacchus*. J Comp Pathol 1980;90:373–8.
[170] Potkay S. Diseases of the callitrichidae: a review. J Med Primatol 1992;21:189–236.
[171] Baskin GB, Wolf RH, Worth CL, Soike K, Gibson SV, Bieri JG. Anemia, steatitis, and muscle necrosis in marmosets (*Saguinus labiatus*). Lab Anim Sci 1983;33:74–80.
[172] Chalmers DT, Murgatroyd LB, Wadsworth PF. A survey of the pathology of marmosets (*Callithrix jacchus*) derived from a marmoset breeding unit. Lab Anim 1983;17:270–9.
[173] Juan-Salles C, Prats N, Resendes A, Domingo M, Hilton D, Ruiz JM, Garner MM, Valls X, Marco AJ. Anemia, myopathy, and pansteatitis in vitamin E-deficient captive marmosets (Callithrix spp.). Vet Pathol 2003;40:540–7.

CHAPTER

15

Viral Diseases of Common Marmosets

Kerstin Mätz-Rensing, Martina Bleyer
German Primate Center, Pathology Unit, Göttingen, Germany

HERPESVIRUSES

Etiology. The family Herpesviridae is divided into the three subfamilies of *Alpha-, Beta-, and Gammaherpesvirinae*. Members of the family Herpesviridae are typical DNA viruses with a complex double-stranded DNA genome embedded in an icosahedral capsid. They are well adapted to their natural hosts and are able to persist in a latent form within their hosts. Some have the potential to cross species barriers and can induce severe zoonotic infections in susceptible species. For common marmosets, the most important herpesviruses are the *Herpes simplex virus* of the *Alphaherpesvirinae* subfamily and the *Callitrichine herpesvirus 3*, a recently recognized lymphocryptovirus of the *Gammaherpesvirinae* subfamily.

Herpes Simplex Virus

Clinical signs. Marmosets are highly susceptible hosts to human *Herpes simplex virus* (HSV-1,2) [1]. Humans are the original host and the reservoir for HSV. In its natural host, HSV-1 and 2 cause only mild facial lesions, including gingivostomatitis or keratitis, or lead to clinically inapparent infections. There is lifelong infection with *Herpes simplex virus* because of viral latency in sensory neurons. Only in neonates or immunocompromised individuals, fatal encephalitis or systemic disease can occur. When immunity is compromised by stress or disease, the virus becomes activated by suppression of the latency-associated transcript gene. At this stage, the virus can be shed even in the absence of visible lesions. Oral, conjunctival infection and meningoencephalitis are most commonly caused by HSV-1, whereas genital and neonatal infection is usually attributed to HSV-2 infection. Primates are susceptible to both serotypes. Transmission occurs from humans to monkeys and vice versa, leading to epizootics with high morbidity and mortality in common marmosets. There is no evidence for the presence of HSV-1,2 genotypes exhibiting a higher virulence for animals [2]. Spontaneous HSV-1 infections are described in wild marmosets (black tufted-ear marmosets, *Callithrix penicillata*) in a natural park in Brazil [3,4], in young pet marmosets (*Callithrix jacchus*) in private husbandries [5–8], and in a semifree-living common marmoset colony in Germany [9]. Most reported cases of herpesvirus infection in callitrichids occurred in animals kept privately, often in close association with the family.

Epizootiology. The most likely source of transmission is direct close contact to humans suffering from oral herpesvirus infection. The disease is spread by contact with infected saliva or via contaminated objects. Animal owners, visitors, students, or caretakers might be able to infect the animals accidentally, for example, by passing partly eaten food into the cages or by practicing mouth-to-mouth feeding of hand-raised offspring [5]. Once brought into a colony, the disease is spread rapidly between the animals by direct contact and by aerogenous transmission.

Pathology. Usually, infection leads to a severe, rapidly progressing and often fatal generalized febrile disease with typical lesions on oral mucous membranes [1,7]. Clinical symptoms depend on the extent and severity of gross alterations. Infected animals show severe apathy, anorexia, weakness, and marked salivation or serous nasal discharge before they become moribund and die [9]. Death occurs within 2–14 days after the onset of the first symptoms. In rare cases, infected animals survive and develop a persistent infection, which can be monitored by serologic examination [5]. Gross pathology is characterized by erosions and ulcers of variable extent on oral or genital mucous membranes and mucocutaneous junctions of the lips, which may become confluent and are covered by a fibrinonecrotic exudate (Fig. 15.1). The lesions are accompanied by severe lymphadenopathy of regional lymph nodes. Areas of necrosis and hemorrhage may be present in other organ

The Common Marmoset in Captivity and Biomedical Research
https://doi.org/10.1016/B978-0-12-811829-0.00015-7

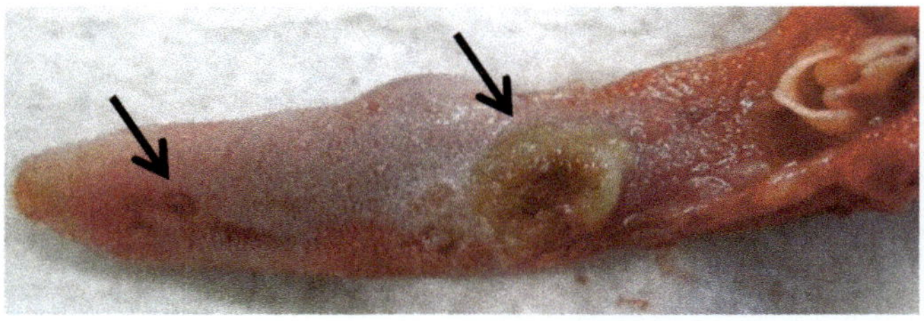

FIGURE 15.1 Common marmoset, tongue, naturally acquired *Herpes simplex virus* infection. Tongue with multifocal to confluent ulcerations (*arrows*).

systems if the disease disseminates. In severe cases, the cerebral cortex may also be involved [5,7]. Histopathologic changes consist of severe vesiculation and ulceration of squamous epithelia of mucous membranes and tongue with acantholysis, parakeratosis, coagulation necrosis, and polykaryocytosis. Necrotic foci may extend from the submucosa to the surface epithelium (Fig. 15.2A). Epithelial cells at the ulcer margins show varying degrees of degeneration and necrosis. Typical intranuclear inclusion bodies are key morphologic features of HSV infection. They are commonly found in epithelial cells at the borders of vesicles and ulcers. These intranuclear inclusions are surrounded by a clear halo and appear red-blue in Feulgen-stained sections. The presence of HSV antigen can be confirmed by immunohistochemistry using commercial antibodies against *Herpes simplex* antigen. Nonsuppurative meningoencephalitis is a frequent finding in HSV-infected callitrichids [5] (Fig. 15.2B). In cases with meningoencephalitis, intranuclear inclusion bodies can be found within neurons and glial cells. The widespread distribution of the virus within neurons and glia cells can be demonstrated by immunohistochemistry (Fig. 15.2C). Using transmission electron microscopy, large numbers of enveloped virions can usually be detected in the cytoplasm and within intercellular spaces (Fig. 15.2D). Intranuclear particles measure 80–100 nm, the enveloped extranuclear particles have an outside diameter of 150 mm or more. Similar gross and histologic lesions are observed in experimentally infected marmosets [10].

Diagnosis. Diagnosis is based on histologic, immunohistologic, and electron microscopic findings, with demonstration of classic intranuclear inclusion bodies. Commercial antibodies used for HSV detection by

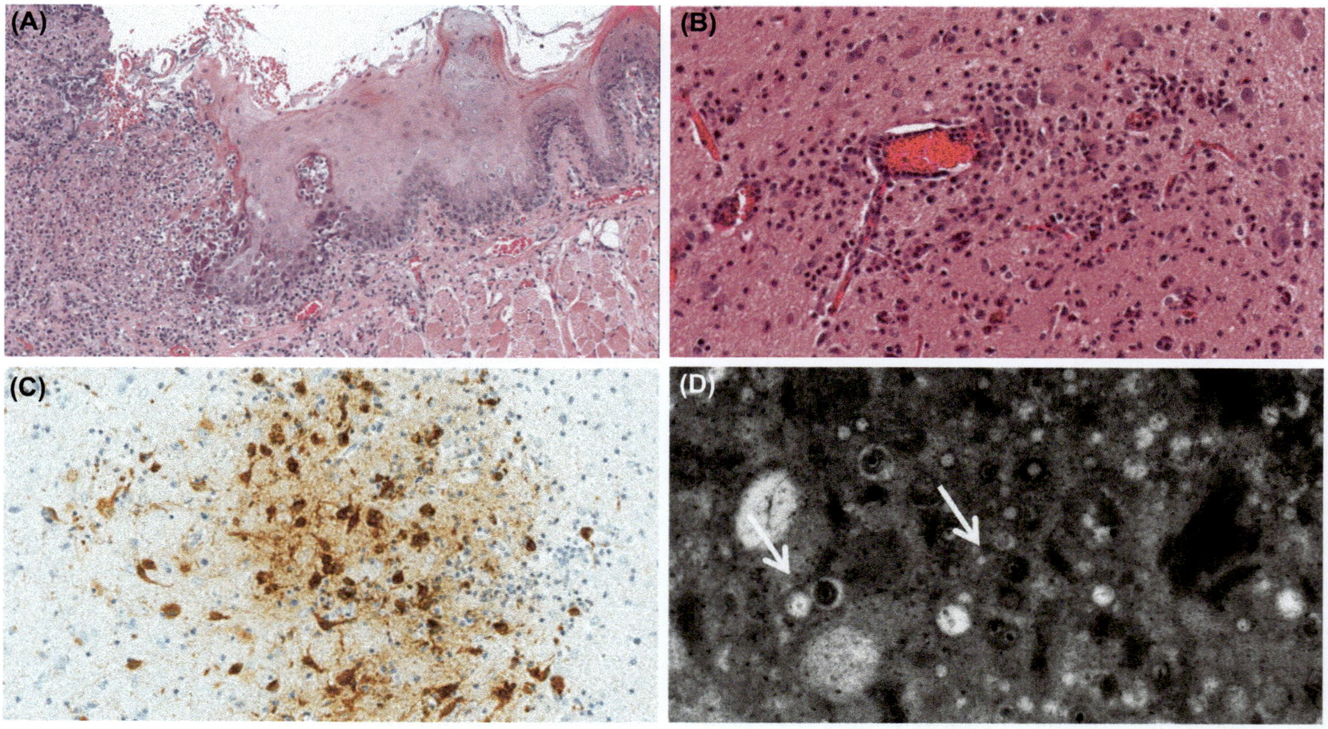

FIGURE 15.2 (A–D) Common marmoset with naturally acquired *Herpes simplex virus* infection. (A) Tongue, severe ulceration and necrosis of the epithelium. Intranuclear inclusion bodies are present in epithelial cells at the ulcer margin, paraffin section stained with hematoxylin-eosin (HE). (B) Moderate nonsuppurative meningoencephalitis with perivascular mononuclear cuffing, paraffin section stained with HE. (C) Widespread evidence of *Herpes simplex* antigen within infected neurons, immunohistochemistry on paraffin sections. (D) Transmission electron microscopy of the ulcerative epithelial lesions reveals the presence of large numbers of enveloped virions with typical herpes morphology (*arrows*).

immunohistochemistry are cross-reactive to other members of the simplexvirus genus. Therefore, definitive diagnosis requires virus isolation and detection of viral sequences using molecular assays.

Prevention and control. Treatment protocols are not available, but application of antiviral drugs in early stages of the disease might be beneficial. Because humans are the natural or reservoir host of the virus, contact with symptomatically and subclinically HSV-1-infected humans should always be avoided. Infected persons can excrete the virus even in the absence of visible lesions. The use of appropriate personal protective clothing and face masks is required.

Saimiriine Herpesvirus 1

Etiology. *Saimiriine herpesvirus 1* (SaHV1), previously known as *Herpesvirus tamarinus* or *Herpes T*, is an alpha herpesvirus, which induces an infection with many similarities to *Herpes simplex virus* infection in New World monkeys. Reservoir hosts are squirrel monkeys, which are naturally infected [11].

Clinical signs. Squirrel monkeys harbor the virus without developing clinical disease. In rare cases, mild oral ulceration indicates SaHV1 infection. Persistence of the virus within sensory ganglia has been documented. The virus may be shed during periods of reactivation with oral secretions. Marmosets and tamarins are highly susceptible to SaHV1 and may develop an acute lethal disease. After an incubation period of 7–10 days, exposed animals show signs of a disseminated infection with necrotizing inflammation in various organs. Clinical signs are vesiculation and ulceration of skin and oral mucous membranes.

Pathology. In addition to vesiculation and ulceration of skin and oral mucous membranes, further gross findings include ulceration and hemorrhage throughout the alimentary tract. Histologically, initial intraepidermal vesicles in the skin progress to full-thickness necrosis. Adnexal structures such as sebaceous glands and hair follicles are spared. Multinucleated giant cells may be present and reveal intranuclear inclusion bodies. Necrotizing inflammation is further noted in liver, spleen, kidney, adrenal gland, and lung. In contrast to *Herpes simplex virus* infection, meningoencephalitis is only minimal [12]. Nevertheless, meningoencephalitis is not a criterion to distinguish between both herpesvirus infections.

Prevention and control. Prevention requires separation of susceptible callitrichids from reservoir hosts such as *Saimiri*, *Ateles*, *Cebus*, and *Lagothrix*. In the past, a live vaccine has been developed, but vaccination frequently failed and vaccinated animals developed vaccine-induced diseases [13].

Cytomegalovirus

Simian cytomegaloviruses (CMVs) are described in a variety of New World and Old World monkey species. CMVs generally have a restricted host range, and infection of the healthy mature host is usually asymptomatic. CMVs are typical opportunistic agents, and infection and disease become apparent only in immunodeficient individuals. Two marmoset CMVs, *Callitrichine herpesvirus 1 and 2*, are known, which are not associated with disease [12].

Lymphocryptovirus

Etiology. Nonhuman primates are naturally infected with species-specific gamma herpesviruses belonging to the genus *Lymphocryptovirus*. More than 50 distinct simian lymphocryptoviruses have been isolated [14]. All lymphocryptoviruses are closely related to *Human herpesvirus 4*, the *Epstein–Barr virus* (EBV). EBV infects B lymphocytes and is indigenous to humans. Infected humans develop lifelong latent infections and most cases remain asymptomatic. Under certain conditions, the virus can cause infectious mononucleosis. Furthermore, there is an association between infection and tumor development. In this context, EBV is involved in the pathogenesis of Burkitt's lymphoma, T-cell lymphoma, and nasopharyngeal carcinoma, as well as in oral hairy leukoplacia and non-Hodgkin's lymphoma in immunocompromised patients. In general, infection is not associated with disease in the natural host. Immunosuppression of latently infected individuals or infection of nonnatural hosts may lead to rapid lymphoproliferation or lymphomatous disease. Spontaneous disease attributed to EBV infection in common marmosets is not reported, but the demonstration of serum antibodies against EBV indicates that marmosets are susceptible hosts. Experimental EBV infection of common marmosets leads to seroconversion, but there is no definite evidence of lymphoproliferative disease or lymphoma [15].

Epizootiology. Infected animals become positive for early antigen, virus capsid antigen, and nuclear antigen at low levels and show an increase in the number of leukocytes [16]. Their responses to experimental EBV infection partly resemble those of humans, and common marmosets may provide a useful model for exploring the potential of cofactors involved in the development of EBV-associated neoplasia [17]. Cotton-top tamarins (*Saguinus oedipus*) experimentally infected with EBV B95-8 virus react in a different manner and develop a fatal lymphoproliferative disease that resembles human Burkitt's lymphoma [18,19]. Similarly, white-lipped marmosets (*Saguinus labiatus*) develop lymphoma following experimental EBV inoculation [20].

Callitrichine Herpesvirus 3

Etiology. Common marmosets are frequently infected with marmoset lymphocryptovirus or *Callitrichine herpesvirus 3* (CalHV3), which was first identified in 2000. The virus may induce lymphoproliferative disease and B-cell lymphoma of the intestinal tract and associated lymph nodes [21,22], but the causative role and oncogenic potential of this virus are not definitive proof. Several outbreaks and individual cases of CalHV3 infection have been documented in zoo and laboratory settings. CalHV3 is a typical lymphocryptovirus with a complex genome that is now completely sequenced. The sequence demonstrates on overall homology with the EBV genome of 43% suggesting that CalHV3 is more closely related to a primitive lymphocryptovirus, from which all other lymphocryptoviruses evolved [23].

Epizootiology. Almost 35%–65% of wild-caught and captive marmosets are seropositive for the virus. The seroprevalence of marmoset lymphocryptovirus infection is not as ubiquitous as infection with EBV in humans or lymphocryptovirus infection in Old World monkeys, in which a prevalence of 100% among adults can be assumed [24]. Most infected marmosets do not show overt signs of clinical disease. Rarely, affected animals develop weight loss, inappetence, and diarrhea.

Pathology. Intestinal obstruction due to intraabdominal masses may occur. Most frequently, the hemogram is nonspecific, although neutrophilia with left shift and elevated liver enzymes may be present. Gross pathology reveals enlarged mesenteric lymph nodes. The digestive tract is frequently dilated, and the mucosa is thickened. Histologically, lymph node architecture is completely obliterated by sheets of infiltrating heterogeneous neoplastic round cells identified as B lymphocytes by immunohistochemistry. Small amounts of reactive T lymphocytes are scattered among neoplastic cells. Similar infiltrates can be found in the colonic mucosa, infiltrating and expanding the lamina propria and less frequently the submucosa, muscularis, and serosa. The mucosa of the small intestine may be similarly altered (Fig. 15.3). Neoplastic cellular infiltrates may also occur in liver, kidney, and lung [22]. The development of spontaneous CalHV3-positive lymphosarcomas in otherwise healthy and immunocompetent marmosets suggests that other cofactors contribute to CalHV3-associated lymphomagenesis [21]. A causative role of CalHV3 in the development of other gastrointestinal tumors, like carcinomas as observed in humans, seems unlikely in this species [25].

Diagnosis, prevention, and control. Serologic tests are available to diagnose animals with latent infections. This offers the possibility to identify and isolate infected animals and build up colonies that are free of this herpesvirus. There is no known zoonotic potential.

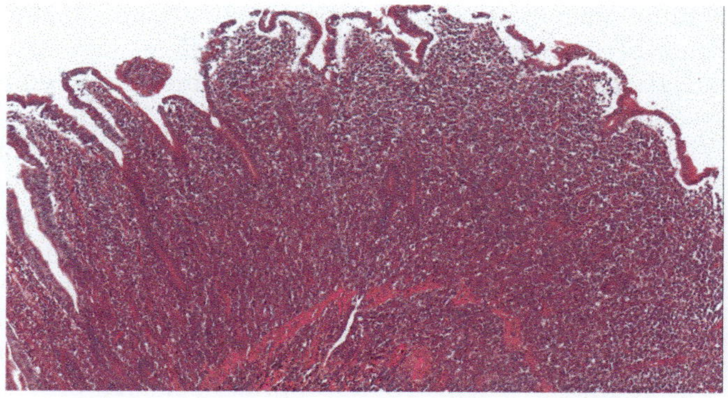

FIGURE 15.3 Common marmoset with natural *Callitrichine herpesvirus 3* infection. The small intestinal mucosa is infiltrated by a heterogenous neoplastic round cell population, paraffin section stained with HE.

Saimiriine Herpesvirus 2

Etiology. *Saimiriine herpesvirus 2*, also known as *Herpesvirus saimiri* (HVS), is a gamma herpesvirus closely related to the *Kaposi's sarcoma-associated herpesvirus* (HHV-8) of humans. The T-lymphotropic virus is naturally found in squirrel monkeys (*Saimiri sciureus*), in which it does not cause disease [26]. Virtually all squirrel monkeys are persistently infected with HVS. Transmission is horizontal, and the infection is acquired via saliva in the first 2 years of life [27]. The virus infects T lymphocytes and persists within sensory ganglia. Infected T lymphocytes are transformed into neoplastic cells by the oncogenic properties of the virus.

Clinical signs. There is no disease in the natural host, but experimental transmission of HVS to common marmosets and other susceptible New World monkeys leads to a fatal acute lymphoproliferative disorder [28]. The clinical outcome of experimental infections largely depends on the virus strain. Three different subgroups of HVA (A, B, and C) are discriminated on the basis of DNA sequence divergence at the left terminus of L-DNA [29]. Subtypes A and C are highly oncogenic, transforming marmoset peripheral blood lymphocytes in vitro and inducing rapidly progressing T-cell lymphomas in a variety of New World primate species [30]. After experimental infection, different disease patterns occur depending on virus strain and host. The time span from infection to lymphoma development may be as short as 3 weeks with a mean survival time of 22–42 days [31,32]. Animals that survive less than 40 days develop aggressive disseminated lymphomas associated with extensive necrosis and replacement of normal tissue structure. Longer survival times are associated with less progressive lymphoma development and/or leukemia.

Pathology. Gross pathology is characterized by severely enlarged peripheral and visceral lymph nodes and splenomegaly. Other characteristic lesions include enlarged and hemorrhagic thymuses, tonsils, and Peyer's patches. The basic histologic lesions are disseminated lymphoblastic infiltrates in almost every organ [33]. The cells are arranged in discrete nodules or, more frequently, in sheets of neoplastic cells exhibiting infiltrative and invasive growth. Neoplastic cells are pleomorphic and round with two large nucleoli in most cells. Leukemia is infrequently present. The neoplastic cells are of T-cell origin.

Prevention and control. HVS is primarily an experimental disease. Accidental natural transmission from squirrel monkeys to callitrichids in zoos is possible. Therefore, squirrel monkeys should not be kept in mixed enclosures together with marmosets and tamarins. There is no known zoonotic potential.

Ateline Herpesvirus

Ateline herpesvirus (AtHV) 2,3 or *Herpesvirus ateles* is another gamma herpesvirus closely related to HVS [34]. The virus is naturally found in spider monkeys (*Ateles* spp.), in which it does not lead to clinical disease. AtHV has similar pathogenic properties as *Herpesvirus saimiri* in other New World primate species. The survival time after experimental infection ranges from 36 to 104 days and is extended when compared with HVS. The lesions induced by both viruses are almost identical [31,35].

POXVIRUSES

Etiology. Poxviruses are large, brick- to ovoid-shaped enveloped DNA viruses. Different members of the *Orthopoxvirus* (OPV) family may infect callitrichids, but infections with members of the genus *Yatapox virus*, including *Yaba monkey tumor virus* and *Yaba-like disease virus*, are not reported to occur among New World monkeys. The most important pathogen among the OPVs is the *Variola virus* (VARV), the causative agent of smallpox, which only naturally infects humans. In contrast, *Vaccinia virus* (VACV) and *Cowpox virus* (CPXV) have a broad host range and are able to infect humans, cattle, cats, rodents, and nonhuman primates. *Vaccinia virus* is a live attenuated orthopoxvirus used in smallpox vaccination programs. Modified Vaccinia virus Ankara (MVA) or Vaccinia virus Lister-Elstree (VACV LE-BN) are classical vaccines used in the smallpox eradication campaigns. MVA is frequently used in different nonhuman primate animal models as a vector for recombinant vaccines. In common marmosets, a typical "vaccine take" arises at the injection site following intradermal injection of VACV LE-BN with the multipuncture method using standard bifurcated needles (Fig. 15.4). A focal poxlike skin lesion develops between days 4 and 28, and the animals remain infectious after vaccination until the skin lesions are completely healed. In vaccinated marmosets, OPV-specific IgM and IgG antibodies are observed between day 11 and day 21 after vaccination. Neutralizing antibodies start rising from day 21 postvaccination onward (own observations).

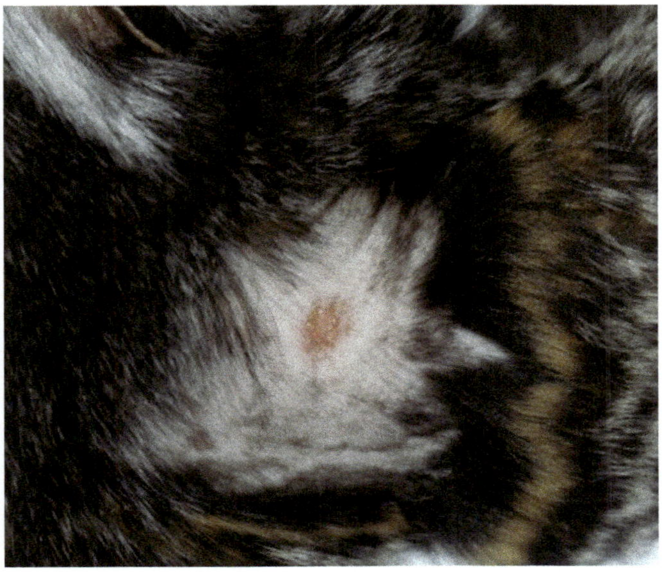

FIGURE 15.4 Common marmoset, skin. Intradermal inoculation of *Vaccinia virus* Lister-Elstree induces a typical pox lesion (vaccine-take) in the skin.

Cowpox Virus

Epizoology. Callitrichids, especially common marmosets, are highly susceptible to *Cowpox virus* (CPXV), another orthopoxvirus that is endemic in rodents. The virus seems to be limited to Europe and Central Asia and is endemic in wild rodents such as bank voles (*Clethrionomys glareolus*), wood mice (*Apodemus sylvaticus*), and rats (*Rattus norvegicus*) [36,37], which serve as vectors. Domestic cats can become infected by contact with infected rodents and are potential vectors responsible for the majority of human CPXV infections [38]. CPXV has also been isolated from a variety of zoo animals including Old World monkeys [39–41].

Clinical signs. Normally, infection is acquired through skin lesions resulting in a local often self-limiting infection at the site of inoculation. A generalized infection with severe often fatal clinical disease can occur in immunodeficient individuals or highly susceptible species. It seems that cowpox infections occur as sporadic outbreaks depending on environmental viral burden and the pathogenetic potential of the virus. In common marmosets, natural and experimental infection

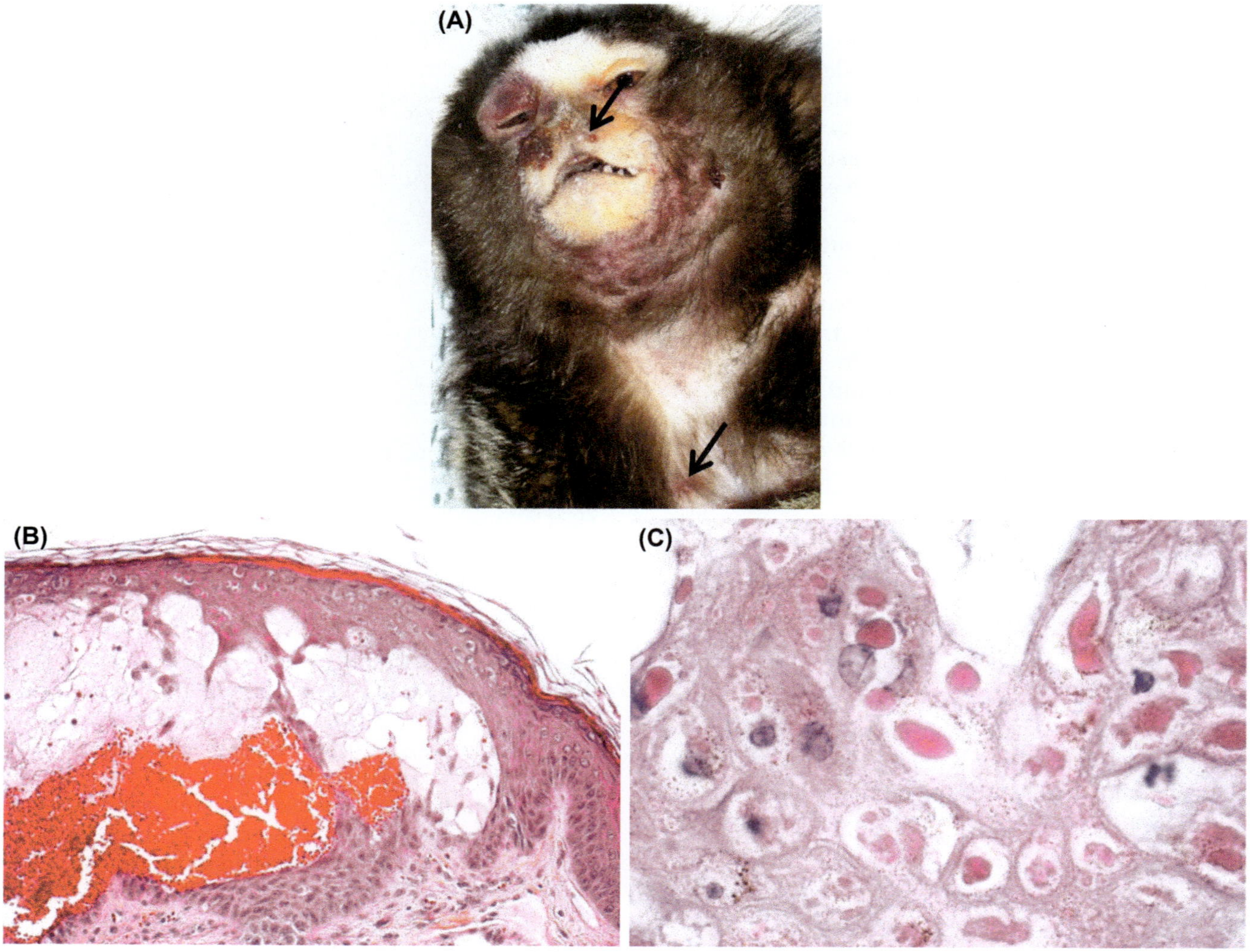

FIGURE 15.5 (A–C) Common marmoset, skin, experimental infection with a New World monkey-adapted cowpox virus named calpox virus, which closely mimics the natural disease. (A) Severe cutaneous edema and hemorrhages and vesicular lesions of different extent around the nose covered by serocellular crusts. More vesicles are randomly distributed on the body skin (*arrow*). (B) Severe focal vesicular dermatitis with intraepidermal hemorrhages, acantholysis, acanthosis, and syncytia formation of basal keratinocytes, paraffin section stained with HE. (C) Accumulation of large Guarnieri bodies in altered epithelial cells, paraffin section stained with HE.

leads to severe vesicular, hemorrhagic dermal lesions, preferentially in the facial skin (Fig. 15.5A), scrotum/labia, and palmar or plantar surfaces. For modeling orthopoxvirus infection, a New World monkey-adapted cowpox virus named calpox virus is used, which was isolated during a natural disease outbreak [42]. Experimentally induced disease is similar to the natural infection [43].

Pathology. Histologically, vesicular skin lesions show acanthosis, vesiculation, hemorrhage, and necrosis with typical orthopoxvirus inclusion bodies (Guanieri bodies) in epithelial cells and syncytia formation of the basal keratinocytes (Fig. 15.5B andC). Lymph nodes are hyperplastic and reveal hemorrhages and necrosis, which are also infrequently found in other organs, especially lung, liver, and spleen. The disease course in marmosets is fatal [41–43]. Similar alterations have been described in common marmosets during an earlier outbreak that was originally attributed to *Yatapoxvirus* [44].

Diagnosis. Diagnosis is based on typical histologic and electron microscopic findings, virus isolation, and molecular biologic identification.

Prevention and control. Cowpox virus infections are zoonotic diseases, and infected animals should be handled with care. Vaccination with modified vaccinia virus Ankara (MVA) or vaccinia virus Lister-Elstree (VACV LE-BN) is protective.

ORTHOMYXOVIRUSES

Etiology. Influenza virus is an enveloped single-stranded RNA virus in the family Orthomyxoviridae. Influenza A viruses are classified according to subtypes based on two surface proteins (hemagglutinin [H] and neuraminidase [N]) [45]. Natural infection of common marmosets with influenza A virus has not been described, but the marmoset is susceptible to experimental infection with pandemic H1N1 influenza virus strains (influenza A/Mexico/InDRE4487/2009 and influenza A/California/07/2009) [46,47]. Animal-to-animal transmission has been demonstrated at least for the influenza A/California/07/2009 virus strain [46]. Infected marmosets develop human-like "flu" symptoms between 5 and 8 days after exposure, including nasal discharge, sneezing, labored breathing, and loss of appetite [46,47]. Total protein levels are elevated in bronchoalveolar lavage fluid of infected animals between days 9 and 27 after exposure, indicating lung damage [46]. Gross pulmonary lesions of marmosets include multifocal hemorrhage, edema, and consolidation [47]. A detailed description of the microscopic lesions in common marmosets is not available.

ADENOVIRUSES

Etiology. Adenoviruses isolated from NHP belong to the genus *Mastadenovirus* in the family Adenoviridae. Adenoviruses are nonenveloped, double-stranded DNA viruses that are associated with mild to moderate respiratory or enteric disease in monkeys and apes. However, many isolates have been obtained from swabs or cell culture derived from clinically healthy animals indicating that subclinical or persistent infections are common in NHP species [12].

Clinical signs. Clinical disease due to natural adenovirus infection is not documented in common marmosets, but evidence of latent infection is supported by the observation of neutralizing antibodies to human and chimpanzee adenovirus serotypes [48] and the detection of adenovirus sequences in liver tissue of captive common marmosets from Central Europe [49]. Experimental infection of common marmosets with *titi monkey adenovirus* (TMAdV), a novel adenovirus previously associated with acute respiratory illness in humans and rapidly fatal hepatitis and pneumonia in titi monkeys [50], leads to a productive infection with the onset of clinical symptoms at days 5–10 after intranasal inoculation, specific neutralizing antibody responses, and detection of the virus in nasal swabs up to 2 weeks postinoculation.

Pathology. Pathological lesions are limited to mild bronchitis, atypical nodular hyperplasia of the liver, and mild enterocolitis. Typical basophilic intranuclear inclusions consistent with active adenovirus infection were not observed in tissues from TMAdV-infected marmosets [51].

Diagnosis. Diagnosis is based on typical histologic and electron microscopic findings, virus isolation, and molecular biologic identification.

Prevention and control. Adenovirus should be considered as a zoonotic agent as there is evidence for cross-species transmission [49,51,52], and infected animals should be handled with care. Wearing protective clothing and face masks when handling marmosets may minimize the risk of transmission.

CORONAVIRUSES

Etiology. Coronaviruses with relevance for NHPs belong to the genus *Betacoronavirus* within the family Coronaviridae. Common marmosets experimentally infected with *severe acute respiratory syndrome coronavirus* (SARS-CoV) and *Middle East respiratory syndrome coronavirus* (MERS-CoV) develop a disease similar to humans and represent a useful animal model for these highly pathogenic coronavirus species. Intratracheal inoculation of marmosets with SARS-CoV (Urbani strain) is associated with mild clinical disease, including an increase in rectal temperature, watery diarrhea, and dyspnea at 4 and 7 days after infection. Grossly, multifocal areas of pulmonary atelectasis and consolidation as well as tracheobronchial lymphadenopathy may be noted. The predominant histological lesion is a multifocal to coalescing interstitial pneumonia with multinucleated syncytial cells, type 2 pneumocyte hyperplasia, edema, and lymphocytic bronchiolitis. Tracheobronchial lymph nodes show parafollicular lymphocytic hyperplasia and histiocytosis within the subcapsular sinuses. Extrapulmonary findings include lymphocytic hepatitis with single cell necrosis, mild diffuse colitis, and mild multifocal lymphocytic myocarditis [53–55]. Common marmosets inoculated with a MERS-CoV strain (HCoV-EMC/2012) develop moderate to severe signs of respiratory disease, including increased respiratory rates, open-mouth breathing, anorexia, decreased levels of activity, and the presence of oral frothy hemorrhagic discharge. A transient decrease in body temperature may also be noted. Radiographic imaging of the lungs shows mild-to-severe bilateral interstitial infiltration. Major gross lesions are multifocal consolidation and dark red discoloration of the lungs. Microscopically, the lungs show multifocal to coalescing, moderate-to-severe acute bronchointerstitial pneumonia, coupled with type 2 pneumocyte hyperplasia and consolidation of pulmonary fibrin [56,57]. Coronavirus-like particles have been detected by electron microscopy in feces of common marmosets and seem to be related to enterocolitis. Particles are characterized by regularly spaced petal-shaped projections from the surface and measure between 100 and 220 nm [58].

Both SARS and MERS are confirmed zoonotic diseases [59].

PARAMYXOVIRUSES

Etiology. Among the single-stranded RNA virus family Paramyxoviridae, parainfluenza virus 1 has previously been associated with epizootics of respiratory disease in common marmosets [60–63]. Infections with parainfluenza I virus result from direct contact or inhalation of aerosols and likely represent anthropozoonoses under conditions of crowding and stress. Clinical disease varies from mild upper respiratory tract disease to severe systemic illness and death and may be more severe in infant animals [61].

Clinical signs. Clinical symptoms include sneezing, ocular and nasal serous or purulent discharge, dyspnea, depression, and anorexia.

Pathology. Gross lesions consist of congestion and/or consolidation of the lungs and alveolar edema [63]. The main histological finding is an acute interstitial pneumonia of variable degree with evidence of multinucleated syncytial cells containing intranuclear and intracytoplasmic inclusion bodies [60,61]. Diagnostic tests include direct electron microscopy of nasal swab and lung specimens, immunofluorescence on lung tissue, and serological detection of neutralizing antibodies [60,61,63].

Prevention and control. As natural infection is usually self-limiting, treatment is largely symptomatic and is aimed to prevent secondary bacterial infections. Vaccines against parainfluenza are not available for NHPs [61]. Wearing protective clothing and face masks when handling marmosets may minimize the risk of transmission from humans.

Measles Virus

Etiology. Measles virus, a morbillivirus of the Paramyxoviridae family, may cause serious epizootics with high morbidity and mortality in marmoset species [61,64]. Measles virus is rapidly spread between animals through contact, fomites, and aerosols, and humans are regarded as the main source of infection [61,65,66]. Measles in NWM exhibit an inconsistent organotropism, whereas in OWM infected with measles virus, the respiratory tract and skin are the common sites of disease manifestation.

Clinical signs. While in mustached marmosets and several other NWM monkey species, the virus frequently targets the gastrointestinal tract and causes a necrotizing enterocolitis with hemorrhagic diarrhea [61,65], the predominant finding in naturally infected common marmosets is the characteristic pneumonia [64,67]. Common marmosets with measles infection become clinically apparent with lethargy, facial edema, and nasal discharge and occasionally develop an exanthema. Death occurs 8–18 h after the onset of the first clinical signs.

Pathology. The characteristic histopathological finding is an interstitial pneumonia with thickened and hyperemic alveolar walls. Syncytial epithelial cells containing eosinophilic intracytoplasmic or intranuclear inclusion bodies are the hallmark of the disease but are not an obligatory finding and are often absent in the skin lesions [64,67]. Secondary bacterial infection occasionally results in patchy bronchopneumonia. Multinucleated syncytial cells with or without inclusion bodies may be found in different organs, including lymph nodes (Warthin–Finkeldey cells), spleen, and colon. The maculopapular skin lesions are histologically characterized by focal hyperemia and hemorrhages in the lamina propria [64]. Experimental intracerebral inoculation of marmosets may cause encephalitis, which is similar to the subacute, sclerosing panencephalitis in humans [67,68]. Measles virus is strongly immunosuppressive, and the clinical picture may be complicated by secondary opportunistic viral or bacterial infections [67].

Diagnosis. Diagnostic tools to confirm measles virus etiology include detection of seroconversion, virus isolation, or immunohistochemical demonstration of viral antigen in tissue sections. Treatment of measles is limited to supportive therapy with fluids and antibiotics to prevent secondary bacterial infections [61].

Prevention and control. Experiences with efficacy and tolerability of commercial vaccines in common marmosets are not reported in the literature. Measles constitutes a zoonotic risk for naive humans [61], and protective clothing and face masks should be used when handling potentially infected marmosets.

Paramyxovirus Saguinus

Etiology. *Paramyxovirus saguinus*, which presumably is a variant of measles virus, is associated with a single outbreak of infectious gastroenteritis in a marmoset colony at the New England Primate Research Center [69]. The virus is described to infect cotton-top tamarins, mustached tamarins, and common marmosets and causes clinical symptoms such as anorexia, diarrhea, and dehydration with rapid progression to death. Gross lesions include congestion and hemorrhage of the gastrointestinal mucosa and enlargement of lymphoid tissues (Peyer's patches, lymph nodes, spleen). The predominant microscopic lesion is a necrotizing and ulcerative typhlocolitis with evidence of multinucleated syncytial cells on the surface epithelium and within crypts. Intranuclear and/or intracytoplasmic inclusions can occasionally be observed in the multinucleated cells. Syncytial cells may also be present in bile and pancreatic duct epithelium, pancreatic acini, hepatic cords, kidney tubules, and in endometrial epithelium. Further histologic findings include cholangitis and necrosis of germinal centers in lymph nodes, spleen, and Peyer's patches [69]. The virus is most likely transmitted via the fecal–oral route. The origin of infection remains unknown. Preventive or curative treatment is not available.

CALLITRICHID HEPATITIS VIRUS

Etiology. Callitrichid hepatitis is an acute fatal infection of New World primates caused by an arenavirus referred to as callitrichid hepatitis virus. The virus is an enveloped RNA virus closely related to lymphocytic choriomeningitis virus (LCMV) [70], for which the house mouse (*Mus musculus*) is the major reservoir. LCMV can cause human disease characterized by mostly self-limiting influenza-like illness, but the disease can progress to acute meningitis. Several outbreaks of hepatitis among captive callitrichids are reported in the United States, which are frequently related to feeding of neonatal mice infected with LCMV [71–73]. The first outbreak in Germany was mostly attributed to transmission by wild mice [74]. In general, transmission occurs through contact with or ingestion of infected rodents.

Clinical signs. Clinical signs include dyspnea, weakness, and jaundice. Serum levels of transaminases and bilirubin are elevated. Infected animals die 7–12 days after the onset of clinical symptoms. Coagulopathy may be apparent.

Pathology. At necropsy, hepatosplenomegaly, pleural and pericardial effusions, jaundice, and subcutaneous and intramuscular hemorrhages are characteristic [72]. Histopathologically, random foci of hepatocellular degeneration and spotty necrosis associated with mononuclear inflammatory cell infiltration occur throughout the liver. LCMV antigen is usually demonstrable within these foci. Round acidophilic structures resembling apoptotic or Councilman-like bodies are commonly present (Fig. 15.6A). They are located free within sinusoids or within Kupffer cells. Necrotic lesions also occur in other organs such as the spleen, lymph nodes, adrenal cortex, and intestine. Nonsuppurative meningitis and encephalitis may accompany the liver lesions (Fig. 15.6B). These extrahepatic lesions are usually milder than the liver alterations [71]. Electron microscopy may reveal enveloped virus-like particles with a diameter of 85–105 nm in the rough endoplasmic reticulum and Golgi complex of hepatocytes. Experimental infection of marmosets is possible and leads to identical clinical and pathological findings [71,75].

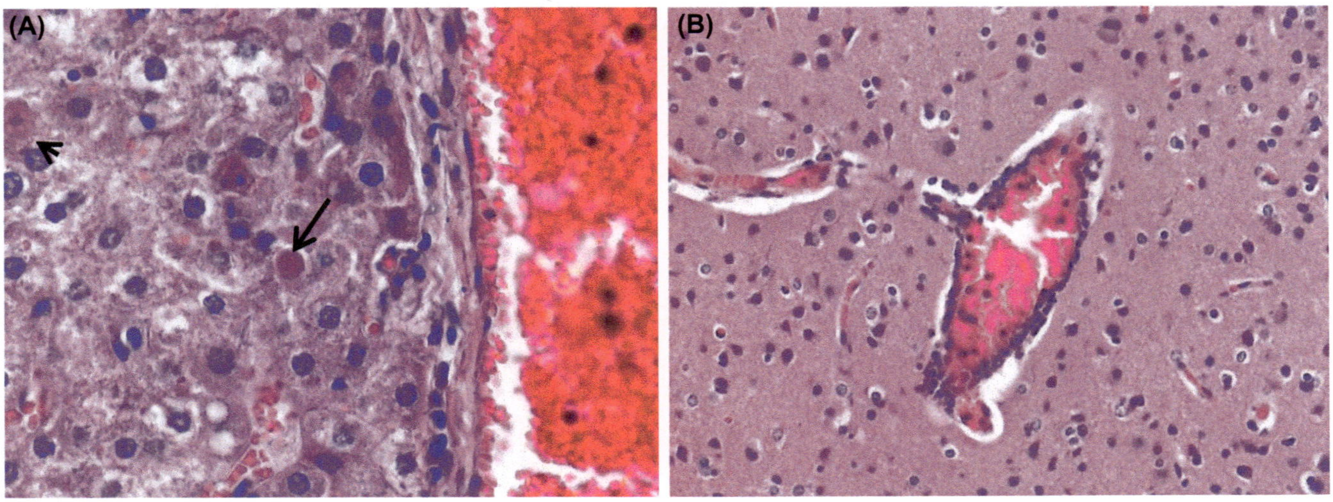

FIGURE 15.6 (A and B) Emperor tamarin (*Saguinus imperator*) with naturally acquired LCMV infection. (A) Liver, hepatocellular degeneration and necrosis with intracytoplasmatic inclusion bodies (*arrows*) and Councilman bodies (*short arrow*), paraffin section stained with HE. (B) Central nervous system, mild encephalitis with perivascular cuffing, paraffin section stained with HE.

Prevention and control. Preventing contact with rodents is essential to avoid this rodent-borne disease. LCMV is a zoonotic agent and may cause disease in humans. Seroconversion is documented in caretakers involved in outbreaks of the disease [71].

HEPATITIS A VIRUS

Etiology. Hepatitis A virus (HAV) is a major cause of acute viral hepatitis in humans and several nonhuman primate species. Common marmosets can also be infected as indicated by serologic surveys [76,77]. The virus is a small RNA virus and belongs to the genus *Hepatovirus* within the family Picornaviridae. Transmission occurs by the fecal–oral route.

Clinical signs. Clinical findings are uncommon, and seroconversion and elevation of liver enzymes are usually the only clinical signs of infection. In the past, common marmosets have been used as an animal model for the disease. They can be infected by the oral route [78], via intragastric infection [79], or by direct inoculation of the virus in the liver [80]. Infected animals develop an acute hepatitis 2 weeks after infection and shed the virus in feces from day 7 onward.

Pathology. The liver is the target organ, and viral replication occurs within hepatocytes. Usually, HAV antigen is not detectable in other organs, indicating that the liver is the only and primary site of virus replication. Shedding of HAV in feces during the late incubation period can be explained by excretion of HAV from the liver with the bile [81]. The end of the incubation period is indicated by an initial increase in serum liver enzymes. Characteristic histopathologic changes such as activation of sinusoidal cells, piecemeal necrosis, and bridging necrosis as well as periportal and parenchymal mononuclear inflammatory cell infiltration are detectable during the acute phase, coinciding with a maximum of transaminase levels and the appearance of anti-HAV antibodies [79]. At the same time, clusters of solid or empty virus-like particles about 27 nm in diameter can be demonstrated by electron microscopy mainly in membrane-bound cytoplasmic vesicles of Kupffer cells and hepatocytes [82]. In the convalescent phase, regeneration of hepatic tissue starts and the transaminase values return to baseline. HAV is a potentially zoonotic agent; monkeys can become infected with human strains, but infection is usually self-limiting. Chronic infections and carrier stages are not described.

HEPATITIS B VIRUS

Etiology. The major cause of human hepatitis is the hepatitis B virus (HBV), which belongs to the genus *Orthohepadnavirus*. Transmission occurs by infected blood, saliva, and semen. HBV can cause persistent infections leading to chronic hepatitis and hepatocellular carcinomas. The virus can be transmitted to chimpanzees and macaques [83,84]. Natural infections among New World monkeys are not observed.

HEPATITIS C VIRUS

Etiology. Hepatitis C virus (HCV) is a small enveloped RNA virus that causes chronic hepatitis in humans worldwide. HCV is a member of the genus *Hepacivirus* within the family Flaviviridae. HCV is not associated with natural disease in nonhuman primates, and experimental infection of several nonhuman primate species failed. Only chimpanzees are susceptible and represent an important animal model for chronic hepatitis. In contrast, GB agent viruses are frequently found in a number of New World monkeys; GBV-A and variants are enzootic viral infections of several NWP species including common marmosets, whereas GBV-B and GBV-C are blood-borne pathogens of man and chimpanzees. GB agent viruses are closely related to HCV of humans. They are single-stranded, positive-sense RNA viruses belonging to the same genus and family as HCV. Several GBV-A variants seem to circulate in wild populations of New World monkeys. No clinical disease is associated with GBV-A and C infection, and diagnosis is confirmed using RT-PCR performed on plasma or serum. In contrast, GBV B can induce an acute hepatitis when inoculated intravenously into several species of New World primates. Common marmosets are susceptible to experimental GBV-B infection and develop a characteristic acute nonsuppurative hepatitis, which is characterized by an infiltration of lymphocytes, mainly CD3+ CD8+ T lymphocytes and CD20+ B lymphocytes, within the first 2 months of primary infection [85]. Experimental GBV-B infection of marmosets can be used as a surrogate model of HCV infection for investigation of pathogenetic pathways and antiviral drug development [86].

FLAVIVIRUSES

Etiology. Yellow fever induced by the yellow fever virus (YFV), another flavivirus belonging to the family Flaviviridae, is a devastating viral infection of free-living New World monkeys. The disease is transmitted by mosquitoes of the genera *Aedes* and *Haemagogus*, which serve as vectors. YFV is endemic in South America, and the disease naturally occurs among wild marmosets and tamarins [87]. Devastating outbreaks are reported among howler monkeys. They frequently coincide with human cases in the geographic vicinity [88,89]. Suspected epizootics among common marmosets in endemic areas could not be confirmed [90]. Experimentally infected callitrichids rapidly succumb to disease, and death occurs within 1 week after infection. Gross lesions are not specific except for fever and jaundice, which are infrequently observed. Histologically, liver lesions are described, including hepatocellular necrosis, fatty change, and hemorrhage. Acidophilic apoptotic (Councilman) bodies are rarely observed [91,92]. As the disease is primarily a problem of wild living primates, it should be considered as a differential diagnosis in cases of hemorrhagic fever occurring in endemic regions or in recently imported monkeys originating from endemic areas. Experimental infection has shown that common marmosets are highly susceptible for other RNA viruses belonging to the Bunyaviridae, Arenaviridae, and Filoviridae families, all inducing hemorrhagic fevers. Common marmosets are used as animal models for Lassa fever induced by Arenaviridae, Argentine hemorrhagic fever induced by Junin virus, and filovirus hemorrhagic fever induced by Ebola and Marburg virus [53].

LYSSAVIRUS

Etiology. Rabies is caused by Lyssaviruses, belonging to the family Rhabdoviridae, encompassing a diverse group of viruses. Rabies is rarely considered as a cause of morbidity and mortality in nonhuman primates. Nevertheless, the disease has been reported in several New World monkey species, among them common marmosets [93]. In endemic regions, wild nonhuman primates may become infected in their natural habitat and represent important vectors of the virus. Several human cases could be traced back to contact with wild cuffed common marmosets raised as pets in a region of Brazil [94]. Here a marmoset-specific Rhabdovirus strain seems to exist.

Epizootiology. Nonhuman primates get infected by contact with reservoir or inadvertent host species such as dogs, bats, or rodents. The virus is spread by scratches or bites from animal to animal or from animal to human. Infection after vaccination with attenuated virus strains is reported [62]. The incubation time is not exactly known and can be long. Initial replication occurs at the site of inoculation followed by spread of the virus to the brain via the peripheral nerves. In the brain, the virus infects exclusively neurons and spreads rapidly. With the onset of clinical symptoms a widespread dissemination of the virus starts.

Clinical signs. Infection leads to disseminated meningoencephalitis inducing clinical furious or paralytic forms of the disease that result in self-mutilation, irritability, and paralysis. The disease can only be diagnosed after the start of symptoms. *Pathology* is characterized by nonsuppurative encephalitis with marked formation of glial nodules and neuronal degeneration. Eosinophilic intracytoplasmatic inclusions called Negri bodies within neurons are pathognomonic.

Diagnosis. The diagnosis is based on typical histologic findings, virus isolation, and molecular biologic identification. Diagnosis can be confirmed by the fluorescent antibody test using a rabies-specific antibody. *Differential diagnosis* in a case of suspected rabies should include any cause of encephalitis, in particular infection with herpesviruses, enteroviruses, or arboviruses.

Prevention and control. Rabies is a significant zoonotic threat, and suspected animals should be handled with care. A rabies vaccination can be included in the preventative health-care regimen. Vaccination of nonhuman primates housed in indoor/outdoor enclosures in rabies enzootic areas is advisable [95]. The use of attenuated vaccines should be avoided.

References

[1] Lefaux B, Duprez R, Tanguy M, Longeart L, Gessain A, Boulanger E. Nonhuman primates might be highly susceptible to cross-species infectivity by human alpha-herpesviruses. Vet Pathol 2004;41:302–4.

[2] Sekulin K, Jankova J, Kolodziejek J, Huemer HP, Gruber A, Meyer J, Nowotny N. Natural zoonotic infections of two marmosets and one domestic rabbit with herpes simplex virus type 1 did not reveal a correlation with a certain gG-, gI- or gE genotype. Clin Microbiol Infect 2010;16:1669–72.

[3] Bruno SF, Liebhold MM, Mätz-Rensing K, Romao MA, Didier A, Brandes F, Bressan AC, Kaup FJ. Herpesvirus infections in free living black tufted ear marmosets (*Callithrix penicillata*, E. Geoffroyi 1812) at the State Park of Serra da Tiririca, Niteroi, Rio de Janeiro, Brazil. Berl Munch Tierarztl Wochenschr 1997;110:427–30.

[4] Longa CS, Bruno SF, Pires AR, Romijn PC, Kimura LS, Costa CH. Human herpesvirus 1 in wild marmosets, Brazil, 2008. Emerg Infect Dis 2011;17:1308–10.

[5] Hatt JM, Grest P, Posthaus H, Bossart W. Serologic survey in a colony of captive common marmosets (*Callithrix jacchus*) after infection with herpes simplex type 1-like virus. J Zoo Wildl Med 2004;35:387–90.

[6] Huemer HP, Larcher C, Czedik-Eysenberg T, Nowotny N, Reifinger M. Fatal infection of a pet monkey with Human herpesvirus. Emerg Infect Dis 2002;8:639–42.

[7] Juan-Salles C, Ramos-Vara JA, Prats N, Sole-Nicolas J, Segales J, Marco AJ. Spontaneous herpes simplex virus infection in common marmosets (*Callithrix jacchus*). J Vet Diagn Investig 1997;9:341–5.

[8] Mello MT, AN R. Surto fatal de infeccao herpetica em pequeno grupo de *Callithrix jacchus*. In: 2nd Congresso Brasileiro de Primatologia. Campinas-S.P. Anais. Campinas. A Primatologia no Brasil; 1985. p. 496.

[9] Mätz-Rensing K, Jentsch KD, Rensing S, Langenhuyzen S, Verschoor E, Niphuis H, Kaup FJ. Fatal Herpes simplex infection in a group of common marmosets (*Callithrix jacchus*). Vet Pathol 2003;40:405–11.

[10] Hunt WL. Responses of rat testes and accessory glands to testosterone, pilocarpine and copulation. Nature 1969;221:669–70.

[11] King NW, Hunt RD, Daniel MD, Melendez LV. Overt herpes-T infection in squirrel monkeys (*Saimiri sciureus*). Lab Anim Care 1967;17:413–23.

[12] Wachtman L, Mansfield K. Viral diseases of non-human primates. In: Abee CR, Mansfield K, Tardif S, Morris T, editors. Nonhuman primates in biomedical research. 2nd ed. Amsterdam, Boston: Elsevier; 2012. p. 1–104.

[13] Daniel MD, Karpas A, Melendez LV, King NW, Hunt RD. Isolation of herpes-T virus from a spontaneous disease in squirrel monkeys (*Saimiri sciureus*). Arch Gesamte Virusforsch 1967;22:324–31.

[14] Ehlers B, Spiess K, Leendertz F, Peeters M, Boesch C, Gatherer D, McGeoch DJ. Lymphocryptovirus phylogeny and the origins of Epstein-Barr virus. J Gen Virol 2010;91:630–42.

[15] de-The G, Dubouch P, Fontaine C, Wedderburn N, Carter RL, Edwards MB, Cohen B. Natural antibodies to EBV-VCA antigens in common marmosets (*Callithrix jacchus*) and response after EBV inoculation. Intervirology 1980;14:284–91.

[16] Felton SC, Hoffmann CC, Kreier JP, Glaser R. Hematologic and immunologic responses in common marmosets (*Callithrix jacchus*) infected with Plasmodium knowlesi and Epstein-Barr virus. Lab Anim Sci 1984;34:164–8.

[17] Wedderburn N, Edwards JM, Desgranges C, Fontaine C, Cohen B, de The G. Infectious mononucleosis-like response in common marmosets infected with Epstein-Barr virus. J Infect Dis 1984; 150:878–82.

[18] Johnson DR, Wolfe LG, Levan G, Klein G, Ernberg I, Aman P. Epstein-Barr virus (EBV)-induced lymphoproliferative disease in cotton-topped marmosets. Int J Cancer 1983;31:91–7.

[19] Neubauer RH, Rabin H, Hopkins 3rd RF, Levy BM. Characteristics of cell lines established from Epstein-Barr virus induced marmoset tumors. Primates Med 1978;10:156–62.

[20] Sundar SK, Levine PH, Ablashi DV, Leiseca SA, Armstrong GR, Cicmanec JL, Parker GA, Nonoyama M. Epstein-Barr virus-induced malignant lymphoma in a white-lipped marmoset. Int J Cancer 1981;27:107–11.

[21] Cho Y, Ramer J, Rivailler P, Quink C, Garber RL, Beier DR, Wang F. An Epstein-Barr-related herpesvirus from marmoset lymphomas. Proc Natl Acad Sci USA 2001;98:1224–9.

[22] Ramer JC, Garber RL, Steele KE, Boyson JF, O'Rourke C, Thomson JA. Fatal lymphoproliferative disease associated with a novel gammaherpesvirus in a captive population of common marmosets. Comp Med 2000;50:59–68.

[23] Rivailler P, Cho YG, Wang F. Complete genomic sequence of an Epstein-Barr virus-related herpesvirus naturally infecting a new world primate: a defining point in the evolution of oncogenic lymphocryptoviruses. J Virol 2002;76:12055–68.

[24] Fogg MH, Carville A, Cameron J, Quink C, Wang F. Reduced prevalence of Epstein-Barr virus-related lymphocryptovirus infection in sera from a new world primate. J Virol 2005;79:10069–72.

[25] Miller AD, Kramer JA, Lin KC, Knight H, Martinot A, Mansfield KG. Small intestinal adenocarcinoma in common marmosets (*Callithrix jacchus*). Vet Pathol 2010;47:969–76.

[26] Melendez LV, Daniel MD, Hunt RD, Garcia FG. An apparently new herpesvirus from primary kidney cultures of the squirrel monkey (*Saimiri sciureus*). Lab Anim Care 1968;18:374–81.

[27] Falk LA, Nigida S, Deinhardt F, Cooper RW, Hernandez-Camacho JI. Oral excretion of Herpesvirus saimiri in captive squirrel monkeys and incidence of infection in feral squirrel monkeys. J Natl Cancer Inst 1973;51:1987–9.

[28] Melendez LV, Hunt RD, Daniel MD, Garcia FG, Fraser CE. Herpesvirus saimiri. II. Experimentally induced malignant lymphoma in primates. Lab Anim Care 1969;19:378–86.

[29] Medveczky P, Szomolanyi E, Desrosiers RC, Mulder C. Classification of herpesvirus saimiri into three groups based on extreme variation in a DNA region required for oncogenicity. J Virol 1984;52:938–44.

[30] Fleckenstein B, Desrosiers R. Herpesvirus saimiri and herpesvirus ateles. In: Roizman B, editor. The herpesviruses. New York: Plenum Press; 1982. p. 253–332.

[31] Hunt RD, Blake BJ. Herpesvirus platyrrhinae infection. In: Jones TC, Mohr U, Hunt RD, editors. Nonhuman primates I. Berlin, Heidelberg (New York): Springer Verlag; 1993. p. 100–3.

[32] Wright J, Falk LA, Wolfe LG, Ogden J, Deinhardt F. Susceptibility of common marmosets (*Callithrix jacchus*) to oncogenic and attenuated strains of *Herpesvirus saimiri*. J Natl Cancer Inst 1977;59: 1475–8.

[33] Hunt RD, Melendez LV, King NW, Gilmore CE, Daniel MD, Williamson ME, Jones TC. Morphology of a disease with features of malignant lymphoma in marmosets and owl monkeys inoculated with *Herpesvirus saimiri*. J Natl Cancer Inst 1970;44: 447–65.

[34] Melendez LV, Castellanos H, Barahona HH, Daniel MD, Hunt RD, Fraser CE, Garcia FG, King NW. Two new herpesviruses from spider monkeys (*Ateles geoffroyi*). J Natl Cancer Inst 1972;49:233–8.

[35] Hunt RD, Melendez LV, Garcia FG, Trum BF. Pathologic features of Herpesvirus ateles lymphoma in cotton-topped marmosets (*Saguinus oedipus*). J Natl Cancer Inst 1972;49:1631–9.

[36] Hazel SM, Bennett M, Chantrey J, Bown K, Cavanagh R, Jones TR, Baxby D, Begon M. A longitudinal study of an endemic disease in its wildlife reservoir: cowpox and wild rodents. Epidemiol Infect 2000;124:551–62.

[37] Wolfs TF, Wagenaar JA, Niesters HG, Osterhaus AD. Rat-to-human transmission of Cowpox infection. Emerg Infect Dis 2002;8: 1495–6.

[38] Baxby D, Bennett M, Getty B. Human cowpox 1969-93: a review based on 54 cases. Br J Dermatol 1994;131:598–607.

[39] Kurth A, Straube M, Kuczka A, Dunsche AJ, Meyer H, Nitsche A. Cowpox virus outbreak in banded mongooses (*Mungos mungo*) and jaguarundis (*Herpailurus yagouaroundi*) with a time-delayed infection to humans. PLoS One 2009;4:e6883.

[40] Kurth A, Wibbelt G, Gerber HP, Petschaelis A, Pauli G, Nitsche A. Rat-to-elephant-to-human transmission of cowpox virus. Emerg Infect Dis 2008;14:670–1.

[41] Martina BE, van Doornum G, Dorrestein GM, Niesters HG, Stittelaar KJ, Wolters MA, van Bolhuis HG, Osterhaus AD. Cowpox virus transmission from rats to monkeys, The Netherlands. Emerg Infect Dis 2006;12:1005–7.

[42] Mätz-Rensing K, Ellerbrok H, Ehlers B, Pauli G, Floto A, Alex M, Czerny CP, Kaup FJ. Fatal poxvirus outbreak in a colony of New World monkeys. Vet Pathol 2006;43:212–8.

[43] Mätz-Rensing K, Stahl-Hennig C, Kramski M, Pauli G, Ellerbrok H, Kaup FJ. The pathology of experimental poxvirus infection in common marmosets (*Callithrix jacchus*): further characterization of a new primate model for orthopoxvirus infections. J Comp Pathol 2012;146:230–42.

[44] Gough AW, Barsoum NJ, Gracon SI, Mitchell L, Sturgess JM. Poxvirus infection in a colony of common marmosets (*Callithrix jacchus*). Lab Anim Sci 1982;32:87–90.

[45] Gamblin SJ, Skehel JJ. Influenza hemagglutinin and neuraminidase membrane glycoproteins. J Biol Chem 2010;285:28403–9.

[46] Moncla LH, Ross TM, Dinis JM, Weinfurter JT, Mortimer TD, Schultz-Darken N, Brunner K, Capuano 3rd SV, Boettcher C, Post J, Johnson M, Bloom CE, Weiler AM, Friedrich TC. A novel nonhuman primate model for influenza transmission. PLoS One 2013;8:e78750.

[47] Mooij P, Koopman G, Mortier D, van Heteren M, Oostermeijer H, Fagrouch Z, de Laat R, Kobinger G, Li Y, Remarque EJ, Kondova I, Verschoor EJ, Bogers WM. Pandemic swine-origin H1N1 influenza virus replicates to higher levels and induces more fever and acute inflammatory cytokines in cynomolgus versus Rhesus Monkeys and can replicate in common marmosets. PLoS One 2015;10:e0126132.

[48] Ersching J, Hernandez MI, Cezarotto FS, Ferreira JD, Martins AB, Switzer WM, Xiang Z, Ertl HC, Zanetti CR, Pinto AR. Neutralizing antibodies to human and simian adenoviruses in humans and New-World monkeys. Virology 2010;407:1−6.

[49] Wevers D, Metzger S, Babweteera F, Bieberbach M, Boesch C, Cameron K, Couacy-Hymann E, Cranfield M, Gray M, Harris LA, Head J, Jeffery K, Knauf S, Lankester F, Leendertz SA, Lonsdorf E, Mugisha L, Nitsche A, Reed P, Robbins M, Travis DA, Zommers Z, Leendertz FH, Ehlers B. Novel adenoviruses in wild primates: a high level of genetic diversity and evidence of zoonotic transmissions. J Virol 2011;85:10774−84.

[50] Chen EC, Yagi S, Kelly KR, Mendoza SP, Tarara RP, Canfield DR, Maninger N, Rosenthal A, Spinner A, Bales KL, Schnurr DP, Lerche NW, Chiu CY. Cross-species transmission of a novel adenovirus associated with a fulminant pneumonia outbreak in a new world monkey colony. PLoS Pathog 2011;7:e1002155.

[51] Yu G, Yagi S, Carrion Jr R, Chen EC, Liu M, Brasky KM, Lanford RE, Kelly KR, Bales KL, Schnurr DP, Canfield DR, Patterson JL, Chiu CY. Experimental cross-species infection of common marmosets by titi monkey adenovirus. PLoS One 2013; 8:e68558.

[52] Roy S, Vandenberghe LH, Kryazhimskiy S, Grant R, Calcedo R, Yuan X, Keough M, Sandhu A, Wang Q, Medina-Jaszek CA, Plotkin JB, Wilson JM. Isolation and characterization of adenoviruses persistently shed from the gastrointestinal tract of non-human primates. PLoS Pathog 2009;5:e1000503.

[53] Carrion Jr R, Patterson JL. An animal model that reflects human disease: the common marmoset (*Callithrix jacchus*). Curr Opin Virol 2012;2:357−62.

[54] Curths C, Knauf S, Kaup F-J. Respiratory animal models in the common marmoset (*Callithrix jacchus*). Vet Sci 2014;1:63.

[55] Greenough TC, Carville A, Coderre J, Somasundaran M, Sullivan JL, Luzuriaga K, Mansfield K. Pneumonitis and multi-organ system disease in common marmosets (*Callithrix jacchus*) infected with the severe acute respiratory syndrome-associated coronavirus. Am J Pathol 2005;167:455−63.

[56] Falzarano D, de Wit E, Feldmann F, Rasmussen AL, Okumura A, Peng X, Thomas MJ, van Doremalen N, Haddock E, Nagy L, LaCasse R, Liu T, Zhu J, McLellan JS, Scott DP, Katze MG, Feldmann H, Munster VJ. Infection with MERS-CoV causes lethal pneumonia in the common marmoset. PLoS Pathog 2014;10: e1004250.

[57] van Doremalen N, Munster VJ. Animal models of Middle East respiratory syndrome coronavirus infection. Antiviral Res 2015;122: 28−38.

[58] Russell RG, Brian DA, Lenhard A, Potgieter LN, Gillespie D, Clapp NK. Coronavirus-like particles and Campylobacter in marmosets with diarrhea and colitis. Dig Dis Sci 1985;30:72S−7S.

[59] Chan JF, Lau SK, To KK, Cheng VC, Woo PC, Yuen KY. Middle East respiratory syndrome coronavirus: another zoonotic betacoronavirus causing SARS-like disease. Clin Microbiol Rev 2015;28: 465−522.

[60] Flecknell PA, Parry R, Needham JR, Ridley RM, Baker HF, Bowes P. Respiratory disease associated with parainfluenza Type I (Sendai) virus in a colony of marmosets (*Callithrix jacchus*). Lab Anim 1983;17:111−3.

[61] Ludlage E, Mansfield K. Clinical care and diseases of the common marmoset (*Callithrix jacchus*). Comp Med 2003;53:369−82.

[62] Potkay S. Diseases of the Callitrichidae: a review. J Med Primatol 1992;21:189−236.

[63] Sutherland SD, Almeida JD, Gardner PS, Skarpa M, Stanton J. Rapid diagnosis and management of parainfluenza I virus infection in common marmosets (*Callithrix jacchus*). Lab Anim 1986; 20:121−6.

[64] Levy BM, Mirkovic RR. An epizootic of measles in a marmoset colony. Lab Anim Sci 1971;21:33−9.

[65] Albrecht P, Lorenz D, Klutch MJ, Vickers JH, Ennis FA. Fatal measles infection in marmosets pathogenesis and prophylaxis. Infect Immun 1980;27:969−78.

[66] van Binnendijk RS, van der Heijden RW, Osterhaus AD. Monkeys in measles research. Curr Top Microbiol Immunol 1995;191: 135−48.

[67] Bailey C, Mansfield K. Emerging and reemerging infectious diseases of nonhuman primates in the laboratory setting. Vet Pathol 2010;47:462−81.

[68] Albrecht P, Lorenz D, Klutch MJ. Encephalitogenicity of measles virus in marmosets. Infect Immun 1981;34:581−7.

[69] Fraser CE, Chalifoux L, Sehgal P, Hunt RD, King NW. A paramyxovirus causing fatal gastroenterocolitis in marmoset monkeys. Primates Med 1978;10:261−70.

[70] Stephensen CB, Park JY, Blount SR. cDNA sequence analysis confirms that the etiologic agent of callitrichid hepatitis is lymphocytic choriomeningitis virus. J Virol 1995;69:1349−52.

[71] Montali RJ, Connolly BM, Armstrong DL, Scanga CA, Holmes KV. Pathology and immunohistochemistry of callitrichid hepatitis, an emerging disease of captive New World primates caused by lymphocytic choriomeningitis virus. Am J Pathol 1995;147:1441−9.

[72] Montali RJ, Scanga CA, Pernikoff D, Wessner DR, Ward R, Holmes KV. A common-source outbreak of callitrichid hepatitis in captive tamarins and marmosets. J Infect Dis 1993;167: 946−50.

[73] Ramsay EC, Montali RJ, Worley M, Stephenson CB, Holmes KV. Callitrichid hepatitis: epizootiology of a fatal hepatitis in zoo tamarins and marmosets. J Zoo Wildl Med 1989;20:178−83.

[74] Asper M, Hofmann P, Osmann C, Funk J, Metzger C, Bruns M, Kaup FJ, Schmitz H, Gunther S. First outbreak of callitrichid hepatitis in Germany: genetic characterization of the causative lymphocytic choriomeningitis virus strains. Virology 2001;284:203−13.

[75] Montali RJ, Ramsay EC, Stephensen CB, Worley M, Davis JA, Holmes KV. A new transmissible viral hepatitis of marmosets and tamarins. J Infect Dis 1989;160:759−65.

[76] Eichberg JW, Kalter SS. Hepatitis A and B: serologic survey of human and nonhuman primate sera. Lab Anim Sci 1980;30:541−3.

[77] Sa-nguanmoo P, Thawornsuk N, Rianthavorn P, Sommanustweechai A, Ratanakorn P, Poovorawan Y. High prevalence of antibodies against hepatitis A virus among captive nonhuman primates. Primates 2010;51:167−70.

[78] Pinto MA, Marchevsky RS, Baptista ML, de Lima MA, Pelajo-Machado M, Vitral CL, Kubelka CF, Pissurno JW, Franca MS, Schatzmayr HG, Gaspar AM. Experimental hepatitis A virus (HAV) infection in *Callithrix jacchus*: early detection of HAV antigen and viral fate. Exp Toxicol Pathol 2002;53:413−20.

[79] Vitral CL, Marchevsky RS, Yoshida CF, Coelho JM, Gaspar AM, Schatzmayr HG. Intragastric infection induced in marmosets (*Callithrix jacchus*) by a Brazilian hepatitis A virus (HAF-203). Braz J Med Biol Res 1995;28:313−21.

[80] Emerson SU, Lewis M, Govindarajan S, Shapiro M, Moskal T, Purcell RH. cDNA clone of hepatitis A virus encoding a virulent virus: induction of viral hepatitis by direct nucleic acid transfection of marmosets. J Virol 1992;66:6649−54.

[81] Mathiesen LR, Moller AM, Purcell RH, London WT, Feinstone SM. Hepatitis A virus in the liver and intestine of marmosets after oral inoculation. Infect Immun 1980;28:45−8.

[82] Shibayama T, Kojima H, Ashida M, Hirose S, Sato A, Kamimura T, Hamada C, Shimizu Y, Suzuki S, Ichida F. Localization of hepatitis A virus in marmoset liver tissue during the acute phase of experimental infection. Gastroenterol Jpn 1985;20:564−72.

[83] Dienstag JL, Popper H, Purcell RH. The pathology of viral hepatitis types A and B in chimpanzees. A comparison. Am J Pathol 1976;85:131–48.
[84] Kornegay RW, Giddens Jr WE, Van Hoosier Jr GL, Morton WR. Subacute nonsuppurative hepatitis associated with hepatitis B virus infection in two cynomolgus monkeys. Lab Anim Sci 1985; 35:400–4.
[85] Jacob JR, Lin KC, Tennant BC, Mansfield KG. GB virus B infection of the common marmoset (*Callithrix jacchus*) and associated liver pathology. J Gen Virol 2004;85:2525–33.
[86] Bright H, Carroll AR, Watts PA, Fenton RJ. Development of a GB virus B marmoset model and its validation with a novel series of hepatitis C virus NS3 protease inhibitors. J Virol 2004;78:2062–71.
[87] Laemmert HW, de Castro Ferreira L. The isolation of yellow fever virus from wild-caught marmosets. Am J Trop Med Hyg 1945; s1–25:231–2.
[88] Moreno ES, Agostini I, Holzmann I, Di Bitetti MS, Oklander LI, Kowalewski MM, Beldomenico PM, Goenaga S, Martinez M, Lestani E, Desbiez AL, Miller P. Yellow fever impact on brown howler monkeys (*Alouatta guariba* clamitans) in Argentina: a metamodelling approach based on population viability analysis and epidemiological dynamics. Mem Inst Oswaldo Cruz 2015; 110:865–76.
[89] Sallis ES, de Barros VL, GarMätz SL, Fighera RA, Graca DL. A case of yellow fever in a brown howler (*Alouatta fusca*) in Southern Brazil. J Vet Diagn Investig 2003;15:574–6.
[90] Coutinho JFV, Lima ID, Silva AMT, do Socorro Borges Freire M, Bonfim WM, Barros VLRS, da Costa Vasconcelos PF, Höfer E, de Queiroz MGL, Araújo GB, Carneiro Muniz JAP, Carvalho RA, da Costa IC, de Souza MdF. Investigação etiológica de uma epizootia em sagüis (*Callithrix jacchus*) numa área indene para febre amarela. A Primatol No Brasil 2007;10:251–68.
[91] Laemmert HW. Susceptibility of marmosets to different strains of yellow fever virus. Am J Trop Med Hyg 1944;s1–24:71–81.
[92] Leal SG, Romano AP, Monteiro RV, Melo CB, Vasconcelos PF, Castro MB. Frequency of histopathological changes in Howler monkeys (Alouatta sp.) naturally infected with yellow fever virus in Brazil. Rev Soc Bras Med Trop 2016;49:29–33.
[93] Fiennes RN. Rabies. In: Fiennes RNTW, editor. Pathology of simian primates. London: Karger; 1972. p. 646–62.
[94] Favoretto SR, de Mattos CC, Morais NB, Alves Araújo FA, de Mattos CA. Rabies in marmosets (*Callithrix jacchus*), Ceará. Brazil Emerg Infect Dis 2001;7:1062–5.
[95] Nieves P, Rodriguez JF, Kessler MJ, Bercovitch F. Subcutaneous rabies vaccination of pigtail macaques. J Med Primatol 1996;25: 14–6.

CHAPTER

16

Bacterial Diseases

Keith G. Mansfield[1], James G. Fox[2]

[1]Discovery and Investigative Pathology, Novartis, Institutes for Biomedical Research, Cambridge, MA, United States; [2]Division of Comparative Medicine, Massachusetts Institute of Technology, Cambridge, MA, United States

INTRODUCTION

Common marmosets are susceptible to a number of bacterial infections, which may be enzootic, causing sporadic but occasionally severe disease, or which may result in epizootics associated with more severe colony morbidity and mortality. The spectrum of these diseases often differs from those observed in macaque species, and veterinarians caring for common marmosets need to be aware of these unique susceptibilities. In formulating differential diagnoses for sick or diseased animals, it should be recognized that diseases once common in imported animals in the 1960s and 1970s are now rare. It is also important to recognize that housing and sanitation conditions can influence exposure to potentially pathogenic bacteria. In a zoological setting where mixed- or free-ranging exhibits are utilized, animals may be exposed to many more potential pathogens, than would be the case in animals raised in a barrier facility.

Additional work is needed to understand the role of bacteria in common marmoset health and disease. We have limited understanding of how the normal microbial flora of the gastrointestinal tract impacts marmoset health issues such as inflammatory bowel disease (IBD) or colonization by pathogenic organisms. Moreover, much of the current literature describing the clinical and epizoological features of bacterial infections in this species is dated, and laboratory animal scientists should be encouraged to study these conditions and publish their results in further detail.

GRAM-POSITIVE COCCI

Staphylococcus spp.

Members of the *Staphylococcus* genus are part of the normal microbial flora of the common marmoset mucosal surfaces [1]. *Staphylococcus albus* and *Staphylococcus faecalis* have been isolated from the throat, conjunctiva, vagina, penis, and rectum of normal marmosets in the absence of disease. *Staphylococcus aureus* was not detected in a survey of normal animals but was identified in a single animal with a purulent otitis externa. While *S. aureus* may be isolated from superficial or deep purulent infections in a variety of primate species, it is relatively uncommon in common marmosets. If identified, antimicrobial selection should be based on culture and sensitivity due to the presence of resistance to betalactam antibiotics. Therapy of 3–6 weeks may be required to eliminate established infections.

Streptococcus pneumoniae

Streptococcus pneumoniae is an important pathogen of nonhuman primates which may cause a rapidly progressive disease resulting from dissemination via the respiratory tract. The disease is characterized by involvement of the meninges, serosal surfaces, or joints and may result in rapid death with few clinical signs. Systemic manifestations may be preceded by mild upper or lower respiratory signs in severely affected or contact animals. The source of epizootics is likely symptomatic or asymptomatic human contacts. While the organism remains highly susceptible to penicillin G, the clinical response in animals with systemic disease is often poor. Treatment of contact animals may help control spread of infection. Previous reviews have suggested that *S. pneumoniae* be considered an important pathogen of callitrichids [2,3]; however, definitive reports of infection of common marmosets with *S. pneumoniae* are lacking and their susceptibility to infection and severe disease remains to be determined.

The Common Marmoset in Captivity and Biomedical Research
https://doi.org/10.1016/B978-0-12-811829-0.00016-9

GRAM-POSITIVE RODS

Erysipelothrix rhusiopathiae

Etiology

Erysipelothrix rhusiopathiae is a Gram-positive, nonmotile bacillus, which is known to infect a variety of domestic and wild animals, and is well described in swine and turkeys in which it causes the disease erysipelas. Septicemia with *E. rhusiopathiae* (*insidiosa*) has been reported in cotton-top tamarins, red-bellied tamarins, and common marmosets [4].

Clinical Signs

In the reported epizootic of *E. rhusiopathiae*, common marmosets were found dead with no antecedent clinical signs.

Epizootiology

E. rhusiopathiae can be shed in the feces by a variety of wild and domestic animals where it may survive in the environment for up to 2 weeks, contaminating fomites, bedding, and food sources. Previous outbreaks in nonhuman primates have been linked to exposure to rodents, birds, and swine. In the reported outbreak observed in common marmosets, animals were housed in a barrier facility and the source of infection was unclear. Worker contact with swine at home was proposed as one potential source. Alternatively, tree branches and foliage used for enrichment may have been contaminated with bird droppings, thereby exposing animals in the barrier facility [4].

Pathology

Grossly at necropsy, hemorrhages may be noted in a number of organs, including the kidneys, adrenal glands, gastrointestinal tract, and brain [4]. These changes may be accompanied by splenomegaly and lymphadenopathy. Acute myocarditis, hepatitis, encephalitis, and enteritis involving the duodenum and jejunum may be recognized histologically.

Diagnosis

Definitive diagnosis requires culture and identification of *E. rhusiopathiae*. The organism can be recovered from blood or tissue, and use of enrichment broth or selective media with sodium azide decreases contamination and improves isolation [5]. Polymerase chain reaction (PCR) assays have also been described [6].

Differential Diagnosis

Other causes of bacterial septicemia and disseminated infection such as *Yersinia* spp., *Pseudomonas* spp., and *Klebsiella* spp. should be considered. *Yersinia* spp. and *Klebsiella* spp. should be visualized on histological evaluation of tissue. None of these organisms are associated with multifocal hemorrhage described with *E. rhusiopathiae* and bacterial culture, and identification would distinguish them. The extensive hemorrhages noted at necropsy are also suggestive of coagulopathy, and rodenticide toxicity should be considered in the differential diagnosis.

Prevention and Control

An epizootic of *E. rhusiopathiae* has been stopped by administration of a commercially available *E. rhusiopathiae* vaccine manufactured for swine to remaining animals [4]. *E. rhusiopathiae* should also be susceptible to penicillins and erythromycin, which could be considered in the treatment of clinically affected or contact animals. As *E. rhusiopathiae* may be carried by rodents, pest control measures should be evaluated.

Research Complications

E. rhusiopathiae is a known zoonotic bacterial infection producing the disease erysipeloid in humans. While described in common marmosets in a single report, risk for this potential infection in laboratory-reared animals appears low. Infection risk may be more of a concern in animals housed in zoological collections, particularly if exposed to other species, including rodents and birds.

Mycobacterium spp.

Mycobacterium spp. are small acid-fast bacilli that can be classified into three groups based on biology and disease causation: (1) *Mycobacterium tuberculosis* complex (MTC), (2) nontuberculous mycobacteria (NTM), and (3) leprosy bacilli. The former two groups may infect and cause disease in common marmosets. As a group, mycobacteria have a complex cell wall high in glycolipids, including mycolic acid that imparts unique staining characteristics and functions as an important virulence factor.

Nontuberculous Mycobacteria

Etiology

More than 170 distinct NTM species have been described, most not associated with human or animal disease [7]. The Runyon classification system for NTM was developed in 1959 and is based on growth rate and pigmentation. The slow-growing mycobacteria include Runyon groups I, II, and III, and the rapid-growing mycobacteria are in group IV. Runyon group I or photochromogens produce a yellow pigment on exposure to light and include *Mycobacterium kansasii, Mycobacterium marinum,* and *Mycobacterium simiae.*

Runyon group II or scotochromogens produce pigment regardless of light exposure and include *Mycobacterium gordonae* and *Mycobacterium scrofulaceum*. Runyon group III or nonchromogens fail to produce pigment and contain the species *Mycobacterium avium* and *Mycobacterium intracellulare*. Runyon group IV or rapid growers fail to produce pigment and contain *M. fortuitum*. More recent phylogenetic analysis of NTM 16S rRNA largely reflects the Runyon classification system [8].

Colonization of common marmosets by NTM has recently been described [9]. In this report, routine screening of marmosets by tuberculin skin testing using old mammalian tuberculin revealed that 3.6% of the colony had equivocal or positive reactions. *M. gordonae* was isolated from the fecal culture of a single animal. Subsequent PCR screening for *M. gordonae* sequences revealed that 66.7% of the tuberculin skin test (TST)-positive animals were colonized by *M. gordonae* compared with 16.7% of the TST-negative controls. Animals with positive TSTs were necropsied, and lymph nodes were cultured. *M. kansasii* was isolated from tissues with morphologic evidence of mycobacterial disease. The authors concluded that detection of *M. gordonae* may serve as an indicator of NTM colonization, whereas *M. kansasii* was associated with tissue pathology [9] (Fig. 16.1).

M. kansasii is a Runyon group I photochromogen and is considered one of the more pathogenic NTM. It may be isolated from immunologically normal human patients or in patients with underlying malignancies such as leukemia or solid tumors. It is most commonly isolated from patients with chronic lung disease but has also been observed to cause disseminated disease, tenosynovitis, and cutaneous infections [10].

FIGURE 16.1 **Tuberculin skin test (TST) in common marmosets with atypical mycobacterial infection.** Grade 5-positive TST demonstrating marked swelling and complete closure of the palpebral fissure.

Clinical Signs

Infected animals are often without clinical signs, and NTM infection is suspected based on equivocal or positive intradermal skin test results. Occasional animals may demonstrate weight loss and peripheral lymphadenopathy [9].

Epizootiology

NTM are found widely in environmental soil, vegetation, and water samples and form biofilms on surfaces that promote survival and persistence of the organism. These films occur commonly in water distribution systems and may serve as a source of infection through ingestion or inhalation. Recirculating hot water distribution systems are often used in buildings such as hospitals and animal facilities and may promote the formation of biofilms. They have been linked to the development of disseminated *M. avium* complex in both immunocompromised humans and rhesus macaques [11].

Because of their ubiquitous nature, exposure to NTM from the environment is nearly constant but disease is rare. Colonization and disease may be dependent on the level of exposure and on poorly defined host factors. In general, cases of NTM arise sporadically, and once an animal is infected, the organisms are not thought to be communicable and therefore do not pose a risk to contact animals or human handlers. One exception to this observation was documented in groups of simian immunodeficiency virus-infected rhesus macaques, in which horizontal transmission between animals occurred [12].

Pathology

At necropsy, multiple gastrointestinal lymph nodes may be enlarged. Less commonly tracheobronchial and/or peripheral lymph nodes may be involved. Histologically, there is granulomatous lymphadenitis with variable perinodal fibrosis. In these granulomas, there are central regions of caseous necrosis surrounded by multinucleated giant cells and variable numbers of epithelioid macrophages and lymphocytes. The remaining lymph nodes often appear hyperplastic. Ziehl–Neelsen stains reveal small numbers of acid-fast bacilli both within regions of caseous necrosis and within multinucleated giant cells. The findings are similar to those previously described for *M. kansasii* infection in squirrel monkeys [13] and differ from those seen with NTM in immunocompromised hosts. It is important to note that these morphological features appear similar or identical to MTC infections of nonhuman primates, and histopathology cannot be used to distinguish the two different infections (Fig. 16.2).

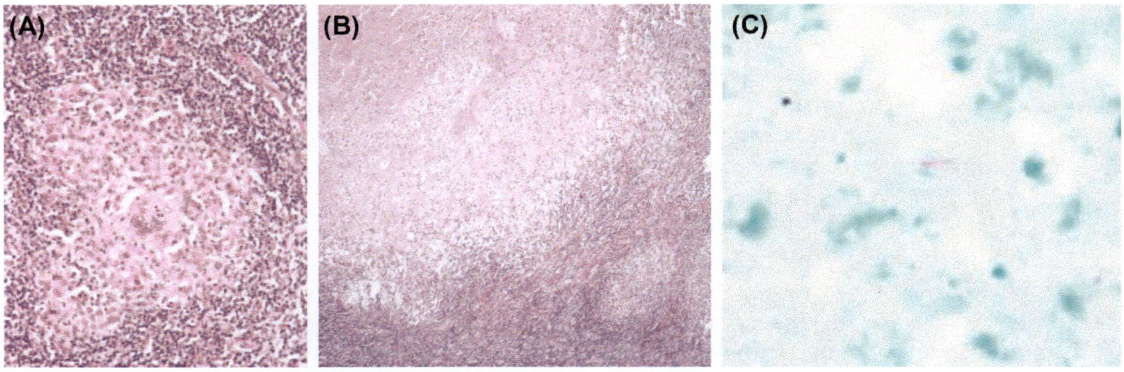

FIGURE 16.2 **Morphologic appearance of atypical mycobacterial infections in common marmosets.** (A) Multinucleated giant cell in the center of a noncaseating granuloma (lymph node, H&E stain, ×200). (B) Large region of caseous necrosis in mesenteric lymph node (lymph node, H&E stain, ×200). (C) Acid-fast bacillus in region of caseous necrosis (lymph node, Ziehl–Neelsen stain, ×500).

Diagnosis

PCR has been widely used to detect mycobacterial nucleic acid in environmental and patient samples and has a number of advantages, including speed and sensitivity. A variety of PCR assays have been described to distinguish MTC from NTM and to further speciate the latter [14,15]. Caution is advised in interpreting PCR results as false positive and negative results are possible. Moreover, the sensitivity of PCR to detect mycobacteria in nonhuman primate tissue or biological samples has not been rigorously assessed. Definitive diagnosis requires bacterial culture and isolation. Most atypical mycobacteria grow well and relatively rapid on Lowenstein–Jensen media.

Differential Diagnosis

NTM must be differentiated from MTC infections; this cannot reliably be done based on clinical and pathological findings. Because suspected cases are first identified by an equivocal or positive TST, other causes of TST reactivity should be considered such as previous administration of complete Freund's adjuvant or iatrogenic bacterial infection at the test site.

Prevention and Control

NTM are common environmental bacteria and can be found in a number of sources. It is likely that most marmosets are constantly exposed to NTM, but the vast majority do not become colonized or develop disease. Host factors that predispose animals to colonization and/or disease are unknown but might include host genetics, concurrent diseases, or alterations in the gastrointestinal microflora. As noted above, NTM form biofilms and can colonize water distribution systems in large facilities. If recurrent or persistent issues are observed, colony water sources should be assessed, and remedial actions to reduce water-borne mycobacterial content should be considered. In immunodeficient macaques, use of autoclaved water sources has been shown to decrease the incidence of NTM infections [16]. Treatment of nonhuman primates for NTM has not been attempted. Treatment of NTM in human patients is difficult, but amikacin has been shown to have the best activity for rapidly growing mycobacteria, and clarithromycin was the most active drug against *M. avium* and other slow growers [17]. *M. kansasii* also responds to rifampin-based therapies, but prolonged treatment may be necessary [18].

Research Complications

Atypical mycobacteria are largely noncommunicable from animal to animal and do not cause zoonotic disease in human handlers. As NTM must be distinguished from MTC infections, colonies should be quarantined until a definitive diagnosis is reached. While this may disrupt ongoing research and breeding programs, it is necessary to protect staff and colony health.

Mycobacterium tuberculosis *Complex*

Etiology

The MTC consists of *Mycobacterium africanum, Mycobacterium bovis, Mycobacterium canettii, Mycobacterium microti,* and *M. tuberculosis* and contains known pathogens of humans and wild and domesticated species of mammals. They are obligate pathogens and the etiologic agents of tuberculosis, so named because of the nodular lesions observed in lungs, which are termed tubercles. These organisms can be visualized using acid-fast stains such as the Ziehl–Neelsen, Fite, and Kinyon stains, which fail to decolorize with weak acids in mycobacteria due to the organism's waxlike cell wall. If a Gram stain is performed, bacilli may be weakly Gram-positive or -negative. *M. tuberculosis* and *M. bovis* are the two most common species in the MTC to be reported in Old World primates and can cause sporadic or epizootic disease of high morbidity and mortality in these species. Infection with other members of the MTC (*M. africanum*) is less commonly reported [19].

PCR surveys of neotropical primates kept as pets or in zoological collections have detected a high rate of *M. tuberculosis* in buccal swabs [20], and mycobacterial cultures of gastric lavage have demonstrated similar results [21]. However, reports of MTC-associated disease in neotropical primates are uncommon. There is a single case report documenting *M. tuberculosis* in a pet common marmoset with likely human-to-animal transmission [22]. In contrast, common marmosets are highly susceptible to experimental inoculation of *M. tuberculosis* by the aerosolized and intratracheal routes [23,24]. The reasons for this disparity are unclear, and efforts should be made to further document the natural history of MTC infection in common marmosets and neotropical primates in general.

Clinical Signs

In the case report, a pet common marmoset was presented for chronic weight loss, and an abdominal mass was palpated [22]. The animal died when a biopsy was attempted, and few additional clinical details were provided. Publications on experimental inoculation of marmosets provide additional insight into disease manifestations in this species [23–25]. Progressive weight loss is the most common clinical sign following experimental infection and is observed as early as 2 weeks following inoculation correlating with disease burden. Other clinical signs recognized less frequently included low-grade pyrexia, dehydration, anorexia, and tachypnea. As in other primate species, coughing is infrequent but may be observed.

Epizootiology

In Old World species of primates, infection most commonly originates from infected human handlers, although ungulates have been implicated in epizootics of *M. bovis*. Transmission in these species most often occurs by aerosolization, but infection has also been documented to occur by ingestion, direct contact, or contact with fomites. Less is known about the transmission in common marmosets, although these routes would all appear relevant. Experimental studies indicated that common marmosets are highly susceptible to infection by the aerosolized route, with as few as 1–12 colony-forming units capable of causing rapidly progressive and fatal disease [24].

Pathology

In the case report of naturally occurring tuberculosis in a pet common marmoset, only a limited necropsy was performed and analysis was restricted to an enlarged mesenteric lymph node, which appeared to be abscessed. Moderate numbers of acid-fast bacilli were visualized in the lymph node, and *M. tuberculosis* was cultured and identified [22].

Experimental studies provide additional details on the distribution and histological appearance of tuberculosis in this species [23–25]. Following inoculation with *M. tuberculosis* or *M. africanum*, marmosets developed progressive weight loss and were euthanized when body weight had decreased by 20% of baseline. Progression rates of clinical disease were dose and strain specific. At necropsy, a range of gross findings and histological features were identified similar to those described in primary human tuberculosis. Pulmonary lesions were most frequent and ranged from complete consolidation of lung lobes to the presence of well-defined nodules. These changes were accompanied by enlargement of the tracheal bronchiolar lymph nodes. Histologically both necrotizing and nonnecrotizing lesions were evident. Typical granulomas consisting of a central region of necrosis surrounded by a cuff of epithelioid macrophages admixed with neutrophils and an outer layer of lymphocytes and histiocytes were observed. Multinucleated giant cells and regions of mineralization characteristic of more chronic infections were not observed. Occasional granulomas appeared cavitated and contained semiliquid caseum. Regions of pulmonary consolidation consisted of large numbers of neutrophils which filled alveolar spaces and effaced the normal architecture, eventually leading to overt necrosis. Discrete granulomas were often lacking in regions of consolidation. Results of acid-fast stains were not described [23–25].

Strain-specific dissemination to the liver and spleen was observed in some animals and recognized at necropsy by hepatosplenomegaly and the formation of discrete miliary to nodular granulomas [23,24]. Histologically these comprise both necrotizing and nonnecrotizing granulomas. Less frequently granulomatous interstitial nephritis was observed.

Diagnosis

Animals are most likely to be first identified through a positive TST or because of chronic weight loss and/or respiratory signs. Definitive antemortem diagnosis of tuberculosis can be difficult in nonhuman primates, and the clinician should rely on a variety of modalities to assess animals. The TST is effective in assessing exposure to mycobacteria on a colony basis, but results can be difficult to interpret in an individual animal. The diagnostic sensitivity and specificity of TST have not been assessed in common marmosets. The diagnostic work up should include a complete review of the animals' housing, medical, and research records. Records from contact animals should also be evaluated to determine the TST status and whether unexpected weight loss has occurred. Thoracic and abdominal radiographs may provide evidence of pulmonary lesions, lymphadenopathy, and/or hepatosplenomegaly. Buccal swabs, gastric lavage, and fecal samples can be utilized to assess for mycobacterial growth or nucleic acids.

Mycobacterial isolation and identification remains the gold standard for the diagnosis of tuberculosis but does have disadvantages. Isolation can be a prolonged process due to the slow-growing nature of MTC organisms and can take weeks to several months to complete. PCR offers the advantage of rapid results, which may be available in a matter of days rather than weeks, and a number of assays have now been developed and are commercially available. Some caution is advised in the interpretation of these results as both false-positive and -negative results are possible. As noted above, detection of environmental NTM sequences does not equate with disease causation and could potentially mask an underlying MTC infection. Moreover, the diagnostic sensitivity and specificity of the various PCR assays have not been rigorously assessed in fluid and tissue samples obtained from clinically affected marmosets. The relative sensitivity of PCR compared with culture in this setting is not defined.

Definitive diagnosis in the living animal may not be possible and euthanasia and necropsy may be required. While this can be a difficult decision, safety of staff and colony health are paramount. Necropsy should be performed with additional respiratory protection such as a fitted N95 respirator or powered air-purifying respirators. At a minimum the lung, spleen, liver, gastrointestinal tract, and tracheobronchial and mesenteric lymph nodes should be collected for formalin fixation, mycobacterial culture, and PCR.

Differential Diagnosis

Progressive weight loss is a common clinical finding in many chronic diseases affecting common marmosets. The clinical and experimental data indicate that mycobacterial infections should be considered in the differential diagnosis in animals presenting with this finding. In animals presenting with a positive TST, differentiating NTM from MTC infections can be difficult but is required due to the different implications for human safety and colony health.

Prevention and Control

Proactive occupational health and safety and colony preventative health programs are critical for tuberculosis prevention and control [26]. Animals should be regularly examined on a quarterly basis; blood obtained for biobanking and 0.05 mL of old mammalian tuberculin should be administered with a 26-gauge needle in the upper eyelid. Purified protein derivative (PPD) used in the human TST is not appropriate. The animals are then visually examined at 24, 48, and 72 h for evidence of redness of swelling. The TST is scored as follows: grade 1, slight bruising of the eyelid; grade 2, erythema of the palpebrum without swelling; grade 3, variable degree of erythema, with minimal swelling; grade 4, obvious swelling, with drooping of the eyelid and erythema; and grade 5, marked swelling and/or necrosis of the eyelid [26]. A grade 3 reaction is considered equivocal, and grades 4 and 5 are considered positive. Due to the risk to colony and human health, in most cases a grade 4 or 5 will result in the euthanasia of the animal. A complete necropsy should be performed with collection of tissue for histopathology, mycobacterial culture, and PCR. While awaiting definitive diagnosis, contact animals should be placed under quarantine and the TST repeated once every 2 weeks until three negative tests are achieved in all remaining animals. Interpretation of equivocal or grade 3 reactions should be made with an assessment of the overall risk of tuberculosis in the individual animal and the colony as a whole. At a minimum, additional diagnostics such as blood work and radiographs should be completed, and the animal and its contacts should be retested by TST. If the risk is deemed high, because of clinical signs such as weight loss or potential recent contact with tuberculosis cases in other animals or humans, euthanasia and necropsy may be warranted.

As most cases of tuberculosis in nonhuman primates are acquired from human handlers, contact with infected individuals should be prevented. Facility staff with contact with nonhuman primates should be assessed for tuberculosis on an annual basis. In general, TST is performed with PPD, and individuals with positive reactions are referred for additional assessment. The QuantiFERON-TB test (manufactured by Cellestis Limited, Carnegie, Victoria, Australia) was approved in 2001 by the Food and Drug Administration as an aid for detecting latent *M. tuberculosis* infection. This in vitro diagnostic test measures a component of cell-mediated immune reactivity to *M. tuberculosis*. Quantification of interferon-gamma (IFN-gamma) released from sensitized lymphocytes is measured in whole blood incubated overnight with PPD from *M. tuberculosis* and control antigens [27,28]. Individuals with signs of upper or lower respiratory tract infections should not work with animals until cleared to do so. Personal protective equipment should be required of all staff while working with nonhuman primates. While a standard face mask does not provide respiratory protection, it does help contain aerosols generated by the wearer when coughing or sneezing. Protection from infected individuals may be more difficult or impossible in zoological settings in which animals are free ranging and may come in contact with visitors. These animals may be at greater risk of contracting tuberculosis as indicated by recent surveys conducted in neotropical primates in Brazil [20,21].

Research Complications

M. tuberculosis is a known zoonotic disease, which has been transmitted from nonhuman primates to humans. Epizootics of *M. tuberculosis* also pose a threat to colony health and may require quarantine and a discontinuation of research efforts while diagnostic and control measures are implemented. Efforts at prevention are extremely important.

ANAEROBIC BACTERIA

Clostridium spp.

Clostridium perfringens—*Clostridial Myonecrosis*

Etiology

Clostridium perfringens (previously named *Clostridium welchii*) is a Gram-positive, rod-shaped, anaerobic, spore-forming pathogenic bacterium, which is found commonly in decaying vegetation and soils. *C. perfringens* isolates can produce more than 17 different bacterial exotoxins that may adversely affect the host. The alpha toxin is a metallophospholipase that causes activation of the arachidonic pathway and production of thromboxane A2 and platelet-activating factor. Coupled with the hemolytic activity of *C. perfringens* alpha toxin, the resulting increase in vascular permeability is responsible for the myonecrosis characteristic of *C. perfringens* in a variety of species. Several other species of *Clostridium* including *Clostridium histolyticum, Clostridium fallax, Clostridium novyi, Clostridium septicum,* and *Clostridium bifermentans* may result in clostridial myonecrosis in other species [29].

Clinical Signs

A single case of *C. perfringens* gas gangrene (clostridial myonecrosis) has been reported in the common marmoset [30]. The animal died acutely with only facial erythema reported the previous day. The animal had a history of diarrhea and weight lost several months prior to death but had recovered.

Epizootiology

Clostridial myonecrosis may be classified as traumatic, postoperative, or spontaneous. *C. perfringens* is commonly found in the environment and may contaminate wounds. In traumatic and postoperative clostridial myonecrosis wound factors such as low oxygen tension play a role in the establishment of tissue infection and in bacterial growth. In human patients, spontaneous clostridial myonecrosis is often associated with underlying malignancy, and it is proposed that initial clostridial colonization of the poorly oxygenated tumor microenvironment allows the infection to become established with subsequent dissemination. There was no evidence of trauma or malignancy in the reported common marmoset case, and it is thought that initial colonization and infection were in the gastrointestinal tract.

Pathology

At necropsy in the reported case, the abdomen was bloated and blood-tinged fluid was observed in the peritoneal cavity. Generalized subcutaneous hemorrhages were present, and the skeletal muscle appeared grayish-yellow and edematous. Histologically, small vacuoles accompanied by bacilli were observed in the skeletal muscle, liver, gastrointestinal tract, and lungs. In skeletal muscle, interstitial edema and foci of rhabdomyolysis and coagulative necrosis of myofibers were observed. Despite tissue damage minimal inflammatory infiltrates were observed.

Diagnosis

Isolation of *C. perfringens* and identification of alpha toxin activity should be coupled with histological examination of tissues demonstrating large numbers of Gram-positive bacilli, scant neutrophilic infiltrates, and the presence of fat globules that result from the breakdown of tissue.

Differential Diagnosis

Differential diagnosis should include death from other causes with postmortem overgrowth of enteric anaerobic bacteria and autolysis.

Prevention and Control

In other species with clostridial myonecrosis, aggressive treatment is required to prevent rapid progression and death. Wound management including debridement and drainage coupled with antibiotic therapy with penicillin G and clindamycin are used.

Clostridium perfringens—*Gastric Dilation*

Gastric dilation and death secondary to overgrowth of *C. perfringens* type A in a group of common marmosets have been described [31]. Shigellosis had been suspected in a colony of 180 marmosets, and animals in the affected housing unit were started on prophylactic treatment with a combination of parenteral gentamicin and oral furoxone administered in the drinking water. Animals subsequently developed signs of gastric distress and were started on gastrointestinal protectants and an anticholinergic antiemetic. Over the next 5 weeks, 29 of 59 animals were found in a moribund state and died with gastrointestinal dilation. *C. perfringens* type A was cultured from gastric contents. It was proposed that antibiotic therapy altered the intestinal microbial flora, allowing clostridial overgrowth and promoting gas accumulation and dilation of the gastrointestinal tract [31].

Clostridium piliforme

Etiology

Clostridium piliforme (previously *Bacillus piliformis*) is a spore-forming, obligate intracellular pathogen and the etiologic agent of Tyzzer's disease. The organism appears filamentous in tissue and cannot be grown on cell-free media. It appears to stain Gram negative unless it is grown under hypoxic conditions in which case it will appear Gram positive as do other members of the *Clostridium* genus. *C. piliforme* can infect a number of species and most often causes disease in neonatal or young animals where it has been recognized in guinea pigs, mice, rats, rabbits, hamsters, foals, birds, cats, dogs, and nonhuman primates [32].

Clinical Signs

Tyzzer's disease has been described in common marmosets [33] and cotton-top tamarins [32] where it was recognized in neonatal animals. The disease is rapidly progressive, and neonatal animals are generally found dead with no antemortem clinical signs recognized.

Epizootiology

The organism is shed in the feces where spores may contaminate water, food, or bedding. Following ingestion, the organism may replicate in the lower gastrointestinal tract and seed the liver and systemic circulation through the portal vein. Rodents may represent natural reservoirs of infection. Once the environment becomes contaminated, the spores may survive for years and serve as a source of infection.

Pathology

C. piliforme infection is most often associated with multifocal hepatic necrosis, which may be recognized both grossly and microscopically. These foci are variable in size and contain karyorrhectic cellular debris and scant inflammatory cells. The organism can often be visualized in the cytoplasm of hepatocytes at the leading edge of necrosis as lightly basophilic filamentous bacteria. Silver or Giemsa stains may highlight the bacteria and are useful diagnostic aids. These morphological findings are diagnostic of Tyzzer's disease. In addition to hepatocellular necrosis, transmural enterocolitis of the lower gastrointestinal tract may be evident. This likely represents the site of bacterial colonization and entry but may be multifocal and missed. Typical organisms may be found in smooth muscle cells adjacent to regions of necrosis. The organism may further disseminate to the myocardium where it can cause myocarditis. Although unusual, involvement of the central nervous system was recognized in the common marmoset where necrosis and neutrophilic infiltration were observed [33] (Fig. 16.3).

Diagnosis

The morphologic findings are suitably distinct to be diagnostic. Recently immunohistochemistry using rabbit anti-*C. piliforme* sera has been described and can be used if routine stains are equivocal [34]. PCR has also been used to detect *C. piliforme* sequences in tissues, and serology can be used in surviving animals to demonstrate seroconversion.

Differential Diagnosis

The triad of intestinal, liver, and heart lesions in young animals is highly suggestive of Tyzzer's disease. Other causes of multifocal hepatic necrosis such as adenovirus, herpes virus simplex 1, and toxoplasmosis may be considered and would have distinct morphologic features on microscopic evaluation.

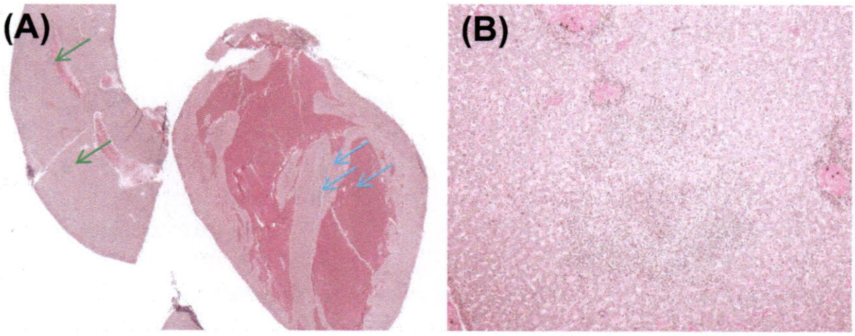

FIGURE 16.3 ***Clostridium piliforme.*** (A) Multifocal necrotizing myocarditis (*blue arrows*) and hepatitis (*green arrows*) (heart, H&E stain, ×4). (B) Focal necrotizing hepatitis (liver, H&E stain, ×100).

Prevention and Control

C. piliforme spores may contaminate the environment and be difficult to eliminate. Ethylene oxide or autoclaving of materials should be effective against the spores and surfaces may be decontaminated with 0.3% hypochlorite. As rodents may serve as a reservoir of infection, additional efforts at rodent control may be warranted.

GRAM-NEGATIVE BACTERIA

Bordetella bronchiseptica

Etiology

Bordetella bronchiseptica is a small, Gram-negative, rod-shaped bacterium and can cause infectious bronchitis in a variety of animal species and rarely humans. *B. bronchiseptica* is closely related to the human pathogen *Bordetella pertussis* to which the marmoset is also susceptible to experimental inoculation.

Clinical Signs

An epizootic of *B. bronchiseptica* has been reported in a colony of common marmosets resulting in pleuropneumonia and deaths [35]. A colony of 156 animals was initially established as specific pathogen free of *B. bronchiseptica*. Despite routine testing and precautions, an epizootic of *B. bronchiseptica* occurred, and the authors were able to demonstrate spread of the organism to 71 individual animals through evaluation of nasal swabs. The most common clinical sign was a purulent nasal discharge, which was observed in animals of all ages. Occasionally these signs were accompanied by an elevated body temperature and dyspnea when the animals were handled; however, affected marmosets never coughed. Sixteen animals died from *B. bronchiseptica* with the vast majority of lethally affected animals being under 1 year of age. In these animals, death was not preceded by premonitory clinical signs and animals were often found dead.

Epizootiology

In the described epizootic, in a specific pathogen-free colony *B. bronchiseptica* spread rapidly, resulting in upper respiratory infection in animals of all ages. While the organism appeared capable of colonizing animals of all ages, severe disease was largely limited to animals under 1 year of age.

Pathology

In animals subjected to necropsy, a purulent bronchopneumonia was observed, most frequently affecting the right cardiac and accessary lung lobes. In 6 of 16 cases the inflammatory process had extended to the pleura resulting in fibrinopurulent pleuritis, and in 2 cases extensive involvement of the pericardium was recognized. Microscopically, bronchi contained plugs of neutrophils admixed with macrophages, fibrin, and sloughed bronchiole epithelial cells. The inflammatory process extended into the surrounding alveolar spaces where edema and destruction of septa were recognized. Blood vessels in affected regions appeared edematous and contained infiltrating neutrophils. Laryngitis and rhinitis were observed in a minority of affected animals (Fig. 16.4).

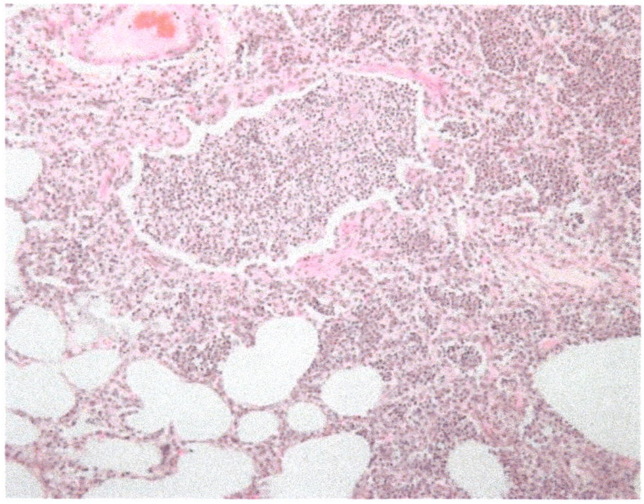

FIGURE 16.4 *Bordetella bronchiseptica*. Suppurative bronchopneumonia (lung, H&E stain, ×200).

Diagnosis

Definitive diagnosis requires bacterial isolation and identification. Antimicrobial profiling should be obtained on all isolates.

Differential Diagnosis

Other viral and bacterial causes of bronchopneumonia such as adenovirus and *Pasteurella* infection may cause similar clinical and/or pathologic findings. In other species, viral coinfection with viral respiratory pathogens such as coronavirus, influenza virus, and adenoviruses may exacerbate *B. bronchiseptica* and should be considered in recognized epizootics.

Prevention and Control

Antimicrobial susceptibility profiles have recently been published for *B. bronchiseptica* isolates from swine and companion animals [36]. *B. bronchiseptica* is generally resistant to macrolide antibiotics and cephalosporins and often carries β-lactamase resistance genes.

Canine isolates have been found to be sensitive to tetracycline, doxycycline, enrofloxacin, and amoxicillin/clavulanic acid [37].

Escherichia coli

Etiology

Six categories of diarrheagenic *Escherichia coli* are defined based on the underlying mechanism of disease pathogenesis, in vivo and in vitro growth characteristics, and the presence of specific genes encoding virulence factors. These include enteropathogenic *E. coli* (EPEC), enteroaggregative *E. coli*, enterotoxigenic *E. coli*, enterohemorrhagic *E. coli*, enteroinvasive *E. coli* (EIEC), and diffuse adherent *E. coli* [38]. EPEC is one of the most common bacterial causes of diarrhea in humans worldwide and is also known to infect a variety of other species including cattle, swine, rabbits, and nonhuman primates, including common marmosets and cotton-top tamarins.

Studies over the past 30 years have revealed the molecular basis of EPEC pathogenesis and virulence. Typical EPEC carry a gene encoding the bundle-forming pilus (bfp), which is responsible for initial attachment of the bacteria to enterocytes. The organisms possess a type 3 secretion system (T3SS), which allows the direct insertion of more than 25 bacterial proteins into the target cell's cytoplasm. Intimin is an outer membrane protein encoded by the eaeA gene and is required for full virulence of the organism. Intimin binds to the translocated intimin receptor (Tir), which has been introduced into the target cell membrane by the T3SS. Tir–intimin interaction initiates full adherence of the organism to the target cell surface, and subsequent phosphorylation of intracytoplasmic Tir domains causes polymerization of actin. It is this effect on actin that results in effacement of the brush border and the formation of an attachment pedestal, both of which can be recognized ultrastructurally. The eventual induction of diarrhea correlates with these morphological changes in enterocytes and is caused by the destruction of the microvilli and tight junctions [39].

Sequencing of the *bfp*A and eae genes from isolates cultured from neotropical primates revealed, respectively, 98.4% and 99.2% identity to those observed in strains obtained from humans [43]. While zoonotic infection of human handlers has not been reported, the close genetic relationship between human and nonhuman primates suggests that care should be exercised when working with infected animals.

Recently, *E. coli* isolates encoding genes for colibactin and cytotoxic necrotizing factor (CNF), both cytotoxins causing DNA damage and cell cycle arrest, have been isolated from the feces of common marmosets [39a]. These *E. coli* strains have been associated with urogenital tract infections, septicemia, and meningitis in humans [40,41]. Their clinical relevance in marmosets requires further investigation. CNF-positive *E. coli* also have been isolated from macaques [42]. In a recent study, the prevalence of cytotoxins encoded by *pks* and *cnf* in *E. coli* cultured from rectal swabs of marmosets housed in three separate colonies was documented. Isolates of *E. coli* were identified and characterized biochemically. Specific PCR for *pks* and *cnf* gene amplification and phylogenetic group identification were performed on all *E. coli* strains. A total of 139 *E. coli* strains were isolated from 120 of the 138 marmosets (87%) in the three colonies. In Colony A, 31 strains of *E. coli* were isolated from 25 of the 31 animals; 55% of the animals had *pks*+ *E. coli*, and 52% had *cnf*+ *E. coli*. In Colony B, 33 strains of *E. coli* were isolated from 32 of the 33 animals, 6% of the animals had *pks*+ *E. coli*, and 9% had *cnf*+ *E. coli*. In Colony C, 75 strains of *E. coli* were isolated from 63 of the 74 animals, 43% of the animals had *pks*+ *E. coli*, and 35% had *cnf*+ *E. coli*. Both *pks*+ and *cnf*+ *E. coli* strains belonged mainly to phylogenetic group B2. Colibactin and CNF cytotoxic activities were confirmed using a HeLa cell cytotoxicity assay. The presence of *pks*+ or *cnf*+ *E. coli* did not correlate with health status. However, given the association of these cytotoxin-producing *E. coli* strains with clinical disease in humans, the colonization of marmosets with these cytotoxin-producing *E. coli* can potentially cause clinical and subclinical diseases that impact marmoset models [39a].

Clinical Signs

EPEC infection presents acutely in juvenile and adult common marmosets [43]. Animals may initially appear lethargic and anorexic and often appear hunched when observed in nest boxes or perches. As signs progress, animals may become anemic and hypotensive and may be found on the floor of large enclosures unable to return to their nest boxes. Diarrhea is voluminous and may appear watery, containing variable amounts of mucus and fresh blood. Feces may have the consistency of strawberry jam, and the animal's perineum may be stained with this material. Clinical disease may present in healthy robust animals who appeared normal 12–24 h previously. Despite the rapidity of disease progression and the severity of diarrhea and apparent blood loss, animals respond dramatically to antibiotics and fluid therapy. With treatment, overall morbidity is low, and animals return to full health within a few days. Failure to respond suggests an alternative diagnosis should be considered or that significant comorbidities may be present (Fig. 16.5).

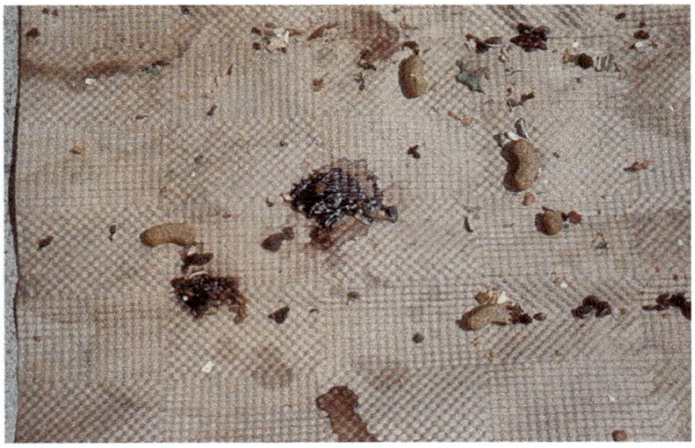

FIGURE 16.5 Enteropathogenic *Escherichia coli* infection. Diarrhea and hematochezia in cage papers.

Epizootiology

The organism is transmitted by the fecal–oral route, and enzootic infections are likely common in most marmoset colonies. As described below, unless specialized diagnostic approaches are used, these infections may go unrecognized and undiagnosed. While enzootic infection is common, clinical signs and more severe disease occur sporadically in colonies. Factors that precipitate clinical disease are unknown. A recent survey demonstrated that 10.2% of clinically healthy marmosets were asymptomatic carriers of EPEC [44].

Pathology

Morphologic findings are characteristic of the attaching and effacing lesion. The colonic surface epithelium appears irregular with bacilli intimately associated with the apical cytoplasmic membrane and accompanied by varying degrees of colonic crypt hyperplasia. Surface epithelium variably has a "cobblestone" or "tombstone" appearance or, when accompanied by necrosis, a flattened or squamous morphology. There are often mild neutrophilic infiltrates and congestion of vessels within the mucosa. Individual epithelial cells with adherent bacilli often appear rounded and vacuolated. Distribution of bacilli is limited to the colonic surface epithelium and varies from diffuse to a locally extensive or focal pattern. Less frequently organisms are noted in the ileum and distal jejunum. In each of these instances, the colon was severely involved. In some cases multiple sections of colon needed to be evaluated to visualize the characteristic findings. Ultrastructurally there is effacement of normal microvillus architecture, and adherent bacilli are attached to the apical cytoplasmic membrane with pedestal formation and rearrangement of the underlying cytoskeleton. Experimental oral inoculation of common marmosets with the EPEC R811 strain produced acute hematochezia in high-dose animals and recapitulated microscopic findings observed following natural infection [44] (Fig. 16.6).

Diagnosis

The morphologic findings of the attaching and effacing lesion described above are pathognomonic for EPEC infection, and biopsy of the colonic mucosa may provide a rapid and reliable means of diagnosis. Bacterial isolation and identification may also be useful, but many clinical bacteriology laboratories do not routinely speciate lactose fermenters. Because nonpathogenic *E. coli* are common commensals of the lower gastrointestinal tract of marmosets, their isolation is not sufficient to diagnose EPEC infection. Additional phenotypic or genotypic characterization is required for definitive identification. Adhesion assays may demonstrate a focal adhesion pattern of EPEC organisms on intestinal epithelial cell lines. Alternatively PCR may be used to

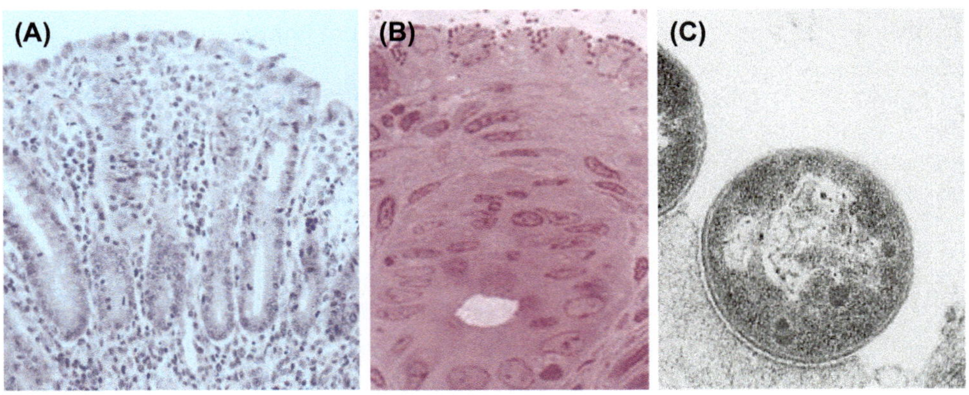

FIGURE 16.6 Morphologic appearance of enteropathogenic *Escherichia coli* infection. (A) Section of colon demonstrating a cobblestone appearance of the surface epithelium, marked hyperplasia within crypts, and minimal inflammatory cell infiltration (colon, H&E stain, 250×). (B) Surface epithelium appears rounded and vacuolated with numerous adherent rod-shaped bacilli (toluidine blue semithin section, colon, 500×). (C) Ultrastructural appearance of EPEC bacillus forming attachment pedestal (electron microscopy, colon).

detect EPEC-specific virulence genes and may be multiplexed to rapidly rule out other common enteric pathogens. Such kits are now commercially available for the diagnosis of human enteric infections but require specialized diagnostic equipment and have not been validated on nonhuman primate isolates.

Differential Diagnosis

EPEC is one of the most common enteric infections of common marmosets but must be distinguished from other enteric pathogens causing acute-onset diarrhea, including most commonly *Campylobacter* spp., *Klebsiella* spp., and *Yersinia* spp. While *Shigella* spp. and *Salmonella* spp. infections may produce similar clinical signs, these are now rare in marmosets raised in a laboratory setting. Adenovirus infection and cryptosporidiosis may be considered as nonbacterial etiologies. Due to the severity of clinical signs, empiric treatment may be required prior to definitive diagnosis, and fluoroquinolones provide adequate coverage against most of the common Gram-negative organisms causing acute enterocolitis in common marmosets.

Prevention and Control

EPEC infection in clinically affected animals responds rapidly to enrofloxacin and supportive therapy such as intravenous fluids and supplemental heat [26]. Diarrhea and clinical signs may resolve within a matter of hours. Previous infection does not prevent subsequent infections or disease months or years later. Whether this is due to the introduction of new strains or absence of an appropriate immune response is not known. The persistence of EPEC in colonies with only episodically recognized disease suggests that an asymptomatic carrier state may be common. Elimination of EPEC from colonies may be difficult due to the ubiquitous nature of *E. coli* commensals and the absence of a rapid diagnostic test.

Research Complications

Sequencing of the intimin gene from both New World primate (NWP) and Old World primate (OWP) isolates indicates a high degree of identity with human isolates, suggesting that primate EPECs may represent a potential zoonotic disease.

Francisella tularensis (Tularemia)

Etiology

Francisella tularensis, a Gram-negative, nonspore-forming, nonmotile, aerobic rod-shaped coccobacillus, is the causative agent of tularemia. Two distinct biovars are recognized: *Francisella tularensis tularensis* (type A), a highly virulent form and found only in North America, and a less virulent *Francisella tularensis holarctica* (type B), which is thought to be endemic throughout the Northern Hemisphere.

F. tularensis can infect a variety of mammals, including wild lagomorphs and rodents, which may serve as important reservoirs. Due to its virulence, ability to spread by the aerosolized route, and highly infectious nature, *F. tularensis* has been designated a select agent by the US government. Colloquially named "rabbit fever" in man, tularemia may take different forms, including ulceroglandular, oculoglandular, typhoidal/oropharyngeal, and pneumonic manifestations. A common marmoset model of pneumonic tularemia has been developed to examine the pathogenesis and antimicrobial therapy [45].

Clinical Signs

Apathy and hyperthermia or hypothermia may initially be observed. However, the disease is rapidly progressive and animals are often found moribund or dead. A survey of 55 contact animals from a German epizootic did not reveal seroreactivity, suggesting that tularemia had been fatal in all animals that had contracted the disease [46].

Epizootiology

Epizootic and sporadic cases of tularemia in common marmosets have been described in Europe with isolation of type B biovar (*F. tularensis holarctica*). In both instances, animals were housed outdoors in either a semifree-living [46] or caged habitat [47]. Increased numbers of ixodid ticks and dead rodents had been observed in the facility prior to the onset of the German epizootic. Rodents were trapped, and PCR identified *F. tularensis* nucleic acid in *Sorex araneus, Apodemus flavicollis*, and *Microtus agrestis* rodents [46]. It is postulated that rodent contamination of food and water sources may have served as the initial source of infection. Once within a colony, the organism may be rapidly spread from animal to animal.

Pathology

At necropsy, multifocal small (<1 mm) white foci may be visible in the liver, spleen, lymph nodes, and kidneys. Histologically, these correspond to foci of acute coagulative necrosis surrounded and infiltrated by variable numbers of neutrophils, macrophages, and lymphocytes. Large intralesional bacterial colonies as observed with yersiniosis are absent. In animals that survive longer, the inflammatory infiltrate may be more granulomatous in nature. Lymph nodes appear hyperplastic and infiltrated by large numbers of neutrophils. Foci of overt lymph node necrosis may be evident. In many cases, a hemorrhagic to necrotizing enteritis may be evident and accompanied by necrotizing lymphadenitis of the submandibular lymph nodes, suggesting an oral route of entry similar to the typhoidal/oropharyngeal form observed in man.

Diagnosis

Several assays proved valuable in the diagnosis of *F. tularensis* in the German epizootic, including bacterial isolation and characterization, antigen detection in the spleen or liver by competitive ELISA, PCR of the *tul4*, *iglC*, and *ISFtu2* genes, and serology [46]. For bacterial isolation 50 μL aliquots of homogenated spleen and liver were plated on cysteine heart agar supplemented with 9% sheep's blood. When plates with antibiotics were used to suppress growth of contaminating bacteria, complete agreement between PCR, competitive ELISA, and culture was observed [46].

Differential Diagnosis

Differential diagnosis should include other rapidly progressive epizootic diseases such as callitrichid hepatitis, *Yersinia enterocolitica*, and *Klebsiella pneumoniae*. In particular, *Y. enterocolitica* can produce similar clinical and gross pathological findings. Histopathologic analysis of tissue samples is helpful in distinguishing these entities and allows a presumptive diagnosis of tularemia.

Prevention and Control

Treatment of affected animals has not been attempted due to the rapidity and severity of the disease course. In human patients, administration of streptomycin, gentamicin, doxycycline, or ciprofloxacin for 10–21 days has been used successfully to control infection and might be considered in nonhuman primates. Levofloxcin 14.5 mg/kg twice daily orally administered 24 h following *F. tularensis* aerosolized challenge protected all common marmosets [48]. Wild animal reservoirs and insect vectors play an important role in epizootics. In the German epizootic, rodent control measures, including the use of clap traps and elimination of rodent habitat, were used to reduce rodent populations. These efforts were coupled with increased hygiene and credited with stemming the epizootic event [46].

Research Complications

Tularemia is a known zoonotic disease that may be transmitted to human handlers by a variety of routes including exposure to open skin wounds and inhalation. One animal handler from the German epizootic was found to be positive by immunoblot and ELISA but had not demonstrated clinical signs [46]. If tularemia is suspected, additional personal protective equipment may be implemented to include N95 and/or powered air-purifying respirators.

Klebsiella pneumoniae

Etiology

K. pneumoniae is a Gram-negative, encapsulated, nonmotile, facultative anaerobic, rod-shaped bacterium, which appears as a mucoid lactose fermenter on MacConkey agar. Lipopolysaccharide and capsular polysaccharide are two of the most important virulence factors of *K. pneumoniae*, promoting invasion and sepsis [49]. A hypermucoviscosity phenotype is increasingly cited as a virulence marker in septic humans and NHPs [50,51]. *Klebsiella* spp. are viewed as opportunistic pathogens and in a hospital setting often infect individuals with concurrent medical issues. Plasmid-mediated antimicrobial resistance is widespread in *K. pneumoniae* isolates making treatment difficult.

Clinical Signs

Clinical signs may be nonspecific such as diarrhea, depression, and anorexia. Animals often lack clinical signs and may be found unexpectedly dead. Cases may occur sporadically over a long period or may cluster as an epizootic. While *Klebsiella* spp. may affect debilitated animals as an opportunistic infection, it may also cause disease in young healthy animals without preexisting conditions.

Epizootiology

K. pneumoniae can be found and isolated from water sources, and these strains possess similar virulence factors such as serum resistance and capsular polysaccharides, pili, and siderophores as found in isolates from clinically affected human patients [49]. The environment may become contaminated with *K. pneumoniae* and serve as a source of infection to colony animals. Fomites and human handlers may spread the infection from the source in the environment to colony animals. *K. pneumoniae* may be nearly ubiquitous in these environments, and host factors that allow marmosets to become colonized and develop disease remain unknown.

Pathology

In animals that die unexpectedly, gross findings at necropsy may be minimal but may be associated with green-yellowish fluid in the gastrointestinal tract. In such animals, large numbers of small Gram-negative encapsulated bacilli may be found disseminated throughout small blood vessels, consistent with fulminant septicemia. The inflammatory reaction to the bacteria may be minimal. In cases in which animals have survived longer, a fibrinopurulent serositis and/or interstitial pneumonia may be evident. The presence of fetid liquid gastrointestinal contents suggests that this may represent the site of bacterial colonization and entry, when interstitial pneumonitis is present. *K. pneumoniae* can also cause a bronchopneumonia similar to that observed with *B. bronchiseptica*. *Klebsiella* spp. can be associated with acute-onset meningitis in several species of NHPs, including rhesus, owl monkeys, and lemurs [50].

Diagnosis

The microscopic findings of myriads of small encapsulated bacilli in blood vessels with scant inflammatory reaction are characteristic. Bacterial isolation and identification is diagnostic. As antibiotic resistance is common, antimicrobial sensitivity profiles should be requested on all isolates. Isolates obtained from animals at necropsy subjected to antibiotic sensitivity assays aid in the empiric choice of antimicrobial therapy in future cases.

Differential Diagnosis

K. pneumoniae should be distinguished from other Gram-negative bacteria such as *E. coli* and *Pseudomonas* spp., which may produce septicemia. If bronchopneumonia is present, *B. bronchiseptica* should also be considered.

Prevention and Control

Treatment of affected animals may be difficult due to the rapidity of disease progression and extensive antimicrobial resistance in *K. pneumoniae*. In the absence of historical antimicrobial resistance data of resident animals, systemic enrofloxacin may be initiated in suspected cases and provides adequate coverage for other Gram-negative bacteria. Careful assessment of contact animals is warranted, and prophylactic treatment may be considered. While *K. pneumoniae* may be difficult to eliminate from the environment, improved levels of sanitation and hygiene may be useful in reducing exposure. A prophylactic autogenous *Klebsiella* vaccine has reportedly been used in a common marmoset colony [3].

Pseudomonas simiae

Pseudomonas simiae is a novel Gram-negative bacillus isolated from a neonatal white-headed marmoset (*Callithrix geoffroyi*) and its dam [52]. The offspring had a bronchointerstitial pneumonia with Gram-negative rods visualized within alveolar spaces, and the dam had glomerulonephritis with uremic pneumonitis. A bacterial isolate was found to belong to the *Pseudomonas fluorescens* intragenic cluster and to be phylogenetically and biochemically distinct from other species.

Salmonella spp.

Salmonella are Gram-negative motile bacilli with peritrichous flagella and fall into two species: *Salmonella enterica* and *Salmonella bongori*. *S. bongori* is restricted to cold-blooded animals, whereas *S. enterica* may infect a variety of warm-blooded animals. The latter is divided into six subspecies and more than 2500 serotypes. Salmonellosis has been described in a number of neotropical primate species, including the common marmoset [53]. Serovars isolated from common marmosets have included *typhimurium, anatum, seftenberg, newington, oranienburg,* and *habana* [3]. Few clinical details are provided and cases were documented in marmosets recently imported from South America in the 1960s and 1970s. Enteritis induced by EPEC and *Campylobacter* spp. should be considered far more common [53]. Clinical disease is likely exacerbated from comorbidities, stress of capture, and shipment. Histologically, enterocolitis with congestion, hemorrhage, and edema was observed and could be accompanied by focal necrosis in the gastrointestinal tract, spleen, and liver. In human patients with noninvasive salmonellosis, symptoms generally resolve in 5–7 days with supportive therapy [54]. Systemic antibiotics are reserved for individuals with invasive disease (enteric fever). Diagnosis of invasive disease can be problematic and relies on clinical findings as well as blood cultures and isolation of the organism. As the natural history of salmonellosis is poorly described in neotropical primates, criteria for initiation of antimicrobial therapy have not been established. In severe cases, enrofloxacin or amoxicillin may be considered empirically until culture and sensitivity results are available and should be used in conjunction with supportive therapy and isolation of affected animals.

Shigella—Bacillary Dysentery

Etiology

Shigella spp. are Gram-negative, facultative anaerobic, nonspore-forming, nonmotile, rod-shaped bacteria closely related to *E. coli* from which they have evolved. *Shigella* spp. are the etiologic agent of bacillary dysentery, a severe and often hemorrhagic infection of the large intestine that may affect humans as well as a number of Old and New World primates. *Shigella* spp. are classified into four serogroups *Shigella dysenteriae* (type A), *Shigella flexneri* (type B), *Shigella boydii* (type C), and *Shigella sonnei* (type D) and must be differentiated from EIEC with which they share a number of characteristics.

Clinical Signs

Shigellosis has rarely been reported in common marmosets [3,55] but can produce signs of bacillary dysentery, including bloody diarrhea, dehydration, anorexia, and depression, which can rapidly progress to death. Animals may initially appear hunched with dried blood adhered to the perineum.

Epizootiology

Shigella spp. are regarded as highly infectious by the fecal–oral route, with fewer than 100 colony-forming units required to produce infection in humans. In nonhuman primates, epizootics may occur when asymptomatic carrier animals are introduced into a naive colony or when animals are inadvertently exposed to waste from infected humans. Once animals are infected, *Shigella* spp. pose a significant zoonotic risk to human handlers.

Pathology

At necropsy, gross findings are typical of typhlocolitis with multifocal punctate ulcers and hemorrhages observed on the cecal and colonic mucosal surfaces and mesenteric lymph nodes. The large intestine may be dilated and contain fresh blood. Occasionally, microhemorrhages may be visible through the serosal surface. Microscopically, mononuclear and neutrophilic infiltrates can be observed in the colon and cecum and accompanied by edema.

Diagnosis

Bacterial culture and isolation can provide definitive diagnosis, but care is required to ensure that samples are collected and handled properly as *Shigella* spp. do not survive well outside of the host. Cary-Blair or Amies and Stuart's transport media is recommended, and cultures should be initiated as soon as possible. Multiplex PCR assays have also been developed and clinically validated to detect common human enteric pathogens in stool, including *Shigella* spp. isogroups. These offer the advantage of detecting other pathogens; however, these assays do not allow assessment of antimicrobial sensitivities achievable with *Shigella* isolates.

Differential Diagnosis

Shigellosis has rarely been recognized in neotropical primates, suggesting there may be some inherent resistance compared with their Old World counterparts who are highly susceptible. EPEC infection is far more common in *Callithrix jacchus* and produces a similar clinical syndrome of hemorrhagic diarrhea. Microscopically EPEC infection is characterized by an attaching and effacing lesion, whereas shigellosis demonstrates erosion and extensive neutrophilic infiltrates. Bacterial culture and isolation and molecular assays will also differentiate EPEC from *Shigella* spp. Initial treatment and control measures are similar for both pathogenic species.

Prevention and Control

In clinically affected animals, treatment should be directed at correcting dehydration and electrolyte imbalances along with initiation of antimicrobial therapy. Both enrofloxacin and trimethoprim–sulfamethoxazole are empiric first-line choices. Antimicrobial resistance has been increasing in recent years, and microbial antibiotic sensitivity should be assessed in any isolates obtained. If an epizootic is recognized, affected and exposed animals should be quarantined from the remainder of the colony. The use of quarantine-dedicated staff and equipment as well as disinfectant foot baths should be considered. Prophylactic treatment of nonaffected contact animals has been of benefit in controlling epizootics in Old World primates.

Strict separation of marmosets from other primates known to harbor *Shigella* and from infectious human waste should prevent exposure and infection. Given the rarity of shigellosis in marmosets, it is unlikely that newly introduced animals pose a significant risk for this infection. However, routine quarantine and health assessment of all new arrivals is warranted and particular attention directed to minimize personnel exposure to *Shigella*, given the documented zoonotic risk of *Shigella* infection in NHPs [56].

Yersinia enterocolitica

Etiology

Yersinia are Gram-negative facultative anaerobes that have peritrichous flagella. Two species have been associated with disease in nonhuman primates (*Yersinia pseudotuberculosis* and *Y. enterocolitica*). In common marmosets, disease (yersiniosis) has been associated with *Y. enterocolitica* [57,58]; however, in other callitrichids *Y. pseudotuberculosis* has produced similar clinical and pathological findings [57].

Clinical Signs

Animals may be found dead without displaying clinical signs. In others cases, animals may develop diarrhea with or without hematochezia and constitutional signs such as anorexia and lethargy.

Epizootiology

Yersinia species are a diverse group of Gram-negative organisms and can be found widely in soils and decaying vegetation. Outbreaks in nonhuman primates often follow a seasonal pattern and increase in the wetter months. Rodents may become infected with *Yersinia* spp. and asymptomatically shed the organism in feces. These animals may represent an important reservoir and initiate epizootics when contaminating food sources or when hunted and eaten by primates.

Pathology

Yersinia infections in nonhuman primates generally present with the gross pathological triad of (1) hepatic or splenic necrosis or abscess formation; (2) mesenteric lymphadenopathy; and (3) ulcerative enterocolitis. It is believed that the gastrointestinal tract represents the point of entry, with subsequent septicemia and seeding of the liver, lymph nodes, and spleen. Grossly micropyogranulomas are recognized as multiple small (1–3 mm) white foci within the liver and spleen and histologically are characterized by necrosuppurative hepatitis, splenitis, and lymphadenitis. An ulcerative enterocolitis, usually visible grossly and microscopically, is characterized by submucosal edema, marked hyperemia, and the presence of discrete masses of bacteria within the mucosa and submucosa. Occasionally, *Yersinia* may disseminate and be associated with pneumonitis, meningitis, and pleuritis. The presence of large colonies of Gram-negative organisms within regions of necrosis is highly suggestive of the diagnosis (Fig. 16.7).

Diagnosis

A presumptive diagnosis may be made based on the characteristic gross and microscopic findings. Definitive diagnosis requires bacterial culture and isolation and allows serotyping of isolates to assist in epizoological investigations.

Differential Diagnosis

Other bacterial gastrointestinal infections such as EPEC and *K. pneumoniae* may produce similar clinical findings. In particular, *F. tularensis*, the cause of tularemia, may produce similar clinical and pathological findings. However, the gross and microscopic findings of yersiniosis should enable a rapid presumptive diagnosis.

Prevention and Control

In clinically affected animals, supportive therapy such as supplemental heat, intravenous fluids, and vitamins may be helpful. Intramuscular treatment of affected and contact animals with enrofloxacin 5 mg/kg once a day for 7–10 days appeared to have some benefit in mildly affected animals [57]. However, in severe cases, animals failed to respond to antimicrobial and supportive therapy.

Control measures should be directed at preventing transmission between exposed and nonexposed animals and reducing potential wild animal reservoirs [58]. Affected and contact animals should be segregated from the remaining colony if possible. Dedicated equipment, personal protective equipment, and disinfectant foot baths should be used.

A *Y. pseudotuberculosis* vaccine (Pseudovac) has been developed and used in common marmosets and tamarins to control seasonal epizootics of yersiniosis observed at the Biomedical Research Center in Rijswijk, the Netherlands [57]. This vaccine contains different serotypes of formalin-killed *Y. pseudotuberculosis* and was developed at the Department of Veterinary Pathology, section Zoo and Exotic Animals, Utrecht University. A dose of 0.25 mL was administered subcutaneously to all colony animals over 7 weeks of age and boosted at 3 weeks. The vaccine has been repeated annually thereafter in the fall months, and no adverse effects have been observed. While a formal efficacy study was

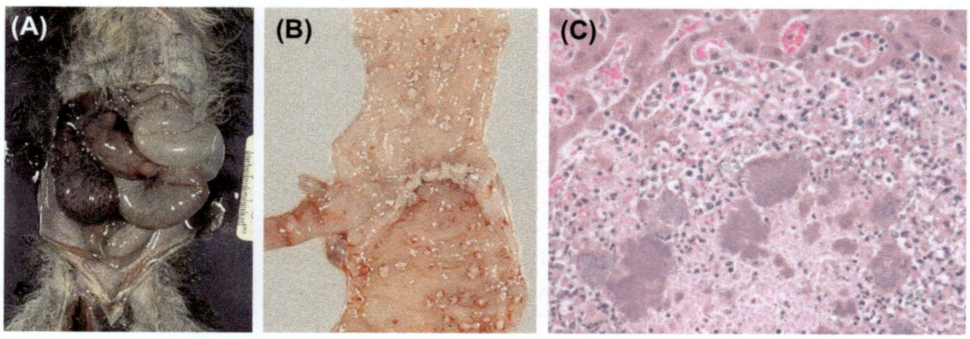

FIGURE 16.7 Morphologic appearance of *Yersinia enterocolitica* infection. (A) Dilatation of the large intestine and cecum. (B) Multifocal necrotizing typhlocolitis. (C) Necrotizing hepatitis (liver, H&E stain, 200×).

not conducted, the investigators conclude that the vaccine was essential in their *Yersinia* spp. control program as no further periods of high mortality were observed [57].

CURVED, SPIRAL, AND FUSIFORM-SHAPED GRAM-NEGATIVE RODS

Campylobacter spp.

Campylobacter is a genus of Gram-negative, curve-shaped bacteria and a frequent cause of foodborne disease in humans. *Campylobacter jejuni* and *Campylobacter coli* are the two most common species and can inhabit the gastrointestinal tract of a variety of warm-blooded animals. Campylobacters can be isolated from rectal swabs of the majority of normal macaques where they are most often not associated with disease. In callitrichids, *Campylobacter* spp. are not considered part of the normal flora and when present are usually associated with disease [26,59]. Histologically infection is associated with neutrophilic colitis with the formation of crypt abscesses. In juvenile cotton-top tamarins, *C. coli* was more frequently associated with diarrhea in juvenile nursery animals, whereas *C. jejuni* was more commonly observed in clinically affected adult animals [59]. Both are susceptible to macrolide and fluoroquinolone antimicrobials, which can effectively eliminate the organisms from infected individuals. *Campylobacter* spp. isolation from rectal swabs of callitrichids should be treated as potential pathogens and steps should be taken to eradicate the organism from the colony. Screening of all colony animals by fecal culture followed by treatment of affected individuals may eliminate infection.

Helicobacter spp.

Etiology

A novel *Helicobacter* species, "*Helicobacter callitrichis*," was isolated from the feces of clinically normal, common marmosets [60]. In addition, an unclassified *Helicobacter* sp. was identified in the stomachs of three species of marmosets, based on immunohistochemistry of paraffin-imbedded tissues using a polyclonal *Helicobacter pylori* antibody [61]. A novel *Helicobacter* species, *Helicobacter jaachi* sp. nov. [62], shared many biochemical properties with those of "*H. callitrichis*," with the exception that *H. jaachi* was urease-positive and alkaline phosphate hydrolysis (PO_4)-negative, whereas "*H. callitrichis*" is urease-negative and PO_4-positive. By 16S rRNA analysis, these two *Helicobacter* spp. are 4% different, which warrants separate species designations [61].

Clinical Signs

In addition to the inflammatory lesions found at necropsy, the colony from which *H. jaachi* was isolated was noted to have clinical gastrointestinal disease as evidenced by sporadic cases of diarrhea and chronic weight loss [63]. The finding of IBD, sometimes in conjunction with colon adenocarcinoma or hepatobiliary inflammatory changes, is reminiscent of findings reported in other nonhuman primates (NHPs) infected with *Helicobacter* spp [64–68]. A novel *Helicobacter* sp., *Helicobacter saguini*, was identified in cotton-top tamarins diagnosed with progressive colitis mimicking features of ulcerative colitis in humans [63,69]. The diarrheic marmosets reflected only the clinical state of the animal at the time of sampling. The importance of *H. jaachi* in eliciting clinical disease needs further study.

Epizootiology

Enterohepatic helicobacters have evolved to colonize and inhabit the mucus of the intestinal crypts of a variety of animals, including humans [70]. *Helicobacter* species that colonize the lower bowel and biliary tract of nonhuman primates have been routinely isolated from cases with chronic inflammation of the lower bowel [60,64–69]. Purpose-bred common marmosets from domestic sources have been reported to have a high incidence of spontaneous IBD and sporadic cholecystitis and cholangiohepatitis. Inflammatory infiltrates increase in incidence and severity with age. Thirty-nine marmosets screened for *Helicobacter* spp. by culture and PCR were frequently detected to be colonized; 48.7% were positive by *Helicobacter* genus-specific PCR, whereas 28.2% of the marmosets were positive for a novel species, *H. jaachi* sp. nov., by culture [62]. *Helicobacter* spp. were also isolated from two liver samples and one gallbladder sample. *Helicobacter* spp. were detected in 36% of the gallbladder samples by PCR, which may be related to the finding that about 10% of the marmosets had cholecystitis or biliary inflammation. Animals aged 6 years and older had higher *Helicobacter*-positive prevalence rates than animals younger than 5 years (39% by culture and 57% by PCR vs. 0% by culture and 27% by PCR). *Helicobacter* species were isolated from animals ingesting a high-fat diet, from primates treated with chemical compounds, and from animals without diarrhea [62].

Pathology

Numerous lesions were noted in this population of marmosets screened for *Helicobacter* spp., and these findings recapitulated the spectrum of survey pathology previously reported [62]. These changes included glomerular and interstitial nephritis, amyloidosis, increased extramedullary hematopoiesis, and metabolic bone disease. In addition, in the recent study, lesions of the hepatobiliary and gastrointestinal systems were recognized and characterized by prominent inflammatory changes in a relatively high percentage of animals [62]. In the liver, the inflammatory lesions consisted of lymphoid infiltrates in hepatic portal regions as well as mild-to-moderate infiltrates in the mucosa and submucosal tissues of large bile ducts and gallbladder. Occasional animals had mild-to-moderate cholangiohepatitis, and several marmosets had cholecystitis with transmural lymphocytic infiltration of the gallbladder. Lymphoid infiltrates increased in incidence and severity with age [62]. Enteric pathology consisted of a continuous spectrum of IBD changes ranging from relatively minimal focal to segmental lymphoid infiltrates in the lamina propria to severe lymphoid cell accumulations with effacement of the normal bowel architecture in both large and small intestines. Severe accumulations of chronic inflammatory cells found in the lamina propria were associated with changes in villous and crypt architecture, with intraepithelial inflammatory cell accumulations. It is not known what role, if any, *H. jaachi* plays in the development of these lesions [62] (Fig. 16.8–16.10).

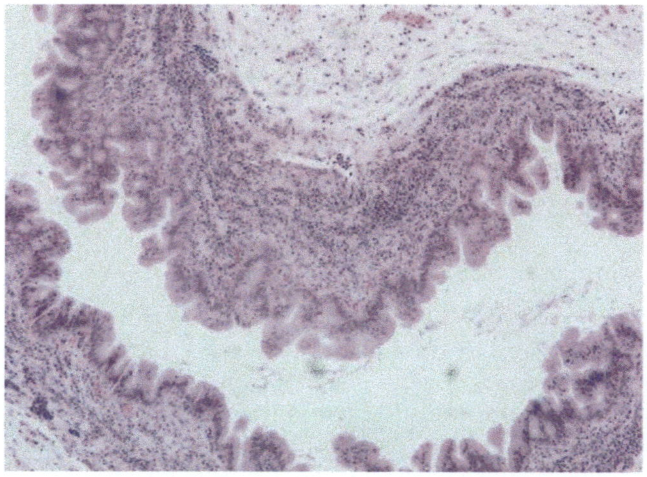

FIGURE 16.9 Chronic cholecystitis characterized by diffuse submucosal infiltrate of lymphoid cells in the wall of the gallbladder (gallbladder, H&E stain, × 200). *Reprinted with permission from Shen Z, Feng Y, Sheh A, Everitt J, Bertram F, Paster BJ, Fox JG. Isolation and characterization of a novel Helicobacter species,* Helicobacter jaachi *sp. nov., from common marmosets* (Callithrix jaachus)*. J Med Microbiol 2015;64: 1063–73.*

Diagnosis

H. jaachi sp. nov. are fastidious, microaerophilic, motile bacteria, which are catalase-, urease-, and oxidase-positive and have fusiform morphology, with periplasmic fibers and multiple bipolar, sheathed flagella. This novel *Helicobacter* species is closely related by 16S rRNA analysis to "*H. saguini*" (97%), a species isolated from cotton-top tamarins. Its 16S rRNA is 96% similar to "*H. callitrichis,*" isolated previously from the feces of common marmosets. The bacterium is Gram-negative and nonsporulating. The organism grows slowly and

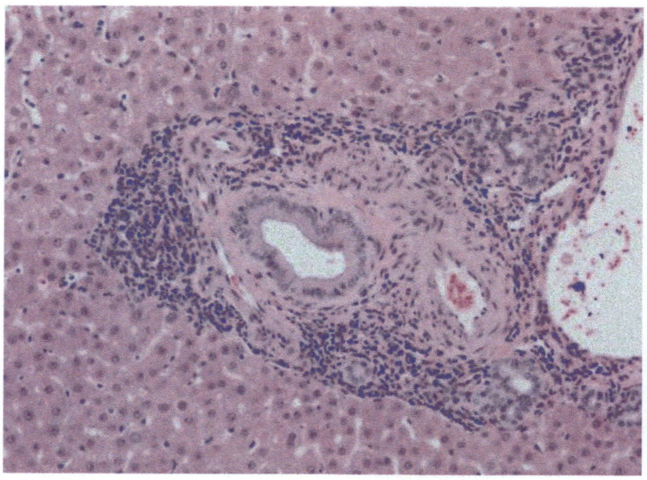

FIGURE 16.8 Lymphohistiocytic inflammatory infiltrate in periportal region of liver from a marmoset with chronic cholangiohepatitis. Inflammatory cells circumscribe the portal triad including the bile duct (liver, H&E stain, × 400). *Reprinted with permission from Shen Z, Feng Y, Sheh A, Everitt J, Bertram F, Paster BJ, Fox JG. Isolation and characterization of a novel Helicobacter species,* Helicobacter jaachi *sp. nov., from common marmosets* (Callithrix jaachus)*. J Med Microbiol 2015;64: 1063–73.*

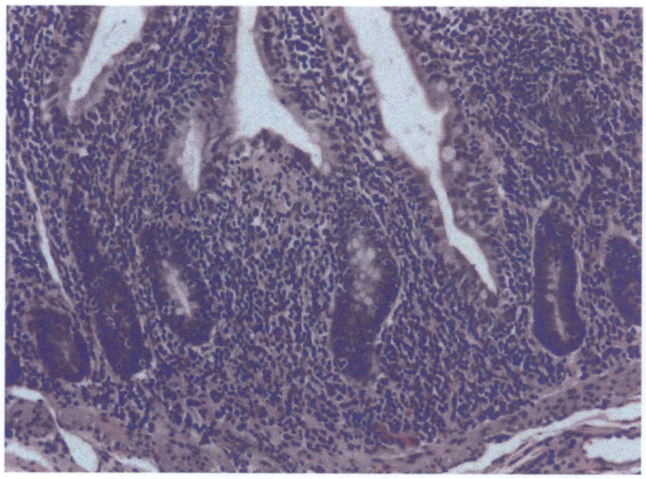

FIGURE 16.10 Section of jejunum from a marmoset with inflammatory bowel disease characterized by loss of crypts and effacement of normal villous structure by severe diffuse lymphohistiocytic infiltrate in the lamina propria (jejunum, H&E stain, × 200). *Reprinted with permission from Shen Z, Feng Y, Sheh A, Everitt J, Bertram F, Paster BJ, Fox JG. Isolation and characterization of a novel Helicobacter species,* Helicobacter jaachi *sp. nov., from common marmosets* (Callithrix jaachus)*. J Med Microbiol 2015;64:1063–73.*

appears on solid agar as a spreading film on the surface. The bacterium grows at 37–42°C, but not at 25°C, under microaerobic conditions, but not aerobically or anaerobically.

Differential Diagnosis

Helicobacter spp. must be distinguished from *Campylobacter* spp., which are also microaerophilic and also colonize the gastrointestinal tract of marmosets.

Prevention and Control

No data exist regarding prevention and control of *Helicobacter* spp. in marmosets. However, triple antibiotic treatment (clarithromycin, amoxicillin, and metronidazole) has successfully eradicated *Helicobacter hepaticus* from endemically infected mouse colonies [71].

Research Complications

Several enteropathogenic *Helicobacter* species (EHS) have zoonotic potential. For example, other EHS isolated from animals and birds, such as *Helicobacter bilis* and *Helicobacter pullorum*, have been identified in patients with cholecystitis and biliary neoplasia [72–74]. Recent studies in humans highlight the ability of enteric helicobacters to translocate across the intestinal epithelia in humans with bacteremia, osteomyelitis, and myositis [75–77]; translocation to the liver is also noted with *H. hepaticus* during the development of hepatitis and hepatocellular carcinoma in A/JCr and B6C3F1 mice [78–80]. The fact that these organisms are routinely cultured from the intestine and biliary tract infers that they may cause inflammation in these tissues and complicate interpretations of studies using marmosets.

Leptospira spp.—Leptospirosis

Etiology

Leptospirosis is a bacterial infection caused by members of the *Leptospira* genus. Over 200 serotypes of *Leptospira* are recognized and may be saprophytic or pathogenic. Morphologically they are thin spiral bacteria 6–20 μm in length and survive in moist or wet environments. Common marmosets are susceptible to experimental infection with pathogenic *Leptospira* serovars, and leptospirosis has been described in Wied's marmosets (*Callithrix kuhlii*). Clinical disease has also been reported in other neotropical primates, including squirrel monkeys (*Saimiri sciureus*) and white-lipped tamarins (*Saguinus labiatus*) [81,82]. Serological surveys of neotropical primates suggest that nonclinical infection with the *icterohemorrhagiae* serogroup may be common in these species, including common marmosets [83].

Clinical Signs

Leptospirosis of two Wied's marmosets has been described in a zoological collection [84]. Both adult animals presented with weight loss and jaundice and succumbed to disease despite antimicrobial and supportive therapy. Complete blood counts on presentation of the first case revealed neutrophilic leukocytosis with a left shift and a regenerative anemia. Serum chemistries demonstrated markedly elevated total and direct bilirubin consistent with intravascular hemolysis. Clinical coagulopathy, a common finding in Weil's disease (icteric leptospirosis) of humans and reported in previous nonhuman primate cases, was not observed. The animal died 17 days into the course of treatment, and necropsy revealed diffuse severe proximal tubular nephrosis in which spirochetes could be demonstrated by Steiner's silver stain. The second case, presented on the same day and while initially recovering with antimicrobial treatment, presented 3 months later with progressive weight loss and jaundice. Histological examination of tissues obtained at necropsy revealed lymphoplasmacytic infiltrates in a number of organs, including the gastrointestinal tract and kidney. Seroconversion to *Leptospira borgpetersenii* serovar *Ballum* was demonstrated in the second case.

Common marmosets are also susceptible to experimental infection with *Leptospira interrogans* serovar *Copenhageni* and develop a severe pulmonary form of leptospirosis (SPFL) [85]. SPFL is recognized in human patients as a rapidly progressive and often fatal form of infection characterized by extensive pulmonary hemorrhage. Marmosets were inoculated by the intraperitoneal route with a clinical isolate obtained from a human patient with SPFL. Animals developed weight loss, dehydration, and jaundice within 6–12 days postinoculation. Serum chemistry results indicated hyperbilirubinemia, increased aminotransferase, creatinine, and blood urea nitrogen. Pathology revealed hemorrhage within the pulmonary parenchyma, and pathological changes were noted in the liver, kidneys, and skeletal muscle. The study was conducted as a time course experiment, so it was not possible to determine if severely affected animals would have recovered.

Epizootiology

Rodents, dogs, swine, and cattle may serve as reservoirs of *Leptospira* spp. and may excrete spirochetes in the urine and contaminate water, food, or bedding. The organisms may survive in stagnant water for several weeks but are susceptible to desiccation. Transmission may then occur through direct oronasal contact, ingestion of infected tissue, or bite wounds. Once a new host is infected, the organism may begin to replicate in the kidneys where it can be shed in large numbers in the urine. Specific serovars may be carried by particular host species and their identification can help in understanding the source of infection in epizootics. The cases in Wied's marmosets were associated with *L. borgpetersenii* serovar *Ballum*, a serovar commonly found in mice [84,86].

Pathology

Necrosis of the proximal convoluted tubules accompanied by variable lymphoplasmacytic infiltrates have been observed with leptospirosis in marmosets. Silver stains, such as Steiner's stain, can be used to visualize the elongated tightly coiled spirochetes within the affected tubules. Additional lesions were observed in the lung, liver, and skeletal muscle of common marmosets with experimentally induced SPFL. In these animals, the alveolar spaces in the lung were filled with erythrocytes and proteinaceous fluid in a multifocal to coalescing pattern. The alveolar septa were thickened by infiltrating macrophages, neutrophils, and lymphocytes. Multifocal myofiber necrosis with interstitial edema and hemorrhage was observed in the skeletal muscle. In the liver, multifocal hepatic necrosis with occasional Councilman bodies was evident. These changes were accompanied by mild interstitial edema, Kupffer cell activation, and foci of parenchymal hemorrhage. Hepatocellular dissociation which may be seen in other species with icteric leptospirosis was not a predominant finding.

Diagnosis

The combination of jaundice, hemolytic anemia, and renal disease is highly suggestive of leptospirosis. Antemortem diagnosis may be achieved by demonstrating seroconversion or detecting the organism in biological samples by bacterial culture/isolation or PCR [87]. Earlier, during the course of infection (<10 days), spirochetes may be visualized in blood by dark-field microscopy and cultured from blood on semisolid media such as Fletcher's Ellinghausen–McCullough–Johnson–Harris media. In later phases of infection, bacteremia may not be evident, but dark-field examination of urine and urine culture may detect the organism. Isolation and serotyping are useful in defining the epizootiology of infection and implementing control programs.

Differential Diagnosis

Jaundice and weight loss appear to be common clinical findings in marmosets naturally infected with *Leptospira* spp. and following experimental inoculation. Callitrichid hepatitis caused by lymphochoriomeningitis virus can also result in these clinical signs and should be considered. Small-intestinal adenocarcinoma can be associated with weight loss and may obstruct the bile duct resulting in jaundice [88]. In more chronic cases, leptospirosis may be associated with lymphoplasmacytic infiltrates and nephrosis. Other causes of renal failure and nephritis such as glomerulonephritis and pyelonephritis may produce similar clinical findings. Silver stains can be used to detect spirochetes in formalin-fixed, paraffin-embedded sections.

Prevention and Control

Identification of serovars responsible for initiating epizootics can provide important information regarding the source of infection. Wild rodents are an important reservoir of leptospirosis and have been linked to infection in a variety of species including marmosets. Rodent control programs should be intensified if this source is suspected. Vaccines can be highly effective in preventing leptospirosis but have not been utilized in nonhuman primates. Vaccination may be considered if repeated epizootics are observed or if the disease becomes enzootic. As immunity is primarily serovar specific, data concerning outbreak-associated isolates would be required to design an effective vaccine program.

Leptospirosis is sensitive to intravenous penicillin which may be attempted in severe acute cases in conjunction with supportive therapy such as intravenous fluids. Penicillins are less effective at eliminating chronic renal infection, and generally doxycycline is administered to animals in the later stage of disease to eliminate this phase of infection.

Research Complications

Leptospirosis is a zoonotic disease and may pose a risk to human contacts. Adequate personal protective equipment to prevent contact with skin and mucous membranes should be provided [86].

ATYPICAL BACTERIA

Ehrlichia canis

Ehrlichia, small Gram-negative coccoid organisms that are obligate intracellular pathogens, are the etiologic agent of ehrlichiosis. *Ehrlichia canis* has recently been detected in free-living primates of the genus callithrix by PCR [89]. Phylogenetic analysis revealed the organism to be closely related to those previously observed in domestic dogs and wild cats. *Ehrlichia* infection in the marmoset has not been associated with clinical disease. In other species, *Ehrlichia* spp. commonly causes fever, lethargy, lymphadenopathy, and splenomegaly. Thrombocytopenia and leukopenias may be present, and morulae may be visible in white bloods during acute infection.

Mycoplasmas and Ureaplasmas

Mycoplasmas are small bacteria that lack a cell wall and are the smallest free-living organism. The lack of a cell wall imparts resistance to antimicrobials such as the beta-lactams, which target the cell wall for activity. *Mycoplasma salivarium* can be isolated from marmosets as part of the normal bacterial flora of the throat and nasal pharynx; its presence, however, has not been associated with disease. Callitrichids have been used as animal models of *Mycoplasma genitalium* (*C. jacchus*) and *Mycoplasma pneumoniae* (*S. labiatus*) infection [90,91]. Marmosets inoculated with a *M. genitalium* isolate, originally recovered from a human patient with nongonococcal urethritis, developed prolonged infections with evidence of salpingitis and antibody-specific responses [90].

Ureaplasmas are commonly isolated from the common marmoset upper respiratory tract and rarely from the genital tract but have not been associated with disease [1,92]. Treatment with minocycline in the drinking water eliminated ureaplasmas, and animals were resistant to reinfection unless in close contact with infected conspecifics. It is believed that marmosets initially acquire infection within 24 h of birth.

References

[1] Hill AC, Turton JA, Bleby J. Bacterial and mycoplasma flora of a laboratory colony of the common marmoset (*Callithrix jacchus*). Vet Rec 1978;103:284–7.

[2] Hunt RD, Anderson MP, Chalifoux LV. Spontaneous infectious diseases of marmosets. Primates Med 1978;10:239–53.

[3] Potkay S. Diseases of the callitrichidae: a review. J Med Primatol 1992;21:189–236.

[4] Brack M, Rensing S, Gatesman TJ. *Erysipelothrix* insidiosa infection in callitrichids kept behind a barrier system. Infect Dis Rev 1999;1:15–9.

[5] Bender JS, Kinyon JM, Kariyawasam S, Halbur PG, Opriessnig T. Comparison of conventional direct and enrichment culture methods for *Erysipelothrix* spp. from experimentally and naturally infected swine. J Vet Diagn Investig 2009;21:863–8.

[6] To H, Koyama T, Nagai S, Tuchiya K, Nunoya T. Development of quantitative real-time polymerase chain reaction for detection of and discrimination between *Erysipelothrix rhusiopathiae* and other *Erysipelothrix* species. J Vet Diagn Investig 2009;21:701–6.

[7] Faria S, Joao I, Jordao L. General overview on nontuberculous mycobacteria, biofilms, and human infection. J Pathog 2015; 2015, 809014.

[8] Stahl DA, Urbance JW. The division between fast- and slow-growing species corresponds to natural relationships among the mycobacteria. J Bacteriol 1990;172:116–24.

[9] Wachtman LM, Miller AD, Xia D, Curran EH, Mansfield KG. Colonization with nontuberculous mycobacteria is associated with positive tuberculin skin test reactions in the common marmoset (*Callithrix jacchus*). Comp Med 2011;61:278–84.

[10] Patel R, Roberts GD, Keating MR, Paya CV. Infections due to nontuberculous mycobacteria in kidney, heart, and liver transplant recipients. Clin Infect Dis 1994;19:263–73.

[11] Mansfield KG, Lackner AA. Simian immunodeficiency virus-inoculated macaques acquire *Mycobacterium avium* from potable water during AIDS. J Infect Dis 1997;175:184–7.

[12] Maslow JN, Brar I, Smith G, Newman GW, Mehta R, Thornton C, Didier P. Latent infection as a source of disseminated disease caused by organisms of the *Mycobacterium avium* complex in simian immunodeficiency virus-infected rhesus macaques. J Infect Dis 2003;187:1748–55.

[13] Brammer DW, O'Rourke CM, Heath LA, Chrisp CE, Peter GK, Hofing GL. *Mycobacterium kansasii* infection in squirrel monkeys (*Saimiri sciureus sciureus*). J Med Primatol 1995;24:231–5.

[14] Wang HY, Kim H, Kim S, Kim DK, Cho SN, Lee H. Performance of a real-time PCR assay for the rapid identification of *Mycobacterium* species. J Microbiol 2015;53:38–46.

[15] Huh HJ, Koh WJ, Song DJ, Ki CS, Lee NY. Evaluation of the Cobas TaqMan MTB test for the detection of *Mycobacterium tuberculosis* complex according to acid-fast-bacillus smear grades in respiratory specimens. J Clin Microbiol 2015;53:696–8.

[16] Mansfield KG, Veazey RS, Hancock A, Carville A, Elliott M, Lin KC, Lackner AA. Induction of disseminated *Mycobacterium avium* in simian AIDS is dependent upon simian immunodeficiency virus strain and defective granuloma formation. Am J Pathol 2001;159:693–702.

[17] Goswami B, Narang P, Mishra PS, Narang R, Narang U, Mendiratta DK. Drug susceptibility of rapid and slow growing non-tuberculous mycobacteria isolated from symptomatics for pulmonary tuberculosis, Central India. Indian J Med Microbiol 2016;34:442–7.

[18] Johnston JC, Chiang L, Elwood K. Mycobacterium kansasii. Microbiol Spectr 2017;5.

[19] Thorel MF. Isolation of *Mycobacterium africanum* from monkeys. Tubercle 1980;61:101–4.

[20] Rosenbaum M, Mendoza P, Ghersi BM, Wilbur AK, Perez-Brumer A, Cavero Yong N, Kasper MR, Montano S, Zunt JR, Jones-Engel L. Detection of *Mycobacterium tuberculosis* complex in New World Monkeys in Peru. EcoHealth 2015;12:288–97.

[21] Alfonso R, Romero RE, Diaz A, Calderon MN, Urdaneta G, Arce J, Patarroyo ME, Patarroyo MA. Isolation and identification of mycobacteria in New World primates maintained in captivity. Vet Microbiol 2004;98:285–95.

[22] Michel AL, Huchzermeyer HF. The zoonotic importance of *Mycobacterium tuberculosis*: transmission from human to monkey. J S Afr Vet Assoc 1998;69:64–5.

[23] Via LE, Weiner DM, Schimel D, Lin PL, Dayao E, Tankersley SL, Cai Y, Coleman MT, Tomko J, Paripati P, Orandle M, Kastenmayer RJ, Tartakovsky M, Rosenthal A, Portevin D, Eum SY, Lahouar S, Gagneux S, Young DB, Flynn JL, Barry 3rd CE. Differential virulence and disease progression following *Mycobacterium tuberculosis* complex infection of the common marmoset (*Callithrix jacchus*). Infect Immun 2013;81: 2909–19.

[24] Cadena AM, Klein EC, White AG, Tomko JA, Chedrick CL, Reed DS, Via LE, Lin PL, Flynn JL. Very low doses of *Mycobacterium tuberculosis* yield diverse host outcomes in common marmosets (*Callithrix jacchus*). Comp Med 2016;66:412–9.

[25] Via LE, England K, Weiner DM, Schimel D, Zimmerman MD, Dayao E, Chen RY, Dodd LE, Richardson M, Robbins KK, Cai Y, Hammoud D, Herscovitch P, Dartois V, Flynn JL, Barry 3rd CE. A sterilizing tuberculosis treatment regimen is associated with faster clearance of bacteria in cavitary lesions in marmosets. Antimicrob Agents Chemother 2015;59:4181–9.

[26] Ludlage E, Mansfield K. Clinical care and diseases of the common marmoset (*Callithrix jacchus*). Comp Med 2003;53:369–82.

[27] Pai M, Riley LW, Colford Jr JM. Interferon-gamma assays in the immunodiagnosis of tuberculosis: a systematic review. Lancet Infect Dis 2004;4:761–76.

[28] Mazurek GH, Villarino ME. Guidelines for using the QuantiFERON-TB test for diagnosing latent *Mycobacterium tuberculosis* infection. Centers for Disease Control and Prevention. MMWR Recomm Rep 2003;52:15–8.
[29] Uzal FA, McClane BA, Cheung JK, Theoret J, Garcia JP, Moore RJ, Rood JI. Animal models to study the pathogenesis of human and animal *Clostridium perfringens* infections. Vet Microbiol 2015;179:23–33.
[30] Yasuda M, Inoue T, Ueno M, Morita H, Hayashimoto N, Kawai K, Itoh T. A case of nontraumatic gas gangrene in a common marmoset (*Callithrix jacchus*). J Vet Med Sci 2016;77:1673–6.
[31] Stein FJ, Lewis DH, Stott GG, Sis RF. Acute gastric dilatation in common marmosets (*Callithrix jacchus*). Lab Anim Sci 1981;31: 522–3.
[32] Sasseville VG, Simon MA, Chalifoux LV, Lin KC, Mansfield KG. Naturally occurring Tyzzer's disease in cotton-top tamarins (*Saguinus oedipus*). Comp Med 2007;57:125–7.
[33] Yoshida K, Nibe K, Nakamura T, Takahashi T, Komatsu M, Ogawa H, Shirota K, Chambers JK, Nakayama H, Uchida K. Spontaneous Tyzzer's disease with the central nerve involvement in a newborn common marmoset. J Vet Med Sci 2013;75: 1119–21.
[34] Artukovic B, Grabarevic Z, Kurilj AG, Beck A, Shirota K, Sakurada Y, Kawamura S, Dzaja P. Clinical and pathological findings of an outbreak of Tyzzers disease in a rabbit colony in Croatia. Vet Arh 2010;80:761–70.
[35] Baskerville A, Dowsett AB, Fitzgeorge RB, Hambleton P, Broster M. Ultrastructure of pulmonary alveoli and macrophages in experimental Legionnaires' disease. J Pathol 1983;140:77–90.
[36] Pruller S, Rensch U, Meemken D, Kaspar H, Kopp PA, Klein G, Kehrenberg C. Antimicrobial susceptibility of *Bordetella bronchiseptica* isolates from swine and companion animals and detection of resistance genes. PLoS One 2015;10:e0135703.
[37] Speakman AJ, Dawson S, Corkill JE, Binns SH, Hart CA, Gaskell RM. Antibiotic susceptibility of canine *Bordetella bronchiseptica* isolates. Vet Microbiol 2000;71:193–200.
[38] Kaper JB, Nataro JP, Mobley HL. Pathogenic *Escherichia coli*. Nat Rev Microbiol 2004;2:123–40.
[39] Carvalho VM, Gyles CL, Ziebell K, Ribeiro MA, Catao-Dias JL, Sinhorini IL, Otman J, Keller R, Trabulsi LR, Pestana de Castro AF. Characterization of monkey enteropathogenic *Escherichia coli* (EPEC) and human typical and atypical EPEC serotype isolates from neotropical nonhuman primates. J Clin Microbiol 2003;41:1225–34.
[39a] McCoy CS, Feng Y, Madden CM, Mannion A, Artim SC, Fox JG. Cytotoxic Escherichia coli strains encoding colibactin and cytotoxic necrotizing factor (CNF) colonize laboratory common marmosets (Callithrix jacchus). Baltimore, MD: AALAS National Meeting; 2018.
[40] Nougayrede JP, Homburg S, Taieb F, Boury M, Brzuszkiewicz E, Gottschalk G, Buchrieser C, Hacker J, Dobrindt U, Oswald E. *Escherichia coli* induces DNA double-strand breaks in eukaryotic cells. Science 2006;313:848–51.
[41] Putze J, Hennequin C, Nougayrede JP, Zhang W, Homburg S, Karch H, Bringer MA, Fayolle C, Carniel E, Rabsch W, Oelschlaeger TA, Oswald E, Forestier C, Hacker J, Dobrindt U. Genetic structure and distribution of the colibactin genomic island among members of the family *Enterobacteriaceae*. Infect Immun 2009;77:4696–703.
[42] Martin HR, Taylor NS, Buckley EM, Marini RP, Patterson MM, Fox JG. Characterization of cytotoxic necrotizing factor 1-producing *Escherichia coli* strains from faeces of healthy macaques. J Med Microbiol 2009;58:1354–8.
[43] Mansfield KG, Lin KC, Xia D, Newman JV, Schauer DB, MacKey J, Lackner AA, Carville A. Enteropathogenic *Escherichia coli* and ulcerative colitis in cotton-top tamarins (*Saguinus oedipus*). J Infect Dis 2001;184:803–7.
[44] Hayashimoto N, Inoue T, Morita H, Yasuda M, Ueno M, Kawai K, Itoh T. Survey and experimental infection of enteropathogenic *Escherichia coli* in common marmosets (*Callithrix jacchus*). PLoS One 2016;11:e0160116.
[45] Nelson M, Lever MS, Dean RE, Savage VL, Salguero FJ, Pearce PC, Stevens DJ, Simpson AJ. Characterization of lethal inhalational infection with *Francisella tularensis* in the common marmoset (*Callithrix jacchus*). J Med Microbiol 2010;59:1107–13.
[46] Splettstoesser WD, Matz-Rensing K, Seibold E, Tomaso H, Al Dahouk S, Grunow R, Essbauer S, Buckendahl A, Finke EJ, Neubauer H. Re-emergence of *Francisella tularensis* in Germany: fatal tularaemia in a colony of semi-free-living marmosets (*Callithrix jacchus*). Epidemiol Infect 2007;135:1256–65.
[47] Posthaus H, Welle M, Morner T, Nicolet J, Kuhnert P. Tularemia in a common marmoset (*Callithrix jacchus*) diagnosed by 16S rRNA sequencing. Vet Microbiol 1998;61:145–50.
[48] Nelson M, Lever MS, Dean RE, Pearce PC, Stevens DJ, Simpson AJ. Bioavailability and efficacy of levofloxacin against *Francisella tularensis* in the common marmoset (*Callithrix jacchus*). Antimicrob Agents Chemother 2010;54:3922–6.
[49] Podschun R, Pietsch S, Holler C, Ullmann U. Incidence of *Klebsiella* species in surface waters and their expression of virulence factors. Appl Environ Microbiol 2001;67:3325–7.
[50] Sasaki E, Tokiwa T, Tsugo K, Higashi Y, Hori H, Une Y. Peracute bacterial meningitis due to Infection with *Klebsiella pneumoniae* in captive-bred ruffed lemurs (*Varecia variegate*). J Comp Pathol 2017;156:281–5.
[51] Shon AS, Bajwa RP, Russo TA. Hypervirulent (hypermucoviscous) *Klebsiella pneumoniae*: a new and dangerous breed. Virulence 2013;4:107–18.
[52] Vela AI, Gutierrez MC, Falsen E, Rollan E, Simarro I, Garcia P, Dominguez L, Ventosa A, Fernandez-Garayzabal JF. *Pseudomonas simiae* sp. nov., isolated from clinical specimens from monkeys (*Callithrix geoffroyi*). Int J Syst Evol Microbiol 2006;56:2671–6.
[53] King NWJ. Synopsis of the pathology of New World monkeys. Scientific Publication, Pan American Health Organization; 1976. p. 169–98. 317.
[54] Andrews JR, Ryan ET. Diagnostics for invasive *Salmonella* infections: current challenges and future directions. Vaccine 2015; 33(Suppl. 3):C8–15.
[55] Cooper JE, Needham JR. An outbreak of shigellosis in laboratory marmosets and tamarins (Family: Callithricidae). J Hyg (Lond) 1976;76:415–24.
[56] Fox JG. Transmissible drug resistance in Shigella and *Salmonella* isolated from pet monkeys and their owners. J Med Primatol 1975;4:165–71.
[57] Bakker J, Kondova I, de Groot C, Remarque EJ, Heidt PJ. A report on *Yersinia*-related mortality in a colony of New World Monkeys. Lab Primate Newsl 2007;463:11–6.
[58] Fredriksson-Ahomaa M, Naglic T, Turk N, Seol B, Grabarevic Z, Bata I, Perkovic D, Stolle A. Yersiniosis in zoo marmosets (*Callitrix jacchuss*) caused by *Yersinia enterocolitica* 4/O:3. Vet Microbiol 2007;121:363–7.
[59] Johnson LD, Ausman LM, Rolland RM, Chalifoux LV, Russell RG. Campylobacter-induced enteritis and diarrhea in captive cotton-top tamarins (*Saguinus oedipus*) during the first year of life. Comp Med 2001;51:257–61.
[60] Won YS, Vandamme P, Yoon JH, Park YH, Hyun BH, Kim HC, Itoh T, Tanioka Y, Choi YK. *Helicobacter callitrichis* sp. nov., a novel *Helicobacter* species isolated from the feces of the common marmoset (*Callithrix jacchus*). FEMS Microbiol Lett 2007;271: 239–44.
[61] de Mello MF, Monteiro AB, Fonseca EC, Pissinatti A, Ferreira AM. Identification of *Helicobacter* sp. in gastric mucosa from captive marmosets (*Callithrix* sp.; callitrichidae, primates). Am J Primatol 2005;66:111–8.

[62] Shen Z, Feng Y, Sheh A, Everitt J, Bertram F, Paster BJ, Fox JG. Isolation and characterization of a novel Helicobacter species, *Helicobacter jaachi* sp. nov., from common marmosets (*Callithrix jaachus*). J Med Microbiol 2015;64:1063–73.
[63] Shen Z, Mannion A, Whary MT, Muthupalani S, Sheh A, Feng Y, Gong G, Vandamme P, Holcombe HR, Paster BJ, Fox JG. *Helicobacter saguini*, a novel helicobacter isolated from cotton-top tamarins with ulcerative colitis, has proinflammatory properties and induces typhlocolitis and dysplasia in gnotobiotic IL-$10^{-/-}$ mice. Infect Immun 2016;84:2307–16.
[64] Fox JG, Handt L, Xu S, Shen Z, Dewhirst FE, Paster BJ, Dangler CA, Lodge K, Motzel S, Klein H. Novel Helicobacter species isolated from rhesus monkeys with chronic idiopathic colitis. J Med Microbiol 2001;50:421–9.
[65] Fox JG, Boutin SR, Handt LK, Taylor NS, Xu S, Rickman B, Marini RP, Dewhirst FE, Paster BJ, Motzel S, Klein HJ. Isolation and characterization of a novel helicobacter species, "*Helicobacter macacae*," from rhesus monkeys with and without chronic idiopathic colitis. J Clin Microbiol 2007;45:4061–3.
[66] Fox JG, Handt L, Sheppard BJ, Xu S, Dewhirst FE, Motzel S, Klein H. Isolation of *Helicobacter cinaedi* from the colon, liver, and mesenteric lymph node of a rhesus monkey with chronic colitis and hepatitis. J Clin Microbiol 2001;39:1580–5.
[67] Lertpiriyapong K, Handt L, Feng Y, Mitchell TW, Lodge KE, Shen Z, Dewhirst FE, Muthupalani S, Fox JG. Pathogenic properties of enterohepatic *Helicobacter* spp. isolated from rhesus macaques with intestinal adenocarcinoma. J Med Microbiol 2014;63:1004–16.
[68] Marini RP, Muthupalani S, Shen Z, Buckley EM, Alvarado C, Taylor NS, Dewhirst FE, Whary MT, Patterson MM, Fox JG. Persistent infection of rhesus monkeys with '*Helicobacter macacae*' and its isolation from an animal with intestinal adenocarcinoma. J Med Microbiol 2010;59:961–9.
[69] Saunders KE, Shen Z, Dewhirst FE, Paster BJ, Dangler CA, Fox JG. Novel intestinal *Helicobacter* species isolated from cotton-top tamarins (*Saguinus oedipus*) with chronic colitis. J Clin Microbiol 1999;37:146–51.
[70] Fox JG. The non-*H pylori* helicobacters: their expanding role in gastrointestinal and systemic diseases. Gut 2002;50:273–83.
[71] Foltz CJ, Fox JG, Yan L, Shames B. Evaluation of antibiotic therapies for eradication of *Helicobacter hepaticus*. Antimicrob Agents Chemother 1995;39:1292–4.
[72] Fox JG, Dewhirst FE, Shen Z, Feng Y, Taylor NS, Paster BJ, Ericson RL, Lau CN, Correa P, Araya JC, Roa I. Hepatic *Helicobacter* species identified in bile and gallbladder tissue from Chileans with chronic cholecystitis. Gastroenterology 1998;114:755–63.
[73] Matsukura N, Yokomuro S, Yamada S, Tajiri T, Sundo T, Hadama T, Kamiya S, Naito Z, Fox JG. Association between *Helicobacter bilis* in bile and biliary tract malignancies: *H. bilis* in bile from Japanese and Thai patients with benign and malignant diseases in the biliary tract. Jpn J Cancer Res 2002;93:842–7.
[74] Segura-López FK, Avilés-Jiménez F, Güitrón-Cantú A, Valdéz-Salazar HA, León-Carballo S, Guerrero-Pérez L, Fox JG, Torres J. Infection with *Helicobacter bilis* but not *Helicobacter hepaticus* was associated with Extrahepatic Cholangiocarcinoma. Helicobacter 2015;20:223–30.
[75] Araoka H, Baba M, Kimura M, Abe M, Inagawa H, Yoneyama A. Clinical characteristics of bacteremia caused by *Helicobacter cinaedi* and time required for blood cultures to become positive. J Clin Microbiol 2014;52:1519–22.
[76] Kitamura T, Kawamura Y, Ohkusu K, Masaki T, Iwashita H, Sawa T, Fujii S, Okamoto T, Akaike T. *Helicobacter cinaedi* cellulitis and bacteremia in immunocompetent hosts after orthopedic surgery. J Clin Microbiol 2007;45:31–8.
[77] Matsumoto T, Goto M, Murakami H, Tanaka T, Nishiyama H, Ono E, Okada C, Sawabe E, Yagoshi M, Yoneyama A, Okuzumi K, Tateda K, Misawa N, Yamaguchi K. Multicenter study to evaluate bloodstream infection by *Helicobacter cinaedi* in Japan. J Clin Microbiol 2007;45:2853–7.
[78] Boutin SR, Rogers AB, Shen Z, Fry RC, Love JA, Nambiar PR, Suerbaum S, Fox JG. Hepatic temporal gene expression profiling in *Helicobacter hepaticus*-infected A/JCr mice. Toxicol Pathol 2004;32:678–93.
[79] Fox JG, Li X, Yan L, Cahill RJ, Hurley R, Lewis R, Murphy JC. Chronic proliferative hepatitis in A/JCr mice associated with persistent *Helicobacter hepaticus* infection: a model of helicobacter-induced carcinogenesis. Infect Immun 1996;64:1548–58.
[80] Hailey JR, Haseman JK, Bucher JR, Radovsky AE, Malarkey DE, Miller RT, Nyska A, Maronpot RR. Impact of *Helicobacter hepaticus* infection in B6C3F1 mice from twelve National Toxicology Program two-year carcinogenesis studies. Toxicol Pathol 1998;26:602–11.
[81] Perolat P, Poingt JP, Vie JC, Jouaneau C, Baranton G, Gysin J. Occurrence of severe leptospirosis in a breeding colony of squirrel monkeys. Am J Trop Med Hyg 1992;46:538–45.
[82] Reid HA, Herron AJ, Hines 2nd ME, Orchard EA, Altman NH. Leptospirosis in a white-lipped tamarin (*Saguinus labiatus*). Lab Anim Sci 1993;43:258–9.
[83] Pinna MH, Martins G, Pinheiro AC, Almeida DS, Oria AP, Lilenbaum W. Detection of anti-Leptospira antibodies in captive nonhuman primates from Salvador, Brazil. Am J Primatol 2012;74:8–11.
[84] Baitchman EJ, Calle PP, James SB, Linn MJ, Raphael BL. Leptospirosis in Wied's marmosets (*Callithrix kuhlii*). J Zoo Wildl Med 2006;37:182–5.
[85] Pereira MM, Da Silva JJ, Pinto MA, Da Silva MF, Machado MP, Lenzi HL, Marchevsky RS. Experimental leptospirosis in marmoset monkeys (*Callithrix jacchus*): a new model for studies of severe pulmonary leptospirosis. Am J Trop Med Hyg 2005;72:13–20.
[86] Fox JG. Selected zoonoses. In: Fox JG, Anderson LC, Otto GM, Pritchett-Corning KR, Whary MT, editors. American College of Laboratory Animal Medicine series. 3rd ed. Elsevier; 2017. p. 1313–70.
[87] Levett PN, Morey RE, Galloway RL, Turner DE, Steigerwalt AG, Mayer LW. Detection of pathogenic leptospires by real-time quantitative PCR. J Med Microbiol 2005;54:45–9.
[88] Miller AD, Kramer JA, Lin KC, Knight H, Martinot A, Mansfield KG. Small intestinal adenocarcinoma in common marmosets (*Callithrix jacchus*). Vet Pathol 2010;47:969–76.
[89] Mafra C, Barcelos RM, Mantovani C, Carrizo J, Soares AC, Moreira HN, Maia NL, da Silva FF, e Silva VHD, Boere V, e Silva IO. Occurrence of *Ehrlichia canis* in free-living primates of the genus *Callithrix*. Rev Bras Parasitol Vet 2015;24:78–81.
[90] Taylor-Robinson D, Furr PM, Tully JG, Barile MF, Moller BR. Animal models of *Mycoplasma genitalium* urogenital infection. Isr J Med Sci 1987;23:561–4.
[91] Mutanda LN, Mufson MA. Experimental *Mycoplasma pneumoniae* infection of marmosets. Lab Anim Sci 1977;27:119.
[92] Furr PM, Hetherington CM, Taylor-Robinson D. Ureaplasmas in the marmoset (*Callithrix jacchus*): transmission and elimination. J Med Primatol 1979;8:321–6.

CHAPTER

17

Parasitic Diseases

David G. Baker

School of Veterinary Medicine, Louisiana State University, Baton Rouge, LA, United States

INTRODUCTION

In the past, when nonhuman primates (NHP), including marmosets, were routinely captured in the wild and shipped to research institutions and zoological collections around the world, parasite infections and infestations were common. Over time, however, as wild populations declined and source country attitudes toward these faunal treasures changed, animals used in research were more often bred in captivity. Standard operating procedures, sanitation, and lack of intermediate hosts greatly reduced the incidence of parasite infection and infestation, particularly for indirectly transmitted parasites [1]. Now, except for a few directly transmitted parasites, infection and infestation are uncommon. However, the occasional movement of animals from the wild into captive populations, particularly in zoological collections, means that veterinarians and biologists working with NHP, including marmosets, should remain aware of these organisms and the conditions caused by them.

This chapter describes the parasites found in or on the common marmoset (*Callithrix jacchus*) and closely related species. The parasite taxonomy used in this chapter is based on current usage. Parasitologists sometimes disagree on specific aspects of parasite classification. Thus, the observant reader will occasionally note that the classification schemes used by various authors may differ. To retain consistency within the current literature, the organization of this chapter will largely follow the structure used by Strait et al. [2] and will serve as an aid to allow the reader to quickly find information on a parasite of interest. It is not intended to be the final word on parasite taxonomy or nomenclature.

PROTOZOAN PARASITES

Protozoan parasites have historically been among the most common parasites found in captive marmosets. For example, older survey reports list the following genera of protozoa identified in marmosets: *Balantidium*, *Eimeria*, *Entamoeba*, *Giardia*, *Isospora*, *Trichomonas*, and *Trypanosoma* [3,4]. Most of these reports were from marmosets that had been captured in the wild and brought into captivity. In this regard, the transition to using captive-bred animals, along with improved housing, sanitation, diagnostics, and treatments, has greatly reduced the prevalence of infection with these organisms.

Flagellates

Marmosets may become infected with flagellated protozoan parasites. For this chapter, discussion has been limited to *Giardia* sp. and *Pentatrichomonas hominis*.

Enteric Flagellates

Giardia sp.

Etiology *Giardia intestinalis* (synonyms *Giardia lamblia*, *Giardia duodenalis*) infects many mammalian hosts, including marmosets. Parasites of the genus *Giardia* are the only parasites known to contain a median body, which appears as a "claw hammer" under light microscopy. The presence of this organelle facilitates identification of the genus. Trophozoites are piriform and bilaterally symmetrical, measure 9–21 μm long by 5–15 μm wide, and bear a sucking disc on the anteroventral side (**Fig. 17.1A**). The dorsal side is convex. There are two anterior nuclei, two slender axostyles, and four pairs of flagella. The infectious cyst stage is oval in shape. Cysts typically measure 8–12 μm long and 7–10 μm wide [5] (Fig. 17.1B). Following binary fission, cysts contain two daughter trophozoites, thereby doubling the number of internal organelles [2,6].

Clinical Signs As with other animals, infection with *Giardia* sp. is most often asymptomatic and self-limiting

The Common Marmoset in Captivity and Biomedical Research
https://doi.org/10.1016/B978-0-12-811829-0.00017-0

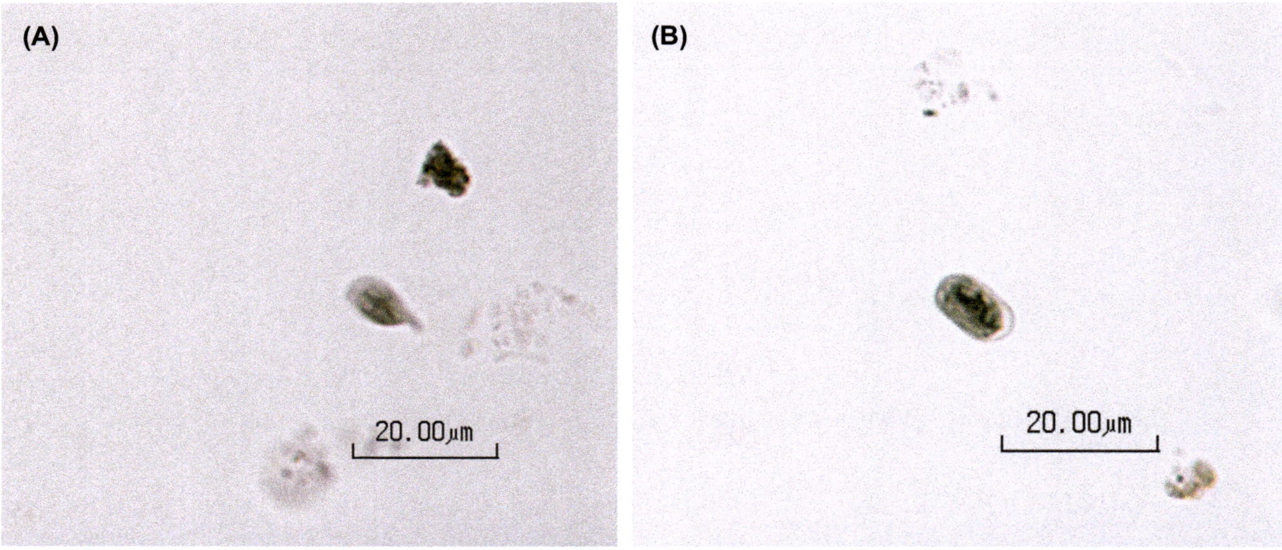

FIGURE 17.1 (A) *Giardia* sp. trophozoite. 600× magnification. (B) *Giardia* sp. cyst. 600× magnification.

in marmosets. However, a chronic carrier state such as that found in some companion animals and humans may be more common than previously recognized [7,8]. Occasionally, clinically affected animals may develop malodorous, fatty stool to watery diarrhea, flatulence, abdominal cramping and distension, anorexia, and weight loss [5,7,9].

Epizootiology Even in well-managed marmoset colonies, giardiasis remains a challenge [1]. Factors that have been associated with *Giardia* sp. infection include situations where large numbers of animals are assembled, exposure to fouled water sources, and in some reports, age [8]. In one survey, 1 of 25 (4%) captive marmosets greater than 1 year of age, and 5 of 25 (20%) captive marmosets less than 1 year of age, were found to be infected, though none of these animals showed clinical signs of infection. In this case, prevalence rates between groups were not significantly different [5].

Pathology *Giardia* sp. causes small intestinal malabsorption of fats and sugars through CD8+ lymphocyte-induced shortening of brush border microvilli [10]. Host–parasite ligand–receptor interactions facilitate attachment to intestinal epithelial cells [11].

Diagnosis Traditionally, diagnosis of *Giardia* sp. infection has been accomplished by light microscopic observation of motile trophozoites or cysts in direct fecal smears by trichrome, fluorescent antibody, or other staining of fecal smears, and more recently, by detection of fecal *Giardia* antigen [2,5] or PCR assay [12]. When traditional methods are employed, multiple fecal samples may be required to diagnose infection due to intermittent shedding [2,5].

Differential Diagnosis A thorough search for other causes of illness along with careful assessment of clinical presentation and response to treatment should all be considered prior to diagnosing clinical giardiasis in an animal shedding cysts [2]. In one comprehensive survey involving 13,385 captive nonhuman primates from several Old and New World species, not including *C. jacchus*, other potential causes of acute diarrhea included bacteria such as *Shigella* sp., *Salmonella* sp., *Campylobacter* sp., and enteropathogenic *Escherichia coli*. Possible viral causes included rotavirus and coronavirus, whereas parasitic causes included *Entamoeba histolytica* and *Strongyloides* sp. [13].

Prevention and Control Prevention of infection with *Giardia* sp. is centered in strict adherence to sanitation practices, including protecting feed and water sources from contamination with infectious cysts. *Giardia* sp. infection in marmosets has been successfully eliminated with tinidazole (150 mg/kg PO followed after 4 days by 77 mg/kg PO, each treatment given once) [7]. Other effective treatments utilized in a variety of nonhuman primates include metronidazole (30–50 mg/kg day PO divided tid for 5–10 days), albendazole (25 mg/kg PO bid for 3–5 days), quinacrine (10 mg/kg/day PO divided tid for 5 days), furazolidone (1.5 mg/kg sid PO for 7 days), and others [2,7].

Research Complications In addition to alterations of the gastrointestinal system associated with the clinical signs of *Giardia* sp. infection, infection of marmosets with *Giardia* sp. may resemble those reported in other host species and include alteration of the lymphoreticular system. For example, infection with *Giardia* sp. has been shown to reduce mouse immune responsiveness to sheep erythrocytes and increase the severity of concurrent infections with other pathogens in athymic mice [14]. Because *Giardia* sp. remains a potential zoonosis, scientific and animal care staff should exercise strict personal hygiene when working with potentially infected marmosets [7].

Pentatrichomonas hominis

Etiology In addition to *Giardia* sp., common marmosets have been found infected with other enteric flagellates, including *Pentatrichomonas hominis*, in the large intestine [15]. Trophozoites of *P. hominis* measure 8–20 μm long by 3–14 μm wide and bear 3–5 anterior flagella. There is no cyst stage [16].

Clinical Signs *P. hominis* infection is considered asymptomatic. In one study, presence or absence of *P. hominis* in fecal samples was not associated with diarrhea [15].

Epizootiology In one study, 58 of 88 (66%) laboratory-bred marmosets were found to be shedding the trichomonad, *P. hominis*. The presence or absence of *P. hominis* in fecal samples was not associated with the age or sex of the host [15].

Pathology, Diagnosis, Prevention, and Control *P. hominis* is considered a nonpathogenic inhabitant of the large intestine. Trophozoites of *P. hominis* may be observed in fresh smears of both normal and diarrheic fecal samples [16]. Because *P. hominis* is nonpathogenic, efforts to eliminate or prevent infection are not usually undertaken.

Research Complications Research complications have not been identified with *P. hominis* infection.

Hemoflagellates

Common marmosets serve as hosts for at least two species of trypanosomes, *Trypanosoma cruzi* and *T. minasense*.

Trypanosoma cruzi

Etiology *Trypanosoma cruzi*, the cause of Chagas disease in humans, infects captive and free-ranging marmosets in Central and South America [16a,16b]. Two forms of the organism are found in marmosets. Early in the infection, the trypomastigote form is found in the blood. Trypomastigotes measure 16–20 μm long, with a pointed posterior end and a curved body. The kinetoplast is subterminal and larger than in other species infecting marmosets. The trypomastigote bears an undulating membrane and a long flagellum [6]. Later in the infection, the nonmotile amastigote stage is found in pseudocysts in skeletal and cardiac muscle and the reticuloendothelial system. In central nervous system infections, organisms may be found in neuroglial cells [2,6].

Clinical Signs Infection of marmosets with *T. cruzi* is most often asymptomatic and may only be detected at necropsy [16]. When present, clinical signs may include generalized edema, anemia, weight loss, and other nonspecific signs. In addition, hepatosplenomegaly and lymphadenitis may develop. Intrauterine fetal death has also been reported [2,17].

Epizootiology Triatomid bugs in the family Reduviidae serve as intermediate hosts and vectors. Other arthropod hosts capable of being infected experimentally include bed bugs, sheep keds, and soft ticks in the genus *Ornithodoros*. These therefore represent additional potential vectors [6]. However, triatomes are clearly most important in transmitting the organism, either through defecation of parasite-laden feces, with subsequent trypomastigote penetration of mucus membranes or entry into the feeding site, or via ingestion of infected triatomes by susceptible hosts. Infection is common in South and Central America. Because of the presence of suitable arthropod vectors in the extreme southern United States, *T. cruzi* has become established in some captive nonhuman primate colonies in Texas, Louisiana, and Georgia. Several species of wild animals serve as reservoirs of the parasite, making eradication of infection difficult once established in a nonhuman primate colony [16,16a].

Pathology As in human infection, the hallmark of *T. cruzi* infection of marmosets is myocarditis. Myocardial fibers become damaged by pseudocysts containing circular- to oval-shaped organisms, the amastigote stage, 1.5–4 μm in diameter. Amastigotes have a central nucleus and a prominent bar-shaped structure, the kinetoplast. Degenerating pseudocysts elicit the host inflammatory response [2].

Diagnosis Traditionally, infection with *T. cruzi* has been diagnosed by finding trypomastigote forms following Giemsa staining of blood smears or blood cultures [16]. The trypomastigote form of *T. cruzi* is smaller and thus easily distinguished from that of *T. minasense*, a nonpathogenic trypanosome also found in marmosets. In addition to direct microscopic examination of blood smears or blood cultures, infection can be diagnosed using serological tests, including ELISA and older complement fixation assays [2]. More recently, quantitative

PCR assay has been used to detect organisms in blood smears (Minuzzi-Souza et al., 2016).

Differential Diagnosis A complete discussion of the differential diagnoses for nonspecific clinical signs is beyond the scope of this chapter. Marmosets exhibiting clinical signs of weight loss, anemia, and generalized edema should receive a complete clinical evaluation.

Prevention and Control Prevention of infection requires exclusion of the arthropod host. There is no effective treatment for *T. cruzi* infection in humans or in NHP, including marmosets. Affected animals should be treated symptomatically. Because Chagas disease occurs in humans, preventing the establishment of *T. cruzi* infection in NHP colonies is critical.

Research Complications Specific research complications involving marmosets infected with *T. cruzi* have not been reported. However, infected animals would be unsuitable for cardiovascular, lymphoreticular, and perhaps other research uses.

Trypanosoma minasense

Etiology *Trypanosoma minasense* (synonyms *T. advieri, T. brirnonti, T. devei, T. escomeli, T. florestali, T. manguinhense, T. mycetae*) is a nonpathogenic trypanosome commonly found in free-ranging marmosets in Central and South America [16–18].

Epizootiology Similar to *T. cruzi*, triatomid bugs in the family Reduviidae serve as intermediate hosts and vectors of *T. minasense*. Natural infection with *T. minasense* has not been reported in laboratory-bred marmosets.

Clinical Signs and Pathology Infection of marmosets with *T. minasense* is asymptomatic as the parasite is nonpathogenic in marmosets. Infection may represent an incidental finding during examination of thin or thick blood films, either ante- or postmortem.

Diagnosis, Prevention, and Control Infection with *T. minasense* may be diagnosed by finding trypomastigote forms following Giemsa staining of blood smears or blood cultures [16]. The trypomastigote form is larger than that of *T. cruzi* and measures 29–46 μm long. The kinetoplast is smaller and is located more anteriorly than that of *T. cruzi* [17a]. Prevention of infection requires exclusion of the arthropod host. There is no need or means to effectively treat *T. minasense* infection in marmosets. *T. minasense* is not considered to be zoonotic.

Research Complications Specific research complications involving marmosets infected with *T. minasense* have not been reported.

Sarcodines: Amoebae

There is scant literature on infections of common marmosets with amoebae. Though rarely reported, it is assumed that, like so many other New World monkeys, marmosets are susceptible to infection with several species of amoebae [16]. Perhaps most important among these is *Entamoeba histolytica*.

Entamoeba histolytica

Etiology

Entamoeba histolytica may infect virtually all species of NHP [19]. Distribution is worldwide. However, compared to Old World monkeys, New World monkeys are less commonly, if not rarely infected. The trophozoite form of *E. histolytica* measures 20–30 μm in diameter (Fig. 17.2). Trophozoites may be observed to contain phagocytized red blood cells. Cysts form in the large intestine and pass out in the fecal stream. Mature cysts contain four nuclei and measure 10–20 μm in diameter. Nuclear morphology is distinctive for the parasite species and serves as a diagnostic aid [16]. When stained with hematoxylin, *E. histolytica* has a small, central endosome and a ring of small peripheral granules interspersed with chromatin granules [6].

Clinical Signs

Many infections with *E. histolytica* are asymptomatic. When present, clinical signs include lethargy, anorexia,

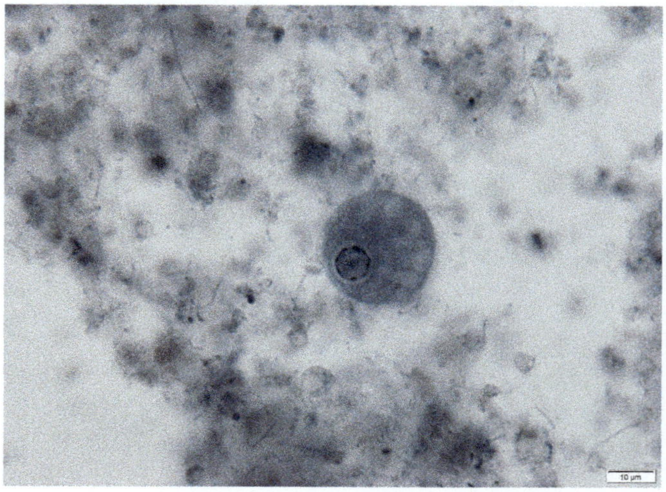

FIGURE 17.2 *Entamoeba histolytica* precystic trophozoite. Fine chromatin granules can be seen lining the nuclear membrane, giving a beaded appearance. Food vacuoles are present in the cytosol. 630× magnification.

weight loss, weakness, dehydration, vomiting, and severe diarrhea, which may be hemorrhagic or catarrhal [16,20,21].

Epizootiology

Cysts of *E. histolytica* are passed in the feces. Prior to encysting, trophozoites round up and become smaller, eliminate food vacuoles, form a cyst wall, and undergo nuclear division twice, resulting in an organism with four distinct nuclei. Following ingestion of mature cysts by a suitable host, the trophozoite will emerge from the cyst and both the nuclei and cytoplasm will divide, resulting in eight daughter trophozoites [6]. New World monkeys, including marmosets, are less commonly infected with *E. histolytica* than are Old World monkeys. However, when infected, clinical disease is generally worse than in Old World monkeys. Other factors influencing severity of disease include the age and nutritional status of the host, with younger animals being more severely affected. In addition, environmental conditions that increase host animal stress and the intestinal microbiota of the host may play a role in disease severity [6,16,22].

Pathology

Early infection with *E. histolytica* is localized to the large intestinal mucosa. Pathologic findings include necro-ulcerative colitis with characteristic flask-shaped lesions. Multiple small lesions may coalesce. Later, ulcerative extension through all layers of the large intestine may result in peritonitis. Pathogen invasion of the vascular or lymphatic system results in dissemination to other organs, most commonly the regional lymph nodes, liver, lungs, and central nervous system [6,20].

Diagnosis

E. histolytica infection can be diagnosed in direct fecal smears, where one can readily observe the motile trophozoites. Alternatively, fecal films fixed in Schaudinn's fluid [6] can be stained with periodic acid Schiff. Other stains, including trichrome, Giemsa, or iron hematoxylin, can be used to evaluate nuclear morphology to allow differentiation of *E. histolytica* from other, nonpathogenic amoebae [6]. Molecular methods of diagnosis and identification have also been developed, but these are not yet in widespread use [22a].

Differential Diagnosis

A complete analysis of the differential diagnoses for diarrhea is beyond the scope of this chapter. The mere finding of *E. histolytica* cysts in diarrheic feces does not prove causality. Other infectious causes should be considered in the clinically affected marmoset. Perhaps most prominent among them are pathogenic bacteria, or bacterial dysbiosis [6].

Prevention and Control

Prevention of infection with *E. histolytica* is grounded in good sanitation practices and maintenance of water supply systems. Routine chlorination procedures do not kill the cyst form. In addition, facility operators should prevent exposure of marmosets to Old World NHP species, a practice that serves many purposes and is uniformly accepted in well-run animal facilities. Because many people are asymptomatic shedders of *E. histolytica*, regular fecal screening of animal facility employees, with treatment of infected personnel, should be standard operating procedure. Persons working with marmosets should practice excellent personal hygiene to prevent transmission of pathogens between themselves and the animals under their care [6,22]. Newly arrived marmosets should be checked for *E. histolytica* infection several times during the first 2 months. Infected marmosets should be treated with metronidazole, whether or not clinical signs of disease are present [2,20].

Research Complications

Infection of marmosets with *E. histolytica* would interfere with research involving the gastrointestinal system and other forms of research in which normal body condition, cognition, and activity levels were required.

Apicomplexa (Coccidians)

A small number of coccidian species have been reported from common marmosets. These include *Cryptosporidium* sp., *Isospora* sp., and *Plasmodium* sp. [2]. It is likely that other species also infect marmosets but may not have been reported. Species such as *Toxoplasma*, which may inadvertently or experimentally infect marmosets but are not considered primary parasites of marmosets, are not discussed. For this chapter, *Cryptosporidium* sp. and *Plasmodium* sp. are presented. Little is known of the *Isospora* sp. reported from marmosets, but the interested reader is referred elsewhere for an excellent review of the available information [23].

Cryptosporidium *sp.*

Etiology

Cryptosporidium sp. is known to infect nearly all species of mammals, including NHP species. The coccidian parasite inhabits an intracellular yet extracytoplasmic location within host intestinal epithelial cells. The life cycle is direct, with schizogony and gametogony occurring in the host. Sporulated, infective oocysts shed in the feces measure 1.5–6 μm in diameter [6,24].

Clinical Signs

Clinical signs consist of acute, self-limiting to chronic, profuse, watery diarrhea resulting in dehydration, anorexia, weight loss, depression, hypothermia, and, occasionally, death [5,24]. Fever is uncommon [24].

Epizootiology

Primary factors that have been associated with *Cryptosporidium* sp. infection include immune deficiency or incompetence, young age, length of time spent in the nursery, and social interactions with other monkeys [1,5,24]. In a survey involving both young and adult marmosets, however, prevalence rates between age groups were not significantly different. In that study, 1 of 25 (4%) captive marmosets greater than 1 year of age, and 4 of 25 (16%) captive marmosets less than 1 year of age, were found to be infected, though none of these animals showed clinical signs of infection. The authors postulated that significant differences may have been uncovered with larger samples [5].

Pathology

Histopathologic changes associated with cryptosporidiosis are found in the small intestine and include mild-to-moderate villus blunting with fusion of villi, variations in height of enterocytes, enterocyte vacuolization or necrosis, and increased numbers of mitotic figures in crypts. Mesenteric lymph nodes are enlarged with reactive lymphoid hyperplasia and histiocytosis [25]. Occasionally, organisms have been found in organ extensions of the gastrointestinal system, including the bile duct, intrahepatic and intrapancreatic ducts, and the gall bladder [26].

Diagnosis

Cryptosporidiosis may be diagnosed using acid-fast or modified acid-fast staining of fixed fecal slides, though this method may have a sensitivity of only 50–70% [5,24]. Immunofluorescent staining methods have been demonstrated to be more specific and easier to read than acid-fast staining [5]. Fecal ELISA has also been used successfully [27].

Differential Diagnosis

There are many potential causes of acute, watery diarrhea in marmosets. Finding *Cryptosporidium* sp. in a fecal sample does not prove causality but necessitates a thorough search for other etiologies, including viral and bacterial pathogens. In one comprehensive survey involving 13,385 captive nonhuman primates from several Old and New World species, not including *C. jacchus*, other potential causes of acute diarrhea included viral agents such as rotavirus and coronavirus, bacteria including *Shigella* sp., *Salmonella* sp., *Campylobacter* sp., and enteropathogenic *Escherichia coli*, and parasitic causes including *Entamoeba histolytica* and *Strongyloides* sp. [13,24].

Prevention and Control

Cryptosporidial oocysts are relatively resistant to dessication, as well as to chemical disinfection, and so may survive in the cage environment for extended periods [2]. Historically, infection of marmosets with *Cryptosporidium* sp. has been refractory to treatment. In a recent report, two marmosets appeared to have been cleared of infection with an extended treatment regiment of the aminoglycoside antibiotic paromomycin (15 mg/kg PO bid for 28 days). The authors of that report acknowledged that the cessation of parasite shedding may not have been due to the treatment, but that the marmosets may have simply and coincidentally "self-cured" during the course of treatment [27]. Because *Cryptosporidium* sp. is known to infect humans, scientific and animal care staff should exercise strict personal hygiene when working with potentially infected marmosets.

Research Complications

Infection of marmosets with *Cryptosporidium* sp. would compromise research involving the gastrointestinal and immune systems, as well as any other system requiring normal hydration, appetite, and physiologic homeostasis.

TREMATODE PARASITES

Several species of trematodes have been reported from all types of marmosets. While trematode infections are common among wild marmosets, the lack of intermediate hosts has greatly reduced the prevalence of these infections in captive marmosets due to an inability to complete the indirect life cycles.

Platynosomum sp.

Etiology

Marmosets are susceptible and have been found infected with several members of the genus *Platynosomum*, including *P. amazonensis, P. illiciens,* and *P. marmoseti* [28] (Fig. 17.3). It should be noted, however, that the taxonomy of species of *Platynosomum* that infect marmosets is not well established. In a recent study, the authors concluded that *P. amazonensis* and *P. marmoseti* should be synonymized with *P. illiciens* [28]. In one study, eggs of *P. illiciens* measured, on average, 40.1 μm long and 26.6 μm wide [29]. Terrestrial snails serve as first intermediate hosts, whereas terrestrial isopods serve as second intermediate hosts. Small lizards may serve as paratenic hosts [30].

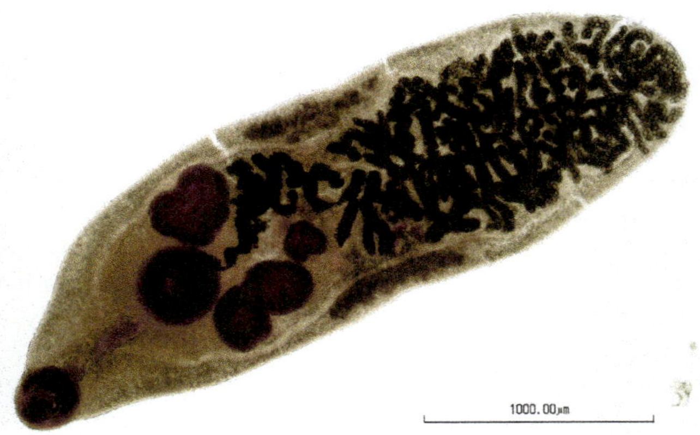

FIGURE 17.3 *Platynosomum* sp. 40× magnification.

Clinical Signs

Light infections are asymptomatic. Clinical signs associated with heavy infection with *Platynosomum* sp. are due to hepatobiliary dysfunction and may include anorexia, weight loss, jaundice, peripheral edema, hepatomegaly, general unthrift, and, occasionally, death [31,32].

Epizootiology

Prevalence of infection of wild marmosets can be high. In captive situations where the intermediate and paratenic hosts are available, thereby facilitating completion of the life cycle, infection rates may remain high [29,33]. Cats also serve as definitive hosts and thus may constitute an important reservoir of infection in some environments [30].

Pathology

Captive black-tufted marmosets (*Callithrix penicillata*) may carry *P. illiciens* in their biliary tree. Infection is associated with mild hepatitis and manifests with hematologic and clinical chemistry perturbations, including eosinophilia, thrombocytopenia, anemia, delayed clotting times, hypoalbuminemia, and elevated serum liver enzymes [31]. Typical findings at necropsy include portal fibrosis, ascending cholangitis, cholestasis, secondary biliary cirrhosis, and thickening of the biliary ducts with traumatic compression of surrounding tissue, resulting in hepatocellular death [31,32].

Diagnosis

Infection with *Platynosomum* sp. can be diagnosed using standard fecal sedimentation procedures. Serial analysis of feces over three consecutive days increases diagnostic sensitivity [33]. Ultrasound examination of the liver may allow for assessment of disease severity and may facilitate diagnosis [31].

Differential Diagnosis

Marmosets displaying clinical signs indicative of hepatobiliary disease should be thoroughly examined for other causes, particularly those associated with protein loss or imbalance.

Prevention and Control

Infection with *Platynosomum* sp. can be prevented with careful exclusion of the intermediate hosts. Infected animals can be cleared of infection with praziquantel, 150 mg/kg PO administered once daily for three consecutive days [31].

Research Complications

Infection of laboratory marmosets with *Platynosomum* sp. would severely compromise research involving the hepatobiliary system and other forms of research requiring normal appetite, behavior, and physiology.

CESTODE PARASITES

Reports of cestode parasitism are surprisingly sparse in the scientific literature and so will not be discussed in detail in this chapter. Adult or larval cestodes found in various species of wild or captive marmosets include *Atriotaenia* sp. (Fig. 17.4), *Bertiella mucronata*, *Raillietina* sp., *Spirometra reptans*, and *Rodentolepis* sp. (Fig. 17.5) [3,4]. It is possible that some cestode species found in other New World monkeys might also infect the common marmoset, but that such infections simply have not yet been reported [2,16].

FIGURE 17.4 *Atriotaenia* sp, anterior end of adult cestode. 20× magnification.

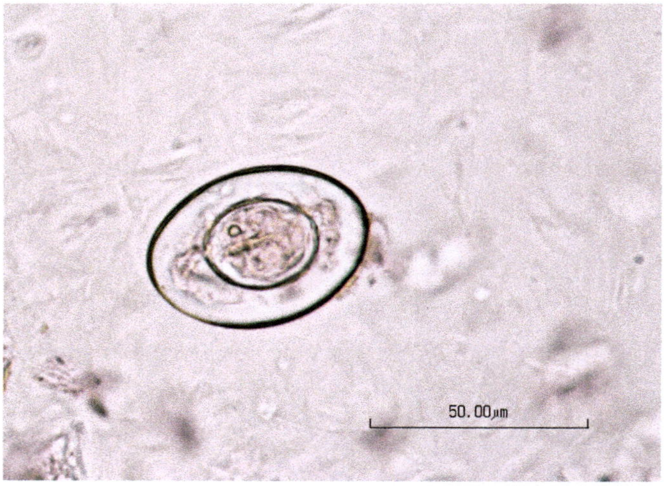

FIGURE 17.5 *Rodentolepis nana* egg. 600× magnification.

NEMATODE PARASITES

Infection with nematodes has historically been common in many species of New World primates. Many, though certainly not all, of these have been found in several varieties of marmosets [2,16]. Those nematode parasitisms that have not been reported in marmosets may not have been so because marmosets are not suitable hosts and thus are not infected, or because infections have simply not yet been reported. For the latter reason, it is likely that the breadth of nematode parasites that may infect marmosets is greater than revealed in the scientific literature. For this reason, the reader should realize that marmosets may be found with parasites that are not on any of the lists of reported marmoset parasitism and so might need to consult the literature on other New World primate species for information regarding specific parasitisms. Furthermore, several nematode superfamilies or species have only infrequently been reported, or have only been reported under somewhat artificial conditions. Uncommonly reported nematode genera parasitizing wild and/or captive marmosets include *Angiostrongylus, Enterobius, Filariopsis, Gongylonema, Physaloptera, Primasubulura, Protospirura, Riticularia, Subulura,* and *Trypanoxyuris* [2,4,16,34–39]. However, relatively little is known of the biology or clinical effects of these parasites. Thus, uncommon or little-known parasitisms such as these are not covered in this chapter. The parasites presented here are those that are reasonably common, and for which there is an established body of scientific literature.

Superfamily Filarioidea

Several species of filarids have been recovered from marmosets and other New World primates [16,40]. Infection with *Dipetalonema gracile* is among the most common, and possesses many of the features common to filarial diseases in New World primates and thus will be covered in this chapter.

Dipetalonema gracile

Etiology

Dipetalonema gracile is known to infect several species of New World monkeys, including the common marmoset [4,41]. Adult worms inhabit the abdominal cavity, where they are often recovered incidentally at necropsy (Fig. 17.6). Adult worms are white and slender. Adult worms of both sexes may vary greatly in their overall length. Male worms measure 8–19 cm in length, whereas females measure 15–60 cm in length [42,43]. The life cycle requires a blood-sucking arthropod intermediate host, typically gnats of the genus *Culicoides* [44]. Microfilaria are found in the blood.

Epizootiology

Infection with filarial worms is common in wild monkeys regularly exposed to arthropod vectors. The incidence of infection should be lower in captive populations where vector control is practiced. Infection should not occur in laboratory colonies of marmosets.

Clinical Signs and Pathology

Clinical signs of abdominal filariasis are usually lacking. However, the presence of abdominal filarids results in parasitic peritonitis and adhesions. Mesenteric fat may contain multiple, perivascular foci of neutrophils and eosinophils [42]. It is likely that heavily infected monkeys may develop nonspecific symptoms, including malaise, anorexia, and weight loss.

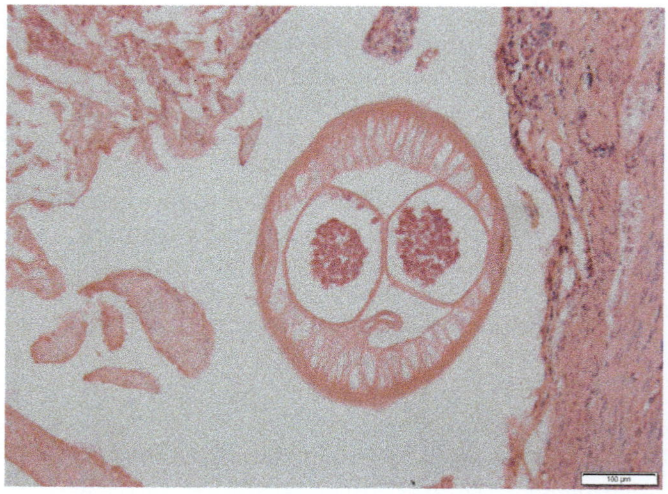

FIGURE 17.6 *Dipetalonema gracile*. Cross-section of adult worm in the abdominal cavity of a common marmoset. 100× magnification.

Diagnosis

Dipetalonemiasis is most commonly diagnosed as an incidental finding at necropsy, when adult worms are found free in the abdominal cavity. However, antemortem diagnosis may be accomplished by finding microfilaria in peripheral blood using standard methods.

Differential Diagnosis

There are many potential causes of general malaise, anorexia, and weight loss in primates, including both infectious and noninfectious diseases, as well as poor husbandry. The primate veterinarian will need to perform a complete clinical workup to exclude causes other than dipetalonemiasis.

Prevention and Control

Prevention of filarial infection relies on vector control. Diethylcarbamazine, given orally at 50 mg/kg for 10 days, has been used to clear both adults and microfilariae of *D. gracile* in squirrel monkeys, and may also be successfully used to treat *D. gracile* in marmosets [45]. While ivermectin would likely be effective against microfilariae and developing larvae, it would likely be ineffective against adult worms [42]. *D. gracile* is not known to be zoonotic and thus poses no direct threat to animal caretakers.

Research Complications

Research complications of abdominal filariasis have not been published. It is expected that animals experiencing general malaise due to abdominal parasitism might not be suitable for research requiring normal appetite, weight, and physiologic balance. The presence of parasitic peritonitis would also complicate research dependent on normal abdominal tissue immune system architecture.

Superfamily Rhabditoidea

Strongyloides cebus

Etiology

Strongyloides cebus is known to infect several species of New World monkeys, including marmosets. Parasitic adult female worms are very small, measuring just 2–5 mm long by 30–80 μm wide [16] (Fig. 17.7). As is the case for other members of the genus, parasitic adult male worms have not been observed. Eggs measure 40–70 μm long by 20–35 μm wide. Eggs are thin-shelled, embryonated when released by the female, and often hatch in the fecal stream, resulting in the finding of first-stage larvae, typical of some other members of the genus [16]. The life cycle is complex and direct. Free-living and parasitic populations exist in the same environment. Following infection, larvae migrate via the bloodstream to the lungs, from where they are brought up and swallowed and so arrive in the small intestine where they complete development to parthenogenic adult female worms.

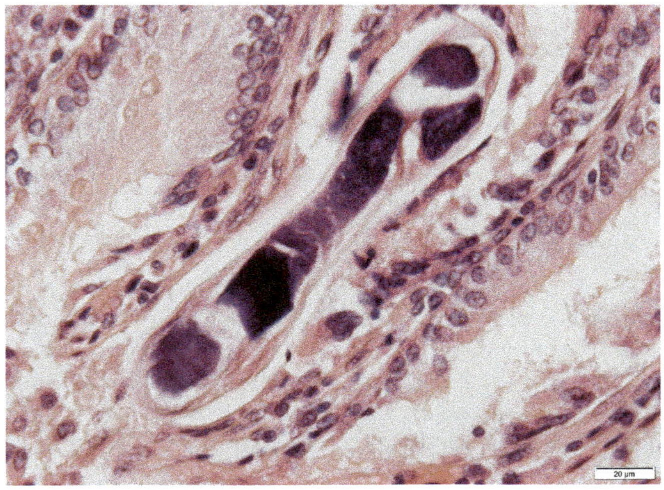

FIGURE 17.7 *Strongyloides cebus*. Longitudinal section of a worm in the intestinal mucosa. 400× magnification.

Clinical Signs and Pathology

Infection with *S. cebus* is usually mild or inapparent. However, clinical signs may develop in heavy infections. These, and the accompanying pathologic changes, can be separated into three distinct phases: invasion, migratory, and intestinal [16]. Each phase is characterized by distinct clinical signs due to the specific location and activities of the worm. The invasion phase may be associated with irritation, erythema, and pruritus as larvae penetrate moist skin or oral mucosa. In the migratory phase, larvae are carried to the cardiopulmonary system and are coughed up and swallowed. Coughing may be sporadic or severe. The presence of larvae in the vasculature and lungs can induce bronchopneumonia, pulmonary hemorrhaging, and death. Finally, in the intestinal phase, worms undergo localized histotrophic migration into the intestinal epithelium where they cause acute enteritis with accompanying diarrhea, listlessness, anorexia, weight loss, reduced growth rate, and, occasionally, death [16].

Epizootiology

Strongyloides spp. are soil-dwelling nematodes. Therefore, moist, warm, and dirty conditions, such as exist in unserviced animal enclosures, facilitate the accumulation of infective forms in the environment and thus exposure of a suitable host to large numbers of infective larvae. In addition, larval development while in transit in the host intestine may result in overwhelming infection ("hyperinfection"). Finally, larvae reaching the infective third stage on arrival at the anus may penetrate the anal or perianal skin, further increasing parasite burden. As for other members of the genus, young monkeys are at greatest risk for patent infection.

Diagnosis

Strongyloidiasis can be diagnosed antemortem by clinical signs and by routine fecal floatation procedures. If a fecal floatation procedure is performed, the technician may find both embryonated eggs and first-stage larvae. Large numbers of very small, translucent adult worms may be found at necropsy in the proximal small intestine [16].

Differential Diagnosis

If only clinical signs are considered, the differential diagnoses for diarrhea are numerous. An aid to diagnosis is found in observing the environmental conditions in which affected animals are housed. Young animals housed in moist, soiled environments are at high risk of infection. Localized damage to the skin, through chronic exposure to moisture and filth, facilitates larval penetration. Thus, the presence of moist, compromised skin on young animals, particularly at points of contact with soiled bedding, is an obvious site to examine for larval activity.

Prevention and Control

Infection with *S. cebus* can be prevented by strict husbandry practices that prevent the accumulation of moist, soiled areas within the animal enclosure. Testing and treatment of new arrivals will help keep environmental contamination to a minimum. Infection can be eliminated with ivermectin (400 μg/kg IM) diluted in sterile propylene glycol. Alternatively, moxidectin (500 μg/kg) can be applied topically [2,16].

Research Complications

Infection of marmosets with *S. cebus* could compromise research involving several body systems. Dermatitis associated with larval skin penetration could compromise skin research, though simply allowing the skin to become chronically moist and erythematous, even without parasite infection, would compromise such research. Cardiopulmonary migration of larvae could compromise research involving the cardiopulmonary and immune systems, whereas intestinal parasitism could compromise research involving this system, as well as any research dependent on normal appetite, activity, and rate of growth.

Superfamily Spiruroidea

Trichospirura leptostoma

Etiology

Trichospirura leptostoma is a spirurid nematode that parasitizes the pancreatic ducts of several species of New World primates, including wild marmosets in Central and South America, and in captive imported populations [1,2,46,47]. Adult male worms measure 15 mm long, whereas female worms measure up to 120 mm long [2] (Fig. 17.8). The eggs are oval, thick-shelled, and embryonated, typical of the superfamily, and measure 40–50 μm long by 15–20 μm wide [48]. The life cycle is indirect, with cockroaches serving as intermediate hosts [49].

Clinical Signs

Light infections with *T. leptostoma* are asymptomatic. Heavy infections result in signs of acute to chronic pancreatitis, pancreatic insufficiency, secondary malnutrition, emaciation in spite of normal appetite, anemia, and, occasionally, biliary disease. Affected monkeys develop increased fecal volume or diarrhea due to pancreatic inflammation and destruction, and jaundice due to biliary outflow obstruction [2,48,50]. Some have suggested *T. leptostoma* as one of the several possible causes of marmoset wasting syndrome [1,48].

Epizootiology

The incidence of infection with *T. leptostoma* is dependent on availability of the cockroach intermediate host. Colony environments infested with cockroaches are prime locations for high levels of infection. The introduction of an infected animal shedding parasite eggs

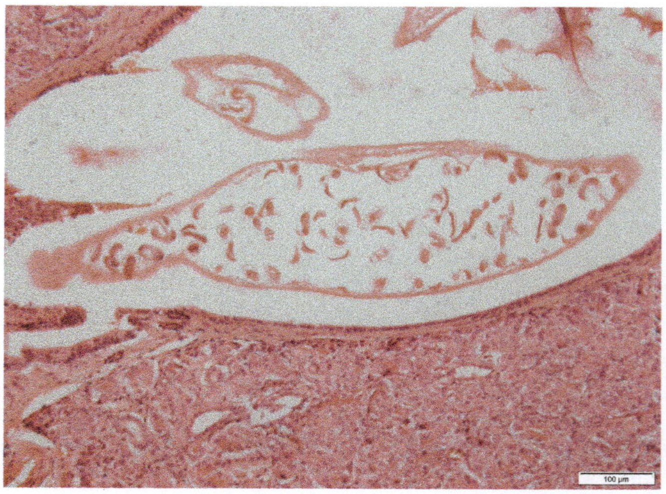

FIGURE 17.8 *Trichospirura leptostoma*. Oblique sections of adult worms in the pancreatic duct of a common marmoset. 100× magnification.

in a facility infested with cockroaches will establish the parasite. Monkeys are fond of catching and eating cockroaches, whereas the roaches are attracted to the plentiful food and water universally found in and around primate caging. Infections are long-lived. In one report, marmosets remained infected for at least 11 months in captivity [47].

Pathology

Infection with *T. leptostoma* results in chronic pancreatitis, with eventual replacement of functional pancreatic parenchyma with fibrotic exocrine tissue [46,47,50,51]. Pathologic changes are proportional to worm burden.

Diagnosis

Infection with *T. leptostoma* is often only made during histopathologic examination of the pancreas, and that with some difficulty [51]. Antemortem, embryonated eggs may be found in fecal preparations as early as 8–9 weeks, and for up to 2.5 years, postinfection [49], though the excretion of eggs is irregular in both timing and quantity [48]. Thus, fecal examinations, repeated at intervals of a few days, may be necessary to make the diagnosis [50]. Formalin–ethyl acetate sedimentation is the method of choice for fecal examination [50].

Differential Diagnosis

Differential diagnoses include other causes of pancreatitis and pancreatic insufficiency. In one report, trichospiruriasis was considered to be a cause of wasting marmoset disease, suggesting that the syndrome should be considered among the differential diagnoses [48].

Prevention and Control

Prevention of *T. leptostoma* depends on exclusion of the cockroach vector and treatment of infected animals [50]. Egg shedding has been greatly reduced with fenbendazole (50 mg/kg PO sid for 14 days). It could not be determined that worms had actually been killed, only that egg shedding ceased in most animals [50].

Research Complications

Infection of marmosets with *T. leptostoma* would compromise research involving the pancreatic and possibly the hepatobiliary systems. No publications are available on this topic.

ACANTHOCEPHALAN PARASITES

Marmosets have been found infected with several species of acanthocephalan parasites, including *Prosthenorchis elegans*, *P. lenti*, *P. sigmoides*, *P. spirula*, and others [4]. Based on literature reviews, it appears that among these, *Prosthenorchis elegans* is most commonly encountered and so will be discussed here [16].

Prosthenorchis elegans

Etiology

Prosthenorchis elegans inhabits the lower intestinal tract of many species of South and Central American primates, including marmosets [4,16,39,52]. Adult male *P. elegans* measure 20–30 mm in length, whereas adult female worms measure 30–50 mm long (Fig. 17.9). The eggs measure 42–53 μm wide and 65–81 μm long [16]. The German cockroach, *Blattella germanica*, serves as an intermediate host. Roaches become infected after ingestion of the parasite egg containing an acanthor larva and passed in the feces of the marmoset definitive host. Marmosets become infected when they ingest roaches harboring the infective cystacanth stage within its coelomic cavity [53].

Clinical Signs

Clinical signs and pathologic lesions are caused by the parasite embedding its spiny proboscis into the intestinal mucosa of the primate definitive host and may be proportional in severity to the worm burden. Clinical signs include watery diarrhea, anorexia, weakness, lethargy, and, if intestinal perforation occurs, death [16,53,54].

Epizootiology

Quite dramatic outbreaks of acanthocephaliasis have been initiated by the entry of infected, egg-shedding animals into a naïve population [53]. Because all acanthocephalans undergo an indirect life cycle, the presence

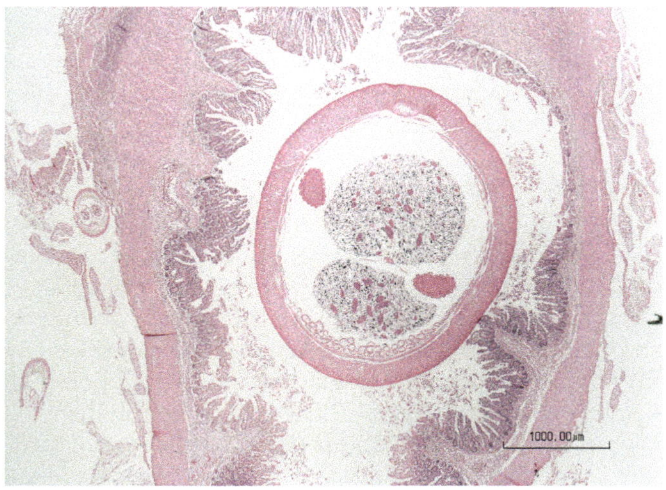

FIGURE 17.9 *Prosthenorchis elegans*. Cross-section of adult worm in the intestine of a common marmoset. An adult *Dipetalonema gracile* in cross-section can be seen intraperitoneally. 20× magnification.

and availability of the intermediate host is an essential component in establishing and maintaining the parasitic life cycle [52]. While the natural range of the parasite is South and Central America, Old World primates are similarly susceptible to infection and thus should be protected from spread of roach vectors from New World to Old World primate colonies.

Pathology

Lesions caused by *P. elegans* include distal small bowel enteritis, colitis, and typhlitis, and intestinal wall perforation with subsequent peritonitis. Grossly, abscesses and granulomata may be observed at the sites of parasite attachment, and white nodules may be seen on the serosal surface. On histologic section, the worm's proboscis can be observed embedded to the level of the submucosal tissue, with surrounding inflammation, ulceration, necrosis, and red blood cell extravasation. Affected animals may become anemic and develop hypoalbuminemia. Occasionally, intussusception or obstruction of the bowel occurs in the region of the ileocecal junction [16,52,53].

Diagnosis

Diagnosis of *P. elegans* infection is based on finding acanthocephalan eggs in the feces of diarrheic animals or adult worms embedded in the lower bowel mucosa. The latter may be found at necropsy or proctoscopically [2]. The infective cystacanth can also be found in the coelomic cavity of infected roach intermediate hosts, though this is rarely done and is not necessary to make a diagnosis [53]. If a fecal examination is performed, formal–ether sedimentation rather than fecal floatation is the method of choice [2].

Differential Diagnosis

Differential diagnosis of chronic diarrhea would include several possible conditions, both infectious and noninfectious. A complete review of causes of diarrhea is beyond the scope of this chapter.

Prevention and Control

Prevention is based on excluding parasite egg shedders from entering the animal facility. This is accomplished through careful quarantine and screening of new arrivals or, preferably, careful screening prior to shipment. One reported outbreak began with the entry into the animal facility of a single infected gibbon [53]. Control of an outbreak will depend on eliminating the cockroach vector population [53,54]. Anthelmintic treatment of acanthocephaliasis is usually unrewarding [53]. However, some success has been reported with oral carbon tetrachloride (0.5 mg/kg) or fenbendazole (20 mg/kg daily for 7 days) [2].

Research Complications

Infection of marmosets with *P. elegans* would render the animals unusable for nearly all types of research. Studies involving the gastrointestinal system would be particularly affected.

ARTHROPOD PARASITES

Reports of natural or experimental infestation of marmosets with arthropod parasites are surprisingly uncommon. However, as was noted for nematode parasitisms, those caring for and working with marmosets should be watchful for external parasitisms, which have been reported in other species of New World primates, as many of these may also, under proper conditions, infest marmosets. Many external parasites are not highly host specific, though they certainly exhibit host preferences, and thus may infest marmosets as conditions provide opportunity. For example, marmosets may be readily infested with the common dog and cat fleas, *Ctenocephalides canis* and *C. felis*, respectively. While these infestations are unlikely to be found in wild marmosets, it is not difficult to envision conditions in which these fleas could move from domestic animals to captive marmosets, especially if housed in outdoor enclosures. Infestations with *Ctenocephalides* sp. should be prevented and managed using approaches similar to those used for domestic animals. For the purposes of this chapter, general descriptions of infestation will be presented for major groups of external parasites, without, for the most part, direct and specific reference to infestations of marmosets.

Acariasis (Mites and Ticks)

Etiology

Many species of mites and very few species of ticks have been reported from New World primates [2]. Mites found on New World primates are in the suborders Trombidiformes or Sarcoptiformes. The trombidiformes are limited to the genus *Demodex*. In contrast, several genera are represented among the sarcoptiformes, including *Dunnalges*, *Rosalialges*, and *Sarcoptes* (Fig. 17.10) [2].

Clinical Signs and Epizootiology

Clinical signs of mite infestation are greatly influenced by whether the infesting parasite is a burrowing or nonburrowing species. As with other hosts, infestation of marmosets with *Sarcoptes scabiei* may cause severe clinical signs, including intense pruritus, anorexia, weakness, and emaciation [2]. It is likely that those individuals capable of mounting a more robust

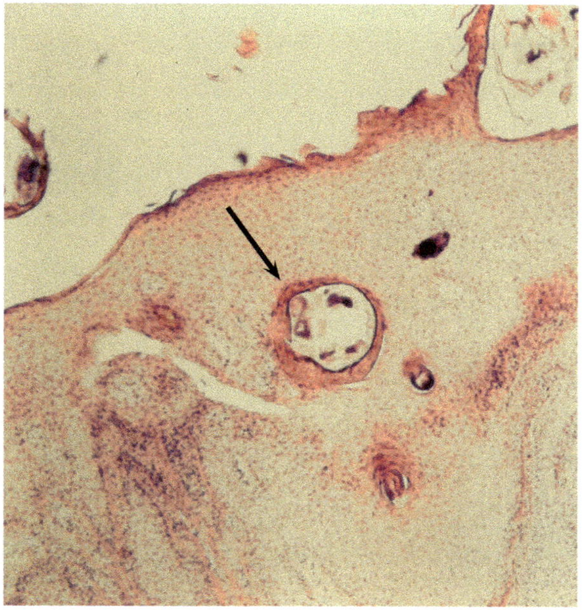

FIGURE 17.10 *Sarcoptes scabiei.* A section of an adult mite (*arrow*) can be seen in the hyperplastic epidermis of a nonhuman primate.

type I hypersensitivity reaction would show the worst signs and harbor the fewest mites, though this has not been reported. In contrast, clinical signs of infestation with nonburrowing species are uncommon. Tick infestations are usually asymptomatic in marmosets and are unlikely to be of concern in captive colonies [2].

Pathology

Mite and tick infestation may lead to dermatologic disease, including hyperkeratosis, anemia, and hypersensitivity [2]. There are no reports of disease transmission to marmosets by mites or ticks.

Diagnosis, Prevention, and Control

Diagnosis of mite or tick infestation is based on visual inspection and the examination of skin scrapings using standard procedures as described for other mammals. Marmosets should be protected from mite and tick infestation through physical barriers to entry and by exclusion of other parasitized primates from close proximity. Infested marmosets can be treated with acaricides commonly used on other veterinary species.

Research Complications

Infestation of marmosets with mites and ticks could compromise research involving the skin and immune system and could compromise research requiring animals exhibiting normal behavior.

Pentastomes

Porocephalus crotali

Etiology

New World primates representing many species, including marmosets, are susceptible to infection with the pentastome *Porocephalus crotali*. Primates, and less commonly other mammals, may serve as intermediate hosts. Adult pentastomes are found in the respiratory tracts of various snakes, which serve as definitive hosts [16]. Parasite eggs containing larvae are passed in the feces of the definitive host. Following ingestion, eggs hatch, releasing larvae, which penetrate the intestinal wall and migrate to several locations within the host, but most commonly to the mesenteric tissues, where they undergo several ecdyses [55], and encyst as the nymphal stage, which is infective to the definitive host.

Clinical Signs, Epizootiology, and Pathology

Clinical signs are usually not observed, though penetration of the intestine by large numbers of larvae may cause fatal peritonitis. Live, encysted nymphs elicit minimal inflammation and so are relatively silent in the intermediate host. In contrast, dead nymphs elicit a strong inflammatory response and can therefore induce nonspecific signs of malaise. Wild-caught monkeys are commonly infected. The replacement of wild-caught, with captive-bred monkeys has virtually eliminated this parasitism from research colonies.

Diagnosis, Prevention, and Control

Diagnosis of *P. crotali* infection is typically made at necropsy or during abdominal surgery. Infections in captive populations are easily prevented by exclusion of the snake definitive host. An effective treatment regimen to eliminate the nymphs has not been reported. However, the common use of ivermectin in captive monkeys may have contributed to the reduction in prevalence of this parasitism as ivermectin has been found to be effective against other pentastomes [56]. While *P. crotali* is a known zoonosis, persons can only become infected by consuming the parasite eggs excreted in the feces of the snake definitive host. Therefore, infected marmosets would not pose a transmission threat to humans.

Research Complications

The presence of nymphal stages in the mesenteric tissues of marmosets is not likely to interfere with most types of research involving monkeys.

References

[1] Ludlage E, Mansfield K. Clinical care and diseases of the common marmoset (*Callithrix jacchus*). Comp Med 2003;53:369–82.

[2] Strait K, Else JG, Eberhard ML. Parasitic diseases of nonhuman primates. In: Abee CR, Mansfield K, Tardif S, Morris T, editors. Nonhuman primates in biomedical research: diseases. 2nd ed., vol. II. Academic Press; 2012. p. 197–297.

[3] Dienhardt F, Holmes AW, Devine J, Deinhardt J. Marmosets as laboratory animals. IV. The microbiology of laboratory kept marmosets. Lab Anim Care 1967;17:48–70.

[4] Pokkay S. Diseases of the callitrichidae: a review. J Med Primatol 1992;21:189–236.

[5] Kalishman J, Paul-Murphy J, Scheffler J, Thomson JA. Survey of *Cryptosporidium* and *Giardia* spp. in a captive population of common marmosets. Lab Anim Sci 1996;46:116–9.

[6] Levine ND. Veterinary protozoology. Iowa State University Press; 1985.

[7] Kramer JA, Hachey AM, Wachtman LM, Mansfield KG. Treatment of giardiasis in common marmosets (*Callithrix jacchus*) with tinidizole. Comp Med 2009;59:174–9.

[8] Levecke B, Dorny P, Geurden T, Vercammen F, Vercruysse J. Gastrointestinal protozoa in non-human primates of four zoological gardens in Belgium. Vet Parasitol 2007;148:236–46.

[9] Ruch TC. Diseases of laboratory primates. W.B. Saunders Company; 1959. p. 140–2.

[10] Cotton JA, Beatty JK, Buret AG. Host parasite interactions and pathophysiology in *Giardia* infections. Int J Parasitol 2011;41: 925–33.

[11] DiGenova BM, Tonelli RR. Infection strategies of intestinal parasite pathogens and host cell responses. Front Microbiol 2016; 7(256):1–16.

[12] Laude A, Valot S, Desoubeaux G, Argy N, Nourrisson C, Pomares C, Machouart M, Le Govic Y, Dalle F, Botterel F, Bourgeois N, Cateau E, Leterrier M, Le Pape P, Levecke B, Dreesen L, Dorny P, Verweij JJ, Vercammen F, Casaert S, Vercruysse J, Geldhof P. Molecular identification of *Entamoeba* spp. in captive nonhuman primates. J Clin Microbiol 2010;48: 2988–90.

[13] Hird D, Anderson JH, Bielitzki JT. Diarrhea in nonhuman primates: a survey of primate colonies for incidence rates and clinical opinion. Lab Anim Sci 1984;34:465–70.

[14] Baker DG. Pathogens of rats and mice. In: Baker DG, editor. Natural pathogens of laboratory animals: their effects on research. ASM Press; 2003.

[15] Inoue T, Hayashimoto N, Yasuda M, Sasaki E, Itoh T. *Pentatrichomonas hominis* in laboratory-bred common marmosets. Exp Anim 2015;64:363–8.

[16] Cogswell F. Parasites of non-human primates. In: Baker DG, editor. Flynn's parasites of laboratory animals. 2nd ed. Blackwell Publishing; 2007.

[16a] Ribeiro RD, Barretto MP, BeldaNeto FM, Correa ST, FerriolliFilho F. Reservoirs and wild vectors of Trypanosomacruzi. LXXV: Natural T. cruzi infection of the monkey Callithrixjacchus (Linnaeus, 1758). Rev Bras Biol 1980;40(1):51–8.

[16b] Minuzzi-Souza TT, Nitz N, Knox MB, Reis F, Hagström L, Cuba CA, Hecht MM, Gurgel-Gonçalves R. Vector-borne transmission of Trypanasoma cruzi among captive Neotropical primates in a Brazilian Zoo. Parasit Vectors 2016 Jan 26;9(39). https://doi.org/10.1186/s13071-016-1334-7.

[17] Lushbaugh CC, Humason G, Gengozian N. Intra-uterine death from congenital Chagas' disease in laboratory-bred marmosets (Saguinus fuscicollis langonatus). Am J Trop Med Hyg 1969; 18(5):662–5.

[17a] Dunn FL, Lambrecht FL, du Plessis R. Trypanosomes of South American monkeys and marmosets. Am J Trop Med Hyg 1963; 12:524–34.

[18] Ziccardi M, Lourenço-De-Oliveira R, Brigido MC, Muniz JA. Trypanosomes of non-human primates from the National Centre of primates, Ananindeua, State of Pará, Brazil. Memórias do Instituto Oswaldo Cruz 2000;95:157–9.

[19] Regan CS, Yon L, Hossain M, Elsheikha HM. Prevalence of *Entamoeba* species in captive primates in zoological gardens in the UK. Peer J 2014;2:e492.

[20] Márquez-Monter H, Fuentes-Orozco R, Correa-Lemas I, Becker I. Invasive amebiasis in a spider monkey (*Ateles geoffroyi*). Case report and a short review of the literature of amebiasis in nonhuman primates. Arch Investig Med (Mexico) 1991;22:75–8.

[21] Verweij JJ, Vermeer J, Brienen EA, Blotkamp C, Laeijendecker D, van Lieshout L, Polderman AM. *Entamoeba histolytica* infections in captive primates. Parasitol Res 2003;90:100–3.

[22] Schuster FL, Visvesvara GS. Amebae and ciliated protozoa as causal agents of waterborne zoonotic disease. Vet Parasitol 2004;126:91–120.

[22a] Levecke B, Dreesen L, Dorny P, Verweij JJ, Vercammen F, Casaert S, Vercruysse J, Geldhof P. Molecular identification of *Entamoeba* spp. in captive nonhuman primates. J Clin Microbiol 2010;48(8):2988–90. https://doi.org/10.1128/JCM.00013-10. Epub 2010 Jun 23.

[23] Lindsey DS, Dubey JP, Blagburn BL. Biology of *Isospora* spp. from humans, nonhuman primates, and domestic animals. Clin Microbiol Rev 1997;10:19–34.

[24] Miller RA, Bronsdon MA, Kuller L, Morton WR. Lab Anim Sci 1990;40:42–6.

[25] Wilson DW, Day PA, Brummer MEG. Diarrhea associated with *Cryptosporidium* spp. in juvenile macaques. Vet Pathol 1984;21: 447–50.

[26] Kovatch RM, White JD. Cryptosporidiosis in two juvenile rhesus monkeys. Vet Pathol 1972;9:426–40.

[27] Hahn NE, Capuano III SV. Successful treatment of cryptosporidiosis in two common marmosets (*Callithrix jacchus*) by using paromomycin. J Assoc Lab Anim Sci 2010;49:873–5.

[28] Pinto HA, Mati VL, Pujoni DF, de Melo AL. *Platynosomum illiciens* (Trematoda: dicrocoeliidae) in captive black-tufted marmoset *Callithrix pencillata* (Primates: Cebidae) from Brazil. A morphometric analysis with taxonomic comments of species of *Platynosomum* from nonhuman primates. J Parasitol 2016. https://doi.org/10.1645/16-1.

[29] Silva KSM, Silva RJ, Pereira WLA. Occurrence of infection by *Platynosomum illiciens* (Braun, 1901) in captive neotropical primates. Primates 2012;53:79–82.

[30] Pinto HA, Mati VL, de Melo AL. New insights into the life cycle of *Platynosomum* (Trematoda: Dicrocoeliidae). Parasitol Res 2014; 113:2701–7.

[31] Mattioli MP, Batista J, dos S, Ferrari M, Paludo GR, Dias CA, Hoppe EG, Monteiro RV. Clinical, hematological, biochemical, and ultrasonographic aspects of *Platynosomum* sp. (Trematoda: Dicrocoeliidae) infection of captive *Callithrix pencillata*. Primates 2016;57:279–87.

[32] Sousa MB, Leão AC, Coutinho JF, de Oliveira Ramos AM. Histopathology findings in common marmosets (*Callithrix jacchus* Linnaeus, 1758) with chronic weight loss associated with bile tract obstruction by infestation with *Platynosomum* (Loos, 1907). Primates 2008;49:283–7.

[33] Mati VL, Pinto HA, de Melo AL. Evaluation of Kato-Katz and spontaneous sedimentation methods for diagnosis of platynosomiasis in neotropical primates. Braz J Vet Parasitol 2015;24: 108–13.

[34] Brack M. Gongylonemiasis in the common marmoset (*Callithrix jacchus*). Lab Anim Sci 1996;46:266–70.
[35] Brack M, Schröpel M. *Angiostrongylus costaricensis* in a black-eared marmoset. Trop Geogr Med 1995;47:136–8.
[36] dos Santos Sales I, Ruiz-Miranda CR, de Paula Santos C. Helminths found in marmosets (*Callithrix pencillata* and *Callithrix jacchus*) introduced to the region of occurrence of golden lion tamarins (*Leontopithicus rosalia*) in Brazil. Vet Parasitol 2010;171:123–9.
[37] Inglis WG. The oxyurid parasites (Nematoda) of primates. J Zool 1960;136:103–22.
[38] Sandosham AA. On *Enterobius vermicularis* (Linnaeus, 1758) and some related species from primates and rodents. J Helminthol 1950;24:171–204.
[39] Tavela Ade O, Fuzessy LF, e Silva VH, da Silva Fde F FR, Junior MC, Silva Ide O, Souza VB. Helminths of wild hybrid marmosets (*Callithrix* sp.) living in an environment with high human activity. Braz J Vet Parasitol 2013;22:391–7.
[40] Notarnicola J, Agustín Jiménez F, Gardner SL. A new species of *Dipetalonema* (Filaroidea: Onchocercidae) from *Ateles chamek* from the Beni of Bolivia. J Parasitol 2007;93:661–7.
[41] Webber WAF. The filarial parasites of primates: a review. I. *Dirofilaria* and *Dipetalonema*. Ann Trop Med Parasitol 1955;49:123–41.
[42] Baker DG, Babcock Jr RE. Diagnostic exercise: abdominal parasites in a spider monkey. Lab Anim Sci 1996;46:338–40.
[43] Webber WAF, Hawking F. The filarial worms *Dipetalonema digitatum* and *D. gracile* in monkeys. Parasitology 1955;45:401–8.
[44] Travi BL, Eberhard ML, Lowrie Jr RC. Development of *Dipetalonema gracile* in the squirrel monkey (*Saimiri sciureus*), with notes on its biology. J Parasitol 1985;71:17–9.
[45] Eberhard ML. Chemotherapy of filariasis in squirrel monkeys (*Saimiri sciureus*). Lab Anim Sci 1982;32:397–400.
[46] Chalmers DT, Murgatroyd LB, Wadsworth PF. A survey of the pathology of marmosets (*Callithrix jaccus*) derived from a marmoset breeding unit. Lab Anim 1983;17:270–9.
[47] de Resende DM, Pereira LH, de Melo AL, Tafuri WL, Moreira NI, de Oliveira CL. Parasitism by *Primasubulura jacchi* (Marcel, 1857) Inglis, 1958 and *Trichospirura leptostoma* Smith and Chitwood, 1967 in *Callithrix penicillata* marmosets, trapped in the wild environment and maintained in captivity. Memórias do Instituto Oswaldo Cruz 1997;89:123–5.
[48] Pfister R, Heider K, Illgen B, Beglinger R. *Trichospirura leptostoma*: a possible cause of wasting disease in the marmoset. Z Veruchstierkunde 1990;33:157–66.
[49] Illgen-Wilcke B, Beglinger R, Pfister R, Heider K. Studies on the developmental cycle of *Trichospirura leptostoma* (Nematoda: Thelaziidae). Experimental infection of the intermediate hosts *Blattella germanica* and *Supella longipalpa* and the definitive host *Callithrix jacchus* and development in the intermediate hosts. Parasitol Res 1992;78:509–12.
[50] Hawkins JV, Clapp NK, Carson RL, Henke MA, McCracken MD, Faulkner CT, Patton S. Diagnosis and treatment of *Trichospirura leptostoma* infection in common marmosets (*Callithrix jacchus*). Contemp Top Lab Anim Med 1997;36:52–5.
[51] Cosgrove GE, Humason G, Lushbaugh CC. *Trichospirura leptostoma*, a nematode of the pancreatic ducts of marmosets (*Saguinus* spp.). J Am Vet Med Assoc 1970;157:696–8.
[52] Catenacci LS, Colosio AC, Oliveira LC, De Vleeschouwer KM, Munhoz AD, Deem SL, Pinto JMS. Occurrence of *Prosthorchis elegans* in free-living primates from the Atlantic forest of Southern Bahia, Brazil. J Wildl Dis 2016;52:365–8.
[53] Moore JG. Epizootic of acanthocephaliasis among primates. J Am Vet Med Assoc 1970;157:699–705.
[54] Benirschke K. Diagnostic exercise. Lab Anim Sci 1979;29:33–4.
[55] Buckle AC, Riley J, Hill GF. The *in vitro* development of the pentastomid *Porocephalus crotali* from the infective instar to the adult stage. Parasitology 1997;115:503–12.
[56] Haugerud RE, Nilssen AC, Rognmo A. On the efficacy of ivermectin against the reindeer sinus worm *Linguatula arctica* (Pentastomida), with a review on ivermectin treatment in reindeer. Rangifer 2010;13:157–62.

CHAPTER

18

Neoplastic Diseases

Andrew D. Miller

Department of Biomedical Sciences, Section of Anatomic Pathology, Cornell University College of Veterinary Medicine, Ithaca, NY, United States

INTRODUCTION

While morbidity and mortality associated with cancer are commonplace in many animal species, the relatively recent rise in the use of the common marmoset in biomedical research has limited our knowledge regarding neoplastic and proliferative diseases of this species compared with other nonhuman primates. Most of the knowledge regarding neoplasia in the common marmoset is derived from short case reports or retrospective studies. The goal of this chapter will be to introduce the reader to basic cancer terminology and then review the known proliferative and neoplastic conditions of the common marmoset.

The English language is replete with terms to describe proliferative conditions from neoplasia, cancer, and tumor. In this chapter, we will use the term neoplasia (new growth) or cancer, rather than the often misused term tumor, which based on its Latin derivation, simply means swelling. When a clinician or pathologist encounters a neoplasm in an animal, various clinical parameters can be assessed prior to pathologic evaluation. While little is published regarding the clinical management of neoplasia in the common marmoset, diagnostic and treatment plans are similar to those described in other species including physical examination, complete blood count, serum chemistry, and urinalysis, radiographs, and ultrasound. Routine assessment of body weight is especially useful as it can aid in recognizing dramatic weight loss associated with cancer cachexia, a progressive debilitation associated with tumor burden and tumor-elaborated cytokines.

Following clinical work-up, if the neoplasm is readily accessible, both aspiration for cytologic examination and biopsy for histologic assessment should be performed. Samples submitted for histopathology are submitted to a reference laboratory in 10% neutral buffered formalin where they will be routinely processed and evaluated by a board-certified pathologist. The clinical management of neoplasia is often dependent on the diagnosis, and a strong working relationship between the clinician and pathologist will aid in the development of an appropriate treatment plan based on the type of cancer diagnosed.

A basic understanding of the terminology used in a pathology report is necessary to fully interpret the findings reported by the pathologist. The field of oncology and pathology is filled with many overlapping terms and the reader is referred to the *Nonhuman Primates in Biomedical Research: Diseases* text for a more in-depth discussion [1]. Herein we will review the basic terms that a clinician should be familiar with when diagnosing and treating cancer in the common marmoset. There are a number of lesions that can mimic neoplasia grossly including hamartomas (abnormal amounts of tissue growing in a normal location), choristomas (normal tissue growing in an abnormal location), and cysts (a cavity or sac lined by epithelium, which may be follicular, dermoid, epidermoid, etc.). Differentiation of these lesions clinically is often not possible and histopathology is necessary for a definitive diagnosis. The differentiation of benign and malignant is made through a combination of observations made through the clinical examination (Is there lymphadenomegaly? Do thoracic radiographs illustrate metastasis?) and the histopathologic examination. Histologically, benign neoplasms are well-demarcated, noninvasive lesions that are usually treated with wide surgical excision, which, depending on the organ, is often curative. Benign neoplasia is typically defined by the suffix—oma. For example, a benign tumor of bone is an osteoma, whereas a benign epithelial tumor would be defined by the location (e.g., bronchial adenoma). In contrast, a malignant neoplasm would include the suffix carcinoma (epithelial origin) or sarcoma (mesenchymal origin). For many epithelial neoplasms, there is often a temporal progression from dysplasia, to carcinoma in situ, to invasive carcinoma

The Common Marmoset in Captivity and Biomedical Research
https://doi.org/10.1016/B978-0-12-811829-0.00018-2

as defined by breaching the basement membrane. While it is seemingly simple to define all cancers based on this terminology, the reality is that over time different verbiage has been defined for cancer types that do not fit into the classic carcinoma/sarcoma classification scheme. For example, a neoplasm of the ovary that contains multiple cell types derived from mesoderm, endoderm, and ectoderm is referred to as a teratoma, which can further be defined as benign or malignant.

Various histologic criteria must be assessed when differentiating benign and malignant neoplasia. One key criterion is the degree of differentiation within the neoplasm. Typically, the more malignant the neoplasm, the less differentiated (or more anaplastic) it appears histologically. The increasing anaplasia within a neoplasm often correlates with other histomorphologic changes, which include pleomorphism (which can be further characterized by cyto/karyomegaly and/or multinucleate neoplastic cells), increased numbers of and often bizarre mitotic figures, and loss of epithelial polarization away from a basement membrane. Although increasing evidence of anaplasia is most indicative of a malignant neoplasm, some benign neoplasms that lack anaplasia can have sinister effects on organ function and patient survival solely based on the location of the neoplasm (e.g., meningiomas are often slow-growing, relatively benign neoplasms that can cause significant clinical disease due to compression of the brain).

All neoplasms grow at different rates; however, typically the more malignant and anaplastic the neoplasm, the faster the growth rate. Cancer growth is governed by how fast the cells replicate (i.e., what is their doubling time?), the rate at which neoplastic cells undergo necrosis/apoptosis, and the number of cells that are mitotically active at any one time. The number of cells that are actively "in the cell cycle" and dividing is an important consideration when treating cancer. The higher the number of actively dividing cells, the better the response to chemotherapy [2]. The growth rate can also be linked to local and distant spread. Local invasion is defined by the localized intrusion of neoplastic cells into surrounding tissue. This is a feature of malignant neoplasm because benign neoplasms typically grow by compressing, not invading, adjacent tissues. While localized invasion is indicative of a malignant neoplasm, the most reliable feature that is used to differentiate benign from malignant neoplasia is metastasis. Metastasis indicates that neoplastic cells have seeded to a location distant from where they originated. This includes seeding of body cavities (either through carcinomatosis or sarcomatosis), invasion into lymphatics, and hematogenous spread. It must be noted that simple lymph node enlargement close to the offending neoplasm does not always indicate metastatic disease. Reactive change secondary to tumor necrosis and draining can cause lymphadenomegaly that will mimic that caused by metastasis. Biopsy or lymph node excision can be used to determine the presence or absence of metastasis. While it would seem that once a neoplasm has entered the blood supply, it would naturally deposit in a stereotypical way into certain organs (such as the lungs or liver); it is clear that at least in human pathology, certain neoplasms have a propensity to metastasize to specific organs preferentially (i.e., prostate carcinoma to bone). Whether the same is true for common marmoset neoplasms is unclear.

The development of cancer involves the complex interplay of numerous mutations in oncogenes that can be growth factors and their receptors, signal-transducing proteins, and transcription factors. Mutations in these oncogenes can occur de novo or can be due to chemical, microbial (viruses and/or bacteria), and radiation carcinogenesis. Very little is known regarding the role that oncogenes have in spontaneous neoplasia in the common marmoset; however, with the increasing use of the common marmoset as a lab animal model, there is ample opportunity to develop novel nonhuman primate models of cancer that are directly applicable to human cancers. Over time, only robust epidemiologic studies and detailed investigative research reports will build our knowledge base of neoplastic diseases of the common marmoset.

Microbial Carcinogenesis

Unlike other nonhuman primate species, where there are numerous viruses known to cause oncogenic transformation, there are only rare examples of viruses causing neoplasia in the common marmoset. Whether this is actually due to a low incidence of microbial carcinogenesis in the common marmoset or because of pathogens yet to be discovered or linked to carcinogenesis remains to be determined. Experimentally, there are several examples in which inoculation with viruses from other nonhuman primate species leads to neoplastic transformation in the marmoset.

Historically, the most widely studied experimental model of neoplasia in common marmosets is the inoculation of *Herpesvirus saimiri* (Saimiriine herpesvirus 1), which results in leukemia and lymphoma. *H. saimiri* is a virus endemic in squirrel monkeys, where disease is typically inapparent; however, direct inoculation into common marmosets is rapidly fatal due to neoplastic transformation [3]. In experimentally infected animals, lymphoma is widespread, affecting many organs including the liver, spleen, and adrenal glands [3,4]. Hematologic changes range from lymphocytosis to leukemia in affected animals [3]. Similarly, *Herpesvirus ateles*, which is endemic in spider monkeys, also causes leukemia and lymphoma when inoculated in common marmosets [5]. Although not typically relevant to the standard modern biomedical research facility, due to the potential for neoplastic transformation when viruses are transferred between species, housing of these nonhuman primates should remain strictly separate.

Other experimental studies have involved infection of common marmosets with Epstein–Barr virus (EBV). Early research indicated that inoculated common marmosets developed a diffuse lymphoproliferative disorder; however, although animals develop an antibody response, transformation to lymphoma was not typical [6,7]. A virus related to EBV, *Callitrichine herpesvirus 3* has been recovered from common marmoset lymphomas. While seroprevalence for CHV3 varies from 35% to 65%, the incidence of lymphoma in common marmoset colonies is varied and generally low [8–10]. Affected animals had multiorgan involvement with lymphoma and neoplastic cells were B cell in origin [10]. The variable seroprevalence may be partly explained by the lack of a significant number of genes in the CHV3 genome when compared with the EBV genome [11]. These lost genes likely explain the variability in transmission and disease development [11]. Sporadic studies have looked for CHV3 genome variations in other spontaneous neoplasms in the common marmoset; however, no other associations between virus and neoplasm development have been reported [12].

Lastly, common marmosets have been used as a model for Kaposi's sarcoma-associated herpesvirus infection [13]. Inoculated animals seroconvert, develop a marked antibody response, and virus can be readily detected in circulating leukocytes. A single animal was reported to develop an atypical mesenchymal proliferation in the skin following inoculation with Kaposi's sarcoma-associated herpesvirus [13]. Due to their small size, high incidence of multiple births, and relative ease of handling, using common marmosets as a model for oncogenic viruses will likely continue in future years.

ORGAN-SPECIFIC NEOPLASIA

Endocrine Neoplasia

As opposed to other New World nonhuman primates, endocrine neoplasia is relatively uncommon in the common marmoset. Thyroid adenomas have been rarely reported and as common marmosets age and proliferative foci of the thyroid follicular cells and C cells can be observed Thyroid adenomas have been rarely reported. As common marmosets age, proliferative foci of the thyroid follicular cells and C cells can be observed [14–16]. Myelolipomas of the adrenal gland are seen uncommonly, typically as an incidental finding [17].

Neoplasia of the Urogenital Tract

Scattered reports of neoplasia of the urogenital tract are present in the literature. There is a single case report of a malignant nephroblastoma arising in the kidney of a 17-month-old common marmoset [18]. This neoplasm had the typical histologic features of primitive renal tissue associated with abundant connective tissue and primitive mesenchyme [18]. Neoplasia of the testis and ovary are similarly uncommon, with only a single report, in a 3-year old common marmoset, of an ovarian teratoma that had cellular differentiation to all three germ cell layers [19].

Integumentary Neoplasia

Neoplasia of the skin is incredibly rare in the common marmoset. There is a single report of a paratrichial (apocrine) sweat gland adenocarcinoma in an 8-year-old male. This neoplasm arose in the right thorax and was composed of lobules of neoplastic epithelial cells forming tubules and cysts [20]. A second, brief report of apocrine gland neoplasia is also present in the literature; however, it is not described in detail [16].

Neoplasia of the Gastrointestinal Tract

The gastrointestinal tract is the most common location for neoplasia in the common marmoset. In the 1980s, a number of reports were produced that described oral and nasopharyngeal carcinomas. Some of these had morphologic features of squamous cell carcinoma and, interestingly, in the largest report, all animals were clustered in the same colony. In these cases, some of the neoplasms arose near salivary glands, whereas others arose from the oral or pharyngeal tissue. The cause of these nasopharyngeal squamous cell carcinomas is unknown, as no cases have been reported in the literature since this cluster in the 1980s. Whether this cluster was due to the effect of an aged colony or another outside factor (i.e., oncogenic virus) is not known [21,22].

Another common gastrointestinal neoplasm of the common marmoset is small intestinal adenocarcinoma [12,23,24]. These neoplasms commonly arise in the proximal small intestine where they begin as carcinoma in situ before becoming transmurally invasive (Fig. 18.1). A single report of an ileal neoplasm of similar histologic appearance is also reported [23]. While carcinomatosis is not typically observed, spread to local lymph nodes is reported [12]. Histologically, signet ring cells are a common feature and in some cases deposition of mucin is also a common feature (Fig. 18.2A and B). Attempts to determine if callitrichid herpesvirus 3 was present in higher amounts in the neoplasm

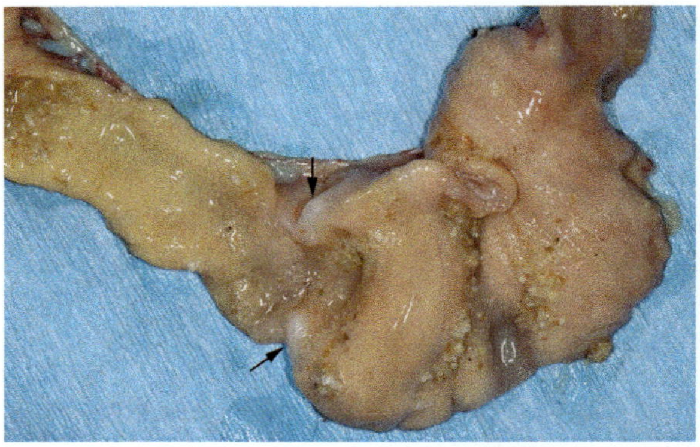

FIGURE 18.1 **Proximal small intestinal carcinoma in a common marmoset (*Callithrix jacchus*).** A *sharp line* of demarcation delineates the site of the carcinoma with distended proximal intestine. At the site of neoplasm the wall is thickened and firm (*arrows*).

compared with other unaffected tissue failed to reveal a causative role for this virus in tumorigenesis [12].

Lastly, there is a single report of a pancreatic adenoma in an aged common marmoset [14].

Musculoskeletal Neoplasia

A single case report exists of an approximately 3-year-old common marmoset that was euthanized for hindlimb paralysis due to the presence of a large, invasive mass in the caudal back muscles. Histologically, this neoplasm had histologic features of a rhabdomyosarcoma and was positive immunohistochemically for desmin, vimentin, and myogenin [25]. No reports of spontaneous neoplasia of the skeletal system exist for common marmosets.

Hematopoietic Neoplasia

T-cell lymphoma, unassociated with an oncogenic viral infection, has been reported in a young male

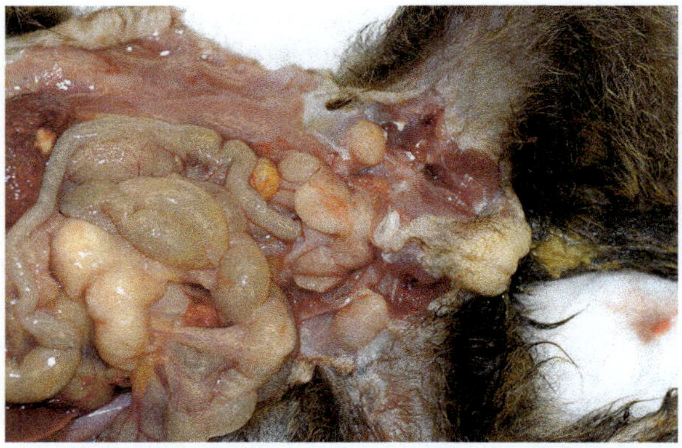

FIGURE 18.3 **Lymphoma in a common marmoset (*Callithrix jacchus*).** The mesenteric and inguinal lymph nodes are enlarged and bulge on the cut section consistent with lymphoma. *Photo courtesy of the Southwest National Primate Center.*

common marmoset that presented with predominately nodal involvement [26]. One large review of common marmoset pathology reports that lymphoma is the most common neoplasm in one colony [14] (Fig. 18.3). Recently, a case of myeloid sarcoma arising in the mediastinum of a 9-month-old common marmoset was reported [27]. This neoplasm extensively replaced the thorax and histologically invaded the heart in addition to replacing the thymus [27]. The neoplastic cells had immunohistochemical features consistent with a hematopoietic origin. The reader is referred to the above microbial carcinogenesis section for a deeper discussion of the virally associated lymphoma incidence in common marmosets.

Cardiovascular and Respiratory Neoplasia

Respiratory neoplasia is uncommon; however, some of the squamous cell carcinomas described above in the gastrointestinal section arose in or invaded the nasal tissue and could have arisen there as a primary site.

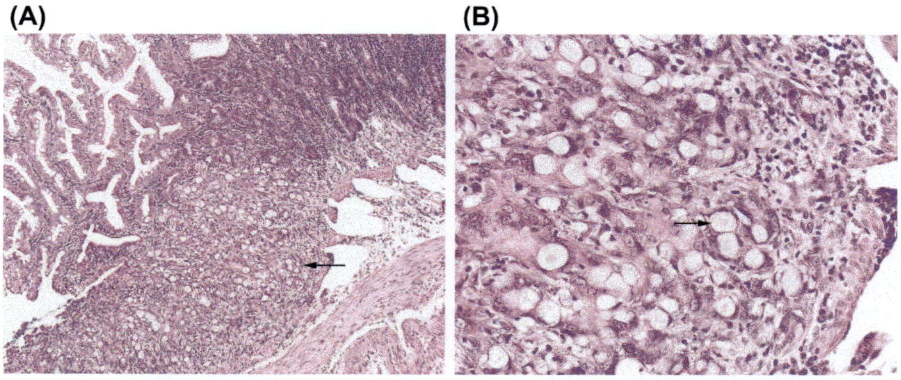

FIGURE 18.2 **Proximal small intestinal carcinoma in a common marmoset (*Callithrix jacchus*).** (A) Neoplastic epithelial cells replace the normal crypts and form vague acinar structures and sheets (*arrow*). (B) A higher magnification illustrates the signet ring cell morphology of the neoplastic cells (*arrow*).

A single report of three common marmosets with undifferentiated carcinomas of the nasal cavity describes a neoplasm that lacked the production of keratin. In this report, all three animals had marked invasion by the neoplasm with metastasis [28]. Lastly, a single bronchial adenoma has been reported in a 2-year-old common marmoset [29].

The organ systems listed above are those in which spontaneous neoplasia reports exist. Neoplasia of the nervous system is not reported in the literature. It must be stressed that neoplasia in common marmosets is relatively uncommon compared with some other larger nonhuman primates. Whether this is due to their relatively short life span, genetic factors, or other, as yet undefined factors, remains to be determined. As common marmosets continue to gain interest for biomedical and aging research, additional reports of neoplasia are certain to come and the field of common marmoset oncology will continue to grow.

References

[1] Miller AD. Nonhuman primates in biomedical research volume 2: Diseases. London: Elsevier; 2012.

[2] Newkirk KM, Kusewitt DF. Pathologic basis of veterinary disease. 6th ed. St. Louis, MO: Elsevier; 2017.

[3] Hunt RD, Melendez LV, King NW, Gilmore CE, Daniel MD, Williamson ME, Jones TC. Morphology of a disease with features of malignant lymphoma in marmosets and owl monkeys inoculated with Herpesvirus saimiri. J Natl Cancer Inst 1970;44:447–65.

[4] Wright J, Falk LA, Wolfe LG, Ogden J, Deinhardt F. Susceptibility of common marmosets (*Callithrix jacchus*) to oncogenic and attenuated strains of Herpesvirus saimiri. J Natl Cancer Inst 1977;59: 1475–8.

[5] Laufs R, Steinke H, Steinke G, Petzold D. Latent infection and malignant lymphoma in marmosets (*Callithrix jacchus*) after infection with two oncogenic herpesviruses from primates. J Natl Cancer Inst 1974;53:195–9.

[6] Emini EA, Luka J, Armstrong ME, Banker FS, Provost PJ, Pearson GR. Establishment and characterization of a chronic infectious mononucleosis-like syndrome in common marmosets. J Med Virol 1986;18:369–79.

[7] Falk L, Deinhardt F, Wolfe L, Johnson D, Hilgers J, de-The G. Epstein-Barr virus: experimental infection of *Callithrix jacchus* marmosets. Int J Cancer 1976;17:785–8.

[8] Cho Y, Ramer J, Rivailler P, Quink C, Garber RL, Beier DR, Wang F. An Epstein-Barr-related herpesvirus from marmoset lymphomas. Proc Natl Acad Sci USA 2001;98:1224–9.

[9] Fogg MH, Carville A, Cameron J, Quink C, Wang F. Reduced prevalence of Epstein-Barr virus-related lymphocryptovirus infection in sera from a new world primate. J Virol 2005;79: 10069–72.

[10] Ramer JC, Garber RL, Steele KE, Boyson JF, O'Rourke C, Thomson JA. Fatal lymphoproliferative disease associated with a novel gammaherpesvirus in a captive population of common marmosets. Comp Med 2000;50:59–68.

[11] Rivailler P, Cho YG, Wang F. Complete genomic sequence of an Epstein-Barr virus-related herpesvirus naturally infecting a new world primate: a defining point in the evolution of oncogenic lymphocryptoviruses. J Virol 2002;76:12055–68.

[12] Miller AD, Kramer JA, Lin KC, Knight H, Martinot A, Mansfield KG. Small intestinal adenocarcinoma in common marmosets (*Callithrix jacchus*). Vet Pathol 2010;47:969–76.

[13] Chang H, Wachtman LM, Pearson CB, Lee JS, Lee HR, Lee SH, Vieira J, Mansfield KG, Jung JU. Non-human primate model of Kaposi's sarcoma-associated herpesvirus infection. PLoS Pathog 2009;5:e1000606.

[14] David JM, Dick Jr EJ, Hubbard GB. Spontaneous pathology of the common marmoset (*Callithrix jacchus*) and tamarins (*Saguinus oedipus, Saguinus mystax*). J Med Primatol 2009;38:347–59.

[15] Kawasako K, Doi T, Kanno T, Wako Y, Hamamura M, Tsuchitani M. Thyroid follicular adenoma with accumulation of collagen type IV in a common marmoset (*Callithrix jacchus*). J Comp Pathol 2014;150:71–4.

[16] Okazaki Y, Kurata Y, Makinodan F, Kidachi F, Yokoyama M, Wako Y, Yamagishi Y, Katsuta O, Takechi M, Tsuchitani M. Spontaneous lesions detected in the common cotton-eared marmosets (*Callithrix jacchus*). J Vet Med Sci 1996;58:181–90.

[17] Kakinuma C, Harada T, Watanabe M, Shibutani Y. Spontaneous adrenal and hepatic myelolipomas in the common marmoset. Toxicol Pathol 1994;22:440–5.

[18] Zoller M, Matz-Rensing K, Fahrion A, Kaup FJ. Malignant nephroblastoma in a common marmoset (*Callithrix jacchus*). Vet Pathol 2008;45:80–4.

[19] Haworth R, Jones S, Sanchez-Morgado J, Pilling A. Ovarian teratoma in a common marmoset (*Callithrix jacchus*). Vet Rec 2003;153: 332–3.

[20] Khan KN, Silverman L, Logan A, Harris RK. Paratrichial sweat gland adenocarcinoma in a marmoset. J Vet Diagn Investig 1999; 11:478–80.

[21] Betton GR. Spontaneous neoplasms of the marmoset (*Callithrix jacchus*). Oral and nasopharyngeal squamous cell carcinomas. Vet Pathol 1984;21:193–7.

[22] McIntosh GH, Giesecke R, Wilson DF, Goss AN. Spontaneous nasopharyngeal malignancies in the common marmoset. Vet Pathol 1985;22:86–8.

[23] Brack M. Gastrointestinal tumors observed in nonhuman primates at the German primate center. J Med Primatol 1998;27: 319–24.

[24] Chalifoux LV. Adenocarcinoma of the small intestine in a common marmoset (*Callithrix jacchus*). Lab Anim Sci 1990;40:210–2.

[25] Tochitani T, Matsumoto I, Hoshino K, Toyosawa K, Kouchi M, Koujitani T, Kimura J, Funabashi H. Spontaneous rhabdomyosarcoma in a common marmoset (*Callithrix jacchus*). J Toxicol Pathol 2013;26:187–91.

[26] Yamaguchi I, Myojo K, Sanada H, Sudo E, Ootsuka S, Okumura H, Takami A, Yoneshige T, Suzuki Y, Imaizumi M, Takada C, Kimoto N, Saeki K, Takaba K. Spontaneous malignant T cell lymphoma in a young male common marmoset (*Callithrix jacchus*). J Toxicol Pathol 2013;26:301–7.

[27] Morosco DT, Cline CR, Owston MA, Kumar S, Dick Jr EJ. Spontaneous mediastinal myeloid sarcoma in a common marmoset (*Callithrix jacchus*) and review of the veterinary literature. J Med Primatol 2017;46:42–7.

[28] Baskerville M, Baskerville A, Manktelow BW. Undifferentiated carcinoma of the nasal tissues in the common marmoset. J Comp Pathol 1984;94:329–38.

[29] Brack M, Schwartz P, Heinrichs T, Schultz M, Fuchs E. Tumors of the respiratory tract observed at the German Primate Center, 1978-1994. J Med Primatol 1996;25:424–34.

SECTION III

RESEARCH APPLICATIONS

CHAPTER

19

The Genome of the Common Marmoset

Eric J. Vallender[1,2]

[1]University of Mississippi Medical Center, Jackson, MS, United States; [2]Tulane National Primate Research Center, Covington, LA, United States

THE SIGNIFICANCE OF THE MARMOSET GENOME

In the early decades of scientific progress, a wide variety of animal models were used as individual scientists focused on what was available and their individual interests. During the second half of the 20th century, a molecular revolution occurred in biology with an increasing focus on genetics. Animal models where genetics could be first controlled and then later manipulated grew in importance. In the last decade, a second molecular revolution has taken place, one that has democratized the powerful tools of genomics.

The marmoset is a unique animal model. Its small size, breeding patterns, and life span as well as its close physiological and genetic relationship to humans make it attractive for biomedical research. Like other primates, however, the marmoset is an outbred species in which animals are genetically individual. While throughout the years marmoset researchers have made targeted discoveries on specific genes or loci, a complete picture of variability and complexity was beyond reach.

At the cusp of the genomic revolution and the introduction of next-generation sequencing technologies, the marmoset genome has now been characterized. With its publication, it is now possible to identify the unique marmoset sequence for a gene or understand the similarities and differences between marmoset and human (or other species) genetics. A number of immediate benefits accrued: Studies that had previously focused on single or few genes could expand to genomic levels, the genic complement of the marmoset could be elucidated, and novel conserved regions across primates could be better characterized. At the same time, the marmoset genome, in concert with new next-generation sequencing technologies, allows for studies that had been previous inaccessible: genomic studies of variation, transcriptomics, and even genetic manipulation.

The marmoset genome is, in many ways, not particularly interesting; it is a fairly standard mammalian, and primate, genome. Its importance lies not in the specific sequence of nucleotides, but rather in the position it occupies allowing for comparative studies and for further developing the common marmoset as a modern animal model for biomedical research.

THE UNIQUE DIFFICULTY OF MARMOSET GENETICS

The cost of genomics work has come down significantly in the past decade as next-generation sequencing technologies become more widespread. This has led to an explosion of genomic work including the sequencing of a number of mammalian genomes [1]. What began as a focus on humans (*Homo sapiens*) and key biomedical research models, mice (*Mus musculus*), rats (*Rattus norvegicus*), and rhesus macaques (*Macaca mulatta*), has expanded to include other evolutionarily significant representatives of diverse clades. At present, genomes with NCBI annotation are available for 24 primate species (Fig. 19.1) with more certain to follow. Due to the unique biology of the marmoset, it is perhaps the most difficult mammalian genome sequenced to date [3].

As discussed extensively in Chapter 8, marmosets are hematopoietic chimeras, harboring myeloid and lymphoid cells derived from their littermates throughout their lives [4–7]. Because DNA is often, though not exclusively, obtained from blood samples, particularly when the organism is still living, genetic work in marmosets has been complicated. Standard molecular methods that do not use clonal approaches do not distinguish between alleles derived from the marmoset or their sibling. Moreover, these alleles are not necessarily in equal numbers. This can result in the appearance of polymorphism or methodological

The Common Marmoset in Captivity and Biomedical Research
https://doi.org/10.1016/B978-0-12-811829-0.00019-4

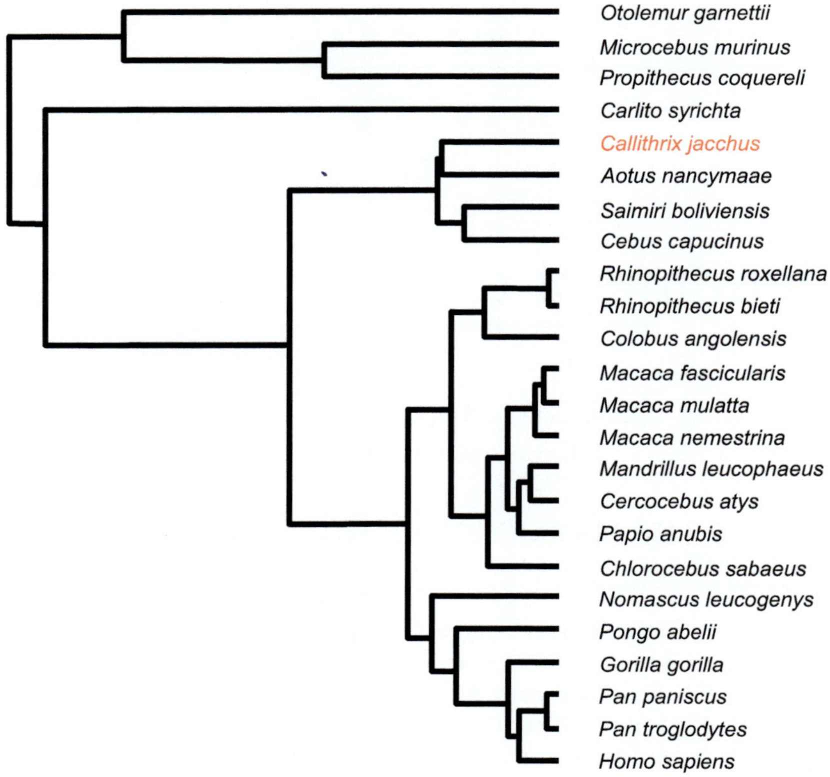

FIGURE 19.1 Phylogeny of currently available primate genomes. The primate genomes currently (12/2017) available and annotated through the NCBI genome annotation pipeline are shown along with their phylogenetic relationships. Branch length is scaled from nuclear DNA mutation rates with nine calibration points derived from paleoanthropological fossil records after [2].

artifacts even when there are none. Avoiding or ameliorating these confounds, then, is a prerequisite for any study of marmoset genetics and genomics.

Chimerism

While chimerism in callitrichids has been recognized for nearly a century, the evolutionary origins remain unclear. There are theories that tie hematopoietic chimerism to reduced major histocompatibility complex (MHC) diversity, but even there the causal relationship between the two is uncertain [8]. It is possible that hematopoietic chimerism necessitated a loss in diversity at the MHC to avoid immune tolerance between dam and offspring during pregnancy, or it is possible that chimerism arose as a way to increase MHC heterogeneity in an individual when population diversity was limited. Of course, it is not unlikely that additional factors are at work that are less appreciated. Regardless, chimerism is a unique factor among callitrichids and one which seemingly pervasively underlies its use as a research model.

During in utero development, dizygotic callitrichid twins have a distinctive placental anastomosis that leads to the sharing of hematopoietic stem cells. These stem cells persist in the adult, particularly in the bone marrow, leading to leukocytes that derive from both the individuals and their in utero siblings. Beginning in the 1930s, karyotyping to detect discordant sex chromosomes in opposite-sex twins firmly established chimerism in lymphatic tissues, but this method was labor-intensive and necessarily limited to relatively small numbers of cells [7]. This approach was, and is, exquisitely specific but lacks sensitivity.

Recent studies using modern molecular genetic techniques, on the other hand, are extremely sensitive but lack specificity. Using sequencing technologies, chimerism has been detected in a variety of somatic tissues beyond the hematopoietic system and even in gametic cells [9]. Increasingly, however, it appears that these low levels of chimerism are the result of cryptic infiltration from lymphatic cells; essentially our modern sequencing technologies are better at detecting these chimeric cells than our ability to isolate pure cell populations [10]. This leads to a central confound of genetic work in marmosets, how is it possible to generate genomic data and interpret it in the context of a "normal" mammalian diploid genome?

It is attractive to attempt to avoid the problem of chimerism entirely by focusing on single births only, but this is not a panacea. Often, if not always, a marmoset born in a single birth will nevertheless appear chimeric; this can result from fetal resorption in utero [11,12]. All marmoset births are believed to begin gestation as dizygotic twins with environmental factors able to modulate litter size throughout pregnancy [11]. It is not possible, then, to obtain a nonchimeric marmoset short of through rather extraordinary surrogacy efforts. Rather, efforts necessarily revolve around identifying and focusing on nonchimeric tissues or in post hoc bioinformatics methods to deconvolute chimeric data.

Sources of Nucleic Acids

Regardless of species, genetic work on living organisms is nearly always derived from blood samples, though human studies using sloughed cheek cells are also common. Unfortunately, marmoset chimerism means that nucleic acids derived from a blood sample can originate from either the animal itself or its dizygotic twin. This inescapable reality makes marmoset studies more difficult than those of other species. One way to approach this difficulty is to use single-cell technologies [13]. This approach would produce pure populations that could then be dissociated. This approach, however, is relatively expensive and only in its infancy. While theoretically possible, it has not been applied to marmoset genomics on a large scale.

The primary approach, then, has been to attempt to use alternative sources of DNA. Because only the hematopoietic system is believed to be chimeric, it is natural to try other tissues. While this ameliorates some of the issues, in practice most tissues are vascularized and there will be contamination by chimeric hematopoietic cells. One common way to avoid this is through the culture of primary cells, usually, though not always, from fibroblasts [14,15]. This has the advantage of creating a pure, nonchimeric, cell population. It does, however, require significantly more effort on the part of the investigator and care must be taken to not allow mutations to accumulate during culturing, usually from repeated passaging.

More recently, a study has been published that suggests an alternative by using DNA derived from the fingernails of marmosets [16]. Developed in humans [17], the advantages in marmosets are particularly meaningful. The DNA derived from fingernail clippings is apparently nonchimeric and not infiltrated by chimeric cells of hematopoietic origins [16]. Moreover, clipping fingernails is noninvasive and minimally stressful to the animals and the processing required by investigators is relatively straightforward and simple. While this approach is relatively novel and remains to be validated across laboratories, its promise is such that it is already being tested in several marmoset colonies.

Bioinformatic Approaches to Chimerism

There is also an argument to be made for simply embracing the chimerism observed in the marmoset blood and resolving it after the fact through bioinformatic approaches. This approach can be traced back to efforts to sequence important alleles, notably the MHC, through molecular cloning [8,18]. With that approach, clonal populations from individual amplifications are sequenced effectively creating multiple instances of individual cells. From there, a complete accounting could be made of the allelic variation observed in an animal and the relative contributions from the animal and its fraternal twin determined. This same approach has been applied to Sanger-based sequencing with some success [9] and can now be taken on a much larger scale using next-generation sequencing.

Next-generation shotgun sequencing at sufficient read densities will produce many individual reads on each nucleotide. In a traditional diploid organism, a heterozygous site would appear as a roughly 50/50 split among the two alleles. Because of the random nature of the process, there can be some deviation from this expected distribution, but adequate read counts and appropriate statistical controls are straightforward and readily incorporated into standard workflows. This work can be extended to marmosets (or to any sample that mixes the DNA from two individual nonchimeric individuals); an even mix of a homozygote and a heterozygote would result in a 75/25 distribution of alleles.

Because the cells derived from the individual are usually found in a greater proportion than the chimeric cells, some further imputation is possible, particularly because the fraction of chimeric cells is the same for each loci. Because of this, a joint estimation approach such as that generated from genomic studies is substantially more informative than approaches that consider only single sites in isolation. Statistically, this becomes more complicated than the relatively straightforward case of a diploid heterozygote, but it is nonetheless tractable. A complete discussion of the statistical methodologies behind this approach is found in the Supplementary Information of the original publication of the marmoset genome [3]. While this approach may require a greater read depth than a more traditional species, it can be relatively easily incorporated into existing workflows for blood samples.

THE MARMOSET GENOME ITSELF

In 2014, the marmoset genome, from the white-tufted marmoset (*Callithrix jacchus*), was first published [3]. It was the first New World monkey (platyrrhine) genome to be sequenced and continues to serve as perhaps the best outgroup genome for catarrhine studies. The marmoset genome was also one of the last mammalian genomes to be sequenced using Sanger sequencing and bacterial artificial chromosome (BAC) libraries; indeed, secondary studies for the publication used a whole-genome shotgun strategy. The BAC library from which the genome was derived was from a male marmoset from the Southwest National Primate Research Center (SNPRC), and the whole-genome shotgun data were from that animal's female co-twin.

A second resequencing effort was completed in 2015 on an animal from the Central Institute for Experimental Animals in Japan [19]. This effort exclusively used next-generation sequencing technology to produce a genome with a coverage depth of roughly 60×. This allowed for the additional mapping of roughly 300 Mbp of unresolved contigs to stabled chromosomes as well as improving the gene annotation from the original study.

At the most basic level, the marmoset genome is 2.62 Gbp in length spread over 22 autosomes and the X and Y sex chromosomes. This is slightly less than catarrhine genomes (e.g., human: 2.99 Gbp, chimpanzee: 2.98 Gbp, rhesus: 3.03 Gbp, baboon: 2.95 Gbp), but comparable to the New World monkey genomes that would follow it (capuchin: 2.72 Gbp, squirrel monkey: 2.61 Gbp, owl monkey: 2.93 Gbp). The chromosomal complement of the marmoset is likewise comparable to other primates, and the sex chromosomes remain similar in length. Understanding the marmoset genome at all levels will be critically important for the work that follows.

The Marmoset Karyotype

Marmoset karyotyping first came to the fore in the context of identifying chimerism [4] but continues to be studied in the context of phylogenetics and genomic evolution [20]. The nomenclature of chromosomes continues to be complicated, however, especially because a number of naming conventions remain in use. The most straightforward convention, and one that is found in nearly all of the assorted annotations, simply numbers the autosomes from largest to smallest. This has the advantage of being simple and obvious, but when comparing between species it lacks information about orthology. A second system, more common within the comparative genomics field and used in Fig. 19.2,

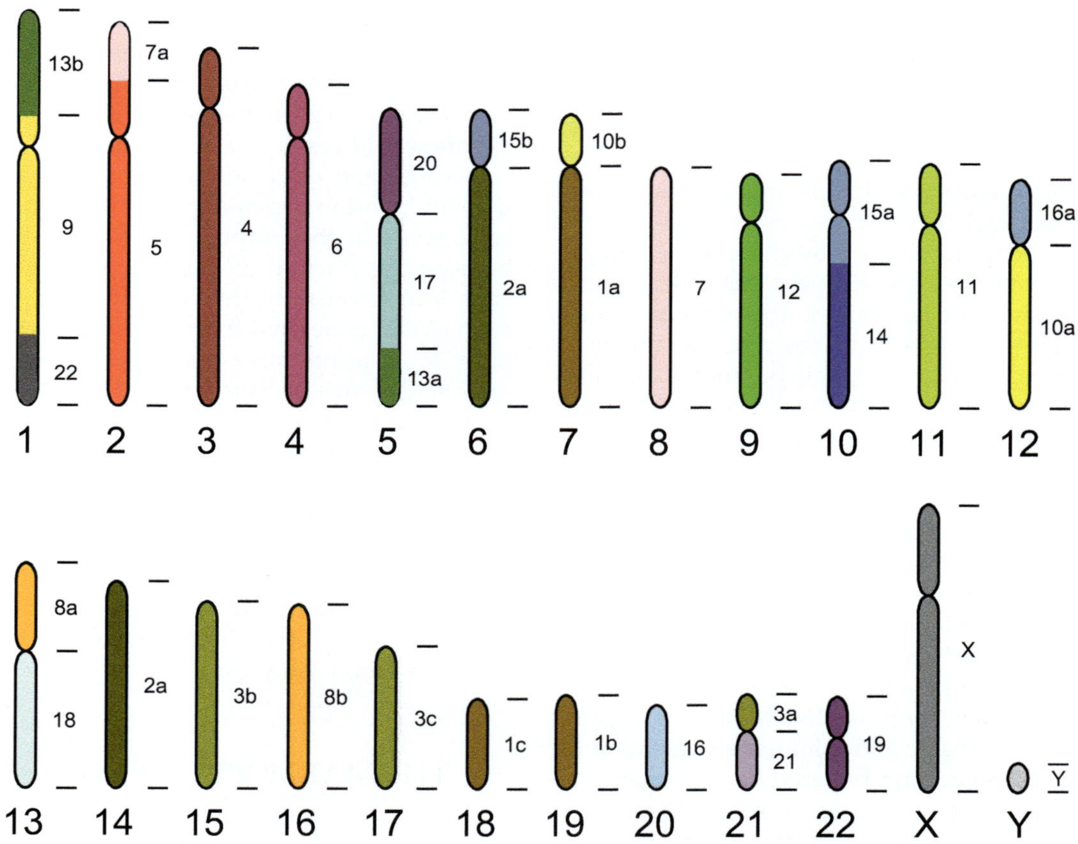

FIGURE 19.2 Idiogram of common marmoset (*Callithrix jacchus*) karyotype with human synteny. The chromosomes of the common marmoset numbered according to the convention used in [3]. Regions of human synteny are indicated by color.

names the nonhuman chromosomes of a species by their human chromosomal counterparts. This more easily demonstrates relationships between species but is meaningful only in that context.

The diploid marmoset (*C. jacchus*) genome contains 22 autosomes in addition to the sex chromosomes (Fig. 19.2). This represents an increase compared with other closely related marmoset genera (*Cebuella* and *Mico*) because of a fission event that created chromosomes 14 (2b) and 20 (16b) from a single chromosome in the ancestral marmosets [21]. The marmosets in turn are distinguished from their tamarin sister taxa (*Saguinus* [22] and *Leontopithecus* [23]) as a result of a fusion creating *C. jacchus* chromosome 7 (1a/10b). The Callitrichinae in turn are separated from the Aotinae and Cebinae by several more complicated fusion events [22]. The ancestral Cebidae karyotype, however, also appears to be the ancestral platyrrhine karyotype.

Elucidating chromosomal evolution not only allows for a better understanding of the evolution of the genome as a whole and speciation events but also helps to true orthology across species. In the context of biomedical models, this is most relevant when identifying marmoset orthologs to human genes. These orthologs share a common ancestor and are more likely to maintain a common functionality. Similarly, these understandings can also aid in the identification of novel gene duplications within a species that may be important for species-specific adaptations.

Large-Scale Duplications and Chromosomal Rearrangements

Large chromosomal rearrangements between species are not uncommon though they are generally few in number except in unique situations such as the gibbon radiation [24]. Smaller-scale duplications, but larger than 1 kB, represent 138 MB of the marmoset genome [3]. Roughly two-thirds of this length is interchromosomal, whereas one-third is intrachromosomal. This is comparable to distributions observed in humans and chimpanzees, though rhesus macaques appear to show a greater percentage of interchromosomal duplications. In general, the patterns of duplications in marmosets and chimpanzees are very similar, but both show an excess of short duplications (<10 kB) relative to humans.

While difficult, comparing the sequences of the duplicated regions across species allows for a general understanding of their age and the role that they may play in speciation and adaptation [25]. Using these approaches it was possible to identify that marmosets have a reduction in the number of more recent duplications compared with humans, possibly the result of demography-influenced population genetic artifacts in humans. In comparing marmoset-only duplications to those shared with rhesus macaques, there does seem to be a greater number with more recent origins [3].

In whole, however, there remain significant difficulties in these analyses both generally and for the marmoset specifically. Chimerism can complicate the inference of segmental duplications, especially recent duplications that might be mistaken for chimeric origin in the absence of sufficient information. At the same time, the divergence time between marmosets and humans can make identifying truly anthropoid basal duplications difficult to identify through identity between or within species. The genomes of additional New World monkey species may allow for some of these issues to be ameliorated, but work is ongoing.

Content and Annotation

For most biomedical researchers, the most important contribution of the marmoset genome is the complete annotation of functional elements. After all, it is these functional elements, whether genes encoding proteins, noncoding RNA genes, or regulatory regions, that are the most common currency of biological understanding. The marmoset, like other historical biomedical animal models, had a piecemeal approach to identifying and characterizing important genetic loci throughout much of the second half of the 20th century. This work proceeded relatively slowly and suffered from unavoidable ascertainment biases. With the generation of the complete marmoset genome, it was possible to more holistically approach these questions.

The end result of a genomic reconstruction is, with the exception of minor gaps, a series of very long strings of nucleic acids. It is only through annotation of these strings that it becomes possible to read meaning into the genome. Annotation can be accomplished by any bioinformatician with the time and inclination, but there are two commonly used and widely accepted pipelines that predominate: NCBI [26] and Ensembl [27]. While the specifics may differ, the underlying process that these two pipelines take is similar. Functional regions are predicted using a combination of conservation and orthology across species, alignment with additional sources (e.g., expressed sequence tag [EST] libraries or others transcriptomics), and known biological mechanisms including splice junctions, transcriptional and translational initiations, and termination.

The most recent Ensembl annotation of the marmoset genome (February 2014) identified 55,116 transcripts of 20,978 protein-coding genes, 9000 noncoding genes, and 2345 pseudogenes. The latest NCBI annotation release (August 2016) has identified 56,158 transcripts of 20,871 protein-coding genes, 6925 noncoding genes, and 8651 pseudogenes. For protein-coding genes the two pipelines closely agree, reflecting the strong understanding of the protein-based central dogma of genetics. Identification of noncoding RNAs differ as there is a less well-established literature. It is also important to note that, particularly among noncoding RNAs, specific definitions may not perfectly concur and remain a moving target even among experts.

Protein-Coding Genes

The total protein-coding gene content identified in the marmoset genome is roughly similar to that seen in humans (Ensembl: 20,441; NCBI: 20,433). At the time of publication of the genome, the annotation relied heavily on known human protein and, to a lesser degree, known proteins from other mammalian species. As RNA-seq technologies and transcriptomics have matured and interest in the marmoset increased, the direct evidence from marmosets themselves has expanded considerably.

Despite a rough similarity in total number of protein-coding genes, the specific constellation present in the marmosets is slightly different from that seen in other primates. Indeed, this is almost always generally true with greater differences in specific gene content seen with greater evolutionary distance. While this work is ongoing, published data have found more than 1200 uniquely marmoset gene gains and 1500 gene losses [3]. It is important to note, however, that this work was done in the absence of any other platyrrhine genomes and thus it cannot be concluded when the last common ancestor of anthropoids these gains/losses occurred. Since the same common ancestor, the lineage leading to humans had gained nearly 2000 genes and lost 800. These numbers, however, can be biased by the completeness of the genomes and accuracy of assemblies. Any impoverishment of quality will necessarily overestimate gene losses and potentially underestimate gene gains. The gain or loss of protein-coding genes is common in genomic evolution and, by itself, does not signify anything particularly remarkable about the marmoset; however, in specific cases, enumerated below, these changes may be related to some unique aspects of marmoset biology that set them apart from other primate species.

Noncoding Genes

In addition to genes that code for proteins, a large number of noncoding genes have been identified. These are loci that are transcribed but whose function occurs at the RNA level. A complete understanding of these noncoding genes is still being developed, but important among what is known are the microRNAs (miRNAs). miRNAs are small noncoding RNAs generally 22 nucleotides in length that mediate posttranscriptional gene regulation through complementary binding to target mRNA sequences.

Unlike mRNAs, little is known about the genomic structure of these loci making de novo prediction difficult. They have typically been identified following reverse transcription and through their conservation across species. In the marmoset, there have been 1164 precursor miRNAs identified. A total of 406, or roughly one-third, of these are perfectly conserved across at least one mammalian species with confirmed expression and another 223 are conserved but with expression remaining to be verified [3]. Another 345, again roughly one-third, have been identified through resequencing, but they do not appear to match miRNAs from other species. This latter group will include those miRNAs unique to marmosets, those unique to platyrrhines, and those that simply remain to be identified in other primate species.

The focus of noncoding genes has been primarily on the brain and placenta, reflecting the importance of the marmoset in these aspects of biology. Unlike protein-coding genes, where there is a fairly clear expectation and where present numbers are unlikely to dramatically change, it is highly likely that miRNA annotation will continue to change as more tissues and developmental stages are included.

Regulatory Regions

Like noncoding RNAs, regulatory regions of the genome tend to be identified either by painstaking basic molecular biology or inferred via conservation across species. This can be accomplished either by comparing regions conserved across distantly related species [28,29] or by comparing regions that have degraded between closely related species [30,31]. Each of these approaches aims to identify regions within the genome that are conserved by evolution, under negative or purifying selection, and thus are likely to be functionally important. This includes, naturally, exons for protein-coding genes, both translated and untranslated regions, but it is particularly meaningful for identifying noncoding RNA genes and regulatory regions. Indeed this comparative genomics revolution has been a driving factor in the development of an appreciation for meaningful functional elements in what was previously considered "junk DNA" [32].

Phylogenetic footprinting uses comparisons of distantly related species to identify regions of shared conservation across a landscape of divergent noise [29]. Early comparisons compared human and mice, incorporating more distantly related species such as chicken and *Xenopus* as they became available. Modern genomic techniques have greatly enlarged taxa availability, and present indices of conservation can be adapted to straightforwardly incorporate additional species as they become available [33]; at present, as many as 100 vertebrate species have been used to identify regions of evolutionary conservation [34]. These techniques are very powerful for identifying regions conserved over long time frames but can have difficulties in identifying more recently emergent functional regions.

Conversely, phylogenetic shadowing uses many similarly related species to recognize conserved regions by identifying regions of divergence across a landscape of highly similar regions [30]. The idea behind this is that for the vast majority of these short lineages there will simply have not been enough evolutionary time to accumulate the mutations needed to deteriorate identity. Because of this, conservation cannot be used to identify functional regions; rather, with enough species the accumulation of divergence can be used to infer nonfunctional regions. This approach can allow for the identification of more evolutionarily recent functional elements but comes at the cost of requiring many species to accumulate the evolutionary branch-time that is needed for identification.

Because of the timing of platyrrhine–catarrhine divergence, average neutral divergence for humans and marmosets is roughly 12%, the marmoset genome has power in phylogenetic footprinting and shadowing approaches. Human–rhesus–marmoset comparisons alone were sufficient for identifying roughly two-thirds of exons present [35]. Incorporating genomes from orangutans, dog, mouse, and horse to these core three anthropoid species allowed for the identification of nearly 24,000 anthropoid-specific constrained elements; 99.7% of which were noncoding [36]. These findings have important implications for not only identifying evolutionary adaptations but also in furthering the general understanding of the regulation of gene expression.

Repeat Families and Mobile Elements

Transposable elements are a common source of genomic variation [37]. They are found in abundance in the genome and can influence gene structure and expression in direct (e.g., exon-insertion or novel noncoding RNA genes) or indirect (e.g., novel regulatory effects) manners. In primates, the most common transposable elements are the long interspersed element (LINE) L1 and the short interspersed element (SINE) *Alu* [38]. In anthropoids transposable elements make up roughly one-half of the genome, true for the marmosets as well as the apes and Old World monkey species [39].

SINEs represent 13.5% of the marmoset genome and include 1.1 million *Alu* elements. *Alu* is a primate-specific SINE that has spawned a number of subfamilies over the course of evolution [40]. These subfamilies can be used to date specific insertion events and for phylogenetic reconstruction. In marmosets, the majority of *Alu* elements derive from the *Alu*TA15 subfamily, a New World monkey-specific derivation [41,42]. LINEs represent roughly 20% of the marmoset genome mostly as L1 elements, a family common across mammals [43]. Over 8000 full-length line elements are identified in the marmoset genome, including six intact line elements that appear to be continuing to propagate [44].

More recently, a novel SINE has been identified in the marmoset genome [45]. Platy-1 appears to have derived from an *AluJ* element and was limited to New World monkeys. A total of 2268 full-length elements were identified in the marmoset genome, representing 62 subfamilies. The oldest of these exists across New World monkey taxa, including Atelidae and Pitheciidae, while 31 are Callitrichinae-specific. These studies demonstrate that transposable elements are not a static feature in the marmoset genome and represent an ongoing opportunity for study of their evolutionary dynamics.

THE MARMOSET TRANSCRIPTOME

The genome is an information carrier for the organism that operates through transcription into RNA; mRNA occupies the key intermediate role between the genomic DNA and the proteins in the cell. Understanding the genome is only the first step in understanding when and where, and under what conditions, specific RNAs are transcribed and how this complex regulation of gene expression leads to the incredible diversity of cellular phenotypes. Transcriptomics is focused on understanding gene expression and bringing this understanding to the biology of the organism.

Just as next-generation sequencing has made it easier to study genomic DNA, so too has it become more straightforward to study the RNA transcriptome. It has become possible to improve older methodologies that required the knowledge of genetic sequence and it has become possible to develop newer technologies that were heretofore inaccessible. This has greatly improved the possibilities for understanding gene expression, including both expanding the breadth of studies away from candidate gene targets and expanding the depth of studies toward ever more specific cellular targets.

Improving Annotation

While not generally the primary focus of transcriptomic studies, sequencing of RNAs is important for identifying genes and improving the annotation of the genome. Previously, the annotation pipelines for genomic sequences were described. While modern genomes can be reasonably well annotated solely through homology with previously annotated genomes, notably human and mouse, originally conservation of identity was unknown and genes could only be identified based on experimental validation from cDNAs. Indeed, even today there is an important divide between experimentally confirmed genic annotations and predicted gene models.

Three major studies have focused on transcriptomic approaches to improving the annotation of the marmoset genome. The first used ESTs generated from RNAs from liver, brain and spinal cord, spleen, testis, and embryonic stem cells sequenced using traditional Sanger methods [46]. A total of nearly 300,000 ESTs were generated that represented 17,232 contigs, which mapped to 13,499 annotated marmoset genes. The meaning contigs included both genes previously annotated in another primate as well as truly novel transcripts.

The second used randomly primed cDNA libraries derived from the liver, intestine, kidney, and brain of three male and three female marmosets sequenced using next-generation sequencing technologies [47]. This approach generated 47,883 transcripts. Although only 29% of these transcripts were previously annotated in the marmoset genome, an additional 32% of the transcripts were found annotated in another primate genome (half of these in humans). This reflects not only the incomplete annotation at the time of the marmoset genome but also a more general lack of knowledge, particularly of alternative splice forms. Beyond its general utility, this study was aimed primarily at identifying the closely related genes of the cytochrome P450 family whose close identity had previously made identification and annotation difficult.

At the same time, another study was underway that used RNA from skeletal muscle, bladder, hippocampus, cerebral cortex, and cerebellum from two marmosets for transcriptomic analysis using next-generation sequencing [48]. This approach yielded 51,163 full-length transcripts from 10,833 unique genes. Roughly three-quarters of these genes had been previously annotated through the Ensembl pipeline, whereas an additional quarter represented previously unannotated human homologs. The variation seen in these studies is not altogether unexpected and likely represents methodological, rather than biological, sources. Nevertheless, these diversities of studies together add information on the marmoset transcriptome that improve annotation of the genome both by experimentally validating specific genes and by the de novo addition of annotation.

Gene Expression

Coincident with the publication of the marmoset genome, there has also been an increase in the number of studies aimed at characterizing gene expression. Partially this is the result simply of the renewed interest in the species as a biomedical model, and partially this results from the greater ease with which these studies can now be done. Regardless, it is clear that these studies are becoming increasingly common and that they are greatly furthering our knowledge of marmoset biology. The focus of these studies reflects the focus of marmoset research generally: neuroscience, immune response, and reproduction.

The first studies of marmoset gene expression in the brain predate the publication of the marmoset genome and the development of next-generation sequencing technologies. These studies used microarrays to characterize gene expression. One microarray was developed targeting roughly 1500 brain transcripts [49] and subsequently used to investigate gene expression in the regions of the hippocampus [50] and changes in the brain associated with serotonin-1A receptor agonist administration [51]. Another marmoset-focused microarray that used more than 7500 brain transcripts was used to characterize gene expression in the frontal lobe, hippocampus, cerebellum, and amygdala [52]. Further elaboration on this array generated a version that targeted nearly 23,000 marmoset transcripts and was used to study synaptic pruning during brain development [53,54].

More recently, next-generation sequencing technologies have been successfully applied to the study of marmoset transcriptomics. This study, which used RNA-Seq shotgun sequencing, compared the transcriptomes of four novel embryonic stem cell lines, a previously derived embryonic stem cell line, an induced pluripotent stem cell line, and six primary skin fibroblasts [55]. The focus of the study was not only in developing tools and resources for future marmoset transgenic and regenerative studies but also serves as an important reflection on the relative ease with which these new technologies can be adapted to existing questions.

MARMOSET GENETIC VARIATION

Unlike the mouse strains that are commonly used in biomedical research, marmosets are an outbred population. This means that, like humans, each individual marmoset is genetically unique. Studies done on outbred animal models control environmental variables but are still subject to individual genetic differences between subjects. For some research this genetic individuality will not matter, but for many studies it will be an important component either to control for or to explicitly test as a variable. It is a consideration that needs must be incorporated into study design. More broadly, however, genetic variation is also important in and of itself. It can help to understand population and demographics and contains valuable information on the evolutionary history of the species.

Appreciation of the genetic variation of a species is a critical next step for the postgenomic era. It will facilitate future studies of genetic association and population health. At present, the tools for assessing genetic variation in marmosets are available and lack only the will for implementation on a large scale. There have been initial forays into this realm, however, that demonstrate the promise and possibility of future work.

Population Genetics

At the time of the publication of the marmoset genome, there were three large academic colonies of *C. jacchus* used for biomedical research in the United States: the New England National Primate Research Center (NEPRC), the SNPRC, and the Wisconsin National Primate Research Center (WNPRC). In addition to the initial whole-genome shotgun sequence and BAC library from twin siblings from the SNPRC, genomic analyses were done on five additional unrelated animals from the SNPRC, two from the NEPRC, and two from the WNPRC; together more than 7.7 million single nucleotide polymorphisms (SNPs) were identified [3]. Additionally, a panel of 24 animals (11 NEPRC, 8 SNPRC, 5 WNPRC) were used for interrogation of *Alu* insertion variation. Together these studies begin to develop a picture of population variation among captive breeding colonies.

Principal component analysis and admixture studies easily separated the three colonies with only two animals (both from Southwest) showing any appreciable amount of intermediacy. Both a clustering methodology for the SNPs and the *Alu* insertions demonstrate some deep similarities between the SNPRC and WNPRC colonies compared with the NEPRC colony, although individual differences between animals still predominate.

Population genetics of natural populations of marmosets are limited. Early studies of the limited MHC diversity in captive tamarin colonies suggested that the genetics of these colonies broadly represented natural populations [56]. Mitochondrial DNA studies have offered some understanding of the genetic relationships among natural marmoset populations [57,58], but these studies are limited to specific interpretations. Genomic-level, microsatellite-based markers have been used with success to study hybridization between natural populations of *C. jacchus* and *Callithrix penicillata*, but again the survey has been limited [59]. In part, this limited focus on the genetics of natural marmoset populations is a result of their population stability; populations that are endangered or threatened have seen more conservation-focused studies (e.g., the golden lion tamarin [*Leontopithecus rosalia*] [60], black-faced lion tamarin [*Leontopithecus caissara*] [61], and pied tamarin [*Saguinus bicolor*] [62]).

Single Nucleotide Variation

At the publication of the common marmoset genome 7.7 million SNPs were identified [3], though dbSNP currently includes more than 16 million [63]. SNPs can be categorized based on their location, with functional importance corresponding imputed (Table 19.1). As expected, the vast majority of identified SNPs in marmoset are intronic or intergenic, and putatively without function. Of increasing interest to biomedical researchers are the variants that may be functionally relevant and whose effects can be predicted in silico, including those that change amino acids in proteins, result in the gain or loss of a stop codon, or affect splicing.

A great deal of work remains on marmoset genetic variation, however. The information on SNPs currently available does not extend to minor allele frequencies, and how common each of those variants are. This is critical information for imputing functional relevance, evolutionary and demographic history, and applicability for downstream analyses. Only relatively common polymorphisms, for instance, will be useful for genetic management decisions or attempts at association studies. This is a pervasive problem, however, that extends beyond marmosets. While this kind of information is readily available for humans, for instance, through the International HapMap Project [64] or the 1000 Genomes

TABLE 19.1 Single Nucleotide Variation in *Callithrix jacchus*

5′ untranslated region (UTR)	4,987
Synonymous	49,385
Nonsynonymous	56,319
Stop gain	1,246
Stop loss	378
Intron	5,153,039
5′ splice site	696
3′ splice site	479
3′ UTR	14,094
Nongenic	10,768,525
Total	16,049,148

Project [65], it largely does not exist for other nonhuman mammal model species, though genetic differences between mouse strains [66], dog [67], or livestock [68] breeds often have been cataloged.

Copy Number Variation

Another finding from the postgenomic era was that it is surprisingly common to find variation in gene copy number between individuals. While pathogenic differences have long been identified, only recently has it become apparent just how common and functionally neutral, or nearly neutral, this variation is. These studies have only recently begun to be expanded into nonhuman primate species [69–75], and they focus almost exclusively on the great apes and macaques.

Copy number variation is interesting, however, because of the potential for similarities between species. SNPs almost never occur at the same spots between species and when they do, except in the most exceptional of cases, they represent fortuitous convergent evolution. Copy number variation, on the other hand, can reflect recurrent mutation at specific hot spots, increasing the likelihood that identical changes will be found between diverse species. This remains an understudied source of genetic variation, however, even in humans.

Functional Genetic Variation

Most of the discussion above has been focused on variation at genomic levels. Indeed, the strength of the next-generation sequencing technologies is that it is increasingly possible and commonplace to focus on variation at this level without succumbing to unintended biases. Most of the work on specific functional variation, however, has focused on candidate genes. These studies revolved around specific targeted questions and hypotheses and included the characterization of genetic variation, often from many individual animals, at a specific gene or locus. The advantage to this work is that it allows for direct tests of function, identifying real biological effects of specific mutations.

These studies are often technically cumbersome and require a significant investment of effort. Because of this, they have been generally reserved to the most generally important genes and functions. As biomedical animal models, there are two categories of genes that are commonly investigated in this way, immune-related genes, where individual variation is important for identifying susceptibility to disease and stratifying animals for studies of prophylaxis and treatment, and cytochrome P450 genes involved in differential drug metabolism. These gene families have both been extensively studied in marmosets and help serve as examples for where studies of genetic variation may proceed in the future.

The Major Histocompatibility Complex and Immunogenetics

Genetic variation in the marmoset MHC genes, class I [8] and class II [76,77], was one of the first to be studied in detail. Across the MHC, marmosets (and callitrichids more broadly) have a significantly reduced genetic variability compared with other mammalian species. This has made them both attractive for studies of infectious diseases and particularly amenable to genetic association studies focused on this region [78]. It is perhaps of interest to note, however, that early studies of marmoset MHC genetics were focused on the classical MHC class I loci, more recent studies have indicated a greater diversity at the nonclassical MHC-G loci in callitrichids compared with other nonhuman primates [79–81].

One early study identified an association between marmoset MHC alleles and the development of experimental autoimmune encephalomyelitis (EAE), a model of multiple sclerosis [82]. Follow-up studies further characterized this association and identified mechanisms by which specific alleles recognized target molecules [83,84]. Chapter 24 contains a greater exploration of EAE in the marmoset and further addresses the role of the MHC as well as other immunogenetic variation.

Marmosets have also historically played a significant role in transplantation research [85–87]. Again this is likely because their reduced diversity at the MHC has led to a reduction in rejection and host-graft incompatibility. Modern molecular studies have confirmed these hypotheses and expanded them, identifying specific MHC alleles that seem to be associated with alloreactive donor–recipient pairs [88,89].

Cytochrome P450 and Metabolism

The marmoset occupies a specific role in toxicology research (Chapter 27), its small body size and phylogenetic similarity to human make it an attractive model for many studies [90,91]. One of the most important aspects of this research is an understanding of the proteins that support drug metabolism, specifically the cytochrome P450 genes. Twenty-two cytochrome P450 genes and typical substrates have been identified in marmosets [47,92]. Differences in these genes between individual animals can lead to differences in drug metabolism and pharmacological responsivity. Understanding these differences is one of the keys underlying the development of personalized medicine in humans and can help greatly improve toxicology studies in marmosets.

To date, little work has been done comparing polymorphic variation in cytochrome P450 genes in marmosets with one exception. Three variants that alter amino acid sequence have been identified in the marmoset cytochrome P450 2C19 (Phe7Leu; Ser254Leu; Ile469Thr); these polymorphisms have been shown to influence the metabolism of warfarin [93] and omeprazole [94]. These differences in metabolism parallel differences seen in humans and can serve as a preclinical model of personalized medicine. While additional work is needed to expand these studies beyond this single gene, variation in marmoset 2C19 operates as a proof-of-concept for future studies interested at incorporating functional genetic variation into toxicology research.

THE MARMOSET IN COMPARATIVE GENETICS

One of the most important roles of the marmoset genome is found in comparing it to genomes from other species. These comparisons allow for a better understanding of both the marmoset itself as well as the other species with which it is compared. At the most basic level, comparative genetics and genomics can help disentangle relationships among taxa, but at higher levels it can help explain the molecular underpinnings that have led to differences between the species and offer insight into the molecular mechanisms by which adaptation occurs.

The marmoset plays an important role in the comparative genomics of primates because of the phylogenetic position that it occupies. New World monkeys are the basal anthropoids and form an outgroup that helps define the catarrhine ancestor, a position that is important in understanding the evolution of apes and humans. As the first New World monkey genome to be completed, and still the most reliable, the marmoset represents this lineage in many comparative genetic studies. As more genomes become available, the marmoset genome is also increasingly useful for understanding the diversification of platyrrhines, including both the Cebidae and the Callitrichinae, as well as the marmoset itself.

Molecular Evolution of Callitrichid Biology

Like any taxa, callitrichids have developed unique biological adaptations that have allowed them to flourish. These adaptations ultimately derive from changes in genetics. In understanding how the mutations at the genetic level drive phenotypic change it is possible to better understand the biology underlying these processes more broadly.

In any species or taxa there are many things that make them distinctive; it is rare that one specific change occurs in isolation. The complexities of marmoset biology are enumerated throughout this book. In some ways, marmosets are very much like other primate species and in other ways they are completely unique. In particular, however, there are four specific traits of marmoset biology that have led to further investigation at the molecular levels. These are qualities where marmosets in particular are vastly different from other species and where an interest exists to better understand human biology or to facilitate the development of the marmoset as an animal model for biomedical research. These include a general reduction in body size, pregnancy-related adaptations, immunogenetics, and social behaviors.

Molecular evolution compares sequences between taxa to identify the changes between genes or loci. During normal evolution, mutations occur regularly like a ticking clock. Selection operates on these mutations either removing them from the populations, negative or purifying selection, driving them to higher frequencies and fixation, positive selection, or ignoring them entirely, neutral selection. By comparing these rates at different loci between species, it is possible to infer the history of selection that acted on them. Couple this with functional information and it is possible to infer, to a first approximation at least, the molecular adaptations that the taxa went through.

When doing comparative and molecular evolution studies, the power of the study is strongly influenced by the relationships among the taxa used as well as by the quality of the genomic sequence available. Most of the initial studies using marmosets used a constellation of taxa including humans, chimpanzees, rhesus macaques, and one or more nonprimate outgroups such as mouse, rat, dog, or cow. With these taxa, the marmoset lineage includes the entirety of the New World monkey evolutionary history. As more data from additional New World monkey species have become available, the possible focus of evolutionary studies has narrowed.

Reduction in Body Size

One of the initial targets of study for molecular evolution was the putative reduction in size that occurred in marmosets [3,95]. The average weight of callitrichid species is roughly an order of magnitude less than most other New and Old World monkeys. While they are certainly not the only taxonomic group to show a smaller body size, it can be inferred by parsimony that the most recent common ancestor shared by anthropoids was likely larger bodied [96,97]; this reduction in body size likely is associated with shifts in feeding behavior and exploitation of unique ecological niches [98,99].

Comparing genes in marmosets to their orthologs in rhesus macaques, chimpanzees, and humans, a number of genes were identified to be under positive selection in the New World monkey lineage [3,95]. This notably included *IGF1R*, an insulin-like growth factor receptor that has previously been associated with size in humans, as well as other growth factors, their receptors, and accessory proteins (*GHSR, IGH2, IGFBP2, IGFBP7, EGFI*). More targeted studies that incorporated additional New World monkey species further identified changes under positive selection in *GDF9, BMP15,* and *BMP4.*

There are also a number of genes identified as under positive selection that are associated with energy production [3]. Specifically, eight genes that are part of mitochondrial respiration were identified: *NDUFA3, NUDFA4, NDUFA8, NDUFA10, NDUFB2, NDUFB8, NDUFS2,* and *NDUFS3.* These changes are proposed to be part of an adaptive response to novel thermoregulatory constraints brought on by the reduction in body size.

While it is yet to be proven that the mutations identified in these genes were directly causative of the reduction of body size in marmosets, a challenge all too common among molecular evolution studies, there is substantial indirect evidence that can drive future hypotheses and more focused studies.

Chimerism and Immunogenetics

One of the earliest roles of marmosets and tamarins in biomedical research was in infectious disease research, largely because they demonstrate a susceptibility to a number of important human diseases [100]. This occurs in part because callitrichids show a significantly reduced MHC diversity at the classical MHC class I genes [101]. *Saguinus oedipus* showed only 11 alleles at three MHC class I loci [102], and, whereas *C. jacchus* may show more [103,104], it still stands in stark comparison to other species such as human with more than 12,021 alleles [105]. Interestingly, the vast majority of callitrichid MHC class I diversity appears to arise from the nonclassical MHC-G loci, suggesting a transfer in function from the classical MHC-A, -B, and -C loci [79–81].

One consequence of this reduced variation is that callitrichids are much more tolerant of transplantation, both intraspecific as well as interspecific [87]. Allografts between species survived an average of 3 weeks, with intraspecific allografts often lasting months. This persistence has been suggested to be a direct consequence of chimerism. In turn, this has led to the suggestion that alloreactivity between chimeric twins may be, in part, driving the reduction in MHC diversity [106]. Regardless of the cause of the reduction in MHC diversity in callitrichids, it is clear that this unique character of the taxa has played a major role in both its evolution and utility as a biomedical model.

Twinning and Pregnancy

Many of the unique features of marmoset biology revolve around pregnancy and gestation. The hematopoietic chimerism is derived from placental anastomosis and it seems likely that there is an association between this chimerism and the reduced diversity in MHC, even if the direction of causation remains unclear [8,107,108]. Body size also has been proposed to be associated with changes in gestation [99], whether through changes in key developmental time points or an overall reduction in prenatal growth rates. This may be further complicated by the fact that callitrichids are unusual among primates in that they typically release multiple ova and regularly give birth to twins or even triplets [109,110]. Again, this entire suite of traits is likely linked in some way even if the directionality is still not understood.

The genetics of both ovulation and gestation are a topic of considerable interest. While a number of specific amino acid changes have been identified in candidate genes associated with twinning in humans or ovulation in other mammalian species, the evidence for a specific role in marmoset reproductive biology remains limited. One particular gene that has generated considerable interest is *WFIKKN1* [3]. This gene has been proposed to play a role in regulating ovulation [95] through regulation of the bone morphogenic protein system [111]. Multiple amino acid changes are observed in callitrichids that are not seen in other New World monkey species, nor catarrhines [3,95]. Perhaps most interestingly, one of these amino acid changes, Thr307Ala, has apparently reverted to the ancestral state in Goeldi's marmoset, the only callitrichid that does not twin.

In addition to genes specifically associated with twinning, there are a number of changes seen in genes associated with reproduction more broadly. This includes the losses of the *KLK2* gene, a protease related to semen viscosity; the *ADAM6* gene, which has a role in fertility; and the duplication of the *ISP2* gene, which has a role in embryo implantation in mice [3,95]. Genes involved in reproduction are commonly among those identified as having undergone meaningful change between species; they are among the most strongly and directly affected by selective pressure and their change may be a proximal factor driving speciation. While it is difficult to ascribe these changes in the marmoset to either a more common effect or uniquely associated with the reproductive differences observed in the species, it is nonetheless meaningful when considering marmoset biology.

Social Behaviors

It has been suggested that the unique gestation and short interbirth intervals of the marmosets have led to unique adaptations in paternal care and affiliative behaviors [95,112,113]. While germ-line chimerism has been offered as a partial explanation for paternal care in marmosets [9], this remains to be validated [114]. On the other hand, there is remarkably strong evidence for the positive selection of the neurohypophysial hormones, vasopressin and oxytocin, and their receptors. Studies of these genes have suggested a mechanism by which paternal care evolved and was reinforced in New World monkeys [115].

AVP, the gene encoding vasopressin, is extremely conserved across mammals generally and in New World monkeys in particular, but its primary receptor, *AVPR1a*, shows significant variability across New World monkey species that is largely absent in catarrhine species [116]. There are 10 amino acid substitutions in *AVPR1a* that separate all New World monkeys from the catarrhines and an additional 33 mutations spread across 22 platyrrhine species. Among these are four mutations that are Callitrichinae-specific. A number of lines of evidence demonstrate that this distribution is not to be expected by chance, suggestive of positive selection. Although specific functional effects cannot yet be attributed to the mutations, these findings are suggestive of a meaningful evolutionary effect.

Oxytocin, *OXT*, and its receptor, *OXTR*, likewise, have seen unexpected changes in New World monkeys. Oxytocin is a nine amino acid peptide that is identical across nearly all mammals, but there have been four mutations in New World monkeys [117]. This, notably, includes the mutation of leucine to proline at the eighth position in the basal Cebidae [118]. This departure from strong conservation across other mammal species has been interpreted as positive selection. Like *AVPR1a*, the oxytocin receptor, *OXTR*, also shows evidence for positive selection in New World monkeys, notably, in the third intracellular loop with a putative effect on β-arrestin signaling [117]. Interestingly, progesterone response elements have been found in the promoter region of *OXTR* uniquely in New World monkeys and they appear to be associated with paternal care [119].

While it is always difficult to ascribe functional meaning to molecular evolutionary changes, it is particularly hard when those changes are purported to be associated with social behaviors. Nevertheless, the current findings on vasopressin and oxytocin suggest the possibility of a molecular preadaptation in Cebidae toward greater affiliative behavior and paternal care that may have allowed for the emergence of twinning and rapid reproduction in the Callitrichinae.

New World Monkey Phylogenetics

One early advantage of the genomic revolution has been in clarifying ambiguous relationships among taxa. Early molecular studies were instrumental in identifying the three crown taxa of Platyrrhini (Atelidae, Cebidae, and Pitheciidae) and in placing Callitrichinae and Aotinae within the Cebidae [120]. It was not until more complete genomic data became available, however, that the specific branching orders, both within the crown taxa and the Cebidae, were more fully resolved [2,121].

Work using a diverse array of nongenic and traditional genomic markers was able to demonstrate that Cebidae and Atelidae are sister taxa and that Aotinae and Callitrichinae should be classified as subfamilies within Cebidae rather that monophyletic families themselves. Within the Cebidae, relationships at the subfamily level are increasingly better understood though still not certain. Saimirinae and Cebinae are certainly sister taxa, but the positioning of Aotinae as a sister taxa to Callitrichinae, while strongly suggested, is still not resolved to a certainty.

The difficulty in ascertaining true phylogenetic relationships occurs when speciation events occur in rapid succession. The closer in time these events occur, the greater the likelihood that individual gene trees will not correspond to the species tree. When only a single or few genetic loci are used for phylogenetic reconstruction there is an increased chance that the history that they reflect will not coincide with the true speciation history. The ability to dissociate these ambiguous relationships is greatly enhanced by genomic data and the *C. jacchus* genome, along with the genomes of *Aotus nancymaae*, *Saimiri boliviensis*, and *Cebus capucinus* may definitively resolve these questions.

The Marmoset as an Anthropoid Outgroup

One of the early justifications for completing the marmoset genome was as an outgroup for anthropoid genetics. The marmoset genome, indeed any New World monkey genome although the marmoset was the first, can be used to root comparisons between Old World monkeys and apes. This allows for genetic changes to be localized to either the lineage leading from the last common catarrhine ancestor to extant Old World monkeys or the lineage leading from that common ancestor to the apes as well as to identify changes unique to catarrhines. This has important implications for studying the molecular evolution of humans by helping with the better understanding of ape and catarrhine origins.

There are numerous studies that make use of the marmoset genome as an outgroup for understanding primate molecular evolution, to include an exhaustive list here would be far beyond the scope of this chapter and counterproductive; rather it is useful to focus on a few meaningful examples. Studies of *GGTA1*, a gene encoding a protein involved in the synthesis of a common mammalian cell surface oligosaccharide, used marmosets as an outgroup to identify a duplication event in the ancestral haplorrhine [122]. While in platyrrhines the pseudogene was lost while the original was maintained, in catarrhines the original gene would eventually be lost as well. These findings have been suggested to demonstrate that catarrhines required a preadaptation before the shift in glycobiology could occur.

A similar study considered the duplication of the relaxin genes, *RLN1* and *RLN2*, which play a role in neuroendocrine regulation [123]. The functional version of these paralogs is only found in apes, but it was initially unclear when the duplication event occurred [124]. Studies using the marmoset were able to demonstrate that the duplication event occurred in the common ancestor of the catarrhines but that while both genes persist in the apes, one of the paralogs was lost in Old World monkeys [123].

Both of these examples elucidate the duplication history of specific genes. While they demonstrate the important role that the marmoset can play genetically, they do not demonstrate the significance of the ability to investigate these questions on the genomic scale. While truly genomic studies remain few, there have been several large-scale attempts to study molecular evolution in primates that incorporate marmosets. One such study included the marmoset among a large cohort of mammals in investigating the molecular evolution of genes involved in neuropsychiatric disorders [125]. Another used the marmoset among other primates to investigate the genes responsible for adrenarche in primates [126]. As more genomic data become available, molecular evolution will rapidly expand from single gene studies into more complex networks of genes.

FRONTIERS OF MARMOSET GENOMICS

As the marmoset reemerges as an important biomedical research model, the genetics and genomics of the marmoset are gaining an increasing importance as well. In part, this is simply the natural outgrowth of advances in genetics itself. The ability to examine genetic variation at a genomic level and with reasonable costs and efforts opens avenues to studying marmoset biology that heretofore had been inaccessible. This is broadly true of any outbred species, but it is particularly useful in the case of common animal models of biomedical research where the breadth and depth of additional knowledge is substantial.

These approaches, however, are simply building on tools that have existed in human genetics for years, taking advantage of technological improvements to bring genetic studies to bear where they were previously too expensive to consider. At the same time, this genetics and genomics revolution has made possible novel approaches. Notable among these is the development of transgenic animals. This technique, an important and emerging avenue of research particularly in neurobiology, is possible because of the presence of a reliable and complete genomic sequence.

In the early to mid-20th century animal models in scientific research were chosen with little regard for genetics or heredity. Indeed, had this consideration been fully appreciated, it is likely that chimeric marmosets would have been passed over. As knowledge grew and genetics came to occupy a more central role in biology, the research community focused on those species for which genetics could be more easily studied, manipulated, or controlled. This, notably, included a shift in focus to rodents generally and mice in particular. In the last decade, however, next-generation sequencing has brought a democratization to the genetics of animal models. It is these tools that have set the stage for a regrowth in marmoset research and placed the marmoset in a central role for future biomedical studies.

Genetic Variation

The most natural next step for marmoset genomics is in better understanding the genetic variation that exists between individuals and subpopulations. This work, as we have seen above, has begun in earnest already; yet it still remains in its infancy. At present, much of the work has been focused on variation from a relatively few animals and/or a relatively small number of loci. The transition from genetic to genomic has only just begun. As more information becomes available, it will be possible not only to continue in the vein of the previously described studies but also to leverage the information into new methodologies.

Colony Health and Genetic Sustainability

As marmoset research develops, it will be increasingly important to understand both the individuality of the animals under study and the genetic health of the populations from which these animals are drawn. In fact, these issues have always been important and researchers across a variety of disciplines have long attempted to make sense of them, but with an increased knowledge of marmoset genomics it is now possible to address these issues much more fully.

Previously, population genetic health was based largely on theoretical modeling and managed through relationships known to or inferred by population managers [127,128]. While aspects of marmoset husbandry such as long-term nonpromiscuous pair bonding make parentage, and thus genetic diversity, somewhat easier to infer than other primate species, these approaches remain limited. Even in situations where parentage can be definitively ascertained (such as paired breeding) often deeper relationships are unknown. This can lead to the underestimation of inbreeding or the development of unwanted genetic substructure. Often population genetic theory refers to effective population size as opposed to total census size. Careful genetic management of population aims to make these numbers as close to equivalent as possible [129].

The most straightforward application of genetic methods to colony management is simply in better determining parentage [130–132]. Especially in naturalistic settings, parentage is not always easily inferred and misattribution can have unwanted, sometimes deleterious, effects with subsequent breeding. The methodologies for genetic parentage testing are well-established and straightforward and can be relatively easily implemented. What is needed is a better understanding of genetic variation in marmosets and the will and effort to genotype the colony individuals.

It is also possible to expand these studies into natural populations [133]. Genetic variation offers a means for identifying structure within populations without a priori knowledge about relationships between individuals. This allows for conclusive identification of genetically isolated or vulnerable groups and a better understanding of their natural ecology. While the marmoset most commonly used in biomedical research, the white-tufted-ear marmoset *C. jacchus*, is listed as of least concern and at a stable population [134], the IUCN Red List shows six callitrichid species as vulnerable, six as endangered, and two as critically endangered. The genetic tools developed for the common marmoset can, with simple modification, be adapted to these species as well.

Association/Linkage Studies

Forward genetics studies have long been used to identify genetic variation associated with phenotypes. In nonhuman mammals this has usually occurred through comparisons between breeds (cows, dogs) or strains (mice, rats). These studies take advantage of known, between group, variation to isolate regions linked to phenotypes of interest. This allows for meaningful information to be gained from smaller numbers and with shallower genotyping densities. In completely outbred populations, including all primate species, this is not possible. Human genetic linkage and association studies require a larger number of markers and, generally, a larger number of individuals, but the studies themselves are still possible with additional effort and expenses.

The techniques that have been applied to human association and linkage studies can likewise be applied to marmosets, and genomic costs have come down to the point that this approach can begin to be realistically considered. Marmosets, like any animal model, have the advantage that environmental variables can be minimized, emphasizing the genetic effects on the phenotype of interest. In addition, studies can be done in marmosets that, because of their invasiveness or other requirements, cannot reasonably be undertaken in humans.

These studies are certainly relevant for understanding marmoset biology itself, asking the question how specific genetic variation in marmosets causes a phenotype. While this is of interest to a specific community of primatologists and geneticists, the studies will be most meaningful if they are translationally relevant to humans. It remains to be seen if the same genotype–phenotype connections exist between species, particularly those as divergent as humans and marmosets. Certainly some broad examples exist, for instance, causative alleles across mammals leading to melanistic coloring [135,136] or the dramatic double muscling phenotype seen in cattle [137] and dogs [138], but the full breadth of associations remain to be explored.

At the far extreme of association studies are targeted studies of genetically unique animals. Every individual, regardless of species or genetic background, harbors some level of genetic load, mildly deleterious mutations that are ultimately under purifying selection [139]. Occasionally, one of these naturally occurring mutations will have a large enough effect or be in an interesting enough location that it will warrant further study. This may include simply targeted phenotyping of an individual animal or it may include additional selective breeding. Often the latter will not be possible in primate species such as marmosets, but it is formally possible.

Gene Expression

Another area of active study focuses on understanding gene expression and transcriptomics. This has led to immediately useful results in improving the quality and annotation of the marmoset genome, but it also has much more long-ranging and varied applications that have only begun to be explored. Targeted studies of biological questions and the molecular basis that underlies them are now possible in marmosets using next-generation sequencing technologies and the information available through the marmoset genome.

Most mammalian molecular studies of gene expression have occurred in rodents, particularly mice, or human samples. Nonhuman primate studies, while their utility was recognized for the physiological and genetic similarities to humans, were largely relegated to specialty labs. The reasons for this are numerous (and not entirely limited to nonhuman primates) but were driven in part by an incomplete understanding of their genomes and a lack of a critical mass to developing standardized testing platforms and resources. Microarrays, in situ hybridization, and quantitative PCR all required a priori knowledge of sequences for targeting and the development of specific targeted probes or primers, a difficult proposition unless there is a strong demand. Next-generation sequencing avoids these requirements by being agnostic to, and independent of, the specific targets under study. This allows the technology to be easily applied to a much more diverse range of species with greater ease.

These studies will play an increasingly important role as research transitions from descriptive to functional [140]. The marmoset can play an important role in this process, sitting at the intersection of an animal model with the increasingly sophisticated control over genetic and molecular manipulations and a close relationship with humans, particularly in relation to understanding the primate brain and behavior.

Transgenics

One of the most exciting developments in marmoset biology in the last decade is the development of transgenic animals [141]. These tools have brought the power of modern genetic manipulation to primates and have spurred the renaissance in the marmoset as a biomedical research model [142]. Chapter 20 elaborates on the methodologies and utility of transgenics, but as a preface it is important to note the role that the marmoset genome has played, and will continue to play, in this work.

The initial transgenic marmoset proof-of-principle studies introduced enhanced green fluorescent protein into marmoset embryos [141]. While this work demonstrated feasibility, it did not fundamentally alter the marmoset genome. Similarly, studies that place specific promoter sequences upstream of reporter genes to identify cell types or regional gene expression localization do not necessarily require marmoset sequence; they in nearly all cases will work equally well with human or rodent promoters [143]. But while these studies are certainly meaningful, there is significant interest in directly manipulating the marmoset genome [144]. As methods including CRISPR/Cas9 [145] and TALENs [146] are adapted to primate transgenics, more traditional methods have been used to introduce changes directly into the β-actin locus [147].

These technologies require a complete understanding of the genome of the host species to be maximally effective [148]. Off-target, unintended, effects not only confound results but can also render the entire experiment inoperable. A high-quality, complete, and reliable genome is necessary for these studies to progress. While many methodological advances were necessary to make transgenic marmosets feasible as a biomedical research model, the completion of the marmoset genome was a vitally important component in taking these next steps.

SUMMATION

The marmoset genome has opened the door to a vast array of studies that were previously inaccessible. Modern genetic tools can now be deployed for biomedical research into aspects of biology for which marmosets are uniquely suitable, including neuroscience and infectious disease. It is also possible to explore molecularly the unique aspects of marmoset biology mainly revolving around adaptation to a unique ecological niche and the behavioral and physiological associations with their unique pregnancies and gestation. Finally, advances in genomics can better help in the understanding of individual differences and population management.

References

[1] Rogers J, Gibbs RA. Comparative primate genomics: emerging patterns of genome content and dynamics. Nat Rev Genet 2014;15:347–59.

[2] Perelman P, Johnson WE, Roos C, Seuanez HN, Horvath JE, Moreira MA, Kessing B, Pontius J, Roelke M, Rumpler Y, Schneider MP, Silva A, O'Brien SJ, Pecon-Slattery J. A molecular phylogeny of living primates. PLoS Genet 2011;7:e1001342.

[3] The Marmoset Genome Sequencing and Analysis Consortium. The common marmoset genome provides insight into primate biology and evolution. Nat Genet 2014;46:850–7.

[4] Benirschke K, Anderson JM, Brownhill LE. Marrow chimerism in marmosets. Science 1962;138:513–5.

[5] Benirschke K, Brownhill LE. Further observations on marrow chimerism in marmosets. Cytogenetics 1962;1:245–57.

[6] Gengozian N, Batson JS, Eide P. Hematologic and cytogenetic evidence for hematopoietic chimerism in the marmoset, Tamarinus nigricollis. Cytogenetics 1964;10:384–93.

[7] Wislocki GB. Observations on twinning in marmosets. Am J Anat 1939;64:445–83.

[8] Watkins DI, Chen ZW, Hughes AL, Hodi FS, Letvin NL. Genetically distinct cell populations in naturally occurring bone marrow-chimeric primates express similar MHC class I gene products. J Immunol 1990;144:3726–35.

[9] Ross CN, French JA, Orti G. Germ-line chimerism and paternal care in marmosets (*Callithrix kuhlii*). Proc Natl Acad Sci USA 2007;104:6278–82.

[10] Sweeney CG, Curran E, Westmoreland SV, Mansfield KG, Vallender EJ. Quantitative molecular assessment of chimerism across tissues in marmosets and tamarins. BMC Genom 2012;13:98.

[11] Jaquish CE, Tardif SD, Toal RL, Carson RL. Patterns of prenatal survival in the common marmoset (*Callithrix jacchus*). J Med Primatol 1996;25:57–63.

[12] Windle CP, Baker HF, Ridley RM, Oerke AK, Martin RD. Unrearable litters and prenatal reduction of litter size in the common marmoset (*Callithrix jacchus*). J Med Primatol 1999;28:73–83.

[13] Gawad C, Koh W, Quake SR. Single-cell genome sequencing: current state of the science. Nat Rev Genet 2016;17:175–88.

[14] Kanda A, Sotomaru Y, Nobukiyo A, Yamaoka E, Hiyama E. Characterization of common marmoset (*Callithrix jacchus*) bone marrow-derived mesenchymal stem cells. Folia Histochem Cytobiol 2013;51:292–9.

[15] Wu Y, Zhang Y, Mishra A, Tardif SD, Hornsby PJ. Generation of induced pluripotent stem cells from newborn marmoset skin fibroblasts. Stem Cell Res 2010;4:180–8.
[16] Takabayashi S, Katoh H. Noninvasive genotyping of common marmoset (*Callithrix jacchus*) by fingernail PCR. Primates 2015; 56:235–40.
[17] Tie J, Uchigasaki S. Detection of short tandem repeat polymorphisms from human nails using direct polymerase chain reaction method. Electrophoresis 2014;35:3188–92.
[18] Trtkova K, Kupfermann H, Grahovac B, Mayer WE, O'HUigin C, Tichy H, Bontrop R, Klein J. Mhc-DRB genes of platyrrhine primates. Immunogenetics 1993;38:210–22.
[19] Sato K, Kuroki Y, Kumita W, Fujiyama A, Toyoda A, Kawai J, Iriki A, Sasaki E, Okano H, Sakakibara Y. Resequencing of the common marmoset genome improves genome assemblies and gene-coding sequence analysis. Sci Rep 2015;5:16894.
[20] de Oliveira EH, Neusser M, Muller S. Chromosome evolution in new world monkeys (Platyrrhini). Cytogenet Genome Res 2012; 137:259–72.
[21] Sherlock JK, Griffin DK, Delhanty JD, Parrington JM. Homologies between human and marmoset (*Callithrix jacchus*) chromosomes revealed by comparative chromosome painting. Genomics 1996;33:214–9.
[22] Neusser M, Stanyon R, Bigoni F, Wienberg J, Muller S. Molecular cytotaxonomy of New World monkeys (Platyrrhini) - comparative analysis of five species by multi-color chromosome painting gives evidence for a classification of *Callimico goeldii* within the family of Callitrichidae. Cytogenet Cell Genet 2001; 94:206–15.
[23] Gerbault-Serreau M, Bonnet-Garnier A, Richard F, Dutrillaux B. Chromosome painting comparison of *Leontopithecus chrysomelas* (Callitrichine, Platyrrhini) with man and its phylogenetic position. Chromosome Res 2004;12:691–701.
[24] Carbone L, Harris RA, Gnerre S, Veeramah KR, Lorente-Galdos B, Huddleston J, Meyer TJ, Herrero J, Roos C, Aken B, Anaclerio F, Archidiacono N, Baker C, Barrell D, Batzer MA, Beal K, Blancher A, Bohrson CL, Brameier M, Campbell MS, Capozzi O, Casola C, Chiatante G, Cree A, Damert A, de Jong PJ, Dumas L, Fernandez-Callejo M, Flicek P, Fuchs NV, Gut I, Gut M, Hahn MW, Hernandez-Rodriguez J, Hillier LW, Hubley R, Ianc B, Izsvak Z, Jablonski NG, Johnstone LM, Karimpour-Fard A, Konkel MK, Kostka D, Lazar NH, Lee SL, Lewis LR, Liu Y, Locke DP, Mallick S, Mendez FL, Muffato M, Nazareth LV, Nevonen KA, O'Bleness M, Ochis C, Odom DT, Pollard KS, Quilez J, Reich D, Rocchi M, Schumann GG, Searle S, Sikela JM, Skollar G, Smit A, Sonmez K, ten Hallers B, Terhune E, Thomas GW, Ullmer B, Ventura M, Walker JA, Wall JD, Walter L, Ward MC, Wheelan SJ, Whelan CW, White S, Wilhelm LJ, Woerner AE, Yandell M, Zhu B, Hammer MF, Marques-Bonet T, Eichler EE, Fulton L, Fronick C, Muzny DM, Warren WC, Worley KC, Rogers J, Wilson RK, Gibbs RA. Gibbon genome and the fast karyotype evolution of small apes. Nature 2014;513:195–201.
[25] Marques-Bonet T, Kidd JM, Ventura M, Graves TA, Cheng Z, Hillier LW, Jiang Z, Baker C, Malfavon-Borja R, Fulton LA, Alkan C, Aksay G, Girirajan S, Siswara P, Chen L, Cardone MF, Navarro A, Mardis ER, Wilson RK, Eichler EE. A burst of segmental duplications in the genome of the African great ape ancestor. Nature 2009;457:877–81.
[26] Pruitt KD, Tatusova T, Brown GR, Maglott DR. NCBI Reference Sequences (RefSeq): current status, new features and genome annotation policy. Nucleic Acids Res 2012;40:D130–5.
[27] Potter SC, Clarke L, Curwen V, Keenan S, Mongin E, Searle SM, Stabenau A, Storey R, Clamp M. The Ensembl analysis pipeline. Genome Res 2004;14:934–41.
[28] Oleksyk TK, Smith MW, O'Brien SJ. Genome-wide scans for footprints of natural selection. Philos Trans R Soc Lond B Biol Sci 2010;365:185–205.
[29] Zhang Z, Gerstein M. Of mice and men: phylogenetic footprinting aids the discovery of regulatory elements. J Biol 2003;2:11.
[30] Boffelli D. Phylogenetic shadowing: sequence comparisons of multiple primate species. Methods Mol Biol 2008;453:217–31.
[31] Boffelli D, McAuliffe J, Ovcharenko D, Lewis KD, Ovcharenko I, Pachter L, Rubin EM. Phylogenetic shadowing of primate sequences to find functional regions of the human genome. Science 2003;299:1391–4.
[32] Gerstein MB, Bruce C, Rozowsky JS, Zheng D, Du J, Korbel JO, Emanuelsson O, Zhang ZD, Weissman S, Snyder M. What is a gene, post-ENCODE? History and updated definition. Genome Res 2007;17:669–81.
[33] Pollard KS, Hubisz MJ, Rosenbloom KR, Siepel A. Detection of nonneutral substitution rates on mammalian phylogenies. Genome Res 2010;20:110–21.
[34] Tyner C, Barber GP, Casper J, Clawson H, Diekhans M, Eisenhart C, Fischer CM, Gibson D, Gonzalez JN, Guruvadoo L, Haeussler M, Heitner S, Hinrichs AS, Karolchik D, Lee BT, Lee CM, Nejad P, Raney BJ, Rosenbloom KR, Speir ML, Villarreal C, Vivian J, Zweig AS, Haussler D, Kuhn RM, Kent WJ. The UCSC Genome Browser database: 2017 update. Nucleic Acids Res 2017;45:D626–d634.
[35] Wang QF, Prabhakar S, Chanan S, Cheng JF, Rubin EM, Boffelli D. Detection of weakly conserved ancestral mammalian regulatory sequences by primate comparisons. Genome Biol 2007;8:R1.
[36] del Rosario RC, Rayan NA, Prabhakar S. Noncoding origins of anthropoid traits and a new null model of transposon functionalization. Genome Res 2014;24:1469–84.
[37] Warren IA, Naville M, Chalopin D, Levin P, Berger CS, Galiana D, Volff JN. Evolutionary impact of transposable elements on genomic diversity and lineage-specific innovation in vertebrates. Chromosome Res 2015;23:505–31.
[38] Konkel MK, Walker JA, Batzer MA. LINEs and SINEs of primate evolution. Evol Anthropol 2010;19:236–49.
[39] Lee HE, Ayarpadikannan S, Kim HS. Role of transposable elements in genomic rearrangement, evolution, gene regulation and epigenetics in primates. Genes Genet Syst 2015;90:245–57.
[40] Deininger P. Alu elements: know the SINEs. Genome Biol 2011; 12:236.
[41] Ray DA, Batzer MA. Tracking Alu evolution in New World primates. BMC Evol Biol 2005;5:51.
[42] Ray DA, Xing J, Hedges DJ, Hall MA, Laborde ME, Anders BA, White BR, Stoilova N, Fowlkes JD, Landry KE, Chemnick LG, Ryder OA, Batzer MA. Alu insertion loci and platyrrhine primate phylogeny. Mol Phylogenet Evol 2005;35:117–26.
[43] Warren WC, Hillier LW, Marshall Graves JA, Birney E, Ponting CP, Grutzner F, Belov K, Miller W, Clarke L, Chinwalla AT, Yang SP, Heger A, Locke DP, Miethke P, Waters PD, Veyrunes F, Fulton L, Fulton B, Graves T, Wallis J, Puente XS, Lopez-Otin C, Ordonez GR, Eichler EE, Chen L, Cheng Z, Deakin JE, Alsop A, Thompson K, Kirby P, Papenfuss AT, Wakefield MJ, Olender T, Lancet D, Huttley GA, Smit AF, Pask A, Temple-Smith P, Batzer MA, Walker JA, Konkel MK, Harris RS, Whittington CM, Wong ES, Gemmell NJ, Buschiazzo E, Vargas Jentzsch IM, Merkel A, Schmitz J, Zemann A, Churakov G, Kriegs JO, Brosius J, Murchison EP, Sachidanandam R, Smith C, Hannon GJ, Tsend-Ayush E, McMillan D, Attenborough R, Rens W, Ferguson-Smith M, Lefevre CM, Sharp JA, Nicholas KR, Ray DA, Kube M, Reinhardt R, Pringle TH, Taylor J, Jones RC, Nixon B, Dacheux JL, Niwa H, Sekita Y, Huang X, Stark A,

Kheradpour P, Kellis M, Flicek P, Chen Y, Webber C, Hardison R, Nelson J, Hallsworth-Pepin K, Delehaunty K, Markovic C, Minx P, Feng Y, Kremitzki C, Mitreva M, Glasscock J, Wylie T, Wohldmann P, Thiru P, Nhan MN, Pohl CS, Smith SM, Hou S, Nefedov M, de Jong PJ, Renfree MB, Mardis ER, Wilson RK. Genome analysis of the platypus reveals unique signatures of evolution. Nature 2008;453:175–83.

[44] Penzkofer T, Jager M, Figlerowicz M, Badge R, Mundlos S, Robinson PN, Zemojtel T. L1Base 2: more retrotransposition-active LINE-1s, more mammalian genomes. Nucleic Acids Res 2017;45:D68–d73.

[45] Konkel MK, Ullmer B, Arceneaux EL, Sanampudi S, Brantley SA, Hubley R, Smit AF, Batzer MA. Discovery of a new repeat family in the *Callithrix jacchus* genome. Genome Res 2016;26: 649–59.

[46] Tatsumoto S, Adati N, Tohtoki Y, Sakaki Y, Boroviak T, Habu S, Okano H, Suemizu H, Sasaki E, Satake M. Development and characterization of cDNA resources for the common marmoset: one of the experimental primate models. DNA Res 2013;20: 255–62.

[47] Shimizu M, Iwano S, Uno Y, Uehara S, Inoue T, Murayama N, Onodera J, Sasaki E, Yamazaki H. Qualitative de novo analysis of full length cDNA and quantitative analysis of gene expression for common marmoset (*Callithrix jacchus*) transcriptomes using parallel long-read technology and short-read sequencing. PLoS One 2014;9:e100936.

[48] Maudhoo MD, Ren D, Gradnigo JS, Gibbs RM, Lubker AC, Moriyama EN, French JA, Norgren Jr RB. De novo assembly of the common marmoset transcriptome from NextGen mRNA sequences. GigaScience 2014;3:14.

[49] Datson NA, Morsink MC, Atanasova S, Armstrong VW, Zischler H, Schlumbohm C, Dutilh BE, Huynen MA, Waegele B, Ruepp A, de Kloet ER, Fuchs E. Development of the first marmoset-specific DNA microarray (EUMAMA): a new genetic tool for large-scale expression profiling in a non-human primate. BMC Genom 2007;8:190.

[50] Datson NA, Morsink MC, Steenbergen PJ, Aubert Y, Schlumbohm C, Fuchs E, de Kloet ER. A molecular blueprint of gene expression in hippocampal subregions CA1, CA3, and DG is conserved in the brain of the common marmoset. Hippocampus 2009;19:739–52.

[51] Aubert Y, Allers KA, Sommer B, de Kloet ER, Abbott DH, Datson NA. Brain region-specific transcriptomic markers of serotonin-1A receptor agonist action mediating sexual rejection and aggression in female marmoset monkeys. J Sex Med 2013; 10:1461–75.

[52] Fukuoka T, Sumida K, Yamada T, Higuchi C, Nakagaki K, Nakamura K, Kohsaka S, Saito K, Oeda K. Gene expression profiles in the common marmoset brain determined using a newly developed common marmoset-specific DNA microarray. Neurosci Res 2010;66:62–85.

[53] Sasaki T, Oga T, Nakagaki K, Sakai K, Sumida K, Hoshino K, Miyawaki I, Saito K, Suto F, Ichinohe N. Developmental expression profiles of axon guidance signaling and the immune system in the marmoset cortex: potential molecular mechanisms of pruning of dendritic spines during primate synapse formation in late infancy and prepuberty (I). Biochem Biophys Res Commun 2014;444:302–6.

[54] Sasaki T, Oga T, Nakagaki K, Sakai K, Sumida K, Hoshino K, Miyawaki I, Saito K, Suto F, Ichinohe N. Developmental genetic profiles of glutamate receptor system, neuromodulator system, protector of normal tissue and mitochondria, and reelin in marmoset cortex: potential molecular mechanisms of pruning phase of spines in primate synaptic formation process during the end of infancy and prepuberty (II). Biochem Biophys Res Commun 2014;444:307–10.

[55] Debowski K, Drummer C, Lentes J, Cors M, Dressel R, Lingner T, Salinas-Riester G, Fuchs S, Sasaki E, Behr R. The transcriptomes of novel marmoset monkey embryonic stem cell lines reflect distinct genomic features. Sci Rep 2016;6:29122.

[56] Gyllensten U, Bergstrom T, Josefsson A, Sundvall M, Savage A, Blumer ES, Giraldo LH, Soto LH, Watkins DI. The cotton-top tamarin revisited: Mhc class I polymorphism of wild tamarins, and polymorphism and allelic diversity of the class II DQA1, DQB1, and DRB loci. Immunogenetics 1994;40:167–76.

[57] Faulkes CG, Arruda MF, Monteiro Da Cruz MA. Matrilineal genetic structure within and among populations of the cooperatively breeding common marmoset, *Callithrix jacchus*. Mol Ecol 2003;12:1101–8.

[58] Malukiewicz J, Boere V, Fuzessy LF, Grativol AD, French JA, de Oliveira e Silva I, Pereira LC, Ruiz-Miranda CR, Valenca YM, Stone AC. Hybridization effects and genetic diversity of the common and black-tufted marmoset (*Callithrix jacchus* and *Callithrix penicillata*) mitochondrial control region. Am J Phys Anthropol 2014;155:522–36.

[59] Malukiewicz J, Boere V, Fuzessy LF, Grativol AD, de Oliveira ESI, Pereira LC, Ruiz-Miranda CR, Valenca YM, Stone AC. Natural and Anthropogenic Hybridization in Two Species of Eastern Brazilian Marmosets (*Callithrix jacchus* and *C. penicillata*). PLoS One 2015;10:e0127268.

[60] Forman L, Kleiman DG, Bush RM, Dietz JM, Ballou JD, Phillips LG, Coimbra-Filho AF, O'Brien SJ. Genetic variation within and among lion tamarins. Am J Phys Anthropol 1986; 71:1–11.

[61] Martins MM, Nascimento AT, Nali C, Velastin GO, Mangini PB, Valladares-Padua CB, Galetti Jr PM. Genetic analysis reveals population structuring and a bottleneck in the black-faced lion tamarin (*Leontopithecus caissara*). Folia Primatol (Basel) 2011;82: 197–211.

[62] Farias IP, Santos WG, Gordo M, Hrbek T. Effects of forest fragmentation on genetic diversity of the critically endangered primate, the pied tamarin (*Saguinus bicolor*): implications for conservation. J Hered 2015;106(Suppl. 1):512–21.

[63] Cao R, Shi Y, Chen S, Ma Y, Chen J, Yang J, Chen G, Shi T. dbSAP: single amino-acid polymorphism database for protein variation detection. Nucleic Acids Res 2017;45:D827–d832.

[64] Frazer KA, Ballinger DG, Cox DR, Hinds DA, Stuve LL, Gibbs RA, Belmont JW, Boudreau A, Hardenbol P, Leal SM, Pasternak S, Wheeler DA, Willis TD, Yu F, Yang H, Zeng C, Gao Y, Hu H, Hu W, Li C, Lin W, Liu S, Pan H, Tang X, Wang J, Wang W, Yu J, Zhang B, Zhang Q, Zhao H, Zhao H, Zhou J, Gabriel SB, Barry R, Blumenstiel B, Camargo A, Defelice M, Faggart M, Goyette M, Gupta S, Moore J, Nguyen H, Onofrio RC, Parkin M, Roy J, Stahl E, Winchester E, Ziaugra L, Altshuler D, Shen Y, Yao Z, Huang W, Chu X, He Y, Jin L, Liu Y, Shen Y, Sun W, Wang H, Wang Y, Wang Y, Xiong X, Xu L, Waye MM, Tsui SK, Xue H, Wong JT, Galver LM, Fan JB, Gunderson K, Murray SS, Oliphant AR, Chee MS, Montpetit A, Chagnon F, Ferretti V, Leboeuf M, Olivier JF, Phillips MS, Roumy S, Sallee C, Verner A, Hudson TJ, Kwok PY, Cai D, Koboldt DC, Miller RD, Pawlikowska L, Taillon-Miller P, Xiao M, Tsui LC, Mak W, Song YQ, Tam PK, Nakamura Y, Kawaguchi T, Kitamoto T, Morizono T, Nagashima A, Ohnishi Y, Sekine A, Tanaka T, Tsunoda T, Deloukas P, Bird CP, Delgado M, Dermitzakis ET, Gwilliam R, Hunt S, Morrison J, Powell D, Stranger BE, Whittaker P, Bentley DR, Daly MJ, de Bakker PI, Barrett J, Chretien YR, Maller J, McCarroll S, Patterson N, Pe'er I, Price A, Purcell S, Richter DJ, Sabeti P, Saxena R, Schaffner SF, Sham PC, Varilly P, Altshuler D, Stein LD, Krishnan L, Smith AV, Tello-Ruiz MK, Thorisson GA, Chakravarti A, Chen PE, Cutler DJ, Kashuk CS, Lin S, Abecasis GR, Guan W,

Li Y, Munro HM, Qin ZS, Thomas DJ, McVean G, Auton A, Bottolo L, Cardin N, Eyheramendy S, Freeman C, Marchini J, Myers S, Spencer C, Stephens M, Donnelly P, Cardon LR, Clarke G, Evans DM, Morris AP, Weir BS, Tsunoda T, Mullikin JC, Sherry ST, Feolo M, Skol A, Zhang H, Zeng C, Zhao H, Matsuda I, Fukushima Y, Macer DR, Suda E, Rotimi CN, Adebamowo CA, Ajayi I, Aniagwu T, Marshall PA, Nkwodimmah C, Royal CD, Leppert MF, Dixon M, Peiffer A, Qiu R, Kent A, Kato K, Niikawa N, Adewole IF, Knoppers BM, Foster MW, Clayton EW, Watkin J, Gibbs RA, Belmont JW, Muzny D, Nazareth L, Sodergren E, Weinstock GM, Wheeler DA, Yakub I, Gabriel SB, Onofrio RC, Richter DJ, Ziaugra L, Birren BW, Daly MJ, Altshuler D, Wilson RK, Fulton LL, Rogers J, Burton J, Carter NP, Clee CM, Griffiths M, Jones MC, McLay K, Plumb RW, Ross MT, Sims SK, Willey DL, Chen Z, Han H, Kang L, Godbout M, Wallenburg JC, L'Archeveque P, Bellemare G, Saeki K, Wang H, An D, Fu H, Li Q, Wang Z, Wang R, Holden AL, Brooks LD, McEwen JE, Guyer MS, Wang VO, Peterson JL, Shi M, Spiegel J, Sung LM, Zacharia LF, Collins FS, Kennedy K, Jamieson R, Stewart J. A second generation human haplotype map of over 3.1 million SNPs. Nature 2007;449:851–61.

[65] Auton A, Brooks LD, Durbin RM, Garrison EP, Kang HM, Korbel JO, Marchini JL, McCarthy S, McVean GA, Abecasis GR. A global reference for human genetic variation. Nature 2015;526:68–74.

[66] Blake JA, Eppig JT, Kadin JA, Richardson JE, Smith CL, Bult CJ. Mouse Genome Database (MGD)-2017: community knowledge resource for the laboratory mouse. Nucleic Acids Res 2017;45: D723–d729.

[67] Vonholdt BM, Pollinger JP, Lohmueller KE, Han E, Parker HG, Quignon P, Degenhardt JD, Boyko AR, Earl DA, Auton A, Reynolds A, Bryc K, Brisbin A, Knowles JC, Mosher DS, Spady TC, Elkahloun A, Geffen E, Pilot M, Jedrzejewski W, Greco C, Randi E, Bannasch D, Wilton A, Shearman J, Musiani M, Cargill M, Jones PG, Qian Z, Huang W, Ding ZL, Zhang YP, Bustamante CD, Ostrander EA, Novembre J, Wayne RK. Genome-wide SNP and haplotype analyses reveal a rich history underlying dog domestication. Nature 2010;464:898–902.

[68] Gurgul A, Semik E, Pawlina K, Szmatola T, Jasielczuk I, Bugno-Poniewierska M. The application of genome-wide SNP genotyping methods in studies on livestock genomes. J Appl Genet 2014; 55:197–208.

[69] Gazave E, Darre F, Morcillo-Suarez C, Petit-Marty N, Carreno A, Marigorta UM, Ryder OA, Blancher A, Rocchi M, Bosch E, Baker C, Marques-Bonet T, Eichler EE, Navarro A. Copy number variation analysis in the great apes reveals species-specific patterns of structural variation. Genome Res 2011;21:1626–39.

[70] Gokcumen O, Babb PL, Iskow RC, Zhu Q, Shi X, Mills RE, Ionita-Laza I, Vallender EJ, Clark AG, Johnson WE, Lee C. Refinement of primate copy number variation hotspots identifies candidate genomic regions evolving under positive selection. Genome Biol 2011;12:R52.

[71] Gokcumen O, Tischler V, Tica J, Zhu Q, Iskow RC, Lee E, Fritz MH, Langdon A, Stutz AM, Pavlidis P, Benes V, Mills RE, Park PJ, Lee C, Korbel JO. Primate genome architecture influences structural variation mechanisms and functional consequences. Proc Natl Acad Sci USA 2013;110:15764–9.

[72] Gschwind AR, Singh A, Certa U, Reymond A, Heckel T. Diversity and regulatory impact of copy number variation in the primate *Macaca fascicularis*. BMC Genom 2017;18:144.

[73] Lee AS, Gutierrez-Arcelus M, Perry GH, Vallender EJ, Johnson WE, Miller GM, Korbel JO, Lee C. Analysis of copy number variation in the rhesus macaque genome identifies candidate loci for evolutionary and human disease studies. Hum Mol Genet 2008;17:1127–36.

[74] Perry GH, Tchinda J, McGrath SD, Zhang J, Picker SR, Caceres AM, Iafrate AJ, Tyler-Smith C, Scherer SW, Eichler EE, Stone AC, Lee C. Hotspots for copy number variation in chimpanzees and humans. Proc Natl Acad Sci USA 2006;103: 8006–11.

[75] Perry GH, Yang F, Marques-Bonet T, Murphy C, Fitzgerald T, Lee AS, Hyland C, Stone AC, Hurles ME, Tyler-Smith C, Eichler EE, Carter NP, Lee C, Redon R. Copy number variation and evolution in humans and chimpanzees. Genome Res 2008; 18:1698–710.

[76] Antunes SG, de Groot NG, Brok H, Doxiadis G, Menezes AA, Otting N, Bontrop RE. The common marmoset: a new world primate species with limited Mhc class II variability. Proc Natl Acad Sci USA 1998;95:11745–50.

[77] Wu MS, Tani K, Sugiyama H, Hibino H, Izawa K, Tanabe T, Nakazaki Y, Ishii H, Ohashi J, Hohjoh H, Iseki T, Tojo A, Nakamura Y, Tanioka Y, Tokunaga K, Asano S. MHC (major histocompatibility complex)-DRB genes and polymorphisms in common marmoset. J Mol Evol 2000;51:214–22.

[78] Carrion Jr R, Patterson JL. An animal model that reflects human disease: the common marmoset (*Callithrix jacchus*). Curr Opin Virol 2012;2:357–62.

[79] Neehus AL, Wistuba J, Ladas N, Eiz-Vesper B, Schlatt S, Müller T. Gene conversion of the major histocompatibility complex class I Caja-G in common marmosets (*Callithrix jacchus*). Immunology 2016;149:343–52.

[80] van der Wiel MK, Otting N, de Groot NG, Doxiadis GGM, Bontrop RE. The repertoire of MHC class I genes in the common marmoset: evidence for functional plasticity. Immunogenetics 2013;65:841–9.

[81] Watkins DI, Chen ZW, Hughes AL, Evans MG, Tedder TF, Letvin NL. Evolution of the MHC class I genes of a New World primate from ancestral homologues of human non-classical genes. Nature 1990;346:60–3.

[82] t Hart BA, van Meurs M, Brok HP, Massacesi L, Bauer J, Boon L, Bontrop RE, Laman JD. A new primate model for multiple sclerosis in the common marmoset. Immunol Today 2000;21: 290–7.

[83] Brok HP, Uccelli A, Kerlero De Rosbo N, Bontrop RE, Roccatagliata L, de Groot NG, Capello E, Laman JD, Nicolay K, Mancardi GL, Ben-Nun A, Hart BA. Myelin/oligodendrocyte glycoprotein-induced autoimmune encephalomyelitis in common marmosets: the encephalitogenic T cell epitope pMOG24-36 is presented by a monomorphic MHC class II molecule. J Immunol 2000;165:1093–101.

[84] Jagessar SA, Heijmans N, Blezer EL, Bauer J, Blokhuis JH, Wubben JA, Drijfhout JW, van den Elsen PJ, Laman JD, Hart BA. Unravelling the T-cell-mediated autoimmune attack on CNS myelin in a new primate EAE model induced with MOG34-56 peptide in incomplete adjuvant. Eur J Immunol 2012;42:217–27.

[85] Crawford DH, Janossy G, Hetherington CM, Francis GE, Edwards AJ, Hoffbrand AV, Prentice HG. Immunological characterization of hemopoietic cells in the common marmoset, rhesus monkey, and man. In search of a model for human marrow transplantation. Transplantation 1981;31:245–50.

[86] Porter RP, Gengozian N. Transplantation in the marmoset (Callithricidae): a technic for orthotopic, fitted skin grafts. Proc Soc Exp Biol Med 1968;127:860–7.

[87] Porter RP, Gengozian N. Immunological tolerance and rejection of skin allografts in the marmoset. Transplantation 1969;8: 653–65.

[88] Prasad S, Humphreys I, Kireta S, Gilchrist RB, Bardy P, Russ GR, Coates PT. MHC Class II DRB genotyping is highly predictive of in-vitro alloreactivity in the common marmoset. J Immunol Methods 2006;314:153–63.

[89] Prasad S, Humphreys I, Kireta S, Gilchrist RB, Bardy P, Russ GR, Coates PT. The common marmoset as a novel preclinical transplant model: identification of new MHC class II DRB alleles and prediction of in vitro alloreactivity. Tissue Antigens 2007; 69(Suppl. 1):72–5.

[90] Smith D, Trennery P, Farningham D, Klapwijk J. The selection of marmoset monkeys (*Callithrix jacchus*) in pharmaceutical toxicology. Lab Anim 2001;35:117–30.

[91] Zuhlke U, Weinbauer G. The common marmoset (*Callithrix jacchus*) as a model in toxicology. Toxicol Pathol 2003;31(Suppl.):123–7.

[92] Uno Y, Uehara S, Yamazaki H. Utility of non-human primates in drug development: comparison of non-human primate and human drug-metabolizing cytochrome P450 enzymes. Biochem Pharmacol 2016;121:1–7.

[93] Uehara S, Uno Y, Inoue T, Kawano M, Shimizu M, Toda A, Utoh M, Sasaki E, Yamazaki H. Individual differences in metabolic clearance of S-warfarin efficiently mediated by polymorphic marmoset cytochrome P450 2C19 in livers. Drug Metab Dispos 2016;44:911–5.

[94] Uehara S, Kawano M, Murayama N, Uno Y, Utoh M, Inoue T, Sasaki E, Yamazaki H. Oxidation of R- and S-omeprazole stereoselectively mediated by liver microsomal cytochrome P450 2C19 enzymes from cynomolgus monkeys and common marmosets. Biochem Pharmacol 2016;120:56–62.

[95] Harris RA, Tardif SD, Vinar T, Wildman DE, Rutherford JN, Rogers J, Worley KC, Aagaard KM. Evolutionary genetics and implications of small size and twinning in callitrichine primates. Proc Natl Acad Sci USA 2014;111:1467–72.

[96] Ford SM. Callitrichids as phyletic dwarfs, and the place of the callitrichidae in platyrrhini. Primates 1980;21:31–43.

[97] Montgomery SH, Mundy NI. Parallel episodes of phyletic dwarfism in callitrichid and cheirogaleid primates. J Evol Biol 2013;26:810–9.

[98] Garber PA. Vertical clinging, small body size, and the evolution of feeding adaptations in the Callitrichinae. Am J Phys Anthropol 1992;88:469–82.

[99] Marroig G, Cheverud JM. Size as a line of least evolutionary resistance: diet and adaptive morphological radiation in New World monkeys. Evolution 2005;59:1128–42.

[100] Mansfield K. Marmoset models commonly used in biomedical research. Comp Med 2003;53:383–92.

[101] Maccari G, Robinson J, Ballingall K, Guethlein LA, Grimholt U, Kaufman J, Ho CS, de Groot NG, Flicek P, Bontrop RE, Hammond JA, Marsh SG. IPD-MHC 2.0: an improved interspecies database for the study of the major histocompatibility complex. Nucleic Acids Res 2017;45:D860–d864.

[102] Watkins DI, Hodi FS, Letvin NL. A primate species with limited major histocompatibility complex class I polymorphism. Proc Natl Acad Sci USA 1988;85:7714–8.

[103] Cao YH, Fan JW, Li AX, Liu HF, Li LR, Zhang CL, Zeng L, Sun ZZ. Identification of MHC I class genes in two Platyrrhini species. Am J Primatol 2015;77:527–34.

[104] Shiina T, Kono A, Westphal N, Suzuki S, Hosomichi K, Kita YF, Roos C, Inoko H, Walter L. Comparative genome analysis of the major histocompatibility complex (MHC) class I B/C segments in primates elucidated by genomic sequencing in common marmoset (*Callithrix jacchus*). Immunogenetics 2011;63:485–99.

[105] Robinson J, Halliwell JA, Hayhurst JD, Flicek P, Parham P, Marsh SG. The IPD and IMGT/HLA database: allele variant databases. Nucleic Acids Res 2015;43:D423–31.

[106] Watkins DI, Garber TL, Chen ZW, Toukatly G, Hughes AL, Letvin NL. Unusually limited nucleotide sequence variation of the expressed major histocompatibility complex class I genes of a New World primate species (*Saguinus oedipus*). Immunogenetics 1991;33:79–89.

[107] Barnhart DD, Gengozian N. An evaluation of the mixed lymphocyte culture reaction in marmosets. Transplantation 1975;20: 107–15.

[108] Picus J, Holley K, Aldrich WR, Griffin JD, Letvin NL. A naturally occurring bone marrow-chimeric primate. II. Environment dictates restriction on cytolytic T lymphocyte-target cell interactions. J Exp Med 1985;162:2035–52.

[109] Tardif SD, Lacker HM, Feuer M. Follicular development and ovulation in the marmoset monkey as determined by repeated laparoscopic examination. Biol Reprod 1993;48:1113–9.

[110] Tardif SD, Smucny DA, Abbott DH, Mansfield K, Schultz-Darken N, Yamamoto ME. Reproduction in captive common marmosets (*Callithrix jacchus*). Comp Med 2003;53:364–8.

[111] Shimasaki S, Moore RK, Otsuka F, Erickson GF. The bone morphogenetic protein system in mammalian reproduction. Endocr Rev 2004;25:72–101.

[112] Goldizen AW. A comparative perspective on the evolution of tamarin and marmoset social systems. Int J Primatol 1990;11: 63–83.

[113] Ross C. Life history patterns of new world monkeys. Int J Primatol 1991;12:481–502.

[114] Gengozian N, Brewen JG, Preston RJ, Batson JS. Presumptive evidence for the absence of functional germ cell chimerism in the marmoset. J Med Primatol 1980;9:9–27.

[115] French JA, Taylor JH, Mustoe AC, Cavanaugh J. Neuropeptide diversity and the regulation of social behavior in New World primates. Front Neuroendocrinol 2016;42:18–39.

[116] Ren D, Chin KR, French JA. Molecular variation in AVP and AVPR1a in New World monkeys (Primates, Platyrrhini): evolution and implications for social monogamy. PLoS One 2014;9: e111638.

[117] Vargas-Pinilla P, Paixao-Cortes VR, Pare P, Tovo-Rodrigues L, Vieira CM, Xavier A, Comas D, Pissinatti A, Sinigaglia M, Rigo MM, Vieira GF, Lucion AB, Salzano FM, Bortolini MC. Evolutionary pattern in the OXT-OXTR system in primates: coevolution and positive selection footprints. Proc Natl Acad Sci USA 2015;112:88–93.

[118] Lee AG, Cool DR, Grunwald Jr WC, Neal DE, Buckmaster CL, Cheng MY, Hyde SA, Lyons DM, Parker KJ. A novel form of oxytocin in New World monkeys. Biol Lett 2011;7:584–7.

[119] Vargas-Pinilla P, Babb P, Nunes L, Pare P, Rosa G, Felkl A, Longo D, Salzano FM, Paixao-Cortes VR, Goncalves GL, Bortolini MC. Progesterone response element variation in the OXTR promoter region and paternal care in New World monkeys. Behav Genet 2017;47:77–87.

[120] Opazo JC, Wildman DE, Prychitko T, Johnson RM, Goodman M. Phylogenetic relationships and divergence times among New World monkeys (Platyrrhini, Primates). Mol Phylogenet Evol 2006;40:274–80.

[121] Wildman DE, Jameson NM, Opazo JC, Yi SV. A fully resolved genus level phylogeny of neotropical primates (Platyrrhini). Mol Phylogenet Evol 2009;53:694–702.

[122] Koike C, Uddin M, Wildman DE, Gray EA, Trucco M, Starzl TE, Goodman M. Functionally important glycosyltransferase gain and loss during catarrhine primate emergence. Proc Natl Acad Sci USA 2007;104:559–64.

[123] Arroyo JI, Hoffmann FG, Opazo JC. Evolution of the relaxin/insulin-like gene family in anthropoid primates. Genome Biol Evol 2014;6:491–9.

[124] Arroyo JI, Hoffmann FG, Opazo JC. Gene turnover and differential retention in the relaxin/insulin-like gene family in primates. Mol Phylogenet Evol 2012;63:768–76.

[125] Ogawa LM, Vallender EJ. Evolutionary conservation in genes underlying human psychiatric disorders. Front Hum Neurosci 2014;8:283.

[126] Bernstein RM, Sterner KN, Wildman DE. Adrenal androgen production in catarrhine primates and the evolution of adrenarche. Am J Phys Anthropol 2012;147:389–400.
[127] Vinson A, Raboin MJ. A practical approach for designing breeding groups to maximize genetic diversity in a large colony of captive rhesus macaques (*Macaca mulatta*). J Am Assoc Lab Anim Sci 2015;54:700–7.
[128] Williams-Blangero S. Research-oriented genetic management of nonhuman primate colonies. Lab Anim Sci 1993;43:535–40.
[129] Amos W, Harwood J. Factors affecting levels of genetic diversity in natural populations. Philos Trans R Soc Lond B Biol Sci 1998; 353:177–86.
[130] Katoh H, Takabayashi S, Itoh T. Development of microsatellite DNA markers and their chromosome assignment in the common marmoset. Am J Primatol 2009;71:912–8.
[131] Nievergelt CM, Mundy NI, Woodruff DS. Microsatellite primers for genotyping common marmosets (*Callithrix jacchus*) and other callitrichids. Mol Ecol 1998;7:1432–4.
[132] Raveendran M, Tardif S, Ross CN, Austad SN, Harris RA, Milosavljevic A, Rogers J. Polymorphic microsatellite loci for the common marmoset (*Callithrix jacchus*) designed using a cost- and time-efficient method. Am J Primatol 2008;70:906–10.
[133] Allendorf FW, Hohenlohe PA, Luikart G. Genomics and the future of conservation genetics. Nat Rev Genet 2010;11:697–709.
[134] Rylands AB, Mittermeier RA, de Oliveira MM, Kierulff MCM. *Callithrix jacchus*. The IUCN Red List of Threatened Species 2008. 2008.
[135] Mundy NI, Kelly J, Theron E, Hawkins K. Evolutionary genetics of the melanocortin-1 receptor in vertebrates. Ann NY Acad Sci 2003;994:307–12.
[136] Jackson IJ. Homologous pigmentation mutations in human, mouse and other model organisms. Hum Mol Genet 1997;6: 1613–24.
[137] McPherron AC, Lee SJ. Double muscling in cattle due to mutations in the myostatin gene. Proc Natl Acad Sci USA 1997;94: 12457–61.
[138] Mosher DS, Quignon P, Bustamante CD, Sutter NB, Mellersh CS, Parker HG, Ostrander EA. A mutation in the myostatin gene increases muscle mass and enhances racing performance in heterozygote dogs. PLoS Genet 2007;3:e79.
[139] de Valles-Ibanez G, Hernandez-Rodriguez J, Prado-Martinez J, Luisi P, Marques-Bonet T, Casals F. Genetic load of loss-of-function polymorphic variants in great apes. Genome Biol Evol 2016; 8:871–7.
[140] Sherwood CC, Duka T. Now that we've got the map, where are we going? Moving from gene candidate lists to function in studies of brain evolution. Brain Behav Evol 2012;80:167–9.
[141] Sasaki E, Suemizu H, Shimada A, Hanazawa K, Oiwa R, Kamioka M, Tomioka I, Sotomaru Y, Hirakawa R, Eto T, Shiozawa S, Maeda T, Ito M, Ito R, Kito C, Yagihashi C, Kawai K, Miyoshi H, Tanioka Y, Tamaoki N, Habu S, Okano H, Nomura T. Generation of transgenic non-human primates with germline transmission. Nature 2009;459:523–7.
[142] Okano H, Mitra P. Brain-mapping projects using the common marmoset. Neurosci Res 2015;93:3–7.
[143] Kvon EZ. Using transgenic reporter assays to functionally characterize enhancers in animals. Genomics 2015;106:185–92.
[144] Kishi N, Sato K, Sasaki E, Okano H. Common marmoset as a new model animal for neuroscience research and genome editing technology. Dev Growth Differ 2014;56:53–62.
[145] Liu Z, Li X, Zhang JT, Cai YJ, Cheng TL, Cheng C, Wang Y, Zhang CC, Nie YH, Chen ZF, Bian WJ, Zhang L, Xiao J, Lu B, Zhang YF, Zhang XD, Sang X, Wu JJ, Xu X, Xiong ZQ, Zhang F, Yu X, Gong N, Zhou WH, Sun Q, Qiu Z. Autism-like behaviours and germline transmission in transgenic monkeys overexpressing MeCP2. Nature 2016;530:98–102.
[146] Ke Q, Li W, Lai X, Chen H, Huang L, Kang Z, Li K, Ren J, Lin X, Zheng H, Huang W, Ma Y, Xu D, Chen Z, Song X, Lin X, Zhuang M, Wang T, Zhuang F, Xi J, Mao FF, Xia H, Lahn BT, Zhou Q, Yang S, Xiang AP. TALEN-based generation of a cynomolgus monkey disease model for human microcephaly. Cell Res 2016;26:1048–61.
[147] Shiozawa S, Kawai K, Okada Y, Tomioka I, Maeda T, Kanda A, Shinohara H, Suemizu H, James Okano H, Sotomaru Y, Sasaki E, Okano H. Gene targeting and subsequent site-specific transgenesis at the beta-actin (ACTB) locus in common marmoset embryonic stem cells. Stem Cells Dev 2011;20: 1587–99.
[148] Naito Y, Hino K, Bono H, Ui-Tei K. CRISPRdirect: software for designing CRISPR/Cas guide RNA with reduced off-target sites. Bioinformatics 2015;31:1120–3.

CHAPTER

20

Creating Genetically Modified Marmosets

Erika Sasaki[1,2,3]

[1]Department of Applied Developmental Biology, Central Institute for Experimental Animals, Kawasaki, Japan; [2]Keio Advanced Research Centers (KARC), Keio University, Tokyo, Japan; [3]Laboratory for Marmoset Neural Architecture, RIKEN Brain Science Institute, Wako, Japan

INTRODUCTION

Laboratory animals serve as intermediaries between basic research and clinical practice, bridging the gap between in vitro studies and clinical medicine. Animal models of disease are necessary to assess the safety and efficacy of newly developed drugs or therapies and understand the molecular mechanisms of disease onset. There are several ways to mimic human diseases in laboratory animals, such as drug administration, active or passive immunization, mechanical injuries, and genetic modification. Although all of these methods can accurately reproduce the symptoms of disease, the advantage to utilizing methods of genetic modification is that the onset and progression of disease may be recapitulated. Genetic engineering technologies can be applied to genetically modified (GM) animal models. In the biomedical sciences, GM mice are the most commonly used model organisms and have contributed to a greater understanding of gene functions and the mechanisms of disease onset.

Many human diseases are caused by genetic disorders or dysfunctions, inherited genetic conditions, and epigenetic dysregulation. Therefore, elucidating the relationships between gene function and phenotype is important for understanding the onset and progression of disease. In classical genetic studies, naturally occurring mutations that gave rise to observable phenotypes have been analyzed in many organisms to identify the essential genes or gene mutations responsible for the mutant phenotypes. This approach is called the "forward genetics approach." In recent decades, mutations have been artificially induced in the mouse genome with chemicals, radiation, or transposons to reveal the biological basis of mutant phenomena [1]. Using these technologies, large-scale mouse mutagenesis projects began in 1997 [2].

In contrast, a "reverse genetics" approach attempts to connect a given genetic sequence with its specific effect on an organism [1]. The reverse genetics approach uncovers individual gene functions by analyzing the hyperactivation or inhibition of the gene function through the engineering of DNA sequences. To understand the effects of individual gene functions on the whole organism, a wide variety of GM model organisms have been utilized, including both invertebrates and vertebrates. In 1974, the world's first GM animals were created by injecting simian virus 40 (SV40) viral DNA into mouse blastocyst stage embryos [3]. The SV40 DNA integrated itself into the mouse genome but was not transmitted to next-generation pups. When early mouse embryos were infected with Moloney leukemia virus (M-MuLV), exogenous M-MuLV was shown to be integrated into the mouse genome and transmitted to the germ line [4]. In 1981, several groups reported that plasmid DNA containing exogenous genes such as herpes thymidine kinase gene, rabbit b-globin gene, or human leukocyte interferon cDNA were injected into single-cell mouse embryos to create transgenic mice, and germline transmission of the injected DNA to next-generation pups was shown for the first time [5–7]. The transgenic mice produced by DNA microinjection were useful for examining gene functions by studying phenotypic changes produced by ectopic expression or overexpression of transgenes, and dominant, or dominantly mutated, functions. This method is still widely used to create transgenic mice; however, the location(s) of insertion and how much exogenous DNA is inserted into the host genome cannot be controlled when introducing transgenes by DNA pronuclear

The Common Marmoset in Captivity and Biomedical Research
https://doi.org/10.1016/B978-0-12-811829-0.00020-0

microinjection. Furthermore, this technique can only add exogenous genes into mice and cannot alter the endogenous host genome.

The limitations of the transgene microinjection method were addressed in 1989 by the creation of target gene knockout (KO) mice with germline transmission [8]. To achieve this, several technological breakthroughs were developed. The first innovation was the establishment of mouse embryonic stem cells (ESCs) [9,10] that produced chimeric mice and contributed to the germ line of the chimeras [11]. Subsequently, gene-targeting methods were developed for introducing exogenous DNA into targeted endogenous gene sequences by homologous recombination [12,13] and for positive–negative selection of gene-targeted ESCs [14]. The combination of homologous recombination and positive–negative selection of gene-targeted ESCs can disrupt endogenous genes in the host genome, exchange exogenous and endogenous genes, and accurately insert mutations into target genes. These technological breakthroughs were achieved by Mario R. Capecchi, Martin Evans, and Oliver Smithies, who were awarded the 2007 Nobel Prize in Physiology or Medicine.

As described above, there are several ways to genetically engineer embryos (Fig. 20.1). Genetic modifications are broadly classified according to two methods: random insertion of an exogenous gene into a host genome, called transgenic technology, and insertion of an exogenous gene into a specific region of a target gene, called gene targeting (Table 20.1). Transgenic techniques result in highly expressed transgenes; therefore, they are suitable for making models by introducing dominant negative-effect genes. Transgenic methods include retroviral transduction, DNA pronuclear injection, and transposons. Retroviral transduction has the advantage of high-efficiency transduction of the transgene, while DNA pronuclear injection is able to introduce large transgenes. Transposons are also suitable to introduce large transgenes but can also excise the transgene. Methods to produce gene-targeted animals include homologous recombination of ESCs delivered by blastocyst injection and homologous recombination of somatic cells delivered by nuclear transfer. Currently, gene-targeting methods work only in the mouse and rat. The reasons for this will be described later in this chapter.

While GM rodent models have significantly contributed to understanding the molecular mechanisms of disease onset and progression, the results from rodent experiments cannot always be extrapolated directly to humans because of the physiological and anatomical

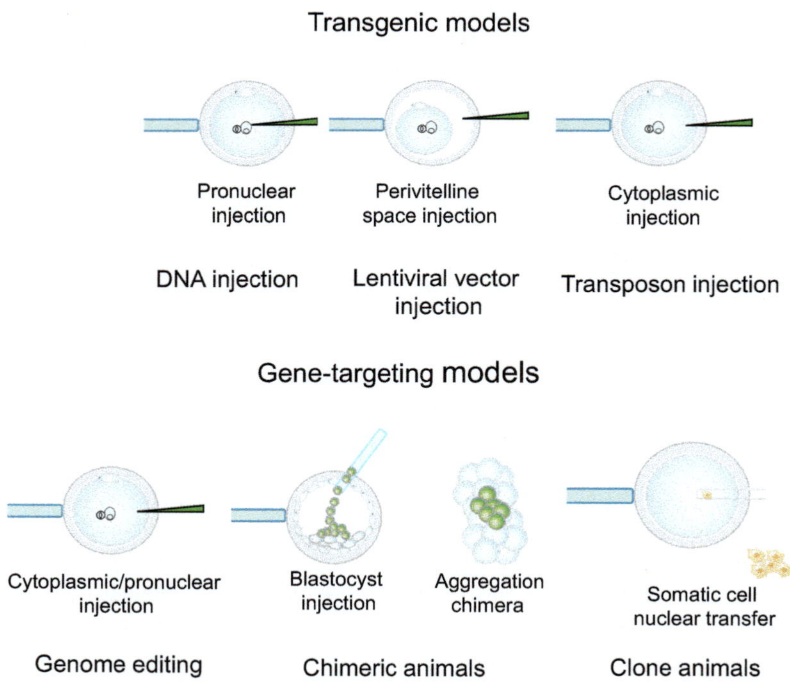

FIGURE 20.1 Methods for producing genetically modified animals. Upper panel: Gene introduction methods and location of the transgene injection. To produce transgenic animals, DNA is injected into pronuclei, lentiviral vector is introduced into the perivitelline space, or a transposon is injected into the cytoplasm or pronuclei. Lower panel: Methods for producing gene-targeted animals. There are three methods for generating target gene KO. In genome editing, genome editing components are introduced into cytoplasm or pronuclei of pronuclear stage embryos. Target gene KO chimeric animals are produced by injection of target gene KO embryonic stem (ES) cells into a blastocyst or by aggregation of eight cell embryos with the KO ES cells to form an aggregation chimera. Somatic cell nuclear transfer can be used for generating KO by nuclear transfer into the oocyte of the target gene KO cells.

TABLE 20.1 Various Methods to Produce Genetically Modified Animals

Type of Animal Model	Method	Features
Transgenic animals	Lentiviral vector	Highly efficient transgenesis
	Pronuclear microinjection	Can introduce a large transgene
	Transposon	Can introduce and remove a large transgene
Gene-targeted animals	Blastocyst injection 8-cell stage embryo aggregation	Precisely modified cells can be selected before injection
	Genome editing of embryos	Can be applied to most species
	Somatic cell nuclear transfer of genetically modified cells	Powerful method for animals that lack embryonic stem cells/ induced pluripotent stem cells, but has low efficiency

differences between mice and humans [15]. Especially, mouse models of high cognitive dysfunction diseases do not always recapitulate the patients' symptoms, while nonhuman primate (NHP) models often show closer alignment to the symptoms of human patients [16]. Therefore, genetically engineered NHP models are expected to play important roles that cannot be accomplished with mouse models. When developing new therapies for neuronal dysfunction diseases, such as Alzheimer's disease or Parkinson's disease, early treatment facilitates improved prognoses; therefore, genetically engineered disease models are essential.

ADVANTAGES OF THE COMMON MARMOSET IN GENETIC ENGINEERING

Marmosets have been widely used in infectious disease research, toxicology, drug development, and neuroscience research. Recent progress in stem cell science and regenerative medicine has increased the importance of NHP models for assessing the safety and efficacy of stem cell therapies, including methods using induced pluripotent stem cells (iPSCs), ESCs, and/or mesenchymal stem cells. To assess the safety and efficacy of these developing therapies in animal models of disease, species-specific research tools are critical to the precise evaluation of the therapeutic effects on the physiologic, genetic, and epigenetic status of the animal model system. To this end, marmoset-specific DNA microarrays and cDNA libraries have been developed and the whole genome sequence of the marmoset has been published [17–20]. Several monoclonal antibodies directed at the marmoset cluster of differentiation molecules have been developed for analysis of the phenotype and function of immune and hematopoietic cells and also to enable differentiation of somatic cells from iPSCs or ESCs [21–23].

Several marmoset ESC and iPSC lines have also been established [24–28]. While there are several published reports describing xenotransplantation of human ESCs or iPS-derived cells in NHP models, these studies required administration of immune-suppressive agents to avoid or prolong the time of immune rejection of the xenotransplanted cells [29,30]. Marmoset ESCs and iPSCs enable allograft or autograft transplantations in the absence of immunosuppressive agents and thus may facilitate a more precise assessment of the safety and efficacy of stem cell transplantation.

Compared to macaque monkeys, the marmoset is a fecund animal that exhibits several unique reproductive characteristics. Details of marmoset reproduction were reviewed by Tardif et al. [31] and are also describe in detail in Chapter 8 of this volume. While the rhesus monkey breeds seasonally, the marmoset is not a seasonal breeder and breeds year-round. Sexual maturation, defined by the onset of ovulation in females and ejaculation of motile spermatozoa in males, begins around 1 year of age, but it is recommended that breeding commences after 1.5 years of age [31]. Marmosets are unusual among primates in having a postpartum ovulation that typically results in conception and successful delivery even when females are lactating. This postpartum ovulation occurs 2 weeks after delivery, allowing common marmosets to deliver twice a year and resulting in a shorter birth interval.

Broad ranges of the length of marmoset gestation periods have been reported. Hearn and Lunn reported that the gestation period of the marmoset was estimated to be about 148 ± 4.3 days [32]. On the other hand, an average gestation period of 143 days was reported by Cashell et al. [33] and 144 days by Hearn [34]. The wide range may be attributed to the variability in identifying the day of ovulation due to the differential sensitivities of radioimmunoassays, enzyme-linked immunoassays, or enzyme immunoassays used to measure serum progesterone. The average gestation period in the Central Institute for Experimental Animals' colony is 141.42 ± 0.44 days.

Twin pregnancies are common but triplets are also frequently observed in captive marmosets; quadruplets are occasionally observed. As a result, females deliver 4–6 infants in a year and 40–80 offspring over their lifetime (Table 20.2). Multizygotic pregnancies are supported by a shared placenta with vascular anastomoses and a

TABLE 20.2 Cross-Species Comparison of Reproductive Traits

	Mouse	Marmoset	Cynomolgus
Sexual maturation	8 weeks	1.5–2 years	3–4 years
Litter size	6–8 pups	2–3 neonates	1 neonates
Gestation period	20 days	145–148 days	175–180 days
Delivery interval	20–28 days	154–157 days	550 days
Number of offspring/year	36–48 pups	4–6 neonates	<1 neonates
Number of deliveries/life	5–6	20–30	10–12
Number of offspring/life	36–48 pups	40–80 neonates	10–12 neonates

fused chorion. This shared placentation results in bone marrow chimerism among multizygotic siblings, which allows for survival of tissue allografts between siblings [35,36].

The marmoset ovulatory cycle is around 28 days. Ovulation cycles can be controlled with prostaglandin F2a analogue (e.g., cloprostenol) when administered greater than 10 days after the onset of the luteal phase. Cloprostenol induces luteolysis of the corpus luteum, which maintains early pregnancies by producing progesterone, estradiol, and inhibin A [26,37,38]. Based on this repertoire of reproductive characteristics, the marmoset is a suitable model animal for studying reproductive biology and creating GM models.

HISTORY OF GENETICALLY MODIFIED NONHUMAN PRIMATES

Development of Transgenic Nonhuman Primates

As human disease models, GM mice have contributed to revealing the functions of genes responsible for diseases; however, many differences exist between humans and rodents. In many instances, the differences in susceptibility to pathogens, neurophysiological function, and drug sensitivity have limited the extrapolation of research outcomes from studies employing mouse models to clinical applications in humans.

NHPs are valuable models for understanding human diseases and developing new therapies due to their genetic, morphological, and physiological similarities to humans. In fact, many NHPs have been utilized to assess safety during drug development, and drug-induced and surgical disease models have been used to evaluate the efficacy of new drugs. However, most diseases are caused by genetic or epigenetic dysfunctions; therefore, drug-induced and surgical disease models have limited applications. In contrast, GM models can mimic diseases caused by genetic dysfunction and also allow for the study of disease induction and progression. Thus, GM NHPs offer tremendous promise as research models for biomedical science.

Over the past 16 years, several transgenic NHPs have been created using rhesus monkeys (*Macaca mulatta*), cynomolgus monkeys (*Macaca fascicularis*), and common marmosets (*Callithrix jacchus*). In 2001, the first transgenic monkeys were created by using retroviral vectors to introduce the green fluorescent protein (GFP) gene into rhesus monkey oocytes, and embryos fertilized by intracytoplasmic sperm injection (ICSI) [39]. Although this transgenic monkey carried the GFP gene, GFP expression was only observed in the hair shafts and toenails of a stillborn male by fluorescence. In a live animal named "ANDi" (for the reverse-transcribed direction of "inserted DNA"), GFP gene was detected by reverse transcription-polymerase chain reaction (RT-PCR) in the placenta and cord blood, but not in the somatic tissues [39]. In the same year, another research group obtained two GFP transgenic rhesus monkeys by injecting lentiviral vectors into blastocyst-stage embryos [40]. In this study, GFP expression was observed in the placenta of the newborn monkeys by direct fluorescence and by immunohistochemical analysis using an anti-GFP monoclonal antibody on frozen tissue sections, but no GFP expression was observed in somatic tissues [40].

After these experiments, transgenic common marmosets were created by injecting lentiviruses encoding GFP into the perivitelline space of preimplantation stage early embryos that ranged from the four-cell stage to the blastocyst stage. Only GFP-expressing embryos were transferred to surrogate mothers. As a result, five transgenic marmosets were obtained, four of which expressed GFP in several somatic tissues. One of these four animals gave birth to a GFP-expressing offspring, thus demonstrating the first germline transmission of a transgene [41]. After publication of this study, two additional female animals also showed germline transmission of the transgene [42].

The first transgenic rhesus monkeys that demonstrated GFP transgene expression in several somatic tissues were produced by simian immunodeficiency virus-based lentiviral vector injection into early cleavage stage embryos [43]. Recently, transgenic cynomolgus monkeys have also been generated [44,45]. To obtain transgenic cynomolgus monkeys that homogeneously expressed GFP in multiple tissues, Seita and colleagues optimized protocols for purifying the lentiviral solution and the protocol for injecting lentiviruses into cynomolgus monkey oocytes [45]. Mature oocytes were subjected to lentivirus injection into the perivitelline space injection prior to ICSI; then, cultured blastocyst-stage embryos were transferred into surrogate mothers [45].

Toward the Application of Transgenic Nonhuman Primates

The first transgenic monkey models of human disease were reported in 2008 [46]. Huntington's disease (HD) models were produced using the human huntingtin (HTT) gene with 84 CAG repeats and GFP-expressing lentivirus vectors, and five HD monkeys carrying 29–88 CAG repeats were obtained. Of these animals, four newborns that carried more than 65–88 CAG repeats exhibited HD patient-like symptoms, such as dystonia, chorea, difficulty swallowing, and respiratory difficulty, but in three of the four animals severe symptoms resulted in early death [46]. Immunohistochemical analysis showed HTT aggregates or inclusions in the striatum and cortex.

Recently, transgenic cynomolgus monkeys expressing human methyl-CpG binding protein 2 (MeCP2) in the brain, under the control of the human synapsin promoter [44], were also generated. These animals exhibited autism-like behaviors, such as a higher frequency of repetitive circular locomotion in their home cages, an increased stress response, and less interaction with wild-type cagemate monkeys. Furthermore, the authors subcutaneously transplanted pieces of testicular tissue into mice, and mature motile sperm were obtained from the xenografts after 10 months. Using these mature motile sperm, next-generation (F1) cynomolgus monkeys carrying MeCP2 transgene were obtained by ICSI [44].

GM NHPs are receiving increasing attention as useful models not only of neurodegenerative diseases, such as Alzheimer's disease, Parkinson's disease, and HD, but also as models for basic neuroscience research aiming to understand how high cognitive functions developed in primates. For this purpose, transgenic marmosets were generated with a genetically encoded calcium indicator, GCaMP, under the control of either cytomegalovirus or the human synapsin promoter [47]. GCaMP visualizes neural activity by intracellular calcium signaling in live animals. These GCaMP transgenic marmosets point toward the possibility of monitoring neural activity and intracellular calcium dynamics in live NHPs with functional confocal and multiphoton optical microscopy and will likely facilitate our understanding of neural circuitry and the development of high cognitive functions in the brain [47].

Recently, a transgenic marmoset model of polyglutamine (polyQ) diseases was reported [48]. PolyQ diseases are inherited neurodegenerative diseases that include HD and spinocerebellar ataxia. The polyQ disease marmosets were produced by introducing the human ataxin 3 gene with 120 CAG repeats encoding an expanded polyQ stretch using lentiviral vectors. Three marmosets with higher transgene expression indicated neurological symptoms of varying degrees such as spontaneous activity and grip strength. Neurological symptoms became evident at 3–4 months after birth, demonstrating a time-dependent disease progression. In this report, protein inclusions were observed in these animals by pathological analyses and revealed neurodegeneration and intranuclear polyQ with gliosis, which recapitulate the neuropathological features of polyQ disease patients. Furthermore, enlargement of the fourth ventricle observed in one living symptomatic marmoset by brain MRI analyses suggests cerebellar atrophy. This report demonstrated the usefulness of transgenic marmosets as human disease models [48]. Moreover, germline transgene transmission in the second generation was confirmed in both the polyQ disease model and the GCaMP transgenic marmoset offspring, indicating that lentiviral vector-mediated transgenesis is a reliable method for establishing transgenic marmoset lines.

PRACTICAL METHODOLOGY FOR PRODUCING TRANSGENIC MARMOSETS

Design and Production of Lentiviral Vectors

The design of the lentiviral vector is the first critical step in producing transgenic marmosets. The lentiviral vector is limited to DNA insertion of up to 8.5 kb, including the promoter region. In our experience, a DNA insert of less than 5.0 kb is optimal for insertion into lentiviral vectors used to introduce exogenous DNA into marmoset embryos.

Due to the long life span of NHPs, the creation of GM NHP models that exhibit observable phenotypes in the founder generation is critically important. Furthermore, crucial ethical considerations require efforts to reduce the number of animals used and avoid euthanasia of failed GM animals after embryonic manipulation. To address these issues, confirmation of transgene insertion into the marmoset genome before embryo transfer to surrogate mothers is an important step.

To confirm transgene insertion, fluorescent protein cDNA, such as GFP or Kusabira Orange genes, should be inserted into individual lentiviral vector constructs, together with the desired transgene, as a transgene integration marker [41]. Unlike plasmid vectors, lentiviral vectors express the encoded gene after it is integrated into host embryo chromosomes [49]. To express the desired transgene and fluorescent protein gene in embryos from a single lentiviral vector construct, 2A peptides are recommended rather than inserting internal ribosomal entry sites due to the small size of the genes [50,51]. These 2A peptides are small "self-cleaving" peptides (18–22 amino acids on average) that were identified in foot-and-mouth disease virus, a member of the picornavirus family [52]. They mediate a ribosome-skipping event that enables multiple gene translations from one mRNA. When a tissue-specific gene promoter is used to drive the transgene, a ubiquitous promoter, such as cytomegalovirus (CMV) or cytomegalovirus (CMV) early enhancer element, chicken β-actin promoter and splice acceptor of the rabbit beta-globin gene (CAG), or an early embryonic development-specific promoter, such as the Early Transposon promoter and Oct-4 and Sox2enhancers (EOS) promoter, is useful for driving the fluorescent protein in preimplantation embryos [53]. This transgene integration marker cassette should be inserted together with the transgene cassette, separated by an insulator to prevent interference of the two promoters. The promoter and the fluorescent protein gene are roughly 1000–1500 base pairs (bp); therefore, an optimal transgene size should be around 3500–4000 bp including the promoter. This transgene integration marker can be detected in the eight-cell stage embryo, corresponding to genome activation in marmoset embryos.

Currently, lentiviral-based gene introduction is the only method for creating transgenic NHPs, including marmosets. This lentiviral vector transgenesis method could be used to produce target gene KO animals using short hairpin RNA expression vectors. However, target gene KO marmosets cannot be produced by this lentiviral method because lentiviral vector can only introduce exogenous gene into chromosomes.

Synchronization of the Ovarian Cycle in the Donor and Recipient

Obtaining and producing in vitro fertilized embryos is critically important for establishing ESCs and creating GM animals. To achieve this, ovarian cycle control is the most important procedure. Many research groups have developed methods of ovarian cycle synchronization, follicular stimulation, and in vitro fertilization (IVF). Recently, Kropp et al. reviewed these methodologies across research groups and compared published protocols [54]. This review makes a valuable contribution to determining the optimal protocols for synchronizing marmoset ovarian cycles and producing IVF embryos (Table 20.3).

To control ovarian cycles, it is necessary to understand the endocrinology of the cycle. As described above, the average marmoset ovarian cycle length is about 28 days, with an approximately 8-day follicular phase and 19-day luteal phase [55]. The follicular and luteal phases can be determined by measurement of serum progesterone concentrations. When serum progesterone levels are higher than 10 ng/mL, this corresponds to the luteal phase, and when serum concentrations are less than 10 ng/mL, this represents the follicular phase [55]. Because progesterone secretion from the corpus luteum can be controlled by PGF2α administration at 10 days after ovulation or later, it is possible to synchronize ovarian cycles between oocyte donors and embryo recipient animals.

By administering PGF2α at 10 days postovulation, the ovarian cycle can be shortened from 28 to 21 days. In the Central Institute for Experimental Animals (CIEA) colony, both donor and recipient animals are divided into three groups. The timing of PGF2α administration to each of these groups is staggered such that members of each of the donor groups ovulate in turn, and time-matched recipients are available for embryo transfer every week on a 3-week cycle. After a single intramuscular administration of PGF2α (0.8 μg) to luteal phase animals, a 0.2 mL of blood sample or urine sample is collected on days −1, 1, 9, 11, and 13 after PGF2α injection. This sample is used to measure serum progesterone concentration by an enzyme immunoassay or enzyme-linked immunoassay. The day of ovulation is defined as the day before which the plasma progesterone concentration exceeds 10 ng/mL (Fig. 20.2). Usually ovulation occurs 9–13 days after PGF2α administration.

Surgical and nonsurgical methods of embryo transfer are available for use in the marmoset. In a study by Marshall et al. using nonsurgical techniques, synchronous transfer of naturally fertilized embryos (performed when both the donor and recipient were 5–8 days postovulation) resulted in a low pregnancy rate (9%) and a 0% birth rate. In contrast, asynchronous embryo transfer (performed when the donor was 5–8 days postovulation and the recipient was 2–4 days postovulation) achieved a 44.4% implantation rate and a 33.3% birth rate [56]. In another study, using surgical techniques, synchronous transfer of naturally fertilized embryos (performed when both the donor and recipient were 6–10 days postovulation) resulted in a 60% implantation rate and a 50% birth rate [57]. Previous work has shown that delivery of in vitro fertilized embryos by synchronous transfer can result in the birth of live neonates [58]. In our experience, both synchronous and asynchronous embryo transfer are successful using nonsurgical methods.

Follicular Stimulation of Donor Animals

Retrieving as many fertile oocytes as possible is important for procuring IVF embryos to create GM marmosets. To obtain large numbers of oocytes, ovaries are stimulated for follicular growth using gonadotropins. Recombinant human follicle-stimulating hormone is delivered daily for 9 days, and human chorionic gonadotropin (hCG) is administered on day 10 to the animals

TABLE 20.3 Published Follicular Stimulation Protocols for Common Marmosets and the Associated Comparative Success Rates of in vitro Embryonic Development

Reference	FSH Delivery Day	FSH Dose IU/Day	FSH Delivery Time(s)	hCG Delivery	hCG Dose IU	hCG Delivery Time	Retrieval Method[a]	Mean Number of Retrieved Oocytes	MII Oocytes (%)	Number of MII Oocytes/ GV Oocytes	Grade A Pronuclei (%)	Grade B Pronuclei (%)	Cleaved (%)	Blastocyst (%)
Lopata 1 [58]	NA	NA	NA	Day 8	75	9:00	Lap	1.4	100	15/15	42	0	22.2	14.3
Lopata 2 [58]	NA	NA	NA	Day 7	75	9:00	Lap	1.8	100	11/11	67[d]	0		
Lopata 3 [58]	NA	NA	NA	Day 7	75	17:00	Lap	2.2	81.3	24/31	70[d]	50		
Wilton [64]	NA	NA	NA	Day 7	75	13:00	Lap	2.7	94.3	83/88	87.3[e]	71.4[e]	42.6	NA
Gilchrist [59]	Day 0–3	3	am/pm	NA	NA	NA	OR	NA	77	421/546	85	85.5	71	4.35
Marshall 1 [101]	NA	NA	NA	Day 7	75	13:00	Lap	2.4	NA	NA	NA	NA	NA	NA
Marshall 2 [101]	Day 1–6	1	am/pm	Day 7	75	15:30	Lap	1.25	60	3/5	0	NA	0	0
Marshall 3 [101]	Day 1–6	10	am/pm	Day 7	75	15:30	Lap	0.5	100	1/1	100	NA	100	0
Marshall 4 [101]	Day 1–6	25	am/pm	Day 7	75	15:30	Lap	1	100	2/2	100	NA	50	0
Marshall 5 [101]	Day 1–5	50	am/pm	Day 6	75	15:30	Lap	5.1	85	109/128	0	NA	0	0
Marshall 6 [101]	Day 1–5	50	am	Day 6	75	pm	Lap				46	NA	90	47
Marshall 7 [101]	Day 1–6	50	am/pm	Day 7	75	15:30	Lap				57	NA	50	20
Grupen 1 [89]	Day 1–6	50	am/pm	NA	NA	NA	OR	50.6[b]	79	NA	67[f]	NA	45	47
Grupen 2 [89]	Day 1–6	50	am/pm	Day 7	500	14:00	OR	60.1[c]	68	NA				
Takahashi [63]	Day 1–9	50	am	Day 10	75	17:00	Lap	15.5	67	NA	85	81	84	49
Park [47]	Day 1–9	50	am	Day 10	75	17:00	Lap	46	79.4	625/787	88	NA	NA	NA

[a]*Retrieval method: Lap = laparotomy and follicular aspiration; OR = ovary removal and follicular aspiration.*
[b]*Total mean of 50.6 includes cumulous–oocyte complexes measuring 0.7–1.5 mm (28.1) and >1.5 mm (22.5).*
[c]*Total mean of 60.1 includes cumulous-oocyte complexes measuring 0.7–1.5 mm (26.8) and >1.5 mm (33.3).*
[d]*Includes polyspermic fertilization.*
[e]*Grade A: Grade I and II, Grade B: Grade III and IV.*
[f]*In vivo mature oocytes.*

For all protocols described below, marmosets were administered PGF2α on day 0. NA indicates that the data are not available. *FSH*, Follicle-stimulating hormone; *hCG*, human chorionic gonadotropin; *MII*, Metaphase II.

This table is adapted from Kropp J, Di Marzo A, Golos T. Assisted reproductive technologies in the common marmoset: an integral species for developing nonhuman primate models of human diseases. Biol Reprod 2017;96:277–87. The numbers that follow the authors in this column indicate multiple protocols reported in the same manuscript.

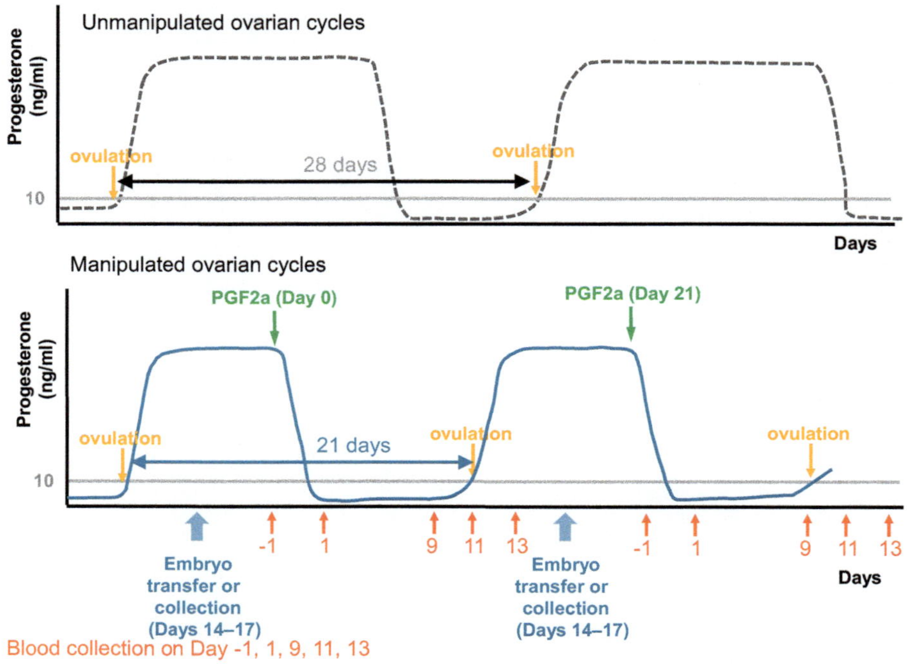

FIGURE 20.2 Unmanipulated and manipulated ovarian cycles. Upper panel: Normal cell cycle is approximately 28 days and consists of an 8-day follicular phase and a 19-day luteal phase. Lower panel: Administration of prostaglandin F2α analogue (PGF2α) later than 10 days after the start of the luteal phase will induce luteolysis. Ovulation occurs 9–13 days after administration. To detect ovulation, blood samples are collected on days −1, 1, 9, 11, and 13 after PGF2α injection. Under natural mating conditions, fertilized embryos can be retrieved or transferred 4–8 days postovulation.

with a confirmed PGF2α-induced reset of the ovarian cycle to the follicular stage (Fig. 20.3). Several studies have reported optimized ovarian stimulation conditions for the retrieval of fertile oocytes (Table 20.3) [47,54,58–64]. From these reports, it seems that stimulating the ovaries to obtain large numbers of oocytes takes more than 6 days and at least 50 IU/day of follicle-stimulating hormone (FSH) priming.

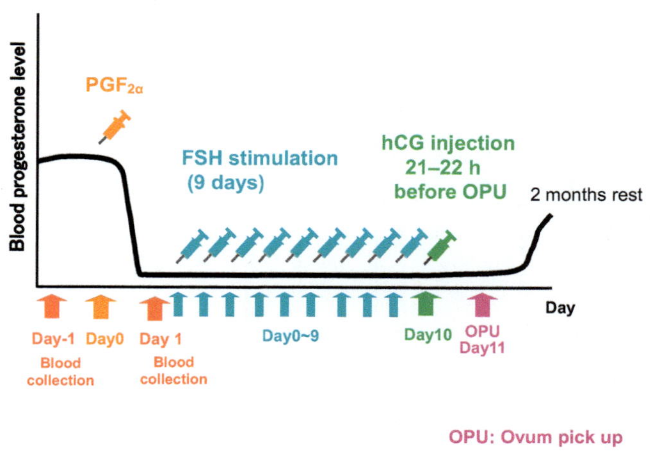

FIGURE 20.3 Schema of ovarian stimulation. After prostaglandin F2a analogue administration and confirmed reset of the luteal phase, 50U of follicle-stimulating hormone (FSH) is administered for 9 days and human chorionic gonadotropin (hCG) is administered on Day 10. Ovum pick-up (OPU) is performed on Day 11 at 18 h after hCG administration.

Administration of 50 IU/day of FSH for 9 days has been used in many previous studies of marmoset transgenesis based on the report by Summers et al., which described an average interval of 10.7 days from PGF2α-induced luteolysis to ovulation [38]. Our preliminary examination indicated that there are no differences in the number of retrieved oocytes between 10 and 9 days of FSH administration [38,41,47,63,65]. However, this protocol may not be the most efficient for retrieving marmoset oocytes, and further improvements of the protocol are still necessary to increase the mean number of retrieved oocytes and achieve much higher success rates of in vitro maturation (IVM) and IVF.

It is reported that exogenous FSH administration affects oocyte quality and reduces the success rates of IVM and IVF [66,67]. One reason is that exogenous gonadotropin administration raises the peripheral blood concentration of inhibin, a heterodimeric glycoprotein hormone secreted by granulosa cells [68]. High inhibin levels decrease the ovulatory response to FSH, consequently reducing the number of developing follicles and fertilization rates or causing a reduction in oocyte and embryo quality [67,69–72]. Neutralization of endogenous inhibin bioactivity by passive or active immunization yields successful multiple ovulations without reduced quality in several species, such as mice, hamsters, rats, guinea pigs, sheep, cows, and mares [73–83]. Inhibin is also secreted by ovarian follicles during the ovarian cycle in the marmoset, and it may also suppress endogenous FSH during ovarian stimulation in marmosets [84]. Thus, neutralization of endogenous inhibin bioactivity in the marmoset by antiinhibin antibody during ovarian stimulation may increase the number and quality of retrieved oocytes.

Retrieval of Oocytes Using the Ovum Pick-Up Procedure

Marmoset oocytes are retrieved by ovum pick-up (OPU) and follicular aspiration. This procedure is performed under general anesthesia. Although several anesthesia protocols for OPU have been reported, our laboratory uses a preanesthetic cocktail comprising medetomidine, midazolam, and butorphanol. General anesthesia is maintained with isoflurane. Heart rate, arterial oxygen saturation, and other physiologic parameters are monitored during the surgery. To prevent infection and dehydration, antibiotics and an electrolyte solution are also administered.

The first OPU was reported by Lopata et al. in 1988 [58]. In our laboratory, a modified protocol for OPU surgery is performed [85]. OPU is performed 18 h after the hCG injection. An approximately 1 cm long surgical incision is made into the abdomen to expose ovaries for midline laparotomy. The stimulated ovarian follicles are aspirated with a 25-G needle attached to a syringe containing heparin and porcine oocyte medium (POM). Collected oocytes are transferred into a Petri dish [86,87]. The muscular, subcutaneous, and skin layers of the incision are sutured and the animal is recovered from anesthesia. According to CIEA guidelines, following recovery from surgery, the donor cannot be used for a subsequent OPU until an interval of more than 2 months has passed, and usually an individual animal can undergo 3–5 OPU procedures.

The collected oocytes are classified according to the stages: germinal vesicle (GV) stage, metaphase I stage, or metaphase II stage. GV stage oocytes are further classified into three groups by visualization of the cumulus–oocyte complexes. The oocytes are classified as follows: GVA, oocytes that are attached to multiple layers of cumulus cells; the GVB; oocytes that are attached to incomplete or few layers of cumulus cells; and GVC, no cumulus cells. These GV stage oocytes are subjected to in vitro maturation (IVM) using POM supplemented with 5% fetal bovine serum, 0.15 IU/mL human follicle-stimulating hormone (hFSH), and 10 IU/mL human chorionic gonadotropin (hCG) at 38°C, with a gas phase of 5% CO_2, 5% O_2, and 90% N_2 for overnight [86–88]. IVM is described in greater detail below.

In Vitro Maturation of Oocytes

In oogenesis, oocyte meiosis is initiated during fetal development but arrested in the diplotene stage of prophase I until puberty. At this stage, oocytes contain a GV, which is a large nucleus. The first meiotic division resumes in the arrested oocyte in response to the luteinizing hormone (LH) surge seen prior to ovulation. Once meiosis has resumed, the GV will undergo dissolution referred to as germinal vesicle breakdown (GVBD). As a result of meiosis I, the primary oocyte develops into the secondary oocyte, and the first polar body is extruded. The oocyte will then initiate another meiotic division giving rise to an oocyte arrested at metaphase II (MII) until fertilization by a sperm. As most of the oocytes obtained by OPU are GV stage, IVM to the MII stage is required.

When a high dose of hCG is administered, GVBD and MII competency varies among OPU oocytes. Although the proportion of oocytes that reach MII is not different when hCG is not administered, treatment with 500 IU of hCG increases the number of expanded cumulous–oocyte complexes from large antral follicles (>1.5 mm) compared with FSH delivered alone [89]. It seems that a high dose of hCG promotes the in vivo maturation of only large antral follicles but does not induce follicular development overall. Usually, an oocyte attached to multiple layers of cumulus cells is classified as a high-quality oocyte that is likely to demonstrate a high competency to mature using IVM protocols, and expansion of those cumulus cells is commonly used as a morphological indicator of oocyte quality following maturation [64,90–93].

For a high success rate of IVF, and to achieve embryo developmental ability, cytoplasmic maturation of MII oocytes is as important as oocyte maturation [63,94]. These two processes are completely independent events [63,94–97]. Cytoplasmic maturation occurs during MII arrest in humans, mice, and cows [98–100]. In marmosets, polar body extrusion was observed between 20 and 26 h of IVM culture, and cytoplasmic maturation appears to occur 2–5 h after polar body extrusion as suggested by the acquisition of the capacity for fertilization [63,64,89,101].

Semen Collection and Sperm Preparation for Producing Embryos

There are several ways to obtain sperm for IVF. In mice, fresh or frozen-thawed epididymal spermatozoa are commonly used for IVF. Usually, to obtain epididymal spermatozoa, mice are sacrificed but orchidectomy also can be utilized. On the other hand, in NHPs, ejaculated semen is collected in a relatively noninvasive way to prevent unnecessary sacrifice. Historically, marmoset semen was collected by rectal probe electro ejaculation (RPE) or by vaginal washing after natural copulation (VW) [102,103]. However, RPE requires anesthesia and VW yields samples contaminated with mucus from the female genital tract, which may influence sperm movement. In 2000, the penile vibrostimulation (PVS) method was reported by Kuederling et al., and this method is currently used in the production of GM marmosets [104]. In their chapter, the authors reported the use of a battery-charged FertiCare personal vibrator for semen collection by the PVS method (Multicept ApS, Rungsted, Denmark). As a substitute, a battery-charged massager or electric toothbrush can also be used to collect marmoset semen. Animals over 1 year old are suitable to be used for collection of semen. To obtain good-quality semen, collection should be performed every week, or precollection should be performed 1 week prior to actual collection. Detailed preparation of sperm for IVF has been described previously [85]. For a discussion of how to perform semen collection by PVS, refer to Chapter 10 of this volume.

In Vitro Fertilization

In the first report of neonates obtained from marmoset IVF embryos, five IVF embryos were transferred into the uteri of three recipients, and two of the recipients gave birth to viable twins and a singleton, respectively [58]. As shown by this report, the marmoset is one of the most useful primates to investigate artificial reproductive technologies. The fertilization rate of mature oocytes in this study was 64%. To improve the IVF rate, various culture conditions have been manipulated, such as the oxygen concentration and type of basal medium and supplements for IVM culture [58,59,63,64,89,101,105–108]. From these reports, it can be speculated that culture conditions improve oocyte quality and IVF rates. However, because each report used different experimental designs and conditions, it is difficult to compare the results. Therefore, optimization of culture conditions for marmoset oocyte IVM is still required. Practical IVM methods for the production of GM marmosets have been described by Kurotaki and Sasaki [85].

Transgenesis by Lentiviral Vector Injection

A method for producing lentiviral vectors was previously reported by Bai et al. [109]. To produce transgenic marmosets, high-titer viral vectors are required; thus the medium containing viral particles is concentrated in an ultracentrifuge, and the viral pellet is resuspended in marmoset embryonic culture medium at 1/1000 volume of the original lentiviral vector supernatant. The biological infectious titer of the lentiviral vector is measured by infecting cultured cell such as 293T cells, HeLa cells, or NIH3T3 cells, and the titer must range from 5.0×10^9 to 5.6×10^{11} transducing units/mL to transduce marmoset embryos. While injecting lentiviral vectors, pronuclear stage marmoset embryos are placed in 0.25 M sucrose supplemented phosphate-buffered (PB1) medium, and the injection is performed to introduce the viral vector solution into the perivitelline space using an Eppendorf FemtoJet Express and a Narishige micromanipulator. The perivitelline space in the early embryonic marmoset is quite small, but the 0.25 M sucrose-supplemented PB1 medium expands the perivitelline space 1.2–7.5-fold. This makes it easy to inject a sufficient amount of lentiviral vector solution into the embryos for gene transfer [41]. After lentiviral vector injection, the embryos are cultured for 3–5 days in a medium prepared for human preimplantation embryos, such as the Origio Sequential Series (Origio, Malov, Denmark), which consists of Cleav medium and Blast medium which are suitable for culturing cleaving stage embryos and blastocyst embryos, until fluorescent protein expression is observed.

A similar method was employed to generate transgenic cynomolgus monkeys. To produce transgenic cynomolgus monkeys, viral vector injection was performed in 0.25 M sucrose/mTALP-HEPES. Injection was directed into the perivitelline space of metaphase II stage oocytes or two-cell stage embryos [45]. The two-cell stage embryos were obtained 24 h after ICSI and the metaphase II stage oocytes were subjected to ICSI after 4 h of viral vector injection. Although transgenic cynomolgus monkeys were obtained from both oocytes and two-cell embryos, only partial GFP expression was observed in the offspring derived from the two-cell stage embryo. In contrast, when lentivirus was injected into cynomolgus monkey oocytes before fertilization, offspring with homogeneous GFP expression were obtained [45].

Nonsurgical Embryo Transfer

The ovarian cycles of both the donor and recipient animals must be synchronized using PGF2α, and ovarian cycles are monitored according to plasma progesterone levels as described above. Before 2014, with the exception of Marshall's work, marmoset embryo transfers were performed surgically [41,56–58,63,88]. More recently, embryo transfers have been carried out nonsurgically to surrogate mothers 3–5 days after ovulation

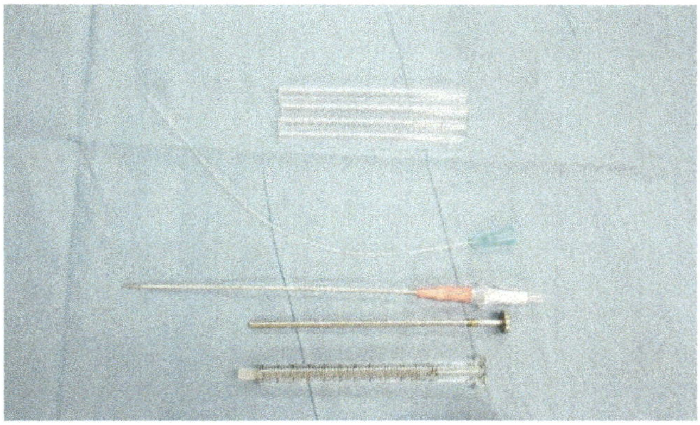

FIGURE 20.4 Devices for nonsurgical embryo transfer. From top to bottom: 7.5-cm long glass tubes of different diameters, polyethylene cannula for embryo transfer (blue connecter), Fluon ETFE cannula (pink connecter) combined with a blunt-ended stainless steel stylet, and 50-μl glass syringe.

[56,63]. The development of new devices to facilitate nonsurgical embryo transfer and a description of their use has been reported [63]. Examples of these devices are shown in Fig. 20.4. To prevent injury to the vagina of the recipient marmoset and to ensure ease of access to the uterus, a 7.5-cm long glass tube is used. Different diameter tubes are available to match the size of the animal's vagina. A Fluon ETFE cannula, a blunt-end stainless steel stylet, and a polyethylene cannula used in conjunction with the appropriately sized glass tube also help to reduce invasiveness and improve the insertion of the cannula into the uterus (Fig. 20.4). Details on the sizes of these devices have been given [63].

The blunt/tapered Fluon ETFE cannula combined with a blunt-ended stainless steel stylet is inserted into the cavitas uteri via the cervix. The Fluon ETFE cannula serves as a guide for the polyethylene cannula through which the embryos are delivered. Before embryo transfer is attempted, appropriate insertion of the cannula should be confirmed by ultrasonography with a linear ultrasound probe. One to three embryos suspended in a small amount of medium are loaded into the polyethylene cannula that is attached to a 50-μL glass syringe (Hamilton Co., Reno, NV, USA). The embryos are then transferred into the uterine lumen by applying gentle pressure to the glass syringe. Gentle pressure is essential to prevent damage to the embryos. Once embryos have been transferred into the cavitas uteri both cannulas are withdrawn. Both cannulas are washed with medium under stereoscopy to confirm that no embryos remain within. Weekly plasma progesterone measurements for the first 4 weeks are performed to check recipient pregnancy, and thereafter, weekly ultrasonography is used to monitor progression of the pregnancy.

TARGETED GENE KNOCKOUT NONHUMAN PRIMATE ANIMALS

Pluripotent Stem Cells of Nonhuman Primates

Targeted gene KO mice are elegant models for understanding the role of unknown functions of a specific gene and the molecular mechanisms underlying disease onset, and DNA regions can be compared between KO and wild-type animals with a similar genetic background. As described above, target gene KO mice have been produced by replacing target gene alleles with the exogenous DNA of interest. Flanking DNA that is identical in sequence to the targeted locus is inserted into ESCs, resulting in disruption of the endogenous target gene by homologous recombination. Usually, this DNA construct carries a drug resistance gene or reporter gene to select successful target gene KO ESCs. Precisely modified ESC selection is followed by ESC injection into host blastocyst stage embryos or aggregates of eight-cell stage embryos with the zona pellucida removed; then, embryos are transferred into recipient mice. Chimeric mice with wild-type and target gene KO cells are obtained from a series of these processes. To analyze gene KO mice, breeding of chimeric mice is required to obtain heterozygous or homozygous KO mice. Some of the chimeric mice will have germ cells derived from ESCs with the gene KO. These chimeric mice are crossbred with wild-type mice from the same genetic background as the ESCs to obtain heterozygous KO mice. Then, these heterozygous offspring are interbred to produce homozygous allelic KO mice. Therefore, chimeric competent ESCs or iPSCs are essential tools for creating targeted gene KO mice.

In marmosets, several lines of ESCs and iPSCs have been established, and these cells can be differentiated into the three germ layers: ectoderm, mesoderm, and endoderm [25–28,110,111]. Furthermore, GFP genes containing heterotypic lox sites of a target gene can be introduced into marmoset ESCs using homologous recombination techniques [112]. However, pluripotent NHP stem cells cannot produce chimeric animals when injected into host blastocysts. This phenomenon is common to most mammalian species; chimeric competent pluripotent stem cells have only been established in mice and rats. A defined ESC culture medium, called 2i, contains the Mek inhibitor PD0325901 to block the Erk pathway, and glycogen synthase kinase 3 inhibition by CHIR99021 can maintain mouse and rat ESCs in a naïve state [113–115]. In 2007, two research groups established new pluripotent stem cells, called epiblast stem cells (EpiSCs), from 5.5 to 6.5 days post coitus post-implanted mouse embryos [116,117]. These EpiSCs can differentiate into the three germ layers, but lack the capacity to produce chimeric animals. NHP and human pluripotent stem cells share several similar characteristics with EpiSCs: colony morphology, tolerance to single cell dissociation, gene expression profiles, X chromosome activation status, and leukemia inhibitory factor responsiveness. From these observations, it is speculated that NHP and human pluripotent cells are more similar to mouse EpiSCs than to ESCs/iPSCs and are termed "primed" pluripotent stem cells [118–120]. Mouse pluripotent stem cells that confer chimeric animals are called "naïve" pluripotent stem cells.

Many reports have been published on the conversion of human pluripotent stem cells into a naive-like phenotype [121–124]. In these reports, converted human naïve pluripotent cells are analyzed based on criteria such as global transcriptional state, naïve transcription factor networks and signaling profiles, epigenetic identity, X chromosome status, and metabolic state, which are known to distinguish mouse naïve cells from primed cells. However, evidence for the chimeric competency of these cells is lacking for ethical reasons. Several studies have reported conversion of the primed status pluripotent stem cells to the (putative) naïve state in nonhuman species, such as pigs, rabbits, and monkeys. Although some NHP putative naïve state cells showed the ability to contribute to normal embryonic and fetal development, chimeric neonates or germline chimeras have not been obtained [125,126]. On the other hand, Tachibana et al. showed that rhesus monkey blastocysts do not incorporate pluripotent stem cells, whether they are naïve or primed, because primitive endodermal cells in rhesus blastocysts prevent uptake of injected cells into host embryos. Consequently, it is difficult to understand why target gene KO NHPs cannot be produced by homologous recombination of pluripotent stem cells, and an understanding of the early development of NHPs is required to solve this issue.

Genome Editing of Nonhuman Primate Embryos

Recent developments in genome editing techniques for disrupting endogenous genes in cells, including early embryos, have changed the situation in the generation of target gene KO animals. Among several kinds of genome editing tools, the three most popular methods are zinc finger nucleases (ZFNs), transcription activator-like effector nucleases (TALENs), and clustered regularly interspaced short palindromic repeat (CRISPR)/CRISPR associated (Cas) 9 systems. While both ZFNs and TALENs use FokI nuclease to digest target genes, ZFNs are recombinant proteins with zinc finger motifs that bind specifically to target DNA sequences [127], and TALENs are chimeric proteins consisting of a transcription activator-like effector DNA binding domain [128]. In the CRISPR/Cas9 system, a small RNA and target DNA sequence guide the Cas9 nuclease so that it digests double-stranded target DNA [129,130] (Fig. 20.5).

All of these tools for genome editing induce double-strand breaks in the specifically targeted gene in the genome and subsequently repair the double-strand breaks via nonhomologous end joining (NHEJ), resulting in frameshifts of the DNA sequence. Consequently, the modified gene is translated into truncated amino acid sequences that disrupt gene function. These genome editing techniques do not require ESCs/iPSCs to produce target gene KO organisms, and injection of the genome editing tools into pronuclei stage embryos can disrupt target genes directly. Using these technologies, targeted gene KO organisms have been produced in many species [131–134].

Recently, target gene KO NHPs have been produced using ZFNs, TALEN, and CRISPR/Cas [65,135,136]. Niu et al. knocked out three genes in cynomolgus monkeys: nuclear receptor subfamily 0 group B member 1, peroxisome proliferator-activated receptor gamma, and recombination activating gene 1 by CRISPR/Cas9 [136]. Liu et al. reported TALEN-mediated MECP2 mutations in rhesus and cynomolgus monkeys [135]. Both papers reported successful mutagenesis of the target genes, with no off-target mutations. When these two papers were published, objective phenotypes had not been observed. Three years later, MeCP2 KO cynomolgus monkeys showed a series of physiological, behavioral, and structural abnormalities resembling the clinical symptoms of Rett syndrome patients [137].

Detection of phenotypes might require longer periods of time, especially in neurodegenerative disease models, although an unaltered phenotype may be observed in the event of mosaicism of the mutated and intact target genes. When mosaic embryos of mutated and intact genes are generated, intact genes rescue the mutant gene phenotypes; therefore, mosaic embryos develop into animals that exhibit wild-type phenotypes. Because NHPs have long life spans, it is critically important to create targeted gene KO NHP models that exhibit altered phenotypes during the founder generation. Furthermore, the production of nonmosaic animals also reduces the number of animals used and avoids euthanasia of failed KO animals after embryonic manipulation for ethical reasons. For this purpose, Sato et al. screened high-efficiency ZFNs and TALENs using marmoset fibroblast cells to generate interleukin-2 receptor subunit gamma (IL2RG) gene KO marmosets [65]. The selected highly efficient ZFNs and TALENs were further analyzed for gene modifications and mosaic rates in marmoset embryos, to avoid obtaining mosaic animals prior to animal production. To estimate mosaicism in the animals to be generated by genome editing, eight-cell stage embryos were dissociated into blastomeres after injection of the genome editing tool,

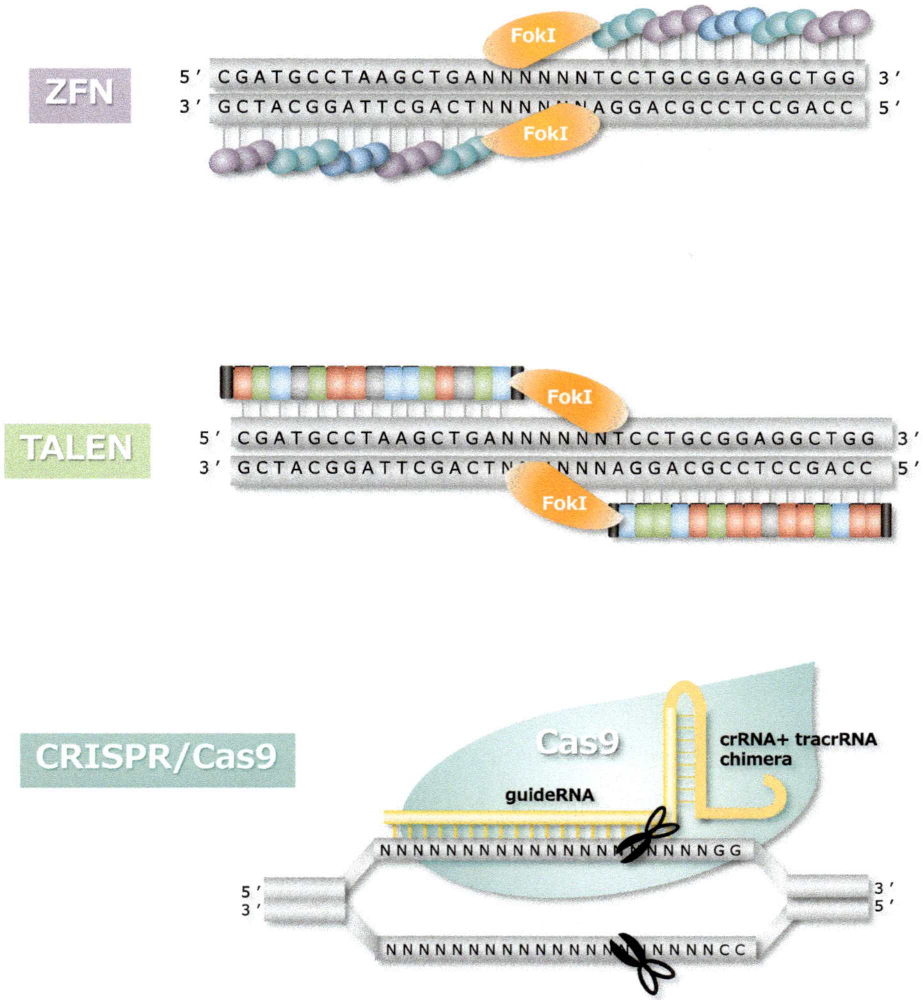

FIGURE 20.5 Genome editing techniques. Upper panel: zinc finger nucleases (ZFNs), middle panel: transcription activator-like effector nucleases (TALENs), lower panel: clustered regularly interspaced short palindromic repeat (CRISPR)/CRISPR associated (Cas) 9 systems. Both ZFNs and TALENs use Fok I nuclease to digest target genes. ZFN comprises recombinant proteins with zinc finger motifs that bind specifically to target DNA sequences and Fok I nuclease. TALEN comprises chimeric proteins including a transcription activator-like effector DNA binding domain and Fok I nuclease. With the CRISPR/Cas9 system, a small guide RNA binds specifically to the target and then a DNA sequence guides the Cas9 nuclease to digest double-stranded target DNA.

and each blastomere was analyzed for genome modification (Fig. 20.6). As a result, the authors obtained nine IL2RG KO marmosets and these founder animals displayed distinct phenotypic and functional changes during lymphocyte development.

FUTURE DIRECTIONS

In the past decade, assisted reproductive technologies in NHPs have developed rapidly, and the production of transgenic models and target gene KO models has been realized; however, several challenges still remain. Currently, only retroviral vectors or lentiviral vectors are available to introduce exogenous genes into NHP embryos. As described above, these vectors cannot introduce exogenous genes longer than 5.0 kb. Transposons can introduce longer DNA; however, this nonintegrating transposon expresses its transgene episomally, which makes it difficult to confirm transgene integration into the embryonic host genome. Therefore, new transgenesis methods for introducing DNA longer than 5.0 kb, that are capable of confirming the integration of the transgene into the host genome, should be developed. Second, new approaches for producing target gene KO NHPs are required. Although genome editing techniques have made it possible to generate target gene KO marmosets, target gene knock-in NHPs have

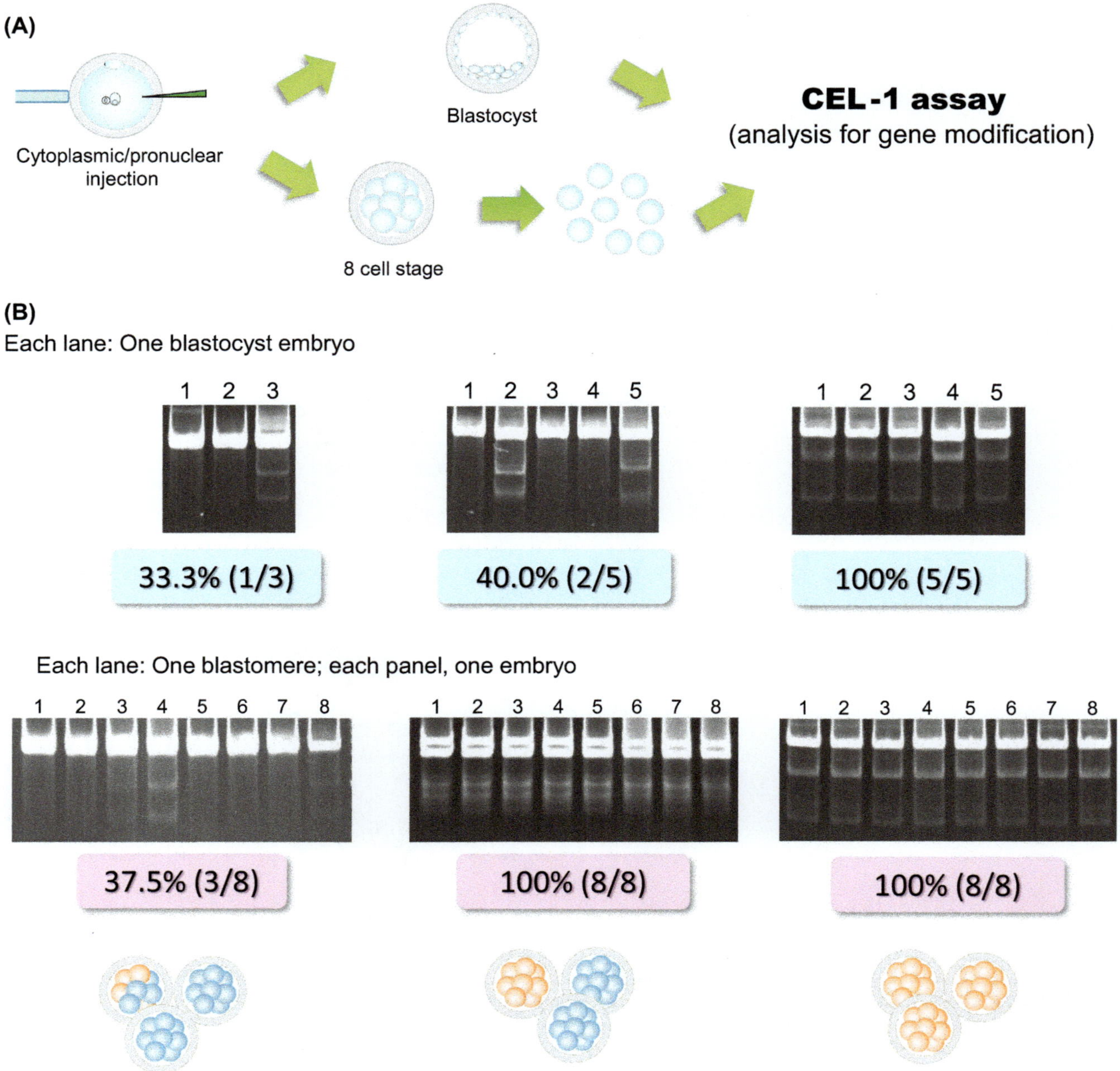

FIGURE 20.6 Strategy for screening high-efficiency genome editing tools in vitro. (A) Gene modification analysis at the blastocyst stage can estimate the gene modification rate of embryos. Gene modification analysis at eight-cell stage embryos can show mosaicism in the embryo. (B) Upper panel shows the results of CEL 1 mutation detection assay in blastocysts. Each lane indicates the result from one embryo. Gene modification rates of each genome editing tool are indicated in blue boxes. Middle panel demonstrates the results of CEL 1 assay of 8-cell stage embryos. Each lane indicates the result of one blastomere and each photo shows mosaicism of the embryos. The mosaic rates are presented in pink boxes. Lower panel indicates synthetic estimation of each genomic tool. Left panel shows that the gene modification rate was 33% and the genome-modified embryo was mosaic after gene modification. Middle panel indicates that the gene modification rate was 40%, but the gene-modified embryo was not mosaic and may exhibit the phenotype when developed to the neonate stage. Right panel indicates that all embryos and blastomeres had target gene modification, suggesting that this genome modification tool is the most effective for producing animal models, and genome modification occurs immediately after embryonic injection.

not yet been produced because homologous recombination to insert donor sequence DNA fragments into a cleaved target site by genome editing has extremely low efficiency in animal embryos [138–140]. The target gene knock-in method is useful for genetic engineering, by substituting DNA sequences to create models of genetic diseases arising from point mutations or inserting a DNA sequence within a locus to monitor its promoter activity in a tissue or during a developmental period. Target gene knock-in NHP models would play crucial roles in understanding the neural control of higher brain functions that are unique to primates. To achieve this, new technologies for target gene knock-in are required. Recently, some studies have reported that NHEJ or microhomology-mediated end joining can integrate long exogenous DNA fragments into the genome at relatively high frequencies [141–145]. In addition, a new genome editing approach, named "base editing," that enables the programmable conversion of one base pair into another without double-stranded DNA cleavage or homologous recombination of the donor genes, has been developed [146]. The authors engineered fusions of CRISPR/Cas9 and a cytidine deaminase enzyme that retains the ability to be programmed with a guide RNA, thereby mediating the direct conversion of cytidine to uridine and effecting C → T (or G → A) substitution [146]. This may be a novel approach for producing NHP models of genetic diseases that arise from point mutations of the responsible genes. To insert DNA sequences at specific loci, selection or confirmation of precise DNA integration is important. However, when embryos are the target cells for knock-in, there is insufficient time for selection or confirmation even if the genome editing tool, such as mRNA of TALEN or protein of Cas9, was injected into pronuclear stage embryos because embryos quickly develop to the blastocyst stage, which must be transferred to a surrogate uterus within 8–10 days. Furthermore, preimplantation genetic diagnosis, such as blastomere extraction, is not a preferred method because confirmation of one blastomere does not confer certainty with respect to the rest of the blastomeres. Establishing stem cells that can develop into animals, after selecting desirable cells such as naïve ESCs/iPSCs or germ stem cells, is the most promising approach to solving this issue. If these kinds of stem cells are established, the bottleneck of GM NHP models, i.e., their long life span for obtaining next-generation animals, may be solved, such that NHPs will contribute greatly to life science research, especially neuroscience.

References

[1] Moresco EM, Li X, Beutler B. Going forward with genetics: recent technological advances and forward genetics in mice. Am J Pathol 2013;182:1462–73.

[2] Gondo Y. Trends in large-scale mouse mutagenesis: from genetics to functional genomics. Nat Rev Genet 2008;9:803–10.

[3] Jaenisch R, Mintz B. Simian virus 40 DNA sequences in DNA of healthy adult mice derived from preimplantation blastocysts injected with viral DNA. Proc Natl Acad Sci USA 1974;71:1250–4.

[4] Jaenisch R. Germ line integration and Mendelian transmission of the exogenous Moloney leukemia virus. Proc Natl Acad Sci USA 1976;73:1260–4.

[5] Brinster RL, Chen HY, Trumbauer M, Senear AW, Warren R, Palmiter RD. Somatic expression of herpes thymidine kinase in mice following injection of a fusion gene into eggs. Cell 1981; 27:223–31.

[6] Costantini F, Lacy E. Introduction of a rabbit beta-globin gene into the mouse germ line. Nature 1981;294:92–4.

[7] Gordon JW, Ruddle FH. Integration and stable germ line transmission of genes injected into mouse pronuclei. Science 1981; 214:1244–6.

[8] Zijlstra M, Li E, Sajjadi F, Subramani S, Jaenisch R. Germ-line transmission of a disrupted beta 2-microglobulin gene produced by homologous recombination in embryonic stem cells. Nature 1989;342:435–8.

[9] Evans MJ, Kaufman MH. Establishment in culture of pluripotential cells from mouse embryos. Nature 1981;292:154–6.

[10] Martin GR. Isolation of a pluripotent cell line from early mouse embryos cultured in medium conditioned by teratocarcinoma stem cells. Proc Natl Acad Sci USA 1981;78:7634–8.

[11] Bradley A, Evans M, Kaufman MH, Robertson E. Formation of germ-line chimaeras from embryo-derived teratocarcinoma cell lines. Nature 1984;309:255–6.

[12] Smithies O, Gregg RG, Boggs SS, Koralewski MA, Kucherlapati RS. Insertion of DNA sequences into the human chromosomal beta-globin locus by homologous recombination. Nature 1985;317:230–4.

[13] Thomas KR, Folger KR, Capecchi MR. High frequency targeting of genes to specific sites in the mammalian genome. Cell 1986;44: 419–28.

[14] Mansour SL, Thomas KR, Capecchi MR. Disruption of the proto-oncogene int-2 in mouse embryo-derived stem cells: a general strategy for targeting mutations to non-selectable genes. Nature 1988;336:348–52.

[15] Sasaki E. Prospects for genetically modified non-human primate models, including the common marmoset. Neurosci Res 2015;93: 110–5.

[16] Kirik D, Annett LE, Burger C, Muzyczka N, Mandel RJ, Bjorklund A. Nigrostriatal alpha-synucleinopathy induced by viral vector-mediated overexpression of human alpha-synuclein: a new primate model of Parkinson's disease. Proc Natl Acad Sci USA 2003;100:2884–9.

[17] Marmoset Genome Sequencing and Analysis Consortium. The common marmoset genome provides insight into primate biology and evolution. Nat Genet 2014;46:850–7.

[18] Datson NA, Morsink MC, Atanasova S, Armstrong VW, Zischler H, Schlumbohm C, Dutilh BE, Huynen MA, Waegele B, Ruepp A, de Kloet ER, Fuchs E. Development of the first marmoset-specific DNA microarray (EUMAMA): a new genetic tool for large-scale expression profiling in a non-human primate. BMC Genom 2007;8:190.

[19] Sato K, Kuroki Y, Kumita W, Fujiyama A, Toyoda A, Kawai J, Iriki A, Sasaki E, Okano H, Sakakibara Y. Resequencing of the common marmoset genome improves genome assemblies and gene-coding sequence analysis. Sci Rep 2015; 5:16894.

[20] Tatsumoto S, Adati N, Tohtoki Y, Sakaki Y, Boroviak T, Habu S, Okano H, Suemizu H, Sasaki E, Satake M. Development and characterization of cDNA resources for the common marmoset: one of the experimental primate models. DNA Res 2013;20: 255–62.

[21] Izawa K, Tani K, Nakazaki Y, Hibino H, Sugiyama H, Kawasaki A, Sasaki E, Nishioka C, Ishii H, Soda Y, Yagita H, Tanioka Y, Tojo A, Asano S. Hematopoietic activity of common marmoset CD34 cells isolated by a novel monoclonal antibody MA24. Exp Hematol 2004;32:843–51.

[22] Kametani Y, Suzuki D, Kohu K, Satake M, Suemizu H, Sasaki E, Ito T, Tamaoki N, Mizushima T, Ozawa M, Tani K, Kito M, Arai H, Koyanagi A, Yagita H, Habu S. Development of monoclonal antibodies for analyzing immune and hematopoietic systems of common marmoset. Exp Hematol 2009;37: 1318–29.

[23] Nunomura S, Shimada S, Kametani Y, Yamada Y, Yoshioka M, Suemizu H, Ozawa M, Itoh T, Kono A, Suzuki R, Tani K, Ando K, Yagita H, Ra C, Habu S, Satake M, Sasaki E. Double expression of CD34 and CD117 on bone marrow progenitors is a hallmark of the development of functional mast cell of *Callithrix jacchus* (common marmoset). Int Immunol 2012;24:593–603.

[24] Debowski K, Drummer C, Lentes J, Cors M, Dressel R, Lingner T, Salinas-Riester G, Fuchs S, Sasaki E, Behr R. The transcriptomes of novel marmoset monkey embryonic stem cell lines reflect distinct genomic features. Sci Rep 2016;6:29122.

[25] Muller T, Fleischmann G, Eildermann K, Matz-Rensing K, Horn PA, Sasaki E, Behr R. A novel embryonic stem cell line derived from the common marmoset monkey (*Callithrix jacchus*) exhibiting germ cell-like characteristics. Hum Reprod 2009;24:1359–72.

[26] Sasaki E, Hanazawa K, Kurita R, Akatsuka A, Yoshizaki T, Ishii H, Tanioka Y, Ohnishi Y, Suemizu H, Sugawara A, Tamaoki N, Izawa K, Nakazaki Y, Hamada H, Suemori H, Asano S, Nakatsuji N, Okano H, Tani K. Establishment of novel embryonic stem cell lines derived from the common marmoset (*Callithrix jacchus*). Stem Cell 2005;23:1304–13.

[27] Tomioka I, Maeda T, Shimada H, Kawai K, Okada Y, Igarashi H, Oiwa R, Iwasaki T, Aoki M, Kimura T, Shiozawa S, Shinohara H, Suemizu H, Sasaki E, Okano H. Generating induced pluripotent stem cells from common marmoset (*Callithrix jacchus*) fetal liver cells using defined factors, including Lin28. Gene Cell 2010;15:959–69.

[28] Wu Y, Zhang Y, Mishra A, Tardif SD, Hornsby PJ. Generation of induced pluripotent stem cells from newborn marmoset skin fibroblasts. Stem Cell Res 2010;4:180–8.

[29] Kikuchi T, Morizane A, Doi D, Magotani H, Onoe H, Hayashi T, Mizuma H, Takara S, Takahashi R, Inoue H, Morita S, Yamamoto M, Okita K, Nakagawa M, Parmar M, Takahashi J. Human iPS cell-derived dopaminergic neurons function in a primate Parkinson's disease model. Nature 2017;548:592–6.

[30] Kobayashi Y, Okada Y, Itakura G, Iwai H, Nishimura S, Yasuda A, Nori S, Hikishima K, Konomi T, Fujiyoshi K, Tsuji O, Toyama Y, Yamanaka S, Nakamura M, Okano H. Pre-evaluated safe human iPSC-derived neural stem cells promote functional recovery after spinal cord injury in common marmoset without tumorigenicity. PLoS One 2012;7:e52787.

[31] Tardif SD, Smucny DA, Abbott DH, Mansfield K, Schultz-Darken N, Yamamoto ME. Reproduction in captive common marmosets (*Callithrix jacchus*). Comp Med 2003;53:364–8.

[32] Hearn JP, Lunn SF. Proceedings: the endocrinology of the ovarian cycle and of pregnancy in the marmoset monkey, *Callithrix jacchus*. J Endocrinol 1975;65:1P.

[33] Jaquish CE, Toal RL, Tardif SD, Carson RL. Use of ultrasound to monitor prenatal growth and development in the common marmoset (*Callithrix jacchus*). Am J Primatol 1995;36:259–79.

[34] Hearn JP. The embryo-maternal dialogue during early pregnancy in primates. J Reprod Fertil 1986;76:809–19.

[35] Yaguchi M, Ohta S, Toyama Y, Kawakami Y, Toda M. Functional recovery after spinal cord injury in mice through activation of microglia and dendritic cells after IL-12 administration. J Neurosci Res 2008;86:1972–80.

[36] Yaguchi M, Tabuse M, Ohta S, Ohkusu-Tsukada K, Takeuchi T, Yamane J, Katoh H, Nakamura M, Matsuzaki Y, Yamada M, Itoh T, Nomura T, Toyama Y, Okano H, Toda M. Transplantation of dendritic cells promotes functional recovery from spinal cord injury in common marmoset. Neurosci Res 2009.

[37] Hearn JP, Webley GE. Regulation of the corpus luteum of early pregnancy in the marmoset monkey: local interactions of luteotrophic and luteolytic hormones in vivo and their effects on the secretion of progesterone. J Endocrinol 1987;114:231–9.

[38] Summers PM, Wennink CJ, Hodges JK. Cloprostenol-induced luteolysis in the marmoset monkey (*Callithrix jacchus*). J Reprod Fertil 1985;73:133–8.

[39] Chan AW, Chong KY, Martinovich C, Simerly C, Schatten G. Transgenic monkeys produced by retroviral gene transfer into mature oocytes. Science 2001;291:309–12.

[40] Wolfgang MJ, Eisele SG, Browne MA, Schotzko ML, Garthwaite MA, Durning M, Ramezani A, Hawley RG, Thomson JA, Golos TG. Rhesus monkey placental transgene expression after lentiviral gene transfer into preimplantation embryos. Proc Natl Acad Sci USA 2001;98:10728–32.

[41] Sasaki E, Suemizu H, Shimada A, Hanazawa K, Oiwa R, Kamioka M, Tomioka I, Sotomaru Y, Hirakawa R, Eto T, Shiozawa S, Maeda T, Ito M, Ito R, Kito C, Yagihashi C, Kawai K, Miyoshi H, Tanioka Y, Tamaoki N, Habu S, Okano H, Nomura T. Generation of transgenic non-human primates with germline transmission. Nature 2009;459:523–7.

[42] Sato K, Sasaki E. Genetic engineering in nonhuman primates for human disease modeling. J Hum Genet 2018;63:125–31.

[43] Niu Y, Yu Y, Bernat A, Yang S, He X, Guo X, Chen D, Chen Y, Ji S, Si W, Lv Y, Tan T, Wei Q, Wang H, Shi L, Guan J, Zhu X, Afanassieff M, Savatier P, Zhang K, Zhou Q, Ji W. Transgenic rhesus monkeys produced by gene transfer into early-cleavage-stage embryos using a simian immunodeficiency virus-based vector. Proc Natl Acad Sci USA 2010;107:17663–7.

[44] Liu Z, Li X, Zhang JT, Cai YJ, Cheng TL, Cheng C, Wang Y, Zhang CC, Nie YH, Chen ZF, Bian WJ, Zhang L, Xiao J, Lu B, Zhang YF, Zhang XD, Sang X, Wu JJ, Xu X, Xiong ZQ, Zhang F, Yu X, Gong N, Zhou WH, Sun Q, Qiu Z. Autism-like behaviours and germline transmission in transgenic monkeys overexpressing MeCP2. Nature 2016;530:98–102.

[45] Seita Y, Tsukiyama T, Iwatani C, Tsuchiya H, Matsushita J, Azami T, Okahara J, Nakamura S, Hayashi Y, Hitoshi S, Itoh Y, Imamura T, Nishimura M, Tooyama I, Miyoshi H, Saitou M, Ogasawara K, Sasaki E, Ema M. Generation of transgenic cynomolgus monkeys that express green fluorescent protein throughout the whole body. Sci Rep 2016;6:24868.

[46] Yang SH, Cheng PH, Banta H, Piotrowska-Nitsche K, Yang JJ, Cheng EC, Snyder B, Larkin K, Liu J, Orkin J, Fang ZH, Smith Y, Bachevalier J, Zola SM, Li SH, Li XJ, Chan AW. Towards a transgenic model of Huntington's disease in a non-human primate. Nature 2008;453:921–4.

[47] Park JE, Zhang XF, Choi SH, Okahara J, Sasaki E, Silva AC. Generation of transgenic marmosets expressing genetically encoded calcium indicators. Sci Rep 2016;6:34931.

[48] Tomioka I, Ishibashi H, Minakawa EN, Motohashi HH, Takayama O, Saito Y, Popiel HA, Puentes S, Owari K, Nakatani T, Nogami N, Yamamoto K, Noguchi S, Yonekawa T, Tanaka Y, Fujita N, Suzuki H, Kikuchi H, Aizawa S, Nagano S, Yamada D, Nishino I, Ichinohe N, Wada K, Kohsaka S, Nagai Y, Seki K. Transgenic monkey model of the polyglutamine diseases recapitulating progressive neurological symptoms. eNeuro 2017;4.

[49] Semenova EA, Marchand C, Pommier Y. HIV-1 integrase inhibitors: update and perspectives. Adv Pharmacol 2008;56:199–228.

[50] Daniels RW, Rossano AJ, Macleod GT, Ganetzky B. Expression of multiple transgenes from a single construct using viral 2A peptides in Drosophila. PLoS One 2014;9:e100637.

[51] Kim JH, Lee SR, Li LH, Park HJ, Park JH, Lee KY, Kim MK, Shin BA, Choi SY. High cleavage efficiency of a 2A peptide derived from porcine teschovirus-1 in human cell lines, zebrafish and mice. PLoS One 2011;6:e18556.

[52] Ryan MD, King AM, Thomas GP. Cleavage of foot-and-mouth disease virus polyprotein is mediated by residues located

within a 19 amino acid sequence. J Gen Virol 1991;72(Pt 11): 2727–32.

[53] Hotta A, Cheung AY, Farra N, Garcha K, Chang WY, Pasceri P, Stanford WL, Ellis J. EOS lentiviral vector selection system for human induced pluripotent stem cells. Nat Protoc 2009;4: 1828–44.

[54] Kropp J, Di Marzo A, Golos T. Assisted reproductive technologies in the common marmoset: an integral species for developing nonhuman primate models of human diseases. Biol Reprod 2017; 96:277–87.

[55] Harlow CR, Gems S, Hodges JK, Hearn JP. The relationship between plasma progesterone and the timing of ovulation and early embryonic development in the marmoset monkey. J Zool 1983;201:273–82.

[56] Marshall VS, Kalishman J, Thomson JA. Nonsurgical embryo transfer in the common marmoset monkey. J Med Primatol 1997;26:241–7.

[57] Summers PM, Shephard AM, Taylor CT, Hearn JP. The effects of cryopreservation and transfer on embryonic development in the common marmoset monkey, *Callithrix jacchus*. J Reprod Fertil 1987;79:241–50.

[58] Lopata A, Summers PM, Hearn JP. Births following the transfer of cultured embryos obtained by in vitro and in vivo fertilization in the marmoset monkey (*Callithrix jacchus*). Fertil Steril 1988;50: 503–9.

[59] Gilchrist RB, Nayudu PL, Hodges JK. Maturation, fertilization, and development of marmoset monkey oocytes in vitro. Biol Reprod 1997;56:238–46.

[60] Gilchrist RB, Nayudu PL, Nowshari MA, Hodges JK. Meiotic competence of marmoset monkey oocytes is related to follicle size and oocyte-somatic cell associations. Biol Reprod 1995;52: 1234–43.

[61] Lopata A. Implantation of the human embryo. Hum Reprod 1996;11(Suppl 1):175–84. Discussion 193–175.

[62] Marshall JW, Duffin KJ, Green AR, Ridley RM. NXY-059, a free radical–trapping agent, substantially lessens the functional disability resulting from cerebral ischemia in a primate species. Stroke 2001;32:190–8.

[63] Takahashi T, Hanazawa K, Inoue T, Sato K, Sedohara A, Okahara J, Suemizu H, Yagihashi C, Yamamoto M, Eto T, Konno Y, Okano H, Suematsu M, Sasaki E. Birth of healthy offspring following ICSI in in vitro-matured common marmoset (*Callithrix jacchus*) oocytes. PLoS One 2014;9.

[64] Wilton LJ, Marshall VS, Piercy EC, Moore HD. In vitro fertilization and embryo development in the marmoset monkey (*Callithrix jacchus*). J Reprod Fertil 1993;97:481–6.

[65] Sato K, Oiwa R, Kumita W, Henry R, Sakuma T, Ito R, Nozu R, Inoue T, Katano I, Okahara N, Okahara J, Shimizu Y, Yamamoto M, Hanazawa K, Kawakami T, Kametani Y, Suzuki R, Takahashi T, Weinstein EJ, Yamamoto T, Sakakibara Y, Habu S, Hata J, Okano H, Sasaki E. Generation of a nonhuman primate model of severe combined immunodeficiency using highly efficient genome editing. Cell Stem Cell 2016;19:127–38.

[66] Santos MA, Kuijk EW, Macklon NS. The impact of ovarian stimulation for IVF on the developing embryo. Reproduction 2010; 139:23–34.

[67] Yan L, Li H, Shi Z. Immunization against inhibin improves in vivo and in vitro embryo production. Anim Reprod Sci 2015;163:1–9.

[68] Kaneko H, Watanabe G, Taya K, Sasamoto S. Changes in peripheral levels of bioactive and immunoreactive inhibin, estradiol-17 beta, progesterone, luteinizing hormone, and follicle-stimulating hormone associated with follicular development in cows induced to superovulate with equine chorionic gonadotropin. Biol Reprod 1992;47:76–82.

[69] De Jong FH. Inhibin. Physiol Rev 1988;68:555–607.

[70] Li DR, Qin GS, Wei YM, Lu FH, Huang QS, Jiang HS, Shi DS, Shi ZD. Immunisation against inhibin enhances follicular development, oocyte maturation and superovulatory response in water buffaloes. Reprod Fertil Dev 2011;23:788–97.

[71] Liu YP, Mao XB, Wei YM, Yu JN, Li H, Chen RA, Shi ZD. Studies on enhancing embryo quantity and quality by immunization against inhibin in repeatedly superovulated Holstein heifers and the associated endocrine mechanisms. Anim Reprod Sci 2013;142:10–8.

[72] Mei C, Li MY, Zhong SQ, Lei Y, Shi ZD. Enhancing embryo yield in superovulated holstein heifers by immunization against inhibin. Reprod Domest Anim 2009;44:735–9.

[73] Akagi S, Kaneko H, Nakanishi Y, Takedomi T, Watanabe G, Taya K. Ovarian response and FSH profile in cows following injection of various doses of inhibin antiserum. J Vet Med Sci 1997; 59:1129–35.

[74] Akagi S, Shi F, Kaneko H, Watanabe G, Tsonis CG, Taya K. Ovarian response and hormonal profiles in heifers after immunization and re-immunization against inhibin α-subunit. J Reprod Dev 2002;48:599–605.

[75] Findlay JK, Doughton B, Robertson DM, Forage RG. Effects of immunization against recombinant bovine inhibin alpha subunit on circulating concentrations of gonadotrophins in ewes. J Endocrinol 1989;120:59–65.

[76] Hasegawa A, Mochida K, Inoue H, Noda Y, Endo T, Watanabe G, Ogura A. High-Yield superovulation in adult mice by anti-inhibin serum treatment combined with estrous cycle synchronization. Biol Reprod 2016;94:21.

[77] Kishi H, Okada T, Otsuka M, Watanabe G, Taya K, Sasamoto S. Induction of superovulation by immunoneutralization of endogenous inhibin through the increase in the secretion of follicle-stimulating hormone in the cyclic golden hamster. J Endocrinol 1996;151:65–75.

[78] Mizumachi M, Voglmayr JK, Washington DW, Chen CL, Bardin CW. Superovulation of ewes immunized against the human recombinant inhibin alpha-subunit associated with increased pre- and postovulatory follicle-stimulating hormone levels. Endocrinology 1990;126:1058–63.

[79] Nambo Y, Kaneko H, Nagata S, Oikawa M, Yoshihara T, Nagamine N, Watanabe G, Taya K. Effect of passive immunization against inhibin on FSH secretion, folliculogenesis and ovulation rate during the follicular phase of the estrous cycle in mares. Theriogenology 1998;50:545–57.

[80] Rivier C, Vale W. Immunoneutralization of endogenous inhibin modifies hormone secretion and ovulation rate in the rat. Endocrinology 1989;125:152–7.

[81] Shi F, Mochida K, Ogura A, Matsuda J, Suzuki O, Watanabe G, Hutz RJ, Tsonis CG, Suzuki AK, Taya K. Follicle selection in cyclic Guinea pigs with active immunization against inhibin alpha-subunit. Life Sci 2000;66:2489–97.

[82] Wang H, Herath CB, Xia G, Watanabe G, Taya K. Superovulation, fertilization and in vitro embryo development in mice after administration of an inhibin-neutralizing antiserum. Reproduction 2001;122:809–16.

[83] Wrathall JH, McLeod BJ, Glencross RG, Knight PG. Effects of active immunization against a synthetic peptide sequence of the inhibin alpha-subunit on plasma gonadotrophin concentrations, ovulation rate and lambing rate in ewes. J Reprod Fertil 1992;95:175–82.

[84] Smith KB, Lunn SF, Fraser HM. Inhibin secretion during the ovulatory cycle and pregnancy in the common marmoset monkey. J Endocrinol 1990;126:489–95.

[85] Kurotaki Y, Sasaki E. Practical reproductive techniques for the common marmoset. J Mammalian Ova Res 2017;34:3–12.

[86] Yoshioka K, Suzuki C, Onishi A. Defined system for in vitro production of porcine embryos using a single basic medium. J Reprod Dev 2008;54:208–13.
[87] Yoshioka K, Suzuki C, Tanaka A, Anas IM, Iwamura S. Birth of piglets derived from porcine zygotes cultured in a chemically defined medium. Biol Reprod 2002;66:112–9.
[88] Tomioka I, Takahashi T, Shimada A, Yoshioka K, Sasaki E. Birth of common marmoset (*Callithrix jacchus*) offspring derived from in vitro-matured oocytes in chemically defined medium. Theriogenology 2012;78:1487–93.
[89] Grupen CG, Gilchrist RB, Nayudu PL, Barry MF, Schulz SJ, Ritter LJ, Armstrong DT. Effects of ovarian stimulation, with and without human chorionic gonadotrophin, on oocyte meiotic and developmental competence in the marmoset monkey (*Callithrix jacchus*). Theriogenology 2007;68:861–72.
[90] Goud PT, Goud AP, Qian C, Laverge H, Van der Elst J, De Sutter P, Dhont M. In-vitro maturation of human germinal vesicle stage oocytes: role of cumulus cells and epidermal growth factor in the culture medium. Hum Reprod 1998;13:1638–44.
[91] Lonergan P, Monaghan P, Rizos D, Boland MP, Gordon I. Effect of follicle size on bovine oocyte quality and developmental competence following maturation, fertilization, and culture in vitro. Mol Reprod Dev 1994;37:48–53.
[92] Madison V, Avery B, Greve T. Selection of immature bovine oocytes for developmental potential in vitro. Anim Reprod Sci 1992;27:1–11.
[93] Shioya Y, Kuwayama M, Fukushima M, Iwasaki S, Hanada A. In vitro fertilization and cleavage capability of bovine follicular oocytes classified by cumulus cells and matured in vitro. Theriogenology 1988;30:489–96.
[94] Eppig JJ. Coordination of nuclear and cytoplasmic oocyte maturation in eutherian mammals. Reprod Fertil Dev 1996;8:485–9.
[95] Albertini DF, Sanfins A, Combelles CM. Origins and manifestations of oocyte maturation competencies. Reprod Biomed Online 2003;6:410–5.
[96] Moor RM, Dai Y, Lee C, Fulka Jr J. Oocyte maturation and embryonic failure. Hum Reprod Update 1998;4:223–36.
[97] Trounson A, Anderiesz C, Jones G. Maturation of human oocytes in vitro and their developmental competence. Reproduction 2001;121:51–75.
[98] Balakier H, Sojecki A, Motamedi G, Librach C. Time-dependent capability of human oocytes for activation and pronuclear formation during metaphase II arrest. Hum Reprod 2004;19:982–7.
[99] Dominko T, First NL. Timing of meiotic progression in bovine oocytes and its effect on early embryo development. Mol Reprod Dev 1997;47:456–67.
[100] Kubiak JZ. Mouse oocytes gradually develop the capacity for activation during the metaphase II arrest. Dev Biol 1989;136:537–45.
[101] Marshall VS, Browne MA, Knowles L, Golos TG, Thomson JA. Ovarian stimulation of marmoset monkeys (*Callithrix jacchus*) using recombinant human follicle stimulating hormone. J Med Primatol 2003;32:57–66.
[102] Kuederling I, Morrell JM, Nayudu PL. Collection of semen from marmoset monkeys (*Callithrix jacchus*) for experimental use by vaginal washing. Lab Anim 1996;30:260–6.
[103] Morrell JM, Kuderling I, Hodges JK. Influence of semen collection method on ejaculate characteristics in the common marmoset, *Callithrix jacchus*. J Androl 1996;17:164–72.
[104] Kuederling I, Schneiders A, Sonksen J, Nayudu PL, Hodges JK. Non-invasive collection of ejaculates from the common marmoset (*Callithrix jacchus*) using penile vibrostimulation. Am J Primatol 2000;52:149–54.
[105] Delimitreva S, Zhivkova R, Isachenko E, Umland N, Nayudu PL. Meiotic abnormalities in in vitro-matured marmoset monkey (*Callithrix jacchus*) oocytes: development of a non-human primate model to investigate causal factors. Hum Reprod 2006;21:240–7.
[106] Gilchrist RB, Wicherek M, Heistermann M, Nayudu PL, Hodges JK. Changes in follicle-stimulating hormone and follicle populations during the ovarian cycle of the common marmoset. Biol Reprod 2001;64:127–35.
[107] Tkachenko OY, Delimitreva S, Heistermann M, Scheerer-Bernhard JU, Wedi E, Nayudu PL. Critical estradiol dose optimization for oocyte in vitro maturation in the common marmoset. Theriogenology 2015;83:1254–63.
[108] Tkachenko OY, Delimitreva S, Isachenko E, Valle RR, Michelmann HW, Berenson A, Nayudu PL. Epidermal growth factor effects on marmoset monkey (*Callithrix jacchus*) oocyte in vitro maturation, IVF and embryo development are altered by gonadotrophin concentration during oocyte maturation. Hum Reprod 2010;25:2047–58.
[109] Bai Y, Soda Y, Izawa K, Tanabe T, Kang X, Tojo A, Hoshino H, Miyoshi H, Asano S, Tani K. Effective transduction and stable transgene expression in human blood cells by a third-generation lentiviral vector. Gene Ther 2003;10:1446–57.
[110] Vermilyea SC, Guthrie S, Meyer M, Smuga-Otto K, Braun K, Howden S, Thomson JA, Zhang SC, Emborg ME, Golos TG. Induced pluripotent stem cell-derived dopaminergic neurons from adult common marmoset fibroblasts. Stem Cells Dev 2017;26:1225–35.
[111] Wiedemann A, Hemmer K, Bernemann I, Gohring G, Pogozhykh O, Figueiredo C, Glage S, Schambach A, Schwamborn JC, Blasczyk R, Muller T. Induced pluripotent stem cells generated from adult bone marrow-derived cells of the nonhuman primate (*Callithrix jacchus*) using a novel quad-cistronic and excisable lentiviral vector. Cell Reprogram 2012;14:485–96.
[112] Shiozawa S, Kawai K, Okada Y, Tomioka I, Maeda T, Kanda A, Shinohara H, Suemizu H, James Okano H, Sotomaru Y, Sasaki E, Okano H. Gene targeting and subsequent site-specific transgenesis at the beta-actin (ACTB) locus in common marmoset embryonic stem cells. Stem Cells Dev 2011;20:1587–99.
[113] Buehr M, Meek S, Blair K, Yang J, Ure J, Silva J, McLay R, Hall J, Ying QL, Smith A. Capture of authentic embryonic stem cells from rat blastocysts. Cell 2008;135:1287–98.
[114] Li P, Tong C, Mehrian-Shai R, Jia L, Wu N, Yan Y, Maxson RE, Schulze EN, Song H, Hsieh CL, Pera MF, Ying QL. Germline competent embryonic stem cells derived from rat blastocysts. Cell 2008;135:1299–310.
[115] Ying QL, Wray J, Nichols J, Batlle-Morera L, Doble B, Woodgett J, Cohen P, Smith A. The ground state of embryonic stem cell self-renewal. Nature 2008;453:519–23.
[116] Brons IG, Smithers LE, Trotter MW, Rugg-Gunn P, Sun B, Chuva de Sousa Lopes SM, Howlett SK, Clarkson A, Ahrlund-Richter L, Pedersen RA, Vallier L. Derivation of pluripotent epiblast stem cells from mammalian embryos. Nature 2007;448:191–5.
[117] Tesar PJ, Chenoweth JG, Brook FA, Davies TJ, Evans EP, Mack DL, Gardner RL, McKay RD. New cell lines from mouse epiblast share defining features with human embryonic stem cells. Nature 2007;448:196–9.
[118] Hackett JA, Surani MA. Regulatory principles of pluripotency: from the ground state up. Cell Stem Cell 2014;15:416–30.
[119] Nichols J, Smith A. Pluripotency in the embryo and in culture. Cold Spring Harb Perspect Biol 2012;4:a008128.
[120] Rossant J. Mouse and human blastocyst-derived stem cells: vive les differences. Development 2015;142:9–12.
[121] Davidson KC, Mason EA, Pera MF. The pluripotent state in mouse and human. Development 2015;142:3090–9.
[122] Hanna J, Cheng AW, Saha K, Kim J, Lengner CJ, Soldner F, Cassady JP, Muffat J, Carey BW, Jaenisch R. Human embryonic stem cells with biological and epigenetic characteristics similar

to those of mouse ESCs. Proc Natl Acad Sci USA 2010;107: 9222–7.

[123] Takashima Y, Guo G, Loos R, Nichols J, Ficz G, Krueger F, Oxley D, Santos F, Clarke J, Mansfield W, Reik W, Bertone P, Smith A. Resetting transcription factor control circuitry toward ground-state pluripotency in human. Cell 2014;158:1254–69.

[124] Theunissen TW, Powell BE, Wang H, Mitalipova M, Faddah DA, Reddy J, Fan ZP, Maetzel D, Ganz K, Shi L, Lungjangwa T, Imsoonthornruksa S, Stelzer Y, Rangarajan S, D'Alessio A, Zhang J, Gao Q, Dawlaty MM, Young RA, Gray NS, Jaenisch R. Systematic identification of culture conditions for induction and maintenance of naive human pluripotency. Cell Stem Cell 2014;15:471–87.

[125] Chen Y, Niu Y, Li Y, Ai Z, Kang Y, Shi H, Xiang Z, Yang Z, Tan T, Si W, Li W, Xia X, Zhou Q, Ji W, Li T. Generation of cynomolgus monkey chimeric fetuses using embryonic stem cells. Cell Stem Cell 2015;17:116–24.

[126] Honda A, Kawano Y, Izu H, Choijookhuu N, Honsho K, Nakamura T, Yabuta Y, Yamamoto T, Takashima Y, Hirose M, Sankai T, Hishikawa Y, Ogura A, Saitou M. Discrimination of stem cell status after subjecting cynomolgus monkey pluripotent stem cells to naive conversion. Sci Rep 2017;7:45285.

[127] Carroll D. Genome engineering with zinc-finger nucleases. Genetics 2011;188:773–82.

[128] Cermak T, Doyle EL, Christian M, Wang L, Zhang Y, Schmidt C, Baller JA, Somia NV, Bogdanove AJ, Voytas DF. Efficient design and assembly of custom TALEN and other TAL effector-based constructs for DNA targeting. Nucleic Acids Res 2011;39:e82.

[129] Cong L, Ran FA, Cox D, Lin S, Barretto R, Habib N, Hsu PD, Wu X, Jiang W, Marraffini LA, Zhang F. Multiplex genome engineering using CRISPR/Cas systems. Science 2013;339:819–23.

[130] Mali P, Aach J, Stranges PB, Esvelt KM, Moosburner M, Kosuri S, Yang L, Church GM. CAS9 transcriptional activators for target specificity screening and paired nickases for cooperative genome engineering. Nat Biotechnol 2013;31:833–8.

[131] Kwon DN, Lee K, Kang MJ, Choi YJ, Park C, Whyte JJ, Brown AN, Kim JH, Samuel M, Mao J, Park KW, Murphy CN, Prather RS. Production of biallelic CMP-Neu5Ac hydroxylase knock-out pigs. Sci Rep 2013;3:1981.

[132] Lillico SG, Proudfoot C, Carlson DF, Stverakova D, Neil C, Blain C, King TJ, Ritchie WA, Tan W, Mileham AJ, McLaren DG, Fahrenkrug SC, Whitelaw CB. Live pigs produced from genome edited zygotes. Sci Rep 2013;3:2847.

[133] Yang D, Zhang J, Xu J, Zhu T, Fan Y, Fan J, Chen YE. Production of apolipoprotein C-III knockout rabbits using zinc finger nucleases. J Vis Exp 2013:e50957.

[134] Zhang C, Wang L, Ren G, Li Z, Ren C, Zhang T, Xu K, Zhang Z. Targeted disruption of the sheep MSTN gene by engineered zinc-finger nucleases. Mol Biol Rep 2014;41:209–15.

[135] Liu H, Chen Y, Niu Y, Zhang K, Kang Y, Ge W, Liu X, Zhao E, Wang C, Lin S, Jing B, Si C, Lin Q, Chen X, Lin H, Pu X, Wang Y, Qin B, Wang F, Wang H, Si W, Zhou J, Tan T, Li T, Ji S, Xue Z, Luo Y, Cheng L, Zhou Q, Li S, Sun YE, Ji W. TALEN-mediated gene mutagenesis in rhesus and cynomolgus monkeys. Cell Stem Cell 2014;14:323–8.

[136] Niu Y, Shen B, Cui Y, Chen Y, Wang J, Wang L, Kang Y, Zhao X, Si W, Li W, Xiang AP, Zhou J, Guo X, Bi Y, Si C, Hu B, Dong G, Wang H, Zhou Z, Li T, Tan T, Pu X, Wang F, Ji S, Zhou Q, Huang X, Ji W, Sha J. Generation of gene-modified cynomolgus monkey via Cas9/RNA-mediated gene targeting in one-cell embryos. Cell 2014;156:836–43.

[137] Chen Y, Yu J, Niu Y, Qin D, Liu H, Li G, Hu Y, Wang J, Lu Y, Kang Y, Jiang Y, Wu K, Li S, Wei J, He J, Liu X, Luo Y, Si C, Bai R, Zhang K, Liu J, Huang S, Chen Z, Wang S, Chen X, Bao X, Zhang Q, Li F, Geng R, Liang A, Shen D, Jiang T, Hu X, Ma Y, Ji W, Sun YE. Modeling Rett Syndrome using TALEN-edited MECP2 mutant cynomolgus monkeys. Cell 2017;169:945–55. e910.

[138] Cox DB, Platt RJ, Zhang F. Therapeutic genome editing: prospects and challenges. Nat Med 2015;21:121–31.

[139] Maruyama T, Dougan SK, Truttmann MC, Bilate AM, Ingram JR, Ploegh HL. Increasing the efficiency of precise genome editing with CRISPR-Cas9 by inhibition of nonhomologous end joining. Nat Biotechnol 2015;33:538–42.

[140] Yang H, Wang H, Shivalila CS, Cheng AW, Shi L, Jaenisch R. One-step generation of mice carrying reporter and conditional alleles by CRISPR/Cas-mediated genome engineering. Cell 2013;154:1370–9.

[141] Cristea S, Freyvert Y, Santiago Y, Holmes MC, Urnov FD, Gregory PD, Cost GJ. In vivo cleavage of transgene donors promotes nuclease-mediated targeted integration. Biotechnol Bioeng 2013;110:871–80.

[142] Maresca M, Lin VG, Guo N, Yang Y. Obligate ligation-gated recombination (ObLiGaRe): custom-designed nuclease-mediated targeted integration through nonhomologous end joining. Genome Res 2013;23:539–46.

[143] Nakade S, Tsubota T, Sakane Y, Kume S, Sakamoto N, Obara M, Daimon T, Sezutsu H, Yamamoto T, Sakuma T, Suzuki KT. Microhomology-mediated end-joining-dependent integration of donor DNA in cells and animals using TALENs and CRISPR/Cas9. Nat Commun 2014;5:5560.

[144] Sakuma T, Nakade S, Sakane Y, Suzuki KT, Yamamoto T. MMEJ-assisted gene knock-in using TALENs and CRISPR-Cas9 with the PITCh systems. Nat Protoc 2016;11:118–33.

[145] Suzuki K, Tsunekawa Y, Hernandez-Benitez R, Wu J, Zhu J, Kim EJ, Hatanaka F, Yamamoto M, Araoka T, Li Z, Kurita M, Hishida T, Li M, Aizawa E, Guo S, Chen S, Goebl A, Soligalla RD, Qu J, Jiang T, Fu X, Jafari M, Esteban CR, Berggren WT, Lajara J, Nunez-Delicado E, Guillen P, Campistol JM, Matsuzaki F, Liu GH, Magistretti P, Zhang K, Callaway EM, Belmonte JC. In vivo genome editing via CRISPR/Cas9 mediated homology-independent targeted integration. Nature 2016;540:144–9.

[146] Komor AC, Kim YB, Packer MS, Zuris JA, Liu DR. Programmable editing of a target base in genomic DNA without double-stranded DNA cleavage. Nature 2016;533:420–4.

CHAPTER

21

Marmosets in Aging Research

Corinna N. Ross

Texas A&M University San Antonio, Department of Science and Mathematics, San Antonio, TX

INTRODUCTION

Aging Research

The demographics of the US population is shifting such that the Center for Disease Control and Prevention (CDC) estimates that by the year 2050 the number of people over the age of 65 will double and with it, the prevalence of age-related disease [1]. This shift in the demographic makeup of the population has been a source of interest, concern, and research for the last two decades [1–11]. This aging population may induce a drastically altered economic burden associated with medical costs, costs of care for the disabled, and the shift in the percentage of the population not working. The CDC Health Statistics estimate that for people over the age of 65 more than 60% suffer from hypertension, roughly 40% are obese, 21% are in fair or poor health, and 7% need personal daily care [12,13]. Due to the dynamic alteration of the human population needs in the near future, researchers focused on studies of longevity and life span are no longer simply interested in extending human life span, they are interested in increasing health span, which is the period of life that is free from chronic disease, thus extending the length of time people can live independently and improving the overall quality of life. The number one complaint given by the elderly is not pain and discomfort, but rather the need for independence as a part of their quality of life [3,9,14]. Independence and health are ultimately tied to disease progression and elderly resilience, which may be related to the ability to be stress resistant and to recover following disease onset, whereas frail individuals are more likely to suffer poor outcomes following an injury or illness [15,16]. Therefore, the focus of much of aging research has now shifted to treating aging as a systemic problem and determining whether an increase in life span alters the ability for that individual to remain disease free. The obvious hope is that the treatment for multiple age-related diseases including diabetes, cardiovascular disease, Alzheimer's, dementia, and sarcopenia could possibly be determined by understanding mechanisms that underlie aging.

The pathways and mechanisms associated with aging are beginning to be more broadly understood, but the connections, triggers, and timing remain unknown. The hallmarks of aging are thought to include stem cell exhaustion, altered intercellular communication, genomic instability, telomere attrition, epigenetic alterations, loss of proteostasis, deregulated nutrient sensing, mitochondrial dysfunction, and cellular senescence [17]. Changes in these hallmarks are related to the diseases associated with aging and may act differentially in organ systems. A systemic approach to evaluating the causes of aging has hypothesized that there are only seven pillars of aging that influence all age-related disease progression: metabolism, macromolecular damage, epigenetics, inflammation, adaptation to stress, proteostasis, and stem cells and regeneration [18]. A great deal of work remains to be done to determine how these pillars interact and ultimately result in aging phenotypes, and most of this work cannot ethically be done on the aging human population.

For decades, research on longevity and aging relied on animal models to evaluate cellular pathways and mechanisms, as well as organismal functional decline. Typical animal models for this research included yeast, *Caenorhabditis elegans*, *Drosophila*, and mice [19]. These models offer a number of advantages including short life span, the ability to control environmental exposure for each individual (or a population), and the ability to manipulate genes of interest. Work with these model organisms has allowed extensive evaluation of conserved biochemical pathways associated with longevity, metabolism, and

The Common Marmoset in Captivity and Biomedical Research
https://doi.org/10.1016/B978-0-12-811829-0.00021-2

development, including the mechanistic target of rapamycin (mTOR), mitochondrial stress pathways, adenosine monophosphate-activated protein kinase, and insulin-like growth factor [20,21]. The rodent aging work has been extensive due to the fact that they are mammals and have a relatively short life span [21,22]. A number of rodent genetic mutant lines that allow direct evaluation of the hereditary basis of age-related disease have been developed [23]. Furthermore, rodents have been extensively examined to evaluate the interventional and therapeutic effects on aging and health span. The gold standard of aging studies is caloric restriction, which has reliably increased life span and health span in specific strains of mice [24–26]. Many new drugs of interest, including rapamycin, have been tested in rodents and continue to reveal details regarding aging and disease progression [27–29]. However, there are many concerns regarding using mice as the only mammalian model of aging because many of the drugs tested in rodents do not translate successfully to humans, and inbred strains of mice often have selected traits that are a by-product of breeding [30]. Investigators have therefore begun to shift to a comparative biology of aging approach in conjunction to the use of model organisms to evaluate life and health span.

The comparative biology of aging approach suggests that we need not only use short-lived model organisms in the laboratory but also need to more broadly examine other organisms that are both short and long lived [31–33]. Animals proposed for further investigation include the naked mole rat, bats, dogs, and nonhuman primates. Naked mole rats live for an extraordinary amount of time in comparison to other rodents of similar size and seem to be resilient to naturally occurring cancers [33–36]. Bats offer the opportunity to compare closely related, exceptionally long-lived species, like the little brown bat *Myotis lucifugus* (34 years), to short-lived species such as the evening bat *Nycticeius humeralis* (6 years) [36a]. Dogs have been proposed as a useful model for studies due to their coevolution with humans as companions, the extensive veterinary knowledge that we have for this species, the interest of pet owners in participating in research to increase their pets' health, and the variability in longevity between different breeds [37]. Finally, nonhuman primates have been proposed due to their close evolutionary history with humans and the variability in longevity between species [32,38–40].

Nonhuman Primate Models

Old World monkeys and apes are the closest living relatives to humans, thus sharing a closer evolutionary history with humans than other mammals, and more similar genetic, biochemical, and phenotypic outcomes. Old World monkeys split from the ancestors of the human lineage around 25 million years ago (mya), and apes, specifically chimpanzees, split as recently as six mya [41,42]. Chimpanzees, our closest living relatives, offer the ability to study aging as they display many of the same diseases as humans (heart disease, cancer, diabetes, etc.) [43–50]. However, they are very long lived, with animals in captivity living up to 60 years, thus making them comparable in difficulty to working with human populations [42]. Additionally, they are endangered animals, and use in the laboratory setting has significantly diminished in many countries, including the US. Other Old World monkeys, including macaques, have been the focus of biomedical translational research including aging studies [32,39,40,51–53]. Macaques live a maximum of 40 years in captivity and their aging phenotypes have been well characterized [54,55]. Of particular interest for aging studies has been the ability to monitor cognitive functional decline in a primate [52]. Caloric restriction is the gold standard for increasing longevity and health span in rodent models, and this technique was translated to macaque models in the 1980s to determine whether it would have the same impact on aging nonhuman primates [32]. While the results have not been the same across the labs using caloric restriction, it does appear that caloric restriction reduces the prevalence of cardiovascular disease, type 2 diabetes, and neoplasias [56–63]. However, these expensive, very lengthy studies are still not complete due to the life span of macaques, and proposals to repeat these types of studies are difficult to justify since one study can outlast the average career life span of a researcher.

To enable shorter studies for examining nonhuman primates, several investigators have suggested shifting work to the development of short-lived, small-bodied nonhuman primates. Several species were suggested for longevity studies including the bush baby (*Galago senegalensis*), the gray mouse lemur (*Microcebus murinus*), and the common marmoset (*Callithrix jacchus*) [38,64]. These species all have relatively shorter life spans (maximum 15–16 years) compared with Old World monkeys and apes and they offer a number of advantages due to their small size and rapid reproduction [38,64]. The only one of these three species to be integrated thoroughly into aging research to date is the common marmoset, which has been emerging as a model for aging-related research [65–70]. This chapter will review what is known so far regarding marmoset aging and phenotypes associated with aging, as well as the types of research that are focused on marmosets as a translational model of aging.

MARMOSET AGING

Life Span

Marmosets typically give birth to fraternal twins following a 5-month gestation period, although litter size can vary from singletons to quadruplets [71–76]. Early development is fairly rapid, with first intake of solid food occurring around day 20, weaning beginning at approximately 30 days, and infants being fully weaned by 90 days of age [77]. Juveniles reach sexual maturity at 12–18 months and adult body length and mass by 24 months [78,79]. This developmental rate produces the comparative rate of aging between human and marmoset years of 1:9 from birth to sexual maturity and 1:12 to reach adult stature, which is much faster than 1:4 and 1:3 calculated for macaques [53]. Most marmoset colonies report a median life span between 5 and 6 years with a maximum life span of around 16 years when mortality associated with experimental protocols are excluded; however, it is not known how participation in some experimental protocols may impact the rate of aging for a marmoset (for example, diet-induced obesity) [64,80,81]. The rate of adult aging between human years and marmoset years would thus be 1:8 to middle age and 1:6 to maximum age (compared with 1:2 and 1:2.5 for macaques). A colony in Japan recently reported dramatically extended life spans in their marmoset colony with a median life span of 9 years for females and 13 years for males, with a maximum reported life span of 22 years [82]. If other colonies can achieve this same slower rate of aging, this would then alter the ratio of aging to 1:4 to middle age and 1:4.5 to maximum age, which is dramatically different than what is seen in most United States and European colonies.

The longevity reported in the Japan production colony was at first a shocking report for those that work with marmosets; however, there are several potential reasons why this colony might have had such dramatically different life spans. First and foremost, the Japanese colony has been a closed colony with no import from other sources since its founding [82]. This ability to prevent the potential transfer of pathogens from other colonies, or have deaths due to the stress of movement and social stress due to exposure to novel conspecifics, may improve life span and has been reported in other instances. In the case of Southwest National Primate Center, mortality increased when the closed colony was open to the import of marmosets from other colonies [70]. Another unique feature of the Japanese colony was the isolation of marmosets from other nonhuman primates that could spread contagion. The final possibility is that in the time the colony was closed, selective breeding of the most robust and healthy animals led to accidental artificial selection of marmosets with extended life span.

One of the difficulties of working with new models of aging is that typically these species have a linear rate of death across middle age. Humans and often used model species, such as mice and rhesus monkeys, in comparison have a plateaued mortality rate during adulthood due to increased health care in the case of humans and better husbandry practices for model species (Fig. 21.1). Most primate species still depict a linear mortality rate even in captivity. Marmosets in standard housing also typically exhibit linear mortality with males typically having slightly better survival in mid adulthood when compared with females (Fig. 21.2) [70,81,82]. These linear declines are thought to not truly represent age-related deaths, but rather deaths associated with infectious disease or trauma. Eliminating or minimizing these deaths should result in a plateaued rate of death during adulthood with increased rates of mortality due to aging in the elderly.

In 2006, a new colony of marmosets was formed at the University of Texas Health Science Center, Barshop Institute for Longevity and Aging Studies, to evaluate marmosets as an aging model. This colony was formed from a founder group of animals selected from the Southwest National Primate Research Center (SNPRC) marmoset colony. These founders were carefully screened for the presence of potential bacterial and viral pathological agents. The colony was different from other marmoset colonies in the United States and Europe as it was to be maintained as an SPF colony under barrier conditions. Specifically, the housing protocols included no presence of other nonhuman primates at the facility, the marmoset rooms had their own HEPA filtered air system, all items entering the room were sterilized, the drinking water was autoclaved and dietary items were irradiated, and these animals were not participating in other experimental protocols. These strict barrier conditions were meant to match barrier conditions maintained for rodent aging models and to decrease the occurrences of death associated with infectious disease.

Maintaining marmosets in barrier conditions significantly increased the survival and plateaued adult survivorship in both males and females when compared with marmosets maintained in conventional housing protocols at the Southwest National Primate Center (Fig. 21.3) [82a]. For marmosets, the first 2 weeks of life present a high risk of neonatal death and are typically removed from longevity analyses. Barrier females had a mean life span of 7.1 years when compared with 4.9 years at the primate center (hazard $= 2.773$, $P = 0.001$). Barrier males had a mean life span of 8 years compared with the 5.4 years at the primate center (hazard $= 3.44$, $P = 0.001$). The plateau effect for adult longevity depicted by the barrier colony is presumably due to the decreased exposure to infectious agents and

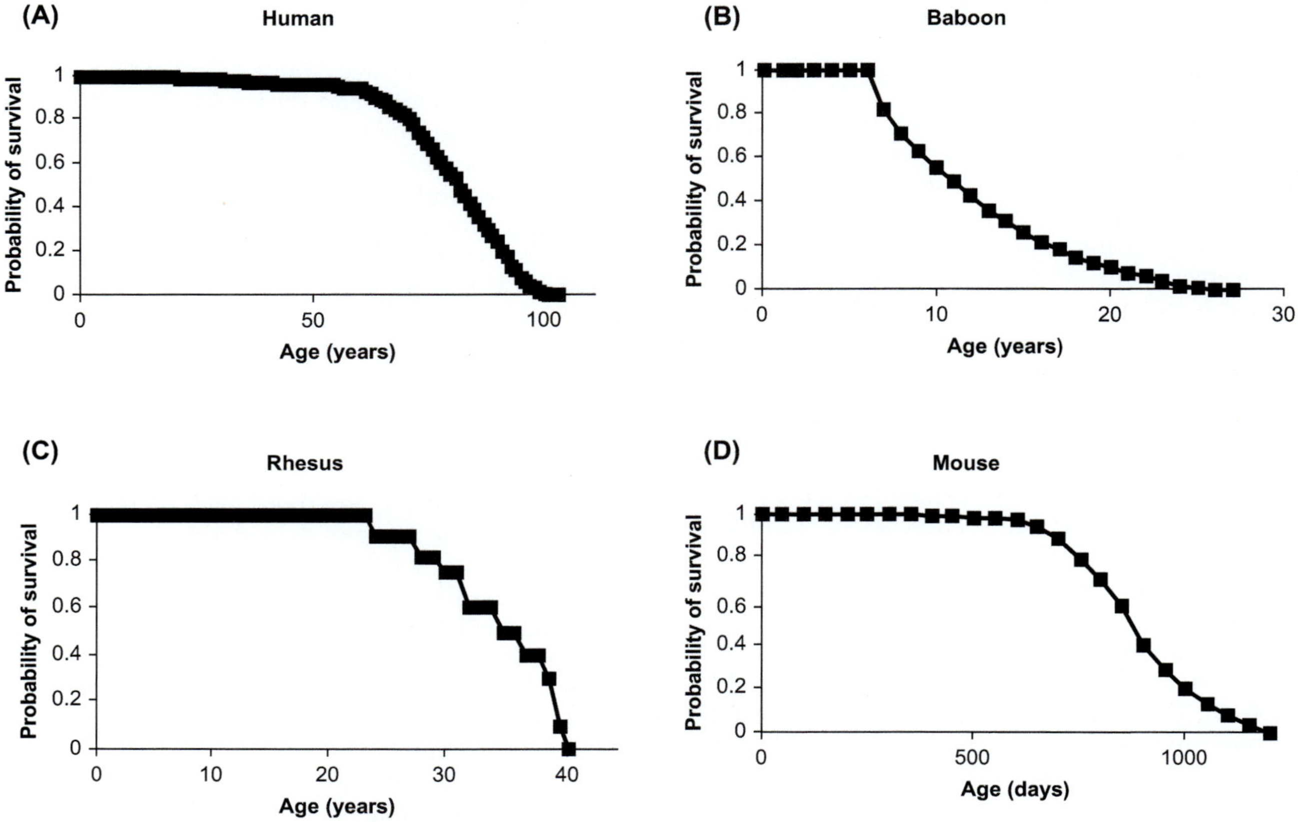

FIGURE 21.1 Representative survival curves from humans and captive maintained model species. (A) Human. (B) Baboon. (C) Rhesus macaque. (D) Mouse. The human, rhesus, and mouse curves all exhibit a plateaued shape in comparison with the baboon linear rate of mortality. *(A) Redrawn from Weon BM. Analysis of trends in human longevity by new model. Mimeo; 2004. http://arxiv.org/ftp/q-bio/papers/0402/0402011.pdf. (B) Redrawn from Martin LJ, Mahaney MC, Bronikowski AM, Carey KD, Dyke B, Comuzzie AG. Lifespan in captive baboons is heritable. Mech Ageing Dev 2002; 123:1461–7. (C) Redrawn from Mattison JA, Colman RJ, Beasley TM, Allison DB, Kemnitz JW, Roth GS, Ingram DK, Weindruch R, de Cabo R, Anderson RM. Caloric restriction improves health and survival of rhesus monkeys. Nat Commun 2017;8:14063. (D) Redrawn from Kanfi Y, Naiman S, Amir G, Peshti V, Zinman G, Nahum L, Bar-Joseph Z, Cohen HY. The sirtuin SIRT6 regulates lifespan in male mice. Nature 2012;483:218–21.*

pathogens rather than genetic effects on longevity because the barrier colony was founded from the SNPRC conventional colony. However, the alteration in mortality could also be correlated to the barrier colony being significantly smaller in size. The colony had few changes in housing or space allocation for the animals, thus increasing the stability of social housing and colony familiarity between animals. It is also possible that the smaller colony size may have increased staff interactions with the individual marmosets, perhaps resulting in increased monitoring and earlier veterinary interventions. The ability to detect differences in life span and mortality rates in association with colony management and husbandry techniques may indeed be valuable for future aging research. It will be of particular importance to determine whether changes in the mortality rates are associated with other phenotypic markers of aging and whether these can also be shifted by altering husbandry.

Causes of Death

Pathologies associated with death have altered greatly over the decades of captive marmoset research. In the 1980s when colonies were a mixture of wild-caught and lab-bred animals, the most common causes of death were pneumonia, otitis media, meningitis, and brain abscesses, all associated with bacterial infections [86]. Anemia, chronic inflammation of the bowels, and colitis were findings at death that were associated with insufficient coverage of dietary needs [86]. A more recent review of spontaneous lesions noted at death for a laboratory-bred colony at Covance laboratories in Germany, reported high incidence of adrenal hematopoiesis (54%), duodenal cystic Brunner's glands (51%), kidney inflammatory cell foci (78.5%), liver hematopoiesis (46.5%), liver inflammatory cell foci (36%), and thyroid C-cell hyperplasia (56%) [87]. These reported pathology findings are not broken down according to age at death.

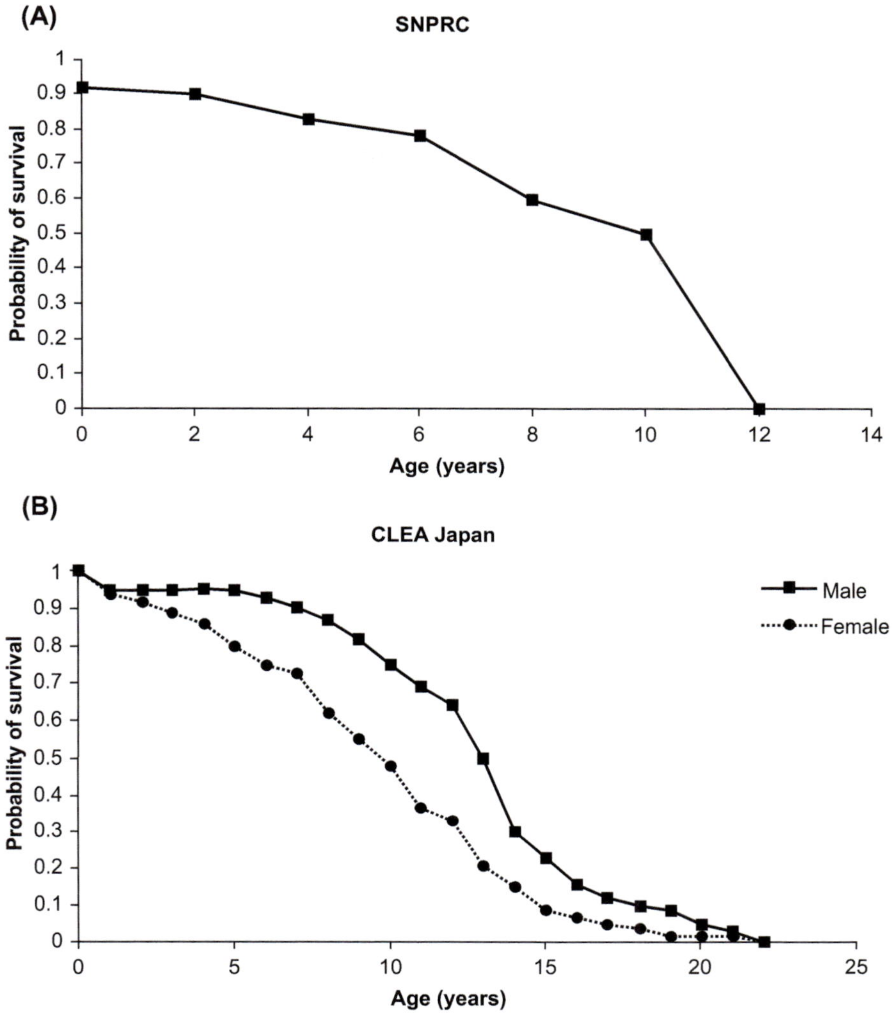

FIGURE 21.2 Survival curves of captive marmoset populations. (A) Southwest National Primate Research Center (SNPRC) (n = 358). (B) Japanese CLEA colony (n = 153) [82]. *(A) Redrawn from Tardif SD, Mansfield KG, Ratnam R, Ross CN, Ziegler TE. The marmoset as a model of aging and age-related disease. Inst Lab Anim Res J 2011;52:54–65.*

An analysis of the pathologies associated with age at the New England National Primate Center revealed dramatic shifts in the associated causes of death as animals aged in the colony. Young adults under the age of 6 years were most likely to die due to injury or idiopathic deaths (41%), whereas deaths associated with infection were the next largest category (39%) [70]. Interestingly, inflammatory bowel disease (IBD) was associated with a large percentage of young adult death (16%), and there were rare reports of neoplasia (2%) and diabetes (2%). In contrast, for animals over the age of six the most likely causes of death were infection (22%), with a decreased association with IBD (13%), and increasing rates of pathologies typically thought of as age-related pathologies in humans and other model organisms, neoplasia (16%), amyloid (13%), cardiac and diabetes (10%), and renal failure (6%) (Fig. 21.4). When marmosets over the age of 10 years are specifically examined, the percentage of deaths due to infection drops to 7%, with leading reported causes being neoplasia at 22% and amyloid at 29% [70].

An assessment of causes of death at the Southwest National Primate Center reported that for animals under the age of six, most of the animals displayed IBD (44%), with death due to infection (19%), amyloid (13%), neoplasias (7%), renal disease (6%), and diabetic/cardiac events (1%) [66] (Fig. 21.4). Animals over the age of six had decreased rates of IBD (31%), infection (14%), neoplasias (4%), and amyloid (12%), but increased occurrences of age-related pathologies including cardiomyopathy (14%) and renal failure (18%). When examining causes of death from both colonies, it appears that aging in marmosets may be associated with increased risk for cardiac disease and renal disease,

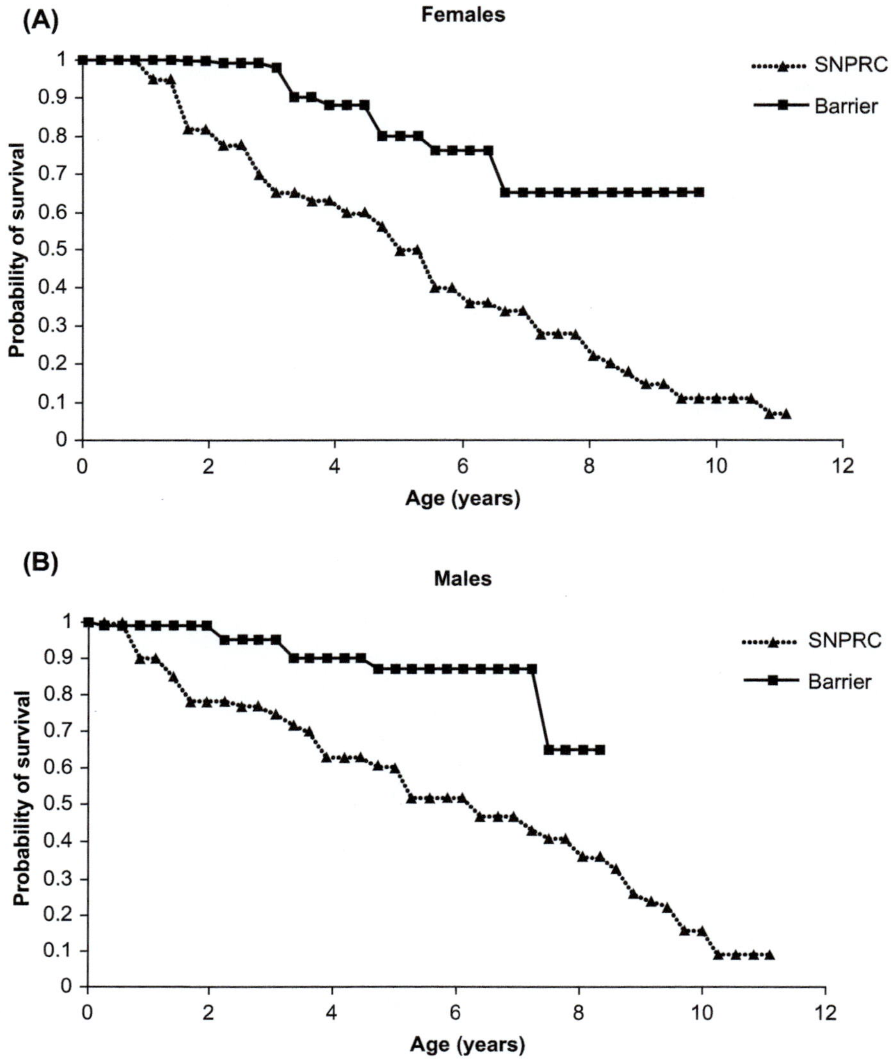

FIGURE 21.3 Comparative analysis of marmosets maintained in the UTHSCSA Barrier colony (n = 247) and the Southwest National Primate Research Center (SNPRC) (n = 370). (A) Females. (B) Males. Redrawn from Ross CN, Austad S, Brasky K, Brown CJ, Forey LJ, Gelfond JA, Lanford R, RIchardson A, Tardif SD. 2017. The development of a specific pathogen free (SPF) barrier colony of marmosets (*Callithrix jacchus*) for aging research. Aging, 9:2544-2558.

with one colony noting higher rates of amyloidosis, and neoplasias with age. It is difficult to determine whether the causes of death truly differ between these two colonies who maintained fairly similar husbandry care at the time of these analyses (2002–11), as pathologists may vary in attributing cause of death and noting dominant pathologies in the necropsy findings. Unfortunately, pathologies and causes of death were not included in the data from the Japanese closed colony with extended life spans, which would have allowed for unique comparisons, although they do note that pneumonia and neoplasias were seen within the colony [82]. The UTHSCSA barrier colony had too few deaths to do a full analysis of causes of death, but pathology did note renal disease in the colony and gallbladder dysfunction in several animals (Dr. Hanes, personal communication). Additionally, there were two deaths with noted metabolic dysfunction prior to death and cardiomyopathy and obesity at necropsy.

Musculoskeletal System

Bone

Degeneration of muscles, joints, and bones is associated with human aging and severe loss is associated with sarcopenia, osteoporosis, and frailty [88–90]. The development of frailty and osteoarthritis causes chronic disability in the elderly, and these diseases are often associated not only with chronic inflammation but also the inability to recover following a stressful event [91–94]. Unfortunately, while treatments exist to relieve

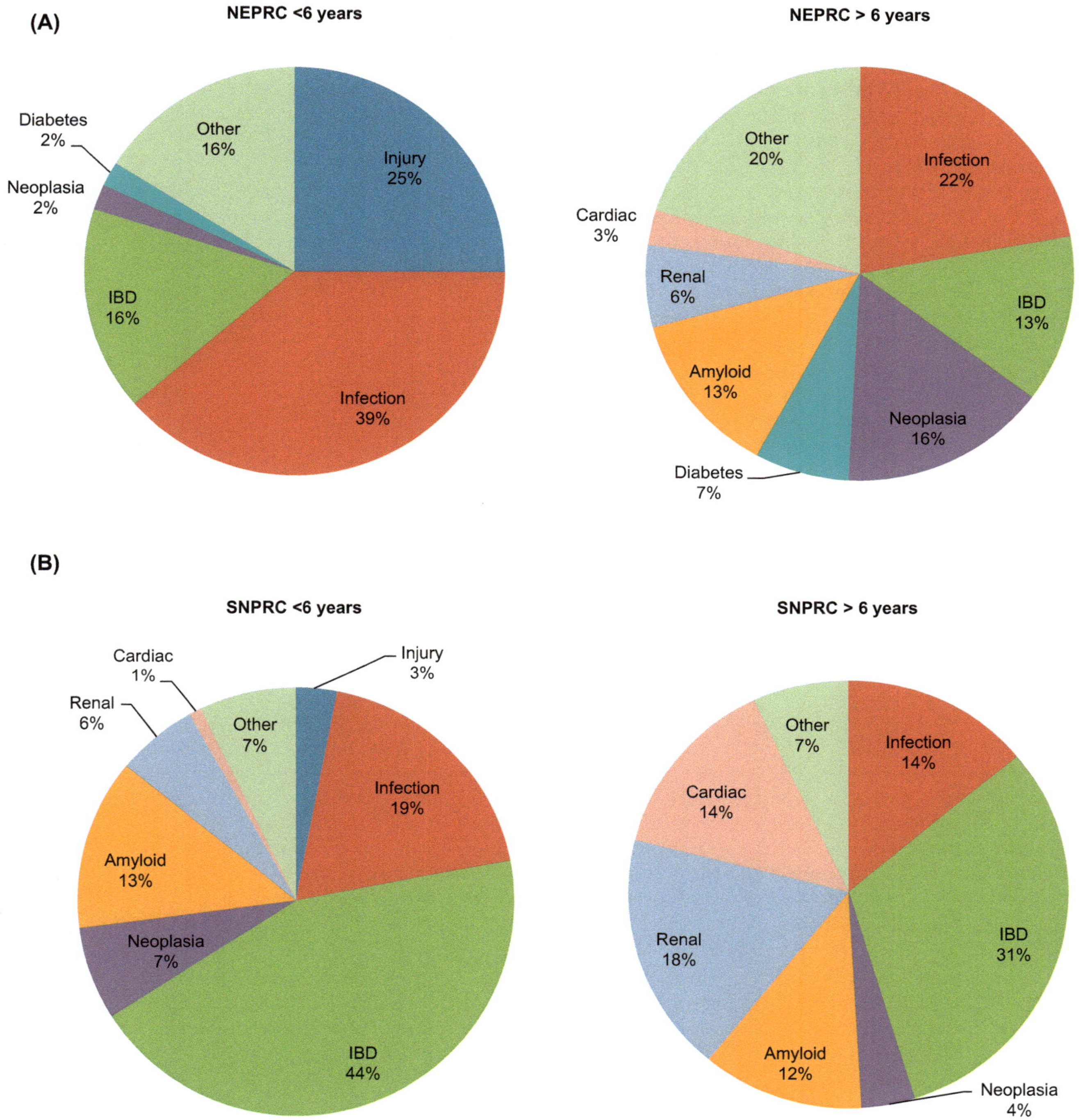

FIGURE 21.4 Common causes of deaths and pathologies reported for common marmoset colonies. (A) New England Primate Research Center (NEPRC) (n = 77). (B) Southwest National Primate Research Center (SNPRC) (n = 150). *(A) Redrawn from Tardif SD, Mansfield KG, Ratnam R, Ross CN, Ziegler TE. The marmoset as a model of aging and age-related disease. Inst Lab Anim Res J 2011;52:54–65. (B) Redrawn from Ross CN, Davis K, Dobek G, Tardif SD. Aging phenotypes of common marmosets (*Callithrix jacchus*). J Aging Res 2012;2012:567143.*

the symptoms of arthritis and osteoporosis, a cure or permanent remediation does not exist. In humans, osteoporosis is caused by a systemic loss of bone mass and deterioration of the trabecular bone structure due to increased absorption of bone by osteoclasts and reduced production by osteoblasts [95–97]. Old World monkeys have been found to display remodeling of cortical bone that is similar to human degeneration and is associated with dysfunction of the reproductive endocrine system [32].

Marmosets offer a unique model for the study of skeletal bone changes because they lack adult growth plates and display similar human age-related structural changes [98]. When treated with a bisphosphate therapy, trabecular number and volume increased in marmosets exhibiting age-related bone loss [98]. However, marmosets appear to be able to maintain bone mass in the absence of gonadal estrogen as females ovariectomized for 6 months exhibited no lumbar spine bone changes, which is not what is found in macaques [58,97,99–101]. Furthermore, while degeneration of bone structure has been associated with aging in bones such as the vertebrae [102], degeneration and softening of the bone in marmosets has also been associated with nutritional deficits and malabsorption of vitamin D [86,103]. It will be particularly important in the future for researchers investigating aging effects on bone to evaluate whether bone disease in the marmoset is a product of aging and degeneration in a way that is similar to human degeneration or whether it is a product of metabolic bone disease which, in the marmosets, is associated with gastrointestinal disease [103] and overexpression of vitamin D response element binding proteins [104]. The ability to tease these effects apart and to focus on age-associated bone decay will depend at least partially on the development of biomarkers for gastrointestinal disease and/or vitamin D malabsorption.

Body Mass

Marmosets have been found to lose weight at an increasing rate following middle age [70,105]. An analysis of body composition of aging animals at the New England National Primate Center supports the data that marmoset weight reaches a maximum by 5 years of age and then begins to decline, and the slope accelerates as aging progresses with animals from four to eight, showing a loss of 7.3 g/year, with animals 8 years old and older losing 14.2 g/year [70]. Furthermore, the maximum weight achieved is significantly associated with the likelihood of survival in older animals. Marmosets with a peak mass of less than 400 g had decreased survival after 8 years of age [70]. Additionally, cross-sectional analyses of body composition for animals ranging in age from approximately 1 to 8 years of age demonstrate an inverted "U"-shaped relationship between fat-free mass (FFM) and age as determined by labeled water dilution analysis [105]. Marmosets exhibit an increase in FFM until reaching a peak mass between two and a half and five years of age, after which the mass steadily declines [105].

To explore the longitudinal changes in body mass and composition associated with aging, 46 marmosets, 19 females age range from 2.2 to 7.3 years old, and 27 males age range from 2.6 to 8.3, were assessed multiple times using quantitative magnetic resonance across a 1–3-year time span. Change in age (aging) was significantly associated with decreasing fat mass, but not lean mass [66]. Further analysis of these data reveals a similar inverted "U"-shaped curve relationship between age and lean mass as was shown in Power et al. [105], with decreasing lean mass in aged marmosets (Fig. 21.5).

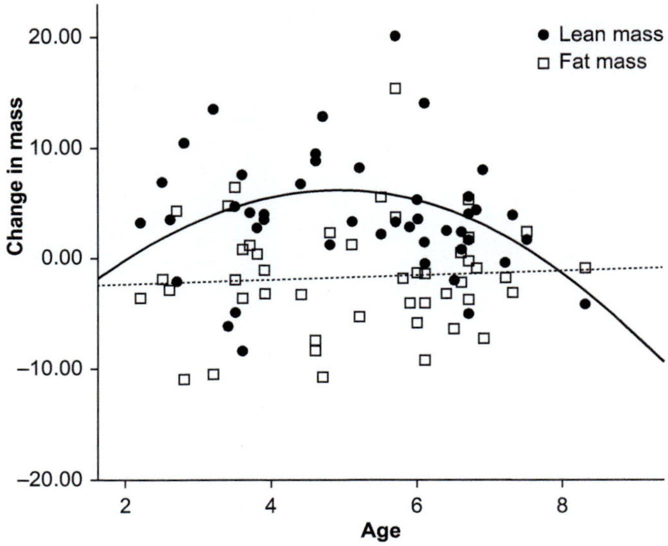

FIGURE 21.5 Changes in lean (*solid*) and fat mass (*open/dashed*) of aging marmosets measured longitudinally at Southwest National Primate Research Center demonstrating loss of lean mass in older marmosets (quadratic linear regression).

Morphometric measures are often substituted for quantitative measures of body composition in human elderly studies. Typically, it has been reported that the elderly have smaller limb circumferences, and this size is associated with loss of muscle mass associated with sarcopenia and frailty [90–92,94]. Power (2001) reported a significant association between FFM and age, and they also reported significant correlations between FFM and body circumferences including abdominal, chest, arm, and thigh [105]. However, the relationship between these measures and age was not reported. In our assessment of aging marmosets, we reported that thigh circumference was associated with age, fat mass, and lean mass; however, abdominal circumference was not associated with age even though it was significantly correlated with fat and lean mass [66]. If marmosets are affected by loss in lean mass associated with age, we would predict significant relationships between activity-related behaviors and age. Males ($n = 30$) ranging in age from 2 to 13.5 were observed for 30 min for behaviors including frequency of jumping, duration of resting, and frequency of a hanging stretch behavior [66]. Contrary to our predictions, the only behavior that was significantly associated with age was the frequency of stretching behaviors. In fact, this behavior, while not a measure of activity, also significantly predicted the risk of death within the following 24 months (hazard = 1.23). It is possible that this behavior represents discomfort and/or pain for the animal, which is similar to what has been noted in other models of aging such as arthritis-related movement changes [106–108].

Strength

To assess frailty in humans, a number of models have been proposed to identify characteristics which predict not only sarcopenia but also the full-spectrum syndrome of frailty [92,93,106]. Muscle mass and weight are often included in these models, as well as measures of strength such as grip strength and walking speed. Modeling these types of activities in animal models has proven difficult due to limitations in equipment or the ability to verify complete compliance of the subjects. Proxies of grip strength have been reported in mouse models in which the investigator has the mouse grasp a small bar with its forefeet and then the animal is pulled backward slowly by the tail. The force that is exerted by the investigator at which the mouse releases the bar is determined to be the subject's grip strength [109,110]. There are a number of reported problems with this measure including consistency between technicians and speed at which the test is done. Adequate proxies of grip strength have not been reported for the marmoset. There is currently a method to evaluate hindlimb strength via vertical jump force by training animals to jump within a tower equipped with force transducers attached to horizontal poles [111,112]. These measures have described the maximum force of the vertical jump to be much higher than forces recorded for humans or bonobos [111]. The difference between human and marmoset hindlimb force generation may be due to marmoset muscle being almost exclusively fast muscle fibers, with 63% of the fibers being IIB, which are not found in human hindlimb muscles [113]. Therefore, while strength measures of hind limbs are still important when investigating functional decline of the marmoset, and the potential risk of an individual to develop frailty symptoms, its direct translation to phenotypic changes in humans may be limited.

In an attempt to model muscle strength in the marmoset, we trained the animals to pull a weighted lever device that was described by Stevens et al. [114] as a device to measure the effects of toxins on strength. The pulley device was used to assess the arm strength of 14 marmosets at the SNPRC ranging from 3.5 to 10 years of age. Animals were trained to pull the lever of the device with one hand and were rewarded for each successful pull. Testing sessions consisted of up to 15 pulls per session in which the weights were added and then decreased in a randomized fashion to reach a maximum weight for the session. Sessions increased by 10 g at a time until no further advancement could be made in three sessions. The maximum weight was achieved by each animal following 25 to 35 sessions. There was a great deal of variance between animals in the amount of weight that could be pulled from a maximum one-handed pull of 130–770 g. Interestingly, the maximum weight pulled by these animals was not associated with age, sex, weight of the animal, or lean mass of the animal. The maximum weight pulled was associated with handedness of the animal, with left-handed individuals pulling significantly more weight than right-handed individuals ($r = 0.715$, $P = .004$) (Fig. 21.6).

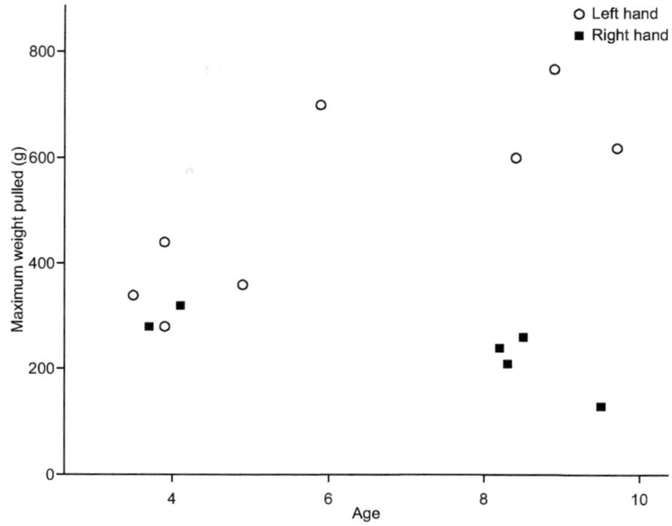

FIGURE 21.6 Maximum weight pulled by marmosets using the strength pulley task. Left-handed marmosets are able to pull significantly more during this task than are right-handed animals [114].

There are several potential explanations for these results. First, while a number of animal studies, including primate studies, have found that the right side of the body is significantly stronger than the left, the right side is used predominantly in supporting body position and may therefore impact how a quadrupedal individual interacts with its environment [115,116]. Marmosets do not develop handedness in relation to early infant nursing and carrying position as seen in many primates [115]. A longitudinal analysis of the development of handedness and supportive feeding postures revealed that animals develop a hand preference early (5–8 months), whereas marmosets between 14 and 25 months readily use both hands and the preference for hand usage stabilizes by 30 months of age [115]. It is unknown whether this preference is maintained through old age. Preferences for feeding posture were correlated to hand preference with right-handed marmosets being more likely to adopt the tripedal posture over a seated or suspended posture [115]. It was hypothesized that animals may choose the left hand for food manipulation to use the stronger right side for stabilized postural control. Thus in our trial, it is possible that those that were manipulating the device with their left hand were able to maintain a more stable body position, whereas those pulling with their right hands were in a more vulnerable body position and thus "weaker." Alternatively, there was a complication during our study regarding the delivery of rewards from the apparatus. Due to the device not functioning properly, the behavioral trainer could not deliver the reward in the center of the device and instead was treating the reward on the right-hand side of the animal. It was not revealed until video analysis that the study had inadvertently introduced a potential hand bias into the testing parameters. There was no difference in the amount of time to retrieval of the reward for animals due to handedness, thus animals that were left handed did not have faster access to the reward. All of the animals let go of the device and actively moved their whole bodies to the side to retrieve the reward, and there were no noted postural differences during those movements for left- or right-handed individuals. There was also no difference in how the animals were manipulating the reward once it was offered because 95% of the retrievals were done by mouth rather than handed manipulation. What did differ between the individuals was the posture adopted during the pull. Both left-handed and right-handed individuals stood on their two hind legs and pulled with a single hand. Animals that were right-handed leaned the left side of their body against the Plexiglas of the device to "brace" themselves while pulling. Animals pulling with their left hands did not have this modified posture. This observation supports the idea that postural differences between right-handed and left-handed individuals resulted in the apparent maximum weight differences rather than those individuals truly being stronger. From these behaviors, we concluded that the hand bias overrode any potential ability to detect differences associated with age. However, all of the older animals were also healthy and had no signs of recent or current weight loss. These data in particular pose a cautionary tale that behavioral testing paradigms can easily produce results with an underlying bias that was not originally being examined. We believe that further strength-based testing needs to be evaluated in the marmosets to determine if there is an accurate, and perhaps easier, method to test strength and relate it not only to age but also body compositional changes.

Walking speed tests have also not been reported in the marmoset as it directly relates to age. A new task has been developed that holds a great deal of promise for this type of study. Marmosets can be trained to walk and run within a large plastic ball on a treadmill [117]. With less than a month of training and habituation time, marmosets were trained to steadily walk in a ball at 0.8–1.2 m/h for 30-min sessions. This allowed the investigators to implement an exercise regime for one group of animals in which they walked on the treadmill three times a week for 3 months. Animals showed few signs of distress, and weight was maintained suggesting there were no large stress-induced weight losses. While the animals were not being exercised to exhaustion and were not tested to a maximum speed, this was an initial report on a new way to examine walking behavior in the marmosets. Future work could readily examine the effects of age on walking speed, maximum speed capabilities, and time to exhaustion.

Neurobehavioral Systems

Neural Degeneration

Marmosets have been used as a model to evaluate a number of neurological diseases including Parkinson's disease, Huntington's disease, Alzheimer's disease, stroke, multiple sclerosis, and spinal cord injury, some of which are discussed elsewhere in this book. Marmosets display a number of common neural changes and degradation that are common to humans and nonhuman primates in aging. They exhibit decreased adult neurogenesis in the dentate gyrus that is evident before the animals would typically be classified as aged [118]. Marmosets aged one and a half to seven years old were injected with BrdU and the number of newly generated cells in the subgranular zone and granule cell layer of the dentate gyrus were counted 3 weeks post injection. Neurogenesis in this region of the brain decreased linearly in association with age, and this substantial decrease occurs before other symptoms of old age in the marmoset brain [118]. As expected, changes in neurogenesis are not limited to just that region of the brain. An evaluation of the lateral ventricle which harbors neural stem cells that create new neurons in rodent brains demonstrated no function of these cells in adult marmoset brains [119]. Brains from marmosets ranging in age from 0 days to 4.7 years were evaluated using immunohistochemistry to mark expression of transcription factors including Dlx2, Pax6, Tbr2, and Ngn2, which are expressed by progenitor cells in the rodent brain. Glutamatergic progenitors were present in the neonates but were absent in the adult lateral ventricle, resulting in lower progenitor diversity in the adult brain [119].

Along with changes in the progenitor cells and neurogenesis, several morphological changes in the aging marmoset brain have been evaluated. Ultrastructural examinations of the frontal cortex and hippocampus for animals more than 12 years of age had widespread accumulation of lipofuscin in the glial cells, perivascular macrophages, and pericytes [120]. Nerve cell nuclei had changes in nuclear morphology including membrane infolding and inclusions. Dystrophic axons with abnormal organelles were also noted, but the investigators found no changes associated with neurofibrillary tangles, neuritic plaques, or amyloid deposits, which have been noted in other studies. The suprachiasmatic nucleus (SCN) is thought to be the central circadian pacemaker for mammals, and many mammals display degraded circadian rhythms as the aging SCN degenerates [121]. Aging female marmosets have significant decreases in the SCN cells that stain with Nissl, NeuN, and CB, but the number of glial cells in the SCN of aged females was significantly increased [121]. These changes in the marmoset SCN suggest compensatory changes due to neuronal loss, and it remains unclear whether aging marmosets have shifted behavioral outcomes associated with changes in the internal rhythm processing.

Another area of focus in the marmoset brain is the basal forebrain and assessment of the loss of calbindin-D [122]. The basal forebrain differs significantly between rodent models, nonhuman primates, and humans. Nonhuman primates and humans have cholinergic neurons that express the calcium-binding protein, calbindin-D. Sixteen marmosets, aged 2 to 15 years, were assessed using quantitative analysis and stereological examinations of the choline acetyltransferase-positive cells, Nissl-stained magnocellular cholinergic neurons, and calbindin cells. There were no differences in the first two cell types, but there was age-associated loss of calbindin cells in all regions of the basal forebrain. These changes may be associated with vulnerability of aged marmosets to insults, which disturb calcium homeostasis, and this remains to be investigated [122].

Marmosets have been evaluated extensively as a model for Parkinson's disease via a number of induced models resulting in dopaminergic cell death, including injection of 1-methyl-4-phenyl-1,2,3,6-tetrahydropyridine (MPTP) and 6-hydroxydopamine. These models have been used to demonstrate the association of dopaminergic cell death with altered motor behavior and increased stress-related behaviors. A new model of synthetic α-synuclein propagation in the marmoset induced Parkinson's disease-like pathologies 3 months after injection, and it is hoped that this model will allow further evaluation of disease progression and disease-modifying drugs [123]. Until recently, reports on MPTP modeling in marmosets did not evaluate social behavioral changes (affect) or cognitive changes [124]. Marmosets are a particularly good model in which to evaluate behavioral changes and social responses during neurodegenerative disease states. Marmosets did not suffer from social isolation or increased rates of receiving aggression, even if physically impaired. Furthermore, a number of behavioral and cognitive tasks could be evaluated for the animals even if they exhibited motor dysfunction.

In addition to induced parkinsonian symptoms, changes associated with the natural aging process in the marmoset are being assessed. A recent study described an increase in α-synuclein aggregations in the olfactory bulb and hippocampus of aged marmosets [125]. The presence of these aggregations in the olfactory bulb of aged marmosets correlates nicely with findings that human parkinsonian patients often express dysfunction in the olfactory bulb prior to other symptomology and suggests that the olfactory bulb is particularly sensitive to degenerative changes. While the presence of Lewy bodies has not been described in marmosets in either the natural aging processes or induced models, many of the symptoms described for Parkinson's patients can be assessed in the marmosets and, more importantly, intervention testing can be done in this model. To further the marmoset as a model for Parkinson's disease, new transgenic marmosets which overexpress α-synuclein are being evaluated [126].

Many aging nonhuman primates including marmosets express increased accumulation of β-amyloid in the brain [70,127−131]. β-Amyloid deposition is associated with Alzheimer's disease progression in humans. However, symptoms of dementia associated with Alzheimer's have not been reported in nonhuman primates. The age at which β-amyloid is detected in the brain of marmosets varies depending on the study [70]. Geula [127] detected β-amyloid deposition in the brain of all animals over the age of seven, and these data are often used as a criterion for defining marmosets as aged when they are 8 years old. However, Ridley [130] did not detect β-amyloid in animals under the age of 10 in their colony and very little deposition in older animals. Brains from adolescent (1.6 year), adult (5.5 years), old (11 years), and aged (18 years) marmosets were evaluated for the presence of amyloid beta markers and tau hyperphosphorylation [132]. Diffuse amyloid plaques were found throughout the cortex of the aged marmosets but not in the younger age groups. Conformational changes in tau were detected in all subjects but increased in frequency with aging. Dystrophic microglia were significantly more likely to exhibit tau hyperphosphorylation than were active microglia. This associated difference in age and presence of β-amyloid suggests environmental differences in rate of aging between these colonies. This again supports the idea that animals at different locations may not be phenotypically aged at the same rate as animals in other environmental conditions, but the results support the continued use of marmosets as a model for this aging disease.

Cognition

Marmosets are capable of learning and performing simple cognitive tasks that can be used to assess disease progression, intervention efficacy, and responses to neurotoxins and pathogens [114,133,134]. Historically these tasks have included examinations of spatial memory, latency retention, stimuli acquisition, working memory, and executive function through the use of manipulatives such as the Wisconsin general test apparatus or computerized touchscreen modules including CANTAB [52,53,133]. While marmosets have been found to successfully learn and test for delayed match to position and reversal learning, they are not capable of performing delayed match to sample tasks [134].

Little work has been done to evaluate cognitive aging in the marmoset. As a model of stroke, marmosets have been found to demonstrate stable measurable changes in cognition on isolateral and contralateral side tasks such as the hill and valley, tube retrieval, and object retrieval task [129,134–137]. Of potential interest for those monitoring health in an aging marmoset colony, it may be particularly useful to train marmosets for simple easily trained tasks such as retrieval of a reward from a set of tubes and then longitudinally assess their performance. Not only would these data be useful for assessment of cognitive changes with aging but it might also be a useful diagnostic tool for animals at risk for stroke. To begin evaluating the effects of aging on cognition in the marmoset, Lacreuse and colleagues [138,139] examined the effects of estrogen therapy on ovariectomized young marmosets (3 years). The marmosets were tested on reversal learning, detoured reaching, and a match to position task. Contrary to predictions, the estrogen-supplemented marmosets performed more poorly on reversal learning and detoured reach tasks in comparison with the ovariectomized females. The investigators suggested that the supplemental estrogens may have been too high for optimal performance but believe that the marmoset is a valuable model for the longitudinal examination of the effects of estrogen on cognition [139].

In our lab, we were interested in the effects of aging on executive function and the ability to plan future actions. To test executive function in marmosets, we modified a task that was designed to examine fine motor control and hand–eye coordination, known as the conveyor task [140]. Typically, this task is used to assess whether marmosets can retrieve a small treat item from a moving conveyor belt where the speed and the direction of the belt can be controlled by the experimenter. To evaluate executive function, the difficulty of the task was increased and the animals were presented with two treats at the same time; a preferred marshmallow reward and an unpreferred apple reward that was cut to match the size of the dehydrated marshmallow reward. If the marmoset was able to functionally control impulsivity and detect the differences between the rewards, then we predicted that the rate of retrieval of the apple would decline after the first presentation and that the successful retrieval of the marshmallow reward would increase after the first presentation. Furthermore, we predicted that older animals would have a delayed switch to the preference of the first treat. We found that older marmosets ($n = 21$, aged $= 8+$) were significantly delayed in their learning of this task when compared with younger individuals ($n = 18$, aged $= 3–7$) (MANOVA repeated: $F = 3.408$, $P = 0.039$) (Fig. 21.7). We also found that there was no difference in their ability to learn the simple random presentation of a single reward, suggesting that the differences between young and old animals were not caused by fine motor skill degeneration.

Hematological Changes With Age

Aging in nonhuman primates and humans is marked by a number of significant changes in hematology and clinical chemistry. Decreased albumin concentrations and hematocrit, as well as increased blood urea nitrogen, have been reported in aging rhesus macaques [63,141–143], chimpanzees [50], and humans [143]. Other markers including decreases in hemoglobin, increased creatinine, and increased globulin have been suggested to be associated with aging in nonhuman primate species, but they are not reported as consistent biomarkers in aging across colonies [141,142,144]. While a number of investigators have reported standard hematological values for marmosets [145–147], the only reported change in marmoset clinical chemistry was a significant decrease in albumin with age [66]. Serum albumin decreases are of particular interest as they are highly predictive of a risk of death in otherwise healthy aging humans [148].

An analysis of the marmosets kept under barrier conditions in comparison to matched controls maintained in conventional housing revealed significant differences between a number of hematological measures and clinical chemistry [82a, Aging]. Marmosets aged six months to six and a half years were matched in age and sex across colonies (barrier $n = 31$). Marmosets maintained in the barrier condition had significantly lower white blood cell, monocyte, and neutrophil counts ($F(1,42) = 14.586$, $p<0.001$), ($F(1,42) = 14.808$, $p<0.001$), and ($F(1,42) = 17.208$, $p<0.001$) respectively. These animals also had significantly higher concentrations of albumin and albumin/globulin ratio ($F(1, 31) = 8.032$, $P = 0.009$; $F = 11.075$, $P = 0.003$). Only the albumin/globulin values were associated with aging when the analyses were controlled for location ($r = -0.507$, $P = 0.003$). Examining

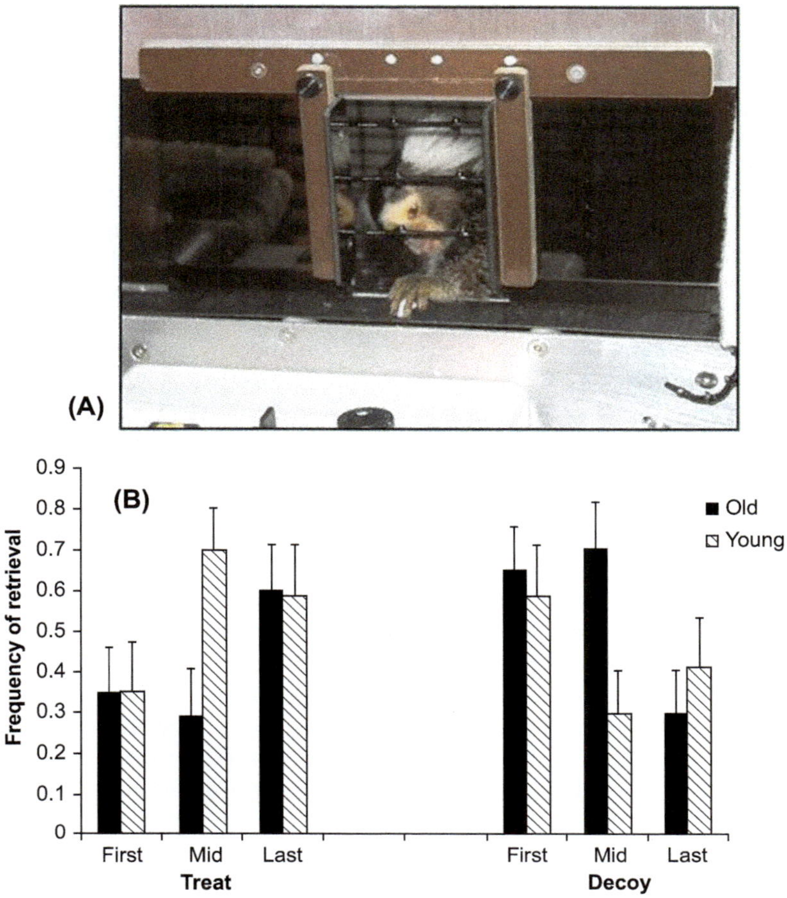

FIGURE 21.7 Executive function tested in marmosets using a conveyor task. (A) Marmoset successfully retrieves a reward from the moving conveyor belt. (B) Marmosets presented with the nonpreferred apple (decoy) followed by the preferred marshmallow tend to retrieve the decoy in the first trial but rapidly learn to wait and retrieve the marshmallow. Older animals exhibit a delay in the learning of this executive function task.

age in relation to other blood counts and chemistry values when location is controlled revealed that lymphocytes ($r = -0.309$, $P = 0.039$), alkaline phosphatase ($r = -0.567$, $P = 0.001$), and BUN/creatinine values ($r = -0.397$, $P = 0.027$) significantly decreased with age, whereas globulin was significantly increased ($r = 0.563$, $P = 0.001$). These results suggest that while some hematological changes may truly be associated with aging affects, a number of values are affected by environmental factors and the colony maintenance of the animals.

This assertion is supported by the fact that marmosets with chronic colitis, also known as chronic lymphocytic enteritis and marmoset wasting syndrome, have markedly decreased serum albumin concentrations, anemia, hypoproteinemia, and elevated alkaline phosphatase [149]. The presence of chronic enteritis is believed to be predictable by assessing serum albumin concentrations in addition to body weight measures, with albumin concentrations below 3.5 g/dL and body weight below 325 g significantly correlated with not only pathological changes in the intestine but also bone disease in the marmosets [103]. If serum albumin is in fact a biomarker for these diseases, it remains to be determined whether its association with aging is truly indicative of aging effects or whether it indicates the presence of nonage-specific underlying diseases or environmental exposures such as infection. An alternative explanation might be that while hypoalbuminemia is associated with intestinal and bone diseases in marmosets, perhaps these animals are depicting early-onset aging phenotypes which can be studied to establish etiology and treatments for similar human dysfunctions such as osteoporosis and sarcopenia. The development of an assay to detect fecal alpha-1-proteinase inhibitor, which is associated with enteric protein loss, may possibly allow further investigations into these questions [150].

Obesity and Metabolism

Obesity and aging are significantly associated with an increased risk of insulin resistance, metabolic dysfunction, metabolic syndrome, and the development of type 2 diabetes in humans and other nonhuman primates [151]. Marmosets have been found to exhibit many of the biomarkers of metabolic dysfunction in association with obesity, including increased insulin resistance, increased serum triglycerides, and increased fasting glucose and HbA1C [152]. Marmosets have been noted to have increased insulin resistance associated with aging in animals weighing more than 400 g [70]. Reported cases of substantiated diabetes phenotypes in captive marmoset colonies are relatively rare, as many of the colonies have differential protocols for determining the presence of diabetic symptoms and prescribing treatment for possibly diabetic animals.

Metabolic changes associated with aging are often complex and integrated in ways that make evaluating single pathways very difficult to interpret. Therefore, it has been proposed that the study of aging in association with metabolism and metabolomics would be best achieved using a systems network approach [153]. This approach suggests that rather than focusing on single markers of metabolic function, shifts in single metabolites or even a series of metabolites associated in a single pathway, the focus needs to become a network analysis which will provide a broad view of the changing associations between metabolites with age. This technique should reveal how the changes in these relationships are linked to functional decline and phenotypic alterations due to aging. A cross-sectional analysis of 60 marmosets ranging in age from 2 to 13 years old identified 2500 metabolites [154,155]. Connectivity between the metabolites was found to decrease with age, and the abundance of several of the metabolites significantly declined with age [153]. A 17-month longitudinal study of marmosets aged 1 to 17 years at the New England Primate Center identified 2100 metabolites that could be consistently isolated and 250 of which were annotated [155]. Forty-six percent of the metabolites assayed were found to increase with age, whereas only 21% decreased. Purine and pyrimidine pathways were significantly enriched with age in both sexes. In the detailed longitudinal examination, the pathways for tryptophan, xenobiotic, and nucleotide metabolism were significantly enriched. This work will highlight specific pathways on which research should focus not only for basic aging science but also for the design of specific interventions to recover from the metabolic dysfunction associated with aging.

Cardiovascular

Increased risk of cardiovascular failure is one of the highest risk factors for mortality in aging humans [53,156,157]. Human aging is characterized by increased prevalence of hypertension, heart arrhythmias, atherosclerosis, and thrombosis [156]. Chimps and other nonhuman primates have also been reported to have increased risk of death associated with cardiovascular disease [39,53]. For marmosets, risk of cardiovascular death increases with age, although reports of this type of death are rare and are often associated with obese animals (see Fig. 21.4). Blood pressure is often used as a biomarker to assess heart health and risk of disease in humans and other mammals [158,159]. Increased systolic, mean arterial, and diastolic pressure may significantly predict mortality, but this may be population- and weight-dependent in humans. In the marmoset, blood pressure measurements have been difficult to assess due to limitations in technology as well as increased stress responses due to handling. Anesthetized values for blood pressure, while offering an opportunity for data, are often skewed due to the anesthesia and the stress of the procedure for many animals including nonhuman primates and domestic pets [159]. Recent technology has increased the ability to assess cardiovascular health in marmosets through the use of blood pressure as measured by oscillometric analyses [159]. Marmosets have significant age effects in blood pressure, with increased mean arterial pressure and diastolic pressure associated with older age, but no age effect on the systolic pressure. Using average values from their colony analysis, they were able to identify hypertension as systolic >163.9, mean arterial >111.8, and diastolic >85.5 mm Hg, whereas hypotension was identified as systolic <122.5, mean arterial <82, and diastolic <58 mm Hg. These values and the ability to evaluate blood pressure will allow researchers to evaluate cardiovascular health of their aging marmosets with a relatively noninvasive technique.

Other Organ Systems

Several other systems have been evaluated in aging marmosets to characterize phenotypic change. Amyloidosis has been reported as affecting multiple organ systems for marmosets and is commonly identified on necropsy. While amyloidosis of the brain is associated with aging [127,128,130,160–162], amyloid deposits are found throughout the organs of individuals of varying ages, and while it may be an age-related cause of death, it remains unclear how the familial or environmental influences affect development in the marmoset [161]. Kidney disease is also quite commonly reported for marmosets (see Fig. 21.4), and its association with aging remains unclear. The presence of IgM mesangial deposits has been noted as early as infancy in some individuals, but there does appear to be a differential form of glomerulosclerosis associated with age without IgM deposits [163]. Mesangial expansions and proteinuria are common in colonies, and while they may coincide with chronic enteritis and weight loss, there appears to be no correlation between the two pathologies [164].

Marmosets have been proposed as a strong model for the evaluation of hearing loss with aging, which is also known as presbycusis [70]. Brainstem auditory-evoked potentials (BAEP) were found to have significantly prolonged latencies in high frequencies for animals 10–12 years of age [165]. BAEP thresholds were elevated and displayed smaller amplitudes in older animals. Furthermore, the auditory brainstem response and the distortion-product otoacoustic emissions were significantly altered in animals over the age of six, but this alteration was idiosyncratic in extent and amplitude of the defects [166]. Specifically, peripheral and cochlear nucleus function is suspected to be altered in older individuals, and central deficits tend to increase with age [166].

While shifts in endocrine function are often associated with age, including complete reproductive dysfunction in older human females, very little is known for marmosets. The function of both the HPA and HPG axes have been shown to decrease with age, with reduction in circulating cortisol, testosterone, and estrogens in humans [167]. Similar descriptions have been reported for chimpanzees, baboons, and macaques [55,168]. However, female marmosets appear to have fully functional reproductive cycles until death, with no severe reduction in the production of estrogens [75,76]. In males, testosterone has been found to decrease in aging marmosets, but they remain capable of reproduction, and no known phenotypic changes are associated with the decreased testosterone [75,169–171]. Nothing is known about the relationship between circulating cortisol and aging in the marmoset.

Care of Elderly Marmosets

To expand our knowledge of elderly marmosets, it is important to develop humane protocols for the care of functionally impaired animals. Few colonies have reported systematic veterinary intervention to extend life span in the marmoset or proposed protocols for the determination of end of life points. While each colony and institution will have specific IACUC and veterinary requirements for determining these protocols, it is important that considerations be given for longevity studies, as these studies differ significantly from other types of research in the effect on the science of ending life via medical intervention. Heiderstadt and Kennett recommend a number of easily implemented protocols for handling geriatric animals [172]. They recommend defining the roles of medical intervention prior to the beginning of the study, including whether antiinflammatories or analgesics can be given to reduce pain. While end-of-life measures are typically literally at the end of life in many studies, it may be more scientifically accurate to define end of life as onset of moribund. Further, they emphasize that to handle aging animals it will be critically necessary to increase monitoring of these animals to evaluate pain and discomfort and train staff carefully to identify these markers.

In our research, we have developed a number of strategies in collaboration with our veterinary staff to identify and handle animals that are displaying signs of disability or discomfort. Animals displaying weight loss, low lean mass, or symptoms of dehydration in our facility are often given probiotics to increase microbiome diversity, fed high-protein high-calorie nutrition drinks, and given extra access to fluids with electrolytes to increase hydration. In cases of severe signs of dehydration, animals are given fluids and then hand fed daily with a mixture of nutrition drinks and the powder, nongelled form of the base diet. This mixture increases fluid delivery as well as caloric delivery. Animals with low mobility can be given softer items to sit on and gather around themselves, such as disposable absorbent pads, but they need to be monitored to ensure they are not shredding the material. Animals can be removed from their home cages and temporarily housed in a temperature-controlled incubator to raise their core body temperature. Aged marmosets can often suffer from pressure sores around the base of the tail as well as on the heel of the hind feet when animals are less active. They seem to respond well to treatment of these sores with triple antibiotic. In some cases, the animal has healed from these sores and recovered to continue life for several months to several years, in other cases it becomes a more chronic condition associated with inactivity. We have seen two instances of deaths associated with identifiable behavioral symptoms of stroke and neurological damage. These symptoms included the inability of the marmoset to right itself, hanging upside down or keeping the head upside down, and eyes that did not track an individual and exhibited rapid left to right movement; these animals were euthanized as quickly as possible. The necropsy identified stroke as the cause of death for these animals, which had shown no sign of disease or altered behavior prior to these events. Increased monitoring of marmosets for behaviors specific to aging disability and disease should increase not only the care for these animals but also the ability to identify precursors for age-related disease.

FUTURE OF AGING RESEARCH

While much remains to be characterized for the aging marmoset, the basis of knowledge currently available allows us to broaden our use of marmosets to evaluate specific aging phenotypes or test specific interventions predicted to impact aging phenotypes. A few studies have already begun to examine target interventions that may be of use to delay aging and increase health span in model species and humans.

Rapamycin Intervention

Rapamycin is an mTOR inhibitor which has been found to increase both mean and maximum life span in mice, even when treatment began later in life (20 months) [173]. Rapamycin has been found to delay the onset of a number of age-related diseases in mouse models including Alzheimer's disease, cardiovascular disease, and various cancers [173–182]. Rapamycin is already FDA approved for use in humans as an immunosuppressant agent during organ transplant. While its cellular actions were once thought to mimic the effects of caloric restriction, it has been determined that the suppression of the mTOR pathway acts differentially to the effects of caloric restriction [67,183,184]. Rapamycin has been associated with a few side effects, including increased insulin resistance and onset of type 2 diabetes in mouse studies and in human patients; however, the interpretations of the data have been contradictory [178,180,185,186]. To evaluate this potential antiaging treatment, encapsulated rapamycin has been given to marmosets to determine drug dose and efficacy [187], evaluate metabolic changes associated with rapamycin dosing [67], and examine its effects on longevity (Salmon, in progress).

Encapsulated rapamycin can be delivered orally in a yogurt vehicle, which is readily consumed by the marmosets within their home cage, with control animals receiving empty eudragit capsules in the yogurt vehicle. Marmosets given rapamycin in the morning at a dose of 1.0 mg/kg day maintain serum concentrations of rapamycin near 6 ng/mL, which is similar to the level reported in human patients and the mice tested on the drug. Marmosets followed for 14 months of rapamycin dosing exhibited no signs of clinical anemia, fibrotic lung changes, mouth ulcers, or slowed wound healing, which are all potential side effects of the drug [187]. Mortality over this short-term study did not appear to be increased due to rapamycin. Treatment with rapamycin had no overall effects on body weight, with only slight decreases in fat mass in the first few months of treatment, after which it recovered to baseline values and was maintained [67]. Interestingly, at the same time that the fat mass of the animals recovered, the rapamycin animals were found to have increased food intake as perhaps a compensation for the effects of fat mass loss. Rapamycin-treated animals also had no changes in daily activity, triglyceride concentrations, fasting glucose, or insulin response to a glucose challenge. While there were no changes in gene expression associated with metabolic genes, there was tissue-specific upregulation of some components of the proteostasis network in animals treated with rapamycin [188]. Overall, we found that there were no negative side effects of rapamycin on the marmosets. While the short-term study also did not note any health benefits for the marmosets, further study is needed to evaluate long-term effects. Recently, a longevity study has begun with marmosets on rapamycin to determine if there are age-related phenotypic changes in treated marmosets and any benefits of treatment. At the same time, pilot clinical examinations have begun to evaluate the efficacy and safety of rapamycin in elderly patients (Kellogg, personal communication). The ability to test drugs and interventions in a small nonhuman primate is invaluable before moving trials into humans, for safety as well as cost reasons.

Genetic Engineering

One of the advantages that the mouse model has maintained over most other animal models, and certainly over other vertebrate models, is the ability to genetically manipulate the subjects to create transgenic and knockout mutants to specifically evaluate target gene effects. The ability to create transgenic nonhuman primates has only recently been achieved [126,189–192]. Beyond the technical difficulties with the embryonic manipulation that plagues this type of work, for nonhuman primates other than marmosets it will be decades before it is known whether the genes can be transmitted through the germline and result in a population of transgenic animals. In other nonhuman primates, it may be decades before the animals are old enough to determine whether the gene affects aging. Marmosets, due to their rapid growth and maturation and production of litters with at least twins, have offered a model for gene manipulation that can be developed more rapidly. Of specific interest is the ability to examine neurological and cognitive disorders that, while modeled in the mice, are not able to be examined through more complex cognitive tasks that nonhuman primates are capable of. Marmoset models of Parkinson's, Alzheimer's, and Rett syndromes are all currently being developed [126]. These models may allow major advances in the biomedical evaluation of age-associated disease and are anxiously awaited by researchers in the field of aging.

CONCLUSIONS

While marmosets are a relatively new model for the study of aging and age-related disease, the characterization of marmosets as biomedical translational models is not new, and the data available regarding age-related phenotypes are expanding rapidly. Marmosets offer a unique nonhuman primate model for studying the dysfunctional phenotypic decline associated with aging and testing interventional treatments to not only increase life span but also to expand health span.

Acknowledgments

The author would like to acknowledge Dr. Suzette Tardif for her continued mentorship and her dedication to marmosets as biomedical models. The author would like to thank the many animal care technicians responsible for care and maintenance of the marmoset colonies at Southwest National Primate Center and the University of Texas Health Science Center, especially Donna Layne-Colon, Aubrey Sills, Theresa Valverde, and Joselyn Artavia. The data reported for these colonies were collected and maintained by a number of individuals including Jenny Spross, Talia Melber, Bryan Rundle, and Alex Greig. The author would like to thank Dr. Arlan Richardson and Dr. Steve Austad for supporting the establishment of the barrier colony of marmosets. The author would like to thank Caitlyn Pleake and Ricki Colman for comments and suggestions for this manuscript. The work presented in this chapter was supported by the following funding sources: NIH P30 AG04427102, NIH R24 OD01093303, Southwest National Primate Center pilot funding, and the Barshop Institute for Longevity and Aging—University of Texas Health Science Center San Antonio.

References

[1] Weir HK, Thompson TD, Soman A, Moller B, Leadbetter S, White MC. Meeting the healthy people 2020 objectives to reduce cancer mortality. Prev Chronic Dis 2015;12:E104.
[2] Anderson WL, Wiener JM, Khatutsky G, Armour BS. Obesity and people with disabilities: the implications for health care expenditures. Obesity (Silver Spring) 2013;21:E798–804.
[3] Caffrey C, Harris-Kojetin L, Rome V, Sengupta M. Operating characteristics of residential care communities, by community bed size: United States, 2012. NCHS Data Brief 2014:1–8.
[4] Centers for Disease Control and Prevention. Prevalence of doctor-diagnosed arthritis and arthritis-attributable activity limitation – United States, 2007-2009. MMWR Morb Mortal Wkly Rep 2010;59:1261–5.
[5] Chambers AJ, Sukits AL, McCrory JL, Cham R. Differences in geriatric anthropometric data between DXA-based subject-specific estimates and non-age-specific traditional regression models. J Appl Biomech 2011;27:197–206.
[6] Curns AT, Holman RC, Sejvar JJ, Owings MF, Schonberger LB. Infectious disease hospitalizations among older adults in the United States from 1990 through 2002. Arch Intern Med 2005; 165:2514–20.
[7] Evenson KR, Morland KB, Wen F, Scanlin K. Physical activity and sedentary behavior among adults 60 years and older: New York City residents compared with a national sample. J Aging Phys Act 2014;22:499–507.
[8] Mokdad AH, Marks JS, Stroup DF, Gerberding JL. Actual causes of death in the United States, 2000. JAMA 2004;291:1238–45.
[9] Stevens JA, Mack KA, Paulozzi LJ, Ballesteros MF. Self-reported falls and fall-related injuries among persons aged>or=65 years—United States, 2006. J Saf Res 2008;39:345–9.
[10] Stevens JA, Rudd RA. The impact of decreasing U.S. hip fracture rates on future hip fracture estimates. Osteoporos Int 2013;24: 2725–8.
[11] Will JC, Valderrama AL, Yoon PW. Preventable hospitalizations for congestive heart failure: establishing a baseline to monitor trends and disparities. Prev Chronic Dis 2012;9:E85.
[12] Bishop KM, Robinson LM, VanLare S. Healthy aging for older adults with intellectual and development disabilities. J Psychosoc Nurs Ment Health Serv 2013;51:15–8.
[13] Centers for Disease Control and Prevention. Trends in aging—United States and worldwide. MMWR Morb Mortal Wkly Rep 2003;52:101–4. 106.
[14] Leveille SG, Fried L, Guralnik JM. Disabling symptoms: what do older women report? J Gen Intern Med 2002;17:766–73.
[15] Fried LP, Tangen CM, Walston J, Newman AB, Hirsch C, Gottdiener J, Seeman T, Tracy R, Kop WJ, Burke G, McBurnie MA, Cardiovascular Health Study Collaborative Research Group. Frailty in older adults: evidence for a phenotype. J Gerontol A Biol Sci Med Sci 2001;56:M146–56.
[16] Hamilton KL, Miller BF. What is the evidence for stress resistance and slowed aging? Exp Gerontol 2016;82:67–72.
[17] Lopez-Otin C, Blasco MA, Partridge L, Serrano M, Kroemer G. The hallmarks of aging. Cell 2013;153:1194–217.
[18] Kennedy BK, Berger SL, Brunet A, Campisi J, Cuervo AM, Epel ES, Franceschi C, Lithgow GJ, Morimoto RI, Pessin JE, Rando TA, Richardson A, Schadt EE, Wyss-Coray T, Sierra F. Geroscience: linking aging to chronic disease. Cell 2014;159:709–13.
[19] Wang AM, Promislow DE, Kaeberlein M. Fertile waters for aging research. Cell 2015;160:814–5.
[20] Bitto A, Wang AM, Bennett CF, Kaeberlein M. Biochemical genetic pathways that modulate aging in multiple species. Cold Spring Harb Perspect Med 2015;5.
[21] Santulli G, Borras C, Bousquet J, Calza L, Cano A, Illario M, Franceschi C, Liotta G, Maggio M, Molloy WD, Montuori N, O'Caoimh R, Orfila F, Rauter AP, Santoro A, Iaccarino G. Models for preclinical studies in aging-related disorders: one is not for all. Transl Med 2015;13:4–12.
[22] Sprott RL. Mouse and rat genotype choices. Exp Gerontol 1997; 32:79–86.
[23] Yuan R, Peters LL, Paigen B. Mice as a mammalian model for research on the genetics of aging. ILAR J 2011;52:4–15.
[24] Koopman JJ, van Heemst D, van Bodegom D, Bonkowski MS, Sun LY, Bartke A. Measuring aging rates of mice subjected to caloric restriction and genetic disruption of growth hormone signaling. Aging (Albany NY) 2016;8:539–46.
[25] Morselli E, Maiuri MC, Markaki M, Megalou E, Pasparaki A, Palikaras K, Criollo A, Galluzzi L, Malik SA, Vitale I, Michaud M, Madeo F, Tavernarakis N, Kroemer G. Caloric restriction and resveratrol promote longevity through the Sirtuin-1-dependent induction of autophagy. Cell Death Dis 2010;1:e10.
[26] Wuttke D, Connor R, Vora C, Craig T, Li Y, Wood S, Vasieva O, Shmookler Reis R, Tang F, de Magalhaes JP. Dissecting the gene network of dietary restriction to identify evolutionarily conserved pathways and new functional genes. PLoS Genet 2012;8:e1002834.
[27] Hasty P, Livi CB, Dodds SG, Jones D, Strong R, Javors M, Fischer KE, Sloane L, Murthy K, Hubbard G, Sun L, Hurez V, Curiel TJ, Sharp ZD. eRapa restores a normal life span in a FAP mouse model. Cancer Prev Res (Phila) 2014;7:169–78.
[28] Kennedy BK, Pennypacker JK. Drugs that modulate aging: the promising yet difficult path ahead. Transl Res 2014;163:456–65.
[29] Sharp ZD, Bartke A. Evidence for down-regulation of phosphoinositide 3-kinase/Akt/mammalian target of rapamycin (PI3K/Akt/mTOR)-dependent translation regulatory signaling pathways in Ames dwarf mice. J Gerontol A Biol Sci Med Sci 2005; 60:293–300.

[30] Zeiss CJ, Allore HG, Beck AP. Established patterns of animal study design undermine translation of disease-modifying therapies for Parkinson's disease. PLoS One 2017;12:e0171790.
[31] Austad SN. Comparative biology of aging. J Gerontol A Biol Sci Med Sci 2009;64:199–201.
[32] Didier ES, MacLean AG, Mohan M, Didier PJ, Lackner AA, Kuroda MJ. Contributions of nonhuman primates to research on aging. Vet Pathol 2016;53:277–90.
[33] Gorbunova V, Bozzella MJ, Seluanov A. Rodents for comparative aging studies: from mice to beavers. Age (Dordr) 2008;30:111–9.
[34] Buffenstein R. The naked mole-rat: a new long-living model for human aging research. J Gerontol A Biol Sci Med Sci 2005;60: 1369–77.
[35] Buffenstein R, Jarvis JU. The naked mole rat–a new record for the oldest living rodent. Sci Aging Knowl Environ 2002; 2002:e7.
[36] Delaney MA, Ward JM, Walsh TF, Chinnadurai SK, Kerns K, Kinsel MJ, Treuting PM. Initial case reports of cancer in naked mole-rats (*Heterocephalus glaber*). Vet Pathol 2016;53:691–6.
[36a] Austad SN. Methusaleh's Zoo: How Nature provides us with Clues for Extending Human Health Span. J. Comp. Path. 2010; 142:S10eS21.
[37] Sprott RL. Introduction: animal models of aging: something old, something new. ILAR J 2011;52:1–3.
[38] Austad SN. Small nonhuman primates as potential models of human aging. Inst Lab Anim Res J 1997;38:142–7.
[39] Lane MA. Nonhuman primate models in biogerontology. Exp Gerontol 2000;35:533–41.
[40] Verdier JM, Acquatella I, Lautier C, Devau G, Trouche S, Lasbleiz C, Mestre-Frances N. Lessons from the analysis of nonhuman primates for understanding human aging and neurodegenerative diseases. Front Neurosci 2015;9:64.
[41] Chatterjee HJ, Ho SYW, Barnes I, Groves C. Estimating the phylogeny and divergence times of primates using a supermatrix approach. BMC Evol Biol 2009;9:259.
[42] Finch C, Austad SN. Primate aging in the mammalian scheme: the puzzle of extreme variation in brain aging. Age 2012;34: 1075–91.
[43] Bellino FL, Wise PM. Nonhuman primate models of menopause workshop. Biol Reprod 2003;68:10–8.
[44] Chilton J, Wilcox A, Lammey M, Meyer D. Characterization of a cardiorenal-like syndrome in aged chimpanzees (*Pan troglodytes*). Vet Pathol 2016;53:417–24.
[45] Ely JJ, Zavaskis T, Lammey ML. Hypertension increases with aging and obesity in chimpanzees (*Pan troglodytes*). Zoo Biol 2013; 32:79–87.
[46] Gilissen EP, Leroy K, Yilmaz Z, Kovari E, Bouras C, Boom A, Poncelet L, Erwin JM, Sherwood CC, Hof PR, Brion JP. A neuronal aging pattern unique to humans and common chimpanzees. Brain Struct Funct 2016;221:647–64.
[47] Lowenstine LJ, McManamon R, Terio KA. Comparative pathology of aging great apes: bonobos, chimpanzees, gorillas, and orangutans. Vet Pathol 2016;53:250–76.
[48] Nunamaker EA, Lee DR, Lammey ML. Chronic diseases in captive geriatric female chimpanzees (*Pan troglodytes*). Comp Med 2012;62:131–6.
[49] Sleeper MM, Drobatz K, Lee DR, Lammey ML. Echocardiography parameters of clinically normal adult captive chimpanzees (*Pan troglodytes*). J Am Vet Med Assoc 2014;244:956–60.
[50] Videan EN, Fritz J, Murphy J. Effects of aging on hematology and serum clinical chemistry in chimpanzees (*Pan troglodytes*). Am J Primatol 2008;70:327–38.
[51] Finch C, Austad S. Primate models for human brain aging and neurological diseases. Annu Rev Gerontol Geriatr 2014;34: 139–70.
[52] Lacreuse A, Herndon JG. Nonhuman primate models of cognitive aging. Anim Models Hum Cogn Aging 2009:1–30.
[53] Mattison JA, Vaughan KL. An overview of nonhuman primates in aging research. Exp Gerontol 2016;94:41–5.
[54] Hoffman CL, Higham JP, Mas-Rivera A, Ayala JE, Maestripieri D. Terminal investment and senescence in rhesus macaques (*Macaca mulatta*) on Cayo Santiago. Behav Ecol 2010;21:972–8.
[55] Maestripieri D, Hoffman CL. Chronic stress, allostatic load, and aging in nonhuman primates. Dev Psychopathol 2011;23: 1187–95.
[56] Colman RJ, Anderson RM. Nonhuman primate calorie restriction. Antioxid Redox Signal 2011;14:229–39.
[57] Colman RJ, Beasley TM, Allison DB, Weindruch R. Attenuation of sarcopenia by dietary restriction in rhesus monkeys. J Gerontol A Biol Sci Med Sci 2008;63:556–9.
[58] Colman RJ, Beasley TM, Allison DB, Weindruch R. Skeletal effects of long-term caloric restriction in rhesus monkeys. Age (Dordr) 2012;34:1133–43.
[59] Colman RJ, Beasley TM, Kemnitz JW, Johnson SC, Weindruch R, Anderson RM. Caloric restriction reduces age-related and all-cause mortality in rhesus monkeys. Nat Commun 2014;5: 3557.
[60] Kemnitz JW. Calorie restriction and aging in nonhuman primates. Inst Lab Anim Res J 2011;52:66–77.
[61] Mattison JA, Colman RJ, Beasley TM, Allison DB, Kemnitz JW, Roth GS, Ingram DK, Weindruch R, de Cabo R, Anderson RM. Caloric restriction improves health and survival of rhesus monkeys. Nat Commun 2017;8:14063.
[62] Mattison JA, Roth GS, Beasley TM, Tilmont EM, Handy AM, Herbert RL, Longo DL, Allison DB, Young JE, Bryant M, Barnard D, Ward WF, Qi W, Ingram DK, de Cabo R. Impact of caloric restriction on health and survival in rhesus monkeys from the NIA study. Nature 2012;489:318–21.
[63] Smucny DA, Allison DB, Ingram DK, Roth GS, Kemnitz JW, Kohama SG, Lane MA. Changes in blood chemistry and hematology variables during aging in captive rhesus macaques (*Macaca mulatta*). J Med Primatol 2001;30:161–73.
[64] Fischer KE, Austad SN. The development of small primate models for aging research. Inst Lab Anim Res J 2011;52:78–88.
[65] Mansfield KG. Marmoset models commonly used in biomedical research. Comp Med 2003;53:383–92.
[66] Ross CN, Davis K, Dobek G, Tardif SD. Aging phenotypes of common marmosets (*Callithrix jacchus*). J Aging Res 2012;2012: 567143.
[67] Ross CN, Salmon AB, Strong R, Fernandez E, Javors M, Richardson A, Tardif SD. Metabolic consequences of long-term rapamycin exposure on common marmoset monkeys (*Callithrix jacchus*). Aging 2015;7:1–10.
[68] Salmon AB. Moving toward 'common' use of the marmoset as a non-human primate aging model. Pathobiol Aging Age Relat Dis 2016;6:32758.
[69] Tardif SD, Abee CR, Mansfield KG. Workshop summary: neoptropical primates in biomedical research. Inst Lab Anim Res J 2011a;52:386–92.
[70] Tardif SD, Mansfield KG, Ratnam R, Ross CN, Ziegler TE. The marmoset as a model of aging and age-related disease. Inst Lab Anim Res J 2011b;52:54–65.
[71] Peters VM, Guerra Mde D. Reproduction and maintenance of two species of marmoset in captivity. Rev Bras Biol 1998;58:169–73.
[72] Phillips IR. The reproductive potential of the common cotton-eared marmoset (*Callithrix jacchus*) in captivity. J Med Primatol 1976;5:49–55.
[73] Poole TB, Evans RG. Reproduction, infant survival and productivity of a colony of common marmosets (*Callithrix jacchus* jacchus). Lab Anim 1982;16:88–97.

[74] Smucny DA, Abbott DH, Mansfield KG, Schultz-Darken NJ, Yamamoto ME, Alencar AI, Tardif SD. Reproductive output, maternal age, and survivorship in captive common marmoset females (*Callithrix jacchus*). Am J Primatol 2004;64:107–21.

[75] Tardif SD, Araujo A, Arruda MF, French JA, Sousa MB, Yamamoto ME. Reproduction and aging in marmosets and tamarins. Interdiscip Top Gerontol 2008;36:29–48.

[76] Tardif SD, Smucny DA, Abbott DH, Mansfield K, Schultz-Darken N, Yamamoto ME. Reproduction in captive common marmosets (*Callithrix jacchus*). Comp Med 2003;53:364–8.

[77] Ross CN, Power ML, Artavia JM, Tardif SD. Relation of food intake behaviors and obesity development in young common marmoset monkeys. Obesity 2013;21:1891–9.

[78] Tardif SD, Bales KL. Relations among birth condition, maternal condition, and postnatal growth in captive common marmoset monkeys (*Callithrix jacchus*). Am J Primatol 2004;62:83–94.

[79] Tardif SD, Power ML, Ross CN, Rutherford JN. Body mass growth in common marmosets: toward a model of pediatric obesity. Am J Phys Anthropol 2013;150:21–8.

[80] Clarke JM. The common marmoset (*Callithrix jacchus*). ANZCCART News 1994;7:1–8.

[81] Ross CN, Fite JE, Jensen H, French JA. Demographic review of a captive colony of callitrichids (*Callithrix kuhlii*). Am J Primatol 2007;69:234–40.

[82] Nishijima K, Saitoh R, Tanaka S, Ohsato-Suzuki M, Ohmo T, Kitajima S. Life span of common marmoset (*Callithrix jacchus*) at CLEA Japan breeding colony. Biogerontology 2012;13:439–43.

[82a] Ross CN, Austad S, Brasky K, Brown CJ, Forey LJ, Gelfond JA, Lanford R, RIchardson A, Tardif SD. The development of a specific pathogen free (SPF) barrier colony of marmosets (*Callithrix jacchus*) for aging research. Aging 2017;9:2544–58.

[83] Weon BM. Analysis of trends in human longevity by new model. Mimeo; 2004. http://arxiv.org/ftp/q-bio/papers/0402/0402011.pdf.

[84] Martin LJ, Mahaney MC, Bronikowski AM, Carey KD, Dyke B, Comuzzie AG. Lifespan in captive baboons is heritable. Mech Ageing Dev 2002;123:1461–7.

[85] Kanfi Y, Naiman S, Amir G, Peshti V, Zinman G, Nahum L, Bar-Joseph Z, Cohen HY. The sirtuin SIRT6 regulates lifespan in male mice. Nature 2012;483:218–21.

[86] Chalmers DT, Murgatroyd LB, Wadsworth PF. A survey of the pathology of marmosets (*Callithrix jacchus*) derived from a marmoset breeding unit. Lab Anim 1983;17.

[87] Kaspareit J, Friderichs-Gromoll S, Buse E, Habermann G. Background pathology of the common marmoset (*Callithrix jacchus*) in toxicological studies. Exp Toxicol Pathol 2006;57:405–10.

[88] Carter CS, Marzetti E, Leeuwenburgh C, Manini T, Foster TC, Groban L, Scarpace PJ, Morgan D. Usefulness of preclinical models for assessing the efficacy of late-life interventions for sarcopenia. J Gerontol A Biol Sci Med Sci 2012;67:17–27.

[89] Longo VD, Antebi A, Bartke A, Barzilai N, Brown-Borg HM, Caruso C, Curiel TJ, de Cabo R, Franceschi C, Gems D, Ingram DK, Johnson TE, Kennedy BK, Kenyon C, Klein S, Kopchick JJ, Lepperdinger G, Madeo F, Mirisola MG, Mitchell JR, Passarino G, Rudolph KL, Sedivy JM, Shadel GS, Sinclair DA, Spindler SR, Suh Y, Vijg J, Vinciguerra M, Fontana L. Interventions to slow aging in humans: are we ready? Aging Cell 2015;14:497–510.

[90] Vinciguerra M, Musaro A, Rosenthal N. Regulation of muscle atrophy in aging and disease. Adv Exp Med Biol 2010;694:211–33.

[91] Collerton J, Martin-Ruiz C, Davies K, Hilkens CM, Isaacs J, Kolenda C, Parker C, Dunn M, Catt M, Jagger C, von Zglinicki T, Kirkwood TB. Frailty and the role of inflammation, immunosenescence and cellular ageing in the very old: cross-sectional findings from the Newcastle 85+ Study. Mech Ageing Dev 2012;133:456–66.

[92] Lipsitz LA. Dynamic models for the study of frailty. Mech Ageing Dev 2008;129:675–6.

[93] Mitnitski A, Collerton J, Martin-Ruiz C, Jagger C, von Zglinicki T, Rockwood K, Kirkwood TB. Age-related frailty and its association with biological markers of ageing. BMC Med 2015;13:161.

[94] Sanchis J, Nunez E, Ruiz V, Bonanad C, Fernandez J, Cauli O, Garcia-Blas S, Mainar L, Valero E, Rodriguez-Borja E, Chorro FJ, Hermenegildo C, Nunez J. Usefulness of clinical data and biomarkers for the identification of frailty after acute coronary syndromes. Can J Cardiol 2015;31:1462–8.

[95] Behrendt AK, Kuhla A, Osterberg A, Polley C, Herlyn P, Fischer DC, Scotland M, Wree A, Histing T, Menger MD, Muller-Hilke B, Mittlmeier T, Vollmar B. Dietary restriction-induced alterations in bone phenotype: effects of lifelong versus short-term caloric restriction on femoral and vertebral bone in C57BL/6 mice. J Bone Miner Res 2016;31:852–63.

[96] Champ JE, Binkley N, Havighurst T, Colman RJ, Kemnitz JW, Roecker EB. The effect of advancing age on bone mineral content of female rhesus monkeys. Bone 1996;19:485–92.

[97] Colman RJ, Lane MA, Binkley N, Wegner FH, Kemnitz JW. Skeletal effects of aging in male rhesus monkeys. Bone 1999;24:17–23.

[98] Bagi CM, Volberg M, Moalli M, Shen V, Olson E, Hanson N, Berryman E, Andresen CJ. Age-related changes in marmoset trabecular and cortical bone and response to alendronate therapy resemble human bone physiology and architecture. Anat Rec (Hoboken) 2007;290:1005–16.

[99] Colman RJ, Kemnitz JW, Lane MA, Abbott DH, Binkley N. Skeletal effects of aging and menopausal status in female rhesus macaques. J Clin Endocrinol Metab 1999;84:4144–8.

[100] Colman RJ, Roecker EB, Ramsey JJ, Kemnitz JW. The effect of dietary restriction on body composition in adult male and female rhesus macaques. Aging (Milano) 1998;10:83–92.

[101] Duncan AE, Colman RJ, Kramer PA. Sex differences in spinal osteoarthritis in humans and rhesus monkeys (*Macaca mulatta*). Spine (Phila Pa 1976) 2012;37:915–22.

[102] Bernick S, Cailliet R, Levy BM. The maturation and aging of the vertebrae of marmosets. Spine 1980;5:519–24.

[103] Baxter VK, Shaw GC, Sotuyo NP, Carlson CS, Olson EJ, Zink MC, Mankowski JL, Adams RJ, Hutchinson E, Metcalf Pate KA. Serum albumin and body weight as biomarkers for the antemortem identification of bone and gastrointestinal disease in the common marmoset. PLoS One 2013;8:e82747.

[104] Abbott DH, Barnett DK, Colman RJ, Yamamoto ME, Schultz-Darken NJ. Aspects of common marmoset basic biology and life history important for biomedical research. Comp Med 2003;53:339–50.

[105] Power RA, Power ML, Layne DG, Jaquish CE, Oftedal OT, Tardif SD. Relations among measures of body composition, age, and sex in the common marmoset monkey (*Callithrix jacchus*). Comp Med 2001;51:281–93.

[106] Andrews JS, Trupin L, Yelin EH, Hough CL, Covinsky KE, Katz PP. Frailty and reduced physical function go hand in hand in adults with rheumatoid arthritis: a US observational cohort study. Clin Rheumatol 2017;36:1031–9.

[107] Greene MA, Loeser RF. Aging-related inflammation in osteoarthritis. Osteoarthr Cartil 2015;23:1966–71.

[108] Tsonga T, Michalopoulou M, Malliou P, Godolias G, Kapetanakis S, Gkasdaris G, Soucacos P. Analyzing the history of falls in patients with severe knee osteoarthritis. Clin Orthop Surg 2015;7:449–56.

[109] Liu H, Graber TG, Ferguson-Stegall L, Thompson LV. Clinically relevant frailty index for mice. J Gerontol A Biol Sci Med Sci 2014;69:1485–91.

[110] Seldeen KL, Pang M, Troen BR. Mouse models of frailty: an emerging field. Curr Osteoporos Rep 2015;13:280–6.

[111] Bobbert MF, Plas RL, Weide G, Clairbois HE, Hofman SO, Jaspers RT, Philippens IH. Mechanical output in jumps of marmosets (*Callithrix jacchus*). J Exp Biol 2014;217:482–8.
[112] Verhave PS, Vanwersch RA, van Helden HP, Smit AB, Philippens IH. Two new test methods to quantify motor deficits in a marmoset model for Parkinson's disease. Behav Brain Res 2009;200:214–9.
[113] Plas RL, Degens H, Meijer JP, de Wit GM, Philippens IH, Bobbert MF, Jaspers RT. Muscle contractile properties as an explanation of the higher mean power output in marmosets than humans during jumping. J Exp Biol 2015;218:2166–73.
[114] Stevens D, Scott EA, Bowditch AP, Griffiths GD, Pearce PC. Multiple vaccine and pyridostigmine interactions: effects on cognition, muscle function and health outcomes in marmosets. Pharmacol Biochem Behav 2006;84:207–18.
[115] Hook MA, Rogers LJ. Development of hand preferences in marmosets (*Callithrix jacchus*) and effects of aging. J Comp Psychol 2000;114:263–71.
[116] Wang Y, Fang Q, Gong N. Motor assessment of developing common marmosets. Neurosci Bull 2014;30:387–93.
[117] Phillips KA, Hambright MK, Hewes K, Schilder BM, Ross CN, Tardif SD. Take the monkey and run. J Neurosci Methods 2015; 248:27–31.
[118] Leuner B, Kozorovitskiy Y, Gross CG, Gould E. Diminished adult neurogenesis in the marmoset brain precedes old age. Proc Natl Acad Sci USA 2007;104:17169–73.
[119] Azim K, Zweifel S, Klaus E, Yoshikawa K, Amrein I, Raineteau O. Early decline in progenitor diversity in the marmoset lateral ventricle. Cerebr Cortex 2013;23:922–31.
[120] Honavar M, Lantos PL. Ultrastructural changes in the frontal cortex and hippocampus in the aging marmoset. Mech Ageing Dev 1987;41:161–75.
[121] Engelberth RCGJ, Silva KDA, Azevedo CVM, Gavioli EC, Santos JR, Soares JG, Nascimento ES, Cavalcante JC, Costa MSMO, Cavalcante JS. Morphological changes in the suprachiasmatic nucleus of aging female marmosets (*Callithrix jacchus*). BioMed Res Int 2014;2014:243825.
[122] Wu C-K, Nagykery N, Hersh LB, Scinto LFM, Geula C. Selective age-related loss of calbindin-D28k from basal forebrain cholinergic neurons in the common marmoset (*Callithrix jacchus*). Neuroscience 2003;120:249–59.
[123] Shimozawa A, Ono M, Takahara D, Tarutani A, Imura S, Masuda-Suzukake M, Higuchi M, Yanai K, Hisanaga SI, Hasegawa M. Propagation of pathological alpha-synuclein in marmoset brain. Acta Neuropathol Commun 2017;5:12.
[124] Phillips KA, Ross CN, Spross J, Cheng CJ, Izquierdo A, Biju KC, Chen C, Li S, Tardif SD. Behavioral phenotypes associated with MPTP induction of partial lesions in common marmosets (*Callithrix jacchus*). Behav Brain Res 2017;325:51–2.
[125] Kobayashi R, Takahashi-Fujigasaki J, Shiozawa S, Hara-Miyauchi C, Inoue T, Okano HJ, Sasaki E, Okano H. alpha-Synuclein aggregation in the olfactory bulb of middle-aged common marmoset. Neurosci Res 2016;106:55–61.
[126] Kishi N, Sato K, Sasaki E, Okano H. Common marmoset as a new model animal for neuroscience research and genome editing technology. Dev Growth Differ 2014;56:53–62.
[127] Geula C, Nagykery N, Wu C-K. Amyloid-B deposits in the cerebral cortex of the aged common marmoset (*Callithrix jacchus*): incidence and chemical composition. Acta Neuropathol 2002; 103:48–58.
[128] Geula C, Wu C-K, Saroff D, Lorenzo A, Yuan M, Yankner B. Aging renders the brain vulnerable to amyloid B- protein neurotoxicity. Nat Med 1998;4:827–31.
[129] Marshall JWB, Ridley RM. Assessment of cognitive and motor deficits in a marmoset model of stroke. Inst Lab Anim Res J 2003;44:153–60.
[130] Ridley RM, Baker HF, Windle CP, Cummings RM. Very long term studies of the seeding of beta-amyloidosis in primates. J Neural Transm (Vienna) 2006;113:1243–51.
[131] Ridley RM, Cummings RM, Leow-Dyke A, Baker HF. Neglect of memory after dopaminergic lesions in monkeys. Behav Brain Res 2006;166:253–62.
[132] Rodriguez-Callejas JD, Fuchs E, Perez-Cruz C. Evidence of Tau hyperphosphorylation and dystrophic microglia in the common marmoset. Front Aging Neurosci 2016;8:315.
[133] Kangas BD, Bergman J, Coyle JT. Touchscreen assays of learning, response inhibition, and motivation in the marmoset (*Callithrix jacchus*). Anim Cognit 2016;19:673–7.
[134] Spinelli S, Pennanen L, Dettling AC, Feldon J, Higgins GA, Pryce CR. Performance of the marmoset monkey on computerized tasks of attention and working memory. Brain Res Cogn Brain Res 2004;19:123–37.
[135] Dias R, Robbins TW, Roberts AC. Dissociation in prefrontal cortex of affective and attentional shifts. Nature 1996a;380:69–72.
[136] Dias R, Robbins TW, Roberts AC. Primate analogue of the Wisconsin Card Sorting Test: effects of excitotoxic lesions of the prefrontal cortex in the marmoset. Behav Neurosci 1996b;110: 872–86.
[137] Dias R, Robbins TW, Roberts AC. Dissociable forms of inhibitory control within prefrontal cortex with an analog of the Wisconsin Card Sort Test: restriction to novel situations and independence from "on-line" processing. J Neurosci 1997;17:9285–97.
[138] Lacreuse A, Chang J, Metevier CM, LaClair M, Meyer JS, Ferris CM. Oestradiol modulation of cognition in adult female marmosets (*Callithrix jacchus*). J Neuroendocrinol 2014;26: 296–309.
[139] Lacreuse A, Mong JA, Hara Y. Neurocognitive effects of estrogens across the adult lifespan in nonhuman primates: state of knowledge and new perspectives. Horm Behav 2015;74:157–66.
[140] D'Mello GD, Duffy EA, Miles SS. A conveyor belt task for assessing visuo-motor coordination in the marmoset (*Callithrix jacchus*): effects of diazepam, chlorpromazine, pentobarbital and d-amphetamine. Psychopharmacology (Berl) 1985;86:125–31.
[141] Nakamura E, Lane MA, Roth GS, Cutler RG, Ingram DK. Evaluating measures of hematology and blood chemistry in male rhesus monkeys as biomarkers of aging. Exp Gerontol 1994;29: 151–77.
[142] Nakamura E, Lane MA, Roth GS, Ingram DK. A strategy for identifying biomarkers of aging: further evaluation of hematology and blood chemistry data from a calorie restriction study in rhesus monkeys. Exp Gerontol 1998;33:421–43.
[143] Nakamura E, Miyao K. A method for identifying biomarkers of aging and constructing an index of biological age in humans. J Gerontol Biol Sci 2007;62A:1096–105.
[144] t Hart BA, Abbott DH, Nakamura K, Fuchs E. The marmoset monkey: a multi-purpose preclinical and translational model of human biology and disease. Drug Discov Today 2012;17:1160–5.
[145] Burns KF, Ferguson FG, Hampton SH. Compendium of normal blood values for baboons chimpanzees and marmosets. Am J Clin Pathol 1967;48:484–94.
[146] Hawkey CM, Hart MG, Jones DM. Clinical hematology of the common marmoset (*Callithrix jacchus*). Am J Primatol 1982;3: 179–99.
[147] McNees DW, Ponzio BJ, Lewis RW, Stein FJ, Sis RF, Levy BM. Hematology of common marmosets (*Callithrix jacchus*). Primates 1982;23:145–50.

[148] Klonoff-Cohen H, Barrett-Connor EL, Edelstein SL. Albumin levels as a predictor of mortality in the healthy elderly. J Clin Epidemiol 1992;45:207–12.
[149] Logan AC, Khan KNM. Clinical pathological changes in two marmosets with wasting syndrome. Toxicol Pathol 1996;24: 707–9.
[150] Parambeth JC, Suchodolski JS, Steiner JM. Purification and partial characterization of alpha1-proteinase inhibitor in the common marmoset (*Callithrix jacchus*). Res Vet Sci 2015;99:17–22.
[151] Vaughan KL, Mattison JA. Obesity and aging in humans and nonhuman primates: a mini-review. Gerontology 2016;62:611–7.
[152] Tardif SD, Power ML, Ross CN, Rutherford JN, Layne-Colon DG, Paulik MA. Characterization of obese phenotypes in a small nonhuman primate, the common marmoset (*Callithrix jacchus*). Obesity (Silver Spring) 2009;17:1499–505.
[153] Soltow QA, Jones DP, Promislow DE. A network perspective on metabolism and aging. Integr Comp Biol 2010;50:844–54.
[154] Go YM, Liang Y, Uppal K, Soltow QA, Promislow DE, Wachtman LM, Jones DP. Metabolic characterization of the common marmoset (*Callithrix jacchus*). PLoS One 2015;10:e0142916.
[155] Hoffman JM, Tran V, Wachtman LM, Green CL, Jones DP, Promislow DE. A longitudinal analysis of the effects of age on the blood plasma metabolome in the common marmoset, *Callithrix jacchus*. Exp Gerontol 2016;76:17–24.
[156] Fontana L, Vinciguerra M, Longo VD. Growth factors, nutrient signaling, and cardiovascular aging. Circ Res 2012;110:1139–50.
[157] Ungvari Z, Buffenstein R, Austad SN, Podlutsky A, Kaley G, Csiszar A. Oxidative stress in vascular senescence: lessons from successfully aging species. Front Biosci 2008;13:5056–70.
[158] Bartholomeusz B, Hardy KJ, Nelson AS, Phillips PA. Modulation of nitric oxide improves cyclosporin A-induced hypertension in rats and primates. J Hum Hypertens 1998;12:839–44.
[159] Mietsch M, Baldauf K, Reitemeier S, Suchowski M, Schoon HA, Einspanier A. Blood pressure as prognostic marker for body condition, cardiovascular, and metabolic diseases in the common marmoset (*Callithrix jacchus*). J Med Primatol 2016;45: 126–38.
[160] Anekonda TS. Resveratrol—a boon for treating Alzheimer's disease? Brain Res Rev 2006;52:316–26.
[161] Ludlage E, Murphy CL, Davern SM, Solomon A, Weiss DT, Glenn-Smith D, Dworkin S, Mansfield KG. Systemic AA amyloidosis in the common marmoset. Vet Pathol 2005;42: 117–24.
[162] Pozueta J, Lefort R, Shelanski ML. Synaptic changes in Alzheimer's disease and its models. Neuroscience 2013;251:51–65.
[163] Brack M. Callitrichid IgM-nephropathy- an old age-related disease? Lab Anim 1993;29:54–8.
[164] Collins MG, Rogers NM, Jesudason S, Kireta S, Brealey J, Coates PT. Spontaneous glomerular mesangial lesions in common marmoset monkeys (*Callitrhix jacchus*): a benign nonprogressive glomerulopathy. J Med Primatol 2014;43:477–87.
[165] Harada T, Tokuriki M, Tanioka Y. Age-related changes in the brainstem auditory evoked potentials of the marmoset. Hear Res 1999;128:119–24.
[166] Valero MD. The common marmoset (*Callithrix jacchus*) as a model for research in age-related hearing loss. ProQuest, University of Texas San Antonio; 2012.
[167] Bjorntorp P. Neuroendocrine ageing. J Intern Med 1995;238: 401–4.
[168] Downs JL, Urbanski HF. Neuroendocrine changes in the aging reproductive axis of female rhesus macaques (*Macaca mulatta*). Biol Reprod 2006;75:539–46.
[169] Kelnar CJ, McKinnell C, Walker M, Morris KD, Wallace WH, Saunders PT, Fraser HM, Sharpe RM. Testicular changes during infantile 'quiescence' in the marmoset and their gonadotrophin dependence: a model for investigating susceptibility of the prepubertal human testis to cancer therapy? Hum Reprod 2002;17: 1367–78.
[170] Kholkute SD, Aitken RJ, Lunn SF. Plasma testosterone response to hCG stimulation in the male marmoset monkey (*Callithrix jacchus jacchus*). J Reprod Fertil 1983;67:457–63.
[171] Nunes S, Fite JE, French JA. Variation in steroid hormones associated with infant care behaviour and experience in male marmosets (*Callithrix kuhlii*). Anim Behav 2000;60:857–65.
[172] Heiderstadt K, Kennett MJ. IACUC issues related to animal models of aging. Inst Lab Anim Res J 2011;52:106–9.
[173] Harrison DE, Strong R, Sharp ZD, Nelson JF, Astle CM, Flurkey K, Nadon NL, Wilkinson JE, Frenkel K, Carter CS, Pahor M, Javors MA, Fernandez E, Miller RA. Rapamycin fed late in life extends lifespan in genetically heterogeneous mice. Nature 2009;460:392–5.
[174] Cox LS, Mattison JA. Increasing longevity through caloric restriction or rapamycin feeding in mammals: common mechanisms for common outcomes? Aging Cell 2009;8:607–13.
[175] Halloran J, Hussong SA, Burbank R, Podlutskaya N, Fischer KE, Sloane LB, Austad SN, Strong R, Richardson A, Hart MJ, Galvan V. Chronic inhibition of mammalian target of rapamycin by rapamycin modulates cognitive and non-cognitive components of behavior throughout lifespan in mice. Neuroscience 2012;223:102–13.
[176] Lamming DW, Ye L, Astle CM, Baur JA, Sabatini DM, Harrison DE. Young and old genetically heterogeneous HET3 mice on a rapamycin diet are glucose intolerant but insulin sensitive. Aging Cell 2013;12:712–8.
[177] Miller RA, Harrison DE, Astle CM, Baur JA, Boyd AR, de Cabo R, Fernandez E, Flurkey K, Javors MA, Nelson JF, Orihuela CJ, Pletcher S, Sharp ZD, Sinclair D, Starnes JW, Wilkinson JE, Nadon NL, Strong R. Rapamycin, but not resveratrol or simvastatin, extends life span of genetically heterogeneous mice. J Gerontol A Biol Sci Med Sci 2011;66: 191–201.
[178] Miller RA, Harrison DE, Astle CM, Fernandez E, Flurkey K, Han M, Javors MA, Li X, Nadon NL, Nelson JF, Pletcher S, Salmon AB, Sharp ZD, Van Roekel S, Winkleman L, Strong R. Rapamycin-mediated lifespan increase in mice is dose and sex dependent and metabolically distinct from dietary restriction. Aging Cell 2014;13:468–77.
[179] Reifsnyder PC, Doty R, Harrison DE. Rapamycin ameliorates nephropathy despite elevating hyperglycemia in a polygenic mouse model of type 2 diabetes, NONcNZO10/LtJ. PLoS One 2014;9:e114324.
[180] Reifsnyder PC, Flurkey K, Te A, Harrison DE. Rapamycin treatment benefits glucose metabolism in mouse models of type 2 diabetes. Aging (Albany NY) 2016;8:3120–30.
[181] Wilkinson JE, Burmeister L, Brooks SV, Chan CC, Friedline S, Harrison DE, Hejtmancik JF, Nadon N, Strong R, Wood LK, Woodward MA, Miller RA. Rapamycin slows aging in mice. Aging Cell 2012;11:675–82.
[182] Ye L, Widlund AL, Sims CA, Lamming DW, Guan Y, Davis JG, Sabatini DM, Harrison DE, Vang O, Baur JA. Rapamycin doses sufficient to extend lifespan do not compromise muscle mitochondrial content or endurance. Aging (Albany NY) 2013;5: 539–50.
[183] Garratt M, Nakagawa S, Simons MJ. Comparative idiosyncrasies in life extension by reduced mTOR signalling and its distinctiveness from dietary restriction. Aging Cell 2016;15: 737–43.
[184] Karunadharma PP, Basisty N, Dai DF, Chiao YA, Quarles EK, Hsieh EJ, Crispin D, Bielas JH, Ericson NG, Beyer RP, MacKay VL, MacCoss MJ, Rabinovitch PS. Subacute calorie

restriction and rapamycin discordantly alter mouse liver proteome homeostasis and reverse aging effects. Aging Cell 2015;14:547–57.

[185] Deepa SS, Walsh ME, Hamilton RT, Pulliam D, Shi Y, Hill S, Li Y, Van Remmen H. Rapamycin modulates markers of mitochondrial biogenesis and fatty acid oxidation in the adipose tissue of db/db mice. J Biochem Pharmacol Res 2013;1:114–23.

[186] Fadini GP, Ceolotto G, Pagnin E, de Kreutzenberg S, Avogaro A. At the crossroads of longevity and metabolism: the metabolic syndrome and lifespan determinant pathways. Aging Cell 2011;10:10–7.

[187] Tardif SD, Ross CN, Bergman P, Fernandez E, Javors M, Salmon AB, Spross JA, Strong R, Richardson A. Testing efficacy of administration of the antiaging drug rapamycin in a nonhuman primate, the common marmoset. J Gerontol Biol Sci 2015;70:577–88.

[188] Lelegren M, Liu Y, Ross CN, Tardif SD, Salmon AB. Pharmaceutical inhibiton of mTOR in the common marmoset: effect of rapamycin on regulators of proteostasis in a nonhuman primate. Pathobiol Aging Age Relat Dis 2016;6:31793.

[189] Gama Sosa MA, De Gasperi R, Elder GA. Modeling human neurodegenerative diseases in transgenic systems. Hum Genet 2012;131:535–63.

[190] Okano H, Hikishima K, Iriki A, Sasaki E. The common marmoset as a novel animal model system for biomedical and neuroscience research applications. Semin Fetal Neonatal Med 2012;17:336–40.

[191] Sasaki E. Prospects for genetically modified non-human primate models, including the common marmoset. Neurosci Res 2015;93:110–5.

[192] Sasaki E, Suemizu H, Shimada A, Hanazawa K, Oiwa R, Kamioka M, Tomioka I, Sotomaru Y, Hirakawa R, Eto T, Shiozawa S, Maeda T, Ito M, Ito R, Kito C, Yagihashi C, Kawai K, Miyoshi H, Tanioka Y, Tamaoki N, Habu S, Okano H, Nomura T. Generation of transgenic non-human primates with germline transmission. Nature 2009;459:523–7.

CHAPTER

22

The Marmoset as a Model for Visual Neuroscience

Jude F. Mitchell[1], *David A. Leopold*[2,3,*]

[1]Brain and Cognitive Sciences Department, Meliora Hall, University of Rochester, Rochester, NY, United States; [2]Section on Cognitive Neurophysiology and Imaging, Laboratory of Neuropsychology, National Institute of Mental Health, National Institutes of Health, Bethesda, MD, United States; [3]Neurophysiology Imaging Facility, National Institute of Mental Health, National Institute of Neurological Disorders and Stroke, National Eye Institute, National Institutes of Health, Bethesda, MD, United States

INTRODUCTION

Vision is central to human cognition and has long been an important model system for studying principles of brain function. Humans, like other primates, depend on vision extensively for navigation, interaction with other individuals, manipulation of objects, and many other aspects of daily life. The coevolution of the eye and visual brain in our distant primate ancestors brought with it many adaptations that benefit diurnal and arboreal living as well as social living in larger extended family groups. These changes are manifested as a pattern of specific features of the visual system that supports unique perceptual and behavioral abilities (for a review, see Kaas [1]). Visual neuroscience has benefitted from decades of comparative studies, as the parallax afforded by studying multiple species has helped to identify traits that are core features of the mammalian brain and other traits that are unique to primates, including humans.

This chapter reviews primate vision from a comparative standpoint and places focus on the common marmoset (*Callithrix jacchus*), an arboreal, small-bodied New World primate. The review is motivated by growing interest in the marmoset as a model species to complement the rhesus monkey (*Macaca mulatta*) and laboratory mouse (*Mus musculus*), which are commonly used to study neural circuits supporting human vision. Humans' most recent common ancestor with the marmoset lived approximately 35–40 million years ago (MYA), before our most recent ancestor with the macaque (25–30 MYA) and long after our most recent ancestor with the mouse (80–90 MYA) [2,3]. Thus, from a purely phylogenetic standpoint, the marmoset offers an intermediate point of comparison between these species (Fig. 22.1). For visual neuroscience, the marmoset also provides a number of distinct experimental advantages over each of these model systems, and we point to areas where a fully developed marmoset animal model promises to cast new light on mechanisms of visual cognition in the human brain.

It makes sense to begin by hailing a success story in neuroscience: the recent flourishing of the mouse model, the facility of its genetic manipulation, and its use as a tool to probe the exquisite detail of the brain's functional circuitry. Advances in the mouse have set new standards for the precision with which animal models can contribute to the investigation of the brain [4–6]. In particular, the development of transgenic lines, such as Cre lines, combined with viral-based optogenetics, has made it possible to express light-sensitive opsins such as channelrhodopsin-2 (ChR2) in highly specific neuronal classes and then causally manipulate their activity with light [7,8]. This approach can be used to study functioning of specific anatomical

The Common Marmoset in Captivity and Biomedical Research
https://doi.org/10.1016/B978-0-12-811829-0.00022-4

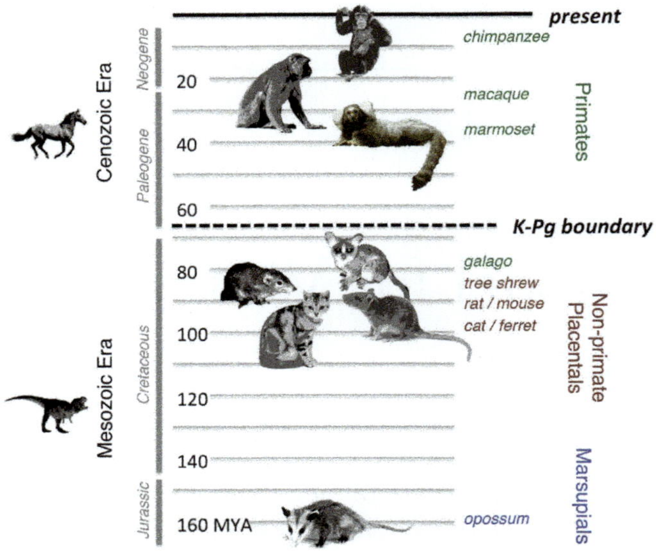

FIGURE 22.1 Phylogenetic portrait of common mammalian experimental models for visual neuroscience, now and from previous scientific generations. Representative species are arranged with respect to human ancestry. The vertical timeline indicates for each species the period at which the most recent common ancestor with humans lived. For the macaque and marmoset, this ancestor lived near the end of the Paleogene Period, long after the so-called K–Pg boundary that marked the end of the Mesozoic Era. However, for other mammalian models, including prosimian primates, the most recent common ancestor lived during the Cretaceous Period, in the Mesozoic Era. *MYA*, million years ago.

pathways and has been used to link activity of specific cell types in the mouse visual cortex to distinct functional roles [9] and to perceptual decisions [10,11]. There is at present a concerted push toward assembling a comprehensive description of visual circuitry in the mouse brain [12].

However, over many decades the Old World rhesus macaque monkey (and closely related macaque species, referred to here collectively as "macaque") has emerged as the species of choice for modeling human vision. The macaque is a logical choice owing to its evolutionary proximity to humans, which is reflected in the similarity of its brain's basic anatomical layout as well as its perceptual capacities and ability to perform complex tasks. As a result, great detail is known about the anatomical connections and electrophysiological response properties of its brain (e.g., Refs. [13,14]). Comparison with the human brain demonstrates that the basic layout of the visual system is similar between the two species [15]. Moreover, because macaques can learn and perform diverse tasks under constrained experimental conditions, researchers have investigated neural responses throughout the brain and have related them to many high-level perceptual and visually guided behaviors, as well as the impact of focal lesions on such behaviors (e.g., Refs. [16,17]).

In some ways, the mouse and macaque models each stand on their own so successfully that they can obscure a very important aspect of neuroscience: comparative studies. As noted by Preuss [18], "the concentration of effort on such a few species would be defensible if cortical organization were basically uniform across mammals, as is commonly believed … phyletic variation in cortical organization is far more extensive than has generally been appreciated or acknowledged." Neuroscience in previous eras was characterized by investigation of a much broader range of species with some research directed toward understanding, for example, the organization of the visual cortex of the cat, ferret, tree shrew, ground squirrel, and many other mammalian and nonmammalian models. Some studies were explicitly comparative and allowed for a contextualization of neural features found in macaque species, such as those related to the anatomical connections between two visual structures [19] or the level of direct cortical control over movement [20]. This breadth of study provided a foundation for describing the pathways of the primate brain from an evolutionary perspective, for example, making the important distinction between ancestral and derived traits. The more recent focus on just two principal species, the mouse and macaque, threatens to diminish this perspective.

In this chapter, we argue that the marmoset provides a strategic choice, based on its specific experimental advantages and phylogenetic relationship to humans, to complement the mouse and the macaque for the study of vision. The marmoset model is already well established for biomedical research in fields such as infectious disease, reproduction, and aging; thus practical issues such as housing and breeding are well understood [21–23]. Some of this research has focused on the central nervous system, including the manipulation of cognition through pharmacological intervention and selective ablation [24]. Marmosets are an important animal model for human audition and active vocal communication [25–27]. Although less studied in vision, a small number of groups have made great progress over the last two decades and built a foundation for understanding the basic organization of their visual system, primarily through visual field mapping and assessment of stimulus selectivity in anesthetized animals (for review, see Ref. [28]). We submit that expanding the role of marmosets in visual neuroscience requires little effort and has the potential to significantly advance our understanding of primate vision. To make our case, we begin in the section The Primate Brain: A Commitment to Vision by reviewing several important features of primate vision. Specifically, we show that the primate eye and brain are adapted in ways that are categorically different from other mammals, with many of the adaptations being consequences of high retinal acuity. In the section Comparing Marmoset and Macaque Vision, we place our focus on the marmoset to compare and contrast features of its eye, brain, and behavior with the macaque. In the section Opportunities Afforded by the Marmoset Model for Visual Neuroscience, we review and discuss experimental aspects of the marmoset model that offer new and exciting opportunities for visual neuroscience. We end by briefly summarizing the main points of the chapter and look ahead to the opportunities and challenges afforded by the marmoset model in the neuroscientific investigation of the human brain.

THE PRIMATE BRAIN: A COMMITMENT TO VISION

Primates in Evolutionary Context

Primates have brains that share much in common with other mammalian species, most notably the prominent cerebral cortex. At the same time, primate brains have unique features that reflect how they interact with the world and with one another [29]. In this section we review the specific adaptations that have made the primate visual system distinct from that of other mammals. This provides the appropriate context for understanding, first, that marmoset vision is highly conserved with other primates and, second, that it differs in important ways from mammals of similar small size. Those readers familiar with the evolution of primate vision may prefer to begin in the section Comparing Marmoset and Macaque Vision, where we focus on the marmoset and its distinctions from other primates.

To best understand the value of the nonhuman primate model it is important to consider the evolutionary context from which the primate lineage emerged. The diversity of the modern mammal was shaped by a volatile period of evolution in the late Mesozoic era, when dinosaurs were still the predominant large animals on Earth (see Fig. 22.1). During this time, the major radiations of mammals diverged, including the Euarchontoglires superorder containing rodents, rabbits, flying lemurs, tree shrews, and primates. The specialization of early primates, which is thought to have been a set of adaptations for a nocturnal arboreal niche, may have strongly shaped the brain of modern extant primates, including humans [30]. A pivotal point in mammalian evolution occurred approximately 66 MYA, commonly termed the Cretaceous–Paleogene (K–Pg) boundary, which marks the time at which a large asteroid impacted the earth and is believed to have led to the extinction of the dinosaurs [31]. Before the K–Pg event, the primate line diverged from other members of Euarchontoglires, followed a few million years later by its division into strepsirrhine (e.g., the prosimian Galago) and haplorhine (e.g., New and Old World monkeys and apes) branches [1,2]. After the K–Pg event, a range of diurnal niches were opened up, which were gradually filled by mammals, including primates. Much more recently, some 35–40 MYA, New World monkeys (e.g., marmosets) diverged from Old World monkeys (e.g., macaques) and apes (e.g., humans). The divergence of New World monkeys has been traced to an incredible example of geographical isolation, where a group of haplorhine monkeys appears to have rafted across the Atlantic Ocean from Africa to South America. This monkey, which is the most recent common ancestor of humans and marmosets, was a small- to medium-sized arboreal fruit and seed eater [32,33]. In Africa, Old World monkeys subsequently diverged from apes and humans approximately 25–30 MYA. When considering the evolutionary relationship between any pair of species, it is important to keep in mind that evolutionary changes can and do occur along both branches emanating from the most recent common ancestor. Thus monkeys and apes should not be treated as evolving from modern prosimians, which have undergone their own evolutionary adaptations. Similarly, humans and other apes did not evolve from rhesus monkeys or any other extant species.

The neural and behavioral evolution of primates is strongly tied to adaptations in the domain of vision, which we describe in detail below. However, it is worth emphasizing here that the importance of vision is in no way unique to primates. It is a vital sense for nearly all vertebrates, with its critical value being that it provides information about the environment from a distance. In many species, vision is important for navigation, interspecies interactions (e.g., predation), intraspecies interactions (e.g., mate selection), and the selection of nesting and foraging sites. In birds, the surviving descendants of dinosaurs whose most recent common ancestor with primates lived approximately 320 MYA, vision is paramount and there are many examples of convergent evolution affecting primate and bird vision, for example, in the specialization of a retinal fovea [33–35] or in aspects of their visual cognition [36]. Among mammals, primates are unusually reliant on vision, as they do not conform to the more typical mammalian pattern of using olfactory cues for social interaction and foraging. Prosimians show less reliance on vision than simians (monkeys, apes, and humans) and may be regarded as intermediate in this respect. The primate commitment to vision has evolved along with specific adaptations of the eye and brain, which we describe in the following sections along with how they uniquely shape primate behaviors.

Adaptations of the Eyes

While cause-and-effect relationships are notoriously difficult to specify in evolutionary biology, it appears evident that a major force driving innovation of the primate brain was changes to the eye. This section identifies three such changes and briefly discusses the importance of each.

Foveal High-Acuity Vision

Perhaps the most distinguishing feature of primate vision is its high spatial acuity or the ability to resolve fine detail. Among mammals, humans rank highest in their visual acuity, which commonly exceeds 50 cycles/degree, and this is followed closely by apes and monkeys [35]. In fact, this aspect of primate vision is unmatched among mammals and only exceeded by a few species of large predatory birds [35].

Primates' unusually high acuity stems from multiple adaptations, including the large size of the eye, its optics, the high density of retinal photoreceptors and ganglion cells in central vision, and the low amount of spatial pooling of photoreceptor signals onto individual ganglion cells [37]. In simian primates (i.e., monkeys and apes), the most unique feature of the retina is the fovea mediating central vision. The fovea is a pit in the inner retina caused by the local absence of cell processes, creating a window of optical clarity for light reaching photoreceptors situated along the outer circumference of the retinal epithelium (Fig. 22.2A). The high density of cone photoreceptors that populate the fovea is, quite remarkably, free of blood vessels (Fig. 22.2B). Across primates, the fovea has a roughly constant size of somewhat less than 0.5 mm despite large variations in eye size. The fact that larger eyes do not have larger foveae may suggest that its size is limited by the diameter within which the overlying cell processes and vasculature can be cleared without harming the photoreceptors themselves [40]. Some prosimian primates such as the nocturnal Galago have a more primitive region of photoreceptor concentration mediating central vision, termed the area centralis. However, the increase in receptor density in this region is notably less than in the simian primate fovea, with only a 2- to 3-fold increase compared with a 20- or more fold increase in New and Old World monkeys [41]. Thus the spatial acuity of the Galago (4.8–6.0 cycles/deg) is much lower than most monkeys and humans [42].

Signals arising from the high density of foveal cones in monkeys and apes are carried by tightly packed midget retinal ganglion cells into the brain. In the foveal region of the retina, midget ganglion cells outnumber the cone photoreceptors and each midget ganglion cell samples from only a single cone photoreceptor via one-to-one connections with midget bipolar cells [43–46]. By comparison, the corresponding classes of small ganglion cells in other species, such as the cat β cells, pool from up to 30 cone photoreceptors [47]. At more peripheral regions of the retina, midget ganglion cells in monkeys show a higher level of spatial pooling from up to 10 cone photoreceptors [47]. Inherent in the spatial pooling of photoreceptor information is a trade-off between light sensitivity and spatial acuity. The demands of night vision seem to have pushed at least two nocturnal species, the Galago and owl monkey, to a level of photoreceptor pooling that is unusually high among primates [48–50], but still much less than in the cat.

The evolution of high acuity in primates may have been gradual. One hypothesis holds that high acuity in early diurnal primates benefitted from the unusually large eyes of their nocturnal ancestors, which was itself an adaptation to low-light conditions [51]. For a given cone density, larger eyes translate directly into higher acuity because a given visual angle subtended on the

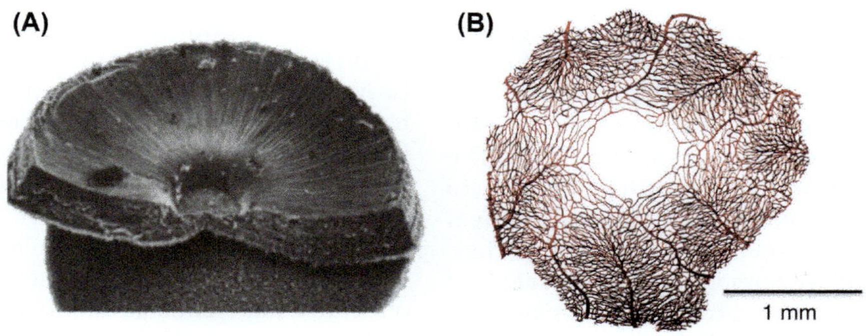

FIGURE 22.2 The unique foveal specialization of the primate retina, which has shaped the organization of the primate brain and much of primate behavior. (A) Scanning electron micrograph of a portion of the macaque retina showing the foveal slope and pit. (B) Pattern of vascularization surrounding the avascular foveal region of the macaque retina. *(A) Adapted from Borwein B. Scanning electron microscopy of monkey foveal photoreceptors. Anat Rec 1983;205(3):363–73. (B) Adapted from Snodderly DM, Weinhaus RS, Choi JC. Neural-vascular relationships in central retina of macaque monkeys* (Macaca fascicularis). *J Neurosci 1992;12(4):1169–93.*

retina is projected onto more photoreceptors. Another set of adaptations led to the intense concentration of photoreceptors and clearing of vasculature at the central part of the retina to create the foveal pit. How and when this came about is far from clear, but it may be linked to the requirements for visually guided insect predation, much like in some insectivorous birds (see Ref. [35]).

For primates, the possession of a fovea has important implications for both near and far vision, both of which critically shape primate behavior. For near vision, the resulting high acuity allows for perception of fine detail in objects and textures, which can be important for manipulating or selecting objects, food, or, in the case of humans, tools. For far vision, high acuity translates to the ability to see details about conspecifics and their movements even at a distance. For primates, this latter capacity is closely related to a unique aspect of primate social behavior: primates constantly use their vision to make judgments about individuals, their actions, emotional state, and attentional focus, to "read" the complex social landscape within a large and hierarchical group.

Trichromacy

Trichromatic color vision is another perceptual capacity that stems directly from adaptations to the retina. Most mammals have two cone types, short (S) and long (L), with the wavelength sensitivity of each cone determined by opsin genes that lead to the selective expression of either S or L photopigment within each cone's outer segment. However, many primates have an additional cone type (medium [M]), offering trichromatic vision that other mammals lack (for review, see Ref. [52]). The expression of these L and M photopigments is highest in the foveal region of the retina, from which ganglion cells carry signals to the brain about the relative stimulation of the different cone types. These internal, comparative signals, often termed red/green opponency, are superimposed on the high-acuity information described above and are thought to be critically important for color perception.

All Old World monkeys and apes are routine trichromats. However, many New World monkey species are marked by genetic polymorphisms that render some individuals as dichromats, similar to other mammals, and others as trichromats, similar to Old World monkeys. In fact, in those species it is only a subset of females that has the capacity for trichromatic vision. This odd relationship between gender and trichromacy can be understood in terms of the underlying molecular genetics. As described above, to be a trichromat one needs three distinct types of cones, each with a different spectral sensitivity. These are the S cones, coded by an autosomal gene that is highly conserved across mammals, and the L and M cones. In Old World monkeys and apes, two separate L and M genes reside on the X chromosome. Because both males and females have at least one X chromosome, all members of those Old World species are trichromats [53]. Humans are the only marginal exception to this rule, where roughly 8% of males have a mutation of one of these genes and are thus dichromats [54]. However, in New World monkeys, only a single opsin gene locus exists on the X chromosome, but there are multiple alleles in the population with varying wavelength sensitivity in the L and M range [55]. As a result, a female, having two X chromosomes, can by chance inherit X chromosome alleles that are sensitive to different wavelengths. In such a case, this female will be a trichromat. Because males from the same species have only one X chromosome, they are obligatory dichromats. As we discuss later, this difference allows for an interesting test of the role played by trichromacy in shaping visual processing.

How did trichromatic vision evolve? Unlike high acuity, primate trichromacy seems to have arisen abruptly. At some point after the divergence of New World and Old World primates (see Fig. 22.1), two genetic events are thought to have occurred in the line leading to extant Old World monkeys and apes. First, there was a gene duplication event leading to two opsin copies on the X chromosome. This was followed by the mutation of one of the genes that led to a shift in its chromatic sensitivity. The result was the routine presence of both M and L opsins on the X chromosome, allowing vision to capitalize on three distinct cone types [56,57]. The stable maintenance of trichromacy among Old World primates is thought to reflect selective advantages. Most often, it is linked to advantages in foraging, as the discrimination of green and red hues can lead to better selection of fruit or leaves [58–60]. However, recent genetic evidence suggests that the emergence of trichromatic vision also affected primate social behavior. Interestingly, this hypothesis stems from a peculiar evolutionary trade-off between photopigment opsin genes and olfactory receptor genes. Most mammals use chemical olfactory signals to convey identity and other social information. To some extent, prosimians and New World monkeys use such signaling mechanisms. However, the genetic potential supporting this mode of social communication virtually disappeared in trichromatic primates, with many olfactory receptor genes mutating into nonfunctional pseudogenes [61]. This deterioration of olfaction is suspiciously coincident with the emergence of trichromacy [62–64], and this linkage may be reflected in the dominance of vision over olfaction for important types of primate social exchange [65], which we will discuss further in the Social Behavior section.

Binocular Visual Field

A third peripheral adaptation is the convergence of the orbits in the primate skull and the resulting high degree of binocular field overlap. These changes are thought to have been part of a larger array of craniofacial changes such as visual acuity, that may have originally benefitted the requirements of nocturnal predation [51]. It has also been suggested that the larger binocular overlap could have developed to meet optical constraints in focusing images for the larger nocturnal eyes of early primates, a feature found not only in prosimians but also in fruit bats and owls (see Ref. [66]). Primates have a higher degree of binocular convergence than other mammals, a condition that affords advantages for vision, including stereoscopic depth perception [67] and redundant sampling to aid perception in a cluttered environment [68]. The requirements placed on the brain for reconciling high-resolution images from the two eyes into a coherent representation appear to have strongly influenced the organization of the visual system. For one, it is hypothesized that such central reconciliation may have eliminated the advantages of feature extraction in the retinal periphery and led to a shift in which such features are first computed in the cortex [69]. While the diversity of retinal cell types found in other mammals remains present in the primate [70], and may support some complex feature extraction, it is notable that what have become the numerically dominant retinal classes in the primate, the parasol and midget cells, lack such feature selectivity. For example, neurons in the rodent retina are readily observed to have directional selectivity, but this is much less common in the primate. In the primate, even the input layers of the primary visual cortex have circularly symmetric, undifferentiated monocular receptive fields [71,72], with higher-order features, such as orientation or directional selectivity, computed subsequently [73]. That this organization can be related to the challenges of binocular overlap is supported by the fact that analogous differences are observed in species of birds with differing levels of binocular overlap [69].

In primates, the need for binocular reconciliation strongly shapes the development and function of the early visual system, for example, in the high degree of crossing at the optic chiasm, the strong lamination of the lateral geniculate nucleus, and the prominent ocular dominance columns in the primary visual cortex. Ultimately it impacts brain function in many other ways, such as in the support of high-resolution stereoscopic vision and the need for precise orienting mechanisms to compensate for the restricted visual field. In the next section we examine more specifically how high-acuity, trichromatic, and binocular vision may have shaped the brain and behavior of primates.

The Visual Brain

The primate brain has a number of unique characteristics that distinguish it from that of other mammals (Fig. 22.3; for review, see Ref. [29]). Most of these features are shared throughout the primate order, despite a wide diversity of sizes and habitats of individual species. Here we describe several brain regions related to vision, focusing on those features that distinguish primates from other mammals.

Superior Colliculus

The most distal target of retinal ganglion cell axons growing in the brain is the superior colliculus (SC) [75]. This structure, which is homologous to the retinorecipient optic tectum in other vertebrates, contains organized maps of visual space derived directly from the layout of the retina. The primate SC receives inputs

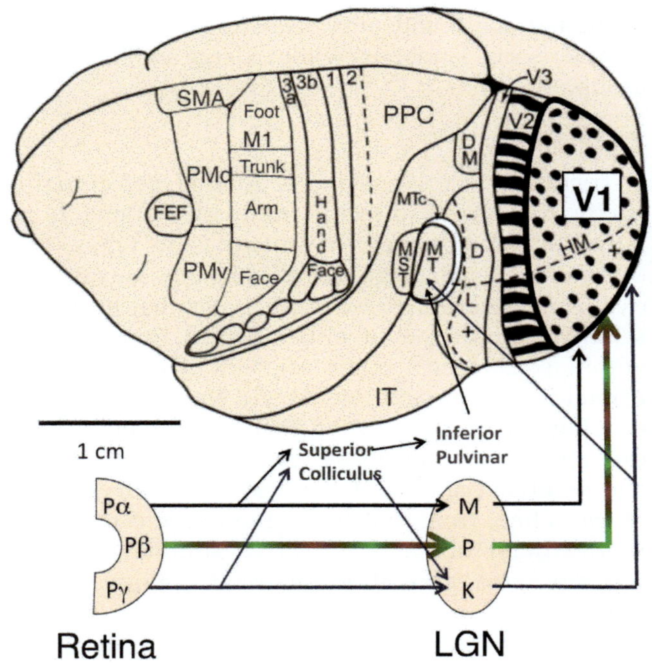

FIGURE 22.3 Schematic drawing of retinal pathways to the cerebral cortex for a prototypical adult primate, shown on the brain of a New World monkey. Parallel streams of visual information leave the retina through three principal classes of projecting ganglion cells (Pα, Pβ, and Pγ), whose lateral geniculate nucleus projection targets are magnocellular (M), parvocellular (P), and koniocellular (K), respectively. The parvocellular pathway dominates, carrying high-resolution foveal information in all primates and also red/green opponent signals in trichromats. In contrast to other mammals, nearly all visual information enters the visual cortex through V1. Only major feedforward projections are depicted. *Adapted from Kaas JH. Evolution of columns, modules, and domains in the neocortex of primates. Proc Natl Acad Sci 2012; 109:10655–60 and Preuss TM. Evolutionary specializations of primate brain organization. In: Ravosa M, Dagosto M, editors. Primate origins. New York: Kluwer; 2007, pp. 625–75.*

that include magnocellular- and koniocellular-type signals from the retina, arising from the parasol (Pα) and diverse types of bistratified (Pγ) ganglion cells, respectively [76,77]. Magnocellular and koniocellular pathways carry visual information of relatively low spatial acuity, but good temporal acuity and sensitivity at low light levels. They can be contrasted with the higher acuity information present in parvocellular signals, which we will discuss in the context of the geniculate pathway shortly.

One notable feature of retinotopic maps in the colliculus that distinguishes primates from other mammals is that input to each colliculus is exclusively from the contralateral visual field. In most other mammals studied to date, visual maps in the colliculus correspond to the complete visual extent of the contralateral eye, including both contralateral and ipsilateral visual fields. The ipsilateral visual field, which is often much smaller, is represented rostrally. The contralateral field occupies the larger portion of the colliculus and extends caudally. The level of orbital convergence determines the level of ipsilateral visual field representation, with more binocular overlap leading to a higher degree of ipsilateral field representation [78]. In primates, however, the retinal input to each SC is very different and derives almost exclusively from the contralateral visual field. Although there must always be a limited overlap near the midline, there is otherwise no rostral ipsilateral field representation despite the large binocular overlap [79]. This strict field segregation in the colliculus is a strong feature that distinguishes primates from all other mammals [80]. In fact, it is such a primate-unique feature that when a somewhat similar organization was found in fruit-eating bats (megachiroptera), it was proposed that these bats must be descended from primates [81]. This proposition has been challenged by subsequent electrophysiological [82] and molecular [83] evidence, though debate continues. The functional implication of this anatomical feature remains unknown but suggests that in our early primate ancestors there was a significant and lasting rerouting of retinal ganglion cell projections to the SC within the optic chiasm [80].

Lateral Geniculate Nucleus

In mammals the lateral geniculate nucleus (LGN) is the principal recipient of retinal ganglion cell afferents. The primate LGN is composed of multiple-layered cells that form parallel maps of the visual world [84] (for a review, see Ref. [85]). These layers are named based on the size of their constituent cells. The large projection neurons in the magnocellular layers ("magno" = large) receive their input primarily from the parasol, or Pα, ganglion cells. The smaller projection neurons in the parvocellular layers ("parvo" = small) receive their input from the midget, or Pβ, ganglion cells. Finally, the smallest projection neurons in the koniocellular layers ("konio" = dust) receive their input from diverse sets of ganglion cells, collectively termed Pγ [86]. In the primate LGN, the inputs from left and right eyes further segregate the cell classes functionally into distinct layers. Thus each LGN contains retinotopically registered maps corresponding to left eye magnocellular, right eye magnocellular, left eye parvocellular, and right eye parvocellular, giving four layers, a number that is shared by all primates. In larger primates, the parvocellular layers are further subdivided into incomplete folds or "subleaflets," giving the appearance of six layers in the macaque, or even eight in the human (see Ref. [87]). Among primates, the parallel projection of retinal ganglion cell types into distinct LGN layers is highly conserved [88], as is the segregated projection of LGN cell classes into the input layers of the visual cortex [89]. It is not until the primary visual cortex that information from the different retinal cell classes, and from the two eyes, is mixed together.

Functional segregation is also a feature of the LGN of other mammals, though it is generally not as strict as in primates. Ocular segregation is common, where retinogeniculate terminations from the contralateral eye show limited overlap with those from the ipsilateral eye [90]. The compartmentalization of physiological types is more varied. In cats and some diurnal rodents (e.g., gray squirrels) LGN cells are layered and receive selective ganglion cell input. Layers of X-like cells (A and A1 layers in the cat) have tonic sustained responses that pool linearly across space, whereas layers of Y-like cells (C and C1 layers in the cat) have transient nonlinear response properties. An additional category of W-like cells is more distributed and carries low-acuity information from diverse bistratified retinal ganglion cells about luminance [91–93]. These classes are physiologically similar to those originating in the Pα, Pβ, and Pγ retinal classes in macaques. Importantly, in mammals where the LGN does not exhibit obvious laminar segregation, similar classes still exist. For example, in some nocturnal rodents, such as rats [94] and mice [95,96], similar physiological classes exist in the LGN (e.g., sustained and transient) despite minimal evidence for their laminar segregation [97–99]. Even in the marsupial opossum, neurons fall into distinct classes that resemble the X, Y, and W responses of other mammals [100], suggesting that this retinogeniculate division of labor evolved before the split between marsupials and placental mammals in the Jurassic period [101] (see Fig. 22.1), and perhaps much earlier. Importantly, comparative studies of the mammalian LGN reveal a clear dissociation between the existence of physiological cell classes and the extent of their laminar compartmentalization. The fact that all extant species evaluated in this way have multiple classes suggests these physiological ganglion cell classes are more fundamental than the compartmentalization of their target neurons within the LGN, which is highly variable across mammals.

In addition to the strict segregation of its physiological cell classes, at least two additional features distinguish the primate LGN, both of which can be traced to retinal specialization. The first feature is the dominance of the parvocellular pathway, which constitutes approximately 90% of the LGN in some primate species. This is the channel for high-resolution vision, which transmits information from the small midget ganglion cells to the primary visual cortex. Parvocellular neurons respond to fine image details and modulate their activity approximately linearly with contrast, particularly in central vision [102]. Although the homology of these classes across species remains unclear, parvocellular neurons have response properties that resemble X-type signals in other mammals, and magnocellular neurons have properties that resemble Y-type signals. This physiological connection is most evident by the similar sustained versus transient nature of the X- and Y-type neural responses, respectively. At the same time, other physiological characteristics of these classes can differ significantly. For example, in rodents the spatial acuity of the sustained X class is no greater than that of transient Y class, and the contrast linearity of response of X class is no greater than Y cells [92,95,103]. In cats, the distinction between these classes lies somewhere in between with regard to both acuity and response linearity [104,105]. Taken together, it appears that the most widely shared property that distinguishes physiological cell classes among mammals is response transience, suggesting that this is a core principle governing parallel transmission through this structure. When there is retinal specialization for high acuity, which is extreme in primates, the parvocellular pathway is much expanded to carry that information.

The second distinguishing feature of the primate LGN is the red/green opponency that is transmitted using the same parvocellular pathway. This feature of the parvocellular system has previously been featured as a possible primate specialization for trichromacy [106], and although given the broad range of dichromatic species that have the same parallel sustained and transient LGN channels, it is unlikely to have evolved specifically for color vision. Experiments in different primate species reveal that the parvocellular channel is most obviously related to conveying high-acuity, sustained visual information, rather than color per se. For example, comparing LGN responses in dichromatic male versus trichromatic female members of the same New World monkey species has revealed essentially identical responses of parvocellular neurons to achromatic stimuli [107]. Similarly, parvocellular neurons in the Galago, a prosimian nocturnal primate that has much higher ratios of rod photoreceptors than cones, were shown to respond in a similar fashion to parvocellular neurons in other primates [49].

Taken together, this comparative analysis points to the order of events in the evolution of the LGN, shaping our understanding of human vision. It is likely that the parallel retinogeniculate pathways evolved first, carrying sustained and transient signals through parallel channels emanating from each position on the retina. The laminar segregation of these channels within the LGN probably came at a later point. Then in some early primate species, regions of exceptionally high photoreceptor density in the retina selectively adopted the sustained pathway for high-resolution signals. Finally, in trichromatic primates, this same pathway was further utilized to relay red/green opponent signals, which were most pronounced in the high-resolution fovea.

Early Retinotopic Visual Cortex

Neurons in the LGN send long-range projections that transmit organized maps of visual space to the cerebral cortex, preserving the topological layout of the retina. This "retinotopy" has analogy in other sensory systems, such as audition and somatosensation, where the layout of the sensory epithelium is also preserved in the cortex. The visual cortex has multiple retinotopic maps. In the early retinotopic maps, a given position responds to input from a unique region of the retina, and a line running along the cortical surface of each map traces out a continuous trajectory of corresponding positions on the retina.

In primates, nearly all LGN projections are directed to area 17, also termed the primary visual cortex or V1. The long thalamocortical axons that run through the optic radiations to V1 remain highly organized until their primary innervation of layer 4C and secondary innervation of layer 6. Parvocellular LGN inputs selectively innervate neurons in layer 4Cβ, and magnocellular inputs selectively innervate neurons in layer 4Cα [108–110], whereas the koniocellular inputs terminate in the supragranular layers [111]. Likewise, the magnocellular pathway provides collateral inputs to layer 6B, whereas the parvocellular pathway provides collateral inputs to layer 6A. Running perpendicular to these cell layers is a different type of functional organization, in which radial units of similar functional properties are assembled into columns. The columnar structure of the early cortex is superimposed on its most conspicuous feature, which is the systematic mapping of visual space. At least one major feature of the tangential organization of the cortex is derived directly from segregation within the LGN. Namely, the ocular dominance columns are a product of the spatially interleaved input, primarily in layer 4C, from the ipsilateral and contralateral eye LGN cell layers [73].

In nonprimate mammals, the organization of projections from the LGN to the cerebral cortex is quite different. First, in most species that have been studied, the LGN projects divergently, not only to area 17 but also significantly to secondary visual areas (e.g., areas 18 and 19) [112,113]. This difference in anatomical projections is obvious following the effects of area 17 lesions, which in primates causes a condition that approximates complete blindness but in other mammals is much less severe [114–116]. Also, the subdivisions within the input layers, and presumed segregation of innervation to sublayers containing different cell classes, vary in other mammals from prominent [117,118] to weak or absent [119]. It is notable that in at least one species, the tree shrew, the geniculostriate terminations in layer 4 convey not transient and sustained responses but rather "on" and "off" responses, which are similarly segregated in the retina and in the LGN [120,121]. Together, these findings suggest that the sublamination of the input layers of the primary visual cortex is closely tied to the segregation of information within the LGN.

Similarly, the columnar organization perpendicular to the V1 input layers is more pronounced in primates than in other mammals. Regarding ocular dominance columns, the tangentially alternating pattern of ocular segregation in layer 4 is less pronounced in the cat [110]. It is also weak or absent in species closely related to macaques such as tree shrews [122] and rodents, including diurnal visual rodents such as squirrels that have laminated LGN [92] (reviewed in Ref. [123]). Ocular dominance columns almost certainly pertain to the reconciliation of images from two frontally positioned eyes. As all primates have considerable binocular overlap, ocular dominance columns are common and have been found in all prosimians and Old World monkeys tested to date. However, it is not a distinguishing feature of the primate brain and is not obviously present in some New World monkeys [124]. At present, the basis of ocular dominance columns and relationship to the LGN and other features of V1 organization remains a puzzle [125].

Another notable specialization of primate V1 is the extremely high density and cellular morphology of its primary geniculorecipient layer, which is almost certainly tied to high visual acuity. While considerable evidence suggests that the areal density of neurons across the cortex in different mammalian species is approximately the same (but see Ref. [126]), a consistent finding is that primate V1 stands out as having the highest neuronal density [126–128]. One factor contributing to this tighter packing is the evolution of a specialized class of pyramidal neurons that have an unusually compact stellate morphology [129]. Such stellate neurons are not found in the input layers of rodent visual cortex [130], although a similar type of cell exists in rodent somatosensory cortex [131]. These distinctions in the input layer structure are also reflected in the local patterns of gene expression in Old World primates, and also appear in New World monkeys but are less pronounced, and are totally absent among rodents [132,133].

In the extragranular layers of primate V1, the parvo-, magno-, and koniocellular information becomes less segregated and is to some extent reorganized to meet the requirements of dorsal and ventral processing streams. For example, information is segregated between the "blobs" and "interblobs" of the superficial layers (see Fig. 22.3), which can be visualized using cytochrome oxidase staining [134]. Neurons in blobs and interblobs have different visual response properties and different projection targets [135]. Cytochrome oxidase blobs are present in primates but absent in other orders of Euarchontoglires, suggesting that this feature evolved early in primate evolution [74]. At the same time, similar blobs have been described in cats [136,137], thus the evolutionary origins of this V1 feature remain unclear. Neurons in the primate interblob regions are organized into columns tuned for a particular orientation with the preferred orientation progressing smoothly across the cortical surface. Orientation columns are also present in certain other mammalian species such as cats [138] and tree shrews [139]. They are, however, curiously absent in rodents [140], including highly visual rodents such as squirrels [92], despite the fact that individual neurons in their area 17 show orientation-tuned responses. The dissociation of orientation tuning and orientation columns is a comparative finding that draws attention to the fact that these two functional characteristics are not fundamentally linked [141].

The segregation of input continues into a second, strongly retinotopic visual area (V2). In area V2, cytochrome oxidase staining reveals a different set of functional zones that take the form of stripes of different intensity and width and run perpendicularly to the V1/V2 border (see Fig. 22.3). The progression of these stripes is superimposed on the map of retinal space and takes the form of a repeated sequence of pale-thick and pale-thin sets of stripes [135,142]. Neurons within the thick and pale stripes, receiving input from the interblob regions, are selective to binocular disparity and orientation. The thin and pale stripes receive afferents from blob regions and are selective to luminance and color. These anatomical features in V1 and V2 are shared across primates, including New World monkeys such as the marmoset [135] and prosimians such as the Galago [74], though the pattern in V2 of prosimians is more patchlike than stripelike [143]. The stripes also appear to project differentially to an array of common extrastriate visual areas that analyze dorsal (dorsomedial [DM], middle temporal/medial superior temporal [MT/MST], fundus of the superior temporal sulcus [FST], posterior parietal cortex [PPC]) and ventral (V3, V4, and inferotemporal [IT]) stream information. The shared projections among different primate species, with thick stripes projecting to visual area MT and the other band projecting to visual area DL (or V4), suggests that this basic organization was present in an early common ancestor [74].

High-Level Visual Cortex

Higher-level visual cortex in primates is typically described as consisting of dorsal and ventral visual pathways, in which different types of visual information are extracted from the retinal input and used for behavior [144,145]. The dorsal stream gains much of its input through projections from area MT. Major projections to subregions of the PPC are critical for guiding visually directed actions to mediate specific behaviors and spatial judgments, whereas other projections carry visual information further to retrosplenial and parahippocampal regions and are thought to be important for navigation [146,147]. The ventral stream is also composed of multiple pathways, in this case passing visual information to different subregions of the inferotemporal cortex. The ventral pathways receive much of their input through area V4, though signals from MT are also important in shaping responses in many ventral stream areas. The ventral pathway is thought to be involved in the processing of the complex structure for recognizing objects and social information [148]. While high-level visual cortex has been most studied in primates, other mammals appear to have an analogous division of labor in their cerebral cortex. Lesion studies in cats reveal a dissociation in the deficits following parietal and temporal lesions that approximate those seen in monkeys [149]. Sheep have abundant face-selective neurons in their temporal cortex that resemble those found in monkeys [150]. In rodents, there is some evidence to suggest that a dedicated dorsal visual stream conducts spatial analysis [151,152], with evidence for a separate ventral visual stream analogue being less obvious [153,154]. Thus, much evidence supports the origination of dorsal and ventral visual pathways in the Mesozoic period, before the first primate (see Fig. 22.1). Nonetheless, there is good reason to believe that the dorsal and ventral systems underwent considerable evolution within the primate radiation. This is because the two most prominent behavioral adaptations related to vision, which we will discuss next, derive directly from parietal and temporal cortex function.

The first prominent adaptation is visually guided reaching and grasping, a behavior at which primates excel and which is unmatched among mammals. This faculty allows for efficient movement through the arboreal environment as well as the manipulation of food and other types of objects, including tools in humans. These abilities are thought to have drawn, at least in part, on expansion and specialization of the PPC, together with the development of motor and premotor cortex [155]. Distinct subnetworks mediate different visually guided behaviors that include reaching, defensive, and grasping movements (for review, see Ref. [146]). These areas have been extensively studied in macaques and include the parietal reach region for reaching movements [156], the lateral intraparietal area for eye movements [157], the anterior intraparietal area for grasping movements [158], and the ventral intraparietal area for defensive movements of the head and arm [159]. Areas with similar functions have been found in the Galago [160,161] and in New World monkeys [162,163], including the marmoset [164–166]. Each subregion of the parietal cortex projects to corresponding subregions of premotor and motor cortex that specialize in similar movements, and thus form distinct networks for particular categories of actions [162,163,167]. Among rodents and the tree shrew the PPC is much smaller and appears to play a less direct role in guiding movement. For example, while the tree shrew has a greatly expanded visual cortex as compared to rodents, most of its visual and somatosensory information still reaches motor cortex through direct projections rather than through the PPC, unlike in primates [146,168]. While comparative physiological studies have been relatively rare in this area, it seems likely that at least some of these PPC regions have no clear homolog in mammals that do not use their vision to guide reaching and grasping.

The specialization of the parietal cortex is particularly important for understanding the human brain. Human bipedalism has fundamentally changed the manner in which extrapersonal space is encoded and has further led to exceedingly complex visually guided actions, and associated brain specializations, which facilitate tool use [169]. Humans have an expansion of the PPC that includes areas that appear to be absent in other primates [170]. This expansion is thought to give rise to a more sophisticated repertoire of motor behaviors including the fine manipulation of tools and other objects [171,172].

The second prominent behavioral adaptation in primates is the use of vision for complex social exchange. As mentioned above, the routine use of vision for individual recognition, sexual selection, and social monitoring is facilitated by specialization of the retina that improved visual acuity and led to trichromatic color vision. The impact of these peripheral specializations on social behavior has been accompanied by massive expansion of the ventral visual cortex, including the large proportion of inferotemporal tissue apparently dedicated to the processing of faces and bodily actions [173].

Ventral visual pathway expansion in the primate appears to be strongly linked with the focus on foveal processing and the use of vision to observe individuals from a distance. Multiple cortical pathways in the inferotemporal cortex project to distinct cortical and subcortical targets and specialize in different aspects of processing, including object recognition, scene recognition, and emotional or affective valence (for review, see Ref. [148]). Along the ventral stream there is a division of labor between cortical areas that lie anterior to, and receive input from, the foveal portions of early visual areas and those which lie anterior to, and receive input from, peripheral field representations. Peripheral regions feed into parahippocampal cortex, where neurons have peripheral receptive fields [173a], and are thought to be involved in the spatial understanding of a scene [147,174,175]. Foveal regions feed into the inferior temporal cortex, which contributes to complex form vision, including the analysis of objects and social stimuli. Nearly all neurons in the inferotemporal cortex respond most strongly to stimuli when they are presented at the fovea [176–179]. As rodents lack a foveal specialization, it may be difficult to identify a homologous, or even analogous, processing pathway [153]. However, a recent study has identified extrastriate visual areas in the mouse whose function appears to map onto dorsal and ventral streams in the primate, with the ventral stream involved in object or landmark recognition [154]. Other work demonstrates that rodents can learn to distinguish between visually complex objects [180,181]. Thus, while there seems to be no clear homolog to inferior temporal cortex, rodents do show some features of visual cognition commonly associated with the ventral processing pathway. There is still much to be learned from comparative research about the origin and evolution of the dorsal and ventral visual pathways that play such a prominent role in understanding the organization of the human brain.

Visually Guided Behaviors

The evolved changes to the primate eye and brain can be linked directly to an array of specialized behaviors. Here we review three visually guided behaviors at which primates excel: the detailed visual exploration of a scene, the perception and interpretation of social information, and the precise guidance of reaching and grasping.

Natural Exploratory Behavior

Monkeys' reputation for being curious is well deserved and is supported by neural systems that promote the efficient acquisition of visual information from the world. More than other mammals, primates have a fast and precise means of directing their gaze from point to point in the form of rapid eye movements called saccades. Saccades, along with smooth pursuit for tracking moving objects, are highly developed in primates and coordinated with actions of the head and body. These movements are tightly coordinated with the processing of the visual input and are controlled by an elaborate network of cortical and subcortical structures (for a review, see Ref. [182]). The purpose of each saccade is to reposition the high-acuity fovea to a new location. Because the eyes are lighter than the head and can be turned more easily, most primates use saccades rather than head movements to scan and orient elements in their environment. This sequential sampling of visual space is used to accumulate information about the structure of a scene. As such, visual exploration can be compared to tactile exploration, such as a person moving their fingertips over an object's surface or a rodent using its whiskers to feel and recognize objects in the dark. In humans, the exploration of visual scenes by eye movements can vary in highly complex ways depending on the context and motivation of the observer [183,184]. With an average of 2–3 saccades issued each second, saccadic eye movements constitute our most frequent overt behavior.

Oculomotor behavior among mammals is determined by the density distribution of photoreceptors on the retina, which is in turn determined by ecology. The small and round foveal pit in the primate retina requires exquisite control over saccadic eye movements to optimally redirect gaze and gather information about a scene. Other species have very different retinal density patterns and correspondingly different eye movement patterns. Mammals such as rabbits and horses with horizontal streaks on their retina tend to reside in open environments, as the higher acuity in the visual streak allows them to monitor the horizon without the need for continual scanning [185]. In rabbits, gaze shifts are infrequent and saccades are strongly coupled with head movements [186]. The cat retina lies somewhere in between, as does its oculomotor behavior. The cat area centralis is much lower in its resolution than the primate fovea and is also elongated horizontally [187]. The saccades of cats are less frequent, slower, and more variable than those of primates [188]. Mice, whose central portion of the retina shows minimal specialization for higher acuity vision, make even fewer saccade-like eye movements and rely primarily on head movements to orient their gaze [189].

It is notable that even closely related species can have very different profiles of retinal receptor density. For example, within the carnivore order, the area centralis found in the domestic cat and other felines is absent in the closely related cheetah, which has a visual streak much like that of a horse. This adaptation has been attributed to living on the open savannah [185]. Similarly, the plains kangaroo has a streak, whereas the tree kangaroo has an area centralis [190]. An important principle in comparative neuroscience is that sensory systems can evolve quickly and that peripheral adaptations can strongly influence the organization of the brain [191]. In primates, the invention of the fovea led to optimization of circuits for directing gaze and processing visual detail. One domain of behavior particularly affected by these features is social interaction, which we will discuss next.

Social Behavior

One feature that sets monkeys and apes apart from other mammals is the complexity and expression of their behavior toward other individuals in the group [192,193]. This widely shared feature of primate life provides good reason to believe that the primate brain is highly adapted to accommodate complex social interaction. All mammals need to engage in certain forms of social interaction, for example, related to the rearing of offspring, mate selection, territorial disputes, cooperation in foraging, and defense. However, primate social interaction involves a higher degree of visual specialization than that of other mammals, starting with their use of high acuity vision to analyze faces, bodies, and actions from a distance [80]. From several meters away, primates are able to observe and monitor the detailed actions of other individuals, giving them an advantage in the social group. However, the rules of the dominance hierarchy of the group also influence how vision can be used, placing certain restrictions on gaze behavior. Because one individual can determine where another individual is looking, dominant males prohibit subordinate animals from looking directly at them. Thus, in the primate competition for power, not all animals are allowed equal access to the social benefits of high-acuity vision.

Another important aspect of social behavior is sexual selection, which, as discussed in the section Trichromacy, has been linked in Old World monkeys to the emergence of trichromatic color [194]. Comparative analysis suggests that the capacity to see reds and yellows may have gradually shaped sexual selection because of the increased perceptual sensitivity to the level of blood perfusion and oxygen saturation in the skin [195]. In the millions of years after genetic mutations that introduced a new cone type in the retina and broadened the capacity to distinguish between different shades of red, Old World monkeys became increasingly barefaced and bore an increasing yellow and red coloration on their face, skin, and pelage. It is suggested that trichromatic females increasingly were able to choose males based on the quality of their red markings, and trichromatic males likewise were able to detect ovulating females on the basis of their red perineal skin swellings [57,196]. Statistical analysis based on extant primate species suggests that not only did color patterns become more prevalent after the advent of routine trichromacy, but also that primate mating became more gregarious [194].

The high level of social monitoring, hierarchical reinforcement, and sexual selection exhibited by primates figures prominently into an aspect of primate social behavior that has been called "Machiavellian intelligence" [192,193,197]. This term refers to the high level of social maneuvering among primates, much of which is based on observing one another and predicting their behavior. Larger groups entail a higher degree of social complexity, as there are more possible pairwise social relationships. It has been suggested that the number of members in a primate social group is ultimately constrained by the capacity of the cerebral cortex, including those regions specialized for high-level vision [198,199], and by brain regions such as the amygdala and hypothalamus, which underlie the expression of social behavior in all mammals [200].

Visually Guided Reaching and Grasping

Primates are also unique among mammals in their use of visually guided reaching and grasping, precision grip, and object manipulation. Humans represent the pinnacle of this behavior in their use of tools or, for example, playing a musical instrument. As mentioned in the section High-Level Visual Cortex, precise visually guided actions are linked to refinement of mammalian dorsal visual stream, with a large expansion of posterior parietal cortex, particularly in humans, and an elaborated domain in the primary motor cortex representing direct cortical control over the hands. Manual manipulation also benefits from stereoscopic vision, along with vergence eye movements that coordinate the two fovea toward targets in depth [201].

The evolution of visually guided reaching and grasping is closely associated with primate ecology and the need to move through the forest canopy. Apes and monkeys that use brachiation rely on vision to select, reach for, and secure their grip onto appropriate branches. These actions are often rapid and coordinated with other self-generated movements that strongly affect the visual input, such as the motion of the body, turning of the head, and frequent saccades. As a miscalculation can have fatal consequences, primates have evolved under a strong selective pressure for accuracy in this domain. The origins of precise manual behavior may stem from specific aspects of the feeding behaviors of early primates. One influential hypothesis holds that early primates were nocturnal predators who were aided by a wide field of stereoscopic vision for catching insects in their hands [30]. An alternative view is that precision reaching and grasping evolved for the purposes of extracting small fruits from terminal branches of angiosperms [202,203].

Rats and other members of Euarchontoglires exhibit some visually guided reaching and grasping, but with much less precision than primates. For example, a rat reaching for a piece of food will not preshape its hand based on visual cues. Instead it will draw on tactile cues for posturing its forepaw, much like the behavior of a blindfolded human subject who is aware of the position of an object but not its shape [204]. Note that much of this deficit is specifically related to the use of vision. Rats have sufficient motor control to preshape their forepaws, as they do so when reaching up to grasp a piece of food already sitting in their mouth. The use of vision to guide reaches and enable manipulation is a difficult computational problem for which special circuits have evolved in primates.

Primate Adaptations: Summary

In this section, we have specified a number of characteristics of the eyes, visual brain, and visual behavior that are shared among primates and in some cases distinguish primates from other mammals. Perhaps the most important are the peripheral adaptations, including the dense packing of photoreceptors in the fovea, the three retinal cone types, and the convergent orbits. These can be linked to many of the specializations of the brain that underlie fundamental primate behaviors, including exploration, social complexity, and precise manual motor behavior.

COMPARING MARMOSET AND MACAQUE VISION

We now turn our attention to one particular primate, the common marmoset (*C. jacchus*), a small New World monkey that has recently attracted much interest as an animal model for the neurosciences. We continue to apply a comparative analysis of vision as above, here placing greater emphasis on comparing the marmoset with the better-studied macaque.

Visual System

The gross appearance of the marmoset brain differs from that of larger monkeys such as the macaque. The most obvious difference is that its surface is nearly free of sulci and in that sense is similar to a rodent. However, visual comparison reveals that in most other respects the marmoset brain more closely resembles the brain of a macaque than that of a rat (Fig. 22.4). For example, the rat telencephalon lacks a Sylvian fissure, has a prominent olfactory bulb and piriform cortex, and bears a different angular relationship to the hindbrain. Beyond its gross anatomy, the detailed anatomy and basic electrophysiology of the marmoset's eye and brain are categorically those of a primate, with many aspects of the visual system nearly indistinguishable from those of the macaque (see Ref. [28]). At the same time, there are notable differences in their eyes and brain that affect their visual behavior. These differences reflect the evolutionary changes that have impacted each species since the time of their most recent common ancestor 35–40 MYA.

Eyes and Retina

In the marmoset, the organization of the retina and placement of the eyes, including the high level of binocular overlap, is typical for a primate. The marmoset's retina has a well-defined fovea with tightly packed cones and dense midget ganglion cells that carry high-resolution visual information out of the eye and into the brain. The estimated visual acuity of marmosets at 30 cycles/deg is somewhat less than the 50 cycles/deg of macaques [35], but this can be largely ascribed to the size of the marmoset eyes, whose axial length is

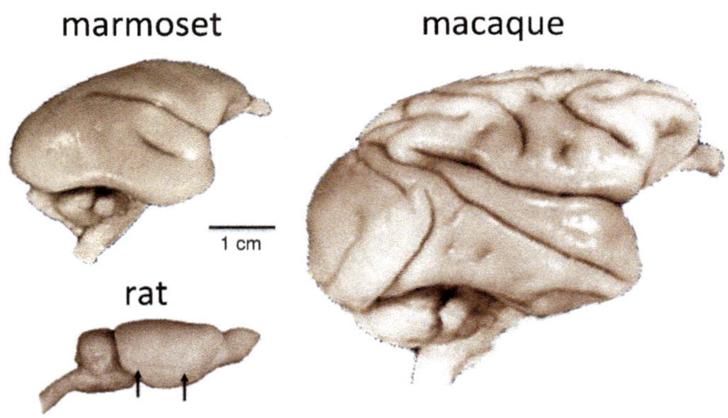

FIGURE 22.4 Comparison of marmoset, rat, and macaque brains, lateral view. Unlike the primate, the rat brain has no Sylvian fissure separating the frontoparietal and occipitotemporal cortical regions. The faint horizontal sulcus in the rat brain (arrows) is the rhinal sulcus, which separates the neocortex from the more primitive piriform (olfactory) cortex. This feature is much smaller in monkeys and not visible from the lateral view. The overall geometry of the marmoset brain is much more similar to the macaque than the rat.

11 mm [205] compared to the macaque's 18 mm [206]. For a given cone density, larger eyes translate to a higher visual acuity as the same arc of visual angle subtends a larger region of retinal epithelium. The peak cone density at the fovea is highly similar between the species (Fig. 22.5A). A more detailed analysis of the marmoset eye further reveals that the visual optics and basic topographic cone density pattern also closely resemble those of macaques and humans [205].

One difference between the marmoset and the other species is a notably higher cone density in the peripheral retina (Fig. 22.5A). However, whether this higher cone density contributes to improved visual acuity depends on several factors. For example, acuity will also depend on the extent of pooling across cone receptors, which in the macaque has been found to be higher for the periphery [208], consistent with the measured behavioral acuities at peripheral eccentricities [209]. A recent study investigated the spatial acuity of the marmoset as a function of retinal eccentricity and found that despite the higher proportion of cones in the periphery, acuity thresholds still scale as roughly half those of macaque or human [210]. These findings are consistent with anatomical data which indicate a higher extent of pooling cone signals in the periphery of marmoset retina [47]. The functional significance of higher cone density in the marmoset peripheral retina remains to be determined but does not appear to enhance peripheral acuity.

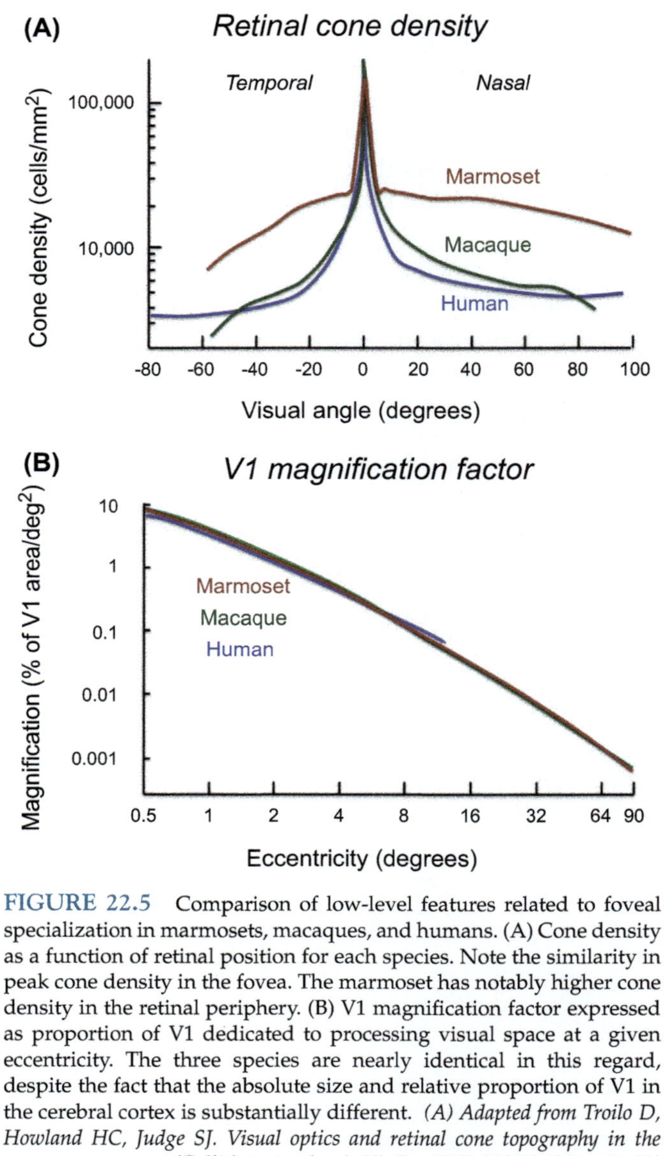

FIGURE 22.5 Comparison of low-level features related to foveal specialization in marmosets, macaques, and humans. (A) Cone density as a function of retinal position for each species. Note the similarity in peak cone density in the fovea. The marmoset has notably higher cone density in the retinal periphery. (B) V1 magnification factor expressed as proportion of V1 dedicated to processing visual space at a given eccentricity. The three species are nearly identical in this regard, despite the fact that the absolute size and relative proportion of V1 in the cerebral cortex is substantially different. *(A) Adapted from Troilo D, Howland HC, Judge SJ. Visual optics and retinal cone topography in the common marmoset (*Callithrix jacchus*). Vis Res 1993;33(10):1301–10. (B) Adapted from Chaplin TA, Yu HH, Rosa MG. Representation of the visual field in the primary visual area of the marmoset monkey: magnification factors, point-image size, and proportionality to retinal ganglion cell density. J Comp Neurol 2013b;521(5):1001–19.*

One cautionary note from studies of acuity in the marmoset is the high prevalence of near-sightedness among marmosets raised in laboratory settings. Two studies have reported average myopia ranging from −1 to −3 diopters correction among those animals raised within indoor colonies with no access to sunlight or open spaces [210,211]. No studies have yet determined if the incidence of myopia is indeed lower among animals raised in the outdoors. However, the robust development of myopia in laboratory settings has made the marmoset valuable as a model of human myopia [205,212]. Proper use of the marmoset in vision and visual neuroscience studies will thus demand consideration of the developmental conditions in which research animals are raised and possibly the correction for myopia.

Color vision in marmosets, like other New World monkeys, is complex and depends on gender. A genetic polymorphism in the longer-wavelength sensitive (L) cone type results in different individuals having different chromatic sensitivities. Moreover, because the corresponding gene is on the X chromosome, this polymorphism differentially affects males and females. Both genders have an autosomal short-wavelength sensitive (S) opsin that is present in most mammals. However, because males have only one X chromosome, they have only one of the L alleles. Thus they are only able to produce two cone types (S + L) and are obligatory dichromats. Females, on the other hand, having two X chromosomes, can possess two different L alleles (for simplicity, we call them here L and M) with somewhat different wavelength sensitivities. For this subset of females, the additional opsin results in three cone types (S + M + L) giving them trichromatic vision similar to that of Old World monkeys. Trichromatic New World monkeys are able to perceive colors differently than their dichromatic conspecifics [213]. However, while there may be distinct survival advantages to the third cone type [59,214], these may actually be matched by advantages to dichromatic vision, such as in foraging under low-light conditions, and these complementary advantages appear to have resulted in a stable genetic balance of dichromats and trichromats in New World monkey populations [215]. This is not the case in Old World monkeys, where both males and females possess fixed L and M genes on each X chromosome and are therefore always trichromats. In fact, among Old World primates, only humans substantially deviate from this pattern, with ~8% of males being dichromats due to a mutation rendering one of their X-linked opsins dysfunctional.

Lateral Geniculate Nucleus

The functional organization of the marmoset LGN closely resembles that of the macaque, with some differences in the layering of the different cell classes. The paired layers of magno-, konio-, and parvocellular neurons fit within the larger pattern of LGN structure that differentiates primates from other mammals [87]. As noted earlier, all primates have four layers in their LGN, although in the macaque the parvocellular layers are folded, giving six layers in cross-section, whereas the marmoset has four in cross-section [87]. One difference with the macaque is that the marmoset has a more defined lamination of the zones containing koniocellular neurons. This latter feature makes the marmoset particularly useful in the study of koniocellular pathways [216–218]. For example, a recent study demonstrated a small proportion of neurons in its koniocellular layers that respond both binocularly and with a relatively strong orientation selectivity [219], two features that are not normally associated with the primate LGN. However, as there have been rare reports of similar phenomena in the macaque (e.g., Ref. [220]), it remains to be determined whether this is a general feature of the primate koniocellular system.

Studying LGN responses in marmosets has also benefited from the coexistence of dichromatic and trichromatic individuals, as mentioned briefly in the Lateral Geniculate Nucleus section. Parvocellular LGN responses of the two groups differ in their red/green opponency, with responses in trichromatic marmosets resembling those in the macaque [221,222]. However, aside from this feature, the spatial and temporal characteristics of the parvocellular LGN were found to be virtually identical in the two groups of marmosets [107] and also similar to the monochromatic Galago [49]. This evidence suggests that the parvocellular specialization did not specifically emerge to carry red/green opponent signals, which is sometimes assumed. Instead, the parvocellular pathway is built on the mammalian X pathway, which expanded and specialized to support high-resolution central vision early in primate evolution [87]. This modified X pathway transmits red/green opponent signals as long as they are provided by the retina [107]. This realization is underscored by recent work showing that a third cone type artificially introduced into the retina of adult male New World monkeys induces trichromatic color vision [223].

In summary, the marmoset LGN differs somewhat from that of macaques and humans, with the largest differences being its layering and the chromatic signals it transmits. However, it bears the key hallmarks of the primate LGN, including most prominently the thick parvocellular layers that transmit high-acuity visual information from the fovea to an expanded region of the primary visual cortex, which we will review in the next section.

Primary Visual Cortex

The organization of the marmoset visual cortex is well understood based on experiments carried out by a relatively small number of laboratories. The detailed anatomy and physiology have been recently reviewed and are in most respects very similar to those of the macaque (for a recent comprehensive review, see Ref. [28]). Area V1 has been studied in great detail [219,224–238]. Electrophysiological mapping has revealed that the visual field layout and basic neural selectivity of this area are similar to those of macaques. Anatomical features, such as the cytochrome oxidase blobs in the supragranular layers, are also present. At the same time, there are some notable differences between the primary visual cortex of the marmoset, macaque, and human.

One difference is the proportion of cortex V1 occupies in the different species. By mass, area V1 is ~14.5% of the cortical gray matter in the marmoset, ~8.8% in the macaque, and only ~1.5% in the human [239]. These values may reflect that an absolute amount of cortical tissue is required to support high-resolution vision and appears to derive directly from the number of neurons providing input to V1 from the LGN [240]. Despite differences in the overall size, the percentage of cortex dedicated to different retinal eccentricities, or magnification factor, is nearly identical between the species (Fig. 22.5B) [207]. That the magnification factor matches so closely may reflect an important constraint for high-resolution central vision, and is in part surprising for eyes of different sizes given the conserved absolute size of the foveal pit [40]. How the marmoset's small eye size, high peripheral cone density, and conserved magnification factor together contribute to the marmoset's perception of visual detail in central and peripheral vision remain an open question.

Another distinguishing feature of marmosets is their lack of strongly defined ocular dominance columns in area V1 [241], which are present in macaques [73] and probably also in humans [242]. Ocular dominance is strongest in the input layer, where afferents from the two eyes remain segregated. Whether the absence of ocular dominance columns has any functional consequence is unclear. Interestingly, marmoset ocular dominance columns do appear transiently in early development, but then nearly disappear in adulthood [243].

A third difference between marmoset and macaque V1 is that, in contrast to the macaque, those neurons in layer 4B that project to area MT have pyramidal rather than stellate morphology [244]. In the macaque, stellate neurons of layer 4B project to MT, whereas neurons of pyramidal morphology project to area V2 [245]. Stellate neurons do have different functional properties, showing greater sensitivity to small inputs and overall higher firing rates [246], and pooling inputs over more limited spatial extents [247]. Thus it is possible that stellate projection neurons in 4B of the macaque might mediate faster or higher precision information to motion-selective area MT than other channels. It is also possible that this species difference stems from modification of the layer 4B developmental trajectory, as cortical stellate cells initially take the form of pyramidal neurons before they lose their apical dendrites as they mature [248]. It is worth emphasizing that the locally projecting stellate cells in the input layers, as well as layer 4B, are abundant in the marmoset, as is typical among primates.

Extrastriate Visual Cortex

The electrophysiological responses of other visual areas in the marmoset have been studied in detail and generally resemble those of the macaque, including V2 [135,224,241,249,250], DM [251–253], V3 [254], and MT [233,252,255–260].

Area MT is of particular interest for comparative studies because there is a question as to whether its evolution is unique to the primate line [261]. It is a compact, motion-selective area containing both upper and lower visual field representations and is very well studied in the macaque because of its conspicuous extraction of direction and stereoscopic depth from retinotopic visual space [262,263]. In the marmoset, the basic topographic organization and response properties appear similar to the macaque, including, for example, the selective neural responses for higher-order, or pattern, motion selectivity [258,260]. Sitting on the cortical surface, area MT can be studied more easily in the marmoset, a fact that has been exploited to address questions that are difficult to study in macaque MT. For example, a recent study used planar surface arrays to measure correlated activity of populations of MT neurons [233]. Another set of studies has focused on the early maturation of MT [264] and the existence of a transient retinopulvinar–MT visual pathway early in development [265]. This putative pathway, which has not been previously described in the macaque, can be well studied in the marmoset due to the surface location of MT and the extended period of postnatal development.

Beyond basic mapping, much less is known about higher-order visual cortex of the temporal and parietal lobes of the marmoset (see Fig. 22.6). This gap in our knowledge is due in part to the paucity of electrophysiological studies in awake marmosets, as activity in these higher visual areas is dependent on behavioral state. While the marmoset visual cortex is generally believed to possess dorsal and ventral visual pathways similar

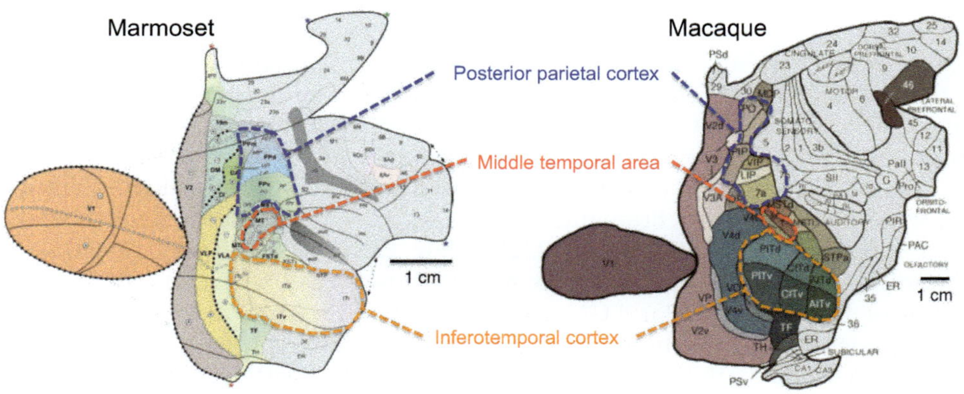

FIGURE 22.6 Flat map comparison between the marmoset and macaque visual cortex. The layout of the visual cortex is generally similar in the two species, including the highlighted features, which are either unique to primates (middle temporal area) or greatly expanded in primates (posterior parietal and inferotemporal cortex). *Adapted from Rosa MG, Palmer SM, Gamberini M, Burman KJ, Yu HH, Reser DH, Bourne JA, Tweedale R, Galletti C. Connections of the dorsomedial visual area: pathways for early integration of dorsal and ventral streams in extrastriate cortex. J Neurosci 2009; 29:4548–63 and Felleman DJ, Van Essen DC. Distributed hierarchical processing in the primate cerebral cortex. Cereb Cortex 1991;1(1):1–47.*

to the macaque, the extent to which the specific functional specializations resemble those of macaques and humans is unknown. This topic is important because, as reviewed in the sections The Visual Brain and Visually Guided Behaviors, these areas are closely associated with primate-specific behaviors, such as visually guided grasping and social perception. Learning whether the physiological responses in the marmoset are similar to those in the macaque may provide insight into the evolution of these behaviors.

Similarly, relatively little is known about responses in the inferotemporal cortex of the marmoset. No previous study has systematically studied IT response properties of the marmoset, though there have been indications that neurons there are selective to faces [266]. Several studies have examined the impact of lesions to IT and found that it causes severe impairments for recognizing visual objects in simple discrimination tasks [267]. A recent study has demonstrated that functional specialization of the marmoset inferotemporal cortex bears striking resemblance to that of the macaque [268]. In the macaque, neurons throughout IT cortex respond with a high degree of selectivity for complex stimuli, and there are a number of regions where neurons are particularly selective for faces (for a review, see Ref. [269]). Marmoset IT cortex is likewise composed of multiple areas that show selective fMRI and electrophysiological responses to faces over other objects [268,270].

The parietal visual cortex in marmosets and its role in visually guided actions are even less explored. Most of the knowledge of parietal cortex function in New World monkeys comes indirectly from microstimulation studies in anesthetized animals [162,163]. Specifically, long-train electrical microstimulation provides a means to study the contribution of an area to particular aspects of behavior [271]. In the posterior parietal cortex, such microstimulation in anesthetized squirrel and owl monkeys leads to coordinated reaching, grasping, and defensive actions. The sites of stimulation, and the pattern of complex motor action, show good correspondence across the prosimian Galago, squirrel and owl monkeys, and also macaques, but not in nonprimate Euarchontoglires species, such as tree shrews or rodents [272]. No studies have examined marmoset parietal cortex using similar microstimulation paradigms, although anatomical evidence suggests similar connectivity between parietal and premotor areas as found in other primates [273]. This frontal–parietal machinery may be unique to the reaching and grasping behavior exhibited by primates, which is under most conditions guided by visual input.

Oculomotor Structures

The cortical and subcortical areas controlling eye movements have been identified in the marmoset and are largely homologous to the macaque. The SC is the main subcortical structure initiating goal-directed eye movements, and it receives cortical inputs from early visual areas (V1, V2, and MT) and from four visuomotor fields which include the frontal eye fields (FEFs), frontal visual area, an area in the region of the supplementary eye fields (SEF), and parietal eye fields [274]. Projections from the SEF itself appear either weak or absent in the marmoset. The FEF, or area 8 aV, was identified in early studies, and microstimulation was shown to evoke either saccade eye movements or slow eye movements of varying speeds [275]. Recent anatomical studies have found that the cortical inputs to FEF in the marmoset resemble those of the macaque [165,276]. A recent fMRI study examined the functional connectivity of frontoparietal networks with the SC and identified similar saccade subnetworks in the marmoset as in macaque and human [277]. No studies have yet examined the projections from the FEF oculomotor nuclei in the marmoset. In other New World species, such as squirrel and owl monkeys, these projections to lower oculomotor control centers appear similar to that in macaques [278].

Visual Behavior

Like other primates, marmosets are skillful in their use of vision. Most of what is known about marmoset visual behavior derives from either natural ethological observations or neuropsychological experiments carried out in a small number of laboratories [279–284, 311]. Recent work has shown that it is also possible to carry out controlled experiments in head-fixed, awake marmosets, where oculomotor behavior and visual stimulation can be precisely determined [268,270,285]. Here we briefly review several aspects of visual cognitive, social, and motor behavior that have been studied in this species.

Visual Cognitive Behavior

In experiments allowing unrestrained movement, marmosets reliably make complex visual discriminations, for example, choosing between pairs of distinct real objects or visual patterns based on a previously established association with food reward [281,286,287]. Several researchers have used the Wisconsin General Test Apparatus in which the animal may use its hand to displace an object it believes to be associated with a food reward into a well underneath. Between trials, the investigator manually changes the choice objects, fills the food well, or sometimes handles the animal, repositioning it to begin the next trial. In this mode of testing, which is often used to study the effect of pharmacological agents or brain lesions on higher-order aspects of cognition, a marmoset will typically perform 40–50 trials in a session. Marmosets can also be trained in more automated testing setups, where they respond to stimuli on visual displays by pushing a touch-sensitive screen or depressing a mechanical lever for receipt of reward [288–291]. Together, these paradigms have been able to tap into aspects of visual perception, for example, by demonstrating that marmosets can distinguish reliably between subtly different colors [213,280] and complex patterns [281,287,292–294]. Moreover, large lesions to the inferotemporal cortex, which are commonly associated with object agnosia in macaques and humans, similarly impair visual discrimination capacity in marmosets [267]. Like macaques, marmosets are able to flexibly use visual context, or the rewarding properties of a visual feature, to alter their response strategy [267,282,290,295–297]. Touchscreen testing has further shown that marmosets can perform visuospatial memory tasks well when the delay is short (e.g., 3 s) but struggle with longer delays (e.g., 12 s) [291]. Attempts to further explore visual memory with delayed match-to-sample and related tasks suggest that marmosets are more challenged by this mode of testing than the macaque [291,298–300], although the reason for this difference is not well understood and could depend on differences in the methods for training and testing the species.

Gaze Behavior and Visual Attention

Like all primates, marmosets use saccadic eye movements to direct their high-resolution fovea to objects of interest within a visual scene. Methods for the monitoring and conditioning of behavior under head restraint have been recently established in the marmoset [301]. One study has begun to use video-based eye tracking to record scan paths and other aspects of saccadic gaze behavior [285]. A second study also examined another category of slow eye movements for pursuit of moving targets [302]. Pursuit eye movements were qualitatively similar to that of macaques and humans, but with an overall lower gain when tracking higher-speed, moving targets. This difference may reflect measurement of marmosets under conditions of head restraint. Like other smaller New World primates such as the squirrel monkey, marmosets typically rely on head movements in orienting and would likely use head movements in pursuit under natural conditions [303]. Nonetheless, it is possible to observe a wide range of interesting eye movement behaviors with head-restrained marmosets. Fig. 22.7A shows

FIGURE 22.7 Marmoset scan paths during the viewing of complex natural images. (A) Scan paths of a marmoset viewing two images for 20s each, with one image depicting a conspecific and the other a macaque. In both cases, gaze is systematically directed toward the body, and particularly toward the face. (B) Comparison of scan paths between a marmoset and macaque viewing the same scene consisting of baboons. Eye position is indicated by yellow lines with red points marking fixations longer than 200 ms. *Adapted from Mitchell JF, Reynolds JH, Miller CT. Active vision in marmosets: a model system for visual neuroscience. J Neurosci 2014;34(4):1183–94.*

typical scan patterns of marmosets viewing a static, natural scene, including the frequent shifts in gaze position induced by saccadic eye movements (yellow lines) punctuated by extended periods of fixation (red circles). The analysis of scan patterns to many such natural stimuli revealed a number of gaze characteristics that closely resemble those demonstrated previously in macaques [285]. First, the general mechanics of the saccades, including, for example, the well-described relationship between peak velocity and amplitude, was closely matched between the species. Second, other aspects of free gaze behavior, such as fixation duration and mean intersaccade interval, were broadly similar, albeit at a somewhat faster pace in the marmoset as compared to the macaque. Third, marmosets, like macaques and humans, target regions of social and biological interest, and particularly faces [285,304].

In one experiment, Mitchell et al. [285] showed both macaques and marmosets the same natural images while recording gaze behavior. Aside from the many similarities, there were also certain differences of comparative interest. One difference relates to the overall spatial distribution of saccades. Because the inertial weight of the smaller marmoset's head is low compared to larger primates, marmosets rely more on head movements to redirect their gaze than do macaques and humans, a feature shared by squirrel monkeys [303,305]. While head-restrained macaques tend to make large saccadic eye movements, head-restrained marmosets seldom make saccades that deviate more than 10° from the head-defined center of view. Similarly restricted saccade behavior has been reported previously in the squirrel monkey [303,305]. Another difference that is apparent in the example in Fig. 22.7B is the reluctance of macaques to look directly at certain faces. Mitchell and colleagues noticed a trend across images that macaque subjects spent less time looking at faces, particularly for those faces in close view or that were directed at the camera. As mentioned in the section Social Behavior, the social rules of primate gaze behavior are strongly influenced by the fact that one animal is able to monitor and enforce the gaze behavior of another. This difference may thus have its origins in deeply ingrained macaque social rules that restrict extended direct gaze toward faces. In macaques, direct gaze can be seen as a challenge or threat toward social dominance, particularly if directed toward an individual higher in the dominance hierarchy [306]. Macaques frequently monitor dominant animals not through direct gaze but instead using covert visual attention. The observation that marmosets are less inhibited to directly foveate the faces of large monkeys in images such as in Fig. 22.7B may in part reflect the fact that gaze aversion and covert attention is not a pronounced feature of marmoset social behavior.

Understanding marmoset gaze behavior under head restraint has practical considerations for electrophysiological and functional imaging experiments. It is clear that marmosets are less willing than macaques to maintain fixation on a small point for many seconds at a time. Nonetheless, Mitchell et al. [285] trained them to reliably perform several tasks involving fixation. In one task, they were required to look repeatedly at a small fixation point for up to 2 seconds while ignoring other stimuli flashed peripherally onto the screen. In a second task, the animals readily performed an orientation discrimination task by directing their eyes to the stimulus having an orientation that differed from an array of identical distracters of the same orientation. This combination of tasks provides the necessary conditions for controlled electrophysiological testing in behaving marmosets, including the initial calibration of video-based eye tracking, the presentation of a range of visual stimuli, and the comparison with behavioral decisions. The last of these is particularly important. In the orientation discrimination task, psychometric functions, along with each animal's orientation discrimination threshold, were reliably measured and appeared similar to data previously reported from macaques [307]. Interestingly, the marmosets' behavior was much more enduring in the discrimination task than the ostensibly simpler visual fixation task, suggesting that marmosets perform better when tasks do not require focused inhibitory control of overt orienting responses.

Although this early behavioral study is likely to represent a lower bound on marmoset task performance that will improve with time, it does appear that marmosets require greater effort to suppress overt motor and orienting responses. This difference from the macaque is likely to be a real one and may reflect one cost of having a small brain. Brain functionality varies with absolute brain size, even among closely related species [308]. It is also the case that of those areas which have expanded most in total brain proportion from New to Old World primates, and then again from Old World primates to humans, include those prefrontal areas that are involved in executive and inhibitory control [170]. It is thus possible that the focused and controlled behavior of macaques during a fixation task stems from a proportionally large number of neurons in the prefrontal cortex or elsewhere, which are able to exert downstream inhibitory control. This same type of control contributing to a fixation task may also explain their natural use of covert attention in the social group. The cerebral cortex of the marmoset, having fewer such projections, may simply be unable to exert sustained inhibitory control for similar periods. In summary, while much still remains unknown about marmoset behavior under constrained conditions, it is clear that they display similarities and differences with the macaque. Their active and exploratory viewing of natural scenes resembles macaques in most respects. They can also be trained for repeated, brief fixations and to perform visual discriminations, albeit for relatively short periods at a time.

Visual Social Behavior

Marmosets are a highly social and hierarchical species in which individuals are continually interacting with members of their family groups [309]. Unlike Old World species, marmosets use cooperative breeding strategies in which the father and older siblings participate in rearing young, which in some ways more closely resemble the socialization in human families [310]. Some of this interaction relies on complex visual analysis, from the reading of faces and bodily postures to the observational learning of adult foraging skills by juvenile observers [311]. While it is now clear that marmosets routinely direct their gaze to faces and bodies, even under artificial experimental conditions [285], the specific types of information that they utilize based on visual inspection of one another remain an open area of research.

An important category of such information is the following of another's orientation and gaze direction, which is a capacity that has been shown in a handful of species [312]. The ability to follow the gaze of one's conspecifics can place one at an advantage for identifying locations, events, and objects of interest, for example, related to food and predators. Marmosets appear to exhibit some gaze-following behavior, but to date this capacity has only been demonstrated with respect to human gaze cues [279,313]. While more research needs to be done on the marmoset in this regard, it is notable that the ring-tailed lemur, a prosimian primate, interprets and uses conspecific gaze direction to orient its own gaze during natural visual behavior [314]. Some known features of marmoset group behavior suggest indirectly that they are able to closely monitor and follow certain types of conspecific orienting. For example, when faced with a visually identified predator, members of a family synchronize their actions through a mobbing behavior that involves facing the threat and engaging in local *tsik* vocalizations [315]. In fact, the eye traces in the right panel of Fig. 22.7A are at least suggestive of gaze following in the marmoset because they appear to show a high density of fixations in the region where the macaque in the picture is looking, though more experiments clearly need to be done.

For animals such as primates that use their vision as their main social sense, faces and bodily actions have the capacity to convey a wide range of important information (for a review, see Ref. [173]). For example, the visual determination of individual identity allows animals living in a hierarchical group to track one anothers' actions and interactions from a distance. Unlike other forms of identification, such as the reciprocal acoustic phee calling used by marmosets [316], visual identification does not require participation of the subject being identified. While it may seem intuitive that most primate species are able to visually recognize their conspecifics as individuals, direct evidence has only been demonstrated in a few species, including Old World macaques and New World capuchin monkeys, each of which is more skilled at discriminating conspecific than heterospecific faces [317]. Visual recognition of individuals has not been tested systematically in the marmoset, though it is now well established that this species recognizes individuals through their vocalizations [316].

It is also an open question whether marmosets systematically use facial expressions for social signaling. Like other primates, marmosets have elaborated facial musculature to support a range of expressions exceeding what is possible in most mammals [284,318,319]. While their use of faces is less conspicuous than that of macaques and humans, some evidence suggests that they produce and perceive certain types of expressions. For example, marmosets and closely related tamarins use their tongues in a sociosexual context as a visual display signal [320]. They also respond to video-recorded facial expressions of fear, disgust, and pleasure of their cage mates [284]. In this sense, marmosets appear similar to other primates [321] and some other nonprimate mammals [150,322] in readily interpreting coarse facial expressions. Whether this ability extends to more nuanced aspects of facial signaling is not known. However, some observations, such as the interest in viewing faces [285] and the apparent specialization of several regions in their inferotemporal cortex for faces [268,270], suggest that there may be more to be discovered. Again the pattern of gaze is informative. Close inspection of Fig. 22.7A reveals that gaze is primarily directed to internal facial features, which are likely to contain information about identity and affect, rather than to species-identifying features such as the ear tufts.

Visually Guided Manual Behavior

As discussed in the section Visually Guided Reaching and Grasping, primates are unique among mammals in their use of vision to guide reaching and grasping. Marmosets often perform such actions from an upright posture, standing on their hindlimbs, and with their eyes binocularly focused on targets (see Fig. 22.8). In experimental testing, marmosets can adapt this behavior to a variety of tasks and conditions, demonstrating great postural control and problem solving. For example, from a variety of bodily postures they learn quickly to steer their hand through holes and around obstacles to

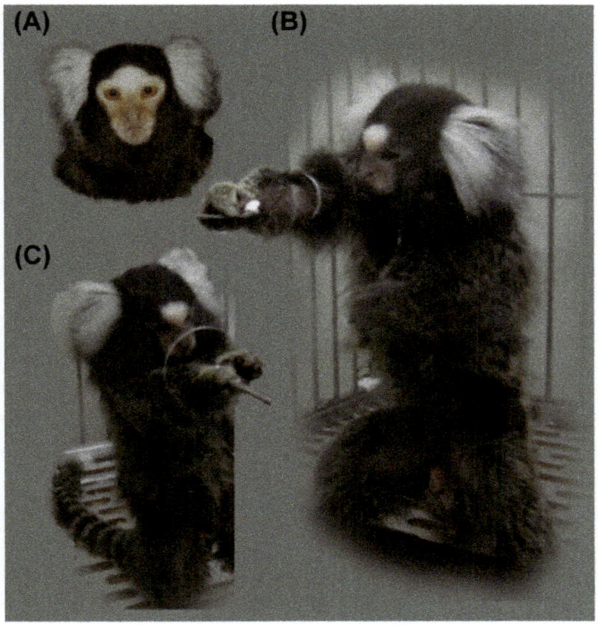

FIGURE 22.8 Visually guided reaching in the marmoset. (A) Frontally positioned eyes allow for stereoscopic near vision typical of primates. (B and C) Reaching movements in marmosets are often performed from an upright posture. Vision is used to guide and preshape the hand or hands based on the features of the target object. *Adapted from Hashimoto T, Yamazaki Y, Iriki A. Hand preference depends on posture in common marmosets. Behav Brain Res 2013;248:144–50.*

grasp food rewards [283]. Their capacity to make detour reaches around visible obstacles is adversely affected by damage to the prefrontal cortex [324,325]. While individual marmosets may show right or left hand preferences, no dominant hand preference is evident across the wider population [283,326,327]. For those with hand preferences, the preferred hand often depends on their postural position [283,323]. Marmosets have also been taught to perform memory-guided sequences of reaching movements to positions on a touchscreen [328] and to perform reaches to eccentric positions while under head and body restraint [329].

Marmosets can also learn to perform sophisticated visually guided actions using tools, such as a special rake with which they can obtain food placed beyond their reach [330]. Marmosets probably preshape their hand to facilitate grasping of objects, though this has not been tested in detail. Hand preshaping in macaques exploits three-dimensional structure, orientation, texture, and movement of a target object based on visual cues [331]. The cotton-top tamarin, a close Callitrichidae relative of the marmoset, preshapes its hand based on the orientation of an object it is about to grasp [332]. However, it is important to point out that despite marmosets' capacity for this type of primate-unique manual behavior, their skills in this domain are much less than macaques, which are in turn less than humans. For example, in learning to use a rake to reach food, marmosets required more than an order of magnitude more trials to achieve the same level of proficiency as Japanese macaques (*Macaca fuscata*) (>7000 trials in marmosets vs. ~500 trials in macaques) [330]. Regarding their dexterity, a key difference from macaques is that marmosets do not have a precision grip. Old World primates have opposable thumbs that allow them to grasp small objects between the thumb and fingers. The precision grip may have been critical for important advances involving fine motor behavior, including tool use among apes and humans [333]. This ability is absent among most New World monkeys, including the marmoset, who make a "power" grip in which all of the fingers close in one sweep.

From a comparative standpoint, it is worth noting that at least one New World monkey, the Cebus or capuchin monkey, has a degree of manual control that includes a precision grip which rivals macaques [334]. This skill is considered to have evolved independently in the Cebus, particularly since the specific changes in hand morphology used to achieve the fine motor actions differ distinctly from those of Old World monkeys [335–338]. This example of convergent evolution in the motor system is thought to have arisen from adaptations to the hand, fitting well with a major theme of this chapter that peripheral adaptations can strongly impact the organization and function of the brain. What is interesting about this example is that changes in the hand, and the capacity for precision grip, profoundly affected the grasping-related regions of Cebus brain. In fact, the posterior parietal cortex of the Cebus bears a much stronger resemblance to that of the macaque than to phylogenetically closer New World monkeys [337]. In addition to a major expansion of parietal and frontal areas related to grasping [337,339,340], the Cebus brain is able to control fine manual movements through direct cortical projections to the ventral spinal cord. For example, like the macaque, but unlike other New World primates that have been studied, Cebus monkeys have large Betz cells in layer 5 of what is sometimes termed "new" M1 that synapse onto primary motor neurons in the lateral ventral horn that controls the digits [341,342]. This control may reflect a more general principle of mammalian brain organization, that as the motor cortex grows relative to the size of the spinal cord, it commands more synaptic territory in the ventral horn [343]. This example also underscores the point that peripheral adaptations, in altering the brain, can also influence cognitive behavior. For example, the Cebus monkey has an ability to use tools in acquiring food, which appears to be much greater than the marmoset [344]. Thus for the study of the neural mechanism of fine motor control, the Cebus is likely to be a considerably better model for human manual control than the marmoset, despite their sharing the same most recent common ancestor with humans.

Summary of Marmoset Versus Macaque Vision

As distinct species of the same mammalian order, the marmoset and macaque clades can be considered rather different variations of a common primate theme. The differences between their brains, and the corresponding behaviors, provide an interesting and important parallax for understanding human brain function. In the case of basic visual perception, the two species are likely very similar. Marmosets have a slightly lower visual acuity, most are dichromats, and their use of vision in social contexts may differ in several subtle ways. Other aspects of their brain and behavior, such as the fine control of the hand, differ considerably from macaques. The most notable physical aspect of the marmoset is its size. For a primate, it is atypically small and in that sense adapted in a manner very different from macaques and particularly from humans. This also affords several new experimental possibilities, which we consider in the following section.

OPPORTUNITIES AFFORDED BY THE MARMOSET MODEL FOR VISUAL NEUROSCIENCE

In this final section, we highlight areas where the marmoset model offers theoretical and practical advantages for visual neuroscience. While there are many useful similarities between the macaque and the marmoset, the two primate species are adapted very differently in their size and in multiple aspects of their behavior. While phylogenetic proximity to humans has made the macaque the model of choice in providing analogy to the human brain, certain features of the marmoset brain, breeding, and behavior offer new opportunities for modern experimental neurobiology.

Comparative Neurobiology

Advancing the marmoset as a second major nonhuman primate model has inherent value simply because it offers an additional perspective on the complex relationship between brain organization and behavior. Throughout this chapter, we have emphasized points of comparison between species that inform our knowledge of brain organization. Since opinions differ widely on the importance of such a comparative perspective, we will discuss this issue briefly here, including why the marmoset is a particularly valuable complement to the macaque.

Though often overlooked in modern neuroscience curricula, it is undeniable that information gained from studying multiple species has always shaped our understanding of the human brain. It may be useful to consider a thought experiment: consider what might have happened if, from the beginning, the macaque was the *only* species studied. In this scenario, a major challenge would be to determine which of the innumerable observable details of macaque's brain are important for understanding its function. For example, how critical is the lamination of the LGN for visual perception? How important is the lamination of the cerebral cortex? Are the specific positions of the sulci important? What about cytochrome oxidase features or orientation columns in V1 or anatomical projections from the visual cortex to the SC and pulvinar? Are alpha rhythms, gamma synchrony, or extraclassical receptive field details essential aspects of neural function? For any species, a comprehensive list of brain features is limitless and continues to expand with the availability of new measurement and analytical methods.

Comparative studies have the capacity to focus researchers' attention on details of brain organization that are most likely to be important for the behavior of the species. For example, owing to comparative studies, we know that the six-layer cerebral cortex is a feature that is present in all mammals and is a very important conserved aspect of the brain; however, the lamination of the LGN is variable even among primates and probably less important. The positions of the sulci do not contribute much to our mechanistic understanding of brain function. For some of the questions raised above, and many more, the answers are still unknown and can actively benefit from comparative research. Sometimes a single observation in a different species is sufficient to call into question long-standing beliefs. We mentioned one such example earlier, in which recordings from the dichromatic marmosets and monochromatic prosimian primates led researchers to conclude that the fundamental function of the LGN parvocellular pathway is not the support of perception of color but rather of detail. We also mentioned a second example, in which the observed absence of V1 orientation columns in nocturnal and diurnal rodent species, despite abundant orientation-tuned neurons, overturned notions that these two features are necessarily coupled through the columnar architecture of the cerebral cortex. These types of comparative observations from decades past, though largely forgotten, have worked their way into our common understanding of what is important in the primate brain.

The marmoset is a strategic counterpoint to the macaque for studying the human brain, in part because these three primate species are adapted so *differently*. The small marmoset brain is on the extreme opposite end of the spectrum compared to humans, while still being closely related. Thus the features that it shares with humans are likely to be essential, shared features of diurnal simian primates, and can be clearly distinguished from other smaller-brained species of comparable size. Similarly, the ways in which marmoset behavior differs from the macaque and human may cast light on specific neural mechanisms critical to support them. For example, differences in visually guided manual behavior or sustained inhibitory control can be linked to specific changes in neural circuits that have taken place along the marmoset, macaque, or human branches since their common ancestry. Finally, in certain domains the behavior of marmosets is quite similar to humans, particularly in aspects of social behavior such as the biparental rearing of offspring [310]. Here, studying the relevant circuitry and its connection to social behavior in macaques and marmosets may reveal areas in which the marmoset is a better animal model for studying the healthy or diseased human brain.

In drawing conclusions from interspecies comparisons, it is also important to recall that since the most recent common ancestor, which in the case of marmosets, humans, and macaques lived more than 35 MYA, each extant species under consideration has undergone profound evolutionary changes. As such, any particular species such as the macaque will have been shaped in a way that is not representative of other primates, including humans. Focusing only on one particular branch of this division, such as that leading to the macaque, thus has the potential to lead researchers astray in their thinking. Studying brain circuits with the appropriate balance of species provides a guard against falling into this trap. Thus the marmoset, as a small New World species with a well-lain path for breeding and housing, would be a logical choice to complement the macaque model for studying the human brain, even if it were not for the several experimental advantages outlined in the next subsections.

Model for Visual System Development

Just as a comparative perspective provides added depth for understanding principles of brain organization, so does neurodevelopment. Marmosets offer an ease of breeding and postnatal maturation timeline that makes them a good model for studying primate brain development. In the past decade, the mouse has opened new vistas for understanding the brain's various developmental events, including the birth and migration of different types of neurons [345], the transient expression of developmental genes [346], the spatial gradients of secreted morphogens [347], and the membrane expression of guidance molecules [348]. Thousands of molecular mapping experiments, and their assembly into databases such as the Allen Brain Atlas [346], now provide an essential reference for understanding how the brain assembles itself and how alterations of this process might lead to brain dysfunction. This work also appears to be leading, quite remarkably, to the overturning of prevailing ideas about the basic axial layout of the brain, which has been in place since it was put forth by neuroembryologists more than a century ago [349]. This revolution is directly relevant for humans, since there is a strong conservation in the basic sequence of neurodevelopmental events among all mammals [350].

At the same time, many features that distinguish the primate brain from that of other mammals are not well modeled by the mouse and need a robust primate model [351]. In vision, for example, our understanding of the areal layout of the primate cortex, its specialization for high-acuity input, or its regions ostensibly devoted to the processing of faces, would all benefit from a neurodevelopmental perspective. Most of what is known about primate visual system development derives from experiments that preceded the modern molecular era [352,353]. Much could be gained from a renewed focus on primate neurodevelopment, taking advantage of the technological and conceptual advances gleaned from work in mice [354].

Marmosets have qualities that make them excellent candidates to study brain development. Notably, they are, like the cat or ferret, born at a comparatively immature developmental stage relative to macaques. While the corresponding time line has not yet been worked out in great detail as it has in the mouse, their caecal period, or the period between conception and eye opening, indicates that their visual system is considerably less developed at birth compared to macaques or humans [265,355]. As a result, there is an unusually wide postnatal window for studying basic development of the visual system. Recent work exploiting this window has, for example, investigated the transition from the retinopulvinar pathway to the retinogeniculate pathway during the first months of life, as mentioned in the Extrastriate Visual Cortex section [265]. Another recent study highlights the importance of the retinopulvinar pathway early in development for establishing reach to grasp circuits in parietal cortex [356]. Other advantages of the marmoset are a gestation period of 4–5 months, a transition to full sexual maturity in approximately 18 months, and the routine production of twins or triplets in each litter, resulting in the highest birth rate of any simian primate [356a]. These features, combined with an increasing availability of molecular and genetic tools in this species [357,358], raise hopes that the marmoset will soon be a principal model to study human neurodevelopment.

Experimental Advantages of a Lissencephalic Brain

The smooth surface of the marmoset lissencephalic brain offers unique opportunities to study the distribution of activity over cortical areas, uninterrupted or obscured by the sulci present in other species. Nearly all of the marmoset visual cortex sits on the brain's lateral surface, as do important oculomotor areas such as the FEFs. By comparison, many areas of visual cortex in macaques lie buried inside sulci, such as V2 and V3 that sit in the lunate sulcus or area MT that sits at the bottom of the superior temporal sulcus (STS). The absence of sulci in the marmoset allows for functional mapping of visual areas that cannot be easily done in the macaque. For some methods, the sulci and gyri of the macaque brain pose insurmountable challenges. For example, optical imaging methods, which have been highly informative in the study of the primary visual cortex [359,360], rely critically on access to the cortical surface. Most optical imaging experiments in the macaque have been carried out in area V1 [361], with a few studies in dorsal area V4 [362], and other cortical regions on the surface [363–365]. Many areas of prime interest for visual neuroscientists, including V2, V3, V3A, MT, LIP, and several areas in the STS, are currently inaccessible. In the marmoset, each of these areas is on the surface and is therefore open to optical investigation.

The continuity of surface cortex is also critical for certain types of electrophysiological investigation, such as large-scale surface recordings using electrocorticography arrays [366,367] or multiple arrays of penetrating electrodes. Only rarely have large electrode arrays been introduced into sulci [368,369]. In macaques, one well-studied primate area that is particularly affected by its sulcal location is the motion-sensitive area MT. While its functional architecture has been carefully charted using single-electrode mapping methods [262,263], and its role in motion perception has been studied in great detail [370,371], its inaccessibility has prohibited investigation with optical methods and large-scale implanted microelectrode arrays. In contrast, area MT of the marmoset sits directly on the cortical surface, and its activity has recently been studied in the anesthetized animal using large, implanted surface arrays [233]. Of course, it remains unknown to what extent marmosets might perform complicated motion perception tasks under head restraint, though recent work indicates that at least simple tasks should be feasible [285].

The convoluted surface of the macaque brain also poses difficulties for the systematic investigation of layer-specific cortical signals, which is of increasing interest to neurophysiologists. The popular method of current source density analysis, which identifies concentrations of synaptic activity in different cortical layers, makes certain assumptions that almost certainly do not hold when a multicontact linear array of electrodes is not inserted perpendicular to the cortical surface [372]. While it is possible to approach sulcal cortex from different angles in the macaque to apply this method [373], such measurements are more straightforward in the marmoset, where a much larger proportion of the visual cortex lends itself to perpendicular penetration.

Of these opportunities in the marmoset that stem from its smooth brain, optical imaging is likely to benefit the most. In addition to the large-scale mapping described above, the flat and compact cortex of the marmoset is ideally suited for spatially restricted methods, such as two-photon imaging. Two-photon calcium imaging has become a powerful method for studying visual processing at the single-cell level in the mouse, but for other species has also been applied to cats [374] and more recently to macaque area V1 [375]. Two-photon microscopy can resolve individual neurons at different depths (though the maximal achievable depth is currently also one of its limitations [376]). It also can reveal aspects of functional architecture expressed over spatial scales less than 1 mm [375]. Finally, the unchanging geometry of the neural positions relative to fixed blood vessels makes it possible to use two-photon imaging to track neural responses not only within a session but also longitudinally across days [377]. This last point may be critical, since it allows one to apply this method to study learning and plasticity within the marmoset visual system. It is not yet known the extent to which other features of the marmoset, including its small size, thin skull, and more transparent dura, offer additional advantages to two-photon imaging. Recent studies have demonstrated the feasibility of two-photon imaging in the species [378,379]. One study introduced methods for reducing the cytotoxicity caused by calcium indicators using tetracycline-inducible expression and was able to track a neural population for over 100 days [378]. There has also been success in building the first transgenic primate lines to express calcium indicators [380]. Several groups are at present pursuing genetic and molecular approaches for the marmoset (see commentary in *Nature*, Ref. [381]), and thus it seems likely that some of these tools will be available to vision scientists soon.

Prospect of Genetic Manipulation

Marmosets appear on the verge of becoming the consensus model for the creation of primate transgenic lines (see commentary in *Nature*, Ref. [382]).

Proof-of-concept experiments in this species have already demonstrated the feasibility of transgenesis germline transmission, using lentiviral infection of embryos [383]. These initial successes have brought with them a cautious optimism that transgenic marmosets might soon play as important a role in neuroscience research as transgenic mice. Several international groups are now moving forward to build genetic models of human mental disease in the marmoset [381], primarily because of the strong overlap with the human in aspects of its primate-specific brain development, as well as its cognitive and social behavior [384].

Lentiviral methods are limited by the size of gene that can be delivered and the lack of control over the insertion point of base pair sequences into the host DNA. For some diseases such as Parkinson's disease, where there is overexpression of a small mutant gene, a lentivirus-based method can be used to create disease models [385]. However, for more precise or extensive genetic manipulations, advanced gene-editing methods are needed. Several such methods have recently been developed in the mouse, allowing for precise targeting of genes in the genome and ultimately making expression in the cell more reliable. These methods, including transcription activator-like effector nucleases (TALEN) and clustered regularly interspaced short palindromic repeats (CRISPR) systems, do not rely on a viral payload capacity and thus are conducive to the insertion of much larger genes [385]. For disease models of schizophrenia and autism, where an extensive manipulation of genes is envisioned, such methods are of particularly high value. For basic neuroscience, these methods may facilitate the creation of marmoset Cre lines that allows for the restricted expression of a target gene in specific neuronal cell types selected by large promotors, the technology that has proven so successful in the mouse.

How might transgenic marmosets be used to study the neurobiology of vision? One can divide the applications into models of disease and models for basic neurobiology research. With respect to schizophrenia and autism, mouse models are impoverished in their capacity to assay some of the most obvious behavioral symptoms in humans. As one example, these afflictions are associated with abnormal active vision and eye movement behavior, particularly for complex stimuli like faces [386–388], as well as deficits in visual perception more generally [389–392]. As mice use their vision less broadly than primates, and attend to different features of the environment, mouse models are not optimal to study important aspects of these diseases. A marmoset model would allow more direct comparison to the human phenotype, especially as tests for marmoset visual perception continue to improve. Even at present, the analysis of marmosets' eye movements, and their targeting of regions of social interest including faces [285], could provide insights into the link between genes, neural circuits, and behavioral deficits in disease. As a strong model for neurodevelopment, the marmoset may further provide concrete information about what processes fail in neuropsychiatric or neurological disorders, for example, relating to the failure of late migrating cells to deliver secreted signals or to confer the proper balance of inhibition and excitation in cortical circuits [393,394]. The same line of reasoning holds for other mental disorders that involve higher-level cognitive or social processing, where a primate transgenic model might mimic a disease state much better than any mouse model. With regard to basic visual neuroscience research, transgenic marmosets promise to be invaluable for studying the functional anatomy of the brain, its cell type-specific neurophysiology, and its circuit principles.

Opportunities and Challenges in Marmoset Behavior

All primate species share aspects of their behavior, such as their inherent curiosity and use of vision to mediate social interaction. However, there is also great diversity in the primate order, and particularly with respect to apparent cognitive behaviors. While the macaque has proven to be a highly valuable research organism for controlled testing, the marmoset is still relatively unexplored. Under freely moving conditions, marmosets can be trained to perform tasks that tap into diverse aspects of their cognition, as reviewed in the Visual Cognitive Behavior section. However, a fully developed marmoset model for studying mechanisms of visual cognition requires physical restraint amenable to various neural measurements and robust behavioral performance by cooperative marmoset participants. What is clear from even a short period working with marmosets is that their "personality" differs from macaques, as does their work ethic. A well-trained macaque can carry out sequences of hundreds or even thousands of consecutive behavioral trials, stopping to take only a few breaks over a period of several hours. They can navigate a large repertoire of psychophysical and behavior paradigms, learn complex rules, and even perform tasks that require a high level of concentration, such as sustaining attentional focus away from their center of gaze during extended fixation. These are key factors that have propelled macaques to such success as a model species for studying the neural basis of perception, cognition, and behavior.

The capacity of macaques for such a wide range of tasks may stem from the fact that they are highly adaptable under natural conditions, making them the most abundant and successful primate species second only to humans [193]. The natural ecology of marmosets is very different and is closely tied to arboreal living in coherent family groups. Marmosets use vocal communication to constantly monitor the positions of their family members in the thick forest canopy [25,395,396]. They are highly emotional and will act against intruders as a group with a mobbing behavior. As codependent individuals, they easily become anxious when they are separated from their family members. This ecology provides a basis for understanding some aspects of their behavior in the laboratory. Mitchell and colleagues [285] found that the marmoset was able to routinely perform sessions of 700–800 trials over the course of 1–2 h. This is approximately one quarter to one half as many trials as a macaque might perform during analogous testing. However, the finding that marmosets can overcome their natural anxiety associated with removal from their social group and work routinely in visual tasks under restrained conditions was an important step, and perhaps the first of many in using and refining the awake marmoset paradigm.

The most successful tasks were those in which the marmosets were required to actively participate in the task, for example, performing a visual discrimination by making an eye movement to a nonmatching stimulus. That they were worse for tasks that involved only prolonged fixation can be understood by the fact that this is the most unnatural of conditions for them. While the work ethic of the marmoset may not ever match that of the macaque, it is important to point out that two visual studies to date have been able to obtain approximately one hour's worth of data from each of their subjects on average [268,270,285]. It also bears mentioning that new methods of neural data collection may place fewer demands on long sessions, since neural activity can be tracked and accumulated across sessions. This possibility was mentioned briefly in the Experimental Advantages of a Lissencephalic Brain section in the context of two-photon imaging with calcium indicators. Similar longitudinal recordings are also possible with chronic microwire electrodes, which can provide the stable isolation of single units across sessions [397].

An important challenge for the future may be to align testing paradigms more closely with marmosets' natural behavioral repertoire. Marmosets are naturally engaged when allowed to explore visual scenes. Their frequent saccades and selection of particular scene content resembles similar active vision processes in humans [183,184]. It might be seen as ironic that, because macaques are so disciplined in their capacity to fixate a small point, thus allowing for the precise visual stimulation of the retina, we know rather little about how the primate brain receives and processes stimuli during active, natural vision. With each shift of gaze, which typically occurs 2–3 times per second, the primate brain redirects the location of its high-acuity fovea to seek new information about the scene. Through these programmed movements, the visual brain must constantly contend with massive changes in retinal stimulation. Our understanding of how the brain reconciles its sensory input and motor actions under natural conditions sorely lags our understanding of concepts of visual processing obtained from fixed-gaze experiments, such as the receptive field, stimulus selectivity, gain fields, and covert attentional orienting. While the marmoset is, at least for now, a weaker candidate model than the macaque for fixed-gaze tasks, it may be just as strong for "active vision" tasks. This may gradually lead to stronger consideration of these topics by a community eager to exploit the many experimental advantages listed above.

Given the rate of emerging technology, studying active vision may soon be possible in animals whose heads, and maybe even bodies, are entirely free. In the rodent, where head restraint is less common than in monkeys, paradigms involving spatial navigation and active whisking have revealed that during active sensing the brain behaves in a way that differs entirely from passive stimulation. For example, certain rhythmic synchronous activity appears only during active exploration of the environment and is now thought to play important roles for sensory coding or memory [398–400]. In the visual domain, recent study in the New World Cebus monkey found that active sensing through eye movements similarly changed the neural dynamics associated with early visual processing, in this case involving synchronous neural activity time-locked to saccadic events [401]. Another recent study examining visual responses in the macaque hippocampus found that during active sensing, oscillatory activity was evident and predictive of subsequent recognition performance [402]. Marmosets seem a natural primate candidate model for studying the brain under natural conditions. This may be particularly useful in the domain of social neuroscience, where activity in the brain can be studied under relatively natural familial contexts. Such experiments may reveal aspects of brain activity never before observed, but highly relevant to understanding neural circuits in the human.

Experimental Opportunities: Summary

For visual neuroscience experiments, the marmoset and macaque have complementary strengths. In the domain of behavior, marmosets fall short of macaques on some counts, such as in the capacity to direct fixation for extended periods or perform the same task continuously for several hours. However, it is important to point out that the experimental behavioral repertoire is, at present, largely unexplored and we might expect considerable improvements in this domain with further research. The marmoset clearly excels in other areas. Its small lissencephalic brain, its suitability for studying neurodevelopment, and the prospect of creating transgenic lines provide unique avenues with which to study the human brain. The similarity of natural viewing behavior in this species compared to that of humans promises to spur creative new directions for understanding brain function. These factors, combined with the comparative benefits of systematically studying the brain and behavior of a New World monkey species, make the marmoset a worthy investment for visual neuroscience.

CONCLUSIONS

In this chapter we considered marmosets as a potential model for visual neuroscience from a broad range of theoretical and practical perspectives. At the heart of this consideration is a deep appreciation of comparative experimental approaches that help navigate the complexity of the brain through recognition of ancestral and derived traits [1]. The marmoset is first and foremost a primate and thus shares many unique traits with other primates, including humans, particularly in the visual and social domains. Moreover, the organization of the primate brain differs in important ways from its closest mammalian relatives.

A critically important visual adaptation in primates is high-acuity foveal vision, which is associated with specializations of the parvocellular pathway in early vision and the emphasis for foveal processing along the ventral stream into the inferior temporal cortex. Foveal vision has strongly shaped visual cognition, giving us the ability to recognize fine detail and read the expressions and intentions of other individuals from a safe distance. It also determines how primates use their vision, with frequent, rapid eye movements serially positioning the fovea on points of interest. The target of each saccade is often the target for a reaching movement or the face of a conspecific, all of which are critical in a primate's life. The frontal placement of eyes in the primate also affords binocular focusing of the left and right fovea toward objects, enabling depth information to guide the precise use of the hands. Together these unique visual and behavioral primate adaptations are foundations on which much of human cognition and intelligence is built. Though we may learn much from the study of mammals such as the mouse, whose adaptations are very different, it is essential to consider those evolved elements in the brain that so strongly shape our own cognition and are most obviously similar in other primates. The marmoset offers a much needed point of comparison to the growing body of circuit-specific knowledge that is now accumulating in the mouse and the array of anatomical, physiological, and behavioral knowledge that has already been obtained from the highly studied macaque.

There are several areas of inquiry where we believe the marmoset will make an important contribution toward furthering our understanding of vision and primate brain organization. Due to its smaller and geometrically tractable brain, the marmoset is ideal for optical imaging, surface electrode arrays, and laminar recordings. Owing to their ease of breeding, inherent fecundity, and rapid maturation, the marmoset is an excellent primate candidate to take advantage of methods developed in the mouse for molecular manipulation. Advances in transgenesis or viral methods will be valuable for developing disease models and enable more precise methods to manipulate neuronal activity. The relative immaturity of the marmoset at birth is also likely to make it important as a model for studying primate brain development, including vision. The natural visual and social behaviors of marmosets are also highly overlapping with macaques and humans, and there is thus great potential to develop behavioral paradigms that exploit this similarity.

In considering whether to turn their attention to the marmoset brain, researchers are presented with trade-offs. This is true for any animal model, including the macaque and the mouse, and we have discussed many of these trade-offs in this chapter. If modern trends continue, marmosets are poised to provide an important experimental bridge by which advances in the mouse community can be applied to the primate brain. Marmosets are at the forefront of this initiative and are likely to play an important and exciting role for understanding our own brains, including how circuits go awry in disease. In this journey, it is vital that neurobiologist researchers retain a comparative and evolutionary perspective, as history teaches us that the deeper insights about the brain over the long term arise from consideration of these points.

Acknowledgments

We thank Drs. James Bourne, Yogita Chudasama, Greg Field, Rebecca Berman, Lindsey Glickfeld, and Sam Nummela for their comments on an earlier version of this manuscript. This work was supported by R21-MH104756 and the Intramural Research Program of the National Institute of Mental Health.

References

[1] Kaas JH. The evolution of brains from early mammals to humans. Wiley Interdiscip Rev Cogn Sci 2013;4(1):33–45.

[2] Janecka JE, Miller W, Pringle TH, Wiens F, Zitzmann A, Helgen KM, Springer MS, Murphy WJ. Molecular and genomic data identify the closest living relative of primates. Science 2007;318(5851):792–4.

[3] Springer MS, Meredith RW, Janecka JE, Murphy WJ. The historical biogeography of Mammalia. Philos Trans R Soc B Biol Sci 2011;366:2478–502.
[4] Bernstein JG, Boyden ES. Optogenetic tools for analyzing the neural circuits of behavior. Trends Cogn Sci 2011;15(12): 592–600.
[5] Callaway EM. A molecular and genetic arsenal for systems neuroscience. Trends Neurosci 2005;28(4):196–201.
[6] Deisseroth K, Feng G, Majewska AK, Miesenböck G, Ting A, Schnitzer MJ. Next-generation optical technologies for illuminating genetically targeted brain circuits. J Neurosci 2006; 26(41):10380–6.
[7] Cardin JA. Dissecting local circuits in vivo: integrated optogenetic and electrophysiology approaches for exploring inhibitory regulation of cortical activity. J Physiol Paris 2012;106:104–11.
[8] Livet J, Weissman TA, Kang H, Draft RW, Lu J, Bennis RA, Sanes JR, Lichtman JW. Transgenic strategies for combinatorial expression of fluorescent proteins in the nervous system. Nature 2007;450:56–62.
[9] Wilson NR, Runyan CA, Wang FL, Sur M. Division and subtraction by distinct cortical inhibitory networks in vivo. Nature 2012; 488(7411):343–8.
[10] Lee SH, Kwan AC, Zhang S, Phoumthipphavong V, Flannery JG, Masmanidis SC, Taniguchi H, Huang ZJ, Zhang F, Boyden ES, Deisseroth K, Dan Y. Activation of specific interneurons improves V1 feature selectivity and visual perception. Nature 2012;488(7411):379–83.
[11] Zhang S, Xu M, Kamigaki T, Phong Hoang Do J, Chang WC, Jenvay S, Miyamich K, Luo L, Dan Y. Long-range and local circuits for top-down modulation of visual cortex processing. Science 2014;345:660–5.
[12] Huberman AD, Niell CM. What can mice tell us about how vision works? Trends Neurosci 2011;34(9):464–73.
[13] Felleman DJ, Van Essen DC. Distributed hierarchical processing in the primate cerebral cortex. Cereb Cortex 1991;1(1):1–47.
[14] Markov NT, Vezoli J, Chameau P, Falchier A, Quilodran R, Huissoud C, Lamy C, Misery P, Giroud P, Ullman S, Barone P, Dehay C, Knoblauch K, Kennedy H. Anatomy of hierarchy: feedforward and feedback pathways in macaque visual cortex. J Comp Neurol 2014;522:225–59.
[15] Orban GA, Van Essen D, Vanduffel W. Comparative mapping of higher visual areas in monkeys and humans. Trends Cogn Sci 2004;8(7):315–24.
[16] Rudebeck PH, Saunders RC, Prescott AT, Chau LS, Murray EA. Prefrontal mechanisms of behavioral flexibility, emotion regulation and value updating. Nat Neurosci 2013;16(8):1140–5.
[17] Shadlen MN, Kiani R. Decision making as a window on cognition. Neuron 2013;80(3):791–806.
[18] Preuss TM. Taking the measure of diversity: comparative alternatives to the model-animal paradigm in cortical neuroscience. Brain Behav Evol 2000;55(6):287–99.
[19] Harting JK, Huerta MF, Hashikawa T, van Lieshout DP. Projection of the mammalian superior colliculus upon the dorsal lateral geniculate nucleus: organization of tectogeniculate pathways in nineteen species. J Comp Neurol 1991;304(2):275–306.
[20] Nudo RJ, Masterton RB. Descending pathways to the spinal cord: a comparative study of 22 mammals. J Comp Neurol 1988; N277(1):53–79.
[21] Carrion Jr R, Patterson JL. An animal model that reflects human disease: the common marmoset (*Callithrix jacchus*). Curr Opin Virol 2012;2(3):357–62.
[22] Mansfield K. Marmoset models commonly used in biomedical research. Comp Med 2003;53(4):383–92.
[23] Tardif SD, Mansfield KG, Ratnam R, Ross CN, Ziegler TE. The marmoset as a model of aging and age-related diseases. ILAR J 2011;52(1):54–65.
[24] Robbins TW, Roberts AC. Differential regulation of fronto-executive function by the monoamines and acetylcholine. Cereb Cortex 2007;17:151–60.
[25] Miller CT, Mandel K, Wang X. The communicative content of the common marmoset phee call during antiphonal calling. Am J Primatol 2010;72(11):974–80.
[26] Takahashi DY, Narayanan DZ, Ghazanfar AA. Coupled oscillator dynamics of vocal turn-taking in monkeys. Curr Biol 2013;23(21): 2162–8.
[27] Wang X. The harmonic organization of auditory cortex. Front Syst Neurosci 2013;17(7):114.
[28] Solomon SG, Rosa MG. A simpler primate brain: the visual system of the marmoset monkey. Front Neural Circ 2014;8(96):1–24.
[29] Preuss TM. Evolutionary specializations of primate brain organization. In: Ravosa M, Dagosto M, editors. Primate origins. New York: Kluwer; 2007. p. 625–75.
[30] Cartmill M. New views on primate origins. Evol Anthropol 1992; 1:105–11.
[31] Renne PR, Deino AL, Hilgen FJ, Kuiper KF, Mark DF, Mitchell WS, Morgan LE, Mundil R, Smit J. Time scales of critical events around the Cretaceous-Paleogene boundary. Science 2013; 339:684–7.
[32] Fleagle JG. Primate evolution and adaptation. N.Y: Academic Press; 1988.
[33] Ross CF. Adaptive explanation for the origins of the Anthropoidea (primates). Am J Primatol 1996;40:205–30.
[34] Fite KV, Rosenfield-Wessels S. A comparative study of deep avian foveas. Brain Behav Evol 1975;12:97–115.
[35] Kirk EC, Kay RF. The evolution of high visual acuity in the Anthropoidea. In: Ross CF, Kay RF, editors. Anthropoid origins: new visions. New York: Kluwer/Plenum Publishing; 2004. p. 539–602.
[36] Emery NJ, Clayton NS. The mentality of crows: convergent evolution of intelligence in corvids and apes. Science 2004;306:1903–7.
[37] Provis JM, Dubis AM, Maddess T, Carroll J. Adaptation of the central retina for high acuity vision: cones, the fovea and the avascular zone. Prog Retin Eye Res 2013;35:63–81.
[38] Borwein B. Scanning electron microscopy of monkey foveal photoreceptors. Anat Rec 1983;205(3):363–73.
[39] Snodderly DM, Weinhaus RS, Choi JC. Neural-vascular relationships in central retina of macaque monkeys (*Macaca fascicularis*). J Neurosci 1992;12(4):1169–93.
[40] Franco ECS, Finlay BL, Silveira LCL, Yamada ES, Crowley JC. Conservation of absolute foveal area in New World monkeys. Brain Behav Evol 2000;56:276–86.
[41] Wikler KC, Rakic P. Distribution of photoreceptor subtypes in the retina of diurnal and nocturnal primates. J Neurosci 1990;10(10): 3390–401.
[42] Langston A, Casagrande VA, Fox R. Spatial resolution of the Galago. Vis Res 1986;26:791–6.
[43] Boycott BB, Dowling JE. Organization of the primate retina: light microscopy. Philos Trans R Soc B (Lond) 1969;255:109–84.
[44] Schein SJ. Anatomy of macaque fovea and spatial densities of neurons in foveal representation. J Comp Neurol 1988;269(4): 479–505.
[45] Wässle H, Grünert U, Röhrenbeck J, Boycott BB. Retinal ganglion cell density and cortical magnification factor in the primate. Vis Res 1990;30(11):1897–911.
[46] Wilder HD, Grünert U, Lee BB, Martin PR. Topography of ganglion cells and photoreceptors in the retina of a New World monkey: the marmoset *Callithrix jacchus*. Vis Neurosci 1996;13(2): 335–52.
[47] Goodchild AK, Ghosh KK, Martin P. A comparison of photoreceptor spatial density and ganglion cell morphology in the retina of human, macaque monkey, cat, and the marmoset *Callithrix jacchus*. J Comp Neurol 1996;366:55–75.

[48] Kilavik BE, Silveira LC, Kremers J. Spatial receptive field properties of lateral geniculate cells in the owl monkey (*Aotus azarae*) at different contrasts: a comparative study. Eur J Neurosci 2007;26(4):992–1006.

[49] Yamada ES, Marshak DW, Silveira LC, Casagrande VA. Morphology of P and M retinal ganglion cells of the bush baby Vis Res 1998;38(21):3345–52.

[50] Yamada ES, Silveira LC, Perry VH, Franco EC. M and P retinal ganglion cells of the owl monkey: morphology, size and photoreceptor convergence. Vis Res 2001;41:119–31.

[51] Ross CF. Into the light: the origin of Anthropoidea. Annu Rev Anthropol 2000:147–94.

[52] Jacobs GH. Primate color vision: a comparative perspective. Vis Neurosci 2008;25:619–33.

[53] Jacobs GH, Deegan JF. Uniformity of colour vision in Old World monkeys. Proc Biol Sci 1999;266:2023–8.

[54] Sharpe LT, Stockman A, Jaegle H, Nathans J. Opsin genes, cone pigments, color vision, and color blindness. In: Gegenfurtner KR, Sharpe LT, editors. Color vision: from genes to perception. Cambridge, UK: Cambridge University Press; 1999.

[55] Jacobs GH, Neitz J, Neitz M. Genetic basis of polymorphism in the color vision of platyrrhine monkeys. Vis Res 1993;33:269–74.

[56] Nathans J, Thomas D, Hogness DS. Molecular genetics of human color vision: the genes encoding blue, green, and red pigments. Science 1986;232:193–202.

[57] Surridge AK, Osorio D, Mundy NI. Evolution and selection of trichromatic vision in primates. Trends Ecol Evol 2003;18: 198–205.

[58] Dominy NJ, Lucas PW. Ecological importance of trichromatic vision to primates. Nature 2001;410:363–6.

[59] Mollon JD. "Tho' she kneel'd in that place where they grew..." the uses and origins of primate colour vision. J Exp Biol 1989; 146:21–38.

[60] Regan BC, Julliot C, Simmen B, Vienot F, Charles-Dominique P, Mollon JD. Fruits, foliage and the evolution of primate colour vision. Philos Trans R Soc Lond B Biol Sci 2001;356:229–83.

[61] Mundy NI. Genetic basis of olfactory communication in primates. Am J Primatol 2006;68:559–67.

[62] Gilad Y, Wiebe V, Przeworski M, Lancet D, Paabo S. Loss of olfactory receptor genes coincides with the acquisition of full trichromatic vision in primates. PLoS Biol 2004;2:E5.

[63] Liman ER, Innan H. Relaxed selective pressure on an essential component of pheromone transduction in primate evolution. Proc Natl Acad Sci USA 2003;100:3328–32.

[64] Zhang J, Webb DM. Evolutionary deterioration of the vomeronasal pheromone transduction pathway in catarrhine primates. Proc Natl Acad Sci USA 2003;100:8337–41.

[65] Liman ER. Use it or lose it: molecular evolution of sensory signaling in primates. Pflugers Arch 2006;453:125–31.

[66] Rosa MG, Schmid LM, Krubitzer LA, Pettigrew JD. Retinotopic organization of the primary visual cortex of flying foxes (*Pteropus poliocephalus* and *Pteropus scapulatus*). J Comp Neurol 1993;335: 55–72.

[67] Parker AJ. Binocular depth perception and the cerebral cortex. Nat Rev Neurosci 2007;8(5):379–91.

[68] Changizi MA, Shimojo S. "X-ray vision" and the evolution of forward-facing eyes. J Theor Biol 2008;254(4):756–67.

[69] Pettigrew JD. Evolution of binocular vision. Vis Neurosci 1986: 208–22.

[70] Dacey DM. 20 Origins of perception: retinal ganglion cell diversity and the creation of parallel visual pathways. Cognit Neurosci 2004;281.

[71] Ringach DL, Shapley RM, Hawken MJ. Orientation selectivity in macaque V1: diversity and laminar dependence. J Neurosci 2002; 22(13):5639–51.

[72] Schiller PH, Finlay BL, Volman SF. Quantitative studies of single-cell properties in monkey striate cortex. II. Orientation specificity and ocular dominance. J Neurophysiol 1976;39(6): 1320–33.

[73] Hubel DH, Wiesel TN. Receptive fields and functional architecture of monkey striate cortex. J Physiol 1968;195:215–43.

[74] Kaas JH. Evolution of columns, modules, and domains in the neocortex of primates. Proc Natl Acad Sci 2012;109:10655–60.

[75] Huerta MF, Harting JK. The mammalian superior colliculus: studies of its morphology and connections. In: Vanegas H, editor. Comparative neurology of the optic tectum. New York: Plenum; 1984. p. 687–773.

[76] Rodieck RW, Watanabe M. Survey of the morphology of macaque retinal ganglion cells that project to the pretectum, superior colliculus, and parvicellular laminae of the lateral geniculate nucleus. J Comp Neurol 1993;338:289–303.

[77] Schiller PH, Malpeli JG. Properties and tectal projections of monkey retinal ganglion cells. J Neurophysiol 1977;40:428–45.

[78] Rosa MG, Schmid LM. Topography and extent of visual-field representation in the superior colliculus of the megachiropteran Pteropus. Vis Neurosci 1994;11:1037–57.

[79] Kaas JH, Huerta MF. The subcortical visual system of primates. In: Steklis HD, Erwin J, editors. Comparative primate biology. Neurosciences, vol. 4. New York: Alan R. Liss; 1988. p. 327–91.

[80] Allman JM. Evolution of the visual system in early primates. In: Sprague JM, Epstein AM, editors. Progress in psychobiology and physiological psychology. New York: Academic Press; 1977. p. 1–53.

[81] Pettigrew JD. Flying primates? Megabats have the advanced pathway from eye to midbrain. Science 1986b;231:1304–6.

[82] Thiele A, Vogelsang M, Hoffmann KP. Pattern of retinotectal projection in the megachiropteran bat Rousettus aegyptiacus. J Comp Neurol 1991;314:671–83.

[83] Murphy WJ, Eizirik E, O'brien SJ, Madsen O, Scally M, Douady CJ, et al. Resolution of the early placental mammal radiation using Bayesian phylogenetics. Science 2001;294:2348–51.

[84] Casagrande VA, Norton TT. Lateral geniculate nucleus: a review of its physiology and function. In: Leventhal A, editor. The neural basis of visual function: vision and visual dysfunction. London: McMillan Press; 1991. p. 41–84.

[85] Casagrande VA, Khaytin I, Boyd J. The evolution of parallel visual pathways in the brains of primates. In: Preuss TM, Kaas J, editors. Evolution of the nervous system, vol. 4; 2006. p. 87–108.

[86] Nassi JJ, Callaway EM. Parallel processing strategies of the primate visual system. Nat Rev Neurosci 2009;10(5):360–72.

[87] Kaas JH, Huerta MF, Weber JT, Harting JK. Patterns of retinal terminations and laminar organization of the lateral geniculate nucleus of primates. J Comp Neurol 1978;182:517–53.

[88] Itoh K, Conley M, Diamond IT. Retinal ganglion cell projections to individual layers of the lateral geniculate body in Galago crassicaudatus. J Comp Neurol 1982;205(3):282–90.

[89] Diamond IT, Conley M, Itoh K, Fitzpatrick D. Laminar organization of geniculocortical projections in *Galago senegalensis* and *Aotus trivirgatus*. J Comp Neurol 1985;242(4):584–610.

[90] Kaas JH, Guillery RW, Allman JM. Some principles of organization in the dorsal lateral geniculate nucleus. Brain Behav Evol 1972;6:253–99.

[91] Enroth-Cugell C, Robson JG. The contrast sensitivity of retinal ganglion cells of the cat. J Physiol 1966;187:517–52.

[92] Van Hooser SD, Heimel JA, Nelson SB. Functional cell classes and functional architecture in the early visual system of a highly visual rodent. Prog Brain Res 2005;2005(149):127–45.

[93] Wilson PD, Rowe MH, Stone J. Properties of relay cells in cat's lateral geniculate nucleus: a comparison of W-cells with X- and Y-cells. J Neurophysiol 1976;39(6):1193–209.

[94] Lennie P, Perry VH. Spatial contrast sensitivity of cells in the lateral geniculate nucleus of the rat. J Physiol 1981;315:69–79.

[95] Grubb MS, Thompson ID. Quantitative characterization of visual response properties in the mouse dorsal lateral geniculate nucleus. J Neurophysiol 2003;90:3594–607.

[96] Piscopo DM, El-Danaf RN, Huberman AD, Niell CM. Diverse visual features encoded in mouse lateral geniculate nucleus. J Neurosci 2013;33:4642–56.

[97] Dräger UC. Autoradiography of tritiated proline and fucose transported transneuronally from the eye to the visual cortex in pigmented and albino mice. Brain Res 1974;82(2):284–92.

[98] Harting JK, Huerta MF. The geniculostriate projection in the grey squirrel: preliminary autoradiographic data. Brain Res 1983; 272(2):341–9.

[99] Hughes HC. Anatomical and neurobehavioral investigations concerning the thalamo-cortical organization of the rat's visual system. J Comp Neurol 1977;175(3):311–36.

[100] Kirby MA, Wilson PD. Receptive field properties and latencies of cells in the lateral geniculate nucleus of the North American opossum (*Didelphis virginiana*). J Neurophysiol 1986;56:907–33.

[101] Luo ZX, Yuan CX, Meng QJ, Ji Q. A Jurassic eutherian mammal and divergence of marsupials and placentals. Nature 2011;476: 442–5.

[102] Malpeli JG, Lee D, Baker FH. Laminar and retinotopic organization of the macaque lateral geniculate nucleus: magnocellular and parvocellular magnification functions. J Comp Neurol 1996;375(3):363–77.

[103] Carcieri SM, Jacobs AL, Nirenberg S. Classification of retinal ganglion cells: a statistical approach. J Neurophysiol 2003;90(3): 1704–13.

[104] Bullier J, Norton TT. Comparison of receptive-field properties of X and Y ganglion cells with X and Y lateral geniculate cells in the cat. J Neurophysiol 1979;42(1):274–91.

[105] Derrington AM, Fuchs AF. Spatial and temporal properties of X and Y cells in the cat lateral geniculate nucleus. J Physiol 1979; 293:347–64.

[106] Shapley R, Hugh Perry V. Cat and monkey retinal ganglion cells and their visual functional roles. Trends Neurosci 1986;9: 229–35.

[107] Martin PR, Blessing EM, Buzás P, Szmajda BA, Forte JD. Transmission of colour and acuity signals by parvocellular cells in marmoset monkeys. J Physiol 2011;589(Pt 11):2795–812.

[108] Blasdel GG, Lund JS. Termination of afferent axons in macaque striate cortex. J Neurosci 1983;3(7):1389–413.

[109] Blasdel GG, Fitzpatrick D. Physiological organization of layer 4 in macaque striate cortex. J Neurosci 1984;4(3):880–95.

[110] Hubel DH, Wiesel TN. Laminar and columnar distribution of geniculocortical fibers in macaque monkey. J Comp Neurol 1972;146:421–50.

[111] Hendry SH, Reid RC. The koniocellular pathway in primate vision. Annu Rev Neurosci 2000;23:127–53.

[112] Dreher B, Cottee LJ. Visual receptive-field properties of cells in area 18 of cat's cerebral cortex before and after acute lesions in area 17. J Neurophysiol 1975;38(4):735–50.

[113] Olavarria J, Torrealba F. The effect of acute lesions of the striate cortex on the retinotopic organization of the lateral peristriate cortex in the rat. Brain Res 1978;151(2):386–91.

[114] Funk AP, Rosa MG. Visual responses of neurones in the second visual area of flying foxes (Pteropus poliocephalus) after lesions of striate cortex. J Physiol 1998;513(Pt 2):507–19.

[115] Killackey H, Snyder M, Diamond IT. Function of striate and temporal cortex in the tree shrew. J Comp Physiol Psychol 1971;74(2): 1–29.

[116] Killackey H, Wilson M, Diamond IT. Further studies of the striate and extrastriate visual cortex in the tree shrew. J Comp Physiol Psychol 1972;81:45–63.

[117] Freund TF, Martin KA, Whitteridge D. Innervation of cat visual areas 17 and 18 by physiologically identified X- and Y- type thalamic afferents. I. Arborization patterns and quantitative distribution of postsynaptic elements. J Comp Neurol 1985;242:263–74.

[118] Wong P, Kaas JH. Architectonic subdivisions of neocortex in the gray squirrel (*Sciurus carolinensis*). Anat Rec (Hoboken) 2008;291: 1301–33.

[119] Wong P, Kaas JH. An architectonic study of the neocortex of the short-tailed opossum (*Monodelphis domestica*). Brain Behav Evol 2009;73:206–28.

[120] Conley M, Fitzpatrick D, Diamond IT. The laminar organization of the lateral geniculate body and the striate cortex in the tree shrew (*Tupaia glis*). J Neurosci 1984;4:171–97.

[121] Van Hooser SD, Roy A, Rhodes HJ, Culp JH, Fitzpatrick D. Transformation of receptive field properties from lateral geniculate nucleus to superficial V1 in the tree shrew. J Neurosci 2013; 33(28):11494–505.

[122] Humphrey AL, Albano JE, Norton TT. Organization of ocular dominance in tree shrew striate cortex. Brain Res 1977;134(2): 225–36.

[123] Horton JC, Hocking DR. Anatomical demonstration of ocular dominance columns in striate cortex of the squirrel monkey. J Neurosci 1996;16:5510–22.

[124] Casagrande VA, Boyd JD. The neural architecture of binocular vision. Eye 1996;10(Pt 2):153–60.

[125] Adams DL, Horton JC. Ocular dominance columns: enigmas and challenges. Neuroscientist 2009;15:62–77.

[126] Collins CE, Airey DC, Young NA, Leitch DB, Kaas JH. Neuron densities vary across and within cortical areas in primates. Proc Natl Acad Sci USA 2010;107(36):15927–32.

[127] Carlo CN, Stevens CF. Structural uniformity of neocortex, revisited. Proc Natl Acad Sci USA 2013;110(4):1488–93.

[128] Rockel AJ, Hiorns RW, Powell TP. The basic uniformity in structure of the neocortex. Brain 1980;103(2):221–44.

[129] Lund JS. Excitatory and inhibitory circuitry and laminar mapping strategies in primary visual cortex of the monkey. In: Edelmann GM, Gall WE, Cowan WM, editors. Signal and sense: local and global order in perceptual maps. New York, NY: John Wiley and Sons; 1990. p. 51–66.

[130] Peters A, Kara DA. The neuronal composition of area 17 of rat visual cortex. I. The pyramidal cells. J Comp Neurol 1985;234(2):218–41.

[131] Woolsey TA, Van der Loos H. The structural organization of layer IV in the somatosensory region (SI) of mouse cerebral cortex. The description of a cortical field composed of discrete cytoarchitectonic units. Brain Res 1970;17(2):205–42.

[132] Takahata T, Komatsu Y, Watakabe A, Hashikawa T, Tochitani S, Yamamori T. Activity-dependent expression of occ1 in excitatory neurons is a characteristic feature of the primate visual cortex. Cereb Cortex 2006;16(7):929–40.

[133] Takahata T, Shukla R, Yamamori T, Kaas JH. Differential expression patterns of striate cortex-enriched genes among Old World, New World, and prosimian primates. Cereb Cortex 2012;22(10): 2313–21.

[134] Horton JC, Hubel DH. Regular patchy distribution of cytochrome oxidase staining in primary visual cortex of macaque monkey. Nature 1981;292(5825):762–4.

[135] Federer F, Ichida JM, Jeffs J, Schiessl I, McLoughlin N, Angelucci A. Four projection streams from primate V1 to the cytochrome oxidase stripes of V2. J Neurosci 2009;29:15455–71.

[136] Boyd JD, Matsubara JA. Laminar and columnar patterns of geniculocortical projections in the cat: relationship to cytochrome oxidase. J Comp Neurol 1996;365:659–82.

[137] Shoham D, Hübener M, Schulze S, Grinvald A, Bonhoeffer T. Spatio-temporal frequency domains and their relation to cytochrome oxidase staining in cat visual cortex. Nature 1997;385: 529–33.

[138] Hubel DH, Wiesel TN. Shape and arrangement of columns in cat's striate cortex. J Physiol 1963;165:559–68.

[139] Humphrey AL, Norton TT. Topographic organization of the orientation column system in the striate cortex of the tree shrew (*Tupaia glis*). I. Microelectrode recording. J Comp Neurol 1980; 192(3):531–47.

[140] Metin C, Godement P, Imbert M. The primary visual cortex in the mouse: receptive field properties and functional organization. Exp Brain Res 1988;69:594–612.

[141] Reid RC. From functional architecture to functional connectomics. Neuron 2012;75(2):209–17.

[142] Tootell RB, Silverman MS, De Valois RL, Jacobs GH. Functional organization of the second cortical visual area in primates. Science 1983;220:737–9.

[143] Collins CE, Stepniewska I, Kaas JH. Topographic patterns of V2 cortical connections in a prosimian primate (*Galago garnetti*). J Comp Neurol 2001;431(2):155–67.

[144] Goodale MA, Milner AD. Separate visual pathways for perception and action. Trends Neurosci 1992;15:20–5.

[145] Ungerleider LG, Mishkin M. Two cortical visual systems. In: Ingle DG, Goodale MA, Mansfield RJQ, editors. Analysis of visual behavior. Cambridge, MA: MIT Press; 1982. p. 549–86.

[146] Kaas JH, Gharbawie OA, Stepniewska I. The organization and evolution of dorsal stream multisensory motor pathways in primates. Front Neuroanat 2011;5:34.

[147] Kravitz DJ, Peng CS, Baker CI. Real-world scene representations in high-level visual cortex: it's the spaces more than the places. J Neurosci 2011;31(20):7322–33.

[148] Kravitz DJ, Saleem KS, Baker CI, Ungerleider LG, Mishkin M. The ventral visual pathway: an expanded neural framework for the processing of object quality. Trends Cogn Sci 2013;17(1):26–49.

[149] Lomber SG, Payne BR, Cornwell P, Long KD. Perceptual and cognitive visual functions of parietal and temporal cortices in the cat. Cereb Cortex 1996;6(5):673–95.

[150] Tate A, Fischer H, Leigh A, Kendrick K. Behavioural and neurophysiological evidence for face identity and face emotion processing in animals. Philos Trans R Soc B Biol Sci 2006;361:2155.

[151] Kolb B. Posterior parietal and temporal association cortex. In: Kolb B, Tees RC, editors. The cerebral cortex of the rat. Cambridge, MA: MIT Press; 1990. p. 459–71.

[152] Reep RL, Chandler HC, King V, Corwin JV. Rat posterior parietal cortex: topography of corticocortical and thalamic connections. Exp Brain Res 1994;100:67–84.

[153] Preuss TM, Goldman-Rakic PS. Architectonics of the parietal and temporal association cortex in the strepsirhine primate Galago compared to the anthropoid primate Macaca. J Comp Neurol 1991;310(4):475–506.

[154] Wang Q, Sporns O, Burkhalter A. Network analysis of corticocortical connections reveals ventral and dorsal processing streams in mouse visual cortex. J Neurosci 2012;32(13):4386–99.

[155] Kaas JH, Stepniewska I, Gharbawie O. Cortical networks subserving upper limb movements in primates. Eur J Phys Rehabil Med 2012;48(2):299–306.

[156] Batista AP, Buneo CA, Snyder LH, Andersen RA. Reach plans in eye-centered coordinates. Science 1999;285:257–60.

[157] Colby CL, Duhamel JR, Goldberg ME. Visual, presaccadic, and cognitive activation of single neurons in monkey lateral intraparietal area. J Neurophysiol 1996;76:2841–52.

[158] Sakata H, Taira M, Murata A, Mine S. Neural mechanisms of visual guidance of hand action in the parietal cortex of the monkey. Cereb Cortex 1995;5:429–38.

[159] Cooke DF, Taylor CR, Moore T, Graziano MSA. Complex movements evoked by microstimulation of the ventral intraparietal area. Proc Natl Acad Sci USA 2003;100:6163–8.

[160] Stepniewska I, Fang PC, Kaas JH. Microstimulation reveals specialized subregions for different complex movements in posterior parietal cortex of prosimian galagos. Proc Natl Acad Sci USA 2005;102:4878–83.

[161] Stepniewska I, Fang PC, Kaas JH. Organization of posterior parietal cortex in galagos: I. Functional zones identified by microstimulation. J Comp Neurol 2009a;517:765–82.

[162] Gharbawie OA, Stepniewska I, Burish MJ, Kaas JH. Thalamocortical connections of functional zones in posterior parietal cortex and frontal cortex motor regions in New World monkeys. Cereb Cortex 2010;20:2391–410.

[163] Gharbawie OA, Stepniewska I, Kaas JH. Cortical connections of functional zones in posterior parietal cortex and frontal cortex motor regions in NewWorld monkeys. Cereb Cortex 2011;21(9): 1981–2002.

[164] Paxinos G, Watson C, Petrides M, Rosa M, Tokuno H. The marmoset brain in stereotaxic coordinates. London: Academic Press, Elsevier; 2012.

[165] Reser DH, Burman KJ, Yu HH, Chaplin TA, Richardson KE, Worthy KH, Rosa MG. Contrasting patterns of cortical input to architectural subdivisions of the area 8 complex: a retrograde tracing study in marmoset monkeys. Cereb Cortex 2013;23(8): 1901–22.

[166] Rosa MG, Palmer SM, Gamberini M, Burman KJ, Yu HH, Reser DH, Bourne JA, Tweedale R, Galletti C. Connections of the dorsomedial visual area: pathways for early integration of dorsal and ventral streams in extrastriate cortex. J Neurosci 2009;29:4548–63.

[167] Stepniewska I, Cerkevich CM, Fang PC, Kaas JH. Organization of posterior parietal cortex in galagos: II. Ipsilateral cortical connections of physiologically identified zones within anterior sensorimotor region. J Comp Neurol 2009b;517:783–807.

[168] Remple MS, Reed JL, Stepniewska I, Lyon DC, Kaas JH. The organization of frontoparietal cortex in the tree shrew (Tupaia belangeri): II. Connectional evidence for a frontal-posterior parietal network. J Comp Neurol 2007:501,121–149.

[169] Orban GA, Caruana F. The neural basis of human tool use. Front Psychol 2014;5:310.

[170] Chaplin TA, Yu HH, Soares JG, Gattass R, Rosa MG. A conserved pattern of differential expansion of cortical areas in simian primates. J Neurosci 2013a;33(38):15120–5.

[171] Orban GA, Claeys K, Nelissen K, Smans R, Sunaert S, Todd JT, Wardak C, Durand JB, Vanduffel W. Mapping the parietal cortex of human and nonhuman primates. Neuropsychologia 2006;44: 2647–67.

[172] Peeters R, Simon L, Nelissen K, Fabbri-Destro M, Vanduffel W, Rizzolatti G, Orban GA. The representation of tool use in humans and monkeys: common and uniquely human features. J Neurosci 2009;29:11523–39.

[173] Leopold DA, Rhodes G. A comparative view of face perception. J Comp Psychol 2010;124:233–51.

[173a] Sato N, Nakamura K. Visual response properties of neurons in the parahippocampal cortex of monkeys. Journal of Neurophysiology 2003;90(2):876–86.

[174] Landis T, Cummings JL, Benson DF. Loss of topographic familiarity. Arch Neurol 1986;43:132–6.

[175] Park S, Brady TF, Greene MR, Oliva A. Disentangling scene content from its spatial boundary: complementary roles for the PPA and LOC in representing real-world scenes. J Neurosci 2011; 31(4):1333–40.

[176] Desimone R, Gross CG. Visual areas in the temporal cortex of the macaque. Brain Res 1979;178:363–80.

[177] Gross CG, Rocha-Miranda CE, Bender DB. Visual properties of neurons in inferotemporal cortex of the macaque. J Neurophysiol 1972;35:96–111.

[178] Op De Beeck H, Vogels R. Spatial sensitivity of macaque inferior temporal neurons. J Comp Neurol 2000;426:505–18.

[179] Tanaka K, Saito H, Fukada Y, Moriya M. Coding visual images of objects in the inferotemporal cortex of the macaque monkey. J Neurophysiol 1991;66:170–89.
[180] Alemi-Neissi A, Rosselli FB, Zoccolan D. Multifeatural shape processing in rats engaged in invariant visual object recognition. J Neurosci 2013;33(14):5939–56.
[181] Zoccolan D, Oertelt N, DiCarlo JJ, Cox DD. A rodent model for the study of invariant visual object recognition. Proc Natl Acad Sci 2009;106(21):8748–53.
[182] Krauzlis RJ. The control of voluntary eye movements: new perspectives. Neuroscientist 2005;11(2):124–37.
[183] Hayhoe M, Ballard D. Eye movements in natural behavior. Trends Cogn Sci 2005;9(4):188–94.
[184] Yarbus AL. In: Haigh B, editor. Eye movements and vision. New York: Plenum Press; 1967.
[185] Ahnelt PK, Schubert C, Kübber-Heiss A, Schiviz A, Anger E. Independent variation of retinal S and M cone photoreceptor topographies: a survey of four families of mammals. Vis Neurosci 2006;23:429–35.
[186] Tegetmeyer H. Influence of visual signals on the coordination of spontaneous eye and head movements in infant rabbits. Brain Reseach Bull 1996;40(5–6):359–62.
[187] Rapaport DH, Stone J. The area centralis of the retina in the cat and other mammals: focal point for function and development of the visual system. Neuroscience 1984;11(2):289–301.
[188] Moeller GU, Kayser C, Knecht F, König P. Interactions between eye movement systems in cats and humans. Exp Brain Res 2004;157(2):215–24.
[189] Sakatani T, Isa T. Quantitative analysis of spontaneous saccade-like rapid eye movements in C57BL/6 mice. Neurosci Res 2007;58(3):324–31.
[190] Hughes A. A comparison of retinal ganglion cell topography in the plains and tree kangaroo. J Physiol 1975;244:61–3.
[191] Krubitzer L, Kaas J. The evolution of the neocortex in mammals: how is phenotypic diversity generated? Curr Opin Neurobiol 2005;15:444–53.
[192] de Waal F, Waal FBM. Chimpanzee politics: power and sex among apes. JHU Press; 2007.
[193] Maestripieri D. Macachiavellian intelligence. University of Chicago Press; 2008.
[194] Fernandez AA, Morris MR. Sexual selection and trichromatic color vision in primates: statistical support for the preexisting-bias hypothesis. Am Nat 2007;170:10–20.
[195] Changizi MA, Zhang Q, Shimojo S. Bare skin, blood and the evolution of primate colour vision. Biol Lett 2006;2:217–21.
[196] Dixson AF. Primate sexuality: comparative studies of the prosimians, monkeys, apes, and humans. Oxford University Press; 2012.
[197] Humphrey N. The social function of intellect. In: Growing points in ethology. Cambridge: Cambridge Univeristy Press; 1976. p. 303–17.
[198] Barton RA. Visual specialization and brain evolution in primates. Proc Biol Sci 1998;265:1933–7.
[199] Dunbar R. Neocortex size as a constraint on group size in primates. J Hum Evol 1992;20:469–93.
[200] Lewis KP, Barton RA. Amygdala size and hypothalamus size predict social play frequency in nonhuman primates: a comparative analysis using independent contrasts. J Comp Psychol 2006; 120:31–7.
[201] Hadjidimitrakis K, Breveglieri R, Bosco A, Fattori P. Three-dimensional eye position signals shape both peripersonal space and arm movement activity in the medial posterior parietal cortex. Front Integr Neurosci 2012;6:37. https://doi.org/10.3389/fnint.2012.00037.
[202] Bloch JI, Boyer DM. Grasping primate origins. Science 2002;298: 1606–10.
[203] Sussman RW, Tab Rasmussen D, Raven PH. Rethinking primate origins again. Am J Primatol 2013;75(2):95–106.
[204] Karl JM, Whishaw IQ. Different evolutionary origins for the reach and the grasp: an explanation for dual visuomotor channels in primate parietofrontal cortex. Front Neurol 2013;4:208.
[205] Troilo D, Howland HC, Judge SJ. Visual optics and retinal cone topography in the common marmoset (*Callithrix jacchus*). Vis Res 1993;33(10):1301–10.
[206] Lapuerta P, Schein SJ. A four-surface schematic eye of macaque monkey obtained by an optical method. Vis Res 1995; 35:2245–54.
[207] Chaplin TA, Yu HH, Rosa MG. Representation of the visual field in the primary visual area of the marmoset monkey: magnification factors, point-image size, and proportionality to retinal ganglion cell density. J Comp Neurol 2013b;521(5):1001–19.
[208] Perry VH, Cowey A. The ganglion cell and cone distributions in the monkey's retina: implications for central magnification factors. Vis Res 1985;25(12):1795–810.
[209] Merigan WH, Katz LM. Spatial resolution across the macaque retina. Vis Res 1990;30(7):985–91.
[210] Nummela SU, Coop SH, Cloherty SL, Boisvert CJ, Leblanc M, Mitchell JF. Psychophysical measurement of marmoset acuity and myopia. Dev Neurobiol 2017;77(3):300–13.
[211] Graham B, Judge SJ. The effects of spectacle wear in infancy on eye growth and refractive error in the marmoset (*Callithrix jacchus*). Vis Res 1999;39(2):189–206.
[212] Troilo D, Nickla DL, Wildsoet CF. Form deprivation myopia in mature common marmosets (*Callithrix jacchus*). Invest Ophthalmol Vis Sci 2000;41(8):2043–9.
[213] Pessoa DM, Tomaz C, Pessoa VF. Color vision in marmosets and tamarins: behavioral evidence. Am J Primatol 2005;67(4):487–95.
[214] Osorio D, Smith AC, Vorobyev M, Buchanan Smith HM. Detection of fruit and the selection of primate visual pigments for color vision. Am Nat 2004;164:696–708.
[215] Caine NG, Osorio D, Mundy NI. A foraging advantage for dichromatic marmosets (*Callithrix geoffroyi*) at low light intensity. Biol Lett 2010;6:36–8.
[216] Cheong SK, Pietersen ANJ. Antidromic latency of magnocellular, parvocellular, and koniocellular (Blue-ON) geniculocortical relay cells in marmosets. Vis Neurosci 2014;31:263–73.
[217] Goodchild AK, Martin PR. The distribution of calcium-binding proteins in the lateral geniculate nucleus and visual cortex of a New World monkey, the marmoset, *Callithrix jacchus*. Vis Neurosci 1998;15:625–42.
[218] White AJ, Solomon SG, Martin PR. Spatial properties of koniocellular cells in the lateral geniculate nucleus of the marmoset *Callithrix jacchus*. J Physiol 2001;533:519–35.
[219] Cheong SK, Tailby C, Solomon SG, Martin PR. Cortical-like receptive fields in the lateral geniculate nucleus of marmoset monkeys. J Neurosci 2013;33:6864–76.
[220] Smith EL, Chino YM, Ridder WH, Kitagawa K, Langston A. Orientation bias of neurons in the lateral geniculate nucleus of macaque monkeys. Vis Neurosci 1990;5:525–45.
[221] Kremers J, Lee BB. Comparative retinal physiology in anthropoids. Vis Res 1998;38(21):3339–44.
[222] White AJ, Wilder HD, Goodchild AK, Sefton AJ, Martin PR. Segregation of receptive field properties in the lateral geniculate nucleus of a New-World monkey, the marmoset *Callithrix jacchus*. J Neurophysiol 1998;80(4):2063–76.
[223] Mancuso K, Hauswirth WW, Li Q, Connor TB, Kuchenbecker JA, Mauck MC, Neitz J, Neitz M. Gene therapy for red-green colour blindness in adult primates. Nature 2009;461(7265):784–7.
[224] Barraclough N, Tinsley C, Webb B, Vincent C, Derrington A. Processing of first-order motion in marmoset visual cortex is influenced by second-order motion. Vis Neurosci 2006;23:815–24.

[225] Bourne JA, Tweedale R, Rosa MGP. Physiological responses of New World monkey V1 neurons to stimuli defined by coherent motion. Cereb Cortex 2002;12:1132–45.

[226] Bourne JA, Lui L, Tweedale R, Rosa MGP. First- and second-order stimulus length selectivity in New World monkey striate cortex. Eur J Neurosci 2004;19:169–80.

[227] Buzás P, Szmajda BA, Hashemi-Nezhad M, Dreher B, Martin PR. Color signals in the primary visual cortex of marmosets. J Vis 2008;8:7. https://doi.org/10.1167/8.10.7.

[228] Forte JD, Hashemi-Nezhad M, Dobbie WJ, Dreher B, Martin PR. Spatial coding and response redundancy in parallel visual pathways of the marmoset *Callithrix jacchus*. Vis Neurosci 2005;22: 479–91.

[229] Guo K, Robertson R, Nevado A, Pulgarin M, Mahmoodi S, Young MP. Primary visual cortex neurons that contribute to resolve the aperture problem. Neuroscience 2006;138:1397–406.

[230] Hashemi-Nezhad M, Blessing EM, Dreher B, Martin PR. Segregation of short-wavelength sensitive ("blue") cone signals among neurons in the lateral geniculate nucleus and striate cortex of marmosets. Vis Res 2008;48:2604–14.

[231] Nowak LG, Barone P. Contrast adaptation contributes to contrast-invariance of orientation tuning of primate V1 cells. PLoS One 2009;4:e4781. https://doi.org/10.1371/journal.pone.0004781.

[232] Sengpiel F, Troilo D, Kind PC, Graham B, Blakemore C. Functional architecture of area 17 in normal and monocularly deprived marmosets (*Callithrix jacchus*). Vis Neurosci 1996;13: 145–60.

[233] Solomon SS, Chen SC, Morley JW, Solomon SG. Local and global correlations between neurons in the middle temporal area of primate visual cortex. Cereb Cortex 2014. https://doi.org/10.1093/cercor/bhu111.

[234] Tinsley CJ, Webb BS, Barraclough NE, Vincent CJ, Parker A, Derrington AM. The nature of V1 neural responses to 2D moving patterns depends on receptive field structure in the marmoset monkey. J Neurophysiol 2003;90:930–7.

[235] Webb BS, Tinsley CJ, Barraclough NE, Parker A, Derrington AM. Gain control from beyond the classical receptive field in primate primary visual cortex. Vis Neurosci 2003;20:221–30.

[236] Yu HH, Verma R, Yang Y, Tibballs HA, Lui LL, Reser DH, Rosa MGP. Spatial and temporal frequency tuning in striate cortex: functional uniformity and specializations related to receptive field eccentricity. Eur J Neurosci 2010;31:1043–62.

[237] Yu HH, Rosa MGP. Uniformity and diversity of response properties of neurons in the primary visual cortex: selectivity for orientation, direction of motion, and stimulus size from center to far periphery. Vis Neurosci 2014;31:85–98.

[238] Zinke W, Roberts MJ, Guo K, McDonald JS, Robertson R, Thiele A. Cholinergic modulation of response properties and orientation tuning of neurons in primary visual cortex of anaesthetized marmoset monkeys. Eur J Neurosci 2006;24:314–28.

[239] Collins CE, Leitch DB, Wong P, Kaas JH, Herculano-Houzel S. Faster scaling of visual neurons in cortical areas relative to subcortical structures in non-human primate brains. Brain Struct Funct 2013;218:805–16.

[240] Stevens CF. An evolutionary scaling law for the primate visual system and its basis in cortical function. Nature 2001; 411:193–5.

[241] Roe AW, Fritsches K, Pettigrew JD. Optical imaging of functional organization of V1 and V2 in marmoset visual cortex. Anat Rec A Discov Mol Cell Evol Biol 2005;287:1213–25.

[242] Yacoub E, Shmuel A, Logothetis N, Ugurbil K. Robust detection of ocular dominance columns in humans using Hahn Spin Echo BOLD functional MRI at 7 Tesla. Neuroimage 2007;37: 1161–77.

[243] Spatz WB. Loss of ocular dominance columns with maturity in the monkey, *Callithrix jacchus*. Brain Res 1989;488:376–80.

[244] Elston GN, Rosa MGP. Ipsilateral corticocortical projections to the primary and middle temporal visual areas of the primate cerebral cortex: area-specific variations in the morphology of connectionally identified pyramidal cells. Eur J Neurosci 2006;23: 3337–45.

[245] Nassi JJ, Callaway EM. Specialized circuits from primary visual cortex to V2 and area MT. Neuron 2007;55:799–808.

[246] Klink R, Alonso A. Ionic mechanisms for the subthreshold oscillations and differential electroresponsiveness of medial entorhinal cortex layer II neurons. J Neurophysiol 1993;70(1): 144–57.

[247] Schubert D, Kötter R, Zilles K, Luhmann HJ, Staiger JF. Cell type-specific circuits of cortical layer IV spiny neurons. J Neurosci 2003;23(7):2961–70.

[248] Callaway EM, Borrell V. Developmental sculpting of dendritic morphology of layer 4 neurons in visual cortex: influence of retinal input. J Neurosci 2011;31(20):7456–70.

[249] Lui LL, Bourne JA, Rosa MGP. Single-unit responses to kinetic stimuli in New World monkey area V2: physiological characteristics of cue-invariant neurones. Exp Brain Res 2005;162:100–8.

[250] Rosa MGP, Fritsches KA, Elston GN. The second visual area in the marmoset monkey: visuotopic organisation, magnification factors, architectonical boundaries, and modularity. J Comp Neurol 1997;387:547–67.

[251] Lui LL, Bourne JA, Rosa MGP. Functional response properties of neurons in the dorsomedial visual area of New World monkeys (*Callithrix jacchus*). Cereb Cortex 2006;16:162–77.

[252] Lui LL, Bourne JA, Rosa MGP. Relationship between size summation properties, contrast sensitivity and response latency in the dorsomedial and middle temporal areas of the primate extrastriate cortex. PLoS One 2013;8:e68276. https://doi.org/10.1371/journal.pone.0068276.

[253] Rosa MGP, Schmid LM. Visual areas in the dorsal and medial extrastriate cortices of the marmoset. J Comp Neurol 1995;359: 272–99.

[254] Rosa MGP, Tweedale R. Visual areas in lateral and ventral extrastriate cortices of the marmoset monkey. J Comp Neurol 2000;422: 621–51.

[255] Lui LL, Bourne JA, Rosa MGP. Spatial and temporal frequency selectivity of neurons in the middle temporal visual area of new world monkeys (*Callithrix jacchus*). Eur J Neurosci 2007a; 25:1780–92.

[256] Lui LL, Bourne JA, Rosa MGP. Spatial summation, end inhibition and side inhibition in the middle temporal visual area (MT). J Neurophysiol 2007b;97:1135–48.

[257] Lui LL, Dobiecki AE, Bourne JA, Rosa MGP. Breaking camouflage: responses of neurons in the middle temporal area to stimuli defined by coherent motion. Eur J Neurosci 2012;36: 2063–76.

[258] McDonald JS, Clifford CW, Solomon SS, Chen SC, Solomon SG. Integration and segregation of multiple motion signals by neurons in area MT of primate. J Neurophysiol 2014;111:369–78.

[259] Rosa MGP, Elston GN. Visuotopic organisation and neuronal response selectivity for direction of motion in visual areas of the caudal temporal lobe of the marmoset monkey (*Callithrix jacchus*): middle temporal area, middle temporal crescent, and surrounding cortex. J Comp Neurol 1998;393:505–27.

[260] Solomon SS, Tailby C, Gharaei S, Camp AJ, Bourne JA, Solomon SG. Visual motion integration by neurons in the middle temporal area of a New World monkey, the marmoset. J Physiol 2011;589:5741–58.

[261] Krubitzer L, Kahn DM. Nature versus nurture revisited: an old idea with a new twist. Prog Neurobiol 2003;70:33–52.

[262] Albright TD, Desimone R, Gross CG. Columnar organization of directionally selective cells in visual area MT of the macaque. J Neurophysiol 1984;51:16–31.

[263] DeAngelis GC, Newsome WT. Organization of disparity-selective neurons in macaque area MT. J Neurosci 1999;19(4):1398–415.

[264] Bourne JA, Rosa MGP. Hierarchical development of the primate visual cortex, as revealed by neurofilament immunoreactivity: early maturation of the middle temporal area (MT). Cereb Cortex 2006;16:405–14.

[265] Warner CE, Kwan WC, Bourne JA. The early maturation of visual cortical area MT is dependent on input from the retinorecipient medial portion of the inferior pulvinar. J Neurosci 2012;32:17073–85.

[266] Tamura H, Fujita I. Complex-image selective visual responses of IT neurons in a new world monkey. In: Proc. 7th IBRO world congr. neurosci., Paris, 304; 2007.

[267] Ridley RM, Warner KA, Maclean CJ, Gaffan D, Baker HF. Visual agnosia and Klüver-Bucy syndrome in marmosets (*Callithrix jacchus*) following ablation of inferotemporal cortex, with additional mnemonic effects of immunotoxic lesions of cholinergic projections to medial temporal areas. Brain Res 2001;898(1):136–51.

[268] Hung CC, Yen CC, Ciuchta JL, Papoti D, Bock NA, Leopold DA, Silva AC. Functional mapping of face-selective regions in the extrastriate visual cortex of the marmoset. J Neurosci 2015a;35(3):1160–72.

[269] Tsao DY, Livingstone MS. Mechanisms of face perception. Annu Rev Neurosci 2008;31:411–37.

[270] Hung CC, Yen CC, Ciuchta JL, Papoti D, Bock NA, Leopold DA, Silva AC. Functional MRI of visual responses in the awake, behaving marmoset. Neuroimage 2015b;120:1–11.

[271] Graziano MSA, Taylor CSR, Moore T. Complex movements evoked by microstimulation of precentral cortex. Neuron 2002;34:841–51.

[272] Kaas JH, Gharbawie OA, Stepniewska I. Cortical networks for ethologically relevant behaviors in primates. Am J Primatol 2013;75:407–14.

[273] Burman KJ, Bakola S, Richardson KE, Reser DH, Rosa MG. Patterns of afferent input to the caudal and rostral areas of the dorsal premotor cortex (6DC and 6DR) in the marmoset monkey. J Comp Neurol 2014;522:3683–716.

[274] Collins CE, Lyon DC, Kaas JH. Distribution across cortical areas of neurons projecting to the superior colliculus in new world monkeys. Anat Rec A Discov Mol Cell Evol Biol 2005;285(1):619–27.

[275] Blum B, Kulikowski JJ, Carden D, Harwood D. Eye movements induced by electrical stimulation of the frontal eye fields of marmosets and squirrel monkeys. Brain Behav Evol 1982;21(1):34–41.

[276] Burman KJ, Palmer SM, Gamberini M, Rosa MG. Cytoarchitectonic subdivisions of the dorsolateral frontal cortex of the marmoset monkey (*Callithrix jacchus*), and their projections to dorsal visual areas. J Comp Neurol 2006;495(2):149–72.

[277] Ghahremani M, Hutchison RM, Menon RS, Everling S. Frontoparietal functional connectivity in the common marmoset. Cerebr Cortex 2016;27(8):3890–905.

[278] Huerta MF, Krubitzer LA, Kaas JH. Frontal eye field as defined by intracortical microstimulation in squirrel monkeys, owl monkeys, and macaque monkeys: I. Subcortical connections. J Comp Neurol 1986;253(4):415–39.

[279] Burkart J, Heschl A. Geometrical gaze following in common marmosets (*Callithrix jacchus*). J Comp Psychol 2006;120(2):120–30.

[280] Derrington AM, Parker A, Barraclough NE, Easton A, Goodson GR, Parker KS, Tinsley CJ, Webb BS. The uses of colour vision: behavioural and physiological distinctiveness of colour stimuli. Philos Trans R Soc Lond B Biol Sci 2002;357(1424):975–85.

[281] Ridley RM, Murray TK, Johnson JA, Baker HF. Learning impairment following lesion of the basal nucleus of Meynert in the marmoset: modification by cholinergic drugs. Brain Res 1986;376(1):108–16.

[282] Roberts AC, Robbins TW, Everitt BJ, Jones GH, Sirkia TE, Wilkinson J, Page K. The effects of excitotoxic lesions of the basal forebrain on the acquisition, retention and serial reversal of visual discriminations in marmosets. Neuroscience 1990;34(2):311–29.

[283] Hook MA, Rogers LJ. Visuospatial reaching preferences of common marmosets (*Callithrix jacchus*): an assessment of individual biases across a variety of tasks. J Comp Psychol 2008;122(1):41–51.

[284] Kemp C, Kaplan G. Facial expressions in common marmosets (*Callithrix jacchus*) and their use by conspecifics. Anim Cogn 2013;16(5):773–88.

[285] Mitchell JF, Reynolds JH, Miller CT. Active vision in marmosets: a model system for visual neuroscience. J Neurosci 2014;34(4):1183–94.

[286] Ridley RM, Haystead TA, Baker HF, Crow TJ. A new approach to the role of noradrenaline in learning: problem-solving in the marmoset after alpha-noradrenergic receptor blockade. Pharmacol Biochem Behav 1981;14(6):849–55.

[287] Ridley RM, Bowes PM, Baker HF, Crow TJ. An involvement of acetylcholine in object discrimination learning and memory in the marmoset. Neuropsychologia 1984;22(3):253–63.

[288] Clarke HF, Dalley JW, Crofts HS, Robbins TW, Roberts AC. Cognitive inflexibility after prefrontal serotonin depletion. Science 2004;304(5672):878–80.

[289] Crofts HS, Muggleton NG, Bowditch AP, Pearce PC, Nutt DJ, Scott EA. Home cage presentation of complex discrimination tasks to marmosets and rhesus monkeys. Lab Anim 1999;33(3):207–14.

[290] Dias R, Robbins TW, Roberts AC. Dissociation in prefrontal cortex of affective and attentional shifts. Nature 1996;380(6569):69–72.

[291] Spinelli S, Pennanen L, Dettling AC, Feldon J, Higgins GA, Pryce CR. Performance of the marmoset monkey on computerized tasks of attention and working memory. Brain Res Cogn Brain Res 2004;19(2):123–37.

[292] Easton A, Parker K, Derrington AM, Parker A. Behaviour of marmoset monkeys in a T-maze: comparison with rats and macaque monkeys on a spatial delayed non-match to sample task. Exp Brain Res 2003;150(1):114–6.

[293] Maclean CJ, Gaffan D, Baker HF, Ridley RM. Visual discrimination learning impairments produced by combined transections of the anterior temporal stem, amygdala and fornix in marmoset monkeys. Brain Res 2001;888(1):34–50.

[294] Ridley RM, Baker HF, Cummings RM, Green ME, Leow-Dyke A. Mild topographical memory impairment following crossed unilateral lesions of the mediodorsal thalamic nucleus and the inferotemporal cortex. Behav Neurosci 2005;119(2):518–25.

[295] Barefoot HC, Baker HF, Ridley RM. Crossed unilateral lesions of temporal lobe structures and cholinergic cell bodies impair visual conditional and object discrimination learning in monkeys. Eur J Neurosci 2002;15(3):507–16.

[296] Barefoot HC, Maclean CJ, Baker HF, Ridley RM. Unilateral hippocampal and inferotemporal cortex lesions in opposite hemispheres impair learning of single-pair visual discriminations as well as visuovisual conditional tasks in monkeys. Behav Brain Res 2003;141(1):51–62.

[297] Roberts AC, Robbins TW, Everitt BJ, Muir JL. A specific form of cognitive rigidity following excitotoxic lesions of the basal forebrain in marmosets. Neuroscience 1992;47(2):251–64.

[298] Ridley RM, Baker HF, Murray TK. Basal nucleus lesions in monkeys: recognition memory impairment or visual agnosia? Psychopharmacology (Berl) 1988;95:289–90.
[299] Ridley RM, Baker HF. A critical evaluation of monkey models of amnesia and dementia. Brain Res Rev 1991;16:15–37.
[300] Ridley RM, Baker HF. Assessing memory in monkeys. In: Sahgal A, editor. Behavioural neuroscience: a practical approach, vol. 1. Oxford: IRL Press at Oxford Univ. Press; 1993. p. 149–63.
[301] Remington ED, Osmanski MS, Wang X. An operant conditioning method for studying auditory behaviors in marmoset monkeys. PLoS One 2012;7(10):e47895.
[302] Mitchell JF, Priebe NJ, Miller CT. Motion dependence of smooth pursuit eye movements in the marmoset. J Neurophysiol 2015; 113(10):3954–60.
[303] Heiney SA, Blazquez PM. Behavioral response of trained squirrel and rhesus monkeys during oculomotor tasks. Exp Brain Res 2011;212(3):409–16.
[304] Keating C, Keating E. Visual scan patterns of rhesus monkeys viewing faces. Perception 1982;11:211–9.
[305] McCrea RA, Gdowski GT. Firing behavior of squirrel monkey eye movement-related vestibular nucleus neurons during gaze saccades. J Physiol 2003;546:207224.
[306] Chance MRA. Attention structure as the basis of primate rank orders. Man 1967;2:503.
[307] Vogels R, Orban GA. How well do response changes of striate neurons signal differences in orientation: a study in the discriminating monkey. J Neurosci 1990;10:3543–58.
[308] Kaas JH. Why is brain size so important: design problems and solutions as neocortex gets bigger or smaller. Brain Mind 2000; 1:7–23.
[309] Stevenson MF, Poole TB. An ethogram of the common marmoset (*Calithrix jacchus jacchus*): general behavioural repertoire. Anim Behav 1976;24(2):428–51.
[310] French JA. The role of androgenic steroids in shaping social phenotypes across the lifespan in male marmosets (Callithrix spp.). Am J Primatol 2013;75(3):212–21.
[311] Schiel N, Huber L. Social influences on the development of foraging behavior in free-living common marmosets (*Callithrix jacchus*). Am J Primatol 2006;68(12):1150–60.
[312] Emery NJ, Clayton NS. Comparative social cognition. Annu Rev Psychol 2009;60:87–113.
[313] Rosati AG, Hare B. Looking past the model species: diversity in gaze-following skills across primates. Curr Opin Neurobiol 2009; 19(1):45–51.
[314] Shepherd SV, Platt ML. Spontaneous social orienting and gaze following in ringtailed lemurs (*Lemur catta*). Anim Cogn 2008; 11(1):13–20.
[315] Clara E, Tommasi L, Rogers LJ. Social mobbing calls in common marmosets (*Callithrix jacchus*): effects of experience and associated cortisol levels. Anim Cogn 2008;11(2):349–58.
[316] Miller CT, Wren Thomas A. Individual recognition during bouts of antiphonal calling in common marmosets. J Comp Physiol A Neuroethol Sens Neural Behav Physiol 2012;198(5):337–46.
[317] Dufour V, Pascalis O, Petit O. Face processing limitation to own species in primates: a comparative study in brown capuchins, Tonkean macaques and humans. Behav Processes 2006;73(1): 107–13.
[318] Preuschoft S, van Hooff JA. Homologizing primate facial displays: a critical review of methods. Folia Primatol 1995;65(3): 121–37.
[319] Burrows AM. The facial expression musculature in primates and its evolutionary significance. Bioessays 2008;30(3):212–25.
[320] Heymann EW, von der Lage F. Brief communication: noninvasive measuring of operational tongue length in callitrichids. Am J Phys Anthropol 2009;139:430–3.
[321] Cook M, Mineka S. Observational conditioning of fear to fear-relevant versus fear-irrelevant stimuli in rhesus monkeys. J Abnormal Psychol 1989;98:448–59.
[322] Langford DJ, Crager SE, Shehzad Z, Smith SB, Sotocinal SG, Levenstadt JS, Chanda ML, Levitin DJ, Mogil JS. Social modulation of pain as evidence for empathy in mice. Science 2006; 312(5782):1967–70.
[323] Hashimoto T, Yamazaki Y, Iriki A. Hand preference depends on posture in common marmosets. Behav Brain Res 2013;248: 144–50.
[324] Wallis JD, Dias R, Robbins TW, Roberts AC. Dissociable contributions of the orbitofrontal and lateral prefrontal cortex of the marmoset to performance on a detour reaching task. Eur J Neurosci 2001;13(9):1797–808.
[325] Walker SC, Mikheenko YP, Argyle LD, Robbins TW, Roberts AC. Selective prefrontal serotonin depletion impairs acquisition of a detour-reaching task. Eur J Neurosci 2006;23(11):3119–23.
[326] Guerra RF, da Silveira NL, Bernardi N, Legal EJ. Hand preference during behavioral tests and spontaneous activity in two species of common marmoset (*Callithrix jacchus* and *Callithrix penicillata*). Rev Bras Biol 1997;57(4):563–70.
[327] Hook-Costigan MA, Rogers LJ. Hand, mouth and eye preferences in the common marmoset (*Callithrix jacchus*). Folia Primatol (Basel) 1995;64(4):180–91.
[328] Nakako T, Murai T, Ikejiri M, Ishiyama T, Taiji M, Ikeda K. Effects of a dopamine D1 agonist on ketamine-induced spatial working memory dysfunction in common marmosets. Behav Brain Res 2013;249:109–15.
[329] Pohlmeyer EA, Geng S, Prins NW, Sanchez JC. Deciding to take action: striatum activation during reaching to targets of varying reward value. Soc Neurosci Abstr 2013.
[330] Yamazaki Y, Echigo C, Saiki M, Inada M, Watanabe S, Iriki A. Tool-use learning by common marmosets (*Callithrix jacchus*). Exp Brain Res 2011;213(1):63–71.
[331] Galletti C, Kutz DF, Gamberini M, Breveglieri R, Fattori P. Role of the medial parieto-occipital cortex in the control of reaching and grasping movements. Exp Brain Res 2003;153(2):158–70.
[332] Weiss DJ, Wark JD, Rosenbaum DA. Monkey see, monkey plan, monkey do: the end-state comfort effect in cotton-top tamarins (*Saguinus oedipus*). Psychol Sci 2007;18(12):1063–8.
[333] Marzke MW. Precision grips, hand morphology, and tools. Am J Phys Anthropol 1997;102(1):91–110.
[334] Fragaszy DM. Preliminary quantitative studies of prehension in squirrel monkeys (*Saimiri sciureus*). Brain Behav Evol 1983; 23(3–4):81–92.
[335] Fleagle JG, Simons EL. Limb skeleton and locomotor adaptations of *Apidium phiomense*, an Oligocene anthropoid from Egypt. Am J Phys Anthropol 1995;97:235–89.
[336] Napier JR, Napier PH. A handbook of living primates. Morphology, ecology and behaviour of nonhuman primates. London: Academic; 1967.
[337] Padberg J, Franca JG, Cooke DF, Soares JG, Rosa MG, Fiorani Jr M, Gattass R, Krubitzer L. Parallel evolution of cortical areas involved in skilled hand use. J Neurosci 2007;27(38): 10106–15.
[338] Rose MD. Functional morphological similarities in the locomotor skeleton of miocene catarrhines and platyrrhine monkeys. Folia Primatol 1996;66:7–14.
[339] Huffman KJ, Krubitzer L. Area 3a: topographic organization and cortical connections in marmoset monkeys. Cereb Cortex 2001; 11(9):849–67.
[340] Padberg J, Disbrow E, Krubitzer L. The organization and connections of anterior and posterior parietal cortex in titi monkeys: do New World monkeys have an area 2? Cereb Cortex 2005;15: 1938–63.

[341] Bortoff GA, Strick PL. Corticospinal terminations in two new-world primates: further evidence that corticoneuronal connections provide part of the neural substrate for manual dexterity. J Neurosci 1993;13:5105–18.
[342] Lemon RN, Griffiths J. Comparing the function of the corticospinal system in different species: organizational differences for motor specialization? Muscle Nerve 2005;32(3):261–79.
[343] Striedter GF. Principles of brain evolution. Sunderland, MA: Sinauer; 2005.
[344] Cummins-Sebree SE, Fragaszy DM. Choosing and using tools: capuchins (*Cebus apella*) use a different metric than tamarins (*Saguinus oedipus*). J Comp Psychol 2005;119(2):210–9.
[345] Marín O, Rubenstein JLR. Cell migration in the forebrain. Annu Rev Neurosci 2003;26:441–83.
[346] Lein ES, Hawrylycz MJ, Ao N, Ayres M, Bensinger A, Bernard A, Boe AF, Boguski MS, Brockway KS, Byrnes EJ, et al. Genome-wide atlas of gene expression in the adult mouse brain. Nature 2007;445:168–76.
[347] Sansom SN, Livesey FJ. Gradients in the brain: the control of the development of form and function in the cerebral cortex. Cold Spring Harb Perspect Biol 2009;1, a002519.
[348] Kolodkin AL, Tessier-Lavigne M. Mechanisms and molecules of neuronal wiring: a primer. Cold Spring Harb Perspect Biol 2011;3(6).
[349] Puelles L, Harrison M, Paxinos G, Watson C. A developmental ontology for the mammalian brain based on the prosomeric model. Trends Neurosci 2013;36(10):570–8.
[350] Workman AD, Charvet CJ, Clancy B, Darlington RB, Finlay BL. Modeling transformations of neurodevelopmental sequences across mammalian species. J Neurosci 2013;33:7368–83.
[351] Homman-Ludiye J, Bourne JA. The guidance molecule Semaphorin3A is differentially involved in the arealization of the mouse and primate neocortex. Cereb Cortex 2013;24(11):2884–98.
[352] Barone P, Dehay C, Berland M, Bullier J, Kennedy H. Developmental remodeling of primate visual cortical pathways. Cereb Cortex 1995;5:22–38.
[353] Rakic P. Prenatal development of the visual system in rhesus monkey. Philos Trans R Soc Lond B Biol Sci 1977;278:245–60.
[354] Homman-Ludiye J, Bourne JA. Mapping arealisation of the visual cortex of non-primate species: lessons for development and evolution. Front Neural Circuits 2014;8:79.
[355] Robinson SR, Dreher B. The visual pathways of eutherian mammals and marsupials develop according to a common timetable. Brain Behav Evol 1990;36:177–95.
[356] Mundinano IC, Fox DM, Kwan WC, Vidaurre D, Teo L, Homman-Ludiye J, Goodale MA, Leopold DA, Bourne JA. Transient visual pathway critical for normal development of primate grasping behavior. Proc Nat Acad Sci 2018;115(6):1364–9.
[356a] Tardif SD, Smucny DA, Abbott DH, Mansfield K, Schultz-Darken N, Yamamoto ME. Reproduction in captive common marmosets (Callithrix jacchus). Comparative medicine 2003; 53(4):364–8.
[357] Goldshmit Y, Homman-Ludiye J, Bourne JA. EphA4 is associated with multiple cell types in the marmoset primary visual cortex throughout the lifespan. Eur J Neurosci 2014;39:1419–28.
[358] Sasaki T, Oga T, Nakagaki K, Sakai K, Sumida K, Hoshino K, Miyawaki I, Saito K, Suto F, Ichinohe N. Developmental expression profiles of axon guidance signaling and the immune system in the marmoset cortex: potential molecular mechanisms of pruning of dendritic spines during primate synapse formation in late infancy and prepuberty (I). Biochem Biophys Res Commun 2014;444:302–6.
[359] Fitzpatrick D. Cortical imaging: capturing the moment. Curr Biol 2000;10(5):R187–90.
[360] Grinvald A. Optical imaging of architecture and function in the living brain sheds new light on cortical mechanisms underlying visual perception. Brain Topogr 1992;5(2):71–5.
[361] Chen Y, Seidemann E. Attentional modulations related to spatial gating but not to allocation of limited resources in primate V1. Neuron 2012;74:557–66.
[362] Tanigawa H, Lu HD, Roe AW. Functional organization for color and orientation in macaque V4. Nat Neurosci 2010;13:1542–8.
[363] Raffi M, Siegel RM. Functional architecture of spatial attention in the parietal cortex of the behaving monkey. J Neurosci 2005;25: 5171–86.
[364] Seidemann E, Arieli A, Grinvald A, Slovin H. Dynamics of depolarization and hyperpolarization in the frontal cortex and saccade goal. Science 2002;295:862–5.
[365] Wang G, Tanaka K, Tanifuji M. Optical imaging of functional organization in the monkey inferotemporal cortex. Science 1996; 272:1665–8.
[366] Rubehn B, Bosman C, Oostenveld R, Fries P, Stieglitz T. A MEMS-based flexible multichannel ECoG-electrode array. J Neural Eng 2009;6, 036003.
[367] Shimoda K, Nagasaka Y, Chao ZC, Fujii N. Decoding continuous three-dimensional hand trajectories from epidural electrocorticographic signals in Japanese macaques. J Neural Eng 2012;9, 036015.
[368] Fukushima M, Saunders RC, Leopold DA, Mishkin M, Averbeck BB. Spontaneous high-gamma band activity reflects functional organization of auditory cortex in the awake macaque. Neuron 2012;74:899–910.
[369] Fukushima M, Saunders RC, Mullarkey M, Doyle AM, Mishkin M, Fujii N. An electrocorticographic electrode array for simultaneous recording from medial, lateral, and intrasulcal surface of the cortex in macaque monkeys. J Neurosci Methods 2014;233:155–65.
[370] Newsome WT, Salzman CD. The neuronal basis of motion perception. Ciba Found Symp 1993;174:217–30.
[371] Shadlen MN, Newsome WT. Noise, neural codes and cortical organization. Curr Opin Neurobiol 1994;4(4):569–79.
[372] Mitzdorf U. Current source-density method and application in cat cerebral cortex: investigation of evoked potentials and EEG phenomena. Physiol Rev 1985;65(1):37–100.
[373] Schroeder CE, Mehta AD, Givre SJ. A spatiotemporal profile of visual system activation revealed by current source density analysis in the awake macaque. Cereb Cortex 1998;8(7):575–92.
[374] Ohki K, Reid RC. In vivo two-photon calcium imaging in the visual system. Cold Spring Harb Protoc 2014;4:402–16.
[375] Nauhaus I, Nielsen KJ, Disney AA, Callaway EM. Orthogonal micro-organization of orientation and spatial frequency in primate primary visual cortex. Nat Neurosci 2012;15(12):1683–90.
[376] Ustione A, Piston DW. A simple introduction to multiphoton microscopy. J Microsc 2011;243(3):221–6.
[377] Huber D, Gutnisky DA, Peron S, O'Connor DH, Wiegert JS, Tian L, Oertner TG, Looger LL, Svoboda K. Multiple dynamic representations in the motor cortex during sensorimotor learning. Nature 2012;484:473–8.
[378] Sadakane O, Masamizu Y, Watakabe A, Terada SI, Ohtsuka M, Takaji M, et al. Long-term two-photon calcium imaging of neuronal populations with subcellular resolution in adult non-human primates. Cell Rep 2015;13(9):1989–99.
[379] Santisakultarm TP, Kersbergen CJ, Bandy DK, Ide DC, Choi SH, Silva AC. Two-photon imaging of cerebral hemodynamics and neural activity in awake and anesthetized marmosets. J Neurosci Meth 2016;271:55–64.
[380] Park JE, Zhang XF, Choi SH, Okahara J, Sasaki E, Silva AC. Generation of transgenic marmosets expressing genetically encoded calcium indicators. Sci Rep 2016;6:34931.
[381] Shen H. Precision gene editing paves way for transgenic monkeys. Nature 2013;503(7474):14–5.
[382] Cyranoski D. Marmosets are stars of Japan's ambitious project. Nature 2014;514:151–2.

[383] Sasaki E, Suemizu H, Shimada A, Hanazawa K, Oiwa R, Kamioka M, Tomioka I, et al. Generation of transgenic non-human primates with germline transmission. Nature 2009;459: 523–7.
[384] Tokuno H, Moriya-Ito K, Tanaka I. Experimental techniques for neuroscience research using common marmosets. Exp Anim 2012;61(4):389–97.
[385] Kishi N, Sato K, Sasaki E, Okano H. Common marmoset as a new model animal for neuroscience research and genome editing technology. Dev Growth Differ 2014;56(1):53–62.
[386] Kennedy DP, Adolphs R. The social brain in psychiatric and neurological disorders. Trends Cogn Sci 2012;16(11):559–72.
[387] Pelphrey KA, Sasson NJ, Reznick JS, Paul G, Goldman BD, Piven J. Visual scanning of faces in autism. J Autism Dev Disord 2002;32(4):249–61.
[388] Williams LM, Loughland CM, Gordon E, Davidson D. Visual scanpaths in schizophrenia: is there a deficit in face recognition? Schizophr Res 1999;40(3):189–99.
[389] Ben-Sasson A, Hen L, Fluss R, Cermak SA, Engel-Yeger B, Gal E. A meta-analysis of sensory modulation symptoms in individuals with autism spectrum disorders. J Autism Dev Disord 2009;39: 1–11.
[390] Foss-Feig JH, Tadin D, Schauder KB, Cascio CJ. A substantial and unexpected enhancement of motion perception in autism. J Neurosci 2013;33(19):8243–9.
[391] Tadin D, Kim J, Doop ML, Gibson C, Lappin JS, Blake R, Park S. Weakened center-surround interactions in visual motion processing in schizophrenia. J Neurosci 2006;26(44):11403–12.
[392] Yang E, Tadin D, Glasser DM, Hong SW, Blake R, Park S. Visual context processing in schizophrenia. Clin Psychol Sci 2013;1(1): 5–15.
[393] Lewis DA. Inhibitory neurons in human cortical circuits: substrate for cognitive dysfunction in schizophrenia. Curr Opin Neurobiol 2014;26:22–6.
[394] Rubenstein JLR, Merzenich MM. Model of autism: increased ratio of excitation/inhibition in key neural systems. Gene Brain Behav 2003;2(5):255–67.
[395] Norcross JL, Newman JD. Context and gender specific differences in the acoustic structure of common marmoset (*Callithrix jacchus*) phee calls. Am J Primatol 1993;30:37–54.
[396] Rylands AB. Marmosets and tamarins: systematics, behavior, and ecology. Oxford University Press; 1993.
[397] McMahon DB, Jones AP, Bondar IV, Leopold DA. Face-selective neurons maintain consistent visual responses across months. Proc Natl Acad Sci USA 2014;111(22):8251–6.
[398] Buzsáki G, Moser EI. Memory, navigation and theta rhythm in the hippocampal-entorhinal system. Nat Neurosci 2013;16(2): 130–8.
[399] Curtis JC, Kleinfeld D. Phase-to-rate transformations encode touch in cortical neurons of a scanning sensorimotor system. Nat Neurosci 2009;12(4):492–501.
[400] Ferezou I, Bolea S, Petersen CC. Visualizing the cortical representation of whisker touch: voltage-sensitive dye imaging in freely moving mice. Neuron 2006;50(4):617–29.
[401] Ito J, Maldonado P, Singer W, Grün S. Saccade-related modulations of neuronal excitability support synchrony of visually elicited spikes. Cereb Cortex 2011;21(11):2482–97.
[402] Jutras MJ, Fries P, Buffalo EA. Oscillatory activity in the monkey hippocampus during visual exploration and memory formation. Proc Natl Acad Sci USA 2013;110(32):13144–9.

Further Reading

[1] Kirk EC. Comparative morphology of the eye in primates. Anat Rec A Discov Mol Cell Evol Biol 2004;281(1):1095–103.
[2] Wright PC. The nocturnal primate niche in the New World. J Human Evol 1989;18(7):635–58.

CHAPTER

23

Marmosets in Neurologic Disease Research: Parkinson's Disease

Ingrid H.C.H.M. Philippens

Animal Science Department, Biomedical Primate Research Centre, Rijswijk, The Netherlands

INTRODUCTION

Parkinson's disease is, next to Alzheimer's disease, the second most common neurodegenerative disorder [1,2]. It is a progressive motor disorder of the central nervous system (CNS). As symptoms develop slowly over time, Parkinson's disease patients suffer from considerable motor disabilities 5–10 years after diagnosis of the disease, with a mean onset age of 60. Although neurodegenerative diseases may appear at any age, the risk increases with age. Presently, about six million patients have been diagnosed worldwide, with a prevalence of 0.3%–1% that increases to 5%–10% at ages above 65 [2,3]. Therefore, the increasing prevalence of neurodegenerative diseases such as Parkinson's disease is an increasing concern for aging societies.

However, Parkinson's disease is not a disease of modern society. An overview of important historical facts is shown in Fig. 23.1. One of the first descriptions dates from 600 BC in ancient Indian medical literature [4]. Around that time (500 BC), there was even a treatment with L-DOPA (levodopa) containing seeds from the *Mucuna pruriens* (velvet bean), which does not differ that much from the treatment that is still the most effective today, synthetic L-DOPA. In Western literature, the first description about the shaking palsy was from the Greek physician and philosopher Galen in 175. But the first detailed medical essay was written by the English surgeon James Parkinson in his Essay on the Shaking Palsy in 1817 (reprint: Parkinson [5]). The observations made by Parkinson are still of clinical relevance for the diagnosis of the disease, 200 years later.

These observations are based on motor symptoms such as resting tremor, rigidity and akinesia (expressed as absence of normal unconscious movements, such as arm swing in walking), bradykinesia (slowness of movement), hypokinesia (reduction in movement amplitude), paucity of normal facial expression, stooped posture, reduced normal postural reflexes, and the inability to begin a voluntary movement such as walking [3,6,7]. These symptoms are caused by specific degeneration of neurons located in the substantia nigra, a nucleus that is involved in control of movement. At the time the patient is diagnosed, at least 50%–60% of these neurons in the substantia nigra are lost [8]. It took more than a century after the publication of the motor symptoms by Parkinson before the role of dopamine in the control of movements was discovered by Carlsson in 1957 [9]. For this important finding, he was rewarded with the 2000 Nobel Prize in Physiology or Medicine together with Paul Greengard who demonstrated the phosphorylation after binding of dopamine on its target receptor. This was also the beginning of the first pharmacological treatment with levodopa (L-DOPA), a precursor of dopamine, in the 1960s.

At that time, the presence of intraneuronal cytoplasmic inclusions, named Lewy bodies after Frederic Lewy (1885–1950), was also discovered [10–12]. The inclusions are aggregated clusters of the protein α-synuclein deposits that are formed from monomers of α-synuclein, which coagulate to oligomers and finally evolve into highly insoluble fibrillar aggregates [13–15]. These protein deposits can increase with age and are classified according to Braak staging [16]. The appearance of Lewy bodies starts in the brain stem and spreads to more frontal regions of the brain over time [14]. Debate is still ongoing as to whether these different aggregates of α-synuclein are toxic to the cell

The Common Marmoset in Captivity and Biomedical Research
https://doi.org/10.1016/B978-0-12-811829-0.00023-6

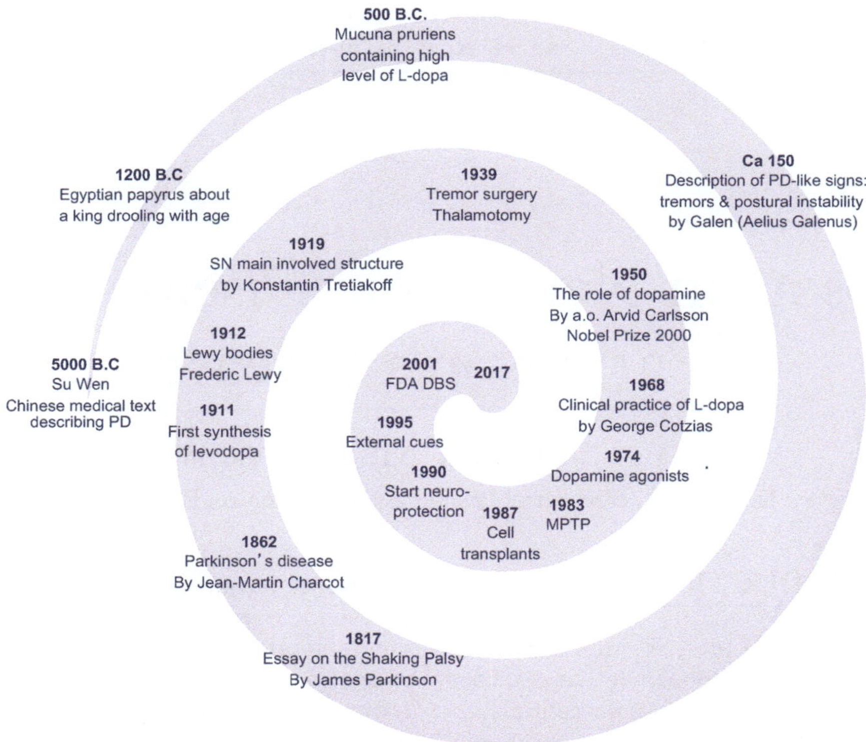

FIGURE 23.1 Historical overview of important insights and therapeutic milestones of Parkinson's disease. *DBS*, deep brain stimulation; *FDA*, Food and Drug Administration; *MPTP*, neurotoxin to induce Parkinson (see section "The MPTP induction model"); *SN*, substantia nigra.

[17]. These typical Lewy bodies are not only exclusively involved in Parkinson's disease but are also found in Lewy body dementia (LBD) in which patients suffer from dementia, but lack the dopaminergic cell loss and accompanying motor symptoms found in Parkinson's disease [12,18]. However, a causal role of Lewy bodies in nonmotor-related symptoms is still evident [17,19].

The nonmotor symptoms associated with PD can often reduce the patient's quality of life even more than motor aspects [20]. One of the nonmotor symptoms, associated with early PD and often presenting in a prodromal phase, is sleep problems which are reported in about 80% of all Parkinson patients [21–23]. Sleep problems can range from reduced sleep efficiency and difficulty in turning in bed to motor problems during rapid eye movement (REM) sleep diagnosed as REM sleep behavior disorder. The latter observation has been proposed as a useful preclinical biomarker for early diagnosis and preclinical research [24–26]. Besides sleep monitoring, live imaging of the dopamine transporter in the brain is introduced as a diagnostic tool, and nonmotor symptoms such as depression and olfactory dysfunction are under current investigation for having diagnostic qualities [27].

Most early nonmotor markers, such as cognitive dysfunction, other sleep abnormalities, dementia, and autonomic dysfunction [6,11,28–30], are often too general for diagnosis of Parkinson's disease but can be used as indicators for developing Parkinson's disease later in life [20]. These early nonmotor markers can also be used in animal models for the validation of early therapeutic strategies to move forward the start of a treatment to limit the neurodegenerative process.

As the current treatment is still heavily dependent on dopamine replacement therapy, which is primarily aimed at symptom control, the progressive loss of nonreplaceable neurons in Parkinson's disease cannot be stopped and reversed. Therefore, it may be a better strategy to focus on early detection of signs of Parkinson's disease and subsequent early intervention in the accumulating neuronal loss. However, to date no disease-modifying therapy has been approved. This shortcoming has been well recognized [30–32], and research efforts to develop disease-modifying treatment strategies are the current focus of both clinicians and basic scientists. However, disease-modifying studies in humans are difficult to perform, and clinical studies of early Parkinson's disease indications are scarce. Animal studies may fulfill this lacuna. Therefore, the availability of adequate animal models would be an excellent asset in current Parkinson's disease research. Cell cultures or invertebrate models are useful [33,34], but they can only model Parkinson's disease to a certain extent. Disease models should ideally mimic the main features of the disease pathology and additionally show the typical parkinsonian syndrome.

ETIOLOGY OF PD

To model a disease, it is important for the construct validity to have some degree of similarity between the mechanisms underlying the behavior in the model. However, models can mimic the pathology and the symptoms of the disease but not easily the etiology [35]. Therefore, an animal model is always limited by the state of knowledge about the disease.

Unfortunately, the cause of most Parkinson's disease cases is unknown, with uncertainty about the role of environmental and genetic factors. It seems to be a complex interplay between different factors.

In 95% of Parkinson's disease cases, there is no genetic linkage. These types of Parkinson's disease are named idiopathic or the sporadic form of Parkinson's disease [36]. The environmental factor theory posits that Parkinson's disease results from exposure to neurotoxins. For example, rural areas have been associated with Parkinson's disease as relations were found with use of herbicides, such as paraquat and pesticides, and rotenone or well water, although results are inconsistent and limited to a small group within the Parkinson's disease population [37,38]. Another example is the finding that people exposed to the neurotoxin MPTP (1-methyl-4-phenyl-1,2,3,6-tetrahydropyridine) developed a syndrome with features similar to Parkinson's disease [39].

The genetically related 5% of Parkinson's disease cases are mostly autosomal dominant or recessive inheritance patterns that result in dopaminergic cell loss [7]. These individuals often develop early-onset Parkinson's disease, which starts before the age of 40 [20]. Single-gene mutations are now identified at 10 loci and 6 genes, including *SNCA* (α-synuclein), *PINK1, PARKIN, DJ-1, MAPT,* and *LRRK2*, which code for proteins involved in protein phosphorylation, protein misfolding, oxidative stress, mitochondrial dysfunction, and impairment of the ubiquitin–proteasome system, processes that are closely interlinked [1,6,40,41].

Then again, it is deemed unlikely that neuronal cell death in Parkinson's disease emerges from one single factor and instead likely results from a multifactorial cascade of deleterious events [42–45]. The different processes involved, besides protein misfolding, are proteasomal dysfunction, excitotoxicity, mitochondrial dysfunction, oxidative stress, and inflammatory responses [6,7,32,45a,46–50]. All of these are intertwined, and dysfunction in one process may lead to disturbances in the others (Fig. 23.2).

Features Involved in the Pathogenesis of Parkinson's Disease

Protein Misfolding

For a healthy protein homeostasis, it is important to clear improperly folded or otherwise dysfunctional proteins [51]. Ubiquitin forms a complex with these proteins and uses activating, conjugating, and ligating enzymes to complete protein degradation [52]. Malfunction of this system leads to protein aggregates, which in the end lead to cell death. These errors are mainly a consequence of malfunctioning of the involved enzymes caused by gene mutations, i.e., PARKIN and UCH-L1 [51–53]. There is still a debate about whether the protein aggregation is neurotoxic or a result of neuroprotective actions of the cell [49].

Proteasomal Dysfunction

The evidence pointing to proteasomal dysfunction in Parkinson's disease comes from analyses of proteasomal enzymatic activity, levels of its subunits, and expression of its activators in the brains of Parkinson's disease patients [54]. Proteasome dysfunction causes misfolding of proteins and protein aggregates, such as the α-synuclein expressing Lewy bodies in Parkinson's disease [55]. The ubiquitin–proteasome system has a critical role in the maintenance of cellular function and viability by clearance of misfolded proteins.

Excitotoxicity

Excitotoxicity occurs when neurons are exposed to high levels of glutamate that causes a persistent activation of the N-methyl-D-aspartate acid (NMDA) and α-amino-3-hydroxy-5-methylisoxazole propionic acid (AMPA) receptors and voltage-gated calcium channels resulting in a lethal influx of extracellular calcium. This overload in calcium leads to cell death via the generation of free radicals, inhibition of protein synthesis, and mitochondrial damage [48,56,57]. The increased glutaminergic drive from the subthalamic nucleus (STN) due to the disturbed balance in the basal ganglia in Parkinson's disease is also suggested to be a source of excitotoxicity [58].

Mitochondrial Dysfunction

The dysfunction of the mitochondria, specifically in complex I of the inner membrane of the electron transport chain, leads to energy failure, increased reactive oxygen species (ROS) production, and a release of apoptosis signaling molecules [59]. The Parkinson gene *PINK1* and the experimental Parkinson's disease inducing toxins (MPTP and rotenone) also specifically affect complex I.

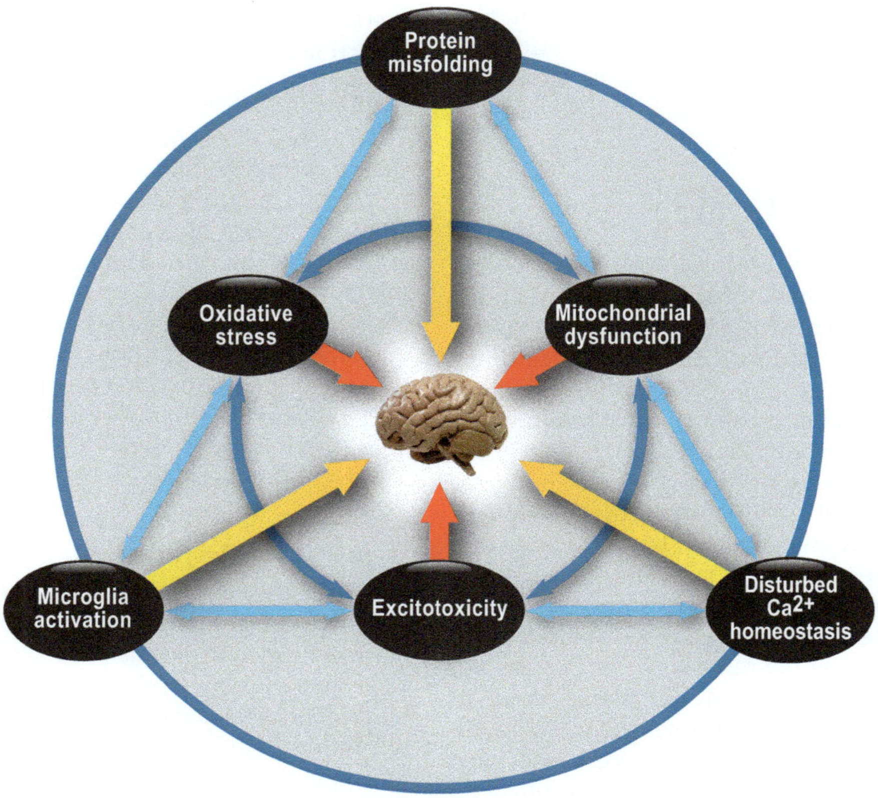

FIGURE 23.2 Schematic diagram depicting several pathogenic features found in a broad range of neurodegenerative diseases. The features in the inner circle are responsible for the initial cellular insult that causes failure of neuronal homeostasis. The features in the outer circle are responsible for the maintenance and progression of disease pathology. In some cases, such as in inherited Parkinson's disease, the pathogenic process may originate from the outer circle. All of these features of disease are potential targets for therapy. *From Philippens IHCHM, 't Hart BA, Torres G. The MPTP marmoset model of Parkinsonism: a multi-purpose non-human primate model for neurodegenerative diseases. Drug Discov Today 2010; 15(23–24):985–90.*

Oxidative Stress

Under normal circumstances, there is a balance between production and removal of ROS formed as by-products of mitochondrial respiration. In Parkinson's disease, brain markers of free radicals are elevated and levels of glutathione, an antioxidant enzyme which is protective against hydrogen peroxide, are decreased [60–62]. Homovanillic acid (HVA), a metabolite of dopamine, is a potential antioxidant [63]. It can thus be envisaged that depletion of dopamine may reduce the antioxidant protection of the neurons. Additionally, inherited mutations in the gene *DJ-1* are directly linked to oxidative stress [64] resulting in Parkinson's disease. Reaction of these molecules with DNA, proteins, and lipids alters their structure and results in cellular damage.

Inflammatory Responses

Postmortem studies revealed elevated levels of inflammatory markers in the Parkinson's disease brain, such as TNF-alpha, IL1-beta, IFN-gamma, and an increase in microglia [6,65]. Additionally, there is also a peripheral immune response observed in the serum of Parkinson's disease patients mediated by several cytokines [66], with the involvement of the gut recently proposed [67]. It is still debated whether the peripheral and CNS immune responses are a cause or a consequence of the disease. However, the immune system may contribute to progressive nigral loss based on increased levels of microglial activation and expression of proinflammatory cytokines [66,68,69].

MODELING PARKINSON'S DISEASE

Introduction

As mentioned before, Parkinson's disease has not one single cause. Most likely, different systems are involved in Parkinson's disease pathology. The multifactorial aspect of Parkinson's disease underscores the need for models that link disease manifestation to the underlying pathology. Studying Parkinson's disease neurobiology in combination with disease manifestation in humans is limited to clinical trials and postmortem material. Therefore, to find new targets for disease-modifying treatments, having adequate animal models available is a great asset in today's research community. Different animal models are available, ranging from fruit flies to monkeys, that mimic Parkinson's disease symptoms and dopaminergic cell loss. Most of the animal models are based on the concept that parkinsonian signs are related to dopaminergic nigral cell loss. However, none mimic the complete range of the complex chronic neurodegenerative features of human Parkinson's disease [70].

The dopamine deficiency observed in Parkinson's disease is the main event underlying the pathophysiology of the motor symptomatology. Particular features of neurodegeneration might be studied in cells or invertebrates [33,34]. However, to investigate neuroprotective therapeutic interventions, animal models should at least feature the typical—preferably progressive—loss of dopamine neurons in the substantia nigra in combination with the associated dopamine reduction in the striatum. In some cases, the presence of the typical Parkinson's disease symptomatology or the Parkinson's disease-specific Lewy body formation can be an important addition to the animal model. Consequently, multiple animal models are available, reflecting different aspects of Parkinson's disease. None of the available models replicates the complexity of Parkinson's disease, but some can be used to gain insight into certain aspects of the disease.

Types of PD Induction Models

Multiple genes have been indicated to play a role in the pathology, onset, mechanism of action, or affected pathway in Parkinson's disease [53]. Genetic models include transgenic, knockout, knockin models or overexpression of genes of particular Parkinson's disease genes. In most cases mice are used as the model organism because of the relatively low costs and the relative ease of genetic modification. Using these models, different genes were found to play a role in the pathology, onset, mechanism of action, or affected pathway in Parkinson's disease [6,49]. Most commonly used genetic variants are DJ-1- and PINK1-mutated animals; however, other genes involved have also been genetically manipulated [40,71]. Besides the genetic models, chemically induced models are available.

Several neurotoxins can be administered to induce Parkinson's disease-like symptoms via different pathways. 6-Hydroxydopamine (6-OHDA) and paraquat cause direct oxidative stress. 6-OHDA was the first neurotoxin discovered to induce selective catecholaminergic cell death [72]. As 6-OHDA is unable to cross the blood–brain barrier, it must be locally administered via injections into the medial forebrain bundle, the substantia nigra pars compacta, or striatum. The severity of the lesion depends on the physical distance toward the region of interest, which makes this model less suitable for studying the molecular mechanisms of neurodegeneration [73]. 6-OHDA is recognized by substantia nigra neurons as dopamine and is taken up by the cell where it then exerts its toxic properties. An advantage of this model is that unilateral 6-OHDA lesions, generally used in rats, have proven to be very reproducible over time, in which the noninfused hemisphere can serve as an interanimal control [74]. Disadvantages are that other neurotransmitter systems are also affected [75], and there is no development of L-DOPA-induced dyskinesias. Other ways to induce Parkinson's disease are by intracerebral injections of lipopolysaccharides (LPS) that cause nitrative stress (caused by reactive nitrogen species) and an innate inflammatory response [6,76]. Proteasomal dysfunction can be induced by Z-Ile-Glu(OBut)-Ala-Leu-al (PSI), epoxomicin, and lactacystin [77], whereas quinolinic and ibotenic acid cause immediate excitotoxicity [6]. Also, repeated systemic administration of agricultural chemicals, such as rotenone and paraquat that easily penetrate the blood–brain barrier, is one of the current alternatives [78,79]. These chemicals affect neurons by disturbing processes such as the glutamate balance, increasing ROS production, affecting the mitochondrial respiration, and the formation of α-synuclein-positive inclusions (reviewed by Bove et al. [73]). Disadvantages of these models are the variable induction effects and lack of specificity [80]. Of the neurotoxins, 1-methyl-4-phenyl-1,2,3,6-tetrahydropyridine (MPTP; Fig. 23.3), that easily crosses the blood–brain barrier with a high specificity for dopamine neurons, is most frequently used in animal models to induce Parkinson's disease [81]. MPTP intoxication is based on mitochondrial dysfunction.

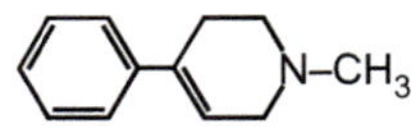

FIGURE 23.3 Structure formula of 1-methyl-4-phenyl-1,2,3,6-tetrahydropyridine (MPTP).

THE MPTP INDUCTION MODEL

Background

In the early 1980s, MPTP (Fig. 23.3) was discovered as a side product formed during the chemical synthesis of meperidine, a synthetic form of heroin [39,82,83] when Californian drug users used the wrong temperature in one of the synthesis steps for the drug meperidine [39].

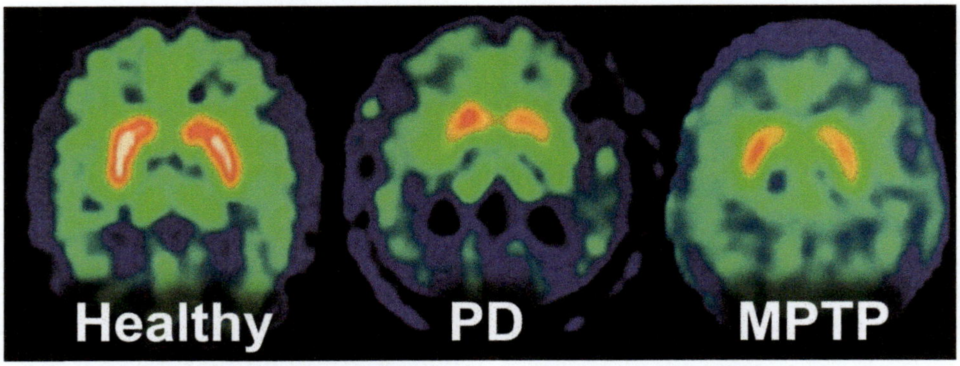

FIGURE 23.4 PET scan of dopamine activity in the striatum of a normal brain, brain from a Parkinson's disease patient, and from a patient exposed to MPTP. *Adapted from Snow BJ, Vingerhoets FJG, Langston JW, Tetrud JW, Sossi V, Calne DB. Pattern of dopaminergic loss in the striatum of humans with MPTP induced parkinsonism. J Neurol Neurosurg Psychiatr 2000;68:313–6.*

Unfortunately, drug users developed Parkinson's disease-like symptoms after the use of this MPTP-polluted synthetic heroin. Positron emission tomography (PET) scans showed similar damage to striatal regions as seen in Parkinson's disease patients, after only mild exposure [39,83] (Fig. 23.4).

In the animal experiments that followed this accidental parkinsonism, it became clear that the MPTP caused selective cell death of dopamine neurons [84], highly specific for the substantia nigra. The highly lipophilic MPTP can easily cross the blood–brain barrier in mice [85], monkeys [86], and humans [87], but not in rats due to clearance of MPTP outside the brain [88]. This observation led to the identification of MPTP as a drug derivative with specific neurotoxic properties. Researchers started to investigate MPTP and its relevance to Parkinson's, leading to the now preferred protocols for modeling Parkinson's disease in animals [49]. MPTP causes dopaminergic cell loss and induces similar clinical and pathological features as observed in Parkinson's disease, which makes the MPTP model a very important and clinically relevant model for studies of Parkinson's disease.

Mechanism of Action

MPTP readily crosses the blood–brain barrier and is metabolized into 1-methyl-2,3-dihydropyridinium ($MPDP^+$) by the enzyme MOA-B and subsequently oxidized into the toxic metabolite 1-methyl-4-phenylpyridinium (MPP^+) in glial cells (e.g., astrocytes) [89] (Fig. 23.5). Thereafter, MPP^+ is released in the extracellular space by unknown mechanisms. MPP^+ is then taken up via the dopamine transporter into dopamine neurons.

Once inside the neuron, MPP^+ is stored in vesicles by the vesicular monoamine transporter [90] or accumulates in the mitochondria where it impairs mitochondrial respiration by inhibiting electron transport enzyme ubiquinone oxidoreductase (complex I) of the electron transport chain and leads to a reduction in cellular adenosine triphosphate (ATP) [91,92]. Additionally, MPP^+ has also been reported to directly inhibit other inner membrane complexes, namely mitochondrial complex III (ferrocytochrome c oxidoreductase) and IV (cytochrome c oxidase) of the electron transport chain [92,93]. Corresponding to these mitochondrial effects of MPP^+, selective reduction of complex I activity has also been found in the substantia nigra and striatum of Parkinson's disease patients.

As already described mitochondria are the energy suppliers of the cell, and mitochondrial dysfunction leads to a multitude of problems inside the cell that ultimately result in cell death. Thus, MPTP intoxication resembles both the dopaminergic cell loss and the behavioral and neurochemical deficits seen in Parkinson's disease patients [76,83]. Although MPP^+ is the metabolite that actually causes these mitochondrial deficits, it is common to refer to MPTP as the toxic agent in this model.

MPTP Protocols

Different dosing regimens of MPTP lead to the induction of different mechanisms of cell death. Acute administration of high dosages of MPTP leads to rapid degeneration through necrotic cell death [94,95], whereas subacute regimens using lower MPTP dosages administered over several days achieve delayed degeneration with apoptotic cell death [96–98]. With these regimens, a 70%–95% loss of tyrosine hydroxylase (TH) mRNA in the substantia nigra pars compacta is found, representing the end stage of Parkinson's disease. By lowering the MPTP dose, a shift can be made from the end stage to an early stage of Parkinson's disease with nonmotor symptoms [99].

For the early stage, a chronic slow progressive model, in which even lower dosages of MPTP are administered over a longer period, is generally regarded as the

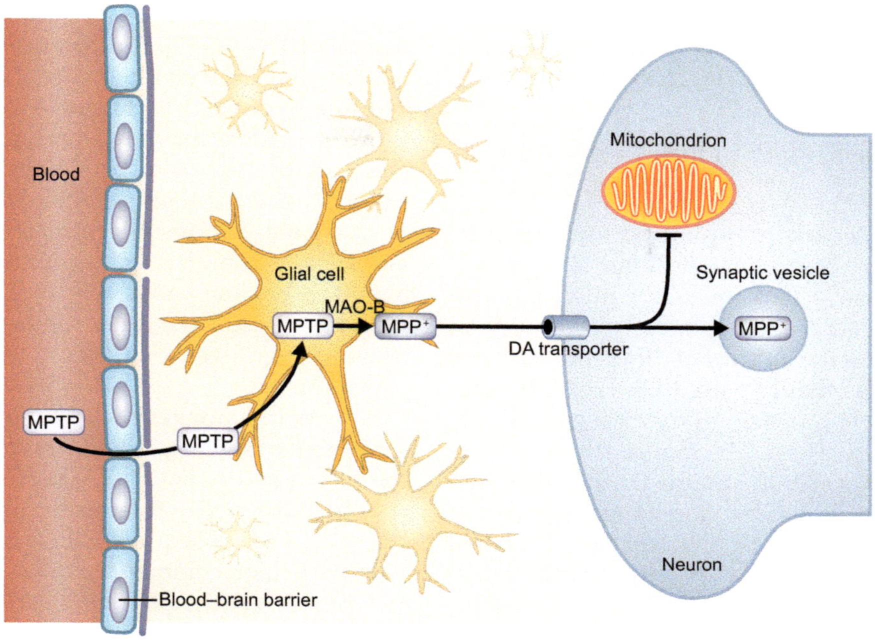

FIGURE 23.5 Schematic overview of MPTP uptake and metabolism in the brain. MPTP crosses the blood–brain barrier and is metabolized by MAO-B in glial cells into $MPDP^+$, which is oxidized into the toxic metabolite MPP^+. After the release in the synaptic cleft, MPP^+ is taken up in dopaminergic neurons by dopamine (DA) transporters. Inside these neurons, MPP^+ is partly stored in the synaptic vesicles. The toxic effect of MPP^+ is its blocking of complex I of the mitochondrial electron transport chain resulting in less ATP production and the formation of free radicals.

preferred animal model of Parkinson's disease [100–102,112]. It best resembles human Parkinson's disease and allows a wider window of opportunity to investigate the early stage of the disease that closely mimics the prodromal and early nonmotor stages of the human counterpart of the disease. A disadvantage of the chronic regimen is the substantial spontaneous behavioral recovery after cessation of the toxin delivery [70,100]. This recovery was not found with a subchronic model for the early stage in which a low dose of MPTP was given for three consecutive days [99]. Not only the dosage but also the time interval between dosing plays an important role in the severity of the brain pathology. In Fig. 23.6, an overview of different induction protocols in the marmoset from the same laboratory is shown. A total dose of 6 mg/kg can result in 24.2% tyrosine hydroxylase (TH)-positive neurons in a subacute protocol [103] and, respectively, 81.0% TH-positive neurons in a chronic induction protocol [100]. Even when the induction takes place within 2 weeks, the interval between the MPTP injections results in differences in brain pathology independent of the dosage used (respectively, 24.2 and 47.8% TH-positive neurons). Therefore, it is important to take the frequency of dosing into account when setting up an induction protocol.

Unfortunately, the MPTP induction model does not include one of the important characteristic features of Parkinson's disease, the Lewy bodies. However, sometimes inclusions resembling Lewy bodies have been described [105,106]. A newly developed combination

week	1					2					3	4	5	6	7	8	9	10	11	12		†	TH+ cells
days	m	t	m	t	f	m	t	m	t	f													
MPTP dose (s.c.)			2	1	1	1			1												(6 mg/kg)	5	24.2 ± 5.3
	2			2		1.5			1.5												(7 mg/kg)	5	47.8 ± 3.0
					0.5					0.5	0.5	0.5	0.5	0.5	0.5	0.5	0.5	0.5	0.5	0.5	(6 mg/kg)	20	81.0 ± 9.0

FIGURE 23.6 Overview of induction protocols in the marmoset from the same research group [100,103,104]. Three protocols are shown. The first two are subacute induction protocols with a total MPTP dose of, respectively, 6 and 7 mg/kg s.c. within a 2-week period (the week days are indicated with the first letter of the day). The third is a chronic induction protocol with a weekly MPTP dose of 0.5 mg/kg s.c. (total dose 6 mg/kg). The time interval between dosing is the determining factor for the severity of brain damage. Brain damage is inversely related to the remaining percentage of TH-positive cells. † indicates the time of sacrifice in weeks starting from the beginning of the MPTP injections.

of α-synuclein viral vectors inducing overexpression of the protein in the MPTP model or continuous infusion of MPTP can cover this lacuna [107,108].

Animal Models for MPTP-Induced Parkinson's Disease

Animal models offer an opportunity to link low-level neurodegeneration to disease manifestation and a possibility to investigate strategies for intervention therapy. Many Parkinson's studies have been performed on different species to test the efficacy of different treatment regimens. MPTP is mainly used in rodents and nonhuman primates [49], although rodents are less sensitive to MPTP toxicity. The C57b6 mouse strain is sensitive for MPTP after systemic administration, but high dosages are needed. Rats are resistant to systemic MPTP due to a different metabolism of MPTP caused by low concentrations of MAO-B [89]. This leads to virtually no MPP^{+} in the brain after MPTP administration [109]. Local intracerebral injections with MPP^{+} in the rat, however, led to similar dopaminergic cell loss as seen in mice and nonhuman primates [110].

As behavioral changes recover very fast in mice after Parkinson's disease induction, the mouse MPTP model is useful in studies investigating neuropathological and neurochemical changes [89]. The monkey MPTP model seems to be more suitable for behavioral tests as they express the same type of long-lasting behavioral disturbances as humans, whereas the behavior of rodents tends to be reversed within a couple of days.

In general, in all mammal species the MPTP model matches all four criteria of validity according to van der Staay et al.: reliability and replicability (internal validity), predictive validity, construct validity, and external validity (i.e., generalizability) [35,111].

THE NONHUMAN PRIMATE AS A MODEL FOR PARKINSON'S DISEASE

Primate research has a high level of validity with respect to many human diseases because nonhuman primates are more similar to humans at all levels of analysis [111]. For Parkinson's disease research, nonhuman primates share with humans a very similar functional organization of the striatal regions [112,113]. As a consequence, monkeys have better developed fine motor skills than rodents, which is important for testing of motor-related dysfunctions in Parkinson's disease. Nonhuman primates have also a similar hand–foot use for measuring complex motor skill such as the hand–eye coordination [114]. In addition, nonhuman primates have a vertical body position, which allows better analysis of postural imbalances characteristic of Parkinson's disease.

Furthermore, monkeys are, like humans, very sensitive to MPTP [86,115]; and because of their genetic, physiologic, and behavioral proximity, they contain many features of clinical Parkinson's disease. Monkeys also show a comparable responsiveness to all dopamine medications known to be effective in Parkinson's disease [76], making them a valuable addition to the range of available Parkinson's disease models. Old World monkeys with high cognitive abilities, such as macaque monkeys [86] and baboons [116], are interesting because they can handle complex behavioral tasks enabling testing cognitive deficits in the late phase of parkinsonism.

New World monkeys, such as squirrel monkeys, capuchin monkeys, and common marmosets, are, although somewhat less sensitive to MPTP, especially useful not only because of the aforementioned abilities but also for their size and consequently their use in the laboratory.

In particular, common marmosets are frequently used for the MPTP model [96,98,112,117,118]. These New World monkeys are genetically approximately 96% identical to humans [119,119a], small in size, easy to handle, and survive well in captivity. The high reproductive efficiency (high availability of marmosets) is another advantage of marmosets for biomedical research. In addition, research with marmosets has proven to reduce costs and safety hazards in comparison with Old World monkeys [76].

The popular marmoset MPTP model offers several advantages in studying disease therapies and neuroprotective methods. Unlike the rodent models [120–122], MPTP-treated marmosets show optimal face validity with a wide range of parkinsonian behaviors [115,123], including levodopa-induced dyskinesia [124], mood [123], hallucinations [125], bladder problems [126], olfactory impairment [99], and sleep-related symptoms [26,127]. The "clinical" condition of the parkinsonian state in this MPTP monkey model is generally focused on the motor symptoms. Motor symptoms reported in marmosets use extensive rating scales [103,127–129] and extensive tests toward locomotor activity [103,123], hand–eye coordination [75,103], jumping behavior and initiation, and turning behavior [127].

PARKINSON'S DISEASE RESEARCH IN MARMOSETS

Parkinsonian behavior of MPTP monkeys is comparable with human Parkinson's disease symptoms [81]. However, like parkinsonian behavior in the clinic, not all types of symptoms are affected in the same manner by the MPTP intoxication. Therefore, a complete behavioral sketch of a parkinsonian marmoset should preferably include a range of motor and nonmotor behavior test systems to create an overview of the full range of impairments. To do so, different behavioral tasks have been developed to monitor several Parkinson's disease-related symptoms, both motor and nonmotor, in the marmoset [26,114,123,127,131]. Nonmotor symptoms can be measured by clinical observations, home-cage activity, observations of fear- and arousal-related behaviors, cognition, and sleep behavior by telemetric electrocencephalogram (EEG). Motor behavior can be examined by investigating jumping behavior, locomotor activity, axial movement, and fine motor skills, such as hand—eye coordination. Together these tests capture multiple features of Parkinson's disease-related motor and nonmotor dysfunction. In the next paragraphs, several methods are described in more detail.

MOTOR-RELATED ASPECTS IN MARMOSETS

As Parkinson's disease is a motor disorder, most of the symptoms are related to motor function. Several test systems and quantification methods have been described for marmoset motor behavior. There are methods to evaluate the "full-body" motor behavior, for example, the general activity tests [114,132]. Some other methods are designed to evaluate unilateral performance, such as the sticky label test, a fine motor task in which the time is measured to remove an adhesive label attached around the ipsilateral foot, or the rotation task, in which the number of the unilateral lesion-induced spontaneous ipsilateral rotations of the animal is measured [75]. And there are methods that focus on more specific motor performances like "arm-reaching" motor behavior such as the staircase test [133] and the hand—eye coordination task [114]. In some test systems it is necessary to motivate the animals to move, by using positive or negative reinforcement, which may influence the outcome of the performance as dopamine is released by reward systems. To minimize the adverse effects of reinforcement on motor behavior, the animals' motivation should be as close to its natural behavioral paradigm as possible.

Clinical Parkinsonian Symptoms

Clinical assessment of Parkinson's disease symptoms is mostly done with the UPDRS (Unified Parkinson's Disease Rating Scale). Adapted rating scales are developed for MPTP monkeys. Key tests to assess Parkinson's disease-related features in marmosets are clinical observations, such as tremor, rigidity, mobility, and general well-being [26,127]. In addition, evoked fear- and arousal-related behaviors can be observed in a human interaction test, in which the observer is positioned in front of the cage [103].

Clinical Score Scale

The "clinical score" is a rating scale, including items for the cardinal parkinsonian symptoms and some marmoset-specific items [103]. The observational tests can be scored on daily base. In most cases the items are rated from 0 (= normal/healthy) to 4 (= severely affected). Clinical scores included the items apathy (no interest in their environment), immobility, muscle rigidity (assessed as stiffness of legs and tail), tremors, and inadequate grooming. The scores are summed to give an overall score that is related to the severity of the disease induction. A mild induction protocol will lead to less severe scores.

Abnormal Involuntary Movement Scale

Another observational rating scale is the abnormal involuntary movement scale [123]. This observational scale is used to quantify dyskinesia in the marmoset. Dyskinesia is an uncontrolled involuntary movement expressed in the arms, legs or head, and trunk, which includes items such as facial behaviors (jaw, facial muscles, tongue, and lips), full-body behaviors (upper, lower, and trunk), and general severity of involuntary movements and incapacitation due to these movements [134,135]. These items are also scored from 0 (= normal/healthy) to 4 (= severely affected). The scores are summed to give an overall score of the involuntary movements. Dyskinesia can be induced by repeated high dosages of oral L-DOPA (12.5 mg/kg, BID) in parkinsonian marmosets for at least 2—3 weeks [136]. The total score is highest after 30 minutes to 1 h after L-DOPA administration.

Human Threat Test

The human threat test is a nonhuman primate putative model of anxiety. It is based on findings that marmosets will exhibit fear-related behavior in the presence of a human observer in front of the home cage [123,137]. Most pronounced behavior would be retreating to the back of the cage and showing characteristic postures. To assess the behavior, the observer made eye contact with the marmoset throughout a 2-min test period. During this period, the movements, behavior, and position of the marmoset in the cage are recorded directly by the observer and by video registration. A range of parameters was obtained according to Carey et al. [137] based on Stevenson and Poole [138]: (1) the number of characteristic postures exhibited: tail posture (tail raise to present the genital region), scent marking (the anal and genital area is pressed against the substrate to be marked with excretion of the glands), arched piloerection (arched back posture with full-body piloerection), slit stare (stare with the eyes half closed in combination with tufts flattened and exposure of the teeth), rearing (upright position with flexed paws), twisting (head and torso movement from side to side); (2) the time spent in the front of the cage; and (3) the number of position changes in the cage.

Home-Cage Activity for General Activity and Circadian Rhythm

Measurement of changes in general activity has direct application to the clinic as hypokinesia and disturbed circadian rhythm are common features of Parkinson's disease. In human Parkinson's disease, activity levels and rhythmic patterns are seriously affected. Telemetric actimeters, connected to the collar of the monkey, are used to measure the 24-h general home-cage activity to assess the circadian rhythm and activity and resting behavior over the day [26,63,138a]. The cumulative activity of 15-s periods is registered. The total 24-h activity or the sum of the daytime and night time can be analyzed. Furthermore, the nap frequency and duration in the afternoon is disturbed in the early stage of Parkinson's disease [138a]. The advantage of this type of measurement is that it gives information of the behavior during the whole day, whereas the other test only measures on specific time points.

FIGURE 23.7 The tower test setup: three successive snapshots of a marmoset performing jumping behavior in the tower. The animal can move freely in the tower setup. The reached-level and the time are noted when the marmoset changes level.

Tower Test for Jumping Behavior

The so-called tower test makes use of the monkey's natural preference to jump in trees. The tower (35 cm × 35 cm × 250 cm) has a transparent Plexiglas front and contains seven levels of horizontal mounted crossbars with an increasing distance varying from 10 to 50 cm (Fig. 23.7). Animals are allowed to move freely in a vertical direction by jumping on the crossbars [127]. A camera is placed in front of the tower for video analysis. A marmoset enters the tower through a sliding door at the bottom of the tower. A trial starts after opening the sliding door. During each test the marmoset can freely move around on the seven levels for 10 min. All marmosets have to be habituated to the tower before testing. The marmoset's location (level) is noted by nonautomated video analysis. A marmoset has reached a certain level when it is observed that the center of gravity of the marmoset is on the crossbar representing that level.

Hourglass Test for Measuring Akinesia and the Righting Reflex

The hourglass test is used to test the marmoset's axial turning ability in a Plexiglas cylinder (ø11 cm, 27 cm) [127]. This method is based on the transportation tubes for marmosets (Fig. 23.8). One trial consists of one manual 180° turn of the cylinder (like turning an hourglass). The time intervals for the quantification of the axial turning are taken between the cylinder turn and the time a marmoset needs to recover its position (head above legs) with a maximum time of 30 s (also for marmosets which do not turn upright at all). A test consists of 10 subsequent trials. After an intertrial interval of 30 s, a new trial is started by turning the cylinder again. The cylinder is always turned after 30 s, keeping the marmoset (when it had not turned upright) no longer than 30 s upside down.

Healthy marmosets quickly correct their posture in normal direction in a reflex-like manner (righting reflex) when placed upside down (within 2 s). When the motor functions are affected the time to recover the normal posture is prolonged, presenting a measure for akinesia (Fig. 23.8).

Bungalow Test for Explorative Locomotor Activity

For measuring the spontaneous explorative locomotor activity of the marmosets a so-called Bungalow test system, testing the activity of the marmoset by compartment changes, has been developed (Fig. 23.9) [114,123,139]. The monkeys are allowed to move freely

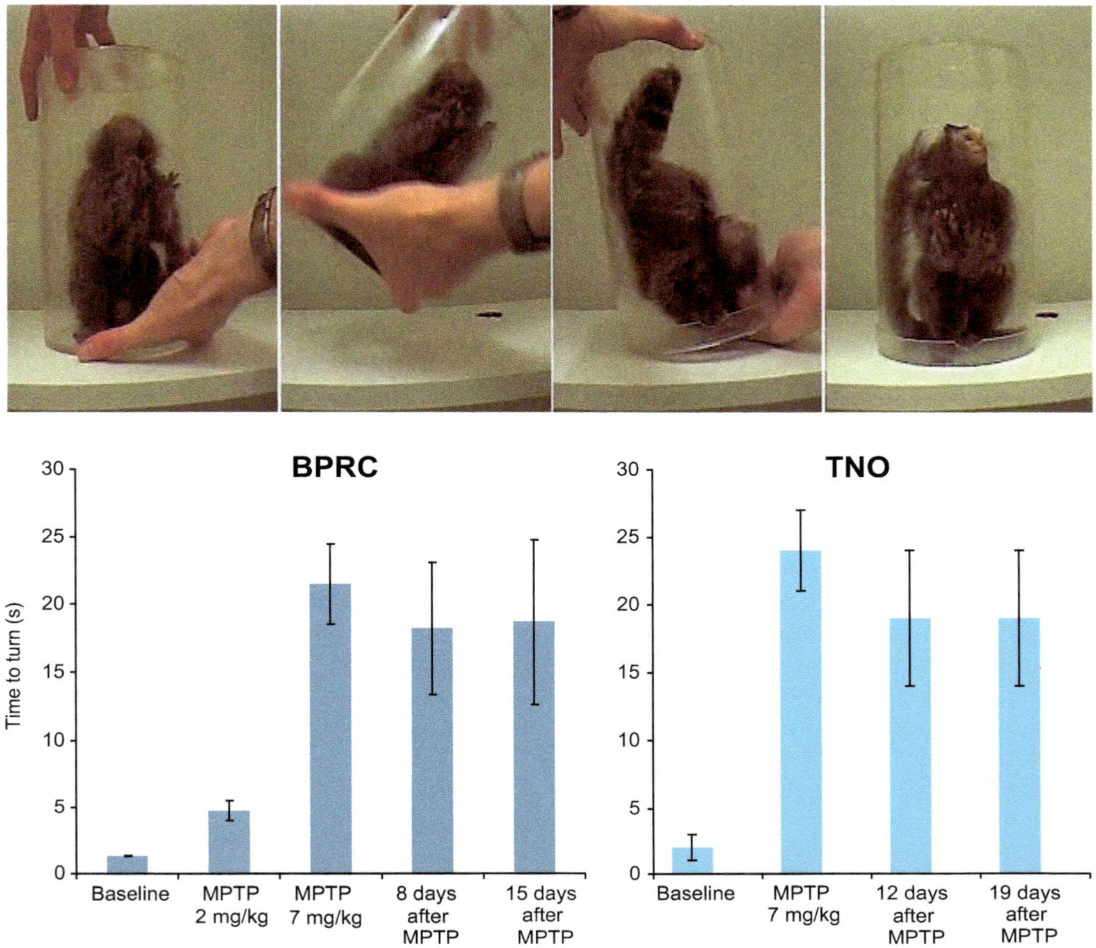

FIGURE 23.8 The hourglass test setup. Top panel: four successive snapshots of a marmoset in the hourglass. The cylinder is turned in one continuous movement and the time it takes for the marmoset to turn upright is noted. Bottom panel: example of external validity of the hourglass method. The average time to turn (s) of MPTP-treated monkeys (7 mg/kg, respectively, n = 5 and n = 6) is shown. Left: a study performed at the BPRC, Rijswijk, The Netherlands [63]. Right: a study performed at TNO Defence, Security and Safety, Rijswijk, The Netherlands [127].

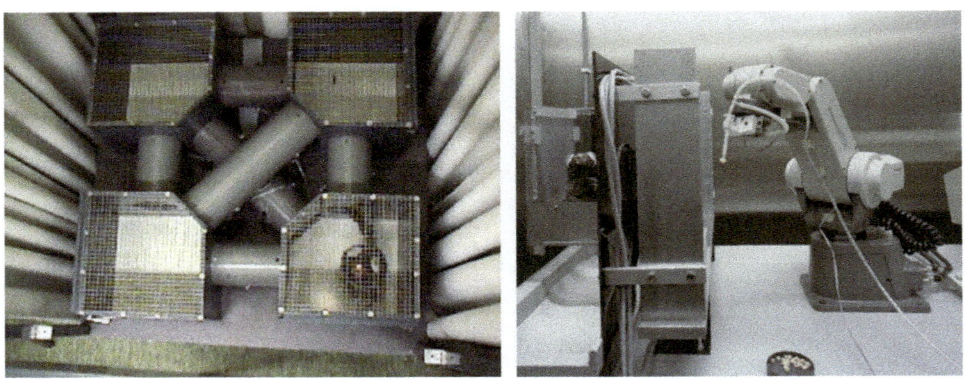

FIGURE 23.9 Pictures of two behavioral test systems. Left: Picture from above of the so-called bungalow test system for explorative locomotor activity. In one of the compartments a marmoset is located. The monkeys can move freely from compartment to compartment. Right: Picture of the robot-guided hand–eye coordination test system. A monkey is sitting in a test cage in front of the testpanel in which the robot arm presents a marshmallow reward.

for about 10 min in a construction consisting of four compartments (23 × 23 × 23 cm) interconnected by nontransparent tubes, which is placed in a homogeneous environment. A video tracking system registers the movement pattern and the position of the monkey in the apparatus. First, the monkeys have to be habituated to the test system for two sessions before the first test session can be recorded. The number of compartment changes is used as a measure of activity. This is a very sensitive method for measuring the progression of parkinsonism in marmosets [114].

Hand–Eye Coordination for Fine Motor Skills

In the clinic, it is noticed that the fine motor skills of patients are reduced due to a combination of tremor, slowness of movement, and disturbed motor planning. In the MPTP-treated monkey fine motor skills can be measured by hand–eye coordination. For measuring these fine motor skills, a reward-related hand–eye coordination test system has been developed (Fig. 23.9). This method makes use of a robot-guided automated noninvasive test setup for learned complex behavior [114,123]. Small marshmallow-like rewards, used as a motivating stimulus, are presented by a robot at different speeds in front of the monkey, challenging the animal to follow and reach out: nonmoving (0.0 m/s for a maximum of 30 s), slow (0.04 m/s), and fast (0.08 m/s) moving. The marmosets are encouraged to display fine motor skill by reward-related grabbing of a moving or static object [114]. All trials start with a brief sound signal to alert the animal. Before the start of the study, all animals have to be trained until they are able to take 75% or more of the presented rewards. The percentage of correct hits is used as a criterion to judge the performance of the animal. It should be noted that test systems that make use of reward-stimulated performance affect the dopamine release that may interfere with the disease-induced reduction of dopamine in the brain.

PREMOTOR PHASE MARKERS OF PARKINSON'S DISEASE

Identification and validation of biomarkers, such as premotor symptoms, are critical to facilitate the early diagnosis of Parkinson's disease [139a]. Examples of premotor indications of early Parkinson's disease include dopamine-related brain imaging [140], olfactory decline, alterations in mood, cognition, and autonomic function, and most notably disturbed sleep [28,30]. For instance, complaints of insomnia are reported in approximately 80% of all Parkinson's disease patients [22,23], and excessive movement during sleep frequently occurs in Parkinson's disease patients [141]. The nonmotor symptoms of Parkinson's disease have only recently started to become of interest to the scientific community [142]. In animal models, this is a very new field so that nonmotor parameters are generally not yet part of the symptom description. However, there are reports on olfactory problems in the MPTP-treated marmoset [143], several studies on sleep problems in nonhuman primates [26,104,144,145], and cognition [146], as well as on constipation [147].

Hence, MPTP-treated nonhuman primates may fill this gap and provide insights into the course of nonmotor effects in relation to the underlying neurodegeneration of the brain. As the focus in Parkinson's disease research has recently shifted from the motor phase to the premotor phase of the disease [26,28,30,32, 104,148–150], it is of particular interest to identify the stages in which premotor symptoms are apparent and precede the motor phase of Parkinson's disease, to investigate early pathological processes of the disease and to develop early treatment approaches. In addition, as there are currently still no blood or cerebrospinal fluid biomarkers of Parkinson's disease, we rely completely on noninvasive markers based on the early premotor symptoms.

Brain Pathology by Imaging Techniques

In animal research, postmortem histopathology is an important method for measuring the efficacy of a disease-modifying treatment at selected time points. Effects on neuropathology are often measured by immunohistochemistry for tyrosine hydroxylase (TH), which is the first and rate-limiting enzyme in the synthesis of catecholamines. This staining is very selective for dopamine-producing neurons and stains only active neurons and is used as a quick and sensitive measurement method for visualization of the surviving dopaminergic neurons [95,151]. Furthermore, determination of the level of dopamine and its metabolites 3,4-dihydroxyphenylacetic acid (DOPAC) and homovanillic acid (HVA), and other monoamines, such as noradrenaline and serotonin, in the striatum, can be used to quantify the severity of neurodegeneration [152].

These neuropathological measures are invasive and static read-out values. Dynamic changes in the brain due to cell death, recovery, and compensation mechanisms cannot be explored with these techniques and for diagnostic purposes it can only be used to confirm the disease after death. Brain imaging techniques, such as positron emission tomography (PET) using radioactive [^{18}F]6-fluoro-DOPA (F-DOPA), single proton emission computed tomography (SPECT) using the dopamine uptake ligand [^{123}I]beta-CIT, and magnetic resonance imaging (MRI), can cover this lacuna as these techniques are noninvasive and can be repeated over time.

In animal studies, brain-imaging techniques can have scientific and ethical benefit as they enable a reduction and refinement of animal studies. The advantage is that these noninvasive techniques can be applied repeatedly over a prolonged period with the opportunity of following the progress of the disease over time. The PET scan quantifies the reduction of dopamine metabolism caused by dopaminergic neuronal death. The SPECT scan targets the dopamine transporter (DaT) on the dopamine neuronal terminals indicating the loss of these axons. DaTSPECT, a technique that has been approved for use in Europe, offers one way of improving the diagnosis of Parkinson's disease. However, it is not clear at what stage neuronal cell loss becomes evident in PET and SPECT imaging. At the onset of the disease manifestation, when more than 50% of the dopaminergic neurons in the substantia nigra are already lost, there is only a 30% reduction of putaminal F-DOPA measured by PET scan [153]. As these techniques involve the exposure of subjects to radiation, they cannot be recommended for identifying individuals at risk for Parkinson's disease. An alternative is magnetic resonance imaging (MRI) and magnetic resonance spectroscopy (MRS). These neuronal imaging techniques examine structural and physiological processes in living organism and are widely used in clinical and experimental research [154]. Therefore, these techniques may be useful for the early detection of Parkinson's disease.

A typical MRI technique used for Parkinson's disease research is T2-weighted imaging in which changes in signal intensities are partly due to the altered water content of a tissue, mostly caused by the presence of extracellular edema [154]. Earlier examinations with T2-weighted imaging on the effect of MPTP intoxication in animals showed changes in relevant brain areas such as the substantia nigra pars compacta, caudate nucleus, and putamen [155–158].

MRS visualizes signals from carbon-bound protons from various metabolites [159]. But with MRS, only a few metabolites can be examined. The most frequently used neuronal marker with MRS is N-acetyl aspartate [160,161]. Reductions of N-acetyl aspartate levels have been observed in the striatum and substantia pars compacta in parkinsonian mice, cats, cynomolgus monkeys, and marmosets [156,157,162,163].

Both MRI and MRS were validated in the marmoset MPTP model using a 4.7 T nuclear magnetic resonance spectrometer (Varian, Palo Alto, California) as a supplement toward disease-modifying research [157]. MPTP treatment resulted in a reduced T2 relaxation time in the substantia nigra [157]. This reduction was in line with the reductions reported in Parkinson's disease patients [140]. The fact that other reports of MPTP studies show the opposite effect [155,156,158] can be explained by the lower dosage and the longer interval time between disease induction and the execution of the brain imaging, preventing extracellular edema, which results in an increase in T2 relaxation time [154]. MPTP intoxication also resulted in a significant reduction of the N-acetyl aspartate/total creatine ratio measured with MRS. A clear correlation between the histopathology and behavioral clinical score was shown in the marmoset [157]. A similar correlation between N-acetyl aspartate levels and immunohistochemistry was found in a mouse MPTP model [162] indicating that these parameters both predict neuronal damage. Decreased N-acetyl aspartate/total creatine ratios have also been found in other MPTP studies: mice show a clear decrease in the absolute N-acetyl aspartate concentration in the substantia nigra both at 2 and 6 days after MPTP intoxication [162], cats show decreased N-acetyl aspartate/total creatine ratios in the striatum 12 h after MPTP intoxication [156], and chronic MPTP-intoxicated cynomolgus monkeys show a persistent reduction in N-acetyl aspartate/total creatine ratio in the caudate and putamen [163]. These data and the strong correlation with clinical motor symptoms indicate the value of brain-imaging markers in research toward neuroprotection. It can be concluded that MRI and MRS are valuable tools for research toward neuroprotection in the MPTP-intoxicated marmoset as correlations indicate a clear relationship between motor functional deficits and measurements of brain damage.

Cognitive Dysfunction

Although it was thought that cognitive decline was a symptom of the late stage of Parkinson's disease [164], patients often complain about loss of memory function even during the early stage of the disease. The cognitive impairment varies from attention, memory, visuospatial skills, to language [165] and varies, not self-evidently related to the severity of the disease, between patients. As an example, dementia with Lewy bodies is clearly linked to the presence of Lewy bodies, which is also one of the hallmarks of Parkinson's disease, whereas language-related and visuomotor aspects do not necessarily depend on the stage of Parkinson's disease and can appear during the early and the late stage of the disease [166]. Although the severity of monoamine transmitter disturbances has been implicated in the cognitive deterioration of Parkinson's disease, it is suggested that the cognitive dysfunctions during the early stage are independent of frontostriatal dopamine deficiency underlying the motor disability and appear to involve nondopaminergic pathology, which was also verified by MRI studies [167]. $^{(18)}$F-fluorodeoxyglucose-PET scans also show that cognitive impairment during the early phase of Parkinson's disease is independent of dopamine-related motor disability but associated with cortical hypometabolism [168]. As reduced cholinergic activity is thought to account for the strongest mechanism in the development of dementia, this transmitter system presumably plays a predominant role in Parkinson's disease-related dementia. This is corroborated by the fact that cholinesterase inhibitors are effective in the treatment of dementia in Parkinson's disease [169]. On the other hand, dementia during the late stage of Parkinson's disease, which covers 15%–30% of the patient population, is mainly due to frontal lobe dysfunction, such as planning, attention, and spatial working memory. The frontal lobe dysfunction together with short-term, working, and learning impairments is related to the severity of the motor symptoms and as a result of the level of neuron degeneration [170]. However, some frontal lobe deficits are also detectable in early Parkinson's disease [166]. As the different types of cognitive symptoms, as mentioned in this paragraph, and their characteristics are presumably related to the stage of the pathological process, it would be of considerable value to learn more about its underlying relation that will help us with a more specific diagnosis of Parkinson's disease and to optimize the individual treatment regimen. As mentioned earlier, brain-imaging studies did lead to more insight into the brain processes during cognitive dysfunction and during different stages of the disease [167,168,171] and can help to gain more insight into the dynamic aspects of the disease progression. The advantage of brain imaging techniques and cognitive measures is that both are noninvasive methods that can be applied to patients and experimental animal models for Parkinson's disease, increasing the predictability of research data to human.

Animal studies have been used to learn more about cognitive function during the early stage of the disease in rodents and nonhuman primates [146,172]. In a subchronic MPTP C57Bl/6 mouse model for Parkinson's disease, leading to activation of apoptotic pathways in striatum, frontal cortex, and hippocampus, the performance in the Morris water maze task for spatial working memory was altered [172]. This study, which combined biochemistry, histopathology, and cognitive measures, shows the relation between the stage of the disease, based on neuronal loss in the substantia nigra, and the alterations in cholinergic brain areas and spatial working memory. Another study in rhesus macaques, in which a slow progressive disease induction with repeated low doses of MPTP (0.2 mg/kg) over several weeks was used, revealed aspects of the dynamics of the disease progression during the presymptomatic analogue to the preclinical nonmotor phase of Parkinson's disease, and the symptomatic motor phase [173]. These kinds of studies offer possibilities to study cognition as an early nonmotor symptom. Aspects of cognition can easily be measured in marmosets using automated test systems with touchscreens for spatial working memory, intra- and extradimensional shifts, reversals, and progressive ratio [131]. This offers the opportunity to shed light on the cause and treatment of cognitive dysfunction in Parkinson's disease, which is a very important aspect of the disease, as cognitive impairment is associated with significant disability for patients and their family members. Experimental research of controlled disease induction in the marmoset in which cognition is combined with motor function and brain pathology related to the disease progression will be of high value to learn more about the underlying pathological processes and understanding of the biological basis of the disabling cognitive dysfunction, which can develop even before the disease manifestation takes place.

Sleep Disorders

Parkinson's disease patients may suffer from different types of sleep disorders, such as difficulties in falling asleep, fragmented sleep, insomnia, and restless leg syndrome [173a]. But the most frequently occurring sleep disorder is REM sleep behavior disorder characterized by a lack of normal muscle atonia during REM sleep [174,175]. In Parkinson's disease, REM sleep muscle tone can be caused by a reduction in nigrostriatal dopamine signaling [176], as clinical conditions affecting dopamine, such as those seen in Parkinson's disease, alter sleep architecture [174] and affect REM sleep [177] and muscle tone during REM sleep [178,179]. On the other hand, the degenerative process in Parkinson's disease is initiated in the medulla and advances to the pons [180], the same brain area in which REM sleep is controlled. Thus, the presence of REM sleep behavior disorder in the preclinical phase of Parkinson's disease might also reflect early involvement of nondopamine medullary and/or pontine REM sleep-related structures [181]. These are important indications for REM sleep behavior disorder during the early onset of the neurodegenerative process of Parkinson's disease and, therefore, of considerable interest for understanding early pathological processes in Parkinson's disease. Indeed, 40% of all REM sleep behavior disorder cases reported eventually go on to develop a neurological disorder [182]. Most of these patients will develop Parkinson's disease as one-third of all REM sleep behavior disorder patients are diagnosed with Parkinson's disease within 10 years [24,183,184]. There is an obvious direct link between REM sleep behavior disorder and Parkinson's disease [24,25,181,183,184]. In general, disturbances in sleep usually begin years before Parkinson's disease is diagnosed [25,30,181]. Therefore, sleep disturbances can serve as a diagnostic tool to identify people at risk, but they can also be used as a marker for the early phase of Parkinson's disease in preclinical studies.

Although rodent studies are frequently used for Parkinson's disease research, the nocturnal nature of these animals' behavior and the short sleep bouts [185,186] make them less suitable as models for sleep in Parkinson's disease. Unlike rodents, the sleep architecture of marmosets resembles that of humans [26,187]. Marmosets are diurnal and, as in humans, their night sleep architecture consists of a recurring pattern of cycles with light sleep, deep sleep, and REM sleep. Because of the striking similarity in sleep macrostructure between marmosets and humans, a demonstration of early abnormalities during sleep in the marmoset MPTP model would be of significant value as potential biomarkers for the early stage of idiopathic Parkinson's disease [26,187].

Indeed, MPTP-treated parkinsonian marmosets showed a significantly increased endogenous muscle tone during REM sleep compared with nontreated healthy monkeys (Fig. 23.10; [26]). Electroencephalogram (EEG) epochs with muscle tone (EMG) were significantly more frequently scored in monkeys after MPTP treatment compared to baseline values before Parkinson's disease induction. Discriminant analysis showed that after MPTP-induced Parkinson's disease, the number of epochs with 10% muscle tone during REM sleep classified 80% of the monkeys correctly, and the epochs with three or more EMG twitches added the remaining 20% of this classification [26]. These two variables together classified the parkinsonian monkeys up to a maximum of 100%, which suggests an REM sleep behavior disorder-like phenomenon. As the neuron degeneration in this marmoset Parkinson's disease induction protocol did lead to a reduction of approximately 50% of dopamine neurons in the substantia nigra [104], which is comparable with a prediagnostic level of neuron degeneration in Parkinson's disease patients, the changes found on REM sleep can be described as a premotor aspect of Parkinson's disease. Therefore, the MPTP-treated marmoset provides an opportunity for quantitative studies on the mechanisms and intervention strategies of REM sleep and the premotor phase of Parkinson's disease, i.e., when patients can be diagnosed with REM sleep behavior disorder but not with Parkinson's disease [24,183,184].

CONCLUDING REMARKS

Today, we still struggle with the fact that there is no cure for Parkinson's disease. As the current dopamine replacement and symptom control regimens do not stop or slow down the progressive neurodegenerative process, research efforts need to change from symptom control toward disease modification. To do so, a shift should be made from diagnosis that is based on motor

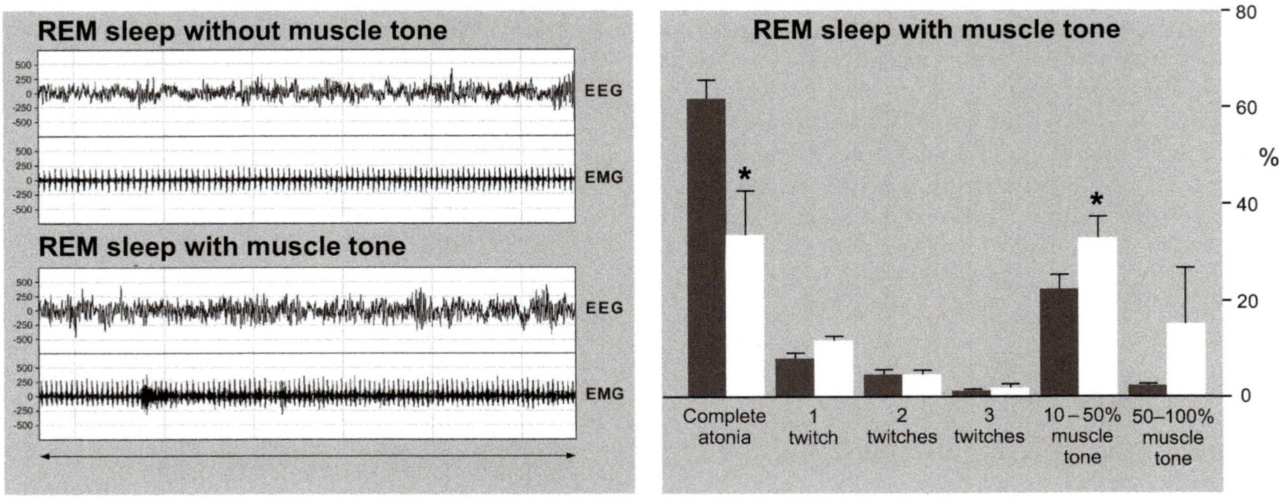

FIGURE 23.10 REM sleep behavior disorder in the marmoset. Left: Example of 30-s EEG and electromyogram (EMG) epoch during REM sleep with and without muscle tone. Upper level shows a trace of a normal healthy marmoset consisting of a random fast EEG, whereas the EMG showed no muscle activity. The bottom level shows a trace of a marmoset suffering from REM sleep behavior disorder consisting of random fast EEG in the presence of 50%–100% muscle tension in the EMG. The regular spikes are artefacts of the heart rate from the arteria carotis. Right: REM sleep epochs with muscle tone as a percentage of REM sleep in marmosets ($n = 5$) before (baseline, black bars) and after MPTP treatment (white bars). *Significant differences between baseline and MPTP treatment ($P < 0.05$ Friedman) [26].

complications [188] to early diagnosis that is supported by premotor symptoms to spot patients or individuals at risk based on early nonmotor indicators. This is crucial because treatment should start at a very early phase of the disease to save as many neurons as possible. It is known that pathological changes in the brain, measured with brain-imaging techniques such as dopamine transporter density, are already present before motor signs appear. Furthermore, sleep disturbances appear years before disease onset.

From this point of view, disease models for preclinical research should also be adapted to these new questions that ask for biomarkers for the early premotor stage of the disease such as sleep and cognition. Deteriorating motor function, sleep quality, and cognition can be related to neuroimaging. But still, the main hurdle is the lack of the availability of valid preclinical models that reflect essential premotor and pathological features of the human counterpart. Animal models should ideally mimic the main features of the disease pathology and additionally show the typical parkinsonian syndrome. The ultimate animal model for Parkinson's disease has not yet been developed; however, there are several experimental models that reproduce Parkinson-like neurodegeneration in combination with motor symptoms. One of these induction models is based on the neurotoxin MPTP that causes selective cell death in the dopamine neurons localized within the substantia nigra. We discussed the perspectives of the MPTP-induced model in the common marmoset as a neurodegenerative disease model to facilitate the development of novel disease-modifying therapies against early neurodegeneration. The MPTP-treated marmoset resembles the human disorder with respect to pathology, biochemistry, symptomatology, and response to treatment with a direct human clinical relevance. Furthermore, the above-mentioned aspects of motor function, sleep, cognition, and brain imaging for early identification of the degenerative process can be measured by neuroimaging techniques, neurophysiological, and behavioral methods in marmosets. This approach will give insight into the dynamic changes in the brain due to neurodegeneration and present the opportunity to monitor the effect of a disease-modifying treatment in the marmoset disease model over time.

Acknowledgments

The authors would like to thank Mrs. F. van Hassel for the artwork in the figures. The studies, described in this chapter, were funded by TNO, The Netherlands, EU transnational access to the research infrastructure PRIMOCID-205 of EUPRIM-Net of the 7th Framework Program and by the US Army Medical Research and Materiel Command, Award No. W81XH-05-1-0517.

References

[1] de Lau LML, Breteler MMB. Epidemiology of Parkinson's disease. Lancet Neurol 2006;5:525–35.

[2] Van Den Eeden SK. Incidence of Parkinson's disease: variation by age, gender, and race/ethnicity. Am J Epidemiol 2003;157:1015–22.

[3] Gillies GE, Pienaar IS, Vohra S, Qamhawi Z. Sex differences in Parkinson's disease. Front Neuroendocrinol 2014;35:370–84. https://doi.org/10.1016/j.yfrne.2014.02.002.

[4] Ovallath S, Deepa P. The history of parkinsonism: descriptions in ancient Indian medical literature. Mov Disord May 2013;28(5):566–8.

[5] Parkinson J. An essay on the shaking palsy. 1817. J Neuropsychiatry Clin Neurosci 2002;14:223–36; discussion 222.

[6] Dexter DT, Jenner P. Parkinson disease: from pathology to molecular disease mechanisms. Free Radical Biol Med 2013;62:132–44.

[7] Mehanna R, Moore S, Hou JG, Sarwar AI, Lai EC. Comparing clinical features of young onset, middle onset and late onset Parkinson's disease. Park Relat Disord 2014;20:530–4.

[8] Jellinger KA. Neuropathological aspects of Alzheimer disease, Parkinson disease and frontotemporal dementia. Neurodegener Dis 2008;5:118–21.

[9] Carlsson A, Lindqvist M, Magnusson T. 3,4-Dihydroxyphenylalanine and 5-hydroxytryptophan as reserpine antagonists. Nature November 30, 1957;180(4596):1200.

[10] Alafuzoff I, Parkkinen L. Staged pathology in Parkinson's disease. Park Relat Disord 2014;20(Suppl. 1):S57–61.

[11] Jenner P, Morris HR, Robbins TW, Goedert M, Hardy J, Ben-Shlomo Y, Bolam P, Burn D, Hindle JV, Brooks D. Parkinson's disease – the debate on the clinical phenomenology, aetiology, pathology and pathogenesis. J Parkinson's Dis 2013;3:1–11.

[12] Mollenhauer B, Förstl H, Deuschl G, Storch A, Oertel W, Trenkwalder C. Lewy body and parkinsonian dementia: common, but often misdiagnosed conditions. Dtsch Arztebl Int 2010;107:684–91.

[13] Eschbach J, Danzer KM. α-synuclein in Parkinson's disease: pathogenic function and translation into animal models. Neurodegener Dis 2014;14:1–17.

[14] Jellinger KA. Formation and development of Lewy pathology: a critical update. J Neurol 2009;256(Suppl. 3):270–9.

[15] Luk KC, Lee VMY. Modeling Lewy pathology propagation in Parkinson's disease. Park Relat Disord 2014;20(Suppl. 1):S85–7.

[16] Braak H, Rub U, Gai WP, Del Tredici K. Idiopathic Parkinson's disease: possible routes by which vulnerable neuronal types may be subject to neuroinvasion by an unknown pathogen. J Neural Transm 2003b;110:517–36.

[17] Schulz-Schaeffer WJ. The synaptic pathology of alpha-synuclein aggregation in dementia with Lewy bodies, Parkinson's disease and Parkinson's disease dementia. Acta Neuropathol 2010;120:131–43.

[18] Gaig C, Valldeoriola F, Gelpi E, Ezquerra M, Llufriu S, Buongiorno M, Rey MJ, Martí MJ, Graus F, Tolosa E. Rapidly progressive diffuse lewy body disease. Mov Disord 2011.

[19] Ulusoy A, Di Monte D. α-synuclein elevation in human neurodegenerative diseases: experimental, pathogenetic, and therapeutic implications. Mol Neurobiol 2013;47:484–94.

[20] Noyce AJ, Bestwick JP, Silveira-Moriyama L, Hawkes CH, Giovannoni G, Lees AJ, Schrag A. Meta-analysis of early nonmotor features and risk factors for Parkinson disease. Ann Neurol 2012;72:893–901. https://doi.org/10.1002/ana.23687.

[21] Garcia-Borreguero D, Larrosa O, Bravo M. Parkinson's disease and sleep. Sleep Med Rev 2003;7:115–29.

[22] Oerlemans WG, de Weerd AW. The prevalence of sleep disorders in patients with Parkinson's disease. A self-reported, community-based survey. Sleep Med 2002;3:47–149.
[23] Tandberg E, Larsen JP, Karlsen K. A community-based study of sleep disorders in patients with Parkinson's disease. Mov Disord 1998;13:895–9.
[24] Iranzo A, Molinuevo JL, Santamaria J, Serradell M, Marti MJ, Valldeoriola F, Tolosa E. Rapid-eye-movement sleep behaviour disorder as an early marker for a neurodegenerative disorder: a descriptive study. Lancet Neurol 2006;5:572–7.
[25] Postuma RB, Lang AE, Massicotte-Marquez J, Montplaisir J. Potential early markers of Parkinson disease in idiopathic REM sleep behavior disorder. Neurology 2006;66:845–51.
[26] Verhave PS, Jongsma MJ, Van den Berg RM, Vis JC, Vanwersch RAP, Smit AB, Someren EJW, Philippens IHCHM. REM sleep behavior disorder in the marmoset MPTP model of early Parkinson disease. Sleep 2011;34(8):1119–44.
[27] Deeb J, Shah M, Muhammed N, Gunasekera R, Gannon K, Findley LJ, Hawkes CH. A basic smell test is as sensitive as a dopamine transporter scan: comparison of olfaction, taste and DaTSCAN in the diagnosis of Parkinson's disease. QJM December 2010;103(12):941–52.
[28] Berg D. Biomarkers for the early detection of Parkinson's and Alzheimer's disease. Neurodegener Dis 2008;5:133–6.
[29] Seibyl J, Russell D, Jennings D, Marek K. Neuroimaging over the course of Parkinson's disease: from early detection of the at-risk patient to improving pharmacotherapy of later-stage disease. Semin Nucl Med 2012;42:406–14.
[30] Tolosa E, Gaig C, Santamaria J, Compta Y. Diagnosis and the premotor phase of Parkinson disease. Neurology 2009;72:S12–20.
[31] Jankovic J. Motor fluctuations and dyskinesias in Parkinson's disease: clinical manifestations. Mov Disord 2005;20(Suppl. 11): S11–6.
[32] Philippens IHCHM, 't Hart BA, Torres G. The MPTP marmoset model of Parkinsonism: a multi-purpose non-human primate model for neurodegenerative diseases. Drug DiscovToday 2010;15(23–24):985–90.
[33] Botella JA, Bayersdorfer F, Gmeiner F, Schneuwly S. Modelling Parkinson's disease in Drosophila. NeuroMolecular Med 2009; 11:268–80.
[34] Schule B, Pera RA, Langston JW. Can cellular models revolutionize drug discovery in Parkinson's disease? Biochim Biophys Acta 2009;1792:1043–51.
[35] van der Staay FJ, Arndt S, Nordquist R. Evaluation of animal models of neurobehavioral disorders. Behav Brain Funct 2009;5:11.
[36] Gasser T. Genetics of Parkinson's disease. Curr Opin Neurol 2005;18:363–9.
[37] Gorrell JM, DiMonte D, Graham D. The role of the environment in Parkinson's disease. Environ Health Perspect 1996;104:652–4.
[38] Semchuk KM, Love EJ, Lee RG. Parkinson's disease and exposure to rural environmental factors: a population based case-control study. Can J Neurol Sci 1991;18:279–86.
[39] Langston JW, Ballard P, Tetrud JW, Irwin I. Chronic Parkinsonism in humans due to a product of meperidine-analog synthesis. Science 1983;219:979–80.
[40] Shulman JM, De Jager PL, Feany MB. Parkinson's disease: genetics and pathogenesis. Annu Rev Pathol Mech Dis 2011;6: 193–222.
[41] Wood-Kaczmar A, Gandhi S, Wood NW. Understanding the molecular causes of Parkinson's disease. Trends Mol Med 2006;12: 521–8.
[42] Chan DK, Woo J, Ho SC, Pang CP, Law LK, Ng PW, Hung WT, Kwok T, Hui E, Orr K, Leung MF, Kay R. Genetic and environmental risk factors for Parkinson's disease in a Chinese population. J Neurol Neurosurg Psychiatry 1998;65:781–4.
[43] Le Couteur DG, Muller M, Yang MC, Mellick GD, McLean AJ. Age-environment and gene-environment interactions in the pathogenesis of Parkinson's disease. Rev Environ Health 2002; 17:51–64.
[44] Migliore L, Coppede F. Genetics, environmental factors and the emerging role of epigenetics in neurodegenerative diseases. Mutat Res 2009;667:82–97.
[45] Schapira AH. Etiology and pathogenesis of Parkinson disease. Neurol Clin 2009;27:583–603.
[45a] Abou-Sleiman PM, Muqit MMK, Wood NW. Expanding insights of mitochondrial dysfunction in Parkinson's disease. Nat Rev Neurosci 2006;7:207–19.
[46] Alexi T, Borlongan CV, Faull RL, Williams CE, Clark RG, Gluckman PD, Hughes PE. Neuroprotective strategies for basal ganglia degeneration: Parkinson's and Huntington's diseases. Prog Neurobiol 2000;60:409–70.
[47] Betarbet R, Sherer TB, MacKenzie G, Garcia-Osuna M, Panov AV, Greenamyre JT. Chronic systemic pesticide exposure reproduces features of Parkinson's disease. Nat Neurosci 2000;3:1301–6.
[48] Caudle WM, Zhang J. Glutamate, excitotoxicity, and programmed cell death in Parkinson disease. Exp Neurol 2009;220:230–3.
[49] Dauer W, Przedborski S. Parkinson's disease: mechanisms and models. Neuron 2003;39:889–909.
[50] Jenner P. The contribution of the MPTP-treated primate model to the development of new treatment strategies for Parkinson's disease. Park Relat Disord 2003;9:131–7.
[51] Jung T, Catalgol B, Grune T. The proteasomal system. Mol Aspect Med 2009;30:191–296.
[52] Hol EM, van Leeuwen FW, Fischer DF. The proteasome in Alzheimer's disease and Parkinson's disease: lessons from ubiquitin B+1. Trends Mol Med 2005;11:488–95.
[53] Cook C, Petrucelli L. A critical evaluation of the ubiquitin–proteasome system in Parkinson's disease. Biochim Biophys Acta 2009;1792(7):664–75.
[54] Olanow CW, McNaught KS. Ubiquitin-proteasome system and Parkinson's disease. Mov Disord 2006;21(11):1806–23.
[55] Cookson MR. alpha-Synuclein and neuronal cell death. Mol Neurodegener 2009;4:9.
[56] Beal MF. Mechanisms of excitotoxicity in neurologic diseases. FASEB J 1992;6:3338–44.
[57] Koutsilieri E, Riederer P. Excitotoxicity and new antiglutamatergic strategies in Parkinson's disease and Alzheimer's disease. Park Relat Disord 2007;13(3):S329–31.
[58] Rodriguez MC, Obeso JA, Olanow CW. Subthalamic nucleus-mediated excitotoxicity in Parkinson's disease: a target for neuroprotection. Ann Neurol 1998;44:S175–88.
[59] Fulda S, Gorman AM, Hori O, Samali A. Cellular stress responses: cell survival and cell death. Int J Cell Biol 2009;2010:214074.
[60] Alam ZI, Jenner A, Daniel SE, Lees AJ, Cairns N, Marsden CD, et al. Oxidative DNA damage in the Parkinsonian brain: an apparent selective increase in 8-hydroxyguanine levels in substantia nigra. J Neurochem 1997;69:1196–203.
[61] Danielson SR, Andersen JK. Oxidative and nitrative protein modifications in Parkinson's disease. Free Radic Biol Med 2008; 44:1787–94.
[62] Pearce RK, Owen A, Daniel S, Jenner P, Marsden CD. Alterations in the distribution of glutathione in the substantia nigra in Parkinson's disease. J Neural Transm 1997;104:661–77.
[63] Philippens IH, Wubben JA, Finsen B, t Hart BA. Oral treatment with the NADPH oxidase antagonist apocynin mitigates clinical and pathological features of parkinsonism in the MPTP marmoset model. J Neuroimmune Pharmacol 2013;8(3):715–26.
[64] Andres-Mateos E, Perier C, Zhang L, Blanchard-Fillion B, Greco TM, Thomas B, Ko HS, Sasaki M, Ischiropoulos H, Przedborski S, Dawson TM, Dawson VL. DJ-1 gene deletion

reveals that DJ-1 is an atypical peroxiredoxin-like peroxidase. Proc Natl Acad Sci USA 2007;104(37):14807–12.
[65] Sawada M, Imamura K, Nagatsu T. Role of cytokines in inflammatory process in Parkinson's disease. J Neural Transm Suppl 2006:373–81.
[66] Su X, Federoff HJ. Immune responses in Parkinson's disease: interplay between central and peripheral immune systems. BioMed Res Int 2014;2014:275178.
[67] Natale G, Pasquali L, Ruggieri S, Paparelli A, Fornai F. Parkinson's disease and the gut: a well known clinical association in need of an effective cure and explanation. Neuro Gastroenterol Motil 2008;20:741–9.
[68] Deleidi M, Gasser T. The role of inflammation in sporadic and familial Parkinson's disease. Cell Mol Life Sci 2013;70:4259–73.
[69] McGeer PL, Itagaki S, Boyes BE, McGeer EG. Reactive microglia are positive for HLA-DR in the substantia nigra of Parkinson's and Alzheimer's disease brains. Neurology 1988;38(8):1285–91.
[70] Emborg ME. Evaluation of animal models of Parkinson's disease for neuroprotective strategies. J Neurosci Meth 2004;139:121–43.
[71] Shah A, Garzon-Muvdi T, Mahajan R, Duenas VJ, Quiñones-Hinojosa A. Animal models of neurological disease. Adv Exp Med Biol 2010;671:23–40.
[72] Senoh S, et al. Enzymatic p-O-methylation by catechol O-methyl transferase. J Am Chem Soc 1959;81:6240–5.
[73] Bove J, Prou D, Perier C, Przedborski S. Toxin-induced models of Parkinson's disease. NeuroRx July 2005;2(3):484–94.
[74] Blandini F, Levandis G, Bazzini E, Nappi G, Armentero MT. Time-course of nigrostriatal damage, basal ganglia metabolic changes and behavioural alterations following intrastriatal injection of 6-hydroxydopamine in the rat: new clues from an old model. Eur J Neurosci 2007;25:397–405.
[75] Annett LE, Martel FL, Rogers DC, Ridley RM, Baker HF, Dunnett SB. Behavioral assessment of the effects of embryonic nigral grafts in marmosets with unilateral 6-OHDA lesions of the nigrostriatal pathway. Exp Neurol 1994;125:228–46.
[76] Eslamboli A. Marmoset monkey models of Parkinson's disease: which model, when and why? Brain Res Bull 2005;68:140–9.
[77] Yong-Kee CJ, Warre R, Monnier PP, Lozano AM, Nash JE. Evidence for synergism between cell death mechanisms in a cellular model of neurodegeneration in Parkinson's disease. Neurotox Res 2012;22:355–64.
[78] Schmidt WJ, Alam M. Controversies on new animal models of Parkinson's disease pro and con: the rotenone model of Parkinson's disease (PD). J Neural Transm Suppl 2006:273–6.
[79] Sherer TB, Kim JH, Betarbet R, Greenamyre JT. Subcutaneous rotenone exposure causes highly selective dopaminergic degeneration and alpha-synuclein aggregation. Exp Neurol 2003;179: 9–16.
[80] Philippens IHCHM. Non-human primate models for Parkinson's disease. Drug Discov Today Dis Model 2008;5(2):105–11.
[81] Przedborski S, Jackson-Lewis V, Naini AB, Jakowec M, Petzinger G, Miller R, Akram M. The parkinsonian toxin 1-methyl-4-phenyl-1,2,3,6-tetrahydropyridine (MPTP): a technical review of its utility and safety. J Neurochem 2001;76:1265–74.
[82] Langston JW, Forno LS, Tetrud J, Reeves AG, Kaplan JA, Karluk D. Evidence of active nerve cell degeneration in the substantia nigra of humans years after 1-methyl-4-phenyl-1,2,3,6-tetrahydropyridine exposure. Ann Neurol 1999;46:598–605.
[83] Snow BJ, Vingerhoets FJG, Langston JW, Tetrud JW, Sossi V, Calne DB. Pattern of dopaminergic loss in the striatum of humans with MPTP induced parkinsonism. J Neurol Neurosurg Psychiatr 2000;68:313–6.
[84] Javitch JA, D'Amato RJ, Strittmatter SM, Snyder SH. Parkinsonism-inducing neurotoxin, N-methyl-4-phenyl-1,2,3,6 -tetrahydropyridine: uptake of the metabolite N-methyl-4-phenylpyridine by dopamine neurons explains selective toxicity. Proc Natl Acad Sci USA 1985;82:2173–7.
[85] Melamed E, Rosenthal J, Globus M, Cohen O, Uzzan A. Suppression of MPTP-induced dopaminergic neurotoxicity in mice by nomifensine and L-DOPA. Brain Res 1985;342:401–4.
[86] Burns RS, Chiueh CC, Markey SP, Ebert MH, Jacobowitz DM, Kopin IJ. A primate model of parkinsonism: selective destruction of dopaminergic neurons in the pars compacta of the substantia nigra by N-methyl-4-phenyl-1,2,3,6-tetrahydropyridine. Proc Natl Acad Sci USA 1983;80:4546–50.
[87] Ballard PA, Tetrud JW, Langston JW. Permanent human parkinsonism due to 1-methyl-4-phenyl-1,2,3,6-tetrahydropyridine (MPTP): seven cases. Neurology 1985;35:949–56.
[88] Kalaria RN, Mitchell MJ, Harik SI. Correlation of 1-methyl-4-phenyl-1,2,3,6-tetrahydropyridine neurotoxicity with blood-brain barrier monoamine oxidase activity. Proc Natl Acad Sci USA 1987;84:3521–5.
[89] Schmidt N, Ferger B. Neurochemical findings in the MPTP model of Parkinson's disease. J Neural Transm 2001;108: 1263–82.
[90] Liu Y, Peter D, Roghani A, Schuldiner S, Prive GG, Eisenberg D, Brecha N, Edwards RH. A cDNA that suppresses MPP+ toxicity encodes a vesicular amine transporter. Cell 1992;70:539–51.
[91] Nicklas WJ, Vyas I, Heikkila RE. Inhibition of NADH-linked oxidation in brain mitochondria by 1-methyl-4-phenyl-pyridine, a metabolite of the neurotoxin, 1-methyl-4-phenyl-1,2,5,6-tetrahydropyridine. Life Sci 1985;36:2503–8.
[92] Smeyne RJ, Jackson-Lewis V. The MPTP model of Parkinson's disease. Brain Res Mol Brain Res 2005;134:57–66.
[93] Mizuno Y, Sone N, Saitoh T. Effects of 1-methyl-4-phenyl-1,2,3,6-tetrahydropyridine and 1-methyl-4-phenylpyridinium ion on activities of the enzymes in the electron transport system in mouse brain. J Neurochem 1987;48:1787–93.
[94] Pain S, Gochard A, Bodard S, Gulhan Z, Prunier-Aesch C, Chalon S. Toxicity of MPTP on neurotransmission in three mouse models of Parkinson's disease. Exp Toxicol Pathol 2013;65: 689–94.
[95] Waters CM, Hunt SP, Jenner P, Marsden CD. An immunohostochemical study of the acute nd long-term effects of 1-methyl-phenyl-1,2,3,6-tetrahydropyridine in the marmoset. Neuroscience 1987;23:1025–39.
[96] Jenner P, Zeng BY, Smith LA, Pearce RK, Tel B, Chancharme L, Moachon G. Antiparkinsonian and neuroprotective effects of modafinil in the mptp-treated common marmoset. Exp Brain Res 2000;133:178–88.
[97] Kuroiwa H, Yokoyama H, Kimoto H, Kato H, Araki T. Biochemical alterations of the striatum in an MPTP-treated mouse model of Parkinson's disease. Metab Brain Dis 2010; 25:177–83.
[98] Rose S, Nomoto M, Jackson EA, Gibb WR, Jaehnig P, Jenner P, Marsden CD. Age-related effects of 1-methyl-4-phenyl-1,2,3,6-tetrahydropyridine treatment of common marmosets. Eur J Pharmacol 1993;230:177–85.
[99] Phillips KA, Ross CN, Spross J, Cheng CJ, Izquierdo A, Biju KC, Chen C, Li S, Tardif SD. Behavioral phenotypes associated with MPTP induction of partial lesions in common marmosets (*Callithrix jacchus*). Behav Brain Res 2017;325(Pt A):51–62.
[100] Franke SK, van Kesteren RE, Wubben JA, Hofman S, Paliukhovich I, van der Schors RC, van Nierop P, Smit AB, Philippens IH. Progression and recovery of Parkinsonism in a chronic progressive MPTP-induction model in the marmoset without persistent molecular and cellular damage. Neuroscience 2016;312:247–59.
[101] Korecka JA, Eggers R, Swaab DF, Bossers K, Verhaagen J. Modeling early Parkinson's disease pathology with chronic

low dose MPTP treatment. Restor Neurol Neurosci 2013;31: 155–67.
[102] Russ H, Mihatsch W, Gerlach M, Riederer P, Przuntek H. Neurochemical and behavioural features induced by chronic low dose treatment with 1-methyl-4-phenyl-1,2,3,6-tetrahydropyridine (MPTP) in the common marmoset: implications for Parkinson's disease? Neurosci Lett 1991;123:115–8.
[103] van Vliet SAM, Vanwersch RAP, Jongsma MJ, van der Gugten J, Olivier B, Philippens IHCHM. Neuroprotective effects of modafinil in a marmoset Parkinson model: behavioral and neurochemical aspects. Behav Pharmacol 2006a;17:453–62.
[104] Verhave PS, Jongsma MJ, Van Den Berg RM, Vanwersch RAP, Smit AB, Philippens IHCHM. Neuroprotective effects of riluzole in early phase Parkinson's disease on clinically relevant parameters in the marmoset MPTP model. Neuropharmacology 2012; 62(4):1700–7.
[105] Forno LS, Langston JW, DeLanney LE, Irwin I, Ricaurte GA. Locus ceruleus lesions and eosinophilic inclusions in MPTP-treated monkeys. Ann Neurol 1986;20:449–55.
[106] Purisai MG, McCormack AL, Langston WJ, Johnston LC, Di Monte DA. α-Synuclein expression in the substantia nigra of MPTP-lesioned non-human primates. Neurobiol Dis 2005;20(3): 898–906.
[107] Fornai F, Schluter OM, Lenzi P, Gesi M, Ruffoli R, Ferrucci M, Lazzeri G, Busceti CL, Pontarelli F, Battaglia G, Pellegrini A, Nicoletti F, Ruggieri S, Paparelli A, Sudhof TC. Parkinson-like syndrome induced by continuous MPTP infusion: convergent roles of the ubiquitin-proteasome system and alpha-synuclein. Proc Natl Acad Sci USA 2005;102:3413–8.
[108] Kirik D, Annett LE, Burger C, Muzyczka N, Mandel RJ, Bjorklund A. Nigrostriatal alpha-synucleinopathy induced by viral vectormediated overexpression of human alpha-synuclein: a new primate model of Parkinson's disease. Proc Natl Acad Sci USA 2003;100:2884–9.
[109] Schober A. Classic toxin-induced animal models of Parkinson's disease: 6-OHDA and MPTP. Cell Tissue Res 2004;318:215–24.
[110] Wu R-M, Mohanakumar KP, Murphy DL, Chiueh CC. Antioxidant mechanism and protection of nigral neurons against MPP+ toxicity by deprenyl (selegiline). Ann NY Acad Sci 1994; 738:214–21.
[111] Philippens IHCHM, Verhave PS. Preclinical solutions for insight in premotor Parkinson. In: Rana AQ, editor. Symptoms of Parkinson's disease. InTech; 2011, ISBN 978-953-307-464-1.
[112] Gnanalingham KK, Milkowski NA, Smith LA, Hunter AJ, Jenner P, Marsden CD. Short- and long-term changes in striatal and extrastriatal dopamine uptake sites in the MPTP-treated common marmoset. Eur J Pharmacol 1995b;277:235–41.
[113] Pifl C, Schingnitz G, Hornykiewicz O. Effect of 1-methyl-4-phenyl-1,2,3,6-tetrahydropyridine on the regional distribution of brain monoamines in the rhesus monkey. Neuroscience 1991;44:591–605.
[114] Philippens IH, Melchers BP, Roeling TA, Bruijnzeel PL. Behavioral test systems in marmoset monkeys. Behav Res Methods Instrum Comput 2000;32:173–9.
[115] Jenner P, Rupniak NM, Rose S, Kelly E, Kilpatrick G, Lees A, Marsden CD. 1-Methyl-4-phenyl-1,2,3,6-tetrahydropyridine-induced parkinsonism in the common marmoset. Neurosci Lett 1984;50:85–90.
[116] Hantraye P, Varastet M, Peschanski M, Riche D, Cesaro P, Willer JC, Maziere M. Stable parkinsonian syndrome and uneven loss of striatal dopamine fibres following chronic MPTP administration in baboons. Neuroscience 1993;53:169–78.
[117] Fox SH, Visanji N, Reyes G, Huot P, Gomez-Ramirez J, Johnston T, Brotchie JM. Neuropsychiatric behaviors in the MPTP marmoset model of Parkinson's disease. Can J Neurol Sci 2010;37:86–95.
[118] Zeng BY, Iravani MM, Lin ST, Irifune M, Kuoppamaki M, Al-Barghouthy G, Smith L, Jackson MJ, Rose S, Medhurst AD, et al. MPTP treatment of common marmosets impairs proteasomal enzyme activity and decreases expression of structural and regulatory elements of the 26S proteasome. Eur J Neurosci 2006;23:1766–74.
[119] Kohu K, Yamabe E, Matsuzawa A, Onda D, Suemizu H, Sasaki E, Tanioka Y, Yagita H, Suzuki D, Kametani Y, Takai T, Toyoda A, Habu S, Satake M. Comparison of 30 immunity-related genes from the common marmoset with orthologues from human and mouse. Tohoku J Exp Med 2008 Jun;215(2):167–80.
[119a] Whitfield LS, Lovell-Badge R, Goodfellow PN. Rapid sequence evolution of the mammalian sex-determining gene SRY. Nature 1993 Aug 19;364(6439):713–5.
[120] Jackson-Lewis V, Jakowec M, Burke RE, Przedborski S. Time course and morphology of dopaminergic neuronal death caused by the neurotoxin 1-methyl-4-phenyl-1,2,3,6-tetrahydropyridine. Neurodegeneration 1995;4:257–69.
[121] Mandel S, Grünblatt E, Riederer P, Gerlach M, Levites Y, Youdim MBH. Neuroprotective strategies in Parkinson's disease: an update on progress. CNS Drugs 2003;17(10):729–62.
[122] Meredith GE, Totterdell S, Potashkin JA, Surmeier DJ. Modeling PD pathogenesis in mice: advantages of a chronic MPTP protocol. Park Relat Disord 2008;14(Suppl. 2):S112–7.
[123] van Vliet SAM, Jongsma MJ, Vanwersch RAP, Olivier B, Philippens IHCHM. Behavioral effects of modafinil in marmoset monkeys. Psychopharmacology (Berlin) 2006b;185(4):433–40.
[124] Visanji NP, Gomez-Ramirez J, Johnston TH, Pires D, Voon V, Brotchie JM, Fox SH. Pharmacological characterization of psychosis-like behavior in the MPTP-lesioned nonhuman primate model of Parkinson's disease. Mov Disord 2006;21: 1879–91.
[125] Fox SH, Visanji NP, Johnston TH, Gomez-Ramirez J, Voon V, Brotchie JM. Dopamine receptor agonists and levodopa and inducing psychosis-like behavior in the MPTP primate model of Parkinson disease. Arch Neurol 2006;63:1343–4.
[126] Albanese A, Jenner P, Marsden CD, Stephenson JD. Bladder hyperreflexia induced in marmosets by 1-methyl-4-phenyl-1,2,3,6-tetrahydropyridine. Neurosci Lett 1988;87:46–50.
[127] Verhave PS, Vanwersch RAP, van Helden HP, Smit AB, Philippens IHCHM. Two new test methods to quantify motor deficits in a marmoset model for Parkinson's disease. Behav Brain Res 2009;200:214–9.
[128] Iravani MM, Jackson MJ, Kuoppamaki M, Smith LA, Jenner P. 3,4-methylenedioxymethamphetamine (ecstasy) inhibits dyskinesia expression and normalizes motor activity in 1-methyl-4-phenyl-1,2,3,6-tetrahydropyridine-treated primates. J Neurosci 2003;23:9107–15.
[129] Pearce RK, Collins P, Jenner P, Emmett C, Marsden CD. Intraventricular infusion of basic fibroblast growth factor (bFGF) in the MPTP-treated common marmoset. Synapse 1996;23:192–200.
[130] Obinu MC, Reibaud M, Blanchard V, Moussaoui S, Imperato A. Neuroprotective effect of riluzole in a primate model of Parkinson's disease: behavioral and histological evidence. Mov Disord 2002;17:13–9.
[131] Spinelli S, Pennanen L, Dettling AC, Feldon J, Higgins GA, Pryce CR. Performance of the marmoset monkey on computerized tasks of attention and working memory. Brain Res Cogn Brain Res 2004;19(2):123–37.
[132] Przybyszewski AW, Sosale S, Chaudhuri A. Quantification of three-dimensional exploration in the cylinder test by the common marmoset (*Callithrix jacchus*). Behav Brain Res 2006;170(1): 62–70.

[133] Marshall JW, Ridley RM. Primate models for studying functional recovery after drug treatment in stroke. Eur Neuropsychopharmacol 1996;9:S191–2.

[134] Guy W. ECDEU assessment manual for psychopharmacology. Washington DC: US Department of Health, Education and Welfare; 1976. p. 534–7.

[135] Di Monte DA, McCormack A, Petzinger G, Janson AM, Quik M, Langston WJ. Relationship among nigrostriatal denervation, parkinsonism, and dyskinesias in the MPTP primate model. Mov Disord 2000;15:459–66.

[136] Pearce RK, Jackson M, Smith L, Jenner P, Marsden CD. Chronic L-DOPA administration induces dyskinesias in the 1-methyl-4-phenyl-1,2,3,6-tetrahydropyridine-treated common marmoset (Callithrix Jacchus). Mov Disord 1995;10(6):731–40.

[137] Carey GJ, Costall B, Domeney AM, Jones DN, Naylor RJ. Behavioural effects of anxiogenic agents in the common marmoset. Pharmacol Biochem Behav 1992;42:143–53.

[138] Stevenson MF, Poole TB. An ethogram of the common marmoset (Calithrix jacchus jacchus): general behavioural repertoire. Anim Behav 1976;24:428–51.

[138a] Franke SK, van Kesteren RE, Wubben JA, Hofman S, Paliukhovich I, van der Schors RC, van Nierop P, Smit AB, Philippens IH. Progression and recovery of Parkinsonism in a chronic progressive MPTP-induction model in the marmoset without persistent molecular and cellular damage. Neuroscience 2016 Jan 15;312:247–59. https://doi.org/10.1016/j.neuroscience.2015.09.065. Epub 2015 Oct 9.

[139] Wolthuis OL, Groen B, Philippens IH. A simple automated test to measure exploratory and motor activity of marmosets. Pharmacol Biochem Behav 1994;47:879–81.

[139a] Chaudhuri KR, Healy DG, Schapira AH, National Institute for Clinical E. Non-motor symptoms of Parkinson's disease: diagnosis and management. Lancet Neurology 2006;5:235–45.

[140] Kosta P, Agyropoulou MI, Markoula S, Konitsiotis S. MRI evaluation of the basal ganglia size and the iron content in patients with Parkinson's disease. J Neurol 2006;253(1):26–32.

[141] van Hilten B, Hoff JI, Middelkoop HA, van der Velde EA, Kerkhof GA, Wauquier A, Kamphuisen HA, Roos RA. Sleep disruption in Parkinson's disease. Assessment by continuous activity monitoring. Arch Neurol 1994;51:922–8.

[142] Park A, Stacy M. Non-motor symptoms in Parkinson's disease. J Neurol 2009;256(Suppl. 3):293–8.

[143] Miwa T, Watanabe A, Mitsumoto Y, Furukawa M, Fukushima N, Moriizumi T. Olfactory impairment and Parkinson's disease-like symptoms observed in the common marmoset following administration of 1-methyl-4-phenyl-1,2,3,6-tetrahydropyridine. Acta Otolaryngol Suppl August 2004;553:80–4.

[144] Almirall H, Pigarev I, de la Calzada MD, Pigareva M, Herrero MT, Sagales T. Nocturnal sleep structure and temperature slope in MPTP treated monkeys. J Neural Transm 1999; 106:1125–34.

[145] Barraud Q, Lambrecq V, Forni C, McGuire S, Hill M, Bioulac B, Balzamo E, Bezard E, Tison F, Ghorayeb I. Sleep disorders in Parkinson's disease: the contribution of the MPTP non-human primate model. Exp Neurol 2009;219:574–82.

[146] Vezoli J, Fifel K, Leviel V, Dehay C, Kennedy H, Cooper HM, Gronfier C, Procyk E. Early presymptomatic and long-term changes of rest activity cycles and cognitive behavior in a MPTP-monkey model of Parkinson's disease. PLoS One 2011; 6(8):e23952.

[147] Anderson G, Noorian AR, Taylor G, Anitha M, Bernhard D, Srinivasan S, Greene JG. Loss of enteric dopaminergic neurons and associated changes in colon motility in an MPTP mouse model of Parkinson's disease. Exp Neurol 2007;207:4–12.

[148] Marek K, Jennings D. Can we image premotor Parkinson disease? Neurology 2009;72:S21–6.

[149] Stephenson R, Siderowf A, Stern MB. Premotor Parkinson's disease: clinical features and detection strategies. Mov Disord 2009; 24(Suppl. 2):S665–70.

[150] Tolosa E, Poewe W. Premotor Parkinson disease. Neurology 2009;72:S1.

[151] Pearson J, Goldstein M, Markey K, Brandeis. Human brainstem catecholamine neuronal anatomy as indicated by immunocytochemistry with antibodies to tyrosine hydroxylase. Neuroscience 1983;8:3–32.

[152] Gerlach M, Riederer P, Przuntek H, Youdim MBH. MPTP mechanisms of neurotoxicity and their implications for Parkinson's disease. Eur J Pharmacol 1991;208:273–86.

[153] Berg, D.. (2006). Marker for a preclinical diagnosis of Parkinson's disease as a basis for neuroprotection, In: Oxidative stress and neuroprotection, J Neural Transm 2006:(71), 123–132, ISBN 0303-6995

[154] Dijkhuizen RM, Nicolay K. Magnetic resonance imaging in experimental models of brain disorders. J Cerebr Blood Flow Metabol 2003;23:1383–402.

[155] Miletich RS, BAnkiewicz KS, Quareantelli M, Plunkett RJ, Frank J, Kopin IJ, Di Chiro G. MRI detects acute degeneration of the nigrostriatal dopamine system after MPTP exposure in hemiparkinsonian monkeys. Ann Neurol 1994;35:689–97.

[156] Podell M, Hadjiconstantinou M, Smith MA, Neff NH. Proton magnetic resonance imaging and spectroscopy identify metabolic changes in the straiatum in the MPTP feline model of parkinsonism. Exp Neurol 2003;179:159–66.

[157] van Vliet SAM, Blezer EL, Jongsma MJ, Vanwersch RAP, Olivier B, Philippens IHCHM. Exploring the neuroprotective effects of modafinil in a marmoset Parkinson model with immunohistochemistry, magnetic resonance imaging and spectroscopy. Brain Res 2008;1189:219–28.

[158] Zhang Z, Zhang M, Ai Y, Avison C, Gash DM. MPTP-Induced pallidal lesions in rhesus monkeys. Exp Neurol 1999;155:140–9.

[159] Kauppinen RA, Williams SR. Nuclear magnetic resonance spectroscopy studies of the brain. Prog Neurobiol 1994;44:87–118.

[160] Castillo M, Kwock L, Mukherji SK. Clinical applications of proton MR spectroscopy. AJNR Am J Neuroradiol January 1996; 17(1):1–15.

[161] Gujar SK, Maheshwari S, Bjorkman-Burtscher I, Sundgren PC. Magnetic resonance spectroscopy. J Neuro Ophthalmol September 2005;25(3):217–26.

[162] Boska MD, Lewis TB, Destache CJ, Benner EJ, Nelson JA, et al. Quantitative ^{1}H magnetic resonance spectroscopic imaging determines therapeutic immunization efficacy in an animals of Parkinson's disease. J Neurosci 2005;25:1691–700.

[163] Brownell AL, Jenkins BG, Elmaleh DR, Deacon TW, et al. Combined PET/MRS brain studies show dynamic and long-term physiological changes in a primate model of Parkinson disease. Nat Med 1998;4:1308–12.

[164] Charcot JM. Leçons sur les maladies du système nerveux faites à la salpétrière. 3rd ed., Tome I. Paris: Recueillies par Bourneville; 1877. p. 155–88.

[165] Watson GS, Leverenz JB. Profile of cognitive impairment in Parkinson's disease. Brain Pathol 2010;20(3):640–5.

[166] Cooper JA, Sagar HJ, Jordan N, Harvey NS, Sullivan EV. Cognitive impairment in early, untreated Parkinson's disease and its relationship to motor disability. Brain 1991;114(5):2095–122.

[167] Ibarretxe-Bilbao N, Junque C, Marti MJ, Tolosa E. Brain structural MRI correlates of cognitive dysfunctions in Parkinson's disease. J Neurol Sci 2011;310(1–2):70–4.

[168] Pappatà S, Santangelo G, Aarsland D, Vicidomini C, Longo K, Bronnick K, Amboni M, Erro R, Vitale C, Caprio MG, Pellecchia MT, Brunetti A, De Michele G, Salvatore M, Barone P. Mild cognitive impairment in drug-naive patients

with PD is associated with cerebral hypometabolism. Neurology 2011;77(14):1357–62.
[169] Caballol N, Martí MJ, Tolosa E. Cognitive dysfunction and dementia in Parkinson disease. Mov Disord 2007;22(17):358–66.
[170] Gracies J-M, Olanow CW. Dementia in Parkinson's disease. In: Charney DS, Nestler EJ, editors. Neurobiology of mental illness. New York: Oxford university press; 2004. p. 896–916. ISBN 0-19-5114962-9.
[171] Weintraub D, Doshi J, Koka D, Davatzikos C, Siderowf AD, Duda JE, Wolk DA, Moberg J, Xie SX, Clark CM. Neurodegeneration across stages of cognitive decline in Parkinson disease. Arch Neurol 2011;68(12):1562–70.
[172] Deguil J, Chavant F, Lafay-Chebassier C, Pérault-Pochat MC, Fauconneau B, Pain S. Neuroprotective effect of PACAP on translational control alteration and cognitive decline in MPTP parkinsonian mice. Neurotox Res 2010;17(2):142–55.
[173] Vezoli J, Dzahini K, Costes N, Wilson CRE, Fifel K, Cooper HM, Kennedy H, Procyk E. Increased DAT binding in the early stage of the dopaminergic lesion: a longitudinal [11C]PE2I binding study in the MPTP-monkey. Neuroimage 2014;102(Part 2): 249–61.
[173a] Dhawan V, Dhoat S, Williams AJ, Dimarco A, Pal S, Forbes A, Tobias A, Martinez-Martin P, Chaudhuri KR. The range and nature of sleep dysfunction in untreated Parkinson's disease (PD). A comparative controlled clinical study using the Parkinson's disease sleep scale and selective polysomnography. J Neurol Sci 2006;248:158–62.
[174] Comella CL, Tanner CM, Ristanovic RK. Polysomnographic sleep measures in Parkinson's disease patients with treatment-induced hallucinations. Ann Neurol 1993;34:710–4.
[175] Ondo WG, Dat Vuong K, Khan H, Atassi F, Kwak C, Jankovic J. Daytime sleepiness and other sleep disorders in Parkinson's disease. Neurology 2001;57:1392–6.
[176] Levey AI, Hersch SM, Rye DB, Sunahara RK, Niznik HB, Kitt CA, Price DL, Maggio R, Brann MR, Ciliax BJ. Localization of D1 and D2 dopamine receptors in brain with subtype-specific antibodies. Proc Natl Acad Sci U S A 1993;90:8861–5.
[177] Dahan L, Astier B, Vautrelle N, Urbain N, Kocsis B, Chouvet G. Prominent burst firing of dopamin-ergic neurons in the ventral tegmental area during paradoxical sleep. Neuropsychopharmacology 2007;32:1232–41.
[178] Fantini ML, Gagnon JF, Filipini D, Montplaisir J. The effects of pramipexole in REM sleep behavior disorder. Neurology 2003; 61:1418–20.
[179] Garcia-Borreguero D, Caminero AB, De La Llave Y, Larrosa O, Barrio S, Granizo JJ, Pareja JA. Decreased phasic EMG activity during rapid eye movement sleep in treatment-naive Parkinson's disease: effects of treatment with levodopa and progression of illness. Mov Disord 2002;17:934–41.
[180] Braak H, Tredici KD, Rüb U, de Vos RAI, Jansen Steur ENH, Braak E. Staging of brain pathology related to sporadic Parkinson's disease. Neurobiol Aging 2003a;24:197–211.
[181] Iranzo A, Santamaria J, Rye DB, Valldeoriola F, Marti MJ, Munoz E, et al. Characteristics of idiopathic REM sleep behavior disorder and that associated with MSA and PD. Neurology 2005; 65:247–52.
[182] Ferini-Strambi L, Zucconi M. REM sleep behavior disorder. Clin Neurophysiol 2000;111:S136–40.
[183] Gagnon JF, Bedard MA, Fantini ML, Petit D, Panisset M, Rompre S, Carrier J, Montplaisir J. REM sleep behavior disorder and REM sleep without atonia in Parkinson's disease. Neurology 2002;59:585–9.
[184] Schenck CH, Bundlie SR, Mahowald MW. Delayed emergence of a parkinsonian disorder in 38% of 29 older men initially diagnosed with idiopathic rapid eye movement sleep behaviour disorder. Neurology 1996;46:388–93.
[185] Monaca C, Laloux C, Jacquesson JM, Gele P, Marechal X, Bordet R, Destee A, Derambure P. Vigilance states in a parkinsonian model, the MPTP mouse. Eur J Neurosci 2004;20:2474–8.
[186] Yi PL, Tsai CH, Lu MK, Liu HJ, Chen YC, Chang FC. Interleukin-1beta mediates sleep alteration in rats with rotenone-induced parkinsonism. Sleep 2007;30:413–25.
[187] Philippens IHCHM, Kersten CJM, Vanwersch RAP, Strijkstra AM. In: Ruigt GSF, et al., editors. Sleep and sleep EEG spectra in marmoset monkeys, vol. 15. Leiden, The Netherlands: Sleep-wake Research in the Netherland; 2004, ISBN 90-73675-13-8. p. 49–51.
[188] Pahwa R, Lyons KE. Early diagnosis of Parkinson's disease: recommendations from diagnostic clinical guidelines. Am J Manag Care 2010;(16 Suppl. Implications):S94–9.

CHAPTER

24

Experimental Autoimmune Encephalomyelitis in the Marmoset as a Translational Model for Multiple Sclerosis

Yolanda S. Kap[1], Jordon Dunham[1,2], Bert A. 't Hart[1,2]

[1]Department of Immunobiology, Biomedical Primate Research Centre, Rijswijk, The Netherlands; [2]University of Groningen, University Medical Center, Department of Neuroscience, Groningen, The Netherlands

BACKGROUND

Multiple sclerosis (MS) is a chronic autoimmune neurological disease affecting the central nervous system (CNS). About 2.5 million people worldwide are diagnosed with MS, which often occurs between the ages of 20 and 40 years. In about 85% of the patients, the disease starts as relapsing–remitting MS (RRMS), defined as periods with symptoms, such as balance disturbances, visual problems, and spasticity, alternating with periods of recovery. A substantial cohort of RRMS patients develops secondary progressive MS (SPMS) in which remissions gradually disappear and progressive worsening of the disease and eventually paralysis may occur. In 15% of MS patients, the disease is primary progressive from the beginning (PPMS) [1,2]. Pathologically, the disease is characterized by lesions in both the white matter (WM) and gray matter (GM) of the CNS, consisting of demyelination and variable degrees of axonal injury and inflammation, i.e., microglia activation and infiltrates of macrophages and lymphocytes [3]. The cause of MS is still unknown, but it is highly likely a complex disease involving the interaction of genetic [4] and environmental factors, such as vitamin D and viral infections [5]. Once established, autoreactive T cells and B cells specific for myelin components may drive evolution of the disease.

Despite an unknown etiology, the availability of effective drugs has tremendously improved during the last two decades, with, at the time of writing, about 14 disease-modifying treatments being approved for RRMS and 1 for SPMS. Some of the current therapies were selected based on efficacy in the animal model for MS, i.e., experimental autoimmune encephalomyelitis (EAE); other therapies were already being used in other human diseases and horizontally translated to MS. However, these treatments are not effective in all RRMS patients, and they can have serious adverse effects, such as progressive multifocal leukoencephalopathy and Graves' disease. More importantly, no treatment for PPMS exists, although some studies show promising results with B-cell targeting therapies [6].

For exploratory and applied translational research into MS, the use of an animal model is indispensable. Although we have learned a considerable amount about MS from rodent EAE models, the marmoset EAE model is the most useful model for translational research into pathogenic mechanisms and therapy development for MS, as we will describe in this chapter.

EXPERIMENTAL AUTOIMMUNE ENCEPHALOMYELITIS

History

There are several animal models for MS that spontaneously develop in transgenic mice [7–10], or that are induced by viruses [11–13], toxins [14,15], active sensitization against myelin, or adoptive transfer of T cells or serum from immunized donor animals to genetically compatible recipients [16]. The most widely used animal model for MS is EAE, which is induced by active immunization with preparations from CNS myelin or myelin

The Common Marmoset in Captivity and Biomedical Research
https://doi.org/10.1016/B978-0-12-811829-0.00024-8

components formulated with an adjuvant. This has been successfully achieved in susceptible strains of rodents, such as mice and rats [16], or nonhuman primates (NHP), such as the Old World monkey species rhesus macaque (*Macaca mulatta*) and cynomolgus macaque (*Macaca fascicularis*) [17], and a New World monkey species, the common marmoset (*Callithrix jacchus*) [18–20].

The idea to induce an MS-like disease by immunizing animals with myelin or myelin-derived components formulated in an adjuvant came from the serendipitous finding that humans developed paralysis after injection with rabies vaccine in 1885. This vaccine was made with spinal cord from rabies-infected rabbits [21]. Based on these findings, Rivers started to investigate the cause of this paralysis by repeated intramuscular injections of rhesus macaques with rabbit brain homogenate. Two of the eight monkeys developed symptoms and pathology after more than 50 injections [22]. However, when aqueous solutions and alcohol–ether extracts of rabbit brains were intramuscularly injected, six of the eight monkeys developed clinical signs and pathology [23]. Next, the procedure was refined to a single injection with brain homogenate in the newly developed complete Freund's adjuvant (CFA) or vaseline oil containing killed *Mycobacterium bovis*. In rhesus macaques, this led to an acute neuroinflammatory disease with paralytic symptoms resembling acute disseminated encephalomyelitis (ADEM) rather than MS [24,25]. In rodents and common marmosets, this led to a clinical course and pathology that mimics certain aspects of MS [16,18–20].

Myelin Oligodendrocyte Glycoprotein

Demyelination is the pathological hallmark of MS and is caused by a combined cellular and humoral autoimmune attack on myelin. The myelin sheath is produced by oligodendrocytes and is wrapped around axons to ensure fast, saltatory pulse conduction, to protect vulnerable axons against inflammatory mediators and to provide trophic support of the high energy-demanding axon. Myelin is composed of 80% lipids and 20% proteins, such as myelin basic protein (MBP), proteolipid protein (PLP), and myelin oligodendrocyte glycoprotein (MOG). MOG is a type I membrane glycoprotein and is located on the outer myelin sheath with the N-terminal amino acids 1 to 125 exposed to the extracellular environment where it is accessible for immune cells and antibodies [26,27]. MOG probably forms homodimers [27a] and is highly conserved among species [28,29].

The exact function of MOG in the CNS is unknown, but it has an essential role in the control of the immune homeostasis of the healthy brain as MOG can bind to DC-SIGN, a C-type lectin receptor expressed on microglia and dendritic cells [30]. Although MOG-deficient C57BL/6 mice do not present clinical or histological abnormalities [31], it has been shown that T cells from MOG-deficient mice display higher reactivity against ex vivo stimulation with MOG [32].

The Refinement of the Marmoset Experimental Autoimmune Encephalomyelitis Model

The marmoset EAE model was first developed by Genain and coworkers [33]. Since then, the model has been continuously refined to improve the relevance for MS (reviewed in Ref. [34]). The first documented EAE model in the marmoset was induced with human myelin emulsified in CFA supplemented with *Bordetella pertussis* particles. The disease was very acute and gave an ADEM-like disease [33,35]. Several large destructive lesions in the WM of the brain and spinal cord were detected in this model. EAE induction with human or mouse myelin in CFA, but without *B. pertussis*, gave a less destructive pathology and a slower, chronic disease progression. In these models, high T- and B-cell responses against various myelin proteins, i.e., MOG, MBP and PLP, were observed.

To investigate the pathogenic significance of MOG, one sibling of five marmoset twins was sensitized against myelin derived from wild-type C57BL/6 mice and the other sibling with myelin derived from MOG-deficient mice of the same strain [36]. While monkeys sensitized against myelin from the wild-type mice developed robust MS-like pathology and disease, animals immunized with MOG-deficient myelin showed only mild clinical EAE symptoms, mild inflammation, and small demyelinated areas in WM. Next, marmosets were immunized with recombinant human MOG (rhMOG) in CFA. This formulation induced a chronic disease and a typical MS-like pathology, characterized by active lesions in the CNS with macrophages containing myelin-degraded products. Moreover, antibody (Ab) formation against the whole MOG protein and specific MOG peptides were observed. In a metaanalysis of T-cell response profiles from about 30 rhMOG-immunized marmosets, we observed that T cells from monkeys developing clinically evident EAE early after immunization (fast progressors) proliferate against a wider range of MOG peptides than T cells from monkeys developing overt disease at a later time point (slow progressors). This was in contrast to mouse EAE models in which T-cell reactivity becomes more diverse when the disease progresses. Therefore, we hypothesized that one of the MOG peptides might have induced highly pathogenic T cells. We observed that T-cell responses against the peptides MOG34-56 and MOG74-96 were confined to fast progressor monkeys [37].

To test the immunogenicity of these peptides, marmosets were immunized with one of the peptides formulated in CFA. It was observed that monkeys immunized with MOG74-96/CFA did not develop clinically evident disease (score ≥ 2), whereas marmosets immunized with MOG34-56/CFA displayed an essentially similar pathology and disease course as the EAE model induced with rhMOG/CFA [37]. In addition, we observed that a single booster immunization of the MOG74-96/CFA-sensitized monkeys with MOG34-56/incomplete Freund's adjuvant (IFA) elicited clinically evident EAE [37]. As T cells against the two peptides do not cross-react, we next tested whether EAE can be induced in naïve monkeys by immunization with MOG34-56/IFA. Hence, animals were subjected to a challenge-boost protocol with MOG34-56 in IFA [38]. Remarkably, although the peptide/IFA inoculum was completely devoid of microbial antigens for innate immune activation, all animals developed clinically evident EAE, which was characterized by marked inflammation and demyelination in WM and GM of brain and spinal cord. In addition, EAE could also be induced by immunization with the rhMOG protein in IFA [39].

The principal difference between CFA and IFA is that with the latter adjuvant mainly antigen-experienced T cells present in the repertoire are activated. An additional advantage of the replacement of CFA by IFA is that the model shows subtle regulatory networks that keep pathogenic T and B cells in check and maintain homeostasis within the CNS. Furthermore, usage of CFA is a major cause of the serious discomfort to animals engaged in the EAE model.

In summary, robust EAE in the marmoset can be induced by active immunization with rhMOG or MOG34-56 emulsified in either CFA or IFA. This emulsion can be injected intracutaneously at the back [37,39] or subcutaneously [40,41]. The induction of EAE by using an adjuvant without mycobacteria seems to be a special aspect of marmoset EAE as neither specific-pathogen free (SPF) rodents (with the exception of Dark Agouti and Lewis rats), nor rhesus monkeys have satisfactory disease induction can be achieved with administration of MOG in IFA [42].

PATHOLOGY OF MULTIPLE SCLEROSIS IS REPLICATED IN MARMOSET EXPERIMENTAL AUTOIMMUNE ENCEPHALOMYELITIS

White Matter Lesions

One of the criteria supporting the clinical diagnosis of MS is the detection of WM lesions in the brain with magnetic resonance imaging (MRI) [43]. The presence of WM lesions disseminating in time and space is the major pathological hallmark of MS. WM lesions are characterized as focal demyelination with a variable degree of inflammation, axonal injury, edema, and blood–brain barrier dysfunction [1,44].

Whereas in mouse EAE models lesions are mostly confined to the spinal cord, lesions in the marmoset are present in WM and GM of brain and spinal cord [45]. An advantage of the marmoset is that the human-like brain architecture allows that WM lesions can be followed over time by in vivo MRI, and that total lesion load in WM and GM can be assessed by postmortem MRI [35,46] (Fig. 24.1). The most commonly applied T2-weighted MRI technique is highly sensitive for altered tissue distribution of water, caused by edema, for example, and can be used for longitudinal imaging of brain lesion development. In recent years, techniques have improved to perform MRI on a 7 T in vivo [47] and a 9 T for postmortem analysis [45].

Similar to MS, WM lesions disseminate in time and space as has been observed by in vivo MRI [35] (Fig. 24.1A). In addition, it has been shown by in vivo MRI that the great majority of focal WM lesions in the marmoset commence around small veins [47]. Furthermore, some MRI findings could be correlated to histological changes. It was shown, for example, that in vivo magnetization transfer ratio values correlated positively with the number of macrophages in the lesion [48]. The search for correlates between MRI and histology can be improved by using custom-fit, 3D-printed brain holders [49]. Brain MRI can be efficiently used as a clinically relevant readout in immunotherapy studies [45,50]. WM brain lesions in marmoset EAE induced by rhMOG or MOG peptide in CFA resemble the MS pattern II lesions with complement/Ab-mediated damage [51] and remain in their active phase containing macrophages and lymphocytes [48] (Fig. 24.1B).

The presence of oxidative damage and iron redistribution in lesions of the MS brain has been suggested to correlate to disease progression [52]. We have shown a similar profound accumulation of oxidative stress markers and iron redistribution in WM and GM lesions of marmosets with EAE. This was observed in all 4 EAE models, induced by rhMOG and MOG34-56 emulsified in either CFA or IFA [52a]. Interestingly, these pathological hallmarks of MS and marmoset EAE cannot be replicated in rodent EAE models [53].

The MS-like pathology of the marmoset EAE model is relevant for preclinical studies aiming at repair by, for example, stem cells. We reported previously on the marked clinical effect of human neural stem cells, which were administered via the intravenous route, although no regeneration of damaged CNS tissue could be observed [54]. A more recently tested approach used

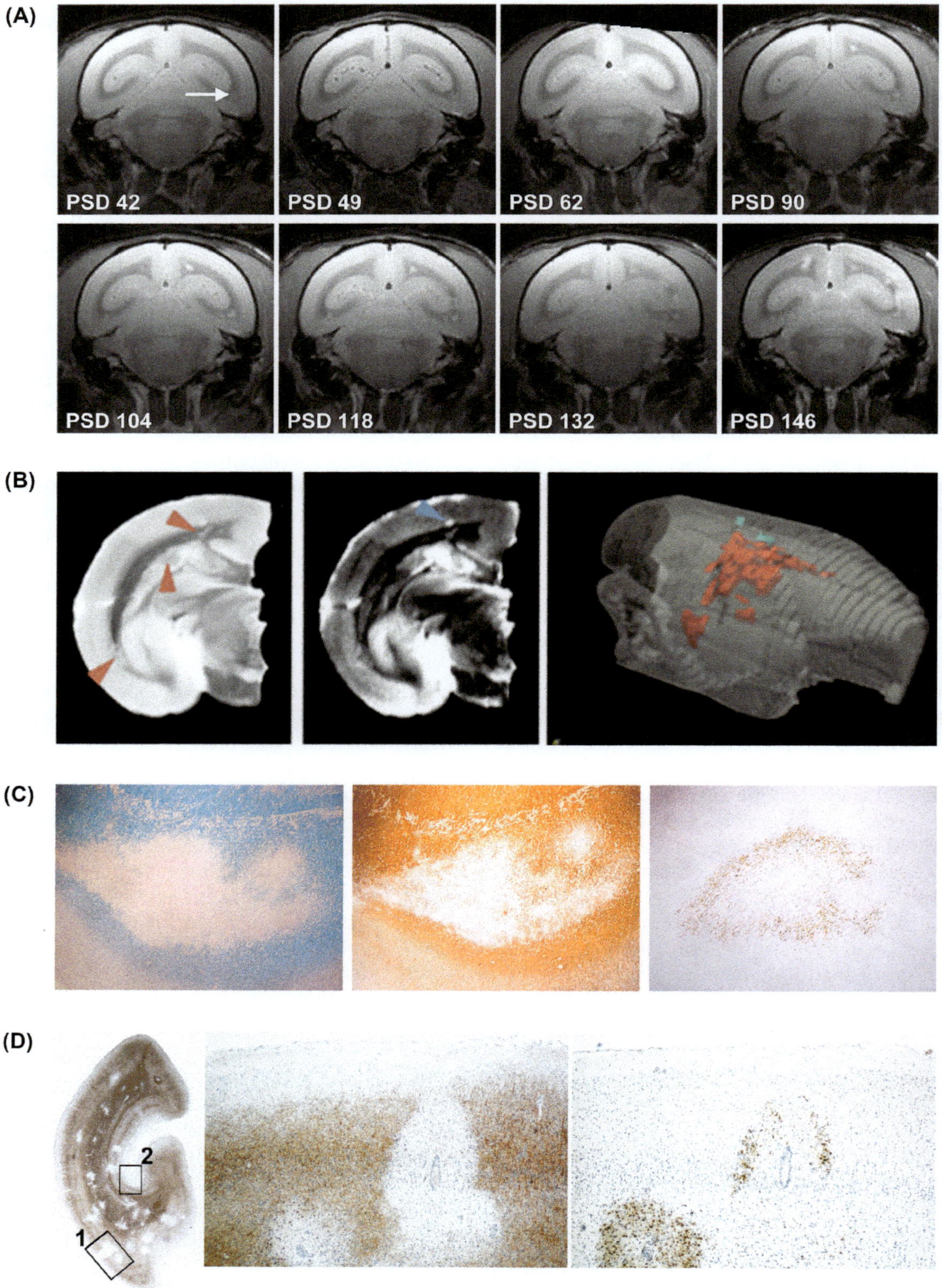

FIGURE 24.1 **Pathology of marmoset experimental autoimmune encephalomyelitis.** Pathology in the brain can be assessed by in vivo magnetic resonance imaging (MRI) (A), post mortem MRI (B), and histology (C and D). (A) The brain of a single marmoset in which EAE was induced with rhMOG/CFA was longitudinally scanned (T2-weighted, 4.7 T) to illustrate the dissemination of lesion formation in time and space. *PSD*, postsensitization day. (B) One hemisphere of a marmoset in which EAE was induced with rhMOG/IFA was scanned postmortem (T2-weighted, 9 T). A representative slide of T2-weighted (T2W) and white matter-attenuated inversion recovery (WAIR) images are depicted, showing white matter lesions (*red arrowheads*) and gray matter lesions (*blue arrowheads*). A three-dimensional reconstruction shows the total outline of white matter (red) and gray matter (green) lesions. (C) A large demyelinated lesion in the white matter of a marmoset in which EAE is induced by MOG34-56/CFA. A Luxol fast blue staining shows myelin in blue (left), a proteolipid protein (PLP) stains myelin in brown (middle), and macrophages and microglia present at the rim of the lesion are stained by myeloid-related protein 14 (MRP-14, right). (D) PLP staining reveals three types of gray matter lesions in a marmoset in which EAE is induced by rhMOG/CFA. A leukocortical and intracortical lesion is indicated by rectangle 1 in the left figure, and a subpial lesion is indicated by rectangle 2 in the left figure. Rectangle 1 is enlarged in the middle and right figures, showing demyelination and macrophages at the rim or throughout the lesion (right, MRP-14).

human oligodendrocyte precursor cells (OPCs) that were differentiated from skin fibroblast-derived induced pluripotent stem cells (iPS) and injected into the corpus callosum of marmosets with EAE. The iPS-derived OPCs migrated toward the lesions where contact with demyelinated axons seemed to induce differentiation of OPCs and production of new myelin [55].

Gray Matter Lesions

MS has been viewed as a mainly WM disease for a long time, but over the past decade it has become clear that lesions are also present in GM and that these play an important role in clinical worsening of patients with progressive MS [56]. GM lesions can be observed not only in the neocortex but also in other GM areas, such as in the hippocampus and the cerebellum. Four types of cortical GM lesions have been described. Subpial lesions (type III and IV), which are extensive and involve several neighboring gyri, are most common. Other lesion types are small, intracortical, and perivascular (type II), or leukocortical (type I), which span both the WM and GM [57].

A major advantage of the marmoset EAE model compared with mouse models is the presence of GM lesions that display the characteristic features of MS pathology. While lesion development in WM can be well visualized and characterized on conventional MR images [35,47,58], this is more problematic for lesions developing in the cortical GM. All types of GM lesions have been found in marmoset EAE, which are comparable with GM lesions observed in MS [41,59–61] (Fig. 24.1). Leukocortical lesions accounted for 57% of the total number of cortical lesions found in the marmoset. Intracortical lesions were found in two of the six marmoset brains examined, whereas subpial lesions accounted for 88% of the total demyelinated cortical area. In the EAE model induced by a subcutaneous injection of rhMOG/CFA with an injection of *B. pertussis*, activated macrophages and microglia were found in leukocortical and intracortical lesions, but the density was lower than in WM lesions, as has also been described for MS. Subpial lesions expressed only marginal signs of inflammation [41]. Furthermore, the cortical thickness was reduced in these marmosets compared to controls, but no differences were observed between demyelinated and myelinated areas [60]. Intracortical, but not subpial, lesions displayed immunoglobulin leakage and complement deposition after subcutaneous injection of rhMOG/CFA [59]. In the marmoset EAE model induced by ratMOG emulsified in IFA and injected subcutaneously, subpial lesions were associated with increased number of T cells and plasma cells in the meninges [62].

As GM lesions have a profound impact on disease progression, it is expected that therapies targeting GM lesion formation have a high impact on MS. The marmoset EAE model can be used to test the effect of candidate treatments as has been shown before: B-cell depletion prevented the development of GM lesions and thereby predicted a positive outcome for progressive MS [45,63]. In line with our prediction, the recently opened trial of the anti-CD20 mAb ocrelizumab showed encouraging clinical benefit in PPMS [6].

AUTOIMMUNITY IN THE MARMOSET EXPERIMENTAL AUTOIMMUNE ENCEPHALOMYELITIS MODEL

Immunoglobulins

In the past, the diagnosis of MS was mainly based on the presence of oligoclonal bands in the cerebrospinal fluid (CSF) [64]. It has been well established that anti-MOG antibodies can amplify demyelination in experimental conditions where demyelination is not or only marginally induced. As an example, marmosets immunized with MBP, PLP, or a chimeric MBP-PLP protein only developed demyelination when anti-MOG antibodies were present [65,66].

In the models induced with rhMOG, emulsified either in CFA or IFA, IgG binding to conformational MOG can be detected in plasma [39]. These antibodies show specific binding with several MOG peptides, including 14–36, 24–46, and 54–76, but not to MOG34-56 [67]. Of note, only preincubation with MOG54-76 blocked the IgG binding to ELISA plate-bound rhMOG [68]. Interestingly, in models induced by MOG34-56, either in CFA or IFA, immunoglobulins specific for MOG24-46, MOG34-56, or rhMOG can be observed [37], but these do not bind to conformationally intact MOG [39]. However, complement factor C9 and IgM were found deposited within demyelinated areas of MOG34-56-induced models, suggesting that antibodies can play a role [37,38]. The difference in binding to conformationally intact MOG shows the great discrepancy between the rhMOG- or MOG34-56-induced models.

T Cells

A significant increase in the number of MOG-specific T cells was reported in peripheral blood mononuclear cells (PBMC) and CSF of MS patients [69]. While half of the 24 MS patients that were tested for proliferation against myelin epitopes showed proliferation against MOG, only five MS patients showed proliferation against MBP and only two against PLP [70]. MOG-reactive T cells are also present in the natural T-cell repertoire of healthy individuals [71,72]. The fact that these autoreactive T cells are not depleted during development of the immune system has been explained by the absence of MOG expression in the thymus [73]. The T-cell response is mainly directed against three domains in the extracellular part of MOG, i.e., residues 1–22, 34–56, and 64–96 [74–76]. The phenotype of these T cells is $CD4^+CD8^-$, and they use αβ T-cell receptors for Ag recognition. MOG-reactive T cells can induce cytolysis of MOG-pulsed PBMC [72,76]. It has long been thought that Th1 cells were the main pathogenic T cells in MS, but more recent literature highlights an important pathogenic contribution of Th17 cells and $CD8^+$ T cells as well [77–85].

The 100% disease incidence after immunization of marmosets with rhMOG in CFA could be mapped to the MHC class II allele Caja-DRB*W1201, which is ubiquitously expressed in marmosets [67,86,87]. In marmosets immunized with rhMOG, Caja-DRB*W1201 presentation of MOG24-46 leads to the activation of MOG24-46-specific Th1 cells in all animals [67]. However, activation of T cells against this peptide alone is not sufficient to induce the full aspect of the disease as shown by immunization [67] or adoptive transfer [88]. This suggests that MOG24-46-specific T cells are involved in the initiation of the disease, but that T cells directed against other MOG epitopes are involved in the progression.

The absence of antibodies binding conformational MOG protein in the MOG34-56-induced models suggests that MOG34-56-specific T cells cause the widespread demyelination that was observed in this model. T-cell lines specific for MOG34-56 were phenotypically characterized as $CD4^+$, $CD8^+$, or CD4/8 double positive, expressing CD56, but not CD16, and capable of lysing MOG34-56-pulsed cells, suggesting they have the phenotype of natural killer-cytotoxic T lymphocytes (NK-CTL) [37]. Interestingly, T cells with a similar phenotype have been associated with demyelination in MS [89–92]. An in-depth analysis of these MOG34-56-specific T cells showed that the cytotoxicity was directed at the epitope MOG40-48 presented by the nonclassical MHC class 1b allele Caja-E, which is an orthologue to HLA-E and is expressed in noninflamed brain [19]. Besides the phenotype of NK-CTL, T cells specific for rhMOG or MGO34-56 produce IFNγ and/or IL17A ex vivo.

B Cells

The pathogenic role of B cells in MS has been thought for a very long time to be only the production of autoantibodies binding and opsonizing myelin Ab. During the clinical trials with the B cell-depleting antibody rituximab, it was observed that the treatment did not affect the production of antibodies [93,94]. Because B cells can have other roles, such as antigen presentation and cytokine production, it is now widely accepted that B cells play a role as antigen-presenting cells (APCs) in MS. This means they may activate pathogenic autoreactive T cells in MS. We could confirm this in the marmoset EAE model. We have shown that B-cell depletion starting 3 weeks after immunization, by an anti-CD20 mAb closely related to ofatumumab, which is clinically effective in RRMS [95], prevented the development of clinical symptoms and pathology, both in the Ab-dependent rhMOG/CFA model [45,63] and in the Ab-independent MOG34-56/IFA model [96]. Reduced T-cell proliferation in secondary lymphoid organs (SLO) was observed after B-cell depletion. In-depth analysis of the SLO showed that B-cell depletion alters the homing of T cells, indicating that T cells may need to be licensed by B cells for exiting the SLO and migrating to the target organ [97].

The Role of an EBV-Related Virus

The marmoset EAE model is used for applied and exploratory research, and the combination of both can lead to new findings. A good example is knowledge gained about the role of Epstein–Barr virus (EBV)-transformed B cells. EBV is a human herpesvirus and the cause of infectious mononucleosis. Considerable evidence suggests a pathogenic role of EBV infection in MS. However, the discrepancy between the high prevalence of EBV infection in the human population (>90%) and the low incidence of MS (±0.1%) is still unexplained [98]. Infection with EBV increases the risk for MS, which is even higher when the infection occurs after the age of 17 years and when the infection is symptomatic [99,100]. All MS patients have serum immunoglobulin (Ig) G antibodies (Ab) against EBV, whereas anti-EBV IgG is detected in 80%–90% of healthy individuals [101]. An increased IgG1 titer against EBV proteins EBNA-1 is predictive of disease exacerbation [102]. In contrast, anti-EBV IgG levels in CSF did not differ between MS patients and non-MS inflammatory CNS disease control patients. Furthermore, EBV RNA could not be detected in the CSF of MS patients [103]. It has been suggested that EBV-infected B cells migrate to the target organ where they become a source of autoantibodies and provide costimulatory signals to T cells [104]. Indeed, EBV-positive B cells were found in the meninges and brain parenchyma of MS patients [83], although others could not confirm this observation [103,105]. Interestingly, EBV seronegative individuals do not develop MS, indicating that being infected with EBV is a prerequisite for MS susceptibility [106].

Recent attention has focused on fundamental differences between the human pathogen-educated immune system and the naïve immune system of mice used in research conditions due to sterile housing conditions of SPF mice [107]. A major advantage of the marmoset is that they are housed conventionally and encounter similar pathogens as the ones that shape the human immune system. NHP are naturally infected with herpesviruses that are closely related to their human counterpart. Thus, in contrast to SPF-bred rodents, conventionally held colonies of NHP have experienced the immune-shaping effect of environmental pathogens throughout their development and may therefore be a good representation of the situation in humans. One example is the natural infection of marmosets with CalHV3, a γ-herpesvirus that shares many similarities with human EBV [108].

The first indication that CalHV3 has a pathogenic role in the marmoset EAE model came from a comparison of the CalHV3 load in SLO of marmosets treated with anti-CD20 Ab and antibodies targeting B-cell growth factors BLyS and APRIL. The CalHV3 load was profoundly lower in anti-CD20 Ab-treated marmosets compared with nontreated animals, but not in marmosets treated with anti-BLyS or anti-APRIL Ab [109]. Interestingly, this coincided with the clinical effect of the treatments: anti-CD20 mAb therapy reduced the clinical score and pathology of marmosets with EAE, whereas mAb targeting the B-cell growth and differentiation factors BLyS or APRIL was only partially effective. This observation suggested that B cells infected with the EBV-related CalHV3 play a crucial role in the disease.

In subsequent studies, we have further analyzed the role of EBV/CalHV3 and we discovered that EBV infection of B cells converts destructive processing of MOG34-56 peptide, which is a potential tolerance mechanism, into productive processing that can lead to cross presentation of the epitope via Caja-E, which is a potential autoimmune mechanism [110]. In addition, coculturing of EBV-transformed B cells with T cells isolated from SLO of marmosets with EAE showed that EBV-transformed B cells alter T cells with respect to activation markers such as CD27, IL-23R, and CD95, and migration markers such as CCR6, CCR7, and the cytokine IL17A [111].

THE MARMOSET AS A TRANSLATIONALLY RELEVANT MODEL IN THERAPY DEVELOPMENT

Because the marmoset EAE model closely resembles MS in its clinical, pathological, and immunopathogenic characteristics, it can very well be used to test new therapies, not only for assessing the efficacy of new compounds but also for investigating underlying mechanisms.

T-Cell Targeting Therapies

IL-12p40

APCs of the myeloid lineage produce IL-12p40, which is a shared subunit between IL-12 (with IL-12p35) and IL-23 (with IL-23p19). IL-12 is a critical cytokine in the induction of Th1 responses, whereas IL-23 is required for Th17 responses. Blocking of IL-12p40 by a mAb starting 14 days after immunization with myelin/CFA completely protected marmosets against EAE [112]. Blocking of IL-12p40 once T2-weighed lesions were present after immunization with rhMOG/CFA suppressed the increase of T2 signal intensity, a measure of inflammation, and the lesion enlargement, a measure of demyelination. However, the onset of clinical signs was only temporarily delayed. Unexpectedly, the same mAb (ustekinumab) lacked clinical efficacy in RRMS [113] but has found a new clinical target in psoriasis [114].

IL17A

Many of the immunopathogenic mechanisms driving EAE that were previously assigned to the IL-12-Th1 pathway are actually mediated via the IL-23-induced activation of Th17 cells [115,116], which exerts its proinflammatory effects via IL17A/F, IL-6, and TNF-α. We have shown that a neutralizing human Ab against human IL17A in the rhMOG/CFA-induced model delayed the onset of EAE but did not protect against the disease [117]. This suggests that IL17A is involved in late-stage disease, rather than in the disease onset.

IFNγ

The signature and immunoactive cytokine of Th1 cells secreted after stimulation with IL-12 is IFNγ. Clinical studies in RRMS showed that administration of IFNγ exacerbated disease activity [118], whereas treatment with anti-IFNγ Ab had only minor clinical effects. In the mouse EAE model both positive and negative effects of IFNγ neutralization have been reported, often depending on the timing of the treatment [119]. For this reason, we have tested whether early (days 0–25) or late (days 56–81) treatment with IFNγ would alter the EAE course in marmosets immunized with MOG34-56/IFA. We observed that neither of the treatment regimens had a positive effect on the disease course, but the disease exacerbation observed in RRMS was not seen. Remarkably, Th1-associated humoral and cellular autoimmune parameters were affected [120].

IL-7 Receptor

The IL-7 receptor (IL-7R) is composed of two subunits, i.e., CD132 and CD127, which are expressed on naïve and memory T cells and may have an association with MS [121]. Blocking CD127 in the marmoset EAE model induced by MOG34-56/IFA was effective only in monkeys with fast-progressing disease, which occurred in three of six marmoset twins; the treated sibling developed clinical symptoms almost 100 days later than the placebo-treated sibling. However, in the three late-responder twins, blockade of CD127 had no effect, suggesting that blocking IL-7 signaling may be a good treatment strategy for a subgroup of MS patients [122].

B-Cell Targeting Therapies

CD20

For many years, it has been thought that B cells only play a role as Ab-producing cells in MS. A clinical trial with the humanized mouse antihuman CD20 mAb rituximab, originally developed for the treatment of lymphoma, showed an unexpectedly profound and long-lasting effect on the annualized relapse rate and the MRI-detectable activity of WM lesions [94]. Currently, fully human CD20^{+} B cell-depleting Ab are available and tested in MS, such as ofatumumab (Arzerra, Novartis, Basel, Switzerland) and ocrelizumab (Ocrevus, Genentech Inc., San Francisco, CA) [123]. A clonal variant of ofatumumab (HuMab 7D8) was tested in the marmoset EAE model and showed similar profound clinical effects when treatment was started 21 days after immunization in both the rhMOG/CFA model and MOG34-56/IFA model [63,96]. Interestingly, depletion of B cells also prevented the development of GM lesions [45], a characteristic of progressive MS. Indeed, it was shown recently that ocrelizumab has a positive effect in PPMS [6]. A major advantage of using the marmoset EAE model for testing the efficacy of new drugs is the availability of SLO. Therewith we could show a possible explanation for the success of depleting antibodies, namely that B-cell depletion prevents the egress of autoaggressive T cells from the SLO to the target organs [97].

CD40

CD40 is a costimulatory molecule that is constitutively expressed on B cells and induced on myeloid APCs on activation. The interaction of CD40 with its ligand CD40L (CD154) induced on activated T cells plays an important role in the activation of T and B cells as well as on macrophage effector functions. A mouse antihuman CD40 Ab (mu5D12) had a strong protective effect in the myelin-induced marmoset EAE model when administered 14 or 25 days after EAE induction, followed by treatment every 2 days for 28 days. However, shortly after the treatment was stopped neurological deficits appeared [124]. On the basis of these promising data, a chimeric version of the Ab was generated, which was further evaluated in the rhMOG/CFA model. In a first experiment, treatment with the chimeric anti-CD40 Ab (ch5D12) was started 1 day prior to disease induction and continued at a frequency of one injection every 2–3 days until the end of a 50-day observation period. The monkeys remained devoid of neurological deficits during the mAb treatment [125]. In the next experiment, we tested the chimeric anti-CD40 mAb in a therapeutic fashion, i.e., starting once brain WM lesions were detected with MRI. To be able to test the effect of the mAb on the lesion activity, several MRI parameters were quantified. The Ab had a beneficial effect on MRI-detectable CNS pathology, but no clear clinical effect [126].

Anti-BLyS/Anti-APRIL mAb

Atacicept (Merck KGaA, Darmstadt, Germany), a fusion protein that blocks the function of BLyS and APRIL, failed in a phase II trial for MS. A reverse translation study was performed in the marmoset in which B cells were depleted by anti-BLyS and anti-APRIL mAbs, but the effect on the clinical scores was minimal. A possible explanation for this effect lies in the depletion of the γ-herpesvirus CalHV3, as described elsewhere in this paper and reviewed in Ref. [127].

CONCLUDING REMARKS

As discussed in this chapter, the marmoset model for MS has many advantages including the presence of both WM and GM lesions, natural infection with an EBV-homologue, and the availability of lymphoid organs. The marmoset EAE model has been refined to two main models induced with either MOGprotein or MOG-peptide 34-56 emulsified in IFA. Furthermore, the marmoset EAE model can very well be used for translational research for therapy development, thereby not only investigating the efficacy of a new therapy but also the underlying mechanisms.

References

[1] Compston A, Coles A. Multiple sclerosis. Lancet 2008;372:1502–17.

[2] Sospedra M, Martin R. Immunology of multiple sclerosis. Annu Rev Immunol 2005;23:683–747.

[3] Lassmann H. Pathology and disease mechanisms in different stages of multiple sclerosis. J Neurol Sci 2013;333:1–4.

[4] International Multiple Sclerosis Genetics Consortium, Wellcome Trust Case Control Consortium, Sawcer S, Hellenthal G, Pirinen M, Spencer CC, Patsopoulos NA, Moutsianas L,

Dilthey A, Su Z, Freeman C, Hunt SE, Edkins S, Gray E, Booth DR, Potter SC, Goris A, Band G, Oturai AB, Strange A, Saarela J, Bellenguez C, Fontaine B, Gillman M, Hemmer B, Gwilliam R, Zipp F, Jayakumar A, Martin R, Leslie S, Hawkins S, Giannoulatou E, D'Alfonso S, Blackburn H, Martinelli Boneschi F, Liddle J, Harbo HF, Perez ML, Spurkland A, Waller MJ, Mycko MP, Ricketts M, Comabella M, Hammond N, Kockum I, McCann OT, Ban M, Whittaker P, Kemppinen A, Weston P, Hawkins C, Widaa S, Zajicek J, Dronov S, Robertson N, Bumpstead SJ, Barcellos LF, Ravindrarajah R, Abraham R, Alfredsson L, Ardlie K, Aubin C, Baker A, Baker K, Baranzini SE, Bergamaschi L, Bergamaschi R, Bernstein A, Berthele A, Boggild M, Bradfield JP, Brassat D, Broadley SA, Buck D, Butzkueven H, Capra R, Carroll WM, Cavalla P, Celius EG, Cepok S, Chiavacci R, Clerget-Darpoux F, Clysters K, Comi G, Cossburn M, Cournu-Rebeix I, Cox MB, Cozen W, Cree BA, Cross AH, Cusi D, Daly MJ, Davis E, de Bakker PI, Debouverie M, D'Hooghe MB, Dixon K, Dobosi R, Dubois B, Ellinghaus D, Elovaara I, Esposito F, Fontenille C, Foote S, Franke A, Galimberti D, Ghezzi A, Glessner J, Gomez R, Gout O, Graham C, Grant SF, Guerini FR, Hakonarson H, Hall P, Hamsten A, Hartung HP, Heard RN, Heath S, Hobart J, Hoshi M, Infante-Duarte C, Ingram G, Ingram W, Islam T, Jagodic M, Kabesch M, Kermode AG, Kilpatrick TJ, Kim C, Klopp N, Koivisto K, Larsson M, Lathrop M, Lechner-Scott JS, Leone MA, Leppa V, Liljedahl U, Bomfim IL, Lincoln RR, Link J, Liu J, Lorentzen AR, Lupoli S, Macciardi F, Mack T, Marriott M, Martinelli V, Mason D, McCauley JL, Mentch F, Mero IL, Mihalova T, Montalban X, Mottershead J, Myhr KM, Naldi P, Ollier W, Page A, Palotie A, Pelletier J, Piccio L, Pickersgill T, Piehl F, Pobywajlo S, Quach HL, Ramsay PP, Reunanen M, Reynolds R, Rioux JD, Rodegher M, Roesner S, Rubio JP, Ruckert IM, Salvetti M, Salvi E, Santaniello A, Schaefer CA, Schreiber S, Schulze C, Scott RJ, Sellebjerg F, Selmaj KW, Sexton D, Shen L, Simms-Acuna B, Skidmore S, Sleiman PM, Smestad C, Sorensen PS, Sondergaard HB, Stankovich J, Strange RC, Sulonen AM, Sundqvist E, Syvanen AC, Taddeo F, Taylor B, Blackwell JM, Tienari P, Bramon E, Tourbah A, Brown MA, Tronczynska E, Casas JP, Tubridy N, Corvin A, Vickery J, Jankowski J, Villoslada P, Markus HS, Wang K, Mathew CG, Wason J, Palmer CN, Wichmann HE, Plomin R, Willoughby E, Rautanen A, Winkelmann J, Wittig M, Trembath RC, Yaouanq J, Viswanathan AC, Zhang H, Wood NW, Zuvich R, Deloukas P, Langford C, Duncanson A, Oksenberg JR, Pericak-Vance MA, Haines JL, Olsson T, Hillert J, Ivinson AJ, De Jager PL, Peltonen L, Stewart GJ, Hafler DA, Hauser SL, McVean G, Donnelly P, Compston A. Genetic risk and a primary role for cell-mediated immune mechanisms in multiple sclerosis. Nature 2011;476:214–9.

[5] Ascherio A, Munger KL. Environmental risk factors for multiple sclerosis. Part II: noninfectious factors. Ann Neurol 2007;61:504–13.

[6] Montalban X, Hauser SL, Kappos L, Arnold DL, Bar-Or A, Comi G, de Seze J, Giovannoni G, Hartung HP, Hemmer B, Lublin F, Rammohan KW, Selmaj K, Traboulsee A, Sauter A, Masterman D, Fontoura P, Belachew S, Garren H, Mairon N, Chin P, Wolinsky JS, Investigators OC. Ocrelizumab versus placebo in primary progressive multiple sclerosis. N Engl J Med 2017;376:209–20.

[7] Bettelli E, Pagany M, Weiner HL, Linington C, Sobel RA, Kuchroo VK. Myelin oligodendrocyte glycoprotein-specific T cell receptor transgenic mice develop spontaneous autoimmune optic neuritis. J Exp Med 2003;197:1073–81.

[8] Ellmerich S, Mycko M, Takacs K, Waldner H, Wahid FN, Boyton RJ, King RH, Smith PA, Amor S, Herlihy AH, Hewitt RE, Jutton M, Price DA, Hafler DA, Kuchroo VK, Altmann DM. High incidence of spontaneous disease in an HLA-DR15 and TCR transgenic multiple sclerosis model. J Immunol 2005;174:1938–46.

[9] Goverman J, Woods A, Larson L, Weiner LP, Hood L, Zaller DM. Transgenic mice that express a myelin basic protein-specific T cell receptor develop spontaneous autoimmunity. Cell 1993;72: 551–60.

[10] Waldner H, Whitters MJ, Sobel RA, Collins M, Kuchroo VK. Fulminant spontaneous autoimmunity of the central nervous system in mice transgenic for the myelin proteolipid protein-specific T cell receptor. Proc Natl Acad Sci USA 2000;97:3412–7.

[11] Oleszak EL, Chang JR, Friedman H, Katsetos CD, Platsoucas CD. Theiler's virus infection: a model for multiple sclerosis. Clin Microbiol Rev 2004;17:174–207.

[12] Smith JP, Morris-Downes M, Brennan FR, Wallace GJ, Amor S. A role for alpha4-integrin in the pathology following Semliki Forest virus infection. J Neuroimmunol 2000;106:60–8.

[13] Tsunoda I, Fujinami RS. Inside-Out versus Outside-In models for virus induced demyelination: axonal damage triggering demyelination. Springer Semin Immunopathol 2002;24:105–25.

[14] Degaonkar MN, Raghunathan P, Jayasundar R, Jagannathan NR. Determination of relaxation characteristics during preacute stage of lysophosphatidyl choline-induced demyelinating lesion in rat brain: an animal model of multiple sclerosis. Magn Reson Imaging 2005;23:69–73.

[15] Lindner M, Fokuhl J, Linsmeier F, Trebst C, Stangel M. Chronic toxic demyelination in the central nervous system leads to axonal damage despite remyelination. Neurosci Lett 2009;453:120–5.

[16] Gold R, Linington C, Lassmann H. Understanding pathogenesis and therapy of multiple sclerosis via animal models: 70 years of merits and culprits in experimental autoimmune encephalomyelitis research. Brain 2006;129:1953–71.

[17] Haanstra KG, Jagessar SA, Bauchet AL, Doussau M, Fovet CM, Heijmans N, Hofman SO, van Lubeek-Veth J, Bajramovic JJ, Kap YS, Laman JD, Touin H, Watroba L, Bauer J, Lachapelle F, Serguera C, 't Hart BA. Induction of experimental autoimmune encephalomyelitis with recombinant human myelin oligodendrocyte glycoprotein in incomplete Freund's adjuvant in three non-human primate species. J Neuroimmune Pharmacol 2013; 8:1251–64.

[18] 't Hart BA, Bauer J, Brok HP, Amor S. Non-human primate models of experimental autoimmune encephalomyelitis: variations on a theme. J Neuroimmunol 2005a;168:1–12.

[19] Jagessar SA, Heijmans N, Blezer EL, Bauer J, Blokhuis JH, Wubben JA, Drijfhout JW, van den Elsen PJ, Laman JD, 't Hart BA. Unravelling the T-cell-mediated autoimmune attack on CNS myelin in a new primate EAE model induced with MOG(34-56) peptide in incomplete adjuvant. Eur J Immunol 2012c;42:217–27.

[20] Kap YS, van Driel N, Arends R, Rouwendal G, Verolin M, Blezer E, Lycke N, 't Hart BA. Immune modulation by a tolerogenic myelin oligodendrocyte glycoprotein (MOG)10-60 containing fusion protein in the marmoset experimental autoimmune encephalomyelitis model. Clin Exp Immunol 2015;180:28–39.

[21] Baxter AG. The origin and application of experimental autoimmune encephalomyelitis. Nat Rev Immunol 2007;7:904–12.

[22] Rivers TM, Sprunt DH, Berry GP. Observations on attempts to produce acute disseminated encephalomyelitis in monkeys. J Exp Med 1933;58:39–53.

[23] Rivers TM, Schwentker FF. Encephalomyelitis accompanied by myelin destruction experimentally produced in monkeys. J Exp Med 1935;61:689–702.

[24] Kabat EA, Wolf A, Bezer AE, Murray JP. Studies on acute disseminated encephalomyelitis produced experimentally in rhesus monkeys. J Exp Med 1951;93:615–33.

[25] Ravkina L, Harib I, Manovitch Z, Deconenko E, Letchinskaja E, Papilova E. Hyperacute experimental allergic encephalomyelitis in rhesus monkeys as a model of acute necrotizing hemorrhagic encephalomyelitis. J Neurol 1979;221:113–25.

[26] Kroepfl JF, Viise LR, Charron AJ, Linington C, Gardinier MV. Investigation of myelin/oligodendrocyte glycoprotein membrane topology. J Neurochem 1996;67:2219–22.

[27] Ohler B, Graf K, Bragg R, Lemons T, Coe R, Genain C, Israelachvili J, Husted C. Role of lipid interactions in autoimmune demyelination. Biochim Biophys Acta 2004;1688:10–7.

[27a] Clements CS, Reid HH, Beddoe T, Tynan FE, Perugini MA, Johns TG, Bernard CC, Rossjohn J. The crystal structure of myelin oligodendrocyte glycoprotein, a key autoantigen in multiple sclerosis. Proc Natl Acad Sci USA 2003;100:11059–64.

[28] Delarasse C, Della Gaspera B, Lu CW, Lachapelle F, Gelot A, Rodriguez D, Dautigny A, Genain C, Pham-Dinh D. Complex alternative splicing of the myelin oligodendrocyte glycoprotein gene is unique to human and non-human primates. J Neurochem 2006;98:1707–17.

[29] Mesleh MF, Belmar N, Lu CW, Krishnan VV, Maxwell RS, Genain CP, Cosman M. Marmoset fine B cell and T cell epitope specificities mapped onto a homology model of the extracellular domain of human myelin oligodendrocyte glycoprotein. Neurobiol Dis 2002;9:160–72.

[30] Garcia-Vallejo JJ, Ilarregui JM, Kalay H, Chamorro S, Koning N, Unger WW, Ambrosini M, Montserrat V, Fernandes RJ, Bruijns SC, van Weering JR, Paauw NJ, O'Toole T, van Horssen J, van der Valk P, Nazmi K, Bolscher JG, Bajramovic J, Dijkstra CD, 't Hart BA, van Kooyk Y. CNS myelin induces regulatory functions of DC-SIGN-expressing, antigen-presenting cells via cognate interaction with MOG. J Exp Med 2014;211: 1465–83.

[31] Delarasse C, Daubas P, Mars LT, Vizler C, Litzenburger T, Iglesias A, Bauer J, Della Gaspera B, Schubart A, Decker L, Dimitri D, Roussel G, Dierich A, Amor S, Dautigny A, Liblau R, Pham-Dinh D. Myelin/oligodendrocyte glycoprotein-deficient (MOG-deficient) mice reveal lack of immune tolerance to MOG in wild-type mice. J Clin Investig 2003;112:544–53.

[32] Linares D, Mana P, Goodyear M, Chow AM, Clavarino C, Huntington ND, Barnett L, Koentgen F, Tomioka R, Bernard CC, Freire-Garabal M, Reid HH. The magnitude and encephalogenic potential of autoimmune response to MOG is enhanced in MOG deficient mice. J Autoimmun 2003;21:339–51.

[33] Massacesi L, Genain CP, Lee-Parritz D, Letvin NL, Canfield D, Hauser SL. Active and passively induced experimental autoimmune encephalomyelitis in common marmosets: a new model for multiple sclerosis. Ann Neurol 1995;37:519–30.

[34] 't Hart BA, van Kooyk Y, Geurts JJG, Gran B. The primate autoimmune encephalomyelitis model; a bridge between mouse and man. Ann Clin Transl Neurol 2015;2:581–93.

[35] 't Hart BA, Bauer J, Muller HJ, Melchers B, Nicolay K, Brok H, Bontrop RE, Lassmann H, Massacesi L. Histopathological characterization of magnetic resonance imaging-detectable brain white matter lesions in a primate model of multiple sclerosis: a correlative study in the experimental autoimmune encephalomyelitis model in common marmosets (*Callithrix jacchus*). Am J Pathol 1998;153:649–63.

[36] Jagessar SA, Smith PA, Blezer E, Delarasse C, Pham-Dinh D, Laman JD, Bauer J, Amor S, 't Hart BA. Autoimmunity against myelin oligodendrocyte glycoprotein is dispensable for the initiation although essential for the progression of chronic encephalomyelitis in common marmosets. J Neuropathol Exp Neurol 2008;67:326–40.

[37] Kap YS, Smith P, Jagessar SA, Remarque E, Blezer E, Strijkers GJ, Laman JD, Hintzen RQ, Bauer J, Brok HP, 't Hart BA. Fast progression of recombinant human myelin/oligodendrocyte glycoprotein (MOG)-induced experimental autoimmune encephalomyelitis in marmosets is associated with the activation of MOG34-56-specific cytotoxic T cells. J Immunol 2008;180: 1326–37.

[38] Jagessar SA, Kap YS, Heijmans N, van Driel N, van Straalen L, Bajramovic JJ, Brok HPM, Blezer ELA, Bauer J, Laman JD, 't Hart BA. Induction of progressive demyelinating autoimmune encephalomyelitis in common marmoset monkeys using MOG34-56 peptide in incomplete Freund adjuvant. J Neuropathol Exp Neurol 2010;69:372–85.

[39] Jagessar SA, Heijmans N, Blezer EL, Bauer J, Weissert R, 't Hart BA. Immune profile of an atypical EAE model in marmoset monkeys immunized with recombinant human myelin oligodendrocyte glycoprotein in incomplete Freund's adjuvant. J Neuroinflammation 2015;12:169.

[40] Boretius S, Schmelting B, Watanabe T, Merkler D, Tammer R, Czeh B, Michaelis T, Frahm J, Fuchs E. Monitoring of EAE onset and progression in the common marmoset monkey by sequential high-resolution 3D MRI. NMR Biomed 2006;19:41–9.

[41] Pomeroy IM, Matthews PM, Frank JA, Jordan EK, Esiri MM. Demyelinated neocortical lesions in marmoset autoimmune encephalomyelitis mimic those in multiple sclerosis. Brain 2005;128:2713–21.

[42] 't Hart BA, Gran B, Weissert R. EAE: imperfect but useful models of multiple sclerosis. Trends Mol Med 2011;17:119–25.

[43] McDonald WI, Compston A, Edan G, Goodkin D, Hartung HP, Lublin FD, McFarland HF, Paty DW, Polman CH, Reingold SC, Sandberg-Wollheim M, Sibley W, Thompson A, van den Noort S, Weinshenker BY, Wolinsky JS. Recommended diagnostic criteria for multiple sclerosis: guidelines from the International Panel on the diagnosis of multiple sclerosis. Ann Neurol 2001;50:121–7.

[44] Compston A, Coles A. Multiple sclerosis. Lancet 2002;359: 1221–31.

[45] Kap YS, Bauer J, Driel N, Bleeker WK, Parren PW, Kooi EJ, Geurts JJ, Laman JD, Craigen JL, Blezer E, 't Hart BA. B-cell depletion attenuates white and gray matter pathology in marmoset experimental autoimmune encephalomyelitis. J Neuropathol Exp Neurol 2011a;70:992–1005.

[46] Maggi P, Sati P, Massacesi L. Magnetic resonance imaging of experimental autoimmune encephalomyelitis in the common marmoset. J Neuroimmunol 2016;304:86–92.

[47] Gaitan MI, Maggi P, Wohler J, Leibovitch E, Sati P, Calandri IL, Merkle H, Massacesi L, Silva AC, Jacobson S, Reich DS. Perivenular brain lesions in a primate multiple sclerosis model at 7-tesla magnetic resonance imaging. Mult Scler 2014;20:64–71.

[48] Blezer EL, Bauer J, Brok HP, Nicolay K, 't Hart BA. Quantitative MRI-pathology correlations of brain white matter lesions developing in a non-human primate model of multiple sclerosis. NMR Biomed 2007;20:90–103.

[49] Guy JR, Sati P, Leibovitch E, Jacobson S, Silva AC, Reich DS. Custom fit 3D-printed brain holders for comparison of histology with MRI in marmosets. J Neurosci Methods 2016;257:55–63.

[50] 't Hart BA, Smith P, Amor S, Strijkers GJ, Blezer EL. MRI-guided immunotherapy development for multiple sclerosis in a primate. Drug Discov Today 2006;11:58–66.

[51] Merkler D, Schmelting B, Czeh B, Fuchs E, Stadelmann C, Bruck W. Myelin oligodendrocyte glycoprotein-induced experimental autoimmune encephalomyelitis in the common marmoset reflects the immunopathology of pattern II multiple sclerosis lesions. Mult Scler 2006b;12:369–74.

[52] Haider L, Fischer MT, Frischer JM, Bauer J, Hoftberger R, Botond G, Esterbauer H, Binder CJ, Witztum JL, Lassmann H. Oxidative damage in multiple sclerosis lesions. Brain 2011;134: 1914–24.

[52a] Dunham J, Bauer J, Campbell GR, Mahad DJ, van Driel N, van der pol SMA, 'tHart BA, Lassmann H, Laman JD, van Horssen J, Kap YS. Oxidative injury and iron redistribution are pathological hallmarks of marmoset experimental autoimmune encephalomyelitis. J Neuropathol Exp Neurol 2017;76:467–78.

[53] Schuh C, Wimmer I, Hametner S, Haider L, Van Dam AM, Liblau RS, Smith KJ, Probert L, Binder CJ, Bauer J, Bradl M, Mahad D, Lassmann H. Oxidative tissue injury in multiple sclerosis is only partly reflected in experimental disease models. Acta Neuropathol 2014;128:247–66.

[54] Pluchino S, Gritti A, Blezer E, Amadio S, Brambilla E, Borsellino G, Cossetti C, Del Carro U, Comi G, 't Hart B, Vescovi A, Martino G. Human neural stem cells ameliorate autoimmune encephalomyelitis in non-human primates. Ann Neurol 2009;66:343–54.

[55] Thiruvalluvan A, Czepiel M, Kap YS, Mantingh-Otter I, Vainchtein I, Kuipers J, Bijlard M, Baron W, Giepmans B, Bruck W, 't Hart BA, Boddeke E, Copray S. Survival and functionality of human induced pluripotent stem cell-derived oligodendrocytes in a nonhuman primate model for multiple sclerosis. Stem Cells Transl Med 2016;5:1550–61.

[56] Bo L, Geurts JJ, Mork SJ, van der Valk P. Grey matter pathology in multiple sclerosis. Acta Neurol Scand Suppl 2006;183:48–50.

[57] Bo L, Vedeler CA, Nyland HI, Trapp BD, Mork SJ. Subpial demyelination in the cerebral cortex of multiple sclerosis patients. J Neuropathol Exp Neurol 2003;62:723–32.

[58] Maggi P, Macri SM, Gaitan MI, Leibovitch E, Wholer JE, Knight HL, Ellis M, Wu T, Silva AC, Massacesi L, Jacobson S, Westmoreland S, Reich DS. The formation of inflammatory demyelinated lesions in cerebral white matter. Ann Neurol 2014;76:594–608.

[59] Merkler D, Boscke R, Schmelting B, Czeh B, Fuchs E, Bruck W, Stadelmann C. Differential macrophage/microglia activation in neocortical EAE lesions in the marmoset monkey. Brain Pathol 2006a;16:117–23.

[60] Pomeroy IM, Jordan EK, Frank JA, Matthews PM, Esiri MM. Diffuse cortical atrophy in a marmoset model of multiple sclerosis. Neurosci Lett 2008;437:121–4.

[61] Pomeroy IM, Jordan EK, Frank JA, Matthews PM, Esiri MM. Focal and diffuse cortical degenerative changes in a marmoset model of multiple sclerosis. Mult Scler 2010;16:537–48.

[62] Kramann N, Neid K, Menken L, Schlumbohm C, Stadelmann C, Fuchs E, Bruck W, Wegner C. Increased meningeal T and plasma cell infiltration is associated with early subpial cortical demyelination in common marmosets with experimental autoimmune encephalomyelitis. Brain Pathol 2015;25:276–86.

[63] Kap YS, van Driel N, Blezer E, Parren PW, Bleeker WK, Laman JD, Craigen JL, 't Hart BA. Late B cell depletion with a human anti-human CD20 IgG1kappa monoclonal antibody halts the development of experimental autoimmune encephalomyelitis in marmosets. J Immunol 2010;185:3990–4003.

[64] Johnson KP, Nelson BJ. Multiple sclerosis: diagnostic usefulness of cerebrospinal fluid. Ann Neurol 1977;2:425–31.

[65] Genain CP, Nguyen MH, Letvin NL, Pearl R, Davis RL, Adelman M, Lees MB, Linington C, Hauser SL. Antibody facilitation of multiple sclerosis-like lesions in a nonhuman primate. J Clin Investig 1995;96:2966–74.

[66] McFarland HI, Lobito AA, Johnson MM, Nyswaner JT, Frank JA, Palardy GR, Tresser N, Genain CP, Mueller JP, Matis LA, Lenardo MJ. Determinant spreading associated with demyelination in a nonhuman primate model of multiple sclerosis. J Immunol 1999;162:2384–90.

[67] Brok HP, Uccelli A, Kerlero De Rosbo N, Bontrop RE, Roccatagliata L, de Groot NG, Capello E, Laman JD, Nicolay K, Mancardi GL, Ben-Nun A, 't Hart BA. Myelin/oligodendrocyte glycoprotein-induced autoimmune encephalomyelitis in common marmosets: the encephalitogenic T cell epitope pMOG24-36 is presented by a monomorphic MHC class II molecule. J Immunol 2000;165:1093–101.

[68] 't Hart BA, Massacesi L. Clinical, pathological, and immunologic aspects of the multiple sclerosis model in common marmosets (*Callithrix jacchus*). J Neuropathol Exp Neurol 2009;68:341–55.

[69] Sun J, Link H, Olsson T, Xiao BG, Andersson G, Ekre HP, Linington C, Diener P. T and B cell responses to myelin-oligodendrocyte glycoprotein in multiple sclerosis. J Immunol 1991;146:1490–5.

[70] Kerlero de Rosbo N, Milo R, Lees MB, Burger D, Bernard CC, Ben-Nun A. Reactivity to myelin antigens in multiple sclerosis. Peripheral blood lymphocytes respond predominantly to myelin oligodendrocyte glycoprotein. J Clin Investig 1993;92:2602–8.

[71] Diaz-Villoslada P, Shih A, Shao L, Genain CP, Hauser SL. Autoreactivity to myelin antigens: myelin/oligodendrocyte glycoprotein is a prevalent autoantigen. J Neuroimmunol 1999;99:36–43.

[72] Koehler NK, Genain CP, Giesser B, Hauser SL. The human T cell response to myelin oligodendrocyte glycoprotein: a multiple sclerosis family-based study. J Immunol 2002;168:5920–7.

[73] Bruno R, Sabater L, Sospedra M, Ferrer-Francesch X, Escudero D, Martinez-Caceres E, Pujol-Borrell R. Multiple sclerosis candidate autoantigens except myelin oligodendrocyte glycoprotein are transcribed in human thymus. Eur J Immunol 2002;32:2737–47.

[74] Bielekova B, Sung MH, Kadom N, Simon R, McFarland H, Martin R. Expansion and functional relevance of high-avidity myelin-specific CD4+ T cells in multiple sclerosis. J Immunol 2004;172:3893–904.

[75] Kerlero de Rosbo N, Hoffman M, Mendel I, Yust I, Kaye J, Bakimer R, Flechter S, Abramsky O, Milo R, Karni A, Ben-Nun A. Predominance of the autoimmune response to myelin oligodendrocyte glycoprotein (MOG) in multiple sclerosis: reactivity to the extracellular domain of MOG is directed against three main regions. Eur J Immunol 1997;27:3059–69.

[76] Van der Aa A, Hellings N, Bernard CC, Raus J, Stinissen P. Functional properties of myelin oligodendrocyte glycoprotein-reactive T cells in multiple sclerosis patients and controls. J Neuroimmunol 2003;137:164–76.

[77] Babbe H, Roers A, Waisman A, Lassmann H, Goebels N, Hohlfeld R, Friese M, Schroder R, Deckert M, Schmidt S, Ravid R, Rajewsky K. Clonal expansions of CD8(+) T cells dominate the T cell infiltrate in active multiple sclerosis lesions as shown by micromanipulation and single cell polymerase chain reaction. J Exp Med 2000;192:393–404.

[78] Brucklacher-Waldert V, Stuerner K, Kolster M, Wolthausen J, Tolosa E. Phenotypical and functional characterization of T helper 17 cells in multiple sclerosis. Brain 2009;132:3329–41.

[79] Ifergan I, Kebir H, Bernard M, Wosik K, Dodelet-Devillers A, Cayrol R, Arbour N, Prat A. The blood-brain barrier induces differentiation of migrating monocytes into Th17-polarizing dendritic cells. Brain 2008;131:785–99.

[80] Kebir H, Kreymborg K, Ifergan I, Dodelet-Devillers A, Cayrol R, Bernard M, Giuliani F, Arbour N, Becher B, Prat A. Human TH17 lymphocytes promote blood-brain barrier disruption and central nervous system inflammation. Nat Med 2007;13:1173–5.

[81] Lock C, Hermans G, Pedotti R, Brendolan A, Schadt E, Garren H, Langer-Gould A, Strober S, Cannella B, Allard J, Klonowski P, Austin A, Lad N, Kaminski N, Galli SJ, Oksenberg JR, Raine CS, Heller R, Steinman L. Gene-microarray analysis of multiple sclerosis lesions yields new targets validated in autoimmune encephalomyelitis. Nat Med 2002;8:500–8.

[82] Matusevicius D, Kivisakk P, He B, Kostulas N, Ozenci V, Fredrikson S, Link H. Interleukin-17 mRNA expression in blood and CSF mononuclear cells is augmented in multiple sclerosis. Mult Scler 1999;5:101–4.

[83] Serafini B, Rosicarelli B, Franciotta D, Magliozzi R, Reynolds R, Cinque P, Andreoni L, Trivedi P, Salvetti M, Faggioni A, Aloisi F. Dysregulated Epstein-Barr virus infection in the multiple sclerosis brain. J Exp Med 2007;204:2899–912.

[84] Skulina C, Schmidt S, Dornmair K, Babbe H, Roers A, Rajewsky K, Wekerle H, Hohlfeld R, Goebels N. Multiple sclerosis: brain-infiltrating CD8+ T cells persist as clonal expansions in the cerebrospinal fluid and blood. Proc Natl Acad Sci USA 2004;101:2428–33.

[85] Tzartos JS, Friese MA, Craner MJ, Palace J, Newcombe J, Esiri MM, Fugger L. Interleukin-17 production in central nervous system-infiltrating T cells and glial cells is associated with active disease in multiple sclerosis. Am J Pathol 2008;172:146–55.

[86] Antunes SG, de Groot NG, Brok H, Doxiadis G, Menezes AA, Otting N, Bontrop RE. The common marmoset: a new world primate species with limited Mhc class II variability. Proc Natl Acad Sci USA 1998;95:11745–50.

[87] Doxiadis GG, van der Wiel MK, Brok HP, de Groot NG, Otting N, t Hart BA, van Rood JJ, Bontrop RE. Reactivation by exon shuffling of a conserved HLA-DR3-like pseudogene segment in a New World primate species. Proc Natl Acad Sci USA 2006;103:5864–8.

[88] Villoslada P, Abel K, Heald N, Goertsches R, Hauser SL, Genain CP. Frequency, heterogeneity and encephalitogenicity of T cells specific for myelin oligodendrocyte glycoprotein in naive outbred primates. Eur J Immunol 2001;31:2942–50.

[89] Antel JP, McCrea E, Ladiwala U, Qin YF, Becher B. Non-MHC-restricted cell-mediated lysis of human oligodendrocytes in vitro: relation with CD56 expression. J Immunol 1998;160:1606–11.

[90] Mazzarino P, Pietra G, Vacca P, Falco M, Colau D, Coulie P, Moretta L, Mingari MC. Identification of effector-memory CMV-specific T lymphocytes that kill CMV-infected target cells in an HLA-E-restricted fashion. Eur J Immunol 2005;35:3240–7.

[91] Vergelli M, Le H, van Noort JM, Dhib-Jalbut S, McFarland H, Martin R. A novel population of CD4+CD56+ myelin-reactive T cells lyses target cells expressing CD56/neural cell adhesion molecule. J Immunol 1996;157:679–88.

[92] Zaguia F, Saikali P, Ludwin S, Newcombe J, Beauseigle D, McCrea E, Duquette P, Prat A, Antel JP, Arbour N. Cytotoxic NKG2C+ CD4 T cells target oligodendrocytes in multiple sclerosis. J Immunol 2013;190:2510–8.

[93] Bar-Or A, Calabresi PA, Arnold D, Markowitz C, Shafer S, Kasper LH, Waubant E, Gazda S, Fox RJ, Panzara M, Sarkar N, Agarwal S, Smith CH. Rituximab in relapsing-remitting multiple sclerosis: a 72-week, open-label, phase I trial. Ann Neurol 2008;63:395–400.

[94] Hauser SL, Waubant E, Arnold DL, Vollmer T, Antel J, Fox RJ, Bar-Or A, Panzara M, Sarkar N, Agarwal S, Langer-Gould A, Smith CH. B-cell depletion with rituximab in relapsing-remitting multiple sclerosis. N Engl J Med 2008;358:676–88.

[95] Sorensen PS, Lisby S, Grove R, Derosier F, Shackelford S, Havrdova E, Drulovic J, Filippi M. Safety and efficacy of ofatumumab in relapsing-remitting multiple sclerosis: a phase 2 study. Neurology 2014;82:573–81.

[96] Jagessar SA, Heijmans N, Bauer J, Blezer EL, Laman JD, Hellings N, 't Hart BA. B-cell depletion abrogates T cell-mediated demyelination in an antibody-nondependent common marmoset experimental autoimmune encephalomyelitis model. J Neuropathol Exp Neurol 2012b;71:716–28.

[97] Kap YS, van Driel N, Laman JD, Tak PP, 't Hart BA. CD20+ B cell depletion alters T cell homing. J Immunol 2014;192:4242–53.

[98] Ascherio A, Munger KL, Lunemann JD. The initiation and prevention of multiple sclerosis. Nat Rev Neurol 2012;8:602–12.

[99] Martyn CN, Cruddas M, Compston DA. Symptomatic Epstein-Barr virus infection and multiple sclerosis. J Neurol Neurosurg Psychiatry 1993;56:167–8.

[100] Thacker EL, Mirzaei F, Ascherio A. Infectious mononucleosis and risk for multiple sclerosis: a meta-analysis. Ann Neurol 2006;59:499–503.

[101] Buljevac D, van Doornum GJ, Flach HZ, Groen J, Osterhaus AD, Hop W, van Doorn PA, van der Meche FG, Hintzen RQ. Epstein-Barr virus and disease activity in multiple sclerosis. J Neurol Neurosurg Psychiatry 2005;76:1377–81.

[102] Ascherio A, Munger KL. EBV and autoimmunity. Curr Top Microbiol Immunol 2015;390:365–85.

[103] Sargsyan SA, Shearer AJ, Ritchie AM, Burgoon MP, Anderson S, Hemmer B, Stadelmann C, Gattenlohner S, Owens GP, Gilden D, Bennett JL. Absence of Epstein-Barr virus in the brain and CSF of patients with multiple sclerosis. Neurology 2010;74:1127–35.

[104] Pender MP. Infection of autoreactive B lymphocytes with EBV, causing chronic autoimmune diseases. Trends Immunol 2003;24:584–8.

[105] Willis SN, Stadelmann C, Rodig SJ, Caron T, Gattenloehner S, Mallozzi SS, Roughan JE, Almendinger SE, Blewett MM, Bruck W, Hafler DA, O'Connor KC. Epstein-Barr virus infection is not a characteristic feature of multiple sclerosis brain. Brain 2009;132:3318–28.

[106] Pakpoor J, Disanto G, Gerber JE, Dobson R, Meier UC, Giovannoni G, Ramagopalan SV. The risk of developing multiple sclerosis in individuals seronegative for Epstein-Barr virus: a meta-analysis. Mult Scler 2013;19:162–6.

[107] Beura LK, Hamilton SE, Bi K, Schenkel JM, Odumade OA, Casey KA, Thompson EA, Fraser KA, Rosato PC, Filali-Mouhim A, Sekaly RP, Jenkins MK, Vezys V, Haining WN, Jameson SC, Masopust D. Normalizing the environment recapitulates adult human immune traits in laboratory mice. Nature 2016;532:512–6.

[108] Cho Y, Ramer J, Rivailler P, Quink C, Garber RL, Beier DR, Wang F. An Epstein-Barr-related herpesvirus from marmoset lymphomas. Proc Natl Acad Sci USA 2001;98:1224–9.

[109] Jagessar SA, Fagrouch Z, Heijmans N, Bauer J, Laman JD, Oh L, Migone T, Verschoor EJ, 't Hart BA. The different clinical effects of anti-BLyS, anti-APRIL and anti-CD20 antibodies point at a critical pathogenic role of gamma-herpesvirus infected B cells in the marmoset EAE model. J Neuroimmune Pharmacol 2013;8:727–38.

[110] Jagessar SA, Holtman IR, Hofman S, Morandi E, Heijmans N, Laman JD, Gran B, Faber BW, van Kasteren SI, Eggen BJ, 't Hart BA. Lymphocryptovirus infection of nonhuman primate B cells converts destructive into productive processing of the pathogenic CD8 T cell epitope in myelin oligodendrocyte glycoprotein. J Immunol 2016;197:1074–88.

[111] Dunham J, van Driel N, Eggen BJL, Paul C, 't Hart BA, Laman JD, Kap YS. Analysis of the cross-talk of Epstein-Barr virus infected B cells with T cells in the marmoset. Clin Transl Immunol 2017;6.

[112] Brok HP, van Meurs M, Blezer E, Schantz A, Peritt D, Treacy G, Laman JD, Bauer J, 't Hart BA. Prevention of experimental autoimmune encephalomyelitis in common marmosets using an anti-IL-12p40 monoclonal antibody. J Immunol 2002;169:6554–63.

[113] Segal BM, Constantinescu CS, Raychaudhuri A, Kim L, Fidelus-Gort R, Kasper LH. Repeated subcutaneous injections of IL12/23 p40 neutralising antibody, ustekinumab, in patients with relapsing-remitting multiple sclerosis: a phase II, double-blind, placebo-controlled, randomised, dose-ranging study. Lancet Neurol 2008;7:796–804.

[114] McInnes IB, Kavanaugh A, Gottlieb AB, Puig L, Rahman P, Ritchlin C, Brodmerkel C, Li S, Wang Y, Mendelsohn AM, Doyle MK, Group PS. Efficacy and safety of ustekinumab in patients with active psoriatic arthritis: 1 year results of the phase 3,

multicentre, double-blind, placebo-controlled PSUMMIT 1 trial. Lancet 2013;382:780–9.

[115] Cua DJ, Sherlock J, Chen Y, Murphy CA, Joyce B, Seymour B, Lucian L, To W, Kwan S, Churakova T, Zurawski S, Wiekowski M, Lira SA, Gorman D, Kastelein RA, Sedgwick JD. Interleukin-23 rather than interleukin-12 is the critical cytokine for autoimmune inflammation of the brain. Nature 2003;421:744–8.

[116] Langrish CL, Chen Y, Blumenschein WM, Mattson J, Basham B, Sedgwick JD, McClanahan T, Kastelein RA, Cua DJ. IL-23 drives a pathogenic T cell population that induces autoimmune inflammation. J Exp Med 2005;201:233–40.

[117] Kap YS, Jagessar SA, van Driel N, Blezer E, Bauer J, van Meurs M, Smith P, Laman JD, 't Hart BA. Effects of early IL-17A neutralization on disease induction in a primate model of experimental autoimmune encephalomyelitis. J Neuroimmune Pharmacol 2011b;6:341–53.

[118] Panitch HS, Hirsch RL, Haley AS, Johnson KP. Exacerbations of multiple sclerosis in patients treated with gamma interferon. Lancet 1987;1:893–5.

[119] Sanvito L, Constantinescu CS, Gran B, 't Hart BA. The multifaceted role of interferon-γ in central nervous system autoimmune demyelination. Open Autoimmun J 2010;2:151–9.

[120] Jagessar SA, Gran B, Heijmans N, Bauer J, Laman JD, t Hart BA, Constantinescu CS. Discrepant effects of human interferon-gamma on clinical and immunological disease parameters in a novel marmoset model for multiple sclerosis. J Neuroimmune Pharmacol 2012a;7:253–65.

[121] Gregory SG, Schmidt S, Seth P, Oksenberg JR, Hart J, Prokop A, Caillier SJ, Ban M, Goris A, Barcellos LF, Lincoln R, McCauley JL, Sawcer SJ, Compston DA, Dubois B, Hauser SL, Garcia-Blanco MA, Pericak-Vance MA, Haines JL. Interleukin 7 receptor alpha chain (IL7R) shows allelic and functional association with multiple sclerosis. Nat Genet 2007;39:1083–91.

[122] Dunham J, Lee LF, van Driel N, Laman JD, Ni I, Zhai W, Tu GH, Lin JC, Bauer J, 't Hart BA, Kap YS. Blockade of CD127 exerts a dichotomous clinical effect in marmoset experimental autoimmune encephalomyelitis. J Neuroimmune Pharmacol 2016;11:73–83.

[123] Lycke J. Monoclonal antibody therapies for the treatment of relapsing-remitting multiple sclerosis: differentiating mechanisms and clinical outcomes. Ther Adv Neurol Disord 2015;8: 274–93.

[124] Laman JD, 't Hart BA, Brok H, Meurs M, Schellekens MM, Kasran A, Boon L, Bauer J, Boer M, Ceuppens J. Protection of marmoset monkeys against EAE by treatment with a murine antibody blocking CD40 (mu5D12). Eur J Immunol 2002;32: 2218–28.

[125] Boon L, Brok HP, Bauer J, Ortiz-Buijsse A, Schellekens MM, Ramdien-Murli S, Blezer E, van Meurs M, Ceuppens J, de Boer M, 't Hart BA, Laman JD. Prevention of experimental autoimmune encephalomyelitis in the common marmoset (*Callithrix jacchus*) using a chimeric antagonist monoclonal antibody against human CD40 is associated with altered B cell responses. J Immunol 2001;167:2942–9.

[126] 't Hart BA, Blezer EL, Brok HP, Boon L, de Boer M, Bauer J, Laman JD. Treatment with chimeric anti-human CD40 antibody suppresses MRI-detectable inflammation and enlargement of pre-existing brain lesions in common marmosets affected by MOG-induced EAE. J Neuroimmunol 2005b;163:31–9.

[127] 't Hart BA, Jagessar SA, Haanstra K, Verschoor E, Laman JD, Kap YS. The primate EAE model points at EBV-infected B cells as a preferential therapy target in multiple sclerosis. Front Immunol 2013;4:145.

CHAPTER

25

Marmosets in Auditory Research

Steven J. Eliades, Joji Tsunada

Auditory and Communications Systems Laboratory, Department of Otorhinolaryngology: Head and Neck Surgery, University of Pennsylvania Perelman School of Medicine, Philadelphia, PA, United States

INTRODUCTION

Over the past few decades, marmosets have engendered a considerable amount of interest as a model system for auditory research. Owing to a rich and diverse vocal repertoire and vocal behaviors, marmosets are becoming an important species for studies of vocal communication and the mechanisms of hearing. Because of their close phylogenetic relationship to humans, nonhuman primates are a logical model for understanding the evolutionary and physiologic processes related to human speech. However, classic primate species used in neurobiological research, notably, macaque monkeys, often fail to produce their vocal behaviors in laboratory and research settings. Marmosets, on the other hand, appear to maintain much of their natural ethologic behaviors in captivity [1,2], making them a natural model for understanding vocal communication (Fig. 25.1). Marmosets can also be easily bred and raised in captivity, often in small university colonies rather than requiring a dedicated primate center. As a result, marmosets provide the potential to understand the full life cycle of auditory and vocal mechanisms, including genetics, development, and aging-related changes.

Because vocal communication requires the production, transmission, and perception of an acoustic signal, understanding these processes requires study of the underlying auditory mechanisms. It is sensory processing in the auditory system that allows for such transduction, processing, and perception of vocalizations and other sounds. Beyond communication sounds, there are also a variety of auditory abilities which are important for marmosets and most other species. These include the ability to localize the origin of sounds, separate out ethologically relevant signals from background noise, and encode many other complex auditory inputs. Unfortunately, there is much that we still do not know about how the auditory system operates, particularly as compared with other sensory systems like vision. In many respects, much of our current understanding of the marmoset auditory system is extrapolated from other species, which may provide insight into shared structure and function, particularly in the early auditory pathway. However, marmosets may also exhibit unique specializations in higher-level auditory functions, particularly as relating to their vocal behaviors. This chapter reviews the current understanding of the auditory perceptual abilities of marmosets, anatomy and physiology of the auditory pathway, methodologic considerations when studying marmoset audition, and finally, select topics in auditory research in which marmosets have provided valuable insights.

AUDITORY PERCEPTION

Hearing Range and Perceptual Thresholds

One of the most fundamental functions of the auditory system is the ability to detect the presence of sound. Such abilities are typically measured by a psychoacoustical threshold, that is, the minimal intensity or sound pressure level (SPL) at which one can perceive the presence of sound on at least half of trials. Because such perceptual thresholds vary across different acoustic frequencies, they can also be used to measure the hearing range of a species. Although earlier studies of the marmoset hearing range using shock avoidance techniques revealed a hearing range of 100 Hz–25 kHz [3], more recent studies using operant conditioning and positive rewards show the upper hearing range to extend as high as 36 kHz [4]. Both studies demonstrate the lowest hearing threshold, the point of maximal sensitivity, to be at 7 kHz in marmosets. The origin of this notch is

The Common Marmoset in Captivity and Biomedical Research
https://doi.org/10.1016/B978-0-12-811829-0.00025-X

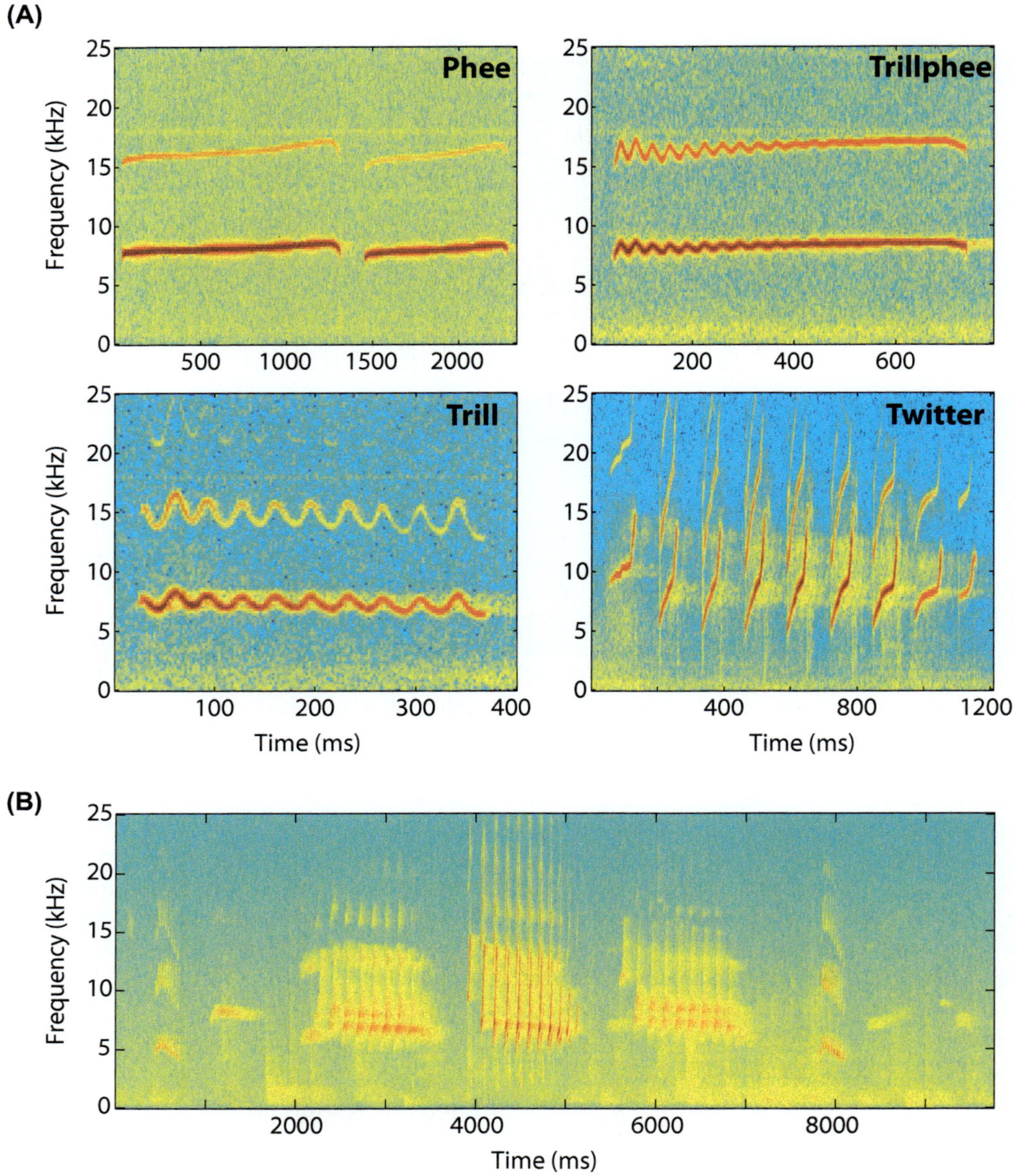

FIGURE 25.1 **Marmoset vocalizations.** (A) Frequency–time spectrogram representations of examples for the four most common marmoset social vocalization types (phee, trillphee, trill, and twitter). (B) Sample of an interactive marmoset vocal "conversation."

unclear, although similar patterns are seen in many nonhuman primate species (see comparison in Ref. [4]) and may be related either to ear canal resonance or to the frequency range of marmoset vocalizations which span 6–10 kHz [1,2]. To what extent this similarity reflects an evolutionary bias to produce vocal sounds that exploit maximal auditory sensitivity, or, alternatively, sensitivity designed to optimally encode such sounds, is not known. Similar ranges of hearing and thresholds have also been estimated based on physiologic thresholds measured from neurons in marmoset auditory cortex (AC) [5].

Frequency Discrimination

Another important psychoacoustical behavior is the ability to discriminate between two different sounds. Measurements of frequency discrimination, for example, reveal the ability of the auditory periphery (inner ear, etc.) to resolve differences between very similar sounds and the ability to use such discrimination to make perceptual decisions. Such abilities are potentially important in the ability of an animal to detect differences between vocalizations produced by other marmosets, which may substantially overlap in frequency [1]. Measurements of frequency discrimination in marmosets reveal an ability to detect differences as small as 0.37 semitones [6], where a semitone represents 1/12th of an octave in frequency change. Such performance is similar to that of other New World primate species, but poorer than that observed in Old World species and humans (0.03–0.3 semitones, see comparison in Ref. [6]). The maximal discrimination sensitivity for marmosets was found for the 3.5–14 kHz frequency range, again overlapping the frequencies of their vocalizations and potentially conveying an advantage in vocal processing. Such overlaps in sensitivity and vocal acoustics have also been observed in humans and other nonhuman primates [7].

Pitch Perception

Pitch, the perceptual location of a sound on a low to high musical scale, is another important perceptual feature of many sounds, including human speech and music. The perception of pitch is attributed to the temporal regularity of periodic sounds (i.e., the period of rapidly repeating sounds or the fundamental frequency of harmonically related sounds). Although perceptual mechanisms of pitch have been extensively studied in humans, it has been unclear to what extent these are shared in some nonhuman animals. Because much of the information necessary to generate a pitch percept is not explicitly coded in the auditory periphery (i.e., in the cochlea and auditory nerve), and must be computed in more central brain regions, understanding pitch processing reveals information about important mechanisms of central auditory processing [8]. Recent work in marmosets has demonstrated that they exhibit many of the key features of pitch perception seen in humans, including use of temporal (timing) information for low-pitch sounds and spectral (frequency) information for higher-pitched sounds [9], as well as a strength of pitch perception based on the resolvability (separability) of harmonics in complex sounds [10,11]. Although specific details of these features differ from humans, possibly related to differences in the hearing range and frequency discrimination abilities, the pitch perceptual mechanisms appear to be generally similar between marmosets and humans.

ANATOMY OF THE MARMOSET AUDITORY PATHWAY

Temporal Bone and Middle Ear Structures

Structures of the auditory periphery are housed within the temporal bone, part of the lateral skull. The lateral portion of the temporal bone, known as the mastoid, consists of a series of mucosalized air cells of unclear function, even for humans. Compared with the temporal bone of humans, the mastoid of marmosets is rather underdeveloped, with little to no mastoid tip [12]. The external ear consists of the pinna, also known as the auricle, and the ear canal itself. The marmoset pinna measures approximately 1.5 by 2 cm and is covered anteriorly by dense tufts [13,14]. As with most other primates, the marmoset pinna appears relatively immobile, or at least lacking the directional movement abilities seen in some predator species. As in most species, the folds of the pinna impart frequency amplification and notching that are important in generating spectral cues for vertical sound localization, and recent measurements in marmosets have revealed similar types of these cues [14].

The middle ear consists of the tympanic membrane (ear drum) and three ossicles: malleus, incus, and stapes (Fig. 25.2). The marmoset tympanic membrane measures approximately 4–4.7 mm in diameter [13,15]. The malleus is approximately 3.2 mm in length, and the incus long process 2.2 mm (short process 1.8 mm). The marmoset stapes is 1.3 mm high and has a footplate measuring about 1.1 mm × 0.6 mm. The sizes of these ossicles do not appear to change much in size between infant and adult marmosets. Other middle ear structures, including the facial nerve and chorda tympani, appear to be anatomically similar in location to humans [13], allowing the potential for classic transmastoid facial recess approaches for cochlear implantation studies in marmosets [15].

Auditory Periphery: Cochlea and Auditory Nerve

Relatively little is known about the marmoset auditory periphery. Imaging and histologic analysis of marmoset cochlea has revealed 2.76–2.84 turns [12,15], similar to the 2.5–3-turn range in humans. The total length of the marmoset cochlea is 16.5 mm, similar to that of guinea pigs, but smaller than that of larger primates such as macaques and humans. Membranous structures in the marmoset cochlea include the typical three fluid compartment organizations: scala vestibuli, scala media, and scala tympani. The cross-sectional area of the scala tympani decreases as one moves from cochlear base to apex, again similar to other species, but if scaled by 2.5× for absolute size, appears to closely match the gradual size progression and tapering seen in the human cochlea [15].

Functionally, little is known about the physiology of the marmoset cochlea or auditory nerve. Measurements of cochlear electrical potentials, using a recording electrode outside the round window membrane of the

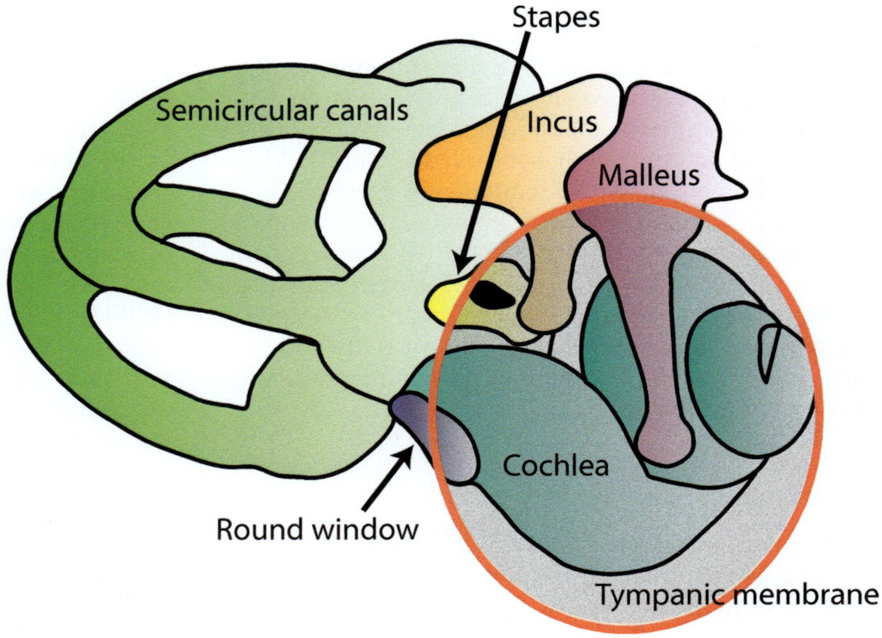

FIGURE 25.2 **Marmoset middle and inner ear.** Illustration of the marmoset auditory periphery for a right ear, including the tympanic membrane, three ossicles (malleus, incus, and stapes), cochlea, and semicircular canals (vestibular system).

cochlea, have revealed sensitivity to sounds up to 100 kHz [16]. This is notably higher than behavioral and neural responses that have been documented, and it is unclear if this reflects a true auditory sensitivity in the ultrasonic range. The presence of a tonotopic organization of the marmoset cochlea has not been explicitly measured. However, measurements of cochlear implant-evoked cortical responses strongly suggest that the marmoset cochlea is arranged tonotopically (high frequencies in the basilar turn near the oval and round windows, low frequencies more distally near the cochlear apex), similar to other mammalian species [17].

Cochlear Nucleus

The cochlear nucleus is the first brain structure of the central auditory pathway (Fig. 25.3). After exiting the cochlea, the auditory nerve enters the brain stem and bifurcates with one branch synapsing in the posteroventral (PVCN) and dorsal (DCN) cochlear nucleus and the other innervating the anteroventral cochlear nucleus (AVCN). This overall structure appears well conserved among primates and other mammals, including marmosets [13,18]. However, the cytoarchitecture of the cochlear nucleus varies widely between mammalian species, with a well-developed lamellar structure and granular cell layer in nonprimates that is reduced in primates and absent in humans [18]. In marmosets, the granular cell layer is thin and external in the DCN and reduced in the AVCN/PVCN, consistent with an intermediary evolutionary position along the primate lineage. Fusiform cells, a prominent feature in cat DCN, lose their radially oriented layer in marmosets and are concentrated instead in central regions without a clear lamellar organization. The origin of these evolutionary differences between primates and nonprimate species is unclear. Other cell types in AVCN/PVCN, including octopus, stellate, and busy cells, appear similar to those in more well-characterized nonprimate species [19].

Physiologically, the marmoset cochlear nucleus also exhibits many similar properties to that of other species. These include the presence of cells with specific temporal response profiles to sound stimulation (chopper, onset, primary-like, pauser, etc.), although the relative numbers of these different cell types vary between species [19]. Marmoset cochlear nucleus neurons also exhibit similar frequency-tuned responses, latencies, and ability to follow time-varying sounds as other species. The presence of a regular tonotopic organization within the marmoset cochlear nucleus has not been demonstrated but is also assumed to exist based on its presence in other animals [20].

Superior Olivary Complex

The superior olivary complex is a series of brain stem structures involved in performing interaural comparisons for use in sound localization [21]. The complex consists of the medial superior olive (MSO) involved in interaural timing difference (ITD) calculation, lateral superior olive (LSO) involved in interaural level difference (ILD) calculations, and the medial nucleus of the

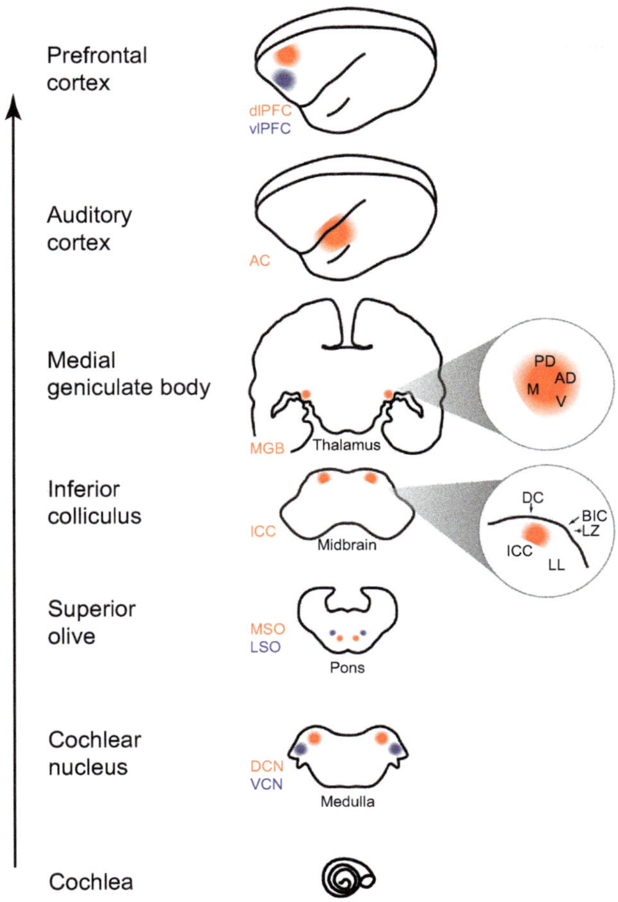

FIGURE 25.3 **Marmoset auditory pathway.** Key structures in the ascending auditory pathway are shown, extending from the cochlea to the prefrontal cortex. *AC*, auditory cortex; *AD*, anteriodorsal MGB; *BIC*, brachium of the inferior colliculus; *DC*, dorsal cortex; *DCN*, dorsal cochlear nucleus; *dlPFC*, dorsolateral prefrontal cortex; *ICC*, central nucleus of the inferior colliculus; *LSO*, lateral superior olive; *LZ*, lateral zone; *M*, medial MGB; *MGB*, medial geniculate body; *MSO*, medial superior olive; *PD*, posteriodorsal MGB; *V*, ventral MGB; *VCN*, ventral cochlear nucleus; *vlPFC*, ventrolateral prefrontal cortex.

trapezoid body. Such sound localization cues are a direct result of the head acting as a barrier to sound transmission and the spacing between the ears, taking slightly longer and being slightly attenuated when going from one side of the head to the other. Information about superior olivary structures is limited in marmosets, consisting only of anatomic studies [22]. Compared with other primates, marmosets have a rather enlarged MSO with an MSO/LSO cell count ratio more similar to nonprimate mammals. The origin of these differences is unclear, but it has been noted that a more prominent LSO might be expected of a species with a small head that is reliant on high-frequency hearing, as relative loudness is a more reliable cue than timing at such frequencies [22].

Inferior Colliculus

The next major structure in the ascending auditory pathway is the inferior colliculus (IC). Anatomically, the IC is dominated by its central nucleus (ICC), which consists of densely packed neurons structured into lamellae thought to reflect the organized tonotopic inputs from the ascending auditory pathway [23]. The IC is considered to be an "obligate" relay station for auditory information, in that all inputs from the core auditory pathway originating in AVCN/PVCN, also known as the lemniscal pathway, synapse within the ICC [24]. Whether or not lemniscal fibers all relay through the ICC, or could bypass to the auditory thalamus, has not been demonstrated in marmosets. However, similar patterns are seen in both cats and closely related squirrel monkeys, suggesting a similar function. In addition to the central nucleus, the marmoset IC also contains a dorsomedially positioned dorsal cortex, a striplike lateral zone, and a small brachial nucleus, most of which are not known to have a clearly defined function.

Based on work in squirrel monkeys, the ICC has been shown to have a tonotopic organization, with low-frequency layers more dorsally and laterally and higher-frequency layers more medially and caudally [25]. Measurements in marmosets have been lacking, but 2-deoxyglucose uptake following sound presentation, which reflects metabolic demands from neural activity, roughly confirms the presence of tonotopic frequency organization [23]. It is presumed that this reflects an evolutionarily conserved lamellar tonotopy within IC. Relatively little is known about the physiology of neurons within the noncentral IC in both marmosets and most other species.

One important function of the IC is the integration of sound localization cues to extract the spatial position of a sound source [26]. Because the IC receives converging inputs from lower auditory centers that each code sound location through a different mechanism (ILD from LSO, ITD from MSO, spectral cues from the DCN), it is a potential site for the reconciliation of this information. Recordings in the marmoset ICC have revealed a mix of neurons encoding sound sources either from unimodal cues (either ITD or ILD) or in combinations of cues, both aligned and conflicting [26]. However, no specific organization or specialization for sound location has been found within ICC. A nearby IC nucleus, the brachium of the IC (BIC), serves as a relay between IC and nonauditory structures, including the superior colliculus. Neurons in the marmoset BIC have been shown to be sensitive to sound location but, unlike the ICC, more strongly respond to aligned or matched location cues rather than misaligned or conflicting cues [27,28]. Such neural responses are consistent with a more specialized role of the BIC in sound localization.

Medial Geniculate

The medial geniculate body (MGB), the auditory thalamic nucleus, receives its ascending auditory inputs from the IC (Fig. 25.3). Anatomically, the MGB is located medially and caudally to the much larger visual lateral geniculate nucleus (LGN). In marmosets, as in other nonhuman primates, the MGB has been subdivided based on cytoarchitecture, myeloarchitectural staining, and calcium-binding proteins [29]. The large ventral division (MGV) is part of the core lemniscal auditory pathway, receiving most of its anatomic input from the central nucleus of the IC. Other subdivisions include the posterodorsal division (MGPD), seen in primates and other mammalian species [30], an anterior-dorsal division (MGAD), which may be unique to primates [31], and finally, medial (MGM) and suprageniculate (SG) areas that receive both auditory and nonauditory inputs in other species [30]. These individual areas exhibit unique thalamocortical connection patterns, with the lemniscal MGV projecting to primary AC, the MGPD projecting to nonprimary AC, and the remaining areas having widespread projections to both primary and nonprimary cortex [29].

As in other areas of the auditory system, the MGB is tonotopically organized. Although such organization has only been demonstrated in squirrel monkeys [32] and other mammalian species, retrograde tracer studies from AC sites of known frequency tuning suggest a similar topography in marmosets [33]. Neurons within the MGV exhibit typical lemniscal-type responses, including strong frequency-tuned responses to narrow-band stimuli, including tones and the ability to synchronize to rapidly time-varying sounds [34–36]. MGAD neurons exhibit less frequency-specificity, responding equally well to tone and noise-based stimuli, but are able to synchronize faithfully to rapidly time-varying sounds, much more than other areas of the MGB. Neurons in MGPD have rather different responses than the other divisions, with many neurons preferring broader-band sounds, such as band-pass or white noise, and a tendency to respond to time-varying stimuli with continuous firing rather than stimulus-synchronized neural activity. Most neurons in marmoset MGB also exhibit nonmonotonic, or tuned, responses to varying sound intensity. Such tuned responses to intensity are distinctly different from responses seen in the auditory nerve and lower auditory structures where firing rates increase monotonically with increasing sound intensity.

Primary Auditory Cortex

One of the important neuroanatomic correlates of the evolutionary process has been the large expansion of the neocortex relative to subcortical structures in both humans and nonhuman primates. This has been particularly evident in the auditory pathway, where the AC has been found to increase both total mass and numbers of neurons much more rapidly than the IC or MGB [37]. It appears that, as total brain mass increases between primate species, the relative mass of AC scales by approximately a power of 1.6 relative to MGB or IC. For example, when compared with the tree shrew, marmosets have approximately the same number of neurons in IC, yet have 1.6 times more neurons in AC. In contrast, primary visual cortex scales approximately equally with LGN [38]. The origin of this relative increase in cortical size is unclear but suggests the need for increasingly complex processing of auditory information.

The AC in marmosets, as in other primates, sits atop the superior temporal gyrus (STG) of the temporal lobe [39,40]. Because the marmoset cortex is relatively lissencephalic, much of the marmoset primary auditory cortex (A1) lies on the surface of the STG and is easily accessible for lateral approaches for electrode recordings [5,33]. The remainder of the more medial A1 extends into the lateral sulcus. Organizationally, recent work in macaque monkeys has put forth the concept of a core-belt arrangement for AC [41,42], a model that appears to be conserved in marmosets [43]. Under this arrangement, the AC consists of core auditory areas, including A1, the rostral field (R), and rostral temporal (RT) field (Fig. 25.4). These core areas receive lemniscal inputs from MGV [29] and generally respond well to narrow-frequency stimuli-like tones [5,34,44]. The differentiation between these areas is based on tonotopically mapped frequency tuning, with A1 exhibiting a caudal to rostral high- to low-frequency gradient. At the A1/R border, this gradient reverses, with more rostral areas showing increasing frequency, followed by yet another reversal at the R/RT border [45]. The auditory core is surrounded by belt areas, including a medial and a lateral belt, with a further parabelt extending laterally from the belt to the superior temporal sulcus (STS), generally considered the boundary between auditory and visual areas of the temporal lobe.

Anatomically, A1 contains a six-layered structure typical of mammalian neocortex [13]. Ascending thalamocortical input fibers synapse on densely packed small cells within the granular layer IV. Upper cortical layers (II–III) consist of small, densely packed pyramidal neurons as well as interneurons, although the boundaries between layers III and IV are often indistinct. Deep layers, V–VI, also consist of pyramidal neurons, more sparsely arranged, and account for half the thickness of marmoset A1. A1 also shows the greatest density of staining for myelin fibers, with greater myelination than other noncore or belt regions. Inputs to A1 come largely from the ventral division of MGB, with a smaller number of inputs from MGAD and MGM, but not MGPD [29,33]. Core field R shows similar patterns of input connections, though with some weaker projections originating throughout the MGB. RT also shows strong MGV inputs. Additionally, core AC contains reciprocal cortico-cortical connections with adjacent medial and lateral belt fields, as well as with the adjacent core fields [43].

Physiologically, A1 and other core AC neurons respond preferentially to tone-based and other narrowband stimuli [45,46]. These areas are tonotopically arranged, as noted above. Earlier studies in anesthetized marmosets documented the strongest responses in the input layer IV [13]; however, more recent work has found strong responses throughout both upper and lower layers [47,48]. These layers also exhibit more complex frequency tuning, including neurons with multiple frequency peaks [49], consistent with their proposed role in processing more complex sounds. Despite many complex frequency-receptive fields, 27% of A1 neurons in awake marmosets have been found to be very narrowly frequency-tuned, even more so than in auditory nerve or in recordings from anesthetized marmoset cortex [34], suggesting sharpening in the auditory pathway. Neurons in R and RT show similar narrowband preferences as A1, but with weaker tone responses, longer response latencies, and poorer ability to synchronize to time-varying sounds [45]. As a result of these observed properties, R and RT have been proposed to be responsible for integrating sound information over longer time windows.

Medial Belt Auditory Cortex

Medial to the core AC is the medial belt, consisting of the caudomedial (CM), middle medial (MM), rostromedial (RM), and rostrotemporal medial fields (Fig. 25.4). These fields are immediately adjacent to and roughly paired with the core fields A1, R, and RT. Anatomically, these structures receive thalamocortical inputs that are distinct from the adjacent core AC [29]. CM, the most well-studied medial field, receives predominant input from MGM and MGAD, with lesser input from MGPD and none from the lemniscal MGV. CM also receives multisensory inputs from the thalamic pulvinar, posterior nucleus, and nucleus limitans. In contrast, RM receives its main inputs from MGPD with minimal inputs from MGM. Both areas also have significant reciprocal cortico-cortical connections [43], including connections with adjacent core structures (i.e., A1 to CM, R to RM), connections to lateral belt, connections between RM, MM, and CM, and connections to contralateral core and medial belt areas. Nonauditory inputs to these areas include amygdala and caudate connections with RM. CM also receives inputs from the entorhinal cortex and has dense reciprocal connections with the retroinsular and posterior parietal cortex, from which CM is thought to receive somatosensory inputs to support multisensory integration [50].

The precise function of the medial belt remains uncertain. Like A1, CM neurons exhibit frequency-tuned responses with a tonotopic arrangement and share a high-frequency border with A1, though with a more irregular arrangement [51,52]. CM neurons exhibit

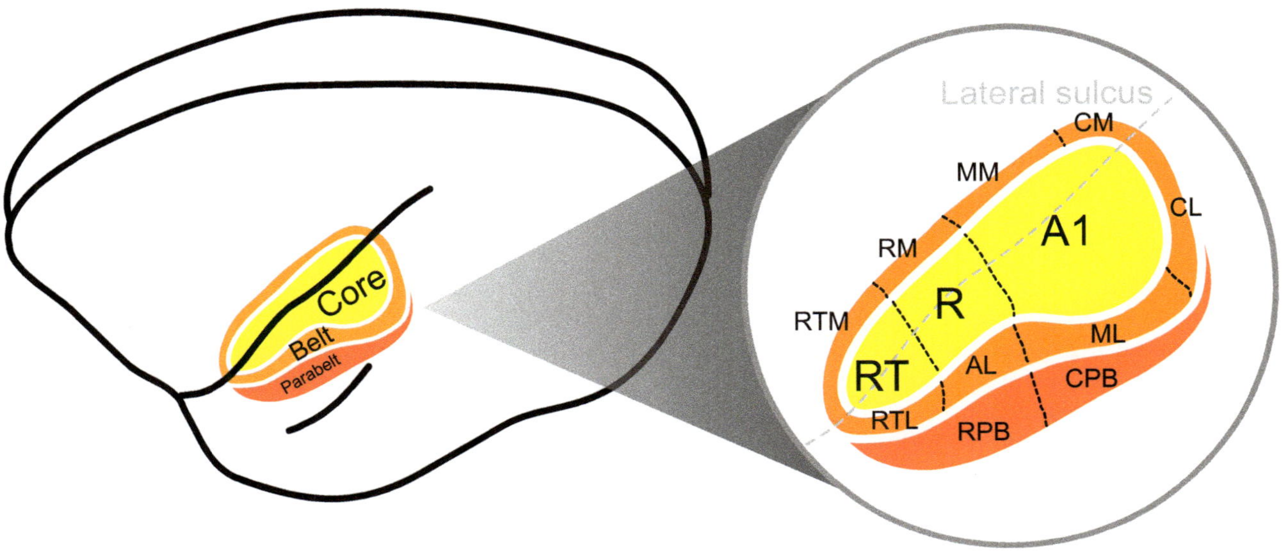

FIGURE 25.4 **Marmoset auditory cortex.** Illustration of the marmoset auditory cortex including core, belt, and parabelt arrangement. *A1*, primary auditory cortex; *AL*, anterolateral; *CL*, caudolateral; *CM*, caudomedial; *CPB*, caudal parabelt; *ML*, middle lateral; *MM*, middle medial; *R*, rostral field; *RM*, rostromedial; *RPB*, rostral parabelt; *RT*, rostrotemporal; *RTL*, rostrotemporal lateral; *RTM*, rostrotemporal medial.

broader frequency tuning, lower sound intensity thresholds, and shorter latencies than adjacent A1. CM neurons also respond in a synchronized, phase-locked pattern to marmoset twitter vocalizations and trains of repeated stimuli, suggesting a possible specialization for encoding rapid timing information [53]. Roles for the somatosensory and other nonauditory connections of the medial belt are unclear.

Lateral Belt Auditory Cortex

Just as there are medial belt areas corresponding to the core AC, there is a large lateral belt consisting of caudolateral (CL), middle lateral (ML), anterolateral (AL), and rostrotemporal lateral (RTL) fields (Fig. 25.4). Histologically, the lateral belt is differentiated from core AC by decreased staining for cytochrome oxidase and acetylcholinesterase compared with A1 and an increase in large layer III pyramidal cells [54]. As with the medial belt, thalamocortical inputs to the lateral belt are distinct from the core and come primarily from the nonlemniscal, or nonprimary, auditory pathway [55]. More caudal areas of the lateral belt (CL, ML) receive their largest thalamic inputs from MGAD, whereas rostral areas (AL, RTL) receive most input from MGPD, and all receive strong inputs from MGM. All lateral belt areas also receive strong inputs from the adjacent core AC, as well as nearby belt areas [54]. Strong interhemispheric connections with core and belt areas of the opposite hemisphere are also present. All these anatomic inputs are consistent with a possible role of the lateral belt in higher-order sensory processing and integration, receiving both direct thalamic input from nonlemniscal nuclei and lemniscal inputs via the core auditory cortical areas, including A1.

The physiologic responses of the lateral belt have been far less studied than in core AC. The lateral belt contains a tonotopic axis with frequency reversals between areas that match those of the adjacent core fields [56], consistent with reciprocal connections to the adjacent core which appear to occur in a frequency-matched fashion [54]. Lateral belt neurons exhibit responses to many auditory stimuli, including tones, band-pass noise, and conspecific vocalizations with sustained firing responses [46]. Unlike A1 neurons, lateral belt neurons exhibit a preference for broader-band sounds, including band-pass noise, responding more strongly than to tone stimuli [46,57], an observation consistent with work in macaque monkeys [58]. These results would suggest that lateral belt neurons are specialized for more spectrally complex sounds, like vocalizations, although conclusive evidence for this role is lacking. Work from macaque monkeys has also suggested that lateral belt neurons exhibit greater correlation with behavioral and perceptual decisions [59], an intriguing hypothesis that may also be applicable to marmosets.

Auditory Parabelt, Superior Temporal Gyrus, and Beyond

Adjacent to the lateral belt is the auditory parabelt that extends laterally out to the STS. The parabelt is broadly divided into rostral and caudal divisions, paralleling the ML to AL transition of the lateral belt. The histologic distinction between the belt and parabelt is weak, with a general decrease in myelin density in the infragranular layers and decreased marker staining, but with a pronounced cellular columnar arrangement [54]. As with the lateral belt, the parabelt receives thalamic inputs from nonlemniscal areas, generally MGM and MGPD [55]. The parabelt, however, also receives dense inputs from nonauditory thalamic areas, including medial pulvinar, nucleus limitans, SG, and the posterior nucleus. Cortico-cortical inputs from the parabelt come primarily from adjacent lateral belt areas, the adjacent parabelt, some medial belt, STS, and similar areas from the contralateral AC, but with relatively few connections from core areas [54]. These strong connections with association areas of thalamus and cortex are consistent with a proposed role of the parabelt in higher auditory and multisensory processing.

Almost nothing is known specifically about the role of the parabelt AC and auditory responsiveness in the remaining STG in marmosets. Based on functional imaging and neural recordings in macaque monkeys, the anterior parabelt and STG have been found to exhibit increasing specialization for species-specific vocalizations [60–62]. A similar pattern has been found by imaging in marmosets [63]. What specific role these areas play remains unclear, as there exists little physiologic recording data in marmosets to support vocal specificity. An anterior specialization for vocalizations and other complex stimuli would be consistent with the dominant working hypothesis for auditory processing in primates and other mammals, the so-called dual-stream or "what" versus "where" model [64]. According to this model, specialization for stimulus identity begins in the anterior lateral belt, refines into the anterior parabelt and the temporal pole, and propagates to the ventrolateral prefrontal cortex (vlPFC). Caudal areas should exhibit increased specialization for spatial location or sensory–motor association and project to more dorsolateral areas of prefrontal cortex (dlPFC).

Prefrontal Cortex and Hearing

The auditory pathway in marmosets and other mammals does not end within the temporal lobe and continues on with projections from nonprimary auditory cortical areas to the prefrontal cortex (PFC) (Fig. 25.3). Similar to other nonhuman primates, several homologous PFC areas have been anatomically delineated in marmosets [65]. These include the premotor cortex (area 6), dorsolateral PFC (dlPFC, area 8), ventrolateral PFC (vlPFC, areas 12/45, 46), and more orbitofrontal areas (areas 9, 10). Area 12/45 is also thought to correspond to the primate homolog of human Broca's area [66,75], a structure involved in the coordination and timing of speech production. Anatomically, many of these areas are connected with auditory areas of the temporal lobe. Area 8Ad (dlPFC) is reciprocally connected with auditory areas and STS [67]. Area 9 has reciprocal connections to parabelt AC. Area 10 also receives inputs from the RT area of AC [68,69]. Area 12/45 has not been as well studied in marmoset, but in macaques, this area receives inputs from rostral AC fields [70,71]. The anatomic arrangement of these temporal to frontal pathways is consistent with the functionally defined dual-stream model of auditory information flow mentioned above [64].

Auditory responsiveness in PFC has been well documented in macaque monkeys [72,73] and is presumed to reflect sensory processing for auditory behavior and decision-making. While there have not yet been any published direct studies of auditory coding in marmoset PFC, labeling of PFC using expression of cFOS, an immediate early gene that is a marker of neural activation, suggests a similar role [74]. Using an interactive vocal paradigm, these studies have shown cFOS expression in vlPFC in animals that failed to produce vocalizations, with lesser expression in more dorsal areas. In contrast, more dorsal expression was noted for vocalizing animals, suggesting a more specific role of ventral areas in vocal auditory processing or perception. Other studies, using another immediate early gene, Egr-1, however, showed similar activation in vlPFC and dlPFC during interactive vocal behaviors, and less expression when hearing alone [75]. As vlPFC is thought to contain a likely Broca's homolog, and might therefore exhibit activity during vocal production, the interpretation of these seemingly conflicting findings is uncertain.

Descending Auditory Pathways

In addition to the ascending auditory pathway, marmosets exhibit a prominent descending efferent pathway similar to that of other mammals. Anatomic tracer studies show that the thalamic and cortical connections are largely reciprocal [29,33]. Additionally, tracer studies show that both A1 (in closely related tamarins [76]), and medial belt [29], exhibit descending corticotectal pathways that bypass the thalamus and connect directly to the IC. These projections are to both central and noncentral IC and appear to label different layers in IC depending on the corresponding cortical location, presumably matching frequency tunings. Descending connections are also more prominent from noncore AC to noncentral IC, possibly reflecting more specialized or dynamic information processing in nonlemniscal pathways. Similar descending pathways have been seen previously in other primates [77,78]. The function of these pathways is unclear, but previous work in bats has shown that descending input can affect sensory tuning in IC to better match that of cortical activation, possibly resulting in more precise encoding of information [79].

METHODS IN MARMOSET AUDITORY RESEARCH

Many methods used in auditory testing and research have been adapted for use in marmosets. Some have required minimal modifications from more common techniques, whereas others have required extensive adaptation to take into account the smaller body size, anatomy, and behavior of the marmoset. Here we summarize some key methods and their adaptations.

Behavioral Methods: Psychoacoustics

Understanding auditory perception and cognition requires behavioral testing using psychoacoustic methods to determine marmosets' abilities to detect or discriminate different sounds. Although early sound perception threshold measurements used a conditioned shock avoidance task [3], use of such aversive conditioning has become ethically questionable for use in nonhuman primates. More recent endeavors have used operant conditioning techniques adapted from other animal species [4,80]. These methods use a conditioned Go/No-Go auditory task wherein the marmoset licks from a spout to receive a reward when it detects a sound. Marmosets rapidly learn this task, in less than 10 sessions for most animals tested, and have a low false-positive lick rate. This technique has also been adapted to test auditory discrimination using a change-detection task, wherein the animal has to lick when hearing a change in stimulus, and not lick when just hearing the reference stimulus repeated [6]. Recent work has also developed an operant conditioning apparatus for freely moving marmosets, using multiple touch sensors or bars rather than a lick response, allowing two (or more) choices rather than a purely Go/No-Go response task [81]. This approach has the potential to allow more complicated discrimination tasks, such as two-alternative, forced-choice methods. On the other hand, the bar press approach may not be as easily paired with neural recordings as increased movement may reduce the stability of single-unit recording methods, particularly for small-bodied animals like marmosets. Further use of implanted electrode techniques, discussed below, may potentially overcome this limitation in the future.

Behavioral Methods: Antiphonal Calling

Although many measures of auditory perceptual abilities in marmosets have focused around traditional training paradigms (shock avoidance, operant conditioning, etc.), recent work on marmoset vocal production has the potential to reveal further information about marmoset auditory abilities. These ethologically inspired behavioral paradigms have proven useful to test auditory cognitive functions, including perception, recognition, and decision-making in marmosets. One such behavior is antiphonal calling, wherein a marmoset will vocally interact with another marmoset when separated by a visual occluder [82]. The antiphonal calling paradigm has allowed behavioral testing of many aspects of vocal communication, including identification of individual callers [82], timing of vocal production to avoid masking noise [83], turn-taking [84], and the role of parental feedback in guiding vocal development [85]. Antiphonal calling can also potentially be used, however, to test other aspects of auditory perceptual abilities. For example, which vocal acoustic features are behaviorally salient remain unclear, despite known variations in such features both among different marmosets and within individuals [1,86]. Because marmosets will also antiphonally interact with a computer that detects vocalizations and then plays back recorded vocal sounds, a computer-controlled "virtual marmoset" [82], there is the potential for investigators to manipulate acoustic aspects of vocal playback to test perceptual boundaries between vocalizations of different classes or different caller identities.

Physiologic Methods: Otoacoustic Emissions

Otoacoustic emissions (OAEs) are a method to measure cochlear function that are commonly used in both humans and animals. Generated by nonlinear mechanical interactions involving the outer hair cells and basilar membrane, OAEs reflect the health and function of the inner ear. The most common method of testing is the distortion-product OAE (DPOAE), generated by simultaneously presenting two tone frequencies, f2 and f1, and then using a microphone in the ear canal to record the resulting sound that is produced by the cochlea (occurring at 2f1-f2). During this sound stimulation, outer hair cell amplification mechanisms result in vibration within the cochlea that is then transmitted along the basilar membrane and out into the middle ear and ear canal, where it can be measured [87]. Marmosets exhibit robust DPOAEs over the frequency range of 3–24 kHz, optimally with an f2/f1 ratio of 1.225 [88]. Repeated measurements show that marmoset DPOAEs are reliable and repeatable, but that this reliability decreases slightly between measurements repeated over longer time periods (minutes apart or longer), particularly for frequencies below 12 kHz [89]. Interestingly, marmosets show stronger OAEs in females and in the right ear, larger than similar differences in humans, the origin of which remains unclear.

Physiologic Methods: Auditory Brain Stem Recordings

Auditory brain stem responses (ABRs), also referred to as brain stem auditory evoked potentials, are another method to measure functions of the cochlea and lower auditory pathways that are commonly used in humans and other animals. Typically measured using either clicks or tone stimuli, ABRs capture electric potentials from subcutaneous or surface electrodes placed behind the ear and the vertex of the head. Typically, ABRs measure a series of individual potentials (referred to as "waves") that are thought to correspond to the serial processing of the auditory pathway including the auditory nerve (waves I and II), cochlear nucleus (wave III), superior olivary complex (wave IV), and IC and nearby structures (wave V), although exact generators of each remain controversial [90]. Marmosets exhibit robust ABRs with all five component waves in the first 10 ms after stimulation, consistent with human results and those of other animals [91]. These ABR potentials exhibit some dependence on tone frequency, with the strongest responses in the center of the marmoset hearing range (~4–8 kHz), but are relatively stable regardless of the exact position of a reference electrode, stimulation rate, etc. The intensity of sound stimulation affects the ABR response strength and latency, as expected from other species, with an optimal click intensity of 80 dB SPL [92]. Marmoset ABRs can be measured up to very high click repetition rates (99 kHz). Marmoset ABR has thus far been used as a relatively simple method to measure age-related [93] and ototoxic [94] changes in marmoset hearing.

Physiologic Methods: Awake Single-Neuron Recordings

Early neural recordings of marmoset AC were performed using classic acute anesthetized preparations, including wide craniotomies and single-electrode methods. While these approaches provide wide access to the entire cortical surface, they suffer from a number of disadvantages, including a limited recording period (a few hours to days at most) and the need for anesthesia with its potential to affect cortical responses. The past few decades have seen increasing use of awake, semichronic recording methods. Adapted from methods developed in macaque monkeys, these approaches use a similar head fixation design, as well as a custom primate chair, to temporarily immobilize the marmoset during physiologic recordings [80,95]. Two decades of use have shown that awake marmosets will sit comfortably in such arrangements for several hours per day and can be recorded for years at a time. Additionally, rather than a large craniotomy and recording chamber requiring frequent cleaning and plagued by tissue growth issues, as in macaques, some recording methods adapted for marmosets have maintained a largely intact cranium covered by a layer of dental acrylic, or similar substance. Typically, small ($\leq$1 mm) microcraniotomies can be made allowing temporary access to cortex for 1–2 weeks (or longer with special care), although larger craniotomies have also been successful. After completion or recording from one opening, the hole can be sealed, and a new one created. When used with high-impedance ($\sim$5 MΩ) sharp microelectrodes, this method yields single-neuron recording of good unit isolation (often >30 dB SNR) and allows stable recording for up to several hours per neuron [95]. This approach has also been successfully adapted for awake intracellular recordings, capable of recording dozens of neurons per animal and holding individual cells for more than an hour [96]. One particular advantage in marmosets for such AC physiology is that A1 is located laterally on the cortical surface, allowing a lateral approach for the electrodes, as opposed to macaques where A1 is largely buried in the lateral sulcus and is generally approached vertically, first passing through parietal lobe brain tissue before reaching auditory areas [97].

Physiologic Methods: Implanted Multielectrode Arrays

Despite the importance of single-electrode methods in marmosets, and other model species, such approaches suffer from several limitations, including the ability to only record from a single location at a time, limited recording duration, and the need for a restrained animal. There has been considerable interest in the use of implanted multielectrode arrays for use in marmosets. Several designs have been developed over the years, largely scaled down adaptations from implants used in other species [98,99]. These implants have been used both for auditory and vocal production studies in marmosets [100,101] and yield reasonable single-unit recording quality (10–30 dB SNR) and stability over hours to days. The principal advantages of such arrays are the ability to simultaneously sample multiple neurons and the ability to perform unrestrained recordings to study more natural behaviors. However, most implants used in auditory work to date have been limited by a small number of electrodes and rather sparse spacing. For example, the $\sim$0.8–1 mm spacing of one frequently used design [101] corresponds to nearly an octave of tonotopy in marmoset A1. Furthermore, the unit isolation quality of these methods is generally poorer than with higher-impedance single electrodes. Future approaches using microwire arrays or other designs may yield better resolution and increased electrode density, but may still have limitations due to the relatively small size of the marmoset brain and limit to the implant weight they can easily carry.

Freely Moving and Wireless Recordings

Studies of neural mechanisms of natural auditory and vocal behaviors, such as marmoset antiphonal calling, require the ability to reduce the restraints typically used during neurophysiologic recordings. Such restraints inhibit natural behaviors like vocal production [102,103]. One potential approach is the use of operant conditioning to train an animal to vocalize on demand, a technique used successfully in macaques, though with a more limited vocal repertoire [104]. Operant training, however, may interfere with many of the normal vocal decision-making processes observed in more natural or interactive contexts. The alternative, reducing restraints to increase willingness to vocalize naturally, requires both implanted electrodes and an ability to relay neural signals from the marmoset to recording hardware. Such relays must maintain signal fidelity while allowing the animal to move with minimal encumbrance. One such approach is the use of tethered recording systems [101]. This tethered approach was adapted from rodent work, where it is common, but is more difficult in the vertically mobile marmoset and required further refinements to allow for control of tether length to reduce entanglement. A second approach is the use of wireless radio telemetry systems, which have also seen use in marmosets [105,106]. Such systems have taken considerable time in their evolution and only in the past few years have achieved sufficiently small size and light weight to be practical in marmosets. However, high channel count systems needed to record multiple neurons or brain areas remain bulky and suffer from limited battery life. Hopefully, the continued evolution in electronics will yield further refinements in devices more appropriate for use in freely moving marmosets.

Imaging in Marmosets

Although microelectrode methods have yielded considerable insights into the physiology of individual neuronal responses to auditory stimuli, such approaches are limited to a single or small group of neurons. Understanding how large areas of cortex are involved in a given auditory response or behavior requires imaging methods to sample a broader area of cortex, including optical and fMRI techniques. Such approaches have been successfully adapted for use in marmosets, including broad measurements of tonotopy [56] and the distribution of responses to vocal stimuli [63]. One limitation of these approaches has been the use of anesthesia in many studies to reduce movement-related artifacts, although with further training it is possible for marmosets to stay sufficiently still during the imaging session [57]. These methods are also limited in that they tend to measure blood flow dynamics as a proxy for neural activity and therefore have poor temporal and spatial resolution as compared with microelectrode studies. This may be overcome with the use of two-photon imaging, which has shown recent use and promise in marmosets [107].

Synthetic Vocal Stimuli

Studies of vocal coding and perception in marmosets and other primates have largely relied on use of vocalization exemplars or simple manipulations such as playing a vocal sound both forward and reversed [108,109]. Although some attempts at stimulus manipulations to change overall frequency or timing of vocal stimuli have been used [109,110], including filtering as well as temporal and spectral scaling/compression, these represent relatively crude signal changes. Real marmoset vocalizations vary over a variety of acoustic parameters, both within and between individual animals [1]. Recent efforts have introduced the use of "virtual vocalizations" using mathematical models of marmoset calls that allow the creation of synthetic vocal sounds to match the statistical structures of real calls [86]. Such synthetic stimuli have the advantage that they can be more precisely parametrically manipulated to test aspects of vocal sensory coding. Synthetic vocalizations have been tested in a limited number of A1 neurons and were found to evoke similar neural responses as their natural vocalization equivalents. Synthetic vocalizations can also be morphed to create artificial vocal stimuli that fall between natural vocalization classes, allowing testing of categorical boundaries between different vocal types. The use of such synthetic stimuli has the potential to allow more systematic testing of both behavioral and physiologic measures of vocal perception and neural coding as an avenue to understand human speech.

TOPICS IN MARMOSET AUDITORY RESEARCH

Studies of neural responsiveness and coding in the marmoset auditory system, and in particular the AC, have played important roles in recent research that has addressed many long-standing questions and controversies in the auditory field. We will review some of these topics in which marmoset auditory research has played a key role.

Sustained Responses and Sound Coding

The ability of the AC to encode long-duration sounds has seen considerable controversy over the years. While early work performed in awake animals demonstrated sustained neural firing in response to sound [111,112], later increased use of anesthetized experimental preparations revealed primarily phasic onset responses to sound, regardless of stimulus duration (i.e., Ref. [113]). As a result, many researchers began to put forth theories that AC was specialized only to respond to brief or time-varying sounds [114]. Other theories also emerged to explain the ability to perceive sustained sounds using elaborate interneuronal connectivity models [115]. However, more recent work in awake marmosets redemonstrated that neurons in AC are, in fact, capable of strong sustained responses even over long-duration stimuli [46]. It is likely that such differences between awake marmosets and anesthetized preparations are the silencing of cortical processing by the anesthesia. Such anesthetic effects may potentially be a result of changes in local processing of the same neurons but may also reflect differences in the neurons sampled in awake and anesthetized preparations. Studies in anesthetized marmosets have found the strongest tone responses in the layer IV input area, with relatively weaker responses in upper and lower layers [5,13]. This is in sharp contrast to the strong responses observed throughout all cortical layers in awake marmosets. To what extent responses in these other layers, which reflect both input and local cortical processing, are different than the input layer IV has not been well studied in marmosets to date.

Another important factor in driving strong sustained responses to sound is the choice of auditory stimulus. Such strong responses only occur when a stimulus is chosen that is "optimal" or preferred for a given neuron, with weaker responses or even onset-only responses when sounds are nonpreferred [46]. For many A1 neurons, this preference can be very narrowly tuned [48] and easily missed when stimulus protocols are fixed and sparse, or not adjusted for tuning characteristics of the neuron being studied. One additional confounder of such stimulus specificity is that many neurons in awake marmoset AC respond nonmonotonically to increasing SPL, responding best to intermediate intensities instead of the loudest levels [116,117] (Fig. 25.5A). This is in sharp contrast to the auditory nerve and anesthetized animals, where the loudest sound level generally evokes the strongest neural response. If loud stimuli are chosen based on those assumptions, one may miss many of the strong auditory responses found in awake cortex.

Temporal Coding and Responses to Time-Varying Sounds

One important feature of many natural communication sounds, including marmoset vocalizations, is the presence of significant, and often rhythmic, changes over time. Marmoset twitter calls contain multiple independent phrases separated by brief periods of quiet [1,2]. Marmoset trills contain both sinusoidally changing frequency (FM) and amplitude (AM) modulations. To what extent these features are behaviorally important is unknown, although they vary between different individual marmosets and may be one feature allowing discrimination of caller identity. Additionally, many other sounds also include time-varying spectral or timing information. It is, therefore, important for neural structures to be capable of detecting and encoding such temporal patterns.

At the level of the auditory nerve, individual neurons can follow the envelope of temporally modulated sounds above 1 kHz in many species [118], but AC is not capable of following such modulations much above 100 Hz in both marmosets and other species [119,120]. Such neurons capable of faithfully following the envelope of a time-varying stimulus are often referred to as synchronized (Fig. 25.5B). In awake marmoset A1, however, higher rates of modulation are encoded in a nonsynchronized fashion, that is, by neurons that increase their firing rates to specific modulation rates without faithfully tracking the envelope of the modulation [120–122]. Lower rates of modulation can still be encoded using synchronized firing, or a mix of synchronized and nonsynchronized responses. Such nonsynchronized neurons can be further divided into those whose firing increases monotonically with modulation rate, those that decrease monotonically with modulation rate, and some that show preference or tuning for specific modulations. At the average population level, marmoset cortex responds most strongly to temporal modulations between 16 and 32 Hz, a range overlapping the modulations seen in marmoset trill calls. There is also a high degree of overlap between modulation preferences for both AM and FM, suggesting common mechanisms. Such dual mechanisms allow flexibility of AC neural populations to encode time-varying sounds over a wide range of modulation rates, including those it cannot faithfully synchronize to, even when the timing [122] or envelope [123] are irregular.

How such a temporal coding (synchronized discharges) to rate coding (nonsynchronized) transformation comes about is unclear. Previous recordings in the IC of awake squirrel monkeys did not demonstrate any nonsynchronized neurons [124]. Studies in awake marmoset MGB, however, have found nonsynchronized responses in approximately 40% of neurons, less than that of A1, as well as neurons with mixed properties of both synchronized and nonsynchronized responses [35]. Additionally, many MGB neurons do synchronize to stimulus modulations and do so at much higher modulation rates than are observed in cortex, consistent with its earlier position in the auditory pathway. To what extent MGB nonsynchronized responses reflect a de novo phenomenon in the ascending pathway, or are a result of cortico-thalamic feedback, is unknown. However, recent intracellular recordings have revealed that even many A1 neurons exhibit subthreshold synchronized potentials at higher rates than for their spiking, demonstrating the presence of a temporally synchronized input to many A1 neurons, and suggesting a temporal-to-rate transformation between thalamus and cortex [96]. This transformation does not appear to end in A1, however, as cortical field R also shows sensitivity to temporal modulations, but with far fewer synchronized responding neurons [45,125]. Such changes compared to A1 demonstrate the presence of an additional temporal-to-rate transformation, rather than just an inherited one, and suggests that R may be integrating information over a longer time scale. Because the ability to synchronize to rapidly time-varying stimuli is gradually degraded along the auditory pathway, with each station exhibiting a lower synchronization limit, the benefits of rate coding transformations may allow the cortex to still encode those sounds to which it cannot faithfully follow by precise, synchronized spike timing. This rate code transformation has also been suggested to play a role in allowing integration of multisensory stimuli, which are processed in other sensory systems at slower time scales than in the auditory pathway [126].

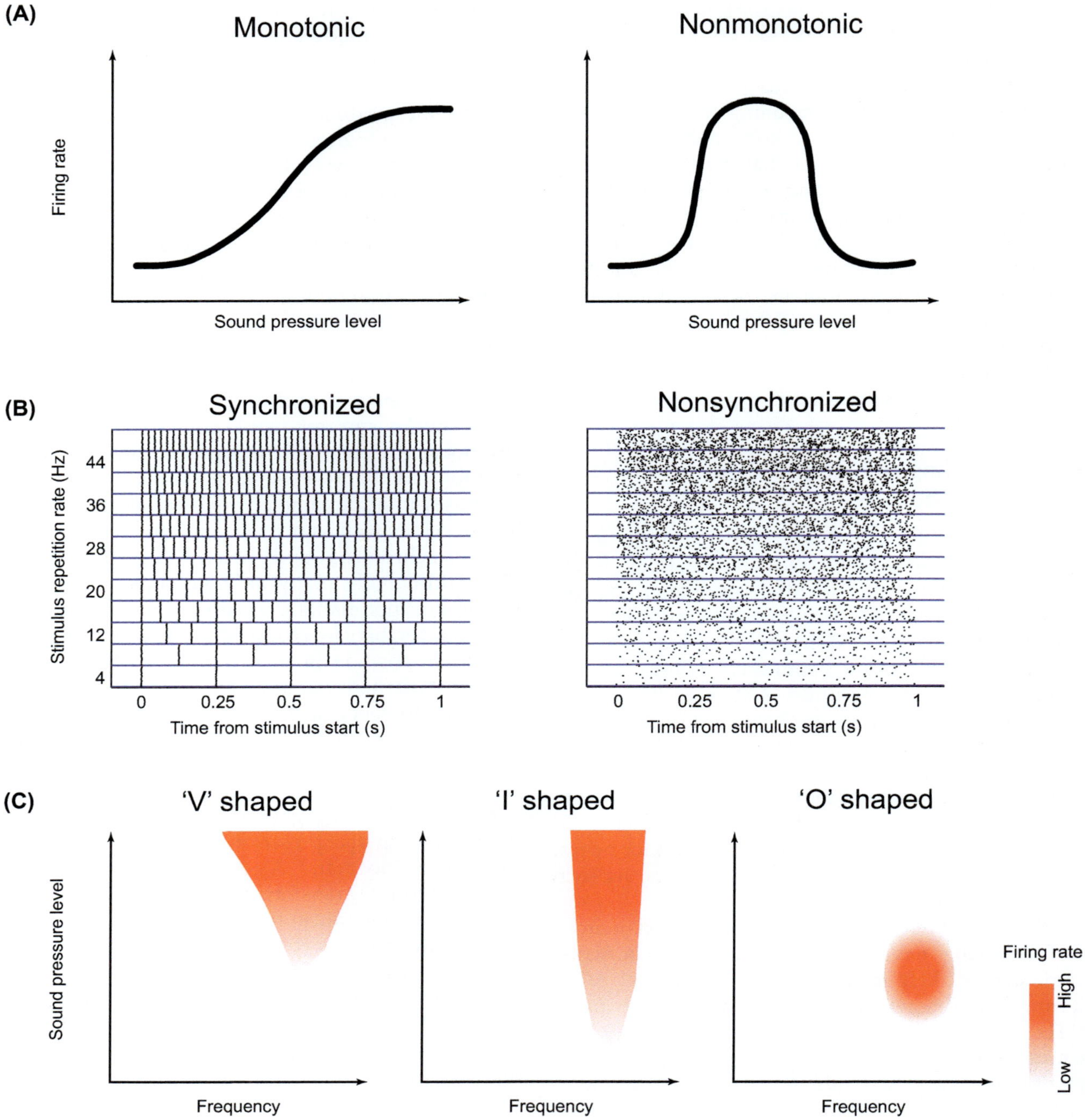

FIGURE 25.5 **Neural coding mechanisms in marmoset auditory cortex.** Illustrations are shown for different coding mechanisms observed for neurons in marmoset auditory cortex. (A) Sound level responses can be divided into monotonic (i.e., increasing with level) or nonmonotonic (tuned to a particular level). (B) Responses to repeated time-varying stimuli can by synchronized (following stimulus timing) or nonsynchronized (not following the stimulus, but sensitive to the timing). (C) Frequency tuning preferences vary with sound level, often either showing "V," "I," or "O" shapes.

Frequency Tuning and Spectral Responses

At the level of the auditory periphery, auditory receptive fields tend to have simple V-shaped frequency tuning. Such receptive fields have narrow tuning at low SPLs and become more broadly tuned at higher sound intensities (Fig. 25.5C). Similar patterns have also been noted in A1 for anesthetized marmosets [5]; however, recordings in awake marmosets have revealed far more complex patterns. Although such V-shaped receptive fields are also present, many neurons show a more I-shaped pattern, with narrow frequency tuning that does not broaden at higher SPLs [116]. These experiments found that such "V" and "I" units only account for 36% of A1 units in marmosets. The remaining units are "O" shaped in that they exhibit selective tuning for both frequency and SPL and do not respond to very loud sounds. Such "O" and "I" neurons may allow the cortex to keep an invariant representation of sounds that does not vary or become less precise at higher sound levels, in contrast to a cortex consisting only of "V" neurons which would not have a level-invariant representation and would exhibit significantly degraded frequency precision when sounds are loud. The mechanisms that create such narrowly tuned O- or I-shaped receptive fields have not been fully elucidated. However, there is evidence to suggest that the presence of sideband inhibition, inhibition occurring at frequencies to either side of the tuning peak, or at higher SPLs, may contribute to help narrow the frequency receptive field [127]. Narrowly tuned neurons receiving V-shaped inputs from the lower auditory pathway also receive inputs from local inhibitory interneurons that, if tuned to frequencies on the edges of the "V" receptive field, would inhibit responses to those edge frequencies, resulting in narrower I-shaped frequency tuning. Similarly, an O-tuned neuron could receive inhibitory input from an interneuron tuned to the same frequency, but only responding to higher SPLs, which would result in the "O" neuron not responding to those sound levels. These inhibitory tuning mechanisms may account for the fine frequency-tuning precision seen in many A1 neurons [34].

In addition to narrow frequency and SPL tuning, many neurons in marmoset AC exhibit more complex frequency-receptive fields, including multiple frequency peaks, although this is by no means exclusive to marmosets [13,128]. A majority of such secondary peaks are harmonically related to the primary frequency of the neuron, i.e., 1.5×, 2×, 2.5×, 3×, 4×, and often only respond to harmonic frequencies at higher SPLs than that for the main frequency peak [49]. Other measurements using a combined two-tone paradigm, playing both a fixed, best frequency tone, and a varying second tone, also reveal secondary peaks clustered around harmonic frequencies [128]. The presence of such harmonically tuned neurons is, perhaps, not surprising, as many natural sounds, including marmoset vocalizations, include a significant number of harmonic components. It stands to reason that the presence of such harmonics must be integrated somewhere within the auditory system. Indeed, many such AC neurons can be considered harmonic template neurons in that they respond to complexes of multiple harmonic tones more strongly than they respond to individual tones alone, or to inharmonically related tones [129]. Harmonically tuned neurons are one possible mechanism by which complex sounds, like vocalizations, can be coded and perceived as a single sound, rather than a group of individual sounds. This is a potentially important intermediary step between encoding sounds by their individual features and grouping multiple features into a single perceptual unit or object.

How such complex frequency receptive fields come about is not clear. While one might expect a simple linearly additive process from multiple frequency-tuned input neurons, evidence suggests that nonlinear processes are also involved. Responses to spectrally complex inputs have been tested in marmosets using random spectrum stimuli (RSS), which consist of random combinations of different tonal components with varying relative loudness [130,131]. Such mixtures have the ability to measure not only the responses to individual frequency elements but also how such frequency responses interact. Many A1 neurons respond quite robustly to RSS, resulting in frequency receptive field maps that closely mimic those measured with tones, including "V," "I," and "O" neurons and the presence of inhibitory sidebands [132]. Such neurons often prefer high degrees of spectral contrast (i.e., a high signal-to-noise ratio between the frequency peak and surrounding frequency energy), accounting for about 60% of A1 neurons [130]. Other neurons actually prefer lower spectral contrasts (~40%), akin to broadband noise preferences observed more prominently in lateral belt areas. Many neurons, however, did not exhibit responses to RSS or did not yield good receptive field measurements using linear estimation methods. This may have been due to the presence of significant nonlinear interactions (i.e., not simply additive or subtractive) between different tone frequencies that were not well captured by this and similar approaches.

Similar complex frequency-tuning properties also extend into the temporal domain. Many natural sounds include complex spectrotemporal modulations, such as frequency modulations wherein the frequency of a sound changes over time. Marmoset vocalizations contain many such frequency modulations including upward sweeps in twitters and phees, downward sweeps in tsiks, and sinusoidal changes in trills. How might such spectrotemporal modulations be encoded by the auditory system? One method that has been tested in marmosets has been the use of short FM sweeps of varying mean frequency and rate of frequency change (modulation rate). Many neurons in CM respond robustly to such FM sweeps in a manner similar to that of twitter calls [53]. Another popular method for estimating such spectrotemporal properties is the spectrotemporal receptive field (STRF), calculated using reverse correlation methods and linear estimations [133]. Such approaches have been popular in the auditory field, have been used in marmoset cortex [134], and are able to explain a modest amount of neural responses, ~18.5% of firing rate variance. Others, however, have found that STRF methods do not yield good estimates in over half of marmoset AC neurons, possibly a result of nonlinearities in the receptive field not well characterized by the linear STRF model [135]. Testing of spectrotemporal responses using two tone pips of varying interval and frequency has found that half of A1 neurons exhibit such nonlinear combination sensitivity, and that such nonlinear sensitivity is predictive of a high degree of selectivity in vocalization responses. On the other hand, the linear half of neurons responds well to pure tones alone and tends to respond nonselectively to multiple vocalizations, suggesting that they were simply encoding the presence or absence of an appropriate frequency component. The presence of such diverse spectrotemporal properties in marmoset AC may be the building blocks by which complex vocalization sensitivity and encoding can be accomplished, but remains unexplored.

Coding of Sound Intensity and Loudness Invariance

Another of the challenges that the auditory system faces is the ability to sensitively respond to sounds of varying intensity that can span five orders of magnitude. For a marmoset, this could range from soft rustling of leaves in the jungle (40–50 dB SPL) to a loud phee call (>105 dB). Additionally, sounds often occur in noisy environments where both the sound of interest and the background have similar SPLs. How might the AC of a marmoset allow them to deal with such problems? Peripheral neurons in the auditory nerve of most species show monotonic rate-level functions to increasing loudness, corresponding to V-shaped frequency responses (Fig. 25.5). In marmoset cortex, however, <50% of the neurons show such monotonic responses, whereas the remaining 50%–60% show nonmonotonic or tuned responses to SPL, corresponding to O-shaped frequency-receptive fields [117]. The presence of a mix of such monotonic and nonmonotonic neurons has been suggested to allow a level-invariant representation of sound across the neural population, one that remains sensitive regardless of overall loudness [116,136]. One additional mechanism contributing to level invariance in marmoset cortex is the ability of cortical neurons, particularly the nonmonotonic ones, to adapt to the overall sound level of the environment. In the presence of ongoing background sounds, such neurons can dynamically shift their sound-level tuning to center around the average loudness of recent sounds [137,138]. This new sound level preference is often at much higher SPLs than the responses of neurons in a quiet background and would allow such neurons to maintain a wide dynamic range for sound intensity that adjusts to maintain sensitivity depending on the acoustic environment (dynamic range adaptation). Such adjustments may also assist in the ability of the AC to encode other sounds, like vocalizations, when embedded in background noise [139].

Pitch and Periodicity

As discussed previously, pitch is an important feature in human speech, music, and other natural sounds. Although most marmoset vocalizations do not operate in a frequency range that would be considered pitch, typically <4 kHz, many ethologically relevant natural sounds contain periodic elements than can generate a perception of pitch, and marmosets exhibit the ability to perceive such pitches [9]. How pitch is coded by the auditory system has long been of interest to many auditory scientists. Recordings in marmoset AC have found "pitch neurons" that are sensitive to the pitch of a sound, regardless of the specific acoustical elements that are used to generate that pitch, including pure tones, harmonic complexes, or periodically repeating noise [140]. In the marmoset cortex, these neurons are localized to the junction between A1, R, and the lateral belt, a location that has been labeled the "pitch area." Similar localization for pitch coding has been suggested in humans (see Ref. [141]). Although many neurons throughout marmoset AC will respond to similar periodic stimuli, such as click trains or AM, many of these neurons are relatively insensitive to small amounts of temporal jitter. In contrast, pitch-selective neurons exhibit weaker responses when faced with such irregular stimuli, a sensitivity that parallels the perception of a reduced pitch strength in humans [142]. Such pitch neurons also exhibit the ability to encode pitch by both spectral (i.e., harmonicity) and temporal (regularity) cues depending on which is most salient, similar to the behavioral performance in marmosets [10,11].

Sound Localization

The ability to localize sounds in space is one of the critical functions of the auditory system. To a marmoset in the wild, such abilities should allow it to maintain spatial cohesion with its family group, localize food sources in response to foraging-related chirp vocalizations, and avoid predation. Surprisingly, relatively little is known about the abilities of marmosets to precisely localize sounds in space, though reflexive head-orienting behaviors suggest that they are capable of doing so at some level [14]. Measurements of head-related transfer functions (HRTFs) for the marmoset head and ears have shown azimuthal ITD and ILD cues that are consistent with other animals of similar head size, though some timings are larger than might be predicted and may be related to the marmoset pinnae. Marmoset HRTFs also exhibit frequency notches, produced by the shape of the pinna, that vary with the elevation of a sound source and can be used as spectral cues in vertical sound localization. Just how accurate marmosets are in using such information to localize a sound source is unknown.

In the marmoset IC, the first major station after extraction of location cues in the superior olivary complex, neurons exhibit sensitivity to ITD and ILD cures, as well as spectral cues [26]. Most low frequency-tuned neurons (<11 kHz) encode information using a single cue, either ITD or ILD, consistent with known dominance of these cues at different frequencies. At higher frequencies (>11 kHz), neurons encode sound location using a mix of cues, generally ILD and spectral cues. However, most neurons in the ICC do not appear to strongly prefer cues that are correctly aligned over those that do not match. On the other hand, neurons in the BIC, a relay nucleus to the superior colliculus, show much stronger preferences for congruent spatial cues [27,28]. This preference is evidence of the BIC as the first site in the marmoset auditory pathway that is specially tuned for sound localization by combining multiple cues in its estimation of a sound source.

At the level of the AC, many neurons have been found to exhibit spatial position sensitivity, either measured based on ILD [143] or with free field speakers [144–146]. Overall, there is a strong preference of AC neurons for the contralateral hemifield (~50%), over the midline (~25%) or ipsilateral (~25%). Results of elevation preferences have been mixed between studies. Interestingly, most AC neurons exhibit broad spatial tuning, but are more precise in CM/CL as compared with A1, consistent with the caudal "where" pathway of the dual-stream model. As with loudness and frequency tuning, the effects of loudness on sound localization are also important. Although some studies have found individual neurons exhibiting degradation in sound localization precision with increasing intensity, population ensemble responses are more level tolerant, consistent with behavioral measurements in humans and other species [145]. Other studies have shown more mixed results for level invariance [144].

Adaptation and Plasticity

Another of the important functions of sensory cortex is the ability to flexibly modify a sensory representation to reflect the situational or behavioral needs of an animal. In some contexts, this may reflect attention- or task-related changes that have been observed in the nonprimate AC [147]. This flexibility may also allow the AC to adapt its responses based on recent stimulus history and the statistics of the acoustic environment. As discussed before, the loudness tuning of many A1 neurons can change to reflect the average loudness of recent sounds (dynamic range adaptation), a mechanism that may help maintain sensitivity to a range of different sound levels even in the face of background noise.

The auditory system exhibits many other context-dependent responses that are sensitive to recent stimulus history. For example, in marmoset A1, neurons often show long-lasting interactions between stimuli, generally with reduced responses to a second sound when it is similar to the preceding sound [148]. Although similar interactions have been observed previously at multiple levels of the auditory pathway, in A1 this effect often persists even when sounds are separated by a second or more. These suppressive interactions in A1 are strongest for two identical sounds, but are also present to a lesser extent for sounds of differing acoustics, and occur regardless of the strength of the neural response to the preceding sound. Additionally, masking is often present when the two sounds are coming from entirely different locations in space [145,146]. Similar masking phenomena have also been observed in the marmoset IC [149,150], possibly suggesting an inherited process.

Another consequence of repetitive sound presentation is a well-described human cortical phenomenon known as mismatch negativity (MMN), which has also been recorded from marmoset cortex [151]. MMN occurs when a series of repeating sounds are presented, with rarer deviant or 'oddball' sounds intermixed, and results in stronger responses to the oddballs than the repeating sounds. This mismatch phenomenon has been suggested to be due to reduced or adapted responses to the repeating sound, termed stimulus-specific adaptation [152]. The potential advantages of such cortical adaptation are that it may facilitate the ability to ignore background or frequent sounds, i.e., routine jungle noise to a wild marmoset, and thereby allow more selective encoding of rare and potentially more important sounds, such as a predator or vocalization from another marmoset. However, to what extent, in any species, there is a measurable perceptual consequence of this cortical adaptation is uncertain.

A second aspect of cortical flexibility is long-term plastic changes in cortical coding which occur during learning and development and may allow the cortex to better represent the sensory environment. For example, auditory deprivation in early development, and subsequent repeated tone exposure, has been found to affect tonotopic map development in rodents [153]. Such cortical plasticity has been relatively poorly studied in marmosets, possibly due to greater expense and both practical and ethical difficulties in performing long-term sensory deprivation or manipulations. However, there is limited evidence for adult cortical plasticity in marmosets. In one study, marmosets underwent a laryngeal procedure to permanently change the frequency contents of their vocalizations, and then subsequently underwent neural recordings after an interval of many months [154]. The ACs of these animals were found to use neurons of different frequency tuning to encode vocalization sounds as compared with normal marmoset AC, but responded normally to tone stimuli. Such changes may reflect plasticity in AC related to the repeated exposure of an animal to its own vocalizations.

Responses to Vocal Communication Sounds

Despite the importance of vocal communication in general, and the interest in marmoset vocalizations specifically, our knowledge of the encoding of such sounds in the auditory system remains relatively limited. At the level of core AC, particularly A1, vocalizations appear to be encoded in a spectrotemporally faithful fashion, with neurons of appropriate frequency tuning responding at those times when the acoustical energy of a vocalization passes through a neuron's frequency receptive field [109]. However, acoustical manipulations of vocal sounds, including time reversal, temporal compression/expansion, and other timing manipulations, also lead to reduced A1 responses [109,110]. Such selectivity for natural vocalizations suggests a more specific sensitivity to the spectrotemporal structure of a vocalization, one that is selective for the relative timing of different components rather than just responding whenever a given frequency is present. Simply playing a time-reversed marmoset twitter, for example, results in reduced responses compared to a natural call, a difference that is not present when forward and reversed twitters are played to another species such as cats [108]. To what extent this selectivity reflects an evolutionary specialization in marmosets for their own communication sounds, or is a result of experience-dependent plasticity due to repeated exposure to natural twitters over an animal's lifetime, is unknown. Observed coding changes in A1 after modifying an animal's own vocal production may be consistent with the latter [154].

Beyond A1, relatively little is known about AC responses to vocal sounds. Medial area CM responds to vocalizations with good phased-locked temporal specificity, similar to its responses to repeated FM stimuli [47,53]. Based on work in macaques, it is presumed that the lateral belt responds to complex sounds like vocalizations with increased stimulus specificity [58]. As discussed above, fMRI in marmosets has shown some increased specificity for vocalizations in more anterior areas of the STG, possibly anterior parabelt areas [63]. To what extent these areas are truly specific for species-specific vocalizations over complex nonvocal sounds remains unknown. Most studies of this sort only compare responses to vocal over nonvocal sounds, or between vocal sounds and time-reversed vocalizations. It is also unclear at what level there is an abstraction in neurons or areas to encode the class or semantic content of a vocalization, regardless of specific acoustical variation or an abstraction to neurons that encode the identity of the producing animal. Previous attempts to find so-called "call-detector" or "grandmother" neurons in squirrel monkeys were unsuccessful (discussed in Ref. [155]), although such experiments may not have extended sufficiently into lateral belt or parabelt areas. It is also possible that such specificity does not exist, and that all vocal sounds continue to be encoded in a more distributed, population fashion.

Sensory—Motor Integration and Mechanisms of Vocal Production

Understanding vocal communication does not end with studying mechanisms of encoding and perception of vocal sound. Marmosets, like other vocal species, both produce and perceive their vocalizations. To ensure accurate communication, both humans and marmosets show evidence that they continually monitor their vocal output to correct for errors, and the auditory system appears to play an important role in this self-monitoring process. Most neurons in the marmoset AC are suppressed during vocal production [102,156], in sharp contrast with more typical excited responses of the same neurons during vocal perception [157]. This suppression begins prior to the onset of vocal production and is thought to represent an efference copy, an internal motor signal predicting the sensory consequences of vocalization. Despite this suppression, these AC neurons are sensitive to changes in vocal feedback, suggesting that they play a role in vocal self-monitoring [103,158]. To what extent these cortical responses allow a marmoset to control its vocalizations, however, remains uncertain.

Where such vocal motor signals originate, or, for that matter, what neural elements are responsible for vocal production in marmosets, is unclear. Is there a specific area that plays a principal role of vocalization in marmosets, a Broca's area homolog? The role of cortex in the initiation and control of vocal production in nonhuman primates has long been an area of controversy, and for many years vocalization in nonhuman primates was viewed as emotional and brain stem controlled (see discussion in Ref. [159]). Recent studies in marmosets, however, have begun to reveal possible roles of frontal cortical areas. Immediate early gene cFOS labeling during antiphonal calling has demonstrated expression in dorsal frontal areas (dlPFC and premotor cortex), greater in antiphonal calling and spontaneously vocalizing animals than for animals that heard vocal sounds but did not respond [74]. Ventrolateral PFC areas, on the other hand, show greater expression during listening than during production conditions. However, other gene expression studies, using Egr-1, show increased vlPFC expression in vocalizing compared to listening-only animals [75]. More recent work using neural recordings during antiphonal calling has found vocal motor-related activity during and preceding vocal production in prefrontal [160] and premotor cortical areas [161]. The vocal activity in these areas is distinct from auditory sensory responses and from nonvocal orofacial somatomotor activity, and therefore attributable specifically as either vocal premotor activity or decision-making. Activity in some of these areas, particularly prefrontal area 8, also showed strong modulation by conversational context, predicting a marmoset's likelihood to respond to a vocalization it hears, even before it hears it, evidence for a potential role in the vocal decision-making process [162]. The question remains, however, to what extent these brain areas are necessary for vocal production, and what specific roles they are playing.

Further supporting a homology between marmosets and humans in the function of auditory and motor areas to support vocal communication, recent studies have also examined the presence of speech-related genes. Brain-wide genetic expression studies have shown that many of the genes implicated in human speech and its disorders, including *FoxP2, FoxP1, ROB01*, and others, are expressed in the marmoset brain [163]. These genes show patterns of overlapping expression in the same marmoset auditory and motor areas that have been found to be involved in vocal production and coding, including AC, motor cortex, and basal ganglia. Such findings are promising for future use of marmosets to study expressive and receptive aspects of vocal communication, particularly as new transgenic marmosets become available to test aspects of human disease.

Sleep

It is generally thought that there is a substantial reduction in sensory processing while sleeping, akin to that seen during anesthesia. However, recordings in naturally sleeping marmoset AC have shown a surprising amount of neural activity and responsiveness to auditory stimuli [164]. Many AC neurons, both in A1 and the lateral belt, exhibit diverse changes in activity with sleep, but overall population activities are similar regardless of wakefulness or the stage of sleep. During slow-wave sleep, many neurons exhibit weaker responses to soft sounds, but preserve their responses to louder sounds [165], possibly explaining why routine environmental sounds are not perceived while sleeping, but loud or novel sounds can awake one from sleep. Slow-wave sleep has also been found to increase synchronization between pairs of AC neurons [166], specifically for stimulus responses, similar to synchronization of slow-wave rhythms during sleep that have been previously observed in other species [167].

Cochlear Implants

A cochlear implant is a neuroprosthetic device used for patients with severe to profound hearing loss. Cochlear implants (CIs) consist of an array of electrodes placed within the scala tympani of the cochlea and function to bypass damaged structures and directly stimulate the auditory nerve. Despite decades of very successful use, we still neither understand how the auditory system makes sense of the impoverished sensory input provided by a CI nor how the brain learns and adapts to process these signals. Marmosets have seen recent use as a model to study CIs and the brain's response to CI stimulation [15]. Many neurons in the AC do not respond well to CI stimulation, particularly those in the ipsilateral hemisphere and those with narrow frequency or loudness tuning [17]. Such narrowly tuned neurons often exhibit inhibitory sidebands, as discussed before, and it is likely that the rather coarse CI signal, which stimulates multiple adjacent frequency regions of the cochlea, may be engaging both the primary frequency receptive field as well as the inhibitory sidebands. CI stimulation using time-varying inputs, however, results in similar types of synchronized and nonsynchronized responses as normal A1 activity, and individual A1 neurons respond with similar patterns both to the CI and normal acoustic inputs, suggesting that CIs preserve normal timing mechanisms [168]. However, these studies have only involved marmosets using a CI in the laboratory setting; the animals did not use the CI on a regular daily basis like human CI recipients. It remains unclear, therefore, whether there is cortical plasticity or plasticity anywhere along the auditory pathway, with implant use. The presence of many AC neurons with poor responses to the CI is one potential mechanism for CI-related, experience-dependent learning and plasticity, an area ripe for future study.

Hearing Loss: Aging, Genetics, and Ototoxicity

Finally, any summary of marmoset audition would not be complete without discussion of potential contributions toward understanding the mechanisms of hearing loss. Marmosets may be a useful model for studying age-related hearing loss. With a maximum life span of 20 years, though typically closer to 10, it is more practical to keep geriatric marmosets than other popular nonhuman primate species such as macaques. Studies of age-related changes in marmosets are sparse but have found changes in ABR signals, including an elevation of response thresholds, decreases in amplitudes, and slightly prolonged latencies in the response to high frequencies, consistent with a moderate age-related hearing loss [93]. To what extent these reflect central deficits or pathologic changes in the inner ear remains to be studied. Marmosets have also been proposed as a model for screening ototoxicity of medications and exhibit ABR changes following administration of ototoxic compounds including ethacrynic acid and neomycin [94].

In addition to acquired forms of deafness and hearing loss, there are numerous genetic causes of deafness. Although mutations in many genes cause deafness in humans, mouse models using similar mutations or knockouts often do not exhibit the same phenotypic hearing loss, making mice a poor model for such genetic deafness. The expression of many of these genes has been examined in the marmoset cochlea and has often been found to be differently expressed than in the mouse. *WFS1*, a gene involved in Wolfram syndrome, shows wide expression in both mouse and marmoset, but only the marmoset shows expression in the basal cells of the stria vascularis [169]. *KIAA1199*, another gene with mutations in deafness, is much more widely expressed in the marmoset cochlea than in mice or rats [170]. Additionally, many ion exchangers important for maintenance of the chemical environment in the cochlea are also differently expressed in marmosets and rodents [171], a primate-specific pattern than may be related to deafness-associated *DFNB4* mutations in human Pendred syndrome. Other deafness-associated genes that are differently expressed in marmosets compared with mice and do not include hearing loss in mice, include *GJB3, CRYM, GRHL2, DFNA5*, and *ATP6B1* [172]. Such genetic expression differences may explain phenotypic differences in hearing loss between humans and rodents and may make the marmoset a better model for understanding aspects of genetic deafness. It should be noted, however, that it is still unclear if marmosets would also exhibit hearing loss with mutations of these genes.

NEW AND UPCOMING RESEARCH APPROACHES

The past several years have seen an explosion of new research methods and approaches in neuroscience. One of these important advances is the introduction of transgenic animals, animals bred with selective genetic modifications. Although these techniques have largely focused on mouse models, recent work has seen the creation of transgenic marmosets [173]. Early work on transgenic marmosets began with simple fluorescence marker proteins, but more recent advances have extended to include the creation of genetic disease models [174]. Such genetic modifications, when coupled with physiologic and behavioral measurements in marmosets, have the potential to revolutionize our understanding of auditory and vocal communication systems. For example, gene mutations that result in hearing loss in humans, but not rodents, can be introduced in transgenic marmosets to better understand mechanisms of genetic hearing loss and rehabilitative approaches. Additionally, as one of the few model species of vocal communication, genetic modification in marmosets of speech-associated protein FoxP2 may yield insights into the neural basis of human speech production and its disorders.

A second recent advance has been the application of imaging techniques to marmosets. While single- and multielectrode recording methods can reveal fine details of neural coding, they are inherently limited to sampling a small number of neural elements at a time. In contrast, imaging approaches can potentially sample larger areas of the brain simultaneously. fMRI, for example, has been successfully used to map vocal sensory responses in the STG [63]. Such methods, simultaneously sampling multiple brain areas, also have the ability to measure the connectivity patterns between brain areas [175], a potential approach to studying the hierarchy of auditory processing. The disadvantages of fMRI, however, include the loss of fine spatial and temporal precision, only yielding signals that represent the average of neural activity over multiple neurons and over time. Recent advances in two-photon optical imaging, which can simultaneously record activities of multiple neurons, have shown promise in marmosets [107]. This approach can capture a broader area of cortex than typically obtained with microelectrode studies, while maintaining the ability to isolate and study individual neurons and their sensory responses. Two-photon imaging also has the ability to repeatedly measure the responses from the same neural population, often over a period of months, which may be applied to studying slowly time-varying phenomena including learning and neural plasticity. The smooth, lissencephalic cortex of marmosets is particularly well suited to such optical approaches.

One final research methodology has been the development of optogenetic techniques to selectively modulate neural activities in vivo, an approach which has revolutionized the study of neural circuits [176], but has largely bypassed nonhuman primates. These techniques use selective viruses to express light-sensitive proteins in neural populations, which can then be used to rapidly activate or inactivate those neurons. Recent work has demonstrated that such viruses express robustly in the marmoset brain [177] and can be used for optogenetic stimulation in awake marmosets [178]. The further development of such approaches will potentially allow more careful study of marmoset neural circuitry, including the contributions of different neural circuits and classes of neurons to auditory sensory processing and behaviors.

CONCLUSIONS

The past few years have seen the emergence of the marmoset as an increasingly popular model in multiple areas of biomedical research. Marmosets have a number of advantages for such research, particularly in neurobiology where their primate lineage may provide better understanding of the human brain than can be inferred from rodents and other nonprimate research models. Because of the prevalence and importance of vocal communication for marmosets, particularly as compared with other captive nonhuman primates, they are a natural model for understanding hearing mechanisms and the auditory basis of vocal perception and production. Here, we have attempted to summarize current understanding of the marmoset auditory system, and how it compares to humans and other species. With the advent of new genetic and physiologic methods that are becoming increasingly available, the use of marmosets in hearing and neuroscience research will only increase. With these new techniques, marmosets will continue to provide valuable insight toward understanding the evolution, normal mechanisms, and disorder of hearing and vocal communication as a way to better understand human audition and speech.

References

[1] Agamaite JA, Chang CJ, Osmanski MS, Wang X. A quantitative acoustic analysis of the vocal repertoire of the common marmoset (*Callithrix jacchus*). J Acoust Soc Am 2015;138:2906–28.

[2] Epple G. Comparative studies on vocalization in marmoset monkeys (Hapalidae). Folia Primatol 1968;8:1–40.

[3] Seiden HR. Auditory acuity of the marmoset monkey (*Hapale jacchus*). Princeton, NJ: Princeton University; 1957.

[4] Osmanski MS, Wang X. Measurement of absolute auditory thresholds in the common marmoset (*Callithrix jacchus*). Hear Res 2011;277:127–33.

[5] Aitkin LM, Merzenich MM, Irvine DR, Clarey JC, Nelson JE. Frequency representation in auditory cortex of the common marmoset (*Callithrix jacchus jacchus*). J Comp Neurol 1986;252: 175–85.

[6] Osmanski MS, Song X, Guo Y, Wang X. Frequency discrimination in the common marmoset (*Callithrix jacchus*). Hear Res 2016;341: 1–8.

[7] Newman JD. Auditory communication and central auditory mechanisms in the squirrel monkey: past and present. In: Ghazanfar AA, editor. Primate audition: ethology and neurobiology. Boca Raton, FL: CRC Press; 2003. p. 198–203.

[8] Plack CJ, Fay RR, Oxenham AJ, Popper AN, editors. Pitch: neural coding and perception. New York: Springer; 2005.

[9] Bendor D, Osmanski MS, Wang X. Dual-pitch processing mechanisms in primate auditory cortex. J Neurosci 2012;32:16149–61.

[10] Osmanski MS, Song X, Wang X. The role of harmonic resolvability in pitch perception in a vocal nonhuman primate, the common marmoset (*Callithrix jacchus*). J Neurosci 2013;33:9161–8.

[11] Song X, Osmanski MS, Guo Y, Wang X. Complex pitch perception mechanisms are shared by humans and a New World monkey. Proc Natl Acad Sci USA 2016;113:781–6.

[12] Borin A, Covolan L, Mello LE, Okada DM, Cruz OL, Testa JR. Anatomical study of a temporal bone from a non-human primate (*Callithrix* sp.). Braz J Otorhinolaryngol 2008;74:370–3.

[13] Aitkin L, Park V. Audition and the auditory pathway of a vocal new world primate, the common marmoset. Prog Neurobiol 1993;41:345–67.

[14] Slee SJ, Young ED. Sound localization cues in the marmoset monkey. Hear Res 2010;260:96–108.

[15] Johnson LA, Della Santina CC, Wang X. Temporal bone characterization and cochlear implant feasibility in the common marmoset (*Callithrix jacchus*). Hear Res 2012;290:37–44.

[16] Wever EG, Vernon JA. Cochlear potentials in the marmoset. Proc Natl Acad Sci USA 1961;47:739–41.

[17] Johnson LA, Della Santina CC, Wang X. Selective neuronal activation by cochlear implant stimulation in auditory cortex of awake Primate. J Neurosci 2016;36:12468–84.

[18] Moore JK. The primate cochlear nuclei: loss of lamination as a phylogenetic process. J Comp Neurol 1980;193:609–29.

[19] Rhode WS, Roth GL, Recio-Spinoso A. Response properties of cochlear nucleus neurons in monkeys. Hear Res 2010;259:1–15.

[20] Clopton BM, Winfield JA, Flammino FJ. Tonotopic organization: review and analysis. Brain Res 1974;76:1–20.

[21] Masterton RB. Role of the central auditory system in hearing: the new direction. Trends Neurosci 1992;15:280–5.

[22] Moore JK, Moore RY. A comparative study of the superior olivary complex in the primate brain. Folia Primatol (Basel) 1971; 16:35–51.

[23] Garey LJ, Webster WR. Functional morphology in the inferior colliculus of the marmoset. Hear Res 1989;38:67–79.

[24] Goldberg JM, Moore RY. Ascending projections of the lateral lemniscus in the cat and monkey. J Comp Neurol 1967;129:143–56.

[25] Fitzpatrick KA. Cellular architecture and topographic organization of the inferior colliculus of the squirrel monkey. J Comp Neurol 1975;164:185–207.

[26] Slee SJ, Young ED. Information conveyed by inferior colliculus neurons about stimuli with aligned and misaligned sound localization cues. J Neurophysiol 2011;106:974–85.

[27] Slee SJ, Young ED. Linear processing of interaural level difference underlies spatial tuning in the nucleus of the brachium of the inferior colliculus. J Neurosci 2013;33:3891–904.

[28] Slee SJ, Young ED. Alignment of sound localization cues in the nucleus of the brachium of the inferior colliculus. J Neurophysiol 2014;111:2624–33.

[29] de la Mothe LA, Blumell S, Kajikawa Y, Hackett TA. Thalamic connections of the auditory cortex in marmoset monkeys: core and medial belt regions. J Comp Neurol 2006;496:72–96.

[30] Hackett TA, Stepniewska I, Kaas JH. Thalamocortical connections of the parabelt auditory cortex in macaque monkeys. J Comp Neurol 1998;400:271–86.

[31] Jones EG. Chemically defined parallel pathways in the monkey auditory system. Ann NY Acad Sci 2003;999:218–33.

[32] Gross NB, Lifschitz WS, Anderson DJ. The tonotopic organization of the auditory thalamus of the squirrel monkey (*Saimiri sciureus*). Brain Res 1974;65:323–32.

[33] Aitkin LM, Kudo M, Irvine DR. Connections of the primary auditory cortex in the common marmoset, *Callithrix jacchus jacchus*. J Comp Neurol 1988;269:235–48.

[34] Bartlett EL, Sadagopan S, Wang X. Fine frequency tuning in monkey auditory cortex and thalamus. J Neurophysiol 2011;106: 849–59.

[35] Bartlett EL, Wang X. Neural representations of temporally modulated signals in the auditory thalamus of awake primates. J Neurophysiol 2007;97:1005–17.

[36] Bartlett EL, Wang X. Correlation of neural response properties with auditory thalamus subdivisions in the awake marmoset. J Neurophysiol 2011;105:2647–67.

[37] Wong P, Peebles JK, Asplund CL, Collins CE, Herculano-Houzel S, Kaas JH. Faster scaling of auditory neurons in cortical areas relative to subcortical structures in primate brains. Brain Behav Evol 2013;81:209–18.

[38] Collins CE, Leitch DB, Wong P, Kaas JH, Herculano-Houzel S. Faster scaling of visual neurons in cortical areas relative to subcortical structures in non-human primate brains. Brain Struct Funct 2013;218:805–16.

[39] Peden JK, Von Bonin G. The neocortex of Hapale. J Comp Neurol 1947;86:37–63.

[40] Woolsey CN. Localization patterns in a lisencephalic primate (*Hapale jacchus*). Am J Physiol 1954;179:686.

[41] Kaas JH, Hackett TA. Subdivisions of auditory cortex and levels of processing in primates. Audiol Neuro Otol 1998;3:73–85.

[42] Rauschecker JP. Parallel processing in the auditory cortex of primates. Audiol Neurotol 1998;3:86–103.

[43] de la Mothe LA, Blumell S, Kajikawa Y, Hackett TA. Cortical connections of the auditory cortex in marmoset monkeys: core and medial belt regions. J Comp Neurol 2006;496:27–71.

[44] Philibert B, Beitel RE, Nagarajan SS, Bonham BH, Schreiner CE, Cheung SW. Functional organization and hemispheric comparison of primary auditory cortex in the common marmoset (*Callithrix jacchus*). J Comp Neurol 2005;487:391–406.

[45] Bendor D, Wang X. Neural response properties of primary, rostral, and rostrotemporal core fields in the auditory cortex of marmoset monkeys. J Neurophysiol 2008;100:888–906.

[46] Wang X, Lu T, Snider RK, Liang L. Sustained firing in auditory cortex evoked by preferred stimuli. Nature 2005;435:341–6.

[47] Rajan R, Dubaj V, Reser DH, Rosa MG. Auditory cortex of the marmoset monkey - complex responses to tones and vocalizations under opiate anaesthesia in core and belt areas. Eur J Neurosci 2013;37:924–41.

[48] Wang X. Neural coding strategies in auditory cortex. Hear Res 2007;229:81–93.

[49] Kadia SC, Wang X. Spectral integration in A1 of awake primates: neurons with single- and multipeaked tuning characteristics. J Neurophysiol 2003;89:1603–22.

[50] Hackett TA, Smiley JF, Ulbert I, Karmos G, Lakatos P, de la Mothe LA, Schroeder CE. Sources of somatosensory input to the caudal belt areas of auditory cortex. Perception 2007;36: 1419–30.

[51] Kajikawa Y, Camalier CR, de la Mothe LA, D'Angelo WR, Sterbing-D'Angelo SJ, Hackett TA. Auditory cortical tuning to band-pass noise in primate A1 and CM: a comparison to pure tones. Neurosci Res 2011;70:401–7.

[52] Kajikawa Y, de La Mothe L, Blumell S, Hackett TA. A comparison of neuron response properties in areas A1 and CM of the marmoset monkey auditory cortex: tones and broadband noise. J Neurophysiol 2005;93:22–34.

[53] Kajikawa Y, de la Mothe LA, Blumell S, Sterbing-D'Angelo SJ, D'Angelo W, Camalier CR, Hackett TA. Coding of FM sweep trains and twitter calls in area CM of marmoset auditory cortex. Hear Res 2008;239:107–25.

[54] de la Mothe LA, Blumell S, Kajikawa Y, Hackett TA. Cortical connections of auditory cortex in marmoset monkeys: lateral belt and parabelt regions. Anat Rec 2012;295:800–21.

[55] de la Mothe LA, Blumell S, Kajikawa Y, Hackett TA. Thalamic connections of auditory cortex in marmoset monkeys: lateral belt and parabelt regions. Anat Rec 2012;295:822–36.

[56] Nishimura M, Takemoto M, Song WJ. Organization of auditory areas in the superior temporal gyrus of marmoset monkeys revealed by real-time optical imaging. Brain Struct Funct 2017. https://doi.org/10.1007/s00429-017-1574-0.

[57] Toarmino CR, Yen CCC, Papoti D, Bock NA, Leopold DA, Miller CT, Silva AC. Functional magnetic resonance imaging of auditory cortical fields in awake marmosets. Neuroimage 2017; 162:86–92.

[58] Rauschecker JP, Tian B, Hauser M. Processing of complex sounds in the macaque nonprimary auditory cortex. Science 1995;268: 111–4.

[59] Tsunada J, Liu AS, Gold JI, Cohen YE. Causal contribution of primate auditory cortex to auditory perceptual decision-making. Nat Neurosci 2016;19:135–42.

[60] Perrodin C, Kayser C, Logothetis NK, Petkov CI. Voice cells in the primate temporal lobe. Curr Biol 2011;21:1408–15.

[61] Petkov CI, Kayser C, Steudel T, Whittingstall K, Augath M, Logothetis NK. A voice region in the monkey brain. Nat Neurosci 2008;11:367–74.

[62] Tian B, Reser D, Durham A, Kustov A, Rauschecker JP. Functional specialization in rhesus monkey auditory cortex. Science 2001;292:290–3.

[63] Sadagopan S, Temiz-Karayol NZ, Voss HU. High-field functional magnetic resonance imaging of vocalization processing in marmosets. Sci Rep 2015;5:10950.

[64] Rauschecker JP, Tian B. Mechanisms and streams for processing of "what" and "where" in auditory cortex. Proc Natl Acad Sci USA 2000;97:11800–6.

[65] Burman KJ, Palmer SM, Gamberini M, Rosa MG. Cytoarchitectonic subdivisions of the dorsolateral frontal cortex of the marmoset monkey (*Callithrix jacchus*), and their projections to dorsal visual areas. J Comp Neurol 2006;495:149–72.

[66] Roberts AC, Tomic DL, Parkinson CH, Roeling TA, Cutter DJ, Robbins TW, Everitt BJ. Forebrain connectivity of the prefrontal cortex in the marmoset monkey (*Callithrix jacchus*): an anterograde and retrograde tract-tracing study. J Comp Neurol 2007; 502:86–112.

[67] Reser DH, Burman KJ, Yu HH, Chaplin TA, Richardson KE, Worthy KH, Rosa MG. Contrasting patterns of cortical input to architectural subdivisions of the area 8 complex: a retrograde tracing study in marmoset monkeys. Cerebr Cortex 2013;23: 1901–22.

[68] Burman KJ, Reser DH, Yu HH, Rosa MG. Cortical input to the frontal pole of the marmoset monkey. Cerebr Cortex 2011;21: 1712–37.

[69] Reser DH, Burman KJ, Richardson KE, Spitzer MW, Rosa MG. Connections of the marmoset rostrotemporal auditory area: express pathways for analysis of affective content in hearing. Eur J Neurosci 2009;30:578–92.

[70] Petrides M, Pandya DN. Dorsolateral prefrontal cortex: comparative cytoarchitectonic analysis in the human and the macaque brain and corticocortical connection patterns. Eur J Neurosci 1999;11:1011–36.

[71] Romanski LM, Bates JF, Goldman-Rakic PS. Auditory belt and parabelt projections to the prefrontal cortex in the rhesus monkey. J Comp Neurol 1999;403:141–57.

[72] Cohen YE, Russ BE, Gifford 3rd GW, Kiringoda R, MacLean KA. Selectivity for the spatial and nonspatial attributes of auditory stimuli in the ventrolateral prefrontal cortex. J Neurosci 2004; 24:11307–16.
[73] Romanski LM, Goldman-Rakic PS. An auditory domain in primate prefrontal cortex. Nat Neurosci 2002;5:15–6.
[74] Miller CT, Dimauro A, Pistorio A, Hendry S, Wang X. Vocalization induced CFos expression in marmoset cortex. Front Integr Neurosci 2010;4:128.
[75] Simoes CS, Vianney PV, de Moura MM, Freire MA, Mello LE, Sameshima K, Araujo JF, Nicolelis MA, Mello CV, Ribeiro S. Activation of frontal neocortical areas by vocal production in marmosets. Front Integr Neurosci 2010;4. https://doi.org/10.3389/fnint.2010.00123.
[76] Luethke LE, Krubitzer LA, Kaas JH. Connections of primary auditory cortex in the New World monkey, Saguinus. J Comp Neurol 1989;285:487–513.
[77] Fitzpatrick KA, Imig TJ. Projections of auditory cortex upon the thalamus and midbrain in the owl monkey. J Comp Neurol 1978; 177:537–55.
[78] Morel A, Kaas JH. Subdivisions and connections of auditory cortex in owl monkeys. J Comp Neurol 1992;318:27–63.
[79] Suga N, Ma X. Multiparametric corticofugal modulation and plasticity in the auditory system. Nat Rev Neurosci 2003;4: 783–y794.
[80] Remington ED, Osmanski MS, Wang X. An operant conditioning method for studying auditory behaviors in marmoset monkeys. PLoS One 2012;7:e47895.
[81] Ribeiro MW, Neto JFR, Morya E, Brasil FL, de Araujo MFP. OBAT: an open-source and low-cost operant box for auditory discriminative tasks. Behav Res Methods 2017. https://doi.org/10.3758/s13428-017-0906-6.
[82] Miller CT, Wang X. Sensory-motor interactions modulate a primate vocal behavior: antiphonal calling in common marmosets. J Comp Physiol 2006;192:27–38.
[83] Roy S, Miller CT, Gottsch D, Wang X. Vocal control by the common marmoset in the presence of interfering noise. J Exp Biol 2011;214:3619–29.
[84] Choi JY, Takahashi DY, Ghazanfar AA. Cooperative vocal control in marmoset monkeys via vocal feedback. J Neurophysiol 2015; 114:274–83.
[85] Takahashi DY, Fenley AR, Teramoto Y, Narayanan DZ, Borjon JI, Holmes P, Ghazanfar AA. The developmental dynamics of marmoset monkey vocal production. Science 2015;349:734–8.
[86] DiMattina C, Wang X. Virtual vocalization stimuli for investigating neural representations of species-specific vocalizations. J Neurophysiol 2006;95:1244–62.
[87] Brown AM, Gaskill SA. Measurement of acoustic distortion reveals underlying similarities between human and rodent mechanical responses. J Acoust Soc Am 1990;88:840–9.
[88] Valero MD, Pasanen EG, McFadden D, Ratnam R. Distortion-product otoacoustic emissions in the common marmoset (*Callithrix jacchus*): parameter optimization. Hear Res 2008;243:57–68.
[89] Valero MD, Ratnam R. Reliability of distortion-product otoacoustic emissions in the common marmoset (*Callithrix jacchus*). Hear Res 2011;282:265–71.
[90] Møller AR, Burgess J. Neural generators of the brain-stem auditory evoked potentials (BAEPs) in the rhesus monkey. Electroencephalogr Clin Neurophysiol 1986;65:361–72.
[91] Harada T, Tokuriki M. Brain-stem auditory evoked potentials in the common marmoset (*Callithrix jacchus*). Electroencephalogr Clin Neurophysiol 1997;104:43–50.
[92] Harada T, Tokuriki M. Effects of click intensity and frequency on the brain-stem auditory evoked potentials in the common marmoset (*Callithrix jacchus*). J Vet Med Sci 1997;59:561–7.
[93] Harada T, Tokuriki M, Tanioka Y. Age-related changes in the brainstem auditory evoked potentials of the marmoset. Hear Res 1999;128:119–24.
[94] Kuzel RA, Smith JM, Trennery PN. Screening procedure for assessment of ototoxicity in the common marmoset. J Pharmacol Methods 1990;24:9–18.
[95] Lu T, Liang L, Wang X. Neural representations of temporally asymmetric stimuli in the auditory cortex of awake primates. J Neurophysiol 2001;85:2364–80.
[96] Gao L, Kostlan K, Wang Y, Wang X. Distinct subthreshold mechanisms underlying rate-coding principles in primate auditory cortex. Neuron 2016;91:905–19.
[97] Pfingst BE, O'Connor TA. A vertical stereotaxic approach to auditory cortex in the unanesthetized monkey. J Neurosci Meth 1980; 2:33–45.
[98] Hoffman KL, McNaughton BL. Coordinated reactivation of distributed memory traces in primate neocortex. Science 2002; 297:2070–3.
[99] Nicolelis MA, Dimitrov D, Carmena JM, Crist R, Lehew G, Kralik JD, Wise SP. Chronic, multisite, multielectrode recordings in macaque monkeys. Proc Natl Acad Sci USA 2003;100:11041–6.
[100] deCharms RC, Blake DT, Merzenich MM. A multielectrode implant device for the cerebral cortex. J Neurosci Meth 1999;93: 27–35.
[101] Eliades SJ, Wang X. Chronic multi-electrode neural recording in free-roaming monkeys. J Neurosci Meth 2008;172:201–14.
[102] Eliades SJ, Wang X. Sensory-motor interaction in the primate auditory cortex during self-initiated vocalizations. J Neurophysiol 2003;89:2194–207.
[103] Eliades SJ, Wang X. Neural substrates of vocalization feedback monitoring in primate auditory cortex. Nature 2008;453:1102–6.
[104] Hage SR, Nieder A. Single neurons in monkey prefrontal cortex encode volitional initiation of vocalizations. Nat Commun 2013; 4:2409.
[105] Mohseni P, Najafi K, Eliades SJ, Wang X. Wireless multichannel biopotential recording using an integrated FM telemetry circuit. IEEE Trans Neural Syst Rehabil Eng 2005;13:263–71.
[106] Roy S, Wang X. Wireless multi-channel single unit recording in freely moving and vocalizing primates. J Neurosci Meth 2012; 203:28–40.
[107] Sadakane O, Masamizu Y, Watakabe A, Terada S, Ohtsuka M, Takaji M, Mizukami H, Ozawa K, Kawasaki H, Matsuzaki M, Yamamori T. Long-term two-photon calcium imaging of neuronal populations with subcellular resolution in adult non-human primates. Cell Rep 2015;13:1989–99.
[108] Wang X, Kadia SC. Differential representation of species-specific primate vocalizations in the auditory cortices of marmoset and cat. J Neurophysiol 2001;86:2616–20.
[109] Wang X, Merzenich MM, Beitel R, Schreiner CE. Representation of a species-specific vocalization in the primary auditory cortex of the common marmoset: temporal and spectral characteristics. J Neurophysiol 1995;74:2685–706.
[110] Nagarajan SS, Cheung SW, Bedenbaugh P, Beitel RE, Schreiner CE, Merzenich MM. Representation of spectral and temporal envelope of twitter vocalizations in common marmoset primary auditory cortex. J Neurophysiol 2002;87:1723–37.
[111] Benson DA, Hienz RD, Goldstein Jr MH. Single-unit activity in the auditory cortex of monkeys actively localizing sound sources: spatial tuning and behavioral dependency. Brain Res 1981;219: 249–67.
[112] Goldstein Jr MH, Hall 2nd JL, Butterfield BO. Single-unit activity in the primary auditory cortex of unanesthetized cats. J Acoust Soc Am 1968;43:444–55.
[113] Heil P. Auditory cortical onset responses revisited. II. Response strength. J Neurophysiol 1997;77:2642–60.

[114] Phillips DP. Neural representation of stimulus times in the primary auditory cortex. Ann NY Acad Sci 1993;682:104–18.
[115] deCharms RC, Merzenich MM. Primary cortical representation of sounds by the coordination of action-potential timing. Nature 1996;381:610–3.
[116] Sadagopan S, Wang X. Level invariant representation of sounds by populations of neurons in primary auditory cortex. J Neurosci 2008;28:3415–26.
[117] Watkins PV, Barbour DL. Rate-level responses in awake marmoset auditory cortex. Hear Res 2011;275:30–42.
[118] Palmer AR. Encoding of rapid amplitude fluctuations by Cochlear-nerve fibres in the Guinea-pig. Arch Oto-Rhino-Laryngol 1982;236:197–202.
[119] Bieser A, Müller-Preuss P. Auditory responsive cortex in the squirrel monkey: neural responses to amplitude-modulated sounds. Exp Brain Res 1996;108:273–84.
[120] Lu T, Liang L, Wang X. Temporal and rate representations of time-varying signals in the auditory cortex of awake primates. Nat Neurosci 2001;4:1131–8.
[121] Liang L, Lu T, Wang X. Neural representations of sinusoidal amplitude and frequency modulations in the primary auditory cortex of awake primates. J Neurophysiol 2002;87:2237–61.
[122] Lu T, Wang X. Information content of auditory cortical responses to time-varying acoustic stimuli. J Neurophysiol 2004;91:301–13.
[123] Zhou Y, Wang X. Cortical processing of dynamic sound envelope transitions. J Neurosci 2010;30:16741–54.
[124] Müller-Preuss P, Flachskamm C, Bieser A. Neural encoding of amplitude modulation within the auditory midbrain of squirrel monkeys. Hear Res 1994;80:197–208.
[125] Bendor D, Wang X. Differential neural coding of acoustic flutter within primate auditory cortex. Nat Neurosci 2007;10:763–71.
[126] Wang X, Lu T, Bendor D, Bartlett E. Neural coding of temporal information in auditory thalamus and cortex. Neuroscience 2008;154:294–303.
[127] Sadagopan S, Wang X. Contribution of inhibition to stimulus selectivity in primary auditory cortex of awake primates. J Neurosci 2010;30:7314–25.
[128] Wang X. The harmonic organization of auditory cortex. Front Syst Neurosci 2013;7:114.
[129] Feng L, Wang X. Harmonic template neurons in primate auditory cortex underlying complex sound processing. Proc Natl Acad Sci USA 2017;114:E840–8.
[130] Barbour DL, Wang X. Contrast tuning in auditory cortex. Science 2003;299:1073–5.
[131] Yu JJ, Young ED. Linear and nonlinear pathways of spectral information transmission in the cochlear nucleus. Proc Natl Acad Sci USA 2000;97:11780–6.
[132] Barbour DL, Wang X. Auditory cortical responses elicited in awake primates by random spectrum stimuli. J Neurosci 2003; 23:7194–206.
[133] Theunissen FE, Sen K, Doupe AJ. Spectral-temporal receptive fields of nonlinear auditory neurons obtained using natural sounds. J Neurosci 2000;20:2315–31.
[134] Luczak A, Hackett TA, Kajikawa Y, Laubach M. Multivariate receptive field mapping in marmoset auditory cortex. J Neurosci Meth 2004;136:77–85.
[135] Sadagopan S, Wang X. Nonlinear spectrotemporal interactions underlying selectivity for complex sounds in auditory cortex. J Neurosci 2009;29:11192–202.
[136] Sun W, Marongelli EN, Watkins PV, Barbour DL. Decoding sound level in the marmoset primary auditory cortex. J Neurophysiol 2017;118:2024–33.
[137] Watkins PV, Barbour DL. Specialized neuronal adaptation for preserving input sensitivity. Nat Neurosci 2008;11:1259–61.
[138] Watkins PV, Barbour DL. Level-tuned neurons in primary auditory cortex adapt differently to loud versus soft sounds. Cerebr Cortex 2011;21:178–90.
[139] Ni R, Bender DA, Shanechi AM, Gamble JR, Barbour DL. Contextual effects of noise on vocalization encoding in primary auditory cortex. J Neurophysiol 2017;117:713–27.
[140] Bendor D, Wang X. The neuronal representation of pitch in primate auditory cortex. Nature 2005;436:1161–5.
[141] Bendor D. Does a pitch center exist in auditory cortex? J Neurophysiol 2012;107:743–6.
[142] Bendor D, Wang X. Neural coding of periodicity in marmoset auditory cortex. J Neurophysiol 2010;103:1809–22.
[143] Lui LL, Mokri Y, Reser DH, Rosa MG, Rajan R. Responses of neurons in the marmoset primary auditory cortex to interaural level differences: comparison of pure tones and vocalizations. Front Neurosci 2015;9:132.
[144] Remington ED, Wang X. Neural representations of the full spatial field in auditory cortex of awake marmoset (*Callithrix jacchus*). Cerebr Cortex 2018. https://doi.org/10.1093/cercor/bhy025 [pmid 29420692].
[145] Zhou Y, Wang X. Level dependence of spatial processing in the primate auditory cortex. J Neurophysiol 2012;108:810–26.
[146] Zhou Y, Wang X. Spatially extended forward suppression in primate auditory cortex. Eur J Neurosci 2014;39:919–33.
[147] Fritz JB, David SV, Radtke-Schuller S, Yin P, Shamma SA. Adaptive, behaviorally gated, persistent encoding of task-relevant auditory information in ferret frontal cortex. Nat Neurosci 2010;13:1011–9.
[148] Bartlett EL, Wang X. Long-lasting modulation by stimulus context in primate auditory cortex. J Neurophysiol 2005;94: 83–104.
[149] Nelson PC, Smith ZM, Young ED. Wide-dynamic-range forward suppression in marmoset inferior colliculus neurons is generated centrally and accounts for perceptual masking. J Neurosci 2009; 29:2553–62.
[150] Nelson PC, Young ED. Neural correlates of context-dependent perceptual enhancement in the inferior colliculus. J Neurosci 2010;30:6577–87.
[151] Komatsu M, Takaura K, Fujii N. Mismatch negativity in common marmosets: whole-cortical recordings with multi-channel electrocorticograms. Sci Rep 2015;5:15006.
[152] Nelken I, Ulanovsky N. Mismatch negativity and stimulus-specific adaptation in animal models. J Psychophysiol 2007;21: 214–23.
[153] Chang EF, Merzenich MM. Environmental noise retards auditory cortical development. Science 2003;300:498–502.
[154] Cheung SW, Nagarajan SS, Schreiner CE, Bedenbaugh PH, Wong A. Plasticity in primary auditory cortex of monkeys with altered vocal production. J Neurosci 2005;25:2490–503.
[155] Symmes D. On the use of natural stimuli in neurophysiological studies of audition. Hear Res 1981;4:203–14.
[156] Eliades SJ, Wang X. Comparison of auditory-vocal interactions across multiple types of vocalizations in marmoset auditory cortex. J Neurophysiol 2013;109:1638–57.
[157] Eliades SJ, Wang X. Contributions of sensory tuning to auditory-vocal interactions in marmoset auditory cortex. Hear Res 2017; 348:98–111.
[158] Eliades SJ, Wang X. Neural correlates of the lombard effect in primate auditory cortex. J Neurosci 2012;32:10737–48.
[159] Hage SR, Nieder A. Dual neural network model for the evolution of speech and language. Trends Neurosci 2016;39:813–29.

[160] Miller CT, Thomas AW, Nummela SU, de la Mothe LA. Responses of primate frontal cortex neurons during natural vocal communication. J Neurophysiol 2015;114:1158–71.
[161] Roy S, Zhao L, Wang X. Distinct neural activities in premotor cortex during natural vocal behaviors in a new world primate, the common marmoset (*Callithrix jacchus*). J Neurosci 2016;36: 12168–79.
[162] Nummela SU, Jovanovic V, de la Mothe L, Miller CT. Social context-dependent activity in marmoset frontal cortex populations during natural conversations. J Neurosci 2017;37: 7036–47.
[163] Kato M, Okanoya K, Koike T, Sasaki E, Okano H, Watanabe S, Iriki A. Human speech- and reading-related genes display partially overlapping expression patterns in the marmoset brain. Brain Lang 2014;133:26–38.
[164] Issa EB, Wang X. Sensory responses during sleep in primate primary and secondary auditory cortex. J Neurosci 2008;28: 14467–80.
[165] Issa EB, Wang X. Altered neural responses to sounds in primate primary auditory cortex during slow-wave sleep. J Neurosci 2011;31:2965–73.
[166] Issa EB, Wang X. Increased neural correlations in primate auditory cortex during slow-wave sleep. J Neurophysiol 2013;109: 2732–8.
[167] Kohn A, Zandvakili A, Smith MA. Correlations and brain states: from electrophysiology to functional imaging. Curr Opin Neurobiol 2009;19:434–8.
[168] Johnson LA, Della Santina CC, Wang X. Representations of time-varying cochlear implant stimulation in auditory cortex of awake marmosets (*Callithrix jacchus*). J Neurosci 2017;37:7008–22.
[169] Suzuki N, Hosoya M, Oishi N, Okano H, Fujioka M, Ogawa K. Expression pattern of wolframin, the WFS1 (Wolfram syndrome-1 gene) product, in common marmoset (*Callithrix jacchus*) cochlea. Neuroreport 2016;27:833–6.
[170] Hosoya M, Fujioka M, Ogawa K, Okano H. Distinct expression patterns of causative genes responsible for hereditary progressive hearing loss in non-human primate cochlea. Sci Rep 2016;6:22250.
[171] Hosoya M, Fujioka M, Kobayashi R, Okano H, Ogawa K. Overlapping expression of anion exchangers in the cochlea of a non-human primate suggests functional compensation. Neurosci Res 2016;110:1–10.
[172] Hosoya M, Fujioka M, Okano H, Ogawa K. Distinct expression pattern of a deafness gene, KIAA1199, in a primate cochlea. Biomed Res Int 2016;2016:1781894.
[173] Sasaki E, Suemizu H, Shimada A, Hanazawa K, Oiwa R, Kamioka M, Tomioka I, Sotomaru Y, Hirakawa R, Eto T, Shiozawa S, Maeda T, Ito M, Ito R, Kito C, Yagihashi C, Kawai K, Miyoshi H, Tanioka Y, Tamaoki N, Habu S, Okano H, Nomura T. Generation of transgenic non-human primates with germline transmission. Nature 2009;459:523–7.
[174] Belmonte J, Callaway EM, Caddick SJ, Churchland P, Feng G, Homatics GE, Lee K, Leopold DA, Miller CT, Mitchell JF, Mitalipov S, Moutri AR, Movschon JA, Okano H, Reynolds J, Ringach D, Sejnowski TJ, Silva AC, Strick PL, Wu J, Zhang F. Brains, genes and primates. Neuron 2015;86:617–31.
[175] Anzellotti S, Coutanche MN. Beyond functional connectivity: investigating networks of multivariate representations. Trends Cogn Sci 2018;22:258–69.
[176] Deisseroth K. Optogenetics: 10 years of microbial opsins in neuroscience. Nat Neurosci 2015;18:1213–25.
[177] Watakabe A, Ohtsuka M, Kinoshita M, Takaji M, Takaji M, Isa K, Mizukami H, Ozawa K, Isa T, Yamamori T. Comparative analyses of adeno-associated viral vector serotypes 1,2,5,8 and 9 in marmoset, mouse and macaque cerebral cortex. Neurosci Res 2015;93:144–57.
[178] MacDougall M, Nummela SU, Coop S, Disney A, Mitchell JF, Miller CT. Optogenetic manipulation of neural circuits in awake marmosets. J Neurophysiol 2016;116:1286–94.

CHAPTER

26

The Marmoset as a Model in Behavioral Neuroscience and Psychiatric Research

Jeffrey A. French

Neuroscience Program and Callitrichid Research Center, University of Nebraska at Omaha, NE, United States

INTRODUCTION

In its mission statement, the US National Institutes of Health provides a clear statement of its focus: "... to seek fundamental knowledge about the nature and *behavior* of living systems and the application of that knowledge to enhance health, lengthen life, and reduce illness and disability" [*emphasis added*, www.nih.gov/about-nih/what-we-do/mission-goals, 2015). Disorders associated with brain or behavioral dysfunction represent the leading disease burden and highest source of lifetime years living with disability on a global basis (YLD: [1]) and together these disorders represent one of the leading contributors to disease-associated mortality worldwide [2]. Clearly, then, there is a premium on understanding both normative behavioral states and their relationship to brain function and the nature of brain dysfunction as is relates to pathological behavioral states.

As in all fields of biomedical science, animal models play an important role in characterizing the basic biology of brain–behavior relationships, in exploring the potential etiology of disorders (genetics, development, environmental perturbations), and in developing and refining treatments in a number of modalities that target dysfunction on brain processing. Nonhuman primates (NHPs) constitute an attractive animal model for behavioral and psychological states for a variety of important reasons. Given the relatively recent common evolutionary origin of primates [3], NHPs share commonalities in complex brain structure and function (homologies), most importantly represented by an increase in brain size and elaboration of the neocortex, especially the prefrontal cortex (PFC) [4]. As a consequence of these morphological adaptations, NHPs are capable of complex cognitions that involve multisensory integration, complex and conditional decision-making, and top-down as well as bottom-up regulation of affect and emotion. Finally, the changes in NHP brain structure and function can facilitate the mediation of challenges associated with group living, including aggression, affiliation, and the establishment and maintenance of long-term complex social relationships that distinguish these species from nonprimate animals [5].

The Utility of Marmosets in Behavioral Models in Neuroscience and Psychiatric Research

From the perspective of a biomedically oriented focus, research on behavioral states (both normative and atypical) is of interest to the extent that it can provide useful information regarding the developmental factors that lead to normative neuropsychological function, as well as the underlying neurobiological properties that contribute to atypical or pathological states. Animal models, primarily rodents, have been widely used in exploring both developmental and neurobiological functions in the behavioral realm [6,7], but progress has been limited with rodent models, relative to primate models, for multiple reasons. First, the complexity of the central nervous system, primarily in cortical regions important for complex thought, cognition, affect, sociality, and decision-making, is not as elaborate in rodents as it is in primates, including humans [8]. The 80 million years since the last common ancestor between rodents and primates has led to divergence in the relative functional morphology and connectivity of the brain, in particular the expansion of the PFC in the primate lineage [9]. Second, the cognitive, social, and behavioral phenotypes of primates are much richer and more complex than those of rodents [10], presumably a result of the elaboration and patterns of connectivity among

The Common Marmoset in Captivity and Biomedical Research
https://doi.org/10.1016/B978-0-12-811829-0.00026-1

cortical regions of the primate brain. As a consequence, there is a premium in developing suitable NHP models from the perspective of face validity (behavioral phenotypes in primates are more likely to be similar to humans) and construct validity (the genetics, ontogeny, and neural architecture are likely to be homologous in primates and humans).

The common marmoset (*Callithrix jacchus*) represents an NHP species of growing interest in biomedical and behavioral research. Among the reasons for the increased use of this species are both practical and management issues. These include their small size (350–500 g), thereby reducing the need for large enclosures and animal facilities. Marmosets are also highly fecund, capable of producing two twin (or larger) litters per year. Relative to cercopithecine models (e.g., rhesus macaques) or hominoid primates (e.g., chimpanzees), the zoonotic risk posed by marmosets is low.

A more compelling case for the marmoset as a model in biobehavioral and neuroscience research derives from the scientific relevance of a host of features associated with the natural history and associated phenotypes of the species. Marmosets (and most other callitrichinae primates) are obligate twinning species, producing dizygotic twins. Twins are dizygotic or fraternal, and because of unique features of the development of placental and shared embryonic vasculature, twin fetuses exchange embryonic stem cells and multiple signaling molecules across these vascular anastomoses. As a consequence, marmoset twins are chimeric in hematopoietically derived tissues [11] and perhaps in other tissues as well [12]. Twinning is thus potentially important from the perspective of experimental design (e.g., differential twin phenotypes that develop in spite of a shared intrauterine environment [13] and in the exploration of immune–brain interactions [14–16]). From a developmental perspective, marmoset life span trajectories are considerably accelerated relative to other NHP models and humans, passing through the infant stage in several months, reaching sexual and social maturity by 18 months of age, and displaying morphological and behavioral signs of senescence within 6–8 years [17,18]. From the perspective of social neuroscience, the marmoset represents an exceptional model for human sociality [19]. The marmoset shares many features in the realm of social phenotypes with humans, including the fundamental social unit (small nuclear or extended family social groups [20]), offspring rearing dynamics (cooperative breeding, which entails shared infant care by mothers, fathers, and older offspring within the family group [21]), and the nature of adult heterosexual social relationships (consisting of many features associated with social monogamy [22]).

From the perspective of biobehavioral model development, there is a wealth of normative information on marmoset brain anatomy and function, including a well-annotated brain atlas that is continuously updated [23], critical information regarding the structure and connectivity of frontal cortices [24,25], a growing knowledge of sensory processing in a number of modalities [26–29], and details on motor control circuitry [30]. In addition, in the past 5 years a remarkable suite of tools in the neurosciences have been developed with and/or applied to the marmoset central nervous system. Silva [31] recently reviewed advances in imaging of the marmoset brain, and the methodologies include high-resolution structural MRI, functional connectivity among multiple brain regions during processing in a host of sensory modalities, and two-photon laser capture microscopy for monitoring neural activity at the level of individual neuronal cells. A host of techniques have been utilized to alter gene function in the marmoset brain (see review in Ref. [32]), including the production of transgenic marmosets relevant for a number of pathologies associated with the brain, lentiviral retrograde vectors for tract tracing in the brain, and shRNA silencing of gene expression in targeted regions of the brain. Finally, the "proof of concept" for an optogenetic preparation in the marmoset, via the induction of channelrhodopsin into multiple cortical regions of the brain [33], anticipates the potential to manipulate neuronal function in the marmoset brain within tight spatial and temporal boundaries. Thus, within the neurosciences the marmoset may well deserve its recent designation as a *"supermodel"* in the biomedical sense of the word [34].

Organization of this Review

This review of marmosets as behavioral models in biomedical and behavioral research is organized in line with the Research Domain Criteria (RDoC) taxonomy. The RDoC movement at the NIH began in light of decreasing rates of morbidity and mortality from a host of disease states (e.g., cardiovascular disease, cancer) as a consequence of investments in and knowledge gained in basic and clinical science, but mortality rates have remained unchanged for mental illness, and depending on the disorder, prevalence rates across recent decades have remained stable or have increased [35]. The RDoC as a research tool represents a move away from the standard clinical diagnostic protocols for psychological disorders [the American Psychiatric Association's DSM-V (www.dsm5.org) and the World Health Organization's ICD-10, www.who.int/whosis/icd10]. These protocols dictate the diagnosis of mental disorders based on clusters of behavioral symptoms, a procedure that likely poses a potential problem for fundamental research and translational insights into the underlying mechanisms of these disorders.

Two major problems have been identified with these approaches [35–38]. First, the breadth of the diagnostic criteria for any given disorder is wide, and any given diagnostic classification includes multiple symptoms, not all of which need be presented by a patient for a diagnostic decision. It is possible therefore that two individuals diagnosed with the same "mental" disorder would share no symptomology in common, and treating the underlying cause therefore becomes problematic. As an analogy to this issue from a nonpsychiatric context, a patient that presents at the clinic with the symptom "shortness of breath" would not be provided with a single treatment regimen without further diagnostic evaluation. A physician would need to identify the underlying cause of dyspnea. Once diagnosed, differential therapeutic treatments would be created, depending on the cause: heart failure, myocardial infarction, lung congestion, a broken rib, or a state of anxiousness. Second, there are multiple symptoms that are common in dramatically different diagnostic states. For instance, dysfunction in social communication constitutes one of the DSM-V diagnostic criteria for schizophrenia spectrum and other psychotic disorders, social anxiety disorder, and autism spectrum disorder. Given that the brain regions and the perceptual and motor circuitry underlying communication in humans is fairly well established, it would seem logical to look across diagnoses for the underlying etiology of the symptom(s) that are common across psychiatric diagnoses.

The RDoC construct is premised on the notion that looking for underlying pathology as a function of the broad DSM and ICD classifications has not proven to be useful. Instead, an inversion of the process may be more fruitful, that is, utilize the knowledge about normative brain function derived from the breadth of modern neuroscience (from genetics to cognition) and identify the degree to which these multiple mechanisms are disrupted in individuals suffering from "brain illness" [37]. While the utility of the RDoC in a diagnostic context remains unresolved and is a matter of considerable debate (e.g., Ref. [39]), there is a clear heuristic value of the RDoC approach in the context of basic research into the neural mechanisms underlying the symptoms associated with psychopathology. Beginning in 2009, the US National Institutes of Mental Health, under the guidance of the director of the institute, Thomas Insel, has developed and elaborated the RDoC construct to guide basic and clinical research in this area. The outline of the RDoC construct is shown in Table 26.1. The five major research "Domains," identified by multiple experts over several years, constitute important, broad areas of psychological and behavioral processes. Within each Domain are multiple constructs that nest within each Domain. What is critical from the RDoC perspective is that each of the constructs can be operationally defined in multiple model systems, can be measured in normative or in perturbed conditions, and can lead to testable, hypothesis-driven science regarding the brain circuitry underlying each construct and can be revised and validated according to standard scientific methods.

The literature review that follows will highlight significant or innovative behavioral protocols that have been developed or modified for use in marmosets and their relevance for RDoC domains. An exhaustive review of the all of the findings in each of the RDoC Domains is beyond the scope of this contribution, and the examples provided within each Domain are meant

TABLE 26.1 RDoC (Research Domain Criteria), NIMH[a]

Negative Valence Domain	**Positive Valence Systems**	**Cognitive Systems**	**Systems for Social Processes**	**Arousal/Modulatory Systems**
Acute threat ("fear")	Approach motivation	Attention	Affiliation and attachment	Arousal
Potential threat ("anxiety")	Initial responsiveness to reward	Perception	Social communication	Biological rhythms
Sustained threat	Sustained responsiveness to reward	Working memory	Perception and understanding of self	Sleep–wake
Frustrative nonreward	Reward learning	Declarative memory	Perception and understanding of others	
	Habit	Language behavior		
		Cognitive control		

Major domains (neurobehavioral systems) are listed in **bold**, and constructs (potentially operationalized and measured features of domains) are listed underneath each domain.
[a]*Cuthbert and Insel [35]; Insel [36]; Insel and Cuthbert [37].*

to highlight the ways in which the unique attributes of the marmoset model can be utilized to explore the neurobiological mechanisms underlying the constructs within each Domain. The RDoC approach is inspired by a recent review by Oikonomidis et al. [38], who applied the RDoC organization to marmoset models in a few of the RDoC Domains (especially Valence and Cognition Domain). This chapter will cover Negative Valence, Social Processes, and Arousal in some detail, and readers interested in a fuller discussion of Positive Valence and Cognitive Systems are referred to Oikonomidis et al. [38], for a more complete review. One final comment: considerable information on normative function in marmosets within each Domain is accessible from observations of routine behavior in undisturbed conditions. However, experimental manipulations in the realm of learning, motivation, cognition, sociality, and arousal and other aspects of marmoset behavioral biology yield greater confidence in the inference of causality, and hence the focus in this review will be primarily on experimental models.

MARMOSET BEHAVIORAL MODELS WITHIN THE CONTEXT OF RDOC

Negative Valence Systems

This domain is characterized by the processing of stimuli and generating responses to aversive stimuli and events, and the antecedents (anticipation) and consequences (long-term function) of these stimuli and events. The constructs within this domain include the activation of *fear* circuits associated with adaptive defensiveness from real or perceived danger and can be elicited by both interoceptive and exteroceptive stimuli. Furthermore, constructs also entail the psychological and physiological processing associated with potential aversive events in which the events are distant, ambiguous, or have a low or uncertain probability of occurring (*anxiety*). Further constructs focus on emotional states that result from a sustained or uninterrupted exposure to contexts or stimuli that are normally avoided (*sustained threat*), situations involving deprivation or separation from a significant physical or social resource (loss), and responses to the lack of positive outcomes in the face of efforts to produce them (*frustrative nonreward*).

Two common behavioral models are utilized in the exploration of the responses of marmosets to fear- and anxiety-inducing stimuli. The first involves exposure of marmosets to actual or simulated models of predators, hereafter referred to as the Marmoset Predator Confrontation Test (MPCT) [40,41]. In the northeastern Brazilian coastal forests, marmosets of the genus *Callithrix* are at predation risk from multiple predators, including aerial raptors, terrestrial felines, and terrestrial and tree-climbing snakes [42]. As such, these classes of predators constitute significant naturalistic threats and reliably elicit a cluster of responses even in captive-house marmosets, including modification in locomotory patterns like movement along substrate or jumps from location to location, piloerection, movement of the upper body back-and-forth on a stable substrate, and spatial proximity to the predator stimulus. A prominent feature of predator exposure is the production of species-specific vocalizations that are observed in wild marmosets [43]. Among those typically elicited by predatory stimuli are vocalizations labeled tsik, egg, peep, and twitter, all of which are carefully characterized and rigorously quantified by Pistorio et al. [44]. The second widely used model for fear and anxiety also involves confronting marmosets with an actual or potential stimulus, in this case a human standing in close proximity to the marmosets' cage (HIT—Human Intruder Test). While less ecologically relevant than the MPCT, this assessment tool relies on the notion that marmosets show fear to the presence of humans, either innately or as a consequence of routine captive husbandry or testing. Similar behavioral measures are employed in the HIT as in the MPCT.

From the perspective of test validation of these common methods, an explicit comparison of the responsivity of marmosets to the HIT and MPCT revealed that the two tests induce similar behavioral responses [45]. This investigation demonstrated a strong, positive correlation between the two methodologies. Furthermore, individual variability among marmosets in their response profiles remained highly stable in independent assessments separated by as much as 3.5 years. Neural and neurochemical correlates of high-responsive (low proximity to stimulus, high rates of tsik and tsik-egg calls) versus low-responsive marmosets were documented. High-responsive marmosets had lower levels of serotonin (5-HT) in the amygdala as measured by microdialysis and smaller structural volume of the anterior cingulate cortex (ACC), an important region for the integration of cognitive and emotional information. While the two methodologies are correlated, another side-by-side comparison of the two protocols revealed greater behavioral responsiveness of marmosets to the HIT than to MPCT [46].

The MPCT has provided insight into both the behavioral components associated with exposure to fear- and anxiety-inducing stimuli and further details on the neural mechanisms that underlie these behavioral elements. A multivariate principal component analysis (PCA) of differential behavioral responding by marmosets when confronted with predator stimuli or human intruders [47] revealed two significant components that account for significant variance: **emotionality**, a PCA dimension that loads heavily on visual avoidance of the threatening stimulus, reduced locomotion, and greater distance from the stimulus, and **coping**, which loads primarily on tsik and tsik-egg vocalizations. Support for the role of predator-induced vocalizations as a coping mechanism for marmoset derives from two findings. First, high rates of tsik vocalizations during predator exposure are associated with low baseline cortisol (as measured by hair cortisol [48]). Second, playback of recorded tsik calls to marmosets during social isolation (a procedure that typically elevates cortisol [49–51]) eliminates separation-induced cortisol responses [52]. A third potential behavioral component relevant to RDoC constructs (**anxiety**) derives from the consequences of repeated exposure to predatory stimuli, which reliably elicits elevated rates, relative to baseline or acute exposure to predatory stimuli, of scratching, self-grooming, and scent-marking behavior [53].

A third paradigm common in this arena is traditional aversive Pavlovian (classical) conditioning, in which an initially neutral conditioned stimulus (CS, e.g., a light, a soft tone, or an environmental context) is paired with an aversive unconditioned stimulus (US, e.g., a loud noise, an exteroceptive shock, a predatory stimulus), and with sufficient associative pairings, the CS acquires the ability to elicit conditioned fear or anxiety response. Marmosets easily acquire conditioned responses under these contexts, as measured by increased autonomic output (heart rate and blood pressure) and enhanced vigilance elicited by the previously neutral or irrelevant CS [54–57].

Additional behavioral models in the marmoset relative to the Negative Valence Domain have not received as much attention as the HIT and MPCT but have the potential to contribute to the development of unique measures of behavioral phenotypes and their underlying neurobiology. Among these methods are open-field testing, a modification of the rodent model in which behavioral indices of fear/anxiety as well as spatial navigation within the open field are assessed [58]. In another modification of a standard rodent model for fear/anxiety (the elevated plus maze), Wang et al. [59] tested location preferences of marmosets in a multiple-chambered testing box, in which some compartments had opaque walls and others had transparent walls. In general, marmosets showed a preference for the transparent chambers, and this technique may prove useful in protocols that induce anxiety-like states, either behaviorally or pharmacologically.

Most of these behavioral protocols have yielded significant insights into the basic neuroscience of Negative Valence constructs in the marmoset, including structural regions of the brain that are important regulators of responsiveness and in the neurotransmitter systems that modulate these brain circuits. Structurally, the orbitofrontal cortex (OFC) and PFC in marmosets appear to independently play an important role in modulating responses to threatening stimuli because bilateral lesions of either area enhanced emotional components and reduced coping components in an MPCT [47]. In the context of aversive Pavlovian conditioning, marmosets with lesions of the OFC and PFC maintain elevated conditioned HR responses during extinction trials (CS only, no US), especially in PFC-lesioned marmosets [54]. In the same study, animals lesioned in either OFC or PFC exhibited elevated emotional responses in the HIT, but reduced coping responses were noted only in the marmosets with PFC lesions, suggesting a structural separation of function for these components of emotional responding. Marmosets that are highly responsive in either HIT or MPCT have reduced volume of the ACC, an important region for the integration of cognitive and emotional information [45]. Finally, localized pharmacological manipulations in restricted brain regions highlight important regulatory circuit nodes. Local deactivation via a GABA agonist of Area 25 in the PFC reduced the strength of aversive Pavlovian conditioning, whereas deactivation of Area 32 enhanced some autonomic measures of conditioned responding [57].

Pharmacological manipulations or measurements in the context of this Domain have also produced findings of basic and potentially therapeutic importance. The $GABA_A$ agonist diazepam, a widely used anxiolytic treatment, clearly impacts behavioral and physiological measures in these paradigms and reveals the importance of the GABA system in regulating fear/anxiety. Marmosets given anxiolytic diazepam in the HIT exhibit reduced threat postures toward the intruder and increased proximity to the intruder, but administration of anxiogenics ($GABA_A$ inverse agonist or amphetamine) does not produce augmented behavioral reactivity in the HIT [60]. Diazepam also reduces behavioral indices of anxiety (phee calls, vigilance) and eliminates the normal preference for spending time in the periphery in the open-field test [58]. Simultaneous testing of marmosets in multiple behavioral paradigms has revealed a differential sensitivity of response characteristics to varying doses of diazepam. High doses of diazepam reduce responsiveness of marmosets in both the HIT and MPCT tests, whereas lower doses of diazepam are effective only in the MPCT model [46]. Treatment of marmosets with the inverse $GABA_A$ agonist FG-7142 increases tsik and tsik-egg vocalizations to comparable levels as those noted in novelty exposure and MPCT [61]. Other neurotransmitters no doubt play a role in the manifestation of fear/anxiety phenotypes in marmosets, including CRH [62], dopamine (DA) [63], and 5-HT [64].

A recent study has provided evidence of a genetic substrate for variation in responding in the behavioral protocols that tap into constructs within the Negative Valence RDoC domain. Santangelo and colleagues [64] genotyped marmosets from three sources—the Cambridge University colony, the NIH colony, and a wild population in Brazil. They documented variable number tandem repeat polymorphisms in several loci of the promoter region of the 5-HT transport allele, *SLC6A4*. On a population basis, a dinucleotide polymorphism in the third repeat and two single nucleotide substitutions in the fourth and 23rd repeat yield the haplotypes AC/C/G in 49.6% of the target colony and CT/C/C in 42.4% of a colony, with roughly similar proportions in the second colony and in wild marmosets. These polymorphisms have meaningful consequences, as qPCR revealed the highest SLC6A4 expression in the CT/C/C haplotype, the lowest in AC/C/G, and intermediate expression in heterozygotes for this polymorphism. Responses in the HIT revealed behavioral consequences of these polymorphisms as well, with AC/C/G marmosets displaying the highest anxiety and lowest coping scores and CT/C/C marmosets exhibiting low anxiety and high coping. Clinical relevance for the use of genetic information for tailoring "personalized" pharmacological treatments for patients on the basis of genotype was demonstrated in the marmoset model by the observation that the selective serotonin uptake inhibitor citalopram enhanced HIT scores in AC/C/G marmosets, but reduced HIT scores in CT/C/C marmosets.

Positive Valence and Cognitive Systems

This section describes behavioral models in two Domains because the distinction among them is not as clear in animal models as they are in the context of human phenotypes. The Positive Valence System addresses constructs associated with unconditioned and learned/adaptive responses to positive motivational stimuli or context. Among the constructs are processes associated with approach tendencies to innate or learned stimuli (*approach motivation*), reward-seeking behavior and hedonic responses to rewards (*initial responsiveness to reward attainment*), and the biobehavioral consequences of attaining reward (*sustained/longer-term responsiveness to reward attainment*). Learning and cognitive processes are reflected by capacities in reinforcement learning and differentiating among stimulus-reinforcement outcomes (*reward learning*), and the persistence of behavioral responses or cognitive processes associated with reward without excessive cognitive resources and/or in the absence of changes in reinforcement outcomes (*habit*). The Cognitive Systems Domain addresses the continuum of processes associated with detecting environmental stimuli (*attention*), peripheral and central computational processing of these stimuli in single and multiple sensory domains (*perception*), the ability to integrate, via cognitive and affective systems, environmental contexts and select adaptive responses to these contexts (*cognitive control*), and the encoding, maintenance, and recall of relevant information for specific tasks or goal outcomes (*working memory*). The constructs of *declarative memory* and *language* (symbolic representation) will not be considered in this review. A common protocol for generating experimental protocols assessing learning and cognition in marmosets, derived from the Cambridge Neuropsychological Test Automated Battery (CANTAB), can be found in Spinelli et al. [65]. Details on training marmosets to provide response selections using touch-sensitive computer screens can be found in Takemoto et al. [66].

Among the most straightforward tasks for assessing the reinforcing property of stimuli for an organism is the conditioned place preference (CPP). Typically a specific environmental context (chamber, room, or other environmental cue) is repeatedly paired with the presentation of a potential rewarding stimulus, whereas a second distinct context does not lead to the stimulus. The subsequent development of a preference for the context that predicts the delivery of the stimulus is a gauge of the positive rewarding property of the stimulus, whereas avoidance of the context is taken as evidence that the stimulus is a negative reinforcer [67]. Marmosets form a CPP for several reinforcing stimuli, including cocaine [68] and sweet/high-fat food (chocolate) [69].

Reversal learning is a common paradigm for assessing generalized discrimination learning, and in particular to assay response persistence in the face of altered reward consequences. Typically marmosets are presented with two stimulus objects, one associated with reward and the other nonrewarded. Once criterion performance (e.g., 75% choice of the correct stimulus) on that set of stimuli is established, reward contingencies are reversed; the responding to the previously unrewarded stimulus now leads to reward, and vice versa. This protocol has identified both structural and neurochemical correlates of performance in the task. Different neurotransmitters in the medial caudate nucleus appear to be critical for successful reversal learning in marmosets because depletion of DA in the caudate disrupts successful reversal learning, whereas depletion of 5-HT does not [70]. Lesions of the medial striatum and OFC also disrupt rapid reversal learning, whereas amygdala lesions do not affect the transition to new stimulus—reward contingencies [71]. Lacreuse and colleagues have addressed the role of sex steroid hormones in modulating learning processes in marmosets in the context of the reversal task. Estradiol appears to disrupt reversal learning performance because relative to untreated ovariectomized (OVX) females, OVX females given estrogen replacement therapy (ERT) took more trials to reach criterion and exhibited a higher number of errors in the task [72]. With repeated exposures to sets of reversed reward contingencies, control females made fewer errors to reach criterion performance, whereas estrogen-treated females made more errors. Hormone replacement appears to have sex-specific effects because untreated and testosterone-treated male marmosets did not differ on reversal learning performance [73].

Marmosets also perform well in the more complex reversal paradigm, probabilistic discrimination learning and reversal. In this task, two stimuli are not associated with reward certainty (e.g., selection of stimulus A leads to reward [or nonreward] 100% of the time). Rather, both stimulus cues are associated with differential probabilities of reward, in which responding to the "correct" stimulus leads to reward in 80% of the trials and nonreward in 20%, and to the "incorrect" stimulus leads to reward in 20% of trials and nonreward in 80% of trials. Performance on this task is related to serotonergic activity as demonstrated by site-specific depletion of 5-HT by the local administration of 5,7-DHT [74]. Reducing 5-HT in either the amygdala or the OFC inhibited the acquisition of the initial task and performance in the reversal phases of the task. Reducing 5-HT in the amygdala altered the marmoset's sensitivity to reward contingency, reducing responding to the "correct"' stimulus and enhancing responding to the incorrect stimulus, which results in overall lower reward density for the marmosets. These data suggest that the amygdala and OFC are critical for learning and adjusting to changing reward outcomes, and that lower 5-HT in the amygdala alone serves to mediate complex probabilistic reinforcement contingencies within the context of reversal learning.

A second common behavioral task within these Domains is the barrier-reach task. In the typical experiment, food is made available to the marmoset that is available by performing a fixed motor reach response (e.g., reaching from the right or left side). This is accomplished in a number of ways, but placing food inside a transparent or opaque cube with a single open face to access the food is common. Once a motor response is established by the marmoset in food acquisition, the reaching response required to access the food is altered by modifying the orientation of the cube (e.g., shifting the open face from the left to the right). This task is particularly important for assessing response flexibility (good performance) and perseverative tendencies (poor performance). Performance on this task in marmosets is reduced by drugs known to affect neurotransmitter systems involved in schizophrenia in humans (the DA D3 receptor agonist PD-128,907 and the NMDA-receptor antagonist ketamine) and enhanced by treatment of marmosets with blonanserin, a pharmacological agent that antagonizes DA D2/3 receptors and blocks 5-HT2A receptors [75] but is insensitive to estrogen replacement in OVX female marmosets [72].

In many cases marmosets are useful for experiments that explore the interstices between Negative, Positive, and Cognitive Domains. Shiba et al. [56] is a good exemplar. Marmosets were initially classified as low versus high reactive to a fear/anxiety stimulus in the standard Pavlovian model described above. Marmosets were then presented with two behavioral tasks: one uses the reward value of a food as a discriminative stimulus with unexpected reward outcomes (an OFC-dependent task) and the second task was the barrier-reach task described earlier in this section, which is a PFC-dependent task. In the first task, marmosets were presented with two transparent boxes, within which were located either an unreachable high- or low-preference food item. The reward contingencies for touching either box yielded reversed reward outcomes—touching the box containing the highly preferred was not rewarded, whereas touching the box with the low-preference food item led to the delivery of highly rewarding syrup bread. Thus, the high probability response of touching the box containing the highly preferred food had to be inhibited to receive a reward. The magnitude of conditioned fear and anxiety in the Pavlovian phase of the experiment predicted performance on the two tasks. More perseverative error in the first cognitive task (touching the high-preference food box) was associated with poorer conditioned vigilance in the Pavlovian task, whereas more perseverative error in the barrier-reach task was associated with lower baseline blood pressure in the Pavlovian trials.

Systems for Social Processes

The scope of this Domain is self-evident from its title and includes multiple components of features associated with conspecific sociality. Among these are the quality and patterning of affiliative social interactions and the consequent development of social bonds (*affiliation and attachment*) and the production, perception, interpretation, and responses to the social stimuli that mediate these interactions and sustain the social bonds (*social communication*). Awareness of one's role in social interactions is reflected in the constructs associated with self-awareness and self-agency (*perception and understanding of self*) and in the ability to recognize the identity of partners, their capacity for agency, and to predict or interpret a partner's mental state (*perception and understanding of others*).

Affiliative processes in marmosets have been explored, including caregiver–offspring attachment and "pair-bonding" in adult males and females. Within the context of caregiver–offspring affiliation, the focus has been on two broad areas: features associated with regulatory processes for caregiver interactions with infants and the consequences of variation in the quality and quantity of caregiving on subsequent offspring development. With regard to the regulation of parental care, correlational studies have implicated a suite of neuroendocrine modulators of parental care, including estradiol, testosterone, glucocorticoids, prolactin, and oxytocin (OT) [76–79]. Experimental models to explore parental motivation include instrumental learning in which the reward state is access to infant stimuli [80] and exposure and response to infants or infant-related stimuli [81–84]. These tasks have revealed that sensitivity to, and interest in, infants and their stimuli are regulated by multiple neuroendocrine systems, and that the behavioral measures are often contingent on age or previous exposure to infants within the social group.

The prominent experimental model to explore the neurobiological and behavioral sequelae to caregiver–infant attachment is the caregiver deprivation model championed by Pryce and colleagues. This model involves short-term, 30–120-min separations of infant marmoset from caregivers from 2 to 28 days of age [85], and the consequence of this disrupted caregiver–infant affiliation is widespread across multiple physiological systems, including behavioral, reward, and cognitive processing, responses to stressors, and gene expression in multiple neurotransmitter systems [86,87]. Less-invasive models include those that examine the impact of normative variation in offspring care on subsequent biobehavioral development. These models demonstrate that "adverse" patterns of early caregiving (e.g., high rates of infant rejection and transfer among caregivers) also produce long-lasting changes in neurobiology and behavior [88–90].

While there is considerable academic argument regarding whether marmosets are strictly "socially monogamous" (e.g., Ref. [91], and articles therein), there is no argument that the close social relationships between adult male and female marmosets, characterized by coordinated activities, high levels of grooming and proximity, cosharing of infant care responsibilities, and joint defense of territories [92], are distinct from other primates considered "nonmonogamous" [53]. Several behavioral protocols have been used to evaluate the establishment, persistence, and strength of this affiliative relationship. The first paradigm is partner preference testing in which marmosets can choose to spend time and/or interact with partners or opposite-sex strangers in a simultaneous choice task [93–95]. With regard to translational potential, the sensitivity of partner preference to environmental and social context should be considered a strength, given the complexities of human social relationships. Partner preference and pair-directed affiliation in marmosets is sensitive to OT manipulations, in that treatment with an OT antagonist reduces affiliative behavior in newly paired marmosets [95] and reduces interest and interactions with an unfamiliar opposite-sex partner in well-established pairs [94]. In addition, correlational studies have revealed that closely bonded family members, including breeding pairs, exhibit highly correlated levels of urinary OT [96], providing additional support for nonapeptide involvement in social affiliation in marmosets.

The second task for assessing pair-related affiliation and attachment involves measuring the behavioral and physiological responses to temporary separation from partners [49,51,97,98]. In heterosexual pairs, partner separation and housing in a novel environment is associated with elevated cortisol, and the presence of the partner [51], but not an opposite-sex stranger [98], can reduce or buffer the stress response in this context. In fact, presentation of phee calls from the partner but not calls from an opposite-sex stranger during separation also buffers the stress response [49]. In long-term same-sex pairs of marmosets, separation stress is buffered by both partners and same-sex strangers in males, but the presence of neither partners nor strangers served to buffer cortisol responses to separation in female marmosets [97]. This would seem an appropriate model system to explore dynamic changes in the socially mediated changes in the HPA axis, and the role of the hippocampus, temporal cortex, and PFC in translating these social buffering effects into physiological and behavioral regulations.

Marmosets have a rich vocal repertoire that mediates a host of intra- and intergroup social interactions. Knowledge of the information content of these social signals and the ways in which they are centrally evaluated in the brain can serve as a useful base to explore normative and atypical vocal communication processes. A standard protocol for assaying information content in marmoset calls involves the playback of recorded marmoset vocalizations and documenting behavioral differences in response to these calls. Information content regarding signalers in marmoset phee calls was experimentally evaluated by Smith et al. [99]. Normative acoustic characteristics of adult phee calls were quantified, and sex differences were noted in numerous acoustic parameters. Synthesized vocalizations that included the important differences in frequency components of the calls were generated, and the responses of marmosets to playback of synthetic and natural calls did not differ; both calls elicited enhanced vigilance, and vigilance was greater for male-typical than female-typical calls. Systematic manipulation of individual components of call structure eliminated differential responding to synthetic phee calls, suggesting that perception and interpretation of the sex of the caller requires a holistic, multivariate assessment of call characteristics. Similar work has been conducted on natural versus synthetic calls in multiple call types, with neural confirmation that these two classes of stimuli are processed similarly [100]. The potential for studying the behavioral and neural processing of acoustic social signals in real time in marmosets has been enhanced by the development of a sophisticated behavioral model of communicative "conversations" [101,102]. In this model, recorded exemplars of marmosets are played back to subjects with varied probabilities and timing, as "virtual marmosets" in an interaction. The patterning of the vocalization by the real marmoset is contingent on these manipulations in calls from the virtual marmosets [103]. Given that vocal exchange patterns appear to be learned during ontogeny in marmosets [104], this behavioral technique has implications for both adult vocal interactions and vocal development.

Several groups have utilized the expression of immediate early gene (IEG) products, signals of early transcriptional activity within neurons that are responsive to stimuli, to identify brain regions associated with the perception and processing of species-specific acoustic signals in marmosets. Marmosets listening to species-specific vocalizations have differential IEG expression in regions of the brain that differ by task (listening to calls, producing calls, and engaging in antiphonal calling). Perception of calls is associated with IEG expression (cFOS) in the ventrolateral PFC and regions of the auditory cortex: producing calls is associated with distinct activation in the premotor cortex, and vocal exchanges activate the perirhinal cortex in addition to the areas previously mentioned [105]. Using a different IEG marker (Egr-1), Simões et al. [106] demonstrated that vocal exchanges also enhanced IEG expression not only in the PFC but also in the ACC. Single-cell recordings in the premotor cortex also support the role of this region in vocal production in marmosets [107]. For a more complete review of the neurobiology of marmoset vocal communication see Refs. [108,109].

Given that marmosets utilize scent-marking behavior extensively in social interactions and have individual- and sex-specific olfactory "signatures" [110,111], the neurobiology underlying olfactory communication would appear ripe for exploitation. fMRI studies have revealed enhanced BOLD signaling in hypothalamic regions of male marmosets exposed to the odors of sexually receptive females [112] and in cortical and subcortical regions associated with reward and emotional processes [113], suggesting a complex integration of information across the brain. The development of a behavioral conditioning model demonstrating that neutral stimuli can be conditioned with sexually arousing female stimuli in male marmosets [114] anticipates future sophisticated experimental assessments of social olfaction and its peripheral and central processing in marmosets.

Evidence for the ability of marmosets to perceive and understand others derives from several behavioral protocols. A fundamental demonstration that marmosets can perceive that other conspecifics have agency and goals derives from a paradigm in which marmosets watch video presentations of one of three actors spending time with and exploring one of two distinct objects [115]. After a short gap, marmosets are presented with a second video, with the actor either interacting with the same object but in a different location (expected outcome) or interacting with the second object that is now in the same location as the object of interest in the first video (unexpected). As with looking time at visual stimuli in human infants, marmosets spend more time looking at the unexpected than the expected object. The critical manipulation in this experiment is the nature of the model in the video: a marmoset, a quadrupedal marmoset-ish robot, or a black box. As expected, marmosets spent more time looking at the unexpected outcome, but only when the actor was a marmoset or the robot. When given access to the actual objects portrayed in the video, marmosets spent more time with the "correct" object but again only when the video actor was a marmoset or robot, demonstrating social learning only when the model possessed "agency."

Marmosets provide an excellent opportunity to study the perception of others in paradigms that require two or more partners to acquire or distribute food rewards, which speaks to the importance of social life in many aspects of the behavioral biology of this species. Several tasks have been employed that require cooperation or altruism among interacting marmosets, in which both marmosets must respond cooperatively to access food items [116] or act "altruistically" to provide food to others, while not receiving rewards themselves, either in a dyadic context [117,118] or in a context in which an individual provides support to multiple group members [119,120]. Marmosets easily learn to perform cooperative tasks, although like many features of marmoset biology, the ease with which cooperation occurs varies as a function of relatedness and status of cooperating partners [116]. Marmosets do exhibit "other-regarding" sharing of food in the altruistic food-sharing paradigms, although again the nature of social relationships between the donors and recipients, and the social role of the "donor" in the family group determines the probability of altruistic responding [117,118,120]. OT is related to these measures of prosociality. The degree of OT synchrony among dyad members predicts levels of prosociality, with higher dyadic synchrony in OT levels correlated with a higher likelihood of provisioning food to the partner [120]. Experimental manipulation of OT paints a different story; OT reduces sharing in both adult males and females with unfamiliar opposite-sex recipients but does not alter altruistic sharing with long-term pairmates [118]. It is certainly the case that OT synchrony measured in urinary samples provides a different index of OT activity than pharmacological manipulation of OT, but nonetheless both recent papers point to an important role for OT in cooperative social behavior. These paradigms have the potential to serve as springboards for more detailed explorations of neuroendocrine mediation of complex social traits such as cooperation and altruism and the cortical decision-making circuits that regulate complex decision-making in a social context.

The ability to perceive and understand self-awareness and agency is a difficult concept to study in pre- and nonverbal organisms, and this certainly applies to marmosets. In spite of its faults (of which there are many, e.g., Ref. [121]), the predominant test in NHP for self-awareness remains the "mirror test" [122]. While no studies have been published regarding the performance of common marmosets on this task, two papers have appeared on other closely related callitrichinae species. Pygmy marmosets with lengthy exposure to mirrors in their home cages showed no evidence of mirror-mediated self-directed behavior, although they appeared to use the mirrors to gain visual access to neighboring groups and exhibited aggressive displays toward individuals in these groups [123]. Thus, while pygmy marmosets show no evidence of self-awareness, they "understand" or at least utilize visual information provided by mirrors. Likewise, cotton-top tamarins with long-term exposure to mirrors spent considerable time looking into the mirrors, displayed repeated actions in front of the mirrors, but showed no evidence of self-directed behavior that appeared to be mirror-guided [124]. These negative findings and absence of findings point to two potential interpretations: (1) like most nonhominoid primates [122], marmosets do not possess the cognitive complexity to express self-awareness or (2) we as investigators lack the cleverness and sophistication to assess this question.

Arousal/Modulatory Systems

The final Domain covers neural systems associated with the coordination of responses to environmental stimuli to achieve a homeostatic state or to effect appropriate responses to environmental change and hence an appropriate allostatic state. Constructs in this Domain include generating an appropriate physiological state for the current interoceptive or exteroceptive context (arousal), maintaining appropriate 24-h patterns across multiple organ systems and behavior (circadian rhythms), and the organization of sleep/wake states in the service of optimizing physiological and behavioral functions (sleep and wakefulness).

Marmosets have proven to be excellent models for photic and nonphotic regulation of the circadian rhythm and represent a model species that is diurnal or crepuscular, as opposed to the commonly used nocturnal rodent models. The free-running cycle (τ) in this species under constant 24-h dim lighting conditions has been estimated in multiple studies (e.g., Refs. [125–127]), and with little variation it has been established as 23.3 h. Like other species, entrainment of the circadian cycle under constant 24-h illumination can be accomplished by photic stimulation with a variety of frequency and timing manipulations [125,126]. What is particularly compelling from an RDoC perspective is the sensitivity of circadian rhythmicity to developmental and social variables. There are important age and sex differences in circadian activity profiles in marmosets, with juveniles more active than adults through the active phase, adults showing earlier activity onset than juveniles, adult males showing the earliest phase-onset and -offset for the active period, and a shifting to later activity onset in postpubertal than prepubertal marmosets [128]. Daily rhythms appear to be exquisitely sensitive to social contexts. Entrainment of activity profiles is more tightly coupled in individuals within a family group than between family groups, and the strongest coupling occurs between sibling twin pairs, followed next most closely by male–female breeding pairs [129]. While direct and preferential social interaction, cofeeding, and coincidental allogrooming could be among the mechanisms by which tight activity coupling is mediated, there is also evidence that vocal communication signals can also play an important role. Rates of phee calling in marmosets exhibit a bimodal distribution during the light phase, with the highest rates occurring shortly after light onset. Entrainment to "light onset" can occur in marmosets housed indoors under constant 24-h dim light conditions, under conditions where they can hear phee calls from marmosets exposed to normal light cycles [127].

Marmosets also appear to be a useful model for exploring the circadian and thermoregulatory changes that are associated with menopause in women [130]. Gervais and colleagues [131] utilized implanted biotelemetry to monitor sleep quality (assessed by EEG) and core body temperature in OVX female marmosets before and after ERT. Both high and low doses of estrogen (6 or 12 μg/kg day) significantly reduced core body temperature during the night phase. EEG analyses revealed fewer nighttime arousals and higher delta wave power, both indices of better-quality sleep.

Circulating plasma cortisol concentrations in marmosets follow the circadian pattern typical for diurnal mammals, with rising concentrations as the light phase approaches, highest concentrations in samples collected shortly after activity onset in the morning, with falling concentrations throughout the day until the nadir 2–4 h after the onset of the dark phase [132]. Similar patterns derive from samples that can be collected noninvasively, including urine [50], saliva [133], and fecal samples [134]. The timing and patterning of circadian cortisol rhythms appear to be sensitive to social context [89,90]. Morning cortisol concentrations in isolated, singly housed marmosets and marmosets in unstable groups with high aggression are higher than those in marmosets housed in stable breeding pairs or family groups. Furthermore, rather than exhibiting the normative reduction in cortisol in the afternoon, cortisol concentrations remain elevated in those marmosets living in unstable or isolated social contexts, and measures of behavioral arousal are strongly correlated with afternoon cortisol levels. In light of the fact that of the multiple parameters that can be measured in the HPA axis, the biggest predictor of risk for depression and other disorders is elevated afternoon cortisol, altered circadian cortisol in marmosets may therefore be a useful proxy measure in models for depression and psychopathology.

Assessing circadian rhythms in marmosets demonstrates the ways in which behavioral models can cut across RDoC domains in meaningful ways, including Arousal and Positive Valence. Marmosets not surprisingly form a CPP for a context and location in which food is available [135]. In an interesting twist, the strength of the CPP was contingent on a match between the time at which associative learning trials were conducted (either in the early morning or late afternoon) and when CPP was assessed. CPP performance was robust when the time of assessment matched the time of training but was effectively eliminated when the times did not match. This result suggests that the phase of the circadian cycle forms an important part of the broad stimulus properties (interoceptive and exteroceptive) to which reward learning in marmosets is sensitive.

SUMMARY

The utility of the marmoset model in behavioral neuroscience is borne out by an NCBI PubMed search on *Marmoset AND (Brain OR Behavior)*, which yields over 1900 citations, 1300 of which have been published since the year 2000. Interpreting this vast literature is a daunting task, but one that is given structure and becomes more heuristically valuable by adopting the RDoC classification matrix, as illustrated by Oikonomidis et al. [38] and this chapter. This review indicates that explorations of the nature of brain–behavior relationships in the marmoset have the potential to speak to all Domains and most constructs in the RDoC framework. This is especially so in the case of the Social Processes Domain, given the analogous similarities in the social systems of the marmoset with those of humans and the homologies between the human and the nonhuman primate brain. The unique position of the marmoset in this regard has been recognized by others [19,136–138].

This review has identified experimental behavioral models that have been employed in the service of understanding the ways in which the brain generates behavioral and social patterns in marmosets and also the ways in which brain structure and function in turn are altered by environmental and social experiences. The degree of sophistication of the neurobiological methods that have been employed vary primarily as a function of the degree to which the neurobiological measures require invasive or restrictive protocols (e.g., single-cell recording, restraint within an fMRI) or can be conducted with minimally or noninvasive protocols (e.g., measurement of neuropeptide metabolites in urine; oral or nasal administration of neuropeptides). Obviously, the degree of resolution with regard to neural function and organization varies as a consequence of the method used, but so, too, does the ability to study complex, interactive behavior in a complex, interactive, and dynamic social group. In a real sense, then, a form of Heisenberg's Uncertainty Principle applies in this context. Simply paraphrased, Heisenberg pointed out that two key components were required to understand some features of atomic physics—a particle's location and its momentum. By measuring either location or momentum, the ability to measure the other was compromised. To complete the analogy as it applies in the current context, methodologies that provide greater precision, resolution, and timing of neural activity that mediate complex affective, cognitive, and social processes require less complexity and sophistication in the behavioral systems that are under study. This conundrum echoes the cogent arguments of Krakauer et al. [139] that generating meaningful basic and translational research in neuroscience requires that experiments be designed and carried out in light of species-specific evolutionary adaptations, with an eye toward species-specific sensory–motor capacities, and an emphasis on appropriately complex social contexts.

The new methodological innovations that were highlighted at the outset of this chapter may provide a way to reduce the Uncertainty Principle as it applies to marmoset brain–behavior relationships. These include the potential to remotely activate selective neural circuits in freely behaving marmosets via optogenetic stimulation or to alter gene expression during development via transgenic models or in adulthood via shRNA silencing. In that sense, the utility of the marmoset model in behavioral research will likely grow in its stature as an important transitional and translational biomedical model in the neurosciences increases—a 'supermodel', indeed!

Acknowledgments

The preparation of this chapter and the research by the author described herein were supported in part by funds from the National Institutes of Health (HD04882, HD089147).

References

[1] Whiteford HA, Degenhardt L, Rehm J, Baxter AJ, Ferrari AJ, Erskine HE, Charlson FJ, Norman RE, Flaxman AD, Johns N, Burstein R, Murray CJ, Vos T. Global burden of disease attributable to mental and substance use disorders: findings from the Global Burden of Disease Study 2010. Lancet 2013;382:1575–86. https://doi.org/10.1016/S0140-6736(13)61611-6.

[2] Walker ER, McGee RE, Druss BG. Mortality in mental disorders and global disease burden implications: a systematic review and meta-analysis. JAMA Psychiatry 2015;72:334. https://doi.org/10.1001/jamapsychiatry.2014.2502.

[3] Perelman P, Johnson WE, Roos C, Seuánez HN, Horvath JE, Moreira MA, Kessing B, Pontius J, Roelke M, Rumpler Y, others. A molecular phylogeny of living primates. PLoS Genet 2011;7:e1001342.

[4] Geschwind DH, Rakic P. Cortical evolution: judge the brain by its cover. Neuron 2013;80:633–47. https://doi.org/10.1016/j.neuron.2013.10.045.

[5] Dunbar RI, Shultz S. Evolution in the social brain. Science 2007; 317:1344–7.

[6] Nestler EJ, Hyman SE. Animal models of neuropsychiatric disorders. Nat Neurosci 2010;13:1161–9. https://doi.org/10.1038/nn.2647.

[7] Stewart AM, Kalueff AV. Developing better and more valid animal models of brain disorders. Behav Brain Res 2015;276:28–31. https://doi.org/10.1016/j.bbr.2013.12.024.

[8] Belmonte JCI, Callaway EM, Caddick SJ, Churchland P, Feng G, Homanics GE, Lee K-F, Leopold DA, Miller CT, Mitchell JF, Mitalipov S, Moutri AR, Movshon JA, Okano H, Reynolds JH, Ringach DL, Sejnowski TJ, Silva AC, Strick PL, Wu J, Zhang F. Brains, genes, and primates. Neuron 2015;86:617–31. https://doi.org/10.1016/j.neuron.2015.03.021.

[9] Wise SP. Forward frontal fields: phylogeny and fundamental function. Trends Neurosci 2008;31:599–608. https://doi.org/10.1016/j.tins.2008.08.008.

[10] Kaiser T, Feng G. Modeling psychiatric disorders for developing effective treatments. Nat Med 2015;21:979–88. https://doi.org/10.1038/nm.3935.

[11] Sweeney CG, Curran E, Westmoreland SV, Mansfield KG, Vallender EJ. Quantitative molecular assessment of chimerism across tissues in marmosets and tamarins. BMC Genom 2012; 13:98.

[12] Ross CN, French JA, Ortí G. Germ-line chimerism and paternal care in marmosets (*Callithrix kuhlii*). Proc Natl Acad Sci 2007; 104:6278–82.

[13] French JA, Smith AS, Gleason AM, Birnie AK, Mustoe A, Korgan A. Stress reactivity in young marmosets (*Callithrix geoffroyi*): ontogeny, stability, and lack of concordance among cotwins. Horm Behav 2012;61:196–203.

[14] Genain CP, Hauser SL. Experimental allergic encephalomyelitis in the New World monkey *Callithrix jacchus*. Immunol Rev 2001;183:159–72.

[15] Neehus A-L, Wistuba J, Ladas N, Eiz-Vesper B, Schlatt S, Müller T. Gene conversion of the major histocompatibility complex class I Caja-G in common marmosets (*Callithrix jacchus*). Immunology 2016;149:343–52. https://doi.org/10.1111/imm.12652.

[16] Rölleke U, Flügge G, Plehm S, Schlumbohm C, Armstrong VW, Dressel R, Uchanska-Ziegler B, Ziegler A, Fuchs E, Czéh B, Walter L. Differential expression of major histocompatibility complex class I molecules in the brain of a New World monkey, the common marmoset (*Callithrix jacchus*). J Neuroimmunol 2006;176:39–50. https://doi.org/10.1016/j.jneuroim.2006.04.015.

[17] Abbott DH, Barnett DK, Colman RJ, Yamamoto ME, Schultz-Darken NJ. Aspects of common marmoset basic biology and life history important for biomedical research. Comp Med 2003;53:339–50.

[18] Schultz-Darken N, Braun KM, Emborg ME. Neurobehavioral development of common marmoset monkeys. Dev Psychobiol 2016;58:141–58.

[19] Burkart JM, Hrdy SB, Van Schaik CP. Cooperative breeding and human cognitive evolution. Evol Anthropol Issues News Rev 2009;18:175–86.

[20] Nievergelt CM, Digby LJ, Ramakrishnan U, Woodruff DS. Genetic analysis of group composition and breeding system in a wild common marmoset (*Callithrix jacchus*) population. Int J Primatol 2000;21:1–20.

[21] Erb WM, Porter LM. Mother's little helpers: what we know (and don't know) about cooperative infant care in callitrichines: ERB and PORTER. Evol Anthropol Issues News Rev 2017;26:25–37. https://doi.org/10.1002/evan.21516.

[22] French JA, Cavanaugh J, Mustoe AC, Carp SB, Womack SL. Social monogamy in nonhuman primates: phylogeny, phenotype, and physiology. J Sex Res 2018;55:410–34 (in press).

[23] Okano H, Mitra P. Brain-mapping projects using the common marmoset. Neurosci Res 2015;9:3–7. https://doi.org/10.1016/j.neures.2014.08.014.

[24] Burman KJ, Rosa MGP. Architectural subdivisions of medial and orbital frontal cortices in the marmoset monkey (*Callithrix jacchus*). J Comp Neurol 2009;514:11–29. https://doi.org/10.1002/cne.21976.

[25] Roberts AC, Tomic DL, Parkinson CH, Roeling TA, Cutter DJ, Robbins TW, Everitt BJ. Forebrain connectivity of the prefrontal cortex in the marmoset monkey (*Callithrix jacchus*): an anterograde and retrograde tract-tracing study. J Comp Neurol 2007; 502:86–112. https://doi.org/10.1002/cne.21300.

[26] Eliades SJ, Wang X. Neural substrates of vocalization feedback monitoring in primate auditory cortex. Nature 2008;453:1102–6.

[27] Mitchell JF, Leopold DA. The marmoset monkey as a model for visual neuroscience. Neurosci Res 2015;93:20–46. https://doi.org/10.1016/j.neures.2015.01.008.

[28] Moriya-Ito K, Tanaka I, Umitsu Y, Ichikawa M, Tokuno H. The olfactory bulb and the number of its glomeruli in the common marmoset (*Callithrix jacchus*). Neurosci Res 2015;93:158–63. https://doi.org/10.1016/j.neures.2014.12.007.

[29] Solomon SG, Rosa MGP. A simpler primate brain: the visual system of the marmoset monkey. Front Neural Circuits 2014;8. https://doi.org/10.3389/fncir.2014.00096.

[30] Bakola S, Burman KJ, Rosa MGP. The cortical motor system of the marmoset monkey (*Callithrix jacchus*). Neurosci Res 2015;93: 72–81. https://doi.org/10.1016/j.neures.2014.11.003.

[31] Silva AC. Anatomical and functional neuroimaging in awake, behaving marmosets: neuroimaging in Awake. Behaving Marmosets Dev Neurobiol 2017;77:373–89. https://doi.org/10.1002/dneu.22456.

[32] Watakabe A, Sadakane O, Hata K, Ohtsuka M, Takaji M, Yamamori T. Application of viral vectors to the study of neural connectivities and neural circuits in the marmoset brain. Dev Neurobiol 2016.

[33] MacDougall M, Nummela SU, Coop S, Disney A, Mitchell JF, Miller CT. Optogenetic manipulation of neural circuits in awake marmosets. J Neurophysiol 2016;116:1286–94.

[34] Cyranoski D. Marmoset model takes centre stage. Nature 2009; 459:492–492. https://doi.org/10.1038/459492a.
[35] Cuthbert BN, Insel TR. Toward the future of psychiatric diagnosis: the seven pillars of RDoC. BMC Med 2013;11. https://doi.org/10.1186/1741-7015-11-126.
[36] Insel T. The NIMH research domain criteria (RDoC) project: precision medicine for psychiatry. Am J Psychiat 2014;171:395–7.
[37] Insel TR, Cuthbert BN. Brain disorders? Precisely Sci 2015;348: 499–500. https://doi.org/10.1126/science.aab2358.
[38] Oikonomidis L, Santangelo A, Shiba Y, Clarke H, Robbins TW, Roberts AC. A dimensional approach to modelling symptoms of neuropsychiatric disorders in the marmoset monkey. Dev Neurobiol 2016.
[39] Lilienfeld SO, Treadway MT. Clashing diagnostic approaches: DSM-ICD versus RDoC. Annu Rev Clin Psychol 2016;12:435–63. https://doi.org/10.1146/annurev-clinpsy-021815-093122.
[40] Barros M, Major RS, Huston JP, Tomaz C. Predatory stress as an experimental strategy to measure fear and anxiety-related behaviors in non-human primates. Rev Neurosci 2008;19:157.
[41] Barros M, Tomaz C. Non-human primate models for investigating fear and anxiety. Neurosci Biobehav Rev 2002;26:187–201.
[42] Ferrari SF, Ferrari MAL. Predator avoidance behaviour in the buffy-headed marmoset, *Callithrix flaviceps*. Primates 1990;31: 323–38. https://doi.org/10.1007/BF02381104.
[43] Bezerra BM, Souto A. Structure and usage of the vocal repertoire of *Callithrix jacchus*. Int J Primatol 2008;29:671–701. https://doi.org/10.1007/s10764-008-9250-0.
[44] Pistorio AL, Vintch B, Wang X. Acoustic analysis of vocal development in a New World primate, the common marmoset (*Callithrix jacchus*). J Acoust Soc Am 2006;120:1655–70.
[45] Mikheenko Y, Shiba Y, Sawiak S, Braesicke K, Cockcroft G, Clarke H, Roberts AC. Serotonergic, brain volume and attentional correlates of trait anxiety in primates. Neuropsychopharmacology 2015;40:1395–404. https://doi.org/10.1038/npp.2014.324.
[46] Cagni P, Gonçalves I, Ziller F, Emile N, Barros M. Humans and natural predators induce different fear/anxiety reactions and response pattern to diazepam in marmoset monkeys. Pharmacol Biochem Behav 2009;93:134–40.
[47] Shiba Y, Kim C, Santangelo AM, Roberts AC. Lesions of either anterior orbitofrontal cortex or ventrolateral prefrontal cortex in marmoset monkeys heighten innate fear and attenuate active coping behaviors to predator threat. Front Syst Neurosci 2015;8. https://doi.org/10.3389/fnsys.2014.00250.
[48] Clara E, Tommasi L, Rogers LJ. Social mobbing calls in common marmosets (*Callithrix jacchus*): effects of experience and associated cortisol levels. Anim Cogn 2008;11:349–58.
[49] Rukstalis M, French JA. Vocal buffering of the stress response: exposure to conspecific vocalizations moderates urinary cortisol excretion in isolated marmosets. Horm Behav 2005;47:1–7.
[50] Smith TE, French JA. Psychosocial stress and urinary cortisol excretion in marmoset monkeys. Physiol Behav 1997;62:225–32.
[51] Smith TE, McGreer-Whitworth B, French JA. Close proximity of the heterosexual partner reduces the physiological and behavioral consequences of novel-cage housing in black tufted-ear marmosets (*Callithrix kuhli*). Horm Behav 1998;34:211–22.
[52] Cross N, Rogers LJ. Mobbing vocalizations as a coping response in the common marmoset. Horm Behav 2006;49:237–46.
[53] Barros M, de Souza Silva MA, Huston JP, Tomaz C. Multibehavioral analysis of fear and anxiety before, during, and after experimentally induced predatory stress in *Callithrix penicillata*. Pharmacol Biochem Behav 2004;78:357–67.
[54] Agustín-Pavón C, Braesicke K, Shiba Y, Santangelo AM, Mikheenko Y, Cockroft G, Asma F, Clarke H, Man M-S, Roberts AC. Lesions of ventrolateral prefrontal or anterior orbitofrontal cortex in primates heighten negative emotion. Biol Psychiatry 2012;72:266–72. https://doi.org/10.1016/j.biopsych.2012.03.007.
[55] Clarke HF, Horst NK, Roberts AC. Regional inactivations of primate ventral prefrontal cortex reveal two distinct mechanisms underlying negative bias in decision making. Proc Natl Acad Sci 2015;112:4176–81. https://doi.org/10.1073/pnas.1422440112.
[56] Shiba Y, Santangelo AM, Braesicke K, AgustÃn-PavÃ3n C, Cockcroft G, Haggard M, Roberts AC. Individual differences in behavioral and cardiovascular reactivity to emotive stimuli and their relationship to cognitive flexibility in a primate model of trait anxiety. Front Behav Neurosci 2014;8. https://doi.org/10.3389/fnbeh.2014.00137.
[57] Wallis CU, Cardinal RN, Alexander L, Roberts AC, Clarke HF. Opposing roles of primate areas 25 and 32 and their putative rodent homologs in the regulation of negative emotion. Proc Natl Acad Sci 2017;114:E4075–84. https://doi.org/10.1073/pnas.1620115114.
[58] Cagni P, Komorowski M, Melo GC, Lima T, Barros M. Diazepam-induced decrease in anxiety-like behaviors of marmoset monkeys exposed to a novel open-field. Pharmacol Biochem Behav 2012;100:518–21.
[59] Wang Y, Fang Q, Gong N. A modified light-dark box test for the common marmoset. Neurosci Bull 2014;30:394–400. https://doi.org/10.1007/s12264-013-1426-8.
[60] Carey GJ, Costall B, Domeney AM, Jones DNC, Naylor RJ. Behavioural effects of anxiogenic agents in the common marmoset. Pharmacol Biochem Behav 1992;42:143–53.
[61] Kato Y, Gokan H, Oh-Nishi A, Suhara T, Watanabe S, Minamimoto T. Vocalizations associated with anxiety and fear in the common marmoset (*Callithrix jacchus*). Behav Brain Res 2014;275:43–52.
[62] French JA, Fite JE, Jensen H, Oparowski K, Rukstalis MR, Fix H, Jones B, Maxwell H, Pacer M, Power ML, others. Treatment with CRH-1 antagonist antalarmin reduces behavioral and endocrine responses to social stressors in marmosets (*Callithrix kuhlii*). Am J Primatol 2007;69:877–89.
[63] Rogers DC, Costall B, Domeney AM, Gerrard PA, Greener M, Kelly ME, Hagan JJ, Hunter AJ. Anxiolytic profile of ropinirole in the rat, mouse and common marmoset. Psychopharmacology (Berl) 2000;151:91–7.
[64] Santangelo AM, Ito M, Shiba Y, Clarke HF, Schut EH, Cockcroft G, Ferguson-Smith AC, Roberts AC. Novel primate model of serotonin transporter genetic polymorphisms associated with gene expression, anxiety and sensitivity to antidepressants. Neuropsychopharmacology 2016;41:2366–76. https://doi.org/10.1038/npp.2016.41.
[65] Spinelli S, Pennanen L, Dettling AC, Feldon J, Higgins GA, Pryce CR. Performance of the marmoset monkey on computerized tasks of attention and working memory. Cogn Brain Res 2004;19:123–37.
[66] Takemoto A, Izumi A, Miwa M, Nakamura K. Development of a compact and general-purpose experimental apparatus with a touch-sensitive screen for use in evaluating cognitive functions in common marmosets. J Neurosci Methods 2011;199:82–6. https://doi.org/10.1016/j.jneumeth.2011.04.029.
[67] Prus AJ, James JR, Rosecrans JA. Conditioned place preference. In: Buccafusco JJ, editor. Methods of behavior analysis in neuroscience, frontiers in neuroscience. Boca Raton (FL): CRC Press/Taylor & Francis; 2009.
[68] Barros M, Dempster EL, Illott N, Chabrawi S, Maior RS, Tomaz C, De Souza Silva MA, Huston JP, Mill J, Müller CP. Decreased methylation of the NK3 receptor coding gene (TACR3) after cocaine-induced place preference in marmoset monkeys: cocaine CPP in monkeys. Addict Biol 2013;18:452–4. https://doi.org/10.1111/j.1369-1600.2011.00409.x.

[69] Duarte RBM, Patrono E, Borges AC, César AAS, Tomaz C, Ventura R, Gasbarri A, Puglisi-Allegra S, Barros M. Consumption of a highly palatable food induces a lasting place-conditioning memory in marmoset monkeys. Behav Processes 2014;107:163–6. https://doi.org/10.1016/j.beproc.2014.08.021.
[70] Clarke HF, Hill GJ, Robbins TW, Roberts AC. Dopamine, but not serotonin, regulates reversal learning in the marmoset caudate nucleus. J Neurosci 2011;31:4290–7. https://doi.org/10.1523/JNEUROSCI.5066-10.2011.
[71] Clarke HF, Robbins TW, Roberts AC. Lesions of the medial striatum in monkeys produce perseverative impairments during reversal learning similar to those produced by lesions of the orbitofrontal cortex. J Neurosci 2008;28:10972–82. https://doi.org/10.1523/JNEUROSCI.1521-08.2008.
[72] Lacreuse A, Chang J, Metevier CM, LaClair M, Meyer JS, Ferris CM. Oestradiol modulation of cognition in adult female marmosets (*Callithrix jacchus*). J Neuroendocrinol 2014;26:296–309.
[73] LaClair M, Lacreuse A. Reversal learning in gonadectomized marmosets with and without hormone replacement: are males more sensitive to punishment? Anim Cogn 2016;19:619–30.
[74] Rygula R, Clarke HF, Cardinal RN, Cockcroft GJ, Xia J, Dalley JW, Robbins TW, Roberts AC. Role of central serotonin in anticipation of rewarding and punishing outcomes: effects of selective amygdala or orbitofrontal 5-HT depletion. Cereb Cortex 2015;25:3064–76. https://doi.org/10.1093/cercor/bhu102.
[75] Kotani M, Enomoto T, Murai T, Nakako T, Iwamura Y, Kiyoshi A, Matsumoto K, Matsumoto A, Ikejiri M, Nakayama T, Ogi Y, Ikeda K. The atypical antipsychotic blonanserin reverses (+)-PD-128907- and ketamine-induced deficit in executive function in common marmosets. Behav Brain Res 2016;305:212–7. https://doi.org/10.1016/j.bbr.2016.02.031.
[76] da Silva Mota MT, Franci CR, de Sousa MBC. Hormonal changes related to paternal and alloparental care in common marmosets (*Callithrix jacchus*). Horm Behav 2006;49:293–302.
[77] Finkenwirth C, Martins E, Deschner T, Burkart JM. Oxytocin is associated with infant-care behavior and motivation in cooperatively breeding marmoset monkeys. Horm Behav 2016;80:10–8.
[78] French JA, Fite JA, Ross CN. Family life in marmosets: causes and consequences of variation in offspring care. In: Bridges RS, editor. The parental brain. New York: Elsevier Press; 2008. p. 461–78.
[79] Storey AE, Ziegler TE. Primate paternal care: interactions between biology and social experience. Horm Behav 2016;77:260–71. https://doi.org/10.1016/j.yhbeh.2015.07.024.
[80] Pryce CR. The regulation of maternal behaviour in marmosets and tamarins. Behav Processes 1993;30:201–24.
[81] Sánchez SM, Ziegler TE, Snowdon CT. Both parents respond equally to infant cues in the cooperatively breeding common marmoset, *Callithrix jacchus*. Anim Behav 2014;97:95–103. https://doi.org/10.1016/j.anbehav.2014.09.002.
[82] Taylor JH, French JA. Oxytocin and vasopressin enhance responsiveness to infant stimuli in adult marmosets. Horm Behav 2015;75:154–9.
[83] Zahed SR, Prudom SL, Snowdon CT, Ziegler TE. Male parenting and response to infant stimuli in the common marmoset (*Callithrix jacchus*). Am J Primatol 2008;70:84–92. https://doi.org/10.1002/ajp.20460.
[84] Ziegler TE, Peterson LJ, Sosa ME, Barnard AM. Differential endocrine responses to infant odors in common marmoset (*Callithrix jacchus*) fathers. Horm Behav 2011;59:265–70. https://doi.org/10.1016/j.yhbeh.2010.12.001.
[85] Dettling AC, Feldon J, Pryce CR. Repeated parental deprivation in the infant common marmoset (*Callithrix jacchus*, primates) and analysis of its effects on early development. Biol Psychiatry 2002;52:1037–46.
[86] French JA, Carp SB. Early-life social adversity and developmental processes in nonhuman primates. Curr Opin Behav Sci 2016;7:40–6. https://doi.org/10.1016/j.cobeha.2015.11.004.
[87] Pryce CR, Rüedi-Bettschen D, Dettling AC, Weston A, Russig H, Ferger B, Feldon J. Long-term effects of early-life environmental manipulations in rodents and primates: potential animal models in depression research. Neurosci Biobehav Rev 2005;29:649–74. https://doi.org/10.1016/j.neubiorev.2005.03.011.
[88] Birnie AK, Taylor JH, Cavanaugh J, French JA. Quality of maternal and paternal care predicts later stress reactivity in the cooperatively-breeding marmoset (*Callithrix geoffroyi*). Psychoneuroendocrinology 2013;38:3003–14.
[89] Johnson EO, Kamilaris TC, Calogero AE, Gold PW, Chrousos GP. Effects of early parenting on growth and development in a small primate. Pediatr Res 1996;39:999–1005. https://doi.org/10.1203/00006450-199606000-00012.
[90] Johnson EO, Kamilaris TC, Carter CS, Calogero AE, Gold PW, Chrousos GP. The biobehavioral consequences of psychogenic stress in a small, social primate (*Callithrix jacchus jacchus*). Biol Psychiatry 1996;40:317–37.
[91] Díaz-Muñoz SL, Bales KL. "Monogamy" in primates: variability, trends, and synthesis. Am J Primatol 2016;78:283–7.
[92] Ågmo A, Smith AS, Birnie AK, French JA. Behavioral characteristics of pair bonding in the black tufted-ear marmoset (*Callithrix penicillata*). Behaviour 2012;149:407–40.
[93] Cavanaugh J, Huffman MC, Harnisch AM, French JA. Marmosets treated with oxytocin are more socially attractive to their long-term mate. Front Behav Neurosci 2015.
[94] Cavanaugh J, Mustoe AC, Taylor JH, French JA. Oxytocin facilitates fidelity in well-established marmoset pairs by reducing sociosexual behavior toward opposite-sex strangers. Psychoneuroendocrinology 2014;49:1–10.
[95] Smith AS, Agmo A, Birnie AK, French JA. Manipulation of the oxytocin system alters social behavior and attraction in pair-bonding primates, *Callithrix penicillata*. Horm Behav 2010;57:255–62. https://doi.org/10.1016/j.yhbeh.2009.12.004.
[96] Finkenwirth C, van Schaik C, Ziegler TE, Burkart JM. Strongly bonded family members in common marmosets show synchronized fluctuations in oxytocin. Physiol Behav 2015;151:246–51.
[97] Galvão-Coelho NL, Silva HPA, De Sousa MBC. The influence of sex and relatedness on stress response in common marmosets (*Callithrix jacchus*). Am J Primatol 2012;74:819–27. https://doi.org/10.1002/ajp.22032.
[98] Norcross JL, Newman JD. Effects of separation and novelty on distress vocalizations and cortisol in the common marmoset (*Callithrix jacchus*). Am J Primatol 1999;47:209–22.
[99] Smith AS, Birnie AK, Lane KR, French JA. Production and perception of sex differences in vocalizations of Wied's black-tufted-ear marmosets (*Callithrix kuhlii*). Am J Primatol 2009;71:324–32. https://doi.org/10.1002/ajp.20656.
[100] DiMattina C. Virtual vocalization stimuli for investigating neural representations of species-specific vocalizations. J Neurophysiol 2005;95:1244–62. https://doi.org/10.1152/jn.00818.2005.
[101] Miller CT, Thomas AW. Individual recognition during bouts of antiphonal calling in common marmosets. J Comp Physiol A 2012;198:337–46.
[102] Miller CT, Wang X. Sensory-motor interactions modulate a primate vocal behavior: antiphonal calling in common marmosets. J Comp Physiol A 2006;192:27–38.
[103] Toarmino CR, Wong L, Miller CT. Audience affects decision-making in a marmoset communication network. Biol Lett 2017;13:20160934. https://doi.org/10.1098/rsbl.2016.0934.

[104] Chow CP, Mitchell JF, Miller CT. Vocal turn-taking in a non-human primate is learned during ontogeny. Proc R Soc B Biol Sci 2015;282:20150069–20150069. https://doi.org/10.1098/rspb.2015.0069.

[105] Miller C, DiMauro A, others. Vocalization induced cFos expression in marmoset cortex. Front Integr Neurosci 2010;4:128.

[106] Simões CS. Activation of frontal neocortical areas by vocal production in marmosets. Front Integr Neurosci 2010;4. https://doi.org/10.3389/fnint.2010.00123.

[107] Roy S, Zhao L, Wang X. Distinct neural activities in premotor cortex during natural vocal behaviors in a new World primate, the common marmoset (*Callithrix jacchus*). J Neurosci 2016;36:12168–79. https://doi.org/10.1523/JNEUROSCI.1646-16.2016.

[108] Eliades SJ, Miller CT. Marmoset vocal communication: behavior and neurobiology: marmoset vocal communication. Dev Neurobiol 2017;77:286–99. https://doi.org/10.1002/dneu.22464.

[109] Ghazanfar AA, Eliades SJ. The neurobiology of primate vocal communication. Curr Opin Neurobiol 2014;28:128–35.

[110] Smith T. Individual olfactory signatures in common marmosets (*Callithrix jacchus*). Am J Primatol 2006;68:585–604. https://doi.org/10.1002/ajp.20254.

[111] Smith TE, Tomlinson AJ, Mlotkiewicz JA, Abbott DH. Female marmoset monkeys (*Callithrix jacchus*) can be identified from the chemical composition of their scent marks. Chem Senses 2001;26:449–58.

[112] Ferris CF, Snowdon CT, King JA, Duong TQ, Ziegler TE, Ugurbil K, Ludwig R, Schultz-Darken NJ, Wu Z, Olson DP, Sullivan Jr JM, Tannenbaum PL, Vaughan JT. Functional imaging of brain activity in conscious monkeys responding to sexually arousing cues. Neuroreport 2001;12:2231–6.

[113] Ferris CF, Snowdon CT, King JA, Sullivan JM, Ziegler TE, Olson DP, Schultz-Darken NJ, Tannenbaum PL, Ludwig R, Wu Z, Einspanier A, Vaughan JT, Duong TQ. Activation of neural pathways associated with sexual arousal in non-human primates. J Magn Reson Imaging 2004;19:168–75. https://doi.org/10.1002/jmri.10456.

[114] Snowdon CT, Tannenbaum PL, Schultz-Darken NJ, Ziegler TE, Ferris CF. Conditioned sexual arousal in a nonhuman primate. Horm Behav 2011;59:696–701. https://doi.org/10.1016/j.yhbeh.2010.10.009.

[115] Burkart J, Kupferberg A, Glasauer S, van Schaik C. Even simple forms of social learning rely on intention attribution in marmoset monkeys (*Callithrix jacchus*). J Comp Psychol 2012;126:129.

[116] Werdenich D, Huber L. Social factors determine cooperation in marmosets. Anim Behav 2002;64:771–81.

[117] Burkart JM, Fehr E, Efferson C, van Schaik CP. Other-regarding preferences in a non-human primate: common marmosets provision food altruistically. Proc Natl Acad Sci 2007;104:19762–6.

[118] Mustoe AC, Cavanaugh J, Harnisch AM, Thompson BE, French JA. Do marmosets care to share? Oxytocin treatment reduces prosocial behavior toward strangers. Horm Behav 2015;71:83–90.

[119] Burkart JM, Allon O, Amici F, Fichtel C, Finkenwirth C, Heschl A, Huber J, Isler K, Kosonen ZK, Martins E, others. The evolutionary origin of human hyper-cooperation. Nat Commun 2014;5:4747.

[120] Finkenwirth C, Burkart JM. Long-term-stability of relationship structure in family groups of common marmosets, and its link to proactive prosociality. Physiol Behav 2017;173:79–86. https://doi.org/10.1016/j.physbeh.2017.01.032.

[121] Heyes CM. Reflections on self-recognition in primates. Anim Behav 1994;47:909–19. https://doi.org/10.1006/anbe.1994.1123.

[122] Anderson JR, Gallup GG. Mirror self-recognition: a review and critique of attempts to promote and engineer self-recognition in primates. Primates 2015;56:317–26. https://doi.org/10.1007/s10329-015-0488-9.

[123] Eglash AR, Snowdon CT. Mirror-image responses in pygmy marmosets (*Cebuella pygmaea*). Am J Primatol 1983;5:211–9. https://doi.org/10.1002/ajp.1350050305.

[124] Inoue-Nakamura N. Mirror self-recognition in nonhuman primates: a phylogenetic approach. Jpn Psychol Res 1997;39:266–75. https://doi.org/10.1111/1468-5884.00059.

[125] Erkert HG. Characteristics of the circadian activity rhythm in common marmosets (*Callithrix j. jacchus*). Am J Primatol 1989;17:271–86. https://doi.org/10.1002/ajp.1350170403.

[126] Glass JD, Tardif SD, Clements R, Mrosovsky N. Photic and nonphotic circadian phase resetting in a diurnal primate, the common marmoset. Am J Physiol Regul Integr Comp Physiol 2001;280:R191–7.

[127] Silva CA da, de Pontes ALB, Cavalcante J de S, de Azevedo CVM. Conspecific vocalisations modulate the circadian activity rhythm of marmosets. Biol Rhythm Res 2014;45:941–54. https://doi.org/10.1080/09291016.2014.939441.

[128] Melo PR, Gonçalves BSB, Menezes AAL, Azevedo CVM. Circadian activity rhythm in pre-pubertal and pubertal marmosets (*Callithrix jacchus*) living in family groups. Physiol Behav 2016;155:242–9. https://doi.org/10.1016/j.physbeh.2015.12.023.

[129] Melo P, Gonçalves B, Menezes A, Azevedo C. Socially adjusted synchrony in the activity profiles of common marmosets in light-dark conditions. Chronobiol Int 2013;30:818–27. https://doi.org/10.3109/07420528.2013.767823.

[130] Pines A. Circadian rhythm and menopause. Climacteric J Int Menopause Soc 2016;19:551–2. https://doi.org/10.1080/13697137.2016.1226608.

[131] Gervais NJ, Viechweg SS, Mong JA, Lacreuse A. The middle-aged ovariectomized marmoset (*Callithrix jacchus*) as a model of menopausal symptoms: preliminary evidence. Neuroscience 2016;337:1–8.

[132] Bertani S, Carboni L, Criado A, Michielin F, Mangiarini L, Vicentini E. Circadian profile of peripheral hormone levels in Sprague-Dawley rats and in common marmosets (*Callithrix jacchus*). Vivo Athens Greece 2010;24:827–36.

[133] Cross N, Rogers LJ. Diurnal cycle in salivary cortisol levels in common marmosets. Dev Psychobiol 2004;45:134–9.

[134] Sousa MBC, Ziegler TE. Diurnal variation on the excretion patterns of fecal steroids in common marmoset (*Callithrix jacchus*) females. Am J Primatol 1998;46:105–17.

[135] Valentinuzzi VS, Neto SPD, Carneiro BTS, Santana KS, Araújo JF, Ralph MR. Memory for time of training modulates performance on a place conditioning task in marmosets. Neurobiol Learn Mem 2008;89:604–7. https://doi.org/10.1016/j.nlm.2007.08.002.

[136] Burkart JM, Finkenwirth C. Marmosets as model species in neuroscience and evolutionary anthropology. Neurosci Res 2015;93:8–19.

[137] Miller CT, Freiwald WA, Leopold DA, Mitchell JF, Silva AC, Wang X. Marmosets: a neuroscientific model of human social behavior. Neuron 2016;90:219–33.

[138] Saito A. The marmoset as a model for the study of primate parental behavior. Neurosci Res 2015.

[139] Krakauer JW, Ghazanfar AA, Gomez-Marin A, MacIver MA, Poeppel D. Neuroscience needs behavior: correcting a reductionist bias. Neuron 2017;93:480–90. https://doi.org/10.1016/j.neuron.2016.12.041.

CHAPTER

27

The Use of the Marmoset in Toxicity Testing and Nonclinical Safety Assessment Studies

Sven Korte[1], *Jeffrey Everitt*[2]

[1]Covance Preclinical Services GmbH, Münster, Germany;

[2]Duke University School of Medicine, Durham, NC, United States

INTRODUCTION

Common marmosets have been used historically in chemical toxicology studies and in nonclinical safety assessment for pharmaceutical development, although at the present time they are not a commonly employed toxicology model. Although they are not commonly used in safety assessment studies relative to macaque species, marmosets have a number of features that make them attractive as nonhuman primate (NHP) models in toxicology studies, as well as having utility as models for human disease, physiology, drug metabolism, and reproductive biology [1,1a]. The extensive use of marmosets as animal models in a number of important therapeutic areas has led to a sizeable amount of background information becoming available over the last decade and has enhanced our ability to use marmosets in toxicology studies [2]. In certain preclinical development programs, particularly when the more commonly utilized cynomolgus (*Macaca fascicularis*) or rhesus (*Macaca mulatta*) macaques do not meet scientific need, marmosets may successfully be used as an alternative NHP species. They are particularly valuable in studies where safety and efficacy are studied together in disease models. It is likely with the advent of new gene editing (e.g., CRISPR-Cas systems) and stem cell technologies that marmosets will become an even more valuable NHP resource and test species in the future.

The small size of the marmoset relative to other NHPs allows for conservation of test substances, especially early in drug development when new chemical entities can be difficult to synthesize, time consuming, and expensive to produce. The relative ease of handling and housing of marmosets, early sexual maturity, and freedom from certain zoonotic agents, such as *Herpesvirus simiae*, make them attractive nonmacaque models when NHP studies are needed. It must be noted though that the advantages of the small size can be offset by the limited ability to sample blood and other biological specimens. In addition, although the common marmoset is easy to handle and house, the complex behaviors and social hierarchies, and husbandry factors, can add features of complexity for the clinical and toxicological evaluation of animals. Marmosets are easily stressed and not as robust as macaques in the laboratory. This limits their practical use in certain toxicology testing situations. The limited availability of animals and laboratories experienced with the use of the common marmoset can also make the use of this species challenging.

CHOICE AND RELEVANCE OF THE SPECIES IN TOXICOLOGY

The common marmoset monkey (Fig. 27.1) has served as an NHP model in nonclinical safety assessment studies for drug development for a variety of small- and large-molecule compounds [2a,2b,3,4]. Marmosets have also been used as disease models in drug discovery for a diverse group of neurologic, inflammatory, metabolic, bone, and infectious diseases, including Parkinson's and Huntington's diseases, multiple sclerosis, rheumatoid arthritis, and diet-induced obesity [5,6,6a]. The utility of the marmoset as a disease model, especially in the areas of neuroscience, immunology, and infectious diseases, makes it a useful NHP for combination studies that assess both the efficacy and toxicity aspects of compound evaluation. Marmosets are particularly

The Common Marmoset in Captivity and Biomedical Research
https://doi.org/10.1016/B978-0-12-811829-0.00027-3

FIGURE 27.1 Pair-housed marmosets.

useful as infectious disease models due to their susceptibility to a wide range of human bacterial, viral, and protozoal pathogens. Their small size and relatively easy housing allow them to be maintained in high-containment housing for infectious agents at much lower cost than other NHP species [7].

Recently there have been several significant advancements and genomic tools that have enhanced our understanding of the biology of *Callithrix jacchus* and have increased the utility of the species for development of biotherapeutic compounds, gene therapy products, and vaccines. These advancements include sequencing of the marmoset genome [8], the development of marmoset DNA microarrays [9], the development of marmoset induced pluripotent stem cells [10], and the production of transgenic marmosets [11].

Due to ethical and societal concerns, the use of NHPs as a nonrodent species in pharmaceutical development has continued to get increasingly stringent and species selection requires careful justification. Some pharmaceutical companies now have company policies that require scientists to specifically deselect other nonrodent species, such as dogs and minipigs, before consideration of NHPs as the potential nonrodent species for studies conducted under International Conference on Harmonization guidelines. Nevertheless, NHPs remain important models for drug development and especially large-molecule development such as monoclonal antibodies. In many instances, they are the only relevant species in which to study target responses. For biopharmaceutical drugs, NHPs are the relevant animal model in a large percentage of pharmaceutical development scenarios [12–14]. Most biopharmaceutical development programs have used and continue to utilize macaque species, especially the cynomolgus monkey (*M. fascicularis*). In a small number of instances, assessment of pharmacological relevance (e.g., binding to the animal ortholog of the human target) can eliminate other NHP species with the marmoset remaining as the only relevant model. Thus, marmosets remain a nontraditional alternative species for nonclinical safety assessment and toxicology studies.

ANIMAL WELFARE, CARE, AND USE IN THE TOXICOLOGY LABORATORY

Most toxicology test facilities that conduct studies using marmosets are fully accredited by the Association for Assessment and Accreditation of Laboratory Animal Care International and thus provide a high standard of animal welfare, with a local institutional animal care and use committee (IACUC) providing internal oversight. All procedures in the toxicology study plan must be in compliance with the local authority's regulations and must be approved by the IACUC before beginning any study. In Europe, studies need to be performed in consideration of the Commission Recommendation 2007/526/EC on guidelines for the accommodation and care of animals used for experimental and other scientific purposes (Appendix A of Convention ETS 123). This European guideline requires, for example, paired or group housing for laboratory primates and a larger cage volume, than that required by regulation in non-European countries. Because different housing and husbandry conditions can have a varying impact on the stress of animals and other toxicology-related endpoints, it is important to consider the test facility conditions and experience when selecting a laboratory to conduct studies with marmosets. Some study parameters cannot be easily followed under group housing conditions, and this has an impact on protocol designs. The standard social housing for marmosets at Covance Preclinical Service GmbH (Muenster, Germany) is according to ETS 123, 2007, as described in Table 27.1 and depicted in

TABLE 27.1 Dimensions of Marmoset Cages as Specified by the New European Guideline Requirements for Housing of Nonhuman Primates (ETS 123, 2007)

	Minimum ETS 123 Guideline Requirements[a]
Floor space (m^2)	0.5
Height (m)	1.5[b]
Volume (m^3)	0.75[c]

[a]*For one or two animals plus offspring up to 5 months old (plus 0.2 m^3 per additional animal over 5 months).*
[b]*Top of enclosure at least 1.8 m above ground.*
[c]*Calculated, not indicated in the guideline.*

FIGURE 27.2 Marmoset group housing under European housing regulations (ETS123) conditions.

Fig 27.2. This laboratory is the source for the toxicology reference data presented in this chapter.

One of the challenges of using common marmosets as a test species for preclinical development is the need to have good reference colony data. Many aspects of marmoset biology are significantly affected by environmental conditions, especially nutrition. Toxicity and toxicologic pathology results can have colony-specific features that need to be considered. Background lesion incidence and severity can vary substantially between sources of marmosets and facilities and are the result of a combination of genetics, husbandry, environment, nutrition, and microbiota. Many of the published reference ranges and lesion incidences in the marmoset literature are outdated and may not be as relevant with current or differing colony husbandry.

JUSTIFICATION FOR USE IN PRECLINICAL TOXICOLOGY STUDIES

When marmosets are selected as the relevant species for use in a study, a clear justification must be provided in the study plan. There are multiple important and diverse reasons why the marmoset may be the species of choice for study. Pharmacologic mechanisms may be shared with humans, metabolism of xenobiotics may be most like human, or previous study experience and/or historical data for the same drug class may be the reason that the marmoset is the species of choice. In some instances, marmosets offer advantages for combining pharmacology or efficacy with compound safety when the marmoset model is the most relevant species with respect to pharmacologic response. Marmosets, by virtue of their propensity to have twins, lend themselves to use in some advantageous experimental designs where compounds can be studied in fraternal siblings that have been shown to be much closer in response than animals of different births [6]. During the development of biotherapeutics, the species-specific cross-reactivity of the test item is often the reason for the choice of NHP. In accordance with optimum 3Rs working practices, the planned study should not duplicate previous studies, and the number of marmosets to be used needs to be chosen to yield meaningful results in the context of the study. This should be obligatory for all animal works but is especially important when using any type of NHP.

CLINICAL EVALUATION ON STUDY

Food and water consumption have always been important parameters for evaluating the general health and condition of test animals on toxicology studies. In contrast to other species, marmosets receive a highly variable diet and receive many differing food items as supplements. Like other NHPs, marmosets often use water supplies as manipulanda, resulting in a certain but highly variable loss of water. Consequently, reduced water intake might be easily overlooked. Food (pellet) intake can be well monitored for animals housed singly by counting or weighing the remaining food. However, some animals also tend to play with pellets, resulting in a more or less pronounced loss outside of the cage. When food supplementation for enrichment is offered, there can be a varying energy content, making calculation of the consumed food in terms of energy even more difficult. When animals are housed under normal group housing conditions, an animal-by-animal assessment of food intake is almost impossible without separating them for a limited period during feeding. The exact amount of food eaten per animal cannot be monitored because the food will be picked at over a long period. In addition to foraging behaviors, the hierarchy between the animals can have a major impact on eating habits. Therefore, the best one can often do is to perform a general overall assessment of food intake per cage. The use of body weight assessment is often a better and more precise measure for the animals' general condition. However, if the pharmacology of a test item is linked to a possible influence on food intake, single housing might be considered justifiable in select studies. Alternatively, animals can be separated, at least during feeding time. Similar approaches might be justified if monitoring feces is of special interest.

BLOOD SAMPLING IN MARMOSET TOXICOLOGY STUDIES

The timing and quantity of blood sampling requires careful consideration due to the small size of the marmoset. Guidelines for blood removal are similar to other species, and depending on analytical requirements, not all parameters for an investigational or pivotal toxicity study can be measured within these required limits. Assuming 300−480 g body weight, the marmoset has a circulating mean blood volume between 19.5 and 31.2 mL (65 mL/kg), whereas the cynomolgus monkey weight 2.5−5.0 kg has a mean blood volume of 175−350 mL (70 mL/kg). Thus, the small size of marmosets necessitates that blood sampling limits be considered much more carefully in comparison with other NHPs such as macaque species. In addition to IACUC and animal welfare requirements, the amount of blood sampled can have confounding effects in tissues as the marmoset is extremely prone to extramedullary hematopoiesis in a variety of tissues evaluated in pathology studies. Fortunately, there are now microsampling technologies that are making the use of small animal subjects increasingly easier to use in studies. One such technology is the use of dried blood spot (DBS) technology in pharmacokinetic studies [14a].

The use of satellite animals for toxicokinetic or pharmacokinetic testing to overcome the limitations of blood sampling issues might also be an option in some study designs, although this adds animals to studies and is not the desired state in working with the 3Rs. In biopharmaceutical testing, there is often a likelihood of antidrug antibody development, which might require an integrated discussion of all data on a single-animal basis. In this instance, the use of satellite animals can result in a suboptimal study design and may confound interpretation of data. To combine good study conduct with optimal animal welfare, the last blood sampling can be performed directly before sedation for necropsy, allowing a larger volume to be collected.

DOSE ADMINISTRATION IN MARMOSET TOXICOLOGY STUDIES

All major routes of compound administration have been utilized in marmoset toxicology studies, including oral, subcutaneous, intraperitoneal, intravascular, intravitreal, and aerosol. Dosing can be accomplished with relatively simple instrumentation (Figs. 27.3 and 27.4). Oral dosing can be easily accomplished with a restraint tube, with the animal's mouth lightly pressed against a thin metal bar. Due to its small size and relative frailty, the use of continuous infusion is more difficult in marmosets than in macaque species but has been utilized. Osmotic pumps and external packs with infusion pumps have been utilized for continuous infusion in marmosets [14b]. Vascular access ports implanted subcutaneously have also been employed as a useful means of repeated dosing in marmosets [15]. Common marmosets have been used in inhalation studies in both infectious disease and chemical toxicity studies. Their small size allows them to be used in whole-body inhalation chambers and within containment facilities [16].

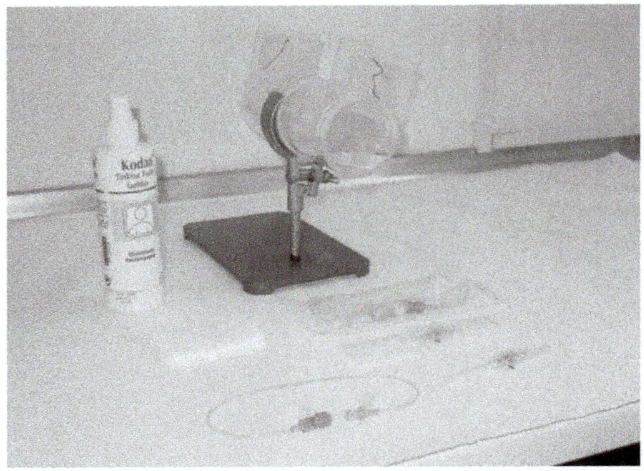

FIGURE 27.3 Restraint tube for intravenous dosing.

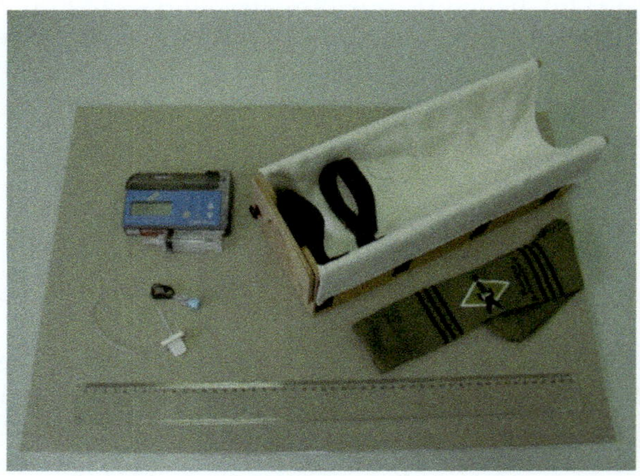

FIGURE 27.4 Restraint bed and infusion pump for prolonged (up to 1 h) intravenous infusion.

CLINICAL PATHOLOGY OF THE COMMON MARMOSET IN TOXICOLOGY STUDIES

Clinical chemistry, urinalysis, coagulation, and hematology parameters are commonly evaluated in toxicology studies including those that utilize marmosets. Clinical chemistry's best practices, including the use of concurrent and historical control data and reference ranges, are performed similarly to that of other nonrodent species where limited numbers of animals are used within experimental groups [17]. Historical reference data for clinical chemistry and clinical hematology parameters from marmosets on toxicology studies are presented from Covance Preclinical Services GmbH (Muenster, Germany) in Tables 27.2–27.5 and have been published in the literature [18–21]. It should be noted that normal clinical pathology parameters vary according to animal gender, age, source, and husbandry factors, and thus experimental protocols in nonclinical safety studies should carefully consider the optimum use of concurrent comparator groups or individual animals as their own internal controls whenever possible. Clinical pathology parameters are often considered during standard health monitoring and screening when assessing suitability of marmosets for assignment to toxicology studies. Studies should always note the conditions of blood withdrawal and note any circumstances of anesthetic use, details of restraint, and vascular access. Like other toxicology models, drugs used for sedation and anesthesia in marmosets can have significant effects on hematology, clinical chemistry, and pathology outcomes that must always be considered when evaluating for compound-induced effects [22].

PHARMACOKINETICS (TOXICOKINETICS) AND METABOLISM STUDIES IN MARMOSETS

The common marmoset has been a useful model for the study of metabolism and kinetics of xenobiotics due to some important shared metabolic characteristics with people. In addition, their small size makes whole-body imaging techniques such as whole-body autoradiography feasible [23]. The availability of new micromethods, such as DBS technology, should further enhance the utility of marmosets for kinetic studies [14a]. Studies have characterized the P450 enzymes, flavin-containing monooxygenases, and other drug-metabolizing enzymes in marmosets [24–26]. These enzymes have been found to be involved in the oxidation of a variety of compounds associated with pharmacological and/or toxicological effects in humans. Uehara et al. [27] recently showed that marmoset P450 3A forms had functional characteristics roughly similar to cynomolgus monkeys and humans in terms of tissue expression patterns and catalytic activities, suggesting that marmosets are suitable animal models for P450 3A-dependent drug metabolism.

Marmosets have also been used to develop physiologically based pharmacokinetic (PB-PK) models that have proven useful for understanding drug effects in neonates and infants. Parrott et al. [28] studied oseltamivir in a PB-PK model in marmosets. Modeled processes included absorption, whole-body distribution, renal clearance, metabolic conversion of a prodrug, permeability-limited hepatic disposition of drug, and age dependencies for all of these processes. Models were refined after comparison of simulations in *C. jacchus* with plasma and liver concentrations measured in adult and newborn marmosets after intravenous and oral dosing. Then simulations with a human model were compared with clinical data taken from intravenous and oral studies in healthy adults and oral studies in infants and neonates to demonstrate the validity of the model [28].

SAFETY PHARMACOLOGY IN NONCLINICAL SAFETY STUDIES

Cardiovascular endpoints such as electrocardiography and blood pressure are usually collected as snapshot measurements in physically restrained marmosets (Table 27.6). If performed, measurements should be taken at least once during the predose phase, at multiple time points in the dosing period, and again around the time of maximum expected postdose plasma compound concentration (T_{max}). Reference data are shown in Table 27.6. Surgically implanted marmoset models have been used for cardiovascular telemetry studies, however, to date, noninvasive jacketed external telemetry available in macaques has not been used in marmosets [29]. Using surgically implanted telemetry devices, Tabo et al. demonstrated that the marmoset was a useful model for detecting QTc prolongation following drug administration [30]. Horii et al. demonstrated that marmosets have a diurnal rhythm heart rate [31]. To account for circadian rhythms, it has been suggested to have time- and vehicle-matched comparator groups in drug studies. The marmoset has been part of a cross-species pharmaceutical industry consortium looking at preclinical QT safety assessment [32]. Blood pressure evaluations are routinely conducted using high-definition oscillometry [33].

Echocardiography has proven useful as a cardiac imaging modality in marmoset cardiotoxicity evaluation [34].

TABLE 27.2 Reference Data/Clinical Chemistry/Group-Housed Predose Females

Parameter	Unit	Number of Animals	Number of Values	Median	Mean	Mean ± 2SD[a]	SD[b]	SE[c]	Min	Max
KONELAB										
Total bilirubin	μmol/L	153	153	4.70	5.14	0.15–10.13	2.50	0.20	2.01	19.50
Creatinine	μmol/L	153	153	49.77	50.10	39.53–60.67	5.29	0.43	39.72	68.70
Blood urea	mmol/L	153	153	6.39	6.48	3.74–9.22	1.37	0.11	3.88	10.78
Glutamate dehydrogenase	U/L	133	133	7.95	10.60	0.00–27.34	8.37	0.73	2.74	51.27
Aspartate aminotransferase	U/L	153	153	128.98	173.73	0.00–630.28	228.28	18.46	61.77	2400.10
Alanine aminotransferase	U/L	153	69	2.54	6.50	0.00–30.49	12.00	1.44	0.01	72.77
Alkaline phosphatase	U/L	153	153	110.71	118.55	30.15–206.95	44.20	3.57	50.44	324.41
Gamma glutamyl transferase	U/L	133	133	9.90	9.83	4.28–15.39	2.78	0.24	2.40	26.85
Glucose	mmol/L	153	153	7.42	7.64	2.83–12.46	2.41	0.19	2.90	24.94
Total cholesterol	mmol/L	153	153	3.93	4.10	1.25–6.96	1.43	0.12	2.07	16.02
Triglycerides	mmol/L	153	153	1.18	1.62	0.00–9.43	3.91	0.32	0.55	49.00
Inorganic phosphorus	mmol/L	153	153	1.25	1.27	0.54–2.00	0.37	0.03	0.57	2.18
Calcium	mmol/L	153	153	2.81	2.79	2.43–3.15	0.18	0.01	2.40	3.20
Sodium	mmol/L	153	153	150.73	152.05	141.05–163.05	5.50	0.44	141.88	171.72
Potassium	mmol/L	153	153	4.32	4.34	2.98–5.69	0.68	0.05	3.03	7.15
Chloride	mmol/L	153	153	112.58	111.47	99.07–123.88	6.20	0.50	93.10	122.72
Total protein	g/L	153	153	72.86	73.20	62.60–83.80	5.30	0.43	56.54	86.83
Albumin	g/L	105	105	39.51	38.98	31.54–46.42	3.72	0.36	27.58	55.64
Globulin	g/L	105	105	34.00	34.29	26.33–42.25	3.98	0.39	24.66	48.59
Albumin/globulin ratio		105	105	1.15	1.15	0.76–1.54	0.20	0.02	0.79	2.26
C-reactive protein	mg/L	42	32	1.02	1.49	0.00–4.09	1.30	0.23	0.02	4.08
Direct bilirubin	μmol/L	22	22	0.81	0.88	0.25–1.52	0.32	0.07	0.51	2.13
Lactate dehydrogenase	U/L	42	42	228.47	271.57	2.91–540.22	134.33	20.73	146.84	788.62
Haptoglobin	g/L	20	20	0.88	0.99	0.36–1.62	0.32	0.07	0.56	1.69
Creatine kinase	U/L	22	22	271.69	1053.88	0.00–7163.49	3054.80	651.29	146.48	14,644.12
Complement factor C3	g/L	20	20	1.04	1.05	0.62–1.48	0.21	0.05	0.62	1.59
Complement factor C4	g/L	20	20	0.27	0.28	0.11–0.44	0.08	0.02	0.17	0.48

[a]*Make up the 95% confidence interval (by nonparametric evaluation).*
[b]*Standard deviation.*
[c]*Standard error.*

TABLE 27.3 Reference Data/Clinical Chemistry Summary/Group-Housed Predose Males

Parameter	Unit	Number of Animals	Number of Values	Median	Mean	Mean ± 2SD[a]	SD[b]	SE[c]	Min	Max
KONELAB										
Total bilirubin	μmol/L	174	171	5.54	6.25	0.00–13.17	3.46	0.26	1.67	25.99
Creatinine	μmol/L	174	171	50.22	50.66	38.94–62.38	5.86	0.45	37.14	68.19
Blood urea	mmol/L	174	171	6.53	6.84	2.70–10.98	2.07	0.16	1.38	17.09
Glutamate dehydrogenase	U/L	154	154	8.68	11.91	0.00–31.98	10.03	0.81	2.65	92.72
Aspartate aminotransferase	U/L	174	171	132.39	143.91	44.10–243.72	49.91	3.82	64.78	374.95
Alanine aminotransferase	U/L	174	89	2.26	4.69	0.00–20.07	7.69	0.82	0.08	56.61
Alkaline phosphatase	U/L	174	171	130.43	137.44	40.16–234.73	48.64	3.72	56.24	313.85
Gamma glutamyl transferase	U/L	154	154	9.83	9.96	5.41–14.51	2.27	0.18	4.29	21.81
Glucose	mmol/L	174	171	6.61	6.79	2.97–10.62	1.91	0.15	3.20	13.28
Total cholesterol	mmol/L	174	171	4.40	4.39	2.42–6.37	0.99	0.08	2.10	6.86
Triglycerides	mmol/L	174	171	1.03	1.21	0.07–2.35	0.57	0.04	0.38	2.94
Inorganic phosphorus	mmol/L	174	171	1.26	1.28	0.45–2.10	0.41	0.03	0.38	3.47
Calcium	mmol/L	174	171	2.75	2.72	2.26–3.17	0.23	0.02	1.90	3.18
Sodium	mmol/L	174	171	151.73	152.84	142.43–163.26	5.21	0.40	139.20	175.48
Potassium	mmol/L	174	171	4.40	4.44	2.94–5.94	0.75	0.06	2.54	6.68
Chloride	mmol/L	174	171	113.94	113.18	103.90–122.45	4.64	0.35	97.31	122.59
Total protein	g/L	174	171	72.51	72.42	60.86–83.98	5.78	0.44	54.04	88.11
Albumin	g/L	126	123	39.96	39.75	32.09–47.40	3.83	0.35	31.18	58.19
Globulin	g/L	106	103	33.09	33.16	25.28–41.04	3.94	0.39	18.08	42.07
Albumin/globulin ratio		106	103	1.20	1.20	0.87–1.54	0.17	0.02	0.90	2.03
C-reactive protein	mg/L	42	32	1.23	1.41	0.00–3.99	1.29	0.23	0.02	4.19
Direct bilirubin	μmol/L	22	22	0.88	0.91	0.53–1.29	0.19	0.04	0.60	1.36
Lactate dehydrogenase	U/L	42	42	257.82	270.78	92.16–449.40	89.31	13.78	132.55	549.48
Haptoglobin	g/L	20	20	1.07	1.12	0.22–2.03	0.45	0.10	0.28	2.24
Creatine kinase	U/L	22	22	158.89	209.28	0.00–474.09	132.41	28.23	54.91	585.35
Complement factor C3	g/L	20	20	1.13	1.12	0.71–1.52	0.20	0.05	0.84	1.59
Complement factor C4	g/L	20	20	0.28	0.29	0.11–0.46	0.09	0.02	0.12	0.51

[a]*Make up the 95% confidence interval (by nonparametric evaluation).*
[b]*Standard deviation.*
[c]*Standard error.*

Marmosets have been shown to be a useful nonrodent toxicology model for cardiac toxicity testing when dogs and other large animal species are particularly sensitive to compounds under study. Marmosets were used to demonstrate the cardiovascular toxicity of minoxidil, a potent vasodilator drug [35].

Robust tests for neurobehavioral evaluation in marmosets in the preclinical setting are not generally conducted, although activity monitoring is possible using implanted telemetry devices [31]. Recently a novel Primate Postnatal Neurobehavioral Assessment Scale for Marmosets (PPNAS-M) was created, which was based on currently available scales for humans and rhesus monkeys [36]. The PPNAS-M consists of 41 noninvasive tests grouped into five testing categories: visual orienting, auditory and spatial orienting, motor responses, righting and body strength, and temperament tests.

TABLE 27.4 Reference Data/Individual Hematology Analysis/Group-Housed Females Predose

Parameter	Unit	Number of Animals	Number of Values	Median	Mean	Mean ± 2SD[a]	SD[b]	SE[c]	Min	Max
SYSMEX K4500										
Red blood cell count	10 E12/L	127	127	7.03	6.94	5.43–8.46	0.76	0.07	4.45	8.83
Hemoglobin concentration	mmol/L	127	127	10.0	9.9	7.9–11.9	1.0	0.1	6.3	11.9
Packed cell volume	%	127	127	51.5	50.8	41.1–60.5	4.8	0.4	33.2	60.9
Mean corpuscular hemoglobin	fmol	127	127	1.427	1.425	1.272–1.578	0.077	0.007	1.217	1.624
Mean corpuscular hemoglobin concentration	mmol/L	127	127	19.5	19.4	18.6–20.3	0.4	0.0	17.9	20.7
Mean corpuscular volume	fL	127	127	73.2	73.3	66.6–80.1	3.4	0.3	65.7	89.2
Platelet count	10 E9/L	127	127	618	631	331–930	150	13	337	1263
Hematocrit	L/L	43	43	0.523	0.517	0.437–0.598	0.040	0.006	0.390	0.600
White blood cell count	10 E9/L	127	127	6.4	6.5	1.7–11.4	2.4	0.2	1.8	17.4
MICROSCOPE										
WBC juvenile cells	%	127	127	0	0	0–0	0	0	0	0
WBC band cells	%	127	127	1	1	0–4	2	0	0	10
WBC segmented cells	%	127	127	56	54	22–86	16	1	17	90
WBC basophil cells	%	127	127	0	1	0–3	1	0	0	5
WBC eosinophil cells	%	127	127	1	2	0–7	3	0	0	12
WBC lymphocytes	%	127	127	39	42	10–73	16	1	8	81
WBC monocytes	%	127	127	0	1	0–3	1	0	0	4
WBC blasts	%	127	127	0	0	0–0	0	0	0	0
WBC diverse cells (normoblasts)	Cells/evaluation area	84	84	0	1	0–8	3	0	0	20
Nucleated red blood cells	–	43	43	0	1	0–5	2	0	0	10

[a]*Make up the 95% confidence interval (by nonparametric evaluation).*
[b]*Standard deviation.*
[c]*Standard error.*

TABLE 27.5 Reference Data/Individual Hematology Analysis/Group-Housed Males Predose

Parameter	Unit	Number of Animals	Number of Values	Median	Mean	Mean ± 2SD[a]	SD[b]	SE[c]	Min	Max
SYSMEX K4500										
Red blood cell count	10 E12/L	122	122	7.18	7.10	5.62–8.58	0.74	0.07	4.16	9.01
Hemoglobin concentration	mmol/L	122	122	10.2	10.1	8.0–12.2	1.0	0.1	6.0	12.4
Packed cell volume	%	122	122	52.6	51.8	42.2–61.4	4.8	0.4	33.4	63.3
Mean corpuscular hemoglobin	fmol	122	122	1.426	1.424	1.284–1.564	0.070	0.006	1.242	1.645
Mean corpuscular hemoglobin concentration	mmol/L	122	122	19.5	19.5	18.5–20.4	0.5	0.0	18.0	21.2
Mean corpuscular volume	fL	122	122	73.2	73.1	67.1–79.2	3.0	0.3	65.4	80.3
Platelet count	10 E9/L	122	122	627	638	348–927	145	13	282	1053
Hematocrit	L/L	40	40	0.544	0.540	0.467–0.612	0.036	0.006	0.458	0.608
White blood cell count	10 E9/L	122	122	6.5	6.8	1.7–11.9	2.6	0.2	2.4	16.7
MICROSCOPE										
WBC juvenile cells	%	122	122	0	0	0–0	0	0	0	0
WBC band cells	%	122	122	1	1	0–4	1	0	0	6
WBC segmented cells	%	122	122	59	57	29–85	14	1	23	85
WBC basophil cells	%	122	122	0	1	0–2	1	0	0	4
WBC eosinophil cells	%	122	122	1	2	0–8	3	0	0	18
WBC lymphocytes	%	122	122	36	39	11–67	14	1	10	74
WBC monocytes	%	122	122	0	1	0–3	1	0	0	5
WBC blasts	%	122	122	0	0	0–0	0	0	0	0
WBC diverse cells (normoblasts)	Cells/evaluation area	82	82	0	1	0–5	2	0	0	17
Nucleated red blood cells	–	40	40	0	1	0–4	1	0	0	6

[a] *Make up the 95% confidence interval (by nonparametric evaluation).*
[b] *Standard deviation.*
[c] *Standard error.*

TABLE 27.6 Reference Data/Individual Electrocardiogram Analysis/Group Housed—Males and Females Predose

Parameter	Unit	Number of Animals	Number of Values	Median	Mean	Mean ± 2SD[a]	SD[b]	SE[c]	Min	Max
Heart rate	beats/min	224	304	408	400	306–493	47	3	221	500
R–R interval	s	224	304	0.15	0.15	0.11–0.20	0.02	0.00	0.12	0.27
QRS COMPLEX										
Width	s	224	304	0.03	0.03	0.02–0.05	0.01	0.00	0.02	0.06
R-wave	mV	224	304	0.66	0.72	0.07–1.36	0.32	0.02	0.15	2.30
S-wave	mV	224	304	0.00	0.07	0.00–0.30	0.12	0.01	0.00	0.75
P–R interval	s	224	268	0.05	0.05	0.03–0.07	0.01	0.00	0.03	0.09
Q–T interval	s	224	268	0.10	0.10	0.08–0.13	0.01	0.00	0.07	0.15
T-wave	mV	224	266	0.22	0.24	0.02–0.46	0.11	0.01	0.05	0.62
QTc	s	224	268	0.19	0.19	0.15–0.22	0.02	0.00	0.13	0.24

[a]*Make up the 95% confidence interval (by nonparametric evaluation).*
[b]*Standard deviation.*
[c]*Standard error.*

USE OF THE MARMOSET IN IMMUNOTOXICOLOGY

There is a high degree of homology between humans and marmosets (*C. jacchus*) with respect to immunoglobulins and T-cell receptors [36a,37]. Many of the surface antigens of leukocytes are well conserved in marmosets, and determination of the cellular phenotypes and quantification in peripheral blood or tissue can be performed with antibodies originally raised against human CD epitopes [36a,38]. White blood cell determinations in preclinical studies for basic cell types in blood can be easily accomplished by measuring their forward and side scatter parameters by flow cytometry. General conformity was noted concerning age-dependent changes in the major subsets (CD4, CD8, CD2, CD20) and the adhesion molecules (CD11a and CD29) on peripheral blood lymphocytes in children and young marmosets [39].

The availability of functional immunotoxicity testing protocols in marmosets is limited when compared with macaques [40]. However, by testing the activity of T-cell dependent antibody responses and B-cell functions, the preclinical safety assessment of immunomodulatory compounds can be assessed. The antibody response specifically mounted to a foreign antigen (e.g., keyhole limpet hemocyanin) can be followed, if the marmoset is the relevant species for a drug which exhibits immunomodulatory activity. Peripheral blood mononuclear cell preparations generally used for functional immune testing also represent a valuable tool in the marmoset, which is obviously limited by blood volume when compared with macaques. However, for many aspects of immunotoxicology and immune regulation, the predictivity of the marmoset primate model to the human remains unknown, and further research is needed.

The shared similarities in immune response with humans recently led to the marmoset being used to evaluate the safety and efficacy of viral vectors in gene therapy regimens for glioblastoma multiforme [41]. The marmoset is a useful preclinical disease model for gaining an understanding of combined safety/efficacy in neurooncology studies as marmosets can be used to model human immune responses to cytokines [42]. For similar reasons, marmosets have contributed importantly to efforts to develop immunotherapy for multiple sclerosis [5].

REPRODUCTIVE AND DEVELOPMENTAL TOXICOLOGY STUDIES IN THE MARMOSET

Marmosets differ substantially from other NHP models such as macaques and from humans with respect to many features of reproductive biology (covered elsewhere in this volume). For instance, females have no menstrual bleeding, each cycle results in multiple ovulations, and postpartum conception occurs. Because macaque reproductive physiology is more similar to that of women, the cynomolgus monkey is the established NHP model for reproductive and developmental toxicology when primate models are required [42a]. Although the marmoset is a useful model for study of comparative reproductive physiology in primates, its relevance for assessment of human reproductive risk needs additional research [1]. There may be several theoretical advantages to conducting developmental toxicity studies using marmosets in some very specific instances. One such instance is when there are issues of species relevance, particularly with biologics, when macaques may be deemed an inappropriate species. There are certain specific teratogenic risks that are more difficult to demonstrate in species with only one offspring, and in those cases, the multiple offspring of marmosets may prove useful. Marmosets have advantages of early sexual maturity and high reproduction efficiency, which are useful characteristics for NHP developmental and reproductive studies.

CHRONIC TOXICITY STUDIES IN THE MARMOSET

The small size, easy handling, and easy to maintain aspects of marmoset husbandry have led to *C. jacchus* being used to evaluate the chronic toxicity and predicted oncogenicity of selected pharmaceutical compounds. One example is the long-term evaluation of lipid-lowering fibrate compounds such as ciprofibrate. In these studies, marmosets were subjected to daily repeated oral gavage dose administration for several years without the development of tumors that were noted in rat oncogenicity bioassays [43]. Comparative studies of fibrate compounds in rats and marmosets were instrumental in helping to assure human safety for this extremely important class of compounds and to show the relationship between peroxisome proliferation and hepatic tumorigenicity for peroxisome proliferator xenobiotics [44]. Ciprofibrate administration in rats was also found to induce gastric carcinoids through a species-specific mechanism (gastric antisecretory activity) that was not operant in marmosets [45]. These studies exemplify the importance of selected nonrodent toxicology studies in the safety assessment of pharmaceutical compounds. Importantly, the use of marmosets in the development of fibrate compounds is a reminder that this NHP species has proven to be crucial for the development of certain classes of drugs in the past. This is a reminder that the same specialized needs can be anticipated in the future as scientific needs dictate.

TOXICOLOGIC PATHOLOGY OF *CALLITHRIX JACCHUS*

One of the most important parameters used in the evaluation of nonclinical safety assessment in regulated toxicity studies is the evaluation of tissue changes following compound administration. During safety assessment in pharmaceutical development, toxicologic pathologists are responsible for evaluating tissues following compound exposure, usually in repeat dose studies. Animals dosed by a variety of routes are administered compounds in several dose levels. Pathologists need to be able to identify when microscopic changes are present in tissues and to determine whether the changes are related to the test article administration or are merely incidental. Morphologic changes not related to the test article are often referred to as "background" or "incidental" changes, and these lesions usually arise spontaneously. The incidence and severity of these background changes in marmosets can be affected by numerous animals, procedural, and environment-related conditions. These lesion-inducing events can occur before animals are exposed to compounds of interest, before animals reach testing facilities, or during the course of study conduct. In addition to age, gender, source, and genetic background of the marmoset itself, a plethora of facility and environmental factors affect lesion development and severity, including diet, microbiome, pathogen status, housing environment, stress, and other factors. Many findings that are typically recognized as background in marmosets can be induced or exacerbated by treatment with pharmaceutical and xenobiotic compounds.

The present situation, where there is infrequent use of the marmoset in toxicology, makes study of this model by the toxicologic pathologist somewhat challenging due to the relative paucity of published reference data for control animals. Relatively few facilities have extensive pathology databases for *C. jacchus*, and background marmoset lesions are often colony dependent. Another challenge relative to the use of macaques in toxicology is that toxicologic pathology terminology is not standardized for callitrichid lesions. The global toxicologic pathology community has an international effort to harmonize nomenclature and diagnostic criteria (INHAND), but the NHP portion of the project is limited to macaque species due to the infrequent use of marmosets in preclinical safety studies [46]. It should be noted, however, that many of the lesions that are noted in *C. jacchus* are not unique and are similar to the kinds of changes noted in other NHPs such as macaque species.

In addition to the use of macroscopic and microscopic evaluation of tissue changes following compound exposure, organ weight changes are often associated with treatment-related effects, and thus selected organs are often weighed at necropsy. The determination of which organs are appropriate to weigh is based on an understanding of the test article and mechanism of action, metabolism and pharmacokinetics, and the physiology of the test species, as well as the cumulative knowledge garnered from previous studies of the same or similar compounds or class of materials. Reference data for common marmoset organ weights are presented in Tables 27.7–27.12. It is a common practice to evaluate organ weights alone and also as ratios to body weight as well as to a reference point such as brain weight that does not change significantly in conjunction with body weight changes. The Society of Toxicologic Pathologists has published best practice guidelines for the use of organ weight data in toxicology studies [47].

There have been a number of useful reports of background lesions in *C. jacchus* from survey pathology compilations of control animals on safety assessment studies and breeding colony data [48–52]. As stated earlier, the type and incidence of background lesions in this species is known to be heavily colony dependent and to change over time as husbandry and dietary practices have been altered. An example of a background

TABLE 27.7 Reference Data/Individual Organ Weight Analysis/Control Group—Housed Females

Parameter	Unit	Number of Animals	Number of Values	Median	Mean	Mean ± 2SD[a]	SD[b]	SE[c]	Min	Max
Body weight at necropsy	g	36	36	334.6	338.1	244.0–432.3	47.1	7.8	224.6	456.0
ADRENALS										
Left	g	36	36	0.070	0.070	0.032–0.108	0.019	0.003	0.040	0.127
Right	g	36	36	0.063	0.067	0.032–0.103	0.018	0.003	0.040	0.108
Total	g	36	36	0.130	0.137	0.067–0.207	0.035	0.006	0.080	0.233
KIDNEYS										
Left	g	36	36	0.975	1.027	0.546–1.509	0.241	0.040	0.620	1.766
Right	g	36	36	0.980	1.031	0.537–1.526	0.247	0.041	0.630	1.950
Total	g	36	36	1.969	2.058	1.091–3.025	0.484	0.081	1.250	3.710
REPRODUCTIVE ORGANS										
OVARIES										
Left	g	33	33	0.080	0.092	0.000–0.196	0.052	0.009	0.020	0.208
Right	g	33	33	0.080	0.086	0.000–0.188	0.051	0.009	0.020	0.269
Total	g	33	33	0.160	0.177	0.000–0.369	0.096	0.017	0.040	0.475
Uterus/cervix	g	10	10	0.239	0.250	0.032–0.467	0.109	0.034	0.120	0.471
UNPAIRED ORGANS										
Liver	g	36	36	13.045	13.895	5.119–22.671	4.388	0.731	8.193	26.797
Lungs	g	6	6	1.710	1.673	1.354–1.992	0.159	0.065	1.478	1.834
Thymus	g	19	19	0.240	0.288	0.000–0.699	0.205	0.047	0.020	0.890
Heart	g	36	36	1.886	1.896	1.139–2.652	0.378	0.063	1.040	2.610
Thyroid (with parathyroids)	g	33	33	0.080	0.092	0.008–0.176	0.042	0.007	0.040	0.190
Brain	g	36	36	8.065	7.941	6.806–9.076	0.567	0.095	6.370	8.760
Pancreas	g	6	6	0.501	0.498	0.234–0.762	0.132	0.054	0.326	0.649
Pituitary	g	29	29	0.013	0.014	0.005–0.022	0.004	0.001	0.007	0.023
Salivary glands, mandibular	g	6	6	0.320	0.371	0.180–0.561	0.095	0.039	0.299	0.524
Spleen	g	32	32	0.510	0.546	0.128–0.965	0.209	0.037	0.254	1.040
Mesenteric lymph nodes	g	6	6	0.280	0.330	0.038–0.622	0.146	0.060	0.210	0.610

[a] *Make up the 95% confidence interval (by nonparametric evaluation).*
[b] *Standard deviation.*
[c] *Standard error.*

TABLE 27.8 Reference Data/Individual Organ Weight/Body Weight Ratio Analysis/Control Group—Housed Females

Parameter	Unit	Number of Animals	Number of Values	Median	Mean	Mean ± 2SD[a]	SD[b]	SE[c]	Min	Max
Body weight at necropsy	g	36	36	334.6	338.1	244.0–432.3	47.1	7.8	224.6	456.0
ADRENALS										
Left	%	36	36	0.020	0.021	0.009–0.034	0.006	0.001	0.010	0.035
Right	%	36	36	0.020	0.020	0.009–0.031	0.006	0.001	0.010	0.033
Total	%	36	36	0.040	0.041	0.018–0.064	0.011	0.002	0.021	0.066
KIDNEYS										
Left	%	36	36	0.300	0.303	0.199–0.408	0.052	0.009	0.209	0.454
Right	%	36	36	0.300	0.305	0.199–0.410	0.053	0.009	0.217	0.502
Total	%	36	36	0.599	0.608	0.401–0.815	0.103	0.017	0.426	0.955
REPRODUCTIVE ORGANS										
Left ovary	%	33	33	0.021	0.027	0.000–0.055	0.014	0.002	0.009	0.055
Right ovary	%	33	33	0.022	0.025	0.000–0.052	0.013	0.002	0.009	0.069
Total	%	33	33	0.045	0.052	0.000–0.103	0.026	0.004	0.018	0.122
Uterus/cervix	%	10	10	0.073	0.073	0.022–0.125	0.026	0.008	0.041	0.121
UNPAIRED ORGANS										
Liver	%	36	36	3.781	4.094	1.980–6.209	1.057	0.176	2.874	6.896
Lungs	%	6	6	0.504	0.507	0.433–0.581	0.037	0.015	0.470	0.569
Thymus	%	19	19	0.074	0.086	0.000–0.194	0.054	0.012	0.008	0.237
Heart	%	36	36	0.550	0.558	0.431–0.684	0.063	0.011	0.430	0.697
Thyroid (with parathyroids)	%	33	33	0.023	0.027	0.002–0.053	0.013	0.002	0.013	0.053
Brain	%	36	36	2.411	2.388	1.711–3.064	0.338	0.056	1.832	3.281
Pancreas	%	6	6	0.151	0.150	0.085–0.214	0.032	0.013	0.104	0.189
Pituitary	%	29	29	0.004	0.004	0.001–0.007	0.001	0.000	0.002	0.008
Salivary glands, mandibular	%	6	6	0.108	0.111	0.077–0.144	0.017	0.007	0.090	0.135
Spleen	%	32	32	0.152	0.162	0.041–0.283	0.061	0.011	0.084	0.315
Mesenteric lymph nodes	%	6	6	0.099	0.114	0.000–0.230	0.058	0.024	0.062	0.222

[a]*Make up the 95% confidence interval (by nonparametric evaluation).*
[b]*Standard deviation.*
[c]*Standard error.*

lesion believed to be due to diet is the finding of a high incidence of hepatic hemosiderosis in some colonies a number of years ago [53]. These lesions were believed to be due to the high iron content of the diets fed. One of the most prevalent background conditions in marmosets historically has been the occurrence of a condition known as marmoset wasting syndrome (MWS). MWS is a clinical condition underpinned by several lesions and etiologies all characterized by the very nonspecific clinical finding of body weight loss and muscle wasting [54,55]. In addition to the nonspecific clinical findings of body weight loss and muscle wasting, there have been various background clinical pathology alterations and tissue lesions associated with the MWS diagnosis. These include anemia, hypoproteinemia, hypoalbuminemia, elevations in muscle and liver enzymes, as well as pansteatitis, hemosiderosis, and skeletal myopathy. Over the years, a variety of

TABLE 27.9 Reference Data/Individual Organ Weight/Brain Weight Ratio Analysis/Control Group—Housed Females

Parameter	Unit	Number of Animals	Number of Values	Median	Mean	Mean ± 2SD[a]	SD[b]	SE[c]	Min	Max
Brain	g	36	36	8.065	7.941	6.806–9.076	0.567	0.095	6.370	8.760
ADRENALS										
Left	%	36	36	0.822	0.886	0.392–1.380	0.247	0.041	0.493	1.535
Right	%	36	36	0.812	0.847	0.401–1.293	0.223	0.037	0.517	1.321
Total	%	36	36	1.673	1.735	0.835–2.635	0.450	0.075	1.034	2.817
KIDNEYS										
Left	%	36	36	12.312	12.963	6.941–18.984	3.011	0.502	8.382	20.781
Right	%	36	36	12.108	13.014	6.928–19.099	3.043	0.507	8.548	22.947
Total	%	36	36	24.363	25.968	13.969–37.968	6.000	1.000	16.961	43.657
REPRODUCTIVE ORGANS										
Left ovary	%	33	33	0.930	1.147	0.000–2.438	0.646	0.112	0.271	2.475
Right ovary	%	33	33	0.926	1.082	0.000–2.336	0.627	0.109	0.271	3.165
Total	%	33	33	1.968	2.220	0.000–4.586	1.183	0.206	0.543	5.590
Uterus/cervix	%	10	10	3.081	3.078	0.589–5.568	1.245	0.394	1.575	5.542
UNPAIRED ORGANS										
Liver	%	36	36	161.435	175.686	60.445–290.926	57.620	9.603	106.636	371.743
Lungs	%	6	6	21.141	21.015	18.115–23.914	1.450	0.592	18.669	22.742
Thymus	%	19	19	2.854	3.614	0.000–8.663	2.525	0.579	0.283	11.056
Heart	%	36	36	23.528	23.902	14.556–33.249	4.673	0.779	14.111	31.937
Thyroid (with parathyroids)	%	33	33	0.994	1.161	0.074–2.248	0.544	0.095	0.474	2.617
Pancreas	%	6	6	6.395	6.226	3.357–9.095	1.435	0.586	4.118	7.846
Pituitary	%	29	29	0.158	0.174	0.072–0.275	0.051	0.009	0.085	0.303
Salivary glands, mandibular	%	6	6	4.278	4.632	2.635–6.629	0.999	0.408	3.581	6.166
Spleen	%	32	32	6.555	6.913	1.473–12.352	2.720	0.481	3.354	16.327
Mesenteric lymph nodes	%	6	6	3.514	4.145	0.546–7.745	1.800	0.735	2.760	7.503

[a] *Make up the 95% confidence interval (by nonparametric evaluation).*
[b] *Standard deviation.*
[c] *Standard error.*

etiologies have been attached with these changes, although the lesions seem to have some relationship to underlying nutrition and feeding practices. Some scientists have speculated that vitamin E deficiency states are responsible for MWS as oxidative injury is common to many of the changes noted [54]. Because a number of diseases processes can result in progressive weight loss and wasting in common marmosets, the term WMS lacks diagnostic specificity, and its use should be discouraged. When such signs are recognized, it is preferable to refer to the condition by the underlying etiologic or morphologic diagnosis.

The classical lesions and clinical findings of MWS found so often 20 years ago are not as common in current marmoset colonies. This is probably the result of increased optimization of callitrichid nutrition and colony management practices. It should be noted, however, that there are several nutritionally influenced conditions

TABLE 27.10 Reference Data/Individual Organ Weight Analysis/Control Group—Housed Males

Parameter	Unit	Number of Animals	Number of Values	Median	Mean	Mean ± 2SD[a]	SD[b]	SE[c]	Min	Max
Body weight at necropsy	g	36	36	347.1	346.2	269.1–423.4	38.6	6.4	267.0	436.0
ADRENALS										
Left	g	36	36	0.050	0.050	0.024–0.075	0.013	0.002	0.030	0.079
Right	g	36	36	0.046	0.047	0.023–0.071	0.012	0.002	0.020	0.071
Total	g	36	36	0.093	0.097	0.051–0.143	0.023	0.004	0.050	0.152
KIDNEYS										
Left	g	36	36	0.885	0.885	0.652–1.119	0.117	0.019	0.630	1.134
Right	g	36	36	0.917	0.913	0.654–1.171	0.129	0.022	0.690	1.210
Total	g	36	36	1.819	1.797	1.317–2.278	0.240	0.040	1.370	2.320
REPRODUCTIVE ORGANS										
Left testis	g	36	36	0.547	0.508	0.252–0.764	0.128	0.021	0.020	0.734
Right testis	g	36	36	0.551	0.540	0.289–0.791	0.125	0.021	0.210	0.784
Total	g	36	36	1.089	1.047	0.597–1.497	0.225	0.037	0.500	1.518
Left epididymis	g	24	24	0.119	0.132	0.037–0.227	0.047	0.010	0.070	0.257
Right epididymis	g	24	24	0.119	0.134	0.036–0.232	0.049	0.010	0.070	0.253
Total	g	24	24	0.235	0.265	0.082–0.448	0.091	0.019	0.140	0.495
Prostate	g	33	33	0.170	0.177	0.101–0.254	0.038	0.007	0.108	0.250
Seminal vesicles	g	23	23	0.121	0.134	0.017–0.250	0.058	0.012	0.050	0.250
Vesicular glands	g	6	6	0.150	0.170	0.020–0.320	0.075	0.031	0.100	0.310
UNPAIRED ORGANS										
Liver	g	36	36	10.915	11.333	7.034–15.632	2.150	0.358	8.010	18.190
Lungs	g	6	6	1.551	1.490	1.201–1.779	0.145	0.059	1.225	1.595
Thymus	g	18	18	0.244	0.309	0.000–0.711	0.201	0.047	0.087	0.905
Heart	g	36	36	2.120	2.099	1.428–2.769	0.335	0.056	1.240	2.920
Thyroid (with parathyroids)	g	33	33	0.089	0.097	0.013–0.181	0.042	0.007	0.047	0.200
Brain	g	36	36	8.085	8.049	6.922–9.176	0.563	0.094	7.090	9.194
Pancreas	g	6	6	0.548	0.578	0.123–1.032	0.227	0.093	0.256	0.851
Pituitary	g	29	29	0.012	0.013	0.006–0.020	0.004	0.001	0.007	0.022
Salivary glands, mandibular	g	6	6	0.374	0.377	0.239–0.514	0.069	0.028	0.294	0.478
Spleen	g	32	32	0.381	0.436	0.027–0.844	0.204	0.036	0.220	1.129
Mesenteric lymph nodes	g	5	5	0.290	0.344	0.000–0.735	0.195	0.087	0.190	0.680

[a]*Make up the 95% confidence interval (by nonparametric evaluation).*
[b]*Standard deviation.*
[c]*Standard error.*

TABLE 27.11 Reference Data/Individual Organ Weight/Body Weight Ratio Analysis/Control Group—Housed Males

Parameter	Unit	Number of Animals	Number of Values	Median	Mean	Mean ± 2SD[a]	SD[b]	SE[c]	Min	Max
Body weight at necropsy	g	36	36	347.1	346.2	269.1–423.4	38.6	6.4	267.0	436.0
Spleen	%	32	32	0.111	0.125	0.010–0.240	0.057	0.010	0.057	0.322
Pancreas	%	6	6	0.166	0.180	0.009–0.350	0.085	0.035	0.083	0.318
ADRENALS										
Left	%	36	36	0.013	0.015	0.006–0.023	0.004	0.001	0.008	0.026
Right	%	36	36	0.014	0.014	0.006–0.021	0.004	0.001	0.007	0.025
Total	%	36	36	0.027	0.028	0.013–0.043	0.008	0.001	0.016	0.048
KIDNEYS										
Left	%	36	36	0.256	0.257	0.195–0.319	0.031	0.005	0.206	0.325
Right	%	36	36	0.262	0.265	0.196–0.334	0.035	0.006	0.212	0.354
Total	%	36	36	0.517	0.521	0.394–0.648	0.064	0.011	0.418	0.678
REPRODUCTIVE ORGANS										
Left testis	%	36	36	0.153	0.147	0.079–0.215	0.034	0.006	0.006	0.197
Right testis	%	36	36	0.157	0.156	0.090–0.222	0.033	0.006	0.075	0.220
Total	%	36	36	0.308	0.302	0.186–0.419	0.058	0.010	0.178	0.415
Left epididymis	%	24	24	0.033	0.037	0.010–0.064	0.013	0.003	0.019	0.066
Right epididymis	%	24	24	0.034	0.038	0.007–0.069	0.015	0.003	0.019	0.082
Total	%	24	24	0.066	0.075	0.020–0.130	0.028	0.006	0.038	0.141
Prostate	%	33	33	0.050	0.051	0.030–0.072	0.011	0.002	0.033	0.081
Seminal vesicles	%	23	23	0.037	0.039	0.007–0.070	0.016	0.003	0.018	0.070
Vesicular glands	%	6	6	0.045	0.044	0.012–0.077	0.016	0.007	0.024	0.071
UNPAIRED ORGANS										
Liver	%	36	36	3.128	3.278	2.251–4.305	0.513	0.086	2.324	4.347
Lungs	%	6	6	0.452	0.454	0.374–0.534	0.040	0.016	0.407	0.519
Thymus	%	18	18	0.080	0.093	0.000–0.200	0.053	0.013	0.028	0.231
Heart	%	36	36	0.603	0.605	0.480–0.730	0.062	0.010	0.464	0.744
Thyroid (with parathyroids)	%	33	33	0.024	0.028	0.004–0.052	0.012	0.002	0.014	0.058
Brain	%	36	36	2.354	2.346	1.845–2.847	0.250	0.042	1.897	3.007
Pituitary	%	29	29	0.003	0.004	0.002–0.006	0.001	0.000	0.002	0.006
Salivary glands, mandibular	%	6	6	0.114	0.114	0.098–0.129	0.008	0.003	0.104	0.122
Mesenteric lymph nodes	%	5	5	0.080	0.111	0.000–0.248	0.069	0.031	0.061	0.227

[a]*Make up the 95% confidence interval (by nonparametric evaluation).*
[b]*Standard deviation.*
[c]*Standard error.*

that are very common as background diseases in marmosets that have been associated with weight loss and muscle wasting clinically [56,57]. These include metabolic bone disease, inflammatory bowel disease, and chronic renal disease, and they increase in incidence and severity with age. Each of these disease states is a complex disorder with many implications for toxicology studies such as effects on pharmacokinetics and complicating lesion interpretation. Laboratory animal veterinarians in toxicology facilities try to screen animals and make colony assessments to minimize the effects of these disorders.

TABLE 27.12 Reference Data/Individual Organ Weight/Brain Weight Ratio Analysis/Control Group—Housed Males

Parameter	Unit	Number of Animals	Number of Values	Median	Mean	Mean ± 2SD[a]	SD[b]	SE[c]	Min	Max
Brain	g	36	36	8.085	8.049	6.922–9.176	0.563	0.094	7.090	9.194
Spleen	%	32	32	4.734	5.453	0.000–10.913	2.730	0.483	2.622	15.122
Pancreas	%	6	6	7.036	7.604	1.360–13.848	3.122	1.275	3.282	11.437
ADRENALS										
Left	%	36	36	0.592	0.622	0.285–0.958	0.168	0.028	0.352	1.058
Right	%	36	36	0.568	0.589	0.275–0.903	0.157	0.026	0.253	0.951
Total	%	36	36	1.138	1.212	0.598–1.825	0.307	0.051	0.632	2.036
KIDNEYS										
Left	%	36	36	10.922	11.030	7.947–14.112	1.541	0.257	8.597	15.189
Right	%	36	36	11.010	11.375	7.947–14.804	1.714	0.286	8.723	15.510
Total	%	36	36	21.858	22.400	16.017–28.783	3.192	0.532	17.320	30.766
REPRODUCTIVE ORGANS										
Left testis	%	36	36	6.634	6.323	3.171–9.474	1.576	0.263	0.245	8.542
Right testis	%	36	36	6.790	6.722	3.639–9.805	1.542	0.257	2.482	9.124
Total	%	36	36	13.290	13.039	7.497–18.580	2.771	0.462	5.910	17.666
Left epididymis	%	24	24	1.404	1.623	0.343–2.904	0.640	0.131	0.830	3.393
Right epididymis	%	24	24	1.440	1.653	0.313–2.993	0.670	0.137	0.847	3.248
Total	%	24	24	2.850	3.265	0.761–5.768	1.252	0.255	1.695	6.536
Prostate	%	33	33	2.145	2.196	1.229–3.162	0.483	0.084	1.306	3.179
Seminal vesicles	%	23	23	1.517	1.644	0.337–2.951	0.653	0.136	0.591	2.917
Vesicular glands	%	6	6	1.804	2.034	0.188–3.880	0.923	0.377	1.175	3.748
UNPAIRED ORGANS										
Liver	%	36	36	134.500	141.743	81.735–201.751	30.004	5.001	88.308	219.952
Lungs	%	6	6	20.196	19.491	15.830–23.153	1.831	0.747	16.463	21.059
Thymus	%	18	18	3.253	4.076	0.000–9.493	2.708	0.638	1.115	11.949
Heart	%	36	36	26.637	26.086	18.399–33.772	3.843	0.641	16.664	34.313
Thyroid (with parathyroids)	%	33	33	1.098	1.204	0.153–2.255	0.525	0.091	0.560	2.478
Pituitary	%	29	29	0.154	0.156	0.065–0.247	0.045	0.008	0.085	0.279
Salivary glands, mandibular	%	6	6	4.826	4.931	3.107–6.755	0.912	0.372	3.951	6.311
Mesenteric lymph nodes	%	5	5	3.901	4.525	0.000–10.117	2.796	1.251	2.464	9.379

[a]*Make up the 95% confidence interval (by nonparametric evaluation).*
[b]*Standard deviation.*
[c]*Standard error.*

Metabolic bone disease is common in many marmoset colonies, and there are several distinct naturally occurring skeletal lesions that have been described and that need to be differentiated from compound-induced effects [58]. Recently, a link has been suggested between the findings of chronic inflammatory bowel disease in marmosets and metabolic bone disease [56]. To further increase the challenges of marmoset background pathology for the toxicologic pathologist, skeletal lesions can be found in association with renal secondary hyperparathyroidism that commonly occurs in marmoset kidney disease (Fig. 27.5). Both glomerular and tubulointerstitial renal diseases are found in relatively high incidence in marmoset colonies, although they tend to be less prevalent in toxicology studies that utilize young animals.

Marmosets have many spontaneous endocrine lesions, including lymphocytic thyroiditis, C-cell hyperplasia, and hyperplasia of the parathyroid glands. These are common background lesions in many colonies and increase in incidence with age. Endocrine lesions can have systemic effects on metabolism creating confounders for studies, but many lesions noted are minor incidental changes. Numerous marmoset organs contain focal infiltrates of lymphocytes, and these must be differentiated from extramedullary hematopoiesis, another extremely common incidental finding that is also believed to vary substantially by animal source, colony husbandry, and history of blood withdrawal on study (Fig. 27.6).

One marmoset organ system with a high incidence of spontaneously occurring disease is the kidney (Fig. 27.7). There are several progressive and nonprogressive renal diseases, including glomerular diseases and tubulointerstitial disease [59–63]. The high incidence of renal disease in marmoset colonies makes routine use of clinical chemistry screening part of periodic physicals performed in many toxicology marmoset colonies. Another common disease is inflammatory bowel disease which is commonly manifested in marmosets as chronic lymphoid enteritis (Fig. 27.8), particularly of the small intestine but also sometimes involving the cecum and/or colon [57]. The high incidence of intestinal inflammatory disease is sometimes associated with changes in absorption and effects on

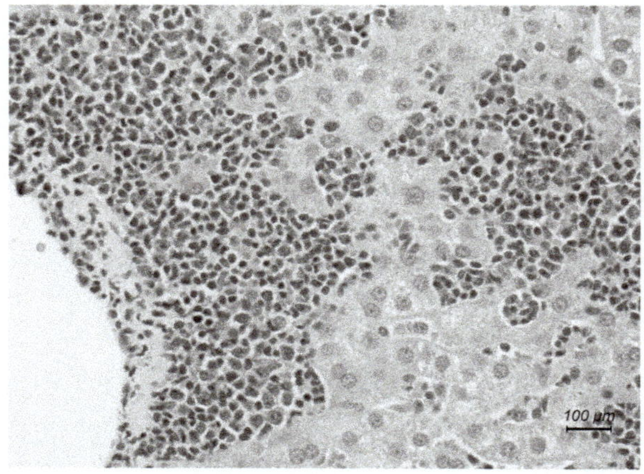

FIGURE 27.6 Incidental extramedullary hematopoiesis noted at euthanasia in the liver of a control animal.

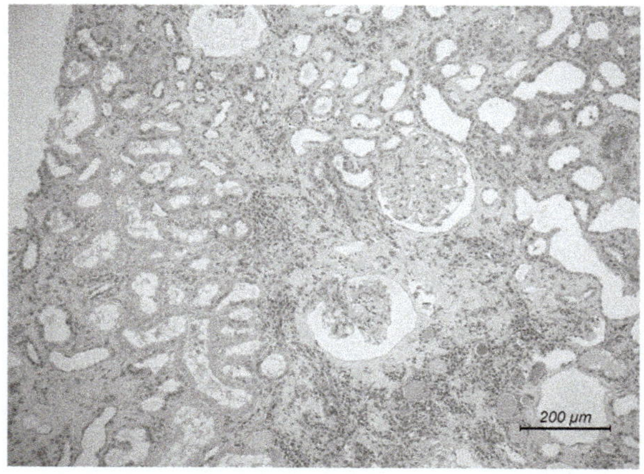

FIGURE 27.7 Chronic renal disease in a control marmoset. There are marked tubule-interstitial and glomerular changes. There is tubular cell atrophy with marked expansion of the interstitium with fibrosis and a lymphoplasmacytic infiltrate. Evidence of mesangial expansion is noted in a single central glomerulus and tuft atrophy and early glomerulosclerosis in remaining glomeruli in the field.

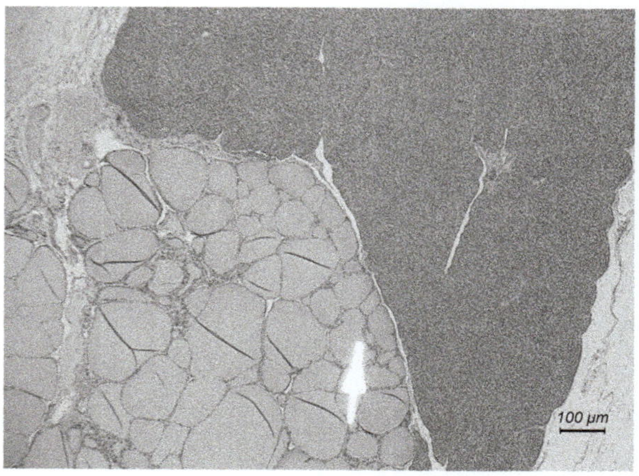

FIGURE 27.5 Marked parathyroid hyperplasia in a nontreated marmoset that had chronic renal disease. This animal also had evidence of widespread fibrous osteodystrophy from renal secondary hyperparathyroidism.

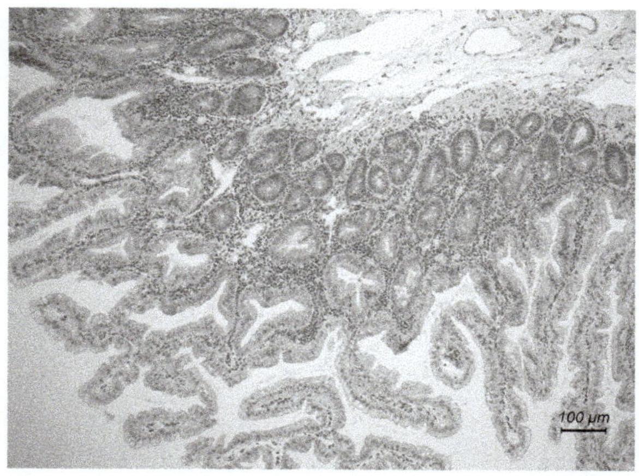

FIGURE 27.8 Moderate diffuse infiltrate of chronic inflammatory cells in the jejunal lamina propria of an untreated marmoset. This is typical of early inflammatory bowel disease in marmosets and progresses to a more severe lesion over time.

other organ systems (i.e., skeletal) and potentially on pharmacokinetic parameters of orally administered compounds.

The toxicologic pathologist that evaluates marmoset tissues following compound administration must carefully consider lesions with full understanding of the colony history, lesion status of concurrent controls, and knowledge of drug and experimental manipulation.

CONCLUSION

Although they are not commonly used in safety assessment studies relative to macaque species, marmosets have a number of features that make them attractive as NHP models in toxicology studies. They have utility as models for human disease and for testing therapeutic efficacy, thus they will continue to be an important alternative NHP toxicology model for select studies.

References

[1] Zühlke U, Weinbauer G. The common marmoset (*Callithrix jacchus*) as a model in toxicology. Toxicol Pathol 2003;31(Suppl.): 123–7.

[1a] Fischer KE, Austad SN. The development of small primate models for aging research. ILAR J 2011;52(1):78–88.

[2] Mätz-rensing K, Korte S. Basic physiology of *Callithrix jacchus*. In: Blümel J, Korte S, Schenck E, Weinbauer GF, editors. The nonhuman primate in nonclinical drug development and safety assessment. Elsevier: Academic Press; 2015. p. 115–33.

[2a] Abbott DH, Barnett DK, Colman RJ, Yamamoto ME, Schultz-Darken NJ. Aspects of common marmoset basic biology and life history important for biomedical research. Comp Med 2003;53(4):339–50.

[2b] Mansfield K. Marmoset models commonly used in biomedical research. Comp Med 2003;53(4):383–92.

[3] Korte S, Zühlke U, Kaspareit J, Friderichs-Gromoll S, Mülle W, Vogel F, Weinbauer GF. Reference control data for the common marmoset (*Callithrix jacchus*): a comparison with macaques. Toxicologist 2004;78:205.

[4] Orsi A, Rees D, Andreini I, Venturella S, Cinelli S, Oberto G. Overview of the marmoset as a model in nonclinical development of pharmaceutical products. Regul Toxicol Pharmacol 2011;59:19–27.

[5] Kap YS, Jagessar SA, Dunham J, 't Hart BA. The common marmoset as an indispensable animal model in immunotherapy development for multiple sclerosis. Drug Discov Today 2016;8:1200–5.

[6] 't Hart BA, Abbott DH, Nakamura K, Fuchs E. The marmoset monkey: a multi-purpose preclinical and translational model of human biology and disease. Drug Discov Today 2012;17:1160–5.

[6a] Vierboom MP, Breedveld E, Kondova I, 't Hart BA. Collagen-induced arthritis in common marmosets: a new nonhuman primate model for chronic arthritis. Arthritis Res Ther 2010;12(5): R200. https://doi.org/10.1186/ar3172.

[7] Carrion Jr R, Patterson JL. An animal model that reflects human disease: the common marmoset (*Callithrix jacchus*). Curr Opin Virol 2012;2:357–62.

[8] Sato K, Kuroki Y, Kumita W, Fujiyama A, Toyoda A, Kawai J, Iriki A, Sasaki E, Okano H, Sakakibara Y. Resequencing of the common marmoset genome improves genome assemblies and gene coding sequence analysis. Sci Rep 2015;5:1–8.

[9] Datson NA, Morsink MC, Atanasova S, Armsrong VW, Zischler H, Schlumbohm C, Dutilh BE, Huynen MA, Waegele B, Ruepp A, de Kloet E/R, Fuchs E. Development of the first marmoset-specific DNA microarray (EUMAMA): a new genetic tool for large-scale expression profiling in a non-human primate. BMC Genom 2007;8:1–9.

[10] Tomioka I, Maeda T, Shimada H, Kawai K, Okada Y, Igarashi H, Oiwa R, Iwasaki T, Aoki M, Kimura T, Shiozawa S, Shinohara H, Suemizu H, Sasaki E, Okano H. Generating induced pluripotent stem cells from common marmoset (*Callithrix jacchus*) fetal liver cells using defined factors, including Lin28. Gene Cells 2010;15: 959–69.

[11] Sasaki E, Suemizu H, Shimada A, Hanazawa K, Oiwa R, Kamioka M, Tomioka I, Sotomaru Y, Hirakawa R, Eto T, Shiozawa S, Maeda T, Ito M, Ito R, Kito C, Yagihashi C, Kawai K, Miyoshi H, Tanioka Y, Tamaoki N, Habu S, Okano H, Nomura T. Generation of transgenic non-human primates with germline transmission. Nature 2009;459:523–7.

[12] Chapman KL, Pullen N, Graham M, Ragan I. Preclinical safety testing of monoclonal antibodies: the significance of species relevance. Nat Rev Drug Discov 2007;6:120–6.

[13] Chapman KL, Pullen N, Coney L, Dempster M, Andrews L, Bajramovic J, Baldrick P, Buckley L, Jacobs A, Hale G, Green C, Ragan I, Robinson V. Preclinical development of monoclonal antibodies: considerations for the use of non-human primates. mAbs 2009;1:505–16.

[14] Chapman KL, Pullen N, Andrews L, Ragan I. The future of non-human primate use in mAb development. Drug Discov Today 2010;15:235–42.

[14a] Wilhelm AJ, den Burger JC, Swart EL. Therapeutic drug monitoring by dried blood spot: progress to date and future directions. Clin Pharmacokinet 2014;53(11):961–73. https://doi.org/10.1007/s40262-014-0177-7.

[14b] Ruiz de Elvira MC, Abbott DH. A backpack system for long-term osmotic minipump infusions into unrestrained marmoset monkeys. Lab Anim 1986;20(4):329–34.

[15] Fitzgerald AL, Dillon LM, Altrogge DM, Bleavins MR, Breider MA. Use of subcutaneous vascular access ports in common marmosets (*Callithrix jacchus*). Contemp Top 1996;35:57–9.

[16] Kurata Y, Takechi M, Toyota N, Tsuchitani M, Katoh M, Rusch GM, Trochimowicz HJ, Shin-Ya S. Four-week repeated inhalation study of HCFC 225ca and HCFC 225cb in the common marmoset. Toxicol Lett 1997;92:209–19.

[17] Tomlinson L, Boone LI, Ramaiah L, Penraat KA, von Beust BR, Ameri M, Poitout-Belissent FM, Weingand K, Workman HC, Aulbach AD, Meyer DJ, Brown DE, MacNeill AL, Bolliger AP, Bounous DI. Best practices for veterinary toxicologic clinical pathology, with emphasis on the pharmaceutical and biotechnology industries. Vet Clin Pathol 2012;42:252–69.

[18] Davy CW, Jackson MR, Walker J. Reference intervals for some clinical chemical parameters in the marmoset (*Callithrix jacchus*): effect of age and sex. Lab Anim 1984;18:135–42.

[19] Hawkey CM, Hart MG, Jones DM. Clinical hematology of the common marmoset *Callithrix jacchus*. Am J Primatol 1982;3: 179–99.

[20] Kuehnel F, Grohmann J, Buchwald U, Koeller G, Teupser D, Einspanier A. Parameters of haematology, clinical chemistry and lipid metabolism in the common marmoset and alterations under stress conditions. J Med Primatol 2012;41: 241–50.

[21] Yarbrough LW, Tollet JL, Montrey RD, Beattie RJ. Serum biochemical, hematological and body measurement data for the common marmosets (*Callithrix jacchus*). Lab Anim Sci 1984; 34:276–80.

[22] Davy CW, Trennery PN, Edmunds JG, Altman JFB, Eichler DA. Local myotoxicity of ketamine hydrochloride in the marmoset. Lab Anim 1987;17:60–7.

[23] Broadmeadow A. The common marmoset (*Callithrix jacchus*) in preclinical research and development of new pharmaceutical products. J Toxicol Sci 2005;30:S123.

[24] Sakuma T, Igarashi T, Hieda M, Ohgiya S, Isogai M, Ninomiya S, Nagata R, Nemoto N, Kamataki T. Marmoset CYP1A2: primary structure and constitutive expression in livers. Carcinogenesis 1997;18:1985–91.

[25] Soars MG, Riley RJ, Burchell B. Evaluation of the marmoset as a model species for drug glucuronidation. Xenobiotica 2001;31: 849–60.

[26] Uno Y, Uehara S, Yamazaki H. Utility of non-human primates in drug development: comparison of non-human primate and human drug-metabolizing cytochrome P450 enzymes. Biochem Pharmacol 2016;121:1–7.

[27] Uehara S, Uno Y, Nakanishi K, Ishii S, Inoue T, Sasaki E, Yamazaki H. Marmoset cytochrome P450 3A4 ortholog expressed in liver and small-intestine tissues efficiently metabolizes midazolam, alprazolam, nifedipine, and testosterone. Drug Metab Dispos 2017;45:457–67.

[28] Parrott N, Davies B, Hoffmann G, Koerner A, Lave T, Prinssen E, Theogarai E, Singer T. Development of a physiologically based model for oseltamivir and simulation of pharmacokinetics in neonates and infants. Clin Pharmacokinet 2011;50: 613–23.

[29] Schnell CR, Wood JM. Measurement of blood pressure and heart rate by telemetry in conscious, unrestrained marmosets. Am J Physiol 1993;264:509–16.

[30] Tabo M, Hara T, Sone S, Shishido N, Kuramoto S, Nakano K, Onodera H, Kimura K, Kobayashi K. Prediction of drug induced QT interval prolongation in telemetered common marmodets. J Toxicol Sci 2008;33:315–25.

[31] Horii I, Kito G, Hamada T, Jikuzono T, Kobayashi K, Hashimoto K. Development of telemetry system in the common marmoset—cardiovascular effects of astemizole and nicardipine. J Toxicol Sci 2002;27:123–30.

[32] Holzgrefe H, Ferber G, Champeroux P, Gill M, Honda M, Greiter-Wilke A, Baird T, Meyer O, Saulnier M. Preclinical QT safety assessment: cross-species comparisons and human translation from an industry consortium. J Pharmacol Toxicol Methods 2014;69:61–101.

[33] Mietsch M, Einspanier A. Non-invasive blood pressure measurement: values, problems and applicability in the common marmoset (*Callithrix jacchus*). Lab Anim 2015;49:241–50.

[34] Hanton G, Eder V, Bonnet P, Rochefort GY. Echocardiography in marmosets: a non-invasive method for the assessment of cardiovascular toxicology and pharmacology. In: Weinbauer GF, Vogel F, editors. Novel approaches towards primate toxicology. Münster (New York): Waxmann Publishing Co; 2006. p. 27–45.

[35] Hanton G, Sobry S, Dagues N, Rochefort GY, Bonnet P, Eder V. Cardiovascular toxicity of minoxidil in the marmoset. Toxicol Lett 2008;180:157–65.

[36] Braun K, Schultz-Darken N, Schneider M, Moore CF, Emborg ME. Development of a novel postnatal neurobehavioral scale for evaluation of common marmoset monkeys. Am J Primatol 2015;77:401–17.

[36a] Neubert E, Gürtler H, Vallentin G. [Effect of restraining growth pigs with snare restraints on plasma levels of catecholamines, cortisol, insulin and metabolic parameters]. Berl Munch Tierarztl Wochenschr 1996;109(11-12):409–13.

[37] Kametani Y, Suzuki D, Kohu K, Satake M, Suemizu H, Sasaki E, Ito T, Tamaoki N, Mizushima T, Ozawa M, Tani K, Kito M, Arai H, Koyanagi A, Yagita H, Habu S. Development of monoclonal antibodies for analyzing immune and hematopoietic systems of common marmoset. Exp Hematol 2009;37:1318–29.

[38] Neumann B, Shi T, Gan LL, Klippert A, Daskalaki M, Stolte-Leeb N, Stahl-Hennig C. Comprehensive panel of cross-reacting monoclonal antibodies for analysis of different immune cells and their distribution in the common marmoset (*Callithrix jacchus*). J Med Primatol 2016;45:139–46.

[39] Foerster M, Delgado I, Abraham K, Gerstmayr S, Neubert R. Comparative study on age-dependent development of surface receptors on peripheral blood lymphocytes in children and young nonhuman primates (marmosets). Life Sci 1997;60: 773–8.

[40] Grote-Wessels S, Frings W, Smith CA, Weinbauer GF. Immunotoxicity testing in non-human primates. Methods Mol Biol 2010;598:341–59.

[41] VanderVeen N, Paran C, Appelhans A, Krasinkiewicz J, Lemons R, Appelman A, Doherty R, Palmer D, Ng P, Lowenstein PR, Castro MG. Marmosets as a preclinical model for testing "off-label" use of doxycycline to turn on Flt3L expression from high-capacity adenovirus vectors. Mol Ther Methods Clin Dev 2014;5:1–9.

[42] Quint DJ, Buckham SP, Bolton EJ, Solari R, Champion BR, Zanders ED. Immunoregulation in the common marmoset, *Calithrix jaccus*: functional properties of T and B lymphocytes and their response to human interleukins 2 and 4. Immunology 1990;69:616–21.

[42a] Chellman GJ, Bussiere JL, Makori N, Martin PL, Ooshima Y, Weinbauer GF. Developmental and reproductive toxicology studies in nonhuman primates. Birth Defects Res B Dev Reprod Toxicol 2009;86(6):446–62. https://doi.org/10.1002/bdrb.20216.

[43] Graham MJ, Wilson SA, Winham MA, Spencer AJ, Rees JA, Old SL, Bonner FW. Lack of peroxisome proliferation in marmoset liver following treatment with ciprofibrate for 3 years. Fundam Appl Toxicol 1994;22:58–64.

[44] Makowska JM, Gibson GG, Bonner FW. Species differences in ciprofibrate induction of hepatic cytochrome P450 4A1 and peroxisome proliferation. J Biochem Toxicol 1992;7:183–91.

[45] Spencer AJ, Barbolt TA, Henry DC, Eason CT, Sauerschell RJ, Bonner FW. Gastric morphological changes including carcinoid tumors in animals treated with a potent hypolipidemic agent, ciprofibrate. Toxicol Pathol 1989;17:7–15.

[46] Keenan CM, Baker JF, Bradley AE, Goodman DG, Harada T, Herbert R, Kaufman W, Kellner R, Mahler B, Meseck E, Nolte T, Rittinghausen S, Vahle J, Yoshizawa K. International harmonization of nomenclature and diagnostic criteria (INHAND) progress to date and future plans. J Toxicol Pathol 2015;28:51–3.

[47] Sellers RS, Morton D, Michael B, Roome N, Johnson JK, Yano BL, Perry R, Schafer K. Society of Toxicologic Pathology position paper: organ weight recommendations for toxicology studies. Toxicol Pathol 2007;35:751–5.

[48] Chalmers DT, Murgatroyd LB, Wadsworth PF. A survey of the pathology of marmosets (*Callithrix jacchus*) derived from a marmoset breeding unit. Lab Anim 1983;17:270–9.

[49] David JM, Dick EJ, Hubbard GB. Spontaneous pathology of the common marmoset (*Callithrix jacchus*) and tamarins (*Saguinus oedipus, Saguinus mystax*). J Med Primatol 2009;38:347–59.

[50] Kaspareit J, Friderichs-Gromoll S, Buse E, Habermann G. Background pathology of the common marmoset (*Callitrix*

jacchus) in toxicological studies. Exp Toxicol Pathol 2006;57: 405–10.

[51] Okasaki Y, Kurata Y, Makinodan F, Kidachi F, Yokoyama M, Wako Y, Yamagishi Y, Katsuta O, Takechi M, Tsuchitani M. Spontaneous lesions detected in the common cotton-eared marmosets (*Callithrix jacchus*). J Vet Med Sci 1996;58:181–90.

[52] Tucker MJ. A survey of the pathology of marmosets (*Callithrix jacchus*) under experiment. Lab Anim 1984;18:351–8.

[53] Miller GF, Barnard DE, Woodward RA, Flynn BM, Bulte JW. Hepatic haemosiderosis in common marmosets, *Callithrix jacchus*: effect of diet on the severity and incidence. Lab Anim Sci 1997; 47:138–42.

[54] Juan-Salles C, Prats N, Resendes A, Domingo M, Hilton D, Ruiz JM, Garner MM, Valls X, Marco AJ. Anemia, myopathy and pansteatitis in vitamin E-deficient captive marmosets (Callithrix spp.). Vet Pathol 2003;40:540–7.

[55] Logan AC, Khan KNM. Clinical pathologic changes in two marmosets with wasting syndrome. Toxicol Pathol 1996;24: 707–9.

[56] Baxter VK, Shaw GC, Sotuyo NP, Carlson CS, Olson EJ, Zink CM, Mankowski JL, Adams RJ, Hutchinson EK, Metcalf Pate KA. Serum albumin and body weight as biomarkers for the antemortem identification of bone and gastrointestinal disease in the common marmoset. PLoS One 2013;8:e827477.

[57] Ludlage E, Mansfield K. Clinical care and diseases of the common marmoset (*Callithrix jacchus*). Comp Med 2003;53:369–82.

[58] Olson EJ, Shaw GC, Hutchinson EK, Schultz-Darken N, Bolton ID, Parker JB, Morrison JM, Baxter VK, Metcalf Pate KA, Mankowski JL, Carlson CS. Bone disease in the common marmoset: radiographic and histological findings. Vet Pathol 2015;52:883–93.

[59] Brack M. IgM-mesangial nephropathy in callithricids. Vet Pathol 1988;25:270–6.

[60] Brack M, Schroeder C, Fooke M, Schlumberger W. IgM/IgA nephropathy in callitrichids: antigen studies. Nephron 1999;82: 221–31.

[61] Collins MG, Rogers NM, Jesudason S, Kireta S, Brealey J, Coates PT. Spontaneous glomerular mesangial lesions in common marmoset monkeys (*Callithrix jacchus*): a benign non-progressive glomerulopathy. J Med Primatol 2014;43: 477–87.

[62] Isobe K, Adachi K, Hayashi S, Ito T, Miyoshi A, Kato A, Suzuki M. Spontaneous glomerular and tubulointerstitial lesions in common marmosets (*Callithrix jacchus*). Vet Pathol 2012;49: 839–45.

[63] Yamada M, Sato J, Kanno T, Wako Y, Tsuchitani M. Morphological study of progressive glomerulonephropathy in common marmosets (*Callithrix jacchus*). Toxicol Pathol 2013;41:1106–15.

Further Reading

[1] Evans GO, Smith DE. Platelet measurements in the common marmoset, *Callithrix jacchus*. J Comp Pathol 1986;96:343–7.

[2] Neubert RT, Webb JR, Neubert D. Feasibility of human trials to assess developmental immunotoxicity, and some comparison with data on New World monkeys. Hum Exp Toxicol 2002;21: 543–67.

[3] Ryffel B, Car BD, Woerly G, Weber M, DiPadova F, Kammuler M, Klug S, Neubert R, Neubert D. Long-term interleukin-6 administration stimulates sustained thrombopoiesis and acute-phase protein synthesis in a small primate-The marmoset. Blood 1994;83: 2093–102.

[4] Zühlke U, Korte S, Niggemann B, Fuchs A, Müller W. Feasibility of long-term continuous intravenous infusion in unrestrained marmoset monkeys. Toxicologist 2003;72(S1-1).

CHAPTER

28

Experimental Infections of the Common Marmoset (*Callithrix jacchus*)

Jean L. Patterson[1], *Robert E. Lanford*[2]

[1]Virology and Immunology, Texas Biomedical Research Institute, San Antonio, TX, United States; [2]Southwest National Primate Research Center, Texas Biomedical Research Institute, San Antonio, TX, United States

Interest in the use of marmosets for experimental infectious disease has dramatically increased in the last decade. These animals are native to the Atlantic coastal forests in northeastern Brazil. The majority of experimental animals come from the National Primate Research Centers and other breeding facilities. They are advantageous because of their relative small size, weighting 350–400 g as adults, their life span is compact compared with other nonhuman primate (NHP), and they produce offspring by 3 years of age. They are free of Herpes B virus and, it is believed, to date, other dangerous human pathogens (Abbot et al., 2003) [1]. We describe here the experimental infections of marmosets to human pathogens. While it is always interesting to compare various NHPs with each other, the importance of an animal model is always in comparing its similarities to human infections.

MARMOSET AS A MODEL FOR VIRAL ENCEPHALITIS

The infection of human Eastern equine encephalitis virus (EEEV) was the first experimental encephalitis in marmosets [2]. EEEV is an alphavirus and a severe human arthropod-borne disease in North America (NA) and is genetically and antigenically distinct from a strain from South America (SA) (strain BeAr436087). South American strains have not been associated with severe human disease or mortality. To confirm this apparent phenotypic difference, the marmoset was infected using the intranasal route and monitored for weight loss, fever, anorexia, depression, and other neurological signs. The NA EEEV-infected animals died or were euthanized within 5 days due to anorexia and other neurological signs. The SA EEEV-infected animals remained healthy throughout the course of the experiment even though they developed viremia titers of $2.8–3.1^{10}$ PFU/mL by days 2–4 after infection. Interestingly, NA EEEV-infected animals showed no detectable viremia; however, virus was detected in the brain, liver, and muscle of NA EEEV-infected animals at the time of death or euthanasia. Additionally, the NA EEEV-infected animals developed meningoencephalitis in the cerebral cortex with some perivascular hemorrhages, which were consistent with brain lesions described in human infections of EEEV. These experiments confirmed previous studies that the SA strain induced a higher level of viremia compared with NA EEEV strains [3], and that NA strains, while showing undetectable or transient levels of viremia, induced encephalitis and subsequently death [4]. In summary, this study showed that the marmoset was a very useful small nonhuman primate (NHP) model of human EEEV.

West Nile virus (WNV), a flavivirus, is another common form of arthropod-borne encephalitis, which occurs worldwide and was responsible for sporadic encephalitis in parts of Africa, Asia, and Europe. In 1999, the virus was detected in New York City and spread rapidly through the United States, the Caribbean, Central America, and Canada. By 2004, there were over 7000 cases of neuroinvasive WNV disease in the United States alone [5]. The enzootic cycle occurs between mosquitoes and birds, whereas the virus can be transmitted, via mosquitoes, to equines and humans. Twenty percent of infected individuals develop a febrile syndrome; however, less than 1% of infected individuals progress to develop meningitis, encephalitis, or paralysis [6].

Experimental infection of the common marmoset with the virulent European WNV strain (WNV-Ita09) resulted in a more productive infection compared

The Common Marmoset in Captivity and Biomedical Research
https://doi.org/10.1016/B978-0-12-811829-0.00028-5

with rhesus monkeys, with higher rates of viral replication and a wider tissue distribution of virus [7]. Neither species showed behavioral changes or impaired kidney function, even though marmosets had detectable WNV RNA in kidney tissues. Both species developed a natural killer response followed by IgM and IgG antibody responses. The common marmoset is at least as susceptible to infection with WNV, if not more susceptible, than macaques, and due to a comparable immune response as that seen in humans, it could serve as a model for WNV disease.

MARMOSET AND DENGUE VIRUS INFECTIONS

Dengue virus (DENV) is an arthropod-borne flavivirus and a major cause of worldwide morbidity and mortality. The four serotypes (DENV 1, 2, 3, and 4) of DENVs are transmitted in tropical countries. All serotypes cause syndromes that are self-limited or severe. DENVs can cause a flu-like disease often referred to as "break-bone fever" or dengue hemorrhagic fever/dengue shock syndrome (DHF/DSS), which is characterized by sudden vascular permeability. It has been hypothesized that DHF/DSS is the result of enhancing and cross-reactive neutralizing antibodies that produce the syndrome with sequential dengue infections of a different strain [8].

Common marmosets developed high levels of viremia after inoculation with clinical isolates of all four serotypes of DENV, generating 1×10^6 genome copies/ml after inoculation with DENV-2. DENV-2 RNA was detected in lymphoid organs including lymph nodes, spleen, and thymus as well as nonlymphoid organs such as lung, liver, kidneys, pancreas, and urinary bladder. When marmosets were inoculated with DENV-2 and then reinfected, it appeared that primary inoculation with DENV-2 induced protective immunity to only DENV-2. More recently [9], it has been shown that after primary infection with DENV-1, 2, and 3, inoculation with heterologous virus produced viremia to that virus as well as high levels of neutralizing antibody titers to the challenge infection not only from serotypes with which they had been inoculated but also to serotypes to which they had not been inoculated. While the primary infection with a serotype elicited lifelong immunity to that serotype, the immunity to other serotypes was short lived. This suggested that DENV infection with any subtype does not prevent reinfection with a different serotype, and that cross-reacting antibodies develop which may complicate the immune response, as seen in human DENV infection. These data suggest that marmosets may prove to be a valuable tool to assess dengue vaccine candidates.

MARMOSET AND CORONAVIRUS INFECTIONS

Severe respiratory infections from coronaviruses (CoVs) have emerged in the last decade. The first of the severe respiratory illnesses to emerge was severe acute respiratory syndrome (SARS). The disease primarily presents with pneumonitis but later can progress to renal, gastrointestinal, and hepatic disease. When common marmosets were inoculated intratracheally with cell supernatant containing the SARS-associated pathogen, SARS-CoV, and evaluated over time up to 7 days after infection [10], all animals developed multifocal mononuclear cell interstitial pneumonitis, multinucleated syncytial cells, and some developed bronchiolitis. At 2, 4, and 7 days after infection, viral RNA was detected from pulmonary extracts and sometimes detected in tracheobronchial lymph node and myocardium. As in human disease, there was a correlation between the expression of the primary functional receptor, ACE-2, SARS-CoV-associated pathology, and the detection of viral RNA [11,12]. This study suggested that ACE-2 expression levels were similar to humans; however, SARS-CoV infection in marmosets did not present many of the severe clinical signs of respiratory disease that were observed in humans.

Middle East respiratory syndrome (MERS) emerged in 2012. The etiological agents of MERS are Middle East respiratory syndrome associated coronavirus (MERS-CoV). WHO estimated that by December 2016 there had been 1841 cases of MERS in 27 countries. The outbreak originated in Saudi Arabia, with the most recent serious outbreak being in South Korea (WHO updates). Marmosets became of interest as a model for MERS-CoV when dipeptidyl peptidase 4 (DPP4), also known as CD26, was shown to be a functional receptor for MERS-CoV [13]. Comparison of marmoset DPP4 showed that it is 96.4% identical to human DPP4 [14]. When marmosets were inoculated via combined intratracheal, intranasal, and/or ocular routes, most marmosets developed a lethal pneumonia. There were extensive lesions evident in the lungs, and viral titers of 1.5×10^6 were detected at 3 days postinfection. MERS-CoV RNA was detected in blood, which suggested a systemic dissemination, and viral RNA was detected in nearly all tissues in infected animals. In another study [15], investigators compared two isolates of MERS-CoV inoculated via the intratracheal route of exposure. This report showed that marmosets developed a mild-to-moderate nonlethal respiratory disease. Chan et al. [16] were able to improve the outcome of disease by treatment with Lopinavir/Ritonavir or interferon-B1b when marmosets were intratracheally delivered MERS-CoV. All these data support that marmosets are a valuable NHP model for MERS-CoV.

HEMORRHAGIC FEVERS AND MARMOSETS

Lassa Fever Virus

Marmosets have been recently evaluated as models for various hemorrhagic fevers. The first hemorrhagic fever examined was Lassa fever, a member of the Arenaviridae family. Lassa virus (LASV) is the causative agent of Lassa fever, with approximately 300,000 cases occurring each year in Western Africa. The fatality rates are between 1% and 2%, with as many as 100,000 hospitalized [17]. Lassa is transmitted from a rodent reservoir, *Mastomys natalensis*, via rodent excreta or by close contact with infected individuals. The disease progression includes nausea, vomiting, diarrhea, and abdominal pain, with hemorrhages occurring in about 20% of cases.

Common marmosets were subcutaneously inoculated with LASV strain Josiah [18]. The inoculation produced systemic disease with clinical features mimicking those of fatal human Lassa infection and included fever, weight loss, high levels of viremia, and viral RNA found in tissues. There were elevated liver enzymes and severe morbidity between days 15 and 20 postinfection. Importantly, there was also multifocal hepatic necrosis with mild inflammation and hepatocyte proliferation, lymphoid depletion, and nephritis. Immunophenotypic alterations in the spleen included reductions in overall numbers of CD20-positive and CD3-positive cells and disturbances with lymphoid follicular architecture. These data showed that in an animal model infected with LASV, infection impaired adaptive immunity as has been noted in LASV infections in humans [19].

The marmoset model was subsequently utilized to evaluate the efficacy of the ML29 reassortant vaccine for LASV [20]. The ML29 vaccine candidate encodes the NP and GPC of LASV and the Z protein and L proteins of Mopeia virus (MOPV). MOPV is an Old World group of arenaviruses related to LASV; however, this virus is nonpathogenic in experimental animals [21]. The virulence for arenaviruses is encoded in the large genome segment, encoding the Z and L proteins. Thus this vaccine has the antigenicity of LASV with the virulence of MOPV. It was determined that a single injection of ML29 reassortant vaccine induced low, transient viremia, and low or moderate levels of ML29 replication in marmoset tissue. This result was expected for a live-attenuated vaccine. The vaccination results in immune responses and the induction of sterilizing cell-mediated immunity. These data support the notion that ML29 is a promising vaccine candidate, and that the marmoset could be used for vaccine and treatment developments.

Filoviridae

Ebola and Marburg viruses (MARV) are members of the Filoviridae and induce a serious hemorrhagic disease. The filoviruses most commonly associated with causing human disease are MARV and Ebola virus (formerly known as Ebola Zaire) (EBOV) [22]. The case fatality rates (CFR) reported, in the various outbreaks, range from 24% to 100%. The average mortality of EBOV disease is estimated at 65% and MARV disease at 54% [23]. Of the know outbreaks, the most recent outbreaks of EBOV in West Africa had a CFR of 47%, the lowest of all of the EBOV outbreaks [24]. It is generally held that the West African outbreaks resulted in lower mortality due to better supportive care that allowed time for the immune system to clear the virus.

Marmosets have been shown to be a model for filovirus-induced disease [25]. The marmosets were inoculated intramuscularly with EBOV and MARV. The infections resulted in systemic fatal disease with clinical features resembling human infections. Animals experienced weight loss, fever, high virus titers in tissues, thrombocytopenia, neutrophilia, high liver transaminases and phosphatases, and disseminated intravascular coagulation. There were some differences in the histopathology between MARV and EBOV. Ebola infection showed widespread intravascular coagulation in multiple organs, including spleen, adrenal gland, kidney, and lung. There was extensive lymphocytic necrosis in medullary and parafollicular regions. However, follicles showed no evidence of necrosis. There was extensive fibrin deposition. MARV-inoculated animals had moderate fibrin deposition in the spleen, whereas intravascular coagulation was not observed. Interestingly, the hepatic lesions were less severe in animals that received a higher dose of virus, which suggested that the disease course developed too rapidly to allow for the development of any hepatic pathology. Also, marmosets did not develop petechial rash, which suggested that rapid progression of disease prevented some symptoms from occurring. These authors suggest that marmosets have a very similar disease course to human disease, and this NHP model may be superior to others in understanding the pathogenesis of the filovirus disease [26].

Development of a vaccine against an air-borne filovirus is a priority, particularly to the military. Filoviruses are select agents and are therefore considered to be a possible biowarfare agent [23]. As filoviruses may be aerosolized, investigators developed marmoset models where MARV was delivered via the respiratory route. Smither et al. [27] showed that low doses of virus, between 4 and 28 $TCID_{50}$, were sufficient to cause a lethal infection. Animals became febrile at days 5 and 6, succumbing to disease between 8 and 11 days post challenge. Typical clinical observations included hemorrhaging in all animals and a transient rash in a small number of animals. The authors suggested that rash may be due to a slightly longer duration of disease, or a slight genetic difference between marmosets from different backgrounds. Virus was isolated from lungs at 3 days post challenge and from liver, spleen, and blood from day 5 post challenge. There was pathology in the kidney and liver on day 3. Those animals with the highest viral titers developed the most severe clinical symptoms.

Marmosets exposed to low doses of EBOV-Kikwit (between 4 and 27 $TCID_{50}$) via the respiratory route [28] developed a lethal infection. Animals developed fever at days 5 and 6 postinfection and succumbed to infection at days 6 and 8 postinfection. Similar results to the MARV marmoset model were observed whereby EBOV was isolated in lungs at day 3 post challenge and from liver, spleen, and blood on day 5 post challenge. Animals that died of the infection had high viral titers in all organs, increased liver function enzymes, increased blood clotting times, decreased levels of platelets, and multifocal moderate-to-severe hepatitis and edema. Marmosets again proved to be excellent models of human disease for another viral family.

OTHER INFECTIOUS AGENTS

As most marmosets are housed indoors and are cared for in Biosafety Level 2 conditions, primarily to protect the animal from acquiring human infections, they can be utilized for a wide variety of Koch's postulates experiments. When prescreened for all clades of enteroviruses only 1 out of 4 showed any significant seroconversion (Patterson and Shigeo, personal communication). Yu et al. [29] were able to use marmosets to test the hypothesis that a novel adenovirus (TMAdV) isolated from a titi monkey (*Callicebus cupreus*) was the cause of a fulminant pneumonia outbreak in a colony of titi monkeys at a national primate center in 2009. This outbreak became more significant when a human researcher at the facility and a household family showed the possibility of a cross-species transmission of the virus. When four marmosets were inoculated with a cell culture-adapted TMAdV three marmosets showed an acute, mild respiratory illness characterized by low-grade fever, reduced activity, anorexia, and sneezing. All monkeys with symptoms developed neutralizing antibodies. This work supported the notion that marmosets could be useful for assessing the zoonotic potential of adenoviruses as they arise in humans and NHPs. Similarly, the marmoset has been shown to be a model for investigations with orthopoxviruses (OPV). Kramski et al. [30] evaluated the infectivity and pathogenicity of calpox virus. In marmosets the minimal infectious dose was 10,000-fold lower than variola virus and monkeypox virus compared with the macaque models. Later, Mucker et al. [31] also showed that marmosets were more susceptible to lower titers of monkeypox virus delivered via the intravenous routes and had incubation periods more characteristic of human smallpox disease compared with macaques. This model overall was deemed to be a suitable primate model by overcoming the limitations of other OPV models, especially the requirement for high viral doses applied by unnatural routes.

Marmosets have been used to model respiratory *Bacillus anthracis* infection [32]. The marmoset was challenged via the respiratory route with small-particle aerosols containing *B. anthracis* Ames strain. The LD_{50} was 1.47×10^3 colony-forming units (CFU) compared with 5.5×10^4 CFU in rhesus macaque and 4.13×10^3 CFU in the cynomolgus macaque. The time to euthanasia was 40–140 h and was not statistically correlated with dosage. Onset of clinical signs, such as dyspnea, disorientation, and reduced response to external stimuli, appeared quickly. Marmosets have been challenged with *Burkholderia mallei* via the intranasal route [33]. Challenge doses of 2.5×10^4 to 2.5×10^5 CFU resulted in an acute lethal infection within 3–4 days postinfection in 83% of marmosets challenged. Those animals challenged with lower infectious doses of bacteria developed mild signs of illness and were culture negative. Those animals challenged with higher infectious doses developed more severe clinical signs of disease, which included lethargy, lack of appetite, conjunctivitis, and mucopurulent and hemorrhagic nasal discharge. Lesions, which resembled those seen in human cases of glanders, were present in tissues.

Marmosets have also been challenged with *Escherichia coli* [34]. These authors showed that *E. coli* can be the cause of diarrhea and hemorrhagic typhlocolitis, seen in captive marmosets. When marmosets were inoculated with three strains of *Mycobacterium tuberculosis*, they developed fulminant disease and it was speculated that marmosets could provide an attractive model for the human disease of *M. tuberculosis* [35].

GBV-B, A HEPATITIS C VIRUS SURROGATE IN NEW WORLD MONKEYS

History of the GB Agent

In the 1960s, studies were conducted in many NHP species in an attempt to find animal models for human hepatitis infections, long before the development of diagnostic assays for HAV, HBV, and HCV. One such study suggested the presence of a new virus when human serum was injected into tamarins (*Saguinus* species). The infection induced hepatitis based on increases in the serum levels of liver enzymes. The serum was derived from a patient with the initials GB. The GB agent was characterized by serial passage in tamarins but was not isolated until modern techniques provided molecular clones of two viruses from tamarins [36,37]. The viral agents, termed GB virus A (GBV-A) and GB virus B (GBV-B), showed similarity to the Flaviviridae. Subsequent studies cloned a related virus from human serum, and it was designated GBV-C. Because GBV-A and GBV-C are not hepatitis viruses and cause no known disease, they will not be discussed. The original studies assumed that the GB agent originated from the serum of the surgeon; however, in retrospect, GBV-B clearly represents a tamarin virus. GBV-B has not been recovered from humans, and GBV-B has a very narrow host range for tamarins and other closely related New World monkeys [38,39]. The fact that it has not been recovered a second time from tamarins may be due to the rapid resolution of the acute infection in tamarins and the limited number of wild-caught tamarins that have been examined immediately on introduction into captivity. Sequence analysis of the GBV-B genome revealed a single ORF of 2864 amino acids [40] with 28% amino acid similarity with HCV [41]. Like HCV, the GBV-B genome encodes a single polyprotein with 10 proteins, possesses a 5′ IRES (internal ribosome entry site) [42], and a unique 3′ NTR (nontranslated region) [43]

Development of the Animal Models for GBV-B

The major advantage of GBV-B as a surrogate for HCV is the small size of the animal model. Although tamarins (*Saguinus* sp.) appear to be the natural host, the virus has been adapted to the closely related marmoset (*Callithrix jacchus*) (see below). An additional advantage is the increased level of viremia, which often reaches 10^9 GE/mL, 1000-fold higher than what is observed in most HCV-infected humans. The basis for higher viremia is not known, but it may involve an increased ability to avoid the innate immune response. The one disadvantage to the GBV-B model is the low frequency of persistent infection. Infection normally resolves within 3 months, but at least four tamarins [44–47] and one marmoset (Lanford, unpublished data) progressed to chronic infection. Insight into the differences between GBV-B and HCV in the development of persistent infections may provide critical insight into vaccine strategies.

Although the natural host for GBV-B appears to be the tamarin, several New World primates are susceptible to infection [38,39]. Marmosets exhibit two distinctive viral profiles when infected with GBV-B, susceptible with rapid escalation of viremia or partially resistant with a clear delay in maximum viremia [48,49]. The different phenotypes are likely to be due to a genetic polymorphism in the marmoset population that affects a critical viral–host interaction. Serial passage of wild-type GBV-B after reaching maximum viremia in resistant animals resulted in adaptation and a uniform susceptible profile in all subsequent infections. The adapted virus contained 21 nucleotide changes and 6 amino acid changes [49]. However, it should be noted that even the adapted virus replicates at levels 10-fold lower in marmosets compared with tamarins.

In Vitro Culture Model

Unlike HCV, GBV-B can be readily cultured in primary hepatocytes from both marmosets and tamarins [50]. Decades of research were conducted on HCV before a single strain was isolated that could replicate in a unique liver cancer cell line [51,52]. Although in vitro models for HCV have become common tools for HCV, the ability to infect primary hepatocytes or to use patient-derived serum remains challenging. GBV-B can move from animal to culture and back to animals without requiring adaptive mutations. The primary hepatocyte culture system for GBV-B [50] has an intact innate immune response [53,54], providing a more authentic environment to evaluate the ability of viral proteins to subvert the innate response.

Molecular Clones and Chimeric Viruses

Infectious cDNA clones of GBV-B have been produced that induced hepatitis on intrahepatic inoculation of tamarins and marmosets with in vitro transcribed RNA [43,55,56]. Chimeric viruses between GBV-B and HCV have been developed with mixed success. Replacing a portion of the GBV-B IRES with domain III of the HCV IRES [57] induced infection in tamarins but only after a prolonged period with no viremia. The virus that emerged had two compensatory mutations in the opposite end of the genome (3′ NTR region) that allowed the chimeric RNA to replicate efficiently [58]. Unlike WT GBV-B, this chimeric virus could not be adapted for efficient infection in marmosets [59], highlighting another difference between the tamarin and marmoset models. Chimeric viruses have also been produced with HCV p7 replacing GBV-B p13 [60] and the HCV E2 hypervariable region [61]. Deletion studies of the 3′ NTR mapped the essential nature of this structure but, surprisingly, deletion of the poly U stretch resulted in one of the few documented chronic infections [45].

ANTIVIRAL STUDIES

The NS3 protease of HCV and GBV-B are interchangeable with regard to processing of the polyprotein [62,63]. This high level of functional homology led to the expectation that antiviral compounds developed for HCV could be studied in the GBV-B animal model. HCV protease inhibitors inhibit GBV-B replication in infected marmosets [48] and infected primary marmoset hepatocytes [39]. The antiviral nature of ribavirin was explored using the GBV-B model as well [64]. The data suggested that incorporation of ribavirin triphosphate by the RNA polymerase induces error-prone replication with a concomitant reduction in infectivity. The induction of error-prone replication with ribavirin has been demonstrated in the HCV replicon model [54].

Marmoset and tamarin interferon alpha (IFNα) and interferon gamma (IFNγ) have potent antiviral activity against GBV-B in vitro [53], similar to what was observed for HCV in vitro [54]. New studies have focused on interferon lambda (IFNλ). The focus on the IFNλ pathway is based on the finding that SNPs in IL28B (IFNλ3) locus in the human population exert an enormous effect on viral clearance during acute infection and during IFNα therapy, yet the mechanism involved is not understood [65]. The IFNλs in humans include IFNλ1 (IL29), IFNλ2 (IL28A), and IFNλ3 (IL28B). The genomic area of the marmoset encoding interferon lambdas was sequenced. Marmosets lack IFNλ2 (IL28A in humans). The transcriptional orientation of IFNλ3 suggests that the human IL28B homologue is present but the human IL28A homologue is absent. Thus, the IL28B gene duplication giving rise to IL28A may have occurred after segregation of the New and Old World primates. In vitro and in vivo studies were conducted to characterize IFNλ3 (IL28B) antiviral activity for GBV-B infection in the marmoset. Exceptionally potent antiviral activity was demonstrated for this class of IFNs in marmosets (Lanford, unpublished data). IFNλ3 is clearly a more potent antiviral agent than IFNα in this model. One animal had undetectable viremia after 3 doses (1 week of therapy). This represents a decline of 7–9 logs of viremia and is greater than the decline observed with potent direct-acting antivirals for HCV.

ANTISENSE KNOCKDOWN OF GBV-B IN MARMOSETS

The marmoset is increasingly becoming the chosen primate model for transgenics with newer technologies that are emerging for efficient gene disruption (e.g., CRISPRs). Methods will be needed to transiently examine phenotypes of gene knockdown in this model before the investment in a transgenic approach. The marmoset model of GBV-B infection has been used to examine the potential for gene knockdown using systemic injection of antisense using the novel Locked Nucleic Acid (LNA) antisense technology. The LNA technology was first used in antiviral studies to demonstrate that sequestration of miR-122 could knockdown HCV replication in chimpanzees [66]. miR-122 is a highly abundant, liver-specific microRNA. HCV has two binding sites for miR-122 in the 5′ NCR, and both are essential for maintaining HCV replication [67,68]. The miR-122 LNA antisense progressed into phase II clinical trials for HCV as Miravirsen and showed highly potent antiviral activity [69]. As a proof of concept that the LNA technology worked as expected in the marmoset model, Miravirsen was used for GBV-B antiviral studies in the marmoset. GBV-B has two binding sites for miR-122 that have identical spacing as the two sites in HCV. In vitro studies confirmed the antiviral effect of Miravirsen using primary marmoset hepatocytes infected with GBV-B. In vivo, pretreatment with Miravirsen rendered marmosets completely resistant to GBV-B infection. These data clearly indicate that miR-122 is a valid antiviral target for GBV-B and LNA technology can be used for gene knockdown in marmosets (Lanford, unpublished data). The marmoset as a small nonhuman model for gene knockdown may be of value in a number of different research fields.

ZIKA VIRUS

The recent emergence of a member of the flavivirus family Zika virus (ZIKV) in the Americas and its role in fetal abnormalities has placed an urgency on the development of animal models for ZIKV. These mosquito-borne infections have been associated with neurological complications, the most severe being microcephaly. The full extent of the adverse events associated with ZIKV infections in pregnant women will not be known for years as we follow infants born to ZIKV-infected pregnant women. Recently, male marmosets experimentally infected with ZIKV reproduced some of the key features of the human disease, an asymptomatic infection, and a short window of detectable virus in the serum of less than a week. There was also virus detected in urine, saliva, semen, and stool for at least 2 weeks following acute infection, and there was a 1-month postinfection persistence in lymph nodes. Only saliva and serum samples produced infectious virus in cell culture. All marmosets developed neutralizing antibody responses with activation of NK cell and B cell subsets. Only circulating cytokines associated with type II interferon signaling were increased. A transcriptome profile showed enrichment of immune responses to an active viral infection and upregulation of type I and II interferon signaling pathway. Studies to examine the role of ZIKV in pregnant marmosets are underway. These authors believe that marmosets may prove to be a valuable model for drug testing, vaccine development, and understanding the pathogenesis of ZIKV [70].

References

[1] Abbott DH, Barnett DK, Colman RJ, Yamamoto ME, Schultz-Darken NJ. Aspects of common marmoset basic biology and life history important for biomedical research. Comp Med 2003;53: 339–50.

[2] Adams AP, Aronson JF, Tardif SD, Patterson JL, Brasky KM, Geiger R, de la Garza M, Carrion Jr R, Weaver SC. Common marmosets (*Callithrix jacchus*) as a nonhuman primate model to assess the virulence of eastern equine encephalitis virus strains. J Virol 2008;82:9035–42.

[3] Aguilar PV, Paessler S, Carrara AS, Baron S, Poast J, Wang E, Moncayo AC, Anishchenko M, Watts D, Tesh RB, Weaver SC. Variation in interferon sensitivity and induction among strains of eastern equine encephalitis virus. J Virol 2005;79:11300–10.

[4] Reed DS, Lackemeyer MG, Garza NL, Norris S, Gamble S, Sullivan LJ, Lind CM, Raymond JL. Severe encephalitis in cynomolgus macaques exposed to aerosolized Eastern equine encephalitis virus. J Infect Dis 2007;196:441–50.

[5] Hayes EB, Komar N, Nasci RS, Montgomery SP, O'Leary DR, Campbell GL. Epidemiology and transmission dynamics of West Nile virus disease. Emerg Infect Dis 2005a;11:1167–73.

[6] Hayes EB, Sejvar JJ, Zaki SR, Lanciotti RS, Bode AV, Campbell GL. Virology, pathology, and clinical manifestations of West Nile virus disease. Emerg Infect Dis 2005b;11:1174–9.

[7] Verstrepen BE, Fagrouch Z, van Heteren M, Buitendijk H, Haaksma T, Beenhakker N, Palù G, Richner JM, Diamond MS, Bogers WM, Barzon L, Chabierski S, Ulbert S, Kondova I, Verschoor EJ. Experimental infection of rhesus macaques and common marmosets with a European strain of West Nile virus. PLOS Negl Trop Dis 2014;8:e2797. https://doi.org/10.1371/journal.pntd.0002797.

[8] Halstead SB. Dengue. Lancet 2007;370:1644–52.

[9] Moi ML, Takasaki T, Saijo M, Kurane I. Determination of antibody concentration as the main parameter in a dengue virus antibody-dependent enhancement assay using FcγR-expressing BHK cells. Arch Virol 2014;159:103–16.

[10] Greenough TC, Carville A, Coderre J, Somasundaran M, Sullivan JL, Luzuriaga K, Mansfield K. Pneumonitis and multi-organ system disease in common marmosets (*Callithrix jacchus*) infected with the severe acute respiratory syndrome-associated coronavirus. Am J Pathol 2005;167:455–63.

[11] Hamming I, Timens W, Bulthuis ML, Lely AT, Navis G, van Goor H. Tissue distribution of ACE2 protein, the functional receptor for SARS coronavirus. A first step in understanding SARS pathogenesis. J Pathol 2004;203:631–7.

[12] To KF, Lo AW. Exploring the pathogenesis of severe acute respiratory syndrome (SARS): the tissue distribution of the coronavirus (SARS-CoV) and its putative receptor, angiotensin-converting enzyme 2 (ACE2). J Pathol 2004;203:740–3.

[13] Raj VS, Mou H, Smits SL, Dekkers DH, Müller MA, Dijkman R, Muth D, Demmers JA, Zaki A, Fouchier RA, Thiel V, Drosten C, Rottier PJ, Osterhaus AD, Bosch BJ, Haagmans BL. Dipeptidyl peptidase 4 is a functional receptor for the emerging human coronavirus-EMC. Nature 2013;495:251–4.

[14] Falzarano D, de Wit E, Feldmann F, Rasmussen AL, Okumura A, Peng X, Thomas MJ, van Doremalen N, Haddock E, Nagy L, LaCasse R, Liu T, Zhu J, McLellan JS, Scott DP, Katze MG, Feldmann H, Munster VJ. Infection with MERS-CoV causes lethal pneumonia in the common marmoset. PLoS Pathog 2014;10: e1004250.

[15] Johnson RF, Via LE, Kumar MR, Cornish JP, Yellayi S, Huzella L, Postnikova E, Oberlander N, Bartos C, Ork BL, Mazur S, Allan C, Holbrook MR, Solomon J, Johnson JC, Pickel J, Hensley LE, Jahrling PB. Intratracheal exposure of common marmosets to MERS-CoV Jordan-n3/2012 or MERS-CoV EMC/2012 isolates does not result in lethal disease. Virology 2015;485:422–30.

[16] Chan JF, Yao Y, Yeung ML, Deng W, Bao L, Jia L, Li F, Xiao C, Gao H, Yu P, Cai JP, Chu H, Zhou J, Chen H, Qin C, Yuen KY. Treatment with Lopinavir/Ritonavir or interferon-β1b improves outcome of MERS-CoV infection in a nonhuman primate model of common marmoset. J Infect Dis 2015;212:1904–13.

[17] McCormick JB, Fisher-Hoch SP. Lassa fever. Curr Top Microbiol Immunol 2002;262:75–109.

[18] Carrion Jr R, Brasky K, Mansfield K, Johnson C, Gonzales M, Ticer A, Lukashevich I, Tardif S, Patterson J. Lassa virus infection in experimentally infected marmosets: liver pathology and immunophenotypic alterations in target tissues. J Virol 2007;81: 6482–90.

[19] Mahanty S, Bausch DG, Thomas RL, Goba A, Bah A, Peters CJ, Rollin PE. Low levels of interleukin-8 and interferon-inducible protein-10 in serum are associated with fatal infections in acute Lassa fever. J Infect Dis 2001;183:1713–21.

[20] Lukashevich IS, Carrion Jr R, Salvato MS, Mansfield K, Brasky K, Zapata J, Cairo C, Goicochea M, Hoosien GE, Ticer A, Bryant J, Davis H, Hammamieh R, Mayda M, Jett M, Patterson J. Safety, immunogenicity, and efficacy of the ML29 reassortant vaccine for Lassa fever in small non-human primates. Vaccine 2008;26: 5246–54.

[21] Lukashevich IS, Patterson J, Carrion R, Moshkoff D, Ticer A, Zapata J, Brasky K, Geiger R, Hubbard GB, Bryant J, Salvato MS. A live attenuated vaccine for Lassa fever made by reassortment of Lassa and Mopeia viruses. J Virol 2005;79: 13934–42.

[22] Kuhn JH, Bào Y, Bavari S, Becker S, Bradfute S, Brauburger K, Rodney, Brister J, Bukreyev AA, Caì Y, Chandran K, Davey RA, Dolnik O, Dye JM, Enterlein S, Gonzalez JP, Formenty P, Freiberg AN, Hensley LE, Hoenen T, Honko AN, Ignatyev GM, Jahrling PB, Johnson KM, Klenk HD, Kobinger G, Lackemeyer MG, Leroy EM, Lever MS, Mühlberger E, Netesov SV, Olinger GG, Palacios G, Patterson JL, Paweska JT, Pitt L, Radoshitzky SR, Ryabchikova EI, Saphire EO, Shestopalov AM, Smither SJ, Sullivan NJ, Swanepoel R, Takada A, Towner JS, van der Groen G, Volchkov VE, Volchkova VA, Wahl-Jensen V, Warren TK, Warfield KL, Weidmann M, Nichol ST. Virus nomenclature below the species level: a standardized nomenclature for filovirus strains and variants rescued from cDNA. Arch Virol 2014;159:1229–37.

[23] Nyakarahuka L, Kankya C, Krontveit R, Mayer B, Mwiine FN, Lutwama J, Skjerve E. How severe and prevalent are Ebola and Marburg viruses? A systematic review and meta-analysis of the case fatality rates and seroprevalence. BMC Infect Dis 2016;16:708.

[24] Petti S, Messano GA, Vingolo EM, Marsella LT, Scully C. The face of Ebola: changing frequency of haemorrhage in the West African compared with Eastern-Central African outbreaks. BMC Infect Dis 2015;15:564.

[25] Carrion Jr R, Patterson JL. An animal model that reflects human disease: the common marmoset (*Callithrix jacchus*). Curr Opin Virol 2012;2:357–62.

[26] Carrion Jr R, Ro Y, Hoosien K, Ticer A, Brasky K, de la Garza M, Mansfield K, Patterson JL. A small nonhuman primate model for filovirus-induced disease. Virology 2011;420:117–24.

[27] Smither SJ, Nelson M, Eastaugh L, Laws TR, Taylor C, Smith SA, Salguero FJ, Lever MS. Experimental respiratory Marburg virus haemorrhagic fever infection in the common marmoset (*Callithrix jacchus*). Int J Exp Pathol 2013;94:156–68.

[28] Smither SJ, Nelson M, Eastaugh L, Nunez A, Salguero FJ, Lever MS. Experimental respiratory infection of marmosets (*Callithrix jacchus*) with Ebola virus Kikwit. J Infect Dis 2015;212(Suppl 2):S336–45.

[29] Yu G, Yagi S, Carrion Jr R, Chen EC, Liu M, Brasky KM, Lanford RE, Kelly KR, Bales KL, Schnurr DP, Canfield DR, Patterson JL, Chiu CY. Experimental cross-species infection of common marmosets by titi monkey adenovirus. PLoS One 2013; 8:e68558. https://doi.org/10.1371/journal.pone.0068558.

[30] Kramski M, Mätz-Rensing K, Stahl-Hennig C, Kaup FJ, Nitsche A, Pauli G, Ellerbrok H. A novel highly reproducible and lethal nonhuman primate model for orthopox virus infection. PLoS One 2010;5:e10412. https://doi.org/10.1371/journal.pone.0010412.

[31] Mucker EM, Chapman J, Huzella LM, Huggins JW, Shamblin J, Robinson CG, Hensley LE. Susceptibility of marmosets (*Callithrix jacchus*) to Monkeypox virus: a low dose prospective model for Monkeypox and smallpox disease. PLoS One 2015;10:e0131742. https://doi.org/10.1371/journal.pone.0131742.

[32] Lever MS, Stagg AJ, Nelson M, Pearce P, Stevens DJ, Scott EA, Simpson AJ, Fulop MJ. Experimental respiratory anthrax infection in the common marmoset (*Callithrix jacchus*). Int J Exp Pathol 2008; 89:171–9.

[33] Jelesijevicc T, Zimmermanc SM, Harvey SB, Mead DG, Shaffer TL, Estes DM, Michel F, Quinn FD, Hogan RJ, Lafontaine ER. Use of the common marmoset to study *Burkholderia mallei* infection. PLoS One 2015;10:e0124181. https://doi.org/10.1371/journal.pone.0124181.

[34] Hayashimoto N, Inoue T, Morita H, Yasuda M, Ueno M, Kawai K, Itoh T. Survey and experimental infection of enteropathogenic *Escherichia coli* in common marmosets (*Callithrix jacchus*). PLoS One 2016;11:e0160116. https://doi.org/10.1371/journal.pone.0160116.

[35] Via LE, Weiner DM, Schimel D, Lin PL, Dayao E, Tankersley SL, Cai Y, Coleman MT, Tomko J, Paripati P, Orandle M, Kastenmayer RJ, Tartakovsky M, Rosenthal A, Portevin D, Eum SY, Lahouar S, Gagneux S, Young DB, Flynn JL, Barry 3rd CE. Differential virulence and disease progression following *Mycobacterium tuberculosis* complex infection of the common marmoset (*Callithrix jacchus*). Infect Immun 2013;81: 2909–19.

[36] Schlauder GG, Dawson GJ, Simons JN, Pilot-Matias TJ, Gutierrez RA, Heynan CA, Kniggo MF, Kurpiewski GS, Buijk SL, Leary TP, Muerhoff AS, Desai SM, Mushahwar IK. Molecular and serologic analysis in the transmission of the GB hepatitis agents. J Med Virol 1995;46:81–90.

[37] Simons JN, Pilot-Matias TJ, Leary TP, Dawson GJ, Desai SM, Schlauder GG, Muerhoff AS, Erker JC, Buijk SL, Chalmers ML, Van Sant CL, Mushahwar IK. Identification of two flavivirus-like genomes in the GB hepatitis agent. Proc Natl Acad Sci USA 1995;92:3401–5.

[38] Bukh J, Apgar CL, Govindarajan S, Purcell RH. Host range studies of GB virus-B hepatitis agent, the closest relative of hepatitis C virus, in New World monkeys and chimpanzees. J Med Virol 2001; 65:694–7.

[39] Lanford RE, Chavez D, Notvall L, Brasky KM. Comparison of tamarins and marmosets as hosts for GBV-B infections and the effect of immunosuppression on duration of viremia. Virology 2003; 311:72–80.

[40] Muerhoff AS, Leary TP, Simons JN, Pilot-Matias TJ, Dawson GJ, Erker JC, Chalmers ML, Schlauder GG, Desai SM, Mushahwar IK. Genomic organization of GB viruses A and B: two new members of the Flaviviridae associated with GB agent hepatitis. J Virol 1995;69:5621–30.

[41] Ohba K, Mizokami M, Lau JYN, Orito E, Ikeo K, Gojobori T. Evolutionary relationship of hepatitis C, pesti-, flavi-, plantviruses, and newly discovered GB hepatitis agents. FEBS Lett 1996;378:232–4.

[42] Grace K, Gartland M, Karayiannis P, McGarvey MJ, Clarke B. The 5′ untranslated region of GB virus B shows functional similarity to the internal ribosome entry site of hepatitis C virus. J Gen Virol 1999;80:2337–41.

[43] Bukh J, Apgar CL, Yanagi M. Toward a surrogate model for hepatitis C virus: an infectious molecular clone of the GB virus-B hepatitis agent. Virology 1999;262:470–8.

[44] Martin A, Rijnbrand R, Yang Y, Cohen L, Goettge K, Bodola F, Lanford RE, Lemon SM. Adaptation of a chimeric GBV-B RNA containing HCV IRES sequence to efficient replication in tamarins. In: Jilbert AR, Grgaci EVL, Vickery K, Burrell CJ, Cossart YE, editors. Viral hepatitis and liver disease. Melbourne: Australian Centre for Hepatitis Virology; 2003b. p. 361–3.

[45] Nam JH, Faulk K, Engle RE, Govindarajan S, St Claire M, Bukh J. In vivo analysis of the 3′ untranslated region of GB virus B after in vitro mutagenesis of an infectious cDNA clone: persistent infection in a transfected tamarin. J Virol 2004;78:9389–99.

[46] Takikawa S, Engle RE, Faulk KN, Emerson SU, Purcell RH, Bukh J. Molecular evolution of GB virus B hepatitis virus during acute resolving and persistent infections in experimentally infected tamarins. J Gen Virol 2010;91:727–33.

[47] Takikawa S, Engle RE, Emerson SU, Purcell RH, St Claire M, Bukh J. Functional analyses of GB virus B p13 protein: development of a recombinant GB virus B hepatitis virus with a p7 protein. Proc Natl Acad Sci USA 2006;103:3345–50.

[48] Bright H, Carroll AR, Watts PA, Fenton RJ. Development of a GB virus B marmoset model and its validation with a novel series of hepatitis C virus NS3 protease inhibitors. J Virol 2004;78:2062–71.

[49] Weatherford T, Chavez D, Brasky KM, Lanford RE. The marmoset model of GB virus B infections: adaptation to host phenotypic variation. J Virol 2009a;83:5806–14.

[50] Beames B, Chavez D, Guerra B, Notvall L, Brasky KM, Lanford RE. Development of a primary tamarin hepatocyte culture system for GB virus-B: a surrogate model for hepatitis C virus. J Virol 2000;74:11764–72.

[51] Wakita T, Pietschmann T, Kato T, Date T, Miyamoto M, Zhao Z, Murthy K, Habermann A, Krausslich HG, Mizokami M, Bartenschlager R, Liang TJ. Production of infectious hepatitis C virus in tissue culture from a cloned viral genome. Nat Med 2005;11: 791–6.

[52] Zhong J, Gastaminza P, Cheng G, Kapadia S, Kato T, Burton DR, Wieland SF, Uprichard SL, Wakita T, Chisari FV. Robust hepatitis C virus infection in vitro. Proc Natl Acad Sci USA 2005;102: 9294–9.

[53] Chavez D, Guerra B, Lanford RE. Antiviral activity and host gene induction by tamarin and marmoset interferon-alpha and interferon-gamma in the GBV-B primary hepatocyte culture model. Virology 2009;390:186–96.

[54] Lanford RE, Guerra B, Lee H, Averett DR, Pfeiffer B, Chavez D, Notvall L, Bigger C. Antiviral effect and virus-host interactions in response to alpha interferon, gamma interferon, poly(I)-poly(C), tumor necrosis factor alpha, and ribavirin in hepatitis C virus subgenomic replicons. J Virol 2003b;77:1092–104.

[55] Martin A, Bodola F, Sangar DV, Goettge K, Popov V, Rijnbrand R, Lanford RE, Lemon SM. Chronic hepatitis associated with GB virus B persistence in a tamarin after intrahepatic inoculation of synthetic viral RNA. Proc Natl Acad Sci USA 2003a;100:9962–7.

[56] Sbardellati A, Scarselli E, Verschoor E, DeTomassi A, Lazzaro D, Traboni C. Generation of infectious and transmissible virions from a GB virus B full-length consensus clone in tamarins. J Gen Virol 2001;82:2437–48.

[57] Rijnbrand R, Yang Y, Beales L, Bodola F, Goettge K, Cohen L, Lanford RE, Lemon SM, Martin A. A chimeric GB virus B with 5′ nontranslated RNA sequence from hepatitis C virus causes hepatitis in tamarins. Hepatology 2005;41:986–94.

[58] Warter L, Cohen L, Benureau Y, Chavez D, Yang Y, Bodola F, Lemon SM, Traboni C, Lanford RE, Martin A. A cooperative interaction between nontranslated RNA sequences and NS5A protein promotes in vivo fitness of a chimeric hepatitis C/GB virus B. PLoS One 2009;4:e4419.

[59] Weatherford T, Chavez D, Brasky KM, Lemon SM, Martin A, Lanford RE. Lack of adaptation of chimeric GB virus B/hepatitis C virus in the marmoset model: possible effects of bottleneck. J Virol 2009b;83:8062–75.

[60] Griffin S, Trowbridge R, Thommes P, Parry N, Rowlands D, Harris M, Bright H. Chimeric GB virus B genomes containing hepatitis C virus p7 are infectious in vivo. J Hepatol 2008;49:908–15.

[61] Haqshenas G, Dong X, Netter H, Torresi J, Gowans EJ. A chimeric GB virus B encoding the hepatitis C virus hypervariable region 1 is infectious in vivo. J Gen Virol 2007;88:895–902.

[62] Butkiewicz N, Yao N, Zhong W, Wright-Minogue J, Ingravallo P, Zhang R, Durkin J, Strandring DN, Baroudy BM, Sangar DV, Lemon SM, Lau JYN, Hong Z. Virus-specific cofactor requirement and chimeric hepatitis C virus/GB virus B nonstructural protein 3. J Virol 2000;74:4291–301.

[63] Scarselli E, Urbani A, Sbardellati A, Tomei L, De Francesco R, Traboni C. GB virus B and hepatitis C virus NS3 serine proteases share substrate specificity. J Virol 1997;71:4985–9.

[64] Lanford RE, Chavez D, Guerra B, Lau JYN, Hong Z, Brasky KM, Beames B. Ribavirin induces error-prone replication of GB virus B in primary tamarin hepatocytes. J Virol 2001;75:8074–81.

[65] Thomas DL, Thio CL, Martin MP, Qi Y, Ge D, O'Huigin C, Kidd J, Kidd K, Khakoo SI, Alexander G, Goedert JJ, Kirk GD, Donfield SM, Rosen HR, Tobler LH, Busch MP, McHutchison JG, Goldstein DB, Carrington M. Genetic variation in IL28B and spontaneous clearance of hepatitis C virus. Nature 2009;461:798–801.

[66] Lanford RE, Hildebrandt-Eriksen ES, Petri A, Persson R, Lindow M, Munk ME, Kauppinen S, Orum H. Therapeutic silencing of microRNA-122 in primates with chronic hepatitis C virus infection. Science 2010;327:198–201.

[67] Jangra RK, Yi M, Lemon SM. Regulation of hepatitis C virus translation and infectious virus production by the microRNA miR-122. J Virol 2010;84:6615–25.

[68] Jopling CL, Schutz S, Sarnow P. Position-dependent function for a tandem microRNA miR-122-binding site located in the hepatitis C virus RNA genome. Cell Host Microbe 2008;4:77–85.

[69] Janssen HL, Reesink HW, Lawitz EJ, Zeuzem S, Rodriguez-Torres M, Patel K, van der Meer AJ, Patick AK, Chen A, Zhou Y, Persson R, King BD, Kauppinen S, Levin AA, Hodges MR. Treatment of HCV infection by targeting microRNA. N Engl J Med 2013;368:1685–94.

[70] Chiu CY, Bouquet J, Li T, Yagi S, Sanchez, San Martin C, Tamhankar M, Hodara VL, Parodi LL, Somasekar S, Yu G, Giavedoni LD, Tardif S, Patterson JL. Experimental Zika virus infection in a new World monkey model reproduces key features of the human disease. BioRxiv 2017 [Preprint posted Jan. 21, 2017] Available from: https://doi.org/10.1101/102145.

CHAPTER

29

Insights Gained From Marmoset Endocrine Research

Marissa Kraynak[1,4], *Jon E. Levine*[2,3,4], *David H. Abbott*[1,3,4]

[1]Department of Obstetrics and Gynecology, University of Wisconsin, Madison, WI, United States; [2]Department of Neuroscience, University of Wisconsin, Madison, WI, United States; [3]Wisconsin National Primate Research Center, University of Wisconsin, Madison, WI, United States; [4]Endocrinology-Reproductive Physiology Training Program, University of Wisconsin, Madison, WI, United States

INTRODUCTION TO MARMOSET-SPECIFIC ENDOCRINOLOGY

The common marmoset monkey, a New World primate (NWP), is a cooperatively breeding, small-bodied, azoonotic, readily tractable nonhuman primate that undergoes rapid maturation into reproductive maturity (~1.5 years) [1] and mimics the endocrine physiology of Old World primates (OWPs), including humans. Typical of NWPs, however, marmosets exhibit a multitude of distinct molecular differences from OWPs in terms of endocrine regulating mechanisms. Such a distinction has contributed to their increased use in biomedical endocrine research since the 1980s, resulting in a tripling of PubMed publications up to 80–100 citations/5 years since the 1990s (Fig. 29.1). Rhesus macaque endocrine-related publications, in contrast, while initially ~25 times more numerous than those of marmosets, underwent ~50% decline in PubMed citations/5 years over the same time period. As marmoset endocrine research has become less constrained by reagent availability and improved understanding, marmosets have become staple biomedical endocrine research models, enabling exploration or exploitation of endocrine variance with OWPs to better rationalize molecular structure to function in primates as models for humans [2–4].

The unique and varied NWP endocrine traits arose during rapid, adaptive selection pressure from undetermined causes, following separation from the OWP lineage, resulting in successive gene changes in a multitude of regulatory systems [5]. One obvious global change includes end-organ resistance to steroid hormones resulting in exaggerated production of NWP steroid hormones driven by increased intracellular concentrations of specific (1) heat shock protein (hsp) chaperones, diminishing steroid hormone binding to intracellular cognate receptors [6–8], and (2) heterogeneous nuclear riboproteins (hnRNPs) that outcompete receptor-bound, steroid hormone transcription complexes for DNA response elements, blocking transcription [9–12]. The generic NWP solution to these global endocrine changes has been compensatory increases in endogenous, bioavailable, circulating steroids [13] sufficient to overwhelm extensive intracellular barriers to biological action in target

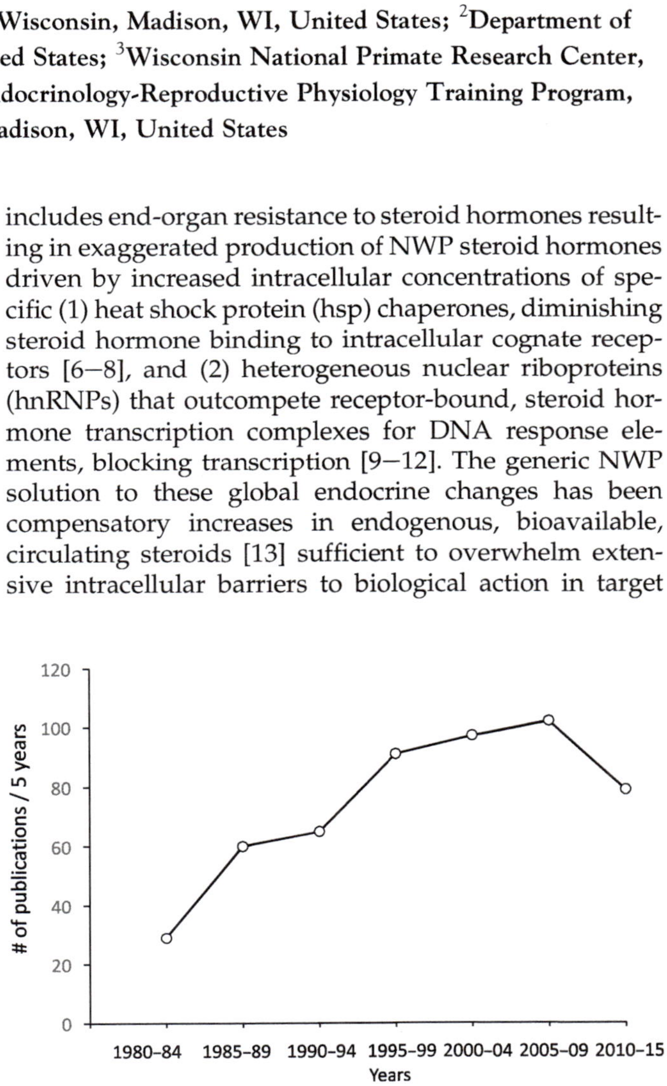

FIGURE 29.1 Numbers of hormone-related publications listed in PubMed per 5-year period since 1980 utilizing marmoset monkeys. Search terms were "marmoset," "hormone."

The Common Marmoset in Captivity and Biomedical Research
https://doi.org/10.1016/B978-0-12-811829-0.00029-7

organs. The original selection pressures may have included environmentally or diet-derived, steroidogenically functional molecules that engaged gene transcription outside homeostatic control [14], as perhaps exhibited today by free-living populations of certain African and Asian monkeys and apes that either increase or diminish fertility, reproductive physiology, and a variety of behaviors because of seasonal increases in dietary phytochemicals that mimic estrogenic and progestogenic actions [15]. The impactful roles of hormonally active phytochemicals are being increasingly recognized as agents of vertebrate evolution [15], particularly with regard to actions that mimic steroid hormones [16].

In this context, this review will provide an overview of the major marmoset endocrine organ systems, providing specific examples of alternative NWP endocrinology to that in humans, and how such NWP findings have led to an improved understanding of endocrine contributions to human disease.

HYPOTHALAMIC–ANTERIOR PITUITARY CONTROL OF SYSTEMIC HORMONE PRODUCTION

Many endocrine systems in marmosets, as in OWPs and humans, are centrally regulated by the hypothalamus and pituitary (HP-axis). Peptide hormones released from the hypothalamus into the hypophyseal portal veins circulate only to the anterior pituitary gland, causing systemic release of pituitary peptide hormones destined to exert systemic endocrine action on the respective target organs (Fig. 29.2). In contrast, hypothalamic peptide hormones systemically released from the posterior pituitary directly stimulate their respective targets. One notable difference in anterior pituitary function between NWPs and OWPs and humans relates to ovarian and testicular regulation. Possibly due to exaggerated expression of a previously rare gene variant of the 12-exon gene coding for primate luteinizing hormone/chorionic gonadotropin receptor type 1 (*LHCGR type 1*) marmosets, typical of NWPs, only express *LHCGR type 2* (affinity for CG binding, alone) that omits exon 10 of *LHCGR type 1* [17]. In addition, their pituitary gonadotropes, while retaining transcriptional responsiveness to gonadotropin-releasing hormone (GnRH), only express the β subunit for CG instead of LH [18]. Marmoset gonadotropes utilize a different promoter sequence than the placenta [19], limiting the addition of a novel exon 1 to placental CG, alone. As reproductive endocrine research continues to provide a rich understanding of alternative molecular signaling regulating marmoset physiology and function [4,20], we will commence with female reproductive endocrinology.

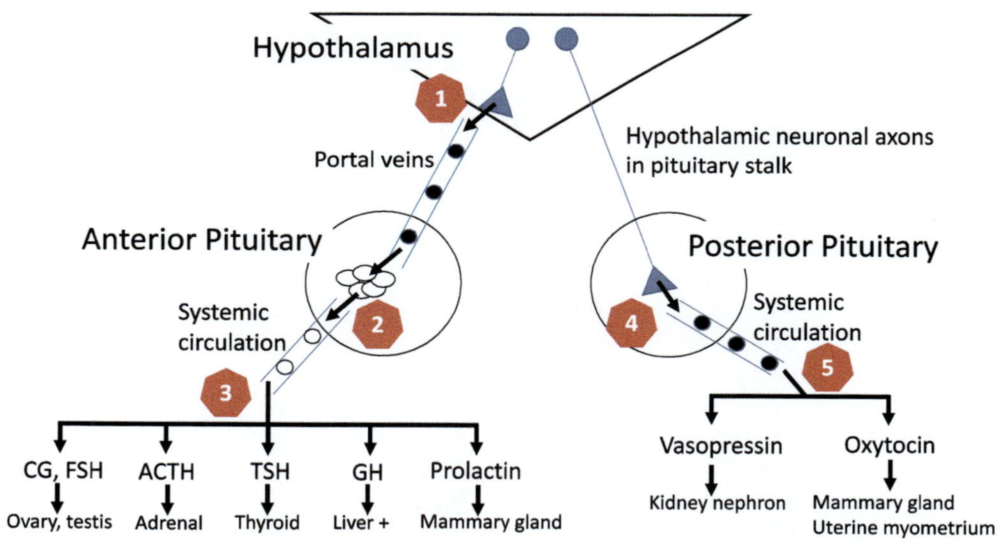

FIGURE 29.2 Diagrammatic summary of marmoset hypothalamic–pituitary hormones regulating peripheral endocrine target organs. (1) Hypothalamic neuropeptides, including a variety of releasing hormones and dopamine (*solid circles*), released from nerve terminals of hypothalamic neurons into the portal venous circulation draining into the anterior pituitary, (2) hypothalamic neuropeptides stimulate specific anterior pituitary cells to release specific pituitary peptides (*open circles*) into the systemic circulation, (3) pituitary peptides stimulate specific peripheral endocrine organs (liver + = liver and a variety of target organs, including bone), (4) hypothalamic neuropeptides released from the nerve terminals of hypothalamic neurons at the posterior pituitary (*solid circles*) into the systemic circulation, and (5) pituitary peptides released from the posterior pituitary stimulate specific target organs (an example of direct, nonsynaptic communication between the brain and peripheral organs). *ACTH*, adrenocorticotropin; *DHEA*, dehydroepiandrosterone; *FSH*, follicle-stimulating hormone; *CG*, chorionic gonadotropin; *TSH*, thyroid stimulating hormone; *GH*, growth hormone.

ENDOCRINE SYSTEMS PRIMARILY REGULATING REPRODUCTION

Regulation of Female Puberty and Ovulatory Cycles

Female marmosets enter puberty about 7–8 months of age with the onset of periodic elevations in circulating levels of estradiol (E_2) [21] and attainment of adult genital pad size [22]. Typical of primates, reproductive neuroendocrine maturation is protracted, with first ovulation usually occurring after the first year of age [20], dependent on a female's social environment [1,23–26].

Within laboratory, and possibly free-living [27,28], social groups of marmosets, complex social mechanisms, together with specialized neuroendocrine and sensory mechanisms, but not stress [29], commonly limit fecundity to one dominant female [26]. Associative learning of dominant female olfactory, visual, and auditory cues contributes to hypothalamic disruption of ovulation [29]. Socially subordinate females or daughters in family groups exhibit varying degrees of hypogonadotropic hypogonadism resulting in anovulation [23,30–32], intermittent ovulation, and luteal phase deficiency (indicative of suboptimal preovulatory follicles and ovulated oocytes) [24,25]. Up to 50% of daughters in family groups exhibit some degree of ovulatory function. This likelihood increases with age beyond 13 months, few or no sisters, and when birthing intervals by the breeding female extend well beyond 6 months [24,25]. Fathers and brothers within families, however, fail to show sexual interest in such ovulating female kin, and ovulating daughters are similarly asexual toward close male kin. Following death or departure of the breeding male, however, and their usually rapid replacement by a biologically unrelated adult male, daughters that are no longer socially subordinate to the breeding female ovulate and mate with this unrelated male, as does the breeding female [25,26,28,33]. Infanticide of subsequent offspring is perpetrated by each of these females when births occur within ~1 month of each other, contributing to social group instability until one female leaves [26–28,33]. Such complex regulation of female reproduction manifests across marmoset and tamarin genera, reflecting their highly specialized cooperative breeding system [26].

Increased sensitivity to E_2-mediated negative feedback on pituitary CG release in socially subordinate females supports underlying neuroendocrine inhibition [29]. The recent demonstration, in ovariectomized (OVX) adult female marmosets, of suprahypergonadotropism (circulating CG in excess of OVX levels), mediated by aromatase inhibition [34], demonstrates the potential for extra-ovarian E_2, likely synthesized within hypothalamic neurons, contributing an E_2-mediated neuroendocrine block.

The hypothalamic–pituitary–gonadal (HPG) axis regulated, 28–30-day marmoset ovarian cycle [35,36], can exhibit unpredictably lengthy, postovulatory luteal phases [35], necessitating administration of prostaglandin-F2alpha (PGF2a) analogue to rapidly induce luteolysis and initiate the next ovarian cycle, providing better predictability for the day of ovulation [37,38]. Unlike OWPs and humans, marmoset corpora lutea (CLs) are uniquely vulnerable to the antigonadotropin action of PGF2a [39,40]. As marmosets fail to exhibit menses, as well as cycle-dependent changes in genital swelling or color, routine endocrine monitoring of circulating progesterone (P_4) levels is required to determine values > 10 ng/mL indicative of the postovulatory luteal phase or early pregnancy [38]. The 5- to 10-fold greater circulating levels of P_4, estrogens, and androgens in female marmosets, compared with those observed in female OWPs, reflect NWP adaptation to selection pressure following their lineage divergence from OWPs (see above); altered expression of intracellular molecular chaperones diminish steroid hormone-mediated transcriptional action, and in compensatory consequence, circulating steroid-binding proteins, such as cortisol-binding globulin (CBG) are absent, resulting in ~100% unbound and bioavailable, circulating P_4 (and cortisol, see below). T and E_2 are exceptions to absent circulating binding globulin, as sex hormone-binding globulin remains expressed, providing primate-typical limited bioavailability of circulating androgens and estrogens [41].

Two major cell types, granulosa and theca cells, work synergistically within ovarian follicles, driven by pituitary follicle-stimulating hormone (FSH) and CG, respectively, to synthesize and release cell-specific hormones throughout the ovarian cycle [42]. Similar to OWP menstrual cycles, marmoset ovarian cycles include a preovulatory, 8- to 10-day follicular phase ($P_4 < 10$ ng/mL), during which FSH drives folliculogenesis, antral follicle granulosa cells to produce E_2 from androgen precursors released from theca cells, as well as insulin-like growth factor-1 (IGF-1) [43] and activin [44] enhancement of granulosa cell aromatase production of E_2. Preovulatory follicle oxytocin and P_4 [45], activin [44], and vascular endothelial growth factor [46] also make significant contributions to ovulatory follicle health, luteinization, and the quality of postovulatory luteal function. The late follicular phase rapid increase in dominant, preovulatory follicle ovarian E_2 production induces a midcycle, ovulation-inducing CG surge via positive feedback on the HP-axis [47], resulting in the ovulation of 2–3 oocytes and follicle conversion into 2–3 predominantly P_4, E_2, estrone, inhibin, and relaxin-secreting corpora lutea [48–50] that provide crucial endocrine support for the uterine endometrium and early pregnancy. Corpora lutea P_4, E_2, and inhibin also engage powerful negative feedback clamps on gonadotropin-driven preovulatory follicle development, preventing ovulation during the post-ovulatory luteal phase [1,36,37]. Unexpectedly, both CG [51] and FSH [48] are effective stimulators of corpus luteum function in marmosets, endowing marmoset FSH with an important postovulatory role, an attribute not found in female OWPs and women. Such dual gonadotropic stimulation of corpus luteum function may contribute to commonly occurring, steroidogenically active, luteinized ovarian follicles in marmosets and other NWPs [52]. While such follicles never ovulate, they contribute increased circulating concentrations of steroid hormones, particularly progesterone, even in postreproductive females, diminishing the use of marmosets as models for postmenopausal health concerns in addition to the absence of female reproductive senescence (menopause) in some aging marmosets [1,20]. The major primate ovarian follicle steroidogenic pathways are illustrated in Fig. 29.3.

Interest in increasing the number of oocyte-containing, preovulatory follicles per cycle and further optimized timing of ovulation in marmosets intensified after these monkeys were used to develop the first primate embryonic stem cells [53] and genetically edited monkeys [54] as novel models for human disease. Research groups have since worked to optimize ovarian hypergonadotropin stimulation protocols for marmosets to once daily injection of 50 IU recombinant human FSH for 9 days, followed by a single injection of 500 IU hCG [37]. Refined endocrine detection of the day of ovulation improves embryo transfer success when female marmoset recipients are within 2–4 days of ovulation [55]. Further improvements in understanding female marmoset reproductive endocrinology will be required to support the artificial reproductive

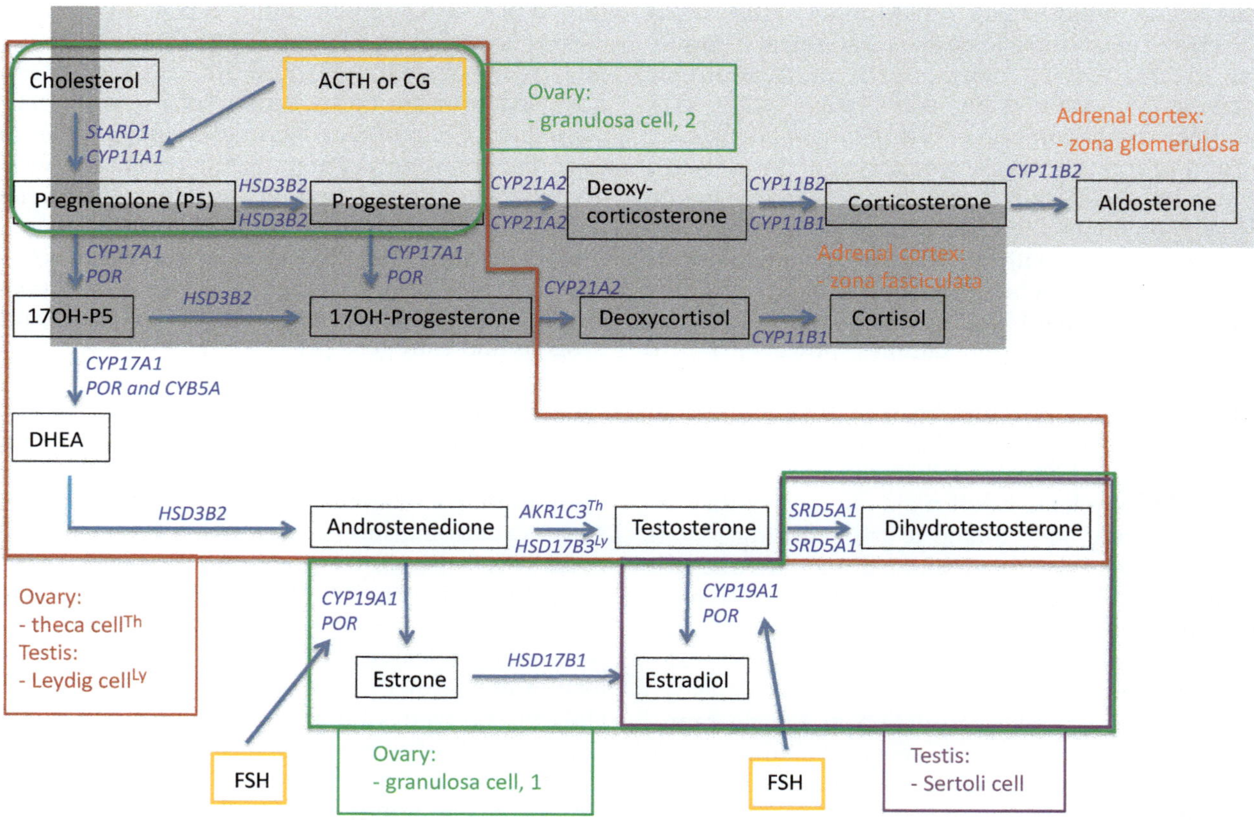

FIGURE 29.3 Diagrammatic representation of the major steroid hormones and steroidogenic enzymatic pathways in the adult marmoset adrenal cortex (shaded gray), ovarian follicle theca cells (red-brown outline), and granulosa cells (green outlines 1 and 2), as well as testicular Leydig cells (red-brown outline) and Sertoli cells (purple outline). Typical of adult marmosets, there is little to no development of an adrenal inner cortical androgenic zone, the zona reticularis [103]. Most of the adult marmoset adrenal cortex is devoted to cortisol biosynthesis. In ovarian granulosa cells, FSH-driven steroidogenesis predominates (granulosa cell, 1) until an ovarian follicle matures into a dominant follicle and its granulosa cells develop LH/CG receptors (granulosa cell, 2) and an ability to synthesize progesterone under CG-driven stimulation. *ACTH*, adrenocorticotropin; *CG*, chorionic gonadotropin; *FSH*, follicle-stimulating hormone; *StARD1*, steroidogenic acute regulatory protein; *CYP11A1*, P450 side chain cleavage; *HSD3B2*, 3beta-hydroxysteroid dehydrogenase type 2; *CYP21A2*, 21-hydroxylase; *CYP11B1*, 11beta-hydroxylase; *CYPB11B2*, aldosterone synthase; *CYP17A1*, 17-hydroxylase and 17,20 lyase; *POR*, P450 oxidoreductase; *CYB5A*, cytochrome b5; *AKR1C3*, aldo-ketoreductase type 1C3; *HSD17B3*, 17-beta hydroxysteroid dehydrogenase type 3; *SRD5A1*, 5-alpha reductase type 1; *CYP19A1*, aromatase.

technologies needed to generate genetically edited monkeys tailored to model human disease.

As in women, increased body fat has profound negative effects on the endocrine function of marmoset ovaries [56]. Folliculogenesis, luteal function, and fecundity are all compromised as marmoset body weight increases and are associated with increased circulating androgen levels and metabolic disease. There is potential here for a marmoset model of the obesity-enhanced, hyperandrogenic polycystic ovary syndrome that increases the risk of infertility and cardiometabolic disease [57].

Compensatory Molecular Responses to Resistance to Ovarian Steroid Hormones

In specific consideration of P_4 and T, increased NWP intracellular expression of the hsp90-bound immunophillin, FKBP51, diminishes high-affinity binding of these steroid hormones and formation of transcription complexes by the respective cytoplasmic or nuclear P_4 or androgen receptors [3,8,58]. Concomitant diminished NWP expression of FKBP52, a molecular chaperone enabling high-affinity receptor binding and transcription complex formation, reinforces this intracellular block. In humans, both P_4 and T upregulate transcription of FKBP51, but not FKBP52, providing rapid, cell-specific ultrashort negative feedback loop on PR- and AR-mediated signaling [59]. Extremely high intracellular concentrations of FKBP51 in humans, associated with pathology, are exploited by malignant cells to support growth and survival [60], enhance DNA demethylation [61], and may promote tissue- or organ-specific P_4 and/or T withdrawal when there is no accompanying dramatic elevation in circulating levels of the respective steroid hormones [58,59]. NWP exploitation of FKBP51 in alternative molecular endocrine regulation of reproduction is thus of more than passing interest [8,62].

Diminished E_2-mediated gene transcription, on the other hand, is accomplished by different, but analogous, intracellular mechanisms. The small heat shock protein hsp27, also called HspB1, is usually upregulated in humans when cells are exposed to altered protein folding, such as during heat shock or malignancy, and increases cell resistance to apoptosis, oxidative stress, or a variety of cell death mechanisms triggered by the immune system [63]. In NWPs, including marmosets, hsp27 has been identified as the intracellular estrogen-binding protein diminishing classical, genomic E_2-mediated action by competing with cytoplasmic estrogen receptor alpha (ERalpha or *ESR1*) receptors for binding to E_2 and thus diminishing translocation of steroid–receptor complexes to the nucleus [7]. There is, however, additional overexpression of another intracellular-binding protein, estrogen response element-binding protein (ERE-BP), also called heterogeneous nuclear ribonucleoprotein C-like 1 (hnRNP C-like 1), an RNA splicing repressor binding microRNAs, that contributes to NWP and marmoset E_2 insensitivity [7,64]. ERE-BP competes with classic, E_2-bound ERalpha and estrogen receptor beta (ERbeta or *ESR2*) assembled transcription complexes for ERE-DNA binding sites around or within target genes. Such transcriptional repression and decoyed cytoplasmic binding are both overcome with elevated levels of E_2 in vitro [7] and by NWPs, including marmosets, in vivo. E_2-ER displacement of ERE-BP from DNA, however, liberates the RNA splice repressor to alter RNA splicing [12] that may include its binding to microRNA, as well as to interact with proteins directly [65]. As analogous posttranscriptional mechanisms regulate apoptosis in cancer [65], translational lessons can be learned from E_2-related research in marmosets.

Endocrinology of Pregnancy

Key endocrine differences separate nonconceptive from conceptive cycles in female marmosets. Similar to OWPs, sustained CL steroid and peptide hormone production support luteotropic function and angiogenesis, as well as maintenance of early pregnancy [41,48]. Relaxin and P_4 enable fallopian tube environments supportive of fertilization and a uterine endometrium receptive to embryo implantation [66,67]. Expression of the relaxin receptor, *RLXFP1*, becomes spatially organized around the endometrial implantation site [68]. Relaxin peaks at midgestation and then declines, independent of pituitary or placental CG levels [69]. The gestational timing of peak relaxin resembles rodent and some ruminant species rather than that observed in OWPs, and the luteal–placental shift for pregnancy support occurs relatively late in marmoset gestation, around 40 days postconception [41,70] instead of 21–30 days observed in rhesus monkeys [71] and humans [72]. Tardy placental endocrine maturity may be in step with tardy marmoset embryo development [73], possibly compensating for postpartum ovulation, conception, and early embryo development coincident with lactation and physically transporting two or more infants [1].

Progressing on to late pregnancy, onset of marmoset parturition may differ from humans. In humans, placental release of corticotropin-releasing hormone (CRH) begins to increase during the second trimester and continues until parturition. CRH-mediated fetal adrenal cortisol and dehydroepiandrosterone (DHEA) responses are thought to contribute to the onset of labor [74]. Additionally, circulating concentrations of CRH-binding protein (CRH-BP) decline in the weeks immediately before parturition onset, enabling cortisol to initiate labor without an obvious decline in P_4 [75].

In marmosets, however, placental CRH rises proportionally to the number of fetuses, peaks at midgestation (~70 days), and then declines [76], without any reduction in circulating levels of CRH-BP [77]. Circulating E_2 levels, nevertheless, rise until parturition [69,70], while unlike humans, P_4 levels fall [70] likely removing P_4-imposed quiescence on myometrial smooth muscle contractions, leading to parturition onset. Such endocrine differences in regulating parturition render marmosets inappropriate for translational studies aimed at ameliorating premature delivery in women.

Endocrine Regulation of Lactation

In primate-typical fashion, pituitary prolactin release increases in the weeks prior to marmoset parturition and remains elevated during postpartum lactation [78], but in contrast to OWPs and women, marmoset ovulation is not suppressed postpartum and conception shortly after birth is the norm [1,79]. One possible explanation for the latter may lie with suckling-induced increases in release of the posterior pituitary hormone, oxytocin, responsible for milk release from the mammary gland. In marmosets, oxytocin stimulates pituitary gonadotropin release, in contrast to its inhibition of gonadotropin in OWPs [80].

Male Reproductive Endocrinology

In contrast to females, male primates (at least in OWPs such as rhesus macaques and humans) exhibit pronounced gonadal endocrine activity during early-to-mid gestation [81,82], when fetal male testosterone levels are elevated well above levels found in fetal females. Sufficient testosterone exposure within a discrete "masculinization programming window (MPW)" is necessary for most, but not all, species- and organ-specific male sexual differentiation [83]. Insufficient testosterone exposure during a species-typical gestational MPW in XY individuals, in contrast, results in deficiencies in male phenotype [83,84]. In XX individuals, the converse applies, and insufficient testosterone exposure during the MPW normally differentiates female phenotypic traits. Excess testosterone exposure in XX individuals during a male-typical MPW, on the other hand, results in varying degrees of male-like differentiation [85,86]. In primates, in contrast to nonprimates, MPW testosterone exposure does not abolish the ability to generate an ovulation-inducing GnRH/LH surge [87–89], and male marmosets are no exception [89a]. Integrating this MPW understanding with marmoset reproductive specializations of fraternal twinning, placental anastomoses, umbilical attachments between all placental discs, and gestation-onset hematopoietic chimerism [1] predicts an increased likelihood for male-like phenotypic traits in XX female marmosets when they share gestation with a male sibling. Almost no evidence for masculinized female outcomes has been found [90], however, in contrast to accumulating evidence for mechanisms protecting female fetuses against masculinization. Genetic modifications in marmoset genes regulating hormonally induced sexual differentiation [90] and a delay in the MPW for marmoset sexually differentiated behaviors from prenatal to neonatal development [30,31] may act in concert to circumvent female marmoset vulnerability to masculinization.

Postnatally, male marmosets display a neonatal rise in circulating T levels, followed by a "quiescent" juvenile phase of mostly undetectable values. Male marmoset neonatal T has been implicated in testicular Sertoli cell proliferation [91] and differentiating minor aspects of male sexual behavior [92]. Episodic increases in T reappear about 7–8 months of age, at the onset of puberty [22,92], reaching adult levels about 18 months of age. Socially subordinate male marmosets, however, exhibit a degree of hypogonadotropic hypogonadism [1,93], with minimal impact on spermatogenesis, which can provide a false indication of unduly protracted endocrine puberty. In adulthood, Sertoli cell proliferation and inhibin production are likely regulated by FSH [91,94], and CG-stimulated Leydig cell production of T initiates and maintains spermatogenesis and adult male behavior [95,96]. Unlike male macaques, male marmosets have proved ineffective models for steroid hormone-based male contraceptives [97]. The intracellular molecular mechanisms engaging and counteracting target organ resistance to T in male marmosets are analogous to those operating in females.

Endocrine Systems Primarily Involved in Regulating Metabolism

Hypothalamic Vasopressin and Oxytocin Released From the Posterior Pituitary

The hypothalamic neuropeptides, vasopressin (antidiuretic hormone, ADH) and oxytocin, synthesized in the magnocellular neurons of the paraventricular (PVN) and supraoptic (SON) nuclei, are released into the systemic circulation from specialized nerve terminals in the posterior pituitary. Both play key roles in regulating behavior [98], lactation, and parturition (oxytocin only, see above), as well as fluid balance [99]. The marmoset vasopressin gene sequence is highly conserved and closely comparable to humans [100], whereas amino acid substitution in marmoset oxytocin likely endows changes in conformation and function [100]. There are marmoset-validated, HPLC-based assays for both neuropeptides [101], and both are readily released from marmoset hypothalamic explants in vitro [102], possibly as a result of dendritic release from PVN and SON magnocellular neurons [99].

Adrenal Gland—Cortex

Cortisol, the major glucocorticoid steroid synthesized and released by the primate adrenal cortex, provides the only negative feedback regulation on hypothalamic–pituitary release of CRH and adrenocorticotropin (ACTH), respectively. The major adrenal steroidogenic pathways are shown in Fig. 29.3, illustrating angiotensin II (see the Renal section) as the major secretagogue for aldosterone released from the outermost cortical zone, zona glomerulosa (ZG), and ACTH as the key secretagogue for the inner zona fasciculata. Marmosets fail to develop an innermost androgenic zona reticularis (ZR) unless the inhibitory influence of gonadally derived E_2 is removed ([103], unpublished results). Typical of primates, and unlike rats and mice, marmoset adrenal cortex synthesizes cortisol and androgens during fetal life, via the delta-5 steroidogenic pathway (Fig. 29.3), resulting in 17-hydroxypregnenolone and not P_4 as the key steroidogenic precursor [103,104]. Fetal and newborn marmoset adrenal cortex, also in common with other primates, exhibits a large androgenic fetal zone (FZ) producing high levels of both DHEA and its sulfoconjugate, DHEAS [103,105,106], in addition to a fetal definitive zone producing cortisol [107]. The FZ regresses during infancy [106].

Typical of NWPs, and unlike their OWP counterparts, marmosets have no effective CBG in circulation [108,109] and effectively exhibit target organ resistance to cortisol due to the same intracellular molecular mechanisms described for P_4 and T, above ([3,8,110]; Scammell and Abbott, unpublished results). Extremely high circulating levels of unbound or "free" cortisol [111,112] and increased brain hippocampal expression of both glucocorticoid type 1 (mineralocorticoid) receptor (MR) and type 2 (glucocorticoid) receptor (GR) [113] compensate for target organ insensitivity so there are no signs or symptoms of Cushing-like hypercortisolemic traits, including hyperglycemia, hyperlipidemia, and osteoporosis [107]. Not surprisingly, nonstressed marmosets and NWPs with relatively high circulating levels of cortisol do not exhibit reduced hippocampal MR and GR expression as observed in chronically hypercortisolemic OWPs [114]. The marmoset blood–brain barrier also only permits entry of $\sim$33% of unbound cortisol in contrast to $\sim$100% in OWPs [114a].

Translational research has examined molecular mechanisms regulating the exaggerated, naturally occurring hypercortisolemia of newborn infant marmosets [107], gestational glucocorticoid excess fetal programming of hippocampal [114b,115], cognitive [116], and metabolic defects [117], as well as gonadal hormone suppression of androgenic ZR development in adolescents and adults [103,117a].

Adrenal Gland—Medulla

As anticipated, marmoset adrenal medullary chromaffin cells predominantly release epinephrine [118] that stimulates lipolysis [119] and cardiac function [120]. Otherwise, little is known concerning marmoset adrenal medulla epinephrine release and its adrenergic actions.

Thyroid Gland

Marmoset thyroid hormones are synthesized by thyrocytes and stored in colloid-filled follicles [121]. Marmoset thyroid is proportioned similarly to humans [122] and concentrates iodide ions to synthesize thyroid hormones for regulation of cell metabolism, development, and growth [121]. While iodine deficiency-induced goiter has not been reported for marmosets, iodine-deficient diets during pregnancy result in anticipated tardy neural and general physical development in the newborn leading to infant abandonment [121]. As expected, thyroxine (T_4) is the major circulating form of thyroid hormone. T_4 demonstrates expected declines in circulating levels following treatment with the commonly used antihyperthyroid drug, methimazole, that inhibits thyroperoxidase-mediated synthesis of bioactive thyroid hormones [122]. Marmoset target tissues express anticipated distributions of deiodinases and thyroid hormone receptors [123], and circulating total thyroxine (T_4) levels are comparable to those in humans with the expected bias in circulating concentrations favoring T_4 rather than the highly bioactive triiodothyronine (T_3). Within target organ cells, T_4 is enzymatically converted into T_3 by d2-deiodinase. As in humans, such a system enables cell-specific regulation of responsiveness to circulating thyroid hormones. Within the hypothalamus and pituitary, local production of T_3 completes hypothalamic–pituitary negative feedback regulation via the beta 2 thyroid hormone receptor at both thyrotropin-releasing hormone neurons and thyroid-stimulating hormone (TSH) releasing anterior pituitary thyrotropes. In marmosets, circulating TSH levels do not exhibit circadian rhythmicity, similar to rhesus monkeys, but in contrast to humans and rats [124].

In contrast to Cebidae NWPs and all OWPs, but similar to rats, marmosets do not express specific thyroid hormone-binding globulins in the circulation

[122,125]. Consequently, their unbound or "free" T_4 and T_3 levels greatly exceed circulating values found in most primates, including humans. As marmosets appear to have expected [126] or relatively low [127] metabolic rates, as well as expected heart rates [128] for their body size, some as yet unknown intracellular mechanism(s) diminish the anticipated elevation in metabolic rate and heart function, possibly in analogous fashion to those regulating NWP P_4, T, cortisol, and E_2.

Such an understanding of thyroid endocrine functioning has enabled exploration of T_3 administration as an ameliorating therapeutic when using the animal model for multiple sclerosis (MS), experimental allergic encephalomyelitis (EAE), in marmosets [123]. T_3 enables multipotent neural stem cell commitment to the oligodendrocyte lineage, stimulating onward oligodendrocyte development and maturation. T_3 therapy administered to EAE, MS-like marmosets not only acts as a protective therapeutic agent combating MS progressive demyelination, but also diminishes inflammation in both central and peripheral nervous systems [123], human-like characteristics not emulated in rodent EAE models. In addition, recent sequencing of marmoset thyroid hormone receptors and deiodinases show high degrees of homology to humans, enabling exploration of thyroid regulation of color vision that is not possible with rodents [129].

Parathyroid Gland

The parathyroid has four discrete locations toward the sides and back of the thyroid gland. The parathyroid(s) synthesize, store, and release parathyroid hormone (PTH) crucial for maintenance of calcium and phosphorus homeostasis, and thus bone mineral density and multiple organ function. Vitamin D, integral to parathyroid function, is diet or sunlight derived and is metabolized first by the liver and then the kidney to generate the bioactive $1,25(OH)_2D_3$ (dihydroxyvitamin D) that emulates PTH function. Unlike the thyroid, however, the parathyroid is not regulated by hypothalamic–pituitary negative feedback. Instead, circulating levels of $1,25(OH)_2D_3$ and calcium ions (ionized calcium) diminish PTH synthesis and release, whereas elevated levels of phosphate ions stimulate PTH release. PTH elevates circulating calcium ion concentrations by stimulating renal nephron resorption, intestinal absorption, and bone resorption (including osteoclast differentiation), while also stimulating renal production of $1,25(OH)_2D_3$.

Marmosets exhibit widely ranging circulating levels of $1,25(OH)_2D_3$ (Table 29.1). It is not known whether they possess a circulating binding protein (DBP) specific for $25(OH)D_3$ and $1,25(OH)_2D_3$, as found in humans and macaques [11], but as the presence or absence of circulating levels of DBP is irrelevant to homeostatic accumulation of $1,25(OH)_2D_3$ in organs and tissues [130], this information is not essential. In common with other NWPs, however, marmosets express very high levels of intracellular binding proteins specific for the vitamin D receptor (VDR-BP), that outcompete all but high levels of $1,25(OH)_2D_3$ [131], as well as intracellular binding proteins specific for vitamin D response elements on chromosomal DNA (VDRE-BP), identified as hnRNPC in humans [6]. Such intracellular hurdles to vitamin D genomic [132] and epigenomic [133,134] actions are overcome through competitive displacement from high circulating levels of $1,25(OH)_2D_3$. Marmoset VDRE-BP also binds microRNA modifying posttranscriptional splicing of mRNA, particularly with regard to diminishing the ubiquitously expressed *CYP24*, the key enzyme degrading $1,25(OH)_2D_3$ or $25(OH)D_3$ into inactive metabolites [135]. Interestingly, when high levels of expression of VDRE-BP occur in humans, they are associated with cancer-related VDRE-mediated DNA transcription and microRNA-modified RNA translocation from transcription sites. Marmoset-derived cells and tissues initially enabled delineation of hnRNPC function, but contemporary research now focuses on human-derived cells [136].

Understanding of NWP-specific vitamin D physiology has enabled laboratories and zoological gardens to supplement NWP diets to eliminate previously all too common rickets, osteopenia, or osteoporosis in captive populations of marmosets and tamarins [137]. Power [138], however, cautions against the negative consequences of oversupplementation with vitamin D in NWP diets. Certainly, elevated circulating levels of PTH accompanying marmoset bone disease indicate secondary hyperparathyroidism in response to low circulating levels of calcium or vitamin D_3 caused by inflammatory gastrointestinal disease impairing absorption [139,140]. As described by Power [138], improved understanding of diet, gastrointestinal pathogens, and microbiome are now required to solve the issue of gastrointestinal-related bone disease in captive marmosets, and likely all NWPs. Current parathyroid (vitamin D)-related research in marmosets focuses on resolving the most accurate and sensitive methods of assessment, such as high-throughput LC-MS/MS, of the most relevant circulating vitamin D metabolite [141].

Parafollicular or C Cells

C cells in the marmoset thyroid gland synthesize, store, and release calcitonin [142]. Calcitonin inhibits

TABLE 29.1 Approximate Ranges Reported for Circulating Hormone Concentrations in Adult Marmosets

Endocrine System	Hormone (Units)	Sex or Reproductive Stage	Circulating concentrations (Minimum—Maximum)
Anterior pituitary	Chorionic gonadotropin (mCG) (ng/mL)	Female, follicular phase:	2–4
		Female, periovulatory:	15–44
		Female, luteal phase:	4–40
		Male:	2–4
	Adrenocorticotropic hormone (ACTH) (pg/mL)	Female:	30–80 (morning)
		Male:	7–20 (morning)
	Thyroid-stimulating hormone (TSH) (ng/mL)	Female:	2–3
		Male:	N/A
	Prolactin (ng/mL)	Female:	<22 (not lactating)
		Female:	50-100 (lactating)
		Male:	0.2–5
Ovary	Estradiol (E_2) (pg/mL)	Follicular phase:	23–200
		Periovulatory:	130–260
		Luteal phase:	130–220
		Early-to-mid gestation	1000–2000
		Mid-to-late gestation	2000–90,000
	Progesterone (P_4) (ng/mL)	Follicular phase:	<10
		Periovulatory:	<10
		Luteal phase:	10–80
		Early-to-mid gestation	50–100
		Mid-to-late gestation	100–300
	Testosterone (ng/mL)	Follicular phase:	N/A
		Periovulatory:	N/A
		Luteal phase:	0.5–7
		Early-to-mid gestation	5–10
		Mid-to-late gestation	4–8
Ovary	Inhibin (mU/L)	Follicular phase:	5–6
		Periovulatory:	N/A
		Luteal phase:	8–13
	Relaxin (ng/mL)	Follicular phase:	<0.6
		Periovulatory	N/A
		Luteal phase:	1–1.5
		Early-to-mid gestation	1–10
		Mid-to-late gestation	4–20
Placenta	Chorionic gonadotropin (mCG) (ng/mL)	Early-to-mid gestation	1–10
		Mid-to-late gestation	4–20
	Corticotropin-releasing hormone (CRH) (pg/mL)	Early-to-mid gestation	200–1000
		Mid-to-late gestation	100–400

Continued

TABLE 29.1 Approximate Ranges Reported for Circulating Hormone Concentrations in Adult Marmosets—cont'd

Endocrine System	Hormone (Units)	Sex or Reproductive Stage	Circulating concentrations (Minimum—Maximum)
Testis	Testosterone (ng/mL)	Male:	10–50
	Dihydrotestosterone (DHT) (ng/mL)	Male:	~0.2
Adrenal	Aldosterone (pg/mL)	Female:	N/A
		Male:	300–1200
	Cortisol (ug/dL)	Female:	300–450
		Male:	73–250
Thyroid	Thyroxin (T_4) (ng/mL)	Female:	135–205 (M + F)
		Male:	80–12
White adipose	Adiponectin (ug/mL)	Female:	N/A
		Male:	4–12 (lean)
	Leptin (ng/mL)	Female:	N/A
		Male:	0.3–1.2 (lean)
	Resistin (pg/mL)	Female:	200–700 (M + F)
		Male:	N/A
	Visfatin (pg/mL)	Female:	100–2000 (M + F)
		Male:	N/A
Gastrointestinal tract	Ghrelin (pg/mL)	Female:	100–380 (M + F)
		Male:	70–130
	Glucagon-like peptide-1 (GLP-1) (pg/mL)	Female:	<10 (M + F)
		Male:	N/A
	Gastric inhibitory polypeptide (GIP) (pg/mL)	Female:	<15 (M + F)
		Male:	N/A
Pancreatic islets	Insulin (uU/mL)	Female:	N/A
		Male:	3–6 (lean)

bone resorption and increases renal calcium excretion, counteracting the effects of PTH and 1,25$(OH)_2D_3$, but has not been a marmoset research target. Marmosets, however, are proving very useful nonhuman primate models for nasal administration of receptor antagonists for calcitonin gene-related peptide (CGRP). CGRP belongs to a family of bioactive peptides that includes calcitonin. Current receptor antagonists are proving highly effective therapeutics for migraine as these drugs lack the cardiovascular liabilities associated with the currently available triptan-based therapeutics [143].

Growth Hormone and Insulin-Like Growth Factor-1

While there are no assays for marmoset growth hormone (GH) or IGF-1, recent whole genome [5] and DNA sequence [144] analyses have demonstrated strong positive selection for accelerated evolution of GH/IGF axis genes after OWP and NWP lineages diverged. Molecular changes were virtually complete before subsequent divergence of marmoset and squirrel monkey lineages within NWPs [144]. GH/IGF axis gene changes, including genes encoding for receptors and IGF-binding proteins, may have played a role in the evolution of diminutive size in marmosets [145], but as biological functions of these molecules appear typical [144], and almost nothing is known of marmoset GH/IGF axis physiology, this area of marmoset endocrine research remains contentious and underexplored.

Cardiac Hormones

Atrial natriuretic peptide (ANP) prohormone is produced and stored in atrial myocardial cells and is released in the form of four endocrinologically active natriuretic peptides: long-acting natriuretic peptide, vessel dilator, kaliuretic peptide, and ANP [146]. All are released in response to stretching of atrial myocardial cells by increased circulation volume. Both forms of ANP stimulate renal sodium ion excretion and glomerular filtration rate, while diminishing renal sodium ion reabsorption, resulting in overall water loss. The major function of the other two peptides is self-evident. All have been shown to have potential anticancer therapies. No ANP endocrine research to date has involved marmosets.

Renal Hormones

Release of renin from juxtaglomerular cells surrounding kidney glomeruli afferent arterioles is driven by beta adrenergic receptor activation from renal sympathetic nerve activity or by falling arterial blood pressure (diminished stretch of vessels). Renin provides the rate-limiting enzymatic conversion of liver-synthesized angiotensin into angiotensin I. This intracirculatory enzymatic action is continued by lung endothelial cell angiotensin-converting enzyme (ACE) conversion of angiotensin I into angiotensin II. Angiotensin II, acting via angiotensin type 1 receptor, stimulates arterial vasoconstriction as well as the release of aldosterone from the outer ZG of the adrenal cortex to stimulate renal sodium ion reabsorption and potassium ion excretion, and thus water retention. Both endocrine actions raise blood pressure. High sodium ion concentrations diminish renin release.

Marmoset renin resembles its human counterpart both in sequence homology and biochemical activity [147]. Not surprisingly, sodium-depleted marmosets figure prominently in preclinical assessments of renin inhibitors as potential hypertensives [148–150]. Renin inhibitors were developed as they do not give rise to persistent cough or angioedema of ACE inhibitors [149]. More recently, however, angiotensin receptor blockers, such as losartan, that do not demonstrate ACE inhibitor side effects, have superseded renin inhibitors in preclinical research [151], and no longer require modeling in marmosets.

Erythropoietin is released from interstitial cells in the renal cortex when oxygen partial pressure in the circulation drops and stimulates bone marrow production of erythrocytes. There are no reports of marmoset involvement in erythropoietin research.

Adipose

Marmoset obesity, naturally occurring [152] or induced by a glucose-enriched diet [153], as well as nonalcoholic steatohepatitis induced by hepatitis C-like infection [154] or naturally occurring [155], have all shown the relevance of marmosets to translational studies of pathological mechanisms underlying obesity-related cardiometabolic disease. In this regard, subcutaneous white adipose depots are now recognized not only as sites of "safe" lipid storage, minimizing inflammatory responses and progression to lipotoxicity of metabolic disease [156], but also as sites of a highly active endocrine organ secreting a variety of hormones or adipokines, such as PAI-1, tumor necrosis factor-α (TNF-α), visfatin, or resistin (potentially pathogenic), and potentially beneficial adipokines, such as adiponectin and leptin [157]. Steroids, including E_2, are also synthesized and released from adipose depots [158]. Such adipose endocrine products are causally implicated in type 2 diabetes mellitus (T2DM), hyperlipidemia, hypertension, and atherosclerosis that comprise metabolic syndrome. Endocrine research in marmosets, however, has focused on only a select number of adipokines, partly due to lack of reagent specificity, insufficient detection limits of techniques, as well as a lack of information about pathophysiology accompanying marmoset metabolic disease.

Circulating leptin, synthesized and released from adipocytes, is usually a reliable biomarker of total body fat. Leptin activates receptors on hypothalamic neurons in the brain to regulate food intake [159,159a]. In pubertal marmosets, as anticipated, leptin is positively correlated with body fat, with the highest circulating levels in obese individuals [160]. A separate study in adult male marmosets, reporting a positive correlation between leptin levels and body weight, supports the expected biomarker potential of leptin levels for degree of marmoset adiposity [161]. As birth weight also positively correlates with pubertal leptin levels in marmosets [160], gestational factors may influence body fat accumulation.

Adiponectin is an anorexigenic peptide that stimulates fatty acid oxidation and glucose uptake in skeletal muscle and adipose tissue, improves whole-body insulin sensitivity, and increases energy expenditure and suppression of hepatic glucose output through activation of AMP-activated protein kinase (AMPK) signaling. Its circulating levels are diminished in obese, insulin- resistant, pubertal male and female marmosets [160,161]. High-molecular-weight adiponectin, considered the most bioactive moiety, showed anticipated negative trends with body fat and weight.

Resistin, visfatin, and PAI-1 are all elevated in the circulation of adult male and female marmosets with liver steatosis-associated hyperlipidemia following infection with GB virus-B and induction of hepatitis C-like disease and metabolic syndrome [154]. In such hyperlipidemic monkeys, resistin also positively correlates with triglyceride levels. Increases in circulating levels of PAI-1 and TNF-α, however, may reflect contributions from the experimentally induced immune response.

Pathophysiological mechanisms engaged by marmoset adipose endocrine secretions in metabolic disease have yet to be explored.

Gastrointestinal Hormones

Ghrelin is synthesized and released by enteroendocrine cells in the mucosal layer of the marmoset stomach [162]. It shares 93% homology with its human counterpart and stimulates appetite and GH secretion. Unlike humans, however, circulating ghrelin levels in either males or females demonstrate no diurnal rhythm [124,161], there is no sex difference, and no difference between fasted and postprandial circulating levels [161]. Low-weight (<300 g) adult marmosets, nevertheless, exhibit higher circulating ghrelin levels than high-weight (>300 g) individuals, potentially indicating appetite-stimulated compensation for compromised gastrointestinal absorption commonly found in laboratory-housed marmosets [139,140].

Meals stimulate the release of two incretins (potent insulin secretagogues) glucagon-like peptide-1 (GLP-1) and gastric inhibitory polypeptide (GIP) from L cells of the lower intestine and colon and K cells of the upper small intestine, respectively [163]. Release is triggered by glucose, amino acid, or fatty acid ingestion. Luminex-detected circulating levels of both GLP-1 and GIP are elevated in marmosets exhibiting a hepatitis C-like, hyperinsulinemic metabolic syndrome following infection with GB virus-B [154]. GIP levels in marmosets demonstrate no diurnal rhythm or sex difference [124].

The potential contributions of marmoset ghrelin and incretins to development of translational models for obesity and insulin resistance in humans have yet to be explored. There are no reports concerning gastrointestinal release of secretin and cholecystokinin and their regulation of digestion.

Pancreatic Islet Hormones

Insulin and glucagon released from pancreatic islet beta and alpha cells, respectively, play classical opposing or antagonistic roles in maintaining glucose homeostasis. The heterogeneous mix of beta and alpha cells within marmoset islets is primate-typical [164,165] in striking contrast to the beta cell core and nonbeta cell mantle of rodents [166]. Marmoset islets provide ~4% fractional area (proportion of islets to entire pancreas) that is comparable to ~2%–4% observed in humans and macaques [167]. With only recent in vitro and in vivo validation of a viable insulin assay enabling accurate quantitation of circulating marmoset insulin levels [161], oral glucose tolerance tests have just started to be used in determining insulin resistance and beta cell defects in marmosets [160,168]. Obese adolescent male marmosets demonstrate insulin resistance even at a relatively early stage in their development of increased body fat and hyperlipidemia [160,161,168], consistent with the hypothesis of adiposity-driven insulin impairment [156,169].

Glucagon has been studied even less, and only one study indicates expected increases in circulating glucagon in a hypoglycemic environment [154]. Somatostatin-producing delta cells have been identified, and the level of mRNA expression is comparable to that in humans [164]. Amylin, or islet amyloid polypeptide, production by marmoset beta cells has yet to be reported. Similar to humans, and in contrast to rodents, glucose transporter GLUT1 predominates in marmoset islet cells, not GLUT2 [164]. Thus regulation of glucose uptake and metabolism, crucial to triggering release of either insulin or glucagon, appears human-like.

Final Considerations

Marmoset endocrinology, while resembling OWP and humans, has distinct functional and molecular differences. Such naturally selected traits provide unique insight to endocrine pathology and human disease. As we have demonstrated, the major distinction of NWP, including marmosets, from OWP and humans is the appearance of end-organ steroid hormone resistance. Marmoset monkeys inherently exhibit higher levels of steroid hormones cortisol, E_2, P_4, androgens, and also vitamin D. Such chronically elevated hormone levels in humans are implicated in disease states, thus it is remarkable that NWPs are able to cope and thrive with these adaptations. Although the fossil record for marmosets is not well characterized, it is likely that natural selection for these NWP changes occurred through rapid evolutionary events related to climatic shift in South America [170,171]. In the event of a drastic climatic event in habitat, being able to thrive in a high-stress, and thus elevated steroid hormone environment, would be advantageous, and may partly explain the selective pressure that created this distinct difference in endocrine phenotype observed in NWPs.

References

[1] Tardif SD. Reproduction, growth and development. In: Marini R, Wachtman L, Tardif S, Mansfield K, Fox J, editors. The common marmoset in captivity and biomedical research. Elsevier Press; 2017. pp. 000–000.

[2] Chen H, Gilbert LC, Lu X, Liu Z, You S, Weitzmann MN, Nanes MS, Adams J. A new regulator of osteoclastogenesis: estrogen response element-binding protein in bone. J Bone Miner Res 2011;26:2537–47.

[3] Fuller PJ, Smith BJ, Rogerson FM. Cortisol resistance in the new world revisited. Trends Endocrinol Metab 2004;15:296–9.

[4] 't Hart BA, Abbott DH, Nakamura K, Fuchs E. The marmoset monkey: a multi-purpose preclinical and translational model of human biology and disease. Drug Discov Today 2012;17:1160–5.

[5] Marmoset Genome Sequencing and Analysis Consortium. The common marmoset genome provides insight into primate biology and evolution. Nat Genet 2014;46:850–7.

[6] Chen H, Hewison M, Adams JS. Functional characterization of heterogeneous nuclear ribonuclear protein C1/C2 in vitamin D resistance: a novel response element-binding protein. J Biol Chem 2006;281:39114–20.

[7] Chen H, Hewison M, Adams JS. Control of estradiol-directed gene transactivation by an intracellular estrogen-binding protein and an estrogen response element-binding protein. Mol Endocrinol 2008;22:559–69.

[8] Westberry JM, Sadosky PW, Hubler TR, Gross KL, Scammell JG. Glucocorticoid resistance in squirrel monkeys results from a combination of a transcriptionally incompetent glucocorticoid receptor and overexpression of the glucocorticoid receptor co-chaperone FKBP51. J Steroid Biochem Mol Biol 2006;100:34–41.

[9] Chen H, Arbelle JE, Gacad MA, Allegretto EA, Adams JS. Vitamin D and gonadal steroid-resistant New World primate cells express an intracellular protein which competes with the estrogen receptor for binding to the estrogen response element. J Clin Investig 1997;99:669–75.

[10] Chen H, Clemens TL, Hewison M, Adams JS. Estradiol and tamoxifen mediate rescue of the dominant-negative effects of estrogen response element-binding protein in vivo and in vitro. Endocrinology 2009;150:2429–35.

[11] Gacad MA, Chen H, Arbelle JE, LeBon T, Adams JS. Functional characterization and purification of an intracellular vitamin D-binding protein in vitamin D-resistant new world primate cells. Amino acid sequence homology with proteins in the hsp-70 family. J Biol Chem 1997;272:8433–40.

[12] Zhou R, Park JW, Chun RF, Lisse TS, Garcia AJ, Zavala K, Sea JL, Lu ZX, Xu J, Adams JS, Xing Y, Hewison M. Concerted effects of heterogeneous nuclear ribonucleoprotein C1/C2 to control vitamin D-directed gene transcription and RNA splicing in human bone cells. Nucleic Acids Res 2017;45:606–18.

[13] Chrousos GP. A new "new" syndrome in the new world: is multiple postreceptor steroid hormone resistance due to a coregulator defect? J Clin Endocrinol Metab 1999;84:4450–3.

[14] Adams JS, Chen H, Chun RF, Nguyen L, Wu S, Ren SY, Barsony J, Gacad MA. Novel regulators of vitamin D action and metabolism: lessons learned at the Los Angeles zoo. J Cell Biochem 2003;88:308–14.

[15] Lambert MR, Edwards TM. Hormonally active phytochemicals and vertebrate evolution. Evol Appl 2017;10:419–32.

[16] Harms MJ, Eick GN, Goswami D, Colucci JK, Griffin PR, Ortlund EA, Thornton JW. Biophysical mechanisms for large-effect mutations in the evolution of steroid hormone receptors. Proc Natl Acad Sci USA 2013;110:11475–80.

[17] Troppmann B, Kleinau G, Krause G, Gromoll J. Structural and functional plasticity of the luteinizing hormone/choriogonadotrophin receptor. Hum Reprod Update 2013;19:583–602.

[18] Müller T, Simoni M, Pekel E, Luetjens CM, Chandolia R, Amato F, Norman RJ, Gromoll J. Chorionic gonadotrophin beta subunit mRNA but not luteinising hormone beta subunit mRNA is expressed in the pituitary of the common marmoset (*Callithrix jacchus*). J Mol Endocrinol 2004;32:115–28.

[19] Adams C, Henke A, Gromoll J. A novel two-promoter-one-gene system of the chorionic gonadotropin β gene enables tissue-specific expression. J Mol Endocrinol 2011;47:285–98.

[20] Abbott DH, Barnett DK, Colman RJ, Yamamoto ME, Schultz-Darken NJ. Aspects of common marmoset basic biology and life history important for biomedical research. Comp Med 2003;53:339–50.

[21] Abbott DH. The physical, hormonal and behavioral development of the common marmoset, *Callithrix jacchus jacchus*. In: Rothe H, Wolters H-J, Hearn JP, editors. Biology and behavior of marmosets. Gottingen: Eigenverlag Hartmut Rothe; 1979. p. 99–106.

[22] Abbott DH, Hearn JP. Physical, hormonal and behavioral aspects of sexual development in the marmoset monkey, *Callithrix jacchus*. J Reprod Fertil 1978;53:155–6.

[23] Abbott DH, McNeilly AS, Lunn SF, Hulme MJ, Burden FJ. Inhibition of ovarian function in subordinate female marmoset monkeys (*Callithrix jacchus jacchus*). J Reprod Fertil 1981;63: 335–45.

[24] Saltzman W, Severin JM, Schultz-Darken NJ, Abbott DH. Behavioral and social correlates of escape from suppression of ovulation in female common marmosets housed with the natal family. Am J Primatol 1997;41:1–21.

[25] Saltzman W, Schultz-Darken NJ, Abbott DH. Familial influences on ovulatory function in common marmosets (*Callithrix jacchus*). Am J Primatol 1997;41:159–77.

[26] Saltzman W, Digby LJ, Abbott DH. Reproductive skew in female common marmosets: what can proximate mechanisms tell us about ultimate causes? Proc Biol Sci 2009;276:389–99.

[27] Arruda MF. Taxonomy and natural history. In: Marini R, Wachtman L, Tardif S, Mansfield K, Fox J, editors. The common marmoset in captivity and biomedical research. Elsevier Press; 2017. pp. 000–000.

[28] Arruda MF, Araújo A, Sousa MB, Albuquerque FS, Albuquerque AC, Yamamoto ME. Two breeding females within free-living groups may not always indicate polygyny: alternative subordinate female strategies in common marmosets (*Callithrix jacchus*). Folia Primatol (Basel) 2005;76:10–20.

[29] Abbott DH, Saltzman W, Schultz-Darken NJ, Smith TE. Specific neuroendocrine mechanisms not involving generalized stress mediate social regulation of female reproduction in cooperatively breeding marmoset monkeys. Ann NY Acad Sci 1997; 807:219–38.

[30] Abbott DH. Behavioral and physiological suppression of fertility in subordinate marmoset monkeys. Am J Primatol 1984;6: 169–86.

[31] Abbott DH. Differentiation of sexual behaviour in female marmoset monkeys: effects of neonatal testosterone or a male co-twin. Prog Brain Res 1984;61:349–58.

[32] Abbott DH, Hodges JK, George LM. Social status controls LH secretion and ovulation in female marmoset monkeys (*Callithrix jacchus*). J Endocrinol 1988;117:329–39.

[33] Saltzman W, Pick RR, Salper OJ, Liedl KJ, Abbott DH. Onset of plural cooperative breeding in common marmoset families following replacement of the breeding male. Anim Behav 2004; 68:59–73.

[34] Kraynak M, Flowers MT, Shapiro RA, Kapoor A, Levine JE, Abbott DH. Extra-ovarian regulation of gonadotropin negative feedback revealed by aromatase inhibition in ovariectomized female marmoset monkeys. Am J Physiol Endocrinol Metab 2017; 313:E507–14.

[35] Harding RD, Hulme MJ, Lunn SF, Henderson C, Aitken RJ. Plasma progesterone levels throughout the ovarian cycle of the common marmoset (*Callithrix jacchus*). J Med Primatol 1982;11: 43–51.
[36] Harlow CR, Hearn JP, Hodges JK. Ovulation in the marmoset monkey: endocrinology, prediction and detection. J Endocrinol 1984;103:17–24.
[37] Kropp J, Di Marzo A, Golos T. Assisted reproductive technologies in the common marmoset: an integral species for developing nonhuman primate models of human diseases. Biol Reprod 2017; 96:277–87.
[38] Summers PM, Wennink CJ, Hodges JK. Cloprostenol-induced luteolysis in the marmoset monkey (*Callithrix jacchus*). J Reprod Fertil 1985;73:133–8.
[39] Abayasekara DR, Michael AE, Webley GE, Flint AP. Mode of action of prostaglandin F2 alpha in human luteinized granulosa cells: role of protein kinase C. Mol Cell Endocrinol 1993;97: 81–91.
[40] Webley GE, Michael AE, Abayasekara DR. The relationship between the production and the anti-gonadotrophic action of prostaglandin F 2 alpha in luteal cells from the marmoset monkey (*Callithrix jacchus*) in the early and mid-luteal phase. Gen Comp Endocrinol 2010;166:436–42.
[41] Hodges JK, Eastman SA, Jenkins N. Sex steroids and their relationship to binding proteins in the serum of the marmoset monkey (*Callithrix jacchus*). J Endocrinol 1983;96:443–50.
[42] Shaw HJ, Hillier SG, Hodges JK. Developmental changes in luteinizing hormone/human chorionic gonadotropin steroidogenic responsiveness in marmoset granulosa cells: effects of follicle-stimulating hormone and androgens. Endocrinology 1989;124: 1669–77.
[43] Shaw HJ, Boddy SE, Hodges JK. Developmental changes in marmoset granulosa cell responsiveness to insulin-like growth factor-I: interactions with follicle-stimulating hormone and estradiol. Biol Reprod 1993;49:568–76.
[44] Hillier SG, Miró F. Local regulation of primate granulosa cell aromatase activity. J Steroid Biochem Mol Biol 1993;44:435–49.
[45] Einspanier A, Ivell R, Hodges JK. Oxytocin: a follicular luteinisation factor in the marmoset monkey. Adv Exp Med Biol 1995;395: 517–22.
[46] Taylor PD, Wilson H, Hillier SG, Wiegand SJ, Fraser HM. Effects of inhibition of vascular endothelial growth factor at time of selection on follicular angiogenesis, expansion, development and atresia in the marmoset. Mol Hum Reprod 2007; 13:729–36.
[47] Hodges JK. Effects of gonadectomy and oestradiol treatment on plasma luteinizing hormone concentrations in the marmoset monkey, *Callithrix jacchus*. J Endocrinol 1978;76:271–81.
[48] Beindorff N, Einspanier A. Luteotrophic effects of relaxin, chorionic gonadotrophin and FSH in common marmoset monkeys (*Callithrix jacchus*). Reproduction 2010;139:923–30.
[49] Fraser HM, Lunn SF, Whitelaw PF, Hillier SG. Induced luteal regression: differential effects on follicular and luteal inhibin/activin subunit mRNAs in the marmoset monkey. J Endocrinol 1995;144:201–8.
[50] Fraser HM, Lunn SF, Cowen GM, Illingworth PJ. Induced luteal regression in the primate: evidence for apoptosis and changes in c-myc protein. J Endocrinol 1995;147:131–7.
[51] Einspanier A, Hodges JK. LH- and chorionic gonadotrophin-stimulated progesterone release in vitro by intact luteal tissue of the marmoset monkey (*Callithrix jacchus*). J Endocrinol 1994; 141:403–9.
[52] Tardif SD, Araujo A, Arruda MF, French JA, Sousa MB, Yamamoto ME. Reproduction and aging in marmosets and tamarins. Interdiscip Top Gerontol 2008;36:29–48.
[53] Thomson JA, Kalishman J, Golos TG, Durning M, Harris CP, Hearn JP. Pluripotent cell lines derived from common marmoset (*Callithrix jacchus*) blastocysts. Biol Reprod 1996;55:254–9.
[54] Sasaki E, Suemizu H, Shimada A, Hanazawa K, Oiwa R, Kamioka M, Tomioka I, Sotomaru Y, Hirakawa R, Eto T, Shiozawa S, Maeda T, Ito M, Ito R, Kito C, Yagihashi C, Kawai K, Miyoshi H, Tanioka Y, Tamaoki N, Habu S, Okano H, Nomura T. Generation of transgenic non-human primates with germline transmission. Nature 2009;459:523–7.
[55] Wilton LJ, Marshall VS, Piercy EC, Moore HD. In vitro fertilization and embryo development in the marmoset monkey (*Callithrix jacchus*). J Reprod Fertil 1993;97:481–6.
[56] Scheerer-Bernhard JU, Tkachenko OY, Heistermann M, Gründker C, Nayudu PL. Body weight-associated differences in ovarian morphology in captive common marmoset (*Callithrix jacchus*). Anim Reprod Sci 2015;157:44–55.
[57] Rosenfield RL, Ehrmann DA. The pathogenesis of polycystic ovary syndrome (PCOS): the hypothesis of PCOS as functional ovarian hyperandrogenism revisited. Endocr Rev 2016;37: 467–520.
[58] Guzeloglu Kayisli O, Kayisli UA, Basar M, Semerci N, Schatz F, Lockwood CJ. Progestins upregulate FKBP51 expression in human endometrial stromal cells to induce functional progesterone and glucocorticoid withdrawal: implications for contraceptive-associated abnormal uterine bleeding. PLoS One 2015;10: e0137855.
[59] Schatz F, Guzeloglu-Kayisli O, Basar M, Buchwalder LF, Ocak N, Guzel E, Guller S, Semerci N, Kayisli UA, Lockwood CJ. Enhanced human decidual cell-expressed FKBP51 may promote labor-related functional progesterone withdrawal. Am J Pathol 2015;185:2402–11.
[60] Russo D, Merolla F, Mascolo M, Ilardi G, Romano S, Varricchio S, Napolitano V, Celetti A, Postiglione L, Di Lorenzo PP, Califano L, Dell'Aversana GO, Astarita F, Romano MF, Staibano S. FKBP51 Immunohistochemical expression: a new prognostic biomarker for OSCC? Int J Mol Sci 2017;18(2). E443.
[61] Gassen NC, Fries GR, Zannas AS, Hartmann J, Zschocke J, Hafner K, Carrillo-Roa T, Steinbacher J, Preißinger SN, Hoeijmakers L, Knop M, Weber F, Kloiber S, Lucas S, Chrousos GP, Carell T, Ising M, Binder EB, Schmidt MV, Rüegg J, Rein T. Chaperoning epigenetics: FKBP51 decreases the activity of DNMT1 and mediates epigenetic effects of the antidepressant paroxetine. Sci Signal 2015;8(404):ra119.
[62] Kadeba PI, Vasauskas AA, Chen H, Wu S, Scammell JG, Cioffi DL. Regulation of store-operated calcium entry by FK506-binding immunophilins. Cell Calcium 2013;53:275–85.
[63] Arrigo AP. Mammalian HspB1 (Hsp27) is a molecular sensor linked to the physiology and environment of the cell. Cell Stress Chaperones 2017;22:517–29. https://doi.org/10.1007/s12192-017-0765-1.
[64] Lisse TS, Hewison M, Adams JS. Hormone response element binding proteins: novel regulators of vitamin D and estrogen signaling. Steroids 2011;76:331–9.
[65] Kedzierska H, Piekiełko-Witkowska A. Splicing factors of SR and hnRNP families as regulators of apoptosis in cancer. Cancer Lett 2017;396:53–65.
[66] Einspanier A, Zarreh-Hoshyari-Khah MR, Balvers M, Kerr L, Fuhrmann K, Ivell R. Local relaxin biosynthesis in the ovary and uterus through the oestrous cycle and early pregnancy in the female marmoset monkey (*Callithrix jacchus*). Hum Reprod 1997;12:1325–37.
[67] Einspanier A, Nubbemeyer R, Schlote S, Schumacher M, Ivell R, Fuhrmann K, Marten A. Relaxin in the marmoset monkey: secretion pattern in the ovarian cycle and early pregnancy. Biol Reprod 1999;61:512–20.

[68] Einspanier A, Lieder K, Husen B, Ebert K, Lier S, Einspanier R, Unemori E, Kemper M. Relaxin supports implantation and early pregnancy in the marmoset monkey. Ann NY Acad Sci 2009;1160: 140–6.

[69] Steinetz BG, Randolph C, Mahoney CJ. Patterns of relaxin and steroids in the reproductive cycle of the common marmoset (*Callithrix jacchus*): effects of prostaglandin F2 alpha on relaxin and progesterone secretion during pregnancy. Biol Reprod 1995;53: 834–9.

[70] Chambers PL, Hearn JP. Peripheral plasma levels of progesterone, oestradiol-17 beta, oestrone, testosterone, androstenedione and chorionic gonadotrophin during pregnancy in the marmoset monkey, *Callithrix jacchus*. J Reprod Fertil 1979;56:23–32.

[71] Tullner WW, Hertz R. Chorionic gonadotropin levels in the rhesus monkey during early pregnancy. Endocrinology 1966;78: 204–7.

[72] Csapo AI, Pulkkinen MO, Wiest WG. Effects of luteectomy and progesterone replacement therapy in early pregnant patients. Am J Obstet Gynecol 1973;115:759–65.

[73] Chambers PL, Hearn JP. Embryonic, foetal and placental development in the common marmet monkey (Callithrix *jacchus*). J Zool Lond 1985;207:545–61.

[74] Markovic D, Bari MF, Lu B, Vatish M, Grammatopoulos DK. Corticotropin-releasing hormone interacts with interleukin-1β to regulate prostaglandin H synthase-2 expression in human myometrium during pregnancy and labor. J Clin Endocrinol Metab 2013;98:2864–75.

[75] Beshay VE, Carr BR, Rainey WE. The human fetal adrenal gland, corticotropin-releasing hormone, and parturition. Semin Reprod Med 2007;25:14–20.

[76] Power ML, Bowman ME, Smith R, Ziegler TE, Layne DG, Schulkin J, Tardif SD. Pattern of maternal serum corticotropin-releasing hormone concentration during pregnancy in the common marmoset (*Callithrix jacchus*). Am J Primatol 2006;68:181–8.

[77] Bowman ME, Lopata A, Jaffe RB, Golos TG, Wickings J, Smith R. Corticotropin-releasing hormone-binding protein in primates. Am J Primatol 2001;53:123–30.

[78] McNeilly AS, Abbott DH, Lunn SF, Chambers PC, Hearn JP. Plasma prolactin concentrations during the ovarian cycle and lactation and their relationship to return of fertility post partum in the common marmoset (*Callithrix jacchus*). J Reprod Fertil 1981; 62:353–60.

[79] Tardif SD, Smucny DA, Abbott DH, Mansfield K, Schultz-Darken N, Yamamoto ME. Reproduction in captive common marmosets (*Callithrix jacchus*). Comp Med 2003;53:364–8.

[80] O'Byrne KT, Lunn SF, Coen CW. Central oxytocin stimulates luteinizing hormone release in the marmoset, a primate which fails to show lactationally-lnduced infertility. J Neuroendocrinol 1990;2:419–21.

[81] Beck-Peccoz P, Padmanabhan V, Baggiani AM, Cortelazzi D, Buscaglia M, Medri G, Marconi AM, Pardi G, Beitins IZ. Maturation of hypothalamic-pituitary-gonadal function in normal human fetuses: circulating levels of gonadotropins, their common alpha-subunit and free testosterone, and discrepancy between immunological and biological activities of circulating follicle-stimulating hormone. J Clin Endocrinol Metab 1991;73:525–32.

[82] Resko JA, Ellinwood WE, Pasztor LM, Huhl AE. Sex steroids in the umbilical circulation of fetal rhesus monkeys from the time of gonadal differentiation. J Clin Endocrinol Metab 1980;50: 900–5.

[83] Dean A, Sharpe RM. Clinical review: anogenital distance or digit length ratio as measures of fetal androgen exposure: relationship to male reproductive development and its disorders. J Clin Endocrinol Metab 2013;98:2230–8.

[84] McFadden D. Masculinization of the mammalian cochlea. Hear Res 2009;252:37–48.

[85] Abbott AD, Colman RJ, Tiefenthaler R, Dumesic DA, Abbott DH. Early-to-mid gestation fetal testosterone increases right hand 2D: 4D finger length ratio in polycystic ovary syndrome-like monkeys. PLoS One 2012;7:e42372.

[86] Goy RW, Resko JA. Gonadal hormones and behavior of normal and pseudohermaphroditic nonhuman female primates. Recent Prog Horm Res 1972;28:707–33.

[87] Dumesic DA, Abbott DH, Eisner JR, Goy RW. Prenatal exposure of female rhesus monkeys to testosterone propionate increases serum luteinizing hormone levels in adulthood. Fertil Steril 1997;67:155–63.

[88] Norman RL, Spies HG. Cyclic ovarian function in a male macaque: additional evidence for a lack of sexual differentiation in the physiological mechanisms that regulate the cyclic release of gonadotropins in primates. Endocrinology 1986;118: 2608–10.

[89] Vargas Trujillo M, Kalil B, Ramaswamy S, Plant TM. Estradiol upregulates Kisspeptin expression in the preoptic area of both the male and female rhesus monkey (Macaca mulatta): implications for the hypothalamic control of ovulation in highly evolved primates. Neuroendocrinology 2017;105:77–89.

[89a] Hodges JK, Hearn JP. A positive feedback effect of oestradiol on LH release in the male marmoset monkey, Callithrix jacchus. J Reprod Fertil 1978;52:83–6.

[90] French JA, Frye B, Cavanaugh J, Ren D, Mustoe AC, Rapaport L, Mickelberg J. Gene changes may minimize masculinizing and defeminizing influences of exposure to male cotwins in female callitrichine primates. Biol Sex Differ 2016;7:28.

[91] Sharpe RM, Walker M, Millar MR, Atanassova N, Morris K, McKinnell C, Saunders PT, Fraser HM. Effect of neonatal gonadotropin-releasing hormone antagonist administration on sertoli cell number and testicular development in the marmoset: comparison with the rat. Biol Reprod 2000;62:1685–93.

[92] Lunn SF, Recio R, Morris K, Fraser HM. Blockade of the neonatal rise in testosterone by a gonadotrophin-releasing hormone antagonist: effects on timing of puberty and sexual behaviour in the male marmoset monkey. J Endocrinol 1994;141:439–47.

[93] Baker JV, Abbott DH, Saltzman W. Social determinants of reproductive failure in male common marmosets housed with their natal family. Anim Behav 1999;58:501–13.

[94] Kulkarni SA, Garde SV, Sheth AR. Immunocytochemical localization of bioregulatory peptides in marmoset testes. Arch Androl 1992;29:87–102.

[95] Dixson AF. Sexual and aggressive behaviour of adult male marmosets (*Callithrix jacchus*) castrated neonatally, prepubertally, or in adulthood. Physiol Behav 1993;354:301–7.

[96] Li LH, Donald JM, Golub MS. Review on testicular development, structure, function, and regulation in common marmoset. Birth Defects Res B Dev Reprod Toxicol 2005;74:450–69.

[97] Wistuba J, Luetjens CM, Ehmcke J, Redmann K, Damm OS, Steinhoff A, Sandhowe-Klaverkamp R, Nieschlag E, Simoni M, Schlatt S. Experimental endocrine manipulation by contraceptive regimen in the male marmoset (*Callithrix jacchus*). Reproduction 2013;145:439–51.

[98] French JA. Marmosets in behavioral research. In: Marini R, Wachtman L, Tardif S, Mansfield K, Fox J, editors. The common marmoset in captivity and biomedical research. Elsevier Press; 2017. pp. 000–000.

[99] Smith JA, Pati D, Wang L, de Kloet AD, Frazier CJ, Krause EG. Hydration and beyond: neuropeptides as mediators of hydromineral balance, anxiety and stress-responsiveness. Front Syst Neurosci 2015;9:46.

[100] Wallis M. Molecular evolution of the neurohypophysial hormone precursors in mammals: comparative genomics reveals novel mammalian oxytocin and vasopressin analogues. Gen Comp Endocrinol 2012;179:313–8.

[101] Seltzer LJ, Ziegler TE. Non-invasive measurement of small peptides in the common marmoset (*Callithrix jacchus*): a radiolabeled clearance study and endogenous excretion under varying social conditions. Horm Behav 2007;51:436–42.

[102] Woller MJ, Sosa ME, Chiang Y, Prudom SL, Keelty P, Moore JE, Ziegler TE. Differential hypothalamic secretion of neurocrines in male common marmosets: parental experience effects? J Neuroendocrinol 2012;24:413–21.

[103] Pattison JC, Abbott DH, Saltzman W, Conley AJ, Bird IM. Plasticity of the zona reticularis in the adult marmoset adrenal cortex: voyages of discovery in the New World. J Endocrinol 2009;203: 313–26.

[104] Conley AJ, Bird IM. The role of cytochrome P450 17 alpha-hydroxylase and 3 beta-hydroxysteroid dehydrogenase in the integration of gonadal and adrenalsteroidogenesis via the delta 5 and delta 4 pathways of steroidogenesis in mammals. Biol Reprod 1997;56:789–99.

[105] Levine J, Wolfe LG, Schiebinger RJ, Loriaux DL, Cutler Jr GB. Rapid regression of fetal adrenal zone and absence of adrenal reticular zone in the marmoset. Endocrinology 1982;111:1797–802.

[106] Pattison JC, Abbott DH, Saltzman W, Nguyen AD, Henderson G, Jing H, Pryce CR, Allen AJ, Conley AJ, Bird IM. Male marmoset monkeys express an adrenal fetal zone at birth, but not a zona reticularis in adulthood. Endocrinology 2005;146:365–74.

[107] Pryce CR, Palme R, Feldon J. Development of pituitary-adrenal endocrine function in the marmoset monkey: infant hypercortisolism is the norm. J Clin Endocrinol Metab 2002;87:691–9.

[108] Robinson PA, Hawkey C, Hammond GL. A phylogenetic study of the structural and functional characteristics of corticosteroid binding globulin in primates. J Endocrinol February 1985; 104(2):251–7.

[109] Yamamoto S, Utsu S, Tanioka Y, Ohsawa N. Extremely high levels of corticosteroids and low levels of corticosteroid binding macromolecule in plasma of marmoset monkeys. Acta Endocrinol (Copenh) 1977;85:398–405.

[110] Scammell JG, Denny WB, Valentine DL, Smith DF. Overexpression of the FK506-binding immunophilin FKBP51 is the common cause of glucocorticoid resistance in three New World primates. Gen Comp Endocrinol 2001;124:152–65.

[111] Brown GM, Grota LJ, Penney DP, Reichlin S. Pituitary-adrenal function in the squirrel monkey. Endocrinology 1970;86:519–59.

[112] Pugeat MM, Chrousos GP, Nisula BC, Loriaux DL, Brandon D, Lipsett MB. Plasma cortisol transport and primate evolution. Endocrinology 1984;115:357–61.

[113] Johnson EO, Brady L, Gold PW, Chrousos GP. Distribution of hippocampal mineralocorticoid and glucocorticoid receptor mRNA in a glucocorticoid resistant nonhuman primate. Steroids 1996;61:69–73.

[114] Pryce CR, Aubert Y, Maier C, Pearce PC, Fuchs E. The developmental impact of prenatal stress, prenatal dexamethasone and postnatal social stress on physiology, behaviour and neuroanatomy of primate offspring: studies in rhesus macaque and common marmoset. Psychopharmacology (Berl) 2011;214:33–53.

[114a] Martensz ND, Herbert J, Stacey PM. Factors regulating levels of cortisol in cerebrospinal fluid of monkeys during acute and chronic hypercortisolemia. Neuroendocrinology 1983;36:39–48.

[114b] Arabadzisz D, Diaz-Heijtz R, Knuesel I, Weber E, Pilloud S, Dettling AC, Feldon J, Law AJ, Harrison PJ, Pryce CR. Primate early life stress leads to long-term mild hippocampal decreases in corticosteroid receptor expression. Biol Psychiatry 2010;67: 1106–9.

[115] Diaz Heijtz R, Fuchs E, Feldon J, Pryce CR, Forssberg H. Effects of antenatal dexamethasone treatment on glucocorticoid receptor and calcyon gene expression in the prefrontal cortex of neonatal and adult common marmoset monkeys. Behav Brain Funct 2010;6:18.

[116] Hauser J, Knapman A, Zürcher NR, Pilloud S, Maier C, Diaz-Heijtz R, Forssberg H, Dettling A, Feldon J, Pryce CR. Effects of prenatal dexamethasone treatment on physical growth, pituitary-adrenal hormones, and performance of motor, motivational, and cognitive tasks in juvenile and adolescent common marmoset monkeys. Endocrinology 2008;149:6343–55.

[117] Nyirenda MJ, Carter R, Tang JI, de Vries A, Schlumbohm C, Hillier SG, Streit F, Oellerich M, Armstrong VW, Fuchs E, Seckl JR. Prenatal programming of metabolic syndrome in the common marmoset is associated with increased expression of 11beta-hydroxysteroid dehydrogenase type 1. Diabetes 2009;58: 2873–9.

[117a] Pattison JC, Saltzman W, Abbott DH, Hogan BK, Nguyen AD, Husen B, Einspanier A, Conley AJ, Bird IM. Gender and gonadal status differences in zona reticularis expression in marmoset monkey adrenals: Cytochrome b5 localization with respect to cytochrome P450 17,20-lyase activity. Mol Cell Endocrinol 2007; 265–266:93–101.

[118] Miraglia T. Histochemical studies on the adrenal medullary cells of the marmoset (*Callithrix jacchus*). J Histochem Cytochem 1965; 13:595–8.

[119] Bousquet-Mélou A, Galitzky J, Lafontan M, Berlan M. Control of lipolysis in,intra-abdominal fat cells of nonhuman primates: comparison with humans. J Lipid Res 1995;36:451–61.

[120] McLennan PL, Barnden LR, Bridle TM, Abeywardena MY, Charnock JS. Dietary fat modulation of left ventricular ejection fraction in the marmoset due to enhanced filling. Cardiovasc Res September 1992;26(9):871–7.

[121] Mano MT, Potter BJ, Belling GB, Chavadej J, Hetzel BS. Fetal brain development in response to iodine deficiency in a primate model (*Callithrix jacchus* jacchus). J Neurol Sci 1987; 79:287–300.

[122] Kurata Y, Wako Y, Tanaka K, Inoue Y, Makinodan F. Thyroid hyperactivity induced by methimazole, spironolactone and phenobarbital in marmosets (*Callithrix jacchus*): histopathology, plasma thyroid hormone levels and hepatic T4 metabolism. J Vet Med Sci 2000;62:607–14.

[123] D'Intino G, Lorenzini L, Fernandez M, Taglioni A, Perretta G, Del Vecchio G, Villoslada P, Giardino L, Calzà L. Triiodothyronine administration ameliorates the demyelination/remyelination ratio in a non-human primate model of multiple sclerosis by correcting tissue hypothyroidism. J Neuroendocrinol 2011;23: 778–90.

[124] Bertani S, Carboni L, Criado A, Michielin F, Mangiarini L, Vicentini E. Circadian profile of peripheral hormone levels in Sprague-Dawley rats and in common marmosets (*Callithrix jacchus*). In Vivo 2010;*24:827–36*.

[125] Seo H, Ando M, Yamauchi K, Matsui N, Takenaka O. Plasma thyroxine-binding proteins and thyroid hormone levels in primate species; is callithricidae thyroid hormone resistant? Endocrinol Jpn 1989;36:665–73.

[126] Lazic SE. Modeling hippocampal neurogenesis across the lifespan in seven species. Neurobiol Aging 2012;33:1664–71.

[127] Genoud M, Martin RD, Glaser D. Rate of metabolism in the smallest simian primate, the pygmy marmoset (*Cebuella pygmaea*). Am J Primatol 1997;41:229–45.

[128] Schnell CR, Wood JM. Measurement of blood pressure and heart rate by telemetry in conscious, unrestrained marmosets. Am J Physiol 1993;264(5 Pt 2):H1509–16.

[129] Iwagawa T, Tanaka Y, Iida A, Itoh T, Watanabe S. Enhancer/promoter activities of the long/middle wavelength-sensitive opsins of vertebrates mediated by thyroid hormone receptor β2 and COUP-TFII. PLoS One August 23, 2013;8(8):e72065.

[130] Chun RF, Peercy BE, Orwoll ES, Nielson CM, Adams JS, Hewison M. Vitamin D and DBP: the free hormone hypothesis revisited. J Steroid Biochem Mol Biol 2014;144(Pt A):132–7.

[131] Adams JS, Chen H, Chun R, Gacad MA, Encinas C, Ren S, Nguyen L, Wu S, Hewison M, Barsony J. Response element binding proteins and intracellular vitamin D binding proteins: novel regulators of vitamin D trafficking, action and metabolism. J Steroid Biochem Mol Biol 2004;89–90:461–5.
[132] Pike JW, Meyer MB, Benkusky NA, Lee SM, St John H, Carlson A, Onal M, Shamsuzzaman S. Genomic determinants of vitamin D-regulated gene expression. Vitam Horm 2016;100:21–44.
[133] Sun M, Guo B. Vitamin D and the epigenetic machinery in colon cancer. Curr Med Chem November 17, 2016 [Epub ahead of print].
[134] Xue J, Schoenrock SA, Valdar W, Tarantino LM, Ideraabdullah FY. Maternal vitamin D depletion alters DNA methylation at imprinted loci in multiple generations. Clin Epigenetics 2016;8:107.
[135] Adams JS, Chen H, Chun R, Ren S, Wu S, Gacad M, Nguyen L, Ride J, Liu P, Modlin R, Hewison M. Substrate and enzyme trafficking as a means of regulating 1,25-dihydroxyvitamin D synthesis and action: the human innate immune response. J Bone Miner Res 2007;22(Suppl. 2):V20–4.
[136] Lisse TS, Vadivel K, Bajaj SP, Chun RF, Hewison M, Adams JS. The heterodimeric structure of heterogeneous nuclear ribonucleoprotein C1/C2 dictates 1,25-dihydroxyvitamin D-directed transcriptional events in osteoblasts. Bone Res 2014;2. 14011.
[137] Olson EJ, Shaw GC, Hutchinson EK, Schultz-Darken N, Bolton ID, Parker JB, Morrison JM, Baxter VK, Pate KA, Mankowski JL, Carlson CS. Bone disease in the common marmoset: radiographic and histological findings. Vet Pathol 2015;52:883–93.
[138] Power ML. Marmoset nutrition and dietary husbandry. In: Marini R, Wachtman L, Tardif S, Mansfield K, Fox J, editors. The common marmoset in captivity and biomedical research. Elsevier Press; 2017. pp. 000–000.
[139] Baxter VK, Shaw GC, Sotuyo NP, Carlson CS, Olson EJ, Zink MC, Mankowski JL, Adams RJ, Hutchinson EK, Metcalf Pate KA. Serum albumin and body weight as biomarkers for the antemortem identification of bone and gastrointestinal disease in the common marmoset. PLoS One December 6, 2013;8(12): e82747.
[140] Jarcho MR, Power ML, Layne-Colon DG, Tardif SD. Digestive efficiency mediated by serum calcium predicts bone mineral density in the common marmoset (*Callithrix jacchus*). Am J Primatol 2013;75:153–60.
[141] Ziegler TE, Kapoor A, Hedman CJ, Binkley N, Kemnitz JW. Measurement of 25-hydroxyvitamin D(2&3) and 1,25-dihydroxyvitamin D(2&3) by tandem mass spectrometry: a primate multispecies comparison. Am J Primatol 2015;77:801–10.
[142] Maile S, Merker HJ. C-cells of marmosets (*Callithrix jacchus*). Ann Anat 1996;178:159–67.
[143] Chaturvedula PV, Mercer SE, Pin SS, Thalody G, Xu C, Conway CM, Keavy D, Signor L, Cantor GH, Mathias N, Moench P, Denton R, Macci R, Schartman R, Whiterock V, Davis C, Macor JE, Dubowchik GM. Discovery of (R)-N-(3-(7-methyl-1H-indazol-5-yl)-1-(4-(1-methylpiperidin-4-yl)-1-oxopropan-2-yl)-4-(2-oxo-1,2-dihydroquinolin-3-yl) piperidine-1-carboxamide (BMS-742413): a potent human CGRP antagonist with superior safety profile for the treatment of migraine through intranasal delivery. Bioorg Med Chem Lett 2013;23:3157–61.
[144] Wallis M. Coevolution of insulin-like growth factors, insulin and their receptors and binding proteins in New World Monkeys. Growth Horm IGF Res 2015;25:158–67.
[145] Harris RA, Tardif SD, Vinar T, Wildman DE, Rutherford JN, Rogers J, Worley KC, Aagaard KM. Evolutionary genetics and implications of small size and twinning in callitrichine primates. Proc Natl Acad Sci USA 2014;111:1467–72.
[146] Vesely DL. Heart peptide hormones: adjunct and primary treatments of cancer. Anticancer Res 2016;36:5693–700.
[147] Valdenaire O, Breu V, Giller T, Bur D, Fischli W. Cloning and characterization of marmoset renin: comparison with human renin. J Cardiovasc Pharmacol 1999;34:893–7.
[148] Göschke R, Stutz S, Rasetti V, Cohen NC, Rahuel J, Rigollier P, Baum HP, Forgiarini P, Schnell CR, Wagner T, Gruetter MG, Fuhrer W, Schilling W, Cumin F, Wood JM, Maibaum J. Novel 2,7-dialkyl-substituted 5(S)-amino-4(S)-hydroxy-8-phenyl-octanecarboxamide transition state peptidomimetics are potent and orally active inhibitors of human renin. J Med Chem 2007;50: 4818–31.
[149] Maibaum J, Stutz S, Göschke R, Rigollier P, Yamaguchi Y, Cumin F, Rahuel J, Baum HP, Cohen NC, Schnell CR, Fuhrer W, Gruetter MG, Schilling W, Wood JM. Structural modification of the P2′ position of 2,7-dialkyl-substituted 5(S)-amino-4(S)-hydroxy-8-phenyl-octanecarboxamides: the discovery of aliskiren, a potent nonpeptide human renin inhibitor active after once daily dosing in marmosets. J Med Chem 2007;50:4832–44.
[150] Wood JM, Schnell CR, Cumin F, Menard J, Webb RL. Aliskiren, a novel, orally effective renin inhibitor, lowers blood pressure in marmosets and spontaneously hypertensive rats. J Hypertens 2005;23:417–26.
[151] White CM. Comparative effectiveness of renin-angiotensin system inhibitors in hypertension. J Comp Eff Res 2012;1:125–7.
[152] Tardif SD, Power ML, Ross CN, Rutherford JN, Layne-Colon DG, Paulik MA. Characterization of obese phenotypes in a small nonhuman primate, the common marmoset (*Callithrix jacchus*). Obesity (Silver Spring) 2009;17:1499–505.
[153] Wachtman LM, Kramer JA, Miller AD, Hachey AM, Curran EH, Mansfield KG. Differential contribution of dietary fat and monosaccharide to metabolic syndrome in the common marmoset (*Callithrix jacchus*). Obesity (Silver Spring) 2011;19:1145–56.
[154] Manickam C, Wachtman L, Martinot AJ, Giavedoni LD, Reeves RK. Metabolic dysregulation in hepacivirus infection of common marmosets (*Callithrix jacchus*). PLoS One 2017;12: e0170240.
[155] Kramer JA, Grindley J, Crowell AM, Makaron L, Kohli R, Kirby M, Mansfield KG, Wachtman LM. The common marmoset as a model for the study of nonalcoholic fatty liver disease and nonalcoholic steatohepatitis. Vet Pathol 2015;52:404–13.
[156] Dumesic DA, Akopians AL, Madrigal VK, Ramirez E, Margolis DJ, Sarma MK, Thomas AM, Grogan TR, Haykal R, Schooler TA, Okeya BL, Abbott DH, Chazenbalk GD. Hyperandrogenism accompanies increased intra-abdominal fat storage in normal weight polycystic ovary syndrome women. J Clin Endocrinol Metab 2016;101:4178–88.
[157] Cohen P, Spiegelman BM. Cell biology of fat storage. Mol Biol Cell 2016;27:2523–7.
[158] Leitner L, Jürets A, Itariu BK, Keck M, Prager G, Langer F, Grablowitz V, Zeyda M, Stulnig TM. Osteopontin promotes aromatase expression and estradiol production in human adipocytes. Breast Cancer Res Treat. 2015;154:63–9.
[159] Kohno D, Koike M, Ninomiya Y, Kojima I, Kitamura T, Yada T. Sweet taste receptor serves to activate glucose- and leptin-responsive neurons in the hypothalamic arcuate nucleus and participates in glucose responsiveness. Front Neurosci 2016; 10:502.
[159a] Cady G, Landeryou T, Garratt M, Kopchick JJ, Qi N, Garcia-Galiano D, Elias CF, Myers Jr MG, Miller RA, Sandoval DA, Sadagurski M. Hypothalamic growth hormone receptor (GHR) controls hepatic glucose production in nutrient-sensing leptin receptor (LepRb) expressing neurons. Mol Metab 2017;6: 393–405.

[160] Power ML, Ross CN, Schulkin J, Ziegler TE, Tardif SD. Metabolic consequences of the early onset of obesity in common marmoset monkeys. Obesity (Silver Spring) 2013;21:E592–8.
[161] Ziegler TE, Colman RJ, Tardif SD, Sosa ME, Wegner FH, Wittwer DJ, Shrestha H. Development of metabolic function biomarkers in the common marmoset, *Callithrix jacchus*. Am J Primatol 2013;75:500–8.
[162] Takemi S, Sakata I, Apu AS, Tsukahara S, Yahashi S, Katsuura G, Iwashige F, Akune A, Inui A, Sakai T. Molecular Cloning of ghrelin and characteristics of ghrelin-producing cells in the gastrointestinal tract of the common marmoset (*Callithrix jacchus*). Zool Sci 2016;33:497–504.
[163] Seino Y, Fukushima M, Yabe D. GIP and GLP-1, the two incretin hormones: similarities and differences. J Diabetes Investig 2010; 1:8–23.
[164] Mohanasundaram D, Drogemuller C, Brealey J, Jessup CF, Milner C, Murgia C, Lang CJ, Milton A, Zalewski PD, Russ GR, Coates PT. Ultrastructural analysis, zinc transporters, glucose transporters and hormones expression in New world primate (*Callithrix jacchus*) and human pancreatic islets. Gen Comp Endocrinol 2011;174:71–9.
[165] Plentz RR, Palagani V, Wiedemann A, Diekmann U, Glage S, Naujok O, Jörns A, Müller T. Islet microarchitecture and glucose transporter expression of the pancreas of the marmoset monkey display similarities to the human. Islets 2012;4:123–9.
[166] Brissova M, Fowler MJ, Nicholson WE, Chu A, Hirshberg B, Harlan DM, Powers AC. Assessment of human pancreatic islet architecture and composition by laser scanning confocal microscopy. J Histochem Cytochem 2005;53:1087–97.
[167] Nicol LE, O'Brien TD, Dumesic DA, Grogan T, Tarantal AF, Abbott DH. Abnormal infant islet morphology precedes insulin resistance in PCOS-like monkeys. PLoS One 2014;9: e106527.
[168] Ziegler TE, Sosa ME, Peterson LJ, Colman RJ. Using snacks high in fat and protein to improve glucoregulatory function in adolescent male marmosets (*Callithrix jacchus*). J Am Assoc Lab Anim Sci 2013;52:756–62.
[169] Keller E, Chazenbalk GD, Aguilera P, Madrigal V, Grogan T, Elashoff D, Dumesic DA, Abbott DH. Impaired preadipocyte differentiation into adipocytes in subcutaneous abdominal adipose of PCOS-like female rhesus monkeys. Endocrinology 2014;155: 2696–703.
[170] Duran A, Meyer AL, Pie MR. Climatic niche evolution in new world monkeys (Platyrrhini). PLoS One 2013;8:e83684.
[171] Kay RF. Anthropology. New World monkey origins. Science 2015;347:1068–9.

Index

'*Note*: Page numbers followed by "f" indicate figures and "t" indicate tables.'

N

CPI Antony Rowe
Chippenham, UK
2019-01-17 19:16